Otoneurology and Vestibular Medicine

Marcello Cherchi

Otoneurology and Vestibular Medicine

A Clinical Handbook

Volume 1

Springer

Marcello Cherchi
Neurology
University of Chicago
Chicago, IL, USA

ISBN 978-3-031-94840-4   ISBN 978-3-031-94841-1 (eBook)
https://doi.org/10.1007/978-3-031-94841-1

This Springer imprint is published by the registered company Springer Nature Switzerland AG
The registered company address is: Gewerbestrasse 11, 6330 Cham, Switzerland

To my family

Preface

Problems with equilibrium and hearing are extraordinarily common in clinical medicine. When these symptoms occur acutely, patients are often first evaluated by emergency room physicians. When these symptoms present in a sub-acute or chronic fashion, patients are often evaluated by internists and family practice physicians. Thus, the vast majority of patients with such symptoms are evaluated by these front-line providers. I am immensely grateful to these physicians for the care they provide to the needs of this enormous population of patients.

After the more straightforward and easily treatable causes of these symptoms have been excluded, patients are often referred to a general otolaryngologist or general neurologist, who frequently have sufficient expertise to identify the cause and provide appropriate treatment.

Patients who pass through those levels of screening and remain undiagnosed and untreated may eventually find their way to a subspecialist. The relevant subspecialists include oto-neurologists and neuro-otologists. There are very few such practitioners, which has always struck me as paradoxical given the immense need for such care.

Programs that train these subspecialists are few, and unfortunately are diminishing. This book will not by itself reverse that trend, but my hope is that it will serve two purposes. First, it should serve as a reference work (for specialists and non-specialists alike) regarding the conditions most commonly encountered in this field, as well as some less common conditions that are instructive regarding pathophysiology, physical and ocular motor examination, otovestibular testing and management. Second, it may serve as framework for a training curriculum.

The present work draws inspiration from several volumes that have stood the test of time—such as Leigh and Zee's definitive reference work on the neurology of eye movements [1], Baloh, Kerber, and Honrubia's work on vestibular neurophysiology [2]; and Jacobson and Shepard's work on vestibular testing [3]—supplemented by recent medical literature, to offer a practical, clinically oriented reference for a broader range of pathologies than those discussed in otherwise similar works by Baloh [4], Bronstein and Lempert [5], Crane and colleagues [6], and Strupp and colleagues [7].

I have been privileged to work with hundreds of wonderful clinicians over the years. I have been lucky to work closely with perhaps a dozen, and very closely with a few. Whatever skill I have as a physician is largely due to what all these people have taught me, and I express my gratitude to all of them. I would like specifically to thank Dr. John Kessler and Dr. Timothy Hain, who went far beyond teaching raw data, by additionally training me to think critically about those data. I am also very fortunate to have had the unwavering support of Dr. Shyam Prabhakaran throughout the process of bringing this work to publication. I thank Gregory Sutorius, Kavitha Mannikannan and their editorial team at SpringerNature for guiding me through the process of assembling the manuscript. Finally, I am grateful to the tens of thousands of patients who entrusted their care to me, and the nearly two decades I spent at Chicago Dizziness and Hearing, where we saw an astonishing breadth of pathology, including nearly every disease discussed in this book (except perhaps all 40+ spinocerebellar ataxias). Any faults in this work are my own.

Chicago, IL, USA Marcello Cherchi

References

1. Leigh RJ, Zee DS. The neurology of eye movements. 5th ed. Oxford/New York: Oxford University Press; 2015. xx, 1109 pages.
2. Baloh RW, Kerber KA. Clinical neurophysiology of the vestibular system. 4th ed. New York: Oxford University Press; 2011. xxi, 455 p.
3. Jacobson GP, Shepard NT. Balance function assessment and management. 2nd ed. San Diego: Plural Publishing Inc.; 2016.
4. Baloh RW. Dizziness, hearing loss, and tinnitus. Philadelphia: F.A. Davis Co.; 1998. xv, 224 p.
5. Bronstein AM, Lempert T. Dizziness : a practical approach to diagnosis and management. 2nd ed. Cambridge: Cambridge University Press; 2017. x, 191 pages.
6. Crane BT, Lustig L, de Souza C. SpringerLink. Disorders of the vestibular system: diagnosis and management. 1st ed. Cham: Springer International Publishing: Imprint: Springer; 2023.
7. Strupp M, Brandt T, Dieterich M. Vertigo and dizziness: common complaints. 3rd ed. London: Springer; 2023. xvii, 405 pages.

Contents of Volume I

Part XV Diagnoses—Neurologically-Mediated Auditory Disorders

Part XVI Diagnoses—Specific Ocular Motor Syndromes

Part XVII Diagnoses—Eye Movements Elicited by Specific Provocative Maneuvers

Part XVIII Diagnoses—Other Central Nervous System Pathologies

Part XIX Diagnoses—Intersections Between Otovestibular Disease, Psychology and Psychiatry

Part XXV Diagnoses—Toxic Causes of Disequilibrium

Part XXVI Diagnoses—The Role of Electromagnetic Fields and Radiation in Disequilibrium

Part XXVII Diagnoses—Presentations with Multiple Potential Causes

Part XXVIII Diagnoses—Otovestibular Symptoms Occurring in the Context of Other Conditions

Part XXIX Diagnoses—Immunologically-Mediated Otovestibular Symptoms

Part XXX Evaluation—Bedside Physical Examination

Contents of Volume II

Contents of Volume III

Abbreviations

ABR	Auditory brainstem-evoked responses
AC	Anterior canal
AICA	Anterior inferior cerebellar artery
ARHL	Age-related hearing loss
bHIT	Bedside head impulse test
BOS	Base of support
BPPV	Benign paroxysmal positional vertigo
BVW	Bilateral vestibular weakness
CANVAS	Cerebellar ataxia, neuropathy, vestibular areflexia syndrome
CDP	Computerized dynamic posturography
CFS	Chronic fatigue syndrome
cGRP	Calcitonin gene-related peptide
CM	Chiari malformation
CN	Congenital nystagmus
CNS	Central nervous system
CO	Carbon monoxide
COG	Center of gravity
COVID	Corona virus disease
CRN	Convergence retraction nystagmus
CT	Computed tomography
CTA	Computed tomographic angiography
CV	Cervicogenic vertigo
CVS	Cyclic vomiting syndrome
cVEMP	Cervical vestibular evoked myogenic potentials
CVS	Cyclic vomiting syndrome
DIE	Dynamic illegible "E" testing
DVA	Dynamic visual acuity testing
EA	Episodic ataxia
ECoG	Electrocochleography
EDS	Ehlers-Danlos syndrome
EHS	Exploding head syndrome
EMF	Electromagnetic field
ENG	Electronystagmography
ETD	Eustachian tube dysfunction
FA	Friedreich ataxia
FAERS	FDA Adverse Events Reporting System
FD	Fahr's disease, fibrous dysplasia
FM	Fibromyalgia
GA	Gluten ataxia
GEN	Gaze-evoked nystagmus
GT	Glomus tympanicum
HAI	High altitude illness

HBOT	Hyperbaric oxygen therapy
HD	Huntington's disease
HIT	Head impulse test
HIV	Human immunodeficiency virus
HSN	Head-shaking nystagmus
HVIN	Hyperventilation-induced nystagmus
HZO	Herpes zoster oticus
INO	Internuclear ophthalmoplegia
IPD	Idiopathic Parkinson's disease
LC	Lateral canal
LN	Latent nystagmus
LOS	Limits of stability
LVA(S)	Large vestibular aqueduct (syndrome)
MAV	Migraine-associated vertigo
MCT	Motor control test
MdDs	Mal de debarquement syndrome
MDS	Motorist disorientation syndrome
MEM	Middle ear myoclonus
MH	Musical hallucination
MM	Multiple myeloma
MRA	Magnetic resonance angiography
MRI	Magnetic resonance imaging
MSA	Multiple system atrophy
NPH	Normal pressure hydrocephalus
OAE	Otoacoustic emissions
OKAN	Optokinetic afternystagmus
OKN	Optokinetic nystagmus
OM	Otitis media
OMIM	Online Mendelian Inheritance in Man website (https://omim.org)
OPT	Oculopalatal tremor
oVEMP	Ocular vestibular evoked myogenic potentials
PAN	Periodic alternating nystagmus
PC	Posterior canal
PCP	Phenyl-cyclohexyl-piperidine hydrochloride (phencyclidine)
PDB	Paget's disease of bone
PE	Pressure equalization
PLF	Perilymphatic fistula
POTS	Postural orthostatic tachycardia syndrome
PPPD	Persistent postural perceptual dizziness
PSP	Progressive supranuclear palsy
RA	Rheumatoid arthritis
RCT	Rotatory chair testing
RD	Residual dizziness
RHS	Ramsay Hunt syndrome
RN	Rebound nystagmus
SCA	Spinocerebellar ataxia
SCD	Superior canal dehiscence
SDBJN	Spontaneous down beat jerk nystagmus
SLE	Systemic lupus erythematosus
SNRI	Serotonin-norepinephrine reuptake inhibitor
SOM	Superior oblique myokymia
SOT	Sensory organization testing
SPS	Stiff person syndrome

SS	Superficial siderosis
SSCD	Superior semicircular canal dehiscence
SSN	See-saw nystagmus
SSNHL	Sudden onset sensorineural hearing loss
SSRI	Serotonin-specific reuptake inhibitor
SUBJN	Spontaneous up beat jerk nystagmus
SWJ	Square wave jerks
TENS	Transcutaneous electrical neurostimulation
TMJ	Temporomandibular joint
TMJD	Temporomandibular joint disorder
TMS	Transcranial magnetic stimulation
TTT	Tilt table testing
VEMP	Vestibular evoked myogenic potentials
vHIT	Video head impulse testing
VIN	Vibration-induced nystagmus
VNG	Videonystagmography
VNS	Vagal nerve stimulation
VOR	Vestibulo-ocular reflex
VS	Visual snow
VSS	Visual snow syndrome
VV	Visual vertigo

Part I

Introductory Materials—The Field and the Diagnostic Process

Otoneurology and Neuro-otology

1

The Field and Its Practitioners

Disequilibrium is an extraordinarily common symptom. Every year, in the United States alone, this symptom prompts millions of emergency room visits and tens of millions of primary care visits [1, 2]—thus, internists, general practitioners, and emergency room physicians end up seeing the vast majority of these patients.

Despite how common disequilibrium is, there are very few medical practitioners who cultivate a specific interest in the population of patients suffering from disturbances of equilibrium.

Neurologists who have done extra fellowship training in diseases of equilibrium and hearing are called "*otoneurologists*." In some settings, they may be referred to as "neuro-otologists," but that designation is usually reserved for surgeons (see below). Otoneurologists specialize in medically managed causes of disequilibrium and auditory problems. Examples of medically managed disorders of equilibrium include migraine-associated vertigo. Examples of medically managed disorders of hearing include autoimmune inner ear disease. Otoneurologists do not currently have a unifying professional organization. As of this writing, there are a few dozen such practitioners in the United States and two active training programs.

Ear-nose-throat surgeons (otolaryngologists) who have done extra fellowship training in diseases of equilibrium and hearing are called "*neuro-otologists*," and among their colleagues are often referred to as "skull base surgeons" because of the anatomical areas on which they operate. Given their training, their expertise is in diseases treated with surgery. Examples of surgically managed disorders of equilibrium include vestibular schwannomas and superior semicircular canal dehiscence. Examples of surgically managed disorders of hearing include otosclerosis and placement of cochlear implants. As of this writing, there are a few hundred such practitioners in the United States and about 25 training programs.

There is considerable overlap in the practice of neuro-otology and otoneurology, and it is common for these specialists to collaborate. For example, a patient diagnosed with Meniere's disease may initially be diagnosed and medically managed by an otoneurologist with dietary sodium restriction and oral medication; as the disease advances, the patient may eventually be referred to a neurotologist to be evaluated for a transtympanic gentamicin injection.

Allied Fields

Since disorders of equilibrium and hearing can involve diseases of the ear and brain, and since the practice of medicine is increasingly subspecialized, it is understandable that the diagnosis and management of these patients is often an interdisciplinary enterprise. Beyond otoneurology and neuro-otology, other relevant subspecialties include:

Audiology Audiologists perform audiologic and vestibular tests. Many audiologists dispense hearing aids to help manage hearing loss. Some audiologists also treat benign paroxysmal positional vertigo (BPPV).

Neuro-ophthalmology Neuro-ophthalmologists are surgeons (ophthalmologists) who have done extra fellowship training in the neurological underpinnings of eye movements and visual processing. Neuro-ophthalmologists are often involved in the surgical management of strabismus (ocular misalignment).

Neuro-optometry Neuro-optometrists are optometrists who have done extra training in disorders of ocular motility and vision processing. They are sometimes referred to as "developmental optometrists" because of their role in managing patients with strabismus (disorders of ocular alignment). Neuro-optometrists often manage patients with diagnoses such as visual vertigo, visual snow, and anomalous retinal correspondence.

Neurosurgery Neurosurgeons are trained to perform surgery on all areas of the nervous system, including the brain,

M. Cherchi, *Otoneurology and Vestibular Medicine*, https://doi.org/10.1007/978-3-031-94841-1_1

spinal cord, and peripheral nerves. Some diseases (e.g., Chiari malformation) require surgery that can only be performed by a neurosurgeon. Some vestibular diseases (such as vestibular schwannoma and superior semicircular canal dehiscence) may require surgical intervention by both an otolaryngologist and a neurosurgeon working in tandem.

Vestibular Rehabilitation Therapy Physical therapists who do extra training in vestibular disorders are sometimes called "vestibular physical therapists." These practitioners play a crucial role in the management of some very common vestibular diseases, such as benign paroxysmal positional vertigo and vestibular neuritis.

The umbrella term "*vestibular medicine*" has been proposed to encompass the various subspecialties involved in the diagnosis and management of these patients [3].

For Pediatric Patients

There are relatively few neuro-otologists, and very few otoneurologists, who take care of adult patients with otologic and vestibular diseases. There are even fewer such practitioners who provide this care to pediatric patients. This is unfortunate, since symptoms of imbalance [4, 5] and hearing deficits are not rare in children, the relative distribution of diagnoses is often different than in adults [6], and responses to treatment may differ from that in adults [7]. As of this writing, there was only one fellowship training program in the United States.

Clinic Organization

A medical facility should be able to provide otoneurologic/neuro-otologic care in a timely fashion, and in such a way, that evaluation (history, examination, diagnostic testing) and treatment occur in proximity, with iterative "course-corrections" as necessary. We have worked in several different settings, with varying equipment, and various combinations of practitioners. In our experience, greater diagnostic certainty, and more appropriately focused therapy, is more likely when the various components of this care can be delivered "under one roof"—colloquially referred to as "one-stop shopping" [8, 9]. Such an arrangement facilitates communication among practitioners and affords internal quality control. Ideally, such an arrangement would include adequate staffing (one or more otoneurologists, a neuro-otologist, an audiologist supported to otologic technicians, a vestibular physical therapist, a neuro-optometrist), infrastructure (appropriate equipment for each specialty), and adequate logistical support.

Why Are There So Few Practitioners If There Are So Many Patients?

In view of the economic tendencies governing supply and demand, it seems paradoxical that there are so many patients needing care for disorders of hearing and balance, yet so few practitioners to provide that care. The reasons for this supply–demand mismatch are unclear. In the author's opinion, there are probably several factors.

One factor is perhaps a byproduct of the *interdisciplinary* nature of the field. While the buzzword "interdisciplinary" may have positive connotations, it has a darker side as well. Specifically, if a medical problem straddles several subspecialties, then there is a risk that a practitioner in one subspecialty will feel uncomfortable taking on the care of a patient whose diagnosis they suspect might be better managed by a clinician in some other subspecialty.

A second factor is a consequence of *lack of exposure during medical training*. Because there are very few otoneurologists and neuro-otologists, it is correspondingly uncommon for a medical trainee to come into contact with these practitioners during their education, so it may never even occur to a student that these disciplines exist as discrete fields.

A third factor is that, given that there are very few practitioners, there is *no pressure* in the design of medical educational curricula at the national level *to make a clinical rotation a standard or mandatory component of a trainee's exposure*, since so few institutions would be able to meet such a requirement.

A fourth factor is the *cost* of setting up a clinic with adequate diagnostic capabilities. In November of 2022, we obtained the following quotes for the minimum new audiologic and vestibular equipment of an otoneurology clinic.

Type of test	Specific model	Company	Price
Binocular otomicroscopy	OPMI Pico	Zeiss	$32,495.12
Video Frenzel goggles for doing videonystagmograms (without caloric testing)	VisualEyes™ 525b	Micromedical by Interacoustics	$37,183.00
Caloric irrigator	Air Fx™ and Aqua Stim™ caloric irrigators	Micromedical by Interacoustics	$8155.83
Video head impulse test	VisualEyes™ EyeSeeCam™	Interacoustics	$16,894.58
Rotatory chair test	Orion™ + VisualEyes™ 525b	Interacoustics	$122,004.17
Tympanometry and otoacoustic emissions	Titan™	Interacoustics	$11,492.17
Otoacoustic emissions	Duet	Natus	$9675.00
Evoked potentials (for VEMP, BAER, ECoG)	Eclipse™ EP25	Interacoustics	$31,447.13
Computerized dynamic posturography	Bertec® CDP/IVR	Bertec	$164,315.00
Soundproof booth	WhisperRoom™ MDL 4872 S	WhisperRoom	$8910.00
Audiometer	Piano Plus™ Clinical	Invetis	$7500.00
Total			**$450,072.00**

So as of this writing, a minimum audiologic and vestibular setup costs >$450k. This high up-front cost is a prohibitive barrier-to-entry for an independent medical practice and will make a hospital system administrator pause when calculating how long the equipment will take to pay for itself. In short, the initial investment (just in equipment start-up costs) makes it financially unattractive to establish an otoneurology practice.

References

1. Cherchi M. Epidemiology of dizziness. J Biol Phys Chem. 2013;13(1):18–29.
2. Newman-Toker DE, Edlow JA. TiTrATE: a novel, evidence-based approach to diagnosing acute dizziness and vertigo. Neurol Clin. 2015;33(3):577–99. viii
3. van de Berg R, Murdin L, Whitney SL, Holmberg J, Bisdorff A. Curriculum for vestibular medicine (VestMed) proposed by the Barany Society. J Vestib Res Equilib Orient. 2022;32(2):89–98.
4. Niemensivu R, Pyykko I, Wiener-Vacher SR, Kentala E. Vertigo and balance problems in children – an epidemiologic study in Finland. Int J Pediatr Otorhinolaryngol. 2006;70(2):259–65.
5. Humphriss RL, Hall AJ. Dizziness in 10 year old children: an epidemiological study. Int J Pediatr Otorhinolaryngol. 2011;75(3):395–400.
6. Gioacchini FM, Alicandri-Ciufelli M, Kaleci S, Magliulo G, Re M. Prevalence and diagnosis of vestibular disorders in children: a review. Int J Pediatr Otorhinolaryngol. 2014;78(5):718–24.
7. Gruber M, Cohen-Kerem R, Kaminer M, Shupak A. Vertigo in children and adolescents: characteristics and outcome. ScientificWorldJournal. 2012;2012:109624.
8. Cook J. The case for a one-stop balance centre. Hosp Med. 2001;62(11):669–72.
9. Mendis D, Irving R. Outcomes of a one-stop balance clinic. The Otorhinolaryngologist. 2015;8(3):132–4.

Symptom Terminology

2

Definition of "Vertigo," "Dizziness," Etc.

Countless articles have been written in an attempt to define what is meant by the terms "vertigo," "dizziness," etc. in the context of clinical medicine [1]. Although these terms and their attempted definitions have been perpetuated in medical training programs for decades, it has been shown repeatedly that there is no one-to-one correspondence between a particular term and a specific diagnosis [2–4]. The 306 practicing otologists who responded to a questionnaire on the definition of the term "vertigo" provided answers that led the authors to conclude that, "The meaning of the word vertigo is imprecise and variable" [5]. Nevertheless, Newman-Toker and Edlow [3] offer the following definitions:

- *Dizziness* is the sensation of disturbed or impaired spatial orientation without a false or distorted sense of motion. This includes sensations sometimes referred to as giddiness, lightheadedness, or nonspecific dizziness but does not include vertigo.
- *Presyncope* (also near-syncope or faintness) is the sensation of impending loss of consciousness. This sensation may or may not be followed by syncope. When patients report "lightheadedness," it should be classified as presyncope, dizziness, or both.
- *Syncope* (also faint) is transient loss of consciousness due to transient global cerebral hypoperfusion characterized by rapid onset, short duration, and spontaneous complete recovery. Syncope usually leads to loss of postural control and falling.
- *Vertigo* is the sensation of self-motion (of head/body) when no self-motion is occurring or the sensation of distorted self-motion during an otherwise normal head movement.
- *Unsteadiness* is the feeling of being unstable while seated, standing, or walking without a particular directional preference. This sensation has previously been called disequilibrium or imbalance.

Despite the lack of a one-to-one relationship between a diagnosis and a subjective symptom description, there have been a number of attempts to develop purely questionnaire-based diagnostic mechanisms [6–8], and while these may serve as helpful screening or triage tools (such as in primary care settings), both patients and their referring physicians expect a more confident diagnosis when they refer to a subspecialty clinician.

Thus, in practice, while we still place importance on taking a good clinical history, we find these descriptive terms in themselves to be of limited utility in arriving at a diagnosis. Our experience is that the symptom's chronology (whether it is constant or episodic; its frequency and duration), its triggers, the factors that worsen or improve the symptom, and any associated symptoms, are much more likely to aid a clinician in narrowing the differential diagnosis, and in this regard, we agree with Dr. Newman-Toker and colleagues [3]. When pressed to use a term, we prefer "*disequilibrium*," as this word is sufficiently general to feel rather neutral for most people.

References

1. Bisdorff A, Von Brevern M, Lempert T, Newman-Toker DE. Classification of vestibular symptoms: towards an international classification of vestibular disorders. J Vestib Res Equilib Orient. 2009;19(1–2):1–13.
2. Kerber KA, Newman-Toker DE. Misdiagnosing dizzy patients: common pitfalls in clinical practice. Neurol Clin. 2015;33(3):565–75. viii

M. Cherchi, *Otoneurology and Vestibular Medicine*, https://doi.org/10.1007/978-3-031-94841-1_2

3. Newman-Toker DE, Edlow JA. TiTrATE: a novel, evidence-based approach to diagnosing acute dizziness and vertigo. Neurol Clin. 2015;33(3):577–99. viii
4. Stanton VA, Hsieh YH, Camargo CA Jr, Edlow JA, Lovett PB, Goldstein JN, et al. Overreliance on symptom quality in diagnosing dizziness: results of a multicenter survey of emergency physicians. Mayo Clin Proc. 2007;82(11):1319–28.
5. Blakley BW, Goebel J. The meaning of the word "vertigo". Otolaryngol Head Neck Surg. 2001;125(3):147–50.
6. Jacobson GP, Piker EG, Hatton K, Watford KE, Trone T, McCaslin DL, et al. Development and preliminary findings of the dizziness symptom profile. Ear Hear. 2019;40(3):568–76.
7. Gavilan C, Gallego J, Gavilan J. 'Carrusel': an expert system for vestibular diagnosis. Acta Otolaryngol. 1990;110(3–4):161–7.
8. Feil K, Feuerecker R, Goldschagg N, Strobl R, Brandt T, von Muller A, et al. Predictive capability of an Ipad-based medical device (medx) for the diagnosis of vertigo and dizziness. Front Neurol. 2018;9:29.

3 Diagnostic Thinking

Introduction

A physician's task of securing a diagnosis can range in difficulty from trivial to extraordinarily challenging. While the public often entertains the notion that physicians can arrive at a diagnosis with something approaching the certainty of mathematics or physics, the reality is that biology, including the pathobiology of disease in humans, is a very messy and imprecise affair.

A practitioner uses a variety of cognitive approaches for this task, and these approaches have been scrutinized by physicians [1] and nonphysicians [2] alike. Physicians generally will not think of these cognitive approaches as distinct, though philosophers may identify several heuristics. Daniel Kahneman defines a heuristic as, "A simple procedure that helps find adequate, though often imperfect, answers to difficult questions" [3].

One (Disease) to One (Symptom): Pathognomonic

A given finding (such as on physical examination or testing) is said to be pathognomonic of a specific disease or condition if there is a one-to-one relationship between that finding and that disease. Such simple correlations are helpful when available, but there are actually relatively few such relations in medicine.

One (Disease) to Many (Symptoms): Occam's Razor

The "principle of parsimony" is a heuristic, a method for approaching a problem. The idea is that if a problem has multiple potential solutions, then the simplest solution is likely to be the correct one. This heuristic is sometimes referred to as Occam's Razor, after the fourteenth-century Franciscan Friar, William of Ockham, even though the idea had been in use for centuries before that. The general application in the practice of medicine is that if a patient presents with three new symptoms, then it is likely that those symptoms are all due to a single underlying diagnosis (sometimes called the "unifying diagnosis"), rather than three separate diseases that just happen to be starting at the same time.

A straightforward example of this from general medicine is that a patient who just developed fever, rust-colored sputum, and a cough is likely to have pneumonia. Is it possible that the patient has the flu (accounting for the fever), lung cancer (accounting for the rust-colored sputum), and congestive heart failure (accounting for the cough)? Yes, it is possible, but it is also less likely than the single unifying diagnosis of pneumonia.

An example of Occam's Razor as applied to neurological symptoms is the triad of gait apraxia, urinary incontinence, and dementia. When these three symptoms occur roughly simultaneously, or in close proximity, then the unifying diagnosis is likely to be normal pressure hydrocephalus, a condition in which the brain ventricles enlarge even though the intracranial pressure is normal. Is it possible that the patient has a severe peripheral sensory neuropathy (accounting for the gait problem), thoracolumbar spinal cord lesion (accounting for the urinary incontinence), and a rapidly progressive variant of Alzheimer's disease? Yes, it is possible, but it is also likely than the single unifying diagnosis of normal pressure hydrocephalus.

In neurology, the logic of Occam's Razor extends beyond mere symptoms (complaints reported by the patient) to the signs (what an observer finds on physical examination). If a patient has, for example, three findings on physical examination, and if you can "place" a single lesion somewhere in the nervous system that would account for all three findings, then the location of that lesion is more likely to be correct; more likely than there being three individual lesions in three separate locations. For instance, if a patient has a small pupil on one side, a droopy eyelid on the same side, and weakness or atrophy of the arm on the same side, then it is likely that the patient has a lesion at the apex of the lung on that side that is eroding into the sympathetic chain (accounting for the small pupil and the droopy eyelid on that side) and the brachial plexus (accounting for the arm weakness on that side). Is it possible that the patient has, for instance, a lesion of the superior branch of the oculomotor nerve (accounting for the droopy

M. Cherchi, *Otoneurology and Vestibular Medicine*, https://doi.org/10.1007/978-3-031-94841-1_3

eyelid), a lesion of the trigeminal nerve (accounting for the small pupil), and a lesion in the brachial plexus? Yes, it is possible, but it is also less likely than the single unifying diagnosis of a lesion at the lung apex—the most common cause of which is cancer involving the lung's superior sulcus (sometimes referred to as a Pancoast tumor). Neurology trainees spend several years, and study thousands of cases, to learn to incorporate this skill of "localization" into their diagnostic thinking, and this approach is basically just Occam's Razor as applied to neuroanatomical pathways, rather than to symptoms.

Many (Diseases) to Many (Symptoms): Hickam's Dictum

Occam's Razor, however, is merely a heuristic. It does not *guarantee* that the simplest solution will be the correct one; it merely says the simplest solution is *more likely* to be the correct one. When one is faced with several different possible solutions, it is also sometimes difficult to figure out which one is "simplest." An obvious dilemma could be: Simplest in what way?

But probably the biggest challenge that medicine poses to Occam's Razor is the reality that people actually do get sick with more than one disease—a situation sometimes referred to as "Hickam's dictum," after Dr. John Bamber Hickam (appointed chair of medicine at the University of Indiana in 1958) who, according to apocryphal accounts, opined that, "A man can have as many diseases as he damn well pleases." Medical literature abounds with case reports of patients presenting with multiple diagnoses, sometimes even multiple rare diagnoses.

Many (Diseases) to One (Symptom)

If three different symptoms are attributable to three individual diseases (in keeping with Hickam's dictum), then that is already somewhat difficult to figure out. Things get even trickier if a patient has one symptom being caused by three different diseases—a problem commonly encountered in the practice of vestibular medicine.

For people who have never experienced problems with their balance, the process of maintaining normal equilibrium seems natural, perhaps inevitable. But maintaining balance on two feet is a complicated process, relying on the integrity of multiple areas of the nervous system, all working in synchrony with one another. Consequently, lesions in a variety of places can manifest with the sole symptom of disturbed equilibrium. These tend to be the cases that find their way into our practice, where we are expected to figure them out.

An example of this would be a patient with diabetic peripheral neuropathy who insidiously develops progressive supranuclear palsy and then additionally is struck with benign paroxysmal positional vertigo. Such a patient may present to our clinic with complaints of falling and feeling unsteady. In order to maintain balance, the brain needs sensory input from the feet, but that input can be impaired if diabetes damages sensory nerves (diabetic peripheral neuropathy). Progressive supranuclear palsy is a neurodegenerative disease that insidiously degrades one's postural reflexes. Benign paroxysmal positional vertigo is an inner ear problem that provokes vertiginous symptoms triggered by changes in position of the head with respect to gravity. To be sure, these three diseases can present with other symptoms as well; for instance, diabetic peripheral neuropathy can cause pain; progressive supranuclear palsy can cause neck stiffness; benign paroxysmal positional vertigo can cause nausea. But each of these diseases individually can provoke a sensation of unsteadiness in isolation. This situation—in which multiple diseases cause a single symptom—eludes both Occam's Razor (one disease causes many symptoms) and Hickam's dictum (many diseases cause many symptoms).

False Positives and False Negatives

In the practice of medicine, everything—any evaluative history, procedure, or technique in medicine—has false positives (incorrect appearance of pathology) and false negatives (incorrect appearance of health). When I say "everything," I am referring to a patient's history, physical examination, blood work, diagnostic imaging, diagnostic testing, pathology—everything. Even tests previously thought to be ultra-precise, such as genetic testing, have proven to be less than perfect.

Breaking Down a Complex Problem

In clinical practice, a physician may hear a patient's history, do a physical examination, and have a "gestalt" that the available information is most compatible with a specific disease. While such thinking often leads to a correct diagnosis, it is also notoriously subject to a wide variety of biases.

Because arriving at a diagnosis is a complex process, it is often helpful to break down that task into smaller problems, each of which is easier. In medicine, this is sometimes done through promulgation of diagnostic criteria, such as, "If you have these two features in the history, and these three features on physical examination, and these two test results, then the likely diagnosis is such-and-such." This approach is sometimes given a fancy name, such as the "mediating assessments protocol" described by Kahneman and colleagues (2). The rationale is that each smaller problem can be evaluated—if possible in isolation from the other problems, before looking at the sum total (of the results of all the little problems)—to arrive at a diagnosis. This approach reduces (though does not eliminate) the possibility of bias and is also helpful because of the issue of false positives and false negatives mentioned earlier.

References

1. Groopman JE. How doctors think. Boston: Houghton Mifflin; 2007. p. 307.
2. Kahneman D, Sibony O, Sunstein CR. Noise: a flaw in human judgment, vol. ix. 1st ed. New York: Little, Brown Spark; 2021. p. 454.
3. Kahneman D. Thinking, fast and slow. 1st ed. New York: Farrar, Straus and Giroux; 2011. p. 499.

Questionnaires for Diagnosing Disequilibrium

4

The Use of Questionnaires for Diagnosing Disequilibrium

For many years, we have used modular intake questionnaires in order to elicit information from a patient, with the goal of constructing a preliminary history. The information is quite detailed, but also rather formulaic.

Clinicians and researchers who notice the formulaic nature of eliciting history have sought to systematize history-taking in such a way as to algorithmize the diagnostic process. In the 1990s, simple computer programs were used for this purpose [1]. More recent attempts have employed more complex algorithms, such as "constellatory reasoning" [2]. Subsequent literature attempted to validate a variety of questionnaire-based approaches [3–6].

These approaches appropriately aim to diagnose some of the most common otoneurological diseases correctly, such as benign paroxysmal positional vertigo, vestibular neuritis, labyrinthitis, Ménière's disease, and migraine-associated vertigo. We believe that these algorithms likely hold promise at the level of triaging patients.

We also think that at their present level, these algorithms are unlikely to be very useful in the setting of a tertiary-care vestibular medicine clinic whose clinicians are expected to recognize less common diseases correctly. It is certainly possible that with accumulated data, and with more advanced techniques (such as various forms of machine learning), it may become feasible to devise algorithms that are sufficiently granular to recognize more esoteric pathology.

References

1. Gavilan C, Gallego J, Gavilan J. 'Carrusel': an expert system for vestibular diagnosis. Acta Otolaryngol. 1990;110(3–4):161–7.
2. Feil K, Feuerecker R, Goldschagg N, Strobl R, Brandt T, von Muller A, et al. Predictive capability of an iPad-based medical device (medx) for the diagnosis of vertigo and dizziness. Front Neurol. 2018;9:29.
3. Jacobson GP, Piker EG, Hatton K, Watford KE, Trone T, McCaslin DL, et al. Development and preliminary findings of the dizziness symptom profile. Ear Hear. 2019;40(3):568–76.
4. Bamiou DE, Kikidis D, Bibas T, Koohi N, Macdonald N, Maurer C, et al. Diagnostic accuracy and usability of the EMBalance decision support system for vestibular disorders in primary care: proof of concept randomised controlled study results. J Neurol. 2022;269(5):2584–98.
5. Yu F, Wu P, Deng H, Wu J, Sun S, Yu H, et al. A questionnaire-based ensemble learning model to predict the diagnosis of vertigo: model development and validation study. J Med Internet Res. 2022;24(8):e34126.
6. Filippopulos FM, Strobl R, Belanovic B, Dunker K, Grill E, Brandt T, et al. Validation of a comprehensive diagnostic algorithm for patients with acute vertigo and dizziness. Eur J Neurol. 2022;29(10):3092–101.

M. Cherchi, *Otoneurology and Vestibular Medicine*, https://doi.org/10.1007/978-3-031-94841-1_4

5 Preliminary Diagnostic Algorithm

Overview

At risk of oversimplification, one can formulate a preliminary differential diagnosis of the symptom of disequilibrium by taking into account the chronology of that symptom and the presence/absence of aural symptoms. In particular:

- *Chronicity*. Is the symptom of disequilibrium *continuous/constant/chronic* (there all the time) or *episodic/intermittent* (starts and stops)?
- *Onset*. Was the onset of symptoms *abrupt* (over seconds to minutes), *intermediate* (over hours to days), or *gradual* (over weeks to months)?
- *Ear symptoms*. Are *ear symptoms (such as hearing loss, ear fullness, ringing in the ears)* absent, sometimes present, or always present? If ear symptoms are present, and are they unilateral or bilateral?

These are presented in the format of Table

Supplementary Information The online version contains supplementary material available at https://doi.org/10.1007/978-3-031-94841-1_5.

M. Cherchi, *Otoneurology and Vestibular Medicine*, https://doi.org/10.1007/978-3-031-94841-1_5

	Abrupt onset (seconds to minutes)	Intermediate onset (hours to days)	Gradual onset (weeks to months)
Episodic/ intermittent	*Abrupt onset and episodic* *Ear symptoms absent*: Benign paroxysmal positional vertigo Otolithic catastrophe of Tumarkin Orthostatic intolerance Cardiac arrhythmia Drop attacks, including vestibular paroxysmia Episodic ataxia Visual vertigo Cervicogenic vertigo Psychological disequilibrium Motion sickness Cybersickness *Ear symptoms present and (usually) unilateral*: Semicircular canal dehiscence Perilymphatic fistula Large vestibular aqueduct *Ear symptoms present and (usually) bilateral*: Intracranial pressure fluctuations	*Intermediate onset and episodic* *Ear symptoms absent*: Migraine Mal de debarquement syndrome Adverse effect of medication Acute alcohol intoxication, recreational drugs Episodic ataxias *Ear symptoms present and (usually) unilateral*: Ménière's disease Cochlear implantation	

	Abrupt onset (seconds to minutes)	Intermediate onset (hours to days)	Gradual onset (weeks to months)
Continuous/ constant/chronic	*Abrupt onset and continuous*	*Intermediate onset and continuous*	*Gradual onset and continuous*
	Ear symptoms absent: Vestibular neuritis Stroke Concussion Epileptic vertigo *Ear symptoms variably present*: Labyrinthine concussion Otic barotrauma *Ear symptoms present and (usually) unilateral*: Labyrinthitis Temporal bone fracture Inner ear ischemia	*Ear symptoms absent*: Persistent postural perceptual dizziness Demyelination Wernicke's encephalopathy (acute B1 deficiency) Other encephalitides *Ear symptoms variably present*: Infection, e.g., COVID-19 Oculopalatal tremor *Ear symptoms present and (usually) unilateral*: Otitis media Ramsay Hunt syndrome	*Ear symptoms absent*: Idiopathic bilateral vestibular weakness Nutritional deficiencies (vitamin B1 deficiency, vitamin B12 deficiency) Spinocerebellar ataxia Other ataxias (e.g., Friedreich's ataxia, gluten ataxia) Normal pressure hydrocephalus Myelopathy CANVAS Basal ganglionic disorders, such as Parkinson's disease, progressive supranuclear palsy, multiple system atrophy Presbyvestibulopathy Space-occupying lesions Chronic alcoholism Paraneoplastic syndromes Chiari malformation Peripheral neuropathy Visual deficits (cataracts, macular degeneration, anisometropia) White matter disease *Ear symptoms variably present*: Ototoxicity Fibromyalgia Radiation *Ear symptoms present and (usually) unilateral*: Vestibular schwannoma Cerebellopontine angle meningioma Labyrinthine ossification Cholesteatoma Glomus tympanicum Fibrous dysplasia Sarcoidosis *Ear symptoms present and (usually) bilateral*: Autoimmune inner ear disease Sympathetic ear disease Superficial siderosis Usher syndrome Otosclerosis Paget's disease of bone Susac syndrome

For people who prefer the logic of a flow chart format, we offer a diagram in Supplementary Figure.

Part II

Introductory Materials — Physiology, Phenomena and Concepts

6 Overview of the Auditory System

Introduction

The auditory system has both afferent and efferent pathways. The afferent auditory pathways detect and interpret acoustic stimuli. The efferent auditory pathways protect the cochleae and serve as an "equalizer" in filtering incoming acoustic stimuli. We briefly review the anatomy and physiology of these systems.

Afferent Auditory System

The main task of the afferent auditory system is to transduce the kinetic energy from the vibration of sound waves into the electrochemical activity that constitutes a neural signal. Acoustic stimuli can present with a range of frequencies and amplitudes, and the auditory system can process this by functioning as a frequency analyzer.

The Middle Ear, the Labyrinth, and the Interfaces Between Them

The bony labyrinth is a convoluted space embedded within the temporal bone. A very small amount of fluid can pass very slowly in or out of the labyrinth via various channels (cochlear aqueduct, endolymphatic duct, vestibular aqueduct), but otherwise the total amount of perilymphatic and endolymphatic fluid remains approximately constant. This fluid is effectively incompressible. The round window and the oval window are two "holes," each covered with a thin membrane; these comprise an interface between the inner ear and the middle ear.

The Tympanic Membrane and Ossicular Chain

As sound enters the external auditory canal, the oscillating pressure waves get transduced into the kinetic energy of the eardrum moving inward and outward, and this motion is transmitted through the ossicular chain. At the end of the ossicular chain, the stapes footplate is attached to the membrane covering the oval window. As sounds make the eardrum vibrate, the stapes footplate will push the membrane of the oval window inward and outward in a piston-like motion. Because endolymphatic and perilymphatic fluid is incompressible, any inward movement of the membrane at the round window will move fluid of a certain volume, and that volume must be exactly offset by outward motion somewhere else, which is usually the round window.

The Cochlea

The alternating inward and outward motions of the membrane of the oval window, and the corresponding alternating outward and inward motion of the membrane of the round window, create fluid waves that mechanically stimulate the cochlea.

Cochlear Hair Cells

For each row of type 1 (inner) cochlear hair cells, there are three rows of type 2 (outer) cochlear hair cells.

Deflection of the Basal and Tectorial Membranes

As a fluid wave travels through a section of the cochlea, the upward and downward motion of the parallel basilar and tectorial membranes results in a change in the relative disposition of their surfaces that face each other. In the deflected phase, the tectorial membrane exerts a shearing force on the stereocilia of the type 1 (inner) cochlear hair cells.

During vibration of the basilar membrane, its upward displacement causes the stereocilia of the type 1 (inner) cochlear hair cells to move against the tectorial membrane; as the ste-

M. Cherchi, *Otoneurology and Vestibular Medicine*, https://doi.org/10.1007/978-3-031-94841-1_6

reocilia move toward the modiolus, mechanically gated ion channels open and current flows in. Conversely, during downward movement of the basilar membrane, the stereocilia to move away from the modiolus, thereby closing the ion channels.

Ion Channel Activity in Type 1 Cochlear Hair Cells

The specific ion channel activity is as follows. When the shearing force deflects the stereocilia toward the kinocilia, mechanically gated ion channels permit influx of K^+ at the cell apex from the endolymph, depolarizing the cell membrane. In turn, that depolarization triggers voltage gated channels at the base of the cell to permit influx of Ca^{2+} into the cell, which in turn causes release of synaptic vesicles into the synaptic cleft. When the shearing force reverses direction, the ion channels close, and the cell repolarizes.

The Cochlea's Tonotopic Mapping Makes It a Frequency Analyzer

The spectrum of vibrations in the fluid causes corresponding vibrations in the tonotopically mapped areas of the basilar membrane. Higher frequencies are mediated toward the proximal (stiffer) end of the basilar membrane, while lower frequencies are mediated toward the distal (floppier) end (apex) of the basilar membrane, as shown in Fig. 6.1 from Li and colleagues [1].

Through this process the cochlea transduces the mechanical energy of vibration (from acoustic stimuli) into electrical voltage. By doing this over a range of pitches, it functions (in engineering terms) as a frequency analyzer. The resulting electrical signals correspond to the spectrum of auditory stimulus originally received.

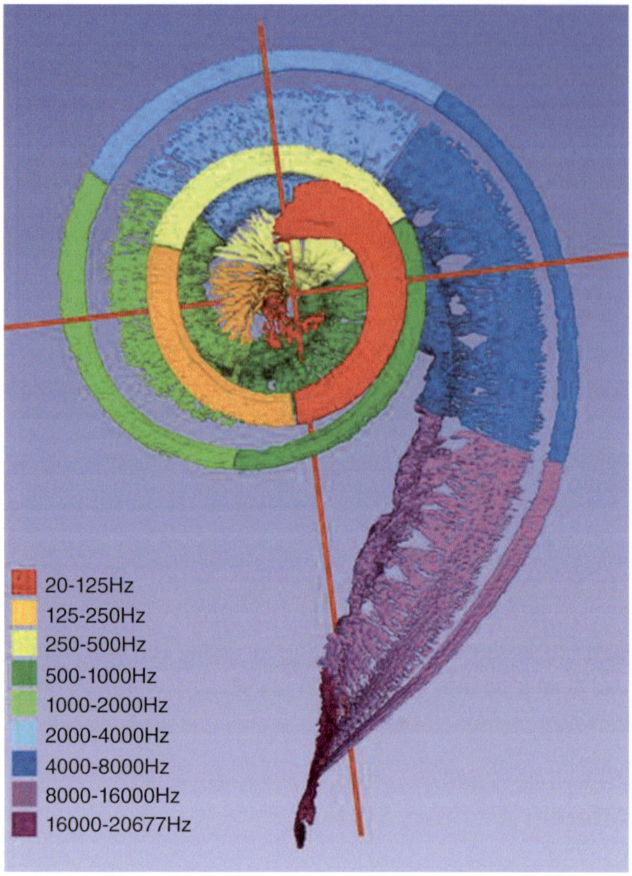

Fig. 6.1 Tonotopic mapping of the cochlea (From Li et al. [1]. Reproduced with written permission from Springer Nature BV)

Cochlear Microphonic, Summation Potential, and Action Potential

As mentioned above, the oscillation of the basilar and tectorial membranes causes an alternating ionic activity of type 1 cochlear hair cells whose frequency parallels that of the auditory stimulus. This alternating current is called the *cochlear microphonic (CM)*.

That same activity (of the inner cochlear hair cells) also induces a *direct* current that reflects the "time-displacement pattern of the cochlear partition in response to the stimulus envelope." This (unidirectional) direct current induces a shift in the cochlear microphonic's electrical baseline, manifesting as the *summation potential (SP)*.

The electrical potential derived from the cochlear microphonic triggers release of neurotransmitters (from the end-organ) into the synaptic cleft between the hair cells and the neurons of the spiral ganglion (the first-order neuron). The post-synaptic potential accumulates and, when it exceeds a specific threshold, triggers depolarization of the post-synaptic neuron. The summed responses of the synchronous firing of several thousand auditory nerve fibers results in an alternating current potential that is termed the *action potential (AP)*. The use of the term "action potential" in this audiologic context has a specific meaning that differs somewhat from what is familiar to neurologists, in that it refers to the *aggregate activity* of thousands of axons, rather than a single axon.

Afferent Auditory Pathways Ascending from the Spiral Ganglion

These signals are transmitted from the primary neurons in the spiral ganglion through the auditory component of the cochleovestibular nerve, and then synapse in the dorsal and ventral cochlear nuclei in the brainstem. From there, some fibers decussate immediately via commissural pathways through the trapezoid body, while other fibers ascend. There are multiple other decussations during this ascent through the afferent auditory system, including at the levels of the nuclei of the lateral lemniscus (via the commissure of Probst), the inferior colliculus (via the commissure of the inferior colliculus) and the auditory cortex (via the corpus callosum). Thus, by the time auditory signals have reached primary auditory cortex, there is significant bilaterality of encoding. (For this reason, a lesion of one primary auditory cortex does not cause deafness, though it can cause a variety of other acoustic perceptual abnormalities.)

A far field electrical potential can be detected for each point in the auditory pathway where a synapse occurs, which is the basis for auditory brainstem evoked responses (ABRs).

The tonotopic arrangement that began in the cochlea is preserved all the way up to the auditory cortex [2].

Ascending Auditory Pathways Terminate in the Primary Auditory Cortex

When these signals reach primary auditory cortex, we experience that neurological activity as the perception of sound. From there, neurons in primary auditory cortex send projections via the arcuate fasciculus to the auditory association cortex. From the auditory association cortex there are projections to many other cortical areas that subserve higher-order functions such as language (e.g., Wernicke's area) and parietal areas that interpret the significance of perceived sounds.

Efferent Auditory System

The efferent auditory system serves several functions, including modulating the activity of the ascending auditory system (through the descending auditory pathways via type 2 cochlear hair cells), and mediating protective acoustic reflexes (the stapedius reflex and the tensor tympani reflex).

Descending Auditory Pathways

Descending pathways from the auditory cortex make reciprocal connections throughout the auditory pathway, forming feedback loops that modulate information arriving from the periphery.

In particular, neurons in the auditory cortex project to the inferior colliculus and synapse there. Neurons in the inferior colliculus in turn project to the lateral and medial periolivary nuclei, and synapse there. Finally, neurons in the periolivary nuclei project to the cochlear nuclei and to the type 2 (outer) cochlear hair cells.

The type 2 (outer) cochlear hair cells serve to modulate the activity of type 1 (inner) cochlear hair cells, effectively serving the function of an auditory equalizer. The activity of type 2 (outer) cochlear hair cells can be assessed by otoacoustic emissions (OAEs).

Protective Reflexes

Protective acoustic reflexes mechanically limit the motion of the ossicular chain in order to protect the cochlea from loud stimuli that might otherwise damage it. The reflexes mechanically limit the ossicular chain at its mechanical "beginning" (dampening the movement of the tympanic membrane via the tensor tympani reflex) and at its mechanical "end" (dampening the movement of the stapes footplate via the stapedial reflex).

Stapedial Reflex

Afferent auditory signals from one cochlear nucleus project to the ipsilateral and contralateral superior olivary nuclei. If that afferent signal is too strong, then the stapedial reflex is triggered. The first-order neuron (whose cell body resides in the superior olivary nucleus) projects to the facial nucleus. The second-order neuron projects through the facial nerve, and the nerve to the stapedius, causing the stapedius muscle to contract, and thereby limit the movement of the stapes footplate on the oval window.

Tensor Tympani Reflex

Each superior olive also sends projections to the ipsilateral trigeminal nucleus. From there, the trigeminal nucleus projects through the trigeminal nerve to the ipsilateral tensor tympani muscle, which contracts and thereby limits the movement of the tympanic membrane.

The pathways for the tensor tympani reflex are depicted in Fig. 6.2, from Niewenhuys and colleagues [3].

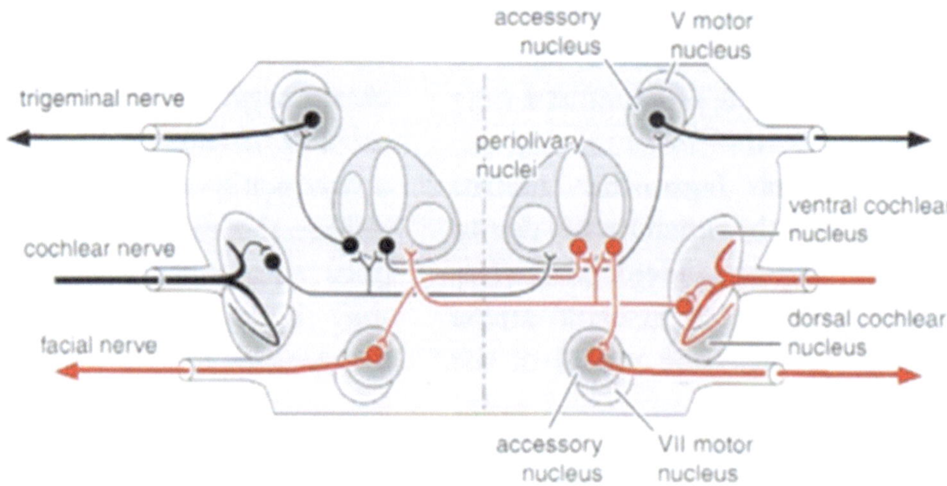

Fig. 6.2 Schematic of the tensor tympani reflex (From Niewenhuys et al. [3]. Reproduced with written permission from Springer Nature BV)

Neurotransmitters in the Auditory System

The auditory system uses a range of excitatory and inhibitory neurotransmitters [3, 4].

- First-order neurons (in the spiral ganglion) project to the cochlear nuclei and release **glutamate** and aspartate (excitatory).
- There are many projections *to* the cochlear nuclei:
  - From the hypothalamus, releasing **histamine**.
  - From the nucleus locus coeruleus, releasing **noradrenaline**.
  - There are also afferent projections that release **dynorphin**.
- There are many projections *to* the inferior colliculus:
  - From the superior olive, releasing **cholecystokinin**.
  - From the nucleus of the lateral lemniscus, releasing dynorphin.
- The olivocochlear bundle contains **enkephalin**.
- The medial superior olive projects ipsilaterally to the dorsal nucleus of the lateral lemniscus and to the inferior colliculus; this is excitatory.
- The lateral superior olive projects bilaterally to the dorsal nucleus of the lateral lemniscus and to the inferior colliculus. The ipsilateral projection is *glycinergic* (inhibitory); the contralateral projection is excitatory.
- The neurons of the nucleus of the trapezoid body release *glycine* (inhibitory).
- Commissural connections between the cochlear nuclei release *glycine* (inhibitory).
- The dorsal nucleus of the lateral lemniscus projects bilaterally to the inferior colliculi and releases *GABA* (inhibitory).
- In the descending auditory pathway, projections from the superior olive to the cochlea release **acetylcholine** (excitatory), **cGRP** (excitatory), and *GABA* (inhibitory).

References

1. Li H, Helpard L, Ekeroot J, Rohani SA, Zhu N, Rask-Andersen H, et al. Three-dimensional tonotopic mapping of the human cochlea based on synchrotron radiation phase-contrast imaging. Sci Rep. 2021;11(1):4437.
2. Saenz M, Langers DR. Tonotopic mapping of human auditory cortex. Hear Res. 2014;307:42–52.
3. Nieuwenhuys R, Voogd J, Huijzen CV. The human central nervous system, vol. xiv. 4th ed. New York: Springer; 2008. p. 967.
4. Haines DE. Fundamental neuroscience. 3rd ed. Philadelphia: Churchill Livingstone; 2006. p. 567.

7 Overview of the Vestibular System

Introduction

The phrase "vestibular system" can have several meanings. In a narrow sense, "vestibular system" refers to the function of the vestibular labyrinth (the inner ear's "gyroscope" function).

In a broader sense, "vestibular system" refers to the arrangement by which multiple sensory inputs (labyrinthine, visual, somatosensory, auditory) are integrated into a coherent perception of orientation and movement, from which motor responses (somatomotor, ocular motor) are orchestrated and executed in order to maintain a desired posture or execute a desired movement. In this section, we will discuss the vestibular system in this broader sense.

Multiple Inputs, Processing, Multiple Outputs

Neurologists tend to apply a "circuitry" approach when thinking about anatomy, physiology and pathology. A simplified circuitry-based schematic of the vestibular system is shown in Fig. 7.1.

This schematic is an oversimplification in many respects. For example, it does not depict feedback loops, nor does it depict the feedforward circuits (efference copies) by which the system compares planned motor output with actual motor output.

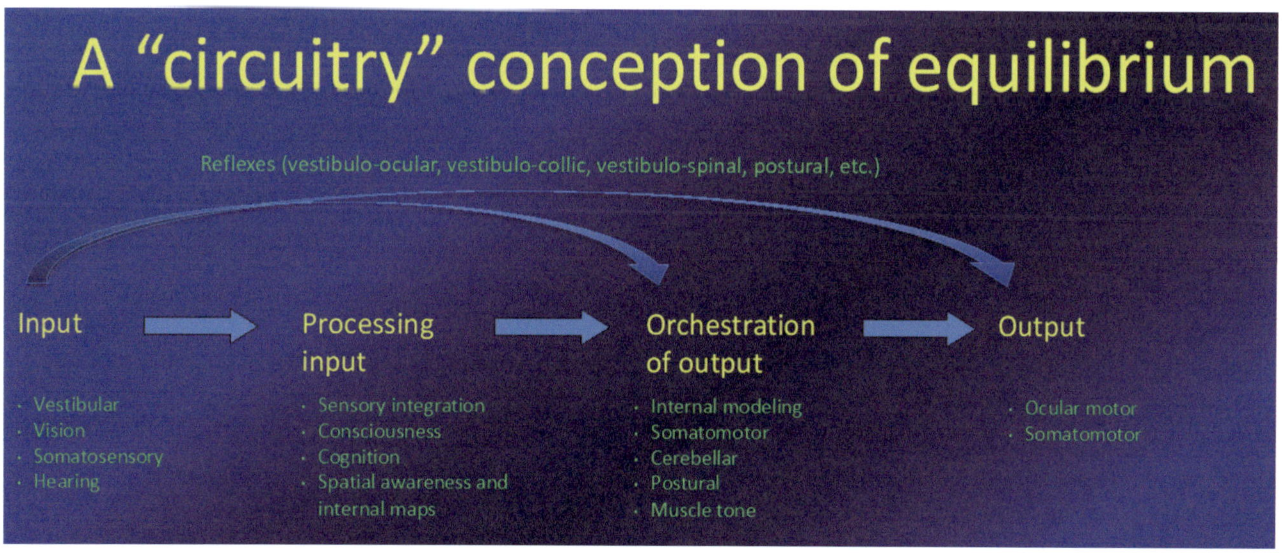

Fig. 7.1 A "circuitry"-based schematic of the vestibular system (By Marcello Cherchi)

M. Cherchi, *Otoneurology and Vestibular Medicine*, https://doi.org/10.1007/978-3-031-94841-1_7

Type 1 Vestibular Hair Cells

The type 1 vestibular hair cells are the fundamental "building blocks" of the labyrinth's gyroscopic capabilities. These cells function as *accelerometers*. They accomplish this task through mechanoelectrical transduction, which is to say that they convert (transduce) the kinetic (mechanical) energy from acceleration into a neurochemical (electrical) signal.

Figure 7. 2, from Soto and colleagues [1], summarizes the sequence through which mechanoelectrical transduction ("MET") occurs.

At baseline, when no acceleration is occurring, each type 1 vestibular hair cell has a tonic (baseline) rate of firing. These cells are morphologically polarized in such a fashion as to make them preferentially sensitive to acceleration in a particular direction; the rate of firing will increase (above the tonic rate) in response to acceleration in that direction, and will decrease (below the tonic rate) in response to acceleration in the opposite direction.

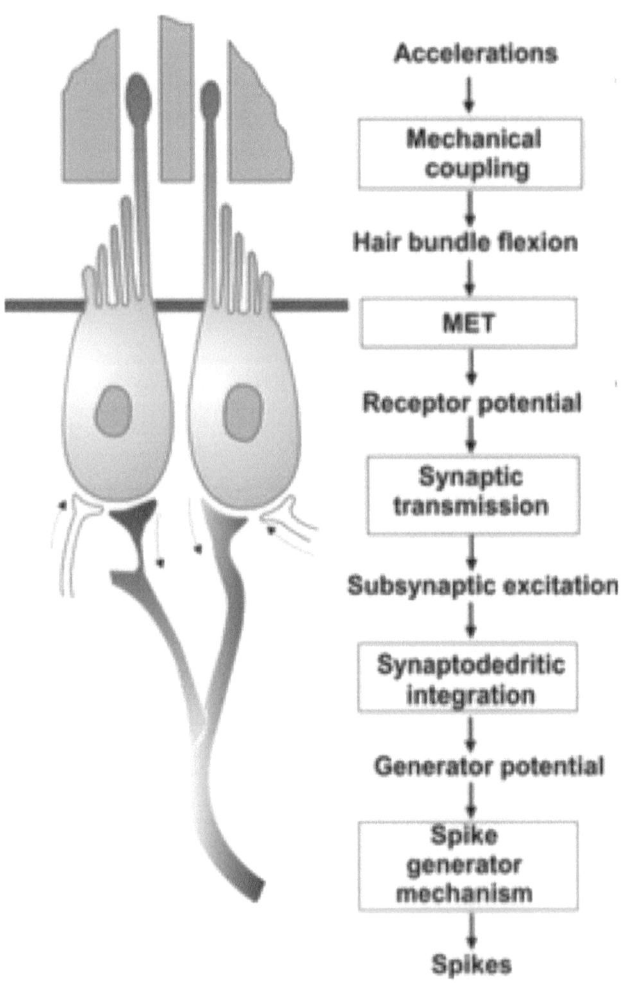

Fig. 7. 2 Summary of the sequence through which the type 1 vestibular hair cells accomplish mechanoelectrical transduction ("MET" in the figure) (From Soto et al. [1]. Reproduced with written permission from IOS Press and from Dr. Enrique Soto)

The morphologic polarization is due to the cellular architecture—specifically, the arrangement of the "hairs" (kinocilium and stereocilia) on each hair cell. Acceleration will deflect these hairs (and we will discuss the mechanics of this shortly); when the deflection is toward the kinocilium one set of chemical processes occurs; when the deflection is away from the kinocilium, a different set of chemical processes occurs.

Vestibular End-Organelles: Semicircular Canals and Otolith Organs

There is one labyrinth on each side of the head, embedded in the temporal bone. Each labyrinth contains five discrete end-organelles, divided into two types:

- Three organelles are the *semicircular canals*, which detect rotational (angular) acceleration. These are as follows:
  - Horizontal semicircular canal, also called the lateral semicircular canal.
  - Anterior semicircular canal, also called the superior semicircular canal.
  - Posterior semicircular canal, sometimes called the inferior semicircular canal.
- Two organelles are the *otolith organs*, which detect translational (linear) acceleration. These are as follows:
  - Utricle.
  - Saccule.

The two types of organelles have different architectures. In the *semicircular canals*, type 1 vestibular hair cells in the ampullary crest extend their hairs (stereocilia and kinocilia) into the gelatinous substance of the cupula. During *rotational acceleration* the inertia of the endolymph in the semicircular canal will deflect the cupula, which will in turn deflect the hairs of the hair cells embedded within it.

In the *otolith organs*, type 1 vestibular hair cells in the macula extend their hairs (stereocilia and kinocilia) into the gelatinous substance of the otolith membrane. Otoliths embedded on the surface of the otolith membrane have a relatively high specific gravity. During *linear acceleration* the inertia that the otoliths impart to the otolith membrane cause the otolith membrane to move (relative to the underlying hair cells) and exert a shearing force on the hairs.

Note that a linear accelerometer cannot distinguish acceleration due to linear movement (such as when a vehicle begins to move) from acceleration due to gravity, as both forces will have the same physical effect. This is illustrated by Fig. 7. 3 from Dakin and Rosenberg [2].

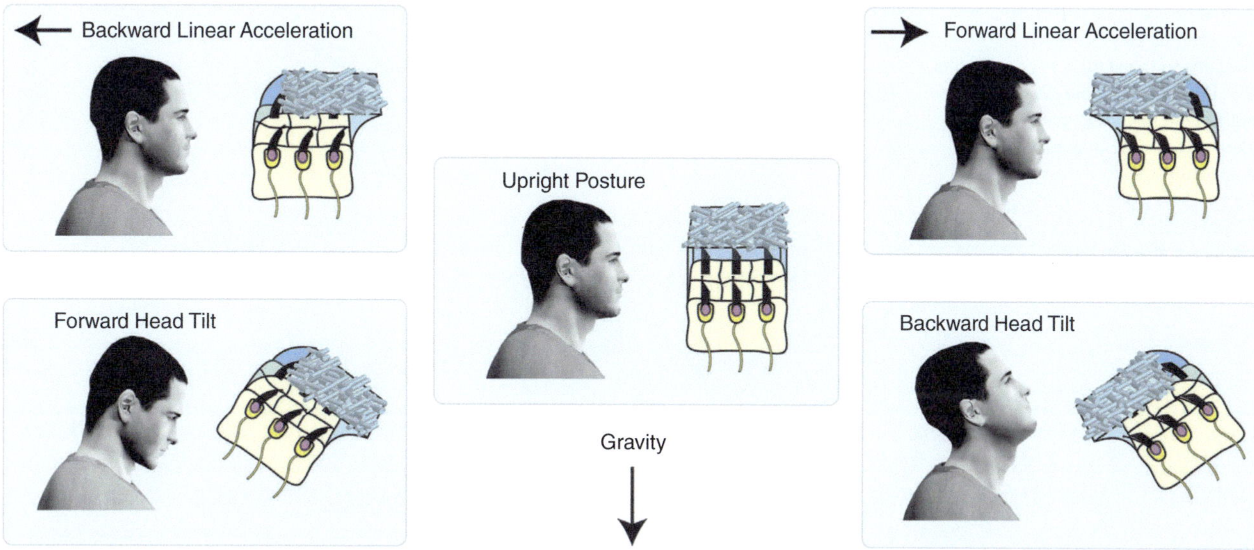

Fig. 7. 3 Einstein's equivalence principle illustrated for the otoliths. During backward linear acceleration (top left), the inertial force acts in the direction opposite to the acceleration, causing the otolithic membrane to lag behind the skull which causes the sensory hair cells to bend. Similarly, tilting the head forward (bottom left) causes the otolithic membrane to sag and the sensory hair cells to bend as they do during backward linear acceleration. The otoliths therefore respond to both linear acceleration of the head and head tilt relative to gravity, and cannot distinguish between the two. The same is true for forward linear accelerations and backward head tilts (right column). Although illustrated here for head pitch, the ambiguity also exists between left–right translations and roll. Graviceptive signals arising from the abdominal viscera suffer from a similar ambiguity (From Dakin and Rosenberg [2]. Reproduced with written permission from Elsevier B.V.)

Vestibular End-Organelles: Contralateral Accelerometers Are Paired in a Push-Pull Relationship

As mentioned earlier, each type 1 vestibular hair cell is morphologically polarized so as to be preferentially sensitive to acceleration in a particular direction. In each end-organelle, collections of type 1 vestibular hair cells are arranged so that the entire zone is preferentially sensitive to acceleration in a particular direction.

A given end-organelle on one side of the head is paired with a contralateral end-organelle such that their preferential directional sensitivities are diametrically opposed. This arrangement means that pairs of end-organelles function in a "push–pull" relationship, in the sense that acceleration (whether rotational or translational) in a given direction will maximally stimulate one end-organelle, and maximally inhibit the corresponding contralateral end-organelle.

The semicircular canals on one side are approximately orthogonal to each other. Semicircular canals that are coplanar are functionally paired. Specifically:

- The horizontal semicircular canals are functionally paired.
- The right anterior canal is functionally paired with the left posterior canal (in the "RALP plane").
- The left anterior canal is functionally paired with the right posterior canal (in the "LARP plane").

These coplanar relationships are depicted in Fig. 7. 4 from Kingma and van de Berg [3].

For a pair of coplanar canals, rotational acceleration in one direction will stimulate one canal and inhibit the other, while rotational acceleration in the other direction will have the opposite effect.

In the anterior and posterior semicircular canals, ampullopetal (toward the ampulla) flow is inhibitory. In the lateral semicircular canals, ampullopetal (toward the ampulla) flow is excitatory. These relationships are depicted in Fig. 7. 5 from Kingma and van de Berg [3].

The otolith organs are also paired, though their morphology is less obvious than that of the semicircular canals, as shown in Fig. 7. 6 from Jaeger and colleagues [4].

The coplanar arrangements of the various zones of the otolith organs are depicted in Fig. 7. 7, from Miller [5].

The saccule detects linear acceleration primarily along the rostro-caudal axis (up and down, "bob"), and partly along the anterior-posterior axis (forward and backward, "surge").

The utricle detects linear acceleration primarily along the interaural axis (side to side, "heave"), and partly along the anterior-posterior axis (forward and backward, "surge").

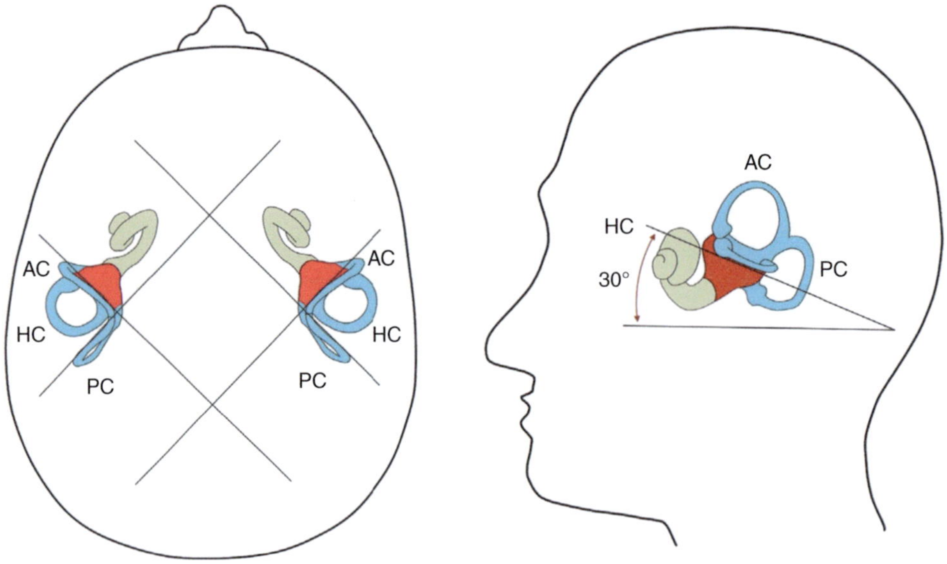

Fig. 7.4 Schematic drawing of the orientation of the two labyrinths in the skull. HC = horizontal (or lateral) canal; PC = posterior (or inferior) canal; AC = anterior (or superior) canal (Reproduced from Kingma and van de Berg [3] with written permission from Elsevier B.V.)

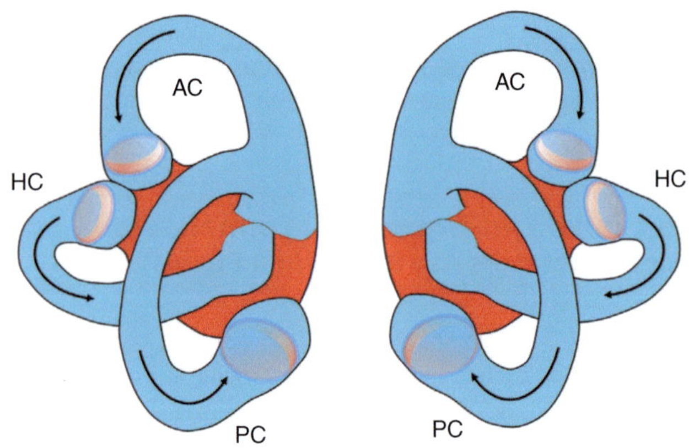

Fig. 7.5 Back view of the two vestibular labyrinths. The arrows indicate the preferred (maximum sensitivity) rotation direction in each canal. HC = horizontal canal; PC = posterior canal; AC = anterior canal. Note that the direction of the sensitivity of the canals is opposite for the HC compared to that of the AC and PC (opposite orientation of the hair cells in the cupulae) (Reproduced from Kingma and van de Berg [3] with written permission from Elsevier B.V.)

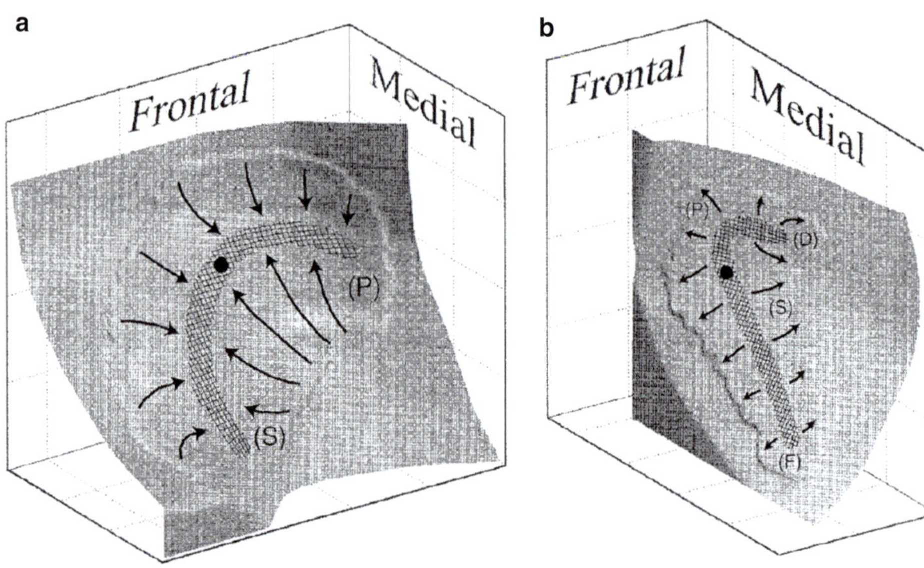

Fig. 7.6 Polarization patterns of utricle (Panel A) and saccule (Panel B), from Jaeger and colleagues [4]. In the modeling study by Jaeger and colleagues, the black dots indicate the locations that were used to calculate simulated displacements; (P) and (S) mark the positions for which polarization during pitch and roll were determined; (F) and D) are the locations considered for the effects of upper layer elasticity (Reproduced with written permission from Elsevier B.V.)

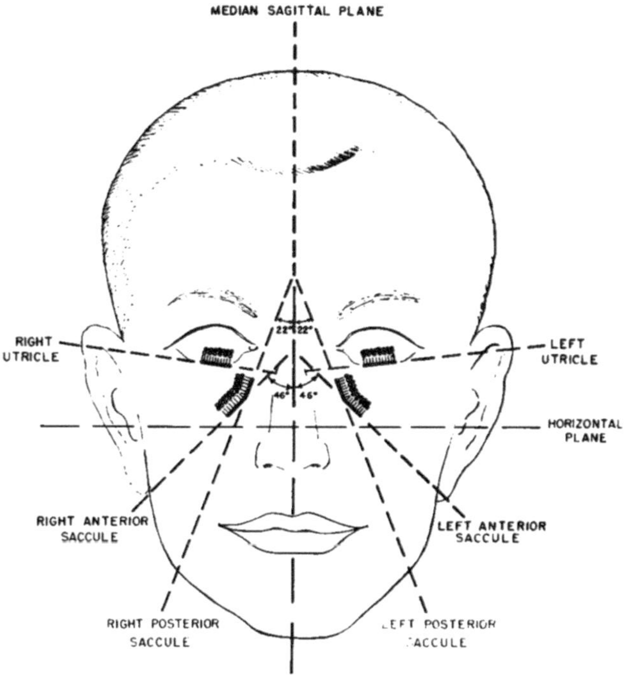

Fig. 7.7 Orientation and planes of the otolith organs (From Miller [5]. Reproduced with written permission from Taylor & Francis, Ltd.)

Neurochemistry of Type 1 Vestibular Hair Cells

At rest, the type 1 vestibular hair cells generate a tonic rate of firing in the first order neuron.

Deflection of the stereocilia *toward* the kinocilium results in opening of K^+ channels at the *apices* of the stereocilia, causing K^+ to enter the cell from the endolymph, depolarizing the cell membrane. This, in turn, causes voltage-gated Ca^{2+} channels at the *base* of the cilia to open, further depolarizing the cell membrane. The influx of Ca^{2+} causes release of synaptic vesicles (containing aspartate or glutamate) into the synaptic cleft, thereby increasing the rate of firing of the first-order neuron.

Deflection of the stereocilia *away from* the kinocilium causes K^+ channels at the *base* of the stereocilia to open, allowing K^+ to exit the cell into the interstitial space, thereby hyperpolarizing the cell membrane and decreasing the rate of firing of the first order neuron.

Pathways Through the Vestibular Nerve

The first order neurons in the vestibular system are located in the vestibular (Scarpa's) ganglion. These are bipolar neurons that receive input from type 1 vestibular hair cells, and project their axons through the vestibular nerve, which has superior and inferior divisions. In summary:

- The superior division of the vestibular nerve transmits afferents from:
  - The anterior semicircular canal
  - The horizontal semicircular canal
  - The utricle
  - Part of the saccule
- The inferior division of the vestibular nerve transmits afferents from:
  - The posterior semicircular canal
  - Part of the saccule

The neuroanatomy is somewhat more complex than what is shown here, due to the presence of other pathways and anastomoses [6].

Afferent Vestibular Projections

After traversing the vestibular nerve, most projections from first order vestibular neurons reach targets in the vestibular nuclei.

A minority of fibers project directly (without synapsing) through the brainstem via the juxtarestiform body into the cerebellum [7]. In fact, the vestibular labyrinth is the only sensory organ in the body that sends direct primary afferent projections to the cerebellar cortex and nuclei [8].

Vestibular Nuclei

The majority of first order vestibular neurons project to vestibular nuclei in the brainstem [9]. The distribution is as follows [8]:

- Afferents from the semicircular canals arrive:
  - Mostly in the superior and medial vestibular nuclei
  - Less in the lateral and inferior vestibular nuclei
- Afferents from the otolith organs (utricle and saccule) arrive:
  - Mostly in the lateral, medial and inferior vestibular nuclei.
  - Some afferents from the saccule arrive in "cell group Y"

It should be noted that the vestibular nuclei receive projections from multiple other structures as well, including the following:

- Cerebral cortex
- Interstitial nucleus of Cajal
- Nucleus prepositus hypoglossi
- Central cervical nucleus
- Dorsal spino-cerebellar tract
- Several areas of the cerebellum

Efferent Projections from the Vestibular Nuclei

Efferent projections from the vestibular nuclei project broadly, with major targets in [8]:

- Oculomotor nuclei
- Vestibulocerebellum
- Contralateral vestibular nuclei
- Spinal cord
- Reticular formation
- Thalamus

Efferent Projections from Vestibular Nuclei to Ocular Motor Nuclei

The vestibular nuclei have substantial projections to the ocular motor nuclei. Among other functions, these projections mediate the vestibulo-ocular reflexes that stabilize the eyes (help maintain foveation of a target) despite the perturbations from head movements. The projections of the vestibulo-ocular reflexes are illustrated in detail in works such as those by Leigh and Zee [10] and Niewenhuys [11].

Vestibular information contributes not only to the vestibulo-ocular reflex, but to other eye movements as well. Slow eye movements (such as smooth pursuit, the slow phase of optokinetic responses, and the slow phase of saccades) rely heavily on vestibular input (either directly from the vestibular nuclei, or indirectly via the cerebellum).

Efferent Projections from Vestibular Nuclei to Cerebellum

The vestibular nuclei also have substantial projections to various areas of the cerebellum, including the following [8]:

- Primary vestibulocerebellar fibers project to the following:
  - Dentate nucleus
  - Nodulus
  - Uvula
  - Flocculus
- Secondary vestibulocerebellar fibers project to the following:
  - Flocculonodular lobe
  - Uvula
  - Parts of the paraflocculus
  - Fastigial nuclei
  - Dentate nuclei

Efferent Projections from Vestibular Nuclei to Contralateral Vestibular Nuclei

The vestibular nuclei on one side have commissural projections to the contralateral vestibular nuclei, some of which are excitatory, others of which are inhibitory [11]. These commissural projections appear to serve at least two purposes:

- They probably comprise the neural substrate for the velocity storage mechanism [12, 13].
- They probably facilitate central vestibular compensation during recovery from a vestibular injury [8].

Efferent Projections from Vestibular Nuclei to Spinal Cord

Projections from the vestibular nuclei reach the spinal cord via the medial and lateral vestibulospinal tracts. These pathways influence muscle tone and mediate reflexive postural adjustments of the head, trunk and limbs. These projections contain both excitatory and inhibitory projections [11].

Neurons mostly in the lateral and inferior vestibular nuclei project through the lateral vestibulospinal tract, and terminate on alpha and gamma motor neurons in the ipsilateral spinal cord, releasing acetylcholine or glutamate.

Neurons in the medial vestibular nuclei (less in the inferior and lateral vestibular nuclei) project through the medial vestibulospinal tract. Some of these efferent projections remain ipsilateral, while others decussate. These projections provide both excitatory and inhibitory influence on cervical flexor and extensor motor neurons, thereby mediating the vestibulocollic reflexes.

Efferent Projections from Vestibular Nuclei to Thalamus

Information about position and motion reaches consciousness through the vestibulo-thalamo-cortical network. Efferent projections from the superior, lateral and inferior vestibular nuclei on one side project bilaterally to two areas in the thalamus: the ventral posterolateral nucleus (VPL), and the ventral posteroinferior nucleus (VPI). Efferent projections from these vestibular nuclei on one side project ipsilaterally to two areas in the temporal lobe (in the base of the intraparietal sulcus and the base of the central sulcus) that comprise the vestibular cortex, which mediates conscious perception of vestibular stimuli.

Type 2 Vestibular Hair Cells

In the auditory system, type 1 cochlear hair cells perform mechanoelectrical transduction from the kinetic energy of oscillating pressure (i.e., sound) into afferent neurochemical signals. Type 2 cochlear hair cells, controlled by descending auditory pathways, modulate the activity of type 1 hair cells, functioning as a type of auditory equalizer.

The auditory and vestibular systems are phylogenetically related [14], so it is unsurprising that the vestibular system also has type 2 vestibular hair cells, controlled by descending vestibular pathways.

Descending vestibular pathways originate primarily from the vestibular nuclei, but there also appear to be contributions from the reticular formation and perhaps other areas. These pathways project to the labyrinth, where they synapse mostly on type 2 vestibular hair cells, and less on type 1 vestibular hair cells.

In some species, the type 2 vestibular hair cells can modulate the activity of type 1 vestibular hair cells in "both directions" (i.e., either augment or diminish the level of activity), analogous to the function of type 2 cochlear hair cells. The type 2 vestibular hair cells are incompletely understood in mammals, but they appear to serve primarily or exclusively an excitatory function [15].

Neurotransmitters in the Vestibular System

A broad array of neurotransmitters is used in the vestibular system, some of which are depicted in Figs. 7. 8 and 7. 9, from Soto and colleagues [1].

A nonexhaustive list of neurotransmitters in the vestibular system is provided below.

- Cerebellar corticovestibular fibers, project to several areas in the vestibular system and, like all output from the cerebellum, are inhibitory and release **GABA** [7].
- Hypothalamic projections to the medial vestibular nucleus release *histamine* [7].
- Other projections to the vestibular nuclei release *dynorphin* [7].
- Vestibular projections to the cerebellum are mostly *cholinergic* [11].
- The nucleus prepositus hypoglossi sends projections to several nuclei in the superior olive where they release GABA and *acetylcholine* [11].
- Efferent projections from the vestibular nuclei through the lateral vestibulospinal tract that terminate on alpha and gamma motor neurons release *glutamate* and *acetylcholine* [8] (Haines 2006, p. 362).
- Among *descending* vestibular pathways:

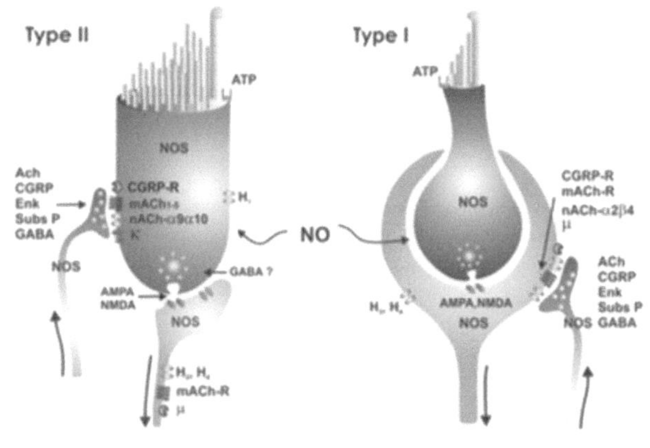

Fig. 7. 8 Neurotransmitters at synapses on vestibular hair cells. See text for details. (From Soto et al. [1]. Reproduced with written permission from IOS Press and from Dr. Enrique Soto)

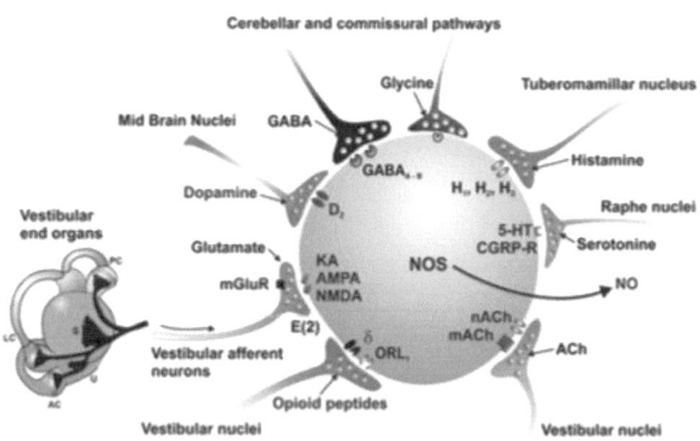

Fig. 7. 9 Neurotransmitters at synapses in the vestibular nuclei. See text for details. (From Soto et al. [1]. Reproduced with written permission from IOS Press and from Dr. Enrique Soto)

  - Animal studies show that these terminate in the labyrinths and release *acetyltransferase* and *cGRP* [11].
  - These have a predominantly excitatory influence of the afferent activity in the vestibular nerve and extend the dynamic range of the afferents during the large accelerations accompanying voluntary head movements [11].

References

1. Soto E, Vega R, Sesena E. Neuropharmacological basis of vestibular system disorder treatment. J Vestib Res Equilib Orient. 2013;23(3):119–37.
2. Dakin CJ, Rosenberg A. Gravity estimation and verticality perception. Handb Clin Neurol. 2018;159:43–59.
3. Kingma H, van de Berg R. Anatomy, physiology, and physics of the peripheral vestibular system. Handb Clin Neurol. 2016;137:1–16.
4. Jaeger R, Takagi A, Haslwanter T. Modeling the relation between head orientations and otolith responses in humans. Hear Res. 2002;173(1-2):29–42.

5. Miller EF 2nd. Counterrolling of the human eyes produced by head tilt with respect to gravity. Acta Otolaryngol. 1962;54:479–501.
6. Cherchi M, Yacovino DA. Histology and neuroanatomy suggest a unified mechanism to explain the distribution of lesion patterns in acute vestibular neuropathy. Exp Brain Res. 2021;239:1395–9.
7. Haines DE. Neuroanatomy: an atlas of structures, sections, and systems. 6th ed. Lippincott, Williams and Wilkins; 2004.
8. Haines DE. Fundamental neuroscience. 3rd ed. Philadelphia: Churchill Livingstone; 2006. 567 p
9. Baloh RW. Dizziness, hearing loss, and tinnitus, vol. xv. Philadelphia: F.A. Davis Co.; 1998. 224 p
10. Leigh RJ, Zee DS. The neurology of eye movements, vol. xx. 5th ed. Oxford/New York: Oxford University Press; 2015. 1109 p
11. Nieuwenhuys R, Voogd J, Huijzen CV. The human central nervous system, vol. xiv. 4th ed. New York: Springer; 2008. 967 p
12. Bhansali SA, Stockwell CW, Bojrab DI. Oscillopsia in patients with loss of vestibular function. Otolaryngol Head Neck Surg. 1993;109(1):120–5.
13. Baloh RW, Kerber KA. Clinical neurophysiology of the vestibular system, vol. xxi. 4th ed. New York: Oxford University Press; 2011. 455 p
14. Manley GA. Cochlear mechanisms from a phylogenetic viewpoint. Proc Natl Acad Sci U S A. 2000;97(22):11736–43.
15. Schneider GT, Lee C, Sinha AK, Jordan PM, Holt JC. The mammalian efferent vestibular system utilizes cholinergic mechanisms to excite primary vestibular afferents. Sci Rep. 2021;11(1):1231.

Nystagmus

8

Nystagmus

Identification and characterization of eye movements comprise a crucial diagnostic tool in the fields of otoneurology, neuro-otology, and neuro-ophthalmology. The interested reader is directed to the foundational compendium by Leigh and Zee [1], an invaluable reference work regarding the neurology, otology, physiology, basic science, and control systems analysis of eye movements. Leigh and Zee [1] organize eye movements into the following broad categories (Table 8.1).

The term "nystagmus" will be used very frequently in our discussions of diagnoses and testing. The term "nystagmus" has been defined in various ways [2] and applied as a general term to a broad variety of eye movements, both normal (physiologic) and abnormal (pathologic). The Barany Society consensus document on this topic [2] offers the following definition:

Table 8.1 Categories of eye movements

Class of eye movement	Main function
Vestibular	Holds images of the visual world steady on the retina during brief head rotations or translations
Visual fixation	Holds the image of a stationary object on the fovea by minimizing ocular drifts
Optokinetic	Holds images of the visual world steady on the retina during sustained head rotation or translation
Smooth pursuit	Holds the image of a small moving target on the fovea; or holds the image of a small target on the retina that is close to the head, during linear motion with; with optokinetic responses, aids gaze stabilization during sustained head rotation
Nystagmus quick phases	Reset the eyes during prolonged rotation and direct gaze toward the ongoing visual scene
Saccades	Bring images of objects of interest onto the fovea
Vergence	Moves the eyes in opposite directions so that images of a single object are placed or held simultaneously on the fovea of each eye

From Leigh and Zee [1]

> *Nystagmus* is an involuntary, rapid, rhythmic, oscillatory eye movement with at least one slow phase. *Jerk nystagmus* is nystagmus with a slow phase and a fast phase. *Pendular nystagmus* is nystagmus with only slow phases. [2]

Eggers and colleagues [2] state that an adequate characterization of nystagmus should include the following (Table 8.2):

Ideally, all of these features of an eye movement should be noted. However, in clinical practice, one may see documentation of only the salient features (abnormal findings, pertinent normal findings).

Table 8.2 Features of eye movements

Characteristic	Notes
Trajectory	Axis or plane of rotation and direction (horizontal, vertical, torsional components)
Binocularity	Is the eye movement binocular or monocular?
Conjugacy	Is the eye movement conjugate or dysconjugate?
Velocity	If the nystagmus includes both fast and slow phases, velocity refers to the slow phase
Waveform	Pendular (only slow phases) or jerk (alternating fast and slow phases)
Frequency	If there is a repetitive pattern, how many cycles per second?
Intensity	A qualitative combination of amplitude and frequency
Effect of direction of gaze	Does eccentric gaze modulate the eye movement?
Effect of convergence	Does convergence modulate the eye movement?
Effect of vision availability	Can the eye movement be modulated by permitting versus blocking visual fixation?
Effect of provocative maneuvers	Such as neck vibration, head shaking, hyperventilation, Valsalva
Temporal profile	Is the eye movement intermittent, continuous, or changing over time
Initial appearance	Patient's age at the first appearance of the eye movement

From Eggers et al. [2]

M. Cherchi, *Otoneurology and Vestibular Medicine*, https://doi.org/10.1007/978-3-031-94841-1_8

References

1. Leigh RJ, Zee DS. The neurology of eye movements, vol. xx. 5th ed. Oxford/New York: Oxford University Press; 2015. 1109 p
2. Eggers SDZ, Bisdorff A, von Brevern M, Zee DS, Kim JS, Perez-Fernandez N, et al. Classification of vestibular signs and examination techniques: nystagmus and nystagmus-like movements. J Vestib Res Equilib Orient. 2019;29(2–3):57–87.

9 Vestibulo-ocular Reflex

Physiology and Neuroanatomy

The function of the vestibulo-ocular reflex (VOR) is to stabilize an image on the retina despite rotational and/or translational movements of the head [1]. The VOR consists of a the rotational VOR (which detects angular acceleration) and the translational VOR (which detects linear acceleration). The rotational VOR (rVOR) can be assessed to some degree with bedside testing, and in greater detail with instrumented otovestibular testing. The translational VOR (tVOR) is more difficult to assess and usually requires instrumentation in a special research laboratory.

Figure 9.1, from Zee [2], illustrates rotational and translational movements.

These are summarized in Table 9.1.

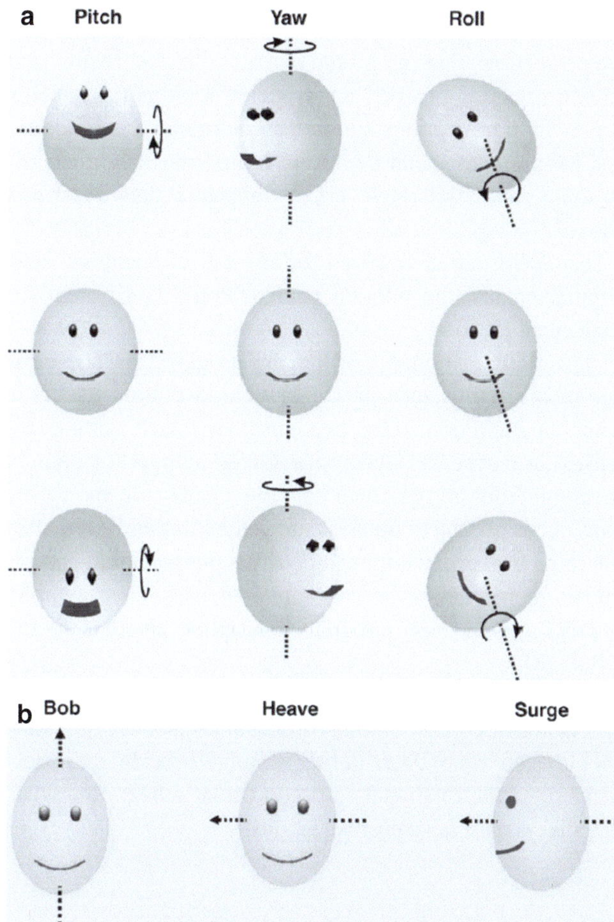

Fig. 9.1 Axes of translation and rotation (From Zee [2]. The columns in Panel (**a**) illustrate rotational movements (pitch, yaw and roll). Panel (**b**) illustrates translational movements (bob, heave and surge). Reproduced with written permission from Wolters Kluwer Health, Inc., and from Dr. Aasifh Shaikh)

The Rotational Vestibulo-ocular Reflex (rVOR)

Ideally, the rotational VOR (rVOR) should rotate the eyes within the orbits by an amount that is equal in magnitude, but opposite in direction, to that of the head, thereby offsetting the head movement and maintaining the point of regard stable on the fovea (in other words, the gain of eye rotation to head rotation is 1.0). The rVOR responds to *pitch* (rotation around the interaural axis), *yaw* (rotation around the rostro-caudal axis), and *roll* (rotation around the naso-occipital axis).

The neuroanatomy of the rVOR is well understood [3]. During rotational acceleration of the head toward one side, there is excitation of the ipsilateral lateral semicircular canal and inhibition of the contralateral lateral semicircular canal. Afferent vestibular excitation from a given lateral semicircular canal passes to the vestibular (Scarpa's) ganglion, through the superior division of the vestibular nerve and into the ipsilateral medial vestibular nucleus (MVN) situated in the mid dorsal portion of the lower pons and upper medulla. From the MVN, there are some projections through the ascending tract of Dieters to the ipsilateral oculomotor nucleus which stimulates contraction of the medial rectus of the ipsilateral eye; there are other projections to the contralateral abducens nucleus which in turn stimulates

M. Cherchi, *Otoneurology and Vestibular Medicine*, https://doi.org/10.1007/978-3-031-94841-1_9

Table 9.1 Axes of translation and rotation

	Rotation around axis (angular movement) When *rotational* acceleration *around* this axis occurs, the ocular motor responses are driven by the *canal-ocular reflex*	*Translation* along axis (linear movement) When *translational* acceleration *along* this axis occurs, the ocular motor responses are driven by the *otolith-ocular reflex*
Interaural axis ("left and right")	Pitch	Heave
Rostro-caudal axis ("up and down")	Yaw	Bob
Naso-occipital axis ("fore and aft")	Roll	Surge

contraction of the lateral rectus of the contralateral eye. In addition, from the MVN, there are projections to the ipsilateral abducens nucleus; from the abducens nucleus, there are projections that inhibit the ipsilateral lateral rectus, and other projections to the contralateral oculomotor nucleus from which projections inhibit the contralateral medial rectus. Afferent vestibular inhibition from a given lateral semicircular canal will have effects opposite that described above.

The VOR can be assessed for any pair of coplanar semicircular canals; the relevant neuroanatomy is different for each canal [3, 4].

The VOR can adapt by altering its gain [5]. The most common circumstance in which such adaptation occurs is when a patient dons refractive lenses (spectacles or contact lenses) to correct for hyperopia (farsightedness) or myopia (nearsightedness) [6]. Such adaptive changes in the rVOR gain occur relatively quickly and can be durable even after the refraction is discontinued [7]. This phenomenon applies mostly to spectacles, because contact lenses move with the eyeball and thus do not induce adaptive changes to the rVOR [8].

Because of this, if a patient habitually uses lenses (whether spectacles or contact lenses) for visual correction, then they should also wear them during testing (bedside or instrumented) of the rotational vestibulo-ocular reflex; otherwise, one may obtain false positive results.

What Tests Can Assess the Rotational Vestibulo-ocular Reflex?

On *bedside testing*, the middle frequency range of the vestibular tuning spectrum can be evaluated with the ophthalmoscope test and with dynamic visual acuity testing, while the high frequency range can be tested with head impulse testing.

On *instrumented otovestibular testing*, the low-frequency range of the vestibular tuning spectrum can be evaluated with caloric testing, the middle-frequency range can be evaluated with rotatory chair testing, and the high-frequency range can be evaluated with video head impulse testing.

The Translational Vestibulo-ocular Reflex (tVOR)

The *translational vestibulo-ocular reflex (tVOR)*, also called the linear VOR or otolith-ocular reflex, maintains an image stable on the retina despite translational movements of the head, specifically, linear acceleration along the *interaural (heave) axis* [9–16], *naso-occipital (surge) axis* [17–19], *rostro-caudal (bob) axis* [20–24], or oblique axes [25]. The tVOR is mediated by the otolith system, which feeds into the neural integrator [26] and can be modulated by viewing distance [10, 21, 27–31] and other visual influences [29, 32–35], and by somatosensory inputs [22]. The tVOR can be adversely affected by various pathologies such as unilateral vestibular weakness [36–39], bilateral vestibular weakness [38], cerebellar disorders [40–42], and skew deviation [43]. Many natural head movements comprise a mix of rotational and translational movements, thereby activating rVOR and tVOR simultaneously [29, 42, 44–54].

What Tests Can Assess the Translational Vestibulo-ocular Reflex?

The tVOR is more difficult to study than the rVOR. Accurate study requires specialized equipment largely restricted to research laboratories, such as human centrifuges, Moog platforms, parallel swings, and linear accelerating sleds.

Although the ocular motor responses of the tVOR are harder to study, it is still possible to assess the otolith system using cervical vestibular evoked myogenic potentials (cVEMPs) and ocular vestibular evoked myogenic potentials (oVEMPs).

Diseases That Affect the Vestibulo-ocular Reflex

Any lesion affecting any of the neuroanatomical pathways described earlier can cause vestibular weakness and thereby influence the vestibulo-ocular reflex—and here, we use the term "lesion" loosely, since not only focal lesions, but also systemic issues (such as pharmacologic vestibular suppression) can cause such affectation.

References

1. Rinaudo CN, Schubert MC, Figtree WVC, Todd CJ, Migliaccio AA. Human vestibulo-ocular reflex adaptation is frequency selective. J Neurophysiol. 2019;122(3):984–93.
2. Zee DS. Pathophysiology of vestibular symptoms and signs: the clinical examination. Continuum. 2006;12(4):13–32.
3. Bronstein AM, Patel M, Arshad Q. A brief review of the clinical anatomy of the vestibular-ocular connections-how much do we know? Eye (Lond). 2015;29(2):163–70.
4. Leigh RJ, Zee DS. The neurology of eye movements, vol. xx. 5th ed. Oxford; New York: Oxford University Press; 2015. 1109 p
5. Watanabe S, Hattori K, Koizuka I. Flexibility of vestibulo-ocular reflex adaptation to modified visual input in human. Auris Nasus Larynx. 2003;30(Suppl):S29–34.
6. Cannon SC, Leigh RJ, Zee DS, Abel LA. The effect of the rotational magnification of corrective spectacles on the quantitative evaluation of the VOR. Acta Otolaryngol. 1985;100(1–2):81–8.
7. Mahfuz MM, Schubert MC, Figtree WVC, Migliaccio AA. Retinal image slip must pass the threshold for human vestibulo-ocular reflex adaptation. J Assoc Res Otolaryngol. 2020;21(3):277–85.
8. Eggers SD, Zee DS. Evaluating the dizzy patient: bedside examination and laboratory assessment of the vestibular system. Semin Neurol. 2003;23(1):47–58.
9. Crane BT, Tian J, Wiest G, Demer JL. Initiation of the human heave linear vestibulo-ocular reflex. Exp Brain Res. 2003;148(2):247–55.
10. King S, Benoit C, Bandealy N, Karmali F. The influence of target distance on perceptual self-motion thresholds and the vestibulo-ocular reflex during interaural translation. Prog Brain Res. 2019;248:197–208.
11. Ramat S, Straumann D, Zee DS. Interaural translational VOR: suppression, enhancement, and cognitive control. J Neurophysiol. 2005;94(4):2391–402.
12. Ramat S, Zee DS. Translational VOR responses to abrupt interaural accelerations in normal humans. Ann N Y Acad Sci. 2002;956:551–4.
13. Ramat S, Zee DS. Ocular motor responses to abrupt interaural head translation in normal humans. J Neurophysiol. 2003;90(2):887–902.
14. Tian JR, Crane BT, Demer JL. Vestibular catch-up saccades augmenting the human transient heave linear vestibulo-ocular reflex. Exp Brain Res. 2003;151(4):435–45.
15. Tian JR, Crane BT, Wiest G, Demer JL. Effect of aging on the human initial interaural linear vestibulo-ocular reflex. Exp Brain Res. 2002;145(2):142–9.
16. Walker MF, Shelhamer M, Zee DS. Eye-position dependence of torsional velocity during interaural translation, horizontal pursuit, and yaw-axis rotation in humans. Vis Res. 2004;44(6):613–20.
17. Seidman SH, Paige GD, Tomko DL. Adaptive plasticity in the naso-occipital linear vestibulo-ocular reflex. Exp Brain Res. 1999;125(4):485–94.
18. Semrau JA, Wei M, Angelaki DE. Scaling of the fore-aft vestibulo-ocular reflex by eye position during smooth pursuit. J Neurophysiol. 2006;96(2):936–40.
19. Tian JR, Mokuno E, Demer JL. Vestibulo-ocular reflex to transient surge translation: complex geometric response ablated by normal aging. J Neurophysiol. 2006;95(4):2042–54.
20. Bush GA, Miles FA. Short-latency compensatory eye movements associated with a brief period of free fall. Exp Brain Res. 1996;108(2):337–40.
21. Cheng RC, Walker MF. Vertical head translation impairs dynamic visual acuity during near viewing. J Vestib Res Equilib Orient. 2016;26(5–6):417–23.
22. Kutz DF, Kolb FP, Glasauer S, Straka H. Somatosensory Influence on Platform-Induced Translational Vestibulo-Ocular Reflex in Vertical Direction in Humans. Frontiers in neurology. 2020;11:332.
23. Liao K, Walker MF, Joshi A, Reschke M, Leigh RJ. Vestibulo-ocular responses to vertical translation in normal human subjects. Exp Brain Res. 2008;185(4):553–62.
24. Liao K, Walker MF, Joshi A, Reschke M, Strupp M, Leigh RJ. The human vertical translational vestibulo-ocular reflex. Normal and abnormal responses. Ann N Y Acad Sci. 2009;1164:68–75.
25. Tomko DL, Paige GD. Linear vestibuloocular reflex during motion along axes between nasooccipital and interaural. Ann N Y Acad Sci. 1992;656:233–41.
26. Hegemann S, Shelhamer M, Kramer PD, Zee DS. Adaptation of the phase of the human linear vestibulo-ocular reflex (LVOR) and effects on the oculomotor neural integrator. J Vestib Res Equilib Orient. 2000;10(4–5):239–47.
27. Busettini C, Miles FA, Schwarz U. Ocular responses to translation and their dependence on viewing distance. II. Motion of the scene. J Neurophysiol. 1991;66(3):865–78.
28. Busettini C, Miles FA, Schwarz U, Carl JR. Human ocular responses to translation of the observer and of the scene: dependence on viewing distance. Exp Brain Res. 1994;100(3):484–94.
29. Paige GD, Telford L, Seidman SH, Barnes GR. Human vestibuloocular reflex and its interactions with vision and fixation distance during linear and angular head movement. J Neurophysiol. 1998;80(5):2391–404.
30. Schwarz U, Miles FA. Ocular responses to translation and their dependence on viewing distance. I. Motion of the observer. J Neurophysiol. 1991;66(3):851–64.
31. Wei M, Angelaki DE. Viewing distance dependence of the vestibulo-ocular reflex during translation: extra-otolith influences. Vis Res. 2004;44(9):933–42.
32. Dits J, King WM, van der Steen J. Scaling of compensatory eye movements during translations: virtual versus real depth. Neuroscience. 2013;246:73–81.
33. Gianna CC, Gresty MA, Bronstein AM. The human linear vestibulo-ocular reflex to transient accelerations: visual modulation of suppression and enhancement. J Vestib Res Equilib Orient. 2000;10(4–5):227–38.
34. Hashiba M, Watanabe S, Watabe H, Matsuoka T, Baba S, Wada Y, et al. Imaginary gaze effects on eye movements induced by linear acceleration: involvement of vestibular induced smooth pursuit eye movement. Acta oto-laryngologica Supplementum. 1995;520(Pt 2):372–6.
35. Liao K, Schneider RM, Yaniglos SS, Bertolini G, Glendinning P, Sawyer RN Jr, et al. Visual and vestibular determinants of the translational vestibulo-ocular reflex. Ann N Y Acad Sci. 2011;1233:263–70.
36. Crane BT, Tian JR, Ishiyama A, Demer JL. Initiation and cancellation of the human heave linear vestibulo-ocular reflex after unilateral vestibular deafferentation. Exp Brain Res. 2005;161(4):519–26.

37. Lempert T, Gianna C, Brookes G, Bronstein A, Gresty M. Horizontal otolith-ocular responses in humans after unilateral vestibular deafferentation. Exp Brain Res. 1998;118(4):533–40.
38. Lempert T, Gresty MA, Bronstein AM. Horizontal linear vestibulo-ocular reflex testing in patients with peripheral vestibular disorders. Ann N Y Acad Sci. 1999;871:232–47.
39. Tian JR, Ishiyama A, Demer JL. Effect of unilateral vestibular deafferentation on the initial human vestibulo-ocular reflex to surge translation. Exp Brain Res. 2007;176(4):575–87.
40. Anastasopoulos D, Haslwanter T, Fetter M, Dichgans J. Smooth pursuit eye movements and otolith-ocular responses are differently impaired in cerebellar ataxia. Brain J Neurol. 1998;121(Pt 8):1497–505.
41. Baloh RW, Yue Q, Demer JL. The linear vestibulo-ocular reflex in normal subjects and patients with vestibular and cerebellar lesions. J Vestib Res Equilib Orient. 1995;5(5):349–61.
42. Crane BT, Tian JR, Demer JL. Initial vestibulo-ocular reflex during transient angular and linear acceleration in human cerebellar dysfunction. Exp Brain Res. 2000;130(4):486–96.
43. Schlenker M, Mirabella G, Goltz HC, Kessler P, Blakeman AW, Wong AM. The linear vestibulo-ocular reflex in patients with skew deviation. Invest Ophthalmol Vis Sci. 2009;50(1):168–74.
44. Anastasopoulos D, Gianna CC, Bronstein AM, Gresty MA. Interaction of linear and angular vestibulo-ocular reflexes of human subjects in response to transient motion. Exp Brain Res. 1996;110(3):465–72.
45. Angelaki DE, Green AM, Dickman JD. Differential sensorimotor processing of vestibulo-ocular signals during rotation and translation. J Neurosci. 2001;21(11):3968–85.
46. Branoner F, Straka H. Semicircular canal influences on the developmental tuning of the translational vestibulo-ocular reflex. Front Neurol. 2018;9:404.
47. Crane BT, Demer JL. Human gaze stabilization during natural activities: translation, rotation, magnification, and target distance effects. J Neurophysiol. 1997;78(4):2129–44.
48. Crane BT, Demer JL. A linear canal-otolith interaction model to describe the human vestibulo-ocular reflex. Biol Cybern. 1999;81(2):109–18.
49. Koizuka I, Katsumi N, Hattori K, Okada T, Kato I. Effect of adaptive plasticity of linear vestibulo-ocular reflex upon angular vestibulo-ocular reflex. Auris Nasus Larynx. 2000;27(2):89–93.
50. Laurens J, Straumann D, Hess BJ. Processing of angular motion and gravity information through an internal model. J Neurophysiol. 2010;104(3):1370–81.
51. McCrea RA, Chen-Huang C. Signal processing related to the vestibulo-ocular reflex during combined angular rotation and linear translation of the head. Ann N Y Acad Sci. 1999;871:65–80.
52. Meng H, Green AM, Dickman JD, Angelaki DE. Pursuit – vestibular interactions in brain stem neurons during rotation and translation. J Neurophysiol. 2005;93(6):3418–33.
53. Ramaioli C, Cuturi LF, Ramat S, Lehnen N, MacNeilage PR. Vestibulo-ocular responses and dynamic visual acuity during horizontal rotation and translation. Front Neurol. 2019;10:321.
54. Schneider R, Liao K, Walker MF, Bronstein A, Leigh RJ. Behavior of the human translational vestibulo-ocular reflex during simultaneous head translation and rotation. J Vestib Res Equilib Orient. 2014;24(5–6):329–33.

Vestibular Tuning Spectrum

10

Introduction

The vestibular component of the labyrinth is phylogenetically older than the auditory component [1]. The auditory component's cochlea evolved a highly developed frequency discriminating capability for detecting and analyzing acoustic stimuli. The vestibular component of the labyrinth also has a frequency discriminating capability, although it is less well-developed than the auditory component in this regard.

Physiology

In the vestibular component of the labyrinth—which essentially functions as an accelerometer for rotational and translational movements—"frequency" refers to the magnitude of acceleration [2] and is homologous to the cochlea [3]. In the cochlea, a particular type 1 hair cell tends to be maximally stimulated by vibration (oscillating acceleration) at a specific frequency. In the vestibular organelles, a particular type 1 hair cell tends to be maximally stimulated by acceleration of a specific magnitude [4–7]. This is the case in various animal species [4, 8–10], including humans [11, 12].

Significance

Although the ability to assess vestibular system function has improved significantly over the past century, there are still significant gaps in what can currently be tested clinically [13]. This means that there are likely pathologies that we are still unable to detect.

Ideally, we would like to have the capability of visualizing the full range of tuning frequencies over the vestibular tuning spectrum for each individual end-organelle of the vestibular system—which some investigators have referred to as a "vestibulogram" [14]. The currently available clinical tools can come closest to this only for the horizontal semicircular canal through the combined use of caloric testing, rotatory chair testing, and video head impulse testing.

References

1. Manley GA, Koppl C. Phylogenetic development of the cochlea and its innervation. Curr Opin Neurobiol. 1998;8(4):468–74.
2. Beraneck M, Straka H. Vestibular signal processing by separate sets of neuronal filters. J Vestib Res Equilib Orient. 2011;21(1):5–19.
3. Ciuman RR. Auditory and vestibular hair cell stereocilia: relationship between functionality and inner ear disease. J Laryngol Otol. 2011;125(10):991–1003.
4. Highstein SM, Rabbitt RD, Holstein GR, Boyle RD. Determinants of spatial and temporal coding by semicircular canal afferents. J Neurophysiol. 2005;93(5):2359–70.
5. Straka H, Lambert FM, Pfanzelt S, Beraneck M. Vestibulo-ocular signal transformation in frequency-tuned channels. Ann N Y Acad Sci. 2009;1164:37–44.
6. Songer JE, Eatock RA. Tuning and timing in mammalian type I hair cells and calyceal synapses. J Neurosci. 2013;33(8):3706–24.
7. Iversen MM, Rabbitt RD. Wave mechanics of the vestibular semicircular canals. Biophys J. 2017;113(5):1133–49.
8. Haque A, Angelaki DE, Dickman JD. Spatial tuning and dynamics of vestibular semicircular canal afferents in rhesus monkeys. Exp Brain Res. 2004;155(1):81–90.
9. Straka H, Dieringer N. Basic organization principles of the VOR: lessons from frogs. Prog Neurobiol. 2004;73(4):259–309.
10. Liu Z, Kimura Y, Higashijima SI, Hildebrand DGC, Morgan JL, Bagnall MW. Central vestibular tuning arises from patterned convergence of otolith afferents. Neuron. 2020;
11. Todd NP, Rosengren SM, Govender S, Colebatch JG. Low-frequency tuning in the human vestibular-ocular projection is determined by both peripheral and central mechanisms. Neurosci Lett. 2009;458(1):43–7.
12. Dumas G, Curthoys IS, Lion A, Perrin P, Schmerber S. The skull vibration-induced nystagmus test of vestibular function – a review. Front Neurol. 2017;8:41.
13. Cherchi M, Yacovino DA. Dysfunction along the continuum of vestibulocochlear anatomy, and the corresponding spectrum of clinical presentation: how little we know, and what else we need to learn. Hear Balance Commun. 2021:1–12.
14. Striteska M, Skoloudik L, Valis M, Mejzlik J, Trnkova K, Chovanec M, et al. Estimated vestibulogram (EVEST) for effective vestibular assessment. Biomed Res Int. 2021;2021:8845943.

M. Cherchi, *Otoneurology and Vestibular Medicine*, https://doi.org/10.1007/978-3-031-94841-1_10

Part III

Introductory Materials—Specific Eye Movements and Their Normative Ranges

Spontaneous Nystagmus

11

Spontaneous Nystagmus

Here, we use the term "*spontaneous nystagmus*" to refer to nystagmus that occurs when an individual is seated upright in the dark, attempting to stare straight ahead. The literature reports a fairly broad range of spontaneous nystagmus in healthy subjects, which in turn raises questions about the diagnostic utility of this finding.

Hajioff and colleagues [1] studied 96 healthy subjects (56 men) age 66–89 years (median age 76 years) using electronystagmography. They reported that 5 out of 96 subjects (5.2%) exhibited spontaneous nystagmus while sitting upright in darkness.

Geisler and colleagues [2] studied 30 healthy subjects age 20–78 with videonystagmography and found 2 (6.6%) exhibited spontaneous nystagmus.

Takahashi and colleagues [3] examined 30 healthy subjects using electronystagmography. They found that 5 (16.7%) subjects exhibited spontaneous nystagmus.

Levo et al. [4] studied 20 healthy subjects using videonystagmography. They reported that spontaneous nystagmus was observed in 20% of individuals while sitting upright in darkness.

Vesterhauge and colleagues [5] studied 50 healthy subjects (25 male) with electronystagmography. They reported that 15 (30%) exhibited spontaneous horizontal nystagmus ranging in amplitude from 1.2 to 8.2 °/s, though of those 15 individuals, only 1 had nystagmus of an amplitude of 8.2 °/s; the remaining 14 had an amplitude of 1.2–5.0 °/s. This same study examined a different subset of 15 patients a second time (24 hours later) and found that 12 (24%) had no spontaneous nystagmus on either the first or second examination; 2 (13.3%) had nystagmus on the first examination but not on the second; none had nystagmus on the second but not the first examination; and 1 (6.6%) had nystagmus on both the first and second examinations; this emphasizes that spontaneous nystagmus can be variable in one and the same healthy subject.

Bisdorff and colleagues [6] studied 40 healthy subjects (20 males) age 21–55 years using videonystagmography. They reported that 16 of 40 subjects (40%) exhibited spontaneous vertical nystagmus while upright in darkness; of these, 4 exhibited downbeat nystagmus (0.7 ± 0.27 °/s) and 12 exhibited upbeat nystagmus (2.58 ± 1.38 °/s).

Taken together, these studies appear to show that there is a very broad range of spontaneous nystagmus (5.2–40%) among different healthy subjects, and one study even shows differences over time in the same individual. Of the studies that report the amplitude of spontaneous nystagmus in healthy individuals, the magnitude appears relatively small, though there are outliers (e.g., 8.2 °/s).

References

1. Hajioff D, Barr-Hamilton RM, Colledge NR, Lewis SJ, Wilson JA. Re-evaluation of normative electronystagmography data in healthy ageing. Clin Otolaryngol Allied Sci. 2000;25(4):249–52.
2. Baba S, Fukumoto A, Aoyagi M, Koizumi Y, Ikezono T, Yagi T. A comparative study on the observation of spontaneous nystagmus with Frenzel glasses and an infrared CCD camera. J Nippon Med Sch. 2004;71(1):25–9.
3. Takahashi J, Kitamura K, Miyata M. Spontaneous nystagmus in normal subjects. ORL J Otorhinolaryngol Relat Spec. 1996;58(1):42–5.
4. Levo H, Aalto H, Petteri Hirvonen T. Nystagmus measured with video-oculography: methodological aspects and normative data. ORL J Otorhinolaryngol Relat Spec. 2004;66(3):101–4.
5. Vesterhauge S, Kildegaard Larsen P. Normal values in a routine ENG test. Acta Otolaryngol. 1977;84(1–2):91–7.
6. Bisdorff AR, Sancovic S, Debatisse D, Bentley C, Gresty MA, Bronstein AM. Positional nystagmus in the dark in normal subjects. Neuro-Ophthalmology. 2000;24:283–90.

M. Cherchi, *Otoneurology and Vestibular Medicine*, https://doi.org/10.1007/978-3-031-94841-1_11

12 Saccades, Normative Ranges

Introduction

A *saccade* is a rapid eye movement during which vision is suppressed. Saccades can be characterized by latency (the delay from the appearance of a target to the time that a saccade is initiated), velocity, and accuracy. Studies vary but overall suggest that the normative ranges for saccade characteristics vary somewhat by age.

Saccades in Children

Salman and colleagues [1] studied horizontal and vertical saccades in 39 healthy children (21 males) aged 8–19 years (mean 13.7 years, SD 3.5 years) using videonystagmography and reported (Table 12.1):

Sinno and colleagues [2] studied 120 healthy children age 5–17 years with videonystagmography. With respect to horizontal saccades, they reported (Table 12.2).

They also observed that saccade velocity and accuracy/precision did not change significantly from ages 5–8 years to ages 15–17 years.

Saccades in Young to Middle Age Adults

Kang and Kim [3] studied 25 healthy Korean subjects (9 men) age 21–38 years using videonystagmography. After manual correction for "errors such as wrong detection of the inflection point, missing movement, and prediction," they found that for saccades at 15° horizontal (Table 12.3):

They found that for saccades at 30° horizontal (Table 12.4):

Table 12.1 Horizontal and vertical saccades in children, mean values for latency, gain, and velocity

Direction and amplitude of saccade	Mean latency in ms (SD)	Mean gain (SD)	Mean asymptotic peak velocity in deg/s (SD)
Rightward 10°	248.7 (20.5)	0.96 (0.08)	521.2 (90.7)
Leftward 10°	253.6 (27.4)	0.96 (0.08)	537.4 (83.6)
Upward 10°	243.3 (29.4)	0.93 (0.15)	466.9 (117.5)
Upward 5°	240.3 (22.6)	0.95 (0.15)	
Downward 10°	247.0 (24.2)	0.90 (0.16)	435.7 (111.5)
Downward 5°	268.3 (23.5)	0.96 (0.16)	

From Salman et al. [1]

Table 12.2 Saccades in children, latency

Age range	Saccadic latency in ms (95% CI)
5–8 year olds	305 ± 48
9–11 year olds	276 ± 22
12–14 year olds	252 ± 34
15–17 year olds	256 ± 33

From Sinno et al. [2]

Table 12.3 Horizontal saccades of 15° in middle age adults, mean peak velocity, latency, and accuracy

	Mean peak velocity in deg/s (95% CI)	Mean latency in ms (95% CI)	Mean accuracy in percent
Left eye, leftward saccade (abduction)	329.5 (281.8–377.1)	208.0 (174.4–241.6)	82.0 (74.5–89.6)
Right eye, leftward saccade (adduction)	311.9 (270.1–353.7)	193.0 (165.7–220.2)	77.8 (69.1–86.4)
Left eye, rightward saccade (adduction)	319.7 (356.5–282.9)	192.8 (166.1–219.5)	75.3 (67.8–82.8)
Right eye, rightward saccade (abduction)	324.0 (352.7–295.3)	190.4 (159.7–221.2)	81.9 (74.7–89.0)

From Kang and Kim [3]

M. Cherchi, *Otoneurology and Vestibular Medicine*, https://doi.org/10.1007/978-3-031-94841-1_12

Table 12.4 Horizontal saccades of 30° in middle age adults, mean peak velocity, latency, and accuracy

	Mean peak velocity in deg/s (95% CI)	Mean latency in ms (95% CI)	Mean accuracy in percent
Left eye, leftward saccade (abduction)	408.2 (336.4–480.1)	233.8 (188.6–279.0)	75.4 (66.8–84.0)
Right eye, leftward saccade (adduction)	337.7 (297.9–377.4)	211.1 (163.0–259.2)	71.4 (63.7–79.0)
Left eye, rightward saccade (adduction)	405.5 (485.2–325.9)	196.3 (163.6–229.1)	76.5 (67.3–85.7)
Right eye, rightward saccade (abduction)	368.1 (406.8–329.4)	191.4 (153.1–229.7)	78.6 (70.9–86.4)

From Kang and Kim [3]

Table 12.5 Horizontal saccades of 20° in middle age and older adults, maximum velocity, latency, and accuracy

	Maximum velocity in deg/s (95% CI)	Latency in ms (95% CI) for both eyes	Dysmetry/ accuracy in percent (95% CI)
Right eye, rightward saccade (abduction)	367 (283–542)	192 (129–255)	98.2 (88–108)
Left eye, rightward saccade (adduction)	458 (336–581)		99.1 (89–110)
Right eye, leftward saccade (adduction)	458 (336–581)		100.1 (89–111)
Left eye, leftward saccade (abduction)	367 (283–542)		98.2 (88–108)

From Van der Stappen et al. [4]

Saccades in the Middle Age and Elderly

Van der Stappen and colleagues [4] studied 40 healthy individuals (15 male) age 35–72 years (average 45 years) using electronystagmography. Regarding saccades at 20°, they reported (Table 12.5):

Saccades in the Elderly

Hajioff and colleagues [5] studied 96 healthy subjects (56 men) age 66–89 years (median age 76 years) using electronystagmography. For random saccades presented at 6–30° jumps, they reported a latency with a lower reference limit of 121.5 ms (95% CI 110.3–132.7 ms) and an upper reference limit of 303.6 ms (95% CI 292.4–314.9 ms). They reported percent accuracy with a lower reference limit of 57.9 (95% CI 52.8–63.0) and an upper reference limit of 120.7 (95% CI 115.6–125.8).

Saccades over a Broad Age Range

Warabi and colleagues [6] used a horizontal saccade protocol in which LEDs were placed at 10° intervals, extending 20° to each side. They state in the text of their article that they studied 8 healthy subjects age 19–26 years, and 24 healthy subjects age 59–82 years using electronystagmography, though in the analysis, they end up breaking the subjects into three groups: (a) age 16 (not 19)–26 years, (b) age 59–82 years, and (c) 67–78 years because, "Four of the elderly subjects exhibited large variations in reaction times (standard deviation more than 110 ms), whereas the variations in the 20 other elderly subjects differed little from those in the young subjects. Therefore, the elderly subjects were classified into groups b and c." They report that the mean reaction times for these groups were (a) 235 ms, (b) 337 ms, and (c) 468 ms (Table 12.6).

Hopf and colleagues [7] studied horizontal and vertical reflexive saccades in 102 healthy subjects (46 male) age 6–76 using videonystagmography. At variance with many prior studies, they concluded that, "We found no relevant effect of age on peak velocity, gain and latency in a fitted linear mixed model." They added that, "However, latency becomes shorter during childhood and adolescence, while in adulthood it is relatively stable with a slight trend to increase in the elderly." They found that the, "Mean peak velocity of saccades was 213 deg/s (±29 deg/s), 352 deg/s (±50 deg/s) and 455 deg/s (±67 deg/s) to a target position 5°, 15° and 30° horizontally, respectively, and 208 deg/s (±36 deg/s), 303 deg/s (±50 deg/s) and 391 deg/s (±71 deg/s) to a target position 5°, 10° and 20° vertically."

Irving and colleagues [8] studied horizontal saccades in 195 healthy subjects (93 male) age 3–86 years using videonystagmography. They observed that, "Mean saccadic latency decreased from 439 ms at 3 years to 172 ms at 14

Table 12.6 Saccades over a broad age range

		Saccade target amplitude		
	Reaction time (latency) in ms	10° amplitude, velocity in deg/s (mean ± SD)	20° amplitude, velocity in deg/s (mean ± SD)	40° amplitude, velocity in deg/s (mean ± SD)
Group a, 19–26 years	235	306 ± 51	408 ± 57	508 ± 76
Group b, 59–82 years	337	300 ± 63	391 ± 65	464 ± 89
Group c, 67–78 years	468			

From Warabi et al. [6]

years, followed by a period of relative stability to age 50 and finally, gradually increased to 264 ms at ≥80 years. For saccadic accuracy (amplitude gain), there was a statistically significant ($P < 0.05$) interaction between saccade size and age. Participants made increasingly hypometric saccades as age and saccade size increased. Average age group saccadic asymptotic peak velocity (Vmax) increased childhood from 446 deg/s at age 3, to a peak of 610 deg/s at 14 years and then gradually declined with age to approximately 345 deg/s for participants ≥80 years."

Seferlis and colleagues [9] studied 250 healthy subjects (87 male) age 18–70 years using videonystagmography. They reported (Tables 12.7 and 12.8):

Spooner and colleagues [10] studied 25 healthy subjects aged 22–64 years (mean 42 years), and 14 healthy subjects aged 50–85 years (mean 65 years) with electronystagmography. The overlap in age between the groups is apparently because the first group "had served as normal controls in our laboratory for several years." They used a horizontal saccade protocol in which, "the target moved in a series of random stepwise jumps from 3 to 36° in amplitude in each direction for a total of 66 jumps." They reported (Table 12.9):

Table 12.7 Horizontal (leftward) saccades across a broad age range

	Age range (years)	Number of subjects	Mean (95% CI)
Saccades latency leftward in ms	Group A, 18–30	50	237.24 (232.06–242.42)
	Group B, 31–40	50	241.44 (234.39–248.49)
	Group C, 41–50	50	255.86 (243.15–268.57)
	Group D, 51–60	50	260.12 (247.18–273.06)
	Group E, 61–70	50	290.42 (275.47–305.37)
	Total	250	257.02 (251.61–262.43)
Saccades velocity leftward in dec/s	Group A, 18–30	50	245.30 (223.07–267.53)
	Group B, 31–40	50	228.28 (209.44–247.12)
	Group C, 41–50	50	236.28 (212.24–260.32)
	Group D, 51–60	50	231.50 (212.06–250.94)
	Group E, 61–70	50	169.52 (152.25–186.79)
	Total	250	222.18 (212.65–231.71)
Saccades accuracy leftward	Group A, 18–30	50	100.28 (95.31–105.25)
	Group B, 31–40	50	91.40 (87.34–95.46)
	Group C, 41–50	50	88.92 (81.87–95.97)
	Group D, 51–60	50	91.24 (87.63–94.85)
	Group E, 61–70	50	70.42 (63.70–77.14)
	Total	250	88.45 (85.78–91.12)

From Seferlis et al. [9]

Table 12.8 Horizontal (rightward) saccades across a broad age range

	Age range (years)	Number of subjects	Mean (95% CI)
Saccades latency rightward in ms	Group A, 18–30	50	265.34 (255.15–275.53)
	Group B, 31–40	50	252.26 (241.56–262.96)
	Group C, 41–50	50	273.82 (259.99–287.65)
	Group D, 51–60	50	264.12 (252.50–275.74)
	Group E, 61–70	50	292.56 (274.72–310.40)
	Total	250	269.62 (263.67–275.57)
Saccades velocity rightward in dec/s	Group A, 18–30	50	237.52 (216.02–259.92)
	Group B, 31–40	50	226.50 (207.44–245.56)
	Group C, 41–50	50	236.28 (212.24–260.32)
	Group D, 51–60	50	233.20 (214.34–251.06)
	Group E, 61–70	50	190.30 (172.03–208.57)
	Total	250	224.76 (215.60–233.92)
Saccades accuracy rightward	Group A, 18–30	50	93.20 (89.26–97.14)
	Group B, 31–40	50	91.40 (87.34–95.46)
	Group C, 41–50	50	89.56 (82.63–96.49)
	Group D, 51–60	50	91.86 (88.49–95.23)
	Group E, 61–70	50	78.22 (73.39–83.05)
	Total	250	88.45 (85.78–91.12)

From Seferlis et al. [9]

Table 12.9 Horizontal saccades across a broad age range

	Maximum velocity in deg/s ± SD	Delay time (latency) in ms
Age 22–64 years	551 ± 65.0	179 ± 27.4
Age 50–85 years	487 ± 85.5	219 ± 32.2

From Spooner et al. [10]

References

1. Salman MS, Sharpe JA, Eizenman M, Lillakas L, Westall C, To T, et al. Saccades in children. Vis Res. 2006;46(8–9):1432–9.
2. Sinno S, Najem F, Smith Abouchacra K, Perrin P, Dumas G. Normative values of saccades and smooth pursuit in children aged 5 to 17 years. J Am Acad Audiol. 2020;31:384–92.
3. Kang S, Kim US. Normative data of videonystagmography in young healthy adults under 40 years old. Korean J Ophthalmol. 2015;29(2):126–30.
4. Van Der Stappen A, Wuyts FL, Van De Heyning PH. Computerized electronystagmography: normative data revisited. Acta Otolaryngol. 2000;120(6):724–30.
5. Hajioff D, Barr-Hamilton RM, Colledge NR, Lewis SJ, Wilson JA. Re-evaluation of normative electronystagmography data in healthy ageing. Clin Otolaryngol Allied Sci. 2000;25(4):249–52.
6. Warabi T, Kase M, Kato T. Effect of aging on the accuracy of visually guided saccadic eye movement. Ann Neurol. 1984;16(4):449–54.
7. Hopf S, Liesenfeld M, Schmidtmann I, Ashayer S, Pitz S. Age dependent normative data of vertical and horizontal reflexive saccades. PLoS One. 2018;13(9):e0204008.
8. Irving EL, Steinbach MJ, Lillakas L, Babu RJ, Hutchings N. Horizontal saccade dynamics across the human life span. Invest Ophthalmol Vis Sci. 2006;47(6):2478–84.
9. Seferlis F, Chimona TS, Papadakis CE, Bizakis J, Triaridis S, Skoulakis C. Age related changes in ocular motor testing in healthy subjects. J Vestib Res Equilib Orient. 2015;25(2):57–66.
10. Spooner JW, Sakala SM, Baloh RW. Effect of aging on eye tracking. Arch Neurol. 1980;37(9):575–6.

13 Eccentric Gaze Nystagmus, Normative Ranges

Eccentric Gaze Nystagmus, Normative Ranges

When the eye is directed to an *eccentric* point of regard and then maintained in that orientation, there may be *nystagmus*. Bárány [1] credits John Hunter as the first to have described such nystagmus in 1786 (apparently referring to the discussion on p. 212 of the chapter entitled, "The Use of the Oblique Muscles") [2].

This nystagmus at eccentric gaze can be physiologic or pathologic, and the literature exhibits some terminological inconsistency in this regard. Some authors use the term "end-point nystagmus" for nonpathologic cases, and "gaze-evoked nystagmus" for pathologic ones. Other authors refer to any nystagmus at eccentric gaze as "gaze-evoked nystagmus." Beyond this terminological inconsistency, the boundary between physiologic and pathologic is not clearly established.

In healthy individuals, when the eyes are horizontally eccentric, one may observe transient, low amplitude nystagmus whose fast phase is centrifugal (i.e., in the same direction of gaze holding, beating away from primary position of gaze) and slow phase is centripetal (i.e., drifting toward primary position of gaze). This phenomenon is sometimes referred to as "*end-point nystagmus*" [3–6]. Some authors [4] characterize end-point nystagmus according to when (after assuming eccentric gaze) it begins and whether it is sustained. For instance, Eizenman and colleagues [5] describe end-point nystagmus that begins immediately or within a few seconds after the eye assumes an eccentric position as "*sustained*" if it persists as long as the eye is maintained in that eccentric position and "*unsustained*" if it does not persist. If the nystagmus "starts after a considerable delay (more than 30 s)" after the eye assumes an eccentric position, they refer to it as "*fatigue*" end-point nystagmus.

The amplitude of end-point nystagmus in healthy subjects appears to be affected by several variables. First, the greater the eccentric position of the eye, the greater the amplitude of the end-point nystagmus; moreover, the amplitude increases linearly only up to 20° of eccentricity, after which the increase in amplitude becomes nonlinear [7]. Second, there is usually some asymmetry in the amplitude of the nystagmus between the abducting and adducting eye [3]. Third, the amplitude of end-point nystagmus varies depending on whether the subject is fixating on a visual target, or simply trying to maintain gaze at a remembered (though not currently visible) target [5].

Given this variability, there does not exist a single "cutoff value" that delimits the normative range. We regard with caution literature that appears simply to stipulate a normative range; for instance, Rett [8] states that gaze-evoked nystagmus >4°/s is pathological.

Perverted nystagmus refers to nystagmus whose fast phase is in some direction other than centrifugal, such as down beat nystagmus on extreme lateral gaze, or torsional nystagmus on upgaze. Perverted nystagmus at eccentric gaze is generally regarded as pathological.

References

1. Bárány R. Über die vom Ohrlabyrinth ausgelöste Gegenrollung der Augen bei Normalhörenden [On the counter-rolling of the eyes triggered by the labyrinth in people with normal hearing]. Arch Ohrenheilk. 1906;68:1–30.
2. Hunter J. Observations on certain parts of the animal oeconomy. London; 1786.
3. Romano F, Tarnutzer AA, Straumann D, Ramat S, Bertolini G. Gaze-evoked nystagmus induced by alcohol intoxication. J Physiol. 2017;595(6):2161–73.
4. Abel LA, Parker L, Daroff RB, Dell'Osso LF. End-point nystagmus. Invest Ophthalmol Vis Sci. 1978;17(6):539–44.
5. Eizenman M, Cheng P, Sharpe JA, Frecker RC. End-point nystagmus and ocular drift: an experimental and theoretical study. Vis Res. 1990;30(6):863–77.
6. Shallo-Hoffmann J, Schwarze H, Simonsz HJ, Muhlendyck H. A reexamination of end-point and rebound nystagmus in normals. Invest Ophthalmol Vis Sci. 1990;31(2):388–92.
7. Bertolini G, Tarnutzer AA, Olasagasti I, Khojasteh E, Weber KP, Bockisch CJ, et al. Gaze holding in healthy subjects. PLoS One. 2013;8(4):e61389.
8. Rett D. Gaze-evoked nystagmus: a case report and literature review. Optometry. 2007;78(9):460–4.

M. Cherchi, *Otoneurology and Vestibular Medicine*, https://doi.org/10.1007/978-3-031-94841-1_13

14 Smooth Pursuit, Normative Ranges

Smooth Pursuit, Normative Ranges

Smooth pursuit refers to a slow eye movement intended to track an object of regard that is moving through the visual field. Characteristics of smooth pursuit include velocity and gain (ratio of eye velocity to target velocity). Smooth pursuit varies by age. Smooth pursuit gain varies by target velocity; gain is better (closer to 1.0) at lower target velocities and poorer at higher target velocities.

Smooth Pursuit in Children

Sinno and colleagues [1] studied 120 healthy children age 5–17 years with videonystagmography. They reported the following regarding smooth pursuit (Table 14.1):

Table 14.1 Smooth pursuit in children, gain

Age range	Smooth pursuit gain
5–8 year olds	0.63
15–17 year olds	0.85

From Sinno et al. [1]

Smooth Pursuit in the Elderly

Hajioff and colleagues [2] studied 96 healthy subjects (56 men) age 66–89 years (median age 76 years) using electronystagmography. For horizontal smooth pursuit (0.4 Hz, amplitude 20 deg/s), they reported gain with a lower reference limit of 0.272 (95% CI 0.197–0.346) and an upper reference limit of 1.13 (95% CI 1.05–1.20).

M. Cherchi, *Otoneurology and Vestibular Medicine*, https://doi.org/10.1007/978-3-031-94841-1_14

Smooth Pursuit over a Broad Age Range

Seferlis and colleagues [3] studied 250 healthy subjects (87 male) age 18–70 years using videonystagmography. To elicit smooth pursuit, they projected a yellow-colored circular target with a width of 1°, moving in a sinusoidal pattern in the horizontal plane, with an amplitude of 10° in each direction, velocity of 10 deg/s, and frequency of 0.6 Hz. They reported the following gains and slow phase velocities stratified by age (Table 14.2):

Table 14.2 Smooth pursuit mean velocity and gain over a broad age range

	Age range (years)	Number of subjects	Mean (95% CI) slow phase velocity (deg/s)	Mean (95% CI) gain (percent)
Smooth pursuit, leftward	Group A, 18–30	50	10.74 (10.12–11.36)	88.92 (85.28–92.56)
	Group B, 31–40	50	10.08 (9.33–10.83)	88.18 (84.50–91.86)
	Group C, 41–50	50	9.60 (8.76–10.44)	85.26 (82.09–88.43)
	Group D, 51–60	50	9.26 (8.80–9.72)	84.92 (81.81–88.03)
	Group E, 61–70	50	8.46 (7.94–8.98)	79.04 (75.10–82.98)
	Total	250	9.63 (9.33–9.93)	85.26 (83.67–86.86)
Smooth pursuit, rightward	Group A, 18–30	50	10.74 (10.19–11.29)	90.24 (86.90–93.58)
	Group B, 31–40	50	11.44 (10.16–12.72)	91.12 (88.31–93.93)
	Group C, 41–50	50	10.52 (9.71–11.33)	88.06 (85.38–90.74)
	Group D, 51–60	50	10.06 (9.28–10.84)	83.24 (79.41–87.07)
	Group E, 61–70	50	8.86 (8.29–9.43)	80.86 (77.51–84.21)
	Total	250	10.32 (9.94–10.70)	86.70 (85.21–88.19)

From Seferlis et al. [3]

Sharpe and colleagues [4] studied 15 healthy subjects (7 men) age 19–32 years and 10 healthy subjects (3 men) age 65–77 years. They studied horizontal smooth pursuit at different target velocities and reported the findings below regarding the slow phase eye velocities (not gain). Each data cell reports the mean velocity ± standard deviation, followed by the maximum individual subject value (Table 14.3).

Sharpe and colleagues [4] also reported the smooth pursuit latencies at several different target velocities as shown below. Each data cell reports the mean latency in milliseconds ± standard deviation, followed by the number of subjects in that trial (Table 14.4).

Spooner and colleagues [5] studied 25 healthy subjects aged 22–64 years (mean 42 years) and 14 healthy subjects age 50–85 years (mean 65 years) with electronystagmography. The overlap in age between the groups is apparently because the first group "had served as normal controls in our laboratory for several years." They used a horizontal smooth pursuit protocol in which the "target moved in fixed-cycle constant velocity ramps (22, 37, 51 deg/s) for ten cycles each at 0.3, 0.5, and 0.7 Hz." In the results displayed below, each data cell shows the mean eye velocity ± standard deviation (Table 14.5).

Table 14.3 Smooth pursuit mean velocity over a broad age range

	Target velocity in deg/s								
	5	10	20	30	40	50	60	80	100
Age 19–32 years	4.9 ± 0.2 5.8	9.6 ± 0.7 12.2	19.6 ± 1.2 25.0	29.3 ± 2.4 43.0	36.0 ± 5.2 47.5	39.4 ± 9.0 56.0	38.4 ± 10.0 61.0	32.2 ± 15.2 75.0	29.1 ± 18.0 70.0
Age 65–77 years	4.7 ± 0.6 5.6	8.7 ± 0.9 11.3	13.8 ± 3.2 22.5	16.2 ± 6.3 32.5	18.4 ± 9.2 42.5	16.0 ± 7.5 30.0	— 25.0	— 15.0	— —

From Sharpe et al. [4]

Table 14.4 Smooth pursuit mean latency over a broad age range

	Target velocity in deg/s		
	5	10	20
Age 19–32 years	131 ± 23 (10 subjects)	138 ± 21 (12 subjects)	134 ± 34 (11 subjects)
Age 65–77 years	176 ± 33 (5 subjects)	183 ± 49 (5 subjects)	169 ± 37 (8 subjects)

From Sharpe et al. [4]

Table 14.5 Smooth pursuit mean eye velocity over a broad age range

	Target velocity in deg/s		
	22	37	51
Age 22–64 years	18.6 ± 1.3	29.0 ± 3.3	37.0 ± 5.6
Age 50–85 years	17.0 ± 2.1	24.7 ± 5.6	29.4 ± 8.7

From Spooner et al. [5]

References

1. Sinno S, Najem F, Smith Abouchacra K, Perrin P, Dumas G. Normative values of saccades and smooth pursuit in children aged 5 to 17 years. J Am Acad Audiol. 2020;31:384–92.
2. Hajioff D, Barr-Hamilton RM, Colledge NR, Lewis SJ, Wilson JA. Re-evaluation of normative electronystagmography data in healthy ageing. Clin Otolaryngol Allied Sci. 2000;25(4):249–52.
3. Seferlis F, Chimona TS, Papadakis CE, Bizakis J, Triaridis S, Skoulakis C. Age related changes in ocular motor testing in healthy subjects. J Vestib Res Equilib Orient. 2015;25(2):57–66.
4. Sharpe JA, Sylvester TO. Effect of aging on horizontal smooth pursuit. Invest Ophthalmol Vis Sci. 1978;17(5):465–8.
5. Spooner JW, Sakala SM, Baloh RW. Effect of aging on eye tracking. Arch Neurol. 1980;37(9):575–6.

15 Optokinetic Nystagmus, Normative Ranges

Horizontal Optokinetic Nystagmus, Normative Ranges

Horizontal optokinetic nystagmus refers to jerk nystagmus that occurs in response to a full-field stimulus moving uniformly at constant velocity in a particular direction; in the laboratory setting, the stimulus is usually a field of vertically oriented evenly spaced alternating black and white stripes that move horizontally (to elicit horizontal optokinetic nystagmus). Most studies report that optokinetic nystagmus varies by age such that the gain (ratio of slow phase eye velocity to optokinetic stimulus velocity) progressively decreases with advancing age.

Horizontal Optokinetic Testing in the Elderly

Hajioff and colleagues [1] studied 96 healthy subjects (56 men) age 66–89 years (median age 76 years) using electronystagmography. For horizontal bidirectional optokinetic testing at a constant velocity of 20°/s, they reported gain with a lower reference range of 0.550 (95% CI 0.30–0.64) and an upper reference range of 1.20 (95% CI 1.05–1.30).

Horizontal Optokinetic Testing over a Broad Age Range

Lynch and colleagues [2] studied 10 healthy subjects age 20–35 years and 10 healthy subjects (4 male) age 69–76 years, using electronystagmography. Of the younger group (age 20–35 years), they reported that, "None showed asymmetry of the optokinetic response," whereas in the older group (age 69–76 years), they reported that "Four… subjects [40%] showed asymmetry of the optokinetic response" (although the table indicates that 5 subjects [50%] had optokinetic asymmetry).

Seferlis and colleagues [3] studied 250 healthy subjects (87 male) age 18–70 years using videonystagmography. They utilized a horizontal optokinetic stimulus moving at 15°/s and reported the optokinetic slow phase velocities and gains stratified by age (Table 15.1):

Table 15.1 Optokinetic nystagmus, mean slow phase velocity and gain over a broad age range

	Age range (years)	Number of subjects	Mean (95% CI) slow phase velocity (°/s)	Mean (95% CI) gain (percent)
Optokinetic nystagmus leftward	Group A, 18–30	50	12.77 (10.91–14.62)	77.64 (72.93–82.34)
	Group B, 31–40	50	11.32 (10.38–12.25)	72.91 (68.89–76.92)
	Group C, 41–50	50	11.20 (10.49–11.90)	73.09 (69.08–77.09)
	Group D, 51–60	50	10.72 (9.74–11.69)	65.91 (60.31–71.51)
	Group E, 61–70	50	10.26 (9.24–11.27)	61.53 (56.59–66.47)
	Total	250	11.25 (10.73–11.77)	70.21 (68.05–72.37)
Optokinetic nystagmus rightward	Group A, 18–30	50	12.53 (11.54–13.51)	80.33 (76.12–84.53)
	Group B, 31–40	50	12.41 (11.52–13.30)	79.39 (75.17–83.61)
	Group C, 41–50	50	11.77 (11.05–12.49)	75.79 (71.92–79.67)
	Group D, 51–60	50	11.41 (10.31–12.51)	67.11 (62.04–72.18)
	Group E, 61–70	50	7.90 (7.37–8.43)	61.04 (56.37–65.71)
	Total	250	11.20 (10.77–11.64)	72.73 (70.59–74.87)

From Seferlis et al. [3]

M. Cherchi, *Otoneurology and Vestibular Medicine*, https://doi.org/10.1007/978-3-031-94841-1_15

Spooner and colleagues [4] studied 25 healthy subjects aged 22–64 years (mean 42 years) and 14 healthy subjects age 50–85 years (mean 65 years) with electronystagmography. The overlap in age between the groups is apparently because the first group "had served as normal controls in our laboratory for several years." They used a horizontal optokinetic protocol with "the subject seated inside a large drum with alternating black and white stripes. The drum rotated at a fixed velocity of 30°/s for 30 s in clockwise and counterclockwise directions. Average slow component velocity over the 30-s period of stimulation was calculated for each group of subjects." They reported that in the group age 22–64 years, the slow component velocity was 25.6 (± SD of 2.8) °/s, and in the group age 50–85 years, the slow component velocity was 19.6 (± SD of 4.6) °/s.

Vertical Optokinetic Nystagmus

Horizontal optokinetic testing is a standard part of most ocular motor testing protocols, whereas assessment of *vertical optokinetic nystagmus* is not.

In healthy subjects, vertical optokinetic nystagmus shows asymmetrical gain in the slow phase velocity; specifically, upward gain (i.e., with the stimulus moving toward the subject's forehead) is higher than downward gain (i.e., with the stimulus moving toward the subject's feet) [5–9]. This asymmetry persists whether the subject is upright (i.e., earth vertical), lying on their side [10], or at some other tilt angle [11]; some investigators report that the asymmetry is actually more pronounced when the subject is lying on their side or is upside-down [12].

In microgravity, this asymmetry initially reverses [13] and then disappears [14]. Most investigators hypothesize that this is related to the influence of otolithic input [6, 14], or perhaps due to microgravity-induced changes in the velocity storage mechanism [15]. These microgravity-induced changes in vertical optokinetic responses disappear when astronauts return to earth [16].

Vertical optokinetic responses may be abnormal in patients with congenital strabismus; specifically, a horizontal component emerges [17], resulting in a diagonally oriented pattern of nystagmus [18].

Abnormalities of vertical optokinetic nystagmus have modest value in neuroanatomical localization. Garbutt and colleagues suggest that the differential diagnosis includes "neurometabolic disease or an abnormality involving the cortex, brainstem, and/or cerebellum. Abnormal VOKN [vertical optokinetic nystagmus] is highly suggestive of a rostral midbrain lesion" [19]. Abnormal vertical optokinetic nystagmus has been reported in progressive supranuclear palsy [20], idiopathic Parkinson's disease [21], and mercury toxicity [22–24].

References

1. Hajioff D, Barr-Hamilton RM, Colledge NR, Lewis SJ, Wilson JA. Re-evaluation of normative electronystagmography data in healthy ageing. Clin Otolaryngol Allied Sci. 2000;25(4):249–52.
2. Lynch SJ, Nayak US, Isaacs B. Positional and optokinetic nystagmus in healthy old people. Age Ageing. 1985;14(2):122–4.
3. Seferlis F, Chimona TS, Papadakis CE, Bizakis J, Triaridis S, Skoulakis C. Age related changes in ocular motor testing in healthy subjects. J Vestib Res Equilib Orient. 2015;25(2):57–66.
4. Spooner JW, Sakala SM, Baloh RW. Effect of aging on eye tracking. Arch Neurol. 1980;37(9):575–6.
5. Murasugi CM, Howard IP. Up-down asymmetry in human vertical optokinetic nystagmus and afternystagmus: contributions of the central and peripheral retinae. Exp Brain Res. 1989;77(1):183–92.
6. Ogino S, Kato I, Sakuma A, Takahashi K, Takeyama I. Vertical optokinetic nystagmus in normal individuals. Acta Otolaryngol Suppl. 1996;522:38–42.
7. Takahashi M, Sakurai S, Kanzaki J. Horizontal and vertical optokinetic nystagmus in man. ORL J Otorhinolaryngol Relat Spec. 1978;40(1):43–52.
8. Wei G, Lafortune SH, Ireland DJ, Jell RM. Stimulus velocity dependence of human vertical optokinetic nystagmus and afternystagmus. J Vestib Res Equilib Orient. 1992;2(2):99–106.
9. Wei G, Lafortune SH, Ireland DJ, Jell RM. Human vertical optokinetic nystagmus and after-response, and their dependence upon head orientation with respect to gravity. J Vestib Res Equilib Orient. 1994;4(1):37–47.
10. Böhmer A, Baloh RW. Vertical optokinetic nystagmus and optokinetic afternystagmus in humans. J Vestib Res Equilib Orient. 1990;1(3):309–15.
11. Correia MJ, Kolev OI, Rupert AH, Guedry FE. Vertical optokinetic nystagmus and after-responses during backward tilt. Aviat Space Environ Med. 1997;68(4):289–95.
12. Clément G, Lathan CE. Effects of static tilt about the roll axis on horizontal and vertical optokinetic nystagmus and optokinetic afternystagmus in humans. Exp Brain Res. 1991;84(2):335–41.
13. Clement G, Vieville T, Lestienne F, Berthoz A. Modifications of gain asymmetry and beating field of vertical optokinetic nystagmus in microgravity. Neurosci Lett. 1986;63(3):271–4.
14. Clément G. A review of the effects of space flight on the asymmetry of vertical optokinetic and vestibulo-ocular reflexes. J Vestib Res Equilib Orient. 2003;13(4–6):255–63.
15. Clement G, Berthoz A. Cross-coupling between horizontal and vertical eye movements during optokinetic nystagmus and optokinetic afternystagmus elicited in microgravity. Acta Otolaryngol. 1990;109(3–4):179–87.
16. Clément G, Popov KE, Berthoz A. Effects of prolonged weightlessness on horizontal and vertical optokinetic nystagmus and optokinetic after-nystagmus in humans. Exp Brain Res. 1993;94(3):456–62.
17. Garbutt S, Han Y, Kumar AN, Harwood M, Rahman R, Leigh RJ. Disorders of vertical optokinetic nystagmus in patients with ocular misalignment. Vis Res. 2003;43(3):347–57.
18. Economides JR, Suh YW, Simmons JB, Adams DL, Horton JC. Vertical optokinetic stimulation induces diagonal eye move-

ments in patients with idiopathic infantile nystagmus. Invest Ophthalmol Vis Sci. 2020;61(6):14.
19. Garbutt S, Harris CM. Abnormal vertical optokinetic nystagmus in infants and children. Br J Ophthalmol. 2000;84(5):451–5.
20. Garbutt S, Riley DE, Kumar AN, Han Y, Harwood MR, Leigh RJ. Abnormalities of optokinetic nystagmus in progressive supranuclear palsy. J Neurol Neurosurg Psychiatry. 2004;75(10):1386–94.
21. Knapp CM, Gottlob I, McLean RJ, Rajabally YA, Abbott RJ, Rafelt S, et al. Vertical optokinetic nystagmus in Parkinson's disease. Mov Disord. 2009;24(10):1533–8.
22. Mizukoshi K, Nagaba M, Ohno Y, Ishikawa K, Aoyagi M, Watanabe Y, et al. Neurotological studies upon intoxication by organic mercury compounds. ORL J Otorhinolaryngol Relat Spec. 1975;37(2):74–87.
23. Mizukoshi K, Watanabe Y, Kobayashi H, Nakano Y, Koide C, Inomata S, et al. Neurotological follow-up studies upon Minamata disease. Acta Otolaryngol Suppl. 1989;468:353–7.
24. Mizukoshi K, Watanabe Y, Shojaku H, Aso S, Asai M, Inukai K, et al. Neurotological long-term follow-up in Minamata disease in Niigata, Japan. Nihon Jibiinkoka Gakkai Kaiho. 2002;105(3):225–31.

Part IV

Diagnoses—Diseases of the Labyrinth

Benign Paroxysmal Positional Vertigo

16

Introduction

Benign paroxysmal positional vertigo (BPPV) is the most common cause of disequilibrium over the lifespan. The mechanism of BPPV is that otoliths (that normally reside in the utricle) break loose and enter the semicircular canals, where their influence on endolymphatic fluid dynamics results in inappropriate stimulation of the ampullar maculae, thereby provoking both nystagmus and symptoms. BPPV usually presents with positionally triggered episodes of disequilibrium that usually, though not always, have a spinning quality. BPPV can occur at any age, though it is less common before the age of about 40 years, after which its incidence increases exponentially. Besides age, additional risk factors include a history of migraine, other ear disease (e.g., vestibular neuritis, labyrinthitis, Ménière's disease), and perhaps vitamin D deficiency. A diagnosis of BPPV is secured when an examiner observes a pattern of eye movements; the specific pattern of eye movements depends on which canal (or combination of canals) is affected by BPPV. Since BPPV is by its nature paroxysmal, these eye movements may not always be evident on examination. The diagnostic eye movements may also be obfuscated by vestibular suppressant medications. Treatment of BPPV consists of appropriately targeted canalith repositioning maneuvers; the specific maneuver will depend on which canal (or combination of canals) is affected. The canalith repositioning maneuvers often have to be repeated. Most literature suggests that sleeping upright (after physical therapy treatment) does not affect treatment outcome. Medications and surgery play no significant role in management of BPPV. The natural history of BPPV is one of recurrence, and there is not yet any firmly established prophylactic intervention.

BPPV Overview

Benign paroxysmal positional vertigo (BPPV) is an inner ear disorder that usually causes positionally triggered episodes of dizziness/imbalance. It is among the most frequently encountered diseases in clinics specializing in evaluating dizzy patients.

Early Descriptions

In 1920, Robert Bárány [1] described a case of a young woman who had been suffering from vertiginous episodes triggered by lying on her right side. Numerous similar cases were described subsequently, and in 1952, Margaret Ruth Dix and Charles Skinner Hallpike [2, 3] described their own cases of "positional nystagmus," noting that symptoms (and simultaneous nystagmus) were induced, "by a critical position of the head in space."

Clinical Presentation

In the majority of cases, BPPV presents as positionally triggered episodes of disequilibrium that patients usually experience as a spinning sensation (sometimes associated with nausea) that can last seconds to minutes, though can be followed by many minutes of unsteadiness. The symptoms can be re-triggered when the patient assumes the same position.

In some instances (usually lateral semicircular canal BPPV, which will be discussed in greater detail below), the sensation of disequilibrium is more vague and may be experienced as a "rocking," "tilting," or "swaying." Despite the fact that the semicircular canals mediate sensations of angu-

M. Cherchi, *Otoneurology and Vestibular Medicine*, https://doi.org/10.1007/978-3-031-94841-1_16

lar acceleration, BPPV can in some cases cause sensation of linear motion as well [4].

If a patient is experiencing these sensations, then BPPV is a reasonable possibility. However, in order to be confident about the diagnosis, a clinician needs to examine a patient's eye movements, as will be discussed below.

Mechanism of Disease

BPPV is the result of otoliths ("crystals" in the inner ear), also called otoconia, coming loose from their normal place in the inner ear (the utricle) and entering an incorrect place (the semicircular canals) in the inner ear. The movement of the otoliths results in stimulation of type 1 vestibular hair cells, which in turn provokes both symptoms (sense of motion perceived by the patient, and oscillopsia) and signs (nystagmus observed by the examiner).

Notice that in order for this disease to manifest, two distinct events need to occur. First, the otoliths need to break loose; this appears to occur spontaneously. Second, the otoliths need to enter one (or more) of the semicircular canals; this simply happens when the head is put in certain orientations with respect to gravity.

Thus, when a person goes to the hairdresser, extends her neck over the edge of the washbasin, becomes dizzy and then gets diagnosed with BPPV, it is easy to fall prey to the misapprehension that "bending my head backward broke the crystals loose and caused BPPV," but that is probably not true; it is far more likely that the otoliths *were already loose* before she went to the hairdresser, and once they were loose, inclining the head backward permitted those (already loose) otoliths to enter one of the semicircular canals, thereby provoking symptoms.

Much of the information available on the internet gives the erroneous impression that BPPV is a matter of there being "one ball bearing that is loose," but actually this is more of a "grains of sand" issue—hence BPPV is a matter of degree, and can be present to a greater or lesser extent.

Epidemiology of BPPV

BPPV is the most common cause of dizziness over the lifespan [5–7]. It is less common before the age of about 40 years, after which its incidence increases exponentially [8]. The recognition of this relationship led to the idea that BPPV is somehow due to "wear and tear" on the inner ear. Electron micrographs comparing otoconia in young healthy individuals [9] versus older healthy individuals [10] support this idea. Given this deterioration of otoliths, it is not hard to imagine fragments of otoliths breaking loose.

Risk Factors for BPPV

Most cases of BPPV appear to occur without any discernible trigger—specifically, there is usually nothing in the patient's history that would explain why the otoliths "broke loose."

Probably the single most important risk factor for BPPV is a history of a prior episode of BPPV. This is because BPPV tends to recur (see below).

Since BPPV occurs more frequently with increasing age (see above), older age is also considered a "risk factor" for BPPV.

There are a few other factors that appear to increase the risk of BPPV, though the literature about these is mixed. A history of migraine appears to increase the risk of BPPV [11, 12]. Low bone density [13] and vitamin D deficiency [14] may increase the risk of BPPV. Finally, the risk of BPPV appears to be increased when a patient has a history of other ear diseases [15], such as vestibular neuritis [16], labyrinthitis [17], and Ménière's disease [18].

BPPV can occur following various dental procedures [19] such as dental extractions [20], placement of dental implants [21–28], removal of dental implants [29], and other procedures [30]. BPPV can also occur after various maxillofacial surgeries, such as nasal septoplasty [31] and sinus floor elevation [32–43].

The literature provides conflicting data on the relationship between head trauma and BPPV. For example, some literature suggests that BPPV following head trauma is less responsive to treatment [44–46]; in contrast, large case series [47] and meta-analyses [48] concluded that the available evidence does *not* show any difference in treatment responsiveness/outcomes between traumatic and nontraumatic BPPV. There is modest literature suggesting that bilateral posterior semicircular canal involvement [49] or (unilateral) anterior semicircular canal involvement [50, 51] are more common following head trauma. From a practical medical perspective, these considerations are irrelevant in the sense that, if a patient has manifest BPPV on examination, treatment should be considered, irrespective of prior history.

Diagnosing BPPV

How Is BPPV Diagnosed?

BPPV can be suspected from the clinical history, but can only be confirmed by observation of a pattern of eye movements [52]. The specific pattern of eye movements depends on which semicircular canal, or combination of the six semicircular canals, is affected by BPPV. The eye movements consist of various combinations of nystagmus.

Tests for BPPV

Observation of eye movements *is the test* for confirming BPPV [52–54].

In a patient who is already known to have BPPV, there may be some circumstances in which otovestibular testing is appropriate. If a patient appears not to be responding well to appropriately targeted treatment, many clinicians may consider checking tests of inner ear function to assess whether the patient has sustained otologic damage by some other disease process, which in turn might have rendered the patient more vulnerable to BPPV. For example, there is emerging evidence that assessment of vestibular evoked myogenic potentials may be useful in this regard [55–64], though not all investigators share this perspective [65].

Examination May Sometimes Not Identify BPPV in Someone Who Truly Has It

There are several reasons why an individual who truly has BPPV may not exhibit the associated eye movements on a particular examination.

The first reason why the diagnostic eye movements might not be seen in a person who truly has BPPV has to do with the behavior of the disease itself. BPPV is an *episodic* problem, which is why it has the word "paroxysmal" in its title. Because of this pattern, the diagnostic eye movements will not necessarily be present every time an examiner looks for them. As an example, the Dix–Hallpike maneuver is considered the "gold standard" for the diagnosis of posterior semicircular canal BPPV [53]; however, the maneuver has a reported specificity of 71% [66] and negative predictive value of 52% [67], meaning that a "negative" Dix–Hallpike maneuver does not entirely exclude the presence of posterior canal BPPV; for this reason, if a patient's history is suggestive of BPPV yet her examination does not confirm this suspicion, it is reasonable to re-examine the patient after some interval. That being said, the more frequently one looks for the abnormal eye movements and fails to detect them, the more skeptical one may become about the diagnosis.

The second reason why the diagnostic eye movements might not be seen in a person who truly has BPPV is that the examiner may be relying on an insufficiently sensitive method of detection; for example, the examiner may not using infrared video Frenzel oculography [68, 69]. If a clinician is only observing a patient's eye movements "face-to-face," or only using Fresnel lenses [70, 71], then some degree of vision is available to the patient; when vision is available, the patient will visually fixate on targets, and such fixation can suppress a variety of eye movement abnormalities [72–81], including many of the eye movement abnormalities that are diagnostically helpful.

The third reason why the diagnostic eye movements might not be seen in a person who truly has BPPV is that the abnormal eye movements (which would otherwise be present) can be suppressed by certain medications, such as vestibular suppressants (e.g., meclizine, dimenhydrinate), or benzodiazepines.

It should also be noted that if a patient suspected of having BPPV has repeatedly normal exams (despite an adequate examination technique, and despite being off vestibular suppressant medications), then the suspected diagnosis may be incorrect. We discuss this further below.

In a Patient with BPPV, Is It Necessary to Get Brain Imaging?

If a person's clinical history and eye movements are compatible with BPPV, and if their physical examination is otherwise normal, then brain imaging is usually unnecessary [53].

Subtypes of BPPV

Although we have been speaking of a "diagnosis of BPPV," there are several subtypes of this disease, depending on the specific semicircular canal or combination of semicircular canals involved.

When BPPV affects a single canal in a single ear, the statistics of involvement are as follows [82]:

- Posterior canal: 81–89% of cases
- Lateral canal: 8–17% of cases
- Anterior canal: 1–3% of cases

These statistics are not uncontroversial; for instance, Bhandari and colleagues [83] report a very different distribution of cases: out of 3975 BPPV patients they identified 47.8% had posterior canal involvement, 46.3% had lateral canal involvement, and 0.7% had anterior canal involvement—in other words, they found that lateral canal involvement comprised a significantly higher proportion of cases than in other studies.

We discuss the specific treatment maneuvers for each type of BPPV elsewhere.

Treatment for BPPV

Is Treatment of BPPV Necessary?

The management of BPPV is "symptom-driven," meaning that if symptoms are infrequent and mild, some patients reasonably elect to refrain from treatment, since spontaneous resolution (i.e., without treatment) can occur. If one initially elects to refrain from treatment but then the symptoms become more aggressive, then treatment can be initiated; it is never "too late" to treat BPPV.

Treatment for BPPV: Physical Therapy

Appropriately targeted vestibular rehabilitation therapy is firmly established as the standard of care for BPPV [53]. The particular series of physical therapy maneuvers is chosen based on which semicircular canal, or combination of semicircular canals, is involved.

Mechanism of Physical Therapy

Note that the physical therapy exercises do not "put the crystals back in place." Rather, by moving the head through certain positions, these maneuvers exploit gravity to "coax" the loose otoliths out of the semicircular canals, and into the utricle, where the otoliths have a higher chance of getting resorbed and phagocytosed by dark cells.

Note that even when BPPV has been correctly diagnosed, and even when the appropriately targeted treatment maneuver has been initiated, the response to therapy may be somewhat uneven; observing symptoms day-to-day, it may appear that on a given day the patient feels worse than the previous day; however, when observing symptoms over a longer span of time (several days in a row), the *overall trajectory* should be one of improvement. If the overall trajectory of symptoms is one of improvement (no matter how slow it seems), then the patient is likely pursuing the correct treatment.

In contrast, if while performing a given exercise the patient finds that symptoms are growing consistently worse with each day of treatment, for several days in a row, then this raises the possibility that the patient has suffered a "canal conversion" (in which the otoliths have exited one semicircular canal but entered a different semicircular canal; see the section below entitled, "Why is treatment for BPPV sometimes unsuccessful?"). In this circumstance, the patient should refrain from further self-treatment, and seek treatment by a physical therapist or physician so that they can examine the patient's eye movements.

Why Do the Physical Therapy Treatment Maneuvers Need to Be Repeated?

The various exercises do help maneuver the loose otoliths into a part of the ear where they have a better chance of getting resorbed, but the exercises *do not cause resorption instantaneously,* they *do not immediately dissolve the crystals,* and they *do not "glue the crystals back into place."* The crystals are *still there,* and while a few of them will get resorbed shortly after the exercises are performed, the remaining otoliths are *still loose,* and in the course of daily activities and head movements, some of those loose crystals will get back into places where they do not belong (the semicircular canals), where they will still be capable of provoking symptoms.

Therefore, where patients get the most benefit out of the treatment exercises is by repeating them over time. Thus, doing the appropriately targeted exercises 30 times in one night is actually *less effective* than doing those exercises 3 times per day over 10 days.

Is It Necessary to Sleep Upright After Doing the Treatment Maneuvers?

In posterior semicircular canal BPPV, lying supine and rolling over in bed often provoke symptoms, so it is reasonable to inquire whether remaining upright, or at least remaining in a partially reclined position (between completely upright and completely supine) might help. The short answer is that remaining upright may avert symptom provocation, but *merely remaining upright does not in itself treat BPPV.*

A related question is whether sleeping upright improves the efficacy of the various treatment maneuvers for BPPV. Most literature on this topic concludes that sleeping upright *does not improve treatment outcomes* for BPPV [84–93]. A few studies suggest that sleeping upright may be helpful [94, 95], while other studies simply conclude that further research is needed on this topic [96].

Treatment for BPPV: Medication

Medications such as vestibular suppressants (meclizine, dimenhydrinate) and benzodiazepines (clonazepam, diazepam) are sometimes prescribed, but their role in treating BPPV is very limited, and guidelines generally recommend *against* their use [53].

It must be kept in mind that these medications may succeed in "dampening" the sensation of dizziness, but they do not actually treat the underlying cause of BPPV—specifically, they do not dissolve otoliths or reattach them in their proper place.

These medications also pose several problems. The first is that by suppressing symptoms, it makes it more difficult to assess whether the disease causing those symptoms is actually improving—in other words, it may be hard to tell whether a given treatment maneuver is effective. The second problem is that these medications also suppress eye movements, thereby making it difficult to verify the diagnosis and monitor treatment responsiveness. The third problem is that these medications tend to be soporific (cause sleepiness).

Despite the discussion above, some investigators have made the case that medication may serve an adjunctive role to physical therapy [97]. We agree that in selected cases this may be appropriate; for example, if a patient's diagnosis of BPPV is secure and confidently localized (meaning that the involved canal is known and has been consistent), but the patient is having difficulty tolerating canalith repositioning maneuvers (e.g., due to significant nausea and vomiting) and thus is refraining from treatment, then premedicating with a vestibular suppressant may make it *feasible* for the patient to engage in appropriately targeted physical therapy.

Treatment for BPPV: Surgery

A variety of essentially destructive procedures have been attempted by otolaryngologists in the management of BPPV.

Of the surgical procedures attempted for management of BPPV the most common is "canal plugging," first reported by Lorne Parnes and Joseph McClure [98]. In this procedure, the surgeon fills the semicircular canal affected by BPPV with either bone pâté or a synthetic substance. The logic of this procedure is that plugging the affected canal "silences" it, because otoliths can no longer enter it. Subsequent studies [99] reported impressive success with this procedure, but it was soon realized [100] that potential complications of the procedure included hearing loss, vestibular deficits on objective testing, and even continued dizziness. The continued dizziness was particularly concerning, since that was the very symptom that the procedure was intended to treat. Among other problems, it was eventually recognized [101] that although plugging one canal might "silence" that canal, there remain five other semicircular canals that are still susceptible to developing BPPV.

A less commonly attempted surgical procedure for management of BPPV is cutting the nerve that transmits vestibular signals from the ear to the brain (which procedure is called "vestibular neurectomy") or, if possible, cutting just the part of that nerve that transmits balance signals from the affected canal (which procedure is called "singular neurectomy" [102]). While the latter procedure (singular neurectomy) appeared promising and minimally destructive, it also turned out to be technically challenging, and few surgeons attempted it.

Taking these factors into consideration, we generally do not recommend surgery as a treatment for patients with BPPV.

Why Is Treatment for BPPV Sometimes Unsuccessful?

There are several reasons why treatment for BPPV may be unsuccessful.

The first is that the specific case of BPPV may be very aggressive. This likely corresponds to a situation in which a very large number of otoliths have broken loose, thereby requiring more extended treatment before resorption is complete.

The second reason by treatment of BPPV may be unsuccessful is what is termed a "canal conversion" [103–109]. Since all parts of the labyrinth are connected, there is nothing to prevent otoliths leaving one semicircular canal and entering another one. This is a problem because the semicircular canals are arranged in different orientations, so a treatment maneuver directed at one semicircular canal will be ineffective as treatment for a different semicircular canal. Clinically this may be evident if a patient who had been improving with treatment suddenly starts to deteriorate.

The third reason why treatment of BPPV may be unsuccessful is that multiple semicircular canals may be involved, and this situation is generally more difficult to treat [110]. In some cases, at the time of diagnosis, it is recognized that multiple canals are involved. In other cases, a patient may begin with involvement of only one canal, but subsequently develop involvement of multiple canals.

A fourth reason why treatment of BPPV may be unsuccessful is that the patient has not only BPPV, but some other vestibular disease as well. This may become evident if, during the course of treatment for BPPV, a patient's eye movements normalize (in other words, the nystagmus that had been diagnostic of BPPV is no longer present on several sequential examinations), yet the patient still complains of symptoms. In this circumstance, it is reasonable to suspect that an additional disease may be at play.

A fifth reason why treatment of BPPV may be unsuccessful is that the patient actually does not have BPPV. Sometimes, even when practitioners are working in good faith and to the best of their abilities, a diagnosis of BPPV may simply be incorrect.

Can Recurrences of BPPV Be Prevented?

Since the treatment for BPPV is a specific series of maneuvers (described earlier), many patients reasonably ask whether they should continue doing these maneuvers in order to "ward off" future episodes. This approach has been studied, and found to be ineffective [111]. The maneuvers only work when BPPV is "active" (i.e., when symptoms are present). When symptoms are absent, that is because there are no more loose otoliths, and in this "quiescent" state, there is nothing for the maneuvers to "act on."

The observation that vitamin D deficiency may increase the risk of BPPV [14] leads to the question of whether supplementation of vitamin D might reduce the risk. There is emerging evidence that, after a first episode of BPPV, taking vitamin D and perhaps calcium may reduce the likelihood of recurrence of BPPV [112]. For instance, a randomized controlled trial [113] reported that supplementation with vitamin D 400 I.U. and calcium carbonate 500 mg twice per day reduces the annual recurrence rate of BPPV. Note that increasing one's level of vitamin D (even without supplementing additional calcium) may increase serum calcium concentrations, which at least in theory can increase the risk of developing kidney stones [114], though it is unclear whether this increased risk is clinically meaningful; one might think that vitamin D supplementation should perhaps be avoided in persons who already have a history of kidney stone formation, but this does not seem to be borne out by the evidence [115], so most patients considering supplementation with vitamin D first consult their primary care physicians.

Residual Disequilibrium Following Apparently Successful Treatment of BPPV

It is well documented that some patients with BPPV who are "successfully treated" (meaning that they no longer have ocular motor evidence for the presence of BPPV) continue to feel disequilibrium for some time [116]. Some literature refers to this as "residual dizziness following successful treatment of BPPV."

There are several plausible mechanisms for "residual dizziness" (RD) following "successful treatment of BPPV."

- Some literature describes persistent otoliths in the semicircular canals [117]. This explanation suggests that although an examiner can no longer see ocular motor evidence of BPPV, there are residual otoliths in the semicircular canals which are sufficient to provoke symptoms, but insufficient to provoke corresponding ocular motor findings.
- Some literature suggests that otolith organ dysfunction accounts for RD. Although the mechanism of BPPV is that otoliths become dislodged from the gelatinous membranes in the otolith organs and enter the semicircular canals, this explanation for RD points to the fact that the otolith organs may not function correctly without sufficient otoliths in their proper place [118]. In support of this idea are published observations of abnormal vestibular evoked myogenic potentials in some patients with BPPV [55–64].
- Since afferent signals from the otolith organs feed in to autonomic function [119, 120], and since autonomic dysfunction can be associated with vertiginous sensations [121, 122], it is possible that in patients with BPPV (whose otolith organs are dysfunctional secondary to the loss of the otoliths) also have some degree of autonomic dysfunction [123] that accounts for RD despite adequate treatment of BPPV.
- Some literature regarding fMRI in patients with BPPV identify functional changes [124–127]. The mechanism of this is poorly understood, but suggests that in some patients, the presence of BPPV induces neuroplastic changes that may persist beyond the presence of BPPV—in other words, a maladaptive response.
- Another possibility is that BPPV triggers persistent postural perceptual dizziness (PPPD) [128, 129].

The main risk factor for RD appears to be older age [130–132]. Other risk factors include longer duration of BPPV [132, 133] and hypovitaminosis D [134].

The prognosis of RD is unclear. Some literature describes it as a "self-limiting" disorder [131] that tends to resolve "in 3 months without specific treatment" [135].

For patients who elect to attempt treatment for RD, the management is unclear. Betahistine has been suggested, but results are mixed, with some studies suggesting efficacy [136] and others not [137]. Some studies suggest that cognitive behavioral therapy (CBT) may play some role [138]. Even in a patient who appears to have RD, it is medically reasonable to reassess periodically to monitor for whether there is true recurrence of BPPV.

Prognosis of BPPV: Recurrence

The natural history of BPPV is one of recurrence. One study [139] of 1105 patients found that after a first episode of BPPV had resolved, recurrence happened anywhere from 3 to 2496 days later; 56% of patients experienced recurrence within the first year, and 76% of patients experienced recurrence within 2 years.

Recurrences usually (but not always) involve the same semicircular canal that was most recently affected. Therefore, if a given vestibular treatment exercise clearly helps a bout of BPPV, but a patient then experiences what feels like a recurrence, it is medically quite reasonable to resume exactly the same exercise that was most recently helpful, because statistically it is likely to be correct. If resuming the previously successful exercise does, in fact, help, then no further intervention is needed. In contrast, if a recurrence fails to respond to a previously successful exercise, then a patient should have her eye movements examined again in order to determine whether the BPPV is "in the same place," or whether the problem is actually still BPPV.

As discussed earlier (see the section above entitled, "Can Recurrences of BPPV Be Prevented?"), there is emerging evidence that supplementation with vitamin D (and perhaps calcium) may reduce the rate of recurrence of BPPV in patients who have vitamin D deficiency; in contrast, in patients whose vitamin D levels are already normal, giving extra vitamin D does not appear to have any influence on the rate of recurrence of BPPV.

There is a modest literature suggesting that BPPV following head trauma is more likely to recur [50], though not all investigators share this viewpoint.

References

1. Bárány E. Diagnose von Krankheitserscheinungen im Bereiche des Otolithenapparates [diagnosis of symptoms in the area of the otolith apparatus]. Acta Otolaryngol. 1920;2(3):434–7.
2. Dix MR, Hallpike CS. The pathology symptomatology and diagnosis of certain common disorders of the vestibular system. Proc R Soc Med. 1952;45(6):341–54.
3. Dix MR, Hallpike CS. The pathology, symptomatology and diagnosis of certain common disorders of the vestibular system. Ann Otol Rhinol Laryngol. 1952;61(4):987–1016.
4. Choi JY, Park YM, Lee SH, Choi J, Hyun SW, Song JM, et al. Linear Vertigo in benign paroxysmal positional Vertigo: prevalence and mechanism. Cerebellum. 2021;20(2):160–8.
5. Parnes LS, Agrawal SK, Atlas J. Diagnosis and management of benign paroxysmal positional vertigo (BPPV). CMAJ. 2003;169(7):681–93.
6. Nedzelski JM, Barber HO, McIlmoyl L. Diagnoses in a dizziness unit. J Otolaryngol. 1986;15(2):101–4.
7. Neuhauser HK. Epidemiology of vertigo. Curr Opin Neurol. 2007;20(1):40–6.
8. von Brevern M, Radtke A, Lezius F, Feldmann M, Ziese T, Lempert T, et al. Epidemiology of benign paroxysmal positional vertigo: a population based study. J Neurol Neurosurg Psychiatry. 2007;78(7):710–5.
9. Otoconia KR. Mimicking a calcite-based functional material of the human body. From basic research to medical aspects. Pure Appl Chem. 2015;87(8):719–36.
10. Walther LE, Westhofen M. Presbyvertigo-aging of otoconia and vestibular sensory cells. J Vestib Res. 2007;17(2–3):89–92.
11. Ishiyama A, Jacobson KM, Baloh RW. Migraine and benign positional vertigo. Ann Otol Rhinol Laryngol. 2000;109(4):377–80.
12. Kim SK, Hong SM, Park IS, Choi HG. Association between migraine and benign paroxysmal positional Vertigo among adults in South Korea. JAMA Otolaryngol Head Neck Surg. 2019;145(4):307–12.
13. Jeong SH, Choi SH, Kim JY, Koo JW, Kim HJ, Kim JS. Osteopenia and osteoporosis in idiopathic benign positional vertigo. Neurology. 2009;72(12):1069–76.
14. Chen J, Zhao W, Yue X, Zhang P. Risk factors for the occurrence of benign paroxysmal positional Vertigo: a systematic review and Meta-analysis. Front Neurol. 2020;11:506.
15. Lee NH, Ban JH, Lee KC, Kim SM. Benign paroxysmal positional vertigo secondary to inner ear disease. Otolaryngol Head Neck Surg. 2010;143(3):413–7.
16. Balatsouras DG, Koukoutsis G, Ganelis P, Economou NC, Moukos A, Aspris A, et al. Benign paroxysmal positional vertigo secondary to vestibular neuritis. Eur Arch Otorrinolaringol. 2014;271(5):919–24.
17. Baloh RW, Honrubia V, Jacobson K. Benign positional vertigo: clinical and oculographic features in 240 cases. Neurology. 1987;37(3):371–8.
18. Kutlubaev MA, Xu Y, Hornibrook J. Benign paroxysmal positional vertigo in Meniere's disease: systematic review and meta-analysis of frequency and clinical characteristics. J Neurol. 2019;268(5):1608–14.
19. Chang TP, Lin YW, Sung PY, Chuang HY, Chung HY, Liao WL. Benign paroxysmal positional Vertigo after dental procedures: a population-based case-control study. PLoS One. 2016;11(4):e0153092.
20. Chiarella G, Leopardi G, De Fazio L, Chiarella R, Cassandro E. Benign paroxysmal positional vertigo after dental surgery. Eur Arch Otorrinolaringol. 2008;265(1):119–22.
21. Wanner L, Manegold-Brauer G, Brauer HU. Review of unusual intraoperative and postoperative complications associated with endosseous implant placement. Quintessence Int. 2013;44(10):773–81.
22. Bennardo F, Barone S, Vocaturo C, Nucci L, Antonelli A, Giudice A. Usefulness of magnetic mallet in oral surgery and implantology: A systematic review. J Pers Med. 2022;12(1):108.
23. Crespi R, Cappare P, Gherlone E. Electrical mallet provides essential advantages in maxillary bone condensing. A prospective clinical study. Clin Implant Dent Relat Res. 2013;15(6):874–82.
24. Crespi R, Cappare P, Gherlone E. A comparison of manual and electrical mallet in maxillary bone condensing for immediately loaded implants: a randomized study. Clin Implant Dent Relat Res. 2014;16(3):374–82.
25. Galli M, Petracca T, Minozzi F, Gallottini L. Complications in implant surgery by summer's technique: benign paroxysmal positional vertigo (BPPV). Minerva Stomatol. 2004;53(9):535–41.
26. Kaplan DM, Attal U, Kraus M. Bilateral benign paroxysmal positional vertigo following a tooth implantation. J Laryngol Otol. 2003;117(4):312–3.
27. Penarrocha-Diago M, Rambla-Ferrer J, Perez V, Perez-Garrigues H. Benign paroxysmal vertigo secondary to placement of maxillary implants using the alveolar expansion technique with osteotomes: a study of 4 cases. Int J Oral Maxillofac Implants. 2008;23(1):129–32.
28. Rodriguez Gutierrez C, Rodriguez GE. Positional vertigo afterwards maxillary dental implant surgery with bone regeneration. Med Oral Patol Oral Cir Bucal. 2007;12(2):E151–3.
29. Messina AM, Marini E. Benign paroxysmal positional Vertigo after Piezosurgical removal of Osseointegrated implants. J Craniofac Surg. 2018;29(8):e739–e40.
30. Brauer HU. Unusual complications associated with third molar surgery: a systematic review. Quintessence Int. 2009;40(7):565–72.
31. Kansu L, Aydin E, Gulsahi K. Benign paroxysmal positional vertigo after nonotologic surgery: case series. J Maxillofac Oral Surg. 2015;14(Suppl 1):113–5.
32. Akcay H, Ulu M, Kelebek S, Aladag I. Benign paroxysmal positional Vertigo following sinus floor elevation in patient with antecedents of Vertigo. J Maxillofac Oral Surg. 2016;15(Suppl 2):351–4.
33. Crespi R, Cappare P, Gherlone E. Sinus floor elevation by osteotome: hand mallet versus electric mallet. A prospective clinical study. Int J Oral Maxillofac Implants. 2012;27(5):1144–50.
34. Damsaz M, Eshghpour M, Grillo R, Jafarpour K, Mazhari K, Samieirad S. Assessment of benign paroxysmal positional Vertigo (BPPV) incidence following closed sinus lift and ridge splitting surgeries: a cohort study. J Maxillofac Oral Surg. 2024;23(1):1–6.
35. Giannini S, Signorini L, Bonanome L, Severino M, Corpaci F, Cielo A. Benign paroxysmal positional vertigo (BPPV): it may occur after dental implantology. A mini topical review. Eur Rev Med Pharmacol Sci. 2015;19(19):3543–7.

36. Hsu YT, Rosen PS, Choksi K, Shih MC, Ninneman S, Lee CT. Complications of sinus floor elevation procedure and management strategies: a systematic review. Clin Implant Dent Relat Res. 2022;24(6):740–65.
37. Molina A, Sanz-Sanchez I, Sanz-Martin I, Ortiz-Vigon A, Sanz M. Complications in sinus lifting procedures: classification and management. Periodontol 2000. 2022;88(1):103–15.
38. Pjetursson BE, Lang NP. Sinus floor elevation utilizing the transalveolar approach. Periodontol 2000. 2014;66(1):59–71.
39. Reddy KS, Shivu ME, Billimaga A. Benign paroxysmal positional vertigo during lateral window sinus lift procedure: a case report and review. Implant Dent. 2015;24(1):106–9.
40. Sammartino G, Mariniello M, Scaravilli MS. Benign paroxysmal positional vertigo following closed sinus floor elevation procedure: mallet osteotomes vs. screwable osteotomes. A triple blind randomized controlled trial. Clin Oral Implants Res. 2011;22(6):669–72.
41. Stacchi C, Bernardello F, Spinato S, Mura R, Perelli M, Lombardi T, et al. Intraoperative complications and early implant failure after transcrestal sinus floor elevation with residual bone height </=5 mm: a retrospective multicenter study. Clin Oral Implants Res. 2022;33(8):783–91.
42. Su GN, Tai PW, Su PT, Chien HH. Protracted benign paroxysmal positional vertigo following osteotome sinus floor elevation: a case report. Int J Oral Maxillofac Implants. 2008;23(5):955–9.
43. Vernamonte S, Mauro V, Vernamonte S, Messina AM. An unusual complication of osteotome sinus floor elevation: benign paroxysmal positional vertigo. Int J Oral Maxillofac Surg. 2011;40(2):216–8.
44. Babac S, Djeric D, Petrovic-Lazic M, Arsovic N, Mikic A. Why do treatment failure and recurrences of benign paroxysmal positional vertigo occur? Otol Neurotol. 2014;35(6):1105–10.
45. Pisani V, Mazzone S, Di Mauro R, Giacomini PG, Di Girolamo S. A survey of the nature of trauma of post-traumatic benign paroxysmal positional vertigo. Int J Audiol. 2015;54(5):329–33.
46. Balatsouras DG, Koukoutsis G, Aspris A, Fassolis A, Moukos A, Economou NC, et al. Benign paroxysmal positional Vertigo secondary to mild head trauma. Ann Otol Rhinol Laryngol. 2017;126(1):54–60.
47. Luryi AL, LaRouere M, Babu S, Bojrab DI, Zappia J, Sargent EW, et al. Traumatic versus idiopathic benign positional Vertigo: analysis of disease, treatment, and outcome characteristics. Otolaryngol Head Neck Surg. 2019;160(1):131–6.
48. Aron M, Lea J, Nakku D, Westerberg BD. Symptom resolution rates of posttraumatic versus nontraumatic benign paroxysmal positional Vertigo: a systematic review. Otolaryngol Head Neck Surg. 2015;153(5):721–30.
49. Katsarkas A. Benign paroxysmal positional vertigo (BPPV): idiopathic versus post-traumatic. Acta Otolaryngol. 1999;119(7):745–9.
50. Dlugaiczyk J, Siebert S, Hecker DJ, Brase C, Schick B. Involvement of the anterior semicircular canal in posttraumatic benign paroxysmal positioning vertigo. Otol Neurotol. 2011;32(8):1285–90.
51. Jackson LE, Morgan B, Fletcher JC Jr, Krueger WW. Anterior canal benign paroxysmal positional vertigo: an underappreciated entity. Otol Neurotol. 2007;28(2):218–22.
52. von Brevern M, Bertholon P, Brandt T, Fife T, Imai T, Nuti D, et al. Benign paroxysmal positional vertigo: Diagnostic criteria. J Vestib Res. 2015;25(3–4):105–17.
53. Bhattacharyya N, Gubbels SP, Schwartz SR, Edlow JA, El-Kashlan H, Fife T, et al. Clinical practice guideline: benign paroxysmal positional Vertigo (update). Otolaryngol Head Neck Surg. 2017;156(3_suppl):S1–S47.
54. von Brevern M, Bertholon P, Brandt T, Fife T, Imai T, Nuti D, et al. Benign paroxysmal positional vertigo: diagnostic criteria consensus document of the Committee for the Classification of vestibular disorders of the Barany society. Acta Otorrinolaringol Esp. 2017;68(6):349–60.
55. Hong SM, Park DC, Yeo SG, Cha CI. Vestibular evoked myogenic potentials in patients with benign paroxysmal positional vertigo involving each semicircular canal. Am J Otolaryngol. 2008;29(3):184–7.
56. Yang WS, Kim SH, Lee JD, Lee WS. Clinical significance of vestibular evoked myogenic potentials in benign paroxysmal positional vertigo. Otol Neurotol. 2008;29(8):1162–6.
57. Yetiser S, Ince D, Gul M. An analysis of vestibular evoked myogenic potentials in patients with benign paroxysmal positional vertigo. Ann Otol Rhinol Laryngol. 2014;123(10):686–95.
58. Hoseinabadi R, Pourbakht A, Yazdani N, Kouhi A, Kamali M. The effects of abnormality of cVEMP and oVEMP on rehabilitation outcomes in patients with idiopathic benign paroxysmal positional vertigo. Eur Arch Otorrinolaringol. 2016;273(3):643–8.
59. Singh NK, Apeksha K. Efficacy of cervical and ocular vestibular-evoked myogenic potentials in evaluation of benign paroxysmal positional vertigo of posterior semicircular canal. Eur Arch Otorrinolaringol. 2016;273(9):2523–32.
60. Lee JD, Park MK, Lee BD, Lee TK, Sung KB, Park JY. Abnormality of cervical vestibular-evoked myogenic potentials and ocular vestibular-evoked myogenic potentials in patients with recurrent benign paroxysmal postitional vertigo. Acta Otolaryngol. 2013;133(2):150–3.
61. Scarpa A, Gioacchini FM, Cassandro E, Tulli M, Ralli M, Re M, et al. Clinical application of cVEMPs and oVEMPs in patients affected by Meniere's disease, vestibular neuritis and benign paroxysmal positional vertigo: a systematic review. Acta Otorhinolaryngol Ital. 2019;39(5):298–307.
62. Xu H, Liang FY, Chen L, Song XC, Tong MC, Thong JF, et al. Evaluation of the utricular and saccular function using oVEMPs and cVEMPs in BPPV patients. J Otolaryngol Head Neck Surg. 2016;45:12.
63. Korres S, Gkoritsa E, Giannakakou-Razelou D, Yiotakis I, Riga M, Nikolpoulos TP. Vestibular evoked myogenic potentials in patients with BPPV. Med Sci Monit. 2011;17(1):CR42-7.
64. Oya R, Imai T, Takenaka Y, Sato T, Oshima K, Ohta Y, et al. Clinical significance of cervical and ocular vestibular evoked myogenic potentials in benign paroxysmal positional vertigo: a meta-analysis. Eur Arch Otorrinolaringol. 2019;276(12):3257–65.
65. Manzari L, Tramontano M. Otolithic and canal functions assessment during the acute phase of benign paroxysmal positional vertigo. Eur Arch Otorrinolaringol. 2023;281(5):2253–7.
66. Lopez-Escamez JA, Lopez-Nevot A, Gamiz MJ, Moreno PM, Bracero F, Castillo JL, et al. Diagnosis of common causes of vertigo using a structured clinical history. Acta Otorrinolaringol Esp. 2000;51(1):25–30.
67. Hanley K, O'Dowd T. Symptoms of vertigo in general practice: a prospective study of diagnosis. Br J Gen Pract. 2002;52(483):809–12.
68. Levo H, Aalto H, Petteri HT. Nystagmus measured with videooculography: methodological aspects and normative data. ORL J Otorhinolaryngol Relat Spec. 2004;66(3):101–4.
69. Frenzel HGW. Pragmatische Systematik in der Hals-Nasen-Ohren-Heilkunde [practical methods of a systematic study of otorhinolaryngology]. Münchener Medizinische Wochenschrift. 1956;98(29):972–5.
70. Fresnel A-J. Mémoire sur un nuveau système d'éclairage des phares. Paris: Imprimerie Royale Lu à L'académie Des Sciences; 1822.
71. Strupp M, Fischer C, Hanss L, Bayer O. The takeaway Frenzel goggles: a Fresnel-based device. Neurology. 2014;83(14):1241–5.
72. Sokolovski A. The influence of mental activity and visual fixation upon caloric-induced nystagmus in normal subjects. Acta Otolaryngol. 1966;61(3):209–20.

73. Hart CW. Ocular fixation and the caloric test. Laryngoscope. 1967;77(12):2103–13.
74. Molnar EM, Torok N. The effect of ocular fixation on the caloric nystagmus. ORL J Otorhinolaryngol Relat Spec. 1974;36(2):76–84.
75. Baloh RW, Solingen L, Sills AW, Honrubia V, Caloric testing. 1. Effect of different conditions of ocular fixation. Ann Otol Rhinol Laryngol Suppl. 1977;86(5 Pt 3 Suppl 43):1–6.
76. Karlsen EA, Goetzinger CP, Hassanein R. Effects of six conditions of ocular fixation on caloric nystagmus. Arch Otolaryngol. 1980;106(8):474–6.
77. Chambers BR, Gresty MA. Effects of fixation and optokinetic stimulation on vestibulo-ocular reflex suppression. J Neurol Neurosurg Psychiatry. 1982;45(11):998–1004.
78. Katsarkas A, Kirkham TH. Failure of suppression of post-caloric nystagmus by fixation. J Otolaryngol. 1982;11(1):57–9.
79. Gauthier GM, Vercher JL. Visual vestibular interaction: vestibulo-ocular reflex suppression with head-fixed target fixation. Exp Brain Res. 1990;81(1):150–60.
80. Gizzi MS, Harper HW. Suppression of the human vestibulo-ocular reflex by visual fixation or forced convergence in the dark, with a model interpretation. Curr Eye Res. 2003;26(5):281–90.
81. Hirvonen TP, Juhola M, Aalto H. Suppression of spontaneous nystagmus during different visual fixation conditions. Eur Arch Otorrinolaringol. 2012;269(7):1759–62.
82. Fife TD. Positional dizziness. Continuum (Minneap Minn). 2012;18(5 Neuro-otology):1060–85.
83. Bhandari R, Bhandari A, Hsieh Y-H, Edlow J, Omron R. Prevalence of Horizontal Canal variant in 3,975 patients with benign paroxysmal positional Vertigo: a cross-sectional study. Neurol Clin Pract. 2023;13:e200191.
84. Simoceli L, Bittar RS, Greters ME. Posture restrictions do not interfere in the results of canalith repositioning maneuver. Braz J Otorhinolaryngol. 2005;71(1):55–9.
85. Roberts RA, Gans RE, DeBoodt JL, Lister JJ. Treatment of benign paroxysmal positional vertigo: necessity of postmaneuver patient restrictions. J Am Acad Audiol. 2005;16(6):357–66.
86. Nuti D, Nati C, Passali D. Treatment of benign paroxysmal positional vertigo: no need for postmaneuver restrictions. Otolaryngol Head Neck Surg. 2000;122(3):440–4.
87. Mostafa BE, Youssef TA, Hamad AS. The necessity of post-maneuver postural restriction in treating benign paroxysmal positional vertigo: a meta-analytic study. Eur Arch Otorrinolaringol. 2013;270(3):849–52.
88. Moon SJ, Bae SH, Kim HD, Kim JH, Cho YB. The effect of postural restrictions in the treatment of benign paroxysmal positional vertigo. Eur Arch Otorrinolaringol. 2005;262(5):408–11.
89. Fyrmpas G, Rachovitsas D, Haidich AB, Constantinidis J, Triaridis S, Vital V, et al. Are postural restrictions after an Epley maneuver unnecessary? First results of a controlled study and review of the literature. Auris Nasus Larynx. 2009;36(6):637–43.
90. De Stefano A, Dispenza F, Citraro L, Petrucci AG, Di Giovanni P, Kulamarva G, et al. Are postural restrictions necessary for management of posterior canal benign paroxysmal positional vertigo? Ann Otol Rhinol Laryngol. 2011;120(7):460–4.
91. Devaiah AK, Andreoli S. Postmaneuver restrictions in benign paroxysmal positional vertigo: an individual patient data meta-analysis. Otolaryngol Head Neck Surg. 2010;142(2):155–9.
92. Casqueiro JC, Ayala A, Monedero G. No more postural restrictions in posterior canal benign paroxysmal positional vertigo. Otol Neurotol. 2008;29(5):706–9.
93. Gordon CR, Gadoth N. Repeated vs single physical maneuver in benign paroxysmal positional vertigo. Acta Neurol Scand. 2004;110(3):166–9.
94. Cakir BO, Ercan I, Cakir ZA, Turgut S. Efficacy of postural restriction in treating benign paroxysmal positional vertigo. Arch Otolaryngol Head Neck Surg. 2006;132(5):501–5.
95. McGinnis PQ, Nebbia M, Saez L, Rudolph K. Retrospective comparison of outcomes for patients with benign paroxysmal positional vertigo based on length of postural restrictions. J Geriatr Phys Ther. 2009;32(4):168–73.
96. Papacharalampous GX, Vlastarakos PV, Kotsis GP, Davilis D, Manolopoulos L. The role of postural restrictions after BPPV treatment: real effect on successful treatment and BPPV'S recurrence rates. Int J Otolaryngol. 2012;2012:932847.
97. Chen J, Liu Z, Xie Y, Jin S. Effects of vestibular rehabilitation training combined with anti-vertigo drugs on vertigo and balance function in patients with vestibular neuronitis: a systematic review and meta-analysis. Front Neurol. 2023;14:1278307.
98. Parnes LS, McClure JA. Posterior semicircular canal occlusion for intractable benign paroxysmal positional vertigo. Ann Otol Rhinol Laryngol. 1990;99(5 Pt 1):330–4.
99. Shaia WT, Zappia JJ, Bojrab DI, LaRouere ML, Sargent EW, Diaz RC. Success of posterior semicircular canal occlusion and application of the dizziness handicap inventory. Otolaryngol Head Neck Surg. 2006;134(3):424–30.
100. Kisilevsky V, Bailie NA, Dutt SN, Rutka JA. Lessons learned from the surgical management of benign paroxysmal positional vertigo: the university health network experience with posterior semicircular canal occlusion surgery (1988-2006). J Otolaryngol Head Neck Surg. 2009;38(2):212–21.
101. Luryi AL, Schutt CA, Bojrab DI, LaRouere M, Zappia J, Sargent EW, et al. Causes of persistent positional Vertigo following posterior Semicircular Canal occlusion for benign paroxysmal positional Vertigo. Otol Neurotol. 2018;39(10):e1078–e83.
102. Gacek RR. Posterior ampullary nerve transection for benign paroxysmal positional vertigo. Trans Pa Acad Ophthalmol Otolaryngol. 1978;31(2):190–3.
103. Lin GC, Basura GJ, Wong HT, Heidenreich KD. Canal switch after canalith repositioning procedure for benign paroxysmal positional vertigo. Laryngoscope. 2012;122(9):2076–8.
104. Babic BB, Jesic SD, Milovanovic JD, Arsovic NA. Unintentional conversion of benign paroxysmal positional vertigo caused by repositioning procedures for canalithiasis: transitional BPPV. Eur Arch Otorrinolaringol. 2014;271(5):967–73.
105. Dispenza F, De Stefano A, Costantino C, Rando D, Giglione M, Stagno R, et al. Canal switch and re-entry phenomenon in benign paroxysmal positional vertigo: difference between immediate and delayed occurrence. Acta Otorhinolaryngol Ital. 2015;35(2):116–20.
106. Lee G, Lee SG, Park HS, Kim BJ, Choi SJ, Choi JW. Clinical characteristics and associated factors of canal switch in benign paroxysmal positional vertigo. J Vestib Res: Equilib Orientat. 2019;29(5):253–60.
107. Anagnostou E, Stamboulis E, Kararizou E. Canal conversion after repositioning procedures: comparison of Semont and Epley maneuver. J Neurol. 2014;261(5):866–9.
108. Foster CA, Zaccaro K, Strong D. Canal conversion and reentry: a risk of Dix-Hallpike during canalith repositioning procedures. Otol Neurotol. 2012;33(2):199–203.
109. Park S, Kim BG, Kim SH, Chu H, Song MY, Kim M. Canal conversion between anterior and posterior semicircular canal in benign paroxysmal positional vertigo. Otol Neurotol. 2013;34(9):1725–8.
110. Shim DB, Song CE, Jung EJ, Ko KM, Park JW, Song MH. Benign paroxysmal positional vertigo with simultaneous involvement of multiple semicircular canals. Korean J Audiol. 2014;18(3):126–30.
111. Helminski JO, Janssen I, Hain TC. Daily exercise does not prevent recurrence of benign paroxysmal positional vertigo. Otol Neurotol. 2008;29(7):976–81.

112. Jeong SH, Lee SU, Kim JS. Prevention of recurrent benign paroxysmal positional vertigo with vitamin D supplementation: a meta-analysis. J Neurol. 2020;269(2):619–26.
113. Jeong SH, Kim JS, Kim HJ, Choi JY, Koo JW, Choi KD, et al. Prevention of benign paroxysmal positional vertigo with vitamin D supplementation: a randomized trial. Neurology. 2020;95(9):e1117–e25.
114. Letavernier E, Daudon M, Vitamin D. Hypercalciuria and kidney stones. Nutrients. 2018;10:3.
115. Ganji MR, Shafii Z, Hakemi MS. Vitamin D supplementation and risk of Hypercalciuria in stone formers. Iran J Kidney Dis. 2019;13(1):27–31.
116. Özgirgin ON, Kingma H, Manzari L, Lacour M. Residual dizziness after BPPV management: exploring pathophysiology and treatment beyond canalith repositioning maneuvers. Front Neurol. 2024;15:1382196.
117. Dispenza F, Mazzucco W, Mazzola S, Martines F. Observational study on risk factors determining residual dizziness after successful benign paroxysmal positional vertigo treatment: the role of subclinical BPPV. Acta Otorhinolaryngol Ital. 2019;39(5):347–52.
118. Ismail NM, Kabil SE, Abdel-Hamid EF. Otolithic functions in patients with residual dizziness after successful repositioning manoeuvres for unilateral posterior canal BPPV. J Int Med Res. 2024;52(5):3000605241249095.
119. Yamamoto KI, Kawasaki Y, Izumi R, Kubo T. Assessment of the otolith-autonomic response using spectral analysis of the R-R interval and blood pressure. Otolaryngol Head Neck Surg. 2000;123(5):582–6.
120. Kim JG, Lee JH, Lee SU, Choi JY, Kim BJ, Kim JS. Utricular dysfunction in patients with orthostatic hypotension. Clin Auton Res. 2022;32(6):431–44.
121. Nakagawa H, Ohashi N, Kanda K, Watanabe Y. Autonomic nervous system disturbance as etiological background of vertigo and dizziness. Acta Otolaryngol Suppl. 1993;504:130–3.
122. Pappas DG Jr. Autonomic related vertigo. Laryngoscope. 2003;113(10):1658–71.
123. Kim HA, Lee H. Autonomic dysfunction as a possible cause of residual dizziness after successful treatment in benign paroxysmal positional vertigo. Clin Neurophysiol. 2014;125(3):608–14.
124. Chen Z, Cai Y, Xiao L, Wei XE, Liu Y, Lin C, et al. Increased functional connectivity between default mode network and visual network potentially correlates with duration of residual dizziness in patients with benign paroxysmal positional vertigo. Front Neurol. 2024;15:1363869.
125. Chen Z, Liu Y, Lin C, Liu D, Xiao L, Liu H, et al. Altered parietal operculum cortex 2 functional connectivity in benign paroxysmal positional vertigo patients with residual dizziness: a resting-state fMRI study. CNS Neurosci Ther. 2024;30(2):e14570.
126. Fu W, Bai Y, He F, Wei D, Wang Y, Shi Y, et al. The association between Precuneus function and residual dizziness in patients with benign paroxysmal positional Vertigo. Front Neurol. 2022;13:828642.
127. Lin C, Liu D, Liu Y, Chen Z, Wei X, Liu H, et al. Altered functional activity of the precuneus and superior temporal gyrus in patients with residual dizziness caused by benign paroxysmal positional vertigo. Front Neurosci. 2023;17:1221579.
128. Casani AP, Ducci N, Lazzerini F, Vernassa N, Bruschini L. Preceding benign paroxysmal positional Vertigo as a trigger for persistent postural-perceptual dizziness: which clinical predictors? Audiol Res. 2023;13(6):942–51.
129. Habs M, Strobl R, Grill E, Dieterich M, Becker-Bense S. Primary or secondary chronic functional dizziness: does it make a difference? A DizzyReg study in 356 patients. J Neurol. 2020;267(Suppl 1):212–22.
130. Fu W, He F, Bai Y, Wang Y, Wei D, Shi Y, et al. Assessment of residual dizziness after successful canalith repositioning maneuvre in benign paroxysmal positional vertigo patients: a questionnaire-based study. Eur Arch Otorrinolaringol. 2023;280(1):137–41.
131. Martellucci S, Pagliuca G, de Vincentiis M, Greco A, De Virgilio A, Nobili Benedetti FM, et al. Features of residual dizziness after Canalith repositioning procedures for benign paroxysmal positional Vertigo. Otolaryngol Head Neck Surg. 2016;154(4):693–701.
132. Teggi R, Giordano L, Bondi S, Fabiano B, Bussi M. Residual dizziness after successful repositioning maneuvers for idiopathic benign paroxysmal positional vertigo in the elderly. Eur Arch Otorrinolaringol. 2011;268(4):507–11.
133. Teggi R, Quaglieri S, Gatti O, Benazzo M, Bussi M. Residual dizziness after successful repositioning maneuvers for idiopathic benign paroxysmal positional vertigo. ORL J Otorhinolaryngol Relat Spec. 2013;75(2):74–81.
134. Wu Y, Han K, Han W, Fan Z, Zhou M, Lu X, et al. Low 25-Hydroxyvitamin D levels are associated with residual dizziness after successful treatment of benign paroxysmal positional Vertigo. Front Neurol. 2022;13:915239.
135. Seok JI, Lee HM, Yoo JH, Lee DK. Residual dizziness after successful repositioning treatment in patients with benign paroxysmal positional vertigo. J Clin Neurol. 2008;4(3):107–10.
136. Jalali MM, Gerami H, Saberi A, Razaghi S. The impact of Betahistine versus Dimenhydrinate in the resolution of residual dizziness in patients with benign paroxysmal positional Vertigo: a randomized clinical trial. Ann Otol Rhinol Laryngol. 2020;129(5):434–40.
137. Acar B, Karasen RM, Buran Y. Efficacy of medical therapy in the prevention of residual dizziness after successful repositioning maneuvers for benign paroxysmal positional Vertigo (BPPV). B-ENT. 2015;11(2):117–21.
138. Wan TJ, Yu YC, Zhao XG, Tang P, Gong YS. Efficacy of betahistine plus cognitive behavioral therapy on residual dizziness after successful canalith repositioning procedure for benign paroxysmal positional vertigo. Neuropsychiatr Dis Treat. 2018;14:2965–71.
139. Luryi AL, Lawrence J, Bojrab DI, LaRouere M, Babu S, Zappia J, et al. Recurrence in benign paroxysmal positional Vertigo: a large, single-institution study. Otol Neurotol. 2018;39(5):622–7.

Benign Paroxysmal Positional Vertigo Involving the Posterior Semicircular Canal

17

Benign Paroxysmal Positional Vertigo Involving the Posterior Semicircular Canal

Posterior semicircular canal benign paroxysmal positional vertigo (BPPV) is by far the most common type, accounting for over 80% of cases.

Posterior semicircular canal BPPV can be triggered in a variety of positions, some of the more common of which are given epithets in the literature; examples include the "top shelf phenomenon" (when a patient inclines the head backward while reaching for an object above her head) [1] and the "under the sink phenomenon" (when an individual leans forward and turns the head horizontally while trying to get the detergent out from under the sink). Other common scenarios include lying on one side, or inclining the head backward (such as in the dentist's chair, or at the hairdresser's).

The diagnosis of posterior semicircular canal BPPV is secured by observing a specific pattern of eye movements when the patient is in a specific position. The diagnostic position is called the Dix–Hallpike maneuver [2–4]. The observed eye movement is a pattern of nystagmus consisting of a combination of nystagmus in which the fast phase of the eye movement is ipsiversive (beating toward the affected ear), ipsitorsional (the top pole of the eye rotates toward the affected ear), and up beat (toward the top of the patient's head).

In cases where the posterior semicircular canal is involved on only one side, treatment typically begins with the Epley maneuver (or a variant thereof) on that side [5]. Multiple trials have established the efficacy of this treatment [6–8]. If an appropriate trial of the Epley maneuver fails, then some physical therapists and clinicians will consider other maneuvers, such as the Semont maneuver [9], the Foster maneuver [10], or the Gans maneuver [11]. The efficacy of these maneuvers has been assessed in meta-analyses [8]. All of these maneuvers are intended to treat posterior semicircular canal BPPV on one side when that side is known to be affected. Note that one cannot "alternate" treatment sides, because treating a given side will "undo" whatever had been accomplished in treating the opposite side. In other words, if an individual is thought to have posterior semicircular canal BPPV on one side but the side is unknown, one cannot "just try treating both sides." Or, if an individual is definitely known to have bilateral posterior semicircular canal involvement, one cannot alternate treatment sides.

In cases where the posterior semicircular canal is known to be involved on both sides, treatment sometimes begins with the Brandt–Daroff maneuver [12].

References

1. Squires TM, Weidman MS, Hain TC, Stone HA. A mathematical model for top-shelf vertigo: the role of sedimenting otoconia in BPPV. J Biomech. 2004;37(8):1137–46.
2. von Brevern M, Bertholon P, Brandt T, Fife T, Imai T, Nuti D, Newman-Toker D. Benign paroxysmal positional vertigo: Diagnostic criteria. J Vestib Res. 2015;25(3–4):105–17.
3. Dix MR, Hallpike CS. The pathology symptomatology and diagnosis of certain common disorders of the vestibular system. Proc R Soc Med. 1952;45(6):341–54.
4. Dix MR, Hallpike CS. The pathology, symptomatology and diagnosis of certain common disorders of the vestibular system. Ann Otol Rhinol Laryngol. 1952;61(4):987–1016.
5. Epley JM. New dimensions of benign paroxysmal positional vertigo. Otolaryngol Head Neck Surg (1979). 1980;88(5):599–605.

M. Cherchi, *Otoneurology and Vestibular Medicine*, https://doi.org/10.1007/978-3-031-94841-1_17

6. Bhattacharyya N, Gubbels SP, Schwartz SR, Edlow JA, El-Kashlan H, Fife T, et al. Clinical practice guideline: benign paroxysmal positional vertigo (update). Otolaryngol Head Neck Surg. 2017;156(3_suppl):S1–S47.
7. Fife TD, Iverson DJ, Lempert T, Furman JM, Baloh RW, Tusa RJ, et al. Practice parameter: therapies for benign paroxysmal positional vertigo (an evidence-based review): report of the quality standards Subcommittee of the American Academy of neurology. Neurology. 2008;70(22):2067–74.
8. Hilton MP, Pinder DK. The Epley (canalith repositioning) manoeuvre for benign paroxysmal positional vertigo. Cochrane Database Syst Rev. 2014;2014(12):CD003162.
9. Semont A, Freyss G, Vitte E. Curing the BPPV with a liberatory maneuver. Adv Otorhinolaryngol. 1988;42:290–3.
10. Foster CA, Ponnapan A, Zaccaro K, Strong D. A comparison of two home exercises for benign positional vertigo: half somersault versus Epley maneuver. Audiol Neurotol Ext. 2012;2(1):16–23.
11. Roberts RA, Gans RE, Montaudo RL. Efficacy of a new treatment maneuver for posterior canal benign paroxysmal positional vertigo. J Am Acad Audiol. 2006;17(8):598–604.
12. Brandt T, Daroff RB. Physical therapy for benign paroxysmal positional vertigo. Arch Otolaryngol. 1980;106(8):484–5.

Benign Paroxysmal Positional Vertigo Involving the Lateral Semicircular Canal

18

Benign Paroxysmal Positional Vertigo Involving the Lateral Semicircular Canal

Lateral (also called horizontal) *semicircular canal benign paroxysmal positional vertigo* (BPPV) is less common than posterior canal BPPV. Estimates in the literature suggest that it accounts for approximately 5–10% of cases of BPPV.

The diagnosis of lateral semicircular canal BPPV is secured by observing a specific pattern of eye movements when the patient is in several different positions. The series of positions is sometimes referred to as the "supine roll test," while some examiners use eponymous designations [1], after the various investigators who provided early descriptions of the phenomenon (e.g., McClure [2], Pagnini [3]). Essentially, when the patient's head is directly on its side (i.e., with the sagittal plane oriented horizontally, in parallel with the surface of the earth), nystagmus that is purely horizontal (with respect to the patient's head) emerges; the nystagmus is either "geotropic" (meaning that the fast phase beats toward the ground) or "apogeotropic" (meaning that the fast phase beats away from the ground), whether the patient is lying on one side or the other. These patterns are sometimes referred to as "direction changing positional nystagmus" [2, 4], because the direction of the nystagmus depends on the position of the patient's head with respect to gravity. The anatomy and physiology of the responses in lateral canal BPPV are discussed by Asprella Libonati [5, 6], and summarized in Table 18.1.

Table 18.1 Tabular summary of lateral canal BPPV canalolithiasis, geotropic, and apogeotropic variants, with patient lying on affected or unaffected side

	Geotropic canalolithiasis BPPV. This variant is attributed to free floating particles in the posterior arm (closer to the utricle) of the lateral canal.	*Apogeotropic canalolithiasis BPPV*. This variant is attributed to free floating particles in the anterior arm (closer to the ampulla) of the lateral canal.
Patient lying on affected ear	When the patient is lying on the affected ear, the otoliths will move toward the ampulla (away from the utricle), resulting in an ampullopetal ***excitatory*** current and causing a nystagmus beating toward the (now undermost) affected ear (i.e., *geo*tropic).	When the patient is lying on the affected ear, the particles will move away from the ampulla (toward the utricle), resulting in an ampullofugal ***inhibitory*** endolymphatic current, causing a nystagmus beating toward the (now uppermost) unaffected ear (i.e., *apogeo*tropic).
Patient lying on unaffected ear	When the patient is lying on the unaffected ear, the otoliths will move away from the ampulla (toward the utricle), resulting in an ampullofugal ***inhibitory*** endolymphatic current, causing a nystagmus beating to the (now undermost) unaffected ear (i.e., *geo*tropic).	When the patient is lying on the unaffected ear, the particles will move toward the ampulla (away from the utricle), resulting in an ampullopetal ***excitatory*** endolymphatic current, causing a nystagmus beating toward the (now uppermost) affected ear (i.e., *apogeo*tropic).

Because both lateral semicircular canals lie in the same plane, either side can produce the same pattern of eye movement abnormalities. Thus, physical examination alone may be able to diagnose lateral semicircular canal BPPV, but may not be able to determine the affected side. Some studies propose methods for determining the affected side [7, 8], such as the "bow and lean test" [8], but practically such determination is often not possible. Consequently, when a physical therapist is confronted with this ambiguity, he or she may simply choose one side and start treating it; if after a few sessions the patient improves, then it is assumed that the originally chosen side was correct; in contrast, if after several sessions a patient fails to improve, then the therapist may switch to the other side, at which point the patient should begin to improve. One *cannot* "alternate" treatment sides, because treating a given side will "undo" whatever had been accomplished by treating the opposite side.

The first maneuver shown to treat lateral semicircular canal BPPV was described by Thomas Lempert [9] as the "barbecue roll;" some clinicians also refer to this as the "log roll." Other maneuvers that have been less thoroughly studied include the Gufoni maneuver [10] and the Appiani maneuver [11].

M. Cherchi, *Otoneurology and Vestibular Medicine*, https://doi.org/10.1007/978-3-031-94841-1_18

As noted earlier, lateral semicircular canal BPPV can cause either geotropic or apogeotropic direction-changing positional nystagmus. Some studies report that both variants respond similarly to the Lempert maneuver and the Gufoni maneuver [12, 13]. In contrast, some studies suggest that the geotropic and apogeotropic variants of lateral semicircular canal BPPV respond differently to different maneuvers; for instance, one randomized trial reported the Lempert maneuver and Gufoni maneuver to be effective specifically for geotropic lateral semicircular canal BPPV [14], while the original paper about the Appiani maneuver described it as efficacious specifically for apogeotropic lateral semicircular canal BPPV [11].

References

1. Bhattacharyya N, Gubbels SP, Schwartz SR, Edlow JA, El-Kashlan H, Fife T, et al. Clinical practice guideline: benign paroxysmal positional vertigo (Update). Otolaryngol Head Neck Surg. 2017;156(3_suppl):S1–S47.
2. McClure JA. Horizontal canal BPV. J Otolaryngol. 1985;14(1):30–5.
3. Pagnini P, Nuti D, Vannucchi P. Benign paroxysmal vertigo of the horizontal canal. ORL J Otorhinolaryngol Relat Spec. 1989;51(3):161–70.
4. Baloh RW, Jacobson K, Honrubia V. Horizontal semicircular canal variant of benign positional vertigo. Neurology. 1993;43(12):2542–9.
5. Asprella LG. Diagnostic and treatment strategy of lateral semicircular canal canalolithiasis. Acta Otorhinolaryngol Ital. 2005;25(5):277–83.
6. Asprella LG. Benign paroxysmal positional vertigo and positional vertigo variants. Otorhinolaryngol Clin. 2012;4:25–40.
7. Choi SY, Oh SW, Kim HJ, Kim JS. Determinants for bedside lateralization of benign paroxysmal positional vertigo involving the horizontal semicircular canal. J Neurol. 2020;267:1709.
8. Choung YH, Shin YR, Kahng H, Park K, Choi SJ. 'Bow and lean test' to determine the affected ear of horizontal canal benign paroxysmal positional vertigo. Laryngoscope. 2006;116(10):1776–81.
9. Lempert T. Horizontal benign positional vertigo. Neurology. 1994;44(11):2213–4.
10. Gufoni M, Mastrosimone L, Di Nasso F. Repositioning maneuver in benign paroxysmal vertigo of horizontal semicircular canal. Acta Otorhinolaryngol Ital. 1998;18(6):363–7.
11. Ciniglio Appiani G, Catania G, Gagliardi M, Cuiuli G. Repositioning maneuver for the treatment of the apogeotropic variant of horizontal canal benign paroxysmal positional vertigo. Otol Neurotol. 2005;26(2):257–60.
12. Fu W, Han J, Chang N, Wei D, Bai Y, Wang Y, et al. Corrigendum to "immediate efficacy of Gufoni maneuver for horizontal canal benign paroxysmal positional vertigo (HC-BPPV): a meta-analysis" [Auris Nasus larynx (2020) 48-54]. Auris Nasus Larynx. 2020;47(1):171–2.
13. Fu W, Han J, Chang N, Wei D, Bai Y, Wang Y, et al. Immediate efficacy of Gufoni maneuver for horizontal canal benign paroxysmal positional vertigo (HC-BPPV): a meta-analysis. Auris Nasus Larynx. 2020;47(1):48–54.
14. Kim JS, Oh SY, Lee SH, Kang JH, Kim DU, Jeong SH, et al. Randomized clinical trial for geotropic horizontal canal benign paroxysmal positional vertigo. Neurology. 2012;79(7):700–7.

19 Benign Paroxysmal Positional Vertigo Involving the Anterior Semicircular Canal

Anterior Semicircular Canal Benign Paroxysmal Positional Vertigo

Anterior (sometimes called superior) *semicircular canal benign paroxysmal positional vertigo* (BPPV) is the rarest form of BPPV. This rarity is plausibly due to the fact that the anterior semicircular canal is the highest of the three canals, and in the course of regular activities, gravity is likely to cause loose otoliths to exit that higher canal and instead enter one of the lower canals (posterior or lateral). Some literature suggest that anterior semicircular canal BPPV is more common after head trauma [1, 2].

Anterior semicircular canal BPPV is diagnosed using the Dix–Hallpike maneuver (like posterior semicircular canal BPPV), but the pattern of observed eye movements is different than that found with posterior semicircular canal BPPV. Specifically, in anterior semicircular canal BPPV, the pattern of observed eye movements consists of a combination of nystagmus in which the fast phase is contraversive (beating toward the unaffected ear), contratorsional (the top pole of the eye rotates toward the unaffected ear), and down beat (toward the patient's feet).

Because anterior semicircular canal BPPV is relatively rare, it has also been more difficult to study. Treatment maneuvers described include the reverse Epley maneuver [3], the Semont maneuver [4], the Casani maneuver [5], the Kim maneuver [6], and the Yacovino maneuver [7].

Given that anterior semicircular canal BPPV is rare, that its pattern of nystagmus is unusual, and that its treatment maneuvers are neither well known nor well demonstrated on internet videos, it is appropriate for patients to consult with a vestibular physical therapist.

References

1. Dlugaiczyk J, Siebert S, Hecker DJ, Brase C, Schick B. Involvement of the anterior semicircular canal in posttraumatic benign paroxysmal positioning vertigo. Otol Neurotol. 2011;32(8):1285–90.
2. Jackson LE, Morgan B, Fletcher JC Jr, Krueger WW. Anterior canal benign paroxysmal positional vertigo: an underappreciated entity. Otol Neurotol. 2007;28(2):218–22.
3. Honrubia V, Baloh RW, Harris MR, Jacobson KM. Paroxysmal positional vertigo syndrome. Am J Otol. 1999;20(4):465–70.
4. Semont A, Freyss G, Vitte E. Curing the BPPV with a liberatory maneuver. Adv Otorhinolaryngol. 1988;42:290–3.
5. Casani AP, Cerchiai N, Dallan I, Sellari-Franceschini S. Anterior canal lithiasis: diagnosis and treatment. Otolaryngol Head Neck Surg. 2011;144(3):412–8.
6. Kim YK, Shin JE, Chung JW. The effect of canalith repositioning for anterior semicircular canal canalithiasis. ORL J Otorhinolaryngol Relat Spec. 2005;67(1):56–60.
7. Yacovino DA, Hain TC, Gualtieri F. New therapeutic maneuver for anterior canal benign paroxysmal positional vertigo. J Neurol. 2009;256(11):1851–5.

M. Cherchi, *Otoneurology and Vestibular Medicine*, https://doi.org/10.1007/978-3-031-94841-1_19

Ménière's Disease

20

Introduction

Ménière's disease is a progressive degenerative disease of the labyrinth whose underlying cause is unknown. The disease presents with a constellation of symptoms consisting of variably frequent episodes of (usually unilateral) aural fullness, hearing loss and tinnitus accompanying or preceding (usually spinning) disequilibrium (often with) nausea and vomiting that can last for several hours, after which the symptom cluster resolves. Patients return to baseline in between episodes, but over time there is cumulative hearing loss in the affected ear. A minority of patients progress to bilateral involvement. In late stages, patients may develop drop attacks, termed "otolithic catastrophes of Tumarkin." Acute management of individual episodes with vestibular suppressants and antiemetics is reasonable. Early in the disease course, prophylaxis (to reduce the frequency of the episodes) is sometimes achievable with daily oral medications (such as a diuretic, verapamil, or betahistine), but in later stages of the disease, invasive interventions (by an otolaryngologist) may be required, such as a transtympanic injection of dexamethasone or gentamicin.

Introduction

Prosper Ménière originally described the constellation of symptoms in 1861 [1] as follows:

1. An auditory organ, otherwise perfectly healthy, may suddenly become the site of a functional disturbance consisting of variable tones, continuous or intermittent, and these noises are soon accompanied by a greater or lesser diminution of hearing.
2. These functional disturbances take place in the internal ear, can cause symptoms [previously] imputed to the brain, such as vertigo, lightheadedness, uncertainty of gait, spins and drops, also accompanied by nausea, vomiting and fainting.
3. These events, which are intermittent, are soon followed by progressive hearing loss, and often hearing is completely abolished.
4. This all leads one to believe that the main lesion of these episodes localizes to the semicircular canals.

In other words, he listed the cardinal symptoms of (1) episodic vertigo, (2) episodic or chronic tinnitus, and (3) episodic worsening of hearing (with an overall trajectory of decline between episodes). He also explicitly articulated a relationship between the symptoms that practitioners had previously "imputed to the brain" (vertigo, impaired gait, nausea, vomiting, fainting) and the aural symptoms (tinnitus and hearing loss), which is what led him to hypothesize that the disease "resides" in the ear—an idea which was novel at the time.

Ménière's original description has basically been carried forward to the current criteria articulated by the American Academy of Otolaryngology in 1995 [2], and the criteria articulated by the Bárány Society [3], which are (1) recurrent, spontaneous episodes of vertigo, (2) hearing loss in at least one ear, and (3) aural fullness, or tinnitus, or both, on the same side as the hearing loss.

Epidemiology

There is considerable controversy regarding the prevalence and incidence of Ménière's disease.

Alexander and Harris [4] reviewed a range of literature on this topic [5–15] and found that reported prevalence rates per 100,000 people ranged from 3.5 to 513—in other words, more than two orders of magnitude.

Alexander and Harris point out several potential sources of variability, including the biases in different study methodologies (e.g., retrospective series, population-based cross-sectional studies). In addition to methodological differences, studies use different methods of data collection and use variably stringent diagnostic criteria.

M. Cherchi, *Otoneurology and Vestibular Medicine*, https://doi.org/10.1007/978-3-031-94841-1_20

To take an example of a study of both prevalence and incidence, Wladislavosky-Wasserman and colleagues [15], reporting on patients seen at Mayo Clinic over a 30-year period, listed a prevalence of 218.2 and an annual incidence of 15.3 per 100,000. In other words, about 0.2% of the population (1 in 500 people) at any given time has Ménière's disease (prevalence); and every year about 0.015% of the population (1 in 6536 people) newly develops Ménière's disease (incidence).

The age distribution of incident Ménière's disease is a separate question. An epidemiologic study by Havia and colleagues [16] noted that the age of symptom onset ranged from 15 to 80 years, with a peak age of incidence in the fifth decade (age 40–50 years), in an approximately bell-curve-like distribution.

Genetics

Exome sequencing in cases of familial clustering of Ménière's disease has implicated numerous genes in connection with Ménière's disease [17], but the role this might play in diagnosing Ménière's disease (whether familial or sporadic) remains unclear.

Pathophysiology: General Considerations

Despite considerable research, the underlying cause of Ménière's disease remains poorly understood. Early autopsy studies of patients with Ménière's disease consistently described finding a dilated inner ear structure; this finding is called "hydrops." This anatomical abnormality is sometimes detectable on high resolution MRI [18], though to be clear, this type of imaging findings is not yet considered a standard diagnostic tool for this disease.

Rauch and colleagues (1989) reported that of 13 patients who in life had been confidently diagnosed with Ménière's disease, at autopsy all 13 had "hydrops." However, autopsies of the general population report that up to 6% of individuals (30 out of 500) have "hydrops," whereas only around 0.2% of people (1 out of 500) have Ménière's disease—a difference greater than one order of magnitude. Consequently, the state of "hydrops" is thought to be *necessary but not sufficient* for Ménière's disease. In other words, something else (in addition to hydrops) must go wrong in order for Ménière's disease to manifest clinically—but exactly what that "something else" is, remains unknown.

Numerous candidate etiologies have been suggested, but none is universally accepted. These candidates have included the following:

- Genetic
- Immunological disorders
  - Autoimmune
  - Allergic
- Infectious
  - Viral
  - Bacterial
- "Hydraulic"
  - Over-production of endolymphatic fluid
  - Under-absorption of endolymphatic fluid (e.g., obstruction of vestibular and cochlear aqueducts)
  - Fluctuations of intracranial pressure

Regarding *genetic* causes, Eppsteiner et al. [19] reported that of individuals affected by Ménière's disease, 7–15% of report having a relative with the disease. Although such numbers are greater than what is likely by chance, it should be kept in mind that family members share more than just their genes; they also share environments and exposures. Genes encoding a range of proteins—including human leukocyte antigens, potassium channels, sodium channels, and others—have been proposed, but none has been widely accepted.

Although *allergies* and Ménière's disease have frequently been noted to be comorbid [20], so far there is little evidence that allergies cause Ménière's disease [21]. Nevertheless, some researchers advocate allergy treatment [22], at least in patients who otherwise have a clear history of allergies (such as to foods or environmental factors).

Although some patients with Ménière's disease also have *autoimmune* diseases, the link between the two remains uncertain [23], and so far there is no firm rationale for attempting to treat Ménière's disease with immunomodulatory strategies.

Although DNA to herpes viruses has been reported in the vestibular ganglion in patients with Ménière's disease [24], this finding does not appear to be higher than in the general population, so no causal link has been firmly established.

The rationale for proposing a "hydraulic" cause of Ménière's disease is that hydrops must be due to either an excess production of fluid (in this case, endolymph), or an insufficient resorption of said fluid, or both. Histopathological studies of the labyrinth in Ménière's disease show a variety of findings, including changes in the cells lining the endolymphatic duct [25], raising the question of whether alteration in the functions of these cells may contribute to the "fluid imbalance" of the labyrinth; this type of finding is often viewed by otolaryngologists as a rationale for attempting a procedure called endolymphatic sac decompression, whose putative goal would be to give any excessively produced endolymph another mechanism for escape (beyond the usual resorption).

Pathophysiology: Can Ménière's Disease Result from Noise Exposure?

There are scattered case reports and case series in the literature of patients who appeared to develop Ménière's disease after noise exposure, typically at a distance of some years from the acoustic insult [26–28]. Ménière's disease is relatively uncommon, whereas noise exposure is extraordinarily common, so this raises the question of whether the association between the two is simply one of chance. This question often comes up in medico-legal cases and disability claims.

Segal and colleagues [29] reviewed and critiqued the analytical flaws in the cases reported by Paparella and Mancini [26], Roitman and colleagues [27], and Ylikoski [28]. They elected an epidemiologically based approach to answer this question:

> We considered that the search for an association between NIHL [noise-induced hearing loss] and subsequent endolymphatic hydrops should be based on epidemiologic research. Our reasoning was based mainly on the small number of reported cases of MD [Ménière's disease] compared with the relatively huge numbers of reported cases of NIHL. Only a significant difference between the incidence of MD among persons with NIHL, in comparison with the incidence seen in the general population, would provide definitive evidence one way or the other [29].

They studied 17,245 military veterans who had all suffered noise exposure and among them identified 11 individuals who met clinical criteria for Ménière's disease, amounting to a prevalence of 1.9 per 100,000 (which is actually *less than* the prevalence in the studies reviewed earlier). They concluded that:

> Our survey showed that the prevalence of MD among patients suffering from acute acoustic trauma or NIHL was 1.9:100000, comparable to that of the general population. We contend that the late appearance of MD in these patients is coincidental [29].

Thus, in this large-scale epidemiologically based study, "No support was found for the hypothesis that Ménière's Disease may be causally related to previous acoustic trauma or noise-induced hearing loss" [29].

Natural History of Ménière's Disease

Although Ménière's disease is not common, it is rare enough that it is difficult to establish what the "natural history" of the disease is—in other words, how the disease progresses over time if untreated. Specific questions include whether Ménière's disease "burns out," and how often it "spreads" from unilateral involvement to bilateral involvement.

For many decades, practitioners were under the impression that Ménière's disease eventually "burns out," in the sense that patients eventually cease experiencing vertiginous episodes. However, more recent longitudinal research suggests that this is probably not true [16].

Most cases of Ménière's disease begin unilaterally. Stahle and colleagues [30] followed a cohort of 167 patients, and found that only 3 (1.7%) presented with bilateral ear symptoms from the outset. Of the cases that begin unilaterally, there is controversy regarding what proportion of cases progresses to bilateral involvement. The cited figures are generally in the range of one-third to one-half of cases. The longitudinal study by Wladislavosky-Wasserman and colleagues [15] reported 34% of patients eventually developed bilateral involvement. Havia and colleagues [16] reported that of their patients who were followed for at least 20 years, 43% developed bilateral ear symptoms. Stahle and colleagues [30] reported that of the 34 patients who were followed for at least 20 years, 47% developed bilateral ear symptoms.

Otolithic Catastrophes of Tumarkin

Some patients, generally at more advanced stages of Ménière's disease, begin experiencing abrupt "drop attacks" without any prodrome of aural symptoms. These attacks are sometimes referred to as "otolithic catastrophes of Tumarkin," after the first investigator to describe them in 1936, Alexis Tumarkin [31]. Patients who have reached this stage of symptoms generally can no longer control the vertiginous symptoms with medical therapy (in the sense of oral medications), and must move on to more invasive procedures for management, such as a transtympanic injection of gentamicin [32].

Diagnostic Testing for Ménière's Disease

When a patient experiences a first-ever episode of disequilibrium and unilateral ear symptoms (hearing loss, fullness, tinnitus), that clinical presentation in itself does not yet meet criteria for Ménière's disease because the disequilibrium has not yet recurred—the presentation could be compatible with, for example, labyrinthitis. When a patient begins to have recurrent attacks, a diagnosis of Ménière's disease becomes more likely. In an attempt to confirm the diagnosis, several diagnostic tests may be useful.

- *Audiometry*. This usually shows sensorineural hearing loss. This hearing loss more commonly begins by involving the lower frequencies. This hearing loss also tends to fluctuate, with an overall course (over years) of gradual decline.
- *Tests of unilateral vestibular function* (vestibular evoked myogenic potentials, caloric testing). These tests often,

though not always, identify a vestibular weakness on the same side as the ear symptoms.

- *Electrocochleography*. There is controversy in the literature regarding the diagnostic utility of this test, which supposedly detects hydrops.
- *MRI of the internal auditory canals without and with contrast*. Although research studies (mentioned earlier) sometimes identify anatomical evidence of hydrops, this is not yet considered a standard test for Ménière's disease. When a diagnosis of Ménière's disease is being considered, the main role of this imaging study is to help exclude competing diagnosis, such as a vestibular schwannoma.

When the clinical history is compatible with Ménière's disease, and audiometry shows fluctuating low frequency sensorineural hearing loss, and electrocochleography shows hydrops, and competing diagnoses have been excluded, then one can diagnose Ménière's disease with a high degree of confidence.

Often the data do not turn out this neatly—in other words, the results do not all "line up" perfectly in support of Ménière's disease. In this circumstance, it is sometimes worthwhile to repeat selected parts of the workup later. For example, since the hearing loss in Ménière's disease fluctuates, if an initial audiogram shows normal hearing, this may be worth repeating later, on the possibility that you will "catch" the fluctuating hearing.

Management of Acute Attacks

The treatment of Ménière's disease consists of both [2] management of acute episodes, and [3] prophylactic strategies.

Management of acute episodes usually consists of medications for disequilibrium and (if that fails) nausea/vomiting.

The medications for acute disequilibrium include *antihistamines* (meclizine, dimenhydrinate) or *benzodiazepines* (such as clonazepam or diazepam). For patients whose nausea/vomiting makes it difficult to ingest medications, they may benefit from an orally disintegrating formulation (such as clonazepam) or suppository formulation (such as diazepam).

The medications for nausea/vomiting include *ondansetron, promethazine,* and *prochlorperazine*. For patients whose nausea/vomiting makes it difficult to ingest medication, they may benefit from either an orally disintegrating formulation (such as clonazepam) or suppository formulation (such as promethazine or prochlorperazine).

If a patient's episodes are very infrequent, then they may be successful in simply treating each attack as it happens.

Longer-Term (Prophylactic) Management, Noninvasive

For many patients, the frequency of the episodes is unacceptably high, and these patients may benefit from prophylactic strategies. The goal of prophylaxis is to decrease the frequency of the episodes. Note that even when a patient is using appropriate prophylactic strategies, this may not completely eliminate the episodes.

A common initial therapy is *restriction of total daily dietary sodium* intake to ≤2000 mg [33].

If dietary sodium restriction fails, a typical first-line medication is a diuretic, such as the combination of hydrochlorothiazide-triamterene. Results of studies on this have been mixed. For instance, van Deelen and colleagues [34] reported that hydrochlorothiazide/triamterene reduced the frequency of attacks. In contrast, a meta-analysis by Thirlwall and colleagues [35] found insufficient evidence of efficacy for diuretics in management of Ménière's.

Another medical approach involves *calcium channel blockers*—such as flunarizine [36], nimodipine [37], or verapamil [38]—or drugs whose actions modulate both calcium channels and histamine receptors—such as cinnarizine [39].

Betahistine is also commonly used oral medication in the prophylaxis for Ménière's disease. This drug is not currently FDA approved in the United States, but is still available through compounding pharmacies. Animal experiments suggest that some of betahistine's catabolites (aminoethylpyridine and hydroxyethylpyridine) promote blood flow in the cochlea [40]. Even if this is also true in humans, it is unclear why this would help Ménière's disease. A meta-analysis by James and colleagues [41] concluded that there is insufficient evidence for efficacy of betahistine in treatment of Ménière's disease.

If a patient has attempted reasonable trials of these approaches (dietary sodium restriction, diuretic, calcium channel blocker, betahistine) without success, then in the interest of avoiding invasive procedures (see below), some practitioners at this stage may provisionally change strategies and attempt treatment for migraine associated vertigo (MAV), on the rationale that there is significant clinical overlap between the disequilibrium caused by MAV and that caused by Ménière's disease. If attempts at treating for migraine also fail, then they may return to the idea of Ménière's disease, and advance to invasive therapies.

Longer-Term (Prophylactic) Management, Invasive

If reasonable attempts at noninvasive prophylactic treatments have failed, then patients are often referred to otolaryngology for further management.

Some otolaryngologists advocate the use of transtympanic injections of *steroids* such as dexamethasone [42]. In favor of this approach is that dexamethasone should not have a destructive effect on the inner ear. Against this idea is the fact that it is difficult to see why this approach would be effective, and even if it provides benefit, that benefit should be transient since the steroids are catabolized and eliminated. In patients who seem to benefit from steroids, they often appear to require repeated steroid injections.

Other otolaryngologists are willing to consider a transtympanic injection of *gentamicin*, an aminoglycoside antibiotic that is preferentially vestibulotoxic. In favor of this approach is well-documented efficacy [43–47]. Against this approach is its destructive nature, and otolaryngologists often voice understandable concern that there is the potential for further hearing loss, though most studies show the risk of additional hearing loss to be very low [44, 45], and hearing preservation may be enhanced if gentamicin is administered concomitantly with dexamethasone [48].

References

1. Méniere P (1861) Sur une forme de surdité grave dépendant d'une lésion de l'oreille interne. Bulletin de l'Adadémie Impérlale de médecine 26:241.
2. American Academy of Otolaryngology-Head and Neck Foundation I. Committee on hearing and equilibrium guidelines for the diagnosis and evaluation of therapy in Meniere's disease. American Academy of Otolaryngology-Head and Neck Foundation, Inc. Otolaryngol Head Neck Surg. 1995;113(3):181–5.
3. Lopez-Escamez JA, Carey J, Chung WH, Goebel JA, Magnusson M, Mandala M, et al. Diagnostic criteria for Meniere's disease. J Vestib Res. 2015;25(1):1–7.
4. Alexander TH, Harris JP. Current epidemiology of Meniere's syndrome. Otolaryngol Clin N Am. 2010;43(5):965–70.
5. Cawthorne T, Hewlett AB. Meniere's disease. Proc R Soc Med. 1954;47(8):663–70.
6. Celestino D, Ralli G. Incidence of Meniere's disease in Italy. Am J Otol. 1991;12(2):135–8.
7. Harris JP, Alexander TH. Current-day prevalence of Meniere's syndrome. Audiol Neurootol. 2010;15(5):318–22.
8. Havia M, Kentala E, Pyykko I. Prevalence of Meniere's disease in general population of southern Finland. Otolaryngol Head Neck Surg. 2005;133(5):762–8.
9. Nakae K, Nitta H, Hattori Y, Maeda K, Watanabe I, Mizukoshi K, et al. The prevalence of Ménière's disease in Japan. Pract Otorhinolaryngol. 1980;73:1023–9.
10. Radtke A, von Brevern M, Feldmann M, Lezius F, Ziese T, Lempert T, et al. Screening for Meniere's disease in the general population – the needle in the haystack. Acta Otolaryngol. 2008;128(3):272–6.
11. Shojaku H, Watanabe Y, Fujisaka M, Tsubota M, Kobayashi K, Yasumura S, et al. Epidemiologic characteristics of definite Meniere's disease in Japan. A long-term survey of Toyama and Niigata prefectures. ORL J Otorhinolaryngol Relat Spec. 2005;67(5):305–9.
12. Stahle J, Stahle C, Arenberg IK. Incidence of Meniere's disease. Arch Otolaryngol. 1978;104(2):99–102.
13. Tokumasu K, Tashiro N, Goto K, Saito A, Kawano R, Hirose Y, et al. Incidence and prevalence of Meniere's disease in Sagamihara city, Kanagawa-ken. Prac Otol (Kyoto). 1982;75((Suppl 3)):1165–73.
14. Watanabe Y, Mizukoshi K, Shojaku H, Watanabe I, Hinoki M, Kitahara M. Epidemiological and clinical characteristics of Meniere's disease in Japan. Acta Otolaryngol Suppl. 1995;519:206–10.
15. Wladislavosky-Waserman P, Facer GW, Mokri B, Kurland LT. Meniere's disease: a 30-year epidemiologic and clinical study in Rochester, Mn, 1951–1980. Laryngoscope. 1984;94(8):1098–102.
16. Havia M, Kentala E. Progression of symptoms of dizziness in Meniere's disease. Arch Otolaryngol Head Neck Surg. 2004;130(4):431–5.
17. Lopez-Escamez JA, Liu Y. Epidemiology and genetics of Meniere's disease. Curr Opin Neurol. 2023;37:88.
18. Gurkov R, Berman A, Dietrich O, Flatz W, Jerin C, Krause E, et al. MR volumetric assessment of endolymphatic hydrops. Eur Radiol. 2015;25(2):585–95.
19. Eppsteiner RW, Smith RJ. Genetic disorders of the vestibular system. Curr Opin Otolaryngol Head Neck Surg. 2011;19(5):397–402.
20. Banks C, McGinness S, Harvey R, Sacks R. Is allergy related to Meniere's disease? Curr Allergy Asthma Rep. 2012;12(3):255–60.
21. Weinreich HM, Agrawal Y. The link between allergy and Meniere's disease. Curr Opin Otolaryngol Head Neck Surg. 2014;22(3):227–30.
22. Derebery MJ, Berliner KI. Allergy and its relation to Meniere's disease. Otolaryngol Clin N Am. 2010;43(5):1047–58.
23. Greco A, Gallo A, Fusconi M, Marinelli C, Macri GF, de Vincentiis M. Meniere's disease might be an autoimmune condition? Autoimmun Rev. 2012;11(10):731–8.
24. Gartner M, Bossart W, Linder T. Herpes virus and Meniere's disease. ORL J Otorhinolaryngol Relat Spec. 2008;70(1):28–31. discussion
25. Michaels L, Soucek S, Linthicum F. The intravestibular source of the vestibular aqueduct: its structure and pathology in Meniere's disease. Acta Otolaryngol. 2009;129(6):592–601.
26. Paparella MM, Mancini F. Trauma and Meniere's syndrome. Laryngoscope. 1983;93(8):1004–12.
27. Roitman R, Talmi YP, Finkelstein Y, Silver S, Sadov R, Zohar Y. Acoustic trauma-induced Meniere's syndrome. ORL J Otorhinolaryngol Relat Spec. 1989;51(4):246–50.
28. Ylikoski J. Delayed endolymphatic hydrops syndrome after heavy exposure to impulse noise. Am J Otol. 1988;9(4):282–5.
29. Segal S, Eviatar E, Berenholz L, Kessler A, Shlamkovitch N. Is there a relation between acoustic trauma or noise-induced hearing loss and a subsequent appearance of Meniere's disease? An epidemiologic study of 17245 cases and a review of the literature. Otol Neurotol. 2003;24(3):387–91.
30. Stahle J, Friberg U, Svedberg A. Long-term progression of Meniere's disease. Acta Otolaryngol Suppl. 1991;485:78–83.
31. Tumarkin A. The Otolithic catastrophe: a new syndrome. Br Med J. 1936;2(3942):175–7.
32. Wu Q, Li X, Sha Y, Dai C. Clinical features and management of Meniere's disease patients with drop attacks. Eur Arch Otorrinolaringol. 2019;276(3):665–72.
33. Santos PM, Hall RA, Snyder JM, Hughes LF, Dobie RA. Diuretic and diet effect on Meniere's disease evaluated by the 1985 committee on hearing and equilibrium guidelines. Otolaryngol Head Neck Surg. 1993;109(4):680–9.

34. van Deelen GW, Huizing EH. Use of a diuretic (Dyazide) in the treatment of Meniere's disease. A double-blind cross-over placebo-controlled study. ORL J Otorhinolaryngol Relat Spec. 1986;48(5):287–92.
35. Thirlwall AS, Kundu S. Diuretics for Meniere's disease or syndrome. Cochrane Database Syst Rev. 2006;2006(3):CD003599.
36. Fraysse B, Bebear JP, Dubreuil C, Berges C, Dauman R. Betahistine dihydrochloride versus flunarizine. A double-blind study on recurrent vertigo with or without cochlear syndrome typical of Meniere's disease. Acta Otolaryngol Suppl. 1991;490:1–10.
37. Lassen LF, Hirsch BE, Kamerer DB. Use of nimodipine in the medical treatment of Meniere's disease: clinical experience. Am J Otol. 1996;17(4):577–80.
38. Kaya I, Eraslan S, Tarhan C, Bilgen C, Kirazli T, Gokcay F, et al. Can verapamil be effective in controlling vertigo and headache attacks in vestibular migraine accompanied with Meniere's disease? A preliminary study. J Neurol. 2019;266(Suppl 1):62–4.
39. Djelilovic-Vranic J, Alajbegovic A, Tiric-Campara M, Volic A, Sarajlic Z, Osmanagic E, et al. Betahistine or Cinnarizine for treatment of Meniere's disease. Med Arch. 2012;66(6):396–8.
40. Bertlich M, Ihler F, Sharaf K, Weiss BG, Strupp M, Canis M. Betahistine metabolites, Aminoethylpyridine, and Hydroxyethylpyridine increase cochlear blood flow in Guinea pigs in vivo. Int J Audiol. 2014;53(10):753–9.
41. James AL, Burton MJ. Betahistine for Meniere's disease or syndrome. Cochrane Database Syst Rev. 2001;2001(1):CD001873.
42. Garduno-Anaya MA, Couthino De Toledo H, Hinojosa-Gonzalez R, Pane-Pianese C, Rios-Castaneda LC. Dexamethasone inner ear perfusion by intratympanic injection in unilateral Meniere's disease: a two-year prospective, placebo-controlled, double-blind, randomized trial. Otolaryngol Head Neck Surg. 2005;133(2):285–94.
43. Martin-Sanz E, Diaz JY, Esteban-Sanchez J, Sanz-Fernandez R, Perez-Fernandez N. Delayed effect and gain restoration after Intratympanic gentamicin for Meniere's disease. Otol Neurotol. 2018;40:79.
44. Casani AP, Piaggi P, Cerchiai N, Seccia V, Franceschini SS, Dallan I. Intratympanic treatment of intractable unilateral Meniere disease: gentamicin or dexamethasone? A randomized controlled trial. Otolaryngol Head Neck Surg. 2012;146(3):430–7.
45. Driscoll CL, Kasperbauer JL, Facer GW, Harner SG, Beatty CW. Low-dose intratympanic gentamicin and the treatment of Meniere's disease: preliminary results. Laryngoscope. 1997;107(1):83–9.
46. Wu Y, Zhou Y, Yu J, Song Z, Dai C, Wang J. Intratympanic gentamicin injection for refractory ménière's disease (MD) has potential effect in preventing contralateral MD occurrence. Acta Otolaryngol. 2021;1-6:657.
47. Webster KE, Harrington-Benton NA, Judd O, Kaski D, Maarsingh OR, MacKeith S, et al. Intratympanic aminoglycosides for Ménière's disease. Cochrane Database Syst Rev. 2021;12:CD015246.
48. Bae S-H, Lee J-M, Lee H-J, Na G, Kim S-H. Effect of dexamethasone combination with gentamicin in chemical Labyrinthectomy on hearing preservation and vertigo control in patients with unilateral Meniere's disease: a randomized controlled clinical trial. Journal of. Clin Med. 2021;10(23):5581.

21 Otolithic Catastrophe of Tumarkin

Otolithic Catastrophe of Tumarkin

In advanced stages of Ménière's disease, approximately 9% of patients [1] develop drop attacks, which came to be known as the "*otolithic catastrophe of Tumarkin*" after the first clinician to describe this, Alexis Tumarkin [2]. Drop attacks have no prodrome, and thus, patients have no warning that an attack will occur. Since such attacks are dangerous, this usually merits more aggressive treatment for Ménière's disease, such as surgical management (gentamicin injection, vestibular neurectomy, etc.), as oral medications are generally ineffective.

Given that an otolithic catastrophe of Tumarkin occurs abruptly (and such occurrences tend to have a very erratic chronology), it is extremely unusual for a clinician to witness, so this is generally diagnosed by history rather than by direct observation. Nevertheless, some cases have been recorded; Chen and colleagues [3] reported a case caught on surveillance camera.

References

1. Wu Q, Li X, Sha Y, Dai C. Clinical features and management of Meniere's disease patients with drop attacks. Eur Arch Otorrinolaringol. 2019;276(3):665–72.
2. Tumarkin A. The otolithic catastrophe: a new syndrome. Br Med J. 1936;2(3942):175–7.
3. Chen Z, Zhang Y, Zhang Q. Tumarkin drop attack recorded by video surveillance. JAMA Neurol. 2020;77:897.

M. Cherchi, *Otoneurology and Vestibular Medicine*, https://doi.org/10.1007/978-3-031-94841-1_21

Posttraumatic Endolymphatic Hydrops

22

Posttraumatic Endolymphatic Hydrops

Around the mid-twentieth century, case reports began to emerge of head trauma followed by fluctuating hearing loss and vertigo. Since the audiologic and vestibular characteristics resembled Ménière's disease, these were sometimes referred to as "posttraumatic Ménière's" [1–3], but we prefer to reserve the term "Ménière's" for the idiopathic syndrome.

Nevertheless, the idea that endolymphatic hydrops could be caused by a traumatic insult was popularized by Dr. Harold Frederick Schuknecht [4, 5], and given his stature in the field [6], eventually the phrase *posttraumatic endolymphatic hydrops* entered the literature [7].

The mechanism of posttraumatic endolymphatic hydrops is unknown. Logically, one would think that trauma causes disordered endolymphatic fluid flow, either from increased production, or decreased resorption, or both; this could in turn be due to alteration in the function of cells that produce/absorb these fluids, or abnormalities in the hydraulic system through which they pass (cochlear duct, endolymphatic duct, and vestibular aqueduct).

Head trauma can result in temporal bone fractures. Similar to cases of head trauma without temporal bone fractures, some of these patients develop hearing loss and disequilibrium. Consequently, hypothesized mechanisms for this similarly include posttraumatic endolymphatic hydrops [8, 9], for which there is some histopathological evidence from temporal bone studies [10].

Audiometry from published cases is compatible with endolymphatic hydrops. For example, Chung and colleagues [11] reported a case of putative posttraumatic endolymphatic hydrops, documenting very clearly the fluctuation in low-frequency hearing. We are not aware of systematic vestibular studies in such patients (such as electrocochleography, or tests for vestibular weakness), so in our view the degree to which posttraumatic endolymphatic hydrops match idiopathic endolymphatic hydrops remains unclear.

Absent better guidance for treatment, it is probably medically reasonable to manage such cases according to their phenotype—meaning, these cases may be treated with strategies similar to Ménière's disease—to a point. Applying relatively low-risk interventions (dietary sodium restriction, oral medications) seems reasonable. In contrast, we would approach ablative therapies (e.g., transtympanic gentamicin injection, vestibular neurectomy) with great caution.

References

1. Gundrum LK. Post-traumatic Meniere's symptom-complex. Eye Ear Nose Throat Mon. 1959;38(5):385–8.
2. Lehrer JF, Poole DC. Post-traumatic Meniere's syndrome. Laryngoscope. 1984;94(1):129.
3. Paparella MM, Mancini F. Trauma and Meniere's syndrome. Laryngoscope. 1983;93(8):1004–12.
4. Gulya AJ, Schuknecht HF. Classification of endolymphatic hydrops. Am J Otolaryngol. 1982;3(5):319–22.
5. Schuknecht HF, Gulya AJ. Endolymphatic hydrops. An overview and classification. Ann Otol Rhinol Laryngol Suppl. 1983;106:1–20.
6. Nadol JB Jr. Harold Frederick Schuknecht. 1917–1996. Audiol Neurootol. 1997;2(3):168–70.
7. DiBiase P, Arriaga MA. Post-traumatic hydrops. Otolaryngol Clin N Am. 1997;30(6):1117–22.
8. Johnson F, Semaan MT, Megerian CA. Temporal bone fracture: evaluation and management in the modern era. Otolaryngol Clin N Am. 2008;41(3):597–618. x
9. Lyos AT, Marsh MA, Jenkins HA, Coker NJ. Progressive hearing loss after transverse temporal bone fracture. Arch Otolaryngol Head Neck Surg. 1995;121(7):795–9.
10. Knoll RM, Ishai R, Lubner RJ, Trakimas DR, Brodsky JR, Jung DH, et al. Peripheral vestibular organ degeneration after temporal bone fracture: a human otopathology study. Laryngoscope. 2020;130(3):752–60.
11. Chung J, Jung HJ, Kim CS, Kim YH. A case of post-traumatic Meniere's disease. Korean J Audiol. 2014;18(1):41–4.

M. Cherchi, *Otoneurology and Vestibular Medicine*, https://doi.org/10.1007/978-3-031-94841-1_22

23 Autoimmune Inner Ear Disease (AIED)

Introduction

Autoimmune inner ear disease (AIED) is an immune-mediated attack on the labyrinth, though the precise underlying autoimmune mechanism is unknown. AIED is rare, though appears more common in patients with other autoimmune diseases. It usually presents with bilateral sensorineural hearing loss that is rapidly progressive (typically over weeks or months); rare unilateral cases have been reported. Vestibular symptoms are inconsistently present. There is no proven biomarker of AIED; the diagnosis is based on a compatible clinical history and on the finding of steroid-responsive hearing loss. Vestibular function testing may show bilateral vestibular weakness, and MRI may show enhancement of the labyrinth. If oral steroids are ineffective, some otolaryngologists may consider transtympanic steroid injections. If AIED is confirmed by steroid-responsive hearing loss, long-term management with an immunomodulatory agent (such as a TNF-alpha blocker) is considered.

One of the earliest descriptions of AIED was in a case report by Dr. Brian F. McCabe [1]. There was initial skepticism about AIED as a distinct clinical entity [2, 3].

Epidemiology

AIED is rare. Its incidence has been difficult to determine [4], though estimates of 5 cases per 100,000 seem plausible [3]. AIED occurs mostly in adults but has been reported in children [5].

AIED seems to occur more commonly in patients with other autoimmune diseases [6] such as Hashimoto's thyroiditis [7], diffuse systemic sclerosis [8], ulcerative colitis [9], other inflammatory bowel disease [10], and several vasculitides [11].

Clinical Presentation

AIED usually presents with bilateral sensorineural hearing loss that is rapidly progressive (typically over weeks to months). It often causes other aural symptoms (such as fullness and tinnitus). Vestibular symptoms are sometimes also present [12–15], with studies estimating that anywhere from 22% [16] to 50% of patients [17] complain of oscillopsia, disequilibrium, etc.

When AIED occurs in the context of malignancy [18, 19], it may be more correct to view the phenomenon as paraneoplastic rather than idiopathic. As with other paraneoplastic syndromes, the logical treatment is that of the underlying cancer.

Rare cases of unilateral AIED have been reported, though often these involve unusual circumstances such as concomitant malignancy [20].

Mechanism of Disease

Although the histopathological findings in AIED have been described [21], the exact immunological mechanism has remained elusive [22–31].

Diagnosis

The diagnosis can be suspected from the clinical history of rapidly progressive, bilateral (symmetrical or asymmetrical) hearing loss. Suspicion is bolstered if audiometry shows the hearing loss to be sensorineural. The diagnosis becomes much more likely if the hearing loss responds (improves) to a course of oral steroids, and if no more compelling alternative diagnosis is discovered.

M. Cherchi, *Otoneurology and Vestibular Medicine*, https://doi.org/10.1007/978-3-031-94841-1_23

There is not yet any widely accepted biomarker of AIED [32, 33]. There was some initial enthusiasm for HSP2 (heat shock protein 2) as a candidate biomarker, but this was subsequently shown to have poor sensitivity and specificity [13].

Vestibular function is often not assessed in AIED patients. However, in patients with vestibular symptoms, formal oto-vestibular testing may show bilateral vestibular weakness [34].

In some cases of AIED, MRI may show contrast enhancement of the labyrinth [35].

Early diagnosis of AIED is desirable [36]. Although the disease tends to progress despite best efforts, early treatment may afford a patient more time with serviceable hearing.

Treatment

Systemic corticosteroids (usually orally administered) are generally regarded as first-line therapy [37]. Patients should respond to this (otherwise the diagnosis becomes questionable), but if the response is suboptimal, then some practitioners may attempt transtympanic steroid injections [37–40].

Steroids, whether systemic or transtympanic, are usually not a long-term solution due to the adverse effects of chronic use. Consequently, patients are often transitioned to immunomodulatory agents. Multiple such agents have been studied, including TNF-alpha blockers (etanercept, adalimumab, infliximab, golimumab), CD20 binders (rituximab), and IL-1 receptor blockers (anakinra, canakinumab), though the studies supporting their efficacy are weak [41, 42]. In practice, we usually see patients treated with etanercept or adalimumab.

The use of plasmapheresis in the management of AIED has also been studied [43] but never gained significant traction in clinical practice.

Prognosis

Prognosis of AIED is poor. Treatment may decelerate the pace of hearing loss, but generally symptoms eventually progress despite vigorous therapy.

References

1. McCabe BF. Autoimmune sensorineural hearing loss. Ann Otol Rhinol Laryngol. 1979;88(5 Pt 1):585–9.
2. Barna BP, Hughes GB. Autoimmune inner ear disease--a real entity? Clin Lab Med. 1997;17(3):581–94.
3. Das S, Bakshi SS, Seepana R. Demystifying autoimmune inner ear disease. Eur Arch Otorrinolaringol. 2019;276(12):3267–74.
4. Ruckenstein MJ. Autoimmune inner ear disease. Curr Opin Otolaryngol Head Neck Surg. 2004;12(5):426–30.
5. Oz RS, Gluth M, Tesher MS. Pediatric autoimmune inner ear disease: a rare, but treatable condition. Pediatr Ann. 2019;48(10):e391–e4.
6. Ralli M, D'Aguanno V, Di Stadio A, De Virgilio A, Croce A, Longo L, et al. Audiovestibular symptoms in systemic autoimmune diseases. J Immunol Res. 2018;2018:5798103.
7. Fayyaz B, Upreti S. Autoimmune inner ear disease secondary to Hashimoto's thyroiditis: a case report. J Community Hosp Intern Med Perspect. 2018;8(4):227–9.
8. Feld J, Shupak A, Zisman D. Diffuse systemic sclerosis presenting as Meniere's disease-like symptoms as part of autoimmune inner ear disease. Isr Med Assoc J. 2015;17(4):263–4.
9. Kariya S, Fukushima K, Kataoka Y, Tominaga S, Nishizaki K. Inner-ear obliteration in ulcerative colitis patients with sensorineural hearing loss. J Laryngol Otol. 2008;122(8):871–4.
10. Fousekis FS, Saridi M, Albani E, Daniel F, Katsanos KH, Kastanioudakis IG, Christodoulou DK. Ear involvement in inflammatory bowel disease: a review of the literature. J Clin Med Res. 2018;10(8):609–14.
11. Rahne T, Plontke S, Keysser G. Vasculitis and the ear: a literature review. Curr Opin Rheumatol. 2020;32(1):47–52.
12. Bovo R, Ciorba A, Martini A. Vertigo and autoimmunity. Eur Arch Otorrinolaringol. 2010;267(1):13–9.
13. Matsuoka AJ, Harris JP. Autoimmune inner ear disease: a retrospective review of forty-seven patients. Audiol Neurootol. 2013;18(4):228–39.
14. Rahman MU, Poe DS, Choi HK. Autoimmune vestibulo-cochlear disorders. Curr Opin Rheumatol. 2001;13(3):184–9.
15. Russo FY, Ralli M, De Seta D, Mancini P, Lambiase A, Artico M, et al. Autoimmune vertigo: an update on vestibular disorders associated with autoimmune mechanisms. Immunol Res. 2018;66:675.
16. Dayal VS, Ellman M, Sweiss N. Autoimmune inner ear disease: clinical and laboratory findings and treatment outcome. J Otolaryngol Head Neck Surg = Le Journal d'oto-rhino-laryngologie et de chirurgie cervico-faciale. 2008;37(4):591–6.
17. Bovo R, Aimoni C, Martini A. Immune-mediated inner ear disease. Acta Otolaryngol. 2006;126(10):1012–21.
18. Greene JJ, Keefe MW, Harris JP, Matsuoka AJ. Paraneoplastic syndrome: a masquerade of autoimmune inner ear disease. Otol Neurotol. 2015;36(1):e3–10.
19. Zibelman M, Pollak N, Olszanski AJ. Autoimmune inner ear disease in a melanoma patient treated with pembrolizumab. J Immunother Cancer. 2016;4:8.
20. Rajapakse A, O'Leary C, Gundelach R, Deva R, O'Byrne K. Unilateral autoimmune inner ear disease in a patient with lung cancer treated with nivolumab. Oxf Med Case Rep. 2020;2020(9):omaa077.
21. Schuknecht HF. Ear pathology in autoimmune disease. Adv Otorhinolaryngol. 1991;46:50–70.
22. Barna BP, Hughes GB. Autoimmunity and otologic disease: clinical and experimental aspects. Clin Lab Med. 1988;8(2):385–98.
23. Baruah P. Cochlin in autoimmune inner ear disease: is the search for an inner ear autoantigen over? Auris Nasus Larynx. 2014;41(6):499–501.
24. Eisner L, Vambutas A, Pathak S. The balance of tissue inhibitor of metalloproteinase-1 and matrix metalloproteinase-9 in the autoimmune inner ear disease patients. J Interf Cytokine Res. 2017;37(8):354–61.
25. Goodall AF, Siddiq MA. Current understanding of the pathogenesis of autoimmune inner ear disease: a review. Clin Otolaryngol. 2015;40(5):412–9.
26. Ianuale C, Cadoni G, De Feo E, Liberati L, Simo RK, Paludetti G, et al. A systematic review and meta-analysis of the diagnostic

accuracy of anti-heat shock protein 70 antibodies for the detection of autoimmune hearing loss. Otol Neurotol. 2013;34(2):214–9.
27. Miwa T, Okano T. Role of inner ear macrophages and autoimmune/autoinflammatory mechanisms in the pathophysiology of inner ear disease. Front Neurol. 2022;13:861992.
28. Pathak S, Stern C, Vambutas A. N-Acetylcysteine attenuates tumor necrosis factor alpha levels in autoimmune inner ear disease patients. Immunol Res. 2015;63(1–3):236–45.
29. Pathak S, Vambutas A. Autoimmune inner ear disease patient-associated 28-kDa proinflammatory IL-1beta fragment results from caspase-7-mediated cleavage in vitro. JCI Insight. 2020;5(3):e130845.
30. Ruckenstein MJ, Harrison RV. Autoimmune inner ear disease: a review of basic mechanisms and clinical correlates. J Otolaryngol. 1991;20(3):196–203.
31. Yoo TJ, Du X, Kwon SS. Molecular mechanism of autoimmune hearing loss. Acta Otolaryngol Suppl. 2002;548:3–9.
32. Ciorba A, Corazzi V, Bianchini C, Aimoni C, Pelucchi S, Skarzynski PH, Hatzopoulos S. Autoimmune inner ear disease (AIED): a diagnostic challenge. Int J Immunopathol Pharmacol. 2018;32:2058738418808680.
33. Shamriz O, Tal Y, Gross M. Autoimmune inner ear disease: immune biomarkers, audiovestibular aspects, and therapeutic modalities of Cogan's syndrome. J Immunol Res. 2018;2018:1498640.
34. Yukawa K, Hagiwara A, Ogawa Y, Nishiyama N, Shimizu S, Kawaguchi S, et al. Bilateral progressive hearing loss and vestibular dysfunction with inner ear antibodies. Auris Nasus Larynx. 2010;37(2):223–8.
35. Lobo D, Tunon M, Villarreal I, Brea B, Garcia-Berrocal JR. Intratympanic gadolinium magnetic resonance imaging supports the role of endolymphatic hydrops in the pathogenesis of immune-mediated inner-ear disease. J Laryngol Otol. 2018;132(6):554–9.
36. Agrup C, Luxon LM. Immune-mediated inner-ear disorders in neuro-otology. Curr Opin Neurol. 2006;19(1):26–32.
37. Breslin NK, Varadarajan VV, Sobel ES, Haberman RS. Autoimmune inner ear disease: a systematic review of management. Laryngoscope Investig Otolaryngol. 2020;5(6):1217–26.
38. Hoffmann KK, Silverstein H. Inner ear perfusion: indications and applications. Curr Opin Otolaryngol Head Neck Surg. 2003;11(5):334–9.
39. Light JP, Silverstein H. Transtympanic perfusion: indications and limitations. Curr Opin Otolaryngol Head Neck Surg. 2004;12(5):378–83.
40. Swan EE, Mescher MJ, Sewell WF, Tao SL, Borenstein JT. Inner ear drug delivery for auditory applications. Adv Drug Deliv Rev. 2008;60(15):1583–99.
41. Balouch B, Meehan R, Suresh A, Zaheer HA, Jabir AR, Qatanani AM, et al. Use of biologics for treatment of autoimmune inner ear disease. Am J Otolaryngol. 2022;43(5):103576.
42. Brant JA, Eliades SJ, Ruckenstein MJ. Systematic review of treatments for autoimmune inner ear disease. Otol Neurotol. 2015;36(10):1585–92.
43. Luetje CM. Theoretical and practical implications for plasmapheresis in autoimmune inner ear disease. Laryngoscope. 1989;99(11):1137–46.

Sympathetic Ear Disease

24

Introduction

Sympathetic ear disease is a rare condition thought to result when antigens (of the labyrinth or vestibular nerve) are exposed to the immune system (secondary to disease, trauma, or an invasive procedure), after which the immune system mounts an attack against the contralateral (previously healthy) ear, typically manifesting as sensorineural hearing loss. Immunosuppression seems like a logical approach in this scenario but has not been formally studied.

Details

Sympathetic ear disease is rare. Sympathetic ear disease can happen when one ear suffers an insult, and then after some delay, the other ear develops symptoms—typically sensorineural hearing loss. Sympathetic ear disease has been reported contralateral to an ear affected by diseases such as vestibular schwannoma [1] and temporal bone fracture [2]. It has also been reported contralateral to an ear that has undergone an invasive procedure, such as after surgery for vestibular schwannoma [3, 4], endolymphatic sac decompression [5], and stapedectomy revision [6].

The hypothesized mechanism of sympathetic ear disease is that trauma or some other disease disrupts the blood–labyrinth barrier [7], which gives the immune system access to labyrinthine antigens, which may in turn trigger the immune system to mount a mis-directed attack at the other (previously healthy) ear. Such immune attacks can happen at some delay after the original insult, ranging from months to years.

Sympathetic ear disease is thought to be the otologic analog of comparable phenomena such as sympathetic ophthalmia [8]. Given the immune-mediated nature of sympathetic ophthalmia, treatment is often attempted with corticosteroids or other immunosuppressants.

While immunosuppression seems logical, there are no trials of this approach in the treatment of sympathetic ear disease.

References

1. Early S, Rinnooy Kan CE, Eggink M, Frijns JHM, Stankovic KM. Progression of contralateral hearing loss in patients with sporadic vestibular schwannoma. Front Neurol. 2020;11:796.
2. ten Cate WJ, Bachor E. Autoimmune-mediated sympathetic hearing loss: a case report. Otol Neurotol. 2005;26(2):161–5.
3. Clemis JD, Mastricola PG, Schuler-Vogler M. Sudden hearing loss in the contralateral ear in postoperative acoustic tumor: three case reports. Laryngoscope. 1982;92(1):76–9.
4. Harris JP, Low NC, House WF. Contralateral hearing loss following inner ear injury: sympathetic cochleolabyrinthitis? Am J Otol. 1985;6(5):371–7.
5. Ochoa VM, Weider DJ. Development of autoimmune sensorineural hearing loss after endolymphatic sac decompression: two case reports. Otol Neurotol. 2003;24(2):279–82.
6. Richards ML, Moorhead JE, Antonelli PJ. Sympathetic cochleolabyrinthitis in revision stapedectomy surgery. Otolaryngol Head Neck Surg. 2002;126(3):273–80.
7. Nyberg S, Abbott NJ, Shi X, Steyger PS, Dabdoub A. Delivery of therapeutics to the inner ear: the challenge of the blood-labyrinth barrier. Sci Transl Med. 2019;11(482):eaao0935.
8. Parchand S, Agrawal D, Ayyadurai N, Agarwal A, Gangwe A, Behera S, et al. Sympathetic ophthalmia: a comprehensive update. Indian J Ophthalmol. 2022;70(6):1931–44.

M. Cherchi, *Otoneurology and Vestibular Medicine*, https://doi.org/10.1007/978-3-031-94841-1_24

Noise-Induced Hearing Loss

25

Overview

The cochlea functions as a frequency analyzing mechano-electrical transducer, converting the changing pattern of kinetic energy from the vibrating motion of the tympanic membrane (driven by sounds of specific frequencies and amplitudes) into discrete neural (electrochemical) signals. As such the cochlea is designed to function optimally within a given range of frequencies and amplitudes. Inappropriately loud sounds (which can occur in many forms of social, recreational, and occupational activities) can result in excessive kinetic energy being transmitted to the cochlea, resulting in structural damage, cellular damage, and their sequelae. There may be some component of genetic predisposition to noise-induced hearing loss. Noise-induced cochlear damage can result in hearing loss, and sometimes tinnitus. The degree of hearing loss can be quantified on audiometry. The best management strategy is primary prevention. After noise-induced hearing loss has occurred, secondary prevention consists of protecting the ear(s) by avoiding loud noises, abrupt pressure changes, and ototoxic medications, and by treating ear infections.

Introduction

The cochlea functions as a frequency analyzing mechano-electrical transducer, effectively converting the changing pattern of kinetic energy from the vibrating motion of the tympanic membrane (driven by sounds of specific frequencies and amplitudes) into discrete neural (electrochemical) signals. As such the cochlea is designed to function optimally within a given range of frequencies and amplitudes. Auditory function can be harmed by sounds carrying excessive kinetic energy.

Forms of Noise Exposure

Noise exposure comes in many forms during social, recreational, and occupational activities [1].

Mechanisms of Noise-Induced Hearing Loss

When the kinetic energy of incoming sounds is excessive, the cochlea can suffer damage through several mechanisms [2]. The first is structural damage (at the cellular level) directly resulting from mechanical forces. The second is a range of cellular insults from the chemical consequences of processing loud sounds, including oxidative stress, calcium overload, and inflammatory responses.

Factors Predisposing to Noise-Induced Hearing Loss

It has been estimated that perhaps as much as 50% of noise-induced hearing loss is made possible due to predisposing genetic factors [1]. Over 200 genes have been implicated as increasing vulnerability to noise-induced hearing loss [2]. The genes encode a variety of proteins, including those involved in the antioxidant system (for processing free radicals), potassium channel proteins, heat shock proteins, and others.

Clinical Manifestations of Noise Exposure

Exposure to loud sounds can induce hearing loss termed a *threshold shift*, meaning that a stimulus of greater amplitude is required to elicit the same perception. This hearing loss can recover in hours to weeks, and in such cases is desig-

M. Cherchi, *Otoneurology and Vestibular Medicine*, https://doi.org/10.1007/978-3-031-94841-1_25

nated a *temporary threshold shift*. The louder the sound, and the greater the time of exposure, the more likely it is that there will be a *permanent threshold shift*, meaning hearing loss from which one does not recover. Often, noise exposure causes a combination of temporary threshold shifts (at some frequencies and amplitudes) and permanent threshold shifts (at other frequencies and amplitudes) [3]. Repeated noise exposure will tend to cause progressive hearing loss [2].

Tinnitus often accompanies hearing loss of any cause [4], and noise-induced hearing loss is no exception [1].

Diagnosis of Noise-Induced Hearing Loss

Hearing loss can be characterized and quantified by audiometry. Most noise-induced hearing loss is sensorineural in nature. Very violent noise exposure, such as from blast injuries, can cause damage to the eardrum or ossicular chain as well, and in such cases, the hearing loss will additionally exhibit a conductive component.

Noise-induced hearing loss tends to affect high frequencies before low frequencies. However, many patients with noise-induced hearing loss exhibit a pattern that audiologists often term a "noise notch" [5] which refers to a dip in hearing around 4–6 kHz, with better hearing below and above that range, irrespective of the particular pitch of sounds to which a person has been exposed. This is apparently attributable to the specific vulnerability conferred by the anatomical properties (with corresponding resonant frequencies) of the human external auditory canal.

Because of the general pattern of high greater than low-frequency sensorineural hearing loss, it is often difficult to distinguish noise-induced hearing loss from presbycusis (age-related hearing loss).

Nonauditory Effects of Noise Exposure

Hearing loss is the perhaps the most obvious effect of noise exposure. But noise exposure has been shown to have a number of nonauditory consequences as well [1], including adverse effects on cognitive function, sleep, and cardiovascular health.

Management: Primary Prevention

When noise induces hearing loss, it is not generally possible to predict whether the threshold shift will be temporary or permanent. Consequently, primary prevention is by far the preferable management strategy.

For individuals who engage in recreational or occupational activities with predictably excessive noise levels, we generally recommend consultation with an audiologist regarding protection with custom ear plugs or noise canceling headphones.

Management: Treatment and Secondary Prevention

Once permanent hearing loss has occurred, management shifts to treatment and secondary prevention.

A reasonable initial treatment of any hearing loss is amplification, often in the form of a hearing aid. Traditionally, this care has been delivered by an audiologist, but in October of 2022, the first over-the-counter hearing aids became available in the United States. In cases of severe bilateral hearing loss, candidacy for cochlear implantation can be assessed by an otolaryngologist.

We also suggest that an individual with hearing loss follow these strategies for protecting their ears:

- Avoid loud noises.
- Avoid abrupt pressure changes. The most common example cited for this is SCUBA diving. However, for most people, airplane travel is the more realistic concern, and for this, we suggest wearing EarPlane ear plugs during airplane ascent and descent. (The EarPlane ear plugs are inexpensive and can be purchased over the counter.)
- Avoid ototoxic medications where medically feasible. If antiplatelet therapy becomes medically warranted (such as for cardiovascular disease), consider avoiding aspirin and instead using a nonototoxic antiplatelet agent (such as clopidogrel).
- Have any ear infections treated promptly.

References

1. Basner M, Babisch W, Davis A, Brink M, Clark C, Janssen S, Stansfeld S. Auditory and non-auditory effects of noise on health. Lancet. 2014;383(9925):1325–32.
2. Ding T, Yan A, Liu K. What is noise-induced hearing loss? Br J Hosp Med (Lond). 2019;80(9):525–9.
3. Ryan AF, Kujawa SG, Hammill T, Le Prell C, Kil J. Temporary and permanent noise-induced threshold shifts: a review of basic and clinical observations. Otol Neurotol. 2016;37(8):e271–5.
4. Lockwood AH, Salvi RJ, Burkard RF. Tinnitus. N Engl J Med. 2002;347(12):904–10.
5. McBride DI, Williams S. Audiometric notch as a sign of noise induced hearing loss. Occup Environ Med. 2001;58(1):46–51.

Noise-Induced Vestibular Damage

26

Overview

It is well recognized that excessively loud sounds can cause hearing loss, but *vestibular damage from noise exposure* is less studied and less well understood. Various studies of individuals who have suffered noise exposure have shown abnormalities in ocular motor function, caloric testing, vestibular evoked myogenic potentials, and video head impulse testing, and it appears that in at least some cases these vestibular deficits can occur independently of hearing loss.

Introduction

At least as far back as the middle of the twentieth century, it was observed that *noise trauma* can produce detectable *deficits in vestibular function* [1–3], though at that time the main clinically available method for assessing vestibular function was only caloric testing.

Examples of Evidence for Vestibular Deficits Associated with Noise-Induced Hearing Loss

Although some investigators regarded that evidence as inconclusive [4], multiple subsequent studies have demonstrated that individuals with noise-induced hearing loss can have objective deficits on otovestibular testing. We review here several representative examples from studies of nystagmography, vestibular evoked myogenic potentials, video head impulse testing, and computerized dynamic posturography.

Oosterveld and colleagues [5] studied 29 airline technicians with an "industrial type of hearing loss" and observed that 18 (62%) exhibited *spontaneous nystagmus*, 24 (83%) exhibited *positional nystagmus,* and 17 (58%) exhibited *cervical nystagmus*. The same research group, apparently studying the same group of 29 airline technicians, subsequently reported [6] that 7 (24%) patients exhibited caloric asymmetry.

Man and colleagues [7] reported an *electronystagmographic* study of 326 patients that identified vestibular abnormalities proportional to the severity of the acoustic trauma.

Golz and colleagues [8] studied 258 male military personnel "exposed to various intense noises" with "complete audiological and *electronystagmographic* evaluation." Of the 258 subjects, "134 had a symmetrical high-tone hearing loss, and 124 had asymmetrical losses" on audiometry. Further, "Each group was divided into 2 subgroups according to the presence or absence of vestibular complaints." The study concluded that, "vestibular damage caused by intense noise exposure might be expressed clinically in subjects with asymmetrical hearing loss," and that, "there was a strong correlation between the subjects' complaints and the results of the vestibular function tests," though, "there was no correlation between the severity of the hearing loss and the vestibular symptomatology and pathology."

Shupak and colleagues [9] studied 22 men with noise-induced hearing loss with *rotatory chair testing* and *videonystagmography*, and reported, "Significantly lower vestibulo-ocular reflex gain (p = 0.05), and a tendency towards decreased caloric responses."

Kumar and colleagues [10] reported that in patients with noise-induced hearing loss the degree of hearing loss was proportional to the prolongation of latencies in *cervical vestibular evoked myogenic potentials.*

Tseng and Young [11] investigated 30 patients with noise-induced hearing loss. They reported that all of the patients had hearing loss on audiometry, 70% exhibited abnormalities on *cervical vestibular evoked myogenic potentials*, 57% exhibited abnormalities on *ocular vestibular evoked myogenic potentials*, and 33% exhibited abnormalities on *caloric testing*.

Cassandro and colleagues [12] took 40 individuals (age 18–26 years) and studied them immediately before and immediately after exposure to loud (128 dB) disco music for

M. Cherchi, *Otoneurology and Vestibular Medicine*, https://doi.org/10.1007/978-3-031-94841-1_26

3 h. They reported that, "post-stimulus recordings have shown a significant increase in the *amplitude of the vestibular evoked myogenic potential* response, thus indicating a possible irritative involvement of the macular receptor" as an acute postexposure consequence of loud noise.

Yilmaz and colleagues [13] used *video head impulse testing* to study 36 men (age 28–55 years) with hearing loss who had worked for ≥4 years "in the steel and metal industry" and compared them to 30 age-matched controls without hearing loss. They found that 20 (56%) of 36 subjects with hearing loss exhibited reduced gain of the vestibulo-ocular reflex (VOR), whereas only 2 (7%) out of 30 normal-hearing subjects exhibited VOR abnormalities.

In addition to the studies cited above, it has also been observed that patients with auditory symptoms and *no complaints of disequilibrium* may still exhibit abnormalities on vestibular testing, suggesting that the labyrinthine insult has caused clinically manifest cochlear dysfunction as well as subclinical vestibular dysfunction. For example, van der Laan [14] reported that tinnitus patients with no symptoms of disequilibrium nevertheless exhibited abnormalities on electronystagmography. Yilkoski and colleagues [15] took 60 patients with firearms-related noise-induced hearing loss who had no complaints of disequilibrium and studied them with posturography, and concluded that the degree of abnormal sway was proportional to the magnitude of hearing loss.

Practical Approach

If a patient with complaints of disequilibrium is found on otovestibular testing to have detectable vestibular deficits, then recognition of a connection with noise-induced hearing loss may not have much practical consequence, as the treatment (vestibular physical therapy) would generally be similar to that of individuals without noise-induced hearing loss.

There may be at least some theoretical interest in potential implications from the public health perspective, though this is speculative. For example, we discussed earlier that some patients with noise-induced hearing loss do not have complaints of disequilibrium yet may have detectable deficits on otovestibular testing; this may be additive to the fall risk from presbyvestibulopathy, so it is conceivable that individuals who suffer noise-induced hearing loss earlier in life should undergo screening later in life for increased fall risk.

References

1. Dickson ED, Chadwick DL. Observations on disturbances of equilibrium and other symptoms induced by jet-engine noise. J Laryngol Otol. 1951;65(3):154–65.
2. Collins EG. Aural trauma caused by gunfire; report on a clinical investigation of 108 soldiers exposed to gunfire who, on some occasion, had complained of injury to their ears. J Laryngol Otol. 1948;62(6):358–90.
3. Dickson ED, Watson NP. A clinical survey into the effects of turbo-jet engine noise on service personnel. J Laryngol Otol. 1949;63(5):276–85.
4. Hinchcliffe R, Coles RR, King PF. Occupational noise induced vestibular malfunction? Br J Ind Med. 1992;49(1):63–5.
5. Oosterveld WJ, Polman AR, Schoonheyt J. Noise-induced hearing loss and vestibular dysfunction. Aviat Space Environ Med. 1980;51(8):823–6.
6. Oosterveld WJ, Polman AR, Schoonheyt J. Vestibular implications of noise-induced hearing loss. Br J Audiol. 1982;16(4):227–32.
7. Man A, Segal S, Naggan L. Vestibular involvement in acoustic trauma. (An electronystagmographic study). J Laryngol Otol. 1980;94(12):1395–400.
8. Golz A, Westerman ST, Westerman LM, Goldenberg D, Netzer A, Wiedmyer T, et al. The effects of noise on the vestibular system. Am J Otolaryngol. 2001;22(3):190–6.
9. Shupak A, Bar-El E, Podoshin L, Spitzer O, Gordon CR, Ben-David J. Vestibular findings associated with chronic noise induced hearing impairment. Acta Otolaryngol. 1994;114(6):579–85.
10. Kumar K, Vivarthini CJ, Bhat JS. Vestibular evoked myogenic potential in noise-induced hearing loss. Noise Health. 2010;12(48):191–4.
11. Tseng CC, Young YH. Sequence of vestibular deficits in patients with noise-induced hearing loss. Eur Arch Otorrinolaringol. 2013;270(7):2021–6.
12. Cassandro E, Chiarella G, Catalano M, Gallo LV, Marcelli V, Nicastri M, Petrolo C. Changes in clinical and instrumental vestibular parameters following acute exposition to auditory stress. Acta Otorhinolaryngol Ital. 2003;23(4):251–6.
13. Yilmaz N, Ila K, Soylemez E, Ozdek A. Evaluation of vestibular system with vHIT in industrial workers with noise-induced hearing loss. Eur Arch Otorrinolaringol. 2018;275(11):2659–65.
14. van der Laan FL. Noise exposure and its effect on the labyrinth, Part II. Int Tinnitus J. 2001;7(2):101–4.
15. Ylikoski J, Juntunen J, Matikainen E, Ylikoski M, Ojala M. Subclinical vestibular pathology in patients with noise-induced hearing loss from intense impulse noise. Acta Otolaryngol. 1988;105(5–6):558–63.

27 Otic Barotrauma

Overview

One of the main functions of the ear is to detect vibrations (minute changes in air pressure) in the external environment. The mechanism for accomplishing this detection is exquisitely sensitive, and therefore also susceptible to damage when the pressure changes in the external environment are very large, or very rapid. Large and/or rapid changes in pressure can traumatize the ear, and such damage is called *otic barotrauma*. The offending pressure changes are often abrupt (such as blast injuries), though more gradual changes can also cause damage when prolonged (such as SCUBA diving). Otic barotrauma can present with auditory symptoms (aural fullness, hearing loss, tinnitus), vestibular symptoms, or both; most literature states that auditory symptoms are more common than vestibular ones.

Introduction

One of the main functions of the ear is to detect vibrations (minute changes in air pressure) in the external environment. The mechanism for accomplishing this detection is exquisitely sensitive, and therefore also susceptible to damage when the pressure changes in the external environment are very large, or very rapid. Large and/or rapid changes in pressure can traumatize the ear, and such damage is called *otic barotrauma* [1, 2].

Consequences of Abrupt Change in Pressure

Common causes of *abrupt* barotrauma include blast injuries [3], or a physical blow to the ear, which result in an extremely abrupt and large *increase* in external pressure. We have also seen patients with avulsion of the stapes footplate after yanking an insert earphone out of their ear—presumably there was an extremely abrupt and large *decrease* in external pressure in this circumstance.

Abrupt pressure changes that are sufficiently large deliver kinetic energy to the ear that can cause physical disruption of its structures, including rupturing the tympanic membrane, dislocating the ossicular chain or fracturing its individual components (malleus, incus, stapes) [4], avulsing the stapes footplate from the oval window, or rupturing the round window [5]. Kinetic energy that traverses the oval and/or round windows can also be transmitted to inner ear structures and cause further damage.

Consequences of More Prolonged Change in Pressure

A pressure change that is more *gradual* yet is sufficiently large can cause symptoms as well, such as remaining in a situation of persistently increased pressure (as in SCUBA diving [6]), or one of persistently decreased pressure (as in air travel [7]).

When the Eustachian tube opens (such as during yawning), it makes the middle ear space continuous with the back of the throat, and therefore with the outside world, and therefore with the outer ear, and thus equilibrates pressure on both sides of the tympanic membrane. If Eustachian tube dysfunction impairs the Eustachian tube's ability to open, then a pressure differential can develop between the sides of the eardrum. As Mirza and Richardson comment, during airplane travel (when cabin pressure is generally lower than at sea level), the following can occur:

> When there is a sustained negative pressure in the middle ear, serous fluid is produced by the mucous glands in the middle ear in an effort to equalize the pressure. Sufficient negative pressure may exist to cause the rupture of blood vessels from the surrounding vasculature, resulting in a haemotympanum, or in severe cases, bulging and perforation of the tympanic membrane [7].

The opposite circumstance occurs when there is unequilibrated elevated external pressure, such as during SCUBA diving [8–10]. Even if an individual dive does not provoke

M. Cherchi, *Otoneurology and Vestibular Medicine*, https://doi.org/10.1007/978-3-031-94841-1_27

symptoms, there is evidence that repeated SCUBA diving may result in cumulative damage from recurrent otic barotrauma [11, 12].

Presentation

Otic barotrauma can present with auditory symptoms (aural fullness, hearing loss, tinnitus), vestibular symptoms, or both. Most literature states that auditory symptoms are more common than vestibular ones.

Diagnosis and Treatment

A patient with a clinical history compatible with otic barotrauma should be examined, including if possible with binocular otomicroscopy (or at least with a standard handheld monocular otoscope). It is also medically reasonable to check a standard audiogram (with tympanometry) and (for age <60 years) otoacoustic emissions. If examination shows a ruptured tympanic membrane, or if there is evidence of conductive hearing loss on audiometry, or if results from tympanometry suggest possible ossicular chain dislocation, then referral to otolaryngology is appropriate.

References

1. King PF. Otic barotrauma. Audiology. 1976;15(4):279–86.
2. Lacey JP, Amedee RG. The otologic manifestations of barotrauma. J La State Med Soc. 2000;152(3):107–11.
3. Darley DS, Kellman RM. Otologic considerations of blast injury. Disaster Med Public Health Prep. 2010;4(2):145–52.
4. Harris JP, Butler D. Recognition of malleus handle fracture in the differential diagnosis of otologic trauma. Laryngoscope. 1985;95(6):665–70.
5. Choo YB. Microscopic characteristics of round window problems in otology. Laryngoscope. 1984;94(1):1–9.
6. Livingstone DM, Smith KA, Lange B. Scuba diving and otology: a systematic review with recommendations on diagnosis, treatment and post-operative care. Diving Hyperb Med. 2017;47(2):97–109.
7. Mirza S, Richardson H. Otic barotrauma from air travel. J Laryngol Otol. 2005;119(5):366–70.
8. Farmer JC Jr. Diving injuries to the inner ear. Ann Otol Rhinol Laryngol Suppl. 1977;86(1 Pt 3 Suppl 36):1–20.
9. Molvaer OI, Albrektsen G. Alternobaric vertigo in professional divers. Undersea Biomed Res. 1988;15(4):271–82.
10. Money KE, Buckingham IP, Calder IM, Johnson WH, King JD, Landolt JP, et al. Damage to the middle ear and the inner ear in underwater divers. Undersea Biomed Res. 1985;12(1):77–84.
11. Molvaer OI, Albrektsen G. Hearing deterioration in professional divers: an epidemiologic study. Undersea Biomed Res. 1990;17(3):231–46.
12. Ramos CC, Rapoport PB, Brito Neto RV. Clinical and tympanometric findings in repeated recreational scuba diving. Travel Med Infect Dis. 2005;3(1):19–25.

Presbycusis

28

Overview

The term *presbycusis* is synonymous with the phrase *age-related hearing loss*. These refer to insidious onset, gradually progressive, bilateral, symmetrical sensorineural hearing loss that is more pronounced in the higher frequencies, resulting from physiologic and pathologic changes in most or all of the structures in the auditory pathway. Presbycusis is extremely common. Risk factors include environmental and probably genetic, though the genetic factors have not yet been clearly defined. Bedside hearing tests may serve the purpose of screening, but formal audiometry is superior for identifying the pattern of hearing loss compatible with presbycusis. Unexpected findings may prompt the search for alternative explanations. There is no curative or arrestive direct therapy for presbycusis. Many patients consult with an audiologist and may eventually consider amplification strategies.

Introduction

Presbycusis, or *age-related hearing loss (ARHL)*, "refers to hearing loss as a result of physiologic and pathologic changes associated with increasing age" [1] and can be "defined as a progressive, bilateral, symmetrical age-related sensorineural hearing loss, which is most pronounced at the higher frequencies" [2]. Hearing loss in general is an extraordinarily common condition. In the United States in 1990–1992, hearing loss was the third most common chronic problem in persons age 65–74 years (after arthritis and hypertension) and the second most common chronic problem in persons age ≥75 (after arthritis) [3].

Epidemiology

Repeated studies have demonstrated an increase in hearing loss with age. For example, a study from Yamasoba and colleagues [4] showed clearly that the prevalence of hearing loss in the United States begins in the fifth decade and increases exponentially thereafter (Fig. 28.1).

M. Cherchi, *Otoneurology and Vestibular Medicine*, https://doi.org/10.1007/978-3-031-94841-1_28

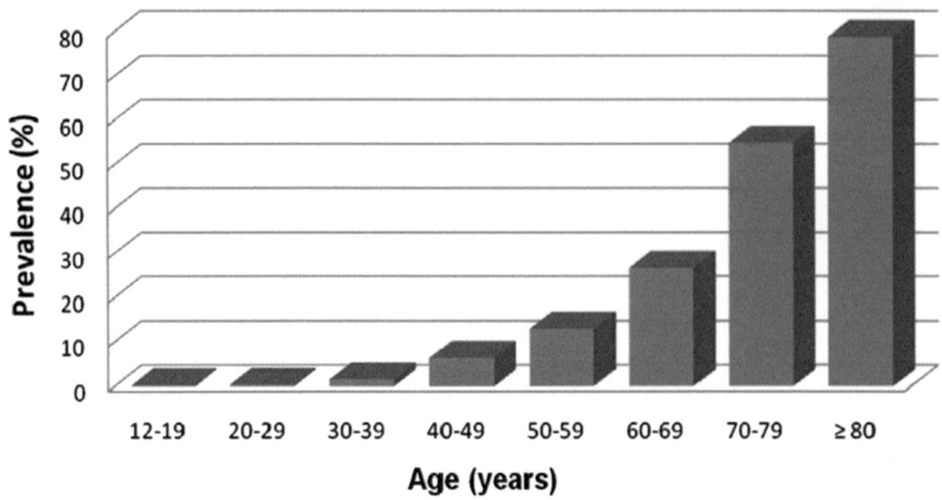

Fig. 28.1 Prevalence of hearing loss by decade in the United States. (From Yamasoba et al. [4]. Reproduced with written permission from Elsevier B.V.)

Genetics

Accumulating studies suggest that "genetic predisposition forms a large and important risk factor for ARHL" [2], but Bowl and Dawson also review the difficulties of identifying candidate genes even using genome-wide association studies. Nevertheless, there have been some advances in identifying likely candidate genes [5].

Pathophysiological Mechanism of Disease

In a general sense, presbycusis reflects age-related attrition of sensoria, but it is likely that multiple mechanisms contribute to this decline [4]. Bowl and Dawson summarize:

> Historical studies of postmortem human temporal bones suggest that ARHL involves a number of auditory structures: degeneration of the mechanotransducing cochlear inner and outer hair cells (sensory presbycusis); reduced function within the stria vascularis (SV) (strial presbycusis, also known as metabolic presbycusis); and degeneration of the auditory nerve (neural presbycusis)… However, in reality, it is likely most people with presbycusis will show a 'mixed' pathology. In addition to peripheral lesions it is also likely that there will be changes occurring in central auditory pathways that contribute to the development and progression of ARHL… As such, ARHL is considered a multifactorial disorder with underlying risk factors that can be divided into several categories: biological age, gender, ethnicity, environment (e.g., noise exposure, ototoxic medications), lifestyle (e.g., smoking, drinking, diet), health comorbidities (e.g., hypertension, diabetes), and genetic predisposition [3].

Elliott and colleagues [6] review the various patterns of degenerative changes occurring at different points along the auditory pathway.

Yamasoba and colleagues [4] discuss the multiple factors that can contribute to presbycusis (Fig. 28.2).

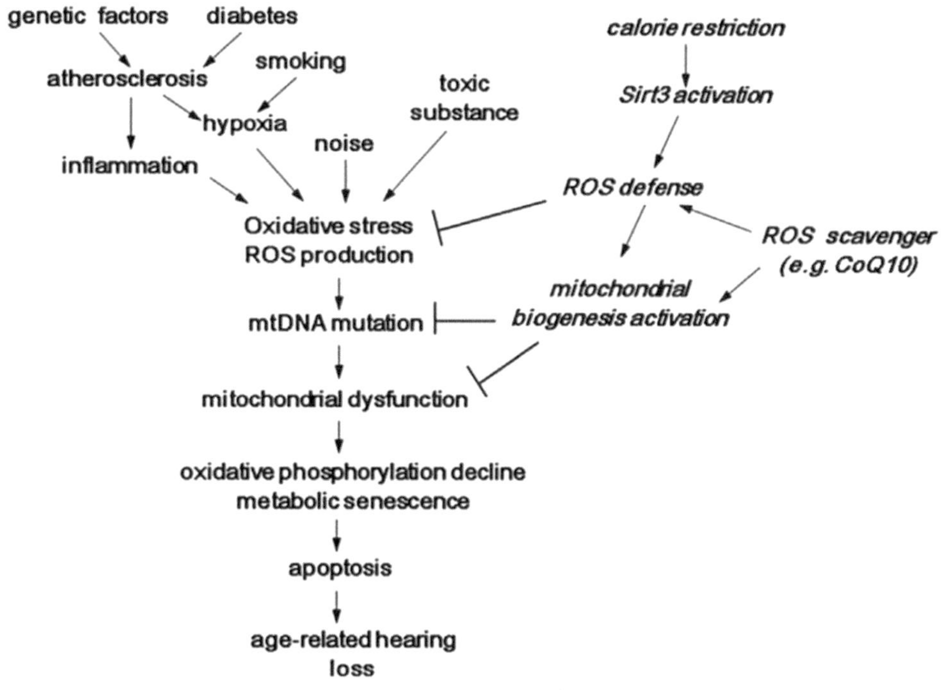

Fig. 28.2 Conceptual model of factors contributing to presbycusis. (From Yamasoba et al. [4]. Reproduced with written permission from Elsevier B.V.)

Clinical Presentation

Presbycusis tends to present insidiously [1], and patients often have difficulty pinpointing when they began noticing difficulty hearing.

Physical Examination

Physical examination can include bedside screening tests for hearing loss [7], such as finger rub, Weber and Rinne tests, but ultimately formal audiometry is a superior method for characterizing the pattern of hearing loss.

Testing: Auditory

Formal audiometry, including pure tone audiometry with word recognition and speech reception thresholds, remains the gold standard for evaluating hearing and supporting a diagnosis of presbycusis.

Presbycusis initially affects predominantly high frequencies, but over time will eventually also involve the middle frequencies, and then the low frequencies.

Testing: Other

As the genetic contributions to presbycusis become clarified, genetic testing may take on an increasing role in diagnosis and informing management decisions.

Imaging

Imaging is not required to secure a diagnosis of presbycusis. However, unexpected findings (such as significant asymmetry in hearing) during a workup for presbycusis may suggest that imaging is indicated.

Histopathology

As mentioned in the section above on the pathophysiologic mechanism of disease, studies (including histopathological works) have demonstrated changes at all levels of the auditory system.

Differential Diagnosis

Not all hearing loss in elderly individuals is due to age alone. Clinicians should have a low threshold for suspecting other diagnoses if the history is atypical (such as abrupt onset, rapidly progressive symptoms, presence of significant disequilibrium, or use of ototoxic medications), or if audiometric findings are incompatible (such as a clear asymmetry in hearing, or hearing loss predominantly in the low frequencies).

Treatment

Patients who are receptive to amplification have several treatment options [7], including hearing aids, osseointegrated auditory implants, and cochlear implants [1], which we discuss elsewhere. By the point at which a patient may be considering these options, care should be supervised by an audiologist.

Other therapeutic modalities being investigated include gene therapy and the use of stem cells. These remain experimental.

Prognosis

Presbycusis is generally slowly progressive. There is no curative or arrestive therapy.

References

1. Patel R, McKinnon BJ. Hearing loss in the elderly. Clin Geriatr Med. 2018;34(2):163–74.
2. Bowl MR, Dawson SJ. Age-related hearing loss. Cold Spring Harb Perspect Med. 2019;9(8):a033217.
3. Collins JG. Prevalence of selected chronic conditions: United States, 1990–1992. Vital Health Stat. 1997;10(194):1–89.
4. Yamasoba T, Lin FR, Someya S, Kashio A, Sakamoto T, Kondo K. Current concepts in age-related hearing loss: epidemiology and mechanistic pathways. Hear Res. 2013;303:30–8.
5. Wells HRR, Newman TA, Williams FMK. Genetics of age-related hearing loss. J Neurosci Res. 2020;98(9):1698–704.
6. Elliott KL, Fritzsch B, Yamoah EN, Zine A. Age-related hearing loss: sensory and neural etiology and their interdependence. Front Aging Neurosci. 2022;14:814528.
7. Michels TC, Duffy MT, Rogers DJ. Hearing loss in adults: differential diagnosis and treatment. Am Fam Physician. 2019;100(2):98–108.

29 Ototoxicity

Introduction

Ototoxicity refers to damage of otologically mediated auditory and/or vestibular function by a chemical compound. Drug-related ototoxicity is nearly always undesirable (with the main exception being purposeful ablation of labyrinthine function by transtympanic gentamicin injection). An extraordinarily broad range of chemical compounds have been documented to have potential ototoxicity, including various antimicrobial agents, antiarrhythmic agents, immunosuppressants, antineoplastic agents, diuretics, and others. Various attempts at monitoring cochlear and vestibular ototoxicity, protecting against ototoxicity, and treating ototoxicity have been attempted, without clear success.

Definition

For purposes of this discussion, we define ototoxicity as damage of otologically mediated auditory and/or vestibular function by a chemical compound.

We use the phrase "otologically mediated" to delimit the topic to damage of the labyrinth and/or vestibulocochlear nerve. We specifically exclude damage of other anatomical structures. Thus, for example, chronic alcoholism can damage the cerebellum; one may say that a chemical (ethanol) has led to imbalance, but the main structure affected by the chemical is the cerebellum, not the labyrinth or vestibular nerve.

We use the phrase "chemical compound" to include prescribed medications (drugs), as well as illegal substances (e.g., marijuana [1]) or misused chemicals (e.g., inhaled toluene). We also use this phrase to excluded nonchemical otologic damage, such as radiation toxicity.

Is Ototoxicity Always Bad?

In short, ototoxicity is nearly always bad, with very few exceptions. The main exception in otoneurology is when gentamicin is intentionally administered transtympanically for chemical ablation of the labyrinth as treatment for Ménière's disease.

Agents Implicated

Hundreds of drugs and other chemicals have been implicated as ototoxic. It is impractical to review these all, but Watts [2] provides a useful starting taxonomy (Fig. 29.1).

The main regulated ototoxic medications encountered in clinical practice include aminoglycoside antibiotics (usually gentamicin) and platinum-based chemotherapeutic agents (usually cisplatin), but there are many other agents for which ototoxicity has been implicated though less well substantiated. As of this writing, a PubMed search with the term "ototoxicity" returns >6000 hits.

Antimicrobial Agents: Aminoglycoside Antibiotics

Aminoglycoside antibiotics are well documented to be ototoxic [3–7], such as ***amikacin*** [8–12], ***gentamicin*** [13, 14], ***kanamycin*** [15–18], ***neomycin*** [19, 20], and ***streptomycin*** [18, 21–23].

M. Cherchi, *Otoneurology and Vestibular Medicine*, https://doi.org/10.1007/978-3-031-94841-1_29

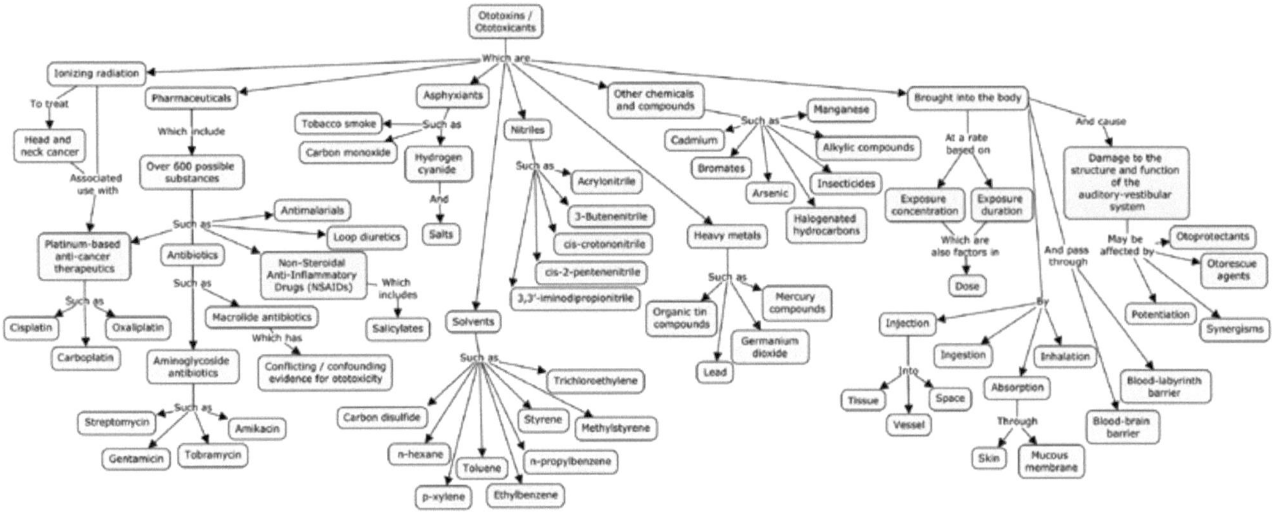

Fig. 29.1 A taxonomy of ototoxins. (From Watts [2]. Reproduced with written permission from Thieme Medical Publishers)

Antimicrobial Agents: Other Antibacterial Drugs

Beyond aminoglycosides [24–26], other *agents used in the treatment of tuberculosis* may also exhibit ototoxicity, such as ***isoniazid*** [27].

Some antibacterial agents from other classes also have documented ototoxicity, including the **macrolide antibiotics *erythromycin*** [28–38] and ***azithromycin*** [39–41], and the *glycopeptide antibiotic* **vancomycin** [19, 42–48].

Antimicrobial Agents: Antifungal Drugs

Some *antifungal agents*, such as ***amphotericin*** [49, 50], have documented ototoxicity.

Antimicrobial Agents: Antiparasitic Drugs

Several *antimalarial medications* have documented ototoxicity [51, 52].

Quinine and Its Derivatives

Quinine-based drugs are used as antimalarial agents [51, 52] and against certain autoimmune diseases such as rheumatoid arthritis; some of these medications have documented ototoxicity. These include ***quinine*** itself [53–55], ***chloroquine*** [56, 57], and ***hydroxychloroquine*** [57].

Antiarrhythmic Agents

Some *antiarrhythmic agents*, such as ***amiodarone*** [58, 59], have been reported to exhibit ototoxic properties.

Immunosuppressants

Several medications and medication classes that influence the immune system have documented ototoxicity. These include a variety of *immunosuppressants* [60] such as ***cyclosporine*** [61].

Antineoplastic Agents: Platinum Derivatives

Platinum-based antineoplastic agents have well-documented ototoxicity. The oldest of these medications, ***cisplatin***, is perhaps the most ototoxic of these [62–73]. Newer platinum-based agents are less so, but still ototoxic, including ***carboplatin*** [65, 74–84] and ***oxaliplatin*** [85–89].

Antineoplastic Agents: Others

Other antineoplastic agents also have documented ototoxicity, including the *taxane*-derived antimitotic agent ***docetaxel*** [90].

More recent antineoplastic agents, such as *immune checkpoint inhibitors*, have documented ototoxicity [91–93].

Loop Diuretics

Several *loop diuretics* have documented ototoxicity, including ***furosemide*** [15, 94–100] and ***bumetanide*** [15, 100].

Other Medications

Unregulated ototoxic medications (not requiring a prescription) include *salicylates* [101–106], the most common of which is ***acetylsalicylic acid (aspirin)***, which often is combined with other medications (such as in Excedrin Migraine).

Iron chelating agents have also documented ototoxicity [107].

Nonmedical Compounds

Various industrial *volatile aromatic solvents* [108, 109] have documented ototoxicity, such as ***styrene*** [110–114] and ***toluene*** [110, 115–117].

Various *heavy metals* have been implicated as ototoxic, including ***arsenic*** [118–121] and ***mercury*** [122].

Marijuana may be ototoxic insofar as it can provoke tinnitus in some patients [1].

Natural History of Ototoxicity

Generally, ototoxicity is dose-dependent, so the larger the dose administered, and the more frequently the drug is administered, the more likely it is that ototoxic effects will occur, and the more pronounced those effects will be.

In contrast to the usual dose-dependent pattern are occasional case reports of dramatic toxicity from a single dose of medication [72, 96, 123–125].

Ototoxicity can be enhanced or potentiated by concomitant issues, such as *renal insufficiency* (which may impede excretion of the offending drug), *coadministration with other agents* [8, 15, 95, 97, 126], *specific genetic mutations* [71, 127–132], *coadministration with radiation therapy* [64, 126, 133, 134], *ischemia* [135], and *noise exposure* [113, 136–141]. Whether these additional factors have a synergistic or merely additive effect is unclear.

Surprisingly, for some ototoxins, there additionally appears to be a diurnal pattern of vulnerability [142, 143].

Some longitudinal studies suggest that long-term hearing loss from cisplatin ototoxicity "dovetails" with age-related hearing loss [144, 145].

Some studies describe some spontaneous improvement from vestibular ototoxicity in some vestibular testing parameters but not others [146].

Can Cochlear Ototoxicity Be Monitored?

Since one of the most common ototoxic clinical manifestations is hearing loss, it is logical to look to assessment of hearing as a method for detecting and monitoring ototoxicity [147–149].

In ototoxicity from aminoglycoside antibiotics and platinum-based chemotherapeutic agents, usually cochleotoxicity initially affects high frequencies and later affects low frequencies [71, 127–132]. Because of this pattern, methods for testing frequencies above those of a standard audiogram have been explored, including high-frequency audiometry [150–156] and otoacoustic emissions [154].

Given the possibility of ototoxic agents directly affecting the cochlear nerve, some studies have explored the utility of using brainstem auditory evoked responses as a detection/monitoring technique [118, 151, 152, 157–159].

Beyond the idea of testing hearing, some studies have explored whether hearing tests can be performed with insert earphones [160], whether they can be performed at home [161, 162] or even by a pharmacist [163] or by telehealth [164].

It has even been suggested that imaging the labyrinth by MRI [165] or monitoring putative serum biomarkers [166] might play a role in detecting/monitoring ototoxicity.

Despite various proposed monitoring recommendations [148, 167], no specific protocol has been widely adopted.

The clinical assessment of ototoxicity, and the broader scientific study of this subject, is sometimes complicated by medico-legal dimensions.

Can Vestibular Ototoxicity Be Monitored?

The vast majority of ototoxicity literature pertains to cochleotoxicity. The literature regarding vestibular ototoxicity is much more limited [13, 14, 108, 168–176].

Can Ototoxicity Be Prevented?

In many instances, the use of an ototoxic medication is unavoidable, such in the chronic management of cystic fibrosis [123, 125, 150, 177, 178], or when a severe infection is only susceptible to an aminoglycoside antibiotic, or when an aggressive malignancy only responds to a platinum-based antineoplastic agent. Faced with likely ototoxicity in such circumstances, investigators have explored whether there are any mechanisms to attenuate ototoxicity [172, 179–186].

Much of this research has been directed at aminoglycoside antibiotics [187–193], specifically at gentamicin [194–196], amikacin [197], and neomycin [198].

Considerable research has also explored otoprotective strategies against cisplatin ototoxicity [6, 68, 111, 184, 188, 199–222].

Relatively little research has been directed against ototoxicity from furosemide [197] and other agents.

Can Ototoxicity Be Treated?

The short answer is that there is not yet any known direct therapy for any form of ototoxicity.

Hearing loss is usually treated with amplification by hearing aids, or (in severe cases) with cochlear implantation.

Imbalance is usually treated with vestibular rehabilitation therapy.

References

1. Qian ZJ, Alyono JC. An association between marijuana use and tinnitus. Am J Otolaryngol. 2020;41(1):102314.
2. Watts KL. Ototoxicity: visualized in concept maps. Semin Hear. 2019;40(2):177–87.
3. Henley CM 3rd, Schacht J. Pharmacokinetics of aminoglycoside antibiotics in blood, inner-ear fluids and tissues and their relationship to ototoxicity. Audiology. 1988;27(3):137–46.
4. Huth ME, Ricci AJ, Cheng AG. Mechanisms of aminoglycoside ototoxicity and targets of hair cell protection. Int J Otolaryngol. 2011;2011:937861.
5. Rizzi MD, Hirose K. Aminoglycoside ototoxicity. Curr Opin Otolaryngol Head Neck Surg. 2007;15(5):352–7.
6. Ruhl D, Du TT, Wagner EL, Choi JH, Li S, Reed R, et al. Necroptosis and apoptosis contribute to cisplatin and aminoglycoside ototoxicity. J Neurosci. 2019;39(15):2951–64.
7. Warchol ME. Cellular mechanisms of aminoglycoside ototoxicity. Curr Opin Otolaryngol Head Neck Surg. 2010;18(5):454–8.
8. Aksoy F, Dogan R, Ozturan O, Yildirim YS, Veyseller B, Yenigun A, et al. Betahistine exacerbates amikacin ototoxicity. Ann Otol Rhinol Laryngol. 2015;124(4):280–7.
9. Beaubien AR, Desjardins S, Ormsby E, Bayne A, Carrier K, Cauchy MJ, et al. Incidence of amikacin ototoxicity: a sigmoid function of total drug exposure independent of plasma levels. Am J Otolaryngol. 1989;10(4):234–43.
10. Beaubien AR, Desjardins S, Ormsby E, Bayne A, Carrier K, Cauchy MJ, et al. Delay in hearing loss following drug administration. A consistent feature of amikacin ototoxicity. Acta Otolaryngol. 1990;109(5-6):345–52.
11. Beaubien AR, Ormsby E, Bayne A, Carrier K, Crossfield G, Downes M, et al. Evidence that amikacin ototoxicity is related to total perilymph area under the concentration-time curve regardless of concentration. Antimicrob Agents Chemother. 1991;35(6):1070–4.
12. Endo A, Hanawa K, Nemoto A, Ishikawa T, Kazama S, Kagami Y, et al. Evaluation of nephrotoxicity and ototoxicity following amikacin administration once daily or every 48 hours in neonates. Medicine (Baltimore). 2022;101(43):e31425.
13. Chatterton S, Wang C, Satyan H, Kotsiou G, Todd CJ, Migliaccio AA, et al. A prospective study on the vestibular toxicity of gentamicin in a clinical setting. Otol Neurotol. 2022;43(9):e1029–e33.
14. Smyth D, Mossman S, Weatherall M, Jolliffe E, Joshi P, Taylor J, et al. Gentamicin vestibulotoxicity with modern systemic dosing regimens: a prospective study using video-oculography. Acta Otolaryngol. 2019;139(9):759–68.
15. Brummett RE, Bendrick T, Himes D. Comparative ototoxicity of bumetanide and furosemide when used in combination with kanamycin. J Clin Pharmacol. 1981;21(11):628–36.
16. Ghafari N, Court R, Chirehwa MT, Wiesner L, Petersen L, Maartens G, et al. Pharmacokinetics and other risk factors for kanamycin-induced hearing loss in patients with multi-drug resistant tuberculosis. Int J Audiol. 2020;59(3):219–23.
17. Mantefardo B, Sisay G. Case report: Kanamycin ototoxicity and MDR-TB treatment regimen. Int Med Case Rep J. 2021;14:815–7.
18. Voogt GR, Schoeman HS. Ototoxicity of aminoglycoside drugs in tuberculosis treatment. S Afr J Commun Disord. 1996;43:3–6.
19. Kavanagh KT, McCabe BF. Ototoxicity of oral neomycin and vancomycin. Laryngoscope. 1983;93(5):649–53.
20. Ward KM, Rounthwaite FJ. Neomycin ototoxicity. Ann Otol Rhinol Laryngol. 1978;87(2 Pt 1):211–5.
21. Adeyemo AA, Oluwatosin O, Omotade OO. Study of streptomycin-induced ototoxicity: protocol for a longitudinal study. Springerplus. 2016;5(1):758.
22. Md Daud MK, Mohamadl H, Haron A, Rahman NA. Ototoxicity screening of patients treated with streptomycin using distortion product otoacoustic emissions. B-ENT. 2014;10(1):53–8.
23. Wanamaker HH, Slepecky NB, Cefaratti LK, Ogata Y. Comparison of vestibular and cochlear ototoxicity from transtympanic streptomycin administration. Am J Otol. 1999;20(4):457–64.
24. Buziashvili M, Mirtskhulava V, Kipiani M, Blumberg HM, Baliashvili D, Magee MJ, et al. Rates and risk factors for nephrotoxicity and ototoxicity among tuberculosis patients in Tbilisi, Georgia. Int J Tuberc Lung Dis. 2019;23(9):1005–11.
25. Owusu E, Amartey BT, Afutu E, Boafo N. Aminoglycoside therapy for tuberculosis: evidence for ototoxicity among tuberculosis patients in Ghana. Diseases. 2022;10(1):10.
26. Stevenson LJ, Biagio-de Jager L, Graham MA, Swanepoel W. Community-based ototoxicity monitoring for drug-resistant tuberculosis in South Africa: an evaluation study. Int J Environ Res Public Health. 2021;18(21):11342.
27. Altiparmak MR, Pamuk ON, Pamuk GE, Ataman R, Serdengecti K. Is isoniazid ototoxic in patients undergoing hemodialysis? Nephron. 2002;92(2):478–80.
28. Agusti C, Ferran F, Gea J, Picado C. Ototoxic reaction to erythromycin. Arch Intern Med. 1991;151(2):380.
29. Bizjak ED, Haug MT 3rd, Schilz RJ, Sarodia BD, Dresing JM. Intravenous azithromycin-induced ototoxicity. Pharmacotherapy. 1999;19(2):245–8.
30. Brummett RE. Ototoxic liability of erythromycin and analogues. Otolaryngol Clin North Am. 1993;26(5):811–9.
31. Levin G, Behrenth E. Irreversible ototoxic effect of erythromycin. Scand Audiol. 1986;15(1):41–2.
32. McGhan LJ, Merchant SN. Erythromycin ototoxicity. Otol Neurotol. 2003;24(4):701–2.
33. Moral A, Navasa M, Rimola A, Garcia-Valdecasas JC, Grande L, Visa J, et al. Erythromycin ototoxicity in liver transplant patients. Transpl Int. 1994;7(1):62–4.
34. Sacristan JA, Angeles De Cos M, Soto J, Zurbano F, Pascual J, Tasis A, et al. Ototoxicity of erythromycin in man: electrophysiologic approach. Am J Otol. 1993;14(2):186–8.
35. Schweitzer VG, Olson NR. Ototoxic effect of erythromycin therapy. Arch Otolaryngol. 1984;110(4):258–60.
36. Swanson DJ, Sung RJ, Fine MJ, Orloff JJ, Chu SY, Yu VL. Erythromycin ototoxicity: prospective assessment with serum concentrations and audiograms in a study of patients with pneumonia. Am J Med. 1992;92(1):61–8.
37. Thompson P, Wood RP 2nd, Bergstrom L. Erythromycin ototoxicity. J Otolaryngol. 1980;9(1):60–2.

38. van Marion WF, van der Meer JW, Kalff MW, Schicht SM. Ototoxicity of erythromycin. Lancet. 1978;2(8082):214–5.
39. Haydon RC, Thelin JW, Davis WE. Erythromycin ototoxicity: analysis and conclusions based on 22 case reports. Otolaryngol Head Neck Surg. 1984;92(6):678–84.
40. Mamikoglu B, Mamikoglu O. Irreversible sensorineural hearing loss as a result of azithromycin ototoxicity. A case report. Ann Otol Rhinol Laryngol. 2001;110(1):102.
41. Wallace MR, Miller LK, Nguyen MT, Shields AR. Ototoxicity with azithromycin. Lancet. 1994;343(8891):241.
42. Gendeh BS, Gibb AG, Aziz NS, Kong N, Zahir ZM. Vancomycin administration in continuous ambulatory peritoneal dialysis: the risk of ototoxicity. Otolaryngol Head Neck Surg. 1998;118(4):551–8.
43. Gomceli U, Vangala S, Zeana C, Kelly PJ, Singh M. An unusual case of ototoxicity with use of oral vancomycin. Case Rep Infect Dis. 2018;2018:2980913.
44. Humphrey C, Veve MP, Walker B, Shorman MA. Long-term vancomycin use had low risk of ototoxicity. PLoS One. 2019;14(11):e0224561.
45. Lestner JM, Hill LF, Heath PT, Sharland M. Vancomycin toxicity in neonates: a review of the evidence. Curr Opin Infect Dis. 2016;29(3):237–47.
46. Marissen J, Fortmann I, Humberg A, Rausch TK, Simon A, Stein A, et al. Vancomycin-induced ototoxicity in very-low-birthweight infants. J Antimicrob Chemother. 2020;75(8):2291–8.
47. Mellor JA, Kingdom J, Cafferkey M, Keane C. Vancomycin ototoxicity in patients with normal renal function. Br J Audiol. 1984;18(3):179–80.
48. Traber PG, Levine DP. Vancomycin Ototoxicity in patient with normal renal function. Ann Intern Med. 1981;95(4):458–60.
49. Ramu R, Sharma B, Karunakara D, Paliwal P, Bansal N, Taneja RS. Liposomal amphotericin B-induced reversible ototoxicity in a patient with disseminated histoplasmosis. Indian J Pharmacol. 2021;53(2):157–9.
50. Singh PK, Sharma V. Ototoxicity, a rare but reversible adverse effect of a commonly used antimicrobial agent. BMJ Case Rep. 2019;12(7):e230251.
51. Dillard LK, Fullerton AM, McMahon CM. Ototoxic hearing loss from antimalarials: a systematic narrative review. Travel Med Infect Dis. 2021;43:102117.
52. Jozefowicz-Korczynska M, Pajor A, Lucas GW. The ototoxicity of antimalarial Drugs-A state of the art review. Front Neurol. 2021;12:661740.
53. Nielsen-Abbring FW, Perenboom RM, van der Hulst RJ. Quinine-induced hearing loss. ORL J Otorhinolaryngol Relat Spec. 1990;52(1):65–8.
54. Semedo MG, Dias-Silva N, Migueis J, Pita JR. Quinine in otology and neurotology: ototoxicity and historic role in therapy. Otol Neurotol. 2021;42(1):145–52.
55. Tange RA, Dreschler WA, Claessen FA, Perenboom RM. Ototoxic reactions of quinine in healthy persons and patients with Plasmodium falciparum infection. Auris Nasus Larynx. 1997;24(2):131–6.
56. Bortoli R, Santiago M. Chloroquine ototoxicity. Clin Rheumatol. 2007;26(11):1809–10.
57. Fernandes CP, Vernier LS, Dallegrave E, Machado MS. The ototoxicity of chloroquine and hydroxychloroquine: a systematic review. Int Arch Otorhinolaryngol. 2022;26(1):e167–e77.
58. Gurkov R. Amiodarone: a newly discovered association with bilateral vestibulopathy. Front Neurol. 2018;9:119.
59. Gurkov R, Manzari L, Blodow A, Wenzel A, Pavlovic D, Luis L. Amiodarone-associated bilateral vestibulopathy. Eur Arch Otorhinolaryngol. 2017;275(3):823–5.
60. Franz L, Frosolini A, Parrino D, Lovato A, de Filippis C, Marioni G. Ototoxicity of immunosuppressant drugs: a systematic review. J Int Adv Otol. 2022;18(2):167–76.
61. Waissbluth S. Is cyclosporine ototoxic? Front Neurol. 2020;11:593917.
62. Brock PR. New insights into cisplatin ototoxicity. Cancer. 2022;128(1):43–6.
63. Hodge SE, Lopez IA, Ishiyama G, Ishiyama A. Cisplatin ototoxicity histopathology. Laryngoscope Investig Otolaryngol. 2021;6(4):852–6.
64. Low WK, Kong SW, Tan MG. Ototoxicity from combined Cisplatin and radiation treatment: an in vitro study. Int J Otolaryngol. 2010;2010:523976.
65. Nitz A, Kontopantelis E, Bielack S, Koscielniak E, Klingebiel T, Langer T, et al. Prospective evaluation of cisplatin- and carboplatin-mediated ototoxicity in paediatric and adult soft tissue and osteosarcoma patients. Oncol Lett. 2013;5(1):311–5.
66. Paken J, Govender CD, Pillay M, Sewram V. Cisplatin-associated ototoxicity: a review for the health professional. J Toxicol. 2016;2016:1809394.
67. Paken J, Govender CD, Pillay M, Sewram V. A review of cisplatin-associated ototoxicity. Semin Hear. 2019;40(2):108–21.
68. Rybak LP, Whitworth CA, Mukherjea D, Ramkumar V. Mechanisms of cisplatin-induced ototoxicity and prevention. Hear Res. 2007;226(1-2):157–67.
69. Santucci NM, Garber B, Ivory R, Kuhn MA, Stephen M, Aizenberg D. Insight into the current practice of ototoxicity monitoring during cisplatin therapy. J Otolaryngol Head Neck Surg. 2021;50(1):19.
70. Sriyapai T, Thongyai K, Phuakpet K, Vathana N, Buaboonnam J, Sanpakit K. Ototoxicity and long-term hearing outcome in pediatric patients receiving cisplatin. Turk J Pediatr. 2022;64(3):531–41.
71. Tserga E, Nandwani T, Edvall NK, Bulla J, Patel P, Canlon B, et al. The genetic vulnerability to cisplatin ototoxicity: a systematic review. Sci Rep. 2019;9(1):3455.
72. Yildirim A, Fidan E, Ozdemir F. Hearing loss after the first low dose administration of cisplatin. J Coll Physicians Surg Pak. 2022;32(9):1216–8.
73. Sanchez VA, Shuey MM, Dinh PC Jr, Monahan PO, Fossa SD, Sesso HD, et al. Patient-reported functional impairment due to hearing loss and tinnitus after cisplatin-based chemotherapy. J Clin Oncol. 2023;41(12):2211–26.
74. Batra A, Thakar A, Bakhshi S. Ototoxicity in retinoblastoma survivors treated with carboplatin based chemotherapy: a cross-sectional study of 116 patients. Pediatr Blood Cancer. 2015;62(11):2060.
75. Bhagat SP, Bass JK, White ST, Qaddoumi I, Wilson MW, Wu J, et al. Monitoring carboplatin ototoxicity with distortion-product otoacoustic emissions in children with retinoblastoma. Int J Pediatr Otorhinolaryngol. 2010;74(10):1156–63.
76. Cavaletti G, Bogliun G, Zincone A, Marzorati L, Melzi P, Frattola L, et al. Neuro- and ototoxicity of high-dose carboplatin treatment in poor prognosis ovarian cancer patients. Anticancer Res. 1998;18(5B):3797–802.
77. Chevreau C, Thomas F, Couteau C, Dalenc F, Mourey L, Chatelut E. Ototoxicity of high-dose carboplatin. J Clin Oncol. 2005;23(15):3649–50. author reply 50-1
78. Geurtsen ML, Kors WA, Moll AC, Smits C. Long-term audiologic follow-up of carboplatin-treated children with retinoblastoma. Ophthalmic Genet. 2017;38(1):74–8.
79. Kennedy IC, Fitzharris BM, Colls BM, Atkinson CH. Carboplatin is ototoxic. Cancer Chemother Pharmacol. 1990;26(3):232–4.
80. Musial-Bright L, Fengler R, Henze G, Hernaiz DP. Carboplatin and ototoxicity: hearing loss rates among survivors of childhood medulloblastoma. Childs Nerv Syst. 2011;27(3):407–13.

81. Parsons SK, Neault MW, Lehmann LE, Brennan LL, Eickhoff CE, Kretschmar CS, et al. Severe ototoxicity following carboplatin-containing conditioning regimen for autologous marrow transplantation for neuroblastoma. Bone Marrow Transplant. 1998;22(7):669–74.
82. Qaddoumi I, Bass JK, Wu J, Billups CA, Wozniak AW, Merchant TE, et al. Carboplatin-associated ototoxicity in children with retinoblastoma. J Clin Oncol. 2012;30(10):1034–41.
83. Waissbluth S, Chuang A, Del Valle A, Cordova M. Long term platinum-induced ototoxicity in pediatric patients. Int J Pediatr Otorhinolaryngol. 2018;107:75–9.
84. Waissbluth S, Peleva E, Daniel SJ. Platinum-induced ototoxicity: a review of prevailing ototoxicity criteria. Eur Arch Otorhinolaryngol. 2017;274(3):1187–96.
85. Guvenc MG, Dizdar D, Dizdar SK, Okutur SK, Demir G. Sudden hearing loss due to oxaliplatin use in a patient with colon cancer. J Chemother. 2016;28(4):341–2.
86. Hijri FZ, Arifi S, Ouattassi N, Mellas N, El Mesbahi O. Oxaliplatin-induced ototoxicity in adjuvant setting for colorectal cancer: unusual side effect. J Gastrointest Cancer. 2014;45(1):106–8.
87. Malhotra NK, Aslam R, Lipman SP, Bilski VJ. Acute ototoxicity from a single infusion of oxaliplatin. Ear Nose Throat J. 2010;89(6):258–61.
88. Oh SY, Wasif N, Garcon MC, Rodriguez G, Saif MW. Ototoxicity associated with oxaliplatin in a patient with pancreatic cancer. JOP. 2013;14(6):676–9.
89. Vietor NO, George BJ. Oxaliplatin-induced hepatocellular injury and ototoxicity: a review of the literature and report of unusual side effects of a commonly used chemotherapeutic agent. J Oncol Pharm Pract. 2012;18(3):355–9.
90. Xuan L, Sun B, Meng X, Liu C, Cong Y, Wu S. Ototoxicity in patients with invasive ductal breast cancer who were treated with docetaxel: report of two cases. Cancer Biol Ther. 2020;21(11):990–3.
91. Hu F, Ye X, Zhai Y, Xu J, Guo X, Guo Z, et al. Ear and labyrinth toxicities induced by immune checkpoint inhibitors: a disproportionality analysis from 2014 to 2019. Immunotherapy. 2020;12(7):531–40.
92. Lemasson J, Cuzzubbo S, Doucet L, Gounant V, Baroudjian B, Herms F, et al. Cochleovestibular toxicity induced by immune checkpoint inhibition: a case series. Eur J Cancer. 2019;117:116–8.
93. Sturmer SH, Lechner A, Berking C. Sudden otovestibular dysfunction in 3 metastatic melanoma patients treated with immune checkpoint inhibitors. J Immunother. 2021;44(5):193–7.
94. Ahmed RM, Hannigan IP, MacDougall HG, Chan RC, Halmagyi GM. Gentamicin ototoxicity: a 23-year selected case series of 103 patients. Med J Aust. 2012;196(11):701–4.
95. Bates DE, Beaumont SJ, Baylis BW. Ototoxicity induced by gentamicin and furosemide. Ann Pharmacother. 2002;36(3):446–51.
96. Halmagyi GM, Curthoys IS. Adverse effects of a single dose of gentamicin. Br J Clin Pharmacol. 2018;84(12):2936.
97. Kaka JS, Lyman C, Kilarski DJ. Tobramycin-furosemide interaction. Drug Intell Clin Pharm. 1984;18(3):235–8.
98. Rybak LP. Furosemide ototoxicity: clinical and experimental aspects. Laryngoscope. 1985;95(9 Pt 2 Suppl 38):1–14.
99. Rybak LP, Whitworth C. Comparative ototoxicity of furosemide and piretanide. Acta Otolaryngol. 1986;101(1-2):59–65.
100. Tuzel IH. Comparison of adverse reactions to bumetanide and furosemide. J Clin Pharmacol. 1981;21(11):615–9.
101. Boettcher FA, Salvi RJ. Salicylate ototoxicity: review and synthesis. Am J Otolaryngol. 1991;12(1):33–47.
102. Brien JA. Ototoxicity associated with salicylates. A brief review. Drug Saf. 1993;9(2):143–8.
103. Day RO, Graham GG, Bieri D, Brown M, Cairns D, Harris G, et al. Concentration-response relationships for salicylate-induced ototoxicity in normal volunteers. Br J Clin Pharmacol. 1989;28(6):695–702.
104. Kim SM, Jo JM, Baek MJ, Jung KH. A case of bilateral sudden hearing loss and tinnitus after salicylate intoxication. Korean J Audiol. 2013;17(1):23–6.
105. Peleg U, Perez R, Freeman S, Sohmer H. Salicylate ototoxicity and its implications for cochlear microphonic potential generation. J Basic Clin Physiol Pharmacol. 2007;18(3):173–88.
106. Stypulkowski PH. Mechanisms of salicylate ototoxicity. Hear Res. 1990;46(1-2):113–45.
107. Derin S, Azik FM, Topal Y, Topal H, Karakus V, Cetinkaya PU, et al. The incidence of ototoxicity in patients using iron chelators. J Int Adv Otol. 2017;13(1):136–9.
108. Fife TD, Robb MJA, Steenerson KK, Saha KC. Bilateral vestibular dysfunction associated with chronic exposure to military jet propellant type-eight jet fuel. Front Neurol. 2018;9:351.
109. Gagnaire F, Langlais C. Relative ototoxicity of 21 aromatic solvents. Arch Toxicol. 2005;79(6):346–54.
110. Hoet P, Lison D. Ototoxicity of toluene and styrene: state of current knowledge. Crit Rev Toxicol. 2008;38(2):127–70.
111. Karasawa T, Steyger PS. An integrated view of cisplatin-induced nephrotoxicity and ototoxicity. Toxicol Lett. 2015;237(3):219–27.
112. Lawton BW, Hoffmann J, Triebig G. The ototoxicity of styrene: a review of occupational investigations. Int Arch Occup Environ Health. 2006;79(2):93–102.
113. Makitie AA, Pirvola U, Pyykko I, Sakakibara H, Riihimaki V, Ylikoski J. The ototoxic interaction of styrene and noise. Hear Res. 2003;179(1-2):9–20.
114. Sass-Kortsak AM, Corey PN, Robertson JM. An investigation of the association between exposure to styrene and hearing loss. Ann Epidemiol. 1995;5(1):15–24.
115. Juarez-Perez CA, Torres-Valenzuela A, Haro-Garcia LC, Borja-Aburto VH, Aguilar-Madrid G. Ototoxicity effects of low exposure to solvent mixture among paint manufacturing workers. Int J Audiol. 2014;53(6):370–6.
116. Schaper M, Demes P, Zupanic M, Blaszkewicz M, Seeber A. Occupational toluene exposure and auditory function: results from a follow-up study. Ann Occup Hyg. 2003;47(6):493–502.
117. Schaper M, Seeber A, van Thriel C. The effects of toluene plus noise on hearing thresholds: an evaluation based on repeated measurements in the German printing industry. Int J Occup Med Environ Health. 2008;21(3):191–200.
118. Ishii N, Mochizuki H, Yamashita M, Yagi K, Shiomi K, Tsuruta K, et al. Auditory brainstem response analysis for long-term central auditory function sequelae in patients with chronic arsenic intoxication: a cross-sectional study. J Neurol Sci. 2019;398:2–3.
119. Kesici GG. Arsenic ototoxicity. J Otol. 2016;11(1):13–7.
120. Kesici GG, Unlu I, Topcu AB, Bal CD, Tutkun E, Yilmaz OH. Arsenic related hearing loss in miners. Am J Otolaryngol. 2016;37(1):6–11.
121. Shokoohi R, Khazaei M, Karami M, Seidmohammadi A, Berijani N, Khotanlou H, et al. The relationship between chronic exposure to arsenic through drinking water and hearing function in exposed population aged 10-49 years: a cross-sectional study. Ecotoxicol Environ Saf. 2021;211:111939.
122. Hoshino AC, Ferreira HP, Malm O, Carvallo RM, Camara VM. A systematic review of mercury ototoxicity. Cad Saude Publica. 2012;28(7):1239–48.
123. Garinis A, Gleser M, Johns A, Larsen E, Vachhani J. Prospective cohort study of ototoxicity in persons with cystic fibrosis following a single course of intravenous tobramycin. J Cyst Fibros. 2021;20(2):278–83.
124. Gooi A, Hochman J, Wellman M, Blakley L, Blakley BW. Ototoxic effects of single-dose versus 19-day daily-dose gentamicin. J Otolaryngol Head Neck Surg. 2008;37(5):664–7.

125. Harruff EE, Kil J, Ortiz MGT, Dorgan D, Jain R, Poth EA, et al. Ototoxicity in cystic fibrosis patients receiving intravenous tobramycin for acute pulmonary exacerbation: ototoxicity following tobramycin treatment. J Cyst Fibros. 2021;20(2):288–94.
126. Driessen CML, Leijendeckers J, Snik A, van der Graaf WTA, de Boer JP, Gelderblom H, et al. Ototoxicity in locally advanced head and neck cancer patients treated with induction chemotherapy followed by intermediate or high-dose cisplatin-based chemoradiotherapy. Head Neck. 2019;41(2):488–94.
127. Steyger PS. Mechanisms of aminoglycoside- and cisplatin-induced ototoxicity. Am J Audiol. 2021;30(3S):887–900.
128. Clemens E, Broer L, Langer T, Uitterlinden AG, de Vries ACH, van Grotel M, et al. Genetic variation of cisplatin-induced ototoxicity in non-cranial-irradiated pediatric patients using a candidate gene approach: the International PanCareLIFE Study. Pharmacogenomics J. 2020;20(2):294–305.
129. Gao Z, Chen Y, Guan MX. Mitochondrial DNA mutations associated with aminoglycoside induced ototoxicity. J Otol. 2017;12(1):1–8.
130. McDermott JH, Mahaveer A, James RA, Booth N, Turner M, Harvey KE, et al. Rapid point-of-care genotyping to avoid aminoglycoside-induced ototoxicity in neonatal intensive care. JAMA Pediatr. 2022;176(5):486–92.
131. Nguyen T, Jeyakumar A. Genetic susceptibility to aminoglycoside ototoxicity. Int J Pediatr Otorhinolaryngol. 2019;120:15–9.
132. Turan C, Kantar M, Aktan C, Kosova B, Orman M, Bilgen C, et al. Cisplatin ototoxicity in children: risk factors and its relationship with polymorphisms of DNA repair genes ERCC1, ERCC2, and XRCC1. Cancer Chemother Pharmacol. 2019;84(6):1333–8.
133. Cohen-Cutler S, Wong K, Mena V, Sianto K, Wright MA, Olch A, et al. Hearing loss risk in pediatric patients treated with cranial irradiation and cisplatin-based chemotherapy. Int J Radiat Oncol Biol Phys. 2021;110(5):1488–95.
134. Low WK, Teng SW, Tan MGK. Synergistic ototoxicity of gentamicin and low-dose irradiation: molecular basis and clinical significance. Audiol Neurootol. 2020;25(3):111–9.
135. Lin CD, Kao MC, Tsai MH, Lai CH, Wei IH, Tsai MH, et al. Transient ischemia/hypoxia enhances gentamicin ototoxicity via caspase-dependent cell death pathway. Lab Invest. 2011;91(7):1092–106.
136. Bhattacharyya TK, Dayal VS. Ototoxicity and noise drug interaction. J Otolaryngol. 1984;13(6):361–6.
137. Brown RD, Penny JE, Henley CM, Hodges KB, Kupetz SA, Glenn DW, et al. Ototoxic drugs and noise. Ciba Found Symp. 1981;85:151–71.
138. Castellanos MJ, Fuente A. The adverse effects of heavy metals with and without noise exposure on the human peripheral and central auditory system: a literature review. Int J Environ Res Public Health. 2016;13(12):1223.
139. Garinis A, Konrad-Martin D, Bramhall N. Ototoxicity and noise damage: from preclinical findings to audiological management. Am J Audiol. 2021;30(3S):797–9.
140. Li H, Steyger PS. Synergistic ototoxicity due to noise exposure and aminoglycoside antibiotics. Noise Health. 2009;11(42):26–32.
141. Steyger PS. Potentiation of chemical ototoxicity by noise. Semin Hear. 2009;30(1):38–46.
142. Blunston MA, Yonovitz A, Woodahl EL, Smolensky MH. Gentamicin-induced ototoxicity and nephrotoxicity vary with circadian time of treatment and entail separate mechanisms. Chronobiol Int. 2015;32(9):1223–32.
143. Tserga E, Moreno-Paublete R, Sarlus H, Bjorn E, Guimaraes E, Goritz C, et al. Circadian vulnerability of cisplatin-induced ototoxicity in the cochlea. FASEB J. 2020;34(10):13978–92.
144. Skalleberg J, Myhrum M, Smastuen MC, Osnes TA, Fossa SD, Bunne M. Speech perception 30 years after cisplatin-based chemotherapy in adults: limited clinical relevance of long-term ototoxicity? Acta Oncol. 2021;60(4):426–33.
145. Skalleberg J, Smastuen MC, Oldenburg J, Osnes T, Fossa SD, Bunne M. The relationship between cisplatin-related and age-related hearing loss during an extended follow-up. Laryngoscope. 2020;130(9):E515–E21.
146. Black FO, Gianna-Poulin C, Pesznecker SC. Recovery from vestibular ototoxicity. Otol Neurotol. 2001;22(5):662–71.
147. Campbell KC, Durrant J. Audiologic monitoring for ototoxicity. Otolaryngol Clin N Am. 1993;26(5):903–14.
148. Le Prell CG, Brewer CC, Campbell KCM. The audiogram: detection of pure-tone stimuli in ototoxicity monitoring and assessments of investigational medicines for the inner ear. J Acoust Soc Am. 2022;152(1):470.
149. Simpson TH, Schwan SA, Rintelmann WF. Audiometric test criteria in the detection of cisplatin ototoxicity. J Am Acad Audiol. 1992;3(3):176–85.
150. Caumo DTM, Geyer LB, Teixeira AR, Barreto SSM. Hearing thresholds at high frequency in patients with cystic fibrosis: a systematic review. Braz J Otorhinolaryngol. 2017;83(4):464–74.
151. Fausti SA, Frey RH, Henry JA, Olson DJ, Schaffer HI. Early detection of ototoxicity using high-frequency, tone-burst-evoked auditory brainstem responses. J Am Acad Audiol. 1992;3(6):397–404.
152. Fausti SA, Frey RH, Henry JA, Olson DJ, Schaffer HI. High-frequency testing techniques and instrumentation for early detection of ototoxicity. J Rehabil Res Dev. 1993;30(3):333–41.
153. Fausti SA, Helt WJ, Phillips DS, Gordon JS, Bratt GW, Sugiura KM, et al. Early detection of ototoxicity using 1/6th-octave steps. J Am Acad Audiol. 2003;14(8):444–50.
154. Jacob LC, Aguiar FP, Tomiasi AA, Tschoeke SN, Bitencourt RF. Auditory monitoring in ototoxicity. Braz J Otorhinolaryngol. 2006;72(6):836–44.
155. Singh Chauhan R, Saxena RK, Varshey S. The role of ultrahigh-frequency audiometry in the early detection of systemic drug-induced hearing loss. Ear Nose Throat J. 2011;90(5):218–22.
156. van der Hulst RJ, Boeschoten EW, Nielsen FW, Struijk DG, Dreschler WD, Tange RA. Ototoxicity monitoring with ultra-high frequency audiometry in peritoneal dialysis patients treated with vancomycin or gentamicin. ORL J Otorhinolaryngol Relat Spec. 1991;53(1):19–22.
157. Bernard P. Alterations of auditory evoked potentials during the course of chloroquine treatment. Acta Otolaryngol. 1985;99(3-4):387–92.
158. Chen Y, Bielefeld EC, Mellott JG, Wang W, Mafi AM, Yamoah EN, et al. Early physiological and cellular indicators of cisplatin-induced ototoxicity. J Assoc Res Otolaryngol. 2021;22(2):107–26.
159. De Lauretis A, De Capua B, Barbieri MT, Bellussi L, Passali D. ABR evaluation of ototoxicity in cancer patients receiving cisplatin or carboplatin. Scand Audiol. 1999;28(3):139–43.
160. Gordon JS, Phillips DS, Helt WJ, Konrad-Martin D, Fausti SA. Evaluation of insert earphones for high-frequency bedside ototoxicity monitoring. J Rehabil Res Dev. 2005;42(3):353–61.
161. Jacobs PG, Silaski G, Wilmington D, Gordon S, Helt W, McMillan G, et al. Development and evaluation of a portable audiometer for high-frequency screening of hearing loss from ototoxicity in homes/clinics. IEEE Trans Bio-med Eng. 2012;59(11):3097–103.
162. Konrad-Martin D, O'Connell Bennett K, Garinis A, McMillan GP. A randomized controlled trial using automated technology for improving ototoxicity monitoring in VA oncology patients. Am J Audiol. 2021;30(3S):870–86.
163. McKinzie CJ, Kam CW, Parry R, Darby M, Tak CR. Pharmacist-administered audiology screening for pediatric cystic fibrosis patients exposed to high-dose aminoglycosides: a pilot study. Pediatr Pulmonol. 2022;57(7):1814–7.

164. Robler SK, Coco L, Krumm M. Telehealth solutions for assessing auditory outcomes related to noise and ototoxic exposures in clinic and research. J Acoust Soc Am. 2022;152(3):1737.
165. Veiga M, Kuhweide R, Demaerel V, De Pauw R, De Foer B, Casselman JW. Labyrinthine enhancement on 3D black blood MR images of the brain as an imaging biomarker for cisplatin ototoxicity in (lung) cancer patients. Neuroradiology. 2021;63(1):81–90.
166. Generotti C, Cox BC, Singh J, Hamilton D, McKenzie E, O'Malley BW Jr, et al. Subclinical diagnosis of cisplatin-induced ototoxicity with biomarkers. Sci Rep. 2022;12(1):18032.
167. Bass JK, Bhagat SP. Challenges in ototoxicity monitoring in the pediatric oncology population. J Am Acad Audiol. 2014;25(8):760–74. quiz 82-3
168. Baguley DM, Prayuenyong P. Looking beyond the audiogram in ototoxicity associated with platinum-based chemotherapy. Cancer Chemother Pharmacol. 2020;85(2):245–50.
169. Black FO, Pesznecker S, Stallings V. Permanent gentamicin vestibulotoxicity. Otol Neurotol. 2004;25(4):559–69.
170. Black FO, Pesznecker SC. Vestibular ototoxicity. Clinical considerations. Otolaryngol Clin N Am. 1993;26(5):713–36.
171. Hsu PC, Cheng PW, Young YH. Ototoxicity from organic solvents assessed by an inner ear test battery. J Vestib Res Equilib Orient. 2015;25(3-4):177–83.
172. Jiang L, Zheng Z, He Y. Progress in protecting vestibular hair cells. Arch Toxicol. 2021;95(8):2613–23.
173. Mount RJ, Takeno S, Wake M, Harrison RV. Carboplatin ototoxicity in the chinchilla: lesions of the vestibular sensory epithelium. Acta oto-laryngologica Supplementum. 1995;519:60–5.
174. Rutka J. Aminoglycoside Vestibulotoxicity. Adv Otorhinolaryngol. 2019;82:101–10.
175. Sedo-Cabezon L, Boadas-Vaello P, Soler-Martin C, Llorens J. Vestibular damage in chronic ototoxicity: a mini-review. Neurotoxicology. 2014;43:21–7.
176. Tsuji K, Velazquez-Villasenor L, Rauch SD, Glynn RJ, Wall C 3rd, Merchant SN. Temporal bone studies of the human peripheral vestibular system. Aminoglycoside ototoxicity. Ann Otol Rhinol Laryngol Suppl. 2000;181:20–5.
177. Dong M, Rodriguez AV, Blankenship CA, McPhail G, Vinks AA, Hunter LL. Pharmacokinetic modelling to predict risk of ototoxicity with intravenous tobramycin treatment in cystic fibrosis. J Antimicrob Chemother. 2021;76(11):2923–31.
178. Elson EC, Meier E, Oermann CM. The implementation of an aminoglycoside induced ototoxicity algorithm for people with cystic fibrosis. J Cyst Fibros. 2021;20(2):284–7.
179. Barbara M, Margani V, Covelli E, Filippi C, Volpini L, El-Borady OM, et al. The use of nanoparticles in otoprotection. Front Neurol. 2022;13:912647.
180. Laurell G. Pharmacological intervention in the field of ototoxicity. HNO. 2019;67(6):434–9.
181. Pham TNM, Jeong SY, Kim DH, Park YH, Lee JS, Lee KW, et al. Protective mechanisms of avocado oil extract against ototoxicity. Nutrients. 2020;12(4):947.
182. Rybak LP, Kelly T. Ototoxicity: bioprotective mechanisms. Curr Opin Otolaryngol Head Neck Surg. 2003;11(5):328–33.
183. Wu P, Wu X, Zhang C, Chen X, Huang Y, Li H. Hair cell protection from ototoxic drugs. Neural Plast. 2021;2021:4909237.
184. Yurtsever KN, Baklaci D, Guler I, Kuzucu I, Kum RO, Ozhamam EU, et al. The protective effect of platelet rich plasma against cisplatin-induced ototoxicity. J Craniofac Surg. 2020;31(5):e506–e9.
185. Zhang X, Yu J. Baicalin attenuates gentamicin-induced cochlear hair cell ototoxicity. J Appl Toxicol. 2019;39(8):1208–14.
186. Zheng Z, Wang Y, Yu H, Li W, Wu J, Cai C, et al. Salvianolic acid B inhibits ototoxic drug-induced ototoxicity by suppression of the mitochondrial apoptosis pathway. J Cell Mol Med. 2020;24(12):6883–97.
187. Kawamoto K, Sha SH, Minoda R, Izumikawa M, Kuriyama H, Schacht J, et al. Antioxidant gene therapy can protect hearing and hair cells from ototoxicity. Mol Ther. 2004;9(2):173–81.
188. Kitcher SR, Kirkwood NK, Camci ED, Wu P, Gibson RM, Redila VA, et al. ORC-13661 protects sensory hair cells from aminoglycoside and cisplatin ototoxicity. JCI Insight. 2019;4(15):e126764.
189. Lopez-Gonzalez MA, Guerrero JM, Torronteras R, Osuna C, Delgado F. Ototoxicity caused by aminoglycosides is ameliorated by melatonin without interfering with the antibiotic capacity of the drugs. J Pineal Res. 2000;28(1):26–33.
190. Momiyama J, Hashimoto T, Matsubara A, Futai K, Namba A, Shinkawa H. Leupeptin, a calpain inhibitor, protects inner ear hair cells from aminoglycoside ototoxicity. Tohoku J Exp Med. 2006;209(2):89–97.
191. Pavlidis P, Maurer J, Apostolidou E, Kekes G, Kouvelas D. Memantine's action against aminoglycoside-induced ototoxicity. Eur Arch Otorhinolaryngol. 2014;271(6):1491–6.
192. Perletti G, Vral A, Patrosso MC, Marras E, Ceriani I, Willems P, et al. Prevention and modulation of aminoglycoside ototoxicity (Review). Mol Med Rep. 2008;1(1):3–13.
193. Zong Y, Chen F, Li S, Zhang H. (-)-Epigallocatechin-3-gallate (EGCG) prevents aminoglycosides-induced ototoxicity via anti-oxidative and anti-apoptotic pathways. Int J Pediatr Otorhinolaryngol. 2021;150:110920.
194. Kucharava K, Sekulic-Jablanovic M, Horvath L, Bodmer D, Petkovic V. Pasireotide protects mammalian cochlear hair cells from gentamicin ototoxicity by activating the PI3K-Akt pathway. Cell Death Dis. 2019;10(2):110.
195. Niu P, Sun Y, Wang S, Li G, Tang X, Sun J, et al. Puerarin alleviates the ototoxicity of gentamicin by inhibiting the mitochondria-dependent apoptosis pathway. Mol Med Rep. 2021;24(6):851.
196. Wen YH, Lin HY, Lin JN, Tseng GF, Hwang CF, Lin CC, et al. 2,3,4′,5-Tetrahydroxystilbene-2-O-beta-D-glucoside ameliorates gentamicin-induced ototoxicity by modulating autophagy via Sesn2/AMPK/mTOR signaling. Int J Mol Med. 2022;49(5):71.
197. Zadrozniak M, Szymanski M, Luszczki JJ. Vitamin C alleviates ototoxic effect caused by coadministration of amikacin and furosemide. Pharmacol Rep. 2019;71(2):351–6.
198. Zheng Z, Tang D, Zhao L, Li W, Han J, Hu B, et al. Liproxstatin-1 protects hair cell-like HEI-OC1 cells and cochlear hair cells against neomycin ototoxicity. Oxid Med Cell Longev. 2020;2020:1782659.
199. Bhatta P, Dhukhwa A, Sheehan K, Al Aameri RFH, Borse V, Ghosh S, et al. Capsaicin protects against cisplatin ototoxicity by changing the STAT3/STAT1 ratio and activating cannabinoid (CB2) receptors in the Cochlea. Sci Rep. 2019;9(1):4131.
200. Drogemoller BI, Wright GEB, Lo C, Le T, Brooks B, Bhavsar AP, et al. Pharmacogenomics of Cisplatin-Induced Ototoxicity: successes, shortcomings, and future avenues of research. Clin Pharmacol Ther. 2019;106(2):350–9.
201. Fang B, Xiao H. Rapamycin alleviates cisplatin-induced ototoxicity in vivo. Biochem Biophys Res Commun. 2014;448(4):443–7.
202. Fernandez KA, Allen P, Campbell M, Page B, Townes T, Li CM, et al. Atorvastatin is associated with reduced cisplatin-induced hearing loss. J Clin Invest. 2021;131(1):e142616.
203. Freyer DR, Brock P, Knight K, Reaman G, Cabral S, Robinson PD, et al. Interventions for cisplatin-induced hearing loss in children and adolescents with cancer. Lancet Child Adolesc Health. 2019;3(8):578–84.
204. Freyer DR, Brock PR, Chang KW, Dupuis LL, Epelman S, Knight K, et al. Prevention of cisplatin-induced ototoxicity in children and adolescents with cancer: a clinical practice guideline. Lancet Child Adolesc Health. 2020;4(2):141–50.

205. Fritzsche S, Strauss C, Scheller C, Leisz S. Nimodipine treatment protects auditory hair cells from cisplatin-induced cell death accompanied by upregulation of LMO4. Int J Mol Sci. 2022;23(10):5780.
206. Guan G, He X, Chen J, Bin L, Tang X. Identifying the mechanisms underlying the protective effect of tetramethylpyrazine against cisplatin-induced in vitro ototoxicity in HEI-OC1 auditory cells using gene expression profiling. Mol Med Rep. 2020;22(6):5053–68.
207. Hu B, Liu Y, Chen X, Zhao J, Han J, Dong H, et al. Ferrostatin-1 protects auditory hair cells from cisplatin-induced ototoxicity in vitro and in vivo. Biochem Biophys Res Commun. 2020;533(4):1442–8.
208. Lee CH, Lee DH, Lee SM, Kim ASY. Otoprotective effects of zingerone on cisplatin-induced ototoxicity. Int J Mol Sci. 2020;21(10):3503.
209. Liang Z, Zhang T, Zhan T, Cheng G, Zhang W, Jia H, et al. Metformin alleviates cisplatin-induced ototoxicity by autophagy induction possibly via the AMPK/FOXO3a pathway. J Neurophysiol. 2021;125(4):1202–12.
210. Liu Y, Wu H, Zhang F, Yang J, He J. Resveratrol upregulates miR-455-5p to antagonize cisplatin ototoxicity via modulating the PTEN-PI3K-AKT axis. Biochem Cell Biol. 2021;99(3):385–95.
211. Lopez-Gonzalez MA, Guerrero JM, Rojas F, Delgado F. Ototoxicity caused by cisplatin is ameliorated by melatonin and other antioxidants. J Pineal Res. 2000;28(2):73–80.
212. Navarro-Ruiz NE. Prevention of platinum-induced hearing loss in children with cancer. Am J Nurs. 2020;120(6):21.
213. Paciello F, Fetoni AR, Mezzogori D, Rolesi R, Di Pino A, Paludetti G, et al. The dual role of curcumin and ferulic acid in counteracting chemoresistance and cisplatin-induced ototoxicity. Sci Rep. 2020;10(1):1063.
214. Santos N, Ferreira RS, Santos ACD. Overview of cisplatin-induced neurotoxicity and ototoxicity, and the protective agents. Food Chem Toxicol. 2020;136:111079.
215. Simsek G, Tas BM, Muluk NB, Azman M, Kilic R. Comparison of the protective efficacy between intratympanic dexamethasone and resveratrol treatments against cisplatin-induced ototoxicity: an experimental study. Eur Arch Otorhinolaryngol. 2019;276(12):3287–93.
216. Tan M, Toplu Y, Varan E, Sapmaz E, Ozhan O, Parlakpinar H, et al. The effect of genistein on cisplatin induced ototoxicity and oxidative stress. Braz J Otorhinolaryngol. 2022;88(1):105–11.
217. Tas BM, Simsek G, Azman M, Kilic R. Efficacy of 2 different intratympanic steroid regimen on prevention of cisplatin ototoxicity: an experimental study. Ear Nose Throat J. 2021;100(6):417–22.
218. Terzi S, Ozgur A, Celiker M, Mercantepe T, Yilmaz A, Tumkaya L, et al. The protective effect of astaxanthin on cisplatin-induced ototoxicity. Adv Clin Exp Med. 2021;30(3):315–21.
219. Wang N, Xu A, Zhang H, Zhang Y, Sui R, Fan X, et al. Melatonin attenuates cisplatin-induced ototoxicity via regulating the cell apoptosis of the inner ear. Comput Math Methods Med. 2022;2022:7160816.
220. Wen Y, Zong S, Liu T, Du P, Li H, Xiao H. Tauroursodeoxycholic acid attenuates cisplatin-induced ototoxicity by inhibiting the accumulation and aggregation of unfolded or misfolded proteins in the endoplasmic reticulum. Toxicology. 2021;453:152736.
221. Zhang W, Xiong H, Pang J, Su Z, Lai L, Lin H, et al. Nrf2 activation protects auditory hair cells from cisplatin-induced ototoxicity independent on mitochondrial ROS production. Toxicol Lett. 2020;331:1–10.
222. Zhang Y, Lv Z, He Q. Agmatine alleviates cisplatin-induced ototoxicity by activating PI3K/AKT signaling pathway. eNeuro. 2022;9(2):ENEURO.0434-21.2022.

Inner Ear Ischemia

30

Overview

Arterial blood supply to the inner ear relies entirely on the labyrinthine artery. Venous drainage is more diffuse, and there are intralabyrinthine anastomoses, but the single arterial supply renders the labyrinth fairly vulnerable to ischemia. That being said, the labyrinthine artery usually branches off the anterior inferior cerebellar artery (AICA) , so usually embolic strokes will result in deficits in the territories of both the labyrinthine artery and the AICA. It is far less common for a stroke to be so focal as to affect only the inner ear, but when it does, the clinical presentation can mimic primary otologic diseases such as vestibular neuritis, labyrinthitis, and sudden sensorineural hearing loss. Most inner ear strokes are ischemic; a minority are hemorrhagic. Management of inner ear strokes is essentially a combination of managing stroke, sensorineural hearing loss (if present), and vestibular weakness (if present). Prognosis is guarded and tends to be poor if the inner ear stroke is hemorrhagic.

Introduction

Arterial blood supply to the inner ear relies entirely on the labyrinthine artery. Venous drainage is more diffuse, and there are intralabyrinthine anastomoses, but the single arterial supply renders the labyrinth fairly vulnerable to ischemia. However, the extent to which vascular disease accounts for otovestibular disorders is unclear.

Epidemiology

Having a discrete acute "stroke" (whether ischemic or hemorrhagic) isolated to the inner ear is uncommon because the vascular territory is so focal. More commonly, inner ear "strokes" occur in association with a stroke in closely related vascular territories (typically the anterior inferior ischemic artery).

Whether gradually progressive age-related microvascular circulatory disturbances cause or contribute to age-related attrition of inner ear function is less clear [1].

Pathophysiological Mechanism of Disease

The vascular anatomy of the inner ear is reasonably well-understood thanks to a number of meticulous temporal bone studies [2, 3], including studies of individuals who in life had known vascular pathology [4, 5]. There are several anastomotic complexes within the inner ear [6, 7]. The inner ear vasculature probably also has some degree of autoregulation in addition to autonomic modulation [8].

Animal studies suggest that temporary interruption of blood supply affects cochlear hair cells sooner than vestibular hair cells [9], though it is unknown whether this pattern also holds in humans.

The arterial supply of the inner ear is tenuous as it relies exclusively on the labyrinthine artery. Most reported cases of "inner ear strokes" are ischemic and presumed to be embolic in origin.

More rarely reported are cases of inner ear hemorrhage [10–14]. Some such cases appear to be spontaneous [11] or presumed secondary to hypertension [13], but many occur in the context of malignancy with presumed hyperviscosity [14].

Clinical Presentation

Clinically, it is difficult to identify cases of inner ear ischemia confidently. Probably, the most convincing cases are those in which the vertebral artery is compromised by dissection (confirmed by imaging), and in close temporal proximity, the patient subsequently develops apoplectic unilateral audiologic and vestibular symptoms [15, 16].

M. Cherchi, *Otoneurology and Vestibular Medicine*, https://doi.org/10.1007/978-3-031-94841-1_30

There are also cases of labyrinthine artery embolization even in the absence of vertebral artery dissection [17, 18]. However, because the labyrinthine artery usually branches off the anterior inferior cerebellar artery (AICA) , it is unusual for embolic events to be restricted to the territory of the labyrinthine artery. More commonly, embolic events leading to ischemia/infarction in the territory of the labyrinthine artery usually are accompanied by ischemia/infarction in territory of the AICA distal to the labyrinthine artery. Lee and colleagues [19] studied a series of 82 patients with AICA strokes and reported that 60% of cases included findings referable to infarction of the cochlea and labyrinth; however, only 5% of cases had isolated vestibular labyrinth infarction (without involvement of the cochlea or cerebellum), only 4% of cases had isolated cochlear involvement (without involvement of the vestibular labyrinth or cerebellum), and only 1% had "isolated audiovestibular loss without central symptoms or signs."

Physical Examination

The audiologic dysfunction of cochlear infarction will manifest as sensorineural hearing loss, and thus could mimic idiopathic sudden onset sensorineural hearing loss.

The vestibular dysfunction of infarction of the vestibular labyrinth will manifest similarly to vestibular neuritis.

If a patient exhibits any physical examination findings of cerebellar or brainstem dysfunction, then a diagnosis of infarction should always be high on the differential diagnosis (as opposed to primary inner ear disease).

Ocular Motor Examination

Ocular motor examination of a patient with an infarct limited to the vestibular labyrinth will appear similar to vestibular neuritis. For example, in the case reported by Nam and colleagues [18] of a patient with a left-sided vestibular labyrinthine infarction, the patient exhibited spontaneous right beat and right torsional nystagmus.

Testing: Auditory

Cochlear infarction will manifest as sensorineural hearing loss. Lee and colleagues [19] provide a well-documented example of this in a patient with an AICA infarction.

Testing: Vestibular

In a patient with labyrinthine vestibular function, otovestibular testing would be expected to show vestibular weakness.

Figure 30.1, from Nam and colleagues [18], shows a video oculographic tracing from a patient with left-sided vestibular labyrinthine infarction. This shows spontaneous nystagmus that is predominantly right beat, with more modest components of right torsional and up beat.

Figure 30.2, from Nam and colleagues [18], shows results of video head impulse testing (vHIT) from a patient with left-sided vestibular labyrinthine infarction. The lower left panel shows vHIT from the left horizontal canal, which exhibits low gain and overt compensatory saccades.

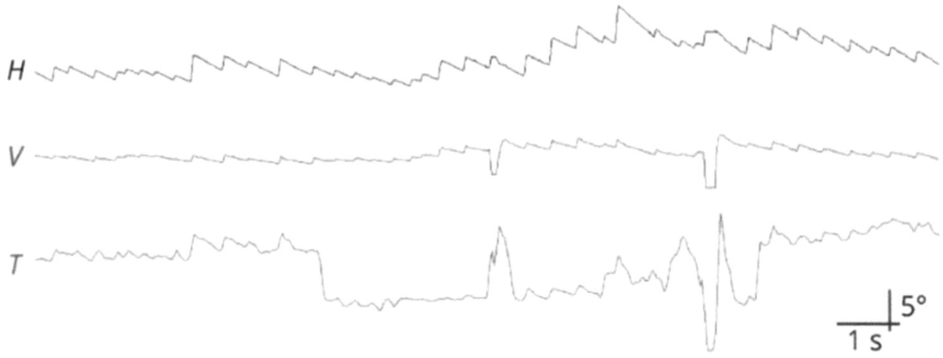

Fig. 30.1 Nystagmus that is predominantly right beat, with more modest components of right torsional and up beat, in a patient with left-sided vestibular labyrinthine infarction. (From Nam et al. [18] Reproduced with written permission from Wolters Kluwer Health Limited)

Fig. 30.2 Video head impulse testing (vHIT) from a patient with left-sided vestibular labyrinthine infarction. The lower left panel shows results from the left horizontal canal, which exhibits low gain and overt compensatory saccades. (From Nam et al. [18] Reproduced with written permission from Wolters Kluwer Health Limited)

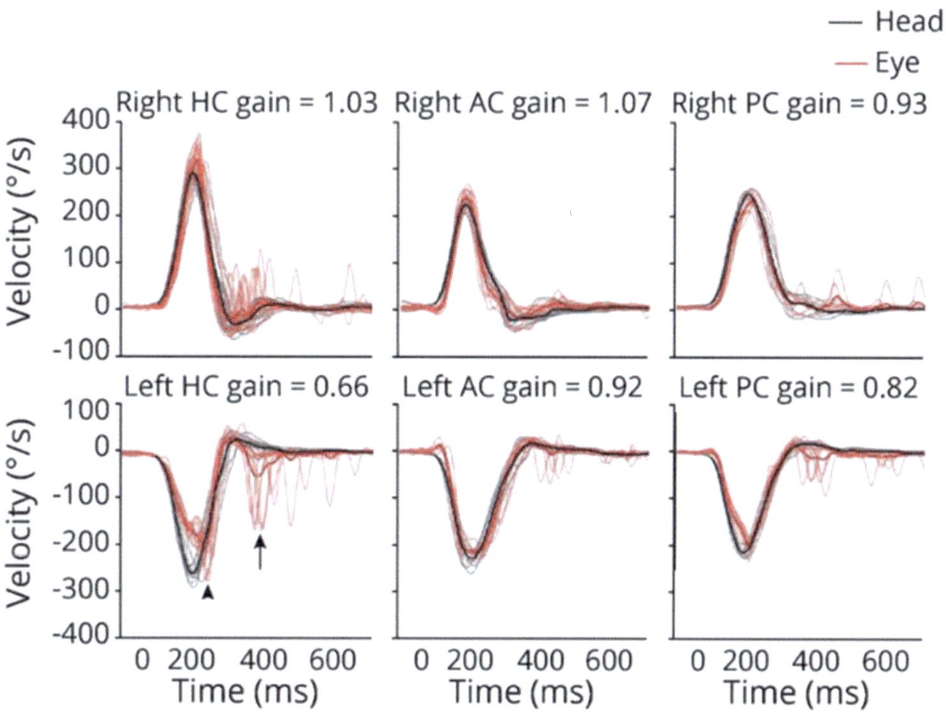

Fig. 30.3 Fundus photographs from a patient with left-sided vestibular labyrinthine infarction showing leftward cyclodeviation. This shows leftward cyclodeviation of both eyes. (From Nam et al. [18] Reproduced with written permission from Wolters Kluwer Health Limited)

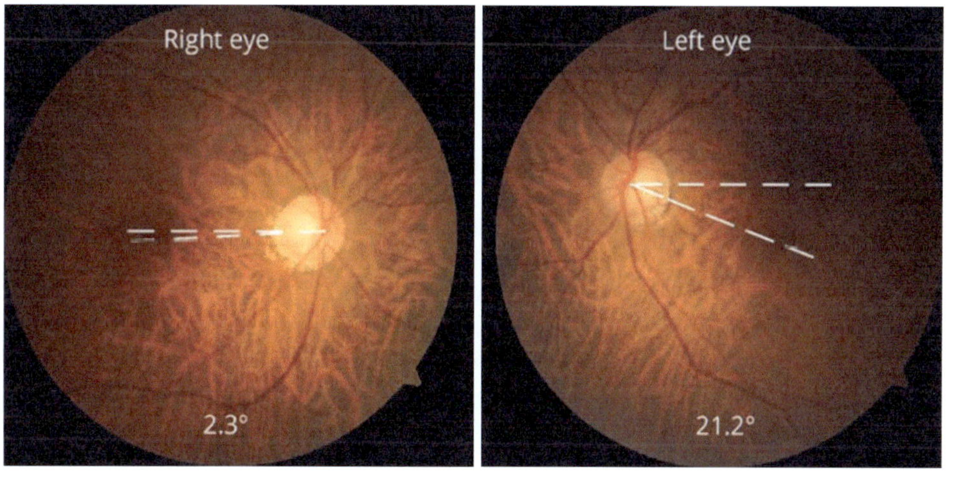

Testing: Other

Figure 30.3, from Nam and colleagues [18], shows results of fundus photography from a patient with left-sided vestibular labyrinthine infarction. This shows leftward cyclodeviation of both eyes.

Imaging

Truly isolated arterial infarction of the labyrinth is rarely detectable on imaging. Venous infarction from hypercoagulable states, such as leukemia, may manifest with labyrinthine hemorrhage [10], which may be more readily detectable.

Histopathology

Temporal bone studies of patients who suffered labyrinthine artery occlusion generally show "severe degenerative changes and ossification of the cochlea and the entire vestibular labyrinth" [4], or "hemorrhages followed by severe degenerative changes that ultimately produce total ossification" [5].

Differential Diagnosis

Practically, because of the vascular anatomy at issue, the majority of patients with audiologic and/or vestibular symptoms secondary to inner ear ischemia will also have signs and symptoms referable to brainstem and cerebellar deficits in the territory of the anterior inferior cerebellar artery [20], which should immediately alert the clinician that the case is not one of primary inner ear disease.

However, the rare cases of infarctions that are so focal as to involve just the inner ear are clinically challenging because they can mimic other diseases; specifically, infarction of the vestibular structures of the labyrinth (in isolation) can mimic vestibular neuritis [18], infarction of the cochlea (in isolation) can mimic idiopathic sudden sensorineural hearing loss, and isolated infarction of the vestibular and cochlear labyrinth (in isolation) can mimic labyrinthitis. The physiologic deficit of infarction of the vestibular labyrinth will be functionally similar to vestibular neuritis, and thus would not be differentiated on HiNTS [18].

Despite the fact that such focal infarctions are possible, it is not practical to obtain advanced imaging in every patient whose presentation is compatible with vestibular neuritis, labyrinthitis, or sudden sensorineural hearing loss. However, the clinical index of suspicion for an inner ear stroke should be higher in patients with known risk factors such as cardiovascular disease, hypercoagulable states, malignancy, or a history suggestive of current or recent vertebral artery dissection.

Treatment

If auditory and vestibular symptoms are truly due to infarction in the territory of the labyrinthine artery, then the source of the infarct should be sought and treated, generally under the guidance of a vascular neurologist. The hearing loss would be treated similarly to other forms of sensorineural hearing loss. For the vestibular deficits, the patient should be offered a trial of vestibular rehabilitation therapy.

Whether the inner ear "component" of stroke merits different or additional management beyond what would normally be offered for stroke is unclear. Animal studies suggest that the rheologic agent pentoxifylline promotes cochlear blood flow [21–24]. Several studies in humans reach similar conclusions [25, 26], but it is unclear whether this has a clinically meaningful effect, so it is similarly unclear whether pentoxifylline has any role in the treatment (or prevention) of inner ear ischemia.

Although aspirin is often used for secondary stroke prophylaxis in patients with atherosclerotic disease, aspirin is also ototoxic, so where medically feasible, it is preferable to use a nonototoxic antiplatelet agent such as clopidogrel.

Prognosis

Infarcted neurons and labyrinthine structures do not regenerate, so if any recovery occurs, it is likely due to compensatory mechanisms. The prognosis in cases of inner ear hemorrhage is generally poor [14].

References

1. Johnsson LG, Hawkins JE Jr. Vascular changes in the human inner ear associated with aging. Ann Otol Rhinol Laryngol. 1972;81(3):364–76.
2. Mazzoni A. The vascular anatomy of the vestibular labyrinth in man. Acta Otolaryngol Suppl. 1990;472:1–83.
3. Mei X, Atturo F, Wadin K, Larsson S, Agrawal S, Ladak HM, et al. Human inner ear blood supply revisited: the Uppsala collection of temporal bone-an international resource of education and collaboration. Ups J Med Sci. 2018;123(3):131–42.
4. Belal A Jr. The effects of vascular occlusion on the human inner ear. J Laryngol Otol. 1979;93(10):955–68.
5. Belal A Jr. Pathology of vascular sensorineural hearing impairment. Laryngoscope. 1980;90(11 Pt 1):1831–9.
6. Hansen CC. Vascular anatomy of the human temporal bone. I. Anastomoses of the membranous labyrinth and its bony capsule. Arch Klin Exp Ohren Nasen Kehlkopfheilkd. 1971;200(2):83–98.
7. Hansen CC. Vascular anatomy of the human temporal bone. II. Anastomoses inside the labyrinthine capsule. Arch Klin Exp Ohren Nasen Kehlkopfheilkd. 1971;200(2):99–114.
8. Lawrence M. Control mechanisms of inner ear microcirculation. Am J Otolaryngol. 1980;1(4):324–33.
9. Kimura RS. Animal models of inner ear vascular disturbances. Am J Otolaryngol. 1986;7(2):130–9.
10. Cherchi M, Huo EJ, Nelson NW, Frankfurt O, Russell EJ, Raizer JJ. Gradual hearing loss with bilateral labyrinthine hemorrhage in chronic myelogenous leukemia. Neurology. 2006;67(1):177–8.
11. Sabatini PR, Kutz JW Jr. Radiology quiz case 1. Intralabyrinthine hemorrhage. Arch Otolaryngol Head Neck Surg. 2009;135(6):612–4.
12. Parikh SK, Frank SA. Clinical reasoning: labyrinthine hemorrhage: an unusual etiology for peripheral vertigo. Neurology. 2017;88(2):e14–e8.
13. Vivas EX, Panella NJ, Baugnon KL. Spontaneous labyrinthine hemorrhage: a case series. Otolaryngol Head Neck Surg. 2018;159(5):908–13.
14. Dumberger LD, Hwa TP, Panara K, Husain S, Yver C, Bigelow DC. Profound sudden sensorineural hearing loss in hematologic malignancy: a case for urgent Cochlear implantation with discussion and systematic review of the literature. Otol Neurotol. 2021;42(7):e815–e24.
15. Eliezer M, Verillaud B, Guichard JP, Kania R, Toupet M, Herman P, et al. Labyrinthine infarction caused by vertebral artery dissection: consideration based on MRI. J Neurol. 2019;266:2575.
16. Choi KD, Chun JU, Han MG, Park SH, Kim JS. Embolic internal auditory artery infarction from vertebral artery dissection. J Neurol Sci. 2006;246(1–2):169–72.
17. Liqun Z, Park KH, Kim HJ, Lee SU, Choi JY, Kim JS. Acute unilateral Audiovestibulopathy due to embolic labyrinthine infarction. Front Neurol. 2018;9:311.
18. Nam HW, Yoo D, Lee SU, Choi JY, Yu S, Kim JS. Pearls & oy-sters: labyrinthine infarction mimicking vestibular neuritis. Neurology. 2021;97(16):787–90.
19. Lee H, Kim JS, Chung EJ, Yi HA, Chung IS, Lee SR, et al. Infarction in the territory of anterior inferior cerebellar artery: spectrum of audiovestibular loss. Stroke. 2009;40(12):3745–51.
20. Adams RD. Occlusion of the anterior inferior cerebellar artery. Arch Neurol Psychiatr. 1943;49(5):765–70.
21. Maass B, Ludwig D. Effect of pentoxifylline ('Trental') on cochlear blood flow as measured by hydrogen wash-out. Curr Med Res Opin. 1984;9(1):52–5.
22. Quirk WS, Dengerink HA, Bademian MJ, Hall KW, Wright JW. The effects of pentoxifylline on cochlear blood flow in normotensive and spontaneously hypertensive rats. Hear Res. 1988;36(2–3):175–80.

23. Coleman JK, Quirk WS, Dengerink HA, Wright JW. Pentoxifylline increases cochlear blood flow while decreasing blood pressure in Guinea pigs. Hear Res. 1990;47(1–2):169–74.
24. LaRouere MJ, Sillman JS, Tsai MT, Nuttall AL. The effect of pentoxifylline on cochlear blood flow. Otolaryngol Head Neck Surg. 1992;106(1):87–91.
25. Cesarone MR, Incandela L, Belcaro G, De Sanctis MT, Nicolaides AN, Griffin M, et al. Treatment of vascular inner ear disease in vascular patients with pentoxifylline: a controlled, randomized trial. Angiology. 2002;53(Suppl 1):S23–6.
26. Incandela L, Cesarone MR, Belcaro G, De Sanctis MT, Nicolaides AN, Griffin M, et al. Treatment of vascular inner ear disease with pentoxifylline: a 4-week, controlled, randomized trial. Angiology. 2002;53(Suppl 1):S19–22.

Labyrinthine Ossification

31

Overview

In response to a variety of labyrinthine insults (infection, trauma, vascular problems, etc.), inflammatory responses may result in scar tissue which, in some cases, leads to abnormal bone formation within the labyrinth called *labyrinthine ossification*. This is a particularly feared complication of meningitis involving the labyrinth [1]. This always manifests with sensorineural hearing loss and less commonly with disequilibrium. The ossification is sometimes detectable by a combination of abnormalities on temporal bone CT and MRI of the internal auditory canals. There is no direct treatment because the end organ (labyrinth) is destroyed. Patients with labyrinthine ossification involving only one ear may benefit from a BiCROS hearing aid or bone-anchored hearing aid.

Introduction

The labyrinth can be damaged by a variety of processes, including infectious, inflammatory, traumatic, vascular, and abnormal bone metabolism (such as Paget's disease) [2, 3]. The body's attempt to heal from these insults often involves inflammatory responses, an unintended consequence of which can be fibrosis [2, 3]—essentially the formation of scar tissue—and formation of new bone [2].

This can manifest in an abnormality variously referred to as *labyrinthine ossification, labyrinthitis ossificans,* or *sclerosis labyrinthitis* [2–4]. This is histopathologically characterized by the "formation of fibrous tissue and new bone within the labyrinthine space" [2] "as the final result of many inflammatory processes" [4] and fibrosis [2, 3]. Cases of labyrinthine ossification always involve hearing loss and can also present with vestibular symptoms [2–4].

Testing: Audiometry

Figure 31.1 shows an audiogram from a patient with left-sided labyrinthine ossification following labyrinthine concussion without temporal bone fracture reported by Schwaber and Tarasidis [2].

M. Cherchi, *Otoneurology and Vestibular Medicine*, https://doi.org/10.1007/978-3-031-94841-1_31

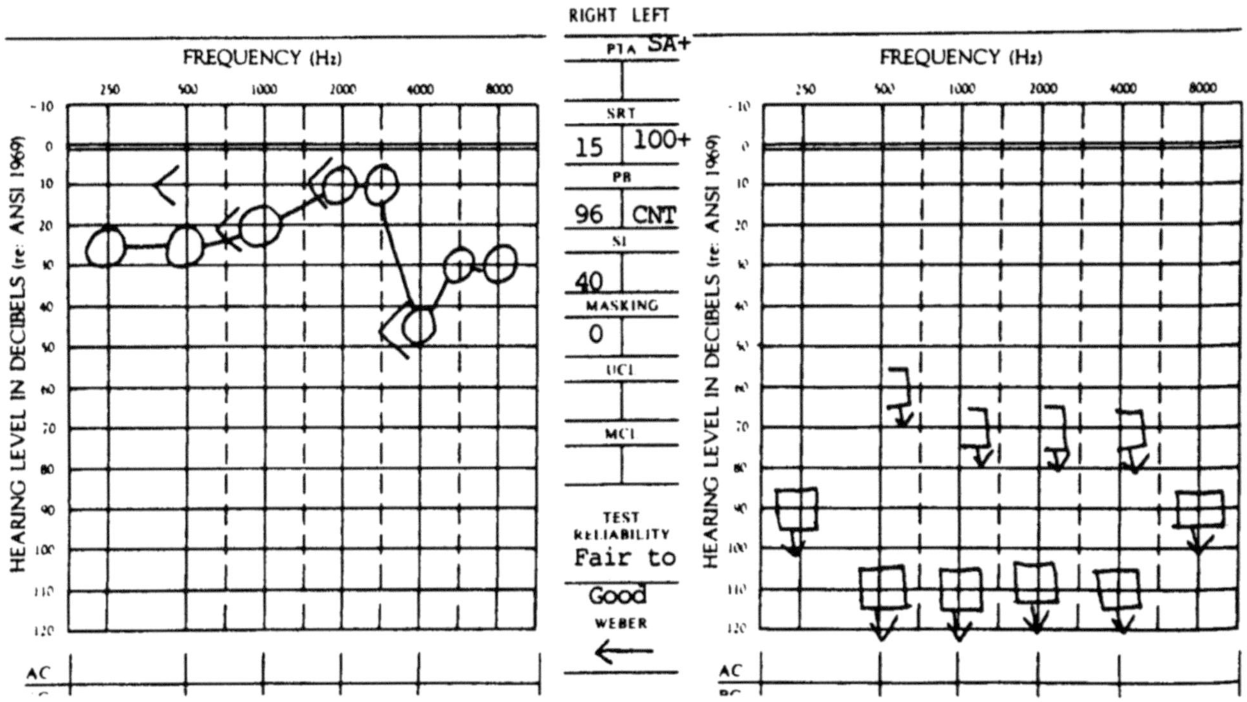

Fig. 31.1 Audiogram from a patient with left-sided labyrinthine ossification following labyrinthine concussion without temporal bone fracture reported by Schwaber and Tarasidis [2]. (Reproduced with written permission from John Wiley & Sons)

Imaging

Labyrinthine ossification can be identified radiographically through a combination of findings on temporal bone CT (which can capture ossification of the bony labyrinth and pneumolabyrinth) and MRI (which better images the membranous labyrinth and vestibulocochlear nerve) [3].

Prognosis

The prognosis of labyrinthine ossification is poor because the end organ is destroyed.

Treatment

There is no known direct therapy for labyrinthine ossification. Since this disease process involves unilateral loss of auditory and vestibular function, patients may benefit from amplification strategies such as a BiCROS (bilateral contralateral routing of signals) hearing aid, or a BAHA (bone attached hearing aid), and from vestibular rehabilitation therapy aimed at unilateral vestibular loss.

References

1. Caye-Thomasen P, Dam MS, Omland SH, Mantoni M. Cochlear ossification in patients with profound hearing loss following bacterial meningitis. Acta Otolaryngol. 2012;132(7):720–5.
2. Schwaber MK, Tarasidis NG. Labyrinthitis ossificans following post-traumatic hearing loss and vertigo: a case report with antemortem histopathology. Otolaryngol Head Neck Surg. 1990;102(1):89–91.
3. Aralasmak A, Dincer E, Arslan G, Cevikol C, Karaali K. Posttraumatic labyrinthitis ossificans with perilymphatic fistulization. Diagn Interv Radiol. 2009;15(4):239–41.
4. Swartz JD, Mandell DM, Faerber EN, Popky GL, Ardito JM, Steinberg SB, et al. Labyrinthine ossification: etiologies and CT findings. Radiology. 1985;157(2):395–8.

32 Audiovestibular Symptoms Following Cochlear Implantation

Overview

A *cochlear implant* (CI) transduces sound into electrical signals that are delivered directly to the spiral ganglion cells in the cochlea, bypassing any malfunctioning structures or cells in the external ear, middle ear, and most of the inner ear. The invasive nature of this device appears to increase the risk of vestibular dysfunction, probably through several distinct mechanisms. Such complications overall seem to be lower in children than in adults. Undertaking a screening otovestibular workup is medically reasonable in a CI patient complaining of disequilibrium. Depending on the underlying mechanism of this symptom, vestibular rehabilitation therapy may be appropriate.

Introduction

In 1961, Dr. William House and Dr. John Doyle successfully placed the first single-channel electrode cochlear implant into a human patient that transduced sound waves into an electrical signal delivered by an electrode through the round window directly to the cochlea [1]. At the time, this was an astounding prosthetic accomplishment, and over the next decades, there were significant advances in this technology [2]. In the United States, the Food and Drug Administration approved cochlear implants for adults in 1984 and for children in 1990.

Of cochlear implants today, "virtually all commercially available devices share a similar design: the external component contains a microphone, battery, sound processor, and transmitting coil, and the internal component includes a radiofrequency receiver coil, microprocessor-based stimulator, and multichannel electrode" [3].

A cochlear implant is usually activated ("turned on") 2–4 weeks after implantation [3].

Epidemiology

As of 2020, it was estimated that about 1.2 million adults in the United States suffered from severe to profound hearing loss [3] and may be candidates for cochlear implantation, yet it has also been estimated that fewer than 6% of these individuals receive this treatment [4].

Pathophysiological Mechanism of Disease

A cochlear implant delivers electrical stimulation directly to spiral ganglion cells of the distal cochlear nerve, thereby bypassing dysfunctional or dead type 1 cochlear hair cells [3]. However, in order for an electrode to reach this location, it must pierce the round window and thus violate the labyrinth. One can imagine that such disruption could result in unintended dysfunction in other areas of the labyrinth, such as the end organelles that mediate vestibular function.

Some individuals experience disequilibrium at some interval after cochlear implantation, and these patients may be referred to otoneurology with the implicit clinical query of assessing whether there is vestibular dysfunction and, if so, whether that dysfunction is the result of cochlear implantation itself.

There are several potential mechanisms by which cochlear implantation may induce vestibular dysfunction.

Occasionally, CI patients are diagnosed with benign paroxysmal positional vertigo (BPPV) [5–7]. This usually occurs ipsilateral to the cochlear implant, but occasionally occurs contralaterally [7]; since BPPV is common, it is possible that chance occurrence of BPPV happens to overlap with cochlear implantation—in other words, this may be coincidental rather than causal; however, the fact that BPPV occurs much more commonly ipsilateral to the CI implicates a causal relationship.

M. Cherchi, *Otoneurology and Vestibular Medicine*, https://doi.org/10.1007/978-3-031-94841-1_32

Another potential mechanism is that the electrode (intended to stimulate the acoustic component of the vestibulocochlear nerve) unintentionally stimulates a vestibular end organ [8], or the presence of the electrode predisposes to ephaptic transmission from auditory to vestibular axons. Some authors conclude that unintentional stimulation of vestibular end organs is rare [9].

There are reports of impaired vestibular function (such as caloric weakness) both ipsilateral and contralateral to a cochlear implant [10]. Since CI necessarily violates the labyrinth (and thus makes labyrinthine proteins accessible to the immune system), it is conceivable that the contralateral deterioration of vestibular function is due to sympathetic ear disease.

Clinical Presentation: Disequilibrium

The onset of disequilibrium in CI patients may occur between implantation and activation, or may occur only following activation [11]. The latency from cochlear implantation to onset of disequilibrium is variable, ranging from as short as "directly postoperatively" [12] or "soon after surgery" [13] or within 2 days [11], to as long as 377 days [14].

Reported rates of disequilibrium following CI are quite variable.

- Carlson [3] reviewed several studies and reported that 102/4664 (***2.2%***) of CI patients experience "persistent vestibular symptoms."
- Farinetti and colleagues [15] studied 168 adult CI patients and 235 pediatric CI patients and reported that 13/168 (7.7%) of adults and 2/235 (0.8%) of children experienced disequilibrium postimplantation, amounting to a pooled rate of 15/403 (***3.7%***).
- Rah and colleagues [6] studied 62 CI patients and reported that 18 (***27%***) experienced postoperative disequilibrium.
- Fina and colleagues [14] report disequilibrium in 29/75 (***39%***) of CI patients. They noted that 4 patients experienced a vertiginous attack less than 24 h after implantation surgery, whereas "the majority, 25 patients, experienced delayed, episodic onset of vertigo," and that "the median (interquartile range) time of delayed onset was 74 (26–377) days after implantation."
- Krause and colleagues [12, 16] reported on 47 adult CI patients and found that 21 (***45%***) experienced disequilibrium after CI.
- Zawawi and colleagues [17] studied 122 patients and reported that 27% complained of disequilibrium preoperatively, and 45.9% complained of disequilibrium postoperatively.
- Kubo and colleagues [13] reported on 94 consecutive adult CI patients and found that 46 (***49%***) experienced disequilibrium after implantation.
- Steenerson and colleagues [18] reported that 35/47 (***74%***) of adult CI patients experienced "vertigo or imbalance."

In short, reported rates of disequilibrium overall following CI vary by over an order of magnitude (2.2–74%).

As far as the symptom of disequilibrium following cochlear implantation is concerned, it is notable that the data from Farinetti and colleagues [15] suggest that this complication is more common in adults (7.7%) than in children (0.8%) by almost an order of magnitude.

A more granular analysis of postimplantation disequilibrium by age may be found in a meta-analysis of CI studies that reported a clear relationship between patient age at the time of implantation and the occurrence of disequilibrium, with a peak in the eighth decade, as shown by Hansel and colleagues [19] (Fig. 32.1).

The duration of vestibular symptoms following cochlear implantation is similarly variable, ranging from brief to "permanent." Katsiari and colleagues [20] studied 20 unilateral CI patients and concluded that, "permanent vertigo was rare." Kubo and colleagues [13] reported on 94 consecutive adult CI patients and found that, "Dizziness of the continuous type, lasting more than 6 months, was a complaint in only two patients" (2%).

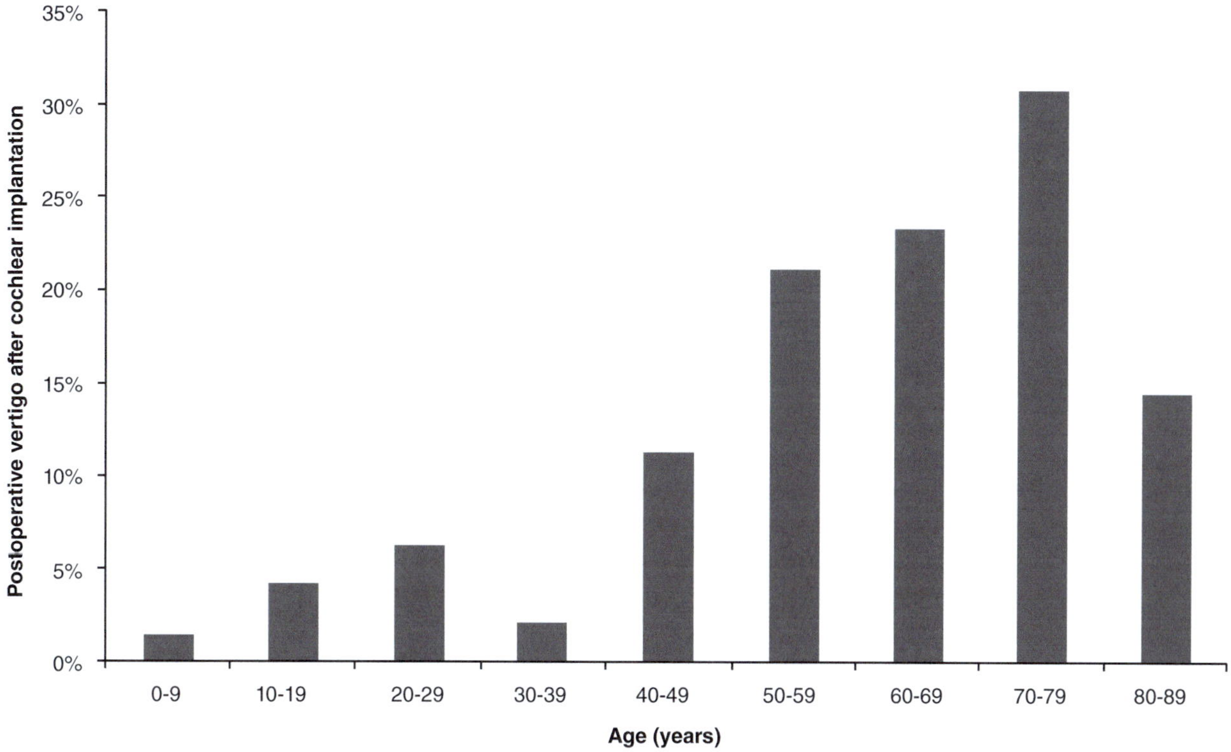

Age group	0-9	10-19	20-29	30-39	40-49	50-59	60-69	70-79	80-89
no vertigo (n)	3122	137	527	1136	1385	2243	306	65	41
vertigo (n)	47	6	35	25	177	604	94	29	7
vertigo (%)	1.5	4.2	6.2	2.2	11.3	21.2	23.5	30.9	14.6

Fig. 32.1 Relationship between patient age at cochlear implantation and occurrence of disequilibrium. (From Hansel et al. [19]. Reproduced with written permission from John Wiley & Sons)

Clinical Presentation: Tinnitus

Farinetti and colleagues [15] note that some CI patients with preimplantation tinnitus experience worsening of that tinnitus postimplantation, while others who had no tinnitus preimplantation develop new tinnitus postimplantation. Tinnitus, similar to vestibular symptoms, appears to be more frequent postimplantation complication in adults than in children; Farinetti and colleagues reported that 14/168 (8.3%) adults and 2/235 (0.9%) children experienced new or worsening of tinnitus following cochlear implantation.

Ocular Motor Examination

Hempel and colleagues [21] described a CI patient with postoperative pneumolabyrinth who under Frenzel glasses had spontaneous and inducible nystagmus.

Testing: Vestibular

Some literature reviews conclude that CI overall has a deleterious effect on the results of instrumented vestibular testing [19], while other reviews emphasize significant variability in the literature [22].

Some literature has reported measurable vestibular dysfunction in all vestibular end organelles following CI [23, 24].

Bonucci and colleagues studied 31 CI patients and reported caloric weakness not only ipsilateral, but also contralateral to a cochlear implant [10].

Bonucci and colleagues [10] studied caloric status preoperatively and postoperatively in implanted and unimplanted ears. In implanted ears, there was preoperative caloric weakness in 58% (11% hyporeflexic, 47% areflexic) and postoperative caloric weakness in 76% (76% areflexic). In unimplanted ears, there was preoperative caloric weakness in 55% (5% hyporeflexic, 50% areflexic) and postoperative caloric weakness in 73% (73% areflexic) (Figs. 32.2 and 32.3).

Fig. 32.2 Preoperative and postoperative caloric responses in an implanted ear. (From Bonucci et al. [10]. Reproduced with written permission from Elsevier B.V.)

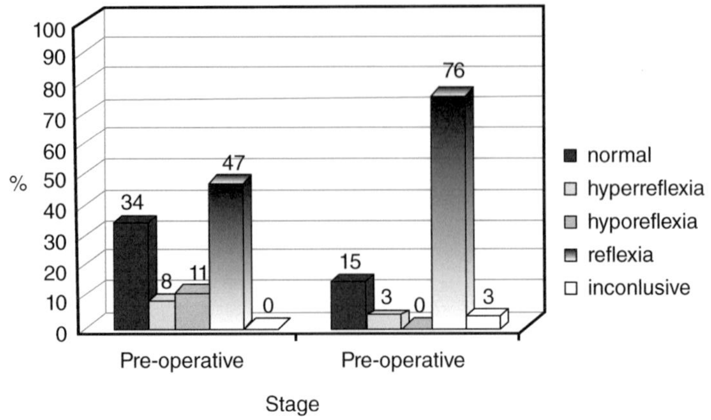

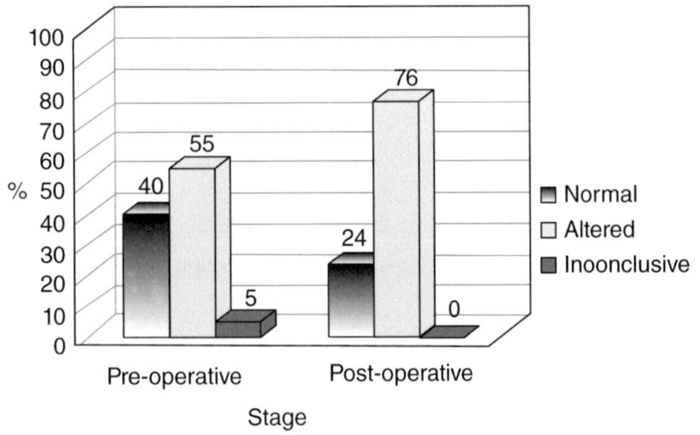

Fig. 32.3 Preoperative and postoperative caloric responses in an unimplanted ear. (From Bonucci et al. [10]. Reproduced with written permission from Elsevier B.V.)

Krause and Louza [16] prospectively studied 47 adult CI patients and reported that, "Thirty-two percent of patients had a substantially reduced hSCC [horizontal semicircular canal] function after CI. Responses of caloric irrigation showed a significant worsening postoperatively in the CI ears."

Chen and colleagues [25] studied 34 patients before and after cochlear implantation; all 34 had pre- and postoperative cervical vestibular evoked myogenic potentials and ocular vestibular evoked myogenic potentials, and 29 of the patients also had caloric testing. They concluded that, "Cochlear implantation can damage the function of the horizontal semicircular canal and utricle, as well as the function of the saccule."

Krause and Louza [26] prospectively studied 32 unilateral CI patients aged 15–83 years and found that, "Cochlear implantation represents a significant risk factor for horizontal semicircular canal impairment ($p<0.001$) and sacculus impairment ($p = 0.047$) in the implanted ear."

Katsiari and colleagues [20] studied 20 unilateral CI patients and reported that a "statistically significant difference was found in the percentages of canal paresis ($p = 0.01$) and the percentages of [c]VEMP waveform absence ($p = 0.002$) between the repeated measurements in the implanted side."

Vibert and colleagues [27] assessed utricular function using the OVAR (off vertical axis rotation) protocol rotatory chair and found no effect of CI on this aspect of otolith function.

Regarding video head impulse testing (vHIT), several studies failed to find any effect of CI on the results [28, 29], while others reported that CI adversely affects vHIT results [23].

Several research groups who studied CI patients observed a poor correlation between vestibular *symptoms* and vestibular *test results* [16, 30, 31].

Imaging

Imaging of vestibularly relevant structures is difficult in CI patients because the implants and their electrodes contain metal, and thus, it will introduce significant artifact into CT and MRI images.

Histopathology

Tien and Linthicum [32] studied 11 pairs of temporal bones from patients with unilateral CI and reported that:

> In the vestibular end organs of the 6 implanted ears that we believe may have shown damage from implantation, no exclusive or dominant findings could be identified. However, the saccule is more susceptible to damage than the utricle or semicircular canals because of its proximity to the pathway of the inserted electrodes, regardless of whether the round window or cochleostomy approach is used [32].

Handzel and colleagues [33] studied 19 temporal bones of CI patients and reported that, "Cochlear implantation does not cause deafferentation of the peripheral vestibular system. Cochlear hydrops accompanied by saccular collapse is common."

Differential Diagnosis

Cochlear implantation is still relatively uncommonly performed, so when an auditory or vestibular symptom appears postimplantation, it is reasonable to undertake a screening assessment for common causes of such symptoms that may be occurring independently of implantation, rather than simply assuming that those symptoms are attributable to the cochlear implant.

As far as workup of vestibular symptoms is concerned, we generally start with bedside infrared video oculography (to assess, for example, for benign paroxysmal positional vertigo). Thereafter, we agree with the view that, "Ideally, all the five vestibular sensors should be tested" [34], and thus, we will usually also consider checking vestibular tests that can detect lateralizing abnormalities, such as cervical vestibular evoked myogenic potentials, ocular vestibular evoked myogenic potentials, video head impulse testing, and videonystagmography with caloric testing. As discussed earlier, it is unsurprising to find vestibular deficits on an implanted ear, but finding such deficits in an unimplanted ear raises suspicion for an independently occurring pathology (such as vestibular neuritis).

Failure of Cochlear Implantation

"Failure" of cochlear implantation is defined in different ways by different studies. Layfield and colleagues [35] conducted a literature review and concluded that the overall CI failure rate was 5.5%.

Liu and colleagues [36] conducted a meta-analysis of 144 articles covering 85,851 initial cochlear implantations and 4276 cochlear reimplantations, and calculated a pooled rate of cochlear reimplantation of 4.7%.

Treatment and Prognosis

Regarding vestibular symptoms thought to be attributable to a cochlear implant, Steenerson and colleagues [18] reported that post-CI disequilibrium responds favorably to vestibular rehabilitation therapy.

In patients experiencing disequilibrium after CI, some surgeons advocate checking whether the electrode needs to be resealed [37].

References

1. Mudry A, Mills M. The early history of the cochlear implant: a retrospective. JAMA Otolaryngol Head Neck Surg. 2013;139(5):446–53.
2. Hainarosie M, Zainea V, Hainarosie R. The evolution of cochlear implant technology and its clinical relevance. J Med Life. 2014;7(Spec Iss 2):1–4.
3. Carlson ML. Cochlear implantation in adults. N Engl J Med. 2020;382(16):1531–42.
4. Sorkin DL. Cochlear implantation in the world's largest medical device market: utilization and awareness of cochlear implants in the United States. Cochlear Implants Int. 2013;14(Suppl 1):S4–12.
5. Di Girolamo S, Fetoni AR, Di Nardo W, Paludetti G. An unusual complication of cochlear implant: benign paroxysmal positional vertigo. J Laryngol Otol. 1999;113(10):922–3.
6. Rah YC, Park JH, Park JH, Choi BY, Koo JW. Dizziness and vestibular function before and after cochlear implantation. Eur Arch Otorrinolaringol. 2016;273(11):3615–21.
7. Viccaro M, Mancini P, La Gamma R, De Seta E, Covelli E, Filipo R. Positional vertigo and cochlear implantation. Otol Neurotol. 2007;28(6):764–7.
8. Coordes A, Basta D, Gotze R, Scholz S, Seidl RO, Ernst A, et al. Sound-induced vertigo after cochlear implantation. Otol Neurotol. 2012;33(3):335–42.
9. Bance ML, O'Driscoll M, Giles E, Ramsden RT. Vestibular stimulation by multichannel cochlear implants. Laryngoscope. 1998;108(2):291–4.
10. Bonucci AS, Costa Filho OA, Mariotto LD, Amantini RC, Alvarenga KF. Vestibular function in cochlear implant users. Braz J Otorhinolaryngol. 2008;74(2):273–8.
11. Filipo R, Patrizi M, La Gamma R, D'Elia C, La Rosa G, Barbara M. Vestibular impairment and cochlear implantation. Acta Otolaryngol. 2006;126(12):1266–74.
12. Krause E, Louza JP, Wechtenbruch J, Hempel JM, Rader T, Gurkov R. Incidence and quality of vertigo symptoms after cochlear implantation. J Laryngol Otol. 2009;123(3):278–82.
13. Kubo T, Yamamoto K, Iwaki T, Doi K, Tamura M. Different forms of dizziness occurring after cochlear implant. Eur Arch Otorrinolaringol. 2001;258(1):9–12.
14. Fina M, Skinner M, Goebel JA, Piccirillo JF, Neely JG, Black O. Vestibular dysfunction after cochlear implantation. Otol Neurotol. 2003;24(2):234–42. discussion 42
15. Farinetti A, Ben Gharbia D, Mancini J, Roman S, Nicollas R, Triglia JM. Cochlear implant complications in 403 patients: comparative study of adults and children and review of the literature. Eur Ann Otorhinolaryngol Head Neck Dis. 2014;131(3):177–82.
16. Krause E, Louza JP, Hempel JM, Wechtenbruch J, Rader T, Gurkov R. Effect of cochlear implantation on horizontal semicircular canal function. Eur Arch Otorrinolaringol. 2009;266(6):811–7.
17. Zawawi F, Alobaid F, Leroux T, Zeitouni AG. Patients reported outcome post-cochlear implantation: how severe is their dizziness? J Otolaryngol Head Neck Surg. 2014;43:49.
18. Steenerson RL, Cronin GW, Gary LB. Vertigo after cochlear implantation. Otol Neurotol. 2001;22(6):842–3.
19. Hansel T, Gauger U, Bernhard N, Behzadi N, Romo Ventura ME, Hofmann V, et al. Meta-analysis of subjective complaints of vertigo and vestibular tests after cochlear implantation. Laryngoscope. 2018;128:2110.
20. Katsiari E, Balatsouras DG, Sengas J, Riga M, Korres GS, Xenelis J. Influence of cochlear implantation on the vestibular function. Eur Arch Otorrinolaringol. 2013;270(2):489–95.
21. Hempel JM, Jager L, Baumann U, Krause E, Rasp G. Labyrinth dysfunction 8 months after cochlear implantation: a case report. Otol Neurotol. 2004;25(5):727–9.
22. Gerdsen M, Jorissen C, Pustjens D, Hof J, Rompaey V, Berg R, et al. Effect of cochlear implantation on vestibular function in children: a scoping review. Front Pediatr. 2022;10:949730.
23. Kwok BYC, Rosengren SM, Kong JHK, Birman CS, Hoskisson E, Young AS, et al. Impact of Cochlear implantation on canal and otolith function. Otol Neurotol. 2022;43(3):304–12.
24. Vaz FC, Petrus L, Martins WR, Silva IMC, Lima JAO, Santos N, et al. The effect of cochlear implant surgery on vestibular function in adults: a meta-analysis study. Front Neurol. 2022;13:947589.

25. Chen X, Chen X, Zhang F, Qin Z. Influence of cochlear implantation on vestibular function. Acta Otolaryngol. 2016;136(7):655–9.
26. Krause E, Louza JP, Wechtenbruch J, Gurkov R. Influence of cochlear implantation on peripheral vestibular receptor function. Otolaryngol Head Neck Surg. 2010;142(6):809–13.
27. Vibert D, Hausler R, Kompis M, Vischer M. Vestibular function in patients with cochlear implantation. Acta Otolaryngol Suppl. 2001;545:29–34.
28. Ajalloueyan M, Saeedi M, Sadeghi M, Zamiri AF. The effects of cochlear implantation on vestibular function in 1-4 years old children. Int J Pediatr Otorhinolaryngol. 2017;94:100–3.
29. Guan R, Wang Y, Wu S, Zhang B, Sun J, Guo X, et al. Vestibular function in children and adults before and after unilateral or sequential bilateral Cochlear implantation. Front Neurol. 2021;12:675502.
30. Stultiens JJA, Kieft HW, Mylanus EAM, Pennings RJE, Terwoert L, Beynon AJ. Impact of cochlear implantation on the function of the three semicircular canals. Int J Audiol. 2020;59(11):843–9.
31. Colin V, Bertholon P, Roy S, Karkas A. Impact of cochlear implantation on peripheral vestibular function in adults. Eur Ann Otorhinolaryngol Head Neck Dis. 2018;135(6):417–20.
32. Tien HC, Linthicum FH Jr. Histopathologic changes in the vestibule after cochlear implantation. Otolaryngol Head Neck Surg. 2002;127(4):260–4.
33. Handzel O, Burgess BJ, Nadol JB Jr. Histopathology of the peripheral vestibular system after cochlear implantation in the human. Otol Neurotol. 2006;27(1):57–64.
34. Abouzayd M, Smith PF, Moreau S, Hitier M. What vestibular tests to choose in symptomatic patients after a cochlear implant? A systematic review and meta-analysis. Eur Arch Otorrinolaringol. 2017;274(1):53–63.
35. Layfield E, Hwa TP, Naples J, Maina I, Brant JA, Eliades SJ, et al. Failure and revision surgery after Cochlear implantation in the adult population: a 10-year single-institution retrospective and systematic review of the literature. Otol Neurotol. 2021;42(3):408–13.
36. Liu H, Yao X, Kong W, Zhang L, Si J, Ding X, et al. Cochlear Reimplantation rate and cause: a 22-year, single-center experience, and a meta-analysis and systematic review. Ear Hear. 2023;44(1):43–52.
37. Karimi D, Mittmann P, Ernst A, Todt I. Surgical treatment of vertigo in cochlear implantees by electrode resealing. Acta Otolaryngol. 2017;137(10):1031–4.

Part V

Diagnoses—Diseases of the Vestibulo-Cochlear Nerve

33 Vestibular Paroxysmia, Cochlear Paroxysmia, and Vestibulocochlear Paroxysmia

The eight cranial nerve (also known as the vestibule-cochlear nerve) transmits auditory (hearing-related) signals and vestibular (balance-related) signals from the ear to the brain. In some circumstances this nerve can become episodically irritable. The term "vestibular paroxysmia" refers to a condition in which only the vestibular portion of this nerve is involved. The term "vestibulocochlear paroxysmia" refers to a condition in which both the vestibular and auditory portions of this nerve are involved. One would think that there should also be a condition corresponding to a situation in which only the cochlear portion of the nerve is involved ("cochlear paroxysmia"); perhaps the closest thing to this is a phenomenon called "typewriter tinnitus." Vestibular paroxysmia is the more common of the three conditions. It usually presents with episodes of disequilibrium that are very abrupt in onset (there is no warning), brief (a fraction of a second), and terminate abruptly. The sense of disequilibrium can be in any direction; some patients describe feeling as if they are thrown to the floor (giving rise to the colloquial term, "drop attack"), but it can also manifest as an abrupt spin or shove in any direction. Presumably, the specific sensation depends on which part of the vestibular nerve is affected. The frequency of attacks is variable, ranging from dozens per day to a few times per year.

Vestibulocochlear paroxysmia is the less common than (purely) vestibular paroxysmia. The vestibular component is similar to that in vestibular paroxysmia, but the episodes are also accompanied by an auditory component—in other words, there is an abrupt-onset, abrupt-offset, and very brief attack of disequilibrium and tinnitus. The tinnitus can present variably, such as a buzzing, humming, or ringing.

The least common of the three conditions is cochlear paroxysmia, in the form of "typewriter tinnitus."

In the majority of cases, no clear cause for these diseases can be discerned. Although the original description of vestibular paroxysmia by neurosurgeon Dr. Peter Jannetta identified cases in which the vestibulocochlear nerve was being impinged upon by a blood vessel, subsequent research brought into question whether this mechanism accounted for most, or even any, of the cases.

Management of these diseases is symptom-driven. In patients who elect to treat, the management is with small doses of antiseizure medications—though to be clear, *this is not a seizure disorder.* When antiseizure medications are used for treating actual seizure disorders, they are prescribed in relatively high doses because seizures are abnormal electrical activity in large portions of the brain. In contrast, in a condition in which only a single nerve is affected (such as vestibular paroxysmia or vestibulocochlear paroxysmia), antiseizure medications can be used in very small doses.

The antiseizure medication with which there is the most experience in treatment of vestibular- and vestibulocochlear paroxysmia is oxcarbazepine. There is less literature regarding other agents, such as carbamazepine, gabapentin, lacosamide, phenytoin, and valproate. There is even modest literature regarding the use of agents other than antiseizure medications, such as betahistine.

Overview

Vestibular paroxysmia, cochlear paroxysmia, and *vestibulocochlear paroxysmia* are conditions in which there appears to be episodic irritability of the vestibular nerve (vestibular paroxysmia), cochlear nerve (cochlear paroxysmia), or both the vestibular and cochlear components of the vestibulocochlear nerve (vestibulocochlear paroxysmia). The presentation consists of recurrent episodes of abrupt-onset, brief (usually seconds), abrupt-offset, symptoms of disequilibrium (for vestibular paroxysmia), tinnitus (for cochlear paroxysmia), or both (vestibulocochlear paroxysmia). The underlying etiology is usually not known. Initial interest in the "vascular cross-compression" theory (the idea that a small artery abuts the vestibulocochlear nerve and wears away its myelin or inappropriately stimulates it) faded when it was recognized that the presence/absence of

M. Cherchi, *Otoneurology and Vestibular Medicine*, https://doi.org/10.1007/978-3-031-94841-1_33

such an artery does not correlate well with the presence/absence of symptoms. Ocular motor examination and infrared video oculography are generally normal. Imaging and otovestibular workup are generally unrevealing, though serve to exclude competing diagnoses. First-line medical therapy is oxcarbazepine; other antiepileptics have been less well studied. If these fail, if imaging shows vascular cross-compression, and (preferably) if otovestibular testing shows abnormalities ipsilateral to the radiographic vascular cross-compression, then referral for neurosurgical evaluation is reasonable.

Introduction

Vestibular paroxysmia, cochlear paroxysmia, and vestibulocochlear paroxysmia are conditions in which there appears to be episodic irritability of the vestibular nerve (vestibular paroxysmia), auditory nerve (cochlear paroxysmia), or vestibular and cochlear components of the vestibulocochlear nerve (vestibulocochlear paroxysmia).

Clinical History

This condition was described by neurosurgeon, Dr. Peter Jannetta [1, 2]. Symptoms can be vestibular alone (vestibular) or vestibular and auditory (vestibulocochlear), and present as episodes with abrupt onset (without prodrome) that are very brief and terminate abruptly.

A study of 22 patients by Ihtijarevic and colleagues [3] listed a mean age of onset of 49 ± 17 years (range 15–80 years), with 15 females (69%) and 7 males (31%). The same study listed the following symptoms:

- Vestibular symptoms:
  - 55% Dizziness
  - 55% Rotational sensations
  - 50% Lightheadedness
  - 45% Drunken feeling
  - 32% Instability
  - 14% Falling without loss of consciousness
  - 10% Tendency to fall
  - 5% Vomiting
  - 5% Lateropulsion
- Nonvestibular symptoms:
  - 77% Cervical pain
  - 59% Headache
  - 45% Tinnitus
  - 36% Hearing loss
  - 32% Aural fullness
  - 29% Nausea.

A study by Hufner and colleagues [4] reported on 32 patients who described the following sensations:

- Vestibular symptoms:
  - 75% Unsteadiness of stance or gait
  - 69% Rotatory vertigo
  - 25% To-and-fro sensation
  - 3% Both rotatory vertigo and a to-and-fro sensation
  - 3% Unable to describe the sensation of disequilibrium
- Nonvestibular symptoms:
  - 41% Nausea or vomiting
  - 28% Tinnitus.

Ihtijarevic and colleagues [3] stated that of their 22 patients:

- 77% of patients reported that symptoms were triggered by positional changes.
  - 55% by head movements
  - 27% by rolling over in bed
- 27% of patients reported that symptoms occurred spontaneously.

Hufner and colleagues [4] reported that of 32 patients:

- 28% said attacks occurred only at rest.
- 22% said attacks were regularly precipitated by a provoking factor.
- 50% said attacks occurred spontaneously at rest or were provoked by a certain action.

Hufner and colleagues [4] reported that of 32 patients who noted provoking factors, those factors were:

- 60% head turn
- 25% forward or backward head movements
- 25% "unspecific" movement
- 22% body turn
- 9% leftward or rightward head movements.

Earlier descriptions characterize the episodes as extremely brief, on the order of a fraction of a second. Subsequent studies of larger number of patients listed durations ranging from seconds to minutes [3].

Cochlear paroxysmia manifests as "typewriter tinnitus" [5], and probably reflects similarly brief bursts of abnormal signaling.

Etiology

The original descriptions by Dr. Peter Jannetta [2] attributed vestibular paroxysmia to "vascular cross compression" of the vestibular nerve by a loop of the anterior inferior cerebellar artery, though subsequent literature described vascular cross compression by other blood vessels as well, such as the posterior inferior cerebellar artery, a vertebral artery or a vein [6], or the superior cerebellar artery [7]. This was thought to be analogous to some cases of trigeminal neuralgia, another condition in which a cranial nerve (the trigeminal nerve) is sometimes found to have vascular impingement [8].

Although some radiologic literature concluded that there was a correlation between vascular cross compression and audiovestibular symptoms [1, 9–16], other literature found no such correlation [17–22].

Since some patients whose clinical history is compatible with vestibular paroxysmia have no evidence of vascular impingement, and since some asymptomatic individuals have vascular impingement, this brought into question the hypothesis of a purely vascular etiology of this disease, and in fact, the Barany Society consensus document states, "The role of imaging in the diagnosis and identification of the affected side is not clear because of the high rate of blood vessels close to the 8th cranial nerve in healthy subjects" [23].

Although there is some logic to the hypothesis of vascular cross compression, there are certainly other neural pathologies that can damage axons and/or Schwann cells (from demyelination, infection, ischemia, and others) that presumably could provoke episodic irritability of the nerve.

Physical Examination

Given the sporadic and brief nature of the symptoms, it is difficult for a clinician to examine a patient during an episode. Young and colleagues [24] were able to observe such a patient, an described a "flurry" of nystagmus (horizontal and torsional) coterminous with the patient's symptoms. Choi and colleagues [25] described a patient with spontaneous baseline horizontal nystagmus that reversed direction during episodes of vestibular paroxysmia.

The more common scenario is for a clinician to be examining a patient interictally. In cases that Ihtijarevic and colleagues [3] classify as "definite vestibular paroxysmia," 50% of patients have "persistent horizontal" nystagmus, 17% have "persistent clockwise/counterclockwise" nystagmus, 8% have "persistent up/downbeating nystagmus," and 25% exhibit no nystagmus. Our own experience is that patients generally have no nystagmus, or minimal horizontal nystagmus.

Testing

In view of the proposed pathophysiology, one would expect to find detectable abnormalities in vestibular and audiologic testing in patients with vestibular and vestibulocochlear paroxysmia. Indeed, a broad variety of audiologic and vestibular tests have been studied in these patients, including audiometry, vestibular evoked myogenic potentials, bedside and video head impulse testing, fundoscopy, static and dynamic subjective visual vertical, caloric testing, video oculography (head-shaking nystagmus, gaze evoked nystagmus, hyperventilation-induced nystagmus), and brainstem auditory evoked potentials [4, 6, 24, 25], and many report lateralizing abnormalities.

However, it should be kept in mind that patients with clinical histories convincing for vestibular and vestibulocochlear paroxysmia can also have entirely normal otovestibular workups [26].

Treatment

Jannetta and colleagues reported many patients in whom the symptoms of vestibular paroxysmia were reduced or eliminated by microvascular decompression [1, 27–30], and other neurosurgeons also attempted this technique [7]. However, as discussed earlier, there are patients with clinical histories compatible with vestibular paroxysmia who have no radiographic evidence of vascular cross compression.

Consequently, most practitioners will first attempt medical (pharmacologic) therapy. The drug with which there is the most clinical experience is oxcarbazepine [4, 31, 32]. There is less experience with carbamazepine [4, 31], gabapentin [31, 33], valproate [31], phenytoin [31], lacosamide [34], and betahistine [35].

In patients who do not respond to reasonable attempts at pharmacologic therapy, who have strong radiographic evidence of vascular impingement on the vestibulocochlear nerve, and (preferably) have objective otovestibular deficits ipsilateral to that vascular impingement, referral to neurosurgery is reasonable.

References

1. Jannetta PJ, Moller MB, Moller AR. Disabling positional vertigo. N Engl J Med. 1984;310(26):1700–5.
2. Jannetta PJ. Neurovascular cross-compression in patients with hyperactive dysfunction symptoms of the eighth cranial nerve. Surg Forum. 1975;26:467–9.
3. Ihtijarevic B, Van Ombergen A, Celis L, Maes LK, Wuyts FL, Van de Heyning PH, et al. Symptoms and signs in 22 patients with vestibular paroxysmia. Clin Otolaryngol. 2019;44(4):682–7.

4. Hufner K, Barresi D, Glaser M, Linn J, Adrion C, Mansmann U, et al. Vestibular paroxysmia: diagnostic features and medical treatment. Neurology. 2008;71(13):1006–14.
5. Levine RA. Typewriter tinnitus: a carbamazepine-responsive syndrome related to auditory nerve vascular compression. ORL J Otorhinolaryngol Relat Spec. 2006;68(1):43–6. discussion 6–7
6. Best C, Gawehn J, Kramer HH, Thomke F, Ibis T, Muller-Forell W, et al. MRI and neurophysiology in vestibular paroxysmia: contradiction and correlation. J Neurol Neurosurg Psychiatry. 2013;84(12):1349–56.
7. Ryu H, Yamamoto S, Sugiyama K, Nozue M. Neurovascular compression syndrome of the eighth cranial nerve. What are the most reliable diagnostic signs? Acta Neurochir. 1998;140(12):1279–86.
8. Gronseth G, Cruccu G, Alksne J, Argoff C, Brainin M, Burchiel K, et al. Practice parameter: the diagnostic evaluation and treatment of trigeminal neuralgia (an evidence-based review): report of the quality standards Subcommittee of the American Academy of neurology and the European Federation of Neurological Societies. Neurology. 2008;71(15):1183–90.
9. Moosa S, Fezeu F, Kesser BW, Ramesh A, Sheehan JP. Sudden unilateral hearing loss and vascular loop in the internal auditory canal: case report and review of literature. J Radiosurg SBRT. 2015;3(3):247–55.
10. De Ridder D, De Ridder L, Nowe V, Thierens H, Van de Heyning P, Moller A. Pulsatile tinnitus and the intrameatal vascular loop: why do we not hear our carotids? Neurosurgery. 2005;57(6):1213–7. discussion –7
11. Herzog JA, Bailey S, Meyer J. Vascular loops of the internal auditory canal: a diagnostic dilemma. Am J Otol. 1997;18(1):26–31.
12. McCabe BF, Harker LA. Vascular loop as a cause of vertigo. Ann Otol Rhinol Laryngol. 1983;92(6 Pt 1):542–3.
13. McDermott AL, Dutt SN, Irving RM, Pahor AL, Chavda SV. Anterior inferior cerebellar artery syndrome: fact or fiction. Clin Otolaryngol Allied Sci. 2003;28(2):75–80.
14. Nowe V, De Ridder D, Van de Heyning PH, Wang XL, Gielen J, Van Goethem J, et al. Does the location of a vascular loop in the cerebellopontine angle explain pulsatile and non-pulsatile tinnitus? Eur Radiol. 2004;14(12):2282–9.
15. Shinn JB, Bush ML, Jones RO. Correlation of central auditory processing deficits and vascular loop syndrome. Ear Nose Throat J. 2009;88(10):E34–7.
16. Wuertenberger CJ, Rosahl SK. Vertigo and tinnitus caused by vascular compression of the vestibulocochlear nerve, not intracanalicular vestibular schwannoma: review and case presentation. Skull Base. 2009;19(6):417–24.
17. Gorrie A, Warren FM 3rd, de la Garza AN, Shelton C, Wiggins RH 3rd. Is there a correlation between vascular loops in the cerebellopontine angle and unexplained unilateral hearing loss? Otol Neurotol. 2010;31(1):48–52.
18. Hoekstra CE, Prijs VF, van Zanten GA. Diagnostic yield of a routine magnetic resonance imaging in tinnitus and clinical relevance of the anterior inferior cerebellar artery loops. Otol Neurotol. 2015;36(2):359–65.
19. Makins AE, Nikolopoulos TP, Ludman C, O'Donoghue GM. Is there a correlation between vascular loops and unilateral auditory symptoms? Laryngoscope. 1998;108(11 Pt 1):1739–42.
20. Parnes LS, Shimotakahara SG, Pelz D, Lee D, Fox AJ. Vascular relationships of the vestibulocochlear nerve on magnetic resonance imaging. Am J Otol. 1990;11(4):278–81.
21. Reisser C, Schuknecht HF. The anterior inferior cerebellar artery in the internal auditory canal. Laryngoscope. 1991;101(7 Pt 1):761–6.
22. van der Steenstraten F, de Ru JA, Witkamp TD. Is microvascular compression of the vestibulocochlear nerve a cause of unilateral hearing loss? Ann Otol Rhinol Laryngol. 2007;116(4):248–52.
23. Strupp M, Lopez-Escamez JA, Kim JS, Straumann D, Jen JC, Carey J, et al. Vestibular paroxysmia: Diagnostic criteria. J Vestib Res. 2016;26(5–6):409–15.
24. Young AS, Jonker B, Welgampola MS. Vestibular paroxysmia presenting with irritative nystagmus. Neurology. 2019;92(15):723–4.
25. Choi SY, Choi JH, Choi KD. The nystagmus of vestibular paroxysmia. J Neurol. 2018;265(7):1711–3.
26. Markowski J, Gierek T, Kluczewska E, Witkowska M. Assessment of vestibulocochlear organ function in patients meeting radiologic criteria of vascular compression syndrome of vestibulocochlear nerve–diagnosis of disabling positional vertigo. Med Sci Monit. 2011;17(3):CR169-73.
27. Moller MB. Results of microvascular decompression of the eighth nerve as treatment for disabling positional vertigo. Ann Otol Rhinol Laryngol. 1990;99(9 Pt 1):724–9.
28. Moller MB. Selection criteria and treatment of patients with disabling positional vertigo. Stereotact Funct Neurosurg. 1997;68(1–4 Pt 1):270–3.
29. Moller MB, Moller AR, Jannetta PJ, Jho HD, Sekhar LN. Microvascular decompression of the eighth nerve in patients with disabling positional vertigo: selection criteria and operative results in 207 patients. Acta Neurochir. 1993;125(1–4):75–82.
30. Moller MB, Moller AR, Jannetta PJ, Sekhar L. Diagnosis and surgical treatment of disabling positional vertigo. J Neurosurg. 1986;64(1):21–8.
31. Strupp M, Dieterich M, Brandt T, Feil K. Therapy of vestibular Paroxysmia, superior oblique Myokymia, and ocular Neuromyotonia. Curr Treat Options Neurol. 2016;18(7):34.
32. Bayer O, Bremova T, Strupp M, Hufner K. A randomized double-blind, placebo-controlled, cross-over trial (Vestparoxy) of the treatment of vestibular paroxysmia with oxcarbazepine. J Neurol. 2018;265(2):291–8.
33. Russell D, Baloh RW. Gabapentin responsive audiovestibular paroxysmia. J Neurol Sci. 2009;281(1–2):99–100.
34. Strupp M, Elger C, Goldschagg N. Treatment of vestibular paroxysmia with lacosamide. Neurol Clin Pract. 2019;9(6):539–41.
35. Xue H, Xiang W, Yu Y, Liu G, Chong Y, Zhou J. Randomized trial of betahistine mesilate tablets as augmentation for oxcarbazepine and carbamazepine in treating vestibular paroxysmia. Drug Des Devel Ther. 2018;12:837–43.

Vestibular Schwannoma

34

Introduction

A *vestibular schwannoma* is an uncommon and (usually) nonmalignant tumor involving the eighth cranial nerve that, depending on its size and location can cause auditory, vestibular symptoms, or symptoms secondary to brainstem compression. The symptoms may be due to a combination of (1) abnormal Schwann cells; (2) mechanical distortion of neural structures; (3) interference with signal propagation; (4) vascular compromise secondary to mass effect; and (5) change in composition of endolymph. Audiometry usually shows gradually progressive, predominantly high-frequency sensorineural hearing loss, and otovestibular testing may show ipsilateral vestibular weakness; but the definitive diagnostic tool is MRI of the internal auditory canals without and with contrast. Depending on the cadence of symptom evolution, management strategies include (1) monitoring; (2) radiosurgery; and (3) open surgery. The radiosurgical and open surgical interventions often result in vestibular weakness (new, or greater than presurgical), so vestibular rehabilitation therapy is often needed postsurgically.

Terminology

A *vestibular schwannoma* is usually a nonmalignant tumor involving the eighth cranial nerve (vestibulocochlear nerve), which contains both acoustic and vestibular afferent fibers.

Older literature referred to this tumor as an *acoustic neuroma* in the mistaken belief that it arose in the cochlear ("acoustic") portion of the nerve, and from actual nerve cells ("neuroma").

Subsequent studies revealed that the tumor more commonly arises in the "vestibular" portion of the nerve [1], and from Schwann cells ("Schwannoma") that ensheathe the nerve [2].

Thus, from the anatomical and histological perspective, the term "vestibular schwannoma" is more correct. Nevertheless, many practitioners and some literature continue to use the term "acoustic neuroma."

Epidemiology

Vestibular schwannomas are relatively uncommon, though there is controversy about the exact numbers. Longitudinal studies typically cite an annual incidence from as low as 5.4 to as high as 10 per million persons, and a lifetime incidence of 1 per thousand persons [3]. Some studies describe a rising incidence of this disease [4], while others claim that this apparent increase is more likely due to increased recognition, as well as the incidental discovery of asymptomatic vestibular schwannomas on imaging (for example, in persons who underwent a brain MRI for some unrelated problem) [5–10].

Vestibular schwannoma is largely a disease of adulthood, with studies describing a mean age at diagnosis of 49–58 years [3]. In children, one study found only 39 cases of nonsyndromic vestibular schwannomas (i.e., those not associated with tumor-predisposing genetic abnormalities such as neurofibromatosis type 2) published in world medical literature [11].

M. Cherchi, *Otoneurology and Vestibular Medicine*, https://doi.org/10.1007/978-3-031-94841-1_34

Presentation

Vestibular schwannomas are generally very slowly growing tumors, so symptoms correspondingly present subtly and are very slowly progressive. Because of this insidious presentation, the median time between symptom onset and diagnosis is 5 years, though the range is very broad (from a few weeks to 40 years) [12].

Given the anatomical structures involved (vestibulocochlear nerve), the disease correspondingly presents with auditory symptoms (hearing loss, sound distortion, tinnitus) and vestibular symptoms.

Other symptoms are also possible, depending largely on whether the tumor is impinging on other structures. For example, *facial weakness* can occur because the facial nerve runs through part of the internal auditory canal and adjacent to the vestibulocochlear nerve for some of its course, so a vestibular schwannoma can compress the facial nerve as well; *clumsiness or incoordination* can result if the vestibular schwannoma is extracanalicular, in the cerebellopontine angle, impinging on the cerebellum; *facial pain* can occur if a vestibular schwannoma in the cerebellopontine angle impinges on the trigeminal nerve. A typical single-institution retrospective study [13] of this patient population (703 individuals) reported the following percentages of symptom presentation:

- 57%: Gradually progressive sensorineural hearing loss
- 14%: Vertigo
- 12%: Tinnitus
- 7%: Sudden deafness
- 4.1%: Trigeminal symptoms, such as facial pain
- 2.8%: Cerebellar symptoms
- 1.3%: Headache
- 1.1%: Facial nerve paresis
- 0.1%: Epilepsy

Anatomical Location of Vestibular Schwannomas

The vestibulocochlear nerve is a relatively long cranial nerve. Vestibular schwannomas can arise anywhere along this structure, including the following:

- Intralabyrinthine, meaning that the tumor lies within the labyrinth itself [12].
- Intracanalicular (intrameatal), meaning that the tumor lies entirely within the internal auditory canal.
- Intra- and extracanalicular (intra- and extrameatal), meaning that the tumor lies within the internal auditory canal, crosses the internal acoustic meatus, and extends into the cerebellopontine angle (Figs. 34.1 and 34.2).

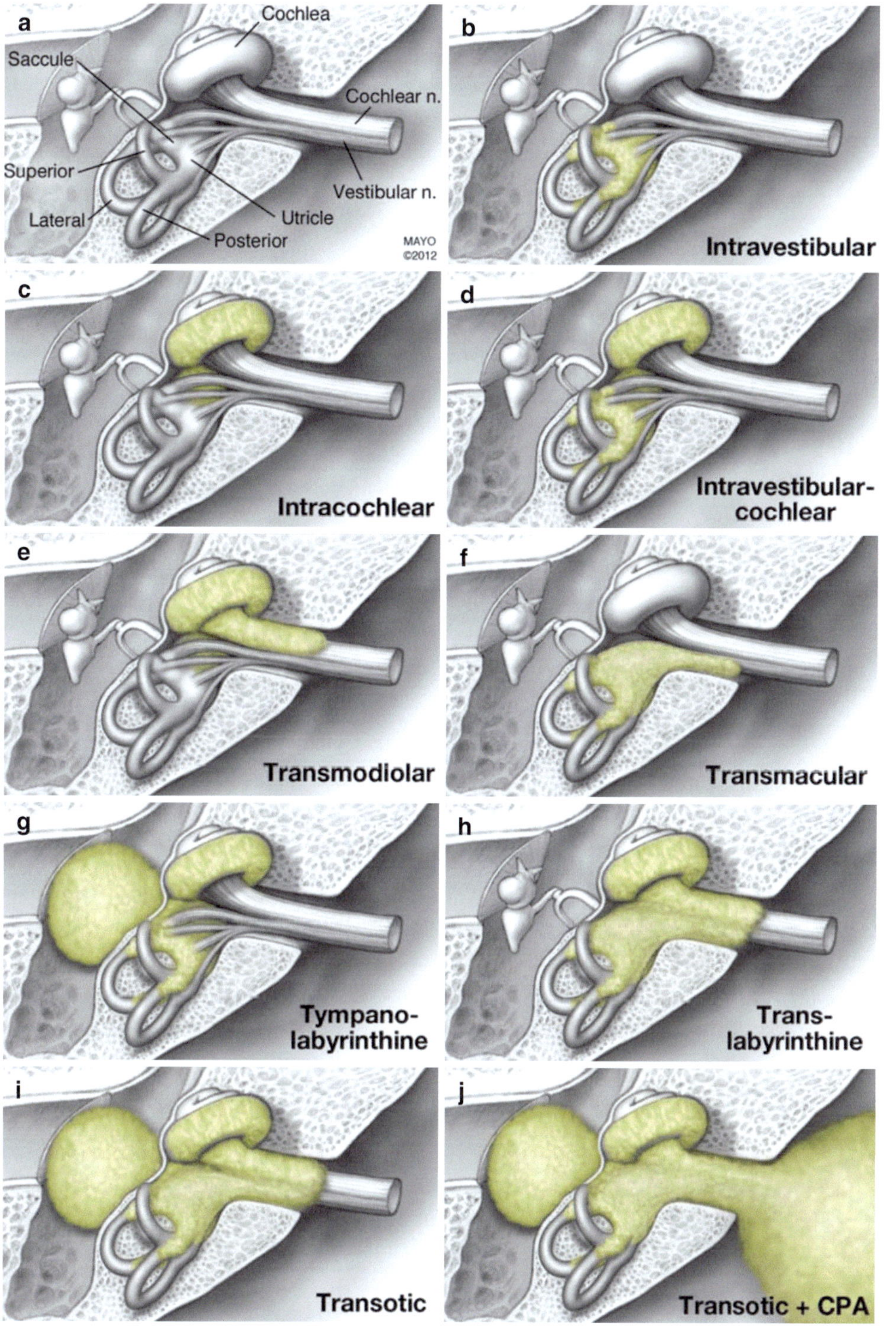

Fig. 34.1 Modified Kennedy classification system for primary inner ear schwannomas. Schematic drawing demonstrating the modified Kennedy classification describing primary inner ear schwannoma location. Panel (**a**): Anatomy of the inner ear and adjacent structures. Panels (**b**–**j**): Extent of tumor involvement for each subtype is depicted in yellow. (From Van Abel et al. [12]. Reproduced with written permission from John Wiley and Sons Limited)

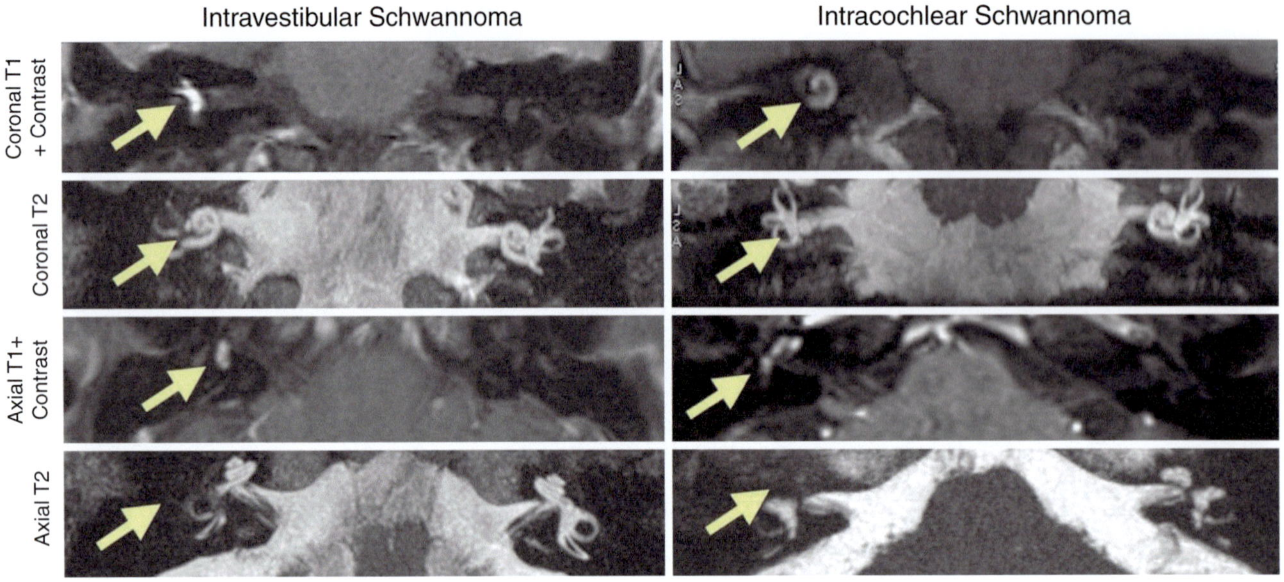

Fig. 34.2 Magnetic resonance imaging depicting an intravestibular and intracochlear primary inner ear schwannoma. Intravestibular and intracochlear tumors as seen on coronal postcontrast T1-weighted coronal reformats from three-dimensional T2-weighted cisternography axial postcontrast T1-weighted, and axial reformats from three-dimensional T2-weighted cisternography. Large arrow indicates tumor location in each image. (From Van Abel et al. [12]. Reproduced with written permission from John Wiley and Sons Limited)

Mechanism of Symptoms

There are probably multiple mechanisms by which vestibular schwannomas can cause auditory and vestibular symptoms, including:

First, vestibular schwannomas may have direct effects on the vestibulocochlear nerve, such as:

- The tumorous Schwann cells (i.e., the Schwann cells that are growing inappropriately) are not normal, and their usual function (namely, facilitating nerve conduction) is disturbed.
- Mechanical distortion of the vestibular nerve, damaging neuronal axons and healthy (nontumorous) Schwann cells. This can disrupt both afferent signals (incoming from type 1 hair cells) and efferent signals (outgoing to type 2 hair cells).
- These result in conduction deficits, including failure to propagate signal; slow propagation; aberrant firing; ephaptic transmission.

Second, vestibular schwannomas may interfere with circulation:

- If intracanalicular: compression of the labyrinthine artery, thereby compromising blood flow to the inner ear.
- Possible change of protein composition in labyrinth's endolymphatic fluid [14].

Finally, vestibular schwannomas that are extracanalicular may exert mass effect on the brainstem or cerebellum.

Diagnosis: Imaging

The cornerstone of diagnosis is MRI of the internal auditory canals without and with contrast, and this is currently considered the criterion standard. However, this is an expensive study, so when the history is suggestive of a vestibular schwannoma, practitioners may consider first checking other (less expensive) tests for screening, including audiometry, brainstem auditory evoked responses (ABRs) and certain vestibular tests. If those screening tests are all normal, then the likelihood of vestibular schwannoma is very low, and imaging may be avoided. In contrast, if those screening tests are abnormal, then seeking confirmation with imaging is appropriate.

Diagnosis: Audiologic Tests

Audiometry. Since the most common presenting symptom is hearing loss, it is logical to assess hearing in greater detail. Although almost any pattern of hearing loss is possible [15], the most common is high-frequency hearing loss.

Auditory brainstem evoked responses (ABRs), also called brainstem auditory evoked responses (BAERs). The sensitivity/specificity of this test for vestibular schwannoma remains controversial, but some clinicians still regard this as a useful screening test [16].

Diagnosis: Vestibular Tests

Videonystagmography (VNG): *Caloric weakness.* Although symptoms of disequilibrium comprise the initial presentation in only a minority of cases, subclinical (i.e., asymptomatic) vestibular deficits can be detected in a larger percentage of patients. For example, a recent single-institution review of 703 patients with vestibular schwannoma [13] found that only 14% of individuals reported dizziness/unsteadiness as the presenting symptom, yet when assessed with videonystagmography, 400 (57%) had an absent caloric response on the affected side, 243 (35%) had a reduced caloric response on the affected side, and only 60 (8%) had a normal caloric response on the affected side; in other words, fully 92% of individuals had an abnormal (reduced or absent) caloric response on the side of the vestibular schwannoma. Other studies report slightly lower numbers, in the range of 77–87% [17].

Videonystagmography: *Hyperventilation-induced nystagmus* (HVIN). Although not in wide use, the finding of hyperventilation-induced nystagmus has been well-studied and is quite sensitive for vestibular schwannomas [18–22]. HVIN is believed to arise from the fact that hyperventilation changes serum acid–base balance, and perhaps also extracellular free calcium concentration, thereby transiently "unmasking" the abnormal neural transmission of the vestibular nerve [23].

Cervical vestibular evoked myogenic potentials (cVEMPs). Cervical VEMPs are often diminished in amplitude ipsilateral to a vestibular schwannoma [24–28]. The deficits detected on cVEMPs do not always correlate with those identified on caloric testing [29], though this is unsurprising since caloric testing assesses afferents from the horizontal semicircular canal (which run through the superior division of the vestibular nerve) whereas cVEMPs test afferents from the saccule (which run through the inferior division of the vestibular nerve).

Ocular vestibular evoked myogenic potentials (oVEMPs). Ocular VEMPs are often diminished in amplitude ipsilateral to a vestibular schwannoma, and this deficit frequently correlates with a caloric weakness [29], which makes sense since caloric testing assesses afferents from the horizontal semicircular canal (which run through the superior division of the vestibular nerve) and oVEMPs assess afferents from the utricle (which also run through the superior division of the vestibular nerve).

Video head impulse testing (vHIT) often shows abnormalities ipsilateral to a vestibular schwannoma [30, 31]. One study [30] reported caloric testing to be abnormal more frequently than video head impulse testing. This discrepancy is likely due to the fact that caloric testing assesses the low end of the vestibular frequency spectrum, while video head impulse testing assesses the high end of the vestibular frequency spectrum, and it may be that these fibers are differentially affected by the lesion.

Galvanic stimulation of the vestibular nerve often elicits asymmetrical ocular motor responses in patients with vestibular schwannomas [32, 33], but this test is generally restricted to research settings.

Practically, since the various parts of the vestibular nerve can be affected in nearly any pattern, our practice has been to check videonystagmography (with caloric testing, hyperventilation and neck vibration), cervical and ocular vestibular evoked myogenic potentials, and video head impulse testing, as these are readily available in most clinical laboratories.

Diagnosis: Differential

Other tumors can manifest with symptoms that, at least initially, resemble those of a vestibular schwannoma. Considerations include meningiomas [34], malignant peripheral nerve sheath tumors [35–37], and tumors metastatic from elsewhere, such as melanoma [38].

Management: Monitoring

If a vestibular schwannoma is suspected by history, if the suspicion is further raised by screening tests and then confirmed on imaging, then the question becomes one of management.

If a vestibular schwannoma is not posing any immediate danger (such as brainstem compression), and if the symptoms for which it is believed to be responsible are manageable, then many practitioners will take a conservative course of watchful waiting, which may include serial imaging.

The rationale behind serial imaging is that the tumor's behavior may be quite variable. Some tumors appear never to change size after they are discovered. Other tumors are found to grow but then plateau and stop growing. Still others continue to grow. Surprisingly, some vestibular schwannomas appear spontaneously to shrink [39]; one study reported that of 1261 patients with vestibular schwannoma, 48 (3.81%) exhibited spontaneous shrinkage of an average of 52.1% during the period of observation [40].

The frequency and interval of serial imaging has remained a matter of debate. For patients who are asymptomatic, or whose symptoms are manageable and not changing, one study [3] suggested that an MRI of the internal auditory canals without and with gadolinium should be performed:

- Every year for years 1–5 after diagnosis.
- Every other year for years 6–10 after diagnosis.
- Once at 15 years after diagnosis.
- At that point, if there has been no growth, then no further imaging is recommended.

In parallel, a patient's symptoms should be monitored clinically. Some clinicians also elect to perform annual audiometry. Interpretation of serial audiometry is made more complicated by the fact that expected age-related hearing loss may be superimposed on whatever component of hearing loss is attributable to the vestibular schwannoma.

Management: Intervention

If symptomatic or radiographic progression warrants action, then an otolaryngologist may consider intervention. The otolaryngologist may take into account several factors, including the tumor's location, size, what (if any) structures are being compressed, and whether any salvageable hearing remains.

Management: Medical

Medical therapy, in the sense of oral medications, does not usually play a significant role in the treatment of vestibular schwannomas. However, there may be some rationale for considering anti-epileptic medications (particularly oxcarbazepine), on the physiologic basis that this approach is thought to stabilize cell membranes and may diminish abnormal signal transmission through the vestibular nerve [41].

Management: Radiosurgery

Radiosurgical modalities include gamma knife and stereotactic radiosurgeries. These approaches have both similarities and differences [42].

Gamma knife surgery directs targeted beams of ionizing radiation such that they hit as much of the tumor as possible, and as little normal tissue as possible [43]. The intent is for the radiation to damage the tumorous Schwann cells so as to retard their growth, or even cause them to shrink [42, 44–51]. Appropriate targeting of the radiation beams is a nontrivial task and requires considerable planning, usually by a team consisting of some combination of an otolaryngologist, neurosurgeon, radiation oncologist and medical physicist. In some cases, the tumor may seem to grow, usually at 3–6 months following gamma knife surgery, but this appears to be due to swelling from tumor necrosis, and thus is sometimes termed "pseudoprogression" [52, 53].

Other types of radiosurgery, including fractionated and unfractionated LINAC (linear particle accelerator) have been explored, though overall have been less studied [42, 54–59] than gamma knife.

Management: Open Surgery

Vestibular schwannomas are located in the skull base, which is not easily accessible. There are several surgical approaches, each with its advantages and disadvantages. The otolaryngologist will decide which approach makes the most sense based on clinical history and tumor location.

Retrosigmoid approach. In this approach, a neurosurgeon performs a suboccipital craniotomy and retracts the cerebellar hemisphere, providing access to the cerebellopontine angle and medial aspect of the temporal bone, from where it is possible to drill and expose the internal auditory canal. Advantages of this approach include that it can be used for tumors of any size, and it offers some possibility of hearing preservation. Disadvantages include that it is quite invasive, and there is a likelihood of damage to the cerebellum during retraction.

Middle fossa approach. In this approach, a neurosurgeon performs a temporal craniotomy and retracts the temporal lobe, giving the otolaryngologist has access to the temporal bone and thus to the vestibular schwannoma. Advantages include that it is sometimes possible to preserve hearing. Disadvantages include that it is limited to small intracanalicular tumors.

Translabyrinthine approach. In this approach, the otolaryngologists enter directly through the temporal bone. Advantages include that it is less invasive than the other approaches (not requiring neurosurgical assist). Disadvantages include that there is complete loss of auditory and vestibular function, as the labyrinth is destroyed.

Management: Transtympanic Gentamicin

For patients whose symptoms are predominantly vestibular, who have no salvageable hearing, and who are failing medical management, some practitioners may consider chemical ablation of vestibular function with a transtympanic gentamicin injection (a technique also used in Ménière's disease) [60]. This approach is controversial because the procedure ablates function of the *inner ear* rather than function of the *vestibular nerve* (where the true abnormality lies).

Outcomes After Surgical Intervention

Some studies report that patients in whom preoperative vestibular testing found a residual caloric response in the ear with the vestibular schwannoma did *worse* (in terms of dizziness) postoperatively than those patients with absent vestibular function preoperatively [61]. At some level, this makes sense, because if a patient has no vestibular function preoperatively, then there is no further vestibular function to lose, so a destructive procedure would cause no new deficit.

With open surgical approaches, postprocedural dizziness and changes in balance are usually fairly immediate.

In contrast, outcomes following gamma knife surgery are different in that some effects (including impact on equilibrium) may be delayed by several months [62].

It is reasonable to ask why there should be new or worsening of symptoms following these interventions. Prior to surgery or radiosurgery the vestibulocochlear nerve was probably already behaving abnormally (due to adverse effects of the vestibular schwannoma on nerve conduction, such as impaired signal propagation, slowed conduction velocities, aberrant firing and ephaptic transmission). Although surgery and radiosurgery may debulk or shrink the tumor, this does not mean that the nerve is restored to its normal function; rather it becomes *abnormal in a different way*, and may therefore continue to behave abnormally. This may be due to mechanisms similar to what happens with demyelinating disease of the central nervous system [63, 64].

Postsurgical Vestibular Rehabilitation

Vestibular rehabilitation is the mainstay of therapy after surgical intervention, though the exact type of therapy, and the timing of therapy, is a matter of debate. One meta-analysis of randomized treatment trials suggested that adaptation exercises are superior to habituation exercises [65], but more research is needed. Therapy involving visual feedback is emerging [66], but also requires further study.

In some cases, despite appropriately-targeted vestibular rehabilitation therapy, outcomes are poor even several years after surgery. One study found that approximately 10% of patients reported postoperative dizziness as "disabling" [67].

It is difficult to predict which patients will have good outcomes, and which will have poor outcomes. Early studies reported correlations such as, "Worse imbalance preoperatively correlates with poorer balance post-operatively," or "Better results on vestibular testing pre-operatively correlates with poorer balance post-operatively," but more recent review found insufficient evidence to support such correlations [68].

Practically, it is probably always better to try *some* sort of formal vestibular rehabilitation therapy rather than no rehabilitation at all. The rehabilitative program should be supervised by a physical therapist with specific expertise in balance disorders. If postoperative disequilibrium is moderate to severe, then a home safety assessment may be warranted, or an occupational therapist should become involved.

Note on Neurofibromatosis Type 2

Neurofibromatosis type 2 (NF2) results from abnormalities in the NF2 gene on chromosome 22 that encodes the synthesis of the protein merlin. These genetic abnormalities can arise from autosomal dominant inheritance (accounting for about half of all cases of NF2), or from de novo mutations (accounting for the other half of all cases of NF2). These mutations result in abnormal production/proliferation of Schwann cells, manifesting with nonmalignant neuroectodermal tumors in multiple areas of the body (central nervous system, peripheral nervous system, skin), including schwannomas, meningiomas, gliomas, and neurofibromas.

The relevance to the present discussion is that NF2 classically causes *bilateral vestibular schwannomas*. Of all people with vestibular schwannomas, approximately 5% of cases are attributable to NF2.

The management of bilateral vestibular schwannomas in the context of NF2 is generally undertaken by a neuro-oncologist.

References

1. Komatsuzaki A, Tsunoda A. Nerve origin of the acoustic neuroma. J Laryngol Otol. 2001;115(5):376–9.
2. Murray MR, Stout AP, Bradley CF. Schwann cell versus fibroblast as the origin of the specific nerve sheath tumor: observations upon normal nerve sheaths and neurilemomas in vitro. Am J Pathol. 1940;16(1):41–60.
3. Stangerup SE, Caye-Thomasen P. Epidemiology and natural history of vestibular schwannomas. Otolaryngol Clin N Am. 2012;45(2):257–68. vii
4. Evans DG, Moran A, King A, Saeed S, Gurusinghe N, Ramsden R. Incidence of vestibular schwannoma and neurofibromatosis 2 in the North West of England over a 10-year period: higher incidence than previously thought. Otol Neurotol. 2005;26(1):93–7.
5. Puls T. Incidental findings of a Vestibular Schwannoma on Brain MRI. B-ENT. 2015;11(4):309–12.
6. Schmidt RF, Boghani Z, Choudhry OJ, Eloy JA, Jyung RW, Liu JK. Incidental vestibular schwannomas: a review of prevalence, growth rate, and management challenges. Neurosurg Focus. 2012;33(3):E4.
7. Hoa M, Drazin D, Hanna G, Schwartz MS, Lekovic GP. The approach to the patient with incidentally diagnosed vestibular schwannoma. Neurosurg Focus. 2012;33(3):E2.

8. Jeyakumar A, Seth R, Brickman TM, Dutcher P. The prevalence and clinical course of patients with 'incidental' acoustic neuromas. Acta Otolaryngol. 2007;127(10):1051–7.
9. Lin D, Hegarty JL, Fischbein NJ, Jackler RK. The prevalence of "incidental" acoustic neuroma. Arch Otolaryngol Head Neck Surg. 2005;131(3):241–4.
10. Selesnick SH, Deora M, Drotman MB, Heier LA. Incidental discovery of acoustic neuromas. Otolaryngol Head Neck Surg. 1999;120(6):815–8.
11. Pothula VB, Lesser T, Mallucci C, May P, Foy P. Vestibular schwannomas in children. Otol Neurotol. 2001;22(6):903–7.
12. Van Abel KM, Carlson ML, Link MJ, Neff BA, Beatty CW, Lohse CM, et al. Primary inner ear schwannomas: a case series and systematic review of the literature. Laryngoscope. 2013;123(8):1957–66.
13. Tos M, Charabi S, Thomsen J. Clinical experience with vestibular schwannomas: epidemiology, symptomatology, diagnosis, and surgical results. Eur Arch Otorrinolaringol. 1998;255(1):1–6.
14. Asthagiri AR, Vasquez RA, Butman JA, Wu T, Morgan K, Brewer CC, et al. Mechanisms of hearing loss in neurofibromatosis type 2. PLoS One. 2012;7(9):e46132.
15. Thomsen J, Terkildsen K, Tos M, Acoustic neuromas. Progression of hearing impairment and function of the eighth cranial nerve. Am J Otol. 1983;5(1):20–33.
16. Koors PD, Thacker LR, Coelho DH. ABR in the diagnosis of vestibular schwannomas: a meta-analysis. Am J Otolaryngol. 2013;34(3):195–204.
17. Hulshof JH, Hilders CG, Baarsma EA. Vestibular investigations in acoustic neuroma. Acta Otolaryngol. 1989;108(1–2):38–44.
18. Bance ML, O'Driscoll M, Patel N, Ramsden RT. Vestibular disease unmasked by hyperventilation. Laryngoscope. 1998;108(4, part 1):610–4.
19. Bradley JP, Hullar TE, Neely JG, Goebel JA. Hyperventilation-induced nystagmus and vertigo after stereotactic radiotherapy for vestibular schwannoma. Otol Neurotol. 2011;32(8):1336–8.
20. Choi KD, Cho HJ, Koo JW, Park SH, Kim JS. Hyperventilation-induced nystagmus in vestibular schwannoma. Neurology. 2005;64(12):2062.
21. Mandala M, Giannuzzi A, Astore S, Trabalzini F, Nuti D. Hyperventilation-induced nystagmus in vestibular schwannoma and unilateral sensorineural hearing loss. Eur Arch Otorrinolaringol. 2013;270(7):2007–11.
22. Minor LB, Haslwanter T, Straumann D, Zee DS. Hyperventilation-induced nystagmus in patients with vestibular schwannoma. Neurology. 1999;53(9):2158–68.
23. Burke D. Microneurography, impulse conduction, and paresthesias. Muscle Nerve. 1993;16(10):1025–32.
24. Murofushi T, Matsuzaki M, Mizuno M. Vestibular evoked myogenic potentials in patients with acoustic neuromas. Arch Otolaryngol Head Neck Surg. 1998;124(5):509–12.
25. Takeichi N, Sakamoto T, Fukuda S, Inuyama Y. Vestibular evoked myogenic potential (VEMP) in patients with acoustic neuromas. Auris Nasus Larynx. 2001;28(Suppplement):S39–41.
26. Suzuki M, Yamada C, Inoue R, Kashio A, Saito Y, Nakanishi W. Analysis of vestibular testing in patients with vestibular schwannoma based on the nerve of origin, the localization, and the size of the tumor. Otol Neurotol. 2008;29(7):1029–33.
27. Ushio M, Iwasaki S, Murofushi T, Sugasawa K, Chihara Y, Fujimoto C, et al. The diagnostic value of vestibular-evoked myogenic potential in patients with vestibular schwannoma. Clin Neurophysiol. 2009;120(6):1149–53.
28. Matsuzaki M, Murofushi T, Mizuno M. Vestibular evoked myogenic potentials in acoustic tumor patients with normal auditory brainstem responses. Eur Arch Otorrinolaringol. 1999;256(1):1–4.
29. Huang CH, Wang SJ, Young YH. Correlation between caloric and ocular vestibular evoked myogenic potential test results. Acta Otolaryngol. 2012;132(2):160–6.
30. Blodow A, Helbig R, Wichmann N, Wenzel A, Walther LE, Bloching MB. Video head impulse test or caloric irrigation?. Contemporary diagnostic tests for vestibular schwannoma. HNO. 2013;61(9):781–5.
31. Aalling M, Skals RK, Abrahamsen ER, Hougaard DD. Comparison of test results from two separate video head impulse test systems in a cohort of patients diagnosed with a unilateral vestibular schwannoma. Eur Arch Otorrinolaringol. 2020;277(11):3185–93.
32. Welgampola MS, Ramsay E, Gleeson MJ, Day BL. Asymmetry of balance responses to monaural galvanic vestibular stimulation in subjects with vestibular schwannoma. Clin Neurophysiol. 2013;124(9):1835–9.
33. Mackenzie SW, Irving R, Monksfield P, Kumar R, Dezso A, Reynolds RF. Ocular torsion responses to electrical vestibular stimulation in vestibular schwannoma. Clin Neurophysiol. 2018;129(11):2350–60.
34. Nakamura M, Roser F, Mirzai S, Matthies C, Vorkapic P, Samii M. Meningiomas of the internal auditory canal. Neurosurgery. 2004;55(1):119–27. discussion 27–8.
35. Gupta G, Mammis A, Maniker A. Malignant peripheral nerve sheath tumors. Neurosurg Clin N Am. 2008;19(4):533–43.
36. Kuzmik GA, Michaelides EM, Chiang VL, Nonaka Y, Fukushima T, Vortmeyer AO, et al. Rapidly progressive epithelioid malignant peripheral nerve sheath tumor of the vestibular nerve. Otol Neurotol. 2013;34(9):1739–42.
37. Wei C, Heman-Ackah SE, Newman K, Zagzag D, Golfinos JG, Roland JT Jr. Temporal bone histopathology case of the month: malignant peripheral nerve sheath tumor arising within vestibular schwannoma. Otol Neurotol. 2012;33(9):e83–4.
38. Lopci E, Monti L, Balzarini L, Chiti A. Cardiac and acoustic metastases in relapsing melanoma. Clin Nucl Med. 2013;38(2):e85–8.
39. Huang X, Caye-Thomasen P, Stangerup SE. Distinct spontaneous shrinkage of a sporadic vestibular schwannoma. Auris Nasus Larynx. 2013;40(2):243–6.
40. Huang X, Caye-Thomasen P, Stangerup SE. Spontaneous tumour shrinkage in 1261 observed patients with sporadic vestibular schwannoma. J Laryngol Otol. 2013;127(8):739–43.
41. Ahle G, Visser W, Juergens A, Schlegel U. Successful treatment of hyperventilation-induced nystagmus in vestibular schwannoma with oxcarbacepine. J Neurol. 2008;255(7):1093.
42. Guadix SW, Tao AJ, An A, Demetres M, Tosi U, Chidambaram S, et al. Assessing the long-term safety and efficacy of gamma knife and linear accelerator radiosurgery for vestibular schwannoma: a systematic review and meta-analysis. Neurooncol Pract. 2021;8(6):639–51.
43. Agarwal P, Natanasabapathi G, Bisht RK, Malhotra RK, Kale SS. Investigation of optimal planning strategy in gamma knife perfexion for vestibular schwannoma tumor using hybrid plan technique. Biomed Phys Eng Express. 2022;8(6):065023.
44. Hafez RFA, Morgan MS, Fahmy OM, Hassan HT. Outcomes of Gamma Knife Surgery retreatment for growing vestibular schwannoma and review of the literature. Clin Neurol Neurosurg. 2020;198:106171.
45. Hwang I, Choi SH, Kim JW, Yeon EK, Lee JY, Yoo RE, et al. Response prediction of vestibular schwannoma after gamma-knife radiosurgery using pretreatment dynamic contrast-enhanced MRI: a prospective study. Eur Radiol. 2022;32(6):3734–43.
46. Kawashima M, Hasegawa H, Shin M, Takahashi W, Shinya Y, Iwasaki S, et al. Long-term outcomes of gamma knife radiosurgery for treating vestibular schwannoma with a lower prescription dose of 12 Gy compared with higher dose treatment. Otol Neurotol. 2020;41(10):e1314–e20.
47. Kieu HD, Vuong DN, Mai KT, Pham PC, Le TD. Long-term outcomes of rotating gamma knife for vestibular schwannoma: a 4-year prospective longitudinal study of 89 consecutive patients in Vietnam. Surg Neurol Int. 2021;12:585.

48. Killeen DE, Tolisano AM, Isaacson B, Kutz JW, Barnett S, Wardak Z, et al. Vestibular Schwannoma tumor size and growth rate predict response with gamma knife stereotactic radiosurgery. J Neurol Surg B Skull Base. 2022;83(1):11–8.
49. Livingston AJ, Espahbodi M, Harvey SA, Albano K, Quinet SA, Bovi JA, et al. Gamma Knife treatment of Vestibular Schwannoma planned with computed tomography cisternography. Adv Radiat Oncol. 2021;6(3):100631.
50. Wage J, Mignano J, Wu J. Tufts medical center experience with long-term follow-up of Vestibular Schwannoma treated with gamma knife stereotactic radiosurgery: novel finding of delayed pseudoprogression. Adv Radiat Oncol. 2021;6(4):100687.
51. Yeole U, Prabhuraj AR, Arivazhagan A, Narasingarao KVL, Vazhayil V, Bhat D, et al. Gamma knife radiosurgery for large Vestibular Schwannoma more than 10 cm (3): a single-center Indian study. J Neurol Surg B Skull Base. 2022;83(Suppl 2):e343–e52.
52. Nasi D, Zunarelli E, Puzzolante A, Moriconi E, Pavesi G. Early life-threating enlargement of a vestibular schwannoma after gamma knife radiosurgery. Acta Neurochir. 2020;162(8):1977–82.
53. Yang HC, Wu CC, Lee CC, Huang HE, Lee WK, Chung WY, et al. Prediction of pseudoprogression and long-term outcome of vestibular schwannoma after Gamma Knife radiosurgery based on preradiosurgical MR radiomics. Radiother Oncol. 2021;155:123–30.
54. Friedman WA. Linear accelerator radiosurgery for vestibular schwannomas. Prog Neurol Surg. 2008;21:228–37.
55. Friedman WA, Foote KD. Linear accelerator-based radiosurgery for vestibular schwannoma. Neurosurg Focus. 2003;14(5):e2.
56. Matsuo T, Okunaga T, Kamada K, Izumo T, Hayashi N, Nagata I. Long-term follow-up results of linear accelerator-based radiosurgery for vestibular schwannoma using serial three-dimensional spoiled gradient-echo MRI. J Clin Neurosci. 2015;22(2):320–5.
57. Okunaga T, Matsuo T, Hayashi N, Hayashi Y, Shabani HK, Kaminogo M, et al. Linear accelerator radiosurgery for vestibular schwannoma: measuring tumor volume changes on serial three-dimensional spoiled gradient-echo magnetic resonance images. J Neurosurg. 2005;103(1):53–8.
58. Spiegelmann R, Lidar Z, Gofman J, Alezra D, Hadani M, Pfeffer R. Linear accelerator radiosurgery for vestibular schwannoma. J Neurosurg. 2001;94(1):7–13.
59. Anderson BM, Khuntia D, Bentzen SM, Geye HM, Hayes LL, Kuo JS, et al. Single institution experience treating 104 vestibular schwannomas with fractionated stereotactic radiation therapy or stereotactic radiosurgery. J Neuro-Oncol. 2014;116(1):187–93.
60. Giannuzzi AL, Merkus P, Falcioni M. The use of intratympanic gentamicin in patients with vestibular schwannoma and disabling vertigo. Otol Neurotol. 2013;34(6):1096–8.
61. Uehara N, Tanimoto H, Nishikawa T, Doi K, Katsunuma S, Kimura H, et al. Vestibular dysfunction and compensation after removal of acoustic neuroma. J Vestib Res. 2011;21(5):289–95.
62. Wackym PA, Hannley MT, Runge-Samuelson CL, Jensen J, Zhu YR. Gamma Knife surgery of vestibular schwannomas: longitudinal changes in vestibular function and measurement of the Dizziness Handicap Inventory. J Neurosurg. 2008;109(Suppl):137–43.
63. Rae-Grant AD. Unusual symptoms and syndromes in multiple sclerosis. Continuum (Minneap Minn). 2013;19(4 Multiple Sclerosis):992–1006.
64. Smith KJ, McDonald WI. The pathophysiology of multiple sclerosis: the mechanisms underlying the production of symptoms and the natural history of the disease. Philos Trans R Soc Lond Ser B Biol Sci. 1999;354(1390):1649–73.
65. Passier L, Doherty D, Smith J, McPhail S. Vestibular rehabilitation following removal of acoustic neuroma: a systematic review of randomised trials. Head Neck Oncol. 2012;4(2):59.
66. Cakrt O, Chovanec M, Funda T, Kalitova P, Betka J, Zverina E, et al. Exercise with visual feedback improves postural stability after vestibular schwannoma surgery. Eur Arch Otorrinolaringol. 2010;267(9):1355–60.
67. Tufarelli D, Meli A, Labini FS, Badaracco C, De Angelis E, Alesii A, et al. Balance impairment after acoustic neuroma surgery. Otol Neurotol. 2007;28(6):814–21.
68. Saman Y, Bamiou DE, Gleeson M. A contemporary review of balance dysfunction following vestibular schwannoma surgery. Laryngoscope. 2009;119(11):2085–93.

35 Auditory Neuropathy

Overview

Auditory neuropathy, sometimes also called "auditory dyssynchrony" or "auditory neuropathy spectrum disorder," refers to hearing loss due to dysfunction of the auditory nerve rather than to the cochlea. In these patients, audiometry shows hearing loss, auditory brainstem evoked responses are abnormal, and yet otoacoustic emissions are present. This is difficult to diagnose in older individuals in whom otoacoustic emissions are absent secondary to normal aging. Auditory neuropathy is uncommon, and when present, is often only one part of a broader neuropathic process. There is emerging evidence that some auditory neuropathies (perhaps as high as 40%) may be associated with specific genetic mutations. First-line treatment is with amplification. Cochlear implantation is sometimes attempted.

Introduction

In the early 1990s, Starr and colleagues [1] and Berlin and colleagues [2] described several patients with subjective hearing loss and a previously unreported pattern of auditory test results. They identified several additional such patients, and in 1996, Starr, Picton, Sininger, Hood, and Berlin [3] wrote:

> We have identified a group of patients with hearing deficits who have preserved otoacoustic emissions and absent or severely abnormal auditory brainstem responses. Most of the patients complain of difficulty understanding speech, particularly in the presence of noise. Using two particular tests of auditory function, otoacoustic emissions and auditory brainstem potentials, we were able to identify that the function of the VIII nerve was disordered, whereas the function of cochlear outer hair cells was normal [3].

On standard audiometry, these patients appeared to have sensorineural hearing loss. In most sensorineural hearing loss (such as presbycusis), the deficit is thought to localize to the cochlear hair cells. But in the patients evaluated by Starr and colleagues, the combination of abnormal auditory brainstem responses yet preserved otoacoustic emissions led these researchers to localize the dysfunction to the *auditory nerve*, rather than the cochlear hair cells. As they later phrased it, "In auditory neuropathy, the hearing impairment commences downstream from mechanoelectrical transduction and cochlear amplification of outer hair cells" [4].

Also of note, some of the patients originally reported additionally had clinical evidence of a sensorimotor neuropathy (such as reduced myotatic reflexes, impaired proprioception, and reduced pallesthesia), and some of the patients also had evidence of vestibular deficits (such as reduced caloric responses). This led the researchers to postulate that, at least in these patients, *auditory neuropathy* was part of a broader neuropathic problem.

Subsequent research on auditory neuropathy "elucidated the wide range of disease mechanisms, which include loss of inner hair cells (IHCs) or IHC synapses, impaired synaptic transmission to spiral ganglion neurons (SGNs), and disrupted propagation of auditory information along the auditory nerve" [4]. In recognition of this range of mechanisms (neuropathy, synaptopathy, spiral ganglionopathy), the literature sometimes refers to this "family" of diseases as "*auditory dyssynchrony*" or "*auditory neuropathy spectrum disorder*."

Demographics

We will discuss below that the auditory test result profile required for a diagnosis of auditory neuropathy often makes it impossible to diagnose in older individuals, so the majority of research on this topic is in younger populations, particularly children with severe hearing impairments. Estimates of the prevalence of auditory neuropathy in this population range over nearly an order of magnitude, from 1.2% [5] to 8.4% [6].

M. Cherchi, *Otoneurology and Vestibular Medicine*, https://doi.org/10.1007/978-3-031-94841-1_35

Diagnosis

In a patient with a clinical history of hearing loss, whose standard audiogram shows poor word recognition and speech perception thresholds out of proportion to performance on pure tone testing, one should check auditory evoked brainstem responses and otoacoustic emissions. If otoacoustic emissions are preserved and brainstem auditory evoked responses are absent or abnormal, then a diagnosis of auditory neuropathy is a reasonable possibility.

Note that this test result profile requires preserved otoacoustic emissions. Since otoacoustic emissions degrade with age, and are often nearly absent by the sixth or seventh decade of life, an older person may have a dysfunctional auditory nerve yet not meet criteria for auditory neuropathy because otoacoustic emissions are absent. This is unfortunate, because some of these patients may go through numerous attempts at amplification with hearing aids and find them all unsatisfactory.

Since these criteria effectively make it impossible to diagnose auditory neuropathy in the elderly, most of the research on this topic is in younger individuals, typically in children with severe hearing loss.

Treatment

Even when the diagnosis seems secure, treatment of auditory neuropathy is difficult [7]. Practically, in a patient with auditory neuropathy, most audiologists will attempt amplification with a hearing aid since this approach has little medical risk, it is noninvasive, and it is reversible. Cochlear implantation is sometimes attempted on the logic that a weakened (but not completely dead) auditory nerve may still be able to transmit the electrically augmented signals driven by a cochlear implant.

However, it is important to recognize that since dysfunction in auditory neuropathy localizes to the auditory nerve itself, interventions at the auditory nerve (such as a cochlear implant) or "upstream" from the auditory nerve (such as hearing aids) are likely to be inadequate if not completely ineffective. Rather, a strategy to bypass the auditory nerve is needed, such as auditory brainstem implants [8], though this approach is certainly much more invasive and risky.

It has been estimated that approximately 40% of cases of auditory neuropathy are attributable to genetic mutations [9], and a broad range of genes has been implicated [10], including over 90 mutations in OTOF which encodes otoferlin; DFNB1 which encodes connexins; locus DFNA25 in the SLC17A8 gene which encodes vesicular glutamate transporter 3; DFNB9 in "deafness, nonsyndromic autosomal recessive 9;" CACNA1D which encodes an alpha-1 subunit on the $Ca_V1.3$ calcium channel; MPZ gene that encodes myelin protein zero in Schwann cells; DFNB59 which encodes pejvakin; DIAPH3. Consequently, the advent of gene therapy may eventually provide opportunities for intervention in these patients.

References

1. Starr A, McPherson D, Patterson J, Don M, Luxford W, Shannon R, et al. Absence of both auditory evoked potentials and auditory percepts dependent on timing cues. Brain. 1991;114(Pt 3):1157–80.
2. Berlin CI, Hood LJ, Cecola RP, Jackson DF, Szabo P. Does type I afferent neuron dysfunction reveal itself through lack of efferent suppression? Hear Res. 1993;65(1–2):40–50.
3. Starr A, Picton TW, Sininger Y, Hood LJ, Berlin CI. Auditory neuropathy. Brain. 1996;119(Pt 3):741–53.
4. Moser T, Starr A. Auditory neuropathy–neural and synaptic mechanisms. Nat Rev Neurol. 2016;12(3):135–49.
5. Penido RC, Isaac ML. Prevalence of auditory neuropathy spectrum disorder in an auditory health care service. Braz J Otorhinolaryngol. 2013;79(4):429–33.
6. Foerst A, Beutner D, Lang-Roth R, Huttenbrink KB, von Wedel H, Walger M. Prevalence of auditory neuropathy/synaptopathy in a population of children with profound hearing loss. Int J Pediatr Otorhinolaryngol. 2006;70(8):1415–22.
7. Giraudet F, Avan P. Auditory neuropathies: understanding their pathogenesis to illuminate intervention strategies. Curr Opin Neurol. 2012;25(1):50–6.
8. Wong K, Kozin ED, Kanumuri VV, Vachicouras N, Miller J, Lacour S, et al. Auditory brainstem implants: recent progress and future perspectives. Front Neurosci. 2019;13:10.
9. Burdo S, Di Berardino F, Bruno G. Is auditory neuropathy an appropriate term? A systematic literature review on its aetiology and pathogenesis. Acta Otorhinolaryngol Ital. 2021;41(6):496–506.
10. Del Castillo FJ, Del Castillo I. Genetics of isolated auditory neuropathies. Front Biosci. 2012;17(4):1251–65.

36 Ramsay Hunt Syndrome

Overview

Ramsay Hunt syndrome (RHS) is a cranial neuropathy mostly of the vestibulocochlear and facial nerves attributed to *Varicella zoster virus* infection. It can occur at any age (though is more common with increasing age), without clear gender predisposition. Presenting symptoms are some combination of facial weakness, aural pain, hearing loss, hyperacusis, tinnitus, disequilibrium, and dysgeusia. Physical examination may show lower motor neuron facial weakness, sensorineural hearing loss, often a vesicular rash in the external auditory canal, and less commonly nystagmus. Audiometry may show hearing loss, which tends to be more pronounced in the high frequencies. Vestibular testing (caloric testing, vestibular evoked myogenic potentials, video head impulse testing) may show evidence of unilateral vestibular hypofunction. MRI of the internal auditory canals without and with contrast may show enhancement of the vestibulocochlear and/or facial nerves. Treatment with steroids (prednisone, methylprednisolone) and antiviral agents (usually valacyclovir) is often attempted, but data supporting efficacy are weak. Prognosis is guarded; the facial weakness is more severe and recovers more poorly than in Bell's palsy; the hearing may recover more poorly than in labyrinthitis; the disequilibrium may recover more poorly than in vestibular neuritis. A patient in whom Ramsay Hunt syndrome is suspected or diagnosed is often referred to an otoneurologist or neuro-otologist.

Fig. 36.1 James Ramsay Hunt (1874–1937). (Reproduced from Wikipedia, public domain)

Introduction

In 1907, the American neurologist, James Ramsay Hunt (1872–1937), published a paper describing varicella zoster affectation of the vestibulocochlear nerve and facial nerve [1] (Fig. 36.1).

This clinical entity is sometimes designated *Herpes zoster oticus*, but that nomenclature implies that only the vestibulocochlear nerve is affected. In fact, patients usually present with some combination of affectation of the vestibulocochlear and facial nerves.

Epidemiology

The incidence of Ramsay Hunt syndrome increases with age, though it can still affect pediatric patients [2, 3]. Figure 36.2, from Wayman and colleagues [4], is a histogram showing the age distribution by decade.

M. Cherchi, *Otoneurology and Vestibular Medicine*, https://doi.org/10.1007/978-3-031-94841-1_36

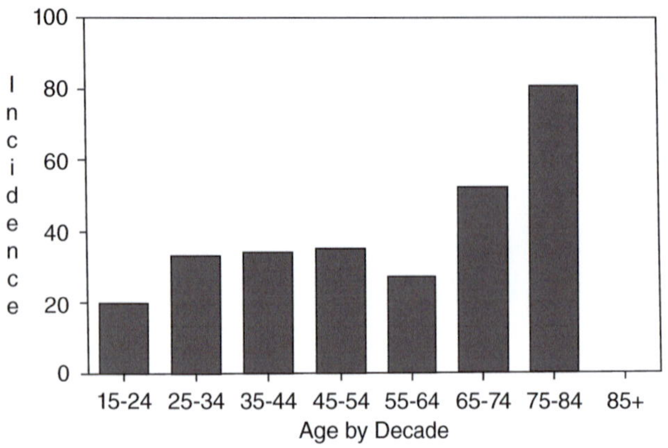

Fig. 36.2 Histogram showing the age distribution by decade of 186 patients with Ramsay Hunt syndrome. (From Wayman et al. [4]. Reproduced with written permission from Cambridge University Press)

Literature reports conflicting data regarding gender predominance of disease distribution.

Some studies report a slight female preponderance. The series of 186 Ramsay Hunt patients studied by Wayman and colleagues [4] included 102 (55%) females and 85 (45%) males. The series of 185 Ramsay Hunt patients studied by Robillard and colleagues [5] reported 101 (54.6%) females and 84 (45.4%) males.

In contrast, other studies report a male preponderance. In the series of 13 Ramsay Hunt patients studied by Martin-Sanz and colleagues [6], 7 (54%) were male and 6 (46%) were female; since these statistics are almost exactly the opposite of the others, one might conclude that it is an artifact of the small sample size (13 patients, rather than 185 and 186 patients in the previously cited 2 studies). However, an investigation with a larger sample size, Coulson and colleagues [7] studied 101 patients with Ramsay Hunt syndrome, and the study population comprised 70 (70%) males and 31 (31%) females.

Clinical Presentation

Patients present with some combination of the following:

- *Vestibulocochlear symptoms* (disequilibrium, hearing loss, tinnitus). These are believed to result from direct affectation of the vestibulocochlear nerve by the Varicella zoster virus (VZV).
- *Ipsilateral facial weakness*. Edema of the vestibulocochlear nerve within the internal auditory canal may compress the facial nerve and cause paresis. Another possibility is passage of infection from the vestibulocochlear nerve to the facial nerve [8]. Since most of the facial nerve supplies facial muscles, the VZV eruptions do not occur on the face (whose sensory innervation would be supplied by the trigeminal nerve).
- *Vesicular lesions in the ipsilateral internal auditory canal*. This may be compatible with VZV involvement of the nervus intermedius—somatosensory fibers that run within the facial nerve to supply "parts of both the medial and lateral surfaces of the auricle, to a strip along the external acoustic meatus, and to a part of the tympanic cavity" [9]—or possibly somatosensory fibers within the vestibulocochlear nerve.

The multiple symptoms may present approximately simultaneously, or serially but in fairly close temporal proximity [10].

In a series of 14 patients with Ramsay Hunt syndrome, Iwasaki and colleagues [10] reported that 14 (100%) had lower motor neuron *facial weakness* (House–Brackman grade IV–VI), 12 (86%) had *disequilibrium*, and 10 (71%) had *hearing loss* ipsilateral to the facial palsy.

In the series of Ramsay Hunt patients studied by Wayman and colleagues [4], of 93 patients with normal audiometry, only 5 (5%) complained of *disequilibrium*, and of 29 patients with abnormal audiometry, 8 (28%) complained of disequilibrium.

Despite the hearing loss, some patients also complain of *hyperacusis*. Two potential contributors to this include loudness recruitment and malfunction of the stapedius (mediated by the nerve to the stapedius, which is from the facial nerve that is paretic in Ramsay Hunt syndrome).

In the series of Ramsay Hunt patients studied by Wayman and colleagues [4], of 93 with normal audiograms, 44 (47%) reported *hyperacusis*, and 45 (48%) reported *hearing loss* and *tinnitus*; of 29 with abnormal audiograms, 5 (31%) reported hyperacusis and 5 (31%) reported hearing loss and tinnitus.

Some patients also complain of *dysgeusia* [5, 11], presumably due to affectation of the gustatory afferents passing through the chorda tympani (gustatory branch of the facial nerve).

Robillard and colleagues [5] reported facial, aural, and auditory symptoms (not disequilibrium) in a series of 102 Ramsay Hunt syndrome patients, as shown in Table 36.1.

Table 36.1 Presenting symptoms by percentage, in decreasing order, of 102 patients with Ramsay Hunt syndrome

Symptom	Ramsay Hunt syndrome ($n = 122$) (%)
Pain	100
Ageusia/dysgeusia	67
Epiphora	55
Hyperacusis	52
Decreased tears	43
Decreased Schirmer test	42
Fifth cranial nerve palsy	34
Hearing loss	21

From Robillard et al. [5]

Table 36.2 Symptoms in 101 patients with Ramsay Hunt syndrome

Symptom	Number affected (*n* = 101)
Bell's pain	90
Decreased hearing	43
Other pain	37
Imbalance	33
Dysgeusia	31
Tinnitus	20
Vertigo	18
Facial numbness	17
Hyperacusis	12
Dry eye	5
Excessive tearing	5
Tongue numbness	5
Swallowing dysfunction	3
Diplopia	2

From Coulson et al. [7]

Table 36.3 Comparison of data from multiple studies regarding symptoms in Ramsay Hunt syndrome

Symptom	Robillard et al. [5]	Coulson et al. [7]	Iwasaki et al. [10]	Wayman et al. [4], patients with *normal* audiometry	Wayman et al. [4], patients with *abnormal* audiometry
Facial weakness	–	–	100%	–	
Aural pain	100%	90%		43%	72%
Hearing loss	21%	43%	71%	48%	31%
Hyperacusis	52%	12%	–	47%	31%
Tinnitus		20%	–	–	48%
Disequilibrium	–	33% ("imbalance") 18% ("vertigo")	86%	5%	28%
Dysgeusia	67%	31%	–	–	–

Notable points include that 100% experienced pain, 67% experienced disturbances of taste, 52% experienced hyperacusis, and 21% experienced hearing loss.

Coulson and colleagues [7] studied findings in 101 patients with Ramsay Hunt syndrome, and reported the findings in Table 36.2; the numbers listed are percentages. Notable points include that 90% of patients complained of ear pain, 43% of hearing loss, 33% of disequilibrium, 31% of gustatory abnormalities, 20% of tinnitus, and 12% of hyperacusis.

Table 36.3 compares data regarding symptoms from the studies discussed earlier.

Physical Examination

A minority of patients lack the vesicular lesions [11], which parallels the "herpete sine herpete" dermatological manifestations of *Varicella zoster* affecting other areas of the body. In the study by Coulson and colleagues [7], of 101 patients with Ramsay Hunt syndrome, ***75%*** presented with vesicular lesions—in other words, 25% of patients lack the vesicular lesions. In the study by Wayman and colleagues [4], of 132 patients with Ramsay Hunt syndrome, 122 had audiometry performed; of those 122 patients, 37/93 (***40%***) with normal hearing had vesicular lesions, and 18/29 (***62%***) with abnormal hearing had vesicular lesions.

In the study by Coulson and colleagues [7], of 101 patients with Ramsay Hunt syndrome, ***5%*** presented with "*nystagmus*."

Lee and colleagues [12] studied 36 patients with Ramsay Hunt syndrome; of 27 patients who complained of disequilibrium, 26 (***96%***) exhibited nystagmus, and of 9 patients who did not complain of disequilibrium, 6 (***67%***) still showed nystagmus. In these patients, they identified direction-fixed nystagmus (either irritative or paretic) and direction-changing positional nystagmus. They reported the following data:

- Ramsay Hunt syndrome (*n* = 36)
  - Ramsay Hunt syndrome with dizziness (*n* = 27)
    - Nystagmus absent (*n* = 1)
    - Nystagmus present (*n* = 26)
      - Positional direction-changing nystagmus (*n* = 12)
      - Direction-fixed nystagmus (*n* = 14)
        - Irritative (*n* = 2)

        - Paretic ($n = 12$)
  - Ramsay Hunt syndrome without dizziness ($n = 9$)
    - Nystagmus absent ($n = 3$)
    - Nystagmus present ($n = 6$)
      - Positional direction-changing nystagmus ($n = 1$)
      - Direction-fixed nystagmus ($n = 5$)
        - Irritative ($n = 2$)
        - Paretic ($n = 3$)

Kim and colleagues [13] studied 28 patients with Ramsay Hunt syndrome who complained of disequilibrium. The patterns of nystagmus they reported were: ***61%*** direction-fixed nystagmus that was contraversive (beating away from the lesion); ***18%*** direction-changing positional nystagmus that was apogeotropic; ***14%*** direction-fixed nystagmus that was ipsiversive (beating away from the lesion); and ***7%*** direction-changing positional nystagmus that was apogeotropic.

Testing: Audiometry

In a series of 14 patients with Ramsay Hunt syndrome, Iwasaki and colleagues [10] reported that audiometry showed sensorineural hearing loss in 10 (***71%***) ipsilateral to the facial palsy.

In a series of 152 patients with Ramsay Hunt syndrome who had audiometry performed, Wayman and colleagues [4] noted that 93 (61%) showed normal hearing, and 29 (***19%***) showed hearing loss attributed to Ramsay Hunt syndrome; the remaining 30 (20%) were deemed to have hearing loss from other causes.

The hearing loss tends to be more severe in the high frequencies. Hearing loss tends to be more severe in patients with disequilibrium. The degree of hearing loss does not correlate significantly with the facial weakness [14].

Testing: Vestibular Assessment with Caloric Testing

In a series of 19 patients with Ramsay Hunt syndrome, Takahashi and colleagues [15] reported that 15 (***79%***) exhibited caloric weakness.

In a series of 14 patients with Ramsay Hunt syndrome, Iwasaki and colleagues [10] reported that 12 (***86%***) had caloric weakness ipsilateral to the facial palsy—the same 12 patients who complained of disequilibrium.

In a series of 10 patients with Ramsay Hunt syndrome, Ozeki and colleagues [16] reported that 10 (***100%***) had caloric weakness.

In a series of 13 patients with Ramsay Hunt syndrome, Martin-Sanz and colleagues [6] reported caloric weakness of 52.5% ± 35.93%.

Testing: Video Head Impulse Testing

Martin-Sanz and colleagues [6] studied video head impulse testing in 18 patients with vestibular neuritis and 13 patients Ramsay Hunt syndrome. In the Ramsay Hunt syndrome patients, they reported reduced vestibulo-ocular reflex gain of 0.63 ± 0.40 on the affected side. They took improvement in vestibulo-ocular reflex gain, and decrease in the latency of compensatory saccades, to reflect vestibular recovery. By these criteria, patients with Ramsay Hunt syndrome recovered vestibular function more poorly than those with vestibular neuritis.

Testing: Other Ocular Motor Testing

Kim and colleagues [17] studied a series of 17 patients with Ramsay Hunt syndrome who complained of disequilibrium, and reported that 94% exhibited vibration-induced nystagmus and 59% exhibited hyperventilation-induced nystagmus.

Testing: Cervical Vestibular Evoked Myogenic Potentials

In a series of 10 patients with Ramsay Hunt syndrome, Ozeki and colleagues [16] reported that 7 (***70%***) exhibited abnormalities in air-conducted cervical vestibular evoked myogenic potentials.

Takahashi and colleagues [15] studied a series of patients with Ramsay Hunt syndrome and reported that 3/18 (***17%***) had abnormal cervical vestibular evoked myogenic potentials.

Testing: Ocular Vestibular Evoked Myogenic Potentials

Takahashi and colleagues [15] studied a series of patients with Ramsay Hunt syndrome and reported that 9/17 (***53%***) had abnormal ocular vestibular evoked myogenic potentials.

Imaging

MRI of the internal auditory canals may show T2/FLAIR signal abnormalities and enhancement of the vestibulocochlear and facial nerves [15, 18, 19].

The series studied by Iwasaki and colleagues [10] reported that of 12 patients with caloric weakness, 11 (***92%***) showed enhancement of the superior division of the ipsilateral *vestibular nerve*. Of all 14 patients with lower motor neuron

Table 36.4 Comparison of test results from multiple studies of patients with Ramsay Hunt syndrome

Test	Iwasaki et al. [10]	Wayman et al. [4]	Takahashi et al. [15]	Ozeki et al. [16]	Kim et al. [14]
Audiometry showing hearing loss	71%	19%			76%
Caloric weakness	86%	–	79%	100%	–
Weakness on cervical VEMP	–	–	17%	70%	–
Weakness on ocular VEMP	–	–	53%	–	–

Table 36.5 Comparison of outcomes of facial paralysis in Ramsay Hunt patients between those treated versus not treated with oral prednisone

Measurement	Prednisone (%)	No prednisone (%)	Statistical significance
Facial paralysis recovery profile (FPRP) <10	19	38	Not statistically significant
Facial paralysis recovery index (FPRI) <10	34	62	$p = 0.05$
Complete denervation	19	25	Not statistically significant
Complete paralysis	20	56	$p = 0.01$
Contracture/synkinesis	22	44	Not statistically significant

From Robillard et al. [5]

facial weakness, only 4 (29%) showed enhancement of the ipsilateral *facial nerve*.

Some studies report enhancement in *brainstem nuclei* [20–22] or *cerebellum* [23–25], suggesting that some cases may involve a component of encephalitis. Sun and colleagues [26] reported a case of Ramsay Hunt syndrome associated with infarction secondary to "vertebral artery vasculopathy" (perhaps VZV vasculitis).

Diagnostic Testing Overall

Table 36.4 summarizes the test results reported in the studies discussed above.

Treatment

Attempted treatments have included oral steroids and antiviral agents. Steroids have included methylprednisolone [10] and prednisone [5]. Antiviral agents have included valacyclovir [10] and acyclovir [27]. Evidence supporting these treatments is weak.

In the series of Ramsay Hunt syndrome patients studied by Robillard and colleagues [5], treatment with oral prednisone modestly improved recovery from facial paralysis, as shown in Table 36.5.

As far as the vestibular symptoms are concerned, since RHS causes vestibular weakness, it is logical to offer vestibular rehabilitation therapy, though there are few reports on outcomes of this treatment specifically for RHS [28, 29].

Prognosis

Despite treatment with oral prednisone, outcomes from *facial paralysis* are still poorer than in patients with Bell's palsy [5, 30].

According to Martin-Sanz and colleagues [6], the presence of facial palsy was a predictor of the severity of the initial *impairment in the vestibulo-ocular reflex* (VOR) of video head impulse testing, but the facial paralysis did *not* correlate with the final VOR (in other words, it did not predict recovery of vestibular function).

A Cochrane review [31] found no difference in outcomes for Ramsay Hunt syndrome patients treated with antiviral agents (acyclovir, valacyclovir, famciclovir, brivudin).

Overall, prognosis in children is worse than in adults [3], though recovery from facial weakness may be better [2].

References

1. Hunt JR. On herpetic inflammations of the geniculate ganglion. A new syndrome and its complicaitons. J Nerv Ment Dis. 1907;34(2):73.
2. Hato N, Kisaki H, Honda N, Gyo K, Murakami S, Yanagihara N. Ramsay Hunt syndrome in children. Ann Neurol. 2000;48(2):254–6.
3. Kansu L, Yilmaz I. Herpes zoster oticus (Ramsay Hunt syndrome) in children: case report and literature review. Int J Pediatr Otorhinolaryngol. 2012;76(6):772–6.
4. Wayman DM, Pham HN, Byl FM, Adour KK. Audiological manifestations of Ramsay Hunt syndrome. J Laryngol Otol. 1990;104(2):104–8.
5. Robillard RB, Hilsinger RL Jr, Adour KK. Ramsay Hunt facial paralysis: clinical analyses of 185 patients. Otolaryngol Head Neck Surg. 1986;95(3 Pt 1):292–7.

6. Martin-Sanz E, Rueda A, Esteban-Sanchez J, Yanes J, Rey-Martinez J, Sanz-Fernandez R. Vestibular restoration and adaptation in vestibular neuritis and Ramsay Hunt syndrome with vertigo. Otol Neurotol. 2017;38(7):e203–e8.
7. Coulson S, Croxson GR, Adams R, Oey V. Prognostic factors in herpes zoster oticus (Ramsay Hunt syndrome). Otol Neurotol. 2011;32(6):1025–30.
8. Kuhweide R, Van de Steene V, Vlaminck S, Casselman JW. Ramsay Hunt syndrome: pathophysiology of cochleovestibular symptoms. J Laryngol Otol. 2002;116(10):844–8.
9. Rhoton AL Jr, Kobayashi S, Hollinshead WH. Nervus intermedius. J Neurosurg. 1968;29(6):609–18.
10. Iwasaki H, Toda N, Takahashi M, Azuma T, Nakamura K, Takao S, et al. Vestibular and cochlear neuritis in patients with Ramsay Hunt syndrome: a Gd-enhanced MRI study. Acta Otolaryngol. 2013;133(4):373–7.
11. Adour KK. Otological complications of herpes zoster. Ann Neurol. 1994;35(Suppl):S62–4.
12. Lee J, Choi B, Noh H, Jeong H, Shin JE, Kim CH. Nystagmus in Ramsay Hunt syndrome with or without dizziness. Neurol Sci. 2021;42(1):193–8.
13. Kim CH, Choi JW, Han KJ, Lee YS, Shin JE. Direction-fixed and direction-changing positional nystagmus in Ramsay Hunt syndrome. Otol Neurotol. 2018;39(3):e209–e13.
14. Kim CH, Choi H, Shin JE. Characteristics of hearing loss in patients with herpes zoster oticus. Medicine (Baltimore). 2016;95(46):e5438.
15. Takahashi M, Sato G, Toda N, Azuma T, Nakamura K, Iwasaki H, et al. Vestibular and cochlear nerve enhancement on MRI and its correlation with vestibulocochlear functional deficits in patients with Ramsay Hunt syndrome. Auris Nasus Larynx. 2021;48(3):347–52.
16. Ozeki H, Iwasaki S, Ushio M, Takeuchi N, Murofushi T. The lesion site of vestibular dysfunction in Ramsay Hunt syndrome: a study by click and galvanic VEMP. J Vestib Res. 2006;16(4–5):217–22.
17. Kim CH, Jeong KH, Ahn SH, Shin DH, Kim YW, Shin JE. Vibration- and hyperventilation-induced nystagmus in patients with Ramsay Hunt syndrome with vertigo. Otolaryngol Head Neck Surg. 2015;152(5):912–8.
18. Chung MS, Lee JH, Kim DY, Lim YM, Ahn JH, Sung YS, et al. The clinical significance of findings obtained on 3D-FLAIR MR imaging in patients with Ramsay-Hunt syndrome. Laryngoscope. 2015;125(4):950–5.
19. Hasuike K, Sekitani T, Imate Y. Enhanced MRI in patients with vestibular neuronitis. Acta Otolaryngol Suppl. 1995;519:272–4.
20. Inagaki A, Toyoda T, Mutou M, Murakami S. Ramsay Hunt syndrome associated with solitary nucleus, spinal trigeminal nucleus and tract, and vestibular nucleus involvement on sequential magnetic resonance imaging. J Neurovirol. 2018;24(6):776–9.
21. Kim JH, Chung PW, Oh S, Hong SB, Chung CS, Jung CW, et al. Ramsay Hunt syndrome complicated by a brainstem lesion. J Clin Virol. 2007;39(4):322–5.
22. Yacovino DA, Perez Akly MS, Ibanez T, Cherchi M. Vestibular nucleus involvement in patients with acute vertigo due to herpes zoster oticus or vestibular neuritis. Neurology. 2023;101:e1461.
23. Letchuman V, Donohoe CD. Brainstem and cerebellar involvement in Ramsay Hunt syndrome. Case Rep Otolaryngol. 2019;2019:7605056.
24. Mizock BA, Bartt R, Agbemazdo B. Herpes zoster oticus with pontine lesion: segmental brain-stem encephalitis. Clin Infect Dis. 2000;30(1):229–30.
25. Sartoretti-Schefer S, Kollias S, Valavanis A. Ramsay Hunt syndrome associated with brain stem enhancement. Am J Neuroradiol. 1999;20(2):278–80.
26. Sun H, You H, Wu H. Brainstem infarction and vertebral artery vasculopathy after Ramsay Hunt syndrome. Neurology. 2022;98(21):890–1.
27. Goswami Y, Gaurkar SS. Ramsay Hunt syndrome: an introduction, signs and symptoms, and treatment. Cureus. 2023;15(1):e33688.
28. Liao WL, Sung PY, Chui WF, Chang TP. Vestibular rehabilitation therapy in a patient with chronic vestibulopathy of ramsay hunt syndrome. Am J Phys Med Rehabil. 2011;90(10):851–5.
29. Gill-Body KM, Krebs DE, Parker SW, Riley PO. Physical therapy management of peripheral vestibular dysfunction: two clinical case reports. Phys Ther. 1994;74(2):129–42.
30. Lee DH, Chae SY, Park YS, Yeo SW. Prognostic value of electroneurography in Bell's palsy and Ramsay-Hunt's syndrome. Clin Otolaryngol. 2006;31(2):144–8.
31. Uscategui T, Doree C, Chamberlain IJ, Burton MJ. Antiviral therapy for Ramsay Hunt syndrome (herpes zoster oticus with facial palsy) in adults. Cochrane Database Syst Rev. 2008;2008(4):CD006851.

37 Multiple Myeloma

Introduction

Multiple myeloma (MM) is the second most common hematologic malignancy and consists of pathological proliferation of monoclonal plasma cells in bone marrow. These plasma cells overproduce immunoglobulins which accumulate in, and interact with, other cells, leading to a variety of problems in multiple organ systems (bone lesions, hypercalcemia, renal failure, anemia, infection) [1].

Vestibulocochlear Manifestations of Multiple Myeloma

It is very unusual for multiple myeloma to cause auditory or vestibular symptoms, and as of this writing, the literature on this topic consists of only case reports. Most such cases occur when a plasmacytoma involves the skull base and impinges either on the vestibulocochlear nerve [2, 3] or on the cerebellum [4, 5].

We have seen many patients who carry a diagnosis of multiple myeloma, but none in whom that disease has been proven as the cause of audiologic or vestibular symptoms.

If the mechanism is compression of the vestibulocochlear nerve by a lytic or space-occupying plasmacytoma in the skull base, then we would anticipate that the auditory symptoms would correspond to sensorineural hearing loss on audiometry, and the disequilibrium would correspond to vestibular weakness. Oddly, one case report [2] instead describes a mixed hearing loss, suggesting possible middle ear involvement. None of the literature reports results of any vestibular testing.

If the mechanism of disequilibrium is due to cerebellar compression by a space-occupying plasmacytoma, then (depending on the location of the lesion) we would anticipate either appendicular or midline cerebellar findings on physical examination. We would also anticipate cerebellar ocular motor findings on oculography, but no ocular motor data were presented in any of the published studies.

References

1. Brigle K, Rogers B. Pathobiology and diagnosis of multiple myeloma. Semin Oncol Nurs. 2017;33(3):225–36.
2. Joshi A, Jiang D, Singh P, Moffat D. Skull base presentation of multiple myeloma. Ear Nose Throat J. 2011;90(1):E6–9.
3. Rakul Nambiar K, Nair SG, Mathew SP. Vertigo and deafness: the sole presenting feature of multiple myeloma. J Egypt Natl Canc Inst. 2017;29(1):57–9.
4. Pika T, Bacovsky J, Vaverka M, Hrbek J, Hubacek J, Spurna D, Scudla V. Unusual manifestation of multiple myeloma: focal affection of central nervous system in a patient with chronic lymphocytic leukaemia. Biomed Pap Med Fac Univ Palacky Olomouc Czech Repub. 2009;153(4):271–3.
5. Tathineni P, Cancarevic I, Malik BH. Uncommon presentations of multiple myeloma. Cureus. 2020;12(6):e8400.

M. Cherchi, *Otoneurology and Vestibular Medicine*, https://doi.org/10.1007/978-3-031-94841-1_37

Part VI

Diagnoses—Diseases Involving the Middle Ear

Middle Ear Myoclonus (MEM)

38

Overview

Middle ear myoclonus (MEM) consists of involuntary rhythmic contractions of the muscle to the stapedius and/or the tensor tympani muscle. Most cases are unilateral and episodic. Most cases are sporadic, though sometimes can occur in the context of another illness, such as multiple sclerosis, facial nerve paralysis, or benign fasciculation syndrome. Binocular otomicroscopy or handheld otoscopy can sometimes reveal contraction of the tympanic membrane synchronous with the patient's perception of tinnitus. Impedance tympanometry may be able to document the abnormal contraction of the tympanic membrane. Treatment includes watchful waiting, physical (zygomatic pressure), audiologic (masking), pharmacologic (such as orphenadrine), by injection (onabotulinum toxin), and surgical (tenotomy of the muscle to the stapedius or the tensor tympani).

Introduction

Middle ear myoclonus (MEM) encompasses the phenomena of *tensor tympani myoclonus* and *stapedial myoclonus* [1].

The *muscle attached to the stapes* is innervated by a branch of the facial nerve. Contraction of the stapedius "pulls the stapes posteriorly with lateral displacement of the posterior tympanic membrane" [2]. In its protective reflex function, "the stapedius contracts in response to intense low frequency acoustic stimuli" [3].

The *tensor tympani muscle* is innervated by the mandibular branch of the trigeminal nerve. When the tensor tympani contracts, it "pulls the malleus medially and anteriorly with inward movement of the tympanic membrane" [2]. In its protective reflex function, "the tensor tympani is believed to contract in response to self-generated noise (chewing, swallowing) and non-auditory stimuli" [3].

In normal circumstances, these muscles mediate protective acoustic reflexes against loud sounds through "increase in the stapes-annular ligament impedance and decreased middle ear sound transmission" [2]. But if the tensor tympani or stapedius pathologically contracts in a rhythmic fashion due to myoclonus, then the ossicular chain will move (as if vibrations were entering the ear from the external environment) and be perceived as sound.

Some authors include within this group the phenomena of "tubal myoclonus" and "peri-tubal myoclonus" (such as tensor veli palatini myoclonus). So-called "tubal myoclonus" [4] actually refers to *palatal tremor*. In this condition, the rhythmic raising and lowering of the soft palate intermittently occludes the pharyngeal orifice of the Eustachian tube, which the patient may perceive as a clicking sound, and can sometimes also be auscultated by the examiner. The *tensor veli palatini muscle*, innervated by the mandibular nerve branch of the trigeminal nerve, mediates some motion of the soft palate. In tensor veli palatini myoclonus [1], unintended contraction of the tensor veli palatini may present with perception of a clicking sound, and this objective tinnitus can sometimes be auscultated by an examiner. In our view, these variants are probably more informatively classified as a "palatal tremor" variant of oculopalatal tremor.

Clinical Presentation

Tensor tympani myoclonus usually presents unilaterally, though rare cases of bilateral involvement have been reported [5]. Patients afflicted with tensor tympani myoclonus often perceive a low- to middle-pitched tinnitus and sometimes experience a "fluttering" sensation in the ear.

Stapedial myoclonus often presents as a "screeching" tinnitus, though in some cases [6], a patient may describe it as low-pitched.

Bento and colleagues [7] presented an unusual case in which middle ear myoclonus caused *continuous* tinnitus. In this case, the investigators simultaneously sectioned both the stapedius and the tensor tympani tendons, so the source of the tinnitus (one or both muscles) was unclear.

M. Cherchi, *Otoneurology and Vestibular Medicine*, https://doi.org/10.1007/978-3-031-94841-1_38

Most instances of tensor tympani myoclonus and stapedial myoclonus occur in isolation, but some cases appear in the context of some other illness. Zipfel and colleagues reported a case of bilateral tensor tympani and stapedial myoclonus occurring in the context of multiple sclerosis [8]. Watanabe and colleagues [9] reported a case of unilateral stapedial myoclonus occurring in the context of facial nerve paralysis. Brigo and colleagues [6] reported a case of unilateral stapedial myoclonus presenting in the context of "benign fasciculation syndrome." The tinnitus spontaneously resolved within 48 h.

Demographics

The epidemiology of middle ear myoclonus has not been well described. A tinnitus prevalence study observed that tensor tympani myoclonus is more common in patients with hyperacusis [10], which seems logical since the tensor tympani reflex serves a protective function against stimuli perceived to be excessively loud.

Pathophysiology

The neuroanatomical pathways mediating the stapedial reflex and tensor tympani reflex are depicted in Figs. 38.1 and 38.2 from Mukerji and colleagues [3].

Although these neuroanatomical pathways are fairly well delineated, most instances of middle ear myoclonus do not have any clear precipitant, and the localization of the underlying lesion or dysfunction is unknown.

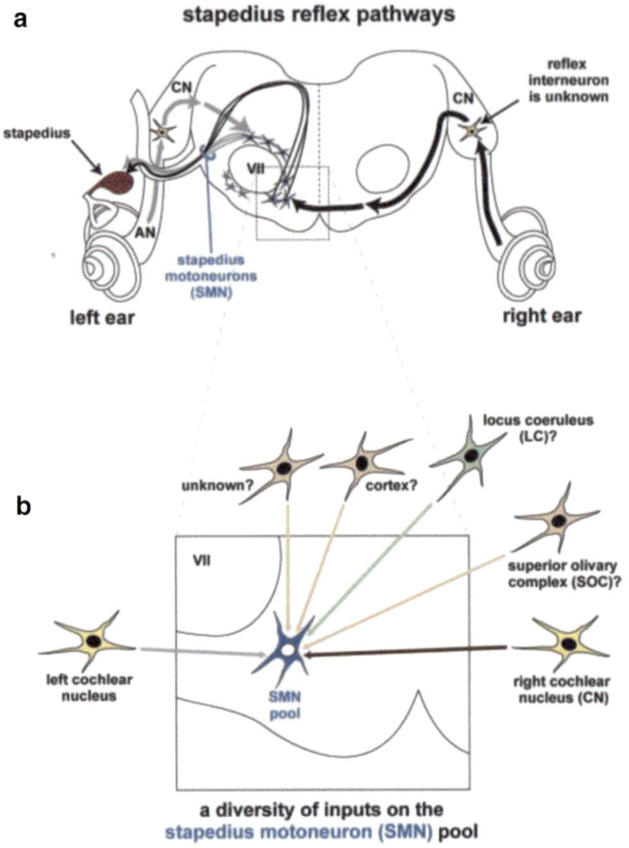

Fig. 38.1 Neuroanatomical pathway mediating the stapedial reflex. Panel A: The uncrossed stapedius reflex (gray arrows) to the left ear is activated by left ear sound and stimulates the subpopulation of stapedius motoneurons (SMNs) that contract the stapedius muscle of the same ear (left ear). The cochlear nucleus (CN) reflex interneuron of the uncrossed pathway is shown by the large yellow star on the left. It is not clear whether it projects directly to the SMNs or whether there is an intervening synapse. The crossed MEM reflex to the left ear (black arrows) is activated by right ear sound and stimulates the subpopulation of SMNs that contract the stapedius muscle of the left ear. The CN reflex interneuron of the crossed pathway is shown by the large yellow star on the right. The crossed reflex pathway ending with contraction of the right stapedius is not shown. Panel B: Magnified schematic showing the possible diversity of inputs on a representative SMN found ventromedial to the facial nucleus (VII). As shown in Panel A, SMNs respond to sound and likely receive input either directly or indirectly from both CN. Some patients can voluntarily contract their stapedius muscle and so these SMNs must also receive input from cortex. Transneuronal tracing using pseudo-rabies virus has shown labeling in the superior olivary complex (SOC), locus coeruleus (LC), and inferior colliculus (IC) after injection into the stapedius muscle in rats. It is unknown whether these regions contain neurons that project directly or indirectly to SMNs. (From Mukerji et al. [3]. Reproduced from Sage Publications under the STM signatory agreement)

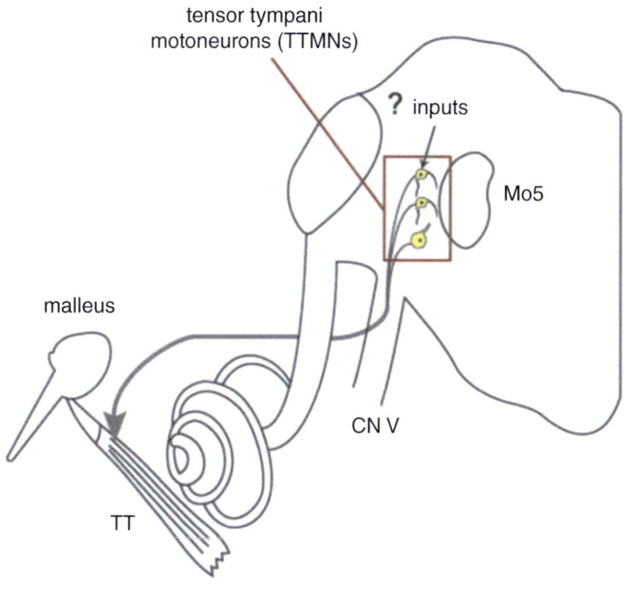

Fig. 38.2 Neuroanatomical pathways mediating the tensor tympani (TT) reflex. Note: The TT reflex to the left ear is activated by left ear sound and by neurons from elsewhere and stimulates the population of tensor tympani motoneurons (TTMNs) that contract the TT muscle of the same ear (left ear). Based on recent pseudorabies viral tracing studies of the TT reflex, the interneurons that project from the cochlear nucleus to TTMNs may be multipolar cells located in the anterior ventral cochlear nuclei (AVCN). From the cochlear nuclei, neurons project either directly or indirectly to TTMN pool in rats. Last, efferent fibers (gray arrow) exit the brainstem to innervate the TT via the "nerve to the TT," a branch of the mandibular division of the trigeminal nerve. A range of inputs from the superior olivary complex, serotoninergic sources and higher brain centers are thought to descend on TTMNs. The variety of inputs may account for the multifunctional roles of this muscle in response to auditory and nonauditory stimuli. The crossed pathway by which the MEM reflex to the left ear is activated by right ear sound is not shown. Abbreviations: Mo5 trigeminal motor nucleus, TT tensor tympani, CN V trigeminal nerve, TTMN tensor tympani motoneurons. (From Mukerji et al. [3]. Reproduced from Sage Publications under the STM signatory agreement)

Physical Examination

In some cases of tensor tympani myoclonus, binocular otomicroscopy or handheld otoscopy may show rhythmic movements of the tympanic membrane synchronous with the patient's report of perceived sound [5, 11].

Testing

In some cases of tensor tympani myoclonus, impedance tympanometry may detect cyclical spikes in eardrum compliance [11].

Figure 38.3 compares tympanometry of the normal left ear, with tympanometry of the abnormal right ear (affected by middle ear myoclonus).

Figure 38.4 shows impedance tympanometry of a patient with middle ear myoclonus (from Zipfel et al. [8]). In this case, it was unclear whether the etiology was tensor tympani myoclonus or stapedial myoclonus.

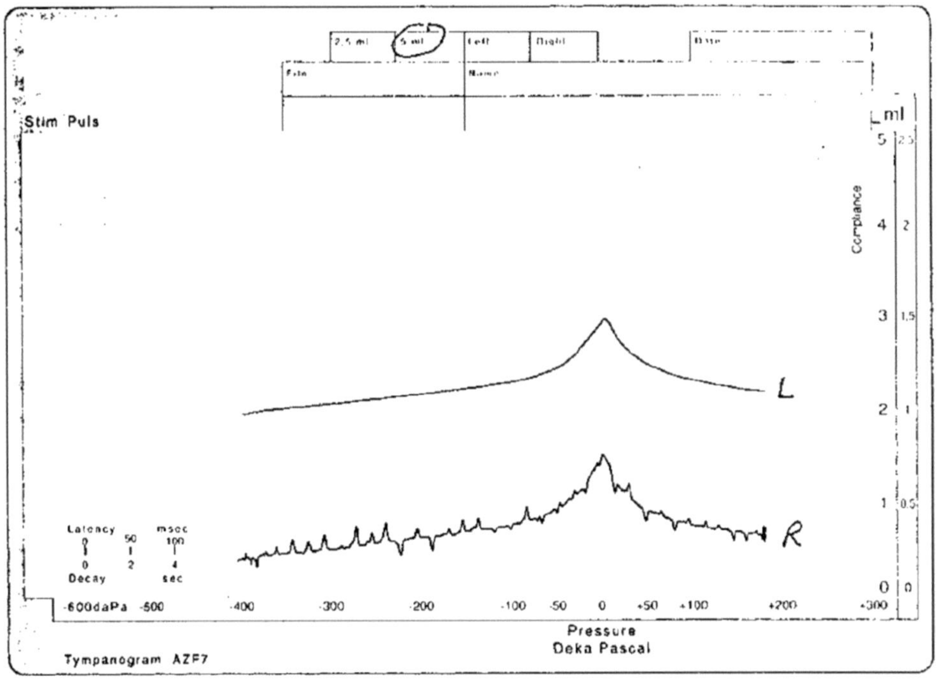

Fig. 38.3 Tympanogram showing right ear compliance changes in a normal ear (upper tracing) and in an ear affected by middle ear myoclonus (lower tracing), from Abdul-Baqi [11]. (Reproduced with written permission from Cambridge University Press)

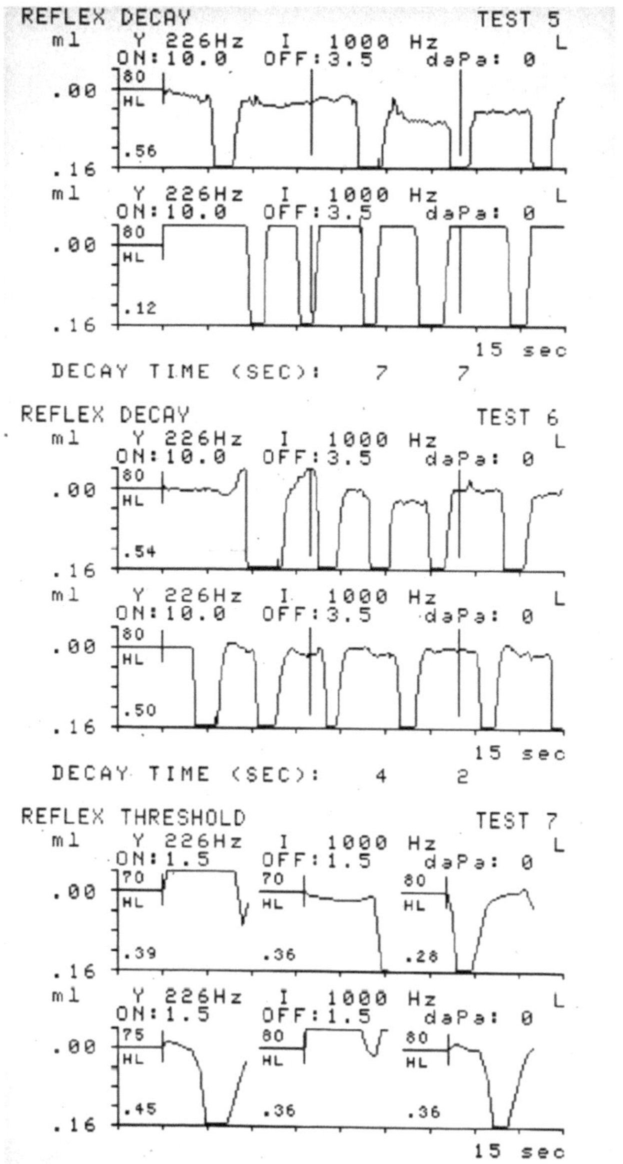

Fig. 38.4 Impedance tympanometry of a patient with middle ear myoclonus. In this case it was unclear whether the etiology was tensor tympani myoclonus or stapedial myoclonus. (From Zipfel et al. [8]. Reproduced with written permission from Cambridge University Press)

Treatment

There are no systematic treatment trials for middle ear myoclonus syndromes. Data are derived from case reports and case series.

Treatment, Physical

Chan and Palaniappan [12] reported two cases of middle ear myoclonus that improved with "non-invasive, self-administered zygomatic pressure."

Treatment, Audiologic

Somewhat surprisingly, East and Hazell [13] report a case series of patients in whom the tinnitus from palatal myoclonus responded to wide band masking.

Treatment, Pharmacologic

Since middle ear myoclonus syndromes are fundamentally myoclonic, some cases may respond to the use of muscle relaxants such as orphenadrine [11].

Treatment, Surgical

Stapedial myoclonus has been treated with injection of onabotulinum toxin [14].

Some cases of tensor tympani or stapedial myoclonus may be treated surgically, typically by resection (tenotomy) of the tendons of the tensor tympani muscle or the muscle to the stapedius [8, 15]. When it is known with confidence that the tinnitus is due to myoclonus of a specific muscle, it is desirable to apply this procedure only to that involved muscle while preserving the unaffected muscle [5, 16]. While these procedures may cure the tinnitus, they also ablate the middle ear protective response that those intact muscles would mediate, leaving the patient more vulnerable to auditory trauma.

References

1. Ellenstein A, Yusuf N, Hallett M. Middle ear myoclonus: two informative cases and a systematic discussion of myogenic tinnitus. Tremor Other Hyperkinet Mov. 2013;3:tre-03.
2. Bhimrao SK, Masterson L, Baguley D. Systematic review of management strategies for middle ear myoclonus. Otolaryngol Head Neck Surg. 2012;146(5):698–706.
3. Mukerji S, Windsor AM, Lee DJ. Auditory brainstem circuits that mediate the middle ear muscle reflex. Trends Amplif. 2010;14(3):170–91.

4. Virtanen H. Objective tubal tinnitus: a report of two cases. J Laryngol Otol. 1983;97(9):857–62.
5. Cohen D, Perez R. Bilateral myoclonus of the tensor tympani: a case report. Otolaryngol Head Neck Surg. 2003;128(3):441.
6. Brigo F, Storti M, Lochner P, Nardone R. Transitory stapedial myoclonus in a patient with benign fasciculation syndrome. J Laryngol Otol. 2013;127(6):605–6.
7. Bento RF, Sanchez TG, Miniti A, Tedesco-Marchesi AJ. Continuous, high-frequency objective tinnitus caused by middle ear myoclonus. Ear Nose Throat J. 1998;77(10):814–8.
8. Zipfel TE, Kaza SR, Greene JS. Middle-ear myoclonus. J Laryngol Otol. 2000;114(3):207–9.
9. Watanabe I, Kumagami H, Tsuda Y. Tinnitus due to abnormal contraction of stapedial muscle. An abnormal phenomenon in the course of facial nerve paralysis and its audiological significance. ORL J Otorhinolaryngol Relat Spec. 1974;36(4):217–26.
10. Westcott M, Sanchez TG, Diges I, Saba C, Dineen R, McNeill C, et al. Tonic tensor tympani syndrome in tinnitus and hyperacusis patients: a multi-clinic prevalence study. Noise Health. 2013;15(63):117–28.
11. Abdul-Baqi KJ. Objective high-frequency tinnitus of middle-ear myoclonus. J Laryngol Otol. 2004;118(3):231–3.
12. Chan C, Palaniappan R. Middle ear myoclonus: a new technique for suppression of spontaneous clicking tinnitus. Int Tinnitus J. 2010;16(1):51–4.
13. East CA, Hazell JW. The suppression of palatal (or intra-tympanic) myoclonus by tinnitus masking devices. A preliminary report. J Laryngol Otol. 1987;101(12):1230–4.
14. Liu HB, Fan JP, Lin SZ, Zhao SW, Lin Z. Botox transient treatment of tinnitus due to stapedius myoclonus: case report. Clin Neurol Neurosurg. 2011;113(1):57–8.
15. Badia L, Parikh A, Brookes GB. Management of middle ear myoclonus. J Laryngol Otol. 1994;108(5):380–2.
16. Hidaka H, Honkura Y, Ota J, Gorai S, Kawase T, Kobayashi T. Middle ear myoclonus cured by selective tenotomy of the tensor tympani: strategies for targeted intervention for middle ear muscles. Otol Neurotol. 2013;34(9):1552–8.

Eustachian Tube Dysfunction (ETD)

39

Overview

The *pharyngotympanic tube*, more commonly referred to as the *Eustachian tube (ET)*, connects the middle ear with the nasopharynx. In the resting state, the ET remains closed; it opens during co-contraction of the tensor veli palatini and the levator palatini muscles, which occurs during deglutition and yawning. Proper functioning of the ET serves to equalize pressure in the middle ear with the outside environment, to clear mucociliary secretions and pathogens from the middle ear, and to protect the middle ear and tympanic membrane. ET dysfunction (ETD) is sometimes classified into subtypes of (1) obstructive, (2) baro-challenge-induced, and (3) patulous. ETD can result from anatomical or functional abnormalities. Sequelae of ETD include middle ear effusions, infections, middle ear atelectasis, tympanic membrane retraction, development of cholesteatoma, and conductive hearing loss. There is no gold-standard test for ETD, so diagnosis usually relies on a combination of history, physical examination and testing (such as tympanometry), less commonly with procedures such as imaging or fiberoptic endoscopy. Treatments may include decongestants (oral or nasal), myringotomy/tympanostomy with or without placement of a grommet, Eustachian tuboplasty (with balloon dilatation or laser), and others, though none of these approaches is firmly supported by evidence-based studies, so prognosis is uncertain.

Introduction

The *pharyngotympanic tube* connecting the middle ear space with the nasopharynx is more commonly referred to as the *Eustachian tube* (ET), after Eustachio Bartolomeo who published a description of it in his anatomical treatise, *Opuscula Anatomica* in 1562 [1].

Anatomy

The Eustachian tube (ET) is situated in the para-pharyngeal region, proximal to the infratemporal fossa. It "acts as a void in fibroelastic cartilage" that "runs from the middle ear's front wall to the sidewall of the nasopharynx" "along the pterygoid plate's posterior edge" [2]. The ET consists of a cartilaginous component (suspended from the skull base) that comprises about two-third of the length of the ET, and an osseous component that comprises about one-third of its length [3]. The Eustachian tube is oriented more horizontally in children than in adults [4, 5].

Demographics

The prevalence of Eustachian tube dysfunction (ETD) is estimated at around 4–20% in the pediatric population. It is less common in adults; "there are an average of 0.77 adult visits for ETD for every pediatric visit" [3]. The higher prevalence of ETD in children is sometimes attributed to the relatively more horizontal orientation of the ET, one result of which is that gravity facilitates drainage less (Fig. 39.1).

Factors in the pediatric population that can predispose to ETD include obstructive sleep apnea and congenital malformations such as cleft palate [3].

M. Cherchi, *Otoneurology and Vestibular Medicine*, https://doi.org/10.1007/978-3-031-94841-1_39

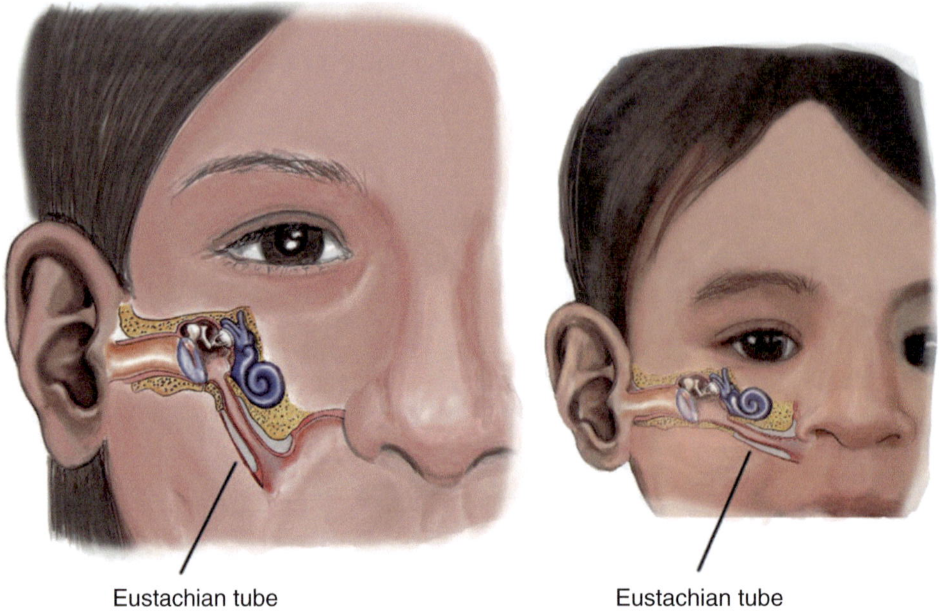

Fig. 39.1 Comparison of normal Eustachian tube anatomy in a healthy adult and in a healthy infant. The Eustachian tube courses more vertically in an adult, allowing gravity to aid the drainage of fluid from the middle ear. (From Vincent et al. [4]. Reproduced with written permission from Springer Nature B.V.)

Physiology

The Eustachian tube "has two primary functions: balancing the pressure between the middle ear and the surroundings, and helping facilitate secretions from the middle ear, as well as protecting the ear from loud and potential sound hazards." Other summaries regard the ET as serving three functions [2, 6]:

- Pressure equalization and ventilation of the middle ear
- Mucociliary clearance of secretions from the middle ear
- Protection of the middle ear and eardrum from sounds, pathogens and secretions

The ET is lined by a "ciliated epithelium that enables mucociliary clearance of middle ear fluid while providing a barrier to the entry of pathogens from the nasopharynx" [3].

The tensor veli palatini muscle, and the levator veli palatini muscle, when simultaneously contracted, dilates the Eustachian tube [2, 3]. At rest, the ET is collapsed; it "opens involuntarily during swallowing, but can also voluntarily open through swallowing or yawning" [3].

Pathophysiology

A variety of processes can interfere with the function of the Eustachian tube, including [2]:

- Anatomical abnormalities, such as:
  - Reduced diameter (stenosis) due to inflammation or wall hypertrophy
  - Enlarged diameter (patulous Eustachian tube) due to slacking of the cartilage around the tube or the tissues
- Elevated mucus production
- Desquamation of the epithelium of the Eustachian tube

There have been various proposals for classifying ET dysfunction (ETD), such as [2, 3, 6]:

- *Obstructive ETD*. Often, this is because the ET is "unable to effectively dilate. The most common reason for the impairment of ET dilation is the inflammation of the mucosa of the cartilaginous ET" from upper respiratory tract infections, reflux, or allergy [3]. Some investigators postulate three subcategories of obstructive ETD [6]:
  - Functional obstruction
  - Dynamic dysfunction (muscular failure)
  - Anatomical obstruction
- *Baro-challenge-induced ETD*, "when there is an ambient pressure change in the environment" [2]. Common scenarios include "deep-sea diving or descent from high altitudes" [3].
- *Patulous (dilatory) ETD*, "characterized by an excessively patent ET that does not sufficiently constrict" [3]. This is less common than obstructive ETD, and "the majority of cases of patulous ETD do not have a clear etiology" [3].

Some common pathophysiologic consequences of ETD include middle ear effusions and infections, atelectasis of the middle ear with tympanic membrane retraction, tympanic membrane damage, development of cholesteatoma, and conductive hearing loss [2, 3].

Clinical History

While each subtype of ETD may have its own characteristics, in general symptoms suggesting ETD include, "pressure in the ear, aural fullness; popping, pain, or discomfort in the ear; muffled hearing; and autophony" [3].

Regarding the characteristics of the various subtypes [3, 6]:

- *Obstructive ETD*: "Can present with aural fullness, pain, altered or muffled hearing, and tinnitus."
- *Baro-challenge-induced ETD*: "Patients may only experience the aural symptoms [aural fullness, pain, muffled hearing] during situations of rapid pressure changes, especially with exposure to high pressures."
- *Patulous ETD*: "can present with positional autophony, audible breathing, pulsatile tinnitus that synchronizes with breathing, a sense of pressure in the ear, and hearing loss."

Diagnosis: By History

Questionnaire-based assessments, such as the Eustachian Tube Dysfunction Questionnaire 7 (ETDF-7), have also been explored [7]. This questionnaire involves summation scoring of questions pertaining to seven parameters of "pressure in ears," "pain in ears," "clicking sound on swallowing," "underwater/clogged sensation," "ringing in ears," "muffled hearing," and "pre-occurring cold/sinusitis," all reported by the patient.

There have also been attempts at developing scoring systems that include both subjective categories and objective testing (from tubomanometry or tympanometry), such as the Eustachian Tube Scoring (ETS-7) [8].

Diagnosis: Physical Examination

Various physical examination maneuvers, such as the *Valsalva maneuver* and the *Toynbee test* (named after Joseph Toynbee, 1815–1866 [9]), have been studied but are not reliable. The Valsalva maneuver involves attempting to exhale against a closed mouth and nose in order to drive air distally through the ET into the middle ear; if during this maneuver the examiner observes movement of the tympanic membrane, then the test is considered "positive." The Toynbee test has the patient swallow while the nostrils are pinched closed; if during this maneuver, the examiner observes movement of the tympanic membrane, then the test is considered "positive." Unfortunately, these are "non-dependable confirmatory diagnostic test[s]" [7].

Diagnosis: Otologic Testing

There is a long history of intermittent enthusiasm for tympanometry in the diagnosis of ETD [10], but this has not been universally accepted, as some studies show it to have very limited diagnostic utility [11].

More technologically elaborate diagnostic methods have included video nasopharyngoscopy [12].

Diagnosis: Imaging

Imaging modalities of CT and MRI have been explored for diagnosing ETD [5], but none has proven reliably diagnostic [3]. Smith and colleagues [5] provide a good example of an MRI of the Eustachian tube of a healthy volunteer at rest, and during Valsalva, demonstrating that Valsalva opens the Eustachian tube. They also provide an example of a CT demonstrating a patulous Eustachian tube.

Diagnosis: Fiberoptic Endoscopy

One study reported video nasopharyngoscopy as a "highly accurate and reliable test for Eustachian tube function as compared to tympanometry" [12].

Diagnosis: Overall

Multiple authors have noted that there does not yet exist a "gold-standard" test for proof-positive diagnosis of ETD [2, 3, 13] and that "more work is required to validate… Eustachian tube function tests" [14].

Absent a definitive test, diagnosis has traditionally relied upon a combination of history, physical examination, otoscopy, and sometimes tympanometry [2].

Treatment

Despite diagnostic uncertainty, various treatments have been attempted.

The most common initial approach usually involves *oral decongestants* or *nasal spray decongestants*. Evidence of efficacy is weak or lacking [15, 16], but the risk is low, so this is a first approach for many clinicians.

Myringotomy/tympanostomy involves making a small incision in the tympanic membrane. Because such incisions usually heal shut relatively rapidly, the otolaryngologist may place a *grommet*, also called a "tympanostomy tube" or "ventilation tube." A meta-analysis reported favorable outcomes [16], though some find the effect to be minimal [17] or zero [18].

Eustachian tuboplasty, which may involve "point laser coagulation of superior and posterior margin of ET nasopharyngeal opening." A meta-analysis of 7 case series (192 patients) of Eustachian laser tuboplasty reported, "Eustachian tuboplasty using various techniques (seven case series, 182 patients) was associated with improvement in symptoms in 36–92% of patients. Improvements in hearing (four studies) were small with limited clinical significance. Three studies documented low rates (13–36%) of conversion to type A tympanogram" [15].

Balloon Eustachian tuboplasty. This involves passing a catheter with a balloon at its end which, when inflated, dilates the ET. A meta-analysis reported favorable outcomes [16].

Microwave ablation. Preliminary studies have explored microwave ablation Eustachian tuboplasty [19], which targets "the ET orifice's hypertrophic tissue" [2].

Research on Eustachian tube *stenting* is conducted in animal models, but the risk of a stent is that it can convert an obstructive dysfunction ET into a patulous dysfunction ET.

For the most part, meta-analyses of interventions for ETD reach disappointing conclusions such as, "It is not possible to draw conclusions regarding the effectiveness of any of the interventions for the treatment of adults with an ETD diagnosis, and there is insufficient evidence to recommend a trial of any particular intervention" [16], and "Eustachian tube dysfunction is a poorly defined condition. Due to the limited and poor-quality evidence, it is inappropriate to make conclusions on the effectiveness of any intervention; the evidence base is insufficient to guide recommendations for a trial of any particular intervention" [15].

Prognosis

A meta-analysis of outcomes from treatment techniques [16] noted that the overall quality of evidence was "limited in quantity and overall was of poor quality," but suggested that:

- "Based on a single RCT, nasal steroids showed no improvement."
- "Single trials fount two pressure equalization devices were each associated with significant short-term improvements in symptoms, middle ear function and/or hearing."
- Eustachian tuboplasty (seven case series) and balloon dilatation (three case series) were associated with improved outcomes."
- "Positive results were also reported for myringotomy (two case series)."

References

1. Eustachio B. Epistola de auditus organis [Letter concerning the hearing organ]. Opuscula Anatomica: Lugduni Batavorum, Apud Johannem vander Linden; 1562. p. 125–42.
2. Bal R, Deshmukh P. Management of Eustachian tube dysfunction: a review. Cureus. 2022;14(11):e31432.
3. Maddineni S, Ahmad I. Updates in Eustachian tube dysfunction. Otolaryngol Clin N Am. 2022;55(6):1151–64.
4. Vincent AG, Rogers DJ, Javia L. Audiology, otology, and airway. In: Swanson JW, editor. Global cleft care in low-resource settings. Cham: Springer International Publishing; 2021. p. 419–32.
5. Smith ME, Scoffings DJ, Tysome JR. Imaging of the Eustachian tube and its function: a systematic review. Neuroradiology. 2016;58(6):543–56.
6. Schilder AG, Bhutta MF, Butler CC, Holy C, Levine LH, Kvaerner KJ, et al. Eustachian tube dysfunction: consensus statement on definition, types, clinical presentation and diagnosis. Clin Otolaryngol. 2015;40(5):407–11.
7. McCoul ED, Anand VK, Christos PJ. Validating the clinical assessment of eustachian tube dysfunction: the Eustachian Tube Dysfunction Questionnaire (ETDQ-7). Laryngoscope. 2012;122(5):1137–41.
8. Schroder S, Lehmann M, Sauzet O, Ebmeyer J, Sudhoff H. A novel diagnostic tool for chronic obstructive eustachian tube dysfunction-the eustachian tube score. Laryngoscope. 2015;125(3):703–8.
9. Mudry A. The making of a career: Joseph Toynbee's first steps in otology. J Laryngol Otol. 2012;126(1):2–7.
10. Givens GD, Seidemann MF. Acoustic immittance testing of the Eustachian tube. Ear Hear. 1984;5(5):297–9.
11. Virtanen H, Marttila T. Middle-ear pressure and eustachian tube function. Arch Otolaryngol. 1982;108(12):766–8.
12. Chauhan B, Chauhan K. A comparative study of eustachian tube functions in normal and diseased ears with tympanometry and videonasopharyngoscopy. Indian J Otolaryngol Head Neck Surg. 2013;65(Suppl 3):468–76.
13. Smith ME, Tysome JR. Tests of Eustachian tube function: a review. Clin Otolaryngol. 2015;40(4):300–11.
14. Smith ME, Bance ML, Tysome JR. Advances in Eustachian tube function testing. World J Otorhinolaryngol Head Neck Surg. 2019;5(3):131–6.
15. Norman G, Llewellyn A, Harden M, Coatesworth A, Kimberling D, Schilder A, et al. Systematic review of the limited evidence base for treatments of Eustachian tube dysfunction: a health technology assessment. Clin Otolaryngol. 2014;39(1):6–21.
16. Llewellyn A, Norman G, Harden M, Coatesworth A, Kimberling D, Schilder A, et al. Interventions for adult Eustachian tube dysfunction: a systematic review. Health Technol Assess. 2014;18(46):1–180. v-vi.
17. Lous J, Burton MJ, Felding JU, Ovesen T, Rovers MM, Williamson I. Grommets (ventilation tubes) for hearing loss associated with otitis media with effusion in children. Cochrane Database Syst Rev. 2005;1:CD001801.
18. Knight LC, Hilger A. The effects of grommet insertion on Eustachian tube function. Clin Otolaryngol Allied Sci. 1993;18(6):459–61.
19. Lou Z, Lou Z, Sun J, Chen Z, Yin S. Microwave ablation eustachian tuboplasty: a preliminary investigation with long-term follow-up. J Otolaryngol Head Neck Surg. 2021;50(1):39.

Middle Ear Effusion, Otitis Media

40

Overview

The middle ear space can be filled with a (noninflammatory, noninfectious) effusion; it can be affected by inflammation (with or without effusion) or infection. Typical symptoms include aural fullness and pain, and possibly hearing loss and disequilibrium. This condition is more frequent in children, probably because of the anatomical orientation of the Eustachian tube. Physical examination under otomicroscopy or otoscopy may show a fluid meniscus behind the tympanic membrane; in some cases, there may be nystagmus, though the patterns of nystagmus are neither specific nor sensitive for these middle ear conditions. The hearing loss, if assessed by audiometry, is usually conductive in nature. Some cases may have detectable abnormalities on otovestibular testing (vestibular evoked myogenic potentials, video head impulse testing, ocular motor testing, caloric testing, rotatory chair testing, posturography). Imaging is usually not necessary with an initial episode. If there is reason to suspect infection, then empiric treatment with antibiotics is reasonable. If that fails, then referral to otolaryngology is appropriate to be evaluated for myringotomy or placement of a grommet.

Terminology

The phrase "*middle ear effusion*" refers to fluid in the middle ear. If the fluid is neither inflammatory nor infectious, then it is usually referred to as a "*serous middle ear effusion*."

The phrase "*otitis media*" refers to an inflammatory process in the middle ear space, irrespective of whether any effusion is also present. If fluid is also present but no infection is involved, then it is referred to as "*serous otitis media*." If otitis media involves infection, then it is designated according to the pathogen (e.g., "bacterial otitis media," "tuberculous otitis media," etc.).

Epidemiology

Middle ear effusion and inflammation are more common in children than in adults, though some studies have also been conducted on adults [1–5].

Risk factors for middle ear effusion and inflammation include Eustachian tube dysfunction.

Pathophysiology

Effusion, inflammation, or infection in the middle ear space has the potential to provoke aural symptoms (pain, fullness), auditory symptoms (hearing loss, tinnitus), and vestibular symptoms (disequilibrium).

Presentation: Hearing Loss

Hearing loss is fairly common in cases of otitis media [2, 3, 6–8]. The associated hearing loss is usually conductive and is due to mechanical interference with the motion of the ossicular chain.

Figure 40.1, from Zaidi and Sinha [9], shows bilateral conductive hearing loss in a case of chronic bilateral suppurative otitis media.

M. Cherchi, *Otoneurology and Vestibular Medicine*, https://doi.org/10.1007/978-3-031-94841-1_40

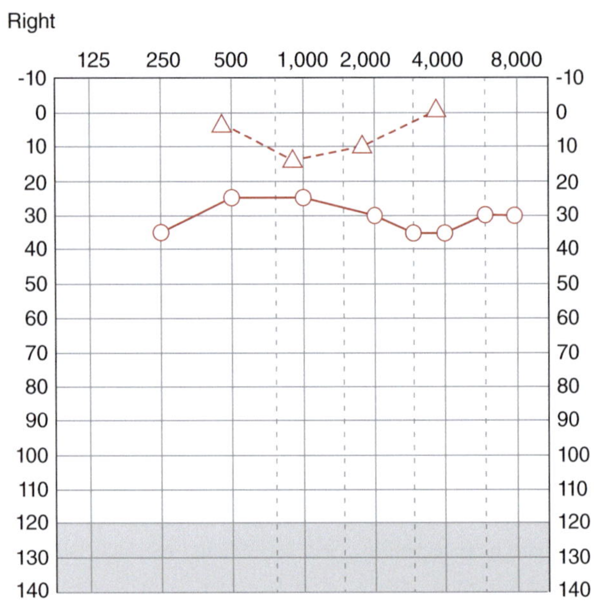

Fig. 40.1 Bilateral conductive hearing loss in a case of bilateral chronic suppurative otitis media. (From Zaidi and Sinha [9]. Reproduced with written permission from Springer Nature B.V.)

Presentation: Disequilibrium

Vestibular symptoms are less common than auditory ones in middle ear effusions and otitis media but do occur [1–3, 5–8, 10–14].

It is certainly possible for otitis media to develop into labyrinthitis [6, 7, 9, 15, 16]. In such cases, presumably pathogens pass from the middle ear to the inner ear by traversing the oval and/or round windows. Labyrinthitis has been well studied, and it is unsurprising that such patients can have vestibular symptoms.

But there are other cases of middle ear effusions and otitis media that do not appear to evolve into labyrinthitis, yet are still associated with vestibular symptoms. In such cases, the mechanism by which the middle ear process provokes vestibular symptoms is unclear. Temporal bone studies of patients who in life had chronic otitis media show reduced density of type I and type II vestibular hair cells and of dark cells [17], suggesting that even if these patients do not clinically present with a discrete episode suggestive of labyrinthitis, the chronic otitis media still provokes pathology in the labyrinth.

Physical Examination: Otomicroscopy/ Otoscopy

A binocular otomicroscopic examination, or at least a monocular otoscopic examination, is important in these patients. A middle ear effusion is often (though not always) visible as a fluid meniscus behind the tympanic membrane [18].

Physical Examination: Nystagmus

Some investigators report nystagmus of various patterns in cases of acute otitis media [3, 7, 15], including direction changing positional nystagmus, and direction-fixed nystagmus (paretic or irritative). One series of patients with acute otitis media *without vestibular complaints* reported nystagmus in 82% (28 out of 34) of patients; we find this statistic impressive and far greater than in our own experience.

Testing

Audiometry often shows hearing loss (usually conductive) in the affected ear [5, 9, 19]. *Tympanometry* often shows reduced compliance (peak immittance).

Vestibular evoked myogenic potentials are sometimes abnormal [4] in a middle ear with effusion or inflammation, but whether this is an artifact of conductive hearing loss or reflects true vestibular dysfunction is not always clear. In this regard, some studies are careful to compare air-conducted and bone-conducted vestibular evoked myogenic potentials [20].

Some patients with middle ear effusion or inflammation exhibit abnormalities on *video head impulse testing*, usually in the form of the presence of corrective saccades, or reduced gain [21]. Comert and colleagues [22] more specifically observed covert compensatory saccades in 32% of the patients and reduced gain in the anterior canal on the affected side.

Several studies have explored ocular motor function with *electronystagmography* or *videonystagmography* in patients with middle ear effusion and/or inflammation. Eliashar and colleagues reported an "abnormal electronystagmogram" in all patients who had vestibular complaints; these "abnormalities" included spontaneous nystagmus and direction changing nystagmus. Gianoli and colleagues reported unilateral or bilateral caloric weakness in 19 (76%) out of 25 patients with chronic suppurative otitis media [10]. Golz and colleagues [23] studied 136 patients with middle ear effusion and reported abnormal electronystagmographic findings in 58% of patients, including 42 (31%) who exhibited spontaneous unidirectional nystagmus and 24 (17.5%) who exhibited positional nystagmus. Lee and colleagues [24] studied 25 patients with unilateral chronic otitis media and reported that 20 (80%) showed some ocular motor abnormality, with 5 (20%) exhibiting unilateral caloric weakness, 12 (48%) exhibiting head shaking nystagmus, and 7 (28%) exhibiting vibration induced nystagmus. Mostafa and colleagues [4] studied 58 adults (mean age 29.5 years) and found 62% had a caloric weakness. Saliba and colleagues [14] studied 15 children with chronic middle ear effusion and found 2 (13%) had abnormal ocular motor findings, 1 with positional nystagmus, and 1 with both spontaneous and positional nystagmus.

Mostafa and colleagues [4] studied 58 adults (mean age 29.5 years) and found 70% of patients *rotatory chair* abnormalities.

Of other vestibular tests, Lee and colleagues [24] studied 25 patients with unilateral chronic otitis media and reported that 5 (20%) had abnormal *subjective visual vertical* (suggesting otolith dysfunction).

Some studies report abnormal findings on static or dynamic *posturography* [12, 21, 25].

Interestingly, of patients with abnormal vestibular test findings, *not all complain of disequilibrium.* For example, Lee and colleagues [24] studied 25 patients with unilateral chronic otitis media and reported that 20 (80%) showed some ocular motor abnormality, yet *none of these patients complained of dizziness or vertigo.* Kim and colleagues [3] used "video Frenzel glasses" to study 34 adult patients with acute otitis media and reported that 28 (82%) exhibited either direction changing positional nystagmus, direction-fixed irritative nystagmus, or direction-fixed paretic nystagmus, yet *none of these patients complained of dizziness or vertigo.* These findings suggest that middle ear effusion and inflammation can cause subclinical vestibular deficits.

Imaging

Imaging is usually not warranted with an initial presentation of middle ear effusion/inflammation/infection. In a patient with recurrent bouts, imaging may be appropriate, though at that point, the patient should probably have already been referred to otolaryngology. Otolaryngologists in that circumstance may consider an MRI of the skull base without and with contrast to evaluate for obstructive causes of Eustachian tube dysfunction, mastoiditis, or an abscess.

Treatment: Medical

Many cases of middle ear effusion resolve spontaneously.

If a middle ear effusion develops a superimposed infection, then antibiotic therapy is appropriate. In a patient without penicillin allergy, a course of amoxicillin or amoxicillin-clavulanate is reasonable. In a patient with a penicillin allergy, consider a course of cefdinir, cefuroxime, or cefpodoxime. If those approaches fail, then (whether or not the patient has a penicillin allergy), consider a course of ceftriaxone. If antibiotic treatment is failing, then the clinician should have a low threshold for referral to otolaryngology.

Treatment: Surgical

For patients with infectious otitis media who have failed empiric courses of antibiotics, referral to otolaryngology is appropriate. Commonly attempted procedures in this circumstance include myringotomy [7] or placement of a grommet [11, 23, 26]. These case series report significant improvement in auditory and vestibular symptoms.

References

1. Aarhus L, Tambs K, Hoffman HJ, Engdahl B. Childhood otitis media is associated with dizziness in adulthood: the HUNT cohort study. Eur Arch Otorrinolaringol. 2016;273(8):2047–54.
2. Kalaycik Ertugay C, Kulekci S, Naiboglu B, Ertugay OC, Kaya KS, Sheidaei S, et al. Clinical evaluation of inner ear changes as result of chronic otitis media and its surgery. Kulak Burun Bogaz Ihtis Derg. 2015;25(1):22–7.
3. Kim CH, Lee J, Choi B, Shin JE. Nystagmus in adult patients with acute otitis media or otitis media with effusion without dizziness. PLoS One. 2021;16(5):e0250357.
4. Mostafa BE, Shafik AG, El Makhzangy AM, Taha H, Abdel Mageed HM. Evaluation of vestibular function in patients with chronic suppurative otitis media. ORL J Otorhinolaryngol Relat Spec. 2013;75(6):357–60.
5. Tailor BV, Phillips JS, Nunney I, Yung MW, Doruk C, Kara H, et al. Presentation of dizziness in individuals with chronic otitis media: data from the multinational collaborative COMQ-12 study. Eur Arch Otorrinolaringol. 2022;279(6):2857–63.
6. Hyden D, Akerlind B, Peebo M. Inner ear and facial nerve complications of acute otitis media with focus on bacteriology and virology. Acta Otolaryngol. 2006;126(5):460–6.
7. Kim CH, Yang YS, Im D, Shin JE. Nystagmus in patients with unilateral acute otitis media complicated by serous labyrinthitis. Acta Otolaryngol. 2016;136(6):559–63.

8. Leskinen K. Complications of acute otitis media in children. Curr Allergy Asthma Rep. 2005;5(4):308–12.
9. Zaidi SH, Sinha A. Chronic suppurative otitis media (CSOM). In: Zaidi SH, Sinha A, editors. Vertigo: a clinical guide. Berlin, Heidelberg: Springer Berlin Heidelberg; 2013. p. 113–8.
10. Gianoli GJ, Soileau JS. Chronic suppurative otitis media, caloric testing, and rotational chair testing. Otol Neurotol. 2008;29(1):13–5.
11. Grace AR, Pfleiderer AG. Dysequilibrium and otitis media with effusion: what is the association? J Laryngol Otol. 1990;104(9):682–4.
12. Monsanto RDC, Kasemodel ALP, Tomaz A, Elias TGA, Paparella MM, Penido NO. Evaluation of vestibular symptoms and postural balance control in patients with chronic otitis media. J Vestib Res. 2020;30(1):35–45.
13. Monsanto RDC, Kasemodel ALP, Tomaz A, Paparella MM, Penido NO. Current evidence of peripheral vestibular symptoms secondary to otitis media. Ann Med. 2018;50(5):391–401.
14. Saliba I, Dagher C, El-Zir E, Yammine FG. A pilot study to assess the vestibular apparatus function with videonystagmography during chronic otitis media with effusion. Curr Pediatr Rev. 2015;11(2):135–40.
15. Choi JW, Han K, Nahm H, Shin JE, Kim CH. Direction-changing positional nystagmus in acute otitis media complicated by serous Labyrinthitis: new insights into positional nystagmus. Otol Neurotol. 2019;40(4):e393–e8.
16. Kitsko DJ, Dohar JE. Inner ear and facial nerve complications of acute otitis media, including vertigo. Curr Allergy Asthma Rep. 2007;7(6):444–50.
17. da Costa MR, Erdil M, Pauna HF, Kwon G, Schachern PA, Tsuprun V, et al. Pathologic changes of the peripheral vestibular system secondary to chronic otitis media. Otolaryngol Head Neck Surg. 2016;155(3):494–500.
18. Bhutta MF, Thornton RB, Kirkham LS, Kerschner JE, Cheeseman MT. Understanding the aetiology and resolution of chronic otitis media from animal and human studies. Dis Model Mech. 2017;10(11):1289–300.
19. Eliashar R, Gross M, Saah D, Elidan J. Vestibular involvement in myringitis bullosa. Acta Otolaryngol. 2004;124(3):249–52.
20. Kolkaila EA, Emara AA, Gabr TA. Vestibular evaluation in children with otitis media with effusion. J Laryngol Otol. 2015;129(4):326–36.
21. Abdelmotaleb H, Sobhy O, Bassiouny M, Elsherif M. Evaluation of postural stability and vestibulo-ocular reflex in adults with chronic suppurative otitis media. Eur Arch Otorrinolaringol. 2023;280(2):897–905.
22. Comert E, Sencan Z, Kocak FM, Simsek G, Muluk NB. Clinical evaluation of the vestibular impairment using video head impulse test in children with acute otitis media. Int J Pediatr Otorhinolaryngol. 2021;141:110568.
23. Golz A, Netzer A, Angel-Yeger B, Westerman ST, Gilbert LM, Joachims HZ. Effects of middle ear effusion on the vestibular system in children. Otolaryngol Head Neck Surg. 1998;119(6):695–9.
24. Lee IS, Park HJ, Shin JE, Jeong YS, Kwak HB, Lee YJ. Results of air caloric and other vestibular tests in patients with chronic otitis media. Clin Exp Otorhinolaryngol. 2009;2(3):145–50.
25. Gawron W, Pospiech L, Orendorz-Fraczkowska K. An evaluation of postural stability and the effects of middle-ear drainage on vestibulo-spinal reflexes of children with chronic otitis media with effusion. Int J Pediatr Otorhinolaryngol. 2004;68(9):1175–9.
26. Golz A, Westerman ST, Gilbert LM, Joachims HZ, Netzer A. Effect of middle ear effusion on the vestibular labyrinth. J Laryngol Otol. 1991;105(12):987–9.

Cholesteatoma

41

Overview

A *cholesteatoma* is a nonmalignant cystic lesion composed of a matrix (consisting of keratinizing squamous epithelium on a stroma of fibrous tissue) and a central white mass (consisting of keratin debris produced by the matrix). Cholesteatomas can arise anywhere in the temporal bone and are locally invasive. They can occur at any age. They usually present to an otolaryngologist when they involve the auditory canal or tympanic membrane. They can cause conductive hearing loss (when they involve the middle ear space), sensorineural hearing loss, or disequilibrium (when they invade the labyrinth). CT and MRI give complementary imaging information to identify and characterize cholesteatoma. Treatment is surgical.

Introduction

A *cholesteatoma* is a nonmalignant cystic lesion composed of a matrix (consisting of keratinizing squamous epithelium on a stroma of fibrous tissue) and a central white mass (consisting of keratin debris produced by the matrix). Cholesteatomas can arise anywhere in the temporal bone and are locally invasive [1].

Classification

There is debate in the literature regarding the classification of cholesteatomas [1–3].

Cholesteatomas can be congenital or acquired.

Congenital cholesteatomas often arise in the middle ear. Less commonly cholesteatomas occur intracranially [4–6], typically at the petrous apex or in the cerebellopontine angle.

Acquired cholesteatomas in turn can either be primary (arising without any clear provoking factor) or secondary (usually arising at a site affected by injury or infection).

Epidemiology

The annual incidence of cholesteatomas is 3 per 100,000 in the pediatric population, and 9.2 per 100,000 in the adult population, with a male predominance of 1.4:1. They are more common among Caucasians [7].

Pathophysiology

Cases presenting to neuro-otology typically involve the tympanic membrane or auditory canal.

Cholesteatomas can occupy the middle ear space (resulting in conductive hearing loss) and erode into the labyrinth (resulting sensorineural hearing loss) [8–10].

When a cholesteatoma invades the labyrinth, it may also cause disequilibrium [10], such as when it gives rise to a labyrinthine fistula [11–13].

In some instances, cholesteatomas co-occur with fibrous dysplasia [14–16], a disorder of bone metabolism that can affect the skull base.

Physical Examination

On otomicroscopy or otoscopy, a cholesteatoma usually appears as a "whitish ovoid or round friable mass with a thin wall that contains pultaceous or a macerated substance," often in the eardrum itself, or sometimes visible behind the eardrum [17].

Figure 41.1, from Mansour and colleagues [18], shows an acquired cholesteatoma.

Tests

Audiometry may detect conductive or sensorineural hearing loss in an ear affected by cholesteatoma. Tympanometry may identify reduced peak immittance.

M. Cherchi, *Otoneurology and Vestibular Medicine*, https://doi.org/10.1007/978-3-031-94841-1_41

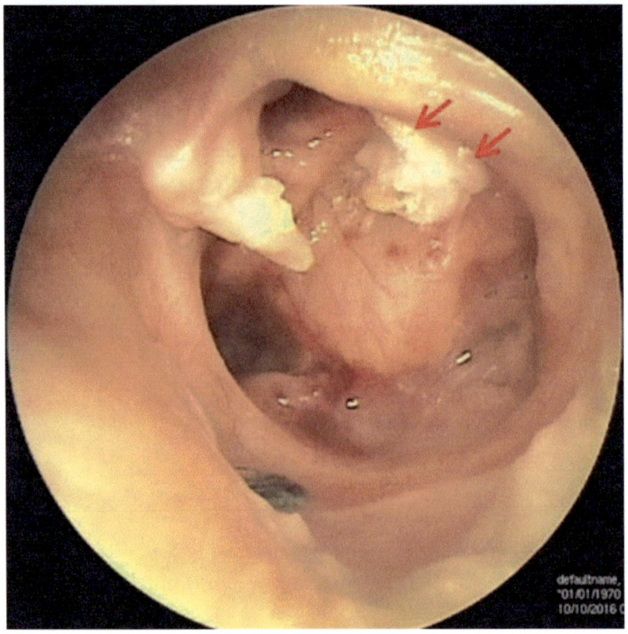

Fig. 41.1 Otoscopy showing left ear with marginal perforation with secondary acquired cholesteatoma due to a migration of the skin epidermis on the edge of the perforation inside the middle ear cavity (red arrows). (From Mansour et al. [18]. Reproduced with written permission from Springer Nature)

A variety of vestibular tests have been investigated in patients with cholesteatoma, including vestibular evoked myogenic potentials [19], caloric testing, and rotatory chair testing [20], but these tests are not considered a standard part of the workup. Video head impulse testing has been studied in cases where cholesteatoma has led to a labyrinthine fistula [12, 13].

Imaging

There is debate regarding the kind of imaging best suited to diagnose cholesteatoma [17, 21–24]. MRI characterizes soft tissue well and thus can image the cholesteatoma itself. A temporal bone CT characterizes bone well, and thus can identify the invasiveness of the lesion. Since the two imaging modalities give complementary information, in practice they are often both checked.

After surgical treatment (see below), there is debate about how frequently to repeat imaging to monitor for recurrence.

Treatment

If a cholesteatoma is identified or suspected, then referral to otolaryngology is appropriate because treatment is surgical. Eradication of the lesion can be attempted with a canal-wall-up or canal-wall-down mastoidectomy. Reconstruction of the tympanic membrane and/or ossicular chain may help address conductive hearing loss.

References

1. Pachpande TG, Singh CV. Diagnosis and treatment modalities of cholesteatomas: a review. Cureus. 2022;14(11):e31153.
2. Yung M, Tono T, Olszewska E, Yamamoto Y, Sudhoff H, Sakagami M, et al. EAONO/JOS Joint consensus statements on the definitions, classification and staging of middle ear cholesteatoma. J Int Adv Otol. 2017;13(1):1–8.
3. Olszewska E, Rutkowska J, Ozgirgin N. Consensus-based recommendations on the definition and classification of cholesteatoma. J Int Adv Otol. 2015;11(1):81–7.
4. Habesoglu TE, Balak N, Habesoglu M, Zemheri E, Isik N, Elmaci I, et al. Intracranial cholesteatoma – case report and critical review. Clin Neuropathol. 2009;28(6):440–4.
5. McHugh TP. Intracranial cholesteatoma: a case report and review. J Emerg Med. 2007;32(4):375–9.
6. Moffat DA, Quaranta N, Baguley DM, Hardy DG, Chang P. Staging and management of primary cerebellopontine cholesteatoma. J Laryngol Otol. 2002;116(5):340–5.
7. Aquino JE, Cruz Filho NA, de Aquino JN. Epidemiology of middle ear and mastoid cholesteatomas: study of 1146 cases. Braz J Otorhinolaryngol. 2011;77(3):341–7.
8. Gulustan F, Yazici ZM, Sayin I, Abakay MA, Gunes S, Akidil AO. Evaluation of the presence of sensorineural hearing loss and the relationship with intraoperative findings in cholesteatoma. Ear Nose Throat J. 2021;100(3_suppl):249S–52S.
9. Stankovic MD. Audiologic results of surgery for cholesteatoma: short- and long-term follow-up of influential factors. Otol Neurotol. 2008;29(7):933–40.
10. Liu YC, Semaan MT, Rayess H, Megerian CA. Hearing and vertigo outcomes after congenital labyrinthine cholesteatoma resection. Am J Otolaryngol. 2014;35(3):417–23.
11. Lim J, Gangal A, Gluth MB. Surgery for cholesteatomatous labyrinthine fistula. Ann Otol Rhinol Laryngol. 2017;126(3):205–15.
12. Covelli E, Talamonti R, Benincasa AT, Filippi C, Marrone V, Tarentini S, et al. Video head impulse test in labyrinthine fistula due to middle ear cholesteatoma. J Int Adv Otol. 2019;15(2):283–8.
13. Fujiwara K, Morita S, Fukuda A, Hoshino K, Nakamaru Y, Homma A. Evaluation of semicircular canal function by video head impulse test in patients with labyrinthine fistula due to cholesteatoma. Otol Neurotol. 2022;43(5):587–93.
14. Liu YH, Chang KP. Fibrous dysplasia of the temporal bone with external auditory canal stenosis and secondary cholesteatoma. J Int Adv Otol. 2016;12(1):125–8.
15. Tochino R, Sunami K, Yamane H. Fibrous dysplasia of the temporal bone with cholesteatoma. Acta Otolaryngol Suppl. 2004;554:47–9.
16. Zanetti D, Gamba P. Cholesteatoma and fibrous dysplasia of the temporal bone: case report and review of the literature. J Otolaryngol. 2007;36(1):59–63.
17. Corrales CE, Blevins NH. Imaging for evaluation of cholesteatoma: current concepts and future directions. Curr Opin Otolaryngol Head Neck Surg. 2013;21(5):461–7.
18. Mansour S, Magnan J, Nicolas K, Haidar H. Cholesteatoma. In: Mansour S, Magnan J, Nicolas K, Haidar H, editors. Middle ear diseases: advances in diagnosis and management. Cham: Springer International Publishing; 2018. p. 311–81.
19. Wang MC, Liu CY, Yu EC, Wu HJ, Lee GS. Vestibular evoked myogenic potentials in chronic otitis media before and after surgery. Acta Otolaryngol. 2009;129(11):1206–11.

20. Mostafa BE, Shafik AG, El Makhzangy AM, Taha H, Abdel Mageed HM. Evaluation of vestibular function in patients with chronic suppurative otitis media. ORL J Otorhinolaryngol Relat Spec. 2013;75(6):357–60.
21. Muzaffar J, Metcalfe C, Colley S, Coulson C. Diffusion-weighted magnetic resonance imaging for residual and recurrent cholesteatoma: a systematic review and meta-analysis. Clin Otolaryngol. 2017;42(3):536–43.
22. Steens S, Venderink W, Kunst D, Meijer A, Mylanus E. Repeated postoperative follow-up diffusion-weighted magnetic resonance imaging to detect residual or recurrent cholesteatoma. Otol Neurotol. 2016;37(4):356–61.
23. Ayache D, Darrouzet V, Dubrulle F, Vincent C, Bobin S, Williams M, et al. Imaging of non-operated cholesteatoma: clinical practice guidelines. Eur Ann Otorhinolaryngol Head Neck Dis. 2012;129(3):148–52.
24. Barath K, Huber AM, Stampfli P, Varga Z, Kollias S. Neuroradiology of cholesteatomas. Am J Neuroradiol. 2011;32(2):221–9.

Glomus Tympanicum

42

Overview

A *glomus tympanicum (GT)* tumor is a paraganglioma arising at various locations in the middle ear. It is highly vascular, and the classic appearance under otomicroscopy/otoscopy is a red pulsatile mass behind the eardrum. It is usually a disease of adulthood and is more common in women. A number of diseases can resemble GTs, and GTs can mimic other lesions. About 15% of GTs secrete catecholamines. The most common presenting symptoms are pulsatile tinnitus and hearing loss. Rarely, GTs can cause disequilibrium and epistaxis. GTs are usually unilateral; they can occur with other paragangliomas anywhere along the sympathetic chain. Audiometry usually shows mixed conductive and sensorineural hearing loss. CT is usually the imaging modality of choice, but MRI and various forms of angiography may also play a role. Treatment is usually surgical, though radiation therapy has become more common. Recurrence is uncommon and occurs more frequently in surgically treated cases. If a clinician observes a red pulsatile mass behind the eardrum, referral to a neuro-otologic otolaryngologist is appropriate.

Introduction

Glomus tympanicum (GT) tumors are a type of paraganglioma.

> "Extra-adrenal paragangliomas (PGOMAs) are rare and can occur anywhere along the sympathetic chain including the carotid body, glomus jugulare, vagal bodies, ganglium tympanicum, larynx, ciliary bodies, organs of Zuckerkandl, urinary bladder, and other locations" [1].

Embryologically, paragangliomas originate from non-chromaffin [2] neural crest cells [3, 4] of the autonomic nervous system [5].

Of particular interest in neuro-otology are paragangliomas arising from the tympanic membrane; these are usually called "glomus tympanicum" (GT) tumors and "are benign arising from paraganglion cells of the tympanic plexus in the middle ear" [6] "located on promontory" [3].

Terminology

The term "glomus" derives from a Latin word usually translated as "ball of yarn," reflecting the vascular nature of these lesions.

Sometimes, paragangliomas are referred to as "glomus tumors," but some authors reserve the term "glomus tumor" to refer to "hyperplasia of glomus bodies" rather than to paragangliomas [7].

Histopathology

Despite the common embryologic origin, GTs differ histopathologically from other paragangliomas [8].

Pathophysiology

GTs cause symptoms by virtue of occupying the middle ear space (where they can impinge on the ossicular chain and cause conductive hearing loss) and invasion into other areas and by virtue of the fact that they are heavily vascularized (and thus can produce pulsatile tinnitus).

A minority of GTs secrete catecholamines. Anand and colleagues report 15% of GTs to be secreting [9]. In such cases, laboratory studies may detect elevated levels of noradrenalin in the serum and vanillylmandelic acid in the urine [10].

Laterality

GTs usually present unilaterally, but cases of bilateral involvement have been reported [11].

M. Cherchi, *Otoneurology and Vestibular Medicine*, https://doi.org/10.1007/978-3-031-94841-1_42

Co-occurrence with Other Paragangliomas

GT can occur with other paragangliomas, such as glomus vagale and glomus jugulare [12], or with glomus of the carotid body [12].

Epidemiology

GT is primarily a disease of adults but has also been reported in children [13–15]. GT are more common in women [16].

Presentation: Tinnitus

Tinnitus is the most common presentation of GT [2, 16–27] and is the primary symptom in over half of cases [24]. The tinnitus is almost always pulsatile, but nonpulsatile tinnitus has been reported [11].

Presentation: Hearing Loss

GT is often associated with hearing loss [2, 17–20, 22, 24, 26, 28, 29]. Hearing loss is the second most common presenting symptom of GT, occurring in about a third of cases [24]. The hearing loss is usually mixed conductive and sensorineural [18].

Presentation: Disequilibrium

Disequilibrium is an uncommon presentation of GT but has been reported [23, 30].

Presentation: Epistaxis

Rarely, GT can present with epistaxis [26, 31].

Physical Examination

A compatible clinical history should prompt examination of the external auditory canal with binocular otomicroscopy, or at least with otoscopy. The classic presentation is a red, sometimes visibly pulsatile, mass behind the eardrum.

Lee and Park [32] recorded a video endoscopy of GT, capturing its visible pulsations.

Glomus Tympanicum Can Be Mimicked by Other Diseases

A number of other pathologies can appear similar to GT, including middle ear adenomatous tumors [33], extramedullary hematopoietic masses [34], carcinoid tumor of the middle ear [35], malignant hemangioendothelioma of the temporal bone [36], metastatic carcinoma of the temporal bone [37], promontory hemangioma [38], endolymphatic sac tumor [39], aberrant petrous carotid artery [40], persistent stapedial artery [41], jugular bulb diverticulum [42], middle ear cholesterol granuloma [43], and middle ear polyp [44].

Glomus Tympanicum Can Mimic Other Diseases

GT can "transgress the tympanic membrane and appear as an inflammatory polyp in the external auditory canal" [16].

The clinical behavior of GT can mimic Ménière's disease [45].

Glomus Tympanicum Can Be Comorbid with Other Otologic Diseases

Unusually, GT can co-occur with other otologic diseases, such as cholesteatoma [46] or otosclerosis [47].

Location

Some GTs arise from the tympanic nerve (Jacobson's nerve) which traverses the middle ear, and in such cases, the lesions may originate from any location along the medial wall of the middle ear [48].

Anatomical Extension

In some cases, GT can extend intracranially [9], including into the middle cranial fossa [49]. GT can also extend into the nasopharynx [50] or even the cervical region [5].

Testing

Audiologic findings are well reported in GT [17–19, 51–56]. While this is interesting, diagnosis of GT does not rely on audiometry.

The role of audiometry is more important when comparing hearing at diagnosis with posttreatment outcomes.

Imaging: CT

Most investigators endorse high-resolution temporal bone CT as the imaging modality of choice in diagnosing GT [57–59].

Imaging: MRI

A few studies recognize complementary roles of CT and MRI in the diagnosis of GT [60].

Imaging: Angiography

Some studies report a potential role for MR angiography, digital subtraction angiography, or transcranial spectral Doppler in the diagnosis of GT [61].

Treatment

Surgery was the mainstay of treatment for decades, and there is a lively debate in the literature regarding optimal surgical approaches [25, 53, 54, 62–73].

More recently, radiation therapy has become an accepted treatment modality [74–86] and is often combined with surgery. As with any radiation therapy, complications include damage of healthy tissue, accelerated atherosclerosis, and subsequent increased risk of malignancy.

Embolization (sometimes followed by surgery) has been another fairly well-studied treatment modality [10, 19, 74, 87, 88], though like any embolization procedure carries a risk of stroke [87, 89].

Less studied treatment modalities include laser coagulation [10, 69, 90–92] and piezoelectric surgery [93, 94].

Recurrence

Recurrence of treated GT is uncommon.

A case series of 8 surgically treated GTs reported no recurrences at up to 15 years [95]. A case series of 16 radiation-treated GTs reported no recurrences at a mean follow-up of 10.5 years (range 4–24 years).

One series of 17 surgically treated cases of GT reported recurrence in 1 patient (6%) after 9 years [16]. Another series of 73 surgically treated cases reported a recurrence rate of 5% [24]. One series noted persistence and/or recurrence of surgically treated GT only in patients in whom the tumor had invaded the central nervous system [96].

Some series report a lower rate of recurrence with radiation-treated GT compared with surgically treated cases [97].

References

1. Snitzer JL, Sheeler LR, Bravo EL, Thacker HL, Lavertu P. A carotid body and glomus jugulare paraganglioma secreting norepinephrine. Endocr Pract. 1995;1(2):82–5.
2. Neto ME, Vuono IM, Souza LR, Testa JR, Pizarro GU, Barros F. Tympanic paragangliomas: case reports. Braz J Otorhinolaryngol. 2005;71(1):97–100.
3. Fountarlis AL, Hajiioannou J, Lachanas V, Tsitiridis I, Saratziotis A, Alagianni A, et al. Endoscopic management of glomus tympanicum tumor: report of three cases and review of the literature. J Audiol Otol. 2022;27:145.
4. Mafee MF, Raofi B, Kumar A, Muscato C. Glomus faciale, glomus jugulare, glomus tympanicum, glomus vagale, carotid body tumors, and simulating lesions. Role of MR imaging. Radiol Clin North Am. 2000;38(5):1059–76.
5. Ma CY, Ji T, Wu YQ, Ow A. A jugulotympanic paraganglioma with craniocervical extension. Int J Oral Maxillofac Surg. 2012;41(2):221–4.
6. Sweeney AD, Carlson ML, Wanna GB, Bennett ML. Glomus tympanicum tumors. Otolaryngol Clin N Am. 2015;48(2):293–304.
7. Wieser ME, Gilley DR, May JG, Rivera AL. A rare case of a middle ear glomangioma. SAGE Open Med Case Rep. 2022;10:2050313X211070520.
8. Branica S, Sprem N, Danilovic Z, Seiwerth S. Immunohistochemistry of glomus jugulare and tympanicum tumors. Acta Med Croatica. 1993;47(3):123–4.
9. Anand VK, Leonetti JP, al-Mefty O. Neurovascular considerations in surgery of glomus tumors with intracranial extensions. Laryngoscope. 1993;103(7):722–8.
10. Kouzaki H, Fukui J, Shimizu T. Management of a catecholamine-secreting tympanicum glomus tumour: case report. J Laryngol Otol. 2008;122(12):1377–80.
11. Fukushima H, Hara H, Paparella MM, Oktay MF, Schachern PA, Cureoglu S. Bilateral glomus tympanicum tumors: human temporalbone study. Clin Pract. 2018;8(3):1035.
12. Balatsouras DG, Eliopoulos PN, Economou CN. Multiple glomus tumours. J Laryngol Otol. 1992;106(6):538–43.
13. Busby DR, Hepp VE. Glomus tympanicum tumor in infancy. Arch Otolaryngol. 1974;99(5):377–8.
14. Jacobs IN, Potsic WP. Glomus tympanicum in infancy. Arch Otolaryngol Head Neck Surg. 1994;120(2):203–5.
15. Yaniv E, Sade J. Glomus tympanicum tumors in a child. Int J Pediatr Otorhinolaryngol. 1983;5(1):93–7.
16. Rohit JY, Caruso A, Russo A, Sanna M. Glomus tympanicum tumour: an alternative surgical technique. J Laryngol Otol. 2003;117(6):462–6.
17. Baguley DM, Irving RM, Hardy DG, Harada T, Moffat DA. Audiological findings in glomus tumours. Br J Audiol. 1994;28(6):291–7.
18. Daneshi A, Asghari A, Mohebbi S, Farhadi M, Farahani F, Mohseni M. Total endoscopic approach in glomus tympanicum surgery. Iran J Otorhinolaryngol. 2017;29(95):305–11.
19. Devuyst L, Defreyne L, Praet M, Geukens S, Dhooge I. Treatment of glomus tympanicum tumors by preoperative embolization and total surgical resection. Am J Otolaryngol. 2016;37(6):544–51.
20. Ghate G, Bhatnagar A, Mukhtar S. Post-embolization excision of glomus tympanicum: a case report. Cureus. 2022;14(1):e21414.

21. Lee SJ, Lee SY, An GS, Lee K, Choi BY, Koo JW, et al. Treatment outcomes of patients with glomus tympanicum tumors presenting with pulsatile tinnitus. J Clin Med. 2021;10(11):2348.
22. Larson TC 3rd, Reese DF, Baker HL Jr, McDonald TJ. Glomus tympanicum chemodectomas: radiographic and clinical characteristics. Radiology. 1987;163(3):801–6.
23. Liscak R, Vladyka V, Simonova G, Vymazal J, Janouskova L. Leksell gamma knife radiosurgery of the tumor glomus jugulare and tympanicum. Stereotact Funct Neurosurg. 1998;70(Suppl 1):152–60.
24. O'Leary MJ, Shelton C, Giddings NA, Kwartler J, Brackmann DE. Glomus tympanicum tumors: a clinical perspective. Laryngoscope. 1991;101(10):1038–43.
25. Ohki M, Kikuchi S. A small glomus tympanicum tumor resected by minimally invasive transcanal endoscopic approach. Case Rep Otolaryngol. 2019;2019:5780161.
26. Tatla T, Savy LE, Wareing MJ. Epistaxis as a rare presenting feature of glomus tympanicum. J Laryngol Otol. 2003;117(7):577–9.
27. Weissman JL, Hirsch BE. Imaging of tinnitus: a review. Radiology. 2000;216(2):342–9.
28. Alshaikhly A, Hamid AM, Azadeh B. Glomus tympanicum chemodectoma: unusual radiological findings. J Laryngol Otol. 1994;108(7):607–9.
29. Kumar G, Andreou Z, Virk JS, Owa A. Unusual presentation of glomus tympanicum tumour: new bone formation in the middle ear. World J Clin Cases. 2014;2(9):463–5.
30. Reddy R. An unusual cause of vertigo and syncope: a case report. Hawaii Med J. 2011;70(1):16–7.
31. Lum C, Keller AM, Kassel E, Blend R, Waldron J, Rutka J. Unusual eustachian tube mass: glomus tympanicum. Am J Neuroradiol. 2001;22(3):508–9.
32. Lee CK, Park KH. Images in clinical medicine. Glomus tympanicum. N Engl J Med. 2010;362(22):e66.
33. Bierry G, Riehm S, Marcellin L, Stierle JL, Veillon F. Middle ear adenomatous tumor: a not so rare glomus tympanicum-mimicking lesion. J Neuroradiol. 2010;37(2):116–21.
34. Chen LN, Espinel A, Reilly B, Luca DC. Extramedullary hematopoiesis presenting as bilateral glomus tympanicum-first case report in a pediatric patient and with bilateral presentation. Pediatr Dev Pathol. 2018;21(3):332–4.
35. Chang H, Silva MA, Torres AA, Weng J, de Lima Guido LP, Velez-Torres J, et al. A carcinoid tumor of the middle ear masquerading as a glomus tympanicum presenting with temporal lobe hemorrhage in a 70-year-old woman: case report and review of the literature. Neurochirurgie. 2022;68(6):654–60.
36. Goldstein WS, Bowen BC, Balkany T. Malignant hemangioendothelioma of the temporal bone masquerading as glomus tympanicum. Ann Otol Rhinol Laryngol. 1994;103(2):156–9.
37. Hellier WP, Crockard HA, Cheesman AD. Metastatic carcinoma of the temporal bone presenting as glomus jugulare and glomus tympanicum tumours: a description of two cases. J Laryngol Otol. 1997;111(10):963–6.
38. Karatas E, Durucu C, Baglam T. Promontory hemangioma mimics glomus tympanicum in the middle ear (promontory hemangioma). Eur Rev Med Pharmacol Sci. 2012;16(Suppl 4):103–5.
39. Muller M, Zammit-Maempel I, Hill J, Wilkins B. An unusual middle-ear mass. J Laryngol Otol. 2010;124(1):108–10.
40. McKiever ME, Carlson ML, Neff BA. Aberrant petrous carotid artery masquerading as a glomus tympanicum. Otol Neurotol. 2014;35(8):e228–30.
41. Sullivan AM, Curtin HD, Moonis G. Arterial anomalies of the middle ear: a pictorial review with clinical-embryologic and imaging correlation. Neuroimaging Clin N Am. 2019;29(1):93–102.
42. Te BC, Ong KP, Zainon IH. A rare cause of red tympanic membrane: jugular bulb diverticulum. Ear Nose Throat J. 2021;102:1455613211031384.
43. Olcott C, Strasnick B. A blue middle ear mass: cholesterol granuloma mimicking a glomus tumor and endolymphatic sac tumor. Am J Otolaryngol. 2017;38(1):100–2.
44. Shew M, Bush J, Tawfik O, Ator G. Middle ear aural polyp mimicking glomus tympanicum in a male adolescent. Otol Neurotol. 2017;38(7):e211–e3.
45. Fitzgerald DC. Glomus tympanicum presenting as Meniere's disease. Ear Nose Throat J. 1985;64(1):68.
46. Gulati K. Unusual presentation of a large glomus tympanicum with a coexisting cholesteatoma. Clin Case Rep. 2020;8(12):2429–32.
47. Yanagisawa E, Principato JJ. Glomus tympanicum tumor in an otosclerotic ear. Arch Otolaryngol. 1970;91(3):299–301.
48. Weissman JL, Hirsch BE. Beyond the promontory: the multifocal origin of glomus tympanicum tumors. Am J Neuroradiol. 1998;19(1):119–22.
49. Carlson RD, Sasaki CT, Friedman SI, Spencer D. Glomus tympanicum tumor with middle cranial fossa extension. Otolaryngol Head Neck Surg. 1987;96(2):186–9.
50. Hirunpat S, Riabroi K, Dechsukhum C, Atchariyasathian V, Tanomkiat W. Nasopharyngeal extension of glomus tympanicum: an unusual clinical and imaging manifestation. Am J Neuroradiol. 2006;27(9):1820–2.
51. Black MJ, Berger H, Tritt RA, Schloss MD. Impedance audiometry: its use in the diagnosis of glomus tympanicum tumors. J Otolaryngol. 1979;8(4):360–7.
52. Medina M, Prasad SC, Patnaik U, Lauda L, Di Lella F, De Donato G, et al. The effects of tympanomastoid paragangliomas on hearing and the audiological outcomes after surgery over a long-term follow-up. Audiol Neurootol. 2014;19(5):342–50.
53. Papaspyrou K, Mewes T, Toth M, Schmidtmann I, Amedee RG, Mann WJ. Hearing results after hypotympanotomy for glomus tympanicum tumors. Otol Neurotol. 2011;32(2):291–6.
54. Patnaik U, Prasad SC, Medina M, Al-Qahtani M, D'Orazio F, Falcioni M, et al. Long term surgical and hearing outcomes in the management of tympanomastoid paragangliomas. Am J Otolaryngol. 2015;36(3):382–9.
55. Thiede O, Stoll W, Schmal F. Clinical and experimental investigations of spontaneous impedance changes of the middle ear. Ann Otol Rhinol Laryngol. 2004;113(7):577–81.
56. Walker DD, Babu S. Temporal bone Paraganglioma: hearing outcomes and rehabilitation. J Neurol Surg B Skull Base. 2019;80(2):209–13.
57. Mafee MF, Valvassori GE, Shugar MA, Yannias DA, Dobben GD. High resolution and dynamic sequential computed tomography. Use in the evaluation of glomus complex tumors. Arch Otolaryngol. 1983;109(10):691–6.
58. Phelps PD, Lloyd GA. Glomus tympanicum tumours: demonstration by high resolution C.T. Clin Otolaryngol Allied Sci. 1983;8(1):15–20.
59. Som PM, Reede DL, Bergeron RT, Parisier SC, Shugar JM, Cohen NL. Computed tomography of glomus tympanicum tumors. J Comput Assist Tomogr. 1983;7(1):14–7.
60. Amin MF, El Ameen NF. Diagnostic efficiency of multidetector computed tomography versus magnetic resonance imaging in differentiation of head and neck paragangliomas from other mimicking vascular lesions: comparison with histopathologic examination. Eur Arch Otorrinolaringol. 2013;270(3):1045–53.
61. De Candia A, Como G, Passon P, Pedace E, Bazzocchi M. Sonographic findings in glomus tympanicum tumor. J Clin Ultrasound. 2002;30(4):236–40.
62. Cannon SJ. Glomus tympanicum removal by transmeatal approach. Laryngoscope. 1962;72:193–200.
63. Alaani A, Chavda SV, Irving RM. The crucial role of imaging in determining the approach to glomus tympanicum tumours. Eur Arch Otorrinolaringol. 2009;266(6):827–31.

64. Chou YL, Huang BR, Wang CH. En bloc surgical removal of an asymptomatic glomus tympanicum tumor. J Chin Med Assoc. 2011;74(11):520–2.
65. Fermi M, Ferri G, Bayoumi Ebaied T, Alicandri-Ciufelli M, Bonali M, Badr El-Dine M, et al. Transcanal endoscopic management of glomus tympanicum: multicentric case series. Otol Neurotol. 2021;42(2):312–8.
66. Farrior JB. Anterior hypotympanic approach for glomus tumor of the infratemporal fossa. Laryngoscope. 1984;94(8):1016–21.
67. Jackson CG. Glomus tympanicum and glomus jugulare tumors. Otolaryngol Clin N Am. 2001;34(5):941–70. vii
68. Kaul VF, Filip P, Schwam ZG, Wanna GB. Nuances in transcanal endoscopic surgical technique for glomus tympanicum tumors. Am J Otolaryngol. 2020;41(5):102562.
69. Noel JE, Sajjadi H. KTP-laser-assisted endoscopic management of glomus tympanicum tumors: a case series. Ear Nose Throat J. 2018;97(12):399–402.
70. Pollak N, Soni RS. Endoscopic excision of a tympanic paraganglioma: training the next generation of ear surgeons. World J Otorhinolaryngol Head Neck Surg. 2017;3(3):160–2.
71. Sillars HA, Fagan PA. The management of multiple paraganglioma of the head and neck. J Laryngol Otol. 1993;107(6):538–42.
72. Sanna M, Fois P, Pasanisi E, Russo A, Bacciu A. Middle ear and mastoid glomus tumors (glomus tympanicum): an algorithm for the surgical management. Auris Nasus Larynx. 2010;37(6):661–8.
73. Smith MF, Shinn JB. Glomus tympanicum–excision by radical mastoidectomy exposure with autograft reconstruction. Laryngoscope. 1976;86(3):431–5.
74. Boyle JO, Shimm DS, Coulthard SW. Radiation therapy for paragangliomas of the temporal bone. Laryngoscope. 1990;100(8):896–901.
75. Fatima N, Pollom E, Soltys S, Chang SD, Meola A. Stereotactic radiosurgery for head and neck paragangliomas: a systematic review and meta-analysis. Neurosurg Rev. 2021;44(2):741–52.
76. Hansen HS, Thomsen KA. Radiotherapy in glomus tumours (paragangliomas). A 25 year-review. Acta Otolaryngol Suppl. 1988;449:151–4.
77. Jansen TTG, Timmers H, Marres HAM, Kaanders J, Kunst HPM. Results of a systematic literature review of treatment modalities for jugulotympanic paraganglioma, stratified per Fisch class. Clin Otolaryngol. 2018;43(2):652–61.
78. Kunzel J, Iro H, Hornung J, Koch M, Brase C, Klautke G, et al. Function-preserving therapy for jugulotympanic paragangliomas: a retrospective analysis from 2000 to 2010. Laryngoscope. 2012;122(7):1545–51.
79. Konefal JB, Pilepich MV, Spector GJ, Perez CA. Radiation therapy in the treatment of chemodectomas. Laryngoscope. 1987;97(11):1331–5.
80. Lee CC, Pan DH, Wu JC, Chung WY, Wu HM, Yang HC, et al. Gamma knife radiosurgery for glomus jugulare and tympanicum. Stereotact Funct Neurosurg. 2011;89(5):291–8.
81. Kalhoro A, Hashim ASM. Glomus jugulare tumors treatment by gamma knife radiosurgery: a single center study. Pak J Med Sci. 2023;39(1):46–9.
82. Liscak R, Urgosik D, Chytka T, Simonova G, Novotny J Jr, Vymazal J, et al. Leksell gamma knife radiosurgery of the jugulotympanic glomus tumor: long-term results. J Neurosurg. 2014;121(Suppl):198–202.
83. Powell S, Peters N, Harmer C. Chemodectoma of the head and neck: results of treatment in 84 patients. Int J Radiat Oncol Biol Phys. 1992;22(5):919–24.
84. Pemberton LS, Swindell R, Sykes AJ. Radical radiotherapy alone for glomus jugulare and tympanicum tumours. Oncol Rep. 2005;14(6):1631–3.
85. Sahyouni R, Mahboubi H, Moshtaghi O, Goshtasbi K, Sahyouni S, Lin HW, et al. Radiosurgery of glomus tumors of temporal bone: a meta-analysis. Otol Neurotol. 2018;39(4):488–93.
86. Scheick SM, Morris CG, Amdur RJ, Bova FJ, Friedman WA, Mendenhall WM. Long-term outcomes after radiosurgery for temporal bone paragangliomas. Am J Clin Oncol. 2018;41(3):223–6.
87. Sawlani V, Browing S, Sawhney IM, Redfern R. Posterior circulation stroke following embolization of glomus tympanicum? Relevance of anatomy and anastomoses of ascending pharyngeal artery. A case report. Interv Neuroradiol. 2009;15(2):229–36.
88. Tasar M, Yetiser S. Glomus tumors: therapeutic role of selective embolization. J Craniofac Surg. 2004;15(3):497–505.
89. Kajtazi NI, Manzoor MU, Ghamdi JA, Zahrani HA, Suwaidan FA, Qahtani SA, et al. Right cerebellar stroke with a right vertebral artery occlusion following an embolization of the right glomus tympanicum tumor: case report with literature review. J Cerebrovasc Endovasc Neurosurg. 2022;24(4):386–92.
90. Durvasula VS, De R, Baguley DM, Moffat DA. Laser excision of glomus tympanicum tumours: long-term results. Eur Arch Otorrinolaringol. 2005;262(4):325–7.
91. Molony NC, Salto-Tellez M, Grant WE. KTP laser assisted excision of glomus tympanicum. J Laryngol Otol. 1998;112(10):956–8.
92. Robinson PJ, Grant HR, Bown SG. NdYAG laser treatment of a glomus tympanicum tumour. J Laryngol Otol. 1993;107(3):236–7.
93. Salami A, Dellepiane M, Proto E, Mora R. Piezosurgery in otologic surgery: four years of experience. Otolaryngol Head Neck Surg. 2009;140(3):412–8.
94. Salami A, Mora R, Dellepiane M. Piezosurgery in the exeresis of glomus tympanicum tumours. Eur Arch Otorrinolaringol. 2008;265(9):1035–8.
95. Cole JM. Panel discussion: glomus jugulare tumors of the temporal bone. Radiation of glomus tumors of the temporal bone. Laryngoscope. 1979;89(10 Pt 1):1623–7.
96. Spector GJ, Sobol S. Surgery for glomus tumors at the skull base. Otolaryngol Head Neck Surg. 1980;88(5):524–30.
97. Contrera KJ, Yong V, Reddy CA, Liu SW, Lorenz RR. Recurrence and progression of head and neck paragangliomas after treatment. Otolaryngol Head Neck Surg. 2020;162(4):504–11.

Part VII

Diagnoses—Diseases Potentially Involving Multiple Otologic Structures

43 Vestibular Weakness

In many instances, it is possible for vestibular testing to identify that vestibular weakness is present, but that identification alone does not reveal the underlying cause. The underlying cause of vestibular weakness can sometimes be determined from other tests or may be suggested by a patient's clinical history.

Most causes of vestibular weakness do not have a direct treatment—in the sense of an intervention that reverses or "undoes" the vestibular weakness.

Almost irrespective of the cause of vestibular weakness, the treatment is usually the same: vestibular physical therapy specifically targeting vestibular weakness. Such therapy usually relies on strategies such as sensory substitution, adaptation, and habituation. While vestibular rehabilitation therapy does not restore inner ear function, it stands some chance of making a patient less uncomfortable, improving the level of function in daily activities, and reducing fall risk.

Overview

We use the term "*vestibular weakness*" to refer to any situation or disease process that reduces the flow of vestibular information traveling from the labyrinth to the vestibular nuclei. Such situations can be unilateral or bilateral, abrupt or gradual in onset, instantaneous or progressive, transient or irreversible. Some examples of each are listed in Table 43.1.

Vestibular weakness may be suspected from the clinical history. The suspicion may increase based on physical examination findings—for example, unilateral weakness may be suspected based on the finding of spontaneous unidirectional horizontal nystagmus on video Frenzel oculography, or bilateral vestibular weakness may be suspected by low retinal gain on ophthalmoscopy or poor performance on dynamic visual acuity testing.

A practical next step is corroboration with objective vestibular testing, such as with vestibular evoked myogenic potentials, video head impulse testing, caloric testing, and rotatory chair testing. In some instances, it may be appropriate to check less common tests, such as fundus photography or optical coherence tomography to identify ocular torsion.

If vestibular weakness is verified, then the differential diagnosis will be narrowed by ascertaining whether the weakness is unilateral or bilateral, by the chronology (acute, subacute, or gradually progressive), by other aspects of the clinical history (e.g., medications), by other otologic testing (e.g., unilateral predominantly high-frequency hearing loss ipsilateral to the vestibular weakness suggests labyrinthitis; unilateral predominantly low-frequency hearing loss ipsilateral to the vestibular weakness suggests Ménière's disease), and by imaging (e.g., vestibular schwannoma).

In some situations, it may be possible to identify on vestibular testing that vestibular weakness is present, yet difficult or impossible to determine the underlying etiology.

If a reversible cause of vestibular weakness is identified (such as pharmacologic vestibular suppression), then this should be treated.

If a treatable cause of vestibular weakness is identified (such as autoimmune inner ear disease), then this should be treated.

If the etiology cannot be identified, then referral to vestibular rehabilitation therapy is appropriate.

M. Cherchi, *Otoneurology and Vestibular Medicine*, https://doi.org/10.1007/978-3-031-94841-1_43

Table 43.1 Etiologies of unilateral and bilateral vestibular weakness

	Usually unilateral	Bilateral versions of usually unilateral disease	Bilateral
Usually acute onset	Vestibular neuritis Labyrinthitis Ménière's disease Temporal bone fracture Labyrinthine ischemia or infarction Congenital otologic malformations Trauma resulting in labyrinthine concussion or a traction injury of the vestibulocochlear nerve Vestibular neurectomy	Vestibular neuritis (such as bilateral sequential) Labyrinthitis Ménière's disease Temporal bone fracture Congenital otologic malformations Trauma resulting in labyrinthine concussion or a traction injury of the vestibulocochlear nerve	Meningitis (with secondary otologic involvement via the vestibular and/or cochlear aqueducts)
Usually subacute onset			Autoimmune inner ear disease Ototoxicity Pharmacologic vestibular suppression
Usually gradual onset	Vestibular schwannoma Other tumors involving the vestibulocochlear nerve (e.g., cerebellopontine angle meningioma) Neurosarcoidosis Neurosyphilis	Vestibular schwannoma (such as in neurofibromatosis type 2) Neurosyphilis	Presbyvestibulopathy CANVAS Superficial siderosis

Bilateral Vestibular Weakness

44

Overview

Reduced or absent vestibular function in both ears is referred to as *bilateral vestibular weakness (BVW)* This condition is almost always the consequence of diseases affecting the labyrinth or vestibular nerve. Patients usually present with imbalance and oscillopsia. BVW is rare. In about half of cases, no cause is found; in the remaining cases, gentamicin ototoxicity is the most common cause, and less common causes include bilateral vestibular neuritis, meningitis, and autoimmune inner ear disease. At the bedside, the diagnosis may be suspected from examination techniques such as the ophthalmoscope test, head impulse test, and dynamic visual acuity testing. The criterion standard for diagnosing BVW is rotatory chair testing; it remains to be established whether video head impulse testing provides comparable diagnostic accuracy. Caloric testing and vestibular evoked myogenic potentials are often abnormal in BVW, but most clinicians rely on these less for diagnosis. There is not yet any restorative treatment, though vestibular rehabilitation therapy, focusing on strategies such as sensory substitution, is nearly always reasonable to attempt. The deficient vestibulo-ocular reflex found in BVW may improve over time, probably due to central compensation. Clinical improvement usually peaks at about 2 years.

Introduction

Passage of vestibular information from the labyrinth through the vestibular nerve to the vestibular nuclei in the brainstem can be diminished or abolished bilaterally if there is damage to both labyrinths, both vestibular nerves, or some combination thereof. Since the vestibular information may be diminished but not entirely absent, we use the term "bilateral vestibular weakness" (BVW) to refer to this condition. Other terminology appears in the literature, such as "bilateral vestibulopathy" [1] or "bilateral vestibular loss." There are several detailed reviews of this topic [2–7], and the present chapter draws heavily on our previous work with Dr. Timothy Hain and Dr. Dario Yacovino [2, 3].

Bilateral vestibular weakness can be subtotal not only in the magnitude of the deficit, but also in the specific pattern of involvement of end-organelles and their afferent pathways. For example, gentamicin ototoxicity affects the entire labyrinth [8] with varying degrees of severity. In contrast, bilateral sequential vestibular neuritis [9–12] tends to affect the superior divisions of the vestibular nerves.

For purposes of diagnosis, treatment, and monitoring, it would be desirable to have a simple grading system for quantifying BVW. Although there have been some attempts at this [13], none has been widely accepted. Here we use the terms, "mild," "moderate" and "severe," while acknowledging that there does not yet exist any universally accepted metric.

Clinical Features and Symptoms of Bilateral Vestibular Weakness

Oscillopsia Almost all patients with BVW experience oscillopsia, which Brickner [14] defined as "oscillating vision." Oscillopsia is a symptom in which head movement is accompanied by the perception that the environment moves. This occurs in BVW because the vestibulo-ocular reflex—which usually moves the eyes an equal amount but opposite in direction of the head movement—is deficient, and the eyes get "dragged along" with head movement. This can certainly occur with large head movements, such as when a car goes over a bump in the road [15], but it can also occur with smaller head movements, such as those that routinely occur during normal ambulation. The "visual" character of this symptom often leads patients to seek ophthalmological consultation.

Imbalance Almost all patients with BVW complain of imbalance, and this symptom tends to be significantly worse if vision or proprioception are challenged. The vestibular

M. Cherchi, *Otoneurology and Vestibular Medicine*, https://doi.org/10.1007/978-3-031-94841-1_44

system calculates movement and position based on several inputs (labyrinthine, visual and somatosensory) and estimates of internal models (from comparisons between perceived motion versus anticipated motion from motor efference copies). If labyrinthine input is diminished or absent (as happens in BVW), then the brain will rely more heavily on visual and somatosensory inputs. If those other inputs are abruptly impaired—for example, if vision is challenged by trying to walk in the dark, or if somatosensory input is challenged by walking on a soft or uneven surface), then the BVW patient will experience greater imbalance.

Auditory Symptoms Auditory symptoms (e.g., tinnitus, hearing loss) are uncommon in BVW. The reason for this is unclear, but one plausible explanation is that the most commonly diagnosed cause of BVW is gentamicin ototoxicity, and that drug is more vestibulotoxic than cochleotoxic [8, 16]. Auditory symptoms are more common in BVW when the cause is a disease known to damage the entire inner ear (e.g., meningitis).

Epidemiology

The prevalence of BVW appears to be low, but is difficult to estimate. Ward and colleagues [17] conducted a survey of more than 21,000 adults complaining of at least one year's worth of ataxia and oscillopsia described as a "big problem," from which they calculated a prevalence of 28/100,000 for "bilateral vestibular weakness," though we would hesitate to draw firm conclusions from this since the survey did not include any instrumented evaluation of vestibular function.

Another method of estimating prevalence is determining its relative frequency of diagnosis. Hain and colleagues [2] report that 213 patients out of a clinic census of about 25,000 were found to have BVW, which amounts to about 0.7%.

By either method of estimation, BVW appears to have a low prevalence.

Etiologies

The etiologies of BVW are usually listed as including ototoxicity, autoimmune inner ear disease, bilateral versions of otherwise unilateral diseases (e.g., tumors, Ménière's disease, vestibular neuritis), and "undetermined" or "idiopathic" [18].

There is disagreement in the literature about the relative frequency of etiologies of BVW. For example, several case series [19–21] report that "idiopathic" or "unknown" comprise the largest category, aminoglycoside ototoxicity is the second most common, and infections (such as meningitis and vestibular neuritis) the third most common.

In contrast, Rinne and colleagues [20] studied a case series of 53 patients with BVW and reported that:

- 39% were associated with neurological disorders (13% cerebellar degeneration, 11% meningitis, and 9% had an association with cranial or peripheral neuropathies)
- 21% were "idiopathic"
- 17% were due to gentamicin ototoxicity
- 10% were due to autoimmune disease
- 8% were attributed to bilateral occurrence of what would usually be unilateral disease (e.g., temporal bone fracture, Ménière's disease)
- 6% were associated with tumors.

Hain and colleagues [2] studied 213 patients whose BVW was confirmed on rotatory chair testing, and reported that:

- 50.7% were "idiopathic"
- 27.7% were due to gentamicin ototoxicity
- 8.9% were due to bilateral vestibular neuritis
- 4.7% were due to tobramycin ototoxicity
- 3.8% were due to head injury
- 3.3% were due to autoimmune inner ear disease
- 1.9% were due to Ménière's disease
- 1.4% were due to streptomycin ototoxicity
- 0.9% were congenital.

Familial bilateral vestibular weakness, with or without hearing loss, is rare and has been reviewed elsewhere [22].

Another genetic syndrome is CANVAS (cerebellar ataxia, neuropathy and vestibular areflexia (BVW) syndrome) may also account for some proportion of cases [23]. This was initially thought to be very rare, but since the causative mutations in the RFC1 gene were identified, more recent work [24] suggests that this may be a more common cause of BVW than originally suspected.

Age-Related Attrition of Vestibular Function

Age In the series reported by Hain and colleagues [2], the median age at diagnosis was 56 years. The older age may reflect that with the passage of time an individual has more exposure to potential vestibular insults.

Problems with balance are well documented to increase with advancing age [25–27]. Temporal bone studies of otherwise healthy individuals show a consistent decrease in the number of vestibular hair cells with normal aging [28]. Histologic studies of the vestibular nerve document that by age 80 the number of fibers in the vestibular nerve declines by 30–50% [29–31].

There is also good evidence for age-related loss of otolith function, as cervical vestibular evoked myogenic potentials and ocular vestibular evoked myogenic potentials responses greatly decrease with age [32].

Ototoxic Causes of Bilateral Vestibular Weakness

Aminoglycoside Antibiotics Excluding "idiopathic" cases of BVW, the aminoglycoside antibiotics (particularly gentamicin and tobramycin), are the most common source of significant BVW. While all aminoglycoside antibiotics have potential ototoxicity, some (such as gentamicin and streptomycin) are preferentially vestibulotoxic [16], while others (e.g., neomycin) are predominantly cochleotoxic.

Gentamicin is the most frequently identified ototoxic medication cases of BVW. Tobramycin is the second most commonly encountered due to its use in the treatment of cystic fibrosis; its risk of ototoxicity is higher when administered intravenously [33–39] and is very low when inhaled [40–43]. Although streptomycin is strongly vestibulotoxic, its use in the United States has greatly diminished, so it is infrequently encountered as a cause of BVW.

The high prevalence of gentamicin ototoxicity in BVW is probably due to a combination of factors. There are several pharmacologic properties of gentamicin that increase its ototoxic potential. First, since gentamicin has relatively less cochleotoxicity, it usually does not cause auditory "warning signs" (tinnitus, hearing loss) that would otherwise alert a patient or treating physician of impending toxicity. Second, even though most aminoglycosides are renally excreted within hours, gentamicin accumulates over months in the inner ear [44], and it is this accumulation that accounts for the drug's ototoxic effects even in patients whose serum concentration has remained within normal limits over the course of treatment. Third, gentamicin is also nephrotoxic; as renal function declines and gentamicin excretion decreases, the drug level (and its ototoxic and nephrotoxic effects) escalates, resulting in a positive feedback loop. Fourth, gentamicin's ototoxicity is potentiated by vancomycin [45], which is commonly coadministered. Besides these pharmacologic properties, another factor is that gentamicin is a relatively inexpensive antibiotic. Finally, certain genes appear to confer particular susceptibility to gentamicin ototoxicity (NOS3, GSTZ1, GSTP1) [46].

Aminoglycoside ototoxicity usually occurs in the context of intravenous administration, less commonly with intraperitoneal administration. However, if aminoglycoside-containing agents are instilled directly into the middle ear (such as through a tympanic membrane perforation), they can diffuse through the round window membrane to the inner ear and cause damage. For this reason, when considering direct aural administration of aminoglycoside-containing agents such as Cortisporin Otic® (which contains neomycin), or gentamicin ophthalmic solution (used off-label as ear drops), one should ensure that no tympanic membrane perforation is present [47–49].

Chemotherapeutic Agents Several chemotherapeutic agents have cochleotoxic potential, though only cisplatin is clearly vestibulotoxic [50]. However, cisplatin can cause numerous other adverse effects which limit its use, so the drug may be discontinued before vestibulotoxicity occurs.

Other Medications There are scattered reports of various medications appearing to cause bilateral vestibular weakness, though isolated case reports comprise weak evidence. The evidence for some of these may be somewhat stronger; for example, Gurkov and colleagues [51] reported a case series in which 15 (12%) of 126 patients taking amiodarone developed BVW.

Nonototoxic Causes of Bilateral Vestibular Weakness

Autoimmune Inner Ear Disease (AIED) AIED and its subtypes are rare causes of BVW. AIED can affect both auditory and vestibular function. This condition generally presents with bilateral sensorineural hearing loss that is rapidly progressive (usually over weeks to months), and the diagnosis is confirmed when the hearing loss significantly improves (or even resolves) after a brief course of high dose oral steroids [52, 53]. Approximately half of cases of AIED also involve vestibular symptoms [54]. There are case reports of AIED presenting exclusively with vestibular symptoms and no hearing loss [55], but absent hearing loss (and thus absent an opportunity to test whether the hearing loss is steroid-responsive), it is unclear how the diagnosis would be secured.

There are several other inner ear conditions that may be immunologically mediated. First, patients who have undergone inner ear surgery on one side may develop auditory and vestibular symptoms in the contralateral (un-operated) ear; this condition is thought to be a "sympathetic autoimmune reaction" [56], analogous to the ocular involvement of Vogt–Koyanagi–Harada syndrome of the eye. Second, Cogan syndrome [57] is similar to AIED but additionally has ocular symptoms; in some respects it resembles postmeningitic hearing loss (see below), as the labyrinth may be occluded with fibrous tissue.

Meningitis Meningitis can damage the entire labyrinth [58], but tends to affect cochlear function more than vestibular function [59]. The meningitic inflammation likely reaches the ear through the vestibular and cochlear aqueducts [60] which are more patent during childhood, and this may in turn be the reason why meningitis is more likely to cause hearing loss in children than in adults [61]. Hearing loss and vestibular deficits often develop immediately in meningitis, though in some cases the vestibular weakness may develop more gradually; such delay is often attributed to the slower process of fibrosis or ossification of the inner ear, which can sometimes be visualized on high-resolution MRI.

Bilateral Vestibular Neuritis Vestibular neuritis can affect any combination of afferent fibers [62–64], and thus can involve the superior or inferior divisions, or both. However, perhaps due to anatomical factors [63, 65, 66], vestibular neuritis more commonly involves only the superior division of the vestibular nerve [62, 67], less commonly involves both the superior and inferior divisions [62], and uncommonly affects only the inferior division [68–70]. Therefore, when vestibular neuritis is bilateral, it tends to involve the superior division on both sides [10, 11]; it is less common for bilateral sequential vestibular neuritis to involve the inferior division on one side and the superior division on the other [71], or to involve bilateral inferior divisions [72]. The tendency of vestibular neuritis to involve the superior division of the vestibular nerve can have diagnostic value in bilateral cases; for instance, if a patient has evidence of bilateral superior division weakness (on caloric testing, rotatory chair testing or video head impulse testing) but preserved inferior division function (with intact cervical vestibular evoked myogenic potentials), then this pattern is more likely to be due to bilateral vestibular neuritis (rather than due to processes that involve the entire labyrinth or the entire vestibular nerve). Loss of caloric function, by itself, is insufficiently specific as caloric testing assesses the horizontal canal alone.

Bilateral Vestibular Schwannomas Neurofibromatosis type 2 is rare, but can manifest with bilateral vestibular schwannomas [73] resulting in BVW.

Bilateral Ménière's Disease [74] Ménière's disease presents with episodes of severe vertigo accompanied by aural symptoms (hearing loss, tinnitus, fullness) with an overall trajectory of progressive hearing loss over years to decades [75], though it is rare for the disease to progress to profound deafness. It is similarly rare to develop the vestibular analog—severe vestibular weakness. In cases of bilateral Ménière's disease, hearing usually remains in the aidable range, and patients generally do not develop severe bilateral vestibular weakness.

Neurosyphilis Despite the increased incidence of syphilis during the AIDS epidemic, neurosyphilis has remained relatively uncommon thanks to the widespread use of antibiotics. When neurosyphilis involves the ear ("otosyphilis"), it generally presents with hearing loss. Vestibular manifestations are less common, but reported in 42% [76] to 52% [77] of cases. Some series report that 80% of patients with vestibular symptoms have electronystagmographic abnormalities [78], some cases of which are bilateral [77, 79]. Given the low incidence of neurosyphilis, testing for this is low yield, but nevertheless some investigators reasonably advocate checking since it is one of the few potentially treatable etiologies of BVW.

Superficial Siderosis Occasional case reports document superficial siderosis as the cause of bilateral vestibular weakness [80, 81]. The mechanism is hypothesized to involve deposition of hemosiderin along the glial segment of the vestibulocochlear nerve [82–84] rather than direct damage of the labyrinth. Superficial siderosis can damage vestibular function, or auditory function, or both.

Vascular Causes It is rare for focal vascular lesions to cause BVW. Vascular supply to the inner ear is via the labyrinthine artery (generally a branch of the anterior inferior cerebellar artery) [85]. Unilateral labyrinthine infarction is very rare [86], so bilateral labyrinthine should be exceedingly rare. If BVW arises from a circulatory disturbance, the etiology is more likely to be a more diffuse vasculopathic/vasculitis process, such as granulomatosis with polyangiitis (Wegener's granulomatosis), or Behçet's disease [87].

Neurosarcoidosis [88] Sarcoidosis is not known to have a specific propensity for the inner ear or vestibular nerve. It is rarely implicated as a cause of unilateral vestibular weakness and therefore should be an exceedingly rare cause of BVW.

Congenital Malformations Malformations of the vestibular end organs occur in several congenital disorders [89], though the relatively low incidence of such diseases makes it difficult to ascertain whether they involve BVW. Aplasia of the semicircular canals [90] occurs in a few rare conditions such as CHARGE (coloboma of the eye, congenital heart defects, choanal atresia, mental and/or growth retardation, genital hypoplasia, ear anomalies and/or deafness) and Mondini dysplasia. Imaging in these cases may demonstrate partial or total absence of the labyrinth; these patients have congenital deafness [91].

Head Trauma If head injuries damage the inner ear, the usual mechanism is thought to be a traction injury of the vestibulocochlear nerve or a labyrinthine concussion [92]. Most, though not all, such cases result in both auditory and vestibular deficits. Since the temporal bone is the hardest bone in the body, any head trauma that is sufficient to cause a temporal bone fracture [93] usually also causes brain injury.

Associations of Other Conditions with Bilateral Vestibular Weakness

Migraine Several case series [94, 95] report an association between migraine and BVW, but this finding is difficult to interpret. Migraine has such a high prevalence that it is unclear whether "migraine plus BVW" is a subcategory migraine [96], or just chance overlap of a common condition (migraine) with a rare one (BVW).

Cerebellar Degeneration Several case series report the combination of BVW and cerebellar degeneration [20, 23, 97–99]. As mentioned earlier, the combination of ataxia, neuropathy and BVW, which has been designated CANVAS (cerebellar ataxia, neuropathy and vestibular areflexia (BVW) syndrome) [23, 97, 98].

Idiopathic We have reviewed literature pertaining to known or suspected causation of bilateral vestibular weakness, but as we mentioned earlier, most clinical series report "idiopathic" to be the most common "diagnosis."

It is possible that at least some patients with "idiopathic" BVW are actually individuals with bilateral vestibular neuritis, as vestibular neuritis is a relatively common inner ear condition, and the bilateral variant of it is well established [10, 11], though this would suggest that the prevalence of vestibular neuritis is higher than generally accepted. Emerging literature suggests that CANVAS may account for a greater proportion of BVW cases than previously suspected [24].

Diagnosis of Bilateral Vestibular Weakness: Clinical Examination

Dynamic visual acuity (DVA) testing [100], also called dynamic illegible E (DIE) testing [101], can be a helpful bedside examination when considering a diagnosis of BVW. This test is performed by comparing visual acuity while the patient's head is stationary, to that when the head is oscillated from side to side [102].

In DVA testing, the examiner passively oscillates the patient's head in a sinusoidal fashion at a frequency of 1 to 2 Hz with an excursion of 15° to 30° to each side. DVA is assessed by having the patient read an eye chart arranged by LogMARs (logarithmic change in the minimum angle of resolution), and comparing performance while the head is stationary to performance while the head is oscillating. Authors have used different criteria for diagnosing BVW by the DVA test; a higher required difference (between the stationary and the oscillating conditions) increases the specificity but decreases the sensitivity of the test.

The bedside *head impulse test* (bHIT), originally recognized as a method for detecting unilateral vestibular weakness [103], is also useful for identifying BVW. The underlying concept is similar to DVA, but the technique differs. The DVA relies on the patient's report of what line on the LogMAR chart he or she is able to read; in contrast, the bedside HIT depends on the examiner's ability to observe a catch-up saccade following a high acceleration, low amplitude rotation of the patient's head while the patient is attempting to maintain his or her gaze fixed on a target; these compensatory saccades (that occur after the head has returned to a stationary position) are called "overt saccades." The sensitivity and specificity of the bHIT depends on patient cooperation, and on the examiner's skill (both in executing the maneuver and observing the compensatory saccades). Patients with cervical pain or limited cervical range of motion may not tolerate this test well. An additional problem is that as patients improve they may learn to produce the compensatory saccades during (rather than after) the head rotation; these are called "covert" saccades and are more difficult for the examiner to observe [104]—in other words, the sensitivity of this test for a given patient may diminish over time. The problem of identifying covert saccades can be addressed by a computerized version of the HIT called video head impulse testing (discussed below).

Ophthalmoscope Test The principle underlying the ophthalmoscope test [102] is similar to DVA, but the technique is different. Whereas the DVA depends on the patient's report of what line on the LogMAR chart he or she is able to read, the ophthalmoscope test instead depends on the examiner's ability to observe (with an ophthalmoscope) the patient's retina during passive oscillation of the patient's head. Keeping the retina in view during movement of the patient's head can be challenging, so the amplitude of oscillation should be relatively small (10°–20°), and the frequency of oscillation should be about 1 Hz [105]. During the passive oscillation of the head in a healthy person, the retina should appear (to the examiner) to remain stationary. In contrast, in a patient with BVW, the retina will appear to oscillate in synchrony with the head oscillation, because the deficient vestibulo-ocular reflex generates insufficient compensatory eye movements to offset the head movements. The ophthalmoscope test should be performed while the patient is wearing any corrective lenses that they usually wear for distance viewing. A patient's performance on the ophthalmoscope test (similar to the bHIT) can improve over time [106]. The ophthalmoscope test is highly specific, but it is not sensitive [107].

General Comments About Bedside Testing for BVW Dynamic visual acuity testing, bedside head impulse testing and the ophthalmoscope test are performed when vision is available, and with high acceleration, and are therefore termed "light, high-frequency" tests. However, BVW is more evident when visual fixation is unavailable, and when the head acceleration is smaller (i.e., in "dark, low-frequency" conditions), so the DVA, bedside HIT, and ophthalmoscope test are not nearly as sensitive as rotatory chair testing (discussed below). Therefore, if the DVA, bHIT and ophthalmoscope test are normal, but one nevertheless has a high index of suspicion for BVW, then one should proceed to rotatory chair testing.

Bilateral Vestibular Weakness: Vestibular Laboratory Testing

Rotatory Chair Rotatory chair testing (RCT) is still regarded as the gold standard test for BVW [108, 109]. This is due to the fact that it measures responses ranging from low to high frequencies (i.e., low to high acceleration), whereas the caloric test (see below) assesses only the very low frequencies, and the vHIT test (see below) assesses only the high frequencies. The responses to the range of stimulus frequencies are plotted during part of the RCT termed slow harmonic acceleration. The deficits in BVW are most evident in the lower frequencies, where one observes low gain and phase lead (if phase can be measured at all). The equipment required for performing adequate RCT is expensive and thus usually only available in major academic medical centers, so there have been attempts to develop less expensive devices, such as the VORTEQ® ("vestibulo-ocular reflex test equipment") [110, 111], but this apparatus is only a "light, high-frequency" test and thus no more sensitive than the bedside tests (DVA, HIT, ophthalmoscope test) described above, so it cannot truly substitute RCT. Furthermore, tests (such as the VORTEQ®) in which the patient actively rotates his or her head (rather than having it passively oscillated by the examiner) reduce the sensitivity of the test, because active performance of head-on-neck movements also enables "preprogramming" of ocular movements that has been shown to augment the vestibulo-ocular reflex [112].

Videonystagmography The caloric portion of videonystagmography (VNG) is sensitive to bilateral vestibular weakness. The thermal stimulus delivered by warm and cool water caloric testing during videonystagmography is generally cited as equivalent to an oscillation frequency of 0.003 Hz [109, 113–115], thus it is a "low frequency test," whence its sensitivity for bilateral vestibular weakness [116]. The total caloric response is the sum of two cool calorics (one in each ear) and two warm calorics (one in each ear). Most studies in healthy individuals state that the average total caloric response is a slow phase velocity of 100°/s [117]. The threshold total caloric response beneath which BVW can be diagnosed is debated; Zapala and colleagues [117] state that, "Fewer than one in 100 otherwise normal subjects demonstrates a T[otal] E[ye] S[peed] of less than 27 degrees/s;" the Bárány Society Consensus document on BVW [1] states that, "the lower limit of the normative data… varies among laboratories from 20 to 25°/s," yet lists the diagnostic criterion (for VNG) as "reduced caloric response (sum of bithermal max[imum] peak S[low] P[hase] V[elocity] on each side <6°/s)," implying that a total caloric response <12°/s is diagnostic. Even if one elects a very stringent criterion such as the ≤10°/s cutoff studied by Furman and colleagues [108], some individuals who on caloric testing appear to have BVW nevertheless exhibit normal responses on RCT [108, 118], showing that caloric testing can render falsely positive results. False positives may be a consequence of a number of factors, including the presence of cerumen, narrow ear canals, or the use of weak stimuli such as balloon irrigation or air caloric stimulation [119]. Conversely, caloric testing can miss moderate BVW, and can thus also render falsely negative results. False negatives are probably due to the fairly wide range of normal responses in healthy controls. The difficulty in interpreting these results is compounded by the fact that many laboratories do not report whether caloric testing was performed with air or water stimulation; since air calorics comprise a weaker thermal stimulus, their results increase the risk of false positives. If the results of caloric testing suggest BVW but the index of suspicion for this diagnosis is low, then the result should be confirmed on rotatory chair testing.

Video Head Impulse Testing (vHIT) The first reports of the clinical utility of the bedside version of the head impulse test emerged in the 1980s [103], but the maneuver can be difficult to execute and the elicited saccades can be difficult to observe. These difficulties have been partially addressed with video head impulse testing (vHIT). The early versions of vHIT were restricted to research settings [120–123], but the technology has evolved to the point that it is more affordable and more readily available. Commercially available products both monitor the movement of the head during the impulse (to ensure that head acceleration is adequate) and process the video of the elicited eye movement, making it easier to calculate the gain and to identify the compensatory saccades (and distinguish whether they are overt or covert). Video head impulse testing appears superior to bedside HIT [124]. Emerging literature [80, 82, 123, 125] suggests that vHIT may play a role in the identification of bilateral vestibular weakness, but its sensitivity and specificity (compared to the gold standard of rotatory chair testing) has yet to be established. Comparison of vHIT with rotatory chair testing is complicated by several factors that can introduce variability into calculation of vestibulo-ocular reflex gain on vHIT; the first reason is physiologic, and has to do with the fact that target viewing distance (which may not be carefully controlled) can have a significant effect on gain both in healthy individuals [126–128] and in patients with peripheral vestibular disease [129]; the second reason is technological, as it has been demonstrated that different devices and different algorithms can render different results in gain [130].

Advantages of the vHIT test include that the equipment is far less expensive than the rotatory chair device, it can be performed rapidly, it can assess all 6 semicircular canals, and the results are generally not influenced by medication. However, the vHIT test is not as capable as the rotatory chair test since it is not designed to detect low-frequency vestibular responses. The predominant frequency of stimulation in the vHIT is 2.5 Hz [131]. As of this writing, vHIT can probably serve as a reasonable screening test for BVW, but one should still seek confirmation of that diagnosis with rotatory chair testing.

Cervical Vestibular Evoked Myogenic Potentials (cVEMPs) Cervical VEMPs are mediated by the saccule and its afferents through the inferior division of the vestibular nerve [132]. Cervical VEMPs require that an adequate vibratory stimulus be delivered to the labyrinth, conductive hearing loss is present, then the stimulus must be delivered by bone vibration. The amplitudes of all VEMPs decline significantly with age [32], so their diagnostic utility diminishes in the elderly. In a young or middle age person with no conductive hearing loss, cVEMPs can help distinguish whether BVW is due to a condition affecting the entire inner ear (such as gentamicin ototoxicity or meningitis, in which the cVEMP should be reduced or absent) or due to a condition with an incomplete lesion (such as bilateral sequential vestibular neuritis, which more commonly involves the superior division of the vestibular nerve and thus will have preserved cVEMPs).

Ocular Vestibular Evoked Myogenic Potentials (oVEMPs) Ocular VEMPs [133, 134] assess the integrity of the utricle and its afferents through the superior division of the vestibular nerve [135, 136], but this test has not yet been studied well in the population of patients with bilateral vestibular weakness. Since oVEMP amplitudes decline significantly with age, and since bilateral loss tends to affect an older population, one would expect that oVEMPs would be far less useful than tests that depend on semicircular canal function, as canal function is less affected by age [137].

Computerized Dynamic Posturography (CDP) Computerized dynamic posturography is sensitive for BVW, but not specific because it cannot distinguish BVW from several important and more common neurological causes of imbalance (e.g., ataxia from cerebellar lesions [138]). On CDP, BVW patients will generally exhibit a low composite score, a "vestibular" pattern on sensory organization testing (SOT), and an ankle dominant sway pattern in conditions 5–6. CDP has some utility in distinguishing malingering of imbalance [139] from BVW; however, patients with severe BVW can be falsely categorized as "aphysiologic" on SOT algorithms [140], so the result should not be interpreted in isolation. These ambiguities pose difficulties in medicolegal situations, as can arise in cases of gentamicin ototoxicity.

Treatment for Bilateral Vestibular Weakness

Most etiologies of BVW cannot be directly treated, so it is important to recognize treatable causes (e.g., treatment of syphilis or autoimmune inner ear disease, stopping an aminoglycoside antibiotic or vestibular suppressant). Vestibular hair cells do not exhibit any significant regenerative capacity [141] in humans, so it is unlikely that any treatment will improve or reverse peripheral vestibular damage. Central compensation, which is probably mediated by plasticity of the commissural connections between the vestibular nuclei [142], requires some minimum residual peripheral vestibular function [143] but even when this mechanism is available, improvement in the vestibulo-ocular reflex (VOR) is limited [144–146], and cannot restore the vestibulo-ocular reflex to its premorbid level. In clinical practice, patients with significant BVW from gentamicin ototoxicity may exhibit no substantial improvement in VOR, reflecting the insufficient compensation despite neural plasticity.

Therefore, functional improvement in BVW is based on substitution. Most BVW patients can improve with physical therapy, though this must be appropriately targeted vestibular rehabilitation therapy [147]. Typically such therapy emphasizes sensory substitution by teaching patients compensatory strategies that rely more heavily on their intact sensoria (vision and somatic sensation) and improving their internal estimates of motion.

Unfortunately, some common practice patterns in medicine can be counterproductive for BVW patients. For example, many of these patients are "diagnosed" with "vertigo" and started on vestibular suppressants such as a benzodiazepine or meclizine, which will further diminish the already deficient peripheral vestibular input and worsen imbalance. More insidious is the use of medications that have the unintended adverse effect of vestibular suppression, and these should be avoided where medically feasible. For example, when treating depression in BVW patients it would be preferable, if deemed appropriate by the treating physician, to avoid medications with antihistaminic or anticholinergic effects (such as tricyclic compounds); when treating anxiety it would be preferable to avoid benzodiazepines.

Vestibular prosthetic devices [148, 149] are being investigated, but remain experimental. Another approach involves technology to improve or restore inner ear function by coaxing regrowth of human inner ear hair cells, similar to some avian species [150, 151].

Prognosis: Natural History of Bilateral Vestibular Weakness

The more extensive the damage in BVW, the more pronounced the symptoms [152]. After 1 to 2 years, patients with mild BVW may have minimal complaints and may function adequately in activities of daily living. Patients with moderate BVW complain of persistent ataxia and oscillopsia. Patients with severe BVW complain not only of ataxia and oscillopsia, but also of functional limitations; for example, severe BVW patients generally refrain from driving. If independently occurring disease imposes other relevant sensory deficits (e.g., proprioceptive impairment from peripheral neuropathy, visual impairment from macular degeneration) or motor deficits (e.g., mechanical limitations from orthopedic problems involving the spine or legs, paresis from stroke), these will interfere with the development of compensatory strategies, and such patients usually have worse outcomes.

Most patients with BVW can improve to some extent, but the degree of improvement depends on the underlying etiology [152, 153]. In cases of gentamicin ototoxicity, the medication's active toxicity continues for some time even after the offending agent is stopped due to some of its pharmacologic properties (see discussion above). Nevertheless, some improvement in the vestibulo-ocular reflex begins within several months and can continue for up to up to 2 years [154]. In moderate to severe BVW cases, patients rarely return to their previous level of function; on physical examination, they will continue to perform poorly on the DVA (see above). Patients with mild BVW, and even some with moderate BVW, may eventually resume driving, though may still avoid driving at night. If a patient's occupation is sedentary or at least does not rely on having good balance, they are usually able to resume work.

Mechanisms for Improvement

Medical improvement usually peaks around 2 years. The main mechanisms of improvement are central compensation, peripheral recovery, and behavioral adaptation.

Central compensation refers to the idea that the brain reprioritizes sensory input by relying more heavily on nonvestibular input (e.g., vision, proprioception) [155]. Results from animal studies suggest that the process of central compensation involves neural plasticity via new synapse formation in the vestibular nuclei of the brainstem [156]. In humans, there is also evidence from fMRI studies that cortical reorganization occurs [157]. Central compensation, defined as increasing the VOR gain, seems to limited to roughly a factor of 2 [158, 159]. Considering that many bilateral patients have lost all or perhaps 90% of their VOR, a factor of 2 is not clinically meaningful.

Evidence from animal studies suggests that vestibular hair cells that are merely damaged (but not dead) may have some capacity to recover function [160]; it is plausible that this also occurs in humans. Similarly, there is sometimes recovery of nerve function in vestibular nerve injuries due to vestibular neuritis; by six months after a vestibular nerve injury, any recovery that is going to occur is probably complete [153, 161].

Behavioral adaptations, whether instinctive or practiced, can play a role in recovering from BVW. Patients tend to avoid activities and situations in which an unanticipated loss of equilibrium would endanger them. Most BVW patients with moderate to severe BVW avoid driving at night, riding bicycles, climbing and standing on ladders. They are also aware of circumstances that would temporarily limit their vision (e.g., walking in the dark) or challenge their proprioception (e.g., walking on a surface that is rough, uneven or moving). We have not personally encountered patients with BVW experiencing problems with swimming; however, swimming imposes proprioceptive and visual challenges, and there are additionally theoretical grounds [162] to suspect that swimming may be difficult for these patients.

References

1. Strupp M, Kim JS, Murofushi T, Straumann D, Jen JC, Rosengren SM, et al. Bilateral vestibulopathy: diagnostic criteria consensus document of the Classification Committee of the Barany Society. J Vestib Res Equilib Orient. 2017;27(4):177–89.
2. Hain TC, Cherchi M, Yacovino DA. Bilateral vestibular weakness. Front Neurol. 2018;9(344):1–14.
3. Hain TC, Cherchi M, Yacovino DA. Bilateral vestibular loss. Semin Neurol. 2013;33(3):195–203.
4. Lee SU, Kim HJ, Kim JS. Bilateral vestibular dysfunction. Semin Neurol. 2020;40(1):40–8.
5. Lucieer F, Vonk P, Guinand N, Stokroos R, Kingma H, van de Berg R. Bilateral vestibular hypofunction: insights in etiologies, clinical subtypes, and diagnostics. Front Neurol. 2016;7:26.
6. Pröpper EJ, van Eijsden HM, Schermer TR, Bruintjes T. Bilateral vestibular hypofunction in a tertiary dizziness center: occurrence and etiology. J Int Adv Otol. 2022;18(4):327–33.
7. van de Berg R, van Tilburg M, Kingma H. Bilateral vestibular hypofunction: challenges in establishing the diagnosis in adults. ORL J Otorhinolaryngol Relat Spec. 2015;77(4):197–218.
8. Ahmed RM, Hannigan IP, MacDougall HG, Chan RC, Halmagyi GM. Gentamicin ototoxicity: a 23-year selected case series of 103 patients. Med J Aust. 2012;196(11):701–4.
9. Comacchio F, Mion M, Armato E, Castellucci A. Sequential vestibular neuritis: report of four cases and literature review. J Audiol Otol. 2021;25(2):89–97.
10. Ogata Y, Sekitani T, Shimogori H, Ikeda T. Bilateral vestibular neuronitis. Acta oto-laryngologica Supplementum. 1993;503:57–60.

11. Schuknecht HF, Witt RL. Acute bilateral sequential vestibular neuritis. Am J Otolaryngol. 1985;6(4):255–7.
12. Young AS, Taylor RL, McGarvie LA, Halmagyi GM, Welgampola MS. Bilateral sequential peripheral vestibulopathy. Neurology. 2016;86(15):1454–6.
13. Hain TC, Cherchi M, Perez-Fernandez N. The gain-time constant product quantifies total vestibular output in bilateral vestibular loss. Front Neurol. 2018;9
14. Brickner RM. Oscillopsia: a new symptom commonly occurring in multiple sclerosis. Arch Neurol Psychiatr. 1936;36(3):586–9.
15. Crawford J. LIVING without a balancing mechanism. N Engl J Med. 1952;246(12):458–60.
16. Selimoglu E. Aminoglycoside-induced ototoxicity. Curr Pharm Des. 2007;13(1):119–26.
17. Ward BK, Agrawal Y, Hoffman HJ, Carey JP, Della Santina CC. Prevalence and impact of bilateral vestibular hypofunction: results from the 2008 US National Health Interview Survey. JAMA Otolaryngol Head Neck Surg. 2013;139(8):803–10.
18. Brandt T. Bilateral vestibulopathy revisited. Eur J Med Res. 1996;1(8):361–8.
19. Zingler VC, Cnyrim C, Jahn K, Weintz E, Fernbacher J, Frenzel C, et al. Causative factors and epidemiology of bilateral vestibulopathy in 255 patients. Ann Neurol. 2007;61(6):524–32.
20. Rinne T, Bronstein AM, Rudge P, Gresty MA, Luxon LM. Bilateral loss of vestibular function: clinical findings in 53 patients. J Neurol. 1998;245(6–7):314–21.
21. Syms CA 3rd, House JW. Idiopathic Dandy's syndrome. Otolaryngol Head Neck Surg. 1997;116(1):75–8.
22. Jen JC. Bilateral vestibulopathy: clinical, diagnostic, and genetic considerations. Semin Neurol. 2009;29(5):528–33.
23. Szmulewicz DJ, Waterston JA, MacDougall HG, Mossman S, Chancellor AM, McLean CA, et al. Cerebellar ataxia, neuropathy, vestibular areflexia syndrome (CANVAS): a review of the clinical features and video-oculographic diagnosis. Ann N Y Acad Sci. 2011;1233:139–47.
24. Traschutz A, Heindl F, Bilal M, Hartmann AM, Dufke C, Riess O, et al. Frequency and phenotype of RFC1 repeat expansions in bilateral vestibulopathy. Neurology. 2023;101(10):e1001–e13.
25. Barin K, Dodson EE. Dizziness in the elderly. Otolaryngol Clin N Am. 2011;44(2):437–54. x
26. Jonsson R, Sixt E, Landahl S, Rosenhall U. Prevalence of dizziness and vertigo in an urban elderly population. J Vestib Res Equilib Orient. 2004;14(1):47–52.
27. Neuhauser HK, von Brevern M, Radtke A, Lezius F, Feldmann M, Ziese T, et al. Epidemiology of vestibular vertigo: a neurotologic survey of the general population. Neurology. 2005;65(6):898–904.
28. Rauch SD, Velazquez-Villasenor L, Dimitri PS, Merchant SN. Decreasing hair cell counts in aging humans. Ann N Y Acad Sci. 2001;942:220–7.
29. Peterka RJ, Black FO, Schoenhoff MB. Age-related changes in human vestibulo-ocular reflexes: sinusoidal rotation and caloric tests. J Vestib Res Equilib Orient. 1990;1(1):49–59.
30. Rosenhall U, Rubin W. Degenerative changes in the human vestibular sensory epithelia. Acta Otolaryngol. 1975;79(1–2):67–80.
31. Ishiyama G, Geiger C, Lopez IA, Ishiyama A. Spiral and vestibular ganglion estimates in archival temporal bones obtained by design based stereology and Abercrombie methods. J Neurosci Methods. 2011;196(1):76–80.
32. Su HC, Huang TW, Young YH, Cheng PW. Aging effect on vestibular evoked myogenic potential. Otol Neurotol. 2004;25(6):977–80.
33. Bates RD, Nahata MC, Jones JW, McCoy K, Young G, Cox S, et al. Pharmacokinetics and safety of tobramycin after once-daily administration in patients with cystic fibrosis. Chest. 1997;112(5):1208–13.
34. Bendush CL, Senior SL, Wooller HO. Evaluation of nephrotoxic and ototoxic effects of tobramycin in worldwide study. Med J Aust. 1977;2(3 Pt 2 Suppl):22–6.
35. Bragonier R, Brown NM. The pharmacokinetics and toxicity of once-daily tobramycin therapy in children with cystic fibrosis. J Antimicrob Chemother. 1998;42(1):103–6.
36. Brogard JM, Conraux C, Collard M, Lavillaureix J. Ototoxicity of tobramycin in humans – influence of renal impairment. Int J Clin Pharmacol Ther Toxicol. 1982;20(9):408–16.
37. Fee WE Jr. Aminoglycoside ototoxicity in the human. Laryngoscope. 1980;90(10 Pt 2 Suppl 24):1–19.
38. Neu HC, Bendush CL. Ototoxicity of tobramycin: a clinical overview. J Infect Dis. 1976;134(Suppl):S206–18.
39. Thomsen J, Friis B. High dosage tobramycin treatment of children with cystic fibrosis. Bacteriological effect and clinical ototoxicity. Int J Pediatr Otorhinolaryngol. 1979;1(1):33–40.
40. Hennig S, McKay K, Vidmar S, O'Brien K, Stacey S, Cheney J, et al. Safety of inhaled (Tobi(R)) and intravenous tobramycin in young children with cystic fibrosis. J Cyst Fibros. 2014;13(4):428–34.
41. Mukhopadhyay S, Baer S, Blanshard J, Coleman M, Carswell F. Assessment of potential ototoxicity following high-dose nebulized tobramycin in patients with cystic fibrosis. J Antimicrob Chemother. 1993;31(3):429–36.
42. Pai VB, Nahata MC. Efficacy and safety of aerosolized tobramycin in cystic fibrosis. Pediatr Pulmonol. 2001;32(4):314–27.
43. Steinkamp G, Tummler B, Gappa M, Albus A, Potel J, Doring G, et al. Long-term tobramycin aerosol therapy in cystic fibrosis. Pediatr Pulmonol. 1989;6(2):91–8.
44. Dulon D, Hiel H, Aurousseau C, Erre JP, Aran JM. Pharmacokinetics of gentamicin in the sensory hair cells of the organ of Corti: rapid uptake and long term persistence. C R Acad Sci III. 1993;316(7):682–7.
45. Brummett RE, Fox KE, Jacobs F, Kempton JB, Stokes Z, Richmond AB. Augmented gentamicin ototoxicity induced by vancomycin in Guinea pigs. Arch Otolaryngol Head Neck Surg. 1990;116(1):61–4.
46. Roth SM, Williams SM, Jiang L, Menon KS, Jeka JJ. Susceptibility genes for gentamicin-induced vestibular dysfunction. J Vestib Res Equilib Orient. 2008;18(1):59–68.
47. Pappas S, Nikolopoulos TP, Korres S, Papacharalampous G, Tzangarulakis A, Ferekidis E. Topical antibiotic ear drops: are they safe? Int J Clin Pract. 2006;60(9):1115–9.
48. Matz G, Rybak L, Roland PS, Hannley M, Friedman R, Manolidis S, et al. Ototoxicity of ototopical antibiotic drops in humans. Otolaryngol Head Neck Surg. 2004;130(3 Suppl):S79–82.
49. Haynes DS. Topical antibiotics: strategies for avoiding ototoxicity. Ear Nose Throat J. 2004;83(1 Suppl):12–4.
50. Schaefer SD, Wright CG, Post JD, Frenkel EP. Cis-platinum vestibular toxicity. Cancer. 1981;47(5):857–9.
51. Gurkov R, Manzari L, Blodow A, Wenzel A, Pavlovic D, Luis L. Amiodarone-associated bilateral vestibulopathy. Eur Arch Otorrinolaringol. 2017;
52. Buniel MC, Geelan-Hansen K, Weber PC, Tuohy VK. Immunosuppressive therapy for autoimmune inner ear disease. Immunotherapy. 2009;1(3):425–34.
53. Ruckenstein MJ. Autoimmune inner ear disease. Curr Opin Otolaryngol Head Neck Surg. 2004;12(5):426–30.
54. Bovo R, Ciorba A, Martini A. Vertigo and autoimmunity. Eur Arch Otorrinolaringol. 2010;267(1):13–9.
55. Dayal VS, Ellman M, Sweiss N. Autoimmune inner ear disease: clinical and laboratory findings and treatment outcome. J Otolaryngol Head Neck Surg Le Journal d'oto-rhino-laryngologie et de chirurgie cervico-faciale. 2008;37(4):591–6.
56. Harris JP, Low NC, House WF. Contralateral hearing loss following inner ear injury: sympathetic cochleolabyrinthitis? Am J Otol. 1985;6(5):371–7.
57. Migliori G, Battisti E, Pari M, Vitelli N, Cingolani C. A shifty diagnosis: Cogan's syndrome. A case report and review of the literature. Acta Otorhinolaryngol Ital. 2009;29(2):108–13.

58. Dichgans M, Jager L, Mayer T, Schorn K, Pfister HW. Bacterial meningitis in adults: demonstration of inner ear involvement using high-resolution MRI. Neurology. 1999;52(5):1003–9.
59. Reeck JB, Lalwani AK. Isolated vestibular ossification after meningitis associated with sensorineural hearing loss. Otol Neurotol. 2003;24(4):576–81.
60. Cushing SL, Papsin BC, Rutka JA, James AL, Blaser SL, Gordon KA. Vestibular end-organ and balance deficits after meningitis and cochlear implantation in children correlate poorly with functional outcome. Otol Neurotol. 2009;30(4):488–95.
61. Wiener-Vacher SR, Obeid R, Abou-Elew M. Vestibular impairment after bacterial meningitis delays infant posturomotor development. J Pediatr. 2012;161(2):246–51 e1.
62. Taylor RL, McGarvie LA, Reid N, Young AS, Halmagyi GM, Welgampola MS. Vestibular neuritis affects both superior and inferior vestibular nerves. Neurology. 2016;87(16):1704–12.
63. Cherchi M, Yacovino DA. Histology and neuroanatomy suggest a unified mechanism to explain the distribution of lesion patterns in acute vestibular neuropathy. Exp Brain Res. 2021;239:1395–9.
64. Yacovino DA, Zanotti E, Cherchi M. The spectrum of acute vestibular neuropathy through modern vestibular testing: a descriptive analysis. Clin Neurophysiol Pract. 2021;
65. Gianoli G, Goebel J, Mowry S, Poomipannit P. Anatomic differences in the lateral vestibular nerve channels and their implications in vestibular neuritis. Otol Neurotol. 2005;26(3):489–94.
66. Goebel JA, O'Mara W, Gianoli G. Anatomic considerations in vestibular neuritis. Otol Neurotol. 2001;22(4):512–8.
67. Fetter M, Dichgans J. Vestibular neuritis spares the inferior division of the vestibular nerve. Brain J Neurol. 1996;119(Pt 3):755–63.
68. Halmagyi GM, Aw ST, Karlberg M, Curthoys IS, Todd MJ. Inferior vestibular neuritis. Ann N Y Acad Sci. 2002;956:306–13.
69. Kim JS, Kim HJ. Inferior vestibular neuritis. J Neurol. 2012;259(8):1553–60.
70. Murofushi T, Halmagyi GM, Yavor RA, Colebatch JG. Absent vestibular evoked myogenic potentials in vestibular neurolabyrinthitis. An indicator of inferior vestibular nerve involvement? Arch Otolaryngol Head Neck Surg. 1996;122(8):845–8.
71. Fujimoto C, Murofushi T, Sugasawa K, Chihara Y, Ushio M, Yamasoba T, et al. Bilateral vestibulopathy with dissociated deficits in the superior and inferior vestibular systems. Ann Otol Rhinol Laryngol. 2012;121(6):383–8.
72. Fujimoto C, Murofushi T, Chihara Y, Suzuki M, Yamasoba T, Iwasaki S. Novel subtype of idiopathic bilateral vestibulopathy: bilateral absence of vestibular evoked myogenic potentials in the presence of normal caloric responses. J Neurol. 2009;256(9):1488–92.
73. Black FO, Brackmann DE, Hitselberger WE, Purdy J. Preservation of auditory and vestibular function after surgical removal of bilateral vestibular schwannomas in a patient with neurofibromatosis type 2. Am J Otol. 1995;16(4):431–43.
74. Nabi S, Parnes LS. Bilateral Meniere's disease. Curr Opin Otolaryngol Head Neck Surg. 2009;17(5):356–62.
75. Huppert D, Strupp M, Brandt T. Long-term course of Meniere's disease revisited. Acta Otolaryngol. 2010;130(6):644–51.
76. Steckelberg JM, McDonald TJ. Otologic involvement in late syphilis. Laryngoscope. 1984;94(6):753–7.
77. Kobayashi H, Mizukoshi K, Watanabe Y, Nagasaki T, Ito M, Aso S. Otoneurological findings in inner ear syphilis. Acta Otolaryngol Suppl. 1991;481:551–5.
78. Wilson WR, Zoller M. Electronystagmography in congenital and acquired syphilitic otitis. Ann Otol Rhinol Laryngol. 1981;90(1 Pt 1):21–4.
79. Durham JS, Longridge NS, Smith JM, Jones H. Clinical manifestations of otological syphilis. J Otolaryngol. 1984;13(3):175–9.
80. Albernaz PL, Cusin FS. The video head impulse test in a case of suspected bilateral loss of vestibular function. Int Arch Otorhinolaryngol. 2016;20(1):84–6.
81. Aran Yoo BS, Kattah JC. Superficial siderosis syndrome with progressive hearing loss and bilateral vestibular failure, 51 years after a neurosurgical procedure: diagnostic value of combined MRI and video head impulse test. J Neurol. 2017;264(2):391–3.
82. Kang KW, Lee C, Kim SH, Cho HH, Lee SH. Bilateral vestibulopathy documented by video head impulse tests in superficial siderosis. Otol Neurotol. 2015;36(10):1683–6.
83. Kumar N, Cohen-Gadol AA, Wright RA, Miller GM, Piepgras DG, Ahlskog JE. Superficial siderosis. Neurology. 2006;66(8):1144–52.
84. Revesz T, Earl CJ, Barnard RO. Superficial siderosis of the central nervous system presenting with longstanding deafness. J R Soc Med. 1988;81(8):479–81.
85. Mazzoni A. The vascular anatomy of the vestibular labyrinth in man. Acta Otolaryngol Suppl. 1990;472:1–83.
86. Lee H, Kim JS, Chung EJ, Yi HA, Chung IS, Lee SR, et al. Infarction in the territory of anterior inferior cerebellar artery: spectrum of audiovestibular loss. Stroke. 2009;40(12):3745–51.
87. Choung YH, Cho MJ, Park K, Choi SJ, Shin YR, Lee ES. Audiovestibular disturbance in patients with Behcet's disease. Laryngoscope. 2006;116(11):1987–90.
88. Colvin IB. Audiovestibular manifestations of sarcoidosis: a review of the literature. Laryngoscope. 2006;116(1):75–82.
89. Sando I, Orita Y, Miura M, Balaban CD. Vestibular abnormalities in congenital disorders. Ann N Y Acad Sci. 2001;942:15–24.
90. Satar B, Mukherji SK, Telian SA. Congenital aplasia of the semicircular canals. Otol Neurotol. 2003;24(3):437–46.
91. Wu CC, Chen YS, Chen PJ, Hsu CJ. Common clinical features of children with enlarged vestibular aqueduct and Mondini dysplasia. Laryngoscope. 2005;115(1):132–7.
92. Feneley MR, Murthy P. Acute bilateral vestibulo-cochlear dysfunction following occipital fracture. J Laryngol Otol. 1994;108(1):54–6.
93. Benitez JT, Bouchard KR, Lane-Szopo D. Pathology of deafness and disequilibrium in head injury: a human temporal bone study. Am J Otol. 1980;1(3):163–7.
94. Cass SP, Furman JM, Ankerstjerne K, Balaban C, Yetiser S, Aydogan B. Migraine-related vestibulopathy. Ann Otol Rhinol Laryngol. 1997;106(3):182–9.
95. Honaker J, Samy RN. Migraine-associated vestibulopathy. Curr Opin Otolaryngol Head Neck Surg. 2008;16(5):412–5.
96. Stewart WF, Lipton RB. Migraine headache: epidemiology and health care utilization. Cephalalgia. 1993;13(Suppl 12):41–6.
97. Szmulewicz DJ, Merchant SN, Halmagyi GM. Cerebellar ataxia with neuropathy and bilateral vestibular areflexia syndrome: a histopathologic case report. Otol Neurotol. 2011;32(8):e63–5.
98. Szmulewicz DJ, Waterston JA, Halmagyi GM, Mossman S, Chancellor AM, McLean CA, et al. Sensory neuropathy as part of the cerebellar ataxia neuropathy vestibular areflexia syndrome. Neurology. 2011;76(22):1903–10.
99. Pothier DD, Rutka JA, Ranalli PJ. Double impairment: clinical identification of 33 cases of cerebellar ataxia with bilateral vestibulopathy. Otolaryngol Head Neck Surg. 2012;146(5):804–8.
100. Demer JL, Honrubia V, Baloh RW. Dynamic visual acuity: a test for oscillopsia and vestibulo-ocular reflex function. Am J Otol. 1994;15(3):340–7.
101. Longridge NS, Mallinson AI. A discussion of the dynamic illegible "E" test: a new method of screening for aminoglycoside vestibulotoxicity. Otolaryngol Head Neck Surg. 1984;92(6):671–7.
102. Hain TC, Cherchi M. Vestib Test. CONTINUUM: Lifelong Learn Neurol. 2021;27(2):330–47.
103. Halmagyi GM, Curthoys IS. A clinical sign of canal paresis. Arch Neurol. 1988;45(7):737–9.

104. Weber KP, Aw ST, Todd MJ, McGarvie LA, Curthoys IS, Halmagyi GM. Horizontal head impulse test detects gentamicin vestibulotoxicity. Neurology. 2009;72(16):1417–24.
105. Zee DS. Ophthalmoscopy in examination of patients with vestibular disorders. Ann Neurol. 1978;3(4):373–4.
106. Atkin A, Bender MB. Ocular stabilization during oscillatory head movements. Arch Neurol. 1968;19(6):559–66.
107. West PD, Sheppard ZA, King EV. Comparison of techniques for identification of peripheral vestibular nystagmus. J Laryngol Otol. 2012;126(12):1209–15.
108. Furman JM, Kamerer DB. Rotational responses in patients with bilateral caloric reduction. Acta Otolaryngol. 1989;108(5–6):355–61.
109. Fife TD, Tusa RJ, Furman JM, Zee DS, Frohman E, Baloh RW, et al. Assessment: vestibular testing techniques in adults and children: report of the Therapeutics and Technology Assessment Subcommittee of the American Academy of Neurology. Neurology. 2000;55(10):1431–41.
110. Hirvonen TP, Aalto H, Pyykko I, Juhola M. Comparison of two head autorotation tests. J Vestib Res Equilib Orient. 1999;9(2):119–25.
111. Tirelli G, Bigarini S, Russolo M, Giacomarra V, Sasso F. Test-retest reliability of the VOR as measured via Vorteq in healthy subjects. Acta Otorhinolaryngol Ital. 2004;24(2):58–62.
112. Della Santina CC, Cremer PD, Carey JP, Minor LB. Comparison of head thrust test with head autorotation test reveals that the vestibulo-ocular reflex is enhanced during voluntary head movements. Arch Otolaryngol Head Neck Surg. 2002;128(9):1044–54.
113. Formby C, Robinson DA. Measurement of vestibular ocular reflex (VOR) time constants with a caloric step stimulus. J Vestib Res Equilib Orient. 2000;10(1):25–39.
114. Goncalves DU, Felipe L, Lima TM. Interpretation and use of caloric testing. Braz J Otorhinolaryngol. 2008;74(3):440–6.
115. Lee SU, Park SH, Kim HJ, Koo JW, Kim JS. Normal caloric responses during acute phase of vestibular neuritis. J Clin Neurol. 2016;12(3):301–7.
116. Jacobson GP, Newman CW, Peterson EL. Interpretation and usefulness of caloric testing. In: Jacobson GP, Newman CW, Kartush JM, editors. Handbook of balance function testing. Clifton Park: Thompson Delmar Learning; 1997. p. 193–233.
117. Zapala DA, Olsholt KF, Lundy LB. A comparison of water and air caloric responses and their ability to distinguish between patients with normal and impaired ears. Ear Hear. 2008;29(4):585–600.
118. Myers SF. Patterns of low-frequency rotational responses in bilateral caloric weakness patients. J Vestib Res Equilib Orient. 1992;2(2):123–31.
119. Karlsen EA, Mikhail HH, Norris CW, Hassanein RS. Comparison of responses to air, water, and closed-loop caloric irrigators. J Speech Hear Res. 1992;35(1):186–91.
120. Aw ST, Halmagyi GM, Haslwanter T, Curthoys IS, Yavor RA, Todd MJ. Three-dimensional vector analysis of the human vestibuloocular reflex in response to high-acceleration head rotations. II. Responses in subjects with unilateral vestibular loss and selective semicircular canal occlusion. J Neurophysiol. 1996;76(6):4021–30.
121. Aw ST, Haslwanter T, Halmagyi GM, Curthoys IS, Yavor RA, Todd MJ. Three-dimensional vector analysis of the human vestibuloocular reflex in response to high-acceleration head rotations. I. Responses in normal subjects. J Neurophysiol. 1996;76(6):4009–20.
122. Bartl K, Lehnen N, Kohlbecher S, Schneider E. Head impulse testing using video-oculography. Ann N Y Acad Sci. 2009;1164:331–3.
123. MacDougall HG, Weber KP, McGarvie LA, Halmagyi GM, Curthoys IS. The video head impulse test: diagnostic accuracy in peripheral vestibulopathy. Neurology. 2009;73(14):1134–41.
124. Yip CW, Glaser M, Frenzel C, Bayer O, Strupp M. Comparison of the bedside head-impulse test with the video head-impulse test in a clinical practice setting: a prospective study of 500 outpatients. Front Neurol. 2016;7:58.
125. Judge PD, Janky KL, Barin K. Can the video head impulse test define severity of bilateral vestibular hypofunction? Otol Neurotol. 2017;38(5):730–6.
126. Clement G, Maciel F. Adjustment of the vestibulo-ocular reflex gain as a function of perceived target distance in humans. Neurosci Lett. 2004;366(2):115–9.
127. Gresty MA, Bronstein AM, Barratt H. Eye movement responses to combined linear and angular head movement. Exp Brain Res. 1987;65(2):377–84.
128. Hine T, Thorn F. Compensatory eye movements during active head rotation for near targets: effects of imagination, rapid head oscillation and vergence. Vis Res. 1987;27(9):1639–57.
129. Crane BT, Demer JL. Human horizontal vestibulo-ocular reflex initiation: effects of acceleration, target distance, and unilateral deafferentation. J Neurophysiol. 1998;80(3):1151–66.
130. Janky KL, Patterson JN, Shepard NT, Thomas MLA, Honaker JA. Effects of device on video head impulse test (vHIT) gain. J Am Acad Audiol. 2017;28(9):778–85.
131. McGarvie LA, MacDougall HG, Halmagyi GM, Burgess AM, Weber KP, Curthoys IS. The video head impulse test (vHIT) of semicircular canal function – age-dependent normative values of VOR gain in healthy subjects. Front Neurol. 2015;6:154.
132. Ferber-Viart C, Dubreuil C, Duclaux R. Vestibular evoked myogenic potentials in humans: a review. Acta Otolaryngol. 1999;119(1):6–15.
133. Todd NP, Rosengren SM, Aw ST, Colebatch JG. Ocular vestibular evoked myogenic potentials (OVEMPs) produced by air- and bone-conducted sound. Clin Neurophysiol. 2007;118(2):381–90.
134. Todd NP, Rosengren SM, Colebatch JG. A short latency vestibular evoked potential (VsEP) produced by bone-conducted acoustic stimulation. J Acoust Soc Am. 2003;114(6 part 1):3264–72.
135. Manzari L, Tedesco A, Burgess AM, Curthoys IS. Ocular vestibular-evoked myogenic potentials to bone-conducted vibration in superior vestibular neuritis show utricular function. Otolaryngol Head Neck Surg. 2010;143(2):274–80.
136. Curthoys IS. A critical review of the neurophysiological evidence underlying clinical vestibular testing using sound, vibration and galvanic stimuli. Clin Neurophysiol. 2010;121(2):132–44.
137. Maes L, Dhooge I, D'Haenens W, Bockstael A, Keppler H, Philips B, et al. The effect of age on the sinusoidal harmonic acceleration test, pseudorandom rotation test, velocity step test, caloric test, and vestibular-evoked myogenic potential test. Ear Hear. 2010;31(1):84–94.
138. Baloh RW, Jacobson KM, Beykirch K, Honrubia V. Static and dynamic posturography in patients with vestibular and cerebellar lesions. Arch Neurol. 1998;55(5):649–54.
139. Cevette MJ, Puetz B, Marion MS, Wertz ML, Muenter MD. Aphysiologic performance on dynamic posturography. Otolaryngol Head Neck Surg. 1995;112(6):676–88.
140. Longridge NS, Mallinson AI. "Across the board" posturography abnormalities in vestibular injury. Otol Neurotol. 2005;26(4):695–8.
141. Staecker H, Praetorius M, Baker K, Brough DE. Vestibular hair cell regeneration and restoration of balance function induced by math1 gene transfer. Otol Neurotol. 2007;28(2):223–31.
142. Graham BP, Dutia MB. Cellular basis of vestibular compensation: analysis and modelling of the role of the commissural inhibitory system. Exp Brain Res. 2001;137(3–4):387–96.
143. Black FO, Wade SW, Nashner LM. What is the minimal vestibular function required for compensation? Am J Otol. 1996;17(3):401–9.
144. McCall AA, Yates BJ. Compensation following bilateral vestibular damage. Front Neurol. 2011;2:88.

145. Herdman SJ. Role of vestibular adaptation in vestibular rehabilitation. Otolaryngol Head Neck Surg. 1998;119(1):49–54.
146. Kasai T, Zee DS. Eye-head coordination in labyrinthine-defective human beings. Brain Res. 1978;144(1):123–41.
147. Porciuncula F, Johnson CC, Glickman LB. The effect of vestibular rehabilitation on adults with bilateral vestibular hypofunction: a systematic review. J Vestib Res Equilib Orient. 2012;22(5–6):283–98.
148. Della Santina CC, Migliaccio AA, Hayden R, Melvin TA, Fridman GY, Chiang B, et al. Current and future management of bilateral loss of vestibular sensation – an update on the Johns Hopkins Multichannel Vestibular Prosthesis Project. Cochlear Implants Int. 2010;11(Suppl 2):2–11.
149. Soto E, Pliego A, Vega R. Vestibular prosthesis: from basic research to clinics. Front Integr Neurosci. 2023;17:1161860.
150. Albu S, Muresanu DF. Vestibular regeneration – experimental models and clinical implications. J Cell Mol Med. 2012;16(9):1970–7.
151. Staecker H, Praetorius M, Brough DE. Development of gene therapy for inner ear disease: using bilateral vestibular hypofunction as a vehicle for translational research. Hear Res. 2011;276(1–2):44–51.
152. Gillespie MB, Minor LB. Prognosis in bilateral vestibular hypofunction. Laryngoscope. 1999;109(1):35–41.
153. Zingler VC, Weintz E, Jahn K, Mike A, Huppert D, Rettinger N, et al. Follow-up of vestibular function in bilateral vestibulopathy. J Neurol Neurosurg Psychiatry. 2008;79(3):284–8.
154. Black FO, Gianna-Poulin C, Pesznecker SC. Recovery from vestibular ototoxicity. Otol Neurotol. 2001;22(5):662–71.
155. Yates BJ, Miller DM. Integration of nonlabyrinthine inputs by the vestibular system: role in compensation following bilateral damage to the inner ear. J Vestib Res Equilib Orient. 2009;19(5–6):183–9.
156. Vibert N, Babalian A, Serafin M, Gasc JP, Muhlethaler M, Vidal PP. Plastic changes underlying vestibular compensation in the Guinea-pig persist in isolated, in vitro whole brain preparations. Neuroscience. 1999;93(2):413–32.
157. Dieterich M, Brandt T. Functional brain imaging of peripheral and central vestibular disorders. Brain J Neurol. 2008;131(Pt 10):2538–52.
158. Demer JL, Porter FI, Goldberg J, Jenkins HA, Schmidt K. Adaptation to telescopic spectacles: vestibulo-ocular reflex plasticity. Invest Ophthalmol Vis Sci. 1989;30(1):159–70.
159. Istl-Lenz Y, Hyden D, Schwarz DW. Response of the human vestibulo-ocular reflex following long-term 2x magnified visual input. Exp Brain Res. 1985;57(3):448–55.
160. Taura A, Kojima K, Ito J, Ohmori H. Recovery of hair cell function after damage induced by gentamicin in organ culture of rat vestibular maculae. Brain Res. 2006;1098(1):33–48.
161. Strupp M, Brandt T. Peripheral vestibular disorders. Curr Opin Neurol. 2013;26(1):81–9.
162. Brandt T, Zwergal A, Glasauer S. 3-D spatial memory and navigation: functions and disorders. Curr Opin Neurol. 2017;30(1):90–7.

45 Vestibular Neuritis

Overview

Vestibular neuritis (VN), also called vestibular neuronitis, refers to damage of the vestibular nerve from a bout of inflammation whose cause is unknown. VN usually presents abruptly, but occasionally can have a more gradual or stuttering onset. Symptoms consist of disequilibrium that often (though not always) involves a spinning sensation, nausea, and vomiting, less commonly aural pain or pressure. VN can happen at any age but peaks in the third to sixth decades. Otovestibular testing (vestibular evoked myogenic potentials, video head impulse testing, caloric testing) shows unilateral vestibular hypofunction. Acute management (in the first 1–3 days) is with antivertiginous medications and antiemetics, but the standard of care for subsequent management is appropriately targeted vestibular rehabilitation therapy. Prognosis is extremely variable in terms of the time required for recovery. The rate of true recurrence is low (about 2% over 10 years).

Introduction

What Is Vestibular Neuritis?

Vestibular neuritis (VN), also called vestibular neuronitis, was originally described by Margaret Ruth Dix and Charles Skinner Hallpike in 1952 [1]. It is currently suspected to be an inflammatory-mediated insult (damage) to the balance-related nerve (vestibular nerve) between the ear and the brain that manifests with abrupt-onset, severe dizziness that lasts days to weeks, and occasionally recurs.

Although vestibular neuritis is usually regarded as a process affecting the vestibular nerve itself, damage restricted to the vestibule (balance components of the inner ear) would manifest clinically in a similar way, and might be termed "vestibulitis," although that term is seldom applied [2]. Thus, distinguishing between "vestibular neuritis" (inflammation of the vestibular nerve) and "vestibulitis" (inflammation of the balance-related components of the inner ear) would be difficult.

Etiologies

The underlying cause of vestibular neuritis is unknown. Viral and vascular causes have been proposed, but reviews of this literature reveal the supporting evidence to be circumstantial at best [3, 4]. Placebo-controlled trials attempting to treat for a presumed viral cause (herpes simplex 1 infection) showed no difference in outcomes [5].

There is a modest literature regarding a possible relationship between ganglioside autoantibodies and vestibular neuritis. Gangliosides are glycosphingolipid structures bearing one or more sialic acid residues; these are most abundantly expressed in the central nervous system [6] and are "believed to play a role in neuroplasticity through influences on neurogenesis, synaptic plasticity, neurotransmission and axonal growth. Furthermore, they also mediate neuronal differentiation, maintenance and repair" [7]. A group of researchers has reported a case [8], a first series of 90 patients [9], and a second series of 105 patients [10] of vestibular neuritis (AUPV, "acute unilateral peripheral vestibulopathy"), in a subset of whom antiganglioside antibodies (anti-*GQ1b* IgG or IgM; anti-*GM1* IgM; anti-*GD1a* IgG) were identified. They reported a positivity rate of 10% [9] to 11% [10]. They conclude that "The causal relationship between antiganglioside antibodies and AUPV may be inferred by the negative seroconversion of antibodies after symptom resolution" [10], yet in their discussion of previously proposed mechanisms (reactivation of type 1 herpes simplex 1 virus, vascular causes), they comment that, "An immune-mediated dysregulation may play a significant role following either cascade of vestibular failure" which (contrary to their postulated causal relationship) suggests that the presence of ganglioside autoantibodies occurs in response to some primary vestibular lesion. Another interpretation of these findings is that in 10–11% of vestibular neuritis patients, axonal or neuronal damage (from some as yet undiscovered etiology) exposes ganglioside antigens against which a maladaptive autoimmune response is mounted. We would be curious to know whether the presence of ganglioside autoantibodies increases

M. Cherchi, *Otoneurology and Vestibular Medicine*, https://doi.org/10.1007/978-3-031-94841-1_45

the risk of subsequently developing recurrent vestibular neuritis (whether ipsilateral or contralateral) or increases the risk of sympathetic ear disease.

Clinical Presentation

Most cases of vestibular neuritis present as a spinning sensation that develops quite rapidly, usually over minutes (less commonly over hours), and is usually associated with nausea and vomiting. Less commonly, the disease may present with more of a "sputtering" onset (with brief bursts of dizziness over several days before the continuous symptoms begin), or the sensation may include some sensation other than spinning (such as wobbling, drifting, tilting, etc.).

In some cases of vestibular neuritis, patients experience pressure, pain, or other discomfort in the affected ear. The mechanism for this sensation is unclear. One possibility is that there are somatosensory fibers (mediating sensations of touch, pain, etc.) in the vestibular nerve, though this has not been described. Another possibility is that the nervus intermedius of Wrisberg, which provides somatosensory innervation to portions of the ear and is neuroanatomically in very close proximity to the vestibular nerve [11], is simultaneously affected by whatever inflammatory process causes vestibular neuritis. Another example of multiple adjacent neural structures being affected is Ramsay Hunt syndrome [12, 13], which affects both the vestibulocochlear nerve and the facial nerve.

Epidemiologic studies [14] suggest that the annual incidence of vestibular neuritis is 3.5 per 100,000 people [3]. Vestibular neuritis is among the most frequently diagnosed conditions in clinics specializing in dizziness. Vestibular neuritis can occur at any age, but peak incidence is in the third to sixth decades, as shown in the histogram in Fig. 45.1 from Sekitani et al. [14].

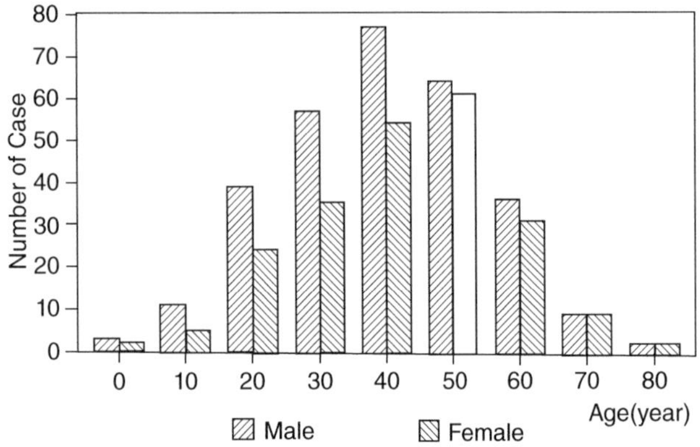

Fig. 45.1 Histogram by age of the incidence of vestibular neuritis. (From Sekitani et al. [14]. Reproduced with written permission from Informa UK Ltd.)

Labyrinthitis is similar to vestibular neuritis in terms of the dizziness/imbalance it causes, but it is different in that labyrinthitis additionally involves auditory symptoms (hearing loss and sometimes tinnitus or ear fullness) on one side; as such, it can be difficult to distinguish labyrinthitis from an initial bout of Ménière's disease.

Diagnosis

How Vestibular Neuritis Diagnosed?

Vestibular neuritis can be suspected based on a compatible clinical history.

Suspicion for vestibular neuritis increases when otovestibular testing (see below) provides evidence of a vestibular deficit on one side, and there is no evidence for a better explanation of the symptoms.

What Tests Help Diagnose Vestibular Neuritis?

In order to detect the vestibular (balance-related) deficits of vestibular neuritis, several tests may be needed. The reason is that there are multiple components of the inner ear that detect movement, each of which sends signals through a different bundle of fibers in the vestibular nerve; there does not currently exist a test to assess these components "all at once," rather, there are individual tests for each component.

Briefly, each labyrinth includes five "organelles" that detect motion—specifically, three semicircular canals (the anterior/superior canal, the lateral/horizontal canal, and the posterior/inferior canal) that detect angular acceleration, and two otolith organs (the saccule and utricle) that detect linear acceleration.

Signals from each of these organelles travel along discrete fibers within the vestibular nerve to reach several vestibular nuclei in the brainstem. The microstructural anatomy of the vestibular nerve is difficult to analyze due to a variety of factors—including neuronal anastomoses with adjacent structures [15], the gradual rotation in the disposition of the fibers along the course of the nerve [16], the variable orientation between individuals [17], variability between genders [18], and the attrition of nerve fibers with age [19, 20]—but a number of painstaking investigations on human specimens have been conducted, such as the study by Lee and Suarez of 3 temporal bones [21], the study by Lopez and colleagues of 13 temporal bones [22], and the study by Bergstrom of 40 human temporal bones [20].

In vestibular neuritis, various subsets of the afferent fibers within the vestibular nerve suffer damage. Some studies [23, 24] oversimplify this (perhaps for didactic purposes) by describing vestibular neuritis that involves either the "superior division" of the vestibular nerve, or the "inferior division" of the vestibular nerve, or both, and conclude with observations that most cases involve either the superior division, or both the superior and inferior divisions [25], whereas it is rare (2.3% [25] to about 5% [26–28]) for only the inferior division to be involved.

However, careful studies show that this pathology can affect nearly any combination of the subsets of fibers [25, 29, 30]. These data suggest that in cases of vestibular neuritis, the horizontal canal (or its afferent pathways) is most commonly affected (90–97.7%), the saccule (or its afferent pathways) is least commonly affected (39.0–44%), whereas the other structures (anterior canal, utricle, posterior canal) exhibit an intermediate level of affectation. The relative ordering of affectation is fairly consistent, but not perfectly so [31].

Given this variability, and given that all tests are fallible (meaning that any test can have false positives and false negatives), we generally try not to base the diagnosis on a single test result in isolation; confidence in the diagnosis increases when multiple independent data points (from separate tests) support the same diagnosis. Tests commonly used in diagnosing vestibular neuritis include cervical vestibular evoked myogenic potentials (cVEMPs), ocular vestibular evoked myogenic potentials (oVEMPs), videonystagmography (VNG) and video head impulse testing (vHIT); this battery is a reasonable selection that assesses many of the inner ear's "balance functions," but it still does not provide a complete picture thereof. Research is exploring other methods of assessing vestibular neuritis [32].

Occasionally, MRI of the brain and internal auditory canals without and with contrast shows enhancement of the vestibular nerve [33–35], though larger studies did not find this [36, 37], suggesting that this finding is uncommon. When enhancement of the vestibular nerve is observed (without any increased "bulk" of the nerve), it can be viewed as compatible with vestibular neuritis, though is not required for the diagnosis.

Treatment, Prognosis, and Recurrence

Treatment: Vestibular Rehabilitation Therapy

During the first few days of vestibular neuritis, management is supportive, sometimes including anti-emetics (such as ondansetron, metoclopramide, or promethazine) and vestibular suppressants (such as meclizine or clonazepam), but these medications play no role beyond the first few days.

Meta-analyses of treatment trials [38] conclude that there is moderate to strong evidence that appropriately targeted vestibular rehabilitation therapy is safe and effective for vestibular neuritis; thus, this treatment modality is generally regarded as the standard of care.

Physical therapy is thought to accelerate recovery by promoting sensory substitution and central compensation for the vestibular deficit [3]. Physical therapy usually is usually initiated with gaze stabilization exercises [39], but evolves depending on an individual's progress.

There is some evidence that the earlier the vestibular rehabilitation therapy is started, the greater the likelihood that it will improve specific physical examination findings in vestibular neuritis [40].

Note that even with appropriately targeted vestibular rehabilitation therapy, recovery may be experienced as "uneven," in the sense that from day to day there may be fluctuation in the intensity of baseline symptoms, though when observing symptoms over a longer time frame (typically weeks) the overall trajectory is one of improvement.

Treatment: Medications

Aside from providing supportive care during the first few days of symptoms (see above), there is not yet any consistent evidence that medications play any role in the long-term management of vestibular neuritis.

There have been several trials of the use of steroids in the treatment of vestibular neuritis. For example, Strupp and colleagues [5] concluded that starting steroids (specifically, methylprednisolone) within 3 days of symptom onset improved outcomes at 12 months—though the outcome measure was the degree of improvement in the caloric response (not in any measure of symptoms or level of patient function). Similarly, a trial by Kim and colleagues [41] reported improvement in caloric response, but no difference in symptomatic recovery as measured by the dizziness handicap inventory. In other words, these trials report that steroids improve a *test result* but do not appear to make any difference in a patient's level of symptoms, or any outcome of function.

Some trials report that treatment with steroids may have some short-term benefit, but no difference in prognosis or long-term outcomes [42, 43]. Studies of vestibular rehabilitation therapy *plus* steroids found that the adjunctive steroids conferred *no* additional long-term benefit [42, 43].

Other trials were unable to detect even an improvement in caloric responses [44]. Consequently, most meta-analyses conclude that on the whole, the evidence does *not* support treatment of vestibular neuritis with steroids [44–46].

The above studies suggest that evidence supporting the use of steroids in the treatment of vestibular neuritis, while not completely absent, is at best weak and conflicting. In view of those conclusions, and in view of the nontrivial risks incurred by the use of steroids, we do not see a compelling rationale for treating vestibular neuritis with steroids.

Prognosis

From a physiologic perspective, it appears that vestibular responses can improve, though may not fully recover. This can be detected on serial measurements of various vestibular tests. For example, on video head impulse testing, McGarvie and colleagues [47] demonstrate very clearly the improvement in gain over time.

Measurable physiologic improvement is interesting, but most patients diagnosed with vestibular neuritis understandably want to know, "When will I get better?" Unfortunately, this is a difficult question to answer because the *symptom* of dizziness is hard to measure and therefore hard to study. In fact, most research of the "prognosis" of vestibular neuritis study an objective (numeric) measure such as a test result, rather than a patient's complaint of a greater or lesser degree of severity of symptoms. For instance, the carefully conducted study by Strupp et al. [5] measured the results of caloric testing rather than patient symptoms.

Many studies of vestibular neuritis comment that symptoms "usually" resolve within days to weeks [48]. However, there are also studies documenting a less favorable prognosis. One study suggested that up to 30–40% of patients experience persistent dizziness after vestibular neuritis [49]. Another study concluded that up to 43.3% of patients remain at least somewhat symptomatic at 5–10 years [50]. The underlying reason for this variability in outcomes is unknown.

Vestibular neuritis does not usually recur (see below). However, an ear affected by vestibular neuritis is at higher risk for developing other vestibular disorders, particularly benign paroxysmal positional vertigo (BPPV). Studies conclude that after a bout of vestibular neuritis, 9.8% [51] to 15.3% [48] developed benign paroxysmal positional vertigo within about 6 years.

Recurrence

Vestibular neuritis can recur, but recurrence is thought to be uncommon. Most studies report recurrence rates of 1.9–2% over 10 years [51, 52]. Other studies cite rates of recurrence as high as 10.7% [48]. In some cases of vestibular neuritis, the recurrence affects the same ear [53]. In other cases of recurrent vestibular neuritis, the recurrence affects "the other ear" (i.e., the originally unaffected ear); this is called "bilateral sequential vestibular neuritis" [54–56].

References

1. Dix MR, Hallpike CS. The pathology symptomatology and diagnosis of certain common disorders of the vestibular system. Proc R Soc Med. 1952;45(6):341–54.
2. Izraeli S, Rachmel A, Nitzan M. Vestibular dysfunction and deafness complicating mumps parotitis. Harefuah. 1989;117(9):243–4.
3. Strupp M, Brandt T. Vestibular neuritis. Semin Neurol. 2009;29(5):509–19.
4. Nadol JB Jr. Vestibular neuritis. Otolaryngol Head Neck Surg. 1995;112(1):162–72.
5. Strupp M, Zingler VC, Arbusow V, Niklas D, Maag KP, Dieterich M, et al. Methylprednisolone, valacyclovir, or the combination for vestibular neuritis. N Engl J Med. 2004;351(4):354–61.
6. Schnaar RL. The biology of gangliosides. Adv Carbohydr Chem Biochem. 2019;76:113–48.
7. Wanleenuwat P, Iwanowski P, Kozubski W. Antiganglioside antibodies in neurological diseases. J Neurol Sci. 2020;408:116576.
8. Lee SU, Kim HJ, Choi JY, Kim JS. Anti-ganglioside antibody-associated acute unilateral peripheral vestibulopathy. J Neurol. 2019;266(1):250–2.
9. Lee SU, Kim HJ, Choi JY, Kim JK, Kim JS. Acute vestibular syndrome associated with anti-GQ1b antibody. Neurology. 2019;93(11):e1085–e92.
10. Kim KT, Park E, Lee SU, Kim B, Kim BJ, Kim JS. Clinical features and neurotologic findings in patients with acute unilateral peripheral vestibulopathy associated with antiganglioside antibody. Neurology. 2023;101(19):e1913–e21.
11. Tubbs RS, Steck DT, Mortazavi MM, Cohen-Gadol AA. The nervus intermedius: a review of its anatomy, function, pathology, and role in neurosurgery. World Neurosurg. 2013;79(5–6):763–7.
12. Hunt JR. On herpetic inflammations of the geniculate ganglion. A new syndrome and its complicaitons. J Nerv Ment Dis. 1907;34(2):73.
13. Kuhweide R, Van de Steene V, Vlaminck S, Casselman JW. Ramsay Hunt syndrome: pathophysiology of cochleovestibular symptoms. J Laryngol Otol. 2002;116(10):844–8.
14. Sekitani T, Imate Y, Noguchi T, Inokuma T. Vestibular neuronitis: epidemiological survey by questionnaire in Japan. Acta Otolaryngol Suppl. 1993;503:9–12.
15. Ozdogmus O, Sezen O, Kubilay U, Saka E, Duman U, San T, et al. Connections between the facial, vestibular and cochlear nerve bundles within the internal auditory canal. J Anat. 2004;205(1):65–75.
16. Silverstein H, Norrell H, Haberkamp T, McDaniel AB. The unrecognized rotation of the vestibular and cochlear nerves from the labyrinth to the brain stem: its implications to surgery of the eighth cranial nerve. Otolaryngol Head Neck Surg. 1986;95(5):543–9.
17. Rasmussen AT. Studies of the VIIIth cranial nerve of man. Laryngoscope. 1940;50:67–83.
18. Moriyama H, Itoh M, Shimada K, Otsuka N. Morphometric analysis of fibers of the human vestibular nerve: sex differences. Eur Arch Otorrinolaringol. 2007;264(5):471–5.
19. Nagai Y, Goto N, Goto J, Kaneko Y, Suzaki H. Morphometric nerve fiber analysis and aging process of the human vestibular nerve. Okajimas Folia Anat Jpn. 1999;76(2–3):95–100.
20. Bergström B. Morphological studies of the vestibular nerve. Uppsala: University of Uppsala; 1973.
21. Lee WS, Suarez C, Honrubia V, Gomez J. Morphological aspects of the human vestibular nerve. Laryngoscope. 1990;100(7):756–64.
22. Lopez I, Ishiyama G, Tang Y, Frank M, Baloh RW, Ishiyama A. Estimation of the number of nerve fibers in the human vestibular endorgans using unbiased stereology and immunohistochemistry. J Neurosci Methods. 2005;145(1–2):37–46.
23. Jeong SH, Kim HJ, Kim JS. Vestibular neuritis. Semin Neurol. 2013;33(3):185–94.
24. Buki B, Hanschek M, Junger H. Vestibular neuritis: involvement and long-term recovery of individual semicircular canals. Auris Nasus Larynx. 2017;44(3):288–93.
25. Taylor RL, McGarvie LA, Reid N, Young AS, Halmagyi GM, Welgampola MS. Vestibular neuritis affects both superior and inferior vestibular nerves. Neurology. 2016;87(16):1704–12.

26. Kim JS, Kim HJ. Inferior vestibular neuritis. J Neurol. 2012;259(8):1553–60.
27. Magliulo G, Gagliardi S, Ciniglio Appiani M, Iannella G, Re M. Vestibular neurolabyrinthitis: a follow-up study with cervical and ocular vestibular evoked myogenic potentials and the video head impulse test. Ann Otol Rhinol Laryngol. 2014;123(3):162–73.
28. Magliulo G, Iannella G, Gagliardi S, Re M. A 1-year follow-up study with C-VEMPs, O-VEMPs and video head impulse testing in vestibular neuritis. Eur Arch Otorrinolaringol. 2015;272(11):3277–81.
29. Yacovino DA, Zanotti E, Cherchi M. The spectrum of acute vestibular neuropathy through modern vestibular testing: a descriptive analysis. Clinical neurophysiology. Practice. 2021;6:137.
30. Navari E, Casani AP. Lesion patterns and possible implications for recovery in acute unilateral vestibulopathy. Otol Neurotol. 2020;41(2):e250–e5.
31. Cherchi M, Yacovino DA. Histology and neuroanatomy suggest a unified mechanism to explain the distribution of lesion patterns in acute vestibular neuropathy. Exp Brain Res. 2021;239:1395–9.
32. Cherchi M. Utricular function in vestibular neuritis: a pilot study of concordance/discordance between ocular vestibular evoked myogenic potentials and ocular cycloposition. Exp Brain Res. 2019;237:1531.
33. Fundakowski CE, Anderson J, Angeli S. Cross-sectional vestibular nerve analysis in vestibular neuritis. Ann Otol Rhinol Laryngol. 2012;121(7):466–70.
34. Karlberg M, Annertz M, Magnusson M. Acute vestibular neuritis visualized by 3-T magnetic resonance imaging with high-dose gadolinium. Arch Otolaryngol Head Neck Surg. 2004;130(2):229–32.
35. Venkatasamy A, Huynh TT, Wohlhuter N, Vuong H, Rohmer D, Charpiot A, et al. Superior vestibular neuritis: improved detection using FLAIR sequence with delayed enhancement (1 h). Eur Arch Otorrinolaringol. 2019;276(12):3309–16.
36. Strupp M, Jager L, Muller-Lisse U, Arbusow V, Reiser M, Brandt T. High resolution Gd-DTPA MR imaging of the inner ear in 60 patients with idiopathic vestibular neuritis: no evidence for contrast enhancement of the labyrinth or vestibular nerve. J Vestib Res Equilib Orient. 1998;8(6):427–33.
37. Hasuike K, Sekitani T, Imate Y. Enhanced MRI in patients with vestibular neuronitis. Acta Otolaryngol Suppl. 1995;519:272–4.
38. McDonnell MN, Hillier SL. Vestibular rehabilitation for unilateral peripheral vestibular dysfunction. Cochrane Database Syst Rev. 2015;1:CD005397.
39. Meldrum D, Jahn K. Gaze stabilisation exercises in vestibular rehabilitation: review of the evidence and recent clinical advances. J Neurol. 2019;266(Suppl 1):11–8.
40. Michel L, Laurent T, Alain T. Rehabilitation of dynamic visual acuity in patients with unilateral vestibular hypofunction: earlier is better. Eur Arch Otorrinolaringol. 2019;
41. Kim G, Seo JH, Lee SJ, Lee DH. Therapeutic effect of steroids on vestibular neuritis: systematic review and meta-analysis. Clin Otolaryngol. 2022;47(1):34–43.
42. Goudakos JK, Markou KD, Psillas G, Vital V, Tsaligopoulos M. Corticosteroids and vestibular exercises in vestibular neuritis. Single-blind randomized clinical trial. JAMA Otolaryngol Head Neck Surg. 2014;140(5):434–40.
43. Ismail EI, Morgan AE, Abdel Rahman AM. Corticosteroids versus vestibular rehabilitation in long-term outcomes in vestibular neuritis. J Vestib Res Equilib Orient. 2019;28:417.
44. Yoo MH, Yang CJ, Kim SA, Park MJ, Ahn JH, Chung JW, et al. Efficacy of steroid therapy based on symptomatic and functional improvement in patients with vestibular neuritis: a prospective randomized controlled trial. Eur Arch Otorrinolaringol. 2017;274(6):2443–51.
45. Fishman JM, Burgess C, Waddell A. Corticosteroids for the treatment of idiopathic acute vestibular dysfunction (vestibular neuritis). Cochrane Database Syst Rev. 2011;(5):CD008607.
46. Leong KJ, Lau T, Stewart V, Canetti EFD. Systematic review and meta-analysis: effectiveness of corticosteroids in treating adults with acute vestibular neuritis. Otolaryngol Head Neck Surg. 2021;165:194599820982910.
47. McGarvie LA, MacDougall HG, Curthoys IS, Halmagyi GM. Spontaneous recovery of the vestibulo-ocular reflex after vestibular neuritis; long-term monitoring with the video head impulse test in a single patient. Front Neurol. 2020;11:732.
48. Kim YH, Kim KS, Kim KJ, Choi H, Choi JS, Hwang IK. Recurrence of vertigo in patients with vestibular neuritis. Acta Otolaryngol. 2011;131(11):1172–7.
49. Neuhauser HK. Epidemiology of vertigo. Curr Opin Neurol. 2007;20(1):40–6.
50. Okinaka Y, Sekitani T, Okazaki H, Miura M, Tahara T. Progress of caloric response of vestibular neuronitis. Acta Otolaryngol Suppl. 1993;503:18–22.
51. Mandala M, Santoro GP, Awrey J, Nuti D. Vestibular neuritis: recurrence and incidence of secondary benign paroxysmal positional vertigo. Acta Otolaryngol. 2010;130(5):565–7.
52. Huppert D, Strupp M, Theil D, Glaser M, Brandt T. Low recurrence rate of vestibular neuritis: a long-term follow-up. Neurology. 2006;67(10):1870–1.
53. Emiliano De Schutter D, Perez FN. Relapsing ipsilateral vestibular neuritis. Case Rep Otolaryngol. 2017;2017:3628402.
54. Yacovino DA, Finlay JB, Jaimes VNU, Verdecchia DH, Schubert MC. Acute bilateral superior branch vestibular neuropathy. Front Neurol. 2018;9(353):6.
55. Schuknecht HF, Witt RL. Acute bilateral sequential vestibular neuritis. Am J Otolaryngol. 1985;6(4):255–7.
56. Ogata Y, Sekitani T, Shimogori H, Ikeda T. Bilateral vestibular neuronitis. Acta oto-laryngologica Supplementum. 1993;503:57–60.

Labyrinthitis

46

Overview

Labyrinthitis is thought to be labyrinthine damage secondary to an inflammatory process whose etiology is unknown. It presents acutely with disequilibrium and unilateral aural symptoms (hearing loss and sometimes tinnitus). Audiometry and otoacoustic emissions show hearing loss on the affected side. Otovestibular testing shows vestibular weakness ipsilateral to the aural symptoms. For the disequilibrium, treatment in the first 1–3 days consists of anti-emetic medications and vestibular suppressants; subsequent management consists of appropriately targeted vestibular rehabilitation therapy. For the hearing loss, treatment includes amplification; some practitioners use steroids (extrapolating from the literature on sudden onset sensorineural hearing loss). Labyrinthitis does not usually recur; when it recurs, other diagnoses (such as Ménière's disease) enter the differential diagnosis.

Introduction

Labyrinthitis, sometimes also called "sudden sensorineural hearing loss with vertigo" in the otolaryngology literature [1–8], is suspected to be due to an inflammatory-mediated insult (damage) to both the "hearing component" (the cochlea) and the "balance component" (the semicircular canals and otolith organs) of the inner ear (labyrinth) itself.

Although labyrinthitis is usually regarded as a process affecting the labyrinth (inner ear) itself, damage to the vestibulocochlear nerve (which transmits hearing and balance signals from the inner ear to the brain) would manifest in a similar fashion clinically; this would be termed "vestibulocochlear neuritis." Thus, distinguishing "labyrinthitis" (inflammation of the entire inner ear) from "vestibulocochlear neuritis" (inflammation of the entire vestibulocochlear nerve) would be difficult.

Etiology

The underlying cause of labyrinthitis is unknown. Viral and vascular causes have been proposed, but reviews of this literature reveal the supporting evidence to be circumstantial at best [9, 10]. Placebo-controlled trials attempting to treat for a presumed viral cause (herpes simplex 1 infection) showed no difference in outcomes [11].

Clinical Presentation

Labyrinthitis usually manifests with severe dizziness (similar to vestibular neuritis) accompanied by ear symptoms on one side (typically hearing loss and tinnitus). These symptoms remain chronic for days to months. In many patients, the symptoms improve; in some, the symptoms completely resolve, while other patients experience some degree of residual symptoms indefinitely. It is less clear whether labyrinthitis recurs, because frequent recurrence is usually clinically more suggestive of Ménière's disease.

Diagnosis

Labyrinthitis can be suspected based on a compatible clinical history. Suspicion for labyrinthitis increases when audiologic testing shows hearing loss on the same side as the vestibular deficits, and there is no evidence for a better explanation of the symptoms.

Otovestibular Testing

In order to detect the vestibular (balance-related) component of labyrinthitis, several tests may be needed. The reason is that there are multiple components of the inner ear that detect movement, each of which sends signals through

M. Cherchi, *Otoneurology and Vestibular Medicine*, https://doi.org/10.1007/978-3-031-94841-1_46

a different bundle of fibers in the vestibular nerve; there does not currently exist a test to assess these components "all at once"; rather, there are individual tests for each component.

Briefly, each labyrinth includes five "organelles" that detect motion—specifically, three semicircular canals (the anterior/superior canal, the lateral/horizontal canal, and the posterior/inferior canal) that detect angular acceleration, and two otolith organs (the saccule and utricle) that detect linear acceleration.

Signals from each of these organelles travel along discrete fibers within the vestibular nerve to reach several vestibular nuclei in the brainstem. The microstructural anatomy of the vestibular nerve is difficult to analyze due to a variety of factors—including neuronal anastomoses with adjacent structures [12], the gradual rotation in the disposition of the fibers along the course of the nerve [13], the variable orientation between individuals [14], variability between genders [15], and the attrition of nerve fibers with age [16, 17]—but a number of painstaking investigations on human specimens have been conducted, such as the study by Lee and Suarez of 3 temporal bones [18], the study by Lopez and colleagues of 13 temporal bones [19], and the study by Bergstrom of 40 human temporal bones [17].

For the auditory part of labyrinthitis (hearing loss, tinnitus), hearing tests are used, including standard audiometry and otoacoustic emissions. Almost any pattern of sensorineural hearing loss can be seen in labyrinthitis, but the most common pattern seen is one of predominantly high frequency sensorineural hearing loss.

We do not usually assess auditory evoked brainstem responses in these patients, as the degree of hearing loss may make the results uninterpretable.

Given this variability, and given that all tests are fallible (meaning that any test can have false positives and false negatives), we generally try not to base the diagnosis on a single test result in isolation; confidence in the diagnosis increases when multiple independent data points (from separate tests) support the same diagnosis. Tests commonly used in diagnosing labyrinthitis include an audiogram, otoacoustic emissions, cervical vestibular evoked myogenic potentials (cVEMPs), ocular vestibular evoked myogenic potentials (oVEMPs), videonystagmography (VNG), and video head impulse testing (vHIT); this battery is a reasonable selection that assesses many of the inner ear's "balance functions," but it still does not provide a complete picture thereof. Ongoing research is exploring other methods of assessing unilateral vestibular weakness [20].

Occasionally, MRI of the brain and internal auditory canals without and with contrast shows enhancement of the labyrinth [7], though the diagnosis does not require this.

Treatment: Vestibular Rehabilitation Therapy

During the first few days of labyrinthitis, management of the symptom of disequilibrium is supportive, sometimes including antiemetics (such as ondansetron, metoclopramide, or promethazine) and vestibular suppressants (such as meclizine or clonazepam), but these medications play no role beyond the first few days.

After the acute phase, physical therapy is the mainstay of treatment. Physical therapy is thought to accelerate recovery by promoting sensory substitution and central compensation for the vestibular deficit [9]. Physical therapy is usually initiated with gaze stabilization exercises [21], but evolves depending on an individual's progress.

Note that even with appropriately targeted vestibular rehabilitation therapy, recovery may be experienced as "uneven," in the sense that from day to day there may be fluctuation in the intensity of baseline symptoms, though when observing symptoms over a longer time frame (typically weeks), the overall trajectory is one of improvement.

The auditory symptoms (hearing loss and tinnitus) may necessitate other treatment. Some practitioners, based on extrapolation from treatment trials for sudden sensorineural hearing loss, attempt a course of oral steroids, even though the data supporting that approach are weak [22, 23]. If that fails, referral to otolaryngology to be evaluated for a transtympanic injection of steroids [24] is sometimes considered, even though the evidence supporting that approach are also weak [25, 26].

Prognosis

The prognosis of the vestibular (dizziness/imbalance) component of labyrinthitis is similarly unpredictable as vestibular neuritis (discussed elsewhere).

The hearing loss from labyrinthitis may improve spontaneously to some degree over about 4–6 months. A common clinical practice is to perform a hearing test as close as possible to the beginning of the disease, and then repeat a hearing test at 4–6 months after symptom onset. If the hearing loss appears to have "plateaued," then consultation with audiology is appropriate in order to discuss amplification options (such as a hearing aid) and perhaps masking strategies (if tinnitus is present).

Recurrence

We have seen patients who appear to have suffered more than one bout of labyrinthitis, though it would be challenging to distinguish recurrent bouts of labyrinthitis from

Ménière's disease. In theory, labyrinthitis should cause predominantly high-frequency hearing loss, whereas Meniere's disease should cause predominantly low-frequency hearing loss, though in reality, either disease can cause any pattern of hearing loss.

References

1. Pogson JM, Taylor RL, Young AS, McGarvie LA, Flanagan S, Halmagyi GM, et al. Vertigo with sudden hearing loss: audio-vestibular characteristics. J Neurol. 2016;263(10):2086–96.
2. Kim CH, Choi HR, Choi S, Lee YS, Shin JE. Patterns of nystagmus conversion in sudden sensorineural hearing loss with vertigo. Medicine (Baltimore). 2018;97(43):e12982.
3. Iwasaki S, Takai Y, Ozeki H, Ito K, Karino S, Murofushi T. Extent of lesions in idiopathic sudden hearing loss with vertigo: study using click and galvanic vestibular evoked myogenic potentials. Arch Otolaryngol Head Neck Surg. 2005;131(10):857–62.
4. Yu H, Li H. Association of Vertigo with hearing outcomes in patients with sudden sensorineural hearing loss: a systematic review and meta-analysis. JAMA Otolaryngol Head Neck Surg. 2018;144(8):677–83.
5. Park HM, Jung SW, Rhee CK. Vestibular diagnosis as prognostic indicator in sudden hearing loss with vertigo. Acta Otolaryngol Suppl. 2001;545:80–3.
6. Mark AS, Seltzer S, Nelson-Drake J, Chapman JC, Fitzgerald DC, Gulya AJ. Labyrinthine enhancement on gadolinium-enhanced magnetic resonance imaging in sudden deafness and vertigo: correlation with audiologic and electronystagmographic studies. Ann Otol Rhinol Laryngol. 1992;101(6):459–64.
7. Seltzer S, Mark AS. Contrast enhancement of the labyrinth on MR scans in patients with sudden hearing loss and vertigo: evidence of labyrinthine disease. Ajnr. 1991;12(1):13–6.
8. Jiang Z, Zhang J, Wang Y, Huang X, Yao Q, Feng Y, et al. Contribution of audiogram classification in evaluating vestibular dysfunction in sudden sensorineural hearing loss with vertigo. Front Neurol. 2021;12:502.
9. Strupp M, Brandt T. Vestibular neuritis. Semin Neurol. 2009;29(5):509–19.
10. Nadol JB Jr. Vestibular neuritis. Otolaryngol Head Neck Surg. 1995;112(1):162–72.
11. Strupp M, Zingler VC, Arbusow V, Niklas D, Maag KP, Dieterich M, et al. Methylprednisolone, valacyclovir, or the combination for vestibular neuritis. N Engl J Med. 2004;351(4):354–61.
12. Ozdogmus O, Sezen O, Kubilay U, Saka E, Duman U, San T, et al. Connections between the facial, vestibular and cochlear nerve bundles within the internal auditory canal. J Anat. 2004;205(1):65–75.
13. Silverstein H, Norrell H, Haberkamp T, McDaniel AB. The unrecognized rotation of the vestibular and cochlear nerves from the labyrinth to the brain stem: its implications to surgery of the eighth cranial nerve. Otolaryngol Head Neck Surg. 1986;95(5):543–9.
14. Rasmussen AT. Studies of the VIIIth cranial nerve of man. Laryngoscope. 1940;50:67–83.
15. Moriyama H, Itoh M, Shimada K, Otsuka N. Morphometric analysis of fibers of the human vestibular nerve: sex differences. Eur Arch Otorrinolaringol. 2007;264(5):471–5.
16. Nagai Y, Goto N, Goto J, Kaneko Y, Suzaki H. Morphometric nerve fiber analysis and aging process of the human vestibular nerve. Okajimas Folia Anat Jpn. 1999;76(2–3):95–100.
17. Bergström B. Morphological studies of the vestibular nerve. Uppsala: University of Uppsala; 1973.
18. Lee WS, Suarez C, Honrubia V, Gomez J. Morphological aspects of the human vestibular nerve. Laryngoscope. 1990;100(7):756–64.
19. Lopez I, Ishiyama G, Tang Y, Frank M, Baloh RW, Ishiyama A. Estimation of the number of nerve fibers in the human vestibular endorgans using unbiased stereology and immunohistochemistry. J Neurosci Methods. 2005;145(1–2):37–46.
20. Cherchi M. Utricular function in vestibular neuritis: a pilot study of concordance/discordance between ocular vestibular evoked myogenic potentials and ocular cycloposition. Exp Brain Res. 2019;237:1531.
21. Meldrum D, Jahn K. Gaze stabilisation exercises in vestibular rehabilitation: review of the evidence and recent clinical advances. J Neurol. 2019;266(Suppl 1):11–8.
22. Nosrati-Zarenoe R, Hultcrantz E. Corticosteroid treatment of idiopathic sudden sensorineural hearing loss: randomized triple-blind placebo-controlled trial. Otol Neurotol. 2012;33(4):523–31.
23. Cinamon U, Bendet E, Kronenberg J. Steroids, carbogen or placebo for sudden hearing loss: a prospective double-blind study. Eur Arch Otorrinolaringol. 2001;258(9):477–80.
24. Garavello W, Galluzzi F, Gaini RM, Zanetti D. Intratympanic steroid treatment for sudden deafness: a meta-analysis of randomized controlled trials. Otol Neurotol. 2012;33(5):724–9.
25. El Sabbagh NG, Sewitch MJ, Bezdjian A, Daniel SJ. Intratympanic dexamethasone in sudden sensorineural hearing loss: a systematic review and meta-analysis. Laryngoscope. 2017;127(8):1897–908.
26. Crane RA, Camilon M, Nguyen S, Meyer TA. Steroids for treatment of sudden sensorineural hearing loss: a meta-analysis of randomized controlled trials. Laryngoscope. 2015;125(1):209–17.

Sudden Sensorineural Hearing Loss (SSNHL)

47

Overview

Idiopathic sudden sensorineural hearing loss (SSNHL) is thought to be due to cochlear inflammation of uncertain etiology. Most cases present with unilateral hearing loss (typically greater in the high frequencies) and sometimes tinnitus. Treatment with oral or transtympanic steroids is often attempted, though the supporting data are weak.

Presentation

Idiopathic sudden sensorineural hearing loss (SSNHL), as its name suggests, is sensorineural hearing loss, presenting acutely (over hours to a couple of days), and generally presenting unilaterally. Bilateral SSNHL appears to be a different disease entity.

Etiology

Idiopathic SSNHL is currently suspected to be due to an inflammatory-mediated insult (damage) to the "hearing component" (cochlea) of the inner ear, which technically should be termed "cochleitis," though that term is seldom used [1–3]. However, damage to the auditory component of the vestibulocochlear nerve could manifest in a similar fashion clinically; this would be termed "cochlear neuritis." Thus, distinguishing between sudden sensorineural hearing loss from "cochleitis" (inflammatory damage of the hearing-related components of the inner ear) and "cochlear neuritis" (inflammation of the hearing-related nerve) would be difficult [4].

Most cases of idiopathic SSNHL present unilaterally. Bilateral SSNHL is much less common and appears to be a different disease entity, with a differential diagnosis including toxic, autoimmune, neoplastic, and vascular conditions [5]. Some research has implicated genetic predisposition to bilateral SSNHL [6].

Demographics of SSNHL

Alexander and Harris [7] report that:

> "During 2006 and 2007, the annual incidence of SSNHL was 27 per 100,000 in the United States. The incidence increased with increasing age, ranging from 11 per 100,000 for patients younger than 18 years to 77 per 100,000 for patients 65 years and older. There was an overall slight male preponderance with a male-to-female ratio of 1.07:1. This was more pronounced in patients 65 years and older, with a ratio of 1.30:1" [7].

The authors also criticize earlier estimates of incidence as suffering from sampling bias. Rausch [8] adds that the peak age of incidence is 43–53 years. Some studies report a slight male preponderance [7], while others report an equal distribution between genders [8].

Clinical Presentation

Idiopathic SSNHL is usually unilateral, and presents acutely, typically over hours to a couple of days. A patient may experience the initial symptoms as aural fullness without perceiving loss of hearing. Tinnitus sometimes accompanies the hearing loss.

Diagnostic Testing

Idiopathic SSNHL is diagnosed based on a compatible clinical history, usually defined as hearing loss occurring over 24–72 h. The criteria for audiometric confirmation usually require 30 dB of hearing loss in at least 3 contiguous frequencies [7]. Note that this requires comparison to a premorbid baseline, and since often no such baseline is available, many practitioners instead look for asymmetry (of the required magnitude) between the two ears, assuming that premorbid hearing was symmetrical [8].

M. Cherchi, *Otoneurology and Vestibular Medicine*, https://doi.org/10.1007/978-3-031-94841-1_47

Treatment

Some practitioners attempt a course of oral steroids, even though the data supporting that approach are weak [9–11]. If that fails, referral to otolaryngology to be evaluated for a transtympanic injection of steroids [12] is sometimes considered, even though the evidence supporting that approach is also weak [13, 14].

If hearing with oral and/or transtympanic steroids does not improve sufficiently after 4–6 months, and if the residual hearing loss is interfering with a patient's level of function, then referral to audiology for amplification is appropriate.

Prognosis

Idiopathic SSNHL may improve spontaneously to some degree over about 4–6 months. A common clinical practice is to perform a hearing test as close as possible to the beginning of the disease, and then repeat a hearing test at 4–6 months after symptom onset. If the hearing loss appears to have "plateaued," then consultation with audiology is appropriate in order to discuss amplification options (such as a hearing aid) and perhaps masking strategies (if tinnitus is present).

The Figure 47.1 below shows a composite audiogram exhibiting spontaneous recovery from idiopathic sudden sensorineural hearing loss over approximately 1 month from Schreiber and colleagues [15].

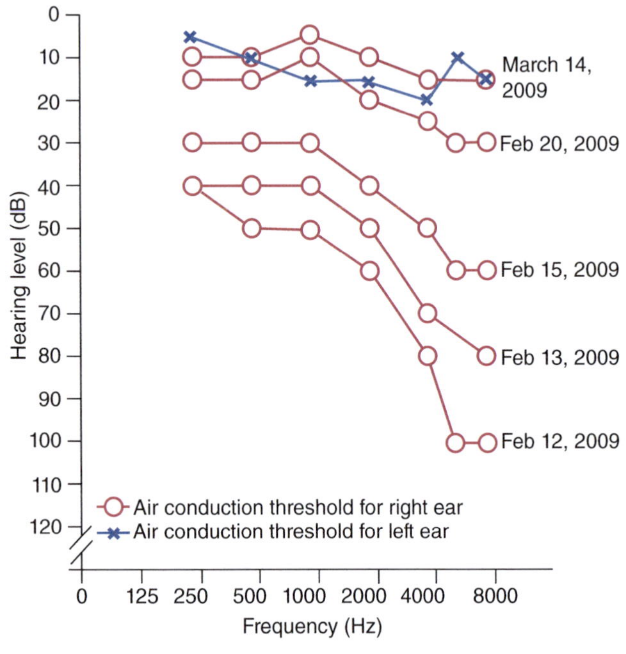

Fig. 47.1 Composite audiogram showing spontaneous improvement from idiopathic sudden sensorineural hearing loss over approximately 1 month. (From Schreiber et al. [15]. Reproduced with written permission from Elsevier B.V.)

References

1. Fitzgerald DC, Mark AS. Viral cochleitis with gadolinium enhancement of the cochlea on magnetic resonance imaging scan. Otolaryngol Head Neck Surg. 1999;121(1):130–2.
2. Luigetti M, Cianfoni A, Modoni A, Conte A, Conti G, Sabatelli M. Teaching NeuroImages: cochleitis: a rare cause of acute deafness in a patient with HCV. Neurology. 2011;77(18):e109.
3. Sataloff J, Vassallo L. Head colds and viral cochleitis. Arch Otolaryngol. 1968;87(1):56–9.
4. Goodale AD, Golub JS, Cornelius RS, Samy RN. Isolated cochlear neuritis from varicella reactivation mimicking a vestibular schwannoma. World J Otorhinolaryngol Head Neck Surg. 2016;2(3):181–4.
5. Sara SA, Teh BM, Friedland P. Bilateral sudden sensorineural hearing loss: review. J Laryngol Otol. 2014;128(Suppl 1):S8–15.
6. Varga L, Jovankovicova A, Huckova M, Demesova L, Gasperikova D, Sebova I, et al. Hereditary bilateral sudden sensorineural hearing loss. Bratisl Lek Listy. 2019;120(9):699–702.
7. Alexander TH, Harris JP. Incidence of sudden sensorineural hearing loss. Otol Neurotol. 2013;34(9):1586–9.
8. Rauch SD. Clinical practice. Idiopathic sudden sensorineural hearing loss. N Engl J Med. 2008;359(8):833–40.
9. Nosrati-Zarenoe R, Hultcrantz E. Corticosteroid treatment of idiopathic sudden sensorineural hearing loss: randomized triple-blind placebo-controlled trial. Otol Neurotol. 2012;33(4):523–31.
10. Cinamon U, Bendet E, Kronenberg J. Steroids, carbogen or placebo for sudden hearing loss: a prospective double-blind study. Eur Arch Otorrinolaringol. 2001;258(9):477–80.
11. Wei BP, Stathopoulos D, O'Leary S. Steroids for idiopathic sudden sensorineural hearing loss. Cochrane Database Syst Rev. 2013;2013(7):CD003998.
12. Garavello W, Galluzzi F, Gaini RM, Zanetti D. Intratympanic steroid treatment for sudden deafness: a meta-analysis of randomized controlled trials. Otol Neurotol. 2012;33(5):724–9.
13. El Sabbagh NG, Sewitch MJ, Bezdjian A, Daniel SJ. Intratympanic dexamethasone in sudden sensorineural hearing loss: a systematic review and meta-analysis. Laryngoscope. 2017;127(8):1897–908.
14. Crane RA, Camilon M, Nguyen S, Meyer TA. Steroids for treatment of sudden sensorineural hearing loss: a meta-analysis of randomized controlled trials. Laryngoscope. 2015;125(1):209–17.
15. Schreiber BE, Agrup C, Haskard DO, Luxon LM. Sudden sensorineural hearing loss. Lancet. 2010;375(9721):1203–11.

48 The Continuum of Vestibular Neuritis, Labyrinthitis, and Sudden Sensorineural Hearing Loss

What Are These Diseases, and How Are They Related?

Discussions in the medical literature use the terms "vestibular neuritis," "labyrinthitis," and "sudden sensorineural hearing loss" under the apparent assumption that these are three discrete diagnoses that are easily distinguishable from one another. While it is true that exemplar cases of each may seem clinically distinct, in real-world clinical practice, one encounters many cases that appear to blur the putative boundaries between these diseases. Consequently, we would argue that these diseases more likely represent a gradation of clinical manifestations that reflect a similar gradation of underlying pathology. We have discussed this in some detail elsewhere [1].

What Is Vestibular Neuritis?

Vestibular neuritis, also called vestibular neuronitis, was originally described by Margaret Ruth Dix and Charles Skinner Hallpike in 1952 [2]. It is currently suspected to be an inflammatory-mediated insult (damage) to the balance-related nerve (vestibular nerve) between the ear and the brain that manifests with abrupt-onset, severe dizziness that lasts days to weeks, and occasionally recurs.

Although vestibular neuritis is usually regarded as a process affecting the vestibular nerve itself, damage restricted to the vestibule (balance components of the inner ear) would manifest clinically in a similar way, and might be termed "vestibulitis," although that term is seldom applied [3]. Thus, distinguishing between "vestibular neuritis" (inflammation of the vestibular nerve) and "vestibulitis" (inflammation of the balance-related components of the inner ear) would be difficult.

What Is Labyrinthitis?

Labyrinthitis is currently suspected to be due to an inflammatory-mediated insult (damage) to both the "hearing component" (the cochlea) and the "balance component" (the semicircular canals and otolith organs) of the inner ear (labyrinth) itself. Labyrinthitis is sometimes also termed "vertigo with sudden hearing loss" [4, 5]—and we will discuss sudden hearing loss further in a moment.

Labyrinthitis usually manifests with severe dizziness (similar to vestibular neuritis) accompanied by ear symptoms on one side (typically hearing loss and tinnitus). It is less clear whether labyrinthitis recurs, because frequent recurrence is usually clinically more suggestive of Ménière's disease.

Although labyrinthitis is usually regarded as a process affecting the labyrinth (inner ear) itself, damage to the vestibulocochlear nerve (which transmits hearing and balance signals from the inner ear to the brain) would manifest in a similar fashion clinically; this would be termed "vestibulocochlear neuritis." Thus, distinguishing "labyrinthitis" (inflammation of the entire inner ear) from "vestibulocochlear neuritis" (inflammation of the entire vestibulocochlear nerve) would be difficult.

What Is Sudden Sensorineural Hearing Loss?

Idiopathic sudden sensorineural hearing loss (SSNHL) is currently suspected to be due to an inflammatory-mediated insult (damage) to the "hearing component" (cochlea) of the inner ear, which technically should be termed "cochleitis," though that term is seldom used [6–8]. However, damage to the auditory component of the vestibulocochlear nerve could manifest in a similar fashion clinically; this would be termed

M. Cherchi, *Otoneurology and Vestibular Medicine*, https://doi.org/10.1007/978-3-031-94841-1_48

Table 48.1 The anatomical and clinical continuum of this group of diseases

		Spectrum of clinical presentation		
		Exclusively vestibular function affected	**Both vestibular and auditory function affected**	**Exclusively auditory function affected**
Spectrum of anatomy	**Exclusively neural structure involved**	"Vestibular neuritis"	"Vestibulocochlear neuritis"	"Cochlear neuritis"
	Both neural structure and inner ear structure involved	"Vestibular neuritis" + "vestibulitis"	"Vestibulocochlear neuritis" + "labyrinthitis"	"Cochlear neuritis" + "cochleitis"
	Exclusively inner ear structure involved	"Vestibulitis"	"Labyrinthitis"	"Cochleitis"

"cochlear neuritis." Thus, distinguishing between sudden sensorineural hearing loss from "cochleitis" (inflammatory damage of the hearing-related components of the inner ear) and "cochlear neuritis" (inflammation of the hearing-related nerve) would be difficult [9].

What Is the Relationship Between Vestibular Neuritis, Labyrinthitis, and Sudden Sensorineural Hearing Loss?

Most textbooks depict vestibular neuritis, labyrinthitis, and sudden sensorineural hearing loss as distinct clinical entities, which may be didactically useful, but is probably an oversimplification. The reality is likely more complex since damage can occur to varying proportions of the vestibulocochlear nerve, varying proportions of the labyrinth, and varying combinations thereof; in other words, the underlying pathobiology probably spans a continuum, with a corresponding range of clinical presentations—with different combinations/degrees of vestibular symptoms (dizziness/imbalance) and ear symptoms (hearing loss, tinnitus).

This "continuum" of disease is likely underappreciated because it is understudied for practical reasons. Specifically, if a patient complains exclusively of hearing symptoms, then she will probably not undergo any balance testing; conversely, if a patient complains exclusively of imbalance, then he will probably not undergo any hearing testing. However, if these patients *do* undergo more comprehensive evaluation, then some surprising results may emerge. Some examples follow.

The literature usually describes vestibular neuritis as presenting purely with vestibular symptoms (dizziness/imbalance) and no auditory symptoms. However, when such patients are studied carefully, it turns out that some of them have detectable hearing deficits on the affected side [10], or abnormalities in other auditory functions such as an elevated stapedius reflex threshold [11, 12]. In other words, "pure" vestibular neuritis may not actually be so pure.

The literature usually describes sudden sensorineural hearing loss as presenting purely with auditory symptoms (hearing loss and sometimes tinnitus) and no vestibular (dizziness/imbalance) symptoms. However, when such patients are studied carefully, it turns out that some of them have detectable vestibular deficits [13–16] on the same side as the hearing loss. In other words, the "pure" sudden sensorineural hearing loss may not actually be so pure.

Thus, these terms ("vestibular neuritis"/"vestibulitis" versus "sudden sensorineural hearing loss"/"cochleitis" versus "labyrinthitis"/"vestibulocochlear neuritis") assume distinct clinical presentations (purely vestibular symptoms, purely auditory symptoms, or a combination thereof, respectively), while other terms ("vestibular neuritis" versus "vestibulitis"; "sudden sensorineural hearing loss" versus "cochleitis"; "cochleovestibular neuritis" versus "labyrinthitis") assume that the underlying disease process is restricted to a discrete anatomical area (the neural pathways versus the inner ear, respectively). But these distinctions, while perhaps terminologically convenient, are flawed, missing what is likely a pathobiological continuum with a corresponding spectrum of clinical presentations. This is illustrated in Table 48.1.

References

1. Cherchi M, Yacovino DA. Dysfunction along the continuum of vestibulocochlear anatomy, and the corresponding spectrum of clinical presentation: how little we know, and what else we need to learn. Hear Balance Commun. 2021;19(4):246–57.
2. Dix MR, Hallpike CS. The pathology symptomatology and diagnosis of certain common disorders of the vestibular system. Proc R Soc Med. 1952;45(6):341–54.
3. Izraeli S, Rachmel A, Nitzan M. Vestibular dysfunction and deafness complicating mumps parotitis. Harefuah. 1989;117(9):243–4.
4. Pogson JM, Taylor RL, Young AS, McGarvie LA, Flanagan S, Halmagyi GM, Welgampola MS. Vertigo with sudden hearing loss: audio-vestibular characteristics. J Neurol. 2016;263(10):2086–96.

5. Kim CH, Choi HR, Choi S, Lee YS, Shin JE. Patterns of nystagmus conversion in sudden sensorineural hearing loss with vertigo. Medicine (Baltimore). 2018;97(43):e12982.
6. Fitzgerald DC, Mark AS. Viral cochleitis with gadolinium enhancement of the cochlea on magnetic resonance imaging scan. Otolaryngol Head Neck Surg. 1999;121(1):130–2.
7. Luigetti M, Cianfoni A, Modoni A, Conte A, Conti G, Sabatelli M. Teaching NeuroImages: cochleitis: a rare cause of acute deafness in a patient with HCV. Neurology. 2011;77(18):e109.
8. Sataloff J, Vassallo L. Head colds and viral cochleitis. Arch Otolaryngol. 1968;87(1):56–9.
9. Goodale AD, Golub JS, Cornelius RS, Samy RN. Isolated cochlear neuritis from varicella reactivation mimicking a vestibular schwannoma. World J Otorhinolaryngol Head Neck Surg. 2016;2(3):181–4.
10. Yao Q, Xu C, Wang H, Shi H, Yu D. Video head impulse test results suggest that different pathomechanisms underlie sudden sensorineural hearing loss with vertigo and vestibular neuritis: our experience in fifty-two patients. Clin Otolaryngol. 2018;43(6):1621–4.
11. Bergenius J, Borg E. Audio-vestibular findings in patients with vestibular neuritis. Acta Otolaryngol. 1983;96(5–6):389–95.
12. Bagger-Sjoback D, Perols O, Bergenius J. Audiovestibular findings in patients with vestibular neuritis: a long-term follow-up study. Acta Otolaryngol Suppl. 1993;503:16–7.
13. Fujimoto C, Egami N, Kinoshita M, Sugasawa K, Yamasoba T, Iwasaki S. Involvement of vestibular organs in idiopathic sudden hearing loss with vertigo: an analysis using oVEMP and cVEMP testing. Clin Neurophysiol. 2015;126(5):1033–8.
14. Hong SM, Byun JY, Park CH, Lee JH, Park MS, Cha CI. Saccular damage in patients with idiopathic sudden sensorineural hearing loss without vertigo. Otolaryngol Head Neck Surg. 2008;139(4):541–5.
15. Zhang Q, Hu J, Xu XD, Chen YF, Zhang Y, Wei JR, et al. Objective evaluation of otolithic end organs in sudden sensorineural hearing loss patients. Zhonghua Er Bi Yan Hou Tou Jing Wai Ke Za Zhi. 2013;48(5):389–93.
16. Korres S, Stamatiou GA, Gkoritsa E, Riga M, Xenelis J. Prognosis of patients with idiopathic sudden hearing loss: role of vestibular assessment. J Laryngol Otol. 2011;125(3):251–7.

Third Window Phenomena

49

Introduction

Third window phenomena [1] are abnormal (or abnormally large) connections between the labyrinth and some other space. Some of these may be congenital, while others are usually acquired. A third window changes hydraulic resistance in the labyrinth and may alter which hair cells get stimulated by vibration and sound. Symptoms often include abnormal auditory perceptions (such as hyperacusis), and sound/vibration-induced oscillopsia, nystagmus, and disequilibrium. Management is surgical.

Anatomy and Normal Physiology

The bony labyrinth is a convoluted space embedded within the temporal bone. A very small amount of fluid can pass very slowly in or out of the labyrinth via various channels (cochlear aqueduct, endolymphatic duct, vestibular aqueduct), but otherwise the total amount of perilymphatic and endolymphatic fluid remains approximately constant. This fluid is effectively incompressible. The ***round window*** and the ***oval window*** are two "holes," each covered with a thin membrane; these comprise an interface between the inner ear and the middle ear.

As sound enters the external auditory canal, the oscillating pressure waves get transduced into the kinetic energy of the eardrum moving inward and outward, and this motion is transmitted through the ossicular chain. At the end of the ossicular chain, the stapes footplate is attached to the membrane covering the oval window. As sounds make the eardrum vibrate, the stapes footplate will push the membrane of the oval window inward and outward in a piston-like motion. Because endolymphatic and perilymphatic fluid are incompressible, any inward movement of the membrane at the round window will move fluid of a certain volume, and that volume must be exactly offset by outward motion somewhere else, which is usually the round window. The alternating inward and outward motion of the membrane of the oval window, and the corresponding alternating outward and inward motion of the membrane of the round window, creates fluid waves that stimulate hair cells in the cochlea that in turn transduce the movement into electrochemical signals that generate action potentials resulting in neural signaling through the cochlear nerve, which we eventually perceive as sound.

Pathology and Pathophysiology

The arrangement described above works when there are two "windows" (the oval window and the round window). If a "***third window***" were to appear somewhere in the labyrinth, there would be two consequences [2].

First, the presence of the third window would lower the total resistance (in the hydrodynamic sense) of the system. The result is that stimulation of inner ear hair cells would be "easier"—which is to say that it would require less input (kinetic energy) to generate the same stimulation of hair cells.

Second, depending on where the third window is placed, it could result in stimulation of different hair cells—specifically, vestibular hair cells [3].

Merchant and Rosowski [4] schematically depict the effects of third window phenomena with air conducted and bone conducted stimuli.

In principle, a third window can be created anywhere in the labyrinth. The main ones are:

- Dehiscence of any of the three semicircular canals.
- Carotid dehiscence.
- Temporal bone fracture involving the labyrinth.

The following phenomena are not "third windows" but are abnormally large passages into the labyrinth which (like third window phenomena) reduce the hydraulic resistance in the labyrinth:

- Large vestibular aqueduct.
- Large cochlear aqueduct.

M. Cherchi, *Otoneurology and Vestibular Medicine*, https://doi.org/10.1007/978-3-031-94841-1_49

Generally, a third window can manifest with clinical correlates of the two consequences described earlier.

First, less energy is required to stimulate inner ear hair cells; clinically, this can manifest with overly sensitive hearing (and conductive hyperacusis on audiometry).

Second, depending on the location of the third window, the path of the fluid waves (generated by sound or vibration) may traverse a region with vestibular hair cells, resulting in some combination of Tullio's phenomenon (sounds cause oscillopsia and disequilibrium) and nystagmus.

Some third window phenomena may be congenital (such as an enlarged vestibular aqueduct), while others may be acquired (such as semicircular canal dehiscence).

Diagnostic Imaging

Depending on the type of third window phenomenon in question, imaging (such as with temporal bone CT) may be confirmatory.

Management

If symptoms from a third window phenomenon are sufficiently intrusive or limiting function, then surgical repair may be contemplated.

References

1. Ota I, Sakagami M, Kitahara T. The third mobile window effects in otology/neurotology. J Int Adv Otol. 2021;17(2):156–61.
2. Iversen MM, Rabbitt RD. Biomechanics of third window syndrome. Front Neurol. 2020;11:891.
3. Iversen MM, Zhu H, Zhou W, Della Santina CC, Carey JP, Rabbitt RD. Sound abnormally stimulates the vestibular system in canal dehiscence syndrome by generating pathological fluid-mechanical waves. Sci Rep. 2018;8(1):10257.
4. Merchant SN, Rosowski JJ. Conductive hearing loss caused by third-window lesions of the inner ear. Otol Neurotol. 2008;29(3):282–9.

50 Semicircular Canal Dehiscence

Overview

In *semicircular canal dehiscence (SCD)*, a bony dehiscence creates an abnormal connection between one of the semicircular canals with another space, and thus is classified as one of several so-called "third window phenomena." Clinically, this can manifest with auditory symptoms (hyperacusis, autophony, pulsatile tinnitus), or vestibular symptoms (Valsalva-induced disequilibrium), or both. The superior semicircular canal is the most commonly involved, though any of the semicircular canals can be affected. Overall, SCD is rare. It can occur spontaneously but has also been documented after head trauma. The diagnosis is suspected based on a combination of a compatible clinical history, results of audiometry (conductive hyperacusis on the affected side), and vestibular evoked myogenic potentials (abnormally large amplitude and abnormally low threshold on the affected side). The diagnosis can be radiographically confirmed on a high-resolution temporal bone CT. Not all cases require treatment; in those that do, surgery can be considered; options include canal plugging and (in the case of the superior semicircular canal) canal re-roofing.

Introduction

Semicircular canal dehiscence (SCD) is one of several so-called "third window phenomena" [1]. In semicircular canal dehiscence, an abnormal hole connects one of the semicircular canals with another space, and this has the two consequences. First, less energy is required to stimulate inner ear hair cells; clinically, this manifests with overly sensitive hearing (and conductive hyperacusis on audiometry). Second, since the hole is within the semicircular canals, fluid waves will propagate from the oval window to the semicircular canal and thereby stimulate vestibular hair cells; clinically, this manifests with oscillopsia from Tullio's phenomenon (loud sounds cause disequilibrium), and ocular motor examination may show corresponding nystagmus.

A "third window" can appear in any of the three semicircular canals. Superior semicircular canal dehiscence (SSCD) is by far the most common variant of this problem, whereas lateral semicircular canal dehiscence (LSCD) and posterior semicircular canal dehiscence (PSCD) are much rarer.

The first kind of SCD described was that involving the superior canal. This was recognized and reported by Minor and colleagues [2]. Carey and colleagues provide excellent postmortem surgical pathology of this condition [3].

SCD is rare. It can occur spontaneously or after trauma. The clinical history includes some combination of vestibular symptoms, hearing loss, hyperacusis, and tinnitus, with exacerbations often caused by Valsalva. On physical examination, Rinne's test may show that bone conduction is superior to air conduction, and ocular motor examination may show nystagmus elicited by loud sounds. Audiometry often shows conductive hyperacusis in the affected ear. On vestibular evoked myogenic potentials (whether cervical or ocular), the affected ear may show an unusually large amplitude and a low threshold. The imaging modality of choice is a temporal bone CT without contrast, with reconstructions in Stenver's and Pöschl's planes. If symptoms are tolerable, then management can be expectant. If symptoms are intolerable, then treatment is surgical, though surgery is not always successful or durable.

Epidemiology

Semicircular canal dehiscence is rare, though the exact incidence and prevalence of SCD is unknown. Part of what makes this difficult to study is the incomplete overlap between the anatomical abnormality and the symptoms.

The phrase "semicircular canal dehiscence" is merely a description of anatomical pathology and is neutral with respect to any clinical manifestations. In contrast, the phrase "semicircular canal dehiscence *syndrome*" refers to a circumstance in which the anatomical abnormality is present, and it is believed to account for a specific set of symptoms.

M. Cherchi, *Otoneurology and Vestibular Medicine*, https://doi.org/10.1007/978-3-031-94841-1_50

This nomenclatural distinction draws attention to the fact that there is incomplete overlap between (1) the group of individuals with radiographic SCD, and (2) the group of individuals with symptoms compatible with SCD syndrome.

Clinical History

Semicircular canal dehiscence can occur after head trauma [4], though it can also occur spontaneously. Some "spontaneously occurring" cases are sometimes suspected to be due to strong Valsalva events, such as coughing or sneezing. The symptoms of SCD are variable.

Naert and colleagues [5] conducted a meta-analysis of symptom frequency reported by 431 patients from 66 articles about superior SCD. The results are summarized in Fig. 50.1.

Our own re-analysis of these data by category of symptom is shown below.

- *Vestibular symptoms*
  - Spontaneous dizziness (51.0%)
  - Sound-induced vertigo (42.7%)
  - Positional vertigo (1.6%)
  - Motion intolerance (1.2%)
  - Drop attack (0.7%)
- *Hearing loss*
  - Hearing loss (39.9%)
- *Hyperacusis*
  - Autophony (42.5%)
  - Hyperacusis to environmental sounds (4.9%)
  - Hyperacusis to bodily sounds (4.6%)
- *Tinnitus*
  - Nonpulsatile tinnitus (17.2%)
  - Spontaneous pulsatile tinnitus (13.7%)
  - Gaze evoked tinnitus (1.2%)
  - Head movement-induced pulsatile tinnitus (0.5%)
  - Tinnitus aggravated by Valsalva (0.2%)
- *Sound distortion* (0.7%)
- *Other aural symptoms*
  - Aural pressure (24.6%)
  - Ear pain (0.7%)
- *Visual symptoms*
  - Spontaneous oscillopsia/tilting (4.2%)
  - Pulsatile oscillopsia (0.9%)
- *Other symptoms*
  - Anxiety (0.2%)
  - Hemifacial numbness (0.2%)

In aggregate, the most common symptom categories of superior SCD are vestibular, hearing loss, hyperacusis, and tinnitus.

In the author's clinical experience, a fair proportion of patients complain of "hearing my eyeballs move," which perhaps corresponds to the "gaze evoked tinnitus" category

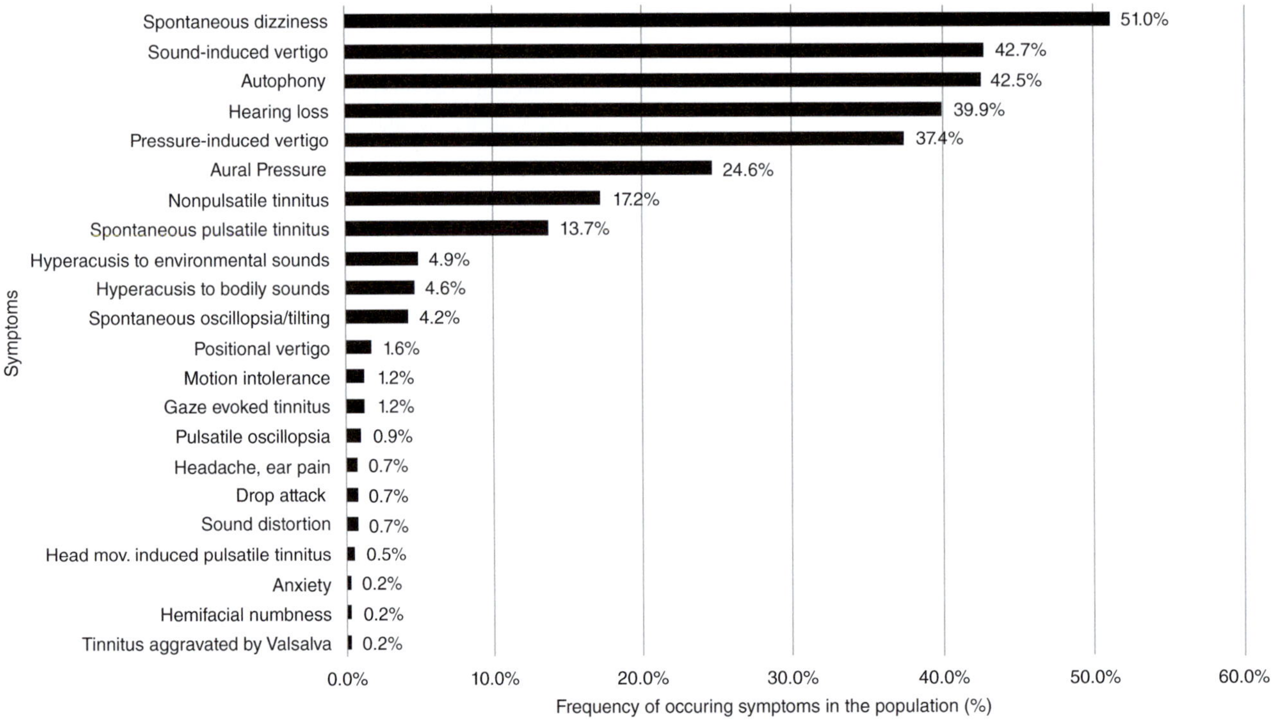

Fig. 50.1 Frequency of symptoms of patients with superior semicircular canal dehiscence. (From Naert et al. [5]. Reproduced with written permission from John Wiley and Sons Limited)

in the analysis of Naert and colleagues [5] above. While it is not frequent, we have found this symptom to be sensitive for superior SCD. Some authors have documented patients also reporting that they can "hear" their eye-blinks [6].

Physical Examination

On physical examination using a tuning for (usually 128 Hz), Rinne's test may show that an ear affected by superior SCD exhibits bone conduction superior to air conduction—in other words, conductive hyperacusis.

Ocular motor examination (especially if done with infrared video oculography) may show oscillatory nystagmus in response to a loud tone (the ocular motor correlate of Tullio's phenomenon). In some cases, there will be a pendular nystagmus (usually with torsional and vertical components) synchronous with the cardiac pulse [7–9].

When an examiner applies the stem of a vibrating 128 Hz tuning fork to either medial malleolus of a patient with unilateral superior semicircular canal dehiscence, the patient may "hear" the tone in the affected ear. This finding is similar to autophony, except that the vibrations in the body are occurring from bone oscillation driven by an externally applied source (the tuning fork). This finding was the basis for "ankle audiometry" [10].

Tests

In SSCD, audiometry usually shows an air–bone gap, and sometimes shows conductive hyperacusis.

Figure 50.2 from Brantberg and colleagues [11] shows audiograms from four patients with SSCD. Each audiogram demonstrates a significant air–bone gap in the affected ear.

On the side affected by SSCD, vestibular evoked myogenic potentials (whether cervical or ocular) often show a combination of an abnormally large amplitude response and a low threshold. Interestingly, the degree of abnormality on vestibular evoked myogenic potentials does not appear to correspond to the degree of symptom severity [12].

Mercado and colleagues [13] provide a clear example from a patient with left-sided superior semicircular canal dehiscence in which cervical vestibular evoked myogenic potentials on the affected side exhibit an unusually large amplitude and unusually low threshold.

In some cases, ocular motor testing may identify rapidly oscillating pendular nystagmus at a specific frequency of stimulation [14–17].

In Fig. 50.3 from Dumas and colleagues [15], a skull vibration frequency of 400 Hz brings out the most prominent nystagmus.

Imaging

The best studied imaging modality to evaluate for SSCD is a temporal bone CT without contrast. The collimation cuts should be 0.6 mm or better. Depending on the radiology service's experience, when there is suspicion for superior canal involvement, it may be helpful to protocol the study with reconstructions in Stenver's and Pöschl's projections.

Stenver's plane, named after Hendrik Willem Stenvers (1889–1973) [18], is an oblique coronal slice (running from anterolateral to posteromedial), which is in the same plane as the superior semicircular canal. Stenver's view acquires images through the plane containing the center of the entire lumen of the semicircular canal.

Pöschl's plane [19], also called the axiopetrosal plane, is an oblique sagittal view (running from anteromedial to posterolateral) that is perpendicular to Stenver's plane. Pöschl's view acquires images through the semicircular canal "end on" (perpendicular to its lumen).

These planes can be imagined by thinking of the so-called RALP-LARP planes. "RALP" refers to the fact that planes of the right anterior semicircular canal and the left posterior canal are parallel. "LARP" refers to the fact that the left anterior semicircular canal and the right posterior semicircular canal are parallel. These are depicted in Fig. 50.4 from Aw and colleagues [20].

For the right superior canal, Stenver's view corresponds to a projection through the RALP plane, while Pöschl's view corresponds to a projection through the LARP plane.

For the left superior canal, Stenver's view corresponds to a projection through the LARP plane, while Pöschl's view corresponds to a projection through the RALP plane.

Thus, the coronal plane, Stenver's plane, and Pöschl's plane provide three different views of the superior semicircular canal. Duman and Dogan [21] juxtapose all three views from the same patient with superior SCD.

Performing a temporal bone CT entails exposure to a significant amount of ionizing radiation, so there is interest in other imaging modalities that could avoid this risk. MRI does not image bone well, but studies are beginning to explore whether high-resolution MRI of the internal auditory canals has utility in the diagnosis of SSCD [22].

While imaging is a helpful tool for securing a diagnosis of SCD, the diagnosis does not rely on radiographic findings alone. As mentioned earlier, there is an incomplete overlap between (1) the group of individuals with radiographic SCD and (2) the group of individuals with symptoms compatible with SCD.

A retrospective study [23] of temporal bone CTs done in 500 consecutive patients with no auditory or vestibular symptoms found that 10 (2%) had radiographic SCD, whereas of 110 patients whose symptoms were compatible with SCD, only 15 (13.6%) had radiographic SCD.

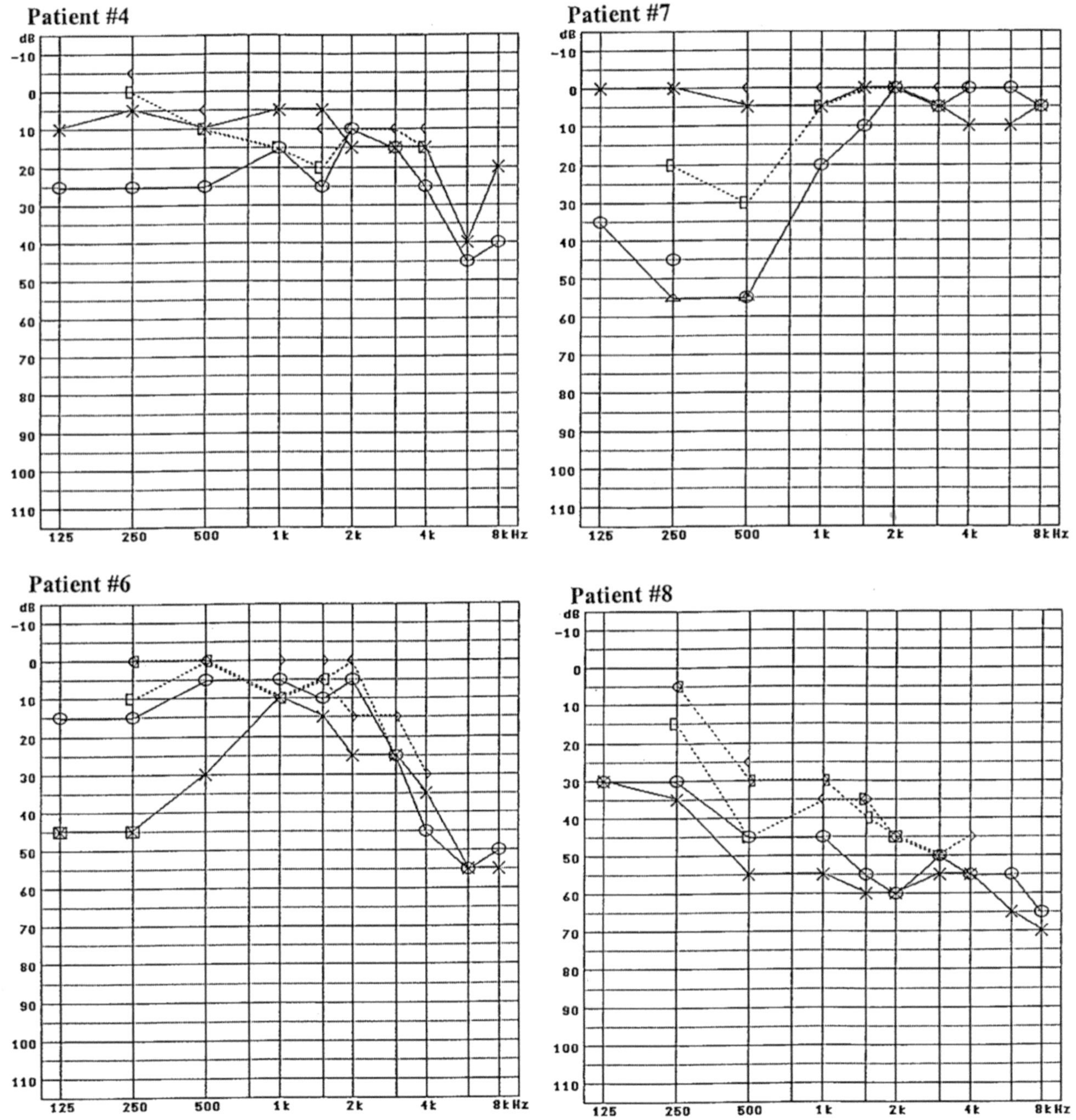

Fig. 50.2 Audiometry in the four patients with asymmetric thresholds. Note low-frequency air–bone gaps with increased thresholds for air-conducted sounds suggesting conductive hearing loss in the symptomatic ear (right ear for patient No. 4 and No. 7, left ear for patient No. 6 and No. 8). (From Brantberg et al. [11]. Reproduced with written permission from Informa UK Limited)

Another study observed that, "When a diagnosis of superior canal dehiscence syndrome (SCDS) was made solely on CT scans, 80% of cases assessed were false positive" [24]. The study retrospectively analyzed CT scans of 82 patients (164 ears), identified that 5 (3%) of 164 ears had radiographic superior SCD, and of those 5 ears, only 1 (0.5%) additionally had a clinical history compatible with superior SCD. The study author concluded that, "This stresses the importance of diagnosing SCDS on the basis of both CT findings and clinical symptoms."

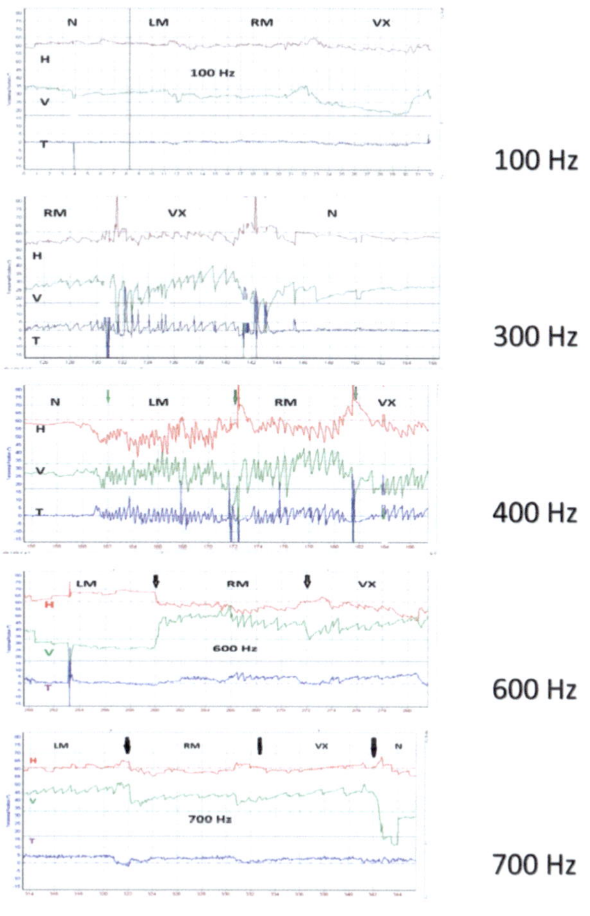

Fig. 50.3 The nystagmus resulting from skull vibration at different frequencies in a patient with right sided superior semicircular canal dehiscence. (From Dumas et al. [15]. Reproduced with written permission from Elsevier)

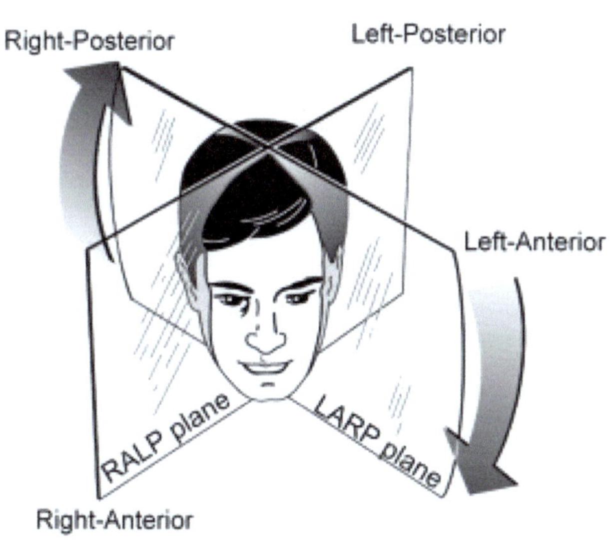

Fig. 50.4 RALP (right anterior and left posterior) and LARP (left anterior and right posterior) planes. (From Aw et al. [20]. Reproduced with written permission from Elsevier)

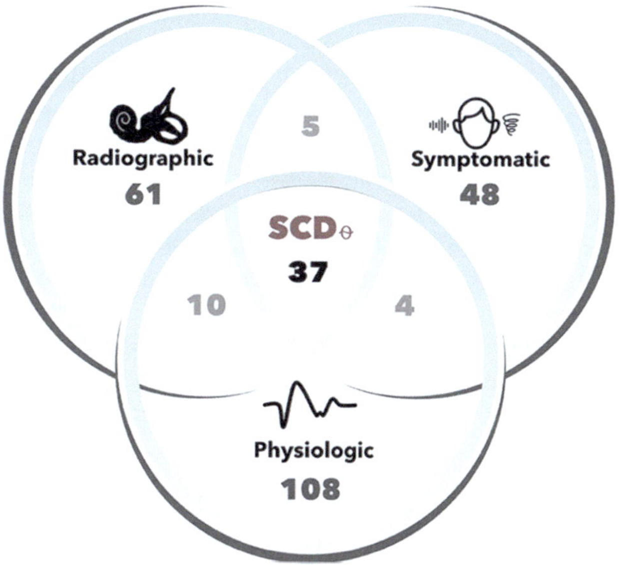

Fig. 50.5 Number of cases for each Ward et al. diagnostic criteria of "SCDθ" which, in the study by Tran et al. [25], refers to the study cohort of clinically significant SCD, as defined by the radiographic, symptomatic, and physiologic diagnostic criteria proposed by Ward et al. [26]. Cases (i.e., ears) were determined to be radiographically positive based upon two-party review of high-resolution (submillimeter) temporal bone-computed tomography (CT) scans. Symptomatic cases included one of the following patient-reported symptoms: hyperacusis/autophony, pulsatile tinnitus, sound-, or pressure-induced vertigo. Physiologic cases were those that had either 500 Hz oVEMP amplitude ≥17 μV, 500 Hz cVEMP threshold ≤75 dB nHL, or negative bone-conduction thresholds. (From Tran et al. [25]. Reproduced from Frontiers Media S.A. under CC-BY 4.0)

Diagnosis

We would go a step further, and add that, when possible, the diagnosis should be supported not only by history and imaging, but also by audiologic and otovestibular testing (such as from audiometry and vestibular evoked myogenic potentials). In other words, the greatest confidence can be achieved when the clinical history, audiologic and otovestibular testing, and imaging all support the diagnosis; in this, we agree with the view espoused by Tran and colleagues [25].

The Venn diagram in Fig. 50.5 from Tran and colleagues [25] schematically depicts the intersection of these types of data; the specific numbers are from their dataset.

Treatment

Depending on whether a patient's symptoms from SSCD are tolerable, one reasonable management strategy is simply watchful waiting.

If a patient's symptoms interfere significantly with function in daily activities, then intervention may be warranted. Treatment of SSCD is surgical. The most common surgical approaches for superior canal involvement are re-roofing the superior canal and plugging the superior canal.

Re-roofing the superior canal is an invasive procedure [27] in which the otolaryngologist usually works in tandem with a neurosurgeon. The neurosurgeon performs a temporal craniotomy and retracts the temporal lobe in order to expose the middle cranial fossa and thus the superior aspect of the temporal bone, after which the otolaryngologist locates the dehiscence and "patches" it, usually with a synthetic substance.

Plugging the superior canal can be performed via a transmastoid surgical approach [11] and is thus less invasive. While plugging the superior canal may eliminate some symptoms from SSCD, it also effectively ablates the function of the superior canal, which can lead to some degree of chronic unsteadiness.

Even if the surgery is initially successful, it is possible for the plugged superior canal to become "un-plugged," or for the re-roofed dehiscence to redehisce.

Prognosis

A multicenter study of 27 patients found that "Patients with larger superior canal dehiscences show significantly more vestibulocochlear symptoms/signs… compared with smaller ones" [28].

Untreated, the clinical course of superior SCD is variable. We have had patients whose symptoms have remained stable for decades, while other patients may deteriorate over months to years.

Variants: Posterior Semicircular Canal Dehiscence (PSCD)

Posterior SCD is rare. One series of 23 patients with PSCD reported that "86% presented with vertigo, 9% with hearing loss or tinnitus" [29]. Some surgeons treat PSCD with canal plugging [30].

Variants: Lateral Semicircular Canal Dehiscence (LSCD)

Lateral SCD is rare. Yaniv and Hacking [31] reported two cases in which each patient had combined round window perilymphatic fistula and lateral SCD. Yaniv and Traub [32] included those patients into a larger series of 15 patients with suspected perilymphatic fistulae, and reported that two patients had both a round window fistula and lateral SCD, and another two patients had isolated lateral SCD.

Differential Diagnosis

The main differential diagnosis is perilymphatic fistula (PLF). Symptoms such as episodic disequilibrium, hearing loss, and tinnitus certainly overlap with other diagnoses, such as Ménière's disease, but in SCD, at least some of the symptoms will tend to have a component of pressure sensitivity.

References

1. Ota I, Sakagami M, Kitahara T. The third mobile window effects in otology/neurotology. J Int Adv Otol. 2021;17(2):156–61.
2. Minor LB, Solomon D, Zinreich JS, Zee DS. Sound- and/or pressure-induced vertigo due to bone dehiscence of the superior semicircular canal. Arch Otolaryngol Head Neck Surg. 1998;124(3):249–58.
3. Carey JP, Minor LB, Nager GT. Dehiscence or thinning of bone overlying the superior semicircular canal in a temporal bone survey. Arch Otolaryngol Head Neck Surg. 2000;126(2):137–47.
4. McCrary HC, Babajanian E, Patel N, Yang S, Kircher M, Carlson ML, et al. Superior semicircular canal dehiscence syndrome following head trauma: a multi-institutional review. Laryngoscope. 2021;131(11):E2810–E8.
5. Naert L, Van de Berg R, Van de Heyning P, Bisdorff A, Sharon JD, Ward BK, et al. Aggregating the symptoms of superior semicircular canal dehiscence syndrome. Laryngoscope. 2018;128(8):1932–8.
6. Bertholon P, Reynard P, Lelonge Y, Peyron R, Vassal F, Karkas A. Hearing eyeball and/or eyelid movements on the side of a unilateral superior semicircular canal dehiscence. Eur Arch Otorrinolaringol. 2018;275(2):629–35.
7. Milenkovic I, Sycha T, Berger-Sieczkowski E, Rommer P, Czerny C, Wiest G. Pulse-synchronous torsional nystagmus. Pract Neurol. 2021;21(5):445–7.
8. Hain TC, Cherchi M. Pulse-synchronous torsional pendular nystagmus in unilateral superior canal dehiscence. Neurology. 2008;70(14):1217–8.
9. Tilikete C, Krolak-Salmon P, Truy E, Vighetto A. Pulse-synchronous eye oscillations revealing bone superior canal dehiscence. Ann Neurol. 2004;56(4):556–60.
10. Verrecchia L, Freden Jansson KJ, Westin M, Velikoselskii A, Reinfeldt S, Hakansson B. Ankle audiometry: a clinical test for the enhanced hearing sensitivity for body sounds in superior canal dehiscence syndrome. Audiol Neurootol. 2023;28(3):219–29.
11. Brantberg K, Bergenius J, Mendel L, Witt H, Tribukait A, Ygge J. Symptoms, findings and treatment in patients with dehiscence of the superior semicircular canal. Acta Otolaryngol. 2001;121(1):68–75.
12. Noij KS, Wong K, Duarte MJ, Masud S, Dewyer NA, Herrmann BS, et al. Audiometric and cVEMP thresholds show little correlation with symptoms in superior semicircular canal dehiscence syndrome. Otol Neurotol. 2018;39(9):1153–62.
13. Mercado VM, Fernández FA, Hernandez CB, Pino CU, Novoa IC, Herrera PC. Dehiscence syndrome superior semicircular canal: a case of dehiscence syndrome of the superior semicircular. J Otolaryngol-ENT Res. 2017;6(6):175–8.
14. Dumas G, Lion A, Karkas A, Perrin P, Perottino F, Schmerber S. Skull vibration-induced nystagmus test in unilateral superior canal dehiscence and otosclerosis: a vestibular Weber test. Acta Otolaryngol. 2014;134(6):588–600.
15. Dumas G, Tan H, Dumas L, Perrin P, Lion A, Schmerber S. Skull vibration induced nystagmus in patients with superior semicircu-

lar canal dehiscence. Eur Ann Otorhinolaryngol Head Neck Dis. 2019;136(4):263–72.
16. Manzari L, Modugno GC, Brandolini C, Pirodda A. Bone vibration-induced nystagmus is useful in diagnosing superior semicircular canal dehiscence. Audiol Neurootol. 2008;13(6):379–87.
17. White JA, Hughes GB, Ruggieri PN. Vibration-induced nystagmus as an office procedure for the diagnosis of superior semicircular canal dehiscence. Otol Neurotol. 2007;28(7):911–6.
18. Stenvers HW. Roentgenology of the os petrosum. Arch Radiol Electrother. 1917;22(4):97–112.
19. Der PM. tomographische Querschnitt durch das Felsenbein [The tomographic cross-section through the petrous bone]. Forschrtschritte auf der Gebiete des Röntgenstrahlen [Adv Radiol]. 1943;68:174–9.
20. Aw ST, Todd MJ, Michael HG. Chapter 12 – Head impulse testing: angular vestibulo-ocular reflex (VOR). Handb Clin Neurophysiol. 2010;9:150–64.
21. Duman IS, Dogan SN. Contribution of reformatted multislice temporal computed tomography images in the planes of Stenvers and Poschl to the diagnosis of superior semicircular canal dehiscence. J Comput Assist Tomogr. 2020;44(1):53–8.
22. Beyazal Celiker F, Ozgur A, Celiker M, Beyazal M, Turan A, Terzi S, et al. The Efficacy of Magnetic Resonance Imaging for the Diagnosis of Superior Semicircular Canal Dehiscence. J Int Adv Otol. 2018;14(1):68–71.
23. Berning AW, Arani K, Branstetter BFt. Prevalence of superior semicircular canal dehiscence on high-resolution CT imaging in patients without vestibular or auditory abnormalities. Ajnr Am J Neuroradiol. 2019;40(4):709–12.
24. Masaki Y. The prevalence of superior canal dehiscence syndrome as assessed by temporal bone computed tomography imaging. Acta Otolaryngol. 2011;131(3):258–62.
25. Tran ED, Swanson A, Sharon JD, Vaisbuch Y, Blevins NH, Fitzgerald MB, et al. Ocular vestibular-evoked myogenic potential amplitudes elicited at 4 kHz optimize detection of superior semicircular canal dehiscence. Front Neurol. 2020;11:879.
26. Ward BK, Carey JP, Minor LB. Superior canal dehiscence syndrome: lessons from the first 20 years. Front Neurol. 2017;8:177.
27. Peterson EC, Lazar DA, Nemecek AN, Duckert L, Rostomily R. Superior semicircular canal dehiscence syndrome: successful treatment with repair of the middle fossa floor: technical case report. Neurosurgery. 2008;63(6):E1207–8. discussion E8.
28. Pfammatter A, Darrouzet V, Gartner M, Somers T, Van Dinther J, Trabalzini F, et al. A superior semicircular canal dehiscence syndrome multicenter study: is there an association between size and symptoms? Otol Neurotol. 2010;31(3):447–54.
29. Krombach GA, DiMartino E, Schmitz-Rode T, Prescher A, Haage P, Kinzel S, et al. Posterior semicircular canal dehiscence: a morphologic cause of vertigo similar to superior semicircular canal dehiscence. Eur Radiol. 2003;13(6):1444–50.
30. Mikulec AA, Poe DS. Operative management of a posterior semicircular canal dehiscence. Laryngoscope. 2006;116(3):375–8.
31. Yaniv E, Hacking DP. Combined perilymphatic fistulas of the round window and lateral semicircular canal. A report of 2 cases. S Afr Med J. 1987;71(6):391–2.
32. Yaniv E, Traub P. Traumatic perilymphatic fistulae of the lateral semicircular canal. J Laryngol Otol. 1988;102(6):521–3.

Perilymphatic Fistula (PLF)

51

Overview

A *perilymphatic fistula (PLF)* is an abnormal connection between the inner ear and the middle ear via a hole in the oval window or round window or both. It is often caused by trauma of various sorts and can occur at any age. It presents with aural, auditory, and vestibular symptoms, often triggered or exacerbated by pressure changes (e.g., Valsalva, diving) or by loud noises. Visualization of the fistula at surgery is the best way to secure the diagnosis. Other testing modalities (imaging, otovestibular testing, assays on middle ear fluid) are inferior to surgical exploration. PLF should be differentiated from labyrinthitis, Ménière's disease, and semicircular canal dehiscence. Treatment of PLF is surgical repair, though it is not always successful. Even when surgical repair is initially successful, PLF can recur.

Surgery, while imperfect, remains the best way to diagnose and treat PLF, but surgery is invasive. Before proceeding to surgery, it is medically reasonable to undertake an otovestibular workup for relevant alternative diagnoses. Specifically:

- On the possibility of labyrinthitis: Check audiometry, otoacoustic emissions, and tests that may detect unilateral vestibular hypofunction (cervical and ocular vestibular evoked myogenic potentials, video head impulse testing, videonystagmography).
- On the possibility of superior semicircular canal dehiscence: Check audiometry, cervical, and ocular vestibular evoked myogenic potentials and (depending on those results) a temporal bone CT.
- On the possibility of Ménière's disease: Check audiometry, electrocochleography, and tests that may detect unilateral vestibular hypofunction (cervical and ocular vestibular evoked myogenic potentials, video head impulse testing, videonystagmography).

If these have been excluded with a reasonable degree of certainty, then referral to otolaryngology is appropriate.

Introduction

Perilymphatic fistula (PLF) is a condition in which an abnormal communication (essentially, a hole) develops in either the oval window, or the round window, or both. This abnormal communication between the inner ear and the middle ear permits vestibular and cochlear hair cells to be stimulated abnormally, manifesting as various combinations of auditory and vestibular symptoms. There are numerous potential causes; it appears that the majority of cases occur following trauma. Diagnosis is difficult; surgical exploration (enabling the surgeon to visualize the middle ear) remains the best way to secure the diagnosis but is obviously invasive. Limited studies suggest that imaging may play some role. Other tests (audiologic, otovestibular, blood tests) are controversial and not widely accepted. Some PLFs heal spontaneously; in other cases, surgery may be curative but is not always successful.

Epidemiology

Because it is difficult to diagnose PLF, the diagnosis is controversial. Consequently, it has been difficult to arrive at useful statistics regarding its incidence and prevalence.

PLF can occur at any age; it can be *congenital* [1–5]; it can occur in the *pediatric* population [6–12], in *young adults* [13], in *middle* and *older* age.

Pathophysiology

An understanding of PLF requires some brief comments on anatomy, hydraulics, and physiology.

The interface between the inner ear and the middle ear consists of the *oval window* (to which the stapes footplate is attached) and the *round window*; both of these "windows" are covered by a soft membrane that can bulge inward or outward. Vibrations (typically from sound) cause an

M. Cherchi, *Otoneurology and Vestibular Medicine*, https://doi.org/10.1007/978-3-031-94841-1_51

oscillating (inward and outward) piston-like motion of the stapes footplate. If the oval window were the only interface, then the stapes would barely be able to move the round window because inner ear fluid is effectively incompressible. However, the round window oscillates in a direction opposite that of the oval window; in other words, when the stapes pushes the oval window inwards, the round window bulges outward, and vice versa. This arrangement makes it possible for vibration of the stapes footplate to be transduced into movement of the perilymph/endolymph and propagate through the labyrinth, usually stimulating hair cells in the cochlea which gets perceived as sound.

If the integrity of the oval window or round window is disrupted, then the overall resistance of the system (in a hydraulic sense) is reduced. There are several consequences to this reduced resistance.

- First, it will be easier to stimulate the inner ear with regular stimuli (such as sound). This can lead to perceiving sounds as louder in the affected ear.
- Second, the labyrinth may be stimulated by phenomena that would not normally do so. This can lead, for example, to auditory and vestibular symptoms triggered by Valsalva (such as during coughing, sneezing, etc.).
- Third, hair cells that are not normally stimulated by sound (specifically, vestibular hair cells) may get stimulated. This can lead to Tullio's phenomenon.

PLF can involve the oval window, or the round window, or both, as illustrated in Fig. 51.1.

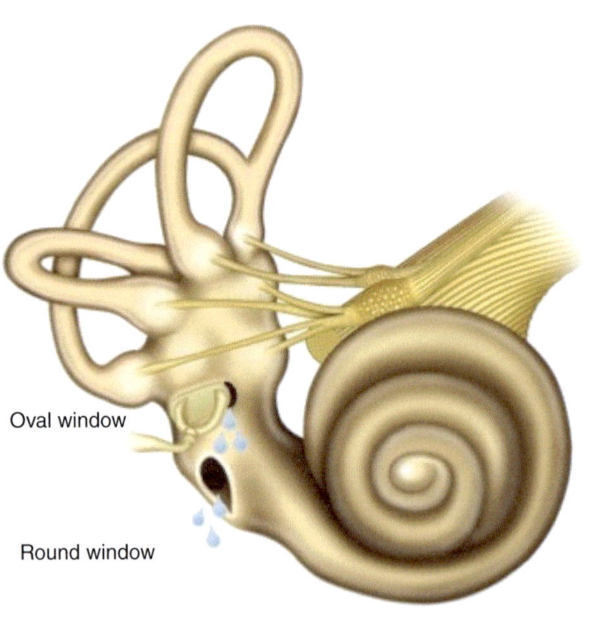

Fig. 51.1 Perilymph fistula at oval and round window (Reproduced with written permission from Dr. Louis Hofmeyr)

Etiology

PLF has been reported in a broad variety of circumstances. The majority of cases reported occur in the context of some sort of trauma [7, 10, 14–27]. A non-exhaustive list of potential etiologies is shown below.

Various sorts of trauma

- Barotrauma [28–32], particularly diving [31–33]
- In the context of the "common cold" [34], during which presumably a person is coughing and/or sneezing
- Sneezing [35]
- Nose blowing [36]
- Vehicular airbag trauma [37]
- Cotton swab use [38]
- Stapes luxation [19, 39]
- Whiplash [40]
- Lightning strike [41, 42]

Pathologies that can potentially erode into the inner ear

- Cholesteatoma [43, 44]
- Granuloma [45]
- Mastoiditis [46]
- Otitis media [47]

Surgical complications

- Complication of stapes surgery [48]
- Complication of surgical drilling elsewhere in the temporal bone [49]
- Complication of transtympanic injection [50]
- Migration of tympanostomy tube [51]

There is a somewhat contentious discussion in the literature regarding whether PLF can also occur spontaneously [11, 52–58].

Clinical Presentation

PLFs usually present with some combination of aural symptoms (e.g., fullness), auditory symptoms (hearing loss, hyperacusis, tinnitus), and vestibular symptoms (disequilibrium).

The symptoms often have episodic exacerbations, sometimes precipitated by pressure changes, whether internally generated (e.g., Valsalva while sneezing, coughing, blowing the nose, straining at stool) or externally generated (e.g., airplane ascent/descent, diving).

In some cases, the symptoms are chronic, though in such cases, the baseline symptoms are generally punctuated by acute exacerbations, as described earlier.

Diagnosis: Physical Examination

Often, physical examination is normal. However, in some cases, it may be possible to observe a bedside Tullio's phenomenon by presenting a loud noise that elicits an ocular motor response (usually an oscillatory nystagmus). In other cases, it may be possible to increase middle ear pressure (either with a pneumatic otoscope or by applying pressure on the tragus) and observe nystagmus [59].

Diagnosis: Imaging

Since PLF involves a very small hole in very small structures, it is difficult for this to be captured on conventional imaging. Nevertheless, a number of studies and case series have collected radiographic observations [8, 60, 61], with some focusing on CT [5, 62, 63] and others focusing on MRI [64–66].

Diagnosis: Audiologic and Vestibular Testing

A variety of otovestibular tests have been explored for the diagnosis of PLF, including:

- Audiometry [67] and "positional audiometry" [68]
- Electrocochleography [69–75] and "positional electrocochleography" [76, 77]
- Otoacoustic emissions [78]
- Vestibular evoked myogenic potentials [79, 80]
- Caloric testing [80, 81]
- Computerized dynamic posturography [82–85]

Ultimately, none of these methods has been widely accepted as diagnostic of PLF.

Diagnosis: Laboratory Tests on Middle Ear Fluid

Although surgery remains considered the best way to diagnose PLF, sometimes the surgeon will not find clear evidence of PLF intraoperatively [74, 86, 87]. On the possibilities that a fistula is very small, or has recently healed, some efforts have been directed at laboratory assays for fluid found in the middle ear at surgery with the goal of confirming/refuting the clinically suspected diagnosis of PLF. The main assays that have been studied are for beta-2-transferrin [12, 88–92], beta trace protein [93, 94], and cochlin-tomoprotein [95–98]. None of these has been widely accepted.

Diagnosis: Surgery

Surgery remains the gold standard for diagnosing PLF [99–108]. In this case, "surgery" means accessing the middle ear space either by myringotomy or by endoscopy, and observing the oval and round windows to detect whether fluid is seeping from them, as shown in Rawal and colleagues [10]. While this remains the gold standard, it is imperfect, in that (1) it is invasive; and (2) many cases of clinically suspected PLF show no evidence of it during surgery [74, 86, 87].

Diagnosis: Differential Diagnosis

A number of diseases besides PLF can cause auditory and vestibular symptoms. While the differential diagnosis can be broad, the main considerations include:

- Chronic symptoms from PLF can be difficult to differentiate from labyrinthitis.
- Episodic symptoms from PLF can be difficult to differentiate from Ménière's disease [109, 110].
- Chronic and episodic symptoms from PLF can be difficult to differentiate from semicircular canal dehiscence [111].
- PLF can coexist with other otologic pathologies, such as lateral semicircular canal dehiscence [112, 113].

Treatment

Surgical repair of PLF remains the mainstay of treatment [10, 28, 114–124]. However, it must be kept in mind that surgery is not always successful [101], and even if surgery is initially successful, PLF can recur [125].

Given that surgery is invasive and not always successful, a number of other treatments have been investigated, including autologous blood patch [126, 127] and ventriculoperitoneal shunt [128, 129]. However, none of these has had the same rate of success as surgery.

Prognosis

Prognosis of PLF, both treated and untreated, is quite variable [26, 123, 130], and often, these patients have long-term follow-up with an otolaryngologist.

References

1. deJong AL. Congenital perilymphatic fistula. Arch Otolaryngol Head Neck Surg. 1998;124(11):1279–81.
2. Reilly JS. Congenital perilymphatic fistula: a prospective study in infants and children. Laryngoscope. 1989;99(4):393–7.
3. Reilly JS, Kenna MA. Congenital perilymphatic fistula: an overlooked diagnosis? Am J Otol. 1989;10(6):496–8.
4. Weber PC, Perez BA, Bluestone CD. Congenital perilymphatic fistula and associated middle ear abnormalities. Laryngoscope. 1993;103(2):160–4.
5. Weissman JL, Weber PC, Bluestone CD. Congenital perilymphatic fistula: computed tomography appearance of middle ear and inner ear anomalies. Otolaryngol Head Neck Surg. 1994;111(3 Pt 1):243–9.
6. Healy GB, Friedman JM, DiTroia J. Ataxia and hearing loss secondary to perilymphatic fistula. Pediatrics. 1978;61(2):238–41.
7. Kim SH, Kazahaya K, Handler SD. Traumatic perilymphatic fistulas in children: etiology, diagnosis and management. Int J Pediatr Otorhinolaryngol. 2001;60(2):147–53.
8. Nakashima T, Sone M, Teranishi MA, Tominaga M, Sugiura M, Naganawa S. Imaging of a congenital perilymphatic fistula. Int J Pediatr Otorhinolaryngol. 2003;67(4):421–5.
9. Pappas DG, Simpson LC, Godwin GH. Perilymphatic fistula in children with preexisting sensorineural hearing loss. Laryngoscope. 1988;98(5):507–10.
10. Rawal R, Zhao X, Lipson S, J RB. Endoscopic repair of traumatic Perilymphatic fistula in children: a case series. J Int Adv Otol. 2021;17(2):182–5.
11. Ruben RJ, Yankelowitz SM. Spontaneous perilymphatic fistula in children. Am J Otol. 1989;10(3):198–207.
12. Weber PC, Kelly RH, Bluestone CD, Bassiouny M. Beta 2-transferrin confirms perilymphatic fistula in children. Otolaryngol Head Neck Surg. 1994;110(4):381–6.
13. Fitzgerald DC. Perilymphatic fistula in teens and young adults: emphasis on preexisting sensorineural hearing loss. Am J Otol. 1996;17(3):397–400.
14. Fee GA. Traumatic perilymphatic fistulas. Arch Otolaryngol. 1968;88(5):477–80.
15. Fife TD, Giza C. Posttraumatic vertigo and dizziness. Semin Neurol. 2013;33(3):238–43.
16. Fitzgerald DC. Persistent dizziness following head trauma and perilymphatic fistula. Arch Phys Med Rehabil. 1995;76(11):1017–20.
17. Glasscock ME 3rd, Hart MJ, Rosdeutscher JD, Bhansali SA. Traumatic perilymphatic fistula: how long can symptoms persist? A follow-up report. Am J Otol. 1992;13(4):333–8.
18. Gunesh RP, Huber AM. Traumatic perilymphatic fistula. Ann Otol Rhinol Laryngol. 2003;112(3):221–2.
19. Khoo LS, Tan TY. Traumatic perilymphatic fistula secondary to stapes luxation into the vestibule: a case report. Ear Nose Throat J. 2011;90(5):E28–31.
20. Kita AE, Kim I, Ishiyama G, Ishiyama A. Perilymphatic fistula after penetrating ear trauma. Clin Pract Cases Emerg Med. 2019;3(2):115–8.
21. Koksal A, Ayyildiz V, Ogul H, Kantarci M. Case report of a patient with posttraumatic Perilymphatic fistula. Ear Nose Throat J. 2022;104:469–71.
22. Lehrer JF, Poole DC, Sigal B. Use of the glycerin test in the diagnosis of post-traumatic perilymphatic fistulas. Am J Otolaryngol. 1980;1(3):207–10.
23. Nishiike S, Hyo Y, Fukushima H. Stapediovestibular dislocation with pneumolabyrinth. J Laryngol Otol. 2008;122(4):419–21.
24. Osetinsky LM, Hamilton GS 3rd, Carlson ML. Sport injuries of the ear and temporal bone. Clin Sports Med. 2017;36(2):315–35.
25. Prisman E, Ramsden JD, Blaser S, Papsin B. Traumatic perilymphatic fistula with pneumolabyrinth: diagnosis and management. Laryngoscope. 2011;121(4):856–9.
26. Tsubota M, Shojaku H, Watanabe Y. Prognosis of inner ear function in pneumolabyrinth: case report and literature review. Am J Otolaryngol. 2009;30(6):423–6.
27. Whitelaw AS, Young I. A case of perilymphatic fistula in blunt head injury. Emerg Med J. 2005;22(12):921.
28. Ahn J, Son SE, Choi JE, Cho YS, Chung WH. Surgical outcomes on hearing and vestibular symptoms in Barotraumatic Perilymphatic fistula. Otol Neurotol. 2019;40(4):e356–e63.
29. McGhee MA, Dornhoffer JL. A case of barotrauma-induced pneumolabyrinth secondary to perilymphatic fistula. Ear Nose Throat J. 2000;79(6):456–9.
30. Pullen FW 2nd. Perilymphatic fistula induced by barotrauma. Am J Otol. 1992;13(3):270–2.
31. Sheridan MF, Hetherington HH, Hull JJ. Inner ear barotrauma from scuba diving. Ear Nose Throat J. 1999;78(3):181. 4, 6-7 passim
32. Shupak A. Recurrent diving-related inner ear barotrauma. Otol Neurotol. 2006;27(8):1193–6.
33. Morvan JB, Gempp E, Riviere D, Louge P, Vallee N, Verdalle P. Perilymphatic fistula after underwater diving: a series of 11 cases. Diving Hyperb Med. 2016;46(2):72–5.
34. Klokker M, Vesterhauge S. Perilymphatic fistula in cabin attendants: an incapacitating consequence of flying with common cold. Aviat Space Environ Med. 2005;76(1):66–8.
35. Comacchio F, Mion M. Sneezing and Perilymphatic fistula of the round window: case report and systematic review of the literature. J Int Adv Otol. 2018;14(1):106–11.
36. Lee JS, Kwon SY, Kim JH, Kim HJ. Bilateral sequential Pneumolabyrinth resulting from nose blowing. J Audiol Otol. 2015;19(3):182–5.
37. Ferber-Viart C, Postec F, Duclaux R, Dubreuil C. Perilymphatic fistula following airbag trauma. Laryngoscope. 1998;108(8 Pt 1):1255–7.
38. Smith M, Darrat I, Seidman M. Otologic complications of cotton swab use: one institution's experience. Laryngoscope. 2012;122(2):409–11.
39. Hatano A, Rikitake M, Komori M, Irie T, Moriyama H. Traumatic perilymphatic fistula with the luxation of the stapes into the vestibule. Auris Nasus Larynx. 2009;36(4):474–8.
40. Markou K, Rachovitsas D, Veros K, Tsiropoulos G, Tsalighopoulos M, Psillas G. Perilymphatic fistula of the round window after whiplash injury: another cause of inner ear conductive hearing loss. Am J Otolaryngol. 2014;35(6):822–5.
41. Kilic E, Genc H, Aydin U, Asik B, Satar B. Variations in otological presentation of lightning strike victims: clinical report of 3 patients. Ulus Travma Acil Cerrahi Derg. 2017;23(2):163–6.
42. Sun GH, Simons JP, Mandell DL. Bilateral perilymphatic fistulas from a lightning strike: a case report. Laryngoscope. 2006;116(6):1039–42.
43. Gormley PK. Surgical management of labyrinthine fistula with cholesteatoma. J Laryngol Otol. 1986;100(10):1115–23.
44. Magliulo G, Terranova G, Varacalli S, Sepe C. Labyrinthine fistula as a complication of cholesteatoma. Am J Otol. 1997;18(6):697–701.
45. Kuhweide R, van de Steene V, Vlaminck S, Casselman JW. Reparative granuloma related to perilymphatic fistula. Adv Otorhinolaryngol. 2007;65:296–9.
46. McCabe BF. Labyrinthine fistula in chronic mastoiditis. Ann Otol Rhinol Laryngol Suppl. 1984;112:138–41.
47. Bluestone CD. Otitis media and congenital perilymphatic fistula as a cause of sensorineural hearing loss in children. Pediatr Infect Dis J. 1988;7(11 Suppl):S141–5.

48. Albera R, Canale A, Lacilla M, Cavalot AL, Ferrero V. Delayed vertigo after stapes surgery. Laryngoscope. 2004;114(5):860–2.
49. Tsunoda A, Anzai T, Matsumoto F, Ikeda K, Tsunoda R. A case of perilymphatic fistula caused by surgical drilling with preserved normal hearing. Am J Otolaryngol. 2021;42(2):102753.
50. Qureshi HA, Zeitler DM. Intratympanic steroid injection complicated by iatrogenic Perilymphatic fistula: a cautionary tale. Laryngoscope. 2021;131(9):2088–90.
51. Hajiioannou JK, Bathala S, Marnane CN. Case of perilymphatic fistula caused by medially displaced tympanostomy tube. J Laryngol Otol. 2009;123(8):928–30.
52. Cole GG. Validity of spontaneous perilymphatic fistula. Am J Otol. 1995;16(6):815–9.
53. Collison PJ, Pons KC. "Spontaneous" perilymph fistula: a case report. Ann Otol Rhinol Laryngol. 2004;113(4):329–34.
54. Gibson WP. Spontaneous perilymphatic fistula: electrophysiologic findings in animals and man. Am J Otol. 1993;14(3):273–7.
55. Kohut RI, Hinojosa R, Thompson JN, Ryu JH. Idiopathic perilymphatic fistulas. A temporal bone histopathologic study with clinical, surgical, and histopathologic correlations. Arch Otolaryngol Head Neck Surg. 1995;121(4):412–20.
56. Kohut RI, Hinojosa R, Thompson JN, Ryu JH. Idiopathic perilymphatic fistulae: a temporal bone histopathological study. Clinical, surgical and histopathological correlations. Acta Otolaryngol Suppl. 1995;520(Pt 1):225–34.
57. Meyerhoff WL. Spontaneous perilymphatic fistula: myth or fact. Am J Otol. 1993;14(5):478–81.
58. Pashley NR, Shapiro R. Spontaneous perilymphatic fistula. J Otolaryngol. 1978;7(2):110–8.
59. Casale M, Errante Y, Sabatino L, Incammisa A, Salvinelli F, Quattrocchi CC. Perilimphatic fistula test: a video clip demonstration. Eur Rev Med Pharmacol Sci. 2014;18(23):3549–50.
60. Ehmer DR Jr, Booth T, Kutz JW Jr, Roland PS. Radiographic diagnosis of trans-stapedial cerebrospinal fluid fistula. Otolaryngol Head Neck Surg. 2010;142(5):694–8.
61. Venkatasamy A, Al Ohraini Z, Karol A, Karch-Georges A, Riehm S, Rohmer D, et al. CT and MRI for the diagnosis of perilymphatic fistula: a study of 17 surgically confirmed patients. Eur Arch Otorrinolaringol. 2020;277(4):1045–51.
62. Bozorg Grayeli A, Bensimon JL, Guyon M, Aho Glele S, Toupet M. Detection of perilymphatic fistula in labyrinthine windows by virtual endoscopy and variation of reconstruction thresholds on CT scan. Acta Otolaryngol. 2020;140(4):270–6.
63. Kvestad E, Kvaerner KJ, Mair IW. Labyrinthine fistula detection: the predictive value of vestibular symptoms and computerized tomography. Acta Otolaryngol. 2001;121(5):622–6.
64. Algin O, Bercin S, Akgunduz G, Turkbey B, Cetin H. Evaluation of labyrinthine fistula by MR cisternography. Emerg Radiol. 2012;19(6):557–60.
65. Dubrulle F, Chaton V, Risoud M, Farah H, Charley Q, Vincent C. The round window sign: a sensitive sign to detect perilymphatic fistulae on delayed postcontrast 3D-FLAIR sequence. Eur Radiol. 2020;30(11):6303–10.
66. Morris MS, Kil J, Carvlin MJ. Magnetic resonance imaging of perilymphatic fistula. Laryngoscope. 1993;103(7):729–33.
67. Fukaya T, Nomura Y. Audiological aspects of idiopathic perilymphatic fistula. Acta Otolaryngol Suppl. 1988;456:68–73.
68. Hazell JW, Fraser JG, Robinson PJ. Positional audiometry in the diagnosis of perilymphatic fistula. Am J Otol. 1992;13(3):263–9.
69. Arenberg IK, Ackley RS, Ferraro J, Muchnik C. ECoG results in perilymphatic fistula: clinical and experimental studies. Otolaryngol Head Neck Surg. 1988;99(5):435–43.
70. Campbell KC, Parnes L. Electrocochleographic recordings in chronic and healed perilymphatic fistula. J Otolaryngol. 1992;21(3):213–7.
71. Campbell KC, Savage MM. Electrocochleographic recordings in acute and healed perilymphatic fistula. Arch Otolaryngol Head Neck Surg. 1992;118(3):301–4.
72. Campbell KC, Savage MM, Harker LA. Electrocochleography in the presence and absence of perilymphatic fistula. Ann Otol Rhinol Laryngol. 1992;101(5):403–7.
73. Gibson WP. Electrocochleography in the diagnosis of perilymphatic fistula: intraoperative observations and assessment of a new diagnostic office procedure. Am J Otol. 1992;13(2):146–51.
74. Morris MS. Perilymphatic fistula with no visible leak of fluid into the middle ear: a new method of intraoperative diagnosis using electrocochleography. Am J Otol. 1994;15(5):695.
75. Sass K, Densert B, Magnusson M. Transtympanic electrocochleography in the assessment of perilymphatic fistulas. Audiol Neurootol. 1997;2(6):391–402.
76. Campbell KC, Abbas PJ. Electrocochleography with postural changes in perilymphatic fistula and Meniere's disease: case reports. J Am Acad Audiol. 1993;4(6):376–83.
77. Campbell KC, Abbas PJ. Electrocochleography with postural changes in perilymphatic fistula. Animal studies. Ann Otol Rhinol Laryngol. 1994;103(6):474–82.
78. Kokesh J, Norton SJ, Duckert LG. Effect of perilymphatic fistulas on evoked otoacoustic emissions in the Guinea pig. Am J Otol. 1994;15(4):466–73.
79. Modugno GC, Magnani G, Brandolini C, Savastio G, Pirodda A. Could vestibular evoked myogenic potentials (VEMPs) also be useful in the diagnosis of perilymphatic fistula? Eur Arch Otorrinolaringol. 2006;263(6):552–5.
80. Suzuki M, Kitajima N, Ushio M, Shintani M, Ishibashi T. Changes in the Tullio phenomenon and the fistula sign in the course of endolymphatic hydrops. ORL J Otorhinolaryngol Relat Spec. 2003;65(2):125–8.
81. Dapsit CP, Churchill D, Linthicum FH Jr. Diagnosis of perilymph fistula using ENG and impedance. Laryngoscope. 1980;90(2):217–23.
82. Black FO, Lilly DJ, Peterka RJ, Shupert C, Hemenway WG, Pesznecker SC. The dynamic posturographic pressure test for the presumptive diagnosis of perilymph fistulas. Neurol Clin. 1990;8(2):361–74.
83. Pyykko I, Selmani Z, Zou J. Low-frequency sound pressure and transtympanic endoscopy of the middle ear in assessment of "spontaneous" perilymphatic fistula. ISRN Otolaryngol. 2012;2012:137623.
84. Selmani Z, Ishizaki H, Pyykko I. Can low frequency sound stimulation during posturography help diagnosing possible perilymphatic fistula in patients with sensorineural hearing loss and/or vertigo? Eur Arch Otorrinolaringol. 2004;261(3):129–32.
85. Shepard NT, Telian SA, Niparko JK, Kemink JL, Fujita S. Platform pressure test in identification of perilymphatic fistula. Am J Otol. 1992;13(1):49–54.
86. Aso S, Gibson WP. Perilymphatic fistula with no visible leak of fluid into the middle ear: a new method of intraoperative diagnosis using electrocochleography. Am J Otol. 1994;15(1):96–100.
87. Haubner F, Rohrmeier C, Koch C, Vielsmeier V, Strutz J, Kleinjung T. Occurence of a round window membrane rupture in patients with sudden sensorineural hearing loss. BMC Ear Nose Throat Disord. 2012;12:14.
88. Bassiouny M, Hirsch BE, Kelly RH, Kamerer DB, Cass SP. Beta 2 transferrin application in otology. Am J Otol. 1992;13(6):552–5.
89. Bluestone CD. Implications of beta-2 transferrin assay as a marker for perilymphatic versus cerebrospinal fluid labyrinthine fistula. Am J Otol. 1999;20(5):701.
90. Delaroche O, Bordure P, Lippert E, Sagniez M. Perilymph detection by beta 2-transferrin immunoblotting assay. Application to the diagnosis of perilymphatic fistulae. Clin Chim Acta. 1996;245(1):93–104.

91. Skedros DG, Cass SP, Hirsch BE, Kelly RH. Sources of error in use of beta-2 transferrin analysis for diagnosing perilymphatic and cerebral spinal fluid leaks. Otolaryngol Head Neck Surg. 1993;109(5):861–4.
92. Weber PC, Bluestone CD, Kenna MA, Kelley RH. Correlation of beta-2 transferrin and middle ear abnormalities in congenital perilymphatic fistula. Am J Otol. 1995;16(3):277–82.
93. Bachmann-Harildstad G, Stenklev NC, Myrvoll E, Jablonski G, Klingenberg O. Beta-trace protein as a diagnostic marker for perilymphatic fluid fistula: a prospective controlled pilot study to test a sample collection technique. Otol Neurotol. 2011;32(1):7–10.
94. Michel O, Petereit H, Klemm E, Walther LE, Bachmann-Harildstad G. First clinical experience with beta-trace protein (prostaglandin D synthase) as a marker for perilymphatic fistula. J Laryngol Otol. 2005;119(10):765–9.
95. Fujita T, Kobayashi T, Saito K, Seo T, Ikezono T, Doi K. Vestibule-middle ear dehiscence tested with perilymph-specific protein Cochlin-Tomoprotein (CTP) detection test. Front Neurol. 2019;10:47.
96. Kataoka Y, Ikezono T, Fukushima K, Yuen K, Maeda Y, Sugaya A, et al. Cochlin-tomoprotein (CTP) detection test identified perilymph leakage preoperatively in revision stapes surgery. Auris Nasus Larynx. 2013;40(4):422–4.
97. Lee K, Ochi N, Yamahara K, Makino K, Ikezono T. A case of Perilymphatic fistula with inner ear anomaly diagnosed preoperatively by the Cochlin-Tomoprotein detection test. Case Rep Otolaryngol. 2020;2020:9476915.
98. Matsuda H, Sakamoto K, Matsumura T, Saito S, Shindo S, Fukushima K, et al. A nationwide multicenter study of the Cochlin Tomo-protein detection test: clinical characteristics of perilymphatic fistula cases. Acta Otolaryngol. 2017;137(sup565):S53–S9.
99. Alzahrani M, Fadous R, Dufour JJ, Saliba I. Perilymphatic fistulas: can we predict the diagnosis? Eur Arch Otorrinolaringol. 2015;272(8):1885–91.
100. Harvey SA, Millen SJ. Absent round window reflex: possible relation to step-wise hearing loss. Am J Otol. 1994;15(2):237–42.
101. Heilen S, Lang CP, Warnecke A, Lenarz T, Durisin M. Exploratory tympanotomy in sudden sensorineural hearing loss for the identification of a perilymphatic fistula - retrospective analysis and review of the literature. J Laryngol Otol. 2020;134(6):501–8.
102. Ogawa K, Kanzaki J, Ogawa S, Tsuchihashi N, Inoue Y, Yamamoto M. Endoscopic diagnosis of idiopathic perilymphatic fistula. Acta Otolaryngol Suppl. 1994;514:63–5.
103. Poe DS, Bottrill ID. Comparison of endoscopic and surgical explorations for perilymphatic fistulas. Am J Otol. 1994;15(6):735–8.
104. Poe DS, Rebeiz EE, Pankratov MM. Evaluation of perilymphatic fistulas by middle ear endoscopy. Am J Otol. 1992;13(6):529–33.
105. Prenzler NK, Schwab B, Kaplan DM, El-Saied S. The role of explorative tympanotomy in patients with sudden sensorineural hearing loss with and without perilymphatic fistula. Am J Otolaryngol. 2018;39(1):46–9.
106. Sekula J, Wlodyka J. The round window in acute hearing loss. Audiology. 1982;21(1):55–60.
107. Selmani Z, Pyykko I, Ishizaki H, Marttila TI. Role of transtympanic endoscopy of the middle ear in the diagnosis of perilymphatic fistula in patients with sensorineural hearing loss or vertigo. ORL J Otorhinolaryngol Relat Spec. 2002;64(5):301–6.
108. Yanagihara N, Nishioka I. Pneumolabyrinth in perilymphatic fistula: report of three cases. Am J Otol. 1987;8(4):313–8.
109. Arenberg IK, May M, Stroud MH. Perilymphatic fistula: an unusual cause of of Meniere's syndrome in a prepubertal child. Laryngoscope. 1974;84(2):243–6.
110. Fitzgerald DC. Perilymphatic fistula and Meniere's disease. Clinical series and literature review. Ann Otol Rhinol Laryngol. 2001;110(5 Pt 1):430–6.
111. Weinreich HM, Carey JP. Perilymphatic fistulas and superior semi-Circular Canal dehiscence syndrome. Adv Otorhinolaryngol. 2019;82:93–100.
112. Yaniv E, Hacking D, Ziv-El L. Perilymphatic fistulas: are they exclusive to the round and oval windows? Laryngoscope. 1986;96(4):360–2.
113. Yaniv E, Hacking DP. Combined perilymphatic fistulas of the round window and lateral semicircular canal. A report of 2 cases. S Afr Med J. 1987;71(6):391–2.
114. Black FO, Pesznecker S, Norton T, Fowler L, Lilly DJ, Shupert C, et al. Surgical management of perilymphatic fistulas: a Portland experience. Am J Otol. 1992;13(3):254–62.
115. Deguine O, Latil d'Albertas D, Fraysse B. Comparison of postoperative results in suspected and confirmed cases of perilymphatic fistula. Rev Laryngol Otol Rhinol (Bord). 1995;116(2):95–8.
116. Fitzgerald DC, Getson P, Brasseux CO. Perilymphatic fistula: a Washington, DC, experience. Ann Otol Rhinol Laryngol. 1997;106(10 Pt 1):830–7.
117. House JW, Morris MS, Kramer SJ, Shasky GL, Coggan BB, Putter JS. Perilymphatic fistula: surgical experience in the United States. Otolaryngol Head Neck Surg. 1991;105(1):51–61.
118. Kimitsuki T, Hara Y, Komune S. Hearing preservation in perilymphatic fistula due to a congenital fistula in an adult. Eur Arch Otorrinolaringol. 2004;261(3):133–5.
119. Lehrer JF, Rubin RC, Poole DC, Hubbard JH, Wille R, Jacobs GB. Perilymphatic fistula–a definitive and curable cause of vertigo following head trauma. West J Med. 1984;141(1):57–60.
120. Maitland CG. Perilymphatic fistula. Curr Neurol Neurosci Rep. 2001;1(5):486–91.
121. Omichi R, Kariya S, Maeda Y, Nishizaki K. Transcanal endoscopic ear surgery for perilymphatic fistula after electric acoustic stimulation. Auris Nasus Larynx. 2018;45(3):657–60.
122. Palva T, Ramsay H. Treatment of labyrinthine fistula. Arch Otolaryngol Head Neck Surg. 1989;115(7):804–6.
123. Weber PC, Bluestone CD, Perez B. Outcome of hearing and vertigo after surgery for congenital perilymphatic fistula in children. Am J Otolaryngol. 2003;24(3):138–42.
124. Weider DJ. Tinnitus: report of ten cases of Perilymphatic fistula and/or endolymphatic hydrops improved by surgery. Int Tinnitus J. 1997;3(1):11–21.
125. Gyo K, Kobayashi T, Yumoto E, Yanagihara N. Postoperative recurrence of perilymphatic fistulas. Acta Otolaryngol Suppl. 1994;514:59–62.
126. Foster PK. Autologous intratympanic blood patch for presumed perilymphatic fistulas. J Laryngol Otol. 2016;130(12):1158–61.
127. Garg R, Djalilian HR. Intratympanic injection of autologous blood for traumatic perilymphatic fistulas. Otolaryngol Head Neck Surg. 2009;141(2):294–5.
128. Lollis SS, Weider DJ, Phillips JM, Roberts DW. Ventriculoperitoneal shunt insertion for the treatment of refractory perilymphatic fistula. J Neurosurg. 2006;105(1):1–5.
129. Weider DJ, Roberts DW, Phillips J. Ventriculoperitoneal shunt as treatment for perilymphatic fistula: a report of six cases. Int Tinnitus J. 2005;11(2):137–45.
130. Kubo T, Kohno M, Naramura H, Itoh M. Clinical characteristics and hearing recovery in perilymphatic fistulas of different etiologies. Acta Otolaryngol. 1993;113(3):307–11.

Tinnitus, Idiopathic

52

Overview

Idiopathic tinnitus is technically an auditory hallucination—the false perception of sound in the absence of any incoming auditory stimulus. Most people have experienced some degree of idiopathic tinnitus, but a percentage of the population experiences distressing tinnitus. The mechanism of idiopathic tinnitus is poorly understood, though it occurs commonly in association with hearing loss. Tinnitus is difficult to measure since it is a purely subjective phenomenon. Treatment of idiopathic tinnitus is difficult.

Demographics

In 2022, Jarach and colleagues [1] published a systematic review and meta-analysis about the global disease burden of tinnitus, which, "estimated that the annual incidence of tinnitus is approximately 1%, with 14% of adults experiencing any tinnitus and 2% experiencing a severe form of it. The prevalence of tinnitus did not differ by sex, but increased prevalence was associated with increasing age, with any tinnitus being present in 10% of young adults, 14% of middle-aged adults, and 24% of older adults."

Definitions

For purposes of this discussion, we focus on *idiopathic tinnitus*, defined as the false perception of sound in the absence of any acoustic stimulus. Tinnitus by this definition is an auditory hallucination (perception in the absence of stimulus).

There are other phenomena that are sometimes classified as tinnitus.

The first is *somatic tinnitus*, which refers to auditory perception of a vibration that is being generated somewhere in the body. For example, some people can "hear" themselves swallow; there are actual vibrations being generated by the musculature in the larynx, and the ear can "hear" these vibrations. Another example of somatic tinnitus is *pulsatile tinnitus*, which is generated by rhythmic turbulent blood flow through vessels. Although somatic tinnitus (such as pulsatile tinnitus) is sometimes classified as "tinnitus," we prefer to regard it as a form of hyperacusis, in the sense that the ear is unusually sensitive to vibrations that would not otherwise be noticeable. We discuss ***somatic tinnitus*** elsewhere.

The second is tinnitus from *middle ear myoclonus*, whose subtypes are tensor tympani myoclonus and stapedial myoclonus. In stapedial myoclonus, the muscle attached to the stapes (innervated by a branch of the facial nerve) rhythmically contracts. In tensor tympani myoclonus, the tensor tympani muscle (innervated by a branch of the trigeminal nerve) rhythmically contracts. In normal circumstances, these muscles mediate a protective auditory reflex against loud sounds, but if the muscles pathologically contract in a rhythmic fashion, then the ossicular chain will move (as if sounds were entering the ear from the external environment) and be perceived as sound. We discuss ***middle ear myoclonus*** elsewhere.

Pathophysiology of Idiopathic Tinnitus

Idiopathic tinnitus, which we will hereafter refer to simply as "tinnitus," is poorly understood. What is known about the neuroanatomy of the ear is that signals generated by the cochlea travel through the cochlear division of the eighth cranial nerve to the brainstem where they synapse in the cochlear nuclei, then ascend through a number of structures (trapezoid bodies, superior olives, lateral lemnisci, inferior colliculi, medial geniculate nuclei, sublenticular limb of the internal capsules) to the auditory cortices in the transverse temporal gyri. Logically, one would think that tinnitus could arise from a lesion anywhere along these pathways.

Many cases of tinnitus likely arise from damage to the cochlea. Suggestive of this mechanism are instances of tinnitus beginning after noise exposure.

M. Cherchi, *Otoneurology and Vestibular Medicine*, https://doi.org/10.1007/978-3-031-94841-1_52

However, it is well reported in the literature [2] that some patients continue to experience tinnitus even after labyrinthectomy and/or vestibulocochlear neurectomy. In cases of labyrinthectomy, there is no longer any cochlea to generate auditory signals; in cases of vestibulocochlear neurectomy, there are no longer any afferent fibers to carry auditory signals from the cochlea; so in patients who continue to experience tinnitus despite having undergone these procedures, it is clear that the tinnitus must be generated somewhere proximal to the vestibulocochlear nerve—either in the cochlear nucleus or anywhere along the ascending pathways up to and including the auditory cortex—even if the tinnitus perhaps originally "began" as a problem in the cochlea.

Studies have shown that tinnitus frequently accompanies hearing loss of any cause [3]. One interpretation of this finding is that the brain does not like a "vacuum," and if the brain is no longer receiving auditory signals it had been accustomed to receive, then it will start "guessing" or "interpolating" what signals "should have" been there. This explanation is somewhat analogous to a phantom limb syndrome, in which a patient may still perceive an amputated limb despite the obvious absence of that limb. On this interpretation, tinnitus following hearing loss is a type of "phantom" perception [4].

Studies have also demonstrated clearly that individuals with normal hearing at least intermittently experience tinnitus [5]—in other words, some degree of tinnitus appears to be physiologic. Taking this observation together with the earlier cited observation that tinnitus often accompanies hearing loss, one possible conclusion is that some degree of tinnitus is "always there" below one's threshold of awareness, but it becomes more noticeable when hearing loss eliminates the distraction that perception of normal environmental sounds would otherwise have provided.

Diagnosis

Simply ascertaining whether or not tinnitus is present is straightforward because it relies entirely on a patient's report of a subjective experience. As such, it is also unverifiable.

A basic rationale for "measuring" tinnitus is that a metric is needed to identify/characterize disease and to evaluate the effect of any interventions [6], and some data suggest that tinnitus management with a masking strategy that is "customized" to a patient's individual tinnitus profile may be more effective than a general (noncustomized) strategy [7].

Despite the subjective nature of tinnitus, there is substantial literature regarding methods for characterizing it. There are several features of tinnitus that researchers have attempted to measure.

The first feature is *tinnitus pitch* [8]. Different pitch matching methods sometimes give different results [9]. Tinnitus pitch matching can be more complex when an individual's tinnitus is present at multiple frequencies [10].

The second feature is *tinnitus loudness* [11]. The reliability of measures tinnitus pitch and loudness is suboptimal [12], which could be due either to variability in the measurement method, or variability in the tinnitus itself, or both. Proposals for objective "biomarkers" of tinnitus [13] have not gained acceptance.

A third set of features comprise the *degree to which tinnitus evokes an emotional response or interferes with an individual's functioning*. This aspect of tinnitus is usually assessed through questionnaire-based investigations, with instruments such as the Tinnitus Handicap Inventory [14], Tinnitus Functional Index [15, 16], and Tinnitus Reaction Questionnaire [15, 16], though the reliability of such instruments has been questioned [17].

Management

Management of idiopathic tinnitus is difficult, and we discuss this topic elsewhere.

References

1. Jarach CM, Lugo A, Scala M, van den Brandt PA, Cederroth CR, Odone A, et al. Global prevalence and incidence of tinnitus: a systematic review and meta-analysis. JAMA Neurol. 2022;79(9):888–900.
2. House JW, Brackmann DE. Tinnitus: surgical treatment. Ciba Found Symp. 1981;85:204–16.
3. Lockwood AH, Salvi RJ, Burkard RF. Tinnitus. N Engl J Med. 2002;347(12):904–10.
4. Muhlnickel W, Elbert T, Taub E, Flor H. Reorganization of auditory cortex in tinnitus. Proc Natl Acad Sci USA. 1998;95(17):10340–3.
5. Heller MF, Bergman M. VII Tinnitus Aurium in normally hearing persons. Ann Otol Rhinol Laryngol. 1953;62(1):73–83.
6. Henry JA. "Measurement" of Tinnitus. Otol Neurotol. 2016;37(8):e276–85.
7. Wang J, Ding J, Song J, Hu L, Cong N, Han Z. A prospective study of the effect of tinnitus sound matching degree on the efficacy of customized sound therapy in patients with chronic tinnitus. ORL J Otorhinolaryngol Relat Spec. 2022;84(3):229–37.
8. McMillan GP, Thielman EJ, Wypych K, Henry JA. A Bayesian perspective on tinnitus pitch matching. Ear Hear. 2014;35(6):687–94.
9. Neff P, Langguth B, Schecklmann M, Hannemann R, Schlee W. Comparing three established methods for tinnitus pitch matching with respect to reliability, matching duration, and subjective satisfaction. Trends Hear. 2019;23:2331216519887247.
10. Tao Y, Chang X, Ye S, Chu G, Guan T, Wang J, Zeng P. Multiple-frequency matching treatment strategy for tinnitus. J Int Adv Otol. 2017;13(2):221–5.
11. Manning C, Grush L, Thielman E, Roberts L, Henry JA. Comparison of tinnitus loudness measures: matching, rating, and scaling. Am J Audiol. 2019;28(1):137–43.
12. Hoare DJ, Edmondson-Jones M, Gander PE, Hall DA. Agreement and reliability of tinnitus loudness matching and pitch likeness rating. PLoS One. 2014;9(12):e114553.

13. Kang DW, Kim SS, Park DC, Kim SH, Yeo SG. Objective and measurable biomarkers in chronic subjective tinnitus. Int J Mol Sci. 2021;22(12):6619.
14. McCombe A, Baguley D, Coles R, McKenna L, McKinney C, Windle-Taylor P, et al. Guidelines for the grading of tinnitus severity: the results of a working group commissioned by the British Association of Otolaryngologists, head and neck surgeons, 1999. Clin Otolaryngol Allied Sci. 2001;26(5):388–93.
15. Meikle MB, Henry JA, Griest SE, Stewart BJ, Abrams HB, McArdle R, et al. The tinnitus functional index: development of a new clinical measure for chronic, intrusive tinnitus. Ear Hear. 2012;33(2):153–76.
16. Henry JA, Griest S, Thielman E, McMillan G, Kaelin C, Carlson KF. Tinnitus Functional Index: development, validation, outcomes research, and clinical application. Hear Res. 2016;334:58–64.
17. Newman CW, Wharton JA, Jacobson GP. Retest stability of the tinnitus handicap questionnaire. Ann Otol Rhinol Laryngol. 1995;104(9 Pt 1):718–23.

53 Exploding Head Syndrome

Overview

Exploding head syndrome (EHS) is a benign parasomnia consisting of a hypnogogic/hypnopompic auditory hallucination in which the patient experiences a very brief, loud sound, accompanied by fear, during the transition into or out of sleep. This can occur at any age, may be slightly more prevalent in females, does not have any clear risk factors, and generally occurs in isolation. Occasional abnormalities in electroencephalography and polysomnography have been described, though these are not needed to secure the diagnosis. Reassurance is generally sufficient for management, as the disease often remits spontaneously. Treatment with medications has been attempted, but none has been studied in randomized controlled trials.

Introduction

Sharpless [1] states that the term *"exploding head syndrome" (EHS)* is usually credited to Pearce [2], but noted that the phenomenon was probably described at least as early as 1920 by Armstrong-Jones [3] under the designation, "snapping of the brain."

EHS is a parasomnia. Although the name implies a painful experience, this is actually a hypnogogic or hypnopompic auditory hallucination [4] often accompanied by fright [5–7]. Patients generally experience this as an abrupt, disturbing sound while they are on the verge of transitioning from the waking state to the sleeping state or from the sleeping state to the waking state.

Most of the literature on this topic consists of case reports and case series [5, 8–13], and there are several reviews [1, 2, 14].

Epidemiology

Aazh and colleagues [15] reported that of 148 consecutive patients attending an audiology clinic for tinnitus/hyperacusis, 12 (8.1%) were diagnosed with EHS. The investigators found no clear correlation between the presence of EHS and demographic factors (age, gender), clinical features (tinnitus/hyperacusis distress), or comorbidities (anxiety, depression, sleep difficulties).

Denis and colleagues [6] conducted two studies on EHS. The first study analyzed 199 female undergraduate students and reported a lifetime prevalence of 37.19%, with 6.54% experiencing at least one episode per month. The second study included a sample of 1683 participants, age range 18–82 years (53.00% female), in which they found a lifetime prevalence of 29.59%, with monthly episodes occurring in 3.89% of participants.

Kirwan and Fortune [16] studied 135 adults and found a lifetime prevalence of EHS of 20.0%. They reported that patients with EHS experienced "more symptoms of anxiety and poorer sleep quality," but found no correlation between the presence of EHS and demographic factors (age, gender) or other comorbidities (depressive symptoms).

Sharpless and colleagues [17] conducted a cross-sectional study of 3286 individuals with EHS and 2954 without EHS. Contrary to other studies, they found a slightly higher prevalence among women.

Pathophysiological Mechanism of Disease

The pathophysiological mechanism of EHS is unknown. Investigators have speculated about etiologies such as dissociative episodes [18], "cranial sensory shock" [19, 20], migraine aura [21], and others.

M. Cherchi, *Otoneurology and Vestibular Medicine*, https://doi.org/10.1007/978-3-031-94841-1_53

Clinical Presentation

EHS presents as the perception of a loud sound, usually lasting less than a second [22], during the transition into or out of sleep, often accompanied by sensations of fear. Some descriptions state that it can be a multisensory phenomenon, with up to 27% of patients reporting associated visual abnormalities [23].

Occasionally, EHS is reported in conjunction with other parasomnias or other diseases. For example, Evans [24] reported the case of a 26-year-old female who would experience EHS followed by sleep paralysis and a migraine headache. Hayreh [25] described his own experience with EHS and noted that the episodes were associated with sick sinus syndrome.

Testing

Some studies of EHS patients have reported electroencephalographic abnormalities [26] or polysomnographic abnormalities [27, 28].

Treatment

Reassurance is the initial management strategy for EHS [29], because after patients understand that this is not a life-threatening condition, the experience is less frightening, and they often do not feel the need for additional intervention.

Lifestyle management strategies to reduce EHS are reviewed by Sharpless and colleagues [17]. These include changes in alcohol intake, sleep position, sleep time, mindfulness and breathing techniques, and changing the number of pillows.

A variety of pharmacologic treatments have been reported, though none has been studied in a randomized controlled trial. These include:

- Amitriptyline [12]
- Clobazam [30]
- Clomipramine [28, 29]
- Nifedipine [31]
- Topiramate [32]

Other attempted treatment modalities include single-pulse transcranial magnetic stimulation [20].

Prognosis

EHS is generally considered a benign disease, and often, the symptoms remint spontaneously [28].

Kitagawa and colleagues [33] reported the appearance of EHS in a patient who subsequently developed mild cognitive impairment that progressed, and the patient eventually was diagnosed with Lewy body disease.

References

1. Sharpless BA. Exploding head syndrome. Sleep Med Rev. 2014;18(6):489–93.
2. Pearce JM. Exploding head syndrome. Lancet. 1988;2(8605):270–1.
3. Armstrong-Jones R. Snapping of the brain. Lancet. 1920;196(5066):720.
4. Blom JD. Auditory hallucinations. Handb Clin Neurol. 2015;129:433–55.
5. Alkhateeb W, Krishnaraj A, Saini V. Single patient multiple explosions: a case report on exploding head syndrome. Cureus. 2023;15(8):e44437.
6. Denis D, Poerio GL, Derveeuw S, Badini I, Gregory AM. Associations between exploding head syndrome and measures of sleep quality and experiences, dissociation, and well-being. Sleep. 2019;42(2):1–11.
7. Sharpless BA. Exploding head syndrome is common in college students. J Sleep Res. 2015;24(4):447–9.
8. Chakravarty A. Exploding head syndrome: report of two new cases. Cephalalgia. 2008;28(4):399–400.
9. Frese A, Summ O, Evers S. Exploding head syndrome: six new cases and review of the literature. Cephalalgia. 2014;34(10):823–7.
10. Ganguly G, Mridha B, Khan A, Rison RA. Exploding head syndrome: a case report. Case Rep Neurol. 2013;5(1):14–7.
11. Pearce JM. Clinical features of the exploding head syndrome. J Neurol Neurosurg Psychiatry. 1989;52(7):907–10.
12. Pirzada AR, Almeneessier AS, BaHammam AS. Exploding head syndrome: a case series of underdiagnosed hypnic parasomnia. Case Rep Neurol. 2020;12(3):348–58.
13. Wang X, Zhang W, Yuan N, Zhang Y, Liu Y. Characteristic symptoms of exploding head syndrome in two male patients. Sleep Med. 2019;57:94–6.
14. Ceriani CEJ, Nahas SJ. Exploding head syndrome: a review. Curr Pain Headache Rep. 2018;22(10):63.
15. Aazh H, Stevens J, Jacquemin L. Exploding head syndrome among patients seeking help for tinnitus and/or hyperacusis at an audiology department in the UK: a preliminary study. J Am Acad Audiol. 2023;34:93–9.
16. Kirwan E, Fortune DG. Exploding head syndrome, chronotype, parasomnias and mental health in young adults. J Sleep Res. 2021;30(2):e13044.
17. Sharpless BA, Denis D, Perach R, French CC, Gregory AM. Exploding head syndrome: clinical features, theories about etiology, and prevention strategies in a large international sample. Sleep Med. 2020;75:251–5.
18. Feketeova E, Buskova J, Skorvanek M, Mudra J, Gdovinova Z. Exploding head syndrome–a rare parasomnia or a dissociative episode? Sleep Med. 2014;15(6):728–30.
19. Goadsby PJ, Sharpless BA. Exploding head syndrome, snapping of the brain or episodic cranial sensory shock? J Neurol Neurosurg Psychiatry. 2016;87(11):1259–60.
20. Puledda F, Moreno-Ajona D, Goadsby PJ. Exploding head syndrome (a.k.a. episodic cranial sensory shock) responds to single-pulse transcranial magnetic stimulation. Eur J Neurol. 2021;28(4):1432–3.
21. Rossi FH, Gonzalez E, Rossi EM, Tsakadze N. Exploding head syndrome as aura of migraine with brainstem aura: a case report. J Oral Facial Pain Headache. 2018;32(2):e34–e6.
22. Khan I, Slowik JM. Exploding head syndrome. Treasure Island (FL): StatPearls; 2023.

23. Sharpless BA. Characteristic symptoms and associated features of exploding head syndrome in undergraduates. Cephalalgia. 2018;38(3):595–9.
24. Evans RW. Exploding head syndrome followed by sleep paralysis: a rare migraine aura. Headache. 2006;46(4):682–3.
25. Hayreh SS. Exploding head syndrome: new observations. Eur J Neurol. 2020;27(11):2333–5.
26. Fotis Sakellariou D, Nesbitt AD, Higgins S, Beniczky S, Rosenzweig J, Drakatos P, et al. Co-activation of rhythms during alpha band oscillations as an interictal biomarker of exploding head syndrome. Cephalalgia. 2020;40(9):949–58.
27. Nakayama M, Nakano N, Mihara T, Arima S, Sato S, Kabaya K, et al. Two cases of exploding head syndrome documented by polysomnography that improved after treatment. J Clin Sleep Med. 2021;17(1):103–6.
28. Sachs C, Svanborg E. The exploding head syndrome: polysomnographic recordings and therapeutic suggestions. Sleep. 1991;14(3):263–6.
29. Green MW. The exploding head syndrome. Curr Pain Headache Rep. 2001;5(3):279–80.
30. Gillis K, Ng MC. Exploding head syndrome in the epilepsy monitoring unit: case report and literature review. Neurodiagn J. 2017;57(2):133–8.
31. Jacome DE. Exploding head syndrome and idiopathic stabbing headache relieved by nifedipine. Cephalalgia. 2001;21(5):617–8.
32. Palikh GM, Vaughn BV. Topiramate responsive exploding head syndrome. J Clin Sleep Med. 2010;6(4):382–3.
33. Kitagawa S, Okamoto N, Ikenouchi A, Igata R, Hashimoto R, Konishi Y, et al. A case of exploding head syndrome preceding the onset of mild cognitive impairment with Lewy bodies. Psychogeriatrics. 2023;23(2):368–70.

54 Pulsatile Tinnitus

Overview

A diagnosis of *pulsatile tinnitus* is suspected when the examiner palpates a patient's pulse, has the patient "sound out" the pulsatile tinnitus that they are perceiving, and finds that the two patterns are synchronous. Pulsatile tinnitus sometimes appears to be associated with vascular abnormalities, though one must keep in mind that some vascular findings on imaging studies may be incidental rather than causal. When vascular phenomena are causal, venous causes are more common than arterial or arteriovenous ones. Some venous cases of pulsatile tinnitus are also associated with increased intracranial pressure.

- *Venous* sources of pulsatile tinnitus include the following:
  - Jugular bulb wall dehiscence
  - Transverse sinus stenosis
  - Sigmoid sinus diverticulum
  - Sigmoid sinus wall dehiscence
  - Enlarged oblique occipital sinus
  - Stenoses of various dural venous sinuses from arachnoid granulations
  - Superficial middle cerebral vein
  - Dilated mastoid emissary vein
- *Arterial* sources of pulsatile tinnitus include the following:
  - Carotid artery stenosis
  - Fibromuscular dysplasia
- *Arteriovenous* sources include the following:
  - Arteriovenous fistula

Practically, in a patient whose physical examination suggests pulsatile tinnitus of a vascular cause, referral to vascular neurosurgery is reasonable, as surgeons may have different recommendations for confirmatory imaging, and different recommendations for intervention (if appropriate).

Introduction

Vascularly mediated *pulsatile tinnitus* is thought to be a hydroacoustic phenomenon that occurs because turbulent blood flow through a vessel produces sufficient vibrations to stimulate the cochlea. The pathway by which the kinetic energy from these vibrations reaches the cochlea is often uncertain [1]. In particular, whether the kinetic energy from the turbulence is transmitted to the cochlea solely via bone conduction, or partly through air conduction via the middle ear space [2], is unclear.

Etiologies

The specific abnormal fluid dynamics required to provoke pulsatile tinnitus have not yet been well characterized.

A study using a novel Doppler technique reported that, "the sensation of PT [pulsatile tinnitus] is closely associated with the flow of kinetic energy rather than the formation of a vortex, whereby the amplitude of PT is correlated to the magnitude of the flow velocity and pressure gradient" [3].

In contrast, another study instead concluded that "vorticity" (production of vortices) is the fluid dynamic mechanism that generates tinnitus [4].

One study reported that unilateral pulsatile tinnitus is usually due to venous etiologies, less commonly to arterial etiologies or arteriovenous shunts [5].

Associated Mood Disorders

Mood disorders are often comorbid with tinnitus in general, and this pattern holds specifically for pulsatile tinnitus as well.

One study of 157 pulsatile tinnitus patients noted mood disorders were common comorbidities, with depression in 25.5% and anxiety in 26.1% of patients [6].

M. Cherchi, *Otoneurology and Vestibular Medicine*, https://doi.org/10.1007/978-3-031-94841-1_54

One study reported that patients with pulsatile tinnitus exhibit a higher rate of anxiety and depression than patients with non-pulsatile tinnitus [7].

Diagnostic Physical Examination

Typically a diagnosis of pulsatile tinnitus is suspected when the examiner palpates a patient's pulse, has the patient "sound out" the pulsatile tinnitus that they are perceiving, and finds that the two patterns are synchronous.

In patients whose pulsatile tinnitus is due to sigmoid sinus dehiscence or diverticulum (see below), compression of the ipsilateral internal jugular vein may suppress the tinnitus [8].

Diagnostic Testing

Some investigators have attempted more technological means of diagnosing pulsatile tinnitus. For example, one group simultaneously recorded (1) sound using a sensitive microphone in the ear, and (2) the cardiac pulse, and then developed an automated detection algorithm to identify synchrony between the two patterns [9].

Imaging Studies

While it seems logical to seek vascular abnormalities in a patient with pulsatile tinnitus, one should keep in mind that some vascular findings on imaging may be incidental rather than actually responsible for the pulsatile tinnitus. This point is made firmly by Walthers and colleagues [10].

Broadly speaking venous sources of pulsatile tinnitus are more common than arterial ones.

Imaging Studies: Jugular Bulb Abnormalities

Investigators performed CT angiography (venography) on patients with unilateral pulsatile tinnitus and identified *jugular bulb wall dehiscence* (JBWD) in 12.1%, and believed the JBWD to be the cause of the tinnitus in 5.0% of cases [11].

A case reported a *terminal plate dehiscence of the jugular bulb* as the source of pulsatile tinnitus [12].

Investigators reported two cases of pulsatile tinnitus due to an "*enlarged oblique occipital sinus*" that, in one case was associated with a sigmoid-jugular wall diverticulum, and in another case was associated with a sigmoid-jugular wall dehiscence [13].

Imaging Studies (Venous): Sigmoid Sinus Diverticulum

Some investigators favor specialized imaging protocols in patients with pulsatile tinnitus. For example, one group favored dynamic volume CT to detect "tinnitus caused by SSD [*sigmoid sinus diverticulum*] with bone defects" [14].

Imaging Studies (Venous): Venous Stenosis

A CT venography study of patients with unilateral pulsatile tinnitus found that "the degree and length of ipsilesional *transverse sinus stenosis* (TSS) are positively correlated with transstenotic pressure gradient (TPG) in unilateral PT [pulsatile tinnitus] patients with SSWA" (sigmoid sinus wall anomalies) [15].

A study of patients whose pulsatile tinnitus was due to *transverse sinus stenosis* noted that, "In white matter regions, the patients with high-degree TSS [transverse sinus stenosis] exhibited decreased CBF [cerebral blood flow]" relative to the healthy control subjects, and concluded that, "TSS in venous PT [pulsatile tinnitus] patients may lead to decreased CBF and cloud-like WM hyperintensity" [16].

Imaging Studies (Venous): Arachnoid Granulations

One study reported a higher incidence of *dural venous sinus arachnoid granulations* in patients with pulsatile tinnitus than in those with non-pulsatile tinnitus [17].

One study of four patients with unilateral pulsatile tinnitus who underwent MR and digital subtraction angiography reported that all patients had "moderate-to-severe *stenoses from large arachnoid granulations within the implicated transverse sinus*" [18].

One study of tinnitus patients found an association between the presence of "brain herniation into arachnoid granulations of the *dural venous sinus*" and the presence of idiopathic intracranial hypertension [19].

Imaging Studies (Venous): Intracranial Hypertension

Investigators compared CT angiography (venography) in one group of idiopathic intracranial hypertension (IIH) patients with unilateral pulsatile tinnitus (PT), and one group of IIH patients without pulsatile tinnitus. In comparison to IIH patients without PT, those IIH patients with PT had a higher proportion of *venous outflow laterality* and *sigmoid sinus diverticula*, and that the sigmoid sinus diverticula were always accompanied with *sigmoid sinus wall dehiscence* [20].

A study of patients with pulsatile tinnitus underwent cerebrospinal fluid manometry and Doppler ultrasound. The investigators concluded that in "Patients with venous PT [pulsatile tinnitus] as the only presenting symptom should be suspected of having increased CSFP [cerebrospinal fluid pressure] when they present with high ITSS [index of transverse sinus stenosis], BMI [body mass index] and low PI [pulsatility index]" [21].

Imaging Studies (Venous): Rarer Etiologies

Pulsatile tinnitus has been reported in individual cases as resulting from a *superficial middle cerebral vein* via a tympanic cavity dehiscence [22], and from a dilated *mastoid emissary vein* [23].

Imaging Studies (Arterial): Carotid Artery Stenosis

A case of a patient with unilateral pulsatile tinnitus attributed to a *stenotic internal carotid artery* reported that the tinnitus resolved after carotid artery stenting [24].

A case of a patient with supraclinoid *stenosis of the internal carotid artery* ipsilateral to unilateral pulsatile tinnitus experienced resolution of the tinnitus after stenting of the stenosis [25].

Imaging Studies (Arterial): Fibromuscular Dysplasia

One study described a patient suffering from medically refractory pulsatile tinnitus who was found to have *fibromuscular dysplasia*. Placement of a Casper stent in the carotid artery eliminated the tinnitus [26].

Imaging Studies: Multiple Etiologies

General discussions of venous pulsatile tinnitus usually cover a differential diagnosis that includes, "sigmoid sinus wall anomalies, transverse and sigmoid sinus stenosis, jugular bulb anomalies, and prominent posterior fossa emissary veins" [27].

A case series of 157 patients with pulsatile tinnitus who underwent CT angiography and venography reported transverse venous sinus stenosis in 59% (bilateral in 39% and unilateral in 20%), osseous etiologies in 10% (superior canal dehiscence or thinning), sigmoid sinus dehiscence in 1 patient, and dural arteriovenous fistula in 2% [6].

Another study concluded that pulsatile tinnitus due to dural arteriovenous fistula is "not rare" [28].

Treatment

Depending on a patient's particular circumstances, the radiographic findings, and the level of confidence that a specific vascular anomaly is the source of pulsatile tinnitus, a vascular neurosurgeon may consider intervention.

Treatment: Venous

One set of investigators reported that, "Two cases (one with *jugular bulb diverticulum* and one with both *sigmoid sinus* and *jugular bulb diverticula*) underwent surgical intervention, and both had immediate resolution of pulsatile tinnitus post-operatively" [29].

A review concluded that venous sinus stenting is effective in treatment of pulsatile tinnitus arising from *venous sinus stenosis* [30].

Treatment of pulsatile tinnitus due to *sigmoid sinus diverticulum* has also been performed with coil embolization [31].

One group reported pulsatile tinnitus in a patient found on CT angiography to have a *sigmoid sinus diverticulum*. This was surgically decompressed and "recontoured" [32].

One group treated idiopathic pulsatile tinnitus (i.e., without any apparent vascular anomalies) with transtemporal sigmoid sinus decompression, with favorable results [33].

In cases where dehiscence (such as venous jugular bulb dehiscence) is believed to be the source of pulsatile tinnitus, there is debate regarding the type of material that should be employed in the surgical repair [34].

Surgical reconstruction of the sigmoid sinus wall in patients with *sigmoid sinus wall dehiscence* or *sigmoid sinus diverticulum* is not always successful in abolishing the pulsatile tinnitus [35].

Treatment: Arterial

A case of a patient with unilateral pulsatile tinnitus attributed to a *stenotic internal carotid artery* reported that the tinnitus resolved after carotid artery stenting [24].

A case of a patient with supraclinoid *stenosis of the internal carotid artery* ipsilateral to unilateral pulsatile tinnitus experienced resolution of the tinnitus after stenting of the stenosis [25].

One study described a patient suffering from medically refractory pulsatile tinnitus who was found to have *fibromuscular dysplasia*. Placement of a Casper stent in the carotid artery eliminated the tinnitus [26].

References

1. Wang X, Hsieh YL, Xu X, Wang W. Influence of sigmoid plate and dura mater on vascular wall displacement, vibroacoustic/hydroacoustic sources characteristics, and frequency-loudness assessments of venous pulsatile tinnitus: a coupled-computational fluid dynamics study combining transcanal recording investigation. Front Bioeng Biotechnol. 2022;10:948230.
2. Hsieh YL, Xu X, Wu Y, Wang W. Evidence of air-conduction transmission pathway and strategized transtemporal operative techniques for venous pulsatile tinnitus: combining water occlusion test and operative sensing applications. Laryngoscope Investig Otolaryngol. 2021;6(6):1436–48.
3. Gao X, Hsieh YL, Wang X, Wang W. Retroauricular/transcranial color-coded doppler ultrasound approach in junction with ipsilateral neck compression on real-time hydroacoustic variation of venous pulsatile tinnitus. Front Hum Neurosci. 2022;16:862420.
4. Hong Z, Liu X, Ding H, Zhao P, Gong S, Wang Z, et al. Flow patterns in the venous sinus of pulsatile tinnitus patients with transverse sinus stenosis and underlying vortical flow as a causative factor. Comput Methods Prog Biomed. 2022;227:107203.
5. Cummins DD, Caton MT, Hemphill K, Lamboy A, Tu-Chan A, Meisel K, et al. Cerebrovascular pulsatile tinnitus: causes, treatments, and outcomes in 164 patients with neuroangiographic correlation. J Neurointerv Surg. 2022;15(10):1014–20.
6. Formeister EJ, Xiao G, Clark J, Carey JP, Hui F, Agrawal Y, et al. Combined arterial and venous phase computed tomographic imaging of the skull base in pulsatile tinnitus. Otol Neurotol. 2022;43(9):1049–55.
7. Williams AP, Gourishetti SC, Flaherty MA, Eisenman DJ. Anxiety, depression, and symptom severity in patients with pulsatile and non-pulsatile tinnitus. Laryngoscope. 2022;133:683–8.
8. Lee SJ, Lee SY, Choi BY, Koo JW, Hong SH, Song JJ. Preoperative Significance of Ipsilateral Manual Neck Compression in Patients With Pulsatile Tinnitus Secondary to Sigmoid Sinus Dehiscences and Diverticula. Front Neurol. 2022;13:869244.
9. Ubbink SWJ, van Dijk JMC, Hofman R, van Dijk P. Performance of an automated detection algorithm to assess objective pulsatile tinnitus. Ear Hear. 2022;44(3):655–60.
10. Walters H, Chung N, Arullendran P. Vascular abnormalities – a true cause of pulsatile tinnitus? J Laryngol Otol. 2022;137(2):138–42.
11. Dai C, Zhao P, Ding H, Lv H, Qiu X, Li X, et al. CT evaluation of unilateral pulsatile tinnitus with jugular bulb wall dehiscence. Eur Radiol. 2023;33(6):4464–71.
12. Guerra-Leal JD, Garcia-Gutierrez YJ, Longoria-Estrada U, Acosta ACG, Garcia EB. Pulsatile tinnitus and imaging diagnosis. A case report. Radiol Case Rep. 2022;17(6):2031–3.
13. Hsieh YL, Wang S, Wang W. Case Report: venous pulsatile tinnitus induced by enlarged oblique occipital sinus and resultant diverticulum/dehiscence of the sigmoid-jugular wall. Front Surg. 2022;9:1014649.
14. Zhang J, Mao QL, Shen FJ, Pan YN, Li AJ. Dynamic volume computed tomography for characterizing pulsatile tinnitus caused by sigmoid sinus diverticulum with bone defects: clinical implications. Curr Med Imaging. 2022;18(14):1479–85.
15. Zhao P, Ding H, Lv H, Li X, Qiu X, Zeng R, et al. CT venography correlate of transverse sinus stenosis and venous transstenotic pressure gradient in unilateral pulsatile tinnitus patients with sigmoid sinus wall anomalies. Eur Radiol. 2021;31(5):2896–902.
16. Li X, Xu N, Meng X, Dai C, Qiu X, Ding H, et al. Transverse sinus stenosis in venous pulsatile tinnitus patients may lead to brain perfusion and white matter changes. Front Neurosci. 2021;15:732113.
17. Bauschard MJ, Reichl K, Socolovsky LD, Sismanis AA, Coelho DH. Dural venous sinus arachnoid granulations in patients with pulsatile tinnitus. Otol Neurotol. 2022;43(7):e787–e90.
18. Gadot R, Hoang AN, Raper DMS, Sweeney AD, Juliano M, Lustrin E, et al. Arachnoid granulation causing unilateral pulsatile tinnitus treated with dural venous sinus stenting. Otol Neurotol. 2023;44(1):86–9.
19. Smith ER, Caton MT, Villanueva-Meyer JE, Remer J, Eisenmenger LB, Baker A, et al. Brain herniation (encephalocele) into arachnoid granulations: prevalence and association with pulsatile tinnitus and idiopathic intracranial hypertension. Neuroradiology. 2022;64(9):1747–54.
20. Zhao P, Jiang C, Lv H, Zhao T, Gong S, Wang Z. Why does unilateral pulsatile tinnitus occur in patients with idiopathic intracranial hypertension? Neuroradiology. 2021;63(2):209–16.
21. Gao X, Hsieh YL, Wang S, Shi S, Wang W. Intracranial pressure, lateral sinus patency, and jugular ultrasound hemodynamics in patients with venous pulsatile tinnitus. Front Neurol. 2022;13:992416.
22. Sing BWH, Lim MLA, See A, Lee SM, Kheok SW. Superficial middle cerebral vein: tympanic cavity dehiscence causing pulsatile tinnitus. Otol Neurotol. 2022;43(5):e602–e4.
23. Yang B, Zhang F, Jiang X. Bilateral pulsatile tinnitus caused by bilateral dilated mastoid emissary vein. Ear Nose Throat J. 2022;103:1455613221077597.
24. Bains NK, Early E, French BR, Siddiq F, Gomez CR, Qureshi AI. Effect of angioplasty and stent placement on sensorineural hearing loss and pulsatile tinnitus in a patient with severe internal carotid artery stenosis. J Stroke Cerebrovasc Dis. 2022;31(11):106718.
25. Capirossi C, Laiso A, Capasso F, Nappini S, Renieri L, Limbucci N. Pulsatile tinnitus due to stenosis of the supraclinoid segment of the internal carotid artery: management with a low-profile self-expanding stent. Interv Neuroradiol. 2022;30(2):293–6.
26. Vanzin JR, Martio AE, Manzato LB, Abud DG. Casper stent in the treatment of pulsatile tinnitus in fibromuscular dysplasia: therapeutic review and case report. Brain Circ. 2021;7(4):281–4.
27. Reardon MA, Raghavan P. Venous abnormalities leading to tinnitus: imaging evaluation. Neuroimaging Clin N Am. 2016;26(2):237–45.
28. Li X, Xu N, Meng X, Dai C, Qiu X, Ding H, et al. Dual-phase contrast-enhanced CT evaluation of dural arteriovenous fistula in patients with pulsatile tinnitus as an initial symptom. Eur J Radiol. 2022;148:110137.
29. Yeo WX, Xu SH, Tan TY, Low YM, Yuen HW. Surgical management of pulsatile tinnitus secondary to jugular bulb or sigmoid sinus diverticulum with review of literature. Am J Otolaryngol. 2018;39(2):247–52.
30. Fiani B, Kondilis A, Doan T, Runnels J, Fiani NJ, Sarno E. Venous sinus stenting for intractable pulsatile tinnitus: a review of indications and outcomes. Surg Neurol Int. 2021;12:81.
31. Liu L, Mu Z, Kang Y, Huang S, Qiu X, Xue X, et al. Hemodynamic mechanism of pulsatile tinnitus caused by venous diverticulum treated with coil embolization. Comput Methods Prog Biomed. 2022;215:106617.
32. Spangler DJ, McElveen JT Jr, McLear PW. Sigmoid sinus diverticulum-a unique case of pulsatile tinnitus. Ear Nose Throat J. 2022;104:78–80.
33. Slater PW, Duhon BH, Feldman DJ. Transtemporal sigmoid sinus decompression: a novel surgical procedure for the treatment of idiopathic pulsatile tinnitus. Otol Neurotol. 2022;43(3):328–36.
34. Hsieh YL, Gao X, Wang X, Hsiang FC, Sun X, Wang W. Therapeutic validation of venous pulsatile tinnitus and biomaterial applications for temporal bone reconstruction surgery using multi-sensing platforms and coupled computational techniques. Front Bioeng Biotechnol. 2021;9:777648.
35. Li D, Wang G, Zeng R, Li W, Chen N, Zhao P, et al. Analysis of revision surgery following surgical reconstruction of the sigmoid sinus wall in patients with pulsatile tinnitus. J Int Adv Otol. 2022;18(4):358–64.

Tinnitus and Temporomandibular Joint Dysfunction (TMJD)

55

Overview

Temporomandibular joint disorder (TMJD)-associated tinnitus can affect any demographic; there may be a slight preponderance of younger women; it seems to have an adverse effect on quality of life. The literature makes clear that the association between TMJD and tinnitus is greater than just overlap by chance, and the few available prospective studies suggest that TMJD may be a risk factor for tinnitus; but these observations do not prove causality. Hypotheses regarding a possible causal relationship remain speculative. Generally TMJD-associated tinnitus tends to be in a higher pitch range. Of the few studies that have sought an anatomical explanation for this problem, attention has focused primarily on the configuration of the petrotympanic fissure, the glenoid fossa, and the temporomandibular joint disc. The most commonly attempted management strategy (and currently the best supported) involves a combination of education, physical therapy and a dental appliance (such as a splint or other jaw-repositioning orthotic).

Medical professionals are generally poorly educated about this problem. Practically, in a patient with tinnitus who additionally has oral problems (poor dentition, bruxism, diagnosed temporomandibular joint dysfunction), referral to a dentist or maxillofacial surgeon is reasonable.

Introduction

The relationship between *temporomandibular joint dysfunction (TMJD)* and *tinnitus* is controversial. Note that the abbreviation "TMJ" is sometimes used in the literature to refer to the "temporomandibular joint" or to "temporomandibular joint disorder," whereas TMD generally refers to the latter. To disambiguate the first abbreviation, in this discussion we use the unconventional abbreviation "TMJD" to refer to "temporomandibular joint disorder/dysfunction."

The Association Between TMJD and Tinnitus

There is substantial, albeit variegated literature demonstrating a greater-than-chance association between TMJD and tinnitus.

- A study reported that the, "Prevalence of tinnitus was found to be 8 times higher in participants with TMD (30 of 82 [36.6%]) than in participants without TMD (38 of 869 [4.4%])" [1].
- A study of 200 tinnitus patients with concomitant temporomandibular joint dysfunction reported that "The severity of tinnitus was significantly associated with the severity of chronic pain," and that "The most prevalent TMD [temporomandibular] subgroups in tinnitus patients… were myofascial pain with limited opening (39.0%), disc displacement with reduction (44.33%), and arthralgia (53.54%)" [2].
- A study comparing 30 patients (13 females, 17 males, age 18–71 years) suffering from TMJD and tinnitus with control subjects reported that, "Increased odds ratios were found for tenderness of the masticatory muscles… TMJ tenderness to dorsal cranial compression," concluded that, "A relationship between tinnitus and TMD was established… Tinnitus patients seem to suffer especially from myofascial and TMJ pain," and recommended that, "A screening for TMD should be included in the diagnostic survey for tinnitus patients" [3].
- A meta-analysis reported that, "For TMD, a bidirectional association with tinnitus was found; odds ratios ranged from 2.3 (95% CI [1.5, 3.6]) for arthrogenous TMD to 6.7 (95%CI [2.4, 18.8]) for unspecified TMD" [4].
- A study of 325 patients with TMJD found that 33 (11.5%) had tinnitus [5].
- A survey of 1032 patients (338 with TMJD and 694 age-matched controls) reported that, "Tinnitus [was] significantly more prevalent in the TMD group than in… the control group" [6].

M. Cherchi, *Otoneurology and Vestibular Medicine*, https://doi.org/10.1007/978-3-031-94841-1_55

- A study of 79 tinnitus patients observed "higher THI [tinnitus handicap inventory scores] was observed in TMD-pain individuals, compared to TMD-free" individuals [7].
- A study of 2482 tinnitus patients (486 with TMJD, 1996 without TMJD), observed that, "The prevalence of TMJ complaints was greater in tinnitus subjects with severe tinnitus (36%) when compared to those with any tinnitus (19%), strongly indicating the contribution of TMJ problems to the severity of tinnitus" [8].
- A study of 100 patients with tinnitus and 100 subjects without tinnitus reported that, "TMD signs and symptoms were detected in 85% of patients with tinnitus and in 55% of controls," and concluded that, "Tinnitus is associated with TMD" [9].
- A study of 11,745 participants with TMJD and dental pain found that, "The prevalence of tinnitus was higher in the subjects with dental pain (21.1%), TMD (22.5%) and both symptoms (31.2%) than subjects without those symptoms (19.6%)" [10].
- A systematic literature review concluded that there is a higher prevalence of tinnitus in patients with TMJD than without [11].
- A study of 250 patients with TMJD found a tinnitus prevalence of 30.4% [12].
- A study of 108 patients with TMJD reported that 33 (30.6%) were experiencing tinnitus [13].
- A study of 20 tinnitus patients reported that, "19 of these individuals had one or more clinical, electromyographic, and radiographic indications of a temporomandibular disorder" [14].
- A systematic literature review concluded that, "In all of the selected studies, the prevalence of tinnitus was higher in patients with TMDs (35.8% to 60.7%) than in patients without TMDs (9.7% to 26.0%). The odds ratio of suffering from tinnitus among patients with TMDs was 4.45 (95% CI 1.64–12.11. $P = 0.003$)" [15].
- A meta-analysis of data extracted from 8 papers found that, "In all of the selected studies, the prevalence of tinnitus was higher in patients with TMDs (35.8% to 60.7%) than in patients without TMDs (9.7% to 26.0%). The odds ratio of suffering from tinnitus among patients with TMDs was 4.45 (95% CI 1.64–12.11. P = 0.003)" [16].
- A controlled study of TMJD patients concluded that tinnitus was, "significantly more prevalent in the TMD group than in [the] control group" [17].
- A systematic literature review of tinnitus and TMJD comorbidity concluded that, "The prevalence of tinnitus in patients with TMD varied from 3.7% to 70% (median 42.3%) whereas the prevalence in control groups without TMD varied between 1.7% and 26% (median 12%)" [18].
- Data from a study of a mixed population of tinnitus patients (some with TMJD, some without it) was, "retrospectively surveyed to determine which traits or attributes of tinnitus could indicate the possibility of temporomandibular joint dysfunction (TMD) as the cause of tinnitus," and found that, "No single benchmark standard was discovered that exclusively indicated tinnitus from TMJ origins. However, a total of 10 "TMJ Indicators" were discovered" [19].
- One study, "compared 30 patients with a confirmed diagnosis of temporomandibular joint dysfunction and tinnitus to a group of 61 patients with tinnitus but without any subjective complaints of TMJ dysfunction with respect to clinical and demographic characteristics," and concluded that, "Classical risk factors for tinnitus (age, male gender, hearing loss) are less relevant in tinnitus patients with TMJ disorder, suggesting a causal role of TMJ pathology in the generation and maintenance of tinnitus" [20]. A subsequent study by Vielsmeier and colleagues reached similar conclusions [21].

A minority of studies failed to identify any relationship between TMJD and tinnitus. For example, Bush [22] found no correlation between the two. Demirkol and colleagues [23] observed that, "gender did show a weak correlation with tinnitus intensity," but TMJD itself did not. Upton and colleagues studied 989 TMJD patients and found the prevalence of tinnitus to be 7.28%, which was less than estimates of tinnitus of the general population at 10–14% [24].

Some authors are careful to articulate that the associative relationship does not prove causality. A study of 224 patients (age 18–76 years) concluded that, "Painful temporomandibular disorders, high levels of depression, and self reported tinnitus are deeply associated," noting, "However, this association does not imply a causal relationship" [25].

Is TMJD a Risk Factor for Developing Tinnitus?

Even if the association between TMJD and tinnitus has not been proven as causal, might TMJD be a risk factor for developing tinnitus? A few prospective studies suggest that this may be.

- A prospective study of 191 patients who on examination had pain on palpation of the TMJ and no tinnitus reported that within 5 years, 142 subjects (5.8%) were diagnosed with tinnitus, and concluded that the presence of TMJD was associated with a risk ratio (adjusted for gender, age, school education and frequent headache) of 2.4 (95% confidence interval, 1.6–3.7) of developing tinnitus [26].
- A study of health insurance claims in Taiwan reported that, "The crude hazard ratio (HR) of tinnitus in the TMJ disorder cohort was 2.73-fold higher than that in the control patients, with an adjusted HR of 2.62 (95 %

CI = 2.29–3.00)," and further that, "We also observed a 3.22-fold significantly higher relative risk of developing tinnitus within the 3-year follow-up period (95 % CI = 2.67–3.89). Patients with TMJ disorder might be at increased risk of tinnitus" [27].

Imaging Studies on TMJD Patients with Tinnitus

Of the many studies that demonstrate an association between TMJD and tinnitus, very few report on imaging findings that might suggest a structural explanation for the association, but below are several such studies. These focus primarily on the configuration of the petrotympanic fissure, the glenoid fossa, and the temporomandibular joint disc, of which the relevant anatomy is shown in Fig. 55.1.

- A cone beam computed tomography study of 100 patients with TMJD (50 with tinnitus, 50 without tinnitus) reported that, "a short, wide petrotympanic fissure (type 1 [wide tubular formation]) might be associated with an increased incidence of tinnitus in patients with TMD" [28].
- An imaging study of 331 patients with TMJD and tinnitus found that 33 of the patients (nearly 10%) at petrotympanic fissure configurations, "characterized by a rear (36.59%) or intracranial-cranial (63.41%) condylar displacement of the temporomandibular joint" [29].
- A study of 82 patients with TMJD (42 patients with tinnitus, 40 patients without tinnitus) found that on CT, "decreased glenoid fossa depth may be associated with an increased incidence of tinnitus" [30].
- An MRI study of 53 patients with TMJD (30 with tinnitus, 23 without tinnitus) reported a higher frequency of disc displacement with reduction in the patients with tinnitus than without tinnitus [31].
- An imaging study of 53 patients with unilateral tinnitus and disk displacement, and 82 individuals with disk displacement but no tinnitus found that, "Disk displacement was found to be present in the ipsilateral joint in all 53 patients with unilateral tinnitus, while the contralateral joint was asymptomatic in 50 (94.3%) of them," and concluded that, "The results of this study revealed a significant correlation between internal derangement of the TMJ and tinnitus" [32].

TMJ Herniation

A few studies, primarily case reports or case series, describe much more dramatic anatomical abnormalities. In these cases, the tinnitus is usually related to mastication with a very disordered temporomandibular joint, typically with herniation.

- A patient with mastication-induced unilateral tinnitus underwent CT which showed that, "the posteromedial bony wall of the glenoid was partially dehiscent" [33].
- A patient with tinnitus was reported to have ipsilateral spontaneous herniation of the temporomandibular joint into the external auditory canal [34].
- Two patients with clicking tinnitus were found to have temporomandibular joint herniation through the foramen of Huschke [35].

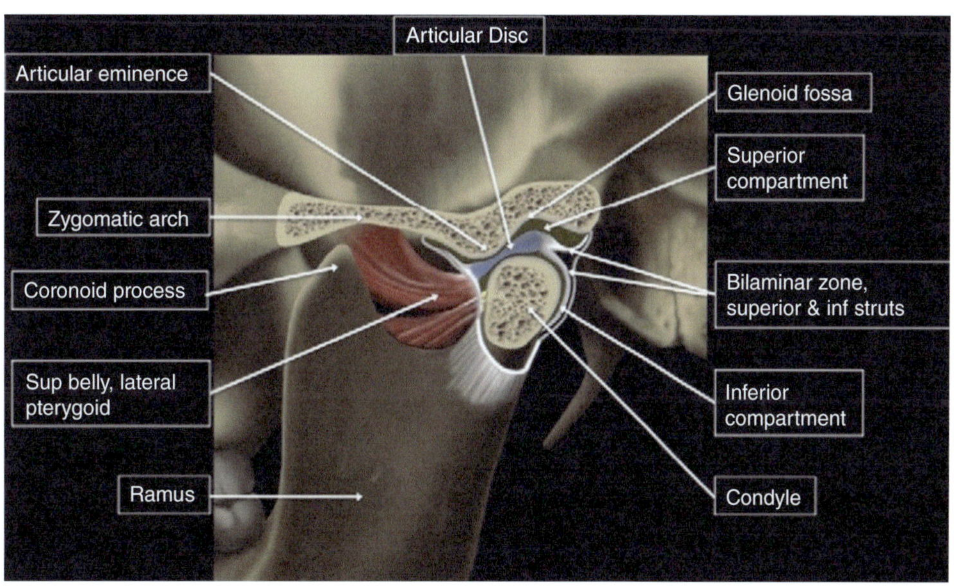

Fig. 55.1 Anatomy relevant to temporomandibular joint dysfunction. (Reproduced with written permission from the artist, Dr. Bruno Bordoni)

Does Tinnitus in TMJD Have Particular Characteristics?

There are very few studies characterizing TMJD-associated tinnitus (such as with tinnitus matching). The few studies that do comment that the tinnitus tends to be in a higher frequency range (6–9 kHz).

- A study of 20 tinnitus patients (10 with TMJD and 10 without TMJD) observed that in the patients with TMJD the tinnitus, "matched to a 6000 Hz tone or noise," and these patients, "associated their tinnitus with a single sound," and concluded that, "tinnitus may occur in patients with TMD, and be of high frequency" [36].
- A study of 20 adults with tinnitus and normal hearing reported that, "the average tinnitus pitch reported by the subjects was 8.6 kHz and the average loudness was 14.1 dBSL," and that "90% of patients had at least one sign or symptom of TMD" [37].

What Is the Mechanism of TMJD-Related Tinnitus?

Even if one embraces the conjecture of a causal relationship between TMJD and tinnitus, this still leaves open the question of the mechanism connecting the two. Literature on this point remains speculative. Salvinelli and colleagues propose:

> Conceivably, TMJD could irritate auriculo-temporal nerve (ATN), triggering a somatosensory pathway-induced disinhibition of dorsal cochlear nucleus (DCN) activity in the auditory pathway. In genetically-predisposed TMJD patients, signals from chronically stimulated DCNs activating specific cortical neuronal networks, could yield plastic neural changes resulting in tinnitus. Based on current evidence of serotoninergic modulation of neural activity and plasticity in sensory pathways, reduced serotoninergic tone could promote plastic changes underlying tinnitus through diminished filtering of incoming signals [38].

Demographics

Tinnitus related to TMJD can occur at any age, though some literature reports a slight preponderance among younger women.

A study of 200 tinnitus patients reported that tinnitus related to temporomandibular joint dysfunction "principally affects a younger population (average fifth decade of life) and mainly women (more than 2/3 of the cases)" [39].

Response to Treatment

A variety of strategies have been attempted to treat TMJD-related tinnitus. Evidence regarding treatment outcomes is weak. A systemic review and meta-analysis reported that, "The eight treatment studies indicated that treatment of TMD symptoms may have a beneficial effect on severity of tinnitus. However, only one treatment study included a control group, meaning that the overall level of evidence is low" [18].

Probably, the most common strategy is some combination of a dental appliance such as a splint, physical therapy and education.

- A study of 55 tinnitus patients aged 18–60 years who had temporomandibular joint dysfunction were evaluated (by the tinnitus handicap inventory and a visual analog scale of temporomandibular joint dysfunction) before and after treatment for TMD with a *splint*. The authors observed that, "The characteristics of tinnitus and the degree of response to treatment confirmed the relationship between tinnitus and TMD," and concluded that, "when the most common causes of tinnitus, such as otologic disorders and neurological diseases are excluded, it is correct to evaluate the functionality of the temporo-mandibular joint and eventually treat its pathology to obtain tinnitus improvement or even resolution" [40].
- A study of 25 patients with tinnitus and ipsilateral TMJD reported symptomatic improvement with "stomatognathic therapy" (oral *splints, physiotherapy*) in 11 of 25 patients (44%) [1].
- A study of, "Sixty-one patients with tinnitus attributed to TMD were randomized into the *physiotherapy and manual therapy* group or physiotherapy alone group," concluded that, "application of cervico-mandibular manual therapies in combination with exercise and education resulted in better outcomes than application of exercise/education alone in individuals with tinnitus attributed to TMD" [41].
- A systematic literature review of 11 studies found, "an overall positive effect of the combination of *splint therapy and exercise* treatment on tinnitus severity and intensity," though noted that the quality of evidence is low [42].
- A secondary analysis study of 61 subjects with TMJD-related tinnitus treated with "cervico-*mandibular manual therapy* into an exercise combined with *education* program" found that, "baseline tinnitus severity and localized PPT over the temporalis muscle were predictive of clinical outcomes in individuals with TMD-related tinnitus following physical therapy" [43].
- A study of 80 patients with TMJD pain and tinnitus were divided into two groups, the first of which underwent "orofacial treatment *physical therapy* complemented with

occlusal *splints* when needed," and the second of which served as a control group that received no treatment. The authors observed that, "A significant total effect of orofacial treatment compared with control on the change in tinnitus functional index (TFI) score was found," and concluded that, "Reduction of TMD pain is a mediating factor in the decrease of tinnitus severity after multidisciplinary orofacial treatment" [44].

- A study of, "Twenty-five patients presenting with a chief complaint of pain around the temporomandibular joints, along with symptoms of internal derangements, i.e., clicking or crepitus, and concomitant vertigo were treated successfully with jaw repositioning *orthotics*" [45].
- A study of US Air Force patients with TMJD and various otologic symptoms, all of whom were treated with "dental *orthotic* and TMD *self-care instructions*," found that of the patients with tinnitus, 64% reported "at least moderate symptom improvement of their tinnitus" after "3 months of orthotic wear" [46].

Less common strategies include injection of anesthetic or steroids, or use of "low level laser therapy."

- An observational study reported that, "In treating patients with temporomandibular joint (TMJ) dysfunction it was noticed that tinnitus and vertigo were common in such patients and there was also muscular tension in jaw and neck. During treatment of these patients it was also noted that *injection of lidocaine in a jaw muscle* (m. pt. lat.) reduced not only their muscular problems but also that the tinnitus was reduced while the local anesthetic was active" [47].
- A study of 70 patients with tinnitus that had "persisted for one month or longer were treated with an intra-articular injection of the temporomandibular joint with steroids," and the investigators reported that, "Relief of tinnitus at seven-week follow-up was achieved in 20% of the patients treated with temporomandibular steroids. At 18 months, 50% of the patients successfully treated with temporomandibular steroids for tinnitus still experienced a benefit" [48].
- A prospective study of 46 patients with "bilateral subjective tinnitus with TMDs" were randomly allocated into three groups, one treated with "Low-level laser therapy (LLLT) with an neodymium-doped yttrium aluminum garnet (Nd:YAG) (1064 nm) laser," a second group treated with "LLLT with a diode laser (810 nm)," and a third group that received placebo treatment," observed that, "There were statistically significant differences in the Nd:YAG laser ($p = 0.001$) and 810 nm diode laser groups ($p = 0.005$), but no difference in the placebo group ($p = 0.065$)," and concluded that, "Both the Nd:YAG and 810 nm diode lasers were effective for the treatment of subjective tinnitus related to TMDs" [49].

More invasive approaches include a variety of surgical techniques.

- A retrospective study of 43 patients who underwent arthroscopic surgery for arthrogenous TMJD reported that symptoms (including tinnitus), "resolved within 24 hours of arthroscopic temporomandibular joint surgery indicating that the temporomandibular joint pathology was the perpetuating force behind, if not the cause of, these symptoms" [50].
- A study compared 120 patients with tinnitus and TMJD who underwent treatment to patients on a waiting list (who received no treatment). The study found that, "Eighty percent of the patients had signs of TMD, most commonly myofascial pain. Forty-three percent of the patients reported that their tinnitus was improved at the 2-year follow-up, 39% that it was unchanged, and 17% that it was impaired compared to before the treatment," whereas, "Twelve percent of the subjects in the control group reported that their tinnitus was improved compared to 2 years previously, 32% that it was unchanged, and 56% that it was impaired. The difference between groups was significant (χ^2: $p<0.001$)," and concluded that, "The results of this study showed that TMD symptoms and signs are frequent in patients with tinnitus and that TMD treatment has a good effect on tinnitus in a long-term perspective, especially in patients with fluctuating tinnitus" [51].

Quality of Life Studies

Most, though not all, studies of TMJD associated tinnitus report that patients suffering with this combination of problems have a lower quality of life.

- A population-based survey of 5786 subjects reported that, "the odds ratio (OR) for lower quality of life was significantly higher in the group with both TMD and tinnitus compared to the group without TMD and tinnitus" [52].
- A quality-of-life study of TMJD patients with tinnitus reported pain (from TMJD) to impair quality of life more than tinnitus [53].
- A quality of life study of 19 patients with TMJD and tinnitus found that, "Tinnitus reported in patients with TMD caused moderate impact on quality of life and can be seen in the presence of background noise although daily activities can still be performed" [54].
- A quality-of-life study of 100 patients with, "somatosensory tinnitus" including 85 with TMJD and 15 without TMJD found that, "Subjects with somatosensory tinnitus… do not have a higher quality of life handicap" [55].

Miscellaneous

A study of 15 patients with painful TMJD without tinnitus, 27 patients with painful TMJD and with tinnitus, and 13 control subjects, identified "dysregulated psychophysiologic interactions" (as measured by several autonomic function tests) in the group of patients with painful TMD and tinnitus compared to the other groups [56].

A case report [57] of a 41-year-old man with, "objective tinnitus complicated with transient low-tone hearing loss coinciding with mouth opening," observed that with the jaw open, audiometry "showed elevations [in hearing thresholds] of approximately 20 dB in the frequencies below 1000 Hz," and "peak pressure on the tympanogram deviated negatively to -200 mm H_2O," which they attributed to the radiographic finding that, "CT showed a connection between the mandibular fossa and middle ear space, as revealed by a gas collection around the joint capsule evaluated in two phases (with and without mouth closing)." The investigators performed a myringotomy and reported that the "ear symptoms resolved." In other words, in this case, jaw opening altered the morphology of the middle ear space, creating negative middle ear pressure resulting in hearing loss and tinnitus.

A study of 310 normal-hearing tinnitus patients with a history of TMJD and/or neck dysfunction and/or a "positive modulation of tinnitus," found a, "strong association between history and modulation for the same regions within the same patients; such an association should always be investigated to improve chances of a correct diagnosis of somatosensory tinnitus" [58].

A study of 226 patients with tinnitus that could be modulated by "somatic maneuvers" concluded that, "Our findings in patients with chronic tinnitus and normal hearing suggest that self-reported history for somatic dysfunction and modulation of tinnitus, when occurring simultaneously in the TMJ region, can be useful to preliminarily identify patients with TMJ disorders" [59].

Based on results from the previous studies, Ralli and colleagues hypothesize that, "In some individuals, tinnitus can be modulated by specific maneuvers of the temporomandibular joint, head and neck, eyes, and limbs. Neuroplasticity seems to play a central role in this capacity for modulation, suggesting that abnormal interactions between the sensory modalities, sensorimotor systems, and neurocognitive and neuroemotional networks may contribute to the development of somatosensory tinnitus" [60].

Medical Professionals Are Often Unaware of a Relationship between TMJD and Tinnitus

Despite the literature support for evidence of an association between TMJD and tinnitus, many medical practitioners are unaware of this.

- A survey of 54 dental surgeons concluded that, "There is a need to inform dentists about the interrelationship between temporomandibular joint disorders and tinnitus" [61].
- A survey of 37 primary health care dentists "indicated insufficient knowledge… on the interrelationship between temporomandibular dysfunction and tinnitus" [62].

References

1. Buergers R, Kleinjung T, Behr M, Vielsmeier V. Is there a link between tinnitus and temporomandibular disorders? J Prosthet Dent. 2014;111(3):222–7.
2. Saldanha AD, Hilgenberg PB, Pinto LM, Conti PC. Are temporomandibular disorders and tinnitus associated? Cranio. 2012;30(3):166–71.
3. Bernhardt O, Gesch D, Schwahn C, Bitter K, Mundt T, Mack F, et al. Signs of temporomandibular disorders in tinnitus patients and in a population-based group of volunteers: results of the study of health in Pomerania. J Oral Rehabil. 2004;31(4):311–9.
4. Bousema EJ, Koops EA, van Dijk P, Dijkstra PU. Association between subjective tinnitus and cervical spine or temporomandibular disorders: a systematic review. Trends Hear. 2018;22:2331216518800640.
5. Cebi AT. Presence of tinnitus and tinnitus-related hearing loss in temporomandibular disorders. Cranio. 2020;41(2):173–7.
6. Chole RA, Parker WS. Tinnitus and vertigo in patients with temporomandibular disorder. Arch Otolaryngol Head Neck Surg. 1992;118(8):817–21.
7. Cimino R, Bucci R, Ambrosio A, Carrabba L, Corrado B, Simeon V, et al. Temporomandibular disorders, neck disability, and oral parafunctions in tinnitus patients: a cross-sectional epidemiological study from southern Italy. Cranio. 2022;40(6):485–93.
8. Edvall NK, Gunan E, Genitsaridi E, Lazar A, Mehraei G, Billing M, et al. Impact of temporomandibular joint complaints on tinnitus-related distress. Front Neurosci. 2019;13:879.
9. Hilgenberg PB, Saldanha AD, Cunha CO, Rubo JH, Conti PC. Temporomandibular disorders, otologic symptoms and depression levels in tinnitus patients. J Oral Rehabil. 2012;39(4):239–44.
10. Kim YH, Park YG, Han KD, Vu D, Cho KH, Lee SY. Prevalence of tinnitus according to temporomandibular joint disorders and dental pain: the Korean National Population-based Study. J Oral Rehabil. 2018;45(3):198–203.
11. Manfredini D. Tinnitus in temporomandibular disorders patients: any clinical implications from research findings? Evid Based Dent. 2019;20(1):30–1.
12. Manfredini D, Olivo M, Ferronato G, Marchese R, Martini A, Guarda-Nardini L. Prevalence of tinnitus in patients with different temporomandibular disorders symptoms. Int Tinnitus J. 2015;19(2):47–51.
13. Mijiritsky E, Winocur E, Emodi-Perlman A, Friedman-Rubin P, Dahar E, Reiter S. Tinnitus in temporomandibular disorders: Axis I and Axis II findings according to the diagnostic criteria for temporomandibular disorders. J Oral Facial Pain Headache. 2020;34(3):265–72.
14. Morgan DH. Tinnitus of TMJ origin: a preliminary report. Cranio. 1992;10(2):124–9.
15. Mottaghi A, Menendez-Diaz I, Cobo JL, Gonzalez-Serrano J, Cobo T. Is there a higher prevalence of tinnitus in patients with temporomandibular disorders? A systematic review and meta-analysis. J Oral Rehabil. 2019;46(1):76–86.

16. Omidvar S, Jafari Z. Association between tinnitus and temporomandibular disorders: a systematic review and meta-analysis. Ann Otol Rhinol Laryngol. 2019;128(7):662–75.
17. Parker WS, Chole RA. Tinnitus, vertigo, and temporomandibular disorders. Am J Orthod Dentofacial Orthop. 1995;107(2):153–8.
18. Skog C, Fjellner J, Ekberg E, Haggman-Henrikson B. Tinnitus as a comorbidity to temporomandibular disorders-a systematic review. J Oral Rehabil. 2019;46(1):87–99.
19. Vernon J, Griest S, Press L. Attributes of tinnitus that may predict temporomandibular joint dysfunction. Cranio. 1992;10(4):282–7. discussion 7–8
20. Vielsmeier V, Kleinjung T, Strutz J, Burgers R, Kreuzer PM, Langguth B. Tinnitus with temporomandibular joint disorders: a specific entity of tinnitus patients? Otolaryngol Head Neck Surg. 2011;145(5):748–52.
21. Vielsmeier V, Strutz J, Kleinjung T, Schecklmann M, Kreuzer PM, Landgrebe M, et al. Temporomandibular joint disorder complaints in tinnitus: further hints for a putative tinnitus subtype. PLoS One. 2012;7(6):e38887.
22. Bush FM. Tinnitus and otalgia in temporomandibular disorders. J Prosthet Dent. 1987;58(4):495–8.
23. Demirkol N, Demirkol M, Usumez A, Sari F, Akcaboy C. The potential etiologic factors influencing tinnitus intensity in patients with temporomandibular disorders. Cranio. 2018;36(6):360–5.
24. Upton LG, Wijeyesakere SJ. The incidence of tinnitus in people with disorders of the temporomandibular joint. Int Tinnitus J. 2004;10(2):174–6.
25. Fernandes G, Goncalves DA, de Siqueira JT, Camparis CM. Painful temporomandibular disorders, self reported tinnitus, and depression are highly associated. Arq Neuropsiquiatr. 2013;71(12):943–7.
26. Bernhardt O, Mundt T, Welk A, Koppl N, Kocher T, Meyer G, et al. Signs and symptoms of temporomandibular disorders and the incidence of tinnitus. J Oral Rehabil. 2011;38(12):891–901.
27. Lee CF, Lin MC, Lin HT, Lin CL, Wang TC, Kao CH. Increased risk of tinnitus in patients with temporomandibular disorder: a retrospective population-based cohort study. Eur Arch Otorrinolaringol. 2016;273(1):203–8.
28. Cakur B, Yasa Y. Correlation between tinnitus and Petrotympanic fissure status among patients with temporomandibular joint dysfunction. J Oral Maxillofac Surg. 2016;74(1):47–52.
29. Kijak E, Szczepek AJ, Margielewicz J. Association between anatomical features of Petrotympanic fissure and tinnitus in patients with temporomandibular joint disorder using CBCT imaging: an exploratory study. Pain Res Manag. 2020;2020:1202751.
30. Koparal M, Sirik M, Yavuz GY, Ege B. Evaluation of the relationship between temporomandibular joint disorders and tinnitus with computed tomography. J Stomatol Oral Maxillofac Surg. 2022;123(4):e199–205.
31. Lavinsky D, Lavinsky J, Setogutti ET, Rehm DDS, Lavinsky L. The role of magnetic resonance imaging of the temporomandibular joint to investigate tinnitus in adults with temporomandibular joint disorder: a comparative study. Int Arch Otorhinolaryngol. 2020;24(1):e68–72.
32. Ren YF, Isberg A. Tinnitus in patients with temporomandibular joint internal derangement. Cranio. 1995;13(2):75–80.
33. Choi JW, Nahm H, Shin JE, Kim CH. Temporomandibular joint herniation into the middle ear: a rare cause of mastication-induced tinnitus. Radiol Case Rep. 2020;15(2):125–7.
34. Ertugrul S. Rare cause of tinnitus: spontaneous temporomandibular joint herniation into the external auditory canal. J Craniofac Surg. 2018;29(5):e521–e2.
35. Lim KH, Jung JY, Rhee J, Choi J. Temporomandibular joint herniation through the foramen of Huschke with clicking tinnitus. Eur Ann Otorhinolaryngol Head Neck Dis. 2019;136(6):497–9.
36. Kanji A, Khoza-Shangase K. Clinical signs and symptoms of tinnitus in temporomandibular joint disorders: a pilot study comparing patients and non-patients. S Afr J Commun Disord. 2013;60:16–20.
37. Morais AA, Gil D. Tinnitus in individuals without hearing loss and its relationship with temporomandibular dysfunction. Braz J Otorhinolaryngol. 2012;78(2):59–65.
38. Salvinelli F, Casale M, Paparo F, Persico AM, Zini C. Subjective tinnitus, temporomandibular joint dysfunction, and serotonin modulation of neural plasticity: causal or casual triad? Med Hypotheses. 2003;61(4):446–8.
39. Algieri GMA, Leonardi A, Arangio P, Vellone V, Paolo CD, Cascone P. Tinnitus in temporomandibular joint disorders: is it a specific somatosensory tinnitus subtype? Int Tinnitus J. 2017;20(2):83–7.
40. Attanasio G, Leonardi A, Arangio P, Minni A, Covelli E, Pucci R, et al. Tinnitus in patients with temporo-mandibular joint disorder: proposal for a new treatment protocol. J Craniomaxillofac Surg. 2015;43(5):724–7.
41. Delgado de la Serna P, Plaza-Manzano G, Cleland J, Fernandez-de-Las-Penas C, Martin-Casas P, Diaz-Arribas MJ. Effects of cervico-mandibular manual therapy in patients with temporomandibular pain disorders and associated somatic tinnitus: a randomized clinical trial. Pain Med. 2020;21(3):613–24.
42. Michiels S, Nieste E, Van de Heyning P, Braem M, Visscher C, Topsakal V, et al. Does conservative temporomandibular therapy affect tinnitus complaints? A systematic review. J Oral Facial Pain Headache. 2019;33(3):308–17.
43. Plaza-Manzano G, Delgado-de-la-Serna P, Diaz-Arribas MJ, Rodrigues-de-Souza DP, Fernandez-de-Las-Penas C, Alburquerque-Sendin F. Influence of clinical, physical, psychological, and psychophysical variables on treatment outcomes in somatic tinnitus associated with temporomandibular pain: evidence from a randomized clinical trial. Pain Pract. 2021;21(1):8–17.
44. van der Wal A, Michiels S, Van de Heyning P, Gilles A, Jacquemin L, Van Rompaey V, et al. Reduction of somatic tinnitus severity is mediated by improvement of temporomandibular disorders. Otol Neurotol. 2022;43(3):e309–e15.
45. Williamson EH. Interrelationship of internal derangements of the temporomandibular joint, headache, vertigo, and tinnitus: a survey of 25 patients. Cranio. 1990;8(4):301–6.
46. Wright EF, Syms CA 3rd, Bifano SL. Tinnitus, dizziness, and non-otologic otalgia improvement through temporomandibular disorder therapy. Mil Med. 2000;165(10):733–6.
47. Bjorne A. Assessment of temporomandibular and cervical spine disorders in tinnitus patients. Prog Brain Res. 2007;166:215–9.
48. Henk K, Mark K. Temporomandibular steroids in patients with tinnitus: only on indication. Int Tinnitus J. 2019;23(1):10–6.
49. Demirkol N, Usumez A, Demirkol M, Sari F, Akcaboy C. Efficacy of low-level laser therapy in subjective tinnitus patients with temporomandibular disorders. Photomed Laser Surg. 2017;35(8):427–31.
50. Steigerwald DP, Verne SV, Young D. A retrospective evaluation of the impact of temporomandibular joint arthroscopy on the symptoms of headache, neck pain, shoulder pain, dizziness, and tinnitus. Cranio. 1996;14(1):46–54.
51. Tullberg M, Ernberg M. Long-term effect on tinnitus by treatment of temporomandibular disorders: a two-year follow-up by questionnaire. Acta Odontol Scand. 2006;64(2):89–96.
52. Kong GS, Lee SH, Park KS, Cho JH, Kim KW, Ha IH. Association of temporomandibular disorders and tinnitus with health-related quality of life: a cross-sectional study using the fifth Korea national health and nutrition examination survey. J Oral Rehabil. 2022;49(3):283–94.
53. Calderon Pdos S, Hilgenberg PB, Rossetti LM, Laurenti JV, Conti PC. Influence of tinnitus on pain severity and quality of life in patients with temporomandibular disorders. J Appl Oral Sci. 2012;20(2):170–3.

54. Lacerda AB, Facco C, Zeigelboim BS, Cristoff K, Stechman JN, Fonseca VR. The impact of tinnitus on the quality of life in patients with temporomandibular dysfunction. Int Tinnitus J. 2016;20(1):24–30.
55. Hilgenberg-Sydney PB, Saldanha ADD, Lopes AC, Conti PCR. Audiological evaluation of patients with somatosensory tinnitus attributed to temporomandibular disorders. Am J Audiol. 2020;29(4):930–4.
56. Kang JH, Song SI. Autonomic and psychologic risk factors for development of tinnitus in patients with chronic temporomandibular disorders. J Oral Facial Pain Headache. 2019;33(4):362–70.
57. Hidaka H, Tatewaki Y, Sakamoto M, Kobayashi T, Yuasa Y, Yuasa R, et al. Transient hearing loss and objective tinnitus induced by mouth opening: a rare connection between the temporomandibular joint and middle ear space. Otol Neurotol. 2016;37(7):e217–21.
58. Ralli M, Altissimi G, Turchetta R, Mazzei F, Salviati M, Cianfrone F, et al. Somatosensory tinnitus: correlation between Cranio-Cervico-mandibular disorder history and somatic modulation. Audiol Neurootol. 2016;21(6):372–82.
59. Ralli M, Greco A, Boccassini A, Altissimi G, Di Paolo C, Falasca V, et al. Subtyping patients with somatic tinnitus: modulation of tinnitus and history for somatic dysfunction help identify tinnitus patients with temporomandibular joint disorders. PLoS One. 2018;13(8):e0202050.
60. Ralli M, Greco A, Turchetta R, Altissimi G, de Vincentiis M, Cianfrone G. Somatosensory tinnitus: current evidence and future perspectives. J Int Med Res. 2017;45(3):933–47.
61. Martins PF, Stechman JN, Marques JM, Martins SK, Cristoff KE, Sampaio RS, et al. Tinnitus and temporomandibular disorders: the knowledge of professionals for primary health care in the city of Curitiba. Int Tinnitus J. 2016;20(1):18–23.
62. Martins PFA, Stechman-Neto J, Cristoff KE, Ravazzi G, Goncalves FM, Sampaio RS. Temporomandibular dysfunction and tinnitus the knowledge of the dental surgeon in primary health care. Int Tinnitus J. 2019;23(2):125–32.

Cervicogenic Tinnitus

56

Overview

In simple terms, the phrase "cervicogenic tinnitus" refers to tinnitus (perception of sound) supposedly originating from the neck, occurring in a patient who also has neck symptoms. In a patient with nonpulsatile tinnitus and neck complaints it is medically reasonable to check screening audiometric testing, including audiometry (with tinnitus matching), otoacoustic emissions and brainstem auditory evoked responses. If these screening tests do not provide evidence for a primary otologic source of tinnitus, then a working diagnosis of *cervicogenic tinnitus* is plausible, and a trial of manual physical therapy for the neck is reasonable. We do not generally recommend neck imaging as there is no evidence (yet) that results of imaging change management. If a reasonable trial of manual physical therapy for the neck is unsuccessful, then referral to a tinnitus audiologist (to be evaluated for tinnitus masking) is reasonable.

Introduction

The phrase "*cervicogenic tinnitus*" refers to a form of somatic tinnitus purportedly originating from neck pathology.

Association of Tinnitus with Cervical Spine Problems

The preliminary evidence for a connection between tinnitus and neck symptoms comes from studies noting an association between the two. For example, a meta-analysis of 22 studies commented that, "patients with tinnitus more frequently reported CSD [cervical spine disease] than subjects without tinnitus. The odds ratio was 2.6 (95% CI [1.1, 6.4])," though the meta-analysis also cautioned that, "Methodological quality was generally limited by a lack of blinding, comparability of groups, and nonvalidated instruments for assessing CSD" [1]. While association does not prove causality, it is a reasonable starting point for studying this phenomenon.

What Is the Mechanism of Cervicogenic Tinnitus?

The topic of cervicogenic tinnitus is controversial for several reasons.

The first difficulty is that many studies of this topic do not clearly define what kind of cervical spine pathology is in question. One research group has contended that cervical spine pathology for this purpose can be adequately assessed by physical examination [2], and on that basis, some of those same investigators conducted a study [3] that compared patients who carried a diagnosis of "cervicogenic somatic tinnitus" (CST) to patients with chronic nonpulsatile tinnitus not believed to be of cervical origin, and tried to identify specific cervical abnormalities on physical examination; the study concluded that the patients had a "positive manual rotation test, 47% a positive AST, 49% a positive score on both, and 81% had positive trigger points," whereas in patients with non-CST, "these percentages were 36, 18, 10, and 50%, respectively." Each group was also surveyed using the Neck Bournemouth Questionnaire [4, 5], and the study found that, "79% of the CST group had a positive NBQ versus 40% in the non-CST group."

The second difficulty is that the mechanism is by which a cervical problem could lead to tinnitus is unclear. There are various theoretical discussions of this [6], which often make the observations such as, "neural connections and integration of the auditory and somatosensory systems of the upper cervical region and head have been confirmed by many studies" [7], but this alone does not explain the mechanism.

One of the more coherent explanations of a potential mechanism comes from Dr. Robert Levine [8]. Based on detailed neuroanatomical studies, human cases and animal studies, Levine hypothesizes, "that craniocervical somatic tinnitus occurs through modulation of the pathway from MSN [medullary somatosensory nuclei] to the ipsilateral DCN [dorsal cochlear nucleus]," with the relevant anatomy as illustrated in Fig. 56.1.

M. Cherchi, *Otoneurology and Vestibular Medicine*, https://doi.org/10.1007/978-3-031-94841-1_56

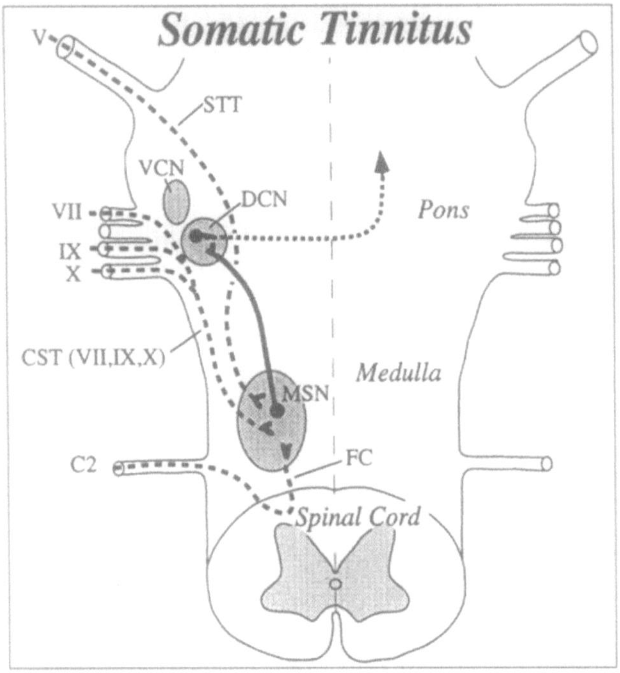

Fig. 56.1 Schematic diagram of brainstem and upper cervical spinal cord, depicting the anatomic basis for the DCN [dorsal cochlear nucleus] hypothesis; both somatic and otic tinnitus occur because of disinhibition of the DCN. In both cases, tinnitus is caused by increased activity in the output of the DCN (curved arrow), which then projects to other centers and eventually leads to activation of the auditory perceptual machinery responsible for tinnitus. For somatic tinnitus (A), sensory inputs (shown as long-dashed lines) from [1] the face (via the trigeminal nerve [V] and the spinal trigeminal tract [STT]), [2] the external and middle ears (via the common spinal tract of the facial, glossopharyngeal, and vagus nerves (CST [VII, IX, X]), and [3] the neck (via the C2 dorsal root and the fasciculus cuneatus [FC]) converge to a common region of the lower medulla, the medullary somatosensory nuclei (MSN) , from which fibers project to the ipsilateral DCN (solid line). Modulation of the activity in the MSN to DCN pathway results in disinhibition of DCN (From Levine [8]. Reproduced with written permission from Elsevier)

How Is Cervicogenic Tinnitus Diagnosed?

Patients who complain of tinnitus and who also complain of neck discomfort or who have suffered a neck injury may be suspected of having cervicogenic tinnitus, though this is merely an observation of association, not proof of a causal relationship. Some investigators suggest that if neck movements can modulate (increase or decrease) the tinnitus, then that supports the idea of a cervical mechanism of tinnitus [9].

How Is Cervicogenic Tinnitus Treated?

Despite the poor understanding of the mechanism of cervicogenic tinnitus, and despite the lack of clear evidence for a causal relationship, numerous attempts at neck therapy have been undertaken.

Some "data" regarding treatment outcomes is simply in the form of case reports. For example, Bechter and colleagues report a case of an individual with cervicogenic tinnitus that remitted after treatment with a cervical collar [10].

Some studies suffer from patient selection problems. For example, the study by Delgado de la Serna and colleagues explored "cervico-mandibular therapies" in patients with either temporomandibular joint dysfunction or cervical problems [11], rather than isolating each group.

One study of cervicogenic tinnitus patients randomized subjects to receive "multimodal cervical physical therapy" immediately or in a delayed fashion, found symptomatic improvements to parallel the timing of the therapy, and concluded that "Cervical physical therapy can have a positive effect on subjective tinnitus complaints in patients with a combination of tinnitus and neck complaints" [12]. In effect, a study of this structure has no control group, and thus no way to exclude a placebo effect.

References

1. Bousema EJ, Koops EA, van Dijk P, Dijkstra PU. Association between subjective tinnitus and cervical spine or temporomandibular disorders: a systematic review. Trends Hear. 2018;22:2331216518800640.
2. Michiels S, Van de Heyning P, Truijen S, De Hertogh W. Diagnostic value of clinical cervical spine tests in patients with cervicogenic somatic tinnitus. Phys Ther. 2015;95(11):1529–35.
3. Michiels S, De Hertogh W, Truijen S, Van de Heyning P. Cervical spine dysfunctions in patients with chronic subjective tinnitus. Otol Neurotol. 2015;36(4):741–5.
4. Bolton JE, Breen AC. The Bournemouth Questionnaire: a short-form comprehensive outcome measure. I. Psychometric properties in back pain patients. J Manip Physiol Ther. 1999;22(8):503–10.
5. Bolton JE, Humphreys BK. The Bournemouth Questionnaire: a short-form comprehensive outcome measure. II. Psychometric properties in neck pain patients. J Manip Physiol Ther. 2002;25(3):141–8.
6. Gelb H, Gelb ML, Wagner ML. The relationship of tinnitus to craniocervical mandibular disorders. Cranio. 1997;15(2):136–43.
7. Oostendorp RA, Bakker I, Elvers H, Mikolajewska E, Michiels S, De Hertogh W, et al. Cervicogenic somatosensory tinnitus: an indication for manual therapy? Part 1: theoretical concept. Man Ther. 2016;23:120–3.
8. Levine RA. Somatic (craniocervical) tinnitus and the dorsal cochlear nucleus hypothesis. Am J Otolaryngol. 1999;20(6):351–62.
9. Ralli M, Altissimi G, Turchetta R, Mazzei F, Salviati M, Cianfrone F, et al. Somatosensory tinnitus: correlation between Cranio-Cervico-mandibular disorder history and somatic modulation. Audiol Neurootol. 2016;21(6):372–82.
10. Bechter K, Wieland M, Hamann GF. Chronic Cervicogenic tinnitus rapidly resolved by intermittent use of cervical collar. Front Psych. 2016;7:43.
11. Delgado de la Serna P, Plaza-Manzano G, Cleland J, Fernandez-de-Las-Penas C, Martin-Casas P, Diaz-Arribas MJ. Effects of cervico-mandibular manual therapy in patients with temporomandibular pain disorders and associated somatic tinnitus: a randomized clinical trial. Pain Med. 2020;21(3):613–24.
12. Michiels S, Van de Heyning P, Truijen S, Hallemans A, De Hertogh W. Does multi-modal cervical physical therapy improve tinnitus in patients with cervicogenic somatic tinnitus? Man Ther. 2016;26:125–31.

Otosclerosis

57

Overview

Otosclerosis is a disease of bone metabolism that can affect the middle ear and inner ear, and cause auditory and vestibular symptoms. The underlying cause of otosclerosis remains unknown. Diagnosis is based primarily on the clinical history and audiologic findings, sometimes supplemented by imaging (mostly to exclude alternative diagnoses). Vestibular impairment is variable. Most, though not all, investigators find high resolution temporal bone CT to be helpful in the diagnosis. There is less research regarding the diagnostic utility of MRI. Hearing loss can sometimes be addressed by amplification, though if this fails, then surgery (such as a stapedectomy) may be considered.

Introduction

Otosclerosis is a primary osteodystrophy [1] restricted to the temporal bone that involves abnormal bone metabolism that can affect the middle ear and inner ear. Depending on the area(s) affected, this can manifest with auditory and (less commonly) vestibular symptoms.

Epidemiology

The incidence of otosclerosis has been estimated at 3 to 4 per 1000 [2]. It is significantly more common in Caucasian populations, and more common in women. Peak age of symptom onset is in the third decade, with a range of the first through the sixth decade [3]. It is relatively uncommon in the pediatric population [4–7].

Genetics

Between 30% and 70% of patients diagnosed with otosclerosis report a family history that is positive for the diagnosis [3]. Some investigators report that familial otosclerosis may present with more rapid onset and with a higher incidence of bilaterality than nonfamilial otosclerosis [3].

Several genetic loci have been associated with otosclerosis, but specific genes have not yet been identified [8]. Inheritance is not always in a clearly Mendelian pattern; it is thought that penetrance is incomplete.

Pathophysiologic Mechanism of Disease

The disorder of bone metabolism in otosclerosis comprises abnormal remodeling of bone. Temporal bone studies recognize several histopathologic stages of otosclerosis:

1. Early phase: Otospongiosis, in which there are histiocytes, osteoblasts, and osteocytes around blood vessels. The affected areas develop abnormal bony morphology with immature "spongy" bone.
2. Transitional phase.
3. Late phase: Otosclerosis. The affected areas evolve into dense sclerotic bone. This process narrows the blood vessels it surrounds.

Otosclerosis affects the following anatomical areas, in descending order of frequency [3]:

- 80%: Stapes footplate
- 30%: Round window
- 21%: Pericochlear region
- 19%: Anterior portion of internal auditory canal
- Less common: malleus, incus, facial canal, semicircular canals, endolymphatic duct.

When otosclerosis involves one or more components of the ossicular chain, this can interfere with the mechanical ability to transmit the kinetic energy of vibrations to the inner ear, thus resulting in a conductive hearing loss.

When otosclerosis involves the bone around the cochlea (pericochlear), sensorineural hearing loss can result. The

M. Cherchi, *Otoneurology and Vestibular Medicine*, https://doi.org/10.1007/978-3-031-94841-1_57

mechanism is not clear, since otosclerosis should only affect the bony labyrinth, not the membranous labyrinth or its contents; for this reason, some investigators have explored whether otosclerosis can affect the membranous labyrinth indirectly, such as by altering the chemical composition of perilymph or endolymph, or in some other way provoking deposition of abnormal material within the membranous labyrinth [9]. Several investigators, such as Harold Schuknecht, dispute even the existence of cochlear otosclerosis [10, 11].

The underlying cause of otosclerosis has not been conclusively identified [12]. Speculations in the literature include an intrinsic abnormality of collagen, autoimmunity, inflammation, measles infection, hormonal influences, oxidative stress and others.

Clinical Presentation

Peak age of symptom onset is in the third decade, with a range of the first through the sixth decade [3]. It is relatively uncommon in the pediatric population [4–7]. Symptoms can include hearing loss, tinnitus [13] or disequilibrium. It is not clear whether otosclerosis is capable of causing purely vestibular deficits and no auditory deficits [14].

Most cases of otosclerosis present with gradually progressive hearing loss that can be conductive, sensorineural or mixed. The hearing loss is bilateral in 70–85% of cases, and often asymmetrical [3].

Esa-Nunez and colleagues studied 40 otosclerotic patients complaining of "dizziness" and reported that, "Dizziness, although frequent in patients with otosclerosis is rarely a cause for specific clinical assessment. There is a lag between the patient's perception of hearing loss and the initiation of vestibular symptoms" [15].

Studies reach different conclusions regarding the frequency of vestibular symptoms among otosclerosis patients.

- Grayeli and colleagues conclude that, "The proportion of patients with otosclerosis among those with balance disorders was similar to the estimated prevalence of otosclerosis in the general population" [16].
- Morales-Garcia studied 202 otosclerotic patients and found that 27.7% reported vestibular symptoms [17].
- Freeman reviewed 438 otosclerotic patients and reported that 165 (37%) had vestibular complaints [18].
- Yetiser reviews literature citing that 25–45% of otosclerotic patients have vestibular complaints [19].

Physical Examination

Bedside hearing tests, such as Rinne's test, may identify conductive hearing loss.

Ocular Motor Examination

Patients with otosclerosis who are otherwise healthy typically have normal ocular motor examinations.

Auditory Testing

Audiometric findings in otosclerosis are variable [20]. Audiometry usually shows bilateral hearing loss that can be conductive, sensorineural or mixed. Conductive hearing loss is more common, and is sometimes the initial type [13]. It may also show a Carhart notch [21], though this finding is inconsistent [22].

Acoustic reflexes are often abnormal in otosclerosis [23, 24], and commonly are inverted [25, 26].

Vestibular Testing

Otosclerotic patients can exhibit a variety of abnormalities on instrumented otovestibular testing.

Vestibular Testing: Caloric Testing

Morales-Garcia studied 202 otosclerotic patients and noted that, "As the age of patients increased there was a parallel increase of deafness and the latter was accompanied by a greater cochleo-vestibular involvement. Those ears with cochlear involvement showed a greater percentage of **caloric** abnormalities than the ears with a purely conductive deafness" [17].

Cody and Baker studied 500 unoperated otosclerotic patients and reported that, "The incidence of vestibular symptoms in patients with otosclerosis increased as the relative and absolute amounts of sensorineural hearing loss increased and was much higher than one would expect in the normal population. In addition, with increasing sensorineural hearing loss, the severity of vestibular symptoms increased along with the incidence of depression in vestibular function determined by the bithermal **caloric** test in patients with vestibular symptoms" [14].

Panda and colleagues [27] studied *caloric* testing in 25 otosclerotic patients before and after undergoing stapes surgery. They reported that, "Patients with otosclerosis demonstrated hypoexcitability compared to controls, which was statistically significant for right cold and hot irrigation… [T] he degree of vestibular dysfunction seems to be inversely proportional to the hearing result" [27].

Vestibular Testing: Vibration-Induced Nystagmus on Video Oculography

Manzari and Modugno studied the effect of bone vibration in otosclerotic patients and found that this had the effect of stimulating the otosclerotic side (i.e., slow phase drifting toward the healthy ear, fast phase beating toward the affected ear) [28]. This pattern is reversed from the vibration induced nystagmus that would normally be seen in static vestibular lesions.

Vestibular Testing: Computerized Dynamic Posturography

Parnes and colleagues [29] studied caloric testing and *computerized dynamic posturography* in 10 otosclerotic patients "before (series 1), 48 hours after (series 2), and between two to four months after (series 3) a stapedectomy." They reported that, "There was no significant change in the electronystagmography (ENG) test findings between series 1 and 3. Posturography, however, demonstrated an uncompensated vestibular pattern in the immediate postoperative period which, after two months, converted to a compensated vestibular pattern," and concluded that, "Posturography serves as a useful quantitative test for the study of patients with balance disorders because the procedure supplements rather than complements the ENG. In addition, posturography detects vestibular reflex abnormalities in stapedectomized patients two months afterwards" [29].

Vestibular Testing: Vestibular Evoked Myogenic Potentials

Akazawa and colleagues studied *cervical vestibular evoked myogenic potentials (cVEMPs)* and *ocular vestibular evoked myogenic potentials (oVEMPs)* of 17 otosclerotic patients both before and after stapes surgery and found no statistically significant change in vestibular evoked myogenic potentials (VEMPs) responses after surgery [30].

Amali and colleagues studied audiometry, air-conducted and bone-conducted *cervical vestibular evoked myogenic potentials* in the affected and unaffected ears of 30 otosclerotic patients and identified that the ears affected by otosclerosis had statistically significantly prolonged p13 latencies and reduced interpeak amplitudes. They concluded, "This study tends to suggest that due to the direct biotoxic effect of the materials released from the otosclerosis foci on saccular receptors, there might be a possibility of vestibular dysfunction in otosclerotic patients" [31].

Catalano and colleagues undertook a prospective case-control study involving 27 otosclerotic patients before and after surgery for otosclerosis using audiometry, *cervical vestibular evoked myogenic potentials* and video head impulse testing. They reported a statistically significantly reduced interpeak amplitude on cervical vestibular evoked myogenic potentials, and no statistically significant difference in gains on video head impulse testing, and concluded, "These findings indicate a probable traumatic saccular impairment in patients with vestibular symptoms" [32].

Lin and Young studied audiometry, caloric testing, *cervical vestibular evoked myogenic potentials* and *ocular vestibular evoked myogenic potentials* in 50 otosclerotic patients—27 with "vertigo" and 23 without vertigo—and observed, "Otosclerosis patients with vertigo have more frequent abnormalities of oVEMPs to impulsive stimulation than do those without, consistent with more frequent abnormalities of the utricle. Abnormalities of oVEMPs and cVEMPs are more frequent than for caloric testing and BC hearing thresholds," which led them to the conclusion that, "The relative frequency of abnormalities may reflect the degree of pathological involvement of the utricle, saccule, semicircular canals and cochlea in otosclerosis patients with vertigo" [33].

Saka and colleagues studied bone-conducted audiometry, caloric testing and *cervical vestibular evoked myogenic potentials* in 25 unoperated otosclerotic patients, of whom 10 had vestibular complaints and 15 did not [34]. They reported that bone-conducted cervical vestibular evoked myogenic potentials were abnormal in 9 out of 10 patients (90%) with vestibular complaints, and in only 1 out of 15 patients (7%) without vestibular complaints.

Tramontani and colleagues conducted a prospective study of audiometry, caloric testing, air-conducted (AC) and bone-conducted (BC) *cervical* and *ocular vestibular evoked myogenic potentials* (VEMPs) in 126 otosclerotic ears before and after stapes surgery and concluded that, "AC and BC VEMPs can be elicited in ears with otosclerosis. AC-VEMP is more vulnerable to conductive hearing loss. Evaluation of AC-VEMP thresholds can be added in the diagnostic work-up of otosclerosis in case of doubt, enhancing differential diagnosis in patients with air-bone gaps," and also somewhat surprisingly that, "Otosclerosis is not a cause of canal paresis or vertigo" [35].

Trivelli and colleagues conducted a randomized prospective trial of air-conducted *vestibular evoked myogenic potentials* in 41 otosclerotic patients before and after stapedotomy and concluded that "VEMPs reduced elicitability, in otosclerosis, is likely due to conductive hearing loss and inner ear impairment" [36].

Yang and Young studied audiometry, air-conducted and bone-conducted *cervical vestibular evoked myogenic potentials* in 21 ears of 15 otosclerotic patients and 20 ears of 10 healthy controls, observed that, "a significant relationship existed among the presence of AC-VEMPs, BC-VEMPs, and magnitude of conductive hearing loss," and concluded, "The presence of an AC-VEMP may indicate an earlier stage of otosclerosis, although absent BC-VEMP infers a later stage. Restated, AC-VEMPs may complement the results obtained with BC-VEMPs to classify the stage of otosclerosis" [37].

Instrumented Vestibular Testing in Otosclerosis Patients: Summary

Results of studies are sometimes discrepant, but generally, otosclerotic patients are more likely than controls to exhibit abnormalities on instrumented vestibular testing such as vestibular evoked myogenic potentials, caloric testing, video head impulse testing, and computerized dynamic posturography. Thus, we would agree with the overall observations of Rajati and colleagues who studied results of an audiovestibular test battery (audiometry, cervical vestibular evoked myogenic potentials, ocular vestibular evoked myogenic potentials, video head impulse testing, caloric testing) in otosclerotic patients and reported that, "the vestibular system even in asymptomatic cases, is affected by otosclerosis. Furthermore, it seems that the otolithic system has a higher chance of involvement, compared to the semicircular canals" [38].

Imaging

Most investigators find CT (usually high resolution temporal bone CT) to be helpful in diagnosing otosclerosis [1, 39–63]. A minority of investigators report that CT findings are not helpful in establishing a diagnosis of otosclerosis [64].

Most investigators report good correlation between CT findings and audiometric findings in otosclerotic patients [45, 46, 51, 55, 60]. A minority of investigators report that CT findings bear little or no correlation with audiometry and have no prognostic value [58].

The application of MRI in cases of otosclerosis has been less studied than CT. MRI findings in otosclerotic patients are often subtle [65, 66], but some investigators still report it to be helpful in diagnosing otosclerosis [67, 68]. In contrast, other investigators find that MRI is not helpful in diagnosing otosclerosis [69]. It may be that high resolution (3 Tesla) MRI without and with gadolinium contrast is more sensitive for detecting otosclerosis [70].

Histopathology

Temporal bone studies report a variety of histopathologic findings of vestibular relevance in otosclerosis.

- Hayashi and colleagues describe "cupular deposits" [9].
- Gussen described various degenerative changes [71] in vestibular structures.
- Sando and colleagues studied four temporal bones of two patients and reported neural degenerative changes distal to, and retrograde from, otosclerotic plaques apposed to the vestibular nerve [72].
- Richter and Schuknecht report, "It seems probable that a loss in vestibular neuronal population caused by involvement of dendritic fibers in the cribrose areas is at least partially responsible for the dysequilibrium or vestibular test abnormalities occurring in some patients with otosclerosis" [73].

Differential Diagnosis

The differential diagnosis of otosclerosis includes other middle ear disorders that can cause conductive hearing loss. If the hearing loss is predominantly sensorineural, then the differential diagnosis includes presbycusis.

Treatment

Treatment Options for Otosclerosis

Treatment of otosclerosis consists of the following:

- Observation.
- Amplification.
- Surgery (usually stapedotomy). If a patient with otosclerosis has conductive hearing loss that interferes with their function, and if the conductive hearing loss is believed to be due to immobilization of the ossicular chain, then an otolaryngologist may consider surgery.
- There has been some literature regarding medical treatment for otosclerosis, though results of such studies are often disappointing. Examples of medical approaches include the following:
  - Fluoride.
  - Bisphosphonates.
  - Bioflavonoids and other anti-oxidants.

If a patient diagnosed with otosclerosis complains of hearing loss, then referral to audiology (to be evaluated for amplification) is appropriate. If that is unsuccessful, then referral to otolaryngology is appropriate.

If an otosclerosis patient complains of vestibular symptoms, management is less clear. Some otolaryngologists maintain that surgery can be beneficial even for vestibular symptoms, but not all investigators agree on this point.

Outcomes of Surgery for Otosclerosis

Investigators disagree on the vestibular outcomes of otosclerotic patients who have undergone surgery. Most, though not all, investigators espouse the position that surgery for otosclerosis causes little, if any, measurable vestibular deficits.

Singbarti and colleagues studied bone-conducted cervical vestibular evoked myogenic potentials in 23 patients (25 ears) with unilateral or bilateral otosclerosis before and after stapedotomy surgery and reported that, "Stapedotomy surgery does not influence VEMPs, implying that the saccular receptors are not injured by surgery" [74].

Winters and colleagues studied bone-conducted ocular vestibular evoked myogenic potentials before and after stapes surgery in 27 otosclerotic patients (compared to 26 healthy controls) and concluded that, "No or undetectably little damage to the utricle is caused by both otosclerotic disease and stapes surgery" [75].

Satar and colleagues conducted a retrospective study of video head impulse testing (vHIT) in 11 otosclerotic patients who had undergone stapedotomy and 30 healthy controls. Out of the 22 ears of the 11 otosclerotic patients, 12 ears had undergone stapedotomy and 10 had not. They concluded that, "Otosclerosis and otosclerosis surgery may have some effects on SCC [semicircular canal] functions and thereby vHIT. Lateral SCC is the most affected SCC in terms of gain" [76].

Kujala and colleagues studied video oculography in 21 otosclerotic patients following oval window surgery and reported that, "The occurrence of nystagmus did not correlate with vestibular symptoms" [77].

Hirvonen and Aalto prospectively studied video oculography in 20 otosclerotic patients who underwent stapes surgery and reported that, "Vertigo was experienced immediately after the operation in five, floating sensation in two, and unspecific dizziness in two patients. Vestibular symptoms were mild or moderate in most patients. The occurrence of nystagmus did not correlate with vestibular symptoms ($p > 0.05$)" [78].

Kujala and colleagues prospectively studied 33 otosclerotic patients before and after stapes surgery and concluded that, "Nystagmus with a low slow-phase velocity can occur in patients with otosclerosis. However, according to the video-oculographic findings and subjective symptoms, significant vestibular dysfunction seems to be rare and temporary after stapes surgery" [79].

Birch and Elbrond studied 925 ears of 722 otosclerotic patients who had undergone stapedectomy and reported that, "at follow-up an average of 15 years after the operation, hearing was somewhat poorer in those having spontaneous nystagmus towards the operated ear. At follow-up 17% had an abnormal caloric test" [80].

De Vilhena and colleagues conducted a prospective observational study of 140 vertiginous patients who underwent stapes surgery and reported that, "vestibular disorders may remain after the immediate postoperative period" [81].

Dybwik studied 108 stapedectomized ears of 96 otosclerosis patients and reported caloric weakness in 6 patients (6%) [82].

Job and colleagues studied 170 otosclerotic patients who underwent surgery and reported that:

> A significant negative influence on bone conduction thresholds, particularly at 2000 Hz, was associated with vestibular symptoms persisting for 7 days after the surgery. Symptoms of impaired bony labyrinth function after stapedotomy, persisting for more than 1 year, were associated with insufficient reduction of the air-bone gap and worse improvement in bone conduction thresholds at 1000 and 2000 Hz. The cause of both problems was related to a prosthesis that was too long or placed too deep in the inner ear during stapedotomy [83].

References

1. Gredilla Molinero J, Mancheño Losa M, Santamaría Guinea N, Arévalo Galeano N, Grande Bárez M. Update on the imaging diagnosis of otosclerosis. Radiologia. 2016;58(4):246–56.
2. Declau F, Van Spaendonck M, Timmermans JP, Michaels L, Liang J, Qiu JP, et al. Prevalence of otosclerosis in an unselected series of temporal bones. Otol Neurotol. 2001;22(5):596–602.
3. Crompton M, Cadge BA, Ziff JL, Mowat AJ, Nash R, Lavy JA, et al. The epidemiology of Otosclerosis in a British cohort. Otol Neurotol. 2019;40(1):22–30.
4. Daniel A, Budiono G, Rao A, Low GK, Ellis MP, Lee J. Juvenile otosclerosis and congenital stapes footplate fixation. A systematic review and meta-analysis of surgical outcomes and management. Int J Pediatr Otorhinolaryngol. 2023;166:111418.
5. Lescanne E, Bakhos D, Metais JP, Robier A, Moriniere S. Otosclerosis in children and adolescents: a clinical and CT-scan survey with review of the literature. Int J Pediatr Otorhinolaryngol. 2008;72(2):147–52.
6. Salomone R, Riskalla PE, Vicente Ade O, Boccalini MC, Chaves AG, Lopes R, et al. Pediatric otosclerosis: case report and literature review. Braz J Otorhinolaryngol. 2008;74(2):303–6.
7. Sobolewska A, Clarós P. Surgical treatment in children with otosclerosis and congenital stapes fixation: our experience and outcome. Otolaryngol Pol. 2018;73(2):23–8.
8. Ealy M, Smith RJ. The genetics of otosclerosis. Hear Res. 2010;266(1–2):70–4.
9. Hayashi H, Cureoglu S, Schachern PA, Oktay MF, Fukushima H, Sone M, et al. Association between cupular deposits and otosclerosis. Arch Otolaryngol Head Neck Surg. 2006;132(12):1331–4.
10. Schuknecht HF. Cochlear otosclerosis. A continuing fantasy. Arch Otorhinolaryngol. 1979;222(2):79–84.
11. Schuknecht HF, Kirchner JC. Cochlear otosclerosis: fact or fantasy. Laryngoscope. 1974;84(5):766–82.
12. Rudic M, Keogh I, Wagner R, Wilkinson E, Kiros N, Ferrary E, et al. The pathophysiology of otosclerosis: review of current research. Hear Res. 2015;330(Pt A):51–6.
13. Danesh AA, Shahnaz N, Hall JW 3rd. The audiology of otosclerosis. Otolaryngol Clin N Am. 2018;51(2):327–42.

14. Cody DT, Baker HL Jr. Otosclerosis: vestibular symptoms and sensorineural hearing loss. Ann Otol Rhinol Laryngol. 1978;87(6 Pt 1):778–96.
15. Eza-Nunez P, Manrique-Rodriguez M, Perez-Fernandez N. Otosclerosis among patients with dizziness. Rev Laryngol Otol Rhinol (Bord). 2010;131(3):199–206.
16. Grayeli AB, Sterkers O, Toupet M. Audiovestibular function in patients with otosclerosis and balance disorders. Otol Neurotol. 2009;30(8):1085–91.
17. Morales-Garcia C. Cochleo-vestibular involvement in otosclerosis. Acta Otolaryngol. 1972;73(6):484–92.
18. Freeman J. Otosclerosis and vestibular dysfunction. Laryngoscope. 1980;90(9):1481–7.
19. Yetiser S. Bilateral Vestibulopathy due to severe Cochlear Otosclerosis: a well-known condition without any favorable solution. Turk Arch Otorhinolaryngol. 2018;56(3):174–6.
20. Hannley MT. Audiologic characteristics of the patient with otosclerosis. Otolaryngol Clin N Am. 1993;26(3):373–87.
21. Carhart R. Effects of stapes fixation on bone-conduction response. In: Ventry IM, Chailkin JB, Dixon RF, editors. Hearing measurement: a book of readings. New York: Appleton-Century-Crofts; 1971. p. 116–29.
22. Wegner I, Bittermann AJ, Hentschel MA, van der Heijden GJ, Grolman W. Pure-tone audiometry in otosclerosis: insufficient evidence for the diagnostic value of the Carhart notch. Otolaryngol Head Neck Surg. 2013;149(4):528–32.
23. Forquer BD, Sheehy JL. Cochlear otosclerosis: acoustic reflex findings. Am J Otol. 1981;2(4):297–300.
24. Forquer BD, Sheehy JL. Cochlear otosclerosis: a review of audiometric findings in 150 cases. Am J Otol. 1987;8(1):1–4.
25. Ried E, Ojeda JP, Agurto M, Ried E, Martinez C. Inverted acoustic reflex in patients with otosclerosis. Acta Otorrinolaringol Esp. 2000;51(6):463–7.
26. Vallejo LA, Herrero D, Sanchez C, Sanchez E, Gil-Carcedo E, Gil-Carcedo LM. Inverted acoustic reflex: an analysis of its morphological characteristics in different physiological and pathological situations. Acta Otorrinolaringol Esp. 2009;60(4):238–52.
27. Panda NK, Saha AK, Gupta AK, Mann SB. Evaluation of vestibular functions in otosclerosis before and after small fenestra stapedotomy. Indian J Otolaryngol Head Neck Surg. 2001;53(1):23–7.
28. Manzari L, Modugno GC. Nystagmus induced by bone (mastoid) vibration in otosclerosis: a new perspective in the study of vestibular function in otosclerosis. Med Sci Monit. 2008;14(10):CR505–10.
29. Parnes S, Black FO, Wall C 3rd, O'Leary DP, Feltyberger E. Vestibular system abnormalities in otosclerotic subjects. Otolaryngology. 1978;86(1):ORL-98–106.
30. Akazawa K, Ohta S, Tsuzuki K, Sakagami M. Bone-conducted vestibular-evoked myogenic potentials before and after stapes surgery. Otol Neurotol. 2018;39(1):6–11.
31. Amali A, Mahdi P, Karimi Yazdi A, Khorsandi Ashtiyani MT, Yazdani N, Vakili V, et al. Saccular function in otosclerosis patients: bone conducted-vestibular evoked myogenic potential analysis. Acta Med Iran. 2014;52(2):111–5.
32. Catalano N, Cammaroto G, Galletti B, Freni F, Nicita RA, Azielli C, et al. The role of cVEMPs and vHIT in the evaluation of otosclerosis and its eventual vestibular impairment: preliminary findings. B-ENT. 2017;13(1 Suppl 27):31–6.
33. Lin KY, Young YH. Role of ocular VEMP test in assessing the occurrence of vertigo in otosclerosis patients. Clin Neurophys. 2015;126(1):187–93.
34. Saka N, Seo T, Fujimori K, Mishiro Y, Sakagami M. Vestibular-evoked myogenic potential in response to bone-conducted sound in patients with otosclerosis. Acta Otolaryngol. 2012;132(11):1155–9.
35. Tramontani O, Gkoritsa E, Ferekidis E, Korres SG. Contribution of Vestibular-Evoked Myogenic Potential (VEMP) testing in the assessment and the differential diagnosis of otosclerosis. Med Sci Monit. 2014;20:205–13.
36. Trivelli M, D'Ascanio L, Pappacena M, Greco F, Salvinelli F. Air- and bone-conducted vestibular evoked myogenic potentials (VEMPs) in otosclerosis: recordings before and after stapes surgery. Acta otorhinolaryngologica Italica. 2010;30(1):5–10.
37. Yang TL, Young YH. Vestibular-evoked myogenic potentials in patients with otosclerosis using air- and bone-conducted tone-burst stimulation. Otol Neurotol. 2007;28(1):1–6.
38. Rajati M, Jafarzadeh S, Javadzadeh R, Disfani MF, Yousefi R. Comprehensive vestibular evaluation in patients with Otosclerosis: a case control study. Indian J Otolaryngol Head Neck Surg. 2022;74(4):582–7.
39. Blakley BW, Hilger PA, Taylor S, Hilger J. Computed tomography in the diagnosis of cochlear otosclerosis. Otolaryngol Head Neck Surg. 1986;94(4):434–8.
40. Brown LA, Mocan BO, Redleaf MI. Diagnostic protocol for detecting Otosclerosis on high-resolution temporal bone CT. Ann Otol Rhinol Laryngol. 2019;128(11):1054–60.
41. Damsma H, de Groot JA, Zonneveld FW, van Waes PF, Huizing EH. CT of cochlear otosclerosis (otospongiosis). Radiol Clin North Am. 1984;22(1):37–43.
42. De Groot JA, Huizing EH, Damsma H, Van Waes PF, Zonneveld FW. Computed tomography of the petrous bone in otosclerosis and Meniere's disease. Acta Otolaryngol Suppl. 1986;434:1–135.
43. Fraysse B. Why do we include CT scanning in the evaluation of otosclerosis patients? ORL J Otorhinolaryngol Relat Spec. 2010;72(3):168. discussion 9
44. Kanzara T, Virk JS. Diagnostic performance of high resolution computed tomography in otosclerosis. World J Clin Cases. 2017;5(7):286–91.
45. Kawase S, Naganawa S, Sone M, Ikeda M, Ishigaki T. Relationship between CT densitometry with a slice thickness of 0.5 mm and audiometry in otosclerosis. Eur Radiol. 2006;16(6):1367–73.
46. Kiyomizu K, Tono T, Yang D, Haruta A, Kodama T, Komune S. Correlation of CT analysis and audiometry in Japanese otosclerosis. Auris Nasus Larynx. 2004;31(2):125–9.
47. Lagleyre S, Sorrentino T, Calmels MN, Shin YJ, Escudé B, Deguine O, et al. Reliability of high-resolution CT scan in diagnosis of otosclerosis. Otol Neurotol. 2009;30(8):1152–9.
48. Lee TC, Aviv RI, Chen JM, Nedzelski JM, Fox AJ, Symons SP. CT grading of otosclerosis. AJNR. 2009;30(7):1435–9.
49. Mafee MF, Henrikson GC, Deitch RL, Norouzi P, Kumar A, Kriz R, et al. Use of CT in stapedial otosclerosis. Radiology. 1985;156(3):709–14.
50. Mafee MF, Valvassori GE, Deitch RL, Norouzi P, Henrikson GC, Capek V, et al. Use of CT in the evaluation of cochlear otosclerosis. Radiology. 1985;156(3):703–8.
51. Marx M, Lagleyre S, Escudé B, Demeslay J, Elhadi T, Deguine O, et al. Correlations between CT scan findings and hearing thresholds in otosclerosis. Acta Otolaryngol. 2011;131(4):351–7.
52. Miura M, Naito Y, Takahashi H, Honjo I. Computed tomographic image analysis of ears with otosclerosis. ORL J Otorhinolaryngol Relat Spec. 1996;58(4):200–3.
53. Purohit B, Hermans R, Op de Beeck K. Imaging in otosclerosis: a pictorial review. Insights. Imaging. 2014;5(2):245–52.
54. Redfors YD, Gröndahl HG, Hellgren J, Lindfors N, Nilsson I, Möller C. Otosclerosis: anatomy and pathology in the temporal bone assessed by multi-slice and cone-beam CT. Otol Neurotol. 2012;33(6):922–7.
55. Swartz JD, Mandell DW, Berman SE, Wolfson RJ, Marlowe FI, Popky GL. Cochlear otosclerosis (otospongiosis): CT analysis with audiometric correlation. Radiology. 1985;155(1):147–50.
56. Valvassori GE. CT densitometry in otosclerosis. Adv Otorhinolaryngol. 1987;37:47–9.

57. Valvassori GE, Dobben GD. CT densitometry of the cochlear capsule in otosclerosis. AJNR. 1985;6(5):661–7.
58. Vartiainen E, Saari T. Value of computed tomography (CT) in the diagnosis of cochlear otosclerosis. Clin Otolaryngol Allied Sci. 1993;18(6):462–4.
59. Vicente Ade O, Yamashita HK, Albernaz PL, Penido NO. Computed tomography in the diagnosis of otosclerosis. Otolaryngol Head Neck Surg. 2006;134(4):685–92.
60. Wilbrand HF. Radioanatomy of cochlear and stapedial otosclerosis. Scand Audiol Suppl. 1988;30:181–3.
61. Wycherly BJ, Berkowitz F, Noone AM, Kim HJ. Computed tomography and otosclerosis: a practical method to correlate the sites affected to hearing loss. Ann Otol Rhinol Laryngol. 2010;119(12):789–94.
62. Yamashita K, Hiwatashi A, Togao O, Kondo M, Kikuchi K, Inoguchi T, et al. Additive value of "otosclerosis-weighted" images for the CT diagnosis of fenestral otosclerosis. Acta Radiol. 2017;58(10):1215–21.
63. Zhu MM, Sha Y, Zhuang PY, Olszewski AE, Jiang JQ, Xu JH, et al. Relationship between high-resolution computed tomography densitometry and audiometry in otosclerosis. Auris Nasus Larynx. 2010;37(6):669–75.
64. Wegner I, van Waes AM, Bittermann AJ, Buitinck SH, Dekker CF, Kurk SA, et al. A systematic review of the diagnostic value of CT imaging in diagnosing Otosclerosis. Otol Neurotol. 2016;37(1):9–15.
65. Goh JP, Chan LL, Tan TY. MRI of cochlear otosclerosis. Br J Radiol. 2002;75(894):502–5.
66. Purohit B, Op de Beeck K, Hermans R. Role of MRI as first-line modality in the detection of previously undiagnosed otosclerosis: a single tertiary institute experience. Insights. Imaging. 2020;11(1):71.
67. de Oliveira Vicente A, Chandrasekhar SS, Yamashita HK, Cruz OL, Barros FA, Penido NO. Magnetic resonance imaging in the evaluation of clinical treatment of otospongiosis: a pilot study. Otolaryngol Head Neck Surg. 2015;152(6):1119–26.
68. Ziyeh S, Berlis A, Ross UH, Reinhardt MJ, Schumacher M. MRI of active otosclerosis. Neuroradiology. 1997;39(6):453–7.
69. Laine J, Hautefort C, Attye A, Guichard JP, Herman P, Houdart E, et al. MRI evaluation of the endolymphatic space in otosclerosis and correlation with clinical findings. Diagn Interv Imaging. 2020;101(9):537–45.
70. Lombardo F, De Cori S, Aghakhanyan G, Montanaro D, De Marchi D, Frijia F, et al. 3D-flair sequence at 3T in cochlear otosclerosis. Eur Radiol. 2016;26(10):3744–51.
71. Gussen R. Otosclerosis and vestibular degeneration. Arch Otolaryngol. 1973;97(6):484–7.
72. Sando I, Hemenway WG, Miller DR, Black FO. Vestibular pathology in otosclerosis temporal bone nistopathological report. Laryngoscope. 1974;84(4):593–605.
73. Richter E, Schuknecht HF. Loss of vestibular neurons in clinical otosclerosis. Arch Otorhinolaryngol. 1982;234(1):1–9.
74. Singbartl F, Basta D, Seidl RO, Ernst A, Todt I. Perioperative recordings of vestibular-evoked myogenic potentials in otosclerosis. Otol Neurotol. 2006;27(8):1070–3.
75. Winters SM, Klis SF, Kool AC, Kraaijenga SA, Tange RA, Grolman W. Perioperative bone-conducted ocular vestibular-evoked myogenic potentials in otosclerosis patients. Otol Neurotol. 2013;34(6):1109–14.
76. Satar B, Karacayli C, Coban VK, Ozdemir S. Do otosclerosis and stapedotomy affect semicircular canal functions? Preliminary results of video head impulse test. Acta Otolaryngol. 2021;141(4):348–53.
77. Kujala J, Aalto H, Hirvonen T. Video-oculography findings and vestibular symptoms on the day of stapes surgery. Eur Arch Otorrinolaringol. 2010;267(2):187–90.
78. Hirvonen TP, Aalto H. Immediate postoperative nystagmus and vestibular symptoms after stapes surgery. Acta Otolaryngol. 2013;133(8):842–5.
79. Kujala J, Aalto H, Hirvonen TP. Video-oculography findings in patients with otosclerosis. Otol Neurotol. 2005;26(6):1134–7.
80. Birch L, Elbrond O. Stapedectomy and vertigo. Clin Otolaryngol Allied Sci. 1985;10(4):217–23.
81. de Vilhena D, Gamboa I, Duarte D, Lopes G. Vestibular disorders after Stapedial surgery in patients with Otosclerosis. Int J Otolaryngol. 2016;2016:6830648.
82. Dybwik H. Vestibular damage in stapedectomy. Acta Otolaryngol. 1966;62(4):292–6.
83. Job K, Wiatr A, Wiatr M. Association between postoperative vertigo and hearing outcomes after stapes surgery for Otosclerosis. Ear Nose Throat J. 2021;1455613211023014:709.

58 Paget's Disease (Osteitis Deformans)

Overview

Paget's disease of bone (PDB) is a disorder of bone metabolism in which abnormal osteoclast activity results in inappropriately accelerated remodeling of bone. PDB is the second most common bone disorder after osteoporosis and tends to become symptomatic in older people. Around 66–70% of PDB patients have involvement of the temporal bone, which can cause sensorineural and/or conductive hearing loss that is usually bilateral and progressive; the mechanism of this hearing loss is incompletely understood. Disequilibrium can also occur. A diagnosis of PDB in general can be secured by plain films and nuclear radioscintigraphy; temporal bone involvement may be better characterized with CT and MRI. Medical treatment with bisphosphonates and calcitonin has had mixed results. In selected cases of hearing loss, surgical treatment with stapedectomy or cochlear implantation has been attempted.

Introduction

Sir James Paget (1814–1899) was an English surgeon and pathologist who published several cases of what he termed "osteitis deformans" [1, 2], currently termed *Paget's disease of bone (PDB)* (Fig. 58.1).

Fig. 58.1 Sir James Paget (1814–1899). (From https://en.wikipedia.org/wiki/James_Paget (public domain))

M. Cherchi, *Otoneurology and Vestibular Medicine*, https://doi.org/10.1007/978-3-031-94841-1_58

Epidemiology

Epidemiologic studies estimate that in the United States the overall prevalence of PDB is 1–2%, with a peak in the range of 65 to 74 years at a rate of 2.32% ± 0.54% [3], making it the second most common bone disorder after osteoporosis [4]. There is a "slight male predominance in the 45- to 74-year age group" [3].

Of all patients affected by PDB, it has been reported that "two-thirds" [5] to 70% have involvement of the skull [6], and 30% have involvement specifically of the temporal bone [7].

Since healthy people exhibit presbycusis, it is reasonable to wonder whether the hearing loss in PDB patients (which tends to afflict older individuals) is due simply to age. Studies that have explored this topic report that hearing loss in temporal bone PDB patients clearly exceeds what could be ascribed to age alone [8–10].

Genetics

The etiology of adult-onset PDB is unknown.

Juvenile onset cases of PDB (OMIM 239000) have been linked to a homozygous or compound heterozygous mutation in the TNFRSF11B (tumor necrosis factor receptor superfamily, member 11B) gene (OMIM 602643) on chromosome 8q24.12, which results in a deficiency of osteoprotegerin.

Pathophysiological Mechanism of Disease

PDB is a disorder of bone metabolism involving inappropriately accelerated bone remodeling, which is thought to be attributable to dysfunctional osteoclasts, whose activity is normally inhibited by osteoprotegerin. PDB can manifest at a single site (monostotic) or multiple sites (polyostotic). Commonly affected bones include the skull, spine, pelvis, and long bones in the legs.

Most cases of PDB are asymptomatic. Symptomatic cases generally result from bone overgrowth/deformation or pathologic fracture.

Below, we discuss that temporal bone involvement by PDB presents with audiologic and vestibular symptoms. Although a number of temporal bone studies have been performed in PDB patients [6, 7, 10–15], the mechanisms by which PDB causes audiologic and vestibular dysfunction remain incompletely understood [12, 15].

PDB involving the skull is well documented to cause multiple compressive cranial neuropathies [4, 16, 17]. When this includes narrowing of the internal auditory canal, it can compress the vestibulocochlear nerve [11, 18]. One could reasonably expect otovestibular symptoms secondary to such compressive neuropathy. Indeed, temporal bone studies have reported, "The histopathology reveals Pagetic bone invading the internal auditory canal and compressing the cochlear division of the VIIIth cranial nerve resulting in severe neural degeneration" [11], with corresponding sensorineural hearing loss.

However, compressive neuropathy does not appear to be the only mechanism underlying hearing loss. It has also been found that:

- PDB patients have *decreased bone mineral density of the otic capsule* [12, 19–21].
- There are cases of PDB patients found with a *fractured stapes footplate* [14].
- Dimitriadis and colleagues stated that, "We hypothesize that sensorineural hearing loss in patients with PDB of the temporal bone may, in some cases, be attributed to obliteration of Cotugno's canal by remodeling pagetoid bone, thus obstructing the venous drainage of the cochlea, with a subsequent effect on the function of stria vascularis and spiral ligament" [14].
- Khetarpal and Schuknecht speculate that, "the hearing losses in Paget's disease are caused by **changes** in bone density, mass, and form **that serve to dampen the finely tuned motion mechanics of the middle and inner ears**" [15].
- Some studies report that PDB patients have normal auditory brainstem evoked responses [22], and others report data reflecting integrity of the auditory component of the vestibulocochlear nerve, reaching conclusions such as, "Findings support the principle that hearing loss in Paget's disease of bone is generally associated with intact auditory nerve function and also support a cochlear site of lesion" [22], and "Studies reviewed show no evidence of auditory nerve dysfunction and confirm a cochlear site of lesion. A loss of bone mineral density in the cochlear capsule is associated with both a high-tone hearing loss and a low-tone air-bone gap" [23].
- Some studies report *degenerative changes of the spiral ligament* [6, 12, 14].

Bahmad and colleagues [12] summarized these findings as follows:

> The pathogenesis for the conductive and sensorineural hearing loss is not known. As illustrated in the present case, histopathologic studies have failed to find consistent correlates for the hearing loss. Clinical studies using audiometric and computed tomographic analysis have demonstrated a loss of otic capsule bone mineral density correlated with both the high-frequency sensorineural hearing loss and the air-bone gap. The air-bone gap has been hypothesized to be the result of the alteration of the mechanics of the ear. The downsloping, sensorineural hearing loss has been hypothesized to represent abnormalities of the

metabolic homeostasis of the cochlea as a result of the liberation of cytokines from the pagetic bone and the dysfunction of the spiral ligament [12].

Clinical Presentation

PDB becomes relevant to otoneurology and neuro-otology when it involves the temporal bone. Depending on the site of temporal bone involvement, PDB can cause various symptoms including:

- *Hearing loss* [4–6, 8–11, 14, 15, 17, 19, 20, 23–32]. Of PDB patients with temporal bone involvement, between 30% and 50% [5] have some form of hearing loss. Deep and colleagues reported that of temporal bone PDB patients with hearing loss, 65% have *sensorineural* hearing loss and 35% have *mixed* hearing loss [10]. The hearing loss is usually *bilateral* and *progressive* [14, 33].
- *Tinnitus* [10, 26, 34]. Deep and colleagues found that tinnitus was reported in 19% of temporal bone PDB patients [8–10].
- *Disequilibrium* [5, 10, 28, 29, 34, 35]. Estimates of this symptom vary. Of PDB patients with temporal bone involvement, 20–25% exhibit vestibular dysfunction [5]. Deep and colleagues report that 52% of temporal bone PDB patients complain of disequilibrium [10].

Ocular Motor Examination

As of this writing, there were no published reports of nystagmus or other ocular motor dysfunction in temporal bone PDB patients.

Testing: Vestibular

As of this writing, there were no published reports of instrumented vestibular testing in temporal bone PDB patients.

Testing: Other

Serum alkaline phosphatase is elevated in PDB patients.

Imaging

For PDB in general, relevant imaging studies include plain films and nuclear radioscintigraphy [24].

CT [21] and MRI [18] may help verify and characterize temporal bone involvement in greater detail.

Histopathology

Histologic studies report that osteoclasts appear abnormal and in high frequency, while affected bone appears disorganized and with abnormal proportions of lamellar and woven bone.

Differential Diagnosis

As far as the audiologic symptoms are concerned, the differential diagnosis includes other causes of bilateral progressive hearing loss.

Since instrumented vestibular testing has not been studied in PDB, such evaluations serve primarily to exclude other diagnoses.

Treatment

Since PDB is an osteometabolic disorder, logical attempts at treatment have attempted to modulate bone metabolism, with mixed results.

Studies of bisphosphonates have reached conflicting conclusions, with some studies demonstrating efficacy [24, 25, 36] and others not [33].

Studies of calcitonin have reached conflicting conclusions, with some studies demonstrating efficacy [25, 37, 38] and others not [27, 39].

Surgical interventions have also been attempted, including stapedectomy [28, 40] and cochlear implantation [10, 33].

As of this writing, treatment of disequilibrium has not been formally studied. Absent any evidence to the contrary, in temporal bone PDB patients complaining of disequilibrium, it is reasonable to hypothesize that there is vestibular dysfunction, and that such dysfunction is probably vestibular weakness (analogous to the hearing loss). On this hypothesis, a trial of vestibular rehabilitation therapy is medically reasonable and incurs no medical risk.

Prognosis

Prognosis of temporal bone PDB is poor. Some studies report that medical treatment with bisphosphonates may retard the progression of hearing loss [24, 25, 36].

References

1. Paget J. On a form of chronic inflammation of bones (Osteitis Deformans). Med Chir Trans. 1877;60:37–63.
2. Paget J. Additional cases of Osteitis Deformans. Med Chir Trans. 1882;65:225–36.
3. Altman RD, Bloch DA, Hochberg MC, Murphy WA. Prevalence of pelvic Paget's disease of bone in the United States. J Bone Miner Res. 2000;15(3):461–5.
4. Ankrom MA, Shapiro JR. Paget's disease of bone (osteitis deformans). J Am Geriatr Soc. 1998;46(8):1025–33.
5. Nager GT. Paget's disease of the temporal bone. Ann Otol Rhinol Laryngol. 1975;84(4 Pt 3 Suppl 22):1–32.
6. Teufert KB, Linthicum F Jr. Paget disease and sensorineural hearing loss associated with spiral ligament degeneration. Otol Neurotol. 2005;26(3):387–91. discussion 91
7. Proops D, Bayley D, Hawke M. Paget's disease and the temporal bone–a clinical and histopathological review of six temporal bones. J Otolaryngol. 1985;14(1):20–9.
8. Amilibia Cabeza E, Holgado Pérez S, Pérez Grau M, Moragues Pastor C, Roca-Ribas Serdà F, Quer Agustí M. Hearing in Paget's disease of bone. Acta Otorrinolaringol Esp (Engl Ed). 2019;70(2):89–96.
9. Baraka ME. Rate of progression of hearing loss in Paget's disease. J Laryngol Otol. 1984;98(6):573–5.
10. Deep NL, Besch-Stokes JG, Lane JI, Driscoll CLW, Carlson ML. Paget's disease of the temporal Bone: a single-institution contemporary review of 27 patients. Otol Neurotol. 2017;38(6):907–15.
11. Applebaum EL, Clemis JD. Temporal bone histopathology of Paget's disease with sensorineural hearing loss and narrowing of the internal auditory canal. Laryngoscope. 1977;87(10 Pt 1):1753–9.
12. Bahmad F Jr, Merchant SN. Paget disease of the temporal bone. Otol Neurotol. 2007;28(8):1157–8.
13. Davies DG. Paget's disease of the temporal bone. A clinical and histopathological survey. Acta Otolaryngol. 1968;Suppl 242:3+.
14. Dimitriadis PA, Bamiou DE, Bibas AG. Hearing loss in Paget's disease: a temporal bone histopathology study. Otol Neurotol. 2012;33(2):142–6.
15. Khetarpal U, Schuknecht HF. In search of pathologic correlates for hearing loss and vertigo in Paget's disease. A clinical and histopathologic study of 26 temporal bones. Ann Otol Rhinol Laryngol Suppl. 1990;145:1–16.
16. Bomfim Fde A, Mendes Wde O, Veras IM, de Freitas JC, Castro AM, de Castro Junior FM. Extramammary Paget's disease affecting the external auditory canal: a case report. J Med Case Rep. 2013;7:250.
17. Hamid M, Touarsa F, Satte A, Bourazza A. Paget's disease of bone presenting with multiple cranial nerve palsies: a case report. Radiol Case Rep. 2022;17(6):1870–3.
18. Ginsberg LE, Elster AD, Moody DM. MRI of Paget disease with temporal bone involvement presenting with sensorineural hearing loss. J Comput Assist Tomogr. 1992;16(2):314–6.
19. Amilibia E, Pérez-Grau M, Callejo À, Cuadras P, Roca-Ribas F, Puyalto P. CT assessment of Otic capsule Bone density in Paget's disease of the temporal Bone and its relationship with hearing loss. Otol Neurotol. 2021;42(6):e788–e94.
20. Monsell EM. The mechanism of hearing loss in Paget's disease of bone. Laryngoscope. 2004;114(4):598–606.
21. Monsell EM, Cody DD, Bone HG, Divine GW, Windham JP, Jacobson GP, et al. Hearing loss in Paget's disease of bone: the relationship between pure-tone thresholds and mineral density of the cochlear capsule. Hear Res. 1995;83(1–2):114–20.
22. Monsell EM, Bone HG, Cody DD, Jacobson GP, Newman CW, Patel SC, Divine GW. Hearing loss in Paget's disease of bone: evidence of auditory nerve integrity. Am J Otol. 1995;16(1):27–33.
23. Monsell EM, Cody DD, Bone HG, Divine GW. Hearing loss as a complication of Paget's disease of bone. J Bone Miner Res. 1999;14(Suppl 2):92–5.
24. Kravets I. Paget's disease of Bone: diagnosis and treatment. Am J Med. 2018;131(11):1298–303.
25. Lando M, Hoover LA, Finerman G. Stabilization of hearing loss in Paget's disease with calcitonin and etidronate. Arch Otolaryngol Head Neck Surg. 1988;114(8):891–4.
26. Mackenzie I, Young C, Fraser WD. Tinnitus and Paget's disease of bone. J Laryngol Otol. 2006;120(11):899–902.
27. Menzies MA, Greenberg PB, Joplin GF. Otological studies in patients with deafness due to Paget's disease before and after treatment with synthetic human calcitonin. Acta Otolaryngol. 1975;79(5–6):378–83.
28. Ramsay HA, Linthicum FH Jr. Cochlear histopathology in Paget's disease. Am J Otolaryngol. 1993;14(1):60–1.
29. Simons RM. Paget's disease in the head and neck. Gerontology. 1980;26(3):155–9.
30. Sparrow NL, Duvall AJ 3rd. Hearing loss and Paget's disease. J Laryngol Otol. 1967;81(6):601–11.
31. Wheeler TT, Alberts MA, Dolan TA, McGorray SP. Dental, visual, auditory and olfactory complications in Paget's disease of bone. J Am Geriatr Soc. 1995;43(12):1384–91.
32. Young CA, Fraser WD, Mackenzie IJ. Detection of hearing impairment and handicap in Paget's disease of bone using a simple scoring system: a case control study. Bone. 2007;40(1):189–93.
33. Aoki M, Tanahashi S, Mizuta K, Kato H. Treatment for progressive hearing loss due to Paget's disease of bone – a case report and literature review. J Int Adv Otol. 2015;11(3):267–70.
34. Harner SG, Rose DE, Facer GW. Paget's disease and hearing loss. Otolaryngology. 1978;86(6 Pt 1):Orl-869–74.
35. Boukas K, Bazoukis G. Vertigo as the first manifestation of Paget's disease: a case report. J Frailty Sarcopenia Falls. 2017;2(1):9–11.
36. Silverman SL. Paget disease of bone: therapeutic options. J Clin Rheumatol. 2008;14(5):299–305.
37. el Sammaa M, Linthicum FH Jr, House HP, House JW. Calcitonin as treatment for hearing loss in Paget's disease. Am J Otol. 1986;7(4):241–3.
38. Solomon LR, Evanson JM, Canty DP, Gill NW. Effect of calcitonin treatment on deafness due to Paget's disease of bone. Br Med J. 1977;2(6085):485–7.
39. Walker GS, Evanson JM, Canty DP, Gill NW. Effect of calcitonin in deafness due to Paget's disease of skull. Br Med J. 1979;2(6186):364–5.
40. Sellars SL, Fine GS. Stapedectomy in the deafness of Paget's disease. S Afr Med J. 1975;49(31):1256–8.

Fibrous Dysplasia

59

Overview

Fibrous dysplasia (FD) is a disorder of bone metabolism in which abnormally functioning osteoblasts produce maturational misdifferentiation of bony mesenchyme resulting in abnormal bone architecture. FD usually presents in the first or second decades, though sometimes as late as the sixth decade. FD can affect any bone and can involve one site or multiple sites. About 12–18% of cases involve the temporal bone, and thus can cause audiologic symptoms; vestibular symptoms are rare. About 29% of temporal bone FD patients experience hearing loss; the majority of these have conductive hearing loss associated with external auditory canal stenosis; but some cases of sensorineural hearing loss have been reported. FD is usually diagnosed and monitored by CT imaging, though MRI may provide complementary information. FD often, though not always, stabilizes after adolescence, so surgical intervention, even if contemplated, is often postponed until then. A patient diagnosed with temporal bone FD should be managed by an otolaryngologist.

Introduction

Fibrous dysplasia (FD) is sometimes also called fibro-osseous dysplasia. According to Adetayo and colleagues [1], FD was first described by Friedrich Daniel von Recklinghausen (1833–1910) in 1891 [2] but named "fibrous dysplasia" by Louis Lichtenstein (1906–1977) in 1938 [3].

Epidemiology

Some studies state that FD becomes symptomatic in the first and second decades [4, 5], while others say that the mean age at diagnosis is 25 years [6]. It can present as late as the fifth [5] or sixth [7] decades.

Of all FD patients, 59% are female; 11% have monostotic forms, and 89% have polyostotic forms [6]. Of FD patients specifically with craniofacial involvement, up to 56% are monostotic and up to 47% are polyostotic [8]. Studies provide discrepant sex ratios, such as a female preponderance of 59% [6] to a male-to-female preponderance of 2:1 [5].

Fibrous dysplasia affects the temporal bone in 12% [8] to 18% of cases [4], and of those, 80% involve the external auditory canal [4].

Rare cases of FD following surgery have been reported [9].

Genetics

Most cases of FD appear to occur sporadically.

Of cases occurring in familial patterns, the inheritance seems to be autosomal dominant with incomplete penetrance. Several genetic loci have been implicated [10–14].

Pathophysiological Mechanism of Disease

Fibrous dysplasia is a disorder in which abnormally functioning osteoblasts [15] produce maturational misdifferentiation of bony mesenchyme [16], resulting in abnormal bone architecture [17].

Clinical Presentation

Hearing loss occurs in 29% of temporal bone FD patients [6].

Symptoms of disequilibrium are rarely reported in association with temporal bone FD [5].

Physical Examination

Depending on the degree of skull involvement, FD patients may exhibit a variety of facial deformities.

M. Cherchi, *Otoneurology and Vestibular Medicine*, https://doi.org/10.1007/978-3-031-94841-1_59

Ocular Motor Examination

As of this writing, there were no published studies on ocular motor examination in FD patients.

Testing: Auditory

In the majority of cases, audiometry will show conductive hearing loss [18–27] and is associated with external auditory canal stenosis [4, 16, 22–25, 28, 29], but sensorineural hearing loss has also been reported in up to 17% of patients [21, 23].

Testing: Vestibular

As of this writing, there was almost no literature regarding vestibular testing in patients with temporal bone FD [30]. In our view, the main role of otovestibular testing is to assess for whether an FD patient complaining of disequilibrium may have co-occurrence of a more common (and possibly treatable) vestibular disorder.

Imaging

CT is the imaging modality of choice for diagnosing, characterizing, and monitoring FD [31, 32]. MRI may also provide helpful information [33].

Differential Diagnosis

FD can co-occur with other diseases such as:

- Cholesteatoma [24, 28, 34–38]. Reported rates vary from 3% [6] to as much as 40% of cases [21].
- Aneurysmal bone cyst [39].
- Posterior semicircular canal dehiscence [40].
- Labyrinthine fistula [39].
- Ménière's disease [41].
- Paget's disease of bone [42].

FD can mimic other diseases, such as:

- Primary malignancy [43].
- Metastatic disease [44].
- Endolymphatic sac tumor [45].
- Osteoma [19].
- Osteochondroma [15].

FD can also be mimicked by other diseases, such as a meningioma [46].

Treatment

Some authors state that FD tends to stabilize after adolescence [47] and thus recommend refraining from surgery if possible until after that point [47, 48]. Various surgical procedures have been attempted depending on the specific loci and complications of temporal bone involvement in FD [26, 36, 48–56].

Prognosis

FD replaces normal bone with abnormal fibrous tissue that cannot truly be restored. The disease often (though not always) stabilizes in adulthood. Surgical treatment may help correct some deficits, but outcomes may be suboptimal.

References

1. Adetayo OA, Salcedo SE, Borad V, Richards SS, Workman AD, Ray AO. Fibrous dysplasia: an overview of disease process, indications for surgical management, and a case report. Eplasty. 2015;15:e6.
2. von Recklinghausen FD. Die fibrose oder deformierende Ostitis, die Osteomalacie und die osteoplastische Carcinose in ihren gegenseitigen Beziehungen [fibrous or deforming osteitis, osteomalacia and osteoplastic carcinosis in their mutual relationships]. In: Festschrift Rudolf Virchow. Berlin: Reimer; 1891.
3. Lichtenstein L. Polyostotic fibrous dysplasia. Arch Surg. 1938;36(5):874–98.
4. Chinski A, Beider B, Cohen D. Fibrous dysplasia of the temporal bone. Int J Pediatr Otorhinolaryngol. 1999;47(3):275–81.
5. Mallina S, Philip R, Chong AW, Gurdeep S. Fibrous dysplasia of the temporal bone. Med J Malaysia. 2007;62(2):160–1.
6. Frisch CD, Carlson ML, Kahue CN, Pelosi S, Haynes DS, Lane JI, et al. Fibrous dysplasia of the temporal bone: a review of 66 cases. Laryngoscope. 2015;125(6):1438–43.
7. Pineker V, Marques Maggio E, Huber A, Pangalu A, Valavanis A, Winklhofer S. Rare appearance, rare location and unusual patient age: report of Osteolytic fibrous dysplasia in the temporal bone of a 52-year-old patient. Clin Neuroradiol. 2017;27(4):521–4.
8. Yang L, Wu H, Lu J, Teng L. Prevalence of different forms and involved bones of craniofacial fibrous dysplasia. J Craniofac Surg. 2017;28(1):21–5.
9. Pardo-Maza A, Lassaletta L, Ruiz-Bravo E, Perez-Mora R, Peñarrocha J, Gavilan J. Fibrous dysplasia of the temporal bone secondary to ear surgery: a case report. J Med Case Rep. 2015;9:129.
10. Albagha OM, Visconti MR, Alonso N, Langston AL, Cundy T, Dargie R, et al. Genome-wide association study identifies variants at CSF1, OPTN and TNFRSF11A as genetic risk factors for Paget's disease of bone. Nat Genet. 2010;42(6):520–4.
11. Albagha OM, Wani SE, Visconti MR, Alonso N, Goodman K, Brandi ML, et al. Genome-wide association identifies three new susceptibility loci for Paget's disease of bone. Nat Genet. 2011;43(7):685–9.
12. Beauregard M, Gagnon E, Guay-Belanger S, Morissette J, Brown JP, Michou L. Identification of rare genetic variants in novel loci associated with Paget's disease of bone. Hum Genet. 2014;133(6):755–68.
13. Chung PY, Beyens G, Boonen S, Papapoulos S, Geusens P, Karperien M, et al. The majority of the genetic risk for Paget's disease of bone is explained by genetic variants close to the

CSF1, OPTN, TM7SF4, and TNFRSF11A genes. Hum Genet. 2010;128(6):615–26.
14. Usategui-Martin R, Garcia-Aparicio J, Corral-Gudino L, Calero-Paniagua I, Del Pino-Montes J, Gonzalez Sarmiento R. Polymorphisms in autophagy genes are associated with Paget disease of bone. PLoS One. 2015;10(6):e0128984.
15. Tambe DA, Patel SK, Sayed SR, Mahajan SR. Fibrous dysplasia of temporal bone presented as a solitary Osteochondroma: a case report. J Orthop Case Rep. 2022;12(6):53–7.
16. Nager GT, Kennedy DW, Kopstein E. Fibrous dysplasia: a review of the disease and its manifestations in the temporal bone. Ann Otol Rhinol Laryngol Suppl. 1982;92:1–52.
17. Neromyliotis E, Sapountzi M, Nikas I, Karydakis P, Mitsios A, Sfakianos G, Themistocleous MS. Fibrous dysplasia of occipital and temporal bone. A case report. Oxf Med Case Rep. 2019;2019(5):omz039.
18. Donnelly MJ, McShane DP, Burns H. Monostotic fibrous dysplasia of the temporal bone with associated lymphadenopathy. Ear Nose Throat J. 1994;73(5):328–30.
19. Kessler A, Wolf M, Ben-Shoshan J. Fibrous dysplasia of the temporal bone presenting as an osteoma of the external auditory canal. Ear Nose Throat J. 1990;69(3):197–9.
20. Liu YH, Chang KP. Fibrous dysplasia of the temporal bone with external auditory canal stenosis and secondary Cholesteatoma. J Int Adv Otol. 2016;12(1):125–8.
21. Megerian CA, Sofferman RA, McKenna MJ, Eavey RD, Nadol JB Jr. Fibrous dysplasia of the temporal bone: ten new cases demonstrating the spectrum of otologic sequelae. Am J Otol. 1995;16(4):408–19.
22. Mohammed S, Jang SH, Moon IS. Fibrous dysplasia of temporal bone. Indian J Otolaryngol Head Neck Surg. 2022;74(Suppl 3):4350–5.
23. Morrissey DD, Talbot JM, Schleuning AJ 2nd. Fibrous dysplasia of the temporal bone: reversal of sensorineural hearing loss after decompression of the internal auditory canal. Laryngoscope. 1997;107(10):1336–40.
24. Pouwels AB, Cremers CW. Fibrous dysplasia of the temporal bone. J Laryngol Otol. 1988;102(2):171–2.
25. Schrimpf R, Karmody CS, Chasin WD, Carter B. Sclerosing lesions of the temporal bone. Laryngoscope. 1982;92(10 Pt 1):1116–9.
26. Yang H, Chen S, Zheng Y, Xu Y, Zhang X, Xiong H, et al. Surgical treatment of monostotic fibrous dysplasia of the temporal bone: a retrospective analysis. Am J Otolaryngol. 2012;33(6):697–701.
27. Yu Hon Wan A, Chi Fai Tong M. Fibrous dysplasia of the temporal bone presenting with facial nerve palsy and conductive hearing loss. Otol Neurotol. 2008;29(7):1039–40.
28. Smouha EE, Edelstein DR, Parisier SC. Fibrous dysplasia involving the temporal bone: report of three new cases. Am J Otol. 1987;8(2):103–7.
29. Talmi YP, Ziv JB, Shimberg R, Finkelstein Y, Zohar Y, Wood BP. Radiological case of the month. Monostotic fibrous dysplasia of the temporal bone. Am J Dis Child. 1989;143(11):1351–2.
30. Brookler KH. History and ENG findings in a patient with fibrous dysplasia of the skull base. Ear Nose Throat J. 2004;83(5):314.
31. Kiroğlu AF, Yuca K, Cankaya H, Kutluhan A, Kara T. Maxillary and temporal fibrous dysplasia: three cases. B-ENT. 2005;1(4):177–80.
32. Lambert PR, Brackmann DE. Fibrous dysplasia of the temporal bone: the use of computerized tomography. Otolaryngol Head Neck Surg. 1984;92(4):461–7.
33. Casselman JW, De Jonge I, Neyt L, De Clercq C, D'Hont G. MRI in craniofacial fibrous dysplasia. Neuroradiology. 1993;35(3):234–7.
34. Djerić D, Stefanović P. Fibrous dysplasia of the temporal bone and maxillofacial region associated with cholesteatoma of the middle ear. Auris Nasus Larynx. 1999;26(1):79–81.
35. Martinez R, Farrior JB. Fibrous dysplasia of the temporal bone complicated by cholesteatoma and thrombophlebitis of the transverse and sigmoid sinuses: a case report. Ear Nose Throat J. 2008;87(2):81–5.
36. Tochino R, Sunami K, Yamane H. Fibrous dysplasia of the temporal bone with cholesteatoma. Acta Otolaryngol Suppl. 2004;554:47–9.
37. Yagoda MR, Selesnick SH. Temporal bone fibrous dysplasia and cholesteatoma leading to the development of a parapharyngeal abscess. J Laryngol Otol. 1994;108(1):51–3.
38. Zanetti D, Gamba P. Cholesteatoma and fibrous dysplasia of the temporal bone: case report and review of the literature. J Otolaryngol. 2007;36(1):59–63.
39. Blanchard M, Abergel A, Williams MT, Ayache D. Aneurysmal bone cyst within fibrous dysplasia causing labyrinthine fistula. Otol Neurotol. 2011;32(2):e11.
40. McCall AA, Curtin HD, McKenna MJ. Posterior semicircular canal dehiscence arising from temporal bone fibrous dysplasia. Otol Neurotol. 2010;31(9):1516–7.
41. Weiss NM, Breitsprecher TM, Dlugaiczyk J, Kämmerer F, Lukas C, Dazert S, et al. Menière's syndrome associated with fibrous dysplasia of the temporal bone. Otol Neurotol. 2023;44(6):e445–e8.
42. Hullar TE, Lustig LR. Paget's disease and fibrous dysplasia. Otolaryngol Clin N Am. 2003;36(4):707–32.
43. Papadakis GZ, Millo C, Sadowski SM, Karantanas AH, Bagci U, Patronas NJ. Fibrous dysplasia mimicking malignancy on 68Ga-DOTATATE PET/CT. Clin Nucl Med. 2017;42(3):209–10.
44. Song Y, Qin C, Liu F, Lan X. Fibrous dysplasia mimicking skeletal metastasis on 68Ga-FAPI PET imaging. Clin Nucl Med. 2021;46(9):774–5.
45. Rawlins KW, Zhan KY, Mattingly JK, Feldman AZ, Adunka OF. Fibrous dysplasia mimicking a pediatric endolymphatic sac tumor. Int J Pediatr Otorhinolaryngol. 2020;129:109785.
46. Mingo K, Sweeney AD, Thompson RC, Rivas A. Hyperostotic En plaque meningioma mimicking fibrous dysplasia of the temporal bone. Otol Neurotol. 2016;37(9):e317–8.
47. Mierzwiński J, Kosowska J, Tyra J, Haber K, Drela M, Paczkowski D, Burduk P. Different clinical presentation and management of temporal bone fibrous dysplasia in children. World J Surg Oncol. 2018;16(1):5.
48. Menon S, Venkatswamy S, Ramu V, Banu K, Ehtaih S, Kashyap VM. Craniofacial fibrous dysplasia: surgery and literature review. Ann Maxillofac Surg. 2013;3(1):66–71.
49. Al-Shawi Y, Alsughayer L, Alradhi A, Alzhrani F. Middle ear implant in a patient with fibrous dysplasia: an alternative for hearing restoration. Ear Nose Throat J. 2021;100(3_suppl):207s–11s.
50. Fandiño M, Bhimrao SK, Saxby AJ, Dar Santos RC, Westerberg BD. Fibrous dysplasia of the temporal bone: systematic review of management and hearing outcomes. Otol Neurotol. 2014;35(10):1698–706.
51. Franzoi G, Bordignon NCT, Marola LHG, Gondak R. Severe craniofacial fibrous dysplasia associated with decreased visual acuity: a case report. Gen Dent. 2023;71(1):50–3.
52. Kim YH, Song JJ, Choi HG, Lee JH, Oh SH, Chang SO, et al. Role of surgical management in temporal bone fibrous dysplasia. Acta Otolaryngol. 2009;129(12):1374–9.
53. Lustig LR, Holliday MJ, McCarthy EF, Nager GT. Fibrous dysplasia involving the skull base and temporal bone. Arch Otolaryngol Head Neck Surg. 2001;127(10):1239–47.
54. Sakamoto Y, Nakajima H, Tamada I, Kishi K, Kawase T. Long-term outcome of fibrous dysplasia: reconstruction with dysplastic bone–case report. Neurol Med Chir (Tokyo). 2011;51(12):857–60.
55. Sataloff RT, Graham MD, Roberts BR. Middle ear surgery in fibrous dysplasia of the temporal bone. Am J Otol. 1985;6(2):153–6.
56. Shakeel M, Steele P, Kamel M, Hussain A. Temporal bone fibrous dysplasia: presentation, resection, and reconstruction. J Otolaryngol Head Neck Surg. 2012;41(6):E58–61.

60 Labyrinthine Concussion

Overview

Labyrinthine concussion refers to labyrinthine damage following head trauma, but without any temporal bone fracture. This damage is usually, though not invariably, ipsilateral to the site of the trauma. The mechanisms of injury may include direct damage from the kinetic energy of a labyrinthine fluid wave, labyrinthine hemorrhage, a traction injury of the vestibulocochlear nerve, circulatory disturbances, impaired production/composition of labyrinthine fluid, and its flow. The clinician should keep in mind that head trauma can cause not only labyrinthine damage but also neurological damage. Cases of labyrinthine concussion can present with any combination of auditory and vestibular symptoms. Hearing loss can usually be identified on bedside examination, though should be quantified on audiometry. Ocular motor examination may identify abnormalities, though these have not been well described in the literature; vestibular testing may find a caloric weakness and/or abnormalities in the vestibulo-ocular reflex. Imaging may be normal, though imaging should still be obtained in order to assess for other sequelae of head trauma, such as temporal bone fracture and intracranial hemorrhage. Rare cases of labyrinthine concussion culminate in labyrinthine ossification. There is no specific treatment for labyrinthine concussion. Prognosis for hearing loss is worse than for symptoms of disequilibrium.

Definition

The phrase "*labyrinthine concussion*" in the literature has come to refer to labyrinthine damage following head trauma, but in the absence of any temporal bone fracture.

The otologic damage from concussion is usually ipsilateral to the site of the trauma, but rare cases describe labyrinthine concussion contralateral to the trauma [1, 2], presumably by a contrecoup mechanism.

Pathophysiology

The usual mechanisms of disease cited in labyrinthine concussion include direct damage from the kinetic energy of a labyrinthine fluid wave, labyrinthine hemorrhage, or a traction injury of the vestibulocochlear nerve [3], and there is some evidence in the literature for each process. Less commonly proposed mechanisms include "circulatory disturbances, impaired production and composition of labyrinthine fluid and problems with its outflow" [4].

The mechanism of hearing loss in labyrinthine concussion is thought to be, "due to basilar membrane shearing, hair cell degeneration and eventual auditory nerve fiber avulsion" [4].

It should also be kept in mind that the actual damage from head trauma need not be restricted to the labyrinth. Since the entire head is concussed, neurological injury should also be considered, and some reports note that in some cases of supposed labyrinthine concussion, the deficits could equally be attributable to "injury to the central auditory pathway (e.g., inferior colliculi or at the upper pontine level)" [5].

Presentation

Labyrinthine concussion can manifest with any combination of auditory disturbances (hearing loss, tinnitus) and vestibular symptoms.

Physical Examination

The hearing loss in labyrinthine concussion is often detectable on physical examination with usual bedside tests of auditory function, including the finger rub test, Weber test, and Rinne test.

Most studies of labyrinthine concussion that report histories of vestibular disturbances do not document an ocular

M. Cherchi, *Otoneurology and Vestibular Medicine*, https://doi.org/10.1007/978-3-031-94841-1_60

motor examination. One study reported "vertical nystagmus appearing in all positions of the head" [4].

Testing: Auditory

Hearing loss after labyrinthine concussion is variable, ranging "from mild to profound sensorineural hearing loss" [5]. The hearing loss tends to affect the higher frequencies preferentially [4], with some authors noting that, "Labyrinthine concussion in most cases involves a SNHL [sensorineural hearing loss] with a notch in the 4–6 kHz resembling acoustic trauma" [1].

Figure 60.1 shows an audiogram from a patient who suffered left-sided labyrinthine concussion in a case reported by Ishai and colleagues [5].

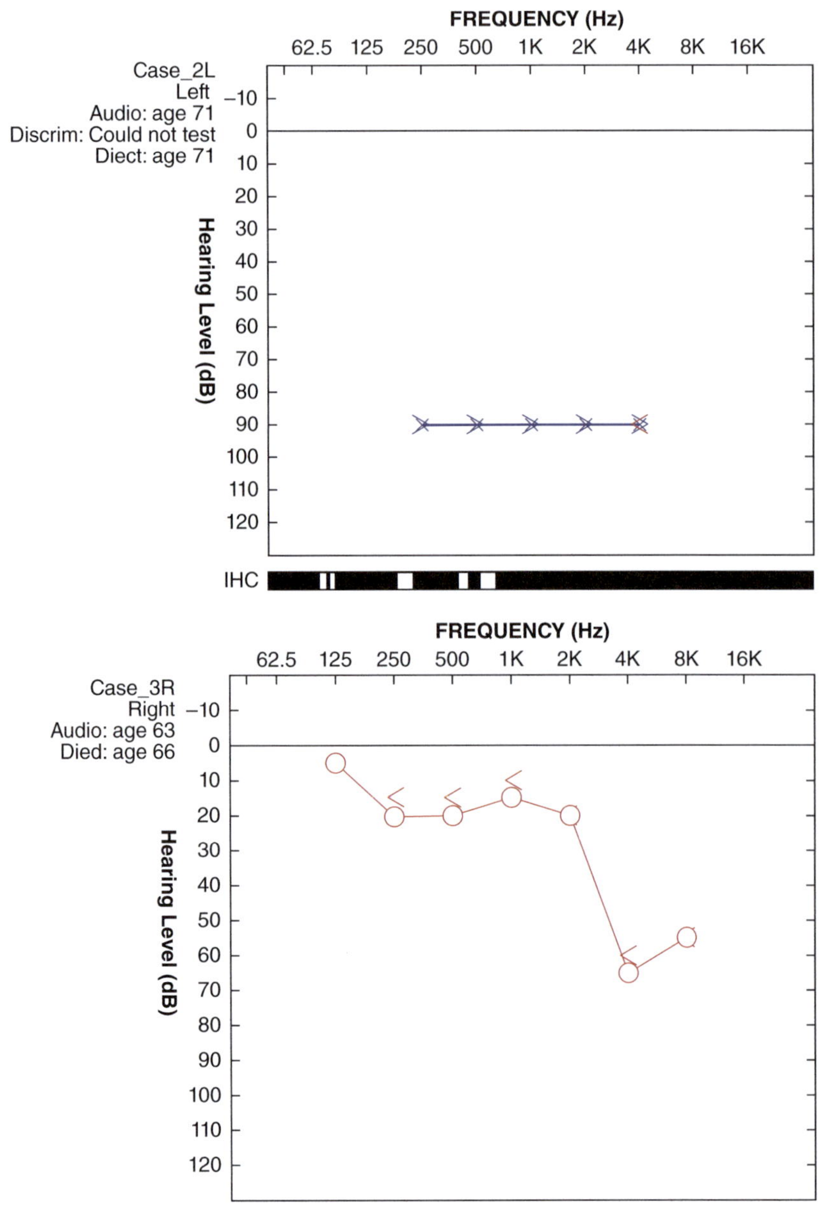

Fig. 60.1 Audiogram from a patient who had suffered a left-sided labyrinthine concussion. (From Ishai and colleagues [5]. Reproduced with written permission from John Wiley and Sons Limited)

Testing: Vestibular

Occasional reports include vestibular testing, which "may demonstrate a reduction in the caloric response, vestibulo-ocular reflex gain abnormality and phase after rotational testing" [4].

Imaging

Some cases of labyrinthine concussion may have detectable imaging correlates, usually on MRI of the internal auditory canals.

Figures 60.2, 60.3, and 60.4, from a case of labyrinthine concussion reported by Chiaramonte and colleagues [4], show right (greater than left)-sided T1 hyperintense signal in the cochlea, and right (greater than left)-sided contrast enhancement of the cochlea.

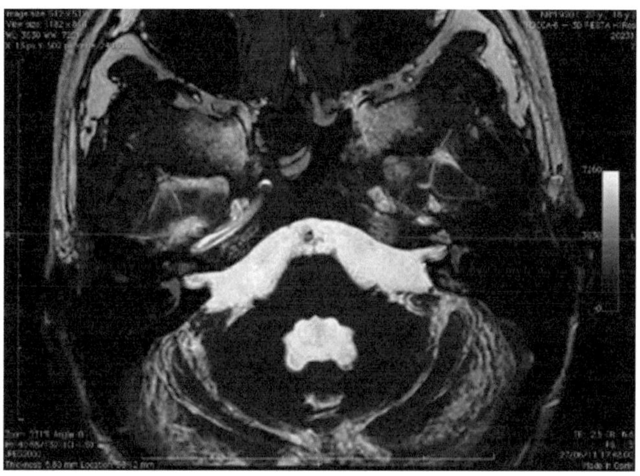

Fig. 60.2 Axial MRI of patient with labyrinthine concussion. MIP image of 3D-FIESTA at the level of the internal auditory canals fails to display the anterior, posterior, and lateral semicircular canals bilaterally. (From Chiaramonte et al. [4]. Reproduced with written permission from Sage Publications)

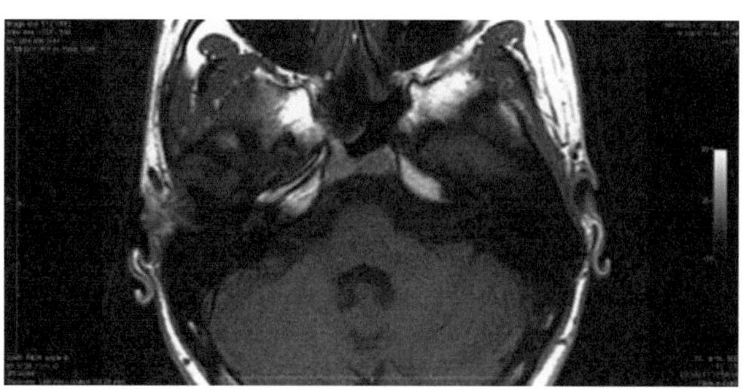

Fig. 60.3 Axial MRI of patient with labyrinthine concussion. SE T1 imaging at the same level showed a hyperintense signal of labyrinth and cochlea, more evident on the right side. (From Chiaramonte et al. [4]. Reproduced with written permission from Sage Publications)

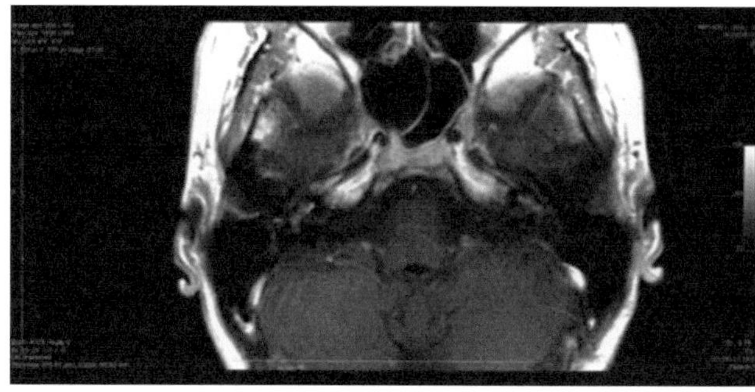

Fig. 60.4 Axial MRI of patient with labyrinthine concussion. 3D-T1 FSPGR after contrast administration revealed enhancement of labyrinth and cochlea, more marked on the right side. (From Chiaramonte et al. [4]. Reproduced with written permission from Sage Publications)

Many cases of apparent labyrinthine concussion have no radiographic correlates. However, given that labyrinthine concussion occurs in the context of head trauma, imaging remains an important component of differential diagnosis in order to identify pathologies such as temporal bone fracture and intracranial hemorrhage.

Labyrinthine Ossification

Some cases of labyrinthine concussion culminate in *labyrinthine ossification* [6–8], though this is relatively rare.

Treatment

Regrettably, "There is no definitive treatment for labyrinthine concussion" [1].

Prognosis

Prognosis of hearing loss after labyrinthine concussion is unfavorable, with most authors concluding that, "the prognosis of hearing recovery is disappointing in most cases" [1], adding that, "the prognosis for hearing function is worse than for vestibular symptoms" [4].

The duration of vestibular symptoms is variable; one paper states that, "The vertigo in these patients can last from several minutes up to several hours" [4]. However, in cases of labyrinthine concussion that result in labyrinthine ossification, the disequilibrium can last longer; in case described by Aralasmak and colleagues, the patient reported "vertigo for the last 20 days" [6].

Treatment

Insofar as labyrinthine concussion may cause permanent (typically unilateral) auditory and vestibular deficits, patients may benefit from amplification strategies such as a BiCROS (bilateral contralateral routing of signals) hearing aid, or a BAHA (bone attached hearing aid), and from vestibular rehabilitation therapy aimed at unilateral vestibular weakness.

References

1. Villarreal IM, Mendez D, Silva JM, Del Alamo PO. Contralateral Cochlear Labyrinthine Concussion without Temporal Bone Fracture: Unusual Posttraumatic Consequence. Case Rep Otolaryngol. 2016;2016:2123182.
2. Toh A, Ho EC, Turner N. Contralateral deafness post head injury without temporal bone fractures. Am J Otolaryngol. 2010;31(1):54–6.
3. Bartholomew RA, Lubner RJ, Knoll RM, Ghanad I, Jung D, Nadol JB Jr, et al. Labyrinthine concussion: Historic otopathologic antecedents of a challenging diagnosis. Laryngoscope Investig Otolaryngol. 2020;5(2):267–77.
4. Chiaramonte R, Bonfiglio M, D'Amore A, Viglianesi A, Cavallaro T, Chiaramonte I. Traumatic labyrinthine concussion in a patient with sensorineural hearing loss. Neuroradiol J. 2013;26(1):52–5.
5. Ishai R, Knoll RM, Chen JX, Wong K, Reinshagen KL, Nadol JB Jr, et al. Otopathologic Changes in the Cochlea following Head Injury without Temporal Bone Fracture. Otolaryngol Head Neck Surg. 2018;159(3):526–34.
6. Aralasmak A, Dincer E, Arslan G, Cevikol C, Karaali K. Posttraumatic labyrinthitis ossificans with perilymphatic fistulization. Diagn Interv Radiol. 2009;15(4):239–41.
7. Schwaber MK, Tarasidis NG. Labyrinthitis ossificans following post-traumatic hearing loss and vertigo: a case report with antemortem histopathology. Otolaryngol Head Neck Surg. 1990;102(1):89–91.
8. Swartz JD, Mandell DM, Faerber EN, Popky GL, Ardito JM, Steinberg SB, et al. Labyrinthine ossification: etiologies and CT findings. Radiology. 1985;157(2):395–8.

Temporal Bone Fracture

61

Overview

Head trauma may result in skull fractures, and a substantial proportion of these may include *temporal bone fractures*. Trauma can occur to anyone at any age, but at highest risk are males in their 20 s, and these demographics are similar for temporal bone fractures. The most common causes are automobile accidents, assaults, and falls. Acute workup includes imaging with high-resolution temporal bone CT and often MRI. There are several classifications of temporal bone fractures, but for otoneurology, an important distinction is whether the fracture violates the otic capsule. Immediate presenting features of a temporal bone fracture may include cerebrospinal fluid leaks, facial paralysis, hearing loss, disequilibrium, and meningitis. Delayed sequelae include cholesteatoma, meningocele, and encephalocele. Acute management often requires involvement of a trauma surgeon, otolaryngologist, and neurosurgeon. Audiologic testing can play a role in characterizing hearing loss and monitoring recovery. If vestibular testing identifies unilateral or bilateral vestibular hypofunction, then vestibular rehabilitation therapy can be considered.

Anatomy

The temporal bone is composed of five parts, the squamous, petrous, mastoid, tympanic, and styloid portions [1].

Figure 61.1, from Gladwell and Viozzi [1], is an illustration of lateral aspect of the left temporal bone with relevant landmarks.

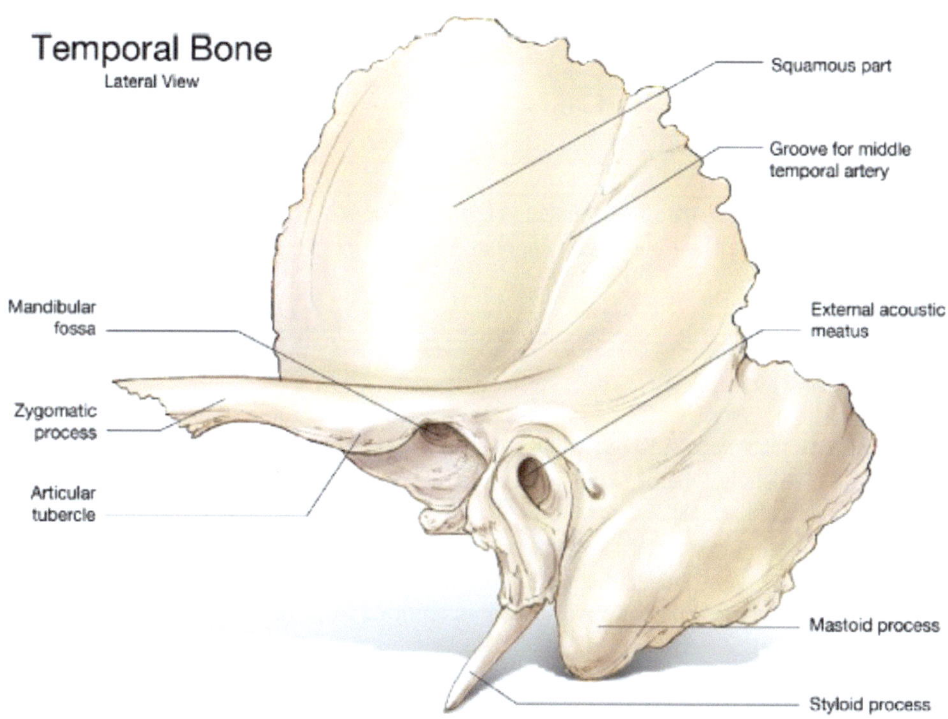

Fig. 61.1 Illustration of the lateral aspect of the left temporal bone with landmarks. (From Gladwell and Viozzi [1]. Reproduced with written permission from Mayo Foundation for Medical Education and Research)

M. Cherchi, *Otoneurology and Vestibular Medicine*, https://doi.org/10.1007/978-3-031-94841-1_61

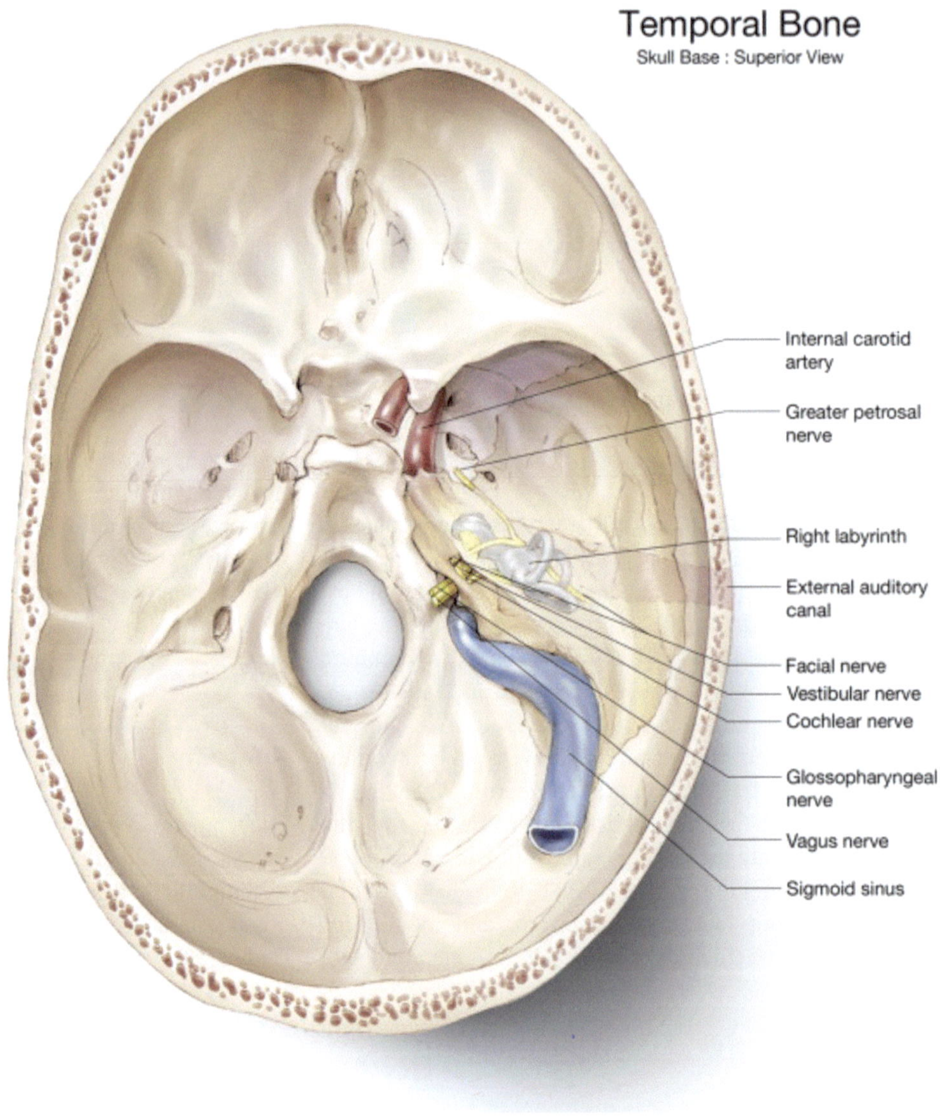

Fig. 61.2 Superior aspect of the skull base with annotated features of the temporal bone. (From Gladwell and Viozzi [1]. Reproduced with written permission from Mayo Foundation for Medical Education and Research)

Figure 61.2, from Gladwell and Viozzi [1], Illustration of superior aspect of the skull base, with annotated features of the temporal bone.

Traumatology

Of all patients who have suffered significant head trauma, 23–66% involve some type of skull fracture and 21% have trauma to the skull base [1].

In patients with any type of skull fracture, from 14% to 22% involve a temporal bone fracture [2]. Of all cases of temporal bone fracture, bilateral involvement is reported in 9–20% of cases [3].

The biomechanics of temporal bone fractures have been studied extensively [4].

Epidemiology

Table 61.1 from Brodie and Thompson [2] reports the etiology of 820 temporal bone fractures in 699 patients.

Similar statistics are reported by other investigators [3, 5].

Table 61.2 from Brodie and Thompson [2], reporting on temporal bone fractures in 699 patients, shows a histogram by age and gender. The peak age at which temporal bone fractures occur is the third decade (age 21–30) and is significantly more common in males than in females at all ages.

Table 61.1 Etiology of temporal bone fractures in 820 temporal bones from 699 patients

Etiology	Number (percent) of 820 temporal bones in 699 patients
Automobile	212 (25.9%)
Assault	118 (14.4%)
Fall	111 (13.5%)
Motorcycle	92 (11.2%)
Pedestrian	67 (8.2%)
Bicycle	44 (5.4%)
Gunshot wound	22 (2.7%)
Miscellaneous	16 (2.0%)
Sports	9 (1.1%)
All-terrain vehicle	7 (0.9%)

Based on data from Brodie and Thompson [2]

Table 61.2 Age distribution by gender in 699 patients with temporal bone fractures

Age range (years)	Males, number (percent) of patients	Females, number (percent) of patients
0–10	53 (7.6%)	24 (3.4%)
11–20	109 (15.6%)	39 (5.6%)
21–30	158 (22.6%)	40 (5.7%)
31–40	116 (16.6%)	36 (5.2%)
41–50	42 (6.0%)	16 (2.3%)
51–60	23 (3.3%)	5 (0.7%)
61–70	16 (2.3%)	3 (0.4%)
71–80	8 (1.1%)	3 (0.4%)
81–90	5 (0.7%)	2 (0.3%)

Based on data from Brodie and Thompson [2]

Similar to other types of trauma, "Risk factors for and causes of temporal bone fractures are similar to those in any patient with head injury: younger age, male gender, motor vehicle accidents, falls, recreational injuries, and assaults" [1].

Presenting Features

Temporal bone fractures commonly manifest with cerebrospinal fluid leaks, facial paralysis, sensorineural hearing loss, conductive hearing loss, disequilibrium, or meningitis [2]. Less common sequelae of temporal bone fractures include cholesteatoma and meningocele/encephalocele [3]. Cholesteatoma may occur at some delay after the original injury, ranging from 2 to 24 years [6].

Brodie and Thompson [2] report that 2.5% of temporal bone fractures involve the otic capsule, and all such cases cause severe to profound *hearing loss*.

Brodie and Thompson [2] report that of all 699 patients in their series, 24% "complained of some degree of hearing loss." In 35 patients (5%), the hearing loss was bilateral. Out of 91 patients who had audiometry, 21% had conductive hearing loss, 57% had sensorineural hearing loss, and 22% had mixed hearing loss.

One study reported *tinnitus* in 41% of patients with temporal bone fractures [7].

Disequilibrium is reported in 24–78% of patients with temporal bone fractures, with mechanisms including otic capsule injury, perilymphatic fistula, labyrinthine concussion, brainstem injury [1], and traction/shearing injury of the vestibular nerve [8, 9]. Other postulated mechanisms include "post-traumatic endolymphatic hydrops" [3], for which there is some histopathological evidence [10]. More mundane causes of disequilibrium following temporal bone fracture include benign paroxysmal positional vertigo [6]. Several investigators state that "vertigo" following temporal bone fractures is "self-limiting" and resolves "spontaneously" in the range of "6 to 12 months" [3, 8], though these do not cite any data in support of that range, and given the variety of mechanisms, we suspect the range may be broader. One series of 20 patients reported that 10 (50%) complained of vestibular symptoms, and that "the vestibular symptoms improved or disappeared in all cases" [11].

Brodie and Thompson [2] report that, of all temporal bone fractures, 7% result in *facial paralysis*. Of temporal bone fractures involving the otic capsule, 48% also caused facial paralysis; in contrast, fractures that did not involve the otic capsule only resulted in facial paralysis in 6% of cases.

Brodie and Thompson [2] report that 122 (17.5%) of 699 patients with temporal bone fracture have a cerebrospinal fluid leak, which included 97 (13.9%) with otorrhea, 16 (2.3%) with rhinorrhea, and 8 (1.1%) with otorrhea.

Brodie and Thompson [2] report that of 122 patients who developed a cerebrospinal fluid leak, 9 (7%) developed meningitis. Of the 578 patients without a cerebrospinal fluid fistula, 5 (1%) developed meningitis.

Classification

A variety of classification systems have been proposed for temporal bone fractures [8]. The most basic one classifies the fractures as either *transverse* or *longitudinal*, "based on the relationship of the fracture line to the axis of the petrous ridge and the otic capsule" [2]. After the orientation of the fracture, the next broad classification pertains to whether or not the fracture involves the otic capsule.

The transverse/longitudinal fracture classification has some utility, but is simplistic. Saraiya and Aygun comment that the, "majority of temporal bone fractures do not fit to this classification system. [The] same anatomic structures may be injured by both types of fractures limiting the clinical value of the classification system in predicting patients' symptoms and outcome" [5].

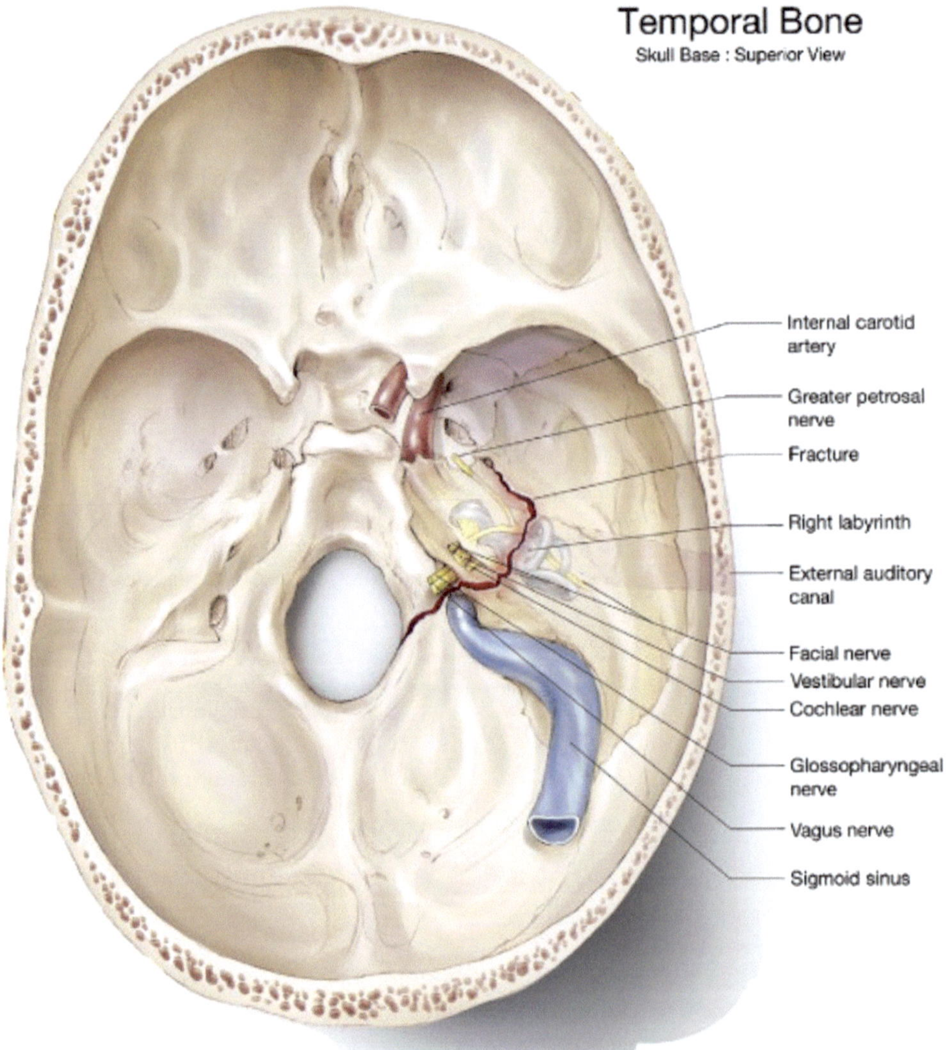

Fig. 61.3 Illustration of a transverse temporal bone fracture involving the otic capsule. (From Gladwell and Viozzi [1]. Reproduced with written permission from Mayo Foundation for Medical Education and Research)

Transverse temporal bone fractures usually result from a blow to the occipital region [12], and comprise 10–30% of temporal bone fractures [13]. Transverse fractures often pass from the foramen magnum across the petrous pyramid and involve the otic capsule and do not usually affect the ossicular chain or the external auditory canal [2].

Figure 61.3 from Gladwell and Viozzi [1] illustrates a transverse temporal bone fracture involving the otic capsule.

Longitudinal temporal bone fractures usually result from a blow to temporoparietal region [12], and comprise 70–90% of temporal bone fractures [13]. Longitudinal fractures lie anterior to the otic capsule (and thus spare the otic capsule) and involve the squamosal portion of the temporal bone and the posterosuperior wall of the external auditory canal, passing through the mastoid air cells, tegmen mastoideum, and tegmen tympani [2].

Figure 61.4 from Gladwell and Viozzi [1] illustrates a longitudinal temporal bone fracture traversing the external auditory canal but sparing the otic capsule.

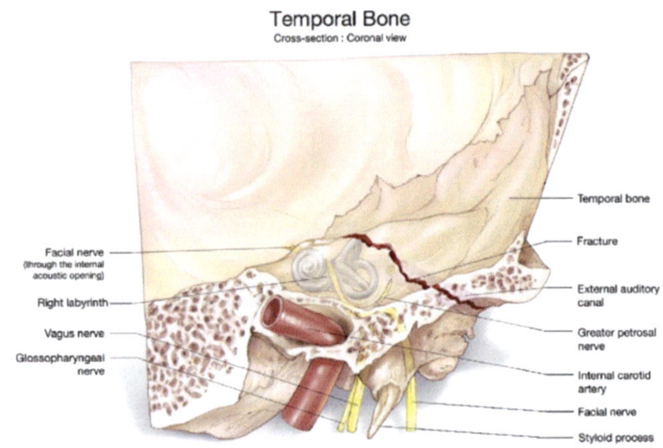

Fig. 61.4 Illustration of a longitudinal temporal bone fracture traversing the external auditory canal but sparing the otic capsule. (From Gladwell and Viozzi [1]. Reproduced with written permission from Mayo Foundation for Medical Education and Research)

Diagnostic Testing: Audiometry

Audiometry is appropriate for any patient with a temporal bone fracture. A longitudinal temporal bone fracture traverses the external auditory canal and can produce conductive hearing loss by several mechanisms (hemotympanum, tympanic, membrane perforation, ossicular chain dislocation). A transverse temporal bone fracture can violate the otic capsule and result in severe sensorineural hearing loss [1].

The incidence of hearing loss reported in temporal bone fractures ranges from 26% to 57% for conductive hearing loss, 14% to 23% for sensorineural hearing loss (including 14% with complete loss), and 20% to 55% for mixed hearing loss. The hearing loss "may be immediate or delayed; may be transient, permanent or progressive" [3]. Hypothesized mechanisms for progressive hearing loss include perilymphatic fistula and post-traumatic endolymphatic hydrops [14].

Given the variability of injury in temporal bone fractures, it is unsurprising that the degree of hearing loss is similarly variable. Suligavi and colleagues [15] studied a series of patients with temporal bone fractures and found that 34% had mild hearing loss and 3% had severe hearing loss.

In cases of temporal bone fracture that violate the otic capsule and cause sensorineural hearing loss, 1–11% of patients eventually develop hearing loss in contralateral (originally uninjured) ear from what is thought to be sympathetic hearing loss [3]. This is thought to occur because violation of the otic capsule disrupts the blood–labyrinth barrier, exposing labyrinthine antigens to the immune system, which may subsequently mount a maladaptive autoimmune response against the contralateral ear.

Diagnostic Testing: Vestibular Testing

There is little literature specifically on vestibular function in patients who have sustained temporal bone fractures. In our experience, vestibular hypofunction can often be identified (such as with cervical and ocular vestibular evoked myogenic potentials, video head impulse testing, videonystagmography).

Imaging

Patients presenting to an emergency room with trauma will generally undergo a noncontrast head CT. In order to identify and characterize a temporal bone fracture, a high-resolution temporal bone CT (with <1 mm collimation) with multiplanar reconstructions is more helpful [16]. High-resolution CT is particularly helpful for smaller structures and is well suited for identifying ossicular fractures and ossicular chain dislocations.

In the context of trauma, an MRI of the brain (and probably also cervical spine) is usually desirable to evaluate nervous system structures and other soft tissues. MRI is also helpful for distinguishing fluid from air and identifying hemorrhage [16].

When reviewing radiology studies of the temporal bone in a patient who has suffered head trauma, it is important to keep in mind that many normal features may mimic fractures [17]. Broadly, "many normal structures are linear and may mimic fracture lines. These mimics include aqueducts, canaliculi, grooves, suture, fissures and other entities" [16].

Histopathology

Individual case reports describe various degenerative changes in the vestibular system [18]. Temporal bone series from patients who suffered temporal bone fractures involving the otic capsule describe distinct degenerative changes in vestibular hair cells, Scarpa's ganglion, and the vestibular nerve [10].

Treatment

Management of temporal bone fractures almost always requires a multidisciplinary approach. Acute management often involves a trauma surgeon, otolaryngologist, neurosurgeon, and intensivist (or neurointensivist). Longer-term management may involve otolaryngology, neurology, audiology, physical therapy, and occupational therapy. Depending on what other injuries occurred, other disciplines may be necessary.

As far as otoneurology and vestibular medicine are concerned, auditory deficits are usually monitored and managed by audiology. Vestibular deficits are usually managed with vestibular physical therapy, though the specific strategies will depend on the lesion(s) involved (e.g., benign paroxysmal positional vertigo, vestibular hypofunction).

References

1. Gladwell M, Viozzi C. Temporal bone fractures: a review for the oral and maxillofacial surgeon. J Oral Maxillofac Surg. 2008;66(3):513–22.
2. Brodie HA, Thompson TC. Management of complications from 820 temporal bone fractures. Am J Otol. 1997;18(2):188–97.
3. Johnson F, Semaan MT, Megerian CA. Temporal bone fracture: evaluation and management in the modern era. Otolaryngol Clin N Am. 2008;41(3):597–618. x

4. Yoganandan N, Pintar FA. Biomechanics of temporo-parietal skull fracture. Clin Biomech (Bristol, Avon). 2004;19(3):225–39.
5. Saraiya PV, Aygun N. Temporal bone fractures. Emerg Radiol. 2009;16(4):255–65.
6. Ishai R, Knoll RM, Chen JX, Wong K, Reinshagen KL, Nadol JB Jr, et al. Otopathologic changes in the cochlea following head injury without temporal bone fracture. Otolaryngol Head Neck Surg. 2018;159(3):526–34.
7. Heid L, Claussen CF, Kersebaum M, Nagy E, Bencze G, Bencsik B. Vertigo, dizziness, and tinnitus after otobasal fractures. Int Tinnitus J. 2004;10(1):94–100.
8. Kennedy TA, Avey GD, Gentry LR. Imaging of temporal bone trauma. Neuroimaging Clin N Am. 2014;24(3):467–86. viii
9. Lantos JE, Leeman K, Weidman EK, Dean KE, Peng T, Pearlman AN. Imaging of temporal bone trauma: a Clinicoradiologic perspective. Neuroimaging Clin N Am. 2019;29(1):129–43.
10. Knoll RM, Ishai R, Lubner RJ, Trakimas DR, Brodsky JR, Jung DH, et al. Peripheral vestibular organ degeneration after temporal bone fracture: a human Otopathology study. Laryngoscope. 2020;130(3):752–60.
11. Wennmo C, Svensson C. Temporal bone fractures. Vestibular and other related ear sequele. Acta Otolaryngol Suppl. 1989;468:379–83.
12. Swartz JD. Temporal bone trauma. Semin Ultrasound CT MR. 2001;22(3):219–28.
13. Collins JM, Krishnamoorthy AK, Kubal WS, Johnson MH, Poon CS. Multidetector CT of temporal bone fractures. Semin Ultrasound CT MR. 2012;33(5):418–31.
14. Lyos AT, Marsh MA, Jenkins HA, Coker NJ. Progressive hearing loss after transverse temporal bone fracture. Arch Otolaryngol Head Neck Surg. 1995;121(7):795–9.
15. Suligavi SS, Saiyad SN, Doddamani SS, Mathri A. Early diagnosis of hidden hearing loss following temporal bone fractures. Indian J Otolaryngol Head Neck Surg. 2022;74(Suppl 3):4336–40.
16. Kurihara YY, Fujikawa A, Tachizawa N, Takaya M, Ikeda H, Starkey J. Temporal bone trauma: typical CT and MRI appearances and important points for evaluation. Radiographics. 2020;40(4):1148–62.
17. Kwong Y, Yu D, Shah J. Fracture mimics on temporal bone CT: a guide for the radiologist. AJR Am J Roentgenol. 2012;199(2):428–34.
18. Benitez JT, Bouchard KR, Lane-Szopo D. Pathology of deafness and disequilibrium in head injury:a human temporal bone study. Am J Otol. 1980;1(3):163–7.

Hyperacusis

62

Overview

There is no universally agreed-upon set of criteria for *hyperacusis*, though most literature relies on definitions including terminology such as "increased sensitivity to sounds," "intolerance to sounds that other individuals would not ordinarily find bothersome," "disorder of loudness perception," etc. Part of the difficulty lies in the fact that hyperacusis involves an objective component (measurable amplitude of a sound stimulus), a subjective percept, and a behavioral response. The epidemiology of hyperacusis has been poorly studied; literature reports prevalence among children of 3.2%, and prevalence among adults as ranging from 8.6% to 15.2%, with differences likely due to different definitions and study methodologies. Etiologies of hyperacusis include peripheral (otologic), central (neurologic), and psychiatric. Testing usually entails audiometry with "loudness discomfort levels (LDL)" or "uncomfortable loudness (UCL)." Ideally, management of hyperacusis should be overseen by an audiologist specializing in this topic.

Introduction

There is no universally agreed-upon set of criteria for *hyperacusis*, though most literature relies on definitions including terminology such as "increased sensitivity to sounds," "intolerance to sounds that other individuals would not ordinarily find bothersome," "disorder of loudness perception," etc.

After conducting a thorough history, at the clinician's discretion, a *workup* for hyperacusis may include:

- Physical examination:
  - Assess for facial paralysis (as can be found in Bell's palsy, Ramsay Hunt syndrome, Lyme disease).
- Audiovestibular workup:
  - Audiometry with loudness discomfort levels. This can also assess for conductive hyperacusis (which can be found with semicircular canal dehiscence), low frequency loss (often associated with Ménière's disease), high-frequency hearing loss (often associated with labyrinthitis), or a "noise notch" (often associated with noise exposure).
  - Otoacoustic emissions.
  - Possibly brainstem auditory evoked responses.
  - Cervical and/or ocular vestibular evoked myogenic potentials with thresholds (to assess for semicircular canal dehiscence).
- Imaging:
  - Depending on the circumstances (e.g., status post head trauma; suspicion for multiple sclerosis; audiovestibular workup suggestive of semicircular canal dehiscence), imaging with a temporal bone CT or MRI of the brain and internal auditory canals without and with contrast may be appropriate.

Management may involve referral to:

- Audiologist experienced with hyperacusis.
- Otolaryngologist, for example, if there is suspicion for perilymphatic fistula or evidence of superior semicircular canal dehiscence.
- Psychologist/psychiatrist if there is evidence for depression or post-traumatic stress disorder.

Definitions of Hyperacusis

Frequently cited definitions of hyperacusis include:

> Increased sensitivity to sound in levels that would not trouble a normal individual. This auditory sensitivity is characterized by a painful or troublesome sensation when the patient is exposed to sound [1].

> Unusual tolerance to ordinary environmental sounds and… consistently exaggerated or inappropriate responses to sounds that are neither threatening nor uncomfortably loud to a typical person [2].

M. Cherchi, *Otoneurology and Vestibular Medicine*, https://doi.org/10.1007/978-3-031-94841-1_62

> Disorder of loudness perception, in which sound intensities that are considered comfortable by most people are perceived unbearably loud [3].

However, it turns out that there is no universally accepted definition of hyperacusis. Tyler and colleagues [4] review a wide variety of definitions that have been proposed for hyperacusis based on different aspects of the phenomenon, which might broadly be grouped into perception (awareness), sensitivity, discomfort, tolerance, response, emotional valence, and auditory function:

- Perception:
  - "Heightened awareness of sounds"
- Sensitivity:
  - "Pathological auditory hypersensitivity"
  - "Increased auditory sensitivity"
  - "Noise sensitivity"
  - "Audiosensitivity"
  - "Soft sound sensitivity"
  - "Select sound sensitivity"
- Discomfort:
  - "Discomfort for sounds that would be acceptable to most normally hearing people"
- Tolerance:
  - "Sound intolerance"
  - "Loudness tolerance problem"
  - "Abnormally low tolerance for sound levels"
- Abnormal response:
  - "Abnormally strong response to moderate sound"
  - "Hyperresponsiveness to sound stimuli"
  - "Hyperresponsiveness to noise"
  - "Aversive response"
- Emotional valence [4]:
  - "Annoyance"
  - "Irritability"
  - "Fear"
- Auditory function:
  - "Disturbed loudness function".

The variety of definitions in turn raises the question of whether each represents a different perspective on one and the same underlying problem or whether there are actually distinct pathologies.

Loudness

At a very basic level, hyperacusis involves sounds being experienced/perceived/responded to as "too loud." But what is loudness?

> Loudness is determined mainly by physical sound intensity (i.e., pressure), although it is also affected by other stimulus attributes such as duration, frequency, band-width, and monaural versus binaural presentation, as well as psychological factors such as mood and emotional exhaustion. Considerable evidence… suggests that the relationship between sound intensity and loudness is not static in time, but can be modified by sound deprivation or stimulation. Deprivation increases sound sensitivity, whereas stimulation decreases it. The neurological bases of those loudness shifts remain largely unknown. A hypothetical central gain process, which regulates supra-threshold sensitivity somewhere in the auditory system, is one of the most commonly proposed mechanisms [5].

In other words, amplitude of sound waves (and thus the energy they contain) is the main, though not the only determinant of loudness.

Epidemiology

There are few studies of the epidemiology of hyperacusis, and their results are difficult to compare because they employed different definitions of hyperacusis and different data collection methodologies, as reviewed by Tyler and colleagues [4].

Andersson and colleagues [6] analyzed the results of a survey of 1147 people (median age 47 years, SD 16.9 years) conducted in Sweden by mail and internet and reported the prevalence of hyperacusis to be 8.6%.

Fabijanska and colleagues [7] analyzed the results of a questionnaire survey from 10,349 adult respondents in Poland and reported the prevalence of hyperacusis to be 15.2% overall (12.5% of women and 17.6% of men).

Coelho and colleagues [8] conducted a questionnaire survey of 506 children aged 5–12 years and reported the prevalence of hyperacusis to be 3.2%.

Etiologies

The differential diagnosis of hyperacusis includes [1, 2, 4]:

- Peripheral etiologies:
  - *Acute acoustic trauma* [9].
  - *Noise-induced hearing loss*. Hyperacusis can occur in individuals with chronic exposure to industrial sounds [10, 11].
  - *Loudness recruitment*. "Converging physiological evidence indicates that intensifying central responsiveness to an existing sound causes loudness recruitment in subjects with hyperacusis" [3].
  - *Ménière's disease* [12–14].
  - *Bell's palsy* (mediated by failure of the stapedial reflex from damage of the facial nerve). Hyperacusis is reported to occur in 29% of patients with Bell's palsy [15].

- *Ramsay Hunt syndrome*. Up to 46% of cases of Ramsay Hunt syndrome are reported to be associated with hyperacusis [16].
- *Stapedectomy* [1].
- *Perilymphatic fistula* [17].
- *Semicircular canal dehiscence* [4]. One series reported that 85% (17 out of 20 patients) with superior semicircular canal dehiscence had hyperacusis [18].

• Central etiologies:
- *Migraine* [19, 20].
- *Williams syndrome*. Up to 95% of patients with Williams syndrome are reported to suffer from hyperacusis [21].
- *Learning disabilities* [22].
- *Dysfunction of auditory efferent system* and its effects on auditory gain [2].
- *Multiple sclerosis* [23, 24].
- *Fibromyalgia* [25, 26].

• Potentially a combination of central and peripheral etiologies:
- *Head trauma* [27]. Although this is usually categorized under "central etiologies" of hyperacusis, head injury can easily cause labyrinthine concussion with cochlear damage, and thus include a peripheral component of hyperacusis.
- *Lyme disease*. In some cases, this may be due to the disease's predilection for causing facial nerve paralysis and thus interference with the stapedial reflex [28], but some cases of hyperacusis in Lyme disease appear to occur independently of facial paralysis [29].

• Hormonal etiologies:
- *Addison's disease* [30].

• Psychiatric etiologies:
- *Depression* [31].
- *Post-traumatic stress disorder* [32].

Loudness Recruitment

Loudness recruitment describes, "an experience commonly associated with cochlear hearing loss and specifically with dysfunction of the outer hair cells of the organ of Corti: with a rising sound level, the perceived loudness increases faster than normal" [2].

Patients who have sustained cochlear damage:

> Conceivably possess the ability of recruitment because weak sounds cannot stimulate the few remaining hair cells and only a minimum number of impulses will be conducted through the acoustic nerve. Strong stimulation, on the other hand, is able to stimulate the remaining intact hair cells (presumably inner hair cells) and the impulses are then conducted through the normal fibers of the acoustic nerve and its connections to the cerebral cortex, which then becomes relatively well saturated [33].

Testing

Contemporary audiologic assessment of hyperacusis often involves characterization of "loudness discomfort levels" (LDL) or "uncomfortable loudness" (UCL), defined as "the lowest sound level judged by the listener to be uncomfortably loud" [4].

Imaging

Few imaging studies have been conducted to investigate presumptive centrally mediated hyperacusis [4].

Management

When initially experienced, hyperacusis can be uncomfortable and frightening, so a natural strategy is one of avoidance, or at least diminution of input. While this may appear sensible and even appropriate in some circumstances, there is some evidence that it may be maladaptive in others.

> For many patients, the first reaction to hyperacusis is to protect themselves with ear plugs, muffs or other devices. There is, however, reason to believe that such strategies to decrease the intensity of sound entering the auditory system may further increase the central gain, exacerbating rather than improving the hyperacusis [2].

After other treatable causes have been excluded, consultation with an audiologist specializing in hyperacusis is medically reasonable.

References

1. Katzenell U, Segal S. Hyperacusis: review and clinical guidelines. Otol Neurotol. 2001;22(3):321–6. discussion 6–7
2. Baguley DM. Hyperacusis. J R Soc Med. 2003;96(12):582–5.
3. Knipper M, Van Dijk P, Nunes I, Ruttiger L, Zimmermann U. Advances in the neurobiology of hearing disorders: recent developments regarding the basis of tinnitus and hyperacusis. Prog Neurobiol. 2013;111:17–33.
4. Tyler RS, Pienkowski M, Roncancio ER, Jun HJ, Brozoski T, Dauman N, et al. A review of hyperacusis and future directions: part I. Definitions and manifestations. Am J Audiol. 2014;23(4):402–19.
5. Fournier P, Schonwiesner M, Hebert S. Loudness modulation after transient and permanent hearing loss: implications for tinnitus and hyperacusis. Neuroscience. 2014;283:64–77.
6. Andersson G, Lindvall N, Hursti T, Carlbring P. Hypersensitivity to sound (hyperacusis): a prevalence study conducted via the Internet and post. Int J Audiol. 2002;41(8):545–54.
7. Fabijanska A, Rogowski M, Bartnik G, Skarzynski H, editors. Epidemiology of tinnitus and hyperacusis in Poland. Proceedings of the Sixth International Tinnitus Seminar; 1999. London 2002.
8. Coelho CB, Sanchez TG, Tyler RS. Hyperacusis, sound annoyance, and loudness hypersensitivity in children. Prog Brain Res. 2007;166:169–78.

9. Axelsson A, Hamernik RP. Acute acoustic trauma. Acta Otolaryngol. 1987;104(3–4):225–33.
10. Niemeyer W. Relations between the discomfort level and the reflex threshold of the middle ear muscles. Audiology. 1971;10(3):172–6.
11. Tzaneva L. Assessment of the discomfort threshold of command board operators in automated productions with respect to its application in professional selection. Cent Eur J Public Health. 1996;4(3):173–5.
12. Brandy William T, Lynn James M. Audiologic Findings in Hyperacusic and Nonhyperacusic Subjects. Am J Audiol. 1995;4(1):46–51.
13. Gordon AG. "Hyperacusis" and origins of lowered sound tolerance. J Neuropsychiatry Clin Neurosci. 2000;12(1):117–9.
14. Vattoth S, Shah R, Cure JK. A compartment-based approach for the imaging evaluation of tinnitus. AJNR. 2010;31(2):211–8.
15. Adour KK, Wingerd J. Idiopathic facial paralysis (Bell's palsy): factors affecting severity and outcome in 446 patients. Neurology. 1974;24(12):1112–6.
16. Bartnik G, Fabijanska A, Rogowski M. Experiences in the treatment of patients with tinnitus and/or hyperacusis using the habituation method. Scand Audiol Suppl. 2001;52:187–90.
17. Fukaya T, Nomura Y. Audiological aspects of idiopathic perilymphatic fistula. Acta Otolaryngol Suppl. 1988;456:68–73.
18. Yuen HW, Eikelboom RH, Atlas MD. Auditory manifestations of superior semicircular canal dehiscence. Otol Neurotol. 2009;30(3):280–5.
19. Vingen JV, Pareja JA, Storen O, White LR, Stovner LJ. Phonophobia in migraine. Cephalalgia. 1998;18(5):243–9.
20. Woodhouse A, Drummond PD. Mechanisms of increased sensitivity to noise and light in migraine headache. Cephalalgia. 1993;13(6):417–21.
21. Nigam A, Samuel PR. Hyperacusis and Williams syndrome. J Laryngol Otol. 1994;108(6):494–6.
22. Gordon AG. Abnormal middle ear muscle reflexes and audiosensitivity. Br J Audiol. 1986;20(2):95–9.
23. Weber H, Pfadenhauer K, Stohr M, Rosler A. Central hyperacusis with phonophobia in multiple sclerosis. Mult Scler. 2002;8(6):505–9.
24. Cohen M, Rudge P, Robinson K, Miller D. The effects on auditory function of damage to the pontine olivo-cochlear bundle in man. Scand Audiol. 1988;17(3):185–9.
25. Geisser ME, Glass JM, Rajcevska LD, Clauw DJ, Williams DA, Kileny PR, Gracely RH. A psychophysical study of auditory and pressure sensitivity in patients with fibromyalgia and healthy controls. J Pain. 2008;9(5):417–22.
26. Staud R, Godfrey MM, Robinson ME. Fibromyalgia patients are not only hypersensitive to painful stimuli but also to acoustic stimuli. J Pain. 2021;22(8):914–25.
27. Waddell PA, Gronwall DM. Sensitivity to light and sound following minor head injury. Acta Neurol Scand. 1984;69(5):270–6.
28. Peltomaa M, Pyykko I, Seppala I, Viljanen M. Lyme borreliosis and facial paralysis–a prospective analysis of risk factors and outcome. Am J Otolaryngol. 2002;23(3):125–32.
29. Nields JA, Fallon BA, Jastreboff PJ. Carbamazepine in the treatment of Lyme disease-induced hyperacusis. J Neuropsychiatry Clin Neurosci. 1999;11(1):97–9.
30. Henkin RI, Daly RL. Auditory detection and perception in normal man and in patients with adrenal cortical insufficiency: effect of adrenal cortical steroids. J Clin Invest. 1968;47(6):1269–80.
31. Carman JS. Imipramine in hyperacusic depression. Am J Psychiatry. 1973;130(8):937.
32. Westcott M, editor. Case study: management of hyperacusis associated with post-traumatic stress disorder. Proceedings of the Seventh International Tinnitus Seminar. Perth: University of Western Australia; 2002.
33. Alexander W. Loudness recruitment and vertigo. Laryngoscope. 1959;69(2):174–88.

Otalgia

63

Overview

Ear pain and discomfort usually result from disease of the ear itself, for which there is a broad differential diagnosis. However, aural symptoms can also be referred from other anatomical locations, with the most common sources being dental, temporomandibular, and cervical. Migraine can also cause aural symptoms. Practically, cases of otalgia are generally referred to an otolaryngologist specializing in otology. If there is evidence of (or clinical suspicion for) dental disease, then a dental evaluation is also reasonable.

Discussion

Hwa and Brant [1] offer the following differential diagnosis of otalgia:

- Congenital.
  - Branchial cleft cyst.
  - Sebaceous cyst.
  - Epidermoid cyst.
- Infectious.
  - Perichondritis.
  - Cellulitis.
  - Acute otitis externa.
  - Herpes zoster oticus.
  - Malignant otitis externa.
  - Acute otitis media—complicated or uncomplicated.
  - Otomastoiditis/petrous apicitis.
  - Suppurative labyrinthitis.
- Traumatic.
  - Auricular trauma/hematoma.
  - External auditory canal laceration.
  - Tympanic membrane perforation (blast or puncture).
  - Ear foreign body.
- Autoimmune/Inflammatory.
  - Relapsing polychondritis.
  - Granulomatosis with polyangiitis (GPA).
  - Sarcoidosis.
- Neurologic.
  - Trigeminal neuralgia.
  - Glossopharyngeal neuralgia.
  - Vagal neuralgia.
  - Migraine.
- Neoplastic.
  - Cutaneous malignancy.
  - Nasopharyngeal carcinoma.
  - Squamous cell carcinoma of the head and neck.
- Other.
  - Chondrodermatitis nodularis helicis.
  - TMJ arthralgia/arthritis.
  - Dental pathology.
  - Eagle syndrome.

The majority of cases of otalgia are attributable to *primary otologic problems*. However, a significant minority of cases of otalgia are actually *referred otalgia*. The work by Taziki and Behnampour [2] reported that of 770 patients with otalgia, 94 patients (12.2%) had referred otalgia. Of the patients with referred otalgia, the pain was referred from dental problems (62.8%), pharyngitis (24.5%), temporomandibular joint dysfunction (8.8%), sinusitis (2.1%), Bell's palsy (1.1%), and pharyngeal abscess (1.1%).

The multiple somatosensory innervations of the ear [2, 3] comprise the mechanism underlying the variety of sources of referred otalgia. The main sensory innervations arise from the auricular branch of the vagus nerve, the auriculotemporal branch of the trigeminal nerve, the lesser occipital nerve, the greater occipital nerve, and the nervus intermedius [4] (Fig. 63.1).

Based on the distributions of sensory innervations depicted in these diagrams, one might suspect that simply identifying the painful area on the ear should enable anatomical localization of the source of otalgia. However, despite what is shown in the figures above, there is actually considerable variation between individuals in the ear's somatosensory innervation [6].

M. Cherchi, *Otoneurology and Vestibular Medicine*, https://doi.org/10.1007/978-3-031-94841-1_63

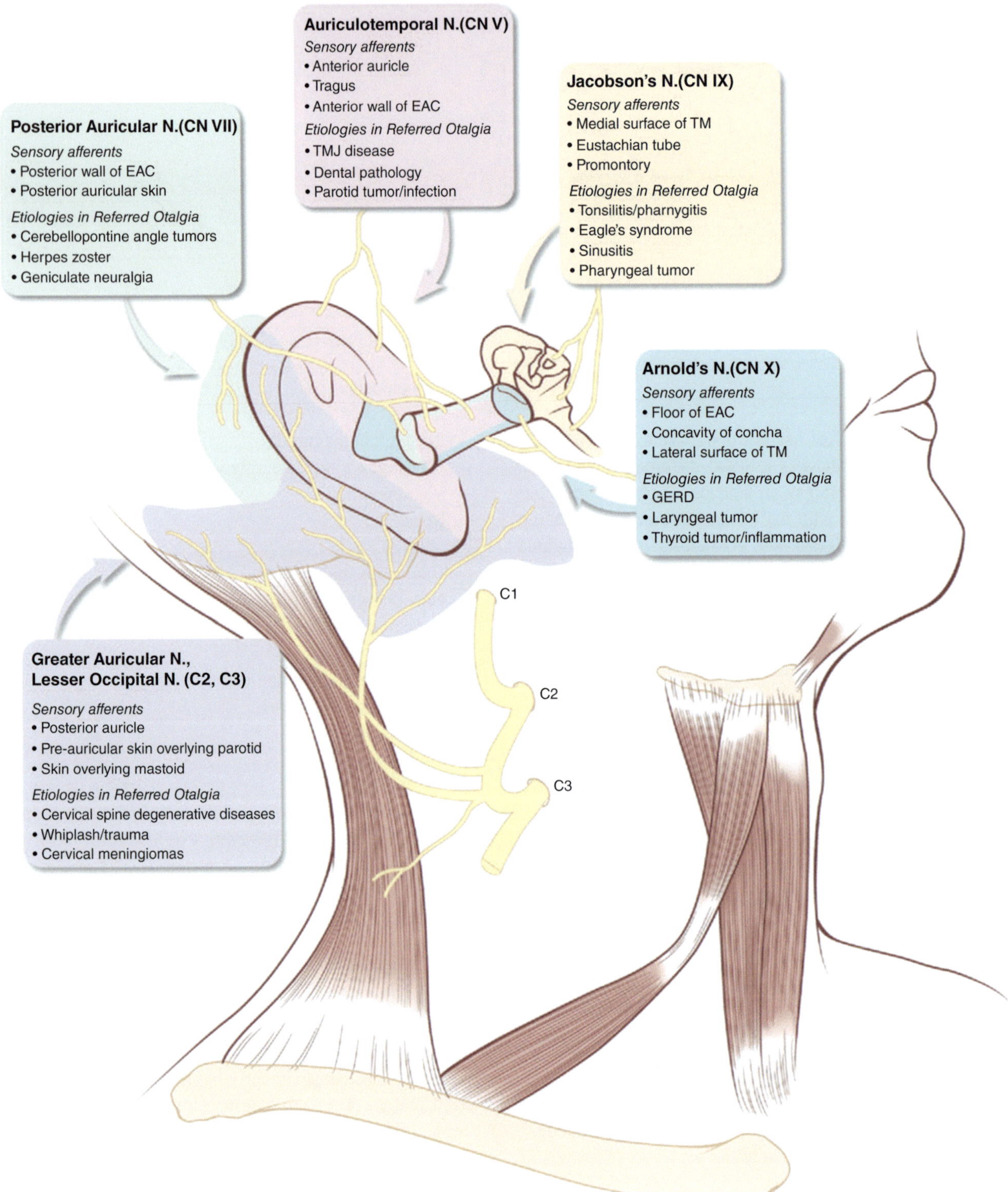

Fig. 63.1 Complex sensory innervation of the ear and periauricular structures and various etiological causes in referred otalgia. (From Jaber et al. [5]. Reproduced with written permission from John Wiley and Sons Limited)

Table 63.1 Somatosensory innervation to the ear, neuroanatomical pathways, and relevant pathologies

Part of ear innervated	Nerve	Frequency in case series [3]	Organs from which referred pain may arise	Pathologies to consider (with percentage frequency in case series [3])
Tragus Anterior auricle Anterior wall of external canal Anterior portion of the lateral tympanic membrane	*Auriculotemporal nerve* (branch of the mandibular division of CN V) and *meningeal nerve* (branch of the maxillary division of CN V)	46%	Temporomandibular joint Dentition Face Mastoid Anterior two-thirds of tongue	Temporomandibular joint disease (28%) Dental disease (9%) Trigeminal neuralgia (3%) Mandibular osteomyelitis/tumor (3%) Parotid infection or tumor (3%)
Posterior wall of external auditory canal Posterior lateral surface of tympanic membrane Posterior skin of auricle	*Posterior auricular nerve* (part of tympanic plexus, chorda tympani branch, and an unnamed branch of CN VII)	2%		Vestibular schwannoma (2%) Herpes zoster affecting facial nerve (1%)
Middle ear Upper Eustachian tube Medial surface of tympanic membrane	*Jacobson's nerve and part of tympanic plexus (CN IX)*	7%	Nasal cavity Nasopharynx Paranasal sinuses Palatine tonsils Tonsillar fossa Soft palate Oropharynx Posterior one-third of tongue Infratemporal fossa	Sinusitis (3%) Tonsillitis/pharyngitis (2%) Pharyngeal tumor (1%) Glossopharyngeal neuroma (1%) A separate case series ($n = 615$) found that 3% of patients with a primary complaint of otalgia had a malignant tumor in the infratemporal fossa [7].
Inferior and posterior aspects of the external auditory canal Concavity of concha Lateral surface of tympanic membrane	*Inferior branch of Arnold's nerve* (auricular branch of CN X)	3%	Thyroid Infratemporal fossa Hypopharynx Larynx Trachea Esophagus Lung Abdominal viscera	The following can irritate the superior division of the laryngeal nerve (a branch of CN X) and be referred to the ear: Gastroesophageal (laryngeal/pharyngeal) reflux (2%) Cricopharyngeal spasm (1%) Vagal stimulator (1%) Others: Thyroiditis Thyroid tumor Laryngeal carcinoma
Posterior auricle Skin overlying the mastoid bone and parotid gland	*Greater auricular nerve* (C2 and C3) and *lesser occipital nerve* (C3) roots of the cervical plexus	42%	Cervical spine	*Cervical spine degenerative disease (37%)*, including: Osteoarthritis Cervical facette disease Spondylosis Disc herniation Spinal stenosis Other: Whiplash (2%) Cervical meningioma Arnold-Chiari type I (1%) Vascular (1%) Fibromyalgia (1%) Cervical nerve root cysts (1%)

Table 63.1 summarizes the anatomy, innervation, and sources of referred pain [3, 5], with pathologies drawn from the statistics of a case series of 123 patients [3].

According to these data, most referred ear pain arises in cervical spine disease [5], followed closely by temporomandibular joint dysfunction [8]. Bruxism, even in the absence of temporomandibular joint dysfunction, can also provoke otalgia [9].

Migraine is not localizing in the anatomical sense but is well known to manifest with otalgia [10, 11] or aural fullness [12–18].

Practically, cases of otalgia are generally referred to an otolaryngologist specializing in otology. If there is evidence of (or clinical suspicion for) dental disease, then a dental evaluation is also reasonable.

References

1. Hwa TP, Brant JA. Evaluation and Management of Otalgia. Med Clin North Am. 2021;105(5):813–26.
2. Taziki MH, Behnampour N. A study of the etiology of referred otalgia. Iran J Otorhinolaryngol. 2012;24(69):171–6.
3. Charlett SD, Coatesworth AP. Referred otalgia: a structured approach to diagnosis and treatment. Int J Clin Pract. 2007;61(6):1015–21.
4. Watanabe K, Tubbs RS, Satoh S, Zomorodi AR, Liedtke W, Labidi M, et al. Isolated deep ear canal pain: possible role of auricular branch of Vagus nerve-case illustrations with cadaveric correlation. World Neurosurg. 2016;96:293–301.
5. Jaber JJ, Leonetti JP, Lawrason AE, Feustel PJ. Cervical spine causes for referred otalgia. Otolaryngol Head Neck Surg. 2008;138(4):479–85.
6. Shah RK, Blevins NH. Otalgia. Otolaryngol Clin N Am. 2003;36(6):1137–51.
7. Leonetti JP, Li J, Smith PG. Otalgia. An isolated symptom of malignant infratemporal tumors. Am J Otol. 1998;19(4):496–8.
8. Hernandez-Nuno de la Rosa MF, Keith DA, Siegel NS, Moreno-Hay I. Is there an association between otologic symptoms and temporomandibular disorders?: an evidence-based review. J Am Dent Assoc. 2022;153(11):1096–103.
9. Kim SH. A case of bruxism-induced otalgia. J Audiol Otol. 2016;20(2):123–6.
10. Sussman S, Zimmerman Z, Chishom T, Reid L, Seyyedi M. Migraine-associated otalgia: an underappreciated entity. J Audiol Otol. 2022;26(2):90–6.
11. Teixido M, Seymour P, Kung B, Lazar S, Sabra O. Otalgia associated with migraine. Otol Neurotol. 2011;32(2):322–5.
12. Celebisoy N, Kisabay Ak A, Ozdemir HN, Gokcay F, Durmaz GS, Karti DT, et al. Vestibular migraine, demographic and clinical features of 415 patients: a multicenter study. Clin Neurol Neurosurg. 2022;215:107201.
13. Danno D, Wolf J, Ishizaki K, Kikui S, Hirata K, Takeshima T. Cranial autonomic symptoms in migraine are related to central sensitization: a prospective study of 164 migraine patients at a tertiary headache center. BMC Neurol. 2022;22(1):89.
14. Karsan N, Nagaraj K, Goadsby PJ. Cranial autonomic symptoms: prevalence, phenotype and laterality in migraine and two potentially new symptoms. J Headache Pain. 2022;23(1):18.
15. Moshtaghi O, Ghavami Y, Mahboubi H, Sahyouni R, Haidar Y, Ziai K, et al. Migraine-related aural fullness: a potential clinical entity. Otolaryngol Head Neck Surg. 2018;158(1):100–2.
16. Risbud A, Muhonen EG, Tsutsumi K, Martin EC, Abouzari M, Djalilian HR. Migraine features in patients with isolated aural fullness and proposal for a new diagnosis. Otol Neurotol. 2021;42:1580.
17. Shi S, Wang D, Ren T, Wang W. Auditory manifestations of vestibular migraine. Front Neurol. 2022;13:944001.
18. Tiwari A, Maurya PK, Qavi A, Kulshreshtha D, Thacker AK, Singh AK. Cranial autonomic symptoms in migraine: an observational study. Ann Indian Acad Neurol. 2022;25(4):654–9.

Part VIII

Diagnoses—Congenital Otologic Malformations

Large Vestibular Aqueduct

64

Overview

Large vestibular aqueduct (LVA) is a genetically mediated developmental disorder in which the vestibular aqueduct is larger than normal. Some epidemiological studies report a slight female preponderance. Cases can be unilateral or bilateral. The pathophysiology remains poorly understood. Hearing loss is the most common presenting symptom; most cases present with hearing loss on the same side as the large vestibular aqueduct; in many cases, the contralateral (radiographically normal) ear eventually also loses hearing. The hearing loss often presents in childhood, and some case reports describe it as precipitated by minor head trauma. The hearing loss is progressive, often in a stepwise fashion. Vestibular symptoms are less common, though the reported incidence is broad (2–71%). Physical examination may show conductive hearing loss on Rinne's test. Ocular motor examination may show nystagmus that is nonlocalizing. Audiologic testing usually shows mixed hearing loss, worse in the high frequencies. Tympanometry may show an abnormally low resonance frequency. A variety of vestibular test result abnormalities may be found, including caloric weakness; VEMPs that are absent or unusually large or with unusually low thresholds; an elevated SP/AP ratio on electrocochleography. Interestingly, the correlation between abnormal vestibular test results and the presence of vestibular symptoms is often poor. Most cases of LVA are attributable to one of hundreds of mutations in the SLC26A4 gene, though mutations in other genes (CEVA, FOXi1, GJB2, KCNJ10, POU3F4) have also been reported. High-resolution temporal bone CT remains the imaging study of choice; MRI may provide complementary information. The clinical course is extremely variable, though the overall trajectory tends to be one of deteriorating hearing. Some patients eventually choose to undergo cochlear implantation. Other treatment modalities (dietary sodium restriction; steroids) have not been well studied. If LVA is suspected, then referral to otolaryngology is appropriate.

Introduction

The phrases *"large vestibular aqueduct" (LVA)* and "enlarged vestibular aqueduct" (EVA) are used interchangeably in the literature, sometimes appended with the word "syndrome" (LVAS, EVAS).

The term "enlarged" implies (incorrectly) that the vestibular aqueduct used to be appropriately small, and subsequently became inappropriately large. In fact, embryologically, the vestibular aqueduct begins as large, and during development, it becomes smaller; so really, this abnormality results from the vestibular aqueduct's "developmental failure to become smaller," thus a more correct term might be, "unshrunken vestibular aqueduct," but such terminology is not used.

For this reason, we prefer the more neutral term, *"large vestibular aqueduct" (LVA)*.

Epidemiology

An epidemiological survey conducted in Japan reported that of all patients diagnosed with LVA, 58.2% were female [1], suggesting female preponderance. Interestingly, although females with LVA tend to have larger vestibular aqueducts than males, their hearing loss on average is not worse [2].

LVA is typically present unilaterally, but bilateral cases are not rare [3–5]. An epidemiological study conducted in Japan surprisingly reported that EVA was bilateral in 91.1% of patients [1].

Pathophysiology

The pathophysiology of LVA remains poorly understood. Several studies suggest that it belongs to the class of "third-window phenomena" [6–10]. Biomechanical analyses have been suggested in modeling LVA [11].

M. Cherchi, *Otoneurology and Vestibular Medicine*, https://doi.org/10.1007/978-3-031-94841-1_64

Presentation: Auditory

The most common presentation of LVA is hearing loss, usually (initially) unilateral. An epidemiological survey conducted in Japan reported that 341/380 (89.7%) reported hearing loss as the initial symptom [1].

Of cases that are unilateral, the hearing loss may remain limited to the side with radiographic LVA [12]. However, it is also well reported that patients with unilateral hearing loss ipsilateral to radiographic LVA eventually develop hearing loss in the contralateral (radiographically normal) ear [13, 14].

Presentation: Vestibular

Studies cite a broad range of the prevalence of vestibular symptoms in LVA patients. For example, an epidemiological survey conducted in Japan reported that 34/380 (***8.9%***) of patients with LVA listed "vertigo/dizziness/imbalance" as the initial symptom [1], compared to 89.7% that presented with hearing loss. A series of 27 children (age 3–12 years) reported that 6 (***22%***) complained of "vertigo" [15]. A series of 28 children with bilateral LVA reported that 11 (***39%***) suffered from vertigo attacks [3]. A retrospective chart review of 11 adult and 21 pediatric patients documented the incidence of vestibular symptoms to be ***46%*** in adults and ***48%*** in children [16]. A series of 22 patients with a median age of 8 years (range 6 months–35 years) reported that 14 (***64%***) complained of "dizziness" [17].

Systematic literature reviews report a similarly broad range of vestibular symptoms. For example, a literature review of 17 studies found that the reported rate of vestibular symptoms in patients with LVA ranged from ***2% to 71%*** [18]. Aside from this apparent variability in incidence, "the degree to which vestibular symptoms impact patients with EVA varies significantly" [18].

One report describes the duration of vestibular symptoms as ranging "minutes to hours," and also points out that the onset of vestibular symptoms may be delayed until adulthood [19].

Presentation: Head Trauma

The literature contains numerous reports of minor head trauma precipitating acute hearing loss in patients with LVA [19, 20], or leading to a diagnosis of LVA.

Some investigators report a fairly consistent relationship between head trauma (including minor head trauma) and acute hearing loss in LVA patients [20–24], while others report no relationship [25, 26]. An epidemiologic survey conducted in Japan reported that only 5.3% of patients with LVA reported hearing loss following head trauma [1]. Since the (putatively) precipitating head trauma can be quite mild, this naturally raises the question of whether patients with LVA should restrict their activities, such as by avoiding contact sports [27, 28].

Although the relationship between head trauma and acute hearing loss in patients with LVA remains debated, since hearing loss can impose significant functional limitations (and is not easily treated), most clinicians recommend that patients with LVA refrain from contact sports out of an abundance of caution.

Physical Examination

Rinne's test may show conductive hearing loss on the side affected by LVA.

Ocular motor examination may reveal nystagmus. One study of three patients reported positional nystagmus, unidirectional horizontal nystagmus, and nystagmus reproduced with head turning to the affected side [29].

Testing: Audiologic

Some, though not all [3], infants with LVA are detected on newborn screening.

Most patients with LVA have hearing loss on audiometry that, on average, tends to slope from better (in the low frequencies) to worse (in the high frequencies), and usually has a conductive component, though often the hearing loss is mixed conductive and sensorineural [30]. In practice, almost any pattern can be seen. Zhou and Gopen [8] analyzed audiograms from a series of studied audiograms from several cases of displays several audiograms from patients with LVA, which showed variable patterns of hearing loss.

Figure 64.1 from Zhou and Gopen [8] displays several audiograms from pediatric patients with LVA.

Tympanometry sometimes shows an abnormally low resonance frequency [30].

One study explored the utility of acoustically evoked short latency negative response (part of the brainstem auditory evoked responses) and suggested that this might be helpful in diagnosis [31]. We are not aware of other literature about this.

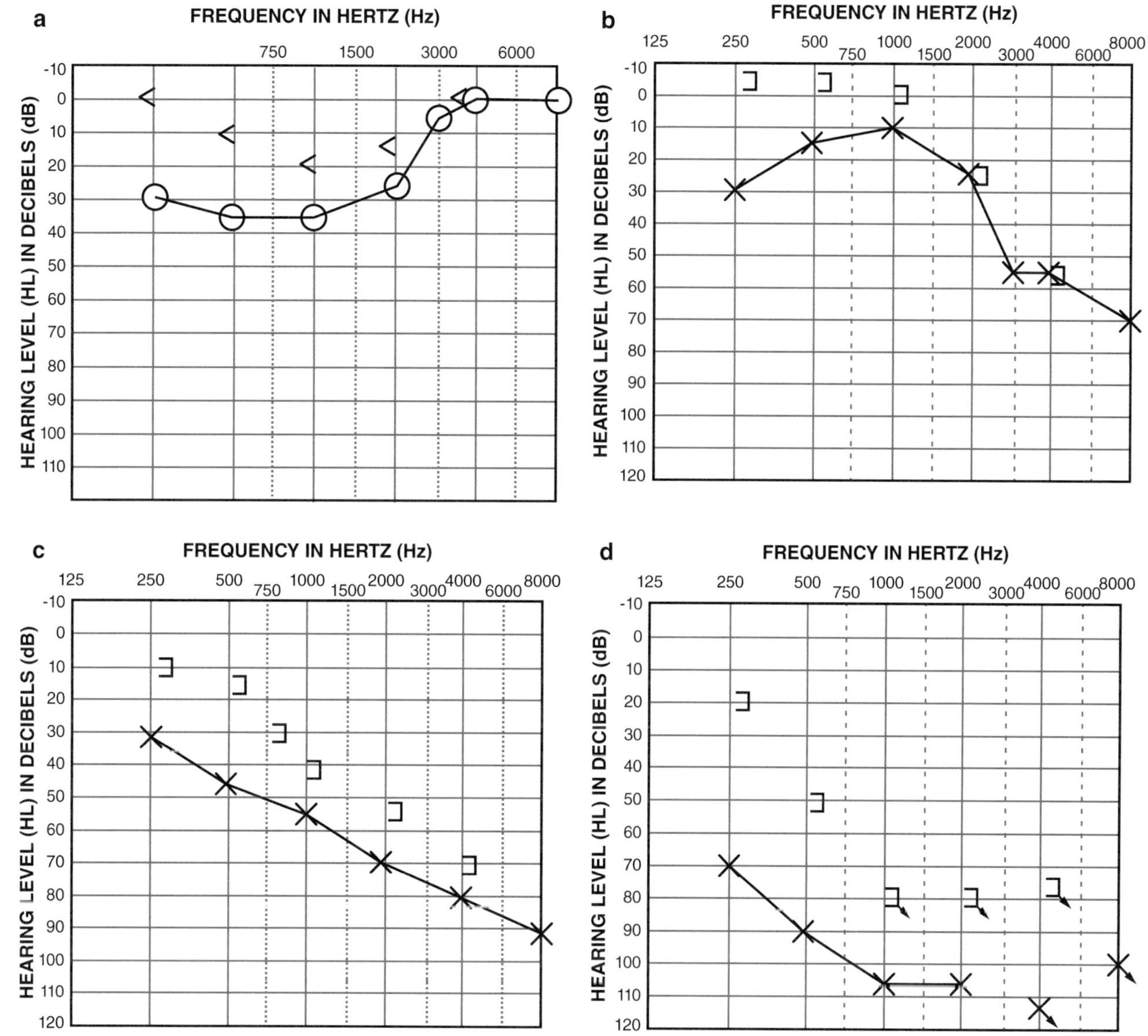

Fig. 64.1 Variations of hearing losses in children with large vestibular aqueduct. Panel (**a**) Low-frequency conductive loss. Panel (**b**) Low-frequency conductive loss plus high-frequency sensorineural loss. Panel (**c**) Sloping mixed hearing loss. Panel (**d**) Profound sensorineural loss with residual hearing at low frequencies. If tested, bone conduction at 250 and 500 Hz usually are still evident. (From Zhou and Gopen [8]. Reproduced with permission from John Wiley and Sons Limited)

Testing: Vestibular

Relatively few studies have specifically explored vestibular function tests in patients with LVA.

One study of 31 LVA patients reported unilateral *caloric weakness* in 45% and bilateral caloric weakness in 6% of patients [32]. Another study of 10 LVA patients reported unilateral caloric weakness in 30% [33].

Several case series studied LVA patients with *rotatory chair testing* (RCT) but did not report any sensitive or specific diagnostic findings [15, 34]. Of 24 patients who underwent RCT, 4 (17%) showed decreased gain and 6 (25%) showed phase lead.

Several investigations have studies *vestibular evoked myogenic potentials* (VEMPs) in LVA patients. One study of vestibular evoked myogenic potentials in 29 bilateral LVA patients reported that in children ocular vestibular evoked myogenic potentials were present in 93% and cervical vestibular evoked myogenic potentials were present in 87%, whereas in adults ocular vestibular evoked myogenic potentials were present in in 83% and cervical vestibular evoked myogenic potentials were present in 75%; the authors concluded that LVA caused "a severer otolithic dysfunction in adults than in children" [5]. One study of children with LVA prior to cochlear implantation reported that compared to controls, the LVA patients exhibited, "shorter N1 latencies, shorter P1 latencies, and higher amplitudes" [35]. One study checked cervical vestibular evoked myogenic potentials on 9 patients and reported abnormalities in 2 (22%); specifically "1 ear from a patient with bilateral EVA had no measurable cVEMP response up to 107 dB nHL, and another ear from a patient with unilateral EVA had an abnormal amplitude symmetry ratio of 51% with a larger cVEMP amplitude in the ear affected with EVA" [34]. A case study of a 42-year-old female with bilateral LVA reported that, "oVEMP reflexes demonstrated pathologically increased amplitudes and reduced thresholds for low-frequency tone bursts. Cervical VEMP amplitudes and thresholds were within normal limits for both ears across all frequencies of stimulation" [36]. A retrospective cohort series studied 25 pediatric patients (37 ears) with cervical vestibular evoked myogenic potentials and found that hearing loss (with an air–bone gap) was present in 36 (97%) of ears; furthermore, *despite the conductive component of hearing loss,* "Abnormally low VEMP thresholds were found in 92% (34/37) of the ears with EVA. In addition, higher amplitudes were also noted in these cases. Statistically, the VEMP thresholds recorded from ears with EVA were significantly lower than that from non-EVA ears or normal ones" [8].

Figure 64.2 from Zhou and Gopen [8] shows cervical vestibular evoked myogenic potentials (cVEMPs) from a patient with left-sided LVA.

One series of 5 pediatric patients studied *electrocochleography* (ECoG) and identified an elevated summation potential (and thus an elevated SP/AP ratio) in 2 (40%).

A few studies have explored *video head impulse testing* (vHIT) in LVA patients. A retrospective case–control study of 16 children (mean age 10.3 ± 4.4 years) with LVA before and after cochlear implantation reported that, "Statistically significant differences were not found in VOR gains of all three semicircular canals before and after surgery" [35].

Another study noted a discrepancy between the results of caloric testing and video head impulse testing. This retrospective case series of 10 patients with LVA reported that, "40% showed unilateral caloric weakness and 30% had recurrent episodes of vertigo. All patients with recurrent vertigo had caloric weakness. However, only 25% of patients with abnormal caloric test results had abnormal vHIT results" [33]. The authors noted that, "The discrepancy between the caloric test and vHIT results was also found in patients with MD." The authors are referring to the fact that studies of vestibular tests in patients with Ménière's disease sometimes report a dissociation between caloric testing (which may show unilateral caloric weakness) and video head impulse testing (which may be normal) [37, 38].

A literature review of 17 studies found that the reported rate of abnormalities among vestibular tests (physical examination maneuvers, caloric testing, electronystagmography, vestibular evoked myogenic potentials) ranged from ***7% to 92%*** [18].

A prospective cross-sectional cohort study evaluated 106 LVA patients with several vestibular tests (videonystagmography, cervical vestibular evoked myogenic potentials and rotatory chair testing) and reported that, "45% had vestibular signs and symptoms, and 44% of tested patients had abnormal VNG test results," further observing that, "An increased number of vestibular signs and symptoms was correlated with the presence of bilateral EVA" [34].

Significantly, some studies report that the correlation between abnormal vestibular test results and the presence of vestibular symptoms is poor. For example, a prospective cohort study of 27 children (age 3–12 years) found abnormalities in vestibular tests (cervical vestibular evoked myogenic potentials, caloric testing, rotatory chair testing and posturography) in 24 (89%), yet only 6 (22%) actually endorsed a complaint of "vertigo," leading the investigators to conclude that, "We found a high rate of vestibular pathology in children with EVA; however, the prevalence of abnormal vestibular testing findings in this patient population was

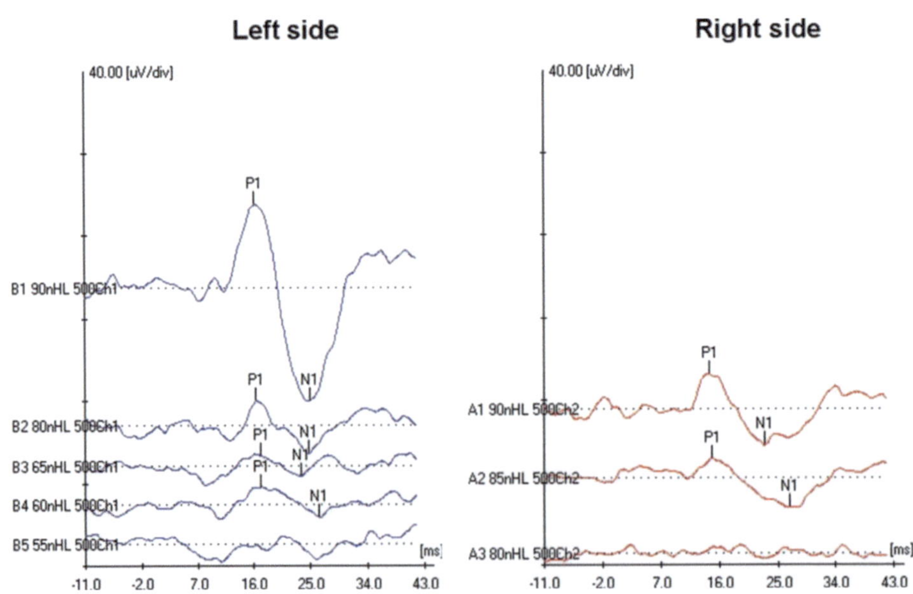

Fig. 64.2 Cervical vestibular evoked myogenic potentials (cVEMP) from a child with left-sided large vestibular aqueduct. The patient's left cVEMP response (shown in the left panel) exhibits a lower threshold (60 dB) and larger amplitude than that of the patients right cVEMP response (shown in the right panel). (From Zhou and Gopen [8]. Reproduced with permission from John Wiley and Sons Limited)

not correlated with vestibular symptoms" [15]. Conversely, a prospective cross-sectional cohort study assessed 106 patients with several vestibular tests (videonystagmography, cervical vestibular evoked myogenic potentials, rotatory chair testing) and observed that "44% of tested patients abnormal VNG test results," but further observed that, "not all patients with vestibular signs and symptoms have abnormal vestibular test results" [34].

Testing: Genetic

Most, though not all [39], cases of LVA have an identifiable genetic abnormality.

Several hundred mutations in the SLC26A4 gene have been identified in patients with LVA. These mutations have been identified in various ethnic groups, including Iranian [40], Czech [41], Austrian [42], and Chinese [43–68].

In other patients with LVA, mutations have been identified in CEVA [69–71], FOXI1 [72, 73], GJB2 [74], KCNJ10 [59, 72, 73], and POU3F4 [74].

Cases in which no genetic mutation is identified are few, and it seems likely that even these will eventually be shown to have identifiable mutations.

Imaging: CT

High-resolution temporal bone CT is usually regarded as the imaging study of choice in identifying and characterizing LVA [75–80]. There are debates regarding the specific CT technique and protocol [81, 82].

Figure 64.3 from Zhou and Gopen [8] shows an axial CT image from a child with a large vestibular aqueduct on the patient's left.

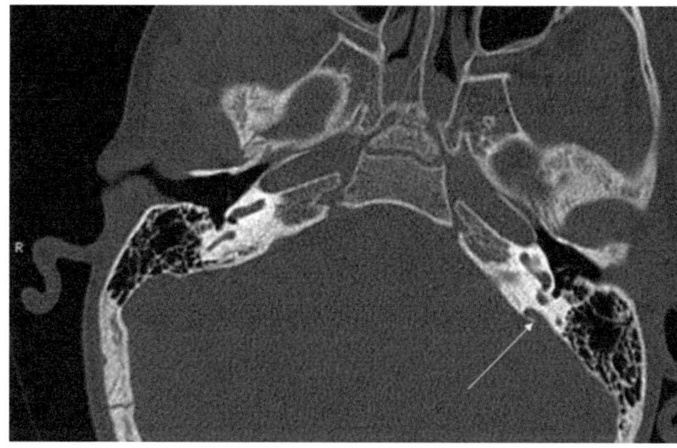

Fig. 64.3 Left large vestibular aqueduct in axial CT of the temporal bone in a child with hearing loss. (From Zhou and Gopen [8]. Reproduced with permission from John Wiley and Sons Limited)

Imaging: MRI

MRI has not (yet) surpassed CT in its ability to identify LVA, but in the interest of avoiding the ionizing radiation exposure of a temporal bone CT, some studies have explored whether MRI might serve as a screening test [83]. Some investigators also advocate the use of MRI because of its superior characterization of soft tissue [84].

There is modest literature regarding whether discrepancies between measurements of the vestibular aqueduct's bony labyrinth and membranous labyrinth might have value in diagnosing LVA [85].

Figure 64.4 from Sarioglu and colleagues [83] juxtaposes CT images (top panels) with MR images (bottom panels).

MRI may also be better than CT at diagnosing other otologic diseases and malformations with which LVA may be comorbid.

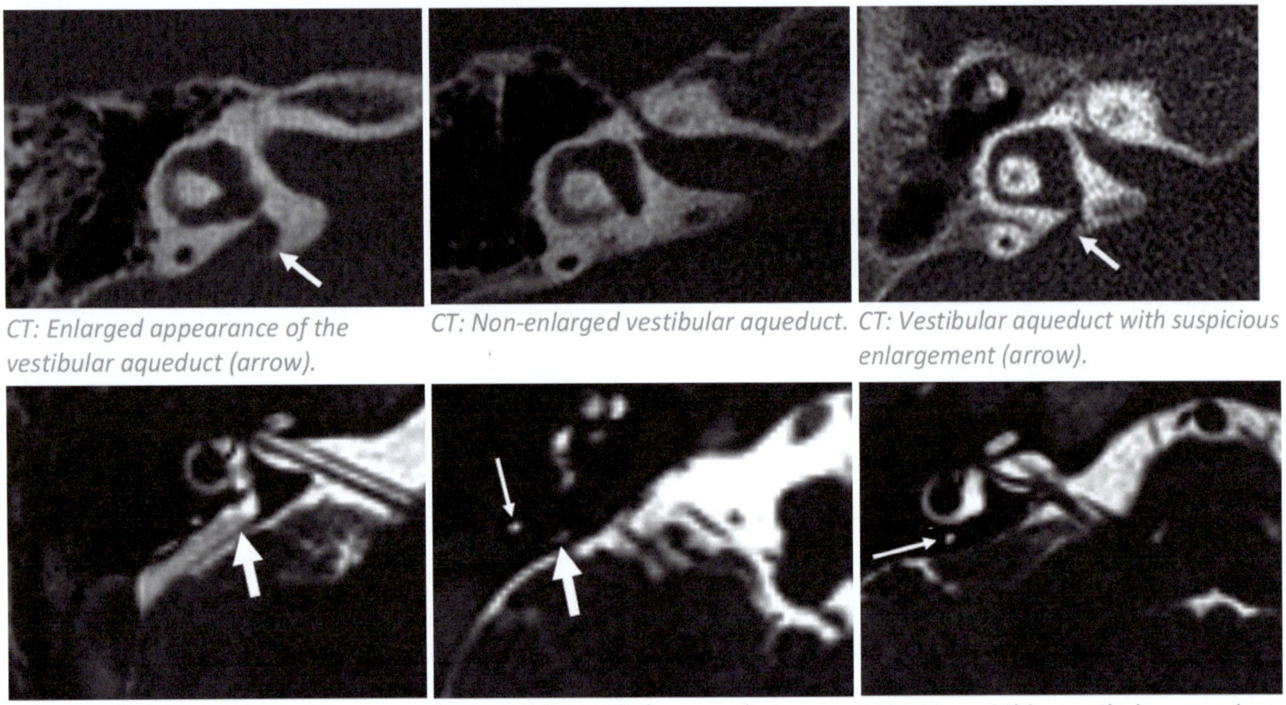

CT: Enlarged appearance of the vestibular aqueduct (arrow).

CT: Non-enlarged vestibular aqueduct.

CT: Vestibular aqueduct with suspicious enlargement (arrow).

MRI: Vestibular aqueduct with enlarged appearance (arrow).

MRI: Visible vestibular aqueduct without an enlarged appearance (thick arrow). Posterior semicircular canal (thin arrow).

MRI: Non-visible vestibular aqueduct. Posterior semicircular canal (arrow).

Fig. 64.4 Comparison of CT (top panel) and MRI (bottom panel) of enlarged and nonenlarged vestibular aqueducts and neighboring structures. (From Sarioglu and colleagues [83]. Reproduced from Galenos Publishing House with permission under CC-BY 4.0)

Correlations Between Workup and Clinical Features

The literature provides discrepant data regarding the correlation between genetics, imaging, audiologic testing and clinical features.

Some studies report some correlation between genotype and degree of hearing loss [86] while others do not [87].

Some studies report a correlation between the genetic heterogeneity and the complexity of inner ear malformations [88].

Most studies report a correlation between various imaging parameters and the degree of hearing loss [7, 26, 89–93], though not all studies reach that conclusion [2, 94].

The clinical course is characterized by significant fluctuations in hearing [95, 96], and this fluctuation may underlie the discrepant results regarding correlation between the genetic, imaging, audiologic and clinical features of the disease. One group concluded, "Our study failed to uncover factors that might be predictive of hearing loss. We conclude that until a better understanding of the natural history and pathophysiologic condition of EVAs is achieved, there is no surgical or other intervention that can be demonstrated as being efficacious" [97].

Differential Diagnosis

Because LVA presents with conductive (or mixed) hearing loss, it is sometimes mistaken for otosclerosis [98, 99].

Comorbidities: LVA Plus Other Otologic Disease

LVA sometimes co-occurs with other inner ear malformations, such as cochlear dysplasia, vestibulocochlear dysplasia, modiolar hypoplasia [100], dilated dysplastic vestibule [101], enlargement of the vestibule [102], shortening of the common crus [103], enlargement of the lateral semicircular canal [104], and superior semicircular canal dehiscence [105]. Whether these additional inner ear malformations are merely associated with LVA, or whether there is a causal relationship, is unclear [106].

Some patients with LVA also develop benign paroxysmal positional vertigo (BPPV). A retrospective review of 26 patients with a radiological diagnosis of LVA reported that 5 (19%) of patients had nystagmus compatible with BPPV [107]. A retrospective review of 22 LVA patients (age 6 months–35 years) reported that 4 (18%) developed BPPV

[17]. Some investigators relate the increased risk of recurrent BPPV to volumetric abnormalities of the vestibular aqueduct on imaging [108].

Comorbidities: LVA Plus Other Nonotologic Disease

A survey of parents of 133 children reported a tenfold greater risk of developing torticollis than the general population [109].

Prognosis

The hearing loss in LVA is quite variable and fluctuates [95 97], though the overall trajectory is usually one of decline [110–112].

Predicting that overall trajectory of decline in hearing has proven difficult. One study reported that "the standard deviation of hearing at 500 Hz was the only prognostic factor for the progression of PTA [pure tone average]" [113]. Another study reported that several features can help predict progression of hearing loss, including male gender, operculum size, hearing at the time of diagnosis, and presence of an incomplete partition type 2 [114].

The clinical course tends to be punctuated by abrupt declines in hearing. Often, these declines occur without any clear precipitant, while others are sometimes suspected to be due to head trauma. The mechanism of the abrupt hearing loss (with or without head trauma) is unknown. Hemorrhage has been suggested [115].

Treatment: Cochlear Implantation

Cochlear implantation is the most commonly performed procedure for treating the hearing loss associated with LVA [116–130].

Treatment: Other Modalities

Dietary sodium restriction has been advocated [131] and incurs no significant medical risk.

Treatment with corticosteroids has been explored. Transtympanic steroids have been tried in LVA patients who suffered acute hearing loss following head trauma [28]. Steroids have also been tried in LVA patients without head trauma [132].

References

1. Noguchi Y, Fukuda S, Fukushima K, Gyo K, Hara A, Nakashima T, et al. A nationwide study on enlargement of the vestibular aqueduct in Japan. Auris Nasus Larynx. 2017;44(1):33–9.
2. Ruthberg J, Ascha MS, Kocharyan A, Gupta A, Murray GS, Megerian CA, et al. Sex-specific enlarged vestibular aqueduct morphology and audiometry. Am J Otolaryngol. 2019;40(4):473–7.
3. Nakano A, Arimoto Y, Mutai H, Nara K, Inoue S, Matsunaga T. Clinical and genetic analysis of children with hearing loss and bilateral enlarged vestibular aqueducts. Int J Pediatr Otorhinolaryngol. 2022;152:110975.
4. Subramaniam S, Tan TY, Yuen HW. Bilateral enlarged vestibular aqueduct with associated bilateral Mondini's dysplasia. Am J Otolaryngol. 2012;33(4):455–6.
5. Zhang Y, Chen Z, Zhang Y, Hu J, Wang J, Xu M, et al. Vestibular-evoked myogenic potentials in patients with large vestibular aqueduct syndrome. Acta Otolaryngol. 2020;140(1):40–5.
6. Riggs WJ, Hiss MM, Varadarajan VV, Mattingly JK, Adunka OF. Enlarged vestibular aqueduct: Intraoperative electrocochleography findings during cochlear implantation. Int J Pediatr Otorhinolaryngol. 2020;134:110065.
7. Seo YJ, Kim J, Choi JY. Correlation of vestibular aqueduct size with air-bone gap in enlarged vestibular aqueduct syndrome. Laryngoscope. 2016;126(7):1633–8.
8. Zhou G, Gopen Q. Characteristics of vestibular evoked myogenic potentials in children with enlarged vestibular aqueduct. Laryngoscope. 2011;121(1):220–5.
9. Mimura T, Sato E, Sugiura M, Yoshino T, Naganawa S, Nakashima T. Hearing loss in patients with enlarged vestibular aqueduct: air-bone gap and audiological Bing test. Int J Audiol. 2005;44(8):466–9.
10. Sheykholeslami K, Schmerber S, Habiby Kermany M, Kaga K. Vestibular-evoked myogenic potentials in three patients with large vestibular aqueduct. Hear Res. 2004;190(1–2):161–8.
11. Han JW, Wang L, Zhao H, Yang SM. Biomechanical analysis of the clinical characteristics of enlarged vestibular aqueduct syndrome with Mondini malformation. Acta Otolaryngol. 2020;140(10):813–7.
12. Macielak RJ, Mattingly JK, Findlen UM, Moberly AC, Malhotra PS, Adunka OF. Audiometric findings in children with unilateral enlarged vestibular aqueduct. Int J Pediatr Otorhinolaryngol. 2019;120:25–9.
13. van Beeck Calkoen EA, Pennings RJE, Smits J, Pegge S, Rotteveel LJC, Merkus P, et al. Contralateral hearing loss in children with a unilateral enlarged vestibular aqueduct. Int J Pediatr Otorhinolaryngol. 2021;150:110891.
14. Greinwald J, DeAlarcon A, Cohen A, Uwiera T, Zhang K, Benton C, et al. Significance of unilateral enlarged vestibular aqueduct. Laryngoscope. 2013;123(6):1537–46.
15. Yang CJ, Lavender V, Meinzen-Derr JK, Cohen AP, Youssif M, Castiglione M, et al. Vestibular pathology in children with enlarged vestibular aqueduct. Laryngoscope. 2016;126(10):2344–50.
16. Grimmer JF, Hedlund G. Vestibular symptoms in children with enlarged vestibular aqueduct anomaly. Int J Pediatr Otorhinolaryngol. 2007;71(2):275–82.
17. Song JJ, Hong SK, Lee SY, Park SJ, Kang SI, An YH, et al. Vestibular manifestations in subjects with enlarged vestibular aqueduct. Otol Neurotol. 2018;39(6):e461–e7.
18. Stahl MC, Otteson T. Systematic review on vestibular symptoms in patients with enlarged vestibular aqueducts. Laryngoscope. 2022;132(4):873–80.

19. Oh AK, Ishiyama A, Baloh RW. Vertigo and the enlarged vestibular aqueduct syndrome. J Neurol. 2001;248(11):971–4.
20. Walsh RM, Ayshford CA, Chavda SV, Proops DW. Large vestibular aqueduct syndrome. ORL J Otorhinolaryngol Relat Spec. 1999;61(1):41–4.
21. Kou B, Macdonald R. Toronto's Hospital for Sick Children study of traumatic sudden sensorineural hearing loss. J Otolaryngol. 1998;27(2):64–8.
22. Noordman BJ, van Beeck Calkoen E, Witte B, Goverts T, Hensen E, Merkus P. Prognostic factors for sudden drops in hearing level after minor head injury in patients with an enlarged vestibular aqueduct: a meta-analysis. Otol Neurotol. 2015;36(1):4–11.
23. Steinbach S, Brockmeier SJ, Kiefer J. The large vestibular aqueduct--case report and review of the literature. Acta Otolaryngol. 2006;126(8):788–95.
24. Wendt WJ, Hashikawa A. Enlarged vestibular aqueduct syndrome: sudden hearing loss in a child with a cerebral shunt. Pediatr Emerg Care. 2019;35(7):e135–e7.
25. Alemi AS, Chan DK. Progressive hearing loss and head trauma in enlarged vestibular aqueduct: a systematic review and meta-analysis. Otolaryngol Head Neck Surg. 2015;153(4):512–7.
26. Madden C, Halsted M, Benton C, Greinwald J, Choo D. Enlarged vestibular aqueduct syndrome in the pediatric population. Otol Neurotol. 2003;24(4):625–32.
27. Brodsky JR, Choi SS. Should children with an enlarged vestibular aqueduct be restricted from playing contact sports? Laryngoscope. 2018;128(10):2219–20.
28. Ralli M, Nola G, Sparvoli L, Ralli G. Unilateral enlarged vestibular aqueduct syndrome and bilateral endolymphatic hydrops. Case Rep Otolaryngol. 2017;2017:6195317.
29. White J, Krakovitz P. Nystagmus in enlarged vestibular aqueduct: a case series. Audiol Res. 2015;5(1):120.
30. Sato E, Nakashima T, Lilly DJ, Fausti SA, Ueda H, Misawa H, et al. Tympanometric findings in patients with enlarged vestibular aqueducts. Laryngoscope. 2002;112(9):1642–6.
31. Liu L, Yang B. Acoustically evoked short latency negative responses in hearing loss patients with enlarged vestibular aqueduct. Acta Neurol Belg. 2013;113(2):157–60.
32. Jung J, Seo YW, Choi JY, Kim SH. Vestibular function is associated with residual low-frequency hearing loss in patients with bi-allelic mutations in the SLC26A4 gene. Hear Res. 2016;335:33–9.
33. Jung J, Suh MJ, Kim SH. Discrepancies between video head impulse and caloric tests in patients with enlarged vestibular aqueduct. Laryngoscope. 2017;127(4):921–6.
34. Zalewski CK, Chien WW, King KA, Muskett JA, Baron RE, Butman JA, et al. Vestibular dysfunction in patients with enlarged vestibular aqueduct. Otolaryngol Head Neck Surg. 2015;153(2):257–62.
35. Wang R, Zhang D, Luo J, Chao X, Xu J, Liu X, et al. Influence of cochlear implantation on vestibular function in children with an enlarged vestibular aqueduct. Front Neurol. 2021;12:663123.
36. Taylor RL, Bradshaw AP, Magnussen JS, Gibson WP, Halmagyi GM, Welgampola MS. Augmented ocular vestibular evoked myogenic potentials to air-conducted sound in large vestibular aqueduct syndrome. Ear Hear. 2012;33(6):768–71.
37. McGarvie LA, Curthoys IS, MacDougall HG, Halmagyi GM. What does the dissociation between the results of video head impulse versus caloric testing reveal about the vestibular dysfunction in Meniere's disease? Acta Otolaryngol. 2015;135(9):859–65.
38. Rubin F, Simon F, Verillaud B, Herman P, Kania R, Hautefort C. Comparison of video head impulse test and caloric reflex test in advanced unilateral definite Meniere's disease. Eur Ann Otorhinolaryngol Head Neck Dis. 2018;135(3):167–9.
39. Claros P, Fokouo JV, Claros A. Cochlear implantation in patients with enlarged vestibular aqueduct. A case series with literature review. Cochlear Implants Int. 2017;18(3):125–9.
40. Yazdanpanahi N, Chaleshtori MH, Tabatabaiefar MA, Noormohammadi Z, Farrokhi E, Najmabadi H, et al. Two novel SLC26A4 mutations in Iranian families with autosomal recessive hearing loss. Int J Pediatr Otorhinolaryngol. 2012;76(6):845–50.
41. Pourova R, Janousek P, Jurovcik M, Dvorakova M, Malikova M, Raskova D, et al. Spectrum and frequency of SLC26A4 mutations among Czech patients with early hearing loss with and without Enlarged Vestibular Aqueduct (EVA). Ann Hum Genet. 2010;74(4):299–307.
42. Roesch S, Bernardinelli E, Nofziger C, Toth M, Patsch W, Rasp G, et al. Functional testing of SLC26A4 variants-clinical and molecular analysis of a cohort with enlarged vestibular aqueduct from Austria. Int J Mol Sci. 2018;19(1):209.
43. Chai Y, Huang Z, Tao Z, Li X, Li L, Li Y, et al. Molecular etiology of hearing impairment associated with nonsyndromic enlarged vestibular aqueduct in East China. Am J Med Genet A. 2013;161A(9):2226–33.
44. Guo YF, Wang YL, Xu BC, Liu XW, Zhu YM, Zhao FF, et al. Identification of two novel mutations, c.232T>C and c.2006A>T, in SLC26A4 in a Chinese family associated with enlarged vestibular aqueduct. Int J Pediatr Otorhinolaryngol. 2010;74(7):831–5.
45. He X, Peng Q, Li S, Zhu P, Wu C, Rao C, et al. A novel missense mutation in the SLC26A4 gene causes nonsyndromic hearing loss and enlarged vestibular aqueduct. Int J Pediatr Otorhinolaryngol. 2017;95:104–8.
46. He X, Zhao S, Shi L, Lu Y, Yang Y, Zhang X. Compound heterozygous variants of the SLC26A4 gene in a Chinese family with enlarged vestibular aqueducts. BMC Med Genet. 2022;15(1):152.
47. Hu H, Wu L, Feng Y, Pan Q, Long Z, Li J, et al. Molecular analysis of hearing loss associated with enlarged vestibular aqueduct in the mainland Chinese: a unique SLC26A4 mutation spectrum. J Hum Genet. 2007;52(6):492–7.
48. Lai R, Hu P, Zhu F, Zhu G, Vivero R, Peng A, et al. Genetic diagnosis and cochlear implantation for patients with nonsyndromic hearing loss and enlarged vestibular aqueduct. J Laryngol Otol. 2012;126(4):349–55.
49. Li Y, Zhu B, Su J, Yin Y, Yu F. Identification of SLC26A4 mutations p.L582LfsX4, p.I188T and p.E704K in a Chinese family with large vestibular aqueduct syndrome (LVAS). Int J Pediatr Otorhinolaryngol. 2016;91:1–5.
50. Liang Y, Peng Q, Wang K, Zhu P, Wu C, Rao C, et al. A novel mutation in the SLC26A4 gene in a Chinese family with non-syndromic hearing loss and enlarged vestibular aqueduct. Int J Pediatr Otorhinolaryngol. 2018;107:97–100.
51. Liu Y, Wang L, Feng Y, He C, Liu D, Cai X, et al. A new genetic diagnostic for enlarged vestibular aqueduct based on next-generation sequencing. PLoS One. 2016;11(12):e0168508.
52. Pang X, Chai Y, Chen P, He L, Wang X, Wu H, et al. Mono-allelic mutations of SLC26A4 is over-presented in deaf patients with non-syndromic enlarged vestibular aqueduct. Int J Pediatr Otorhinolaryngol. 2015;79(8):1351–3.
53. Pang X, Chai Y, He L, Chen P, Wang X, Li L, et al. A 7666-bp genomic deletion is frequent in Chinese Han deaf patients with non-syndromic enlarged vestibular aqueduct but without bi-allelic SLC26A4 mutations. Int J Pediatr Otorhinolaryngol. 2015;79(12):2248–52.
54. Tian Y, Xu H, Liu D, Zhang J, Yang Z, Zhang S, et al. Increased diagnosis of enlarged vestibular aqueduct by multiplex PCR enrichment and next-generation sequencing of the SLC26A4 gene. Mol Genet Genomic Med. 2021;9(8):e1734.
55. Wang M, Zhang F, Xu L, Xiao Y, Li J, Fan Z, et al. Novel compound heterozygous mutations in SLC26A4 gene in a Chinese Han family with enlarged vestibular aqueduct. Int J Pediatr Otorhinolaryngol. 2016;90:170–4.

56. Wang QJ, Zhao YL, Rao SQ, Guo YF, Yuan H, Zong L, et al. A distinct spectrum of SLC26A4 mutations in patients with enlarged vestibular aqueduct in China. Clin Genet. 2007;72(3):245–54.
57. Wen C, Wang S, Zhao X, Wang X, Wang X, Cheng X, et al. Mutation analysis of the SLC26A4 gene in three Chinese families. Biosci Trends. 2019;13(5):441–7.
58. Wu T, Cui L, Mou Y, Guo W, Liu D, Qiu J, et al. A newly identified mutation (c.2029 C > T) in SLC26A4 gene is associated with enlarged vestibular aqueducts in a Chinese family. BMC Med Genet. 2022;15(1):49.
59. Yang T, Gurrola JG 2nd, Wu H, Chiu SM, Wangemann P, Snyder PM, et al. Mutations of KCNJ10 together with mutations of SLC26A4 cause digenic nonsyndromic hearing loss associated with enlarged vestibular aqueduct syndrome. Am J Hum Genet. 2009;84(5):651–7.
60. Yao G, Li S, Chen D, Wang H, Zhang J, Feng Z, et al. Compound heterozygous mutations of SLC26A4 in 4 Chinese families with enlarged vestibular aqueduct. Int J Pediatr Otorhinolaryngol. 2013;77(4):544–9.
61. Yao J, Qian X, Bao J, Wei Q, Lu Y, Zheng H, et al. Probing the effect of two heterozygous mutations in codon 723 of SLC26A4 on deafness phenotype based on molecular dynamics simulations. Sci Rep. 2015;5:10831.
62. Yu Y, Yang Y, Lu J, Jin Y, Yang Y, Hong E, et al. Two compound heterozygous were identified in SLC26A4 gene in two Chinese families with enlarged vestibular aqueduct. Clin Exp Otorhinolaryngol. 2019;12(1):50–7.
63. Yuan Y, Guo W, Tang J, Zhang G, Wang G, Han M, et al. Molecular epidemiology and functional assessment of novel allelic variants of SLC26A4 in non-syndromic hearing loss patients with enlarged vestibular aqueduct in China. PLoS One. 2012;7(11):e49984.
64. Zhang F, Bai X, Xiao Y, Zhang X, Zhang G, Li J, et al. Identification of a novel mutation in SLC26A4 gene in a Chinese family with enlarged vestibular aqueduct syndrome. Int J Pediatr Otorhinolaryngol. 2016;85:75–9.
65. Zhao J, Yuan Y, Chen J, Huang S, Wang G, Han D, et al. SLC26A4 gene copy number variations in Chinese patients with non-syndromic enlarged vestibular aqueduct. J Transl Med. 2012;10:82.
66. Zhao X, Cheng X, Huang L, Wang X, Wen C, Wang X. Novel compound heterozygous mutations in SLC26A4 gene in a Chinese family with enlarged vestibular aqueduct. Biosci Trends. 2018;12(5):502–6.
67. Zhou K, Huang L, Feng M, Li X, Zhao Y, Liu F, et al. A novel SLC26A4 splicing mutation identified in two deaf Chinese twin sisters with enlarged vestibular aqueducts. Mol Genet Genomic Med. 2020;8(10):e1447.
68. Zhu GJ, Shi LS, Zhou H, Yang Y, Chen J, Gao X. A novel compound heterozygous mutation of SLC26A4 in two Chinese families with nonsyndromic hearing loss and enlarged vestibular aqueducts. Mol Med Rep. 2017;16(6):9011–6.
69. Chao JR, Chattaraj P, Munjal T, Honda K, King KA, Zalewski CK, et al. SLC26A4-linked CEVA haplotype correlates with phenotype in patients with enlargement of the vestibular aqueduct. BMC Med Genet. 2019;20(1):118.
70. Chattaraj P, Munjal T, Honda K, Rendtorff ND, Ratay JS, Muskett JA, et al. A common SLC26A4-linked haplotype underlying nonsyndromic hearing loss with enlargement of the vestibular aqueduct. J Med Genet. 2017;54(10):665–73.
71. Smits JJ, de Bruijn SE, Lanting CP, Oostrik J, O'Gorman L, Mantere T, et al. Exploring the missing heritability in subjects with hearing loss, enlarged vestibular aqueducts, and a single or no pathogenic SLC26A4 variant. Hum Genet. 2022;141(3–4):465–84.
72. Liu Y, Wen J, Sang S, Mei L, He C, Jiang L, et al. Next-generation sequencing-based mutation analysis of genes associated with enlarged vestibular aqueduct in Chinese families. Eur Arch Otorrinolaringol. 2020;277(12):3331–9.
73. Song MH, Shin JW, Park HJ, Lee KA, Kim Y, Kim UK, et al. Intrafamilial phenotypic variability in families with biallelic SLC26A4 mutations. Laryngoscope. 2014;124(5):E194–202.
74. Roesch S, Rasp G, Sarikas A, Dossena S. Genetic determinants of non-syndromic enlarged vestibular aqueduct: a review. Audiol Res. 2021;11(3):423–42.
75. Chen JX, Kachniarz B, Shin JJ. Diagnostic yield of computed tomography scan for pediatric hearing loss: a systematic review. Otolaryngol Head Neck Surg. 2014;151(5):718–39.
76. Gao B, Jiang Y, Huang S, Wang Q, Gao X, Cai J, et al. Application of multiplanar reconstruction of spiral CT in the diagnosis and treatment of enlarged vestibular aqueducts. Acta Otolaryngol. 2019;139(8):665–70.
77. Ma H, Han P, Liang B, Tian ZL, Lei ZQ, Kong WJ, et al. Multislice spiral computed tomography imaging in congenital inner ear malformations. J Comput Assist Tomogr. 2008;32(1):146–50.
78. Maiolo V, Savastio G, Modugno GC, Barozzi L. Relationship between multidetector CT imaging of the vestibular aqueduct and inner ear pathologies. Neuroradiol J. 2013;26(6):683–92.
79. Swartz JD, Yussen PS, Mandell DW, Mikaelian DO, Berger AS, Wolfson RJ. The vestibular aqueduct syndrome: computed tomographic appearance. Clin Radiol. 1985;36(3):241–3.
80. Wang L, Qin Y, Zhu L, Li X, Chen Y, Zhang L. Auditory and imaging markers of atypical enlarged vestibular aqueduct. Eur Arch Otorrinolaringol. 2022;279(2):695–702.
81. Hwang M, Marovich R, Shin SS, Chi D, Branstetter BF. Optimizing CT for the evaluation of vestibular aqueduct enlargement: inter-rater reproducibility and predictive value of reformatted CT measurements. J Otol. 2015;10(1):13–7.
82. Weiss NM, Breitsprecher TM, Pscheidl A, Bachinger D, Volkenstein S, Dazert S, et al. Volumetry improves the assessment of the vestibular aqueduct size in inner ear malformation. Eur Arch Otorrinolaringol. 2023;280:2155–63.
83. Sarioglu FC, Cetin AC, Guleryuz H, Guneri EA. The diagnostic efficacy of MRI in the evaluation of the enlarged vestibular aqueduct in children with hearing loss. Turk Arch Otorhinolaryngol. 2020;58(4):220–6.
84. Yang L, Liu J. Comparative analysis of CT and MRI diagnosis of large vestibular aqueduct syndrome (LVAS) in children. J Coll Physicians Surg Pak. 2019;29(8):753–6.
85. Sun W, Liang Q, Kuang S, Zhou S, Wang W. 3D-real IR MRI detects serendipity of inner ear in enlarged vestibular aqueduct syndrome. Acta Otolaryngol. 2019;139(3):233–7.
86. Zhao FF, Lan L, Wang DY, Han B, Qi Y, Zhao Y, et al. Correlation analysis of genotypes, auditory function, and vestibular size in Chinese children with enlarged vestibular aqueduct syndrome. Acta Otolaryngol. 2013;133(12):1242–9.
87. Aimoni C, Ciorba A, Cerritelli L, Ceruti S, Skarzynski PH, Hatzopoulos S. Enlarged vestibular aqueduct: audiological and genetical features in children and adolescents. Int J Pediatr Otorhinolaryngol. 2017;101:254–8.
88. Azaiez H, Yang T, Prasad S, Sorensen JL, Nishimura CJ, Kimberling WJ, et al. Genotype-phenotype correlations for SLC26A4-related deafness. Hum Genet. 2007;122(5):451–7.
89. Anand R, Stahl M, Hicks K, Murray G, Patel N, Gupta A, et al. Assessing the clinical utility of volumetric HRCT in pediatric enlarged vestibular aqueduct related hearing loss. Int J Pediatr Otorhinolaryngol. 2022;155:111067.
90. Archibald HD, Ascha M, Gupta A, Megerian C, Otteson T. Hearing loss in unilateral and bilateral enlarged vestibular aqueduct syndrome. Int J Pediatr Otorhinolaryngol. 2019;118:147–51.
91. Ascha MS, Manzoor N, Gupta A, Semaan M, Megerian C, Otteson TD. Vestibular aqueduct midpoint width and hearing loss in

patients with an enlarged vestibular aqueduct. JAMA Otolaryngol Head Neck Surg. 2017;143(6):601–8.
92. Bouhadjer K, Tissera K, Farris CW, Juliano AF, Cunnane ME, Curtin HD, et al. Retrospective review of midpoint vestibular aqueduct size in the 45 degrees oblique (Poschl) plane and correlation with hearing loss in patients with enlarged vestibular aqueduct. Am J Neuroradiol. 2021;42(12):2215–21.
93. Van Beck J, Chinnadurai S, Morrison AK, Zuniga MG, Smith B, Lohse CM, et al. Correlation of air-bone gap and size of enlarged vestibular aqueduct in children. Int J Pediatr Otorhinolaryngol. 2020;132:109909.
94. Abou-Elew M, El-Khousht M, El-Minawi MS, Selim M, Kamel AI. Enlarged vestibular aqueduct in congenital non-syndromic sensorineural hearing loss in Egypt. Indian J Otolaryngol Head Neck Surg. 2014;66(Suppl 1):88–94.
95. Berrettini S, Forli F, Bogazzi F, Neri E, Salvatori L, Casani AP, et al. Large vestibular aqueduct syndrome: audiological, radiological, clinical, and genetic features. Am J Otolaryngol. 2005;26(6):363–71.
96. Ruthberg JS, Kocharyan A, Farrokhian N, Stahl MC, Hicks K, Scarborough J, et al. Hearing loss patterns in enlarged vestibular aqueduct syndrome: do fluctuations have clinical significance? Int J Pediatr Otorhinolaryngol. 2022;156:111072.
97. Zalzal GH, Tomaski SM, Vezina LG, Bjornsti P, Grundfast KM. Enlarged vestibular aqueduct and sensorineural hearing loss in childhood. Arch Otolaryngol Head Neck Surg. 1995;121(1):23–8.
98. Tavora-Vieira D, Miller S. Misdiagnosis of otosclerosis in a patient with enlarged vestibular aqueduct syndrome: a case report. J Med Case Rep. 2012;6:178.
99. Wieczorek SS, Anderson ME Jr, Harris DA, Mikulec AA. Enlarged vestibular aqueduct syndrome mimicking otosclerosis in adults. Am J Otolaryngol. 2013;34(6):619–25.
100. Atkin JS, Grimmer JF, Hedlund G, Park AH. Cochlear abnormalities associated with enlarged vestibular aqueduct anomaly. Int J Pediatr Otorhinolaryngol. 2009;73(12):1682–5.
101. Emmrich JV, Fatterpekar GM. Dilated dysplastic vestibule: a new computed tomographic finding in patients with large vestibular aqueduct syndrome. J Comput Assist Tomogr. 2011;35(6):674–8.
102. Lyu H, Chen K, Xie Y, Yang L, Zhang T, Dai P. Morphometric study of the vestibular aqueduct in patients with enlarged vestibular aqueduct. J Comput Assist Tomogr. 2017;41(3):467–71.
103. Lyu H, Hong J, Yin D, Chen K, Li J, Yang L, et al. Position of the internal aperture of vestibular aqueduct in patients with enlarged vestibular aqueduct. Otol Neurotol. 2017;38(8):1198–204.
104. Suzuki M, Ota Y, Tanaka T, Ota Y. Lateral semicircular canal-enlarged vestibular aqueduct fistula associated with paroxysmal positional nystagmus. Otol Neurotol. 2016;37(6):e192–3.
105. Brandolini C, Modugno GC. Superior semicircular canal dehiscence and enlarged vestibular aqueduct. Int J Pediatr Otorhinolaryngol. 2011;75(6):861–3.
106. Grover M. Enlarged vestibular aqueduct and cochlear anomalies: just an association or a causal relationship? Cochlear Implants Int. 2021;22(3):182.
107. Song JJ, Hong SK, Kim JS, Koo JW. Enlarged vestibular aqueduct may precipitate benign paroxysmal positional vertigo in children. Acta Otolaryngol. 2012;132(Suppl 1):S109–17.
108. Manzari L. Enlarged vestibular aqueduct (EVA) related with recurrent benign paroxysmal positional vertigo (BPPV). Med Hypotheses. 2008;70(1):61–5.
109. Brodsky JR, Kaur K, Shoshany T, Manganella J, Barrett D, Kawai K, et al. Torticollis in children with enlarged vestibular aqueducts. Int J Pediatr Otorhinolaryngol. 2020;131:109862.
110. Govaerts PJ, Casselman J, Daemers K, De Ceulaer G, Somers T, Offeciers FE. Audiological findings in large vestibular aqueduct syndrome. Int J Pediatr Otorhinolaryngol. 1999;51(3):157–64.
111. Mey K, Bille M, Rye Rasmussen SH, Tranebjaerg L, Caye-Thomasen P. The natural history of hearing loss in pendred syndrome and non-syndromic enlarged vestibular aqueduct. Otol Neurotol. 2019;40(3):e178–e85.
112. Mori T, Westerberg BD, Atashband S, Kozak FK. Natural history of hearing loss in children with enlarged vestibular aqueduct syndrome. J Otolaryngol Head Neck Surg = Le Journal d'oto-rhino-laryngologie et de chirurgie cervico-faciale. 2008;37(1):112–8.
113. Lai CC, Shiao AS. Chronological changes of hearing in pediatric patients with large vestibular aqueduct syndrome. Laryngoscope. 2004;114(5):832–8.
114. Farrokhian N, Kocharyan A, Ruthberg J, Piper R, Rivas A, Semaan M, et al. Predictive modeling and risk stratification of patients with enlarged vestibular aqueduct. Laryngoscope. 2022;132(7):1439–45.
115. Kim M, Kim J, Kim SH, Kim SC, Jeon JH, Lee WS, et al. Hemorrhage in the endolymphatic sac: a cause of hearing fluctuation in enlarged vestibular aqueduct. Int J Pediatr Otorhinolaryngol. 2011;75(12):1538–44.
116. Hansen MU, Rye Rasmussen E, Caye-Thomasen P, Mey K. Cochlear implantation in children with enlarged vestibular aqueduct: a systematic review of surgical implications and outcomes. Ear Hear. 2023;44:440–7.
117. Li J, Kang S, Du H, Wang S, Wang D, Liu M, et al. Analysis of cochlear parameters in paediatric inner ears with enlarged vestibular aqueduct and patent cochlea. J Pers Med. 2022;12(10):1666.
118. Alahmadi A, Abdelsamad Y, Salamah M, Alenzi S, Badr KM, Alghamdi S, et al. Cochlear implantation in adults and pediatrics with enlarged vestibular aqueduct: a systematic review on the surgical findings and patients' performance. Eur Arch Otorrinolaringol. 2022;279(12):5497–509.
119. Saeed HS, Rajai A, Nash R, Saeed SR, Stivaros SM, Black G, et al. Enlarged vestibular aqueduct: disease characterization and exploration of potential prognostic factors for cochlear implantation. Otol Neurotol. 2022;43(5):e563–e70.
120. Benchetrit L, Jabbour N, Appachi S, Liu YC, Cohen MS, Anne S. Cochlear implantation in pediatric patients with enlarged vestibular aqueduct: a systematic review. Laryngoscope. 2022;132(7):1459–72.
121. Hodge SE, Thompson NJ, Park LR, Brown KD. Enlarged vestibular aqueduct: hearing progression and cochlear implant candidacy in pediatric patients. Otol Neurotol. 2021;42(1):203–6.
122. Reynard P, Ionescu E, Joly CA, Ltaief-Boudrigua A, Coudert A, Thai-Van H. Vestibular impairment in cochlear implanted children presenting enlarged vestibular aqueduct and enlarged endolymphatic sac. Int J Pediatr Otorhinolaryngol. 2021;141:110557.
123. Jahn KN, Bergan MD, Arenberg JG. Auditory detection thresholds and cochlear resistivity differ between pediatric cochlear implant listeners with enlarged vestibular aqueduct and those with connexin-26 mutations. Am J Audiol. 2020;29(1):23–34.
124. Patel ND, Ascha MS, Manzoor NF, Gupta A, Semaan M, Megerian C, et al. Morphology and cochlear implantation in enlarged vestibular aqueduct. Am J Otolaryngol. 2018;39(6):657–63.
125. Bostic K, Lewis RM, Chai B, Manganella JL, Barrett DL, Kawai K, et al. Enlarged vestibular aqueduct and cochlear implants: the

effect of early counseling on the length of time between candidacy and implantation. Otol Neurotol. 2018;39(2):e90–e5.
126. Manzoor NF, Wick CC, Wahba M, Gupta A, Piper R, Murray GS, et al. Bilateral Sequential Cochlear Implantation in Patients With Enlarged Vestibular Aqueduct (EVA) Syndrome. Otol Neurotol. 2016;37(2):e96–103.
127. Powell HR, Birman CS. Large vestibular aqueduct syndrome: Impedance changes over time with different cochlear implant electrode arrays. Cochlear Implants Int. 2015;16(6):326–30.
128. Pritchett C, Zwolan T, Huq F, Phillips A, Parmar H, Ibrahim M, et al. Variations in the cochlear implant experience in children with enlarged vestibular aqueduct. Laryngoscope. 2015;125(9):2169–74.
129. Lee KH, Lee J, Isaacson B, Kutz JW, Roland PS. Cochlear implantation in children with enlarged vestibular aqueduct. Laryngoscope. 2010;120(8):1675–81.
130. Dewan K, Wippold FJ 2nd, Lieu JE. Enlarged vestibular aqueduct in pediatric sensorineural hearing loss. Otolaryngol Head Neck Surg. 2009;140(4):552–8.
131. Hill JH, Freint AJ, Mafee MF. Enlargement of the vestibular aqueduct. Am J Otolaryngol. 1984;5(6):411–4.
132. Grimmer JF, Hedlund G, Park A. Steroid treatment of hearing loss in enlarged vestibular aqueduct anomaly. Int J Pediatr Otorhinolaryngol. 2008;72(11):1711–5.

65 Large Cochlear Aqueduct

Overview

Literature suggests that the normative range of sizes and configurations of the cochlear aqueduct is variable, bringing into question what constitutes a "large" cochlear aqueduct (CA). Even if a high-resolution temporal bone CT is interpreted as showing a large CA, it is presently unclear whether this explains any symptoms and whether it predicts surgical complications during cochlear implantation. If a "large" CA is incidentally discovered on imaging, then referral to an otolaryngologist is reasonable.

Introduction

When a neuro-otologic surgeon is placing a cochlear implant, some patients exhibit an unexpectedly brisk flow of cerebrospinal fluid—colloquially referred to as "gushers" and "oozers" [1]. In the interest of being able to predict (and avoid) this surgical complication, investigators have explored whether the underlying cause might be an (inappropriately) *large cochlear aqueduct* (CA).

Some investigators have also postulated that large CAs may be the etiology of otopathologies such as sensorineural hearing loss and perilymphatic fistula [2].

However, a large CA is an unusual finding. Some authors have expressed skepticism about its existence [3, 4], let alone whether it causes or explains any disease.

Anatomy

The cochlear aqueduct (CA) is a small canal in the temporal bone connecting the cochlea to the subarachnoid space [5]. The CA contains the cochlear canaliculus [6] and perilymphatic (periotic) duct [2, 5], which connects the perilymphatic compartment of the inner ear (scala tympani) with the cerebrospinal fluid space of the posterior cranial fossa [2, 4].

Multiple authors [2, 5, 6], citing Jackler and Hwang [4], recognize four segments of the CA, proceeding from lateral to medial:

1. The *lateral (cochlear) orifice/aperture*, also called the *medial aperture*, which opens into the scala tympani at the basal turn of the cochlea, adjacent to the round window membrane.
2. The *otic capsule segment*, which runs medially through the temporal bone. In normal individuals, the narrowest portion (isthmus) of the cochlear aqueduct lies within the otic capsule segment [6].
3. The *petrous apex segment*.
4. The *medial (cranial) orifice/aperture*, which opens into the subarachnoid space in the posterior cranial fossa, adjacent to the pars nervosa of the jugular foramen.

Histologically, "Unlike the vestibular aqueduct, the cochlear aqueduct does not contain a true epithelium-lined duct," but rather is "filled with a loose mesh of connective tissue that, although permeable to fluid, limits the patency" of the CA [5].

Migirof and Kronenberg [6] studied high-resolution CT images of 488 normal patients. They reported the cochlear aqueducts to be "asymmetric… in 10.6% of the images," suggesting that 89.4% of individuals have symmetric CAs.

Embryology

In the search for possible explanations of large CA, some authors begin by focusing on the embryonic development.

> The embryology of the CA [cochlear aqueduct] differs substantially from that of the other components of the inner ear, both in timing and tissue of origin… The CA… is a secondary formation created by enclosure of a soft tissue mesh that is entrapped in a channel within the developing otic capsule [4].

More specifically, there are:

M. Cherchi, *Otoneurology and Vestibular Medicine*, https://doi.org/10.1007/978-3-031-94841-1_65

> Separate embryological origins of the cochlear aqueduct and membranous labyrinth. The membranous labyrinth is derived from the otic vesicle. The perilymphatic duct initially appears as an outpouching of the subarachnoid space. The cochlear aqueduct forms in the mesoderm surrounding the perilymphatic duct and gradually ossified during embryogenesis [1].

In view of the embryonic development, some authors hypothesize that a large CA reflects developmental arrest.

> The development of the periotic duct and its adjacent structures in the human fetus occur over a 6-week period (20–26 weeks). At 22 weeks of gestation, the petrous apex ossifies and forms the medial wall of the CA [cochlear aqueduct], and the progressive otic capsule ossification leads to the formation of the lateral wall of the CA. Subsequently, the width of the CA diminishes with gestational age. Therefore, an enlarged CA may result from the developmental arrest of the primitive CA [7].

Postnatally, the CA appears to change to some degree. "During the neonatal period the CA [cochlear aqueduct] is shorter, straighter, wider, and more free of arachnoid mesh than in adults" [4]. However, as far as the narrowest point of the CA is concerned, the temporal bone study by Gopen and colleagues showed that, "there were no significant changes in this diameter with increasing [post-natal] age" [8].

Hydraulics

Figure 65.1, from Bianchin and colleagues [5], is a schematic depiction of the relative dispositions of the cochlear aqueduct and nearby structures.

Figure 65.2, from Gopen and colleagues [8], schematically depicts the hydraulic arrangement of the stapes (in the oval window), the cochlea, the round window, and the cochlear aqueduct.

The series of 101 temporal bones studied by Gopen and colleagues revealed that, "The narrowest point of the [cochlear] aqueduct as defined by its bony wall was found to lie 200–300 micrometers from its cochlear end. The mean value (±1 SD) for this smallest diameter was 138 (±58) micrometers" [8].

Allen [9] used the Poiseuille–Hagen equation to model the flow of fluid through the cochlear aqueduct.

Jackler and Hwang commented that, "Allen has argued that because fluid flow through a tube varies with the fourth power of its diameter, even slight enlargements of the CA [cochlear aqueduct] may have a substantial impact on perilymphatic flows into the vestibule" [4]. Stimmer

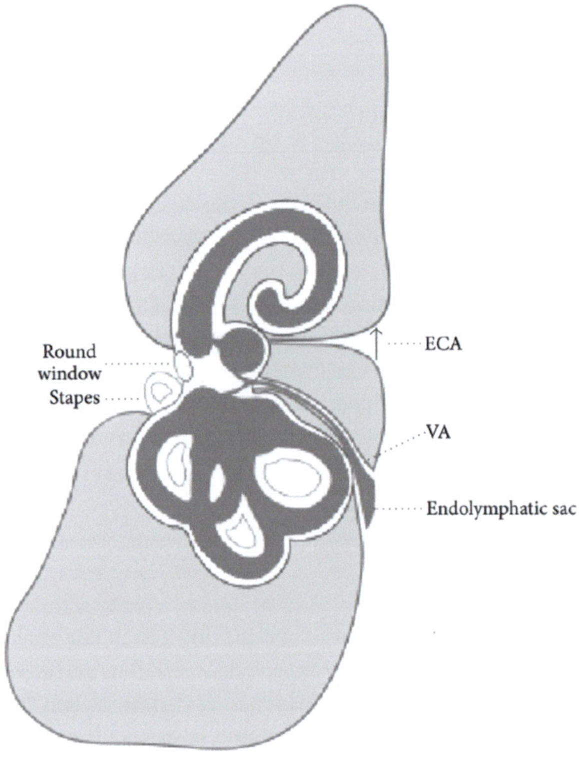

Fig. 65.1 Schematic depiction of the relative disposition of the cochlear aqueduct and nearby structures. Abbreviations: ECA enlarged cochlear aqueduct, VA vestibular aqueduct. (From Bianchin et al. [5]. Reproduced from Hindawi Limited with permission under the CC-BY 4.0)

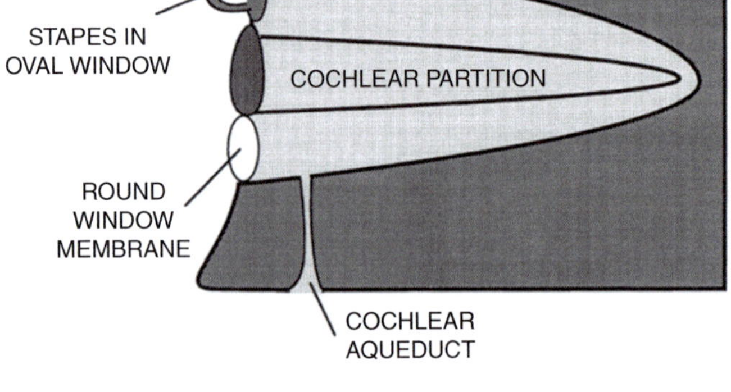

Fig. 65.2 Schematic depiction to the hydraulic arrangement of the stapes (in the oval window), the cochlea, the round window, and the cochlear aqueduct. (From Gopen et al. [8]. Reproduced from Elsevier B.V. and with permission under STM signatory guidelines)

similarly commented that, "The flow of liquids through tubes is a linear function of pressure, viscosity and length of the tube and in fourth degree power function of the radius of the lumen. Therefore, the radius, i.e. the diameter is the most critical factor for the flow rate; small variations in size of the CA [cochlear aqueduct] cause large variations in flow rate" [10].

However, Jackler and Hwang also critiqued this model, noting that, "This simple model of fluid flow through a tube, while conceptually useful, does not take into account the obstructing effect of the membranous mesh that fills the aqueduct. Deficiencies in these membrane baffles theoretically may result in increased flows through even a normal-sized aqueduct" [4].

Significance of Patency/Occlusion of the Cochlear Aqueduct

Gopen and colleagues [8] studied 101 temporal bones from patients aged 0 to 100 years and observed four categories of CA:

1. Patent lumen throughout
2. Lumen filled with loose connective tissue
3. Lumen occluded by bone
4. Obliteration of the aqueduct

Gopen and colleagues [8] found the distribution of these categories by decade of life to be as in Table 65.1.

Overall, Gopen and colleagues found that, "The distribution of patency of the lumen in 101 normal temporal bones was: patent lumen throughout aqueduct 34%, lumen filled with loose connective tissue 59%, lumen blocked by bone 4%, and obliteration of the aqueduct 3%," and they specifically commented that, "Normal ears may have cochlear aqueducts that are completely occluded or obliterated" [8]. These findings led Gopen and colleagues to conclude that, "occlusion of the cochlear aqueduct in the human causes no apparent ill effect on the normal inner ear" [8].

Table 65.1 Patency of the lumen of the cochlear aqueduct by age

	Number of bones				
Age (years)	Total	With patent lumen	With lumen filled with loose connective tissue	With lumen blocked by bone	With aqueduct obliterated
0–1	19	4	13	0	2
1–9	5	2	3	0	0
10–19	3	3	0	0	0
20–29	9	2	6	1	0
30–39	6	5	0	1	0
40–49	10	3	7	0	0
50–59	10	4	6	0	0
60–69	10	3	7	0	0
70–79	9	2	6	0	1
80–89	10	2	8	0	0
90–99	10	4	4	2	0
Total	101	34	60	4	3

From Gopen et al. [8]

Stimmer [10] studied 400 high-resolution temporal bone CTs of patients aged 1–94 years (average 43 years) and reported, "a diameter of more than 1 mm in the otic capsule portion could not be found in any of the 400 cases we examined," and "we did not find a single CA [cochlear aqueduct] that was completely visible in all segments with a diameter of more than 1 mm in the otic capsule segment," and therefore, "we concluded that not a single case of pathologic CA enlargement could be found using highly developed multislice CT -technique. According to these findings CA-enlargement could not be used as an explanation for clinical inner ear deficits."

Risk Factors

Kim and colleagues [7] studied 35 patients (70 ears) to determine whether the presence of a large vestibular aqueduct correlates in any way with a large cochlear aqueduct. They reported that, "The findings indicated that the mean CA [cochlear aqueduct] size was significantly larger in patients with EVA [enlarged vestibular aqueduct]," yet, "linear regression analysis revealed no statistically significant correlation between CA and VA [vestibular aqueduct] size." In other words, patients with large vestibular aqueducts also have large cochlear aqueducts, but beyond simply being "large," there is otherwise no correlation between the degree of oversizing of the two structures.

Imaging

High-resolution temporal bone CT is the imaging modality of choice for diagnosing and characterizing the cochlear aqueduct.

Figures 65.3 and 65.4, from Mukherji and colleagues [2], compare images of a large cochlear aqueduct (top three panels) with those of a normal cochlear aqueduct (bottom 2 panels).

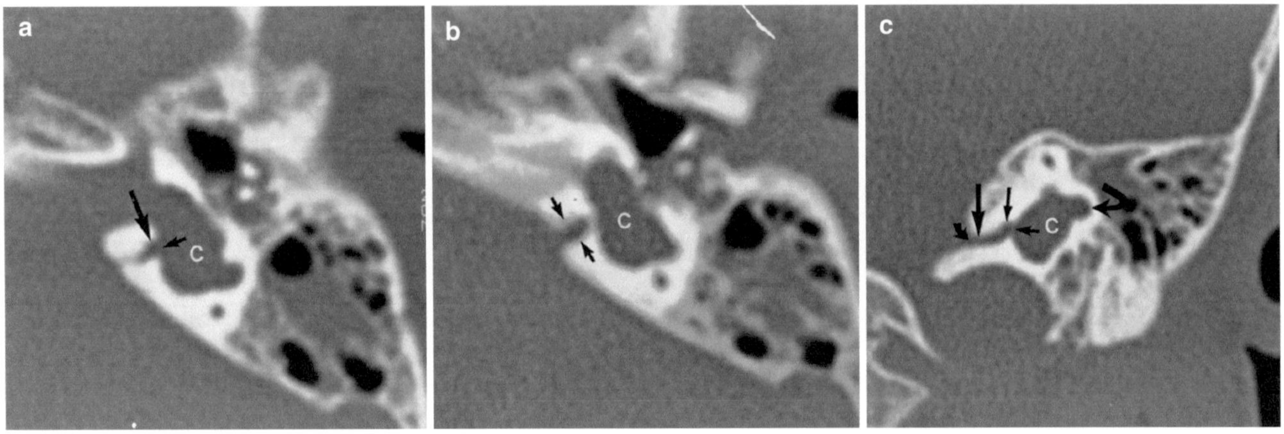

Fig. 65.3 CT findings of enlarged cochlear aqueduct in an 18-month-old boy. Panel (**a**) Axial scan obtained through the temporal bone shows dilatation of the lateral orifice (small arrow) and otic segment (large arrow) of the cochlear aqueduct. Note the associated common cavity malformation (C) of the inner ear. On otoscopic examination, the soft-tissue attenuation surrounding the ossicles was thought to represent fluid in the middle ear cavity rather than cholesteatoma. Panel (**b**) Axial scan obtained 1 mm inferior to (**a**) shows the dilated duct is continuous with the funnel-shaped medial orifice (arrows) that opens into the region of the jugular foramen. The jugular foramen and the pars nervosa were not malformed (not shown) (C indicates common cavity malformation). Panel (**c**) Coronal scan shows diffuse dilatation throughout the course of the cochlear aqueduct (smallest straight arrow indicates lateral orifice; medium-sized straight arrow, otic segment; large straight arrow, petrous apex segment; small curved arrow, medial orifice; and large curved arrow, dysplastic semicircular canal). The diameter of the cochlear aqueduct at the midotic segment measures 1.2 mm. (From Mukherji et al. [2]. Reproduced with written permission from Williams & Wilkins Co.)

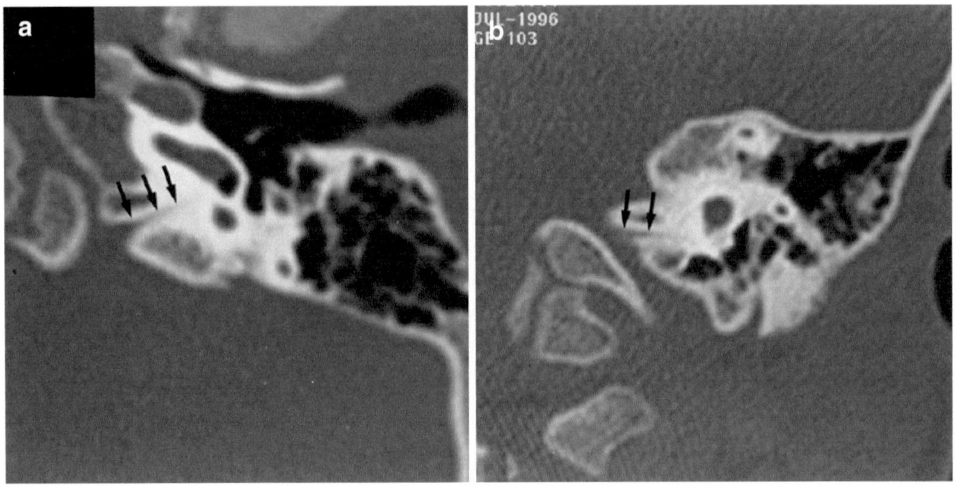

Fig. 65.4 CT appearance of normal cochlear aqueduct. Axial (Panel **a**) and coronal (Panel **b**) scans show the normal appearance of the cochlear aqueduct (arrows) in a 20-month-old patient examined for cholesteatoma of the contralateral ear (not shown). Note the gradual widening of the cochlear aqueduct on the axial scan as the duct courses medially. (From Mukherji et al. [2]. Reproduced with written permission from Williams & Wilkins Co.)

References

1. Schuknecht HF, Reisser C. The morphologic basis for perilymphatic gushers and oozers. Adv Otorhinolaryngol. 1988;39:1–12.
2. Mukherji SK, Baggett HC, Alley J, Carrasco VH. Enlarged cochlear aqueduct. Am J Neuroradiol. 1998;19(2):330–2.
3. Daly CA, Donnelly MJ. The enlarged cochlear aqueduct--a radiological flying saucer? Clin Radiol. 1996;51(11):821.
4. Jackler RK, Hwang PH. Enlargement of the cochlear aqueduct: fact or fiction? Otolaryngol Head Neck Surg. 1993;109(1):14–25.
5. Bianchin G, Polizzi V, Formigoni P, Russo C, Tribi L. Cerebrospinal fluid leak in cochlear implantation: enlarged cochlear versus enlarged vestibular aqueduct (common cavity excluded). Int J Otolaryngol. 2016;2016:6591684.
6. Migirov L, Kronenberg J. Radiology of the cochlear aqueduct. Ann Otol Rhinol Laryngol. 2005;114(11):863–6.
7. Kim BG, Sim NS, Kim SH, Kim UK, Kim S, Choi JY. Enlarged cochlear aqueducts: a potential route for CSF gushers in patients with enlarged vestibular aqueducts. Otol Neurotol. 2013;34(9):1660–5.
8. Gopen Q, Rosowski JJ, Merchant SN. Anatomy of the normal human cochlear aqueduct with functional implications. Hear Res. 1997;107(1–2):9–22.
9. Allen GW. Fluid flow in the cochlear aqueduct and cochlea-hydrodynamic considerations in perilymph fistula, stapes gusher, and secondary endolymphatic hydrops. Am J Otol. 1987;8(4):319–22.
10. Stimmer H. Enlargement of the cochlear aqueduct: does it exist? Eur Arch Otorrinolaringol. 2011;268(11):1655–61.

Mondini Dysplasia

66

Overview

Mondini dysplasia is a congenital inner ear anomaly in which the cochlea has only 1.5 turns, the vestibule is enlarged, and the vestibular aqueduct is large. Mondini dysplasia accounts for over 50% of all congenital inner ear anomalies. It can be unilateral or bilateral and can occur in syndromic contexts (Klippel–Feil syndrome, Pendred syndrome, DiGeorge syndrome, and Down syndrome). It usually presents with severe to profound hearing loss, though some patients have relatively preserved high frequency hearing. Some patients also complain of disequilibrium. Mondini dysplasia is usually identified on a high-resolution temporal bone CT, though MRI of the internal auditory canals may also be informative. Some patients benefit from cochlear implantation. A patient with diagnosed (or suspected) Mondini dysplasia should be referred to otolaryngology for evaluation.

Introduction

Carlo Mondini (1729–1803) was an Italian physician and anatomist who served as the chair of anatomy at the University of Bologna beginning in 1782. In 1791, he published a report of a dissection of a congenitally deaf 8-year-old boy [1, 2]. The findings he described included [3]:

- The *cochlea consisted of 1.5 turns* (instead of the normal 2.5 or 2.75 turns). Subsequent authors also comment on a cystic apex [4] and incomplete cochlear partition type 2 (referring to partial absence of the interscalar septum).
- The *vestibule was enlarged*, but the *semicircular canals were normal*.
- The *vestibular aqueduct was enlarged*, and the *endolymphatic sac was dilated*.

Some authors are careful to invoke the eponymous designation to refer to the specific set of abnormalities that Mondini originally described. Some commentators [3, 5] have noted that much otologic literature unfortunately uses the phrases "Mondini dysplasia," "Mondini defect," "Mondini anomaly," and "Mondini malformation" to refer to almost any inner ear anomaly, which has made diagnosis, classification, and analysis more difficult.

Epidemiology

Mondini dysplasia accounts for >50% of all congenital inner ear malformations [6]. It can occur unilaterally or bilaterally [7, 8].

Embryology

The embryologically based classification of congenital inner ear anomalies proposed by Jackler and colleagues [5] comments that an intrauterine insult around the seventh week of gestation can lead to the incomplete cochlear partition characteristic of the "classical Mondini's" dysplasia.

Histopathology

Kaya and colleagues reported that patients with Mondini dysplasia exhibited a statistically significant reduction in density of type 1 and type 2 hair cells and reduced total number of hair cells, in the cristae of all the semicircular canals, compared to controls [6].

Clinical Presentation: Hearing Loss

All patients have hearing loss, which can range from severe to profound. Some patients have relatively preserved high-frequency hearing [9], which is usually attributed to the fact that the base of the cochlea (that mediates high frequency hearing) is still present in Mondini dysplasia [5].

M. Cherchi, *Otoneurology and Vestibular Medicine*, https://doi.org/10.1007/978-3-031-94841-1_66

Clinical Presentation: Vestibular Symptoms

Some literature states that patients occasionally have vestibular symptoms [6].

Comorbidities

Mondini dysplasia can occur in association with other congenital abnormalities, such as Klippel–Feil syndrome, Pendred syndrome, DiGeorge syndrome, and Down syndrome [6].

Imaging

High-resolution temporal bone CT is the imaging study of choice, though MRI of the internal auditory canals can provide complementary information.

Figure 66.1, from Kitazawa and colleagues [8], shows bilateral Mondini dysplasia, with corresponding schematics.

The case reported by Al-Rawy and colleagues [10] also provides an instructive example.

Treatment

Some patients elect cochlear implantation to manage the hearing loss.

References

1. Mondini C. Minor works of Carlo Mondini: the anatomical section of a boy born deaf. Am J Otol. 1997;18(3):288–93.
2. Mondini C. Anatomica surdi natisectio. De Bononiensi scientarum et artium institute atque academia commentarii. Bononiae. 1791;7:419–31.
3. Lo WW. What is a 'Mondini' and what difference does a name make? Am J Neuroradiol. 1999;20(8):1442–4.
4. Sennaroglu L, Saatci I. Unpartitioned versus incompletely partitioned cochleae: radiologic differentiation. Otol Neurotol. 2004;25(4):520–9; discussion 9.
5. Jackler RK, Luxford WM, House WF. Congenital malformations of the inner ear: a classification based on embryogenesis. Laryngoscope. 1987;97(3 Pt 2 Suppl 40):2–14.
6. Kaya S, Hizli O, Kaya FK, Monsanto RD, Paparella MM, Cureoglu S. Peripheral vestibular pathology in Mondini dysplasia. Laryngoscope. 2017;127(1):206–9.
7. Schuknecht HF. Mondini dysplasia; a clinical and pathological study. Ann Otol Rhinol Laryngol Suppl. 1980;89(1 Pt 2):1–23.

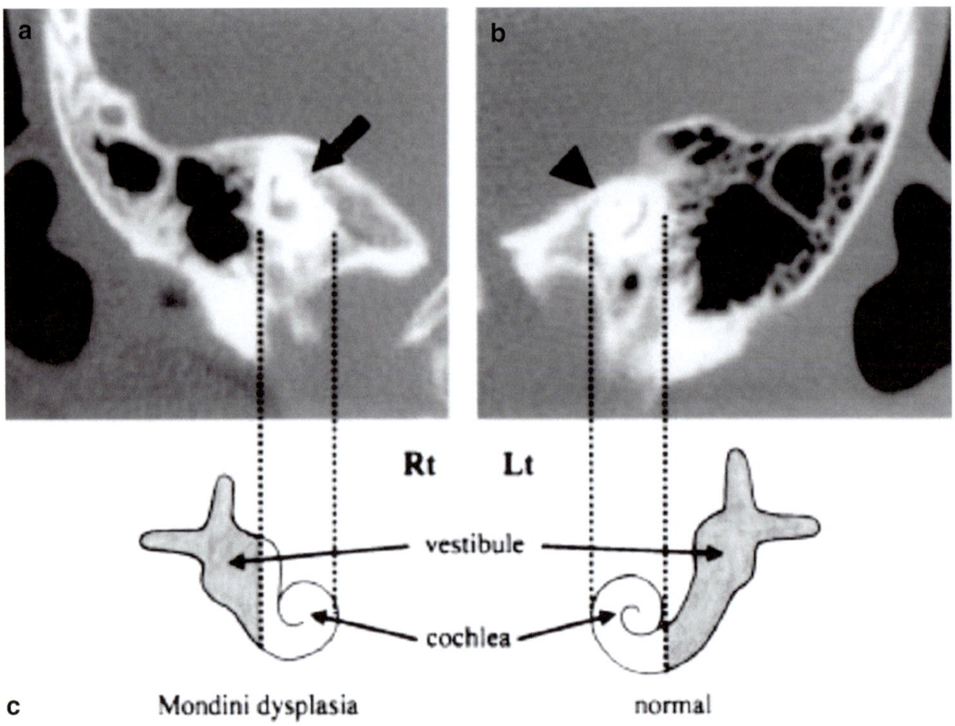

Fig. 66.1 Coronal CT images of bilateral Mondini dysplasia with corresponding schematic illustrations. Panel (**a**) An incomplete partition of the cochlea (arrow) in the right ear demonstrated by coronal high resolution computed tomography. Panel (**b**) A normal cochlea (arrowhead) in the left ear. Panel (**c**) Schematic diagram of the cochleovestibular anatomy in Mondini dysplasia and normal ear. The darkened vestibules illustrated are not visible on the CT slices (panels **a** and **b**). (From Kitazawa et al. [8]. Reproduced with written permission from John Wiley and Sons Limited)

8. Kitazawa K, Matsumoto M, Senda M, Honda A, Morimoto N, Kawashiro N, et al. Mondini dysplasia and recurrent bacterial meningitis in a girl with relapsing Langerhans cell histiocytosis. Pediatr Blood Cancer. 2004;43(1):85–7.
9. Feghali JG, Leone CA, Linthicum FH Jr. Residual high-frequency hearing in a patient with Mondini's deformity: clinical implications. Am J Otol. 1985;6(4):336–7.
10. Al-Rawy A, Al-Shatouri M, Tabbakh M, Gad A. Systematic approach by computed tomography and magnetic resonance imaging in cochlear implantation candidates in Suez Canal University Hospital. Egypt J Radiol Nucl Med. 2018;48:877–84.

67 Pendred Syndrome

Overview

Pendred syndrome (OMIM 274600) is a genetic condition transmitted in an autosomal recessive pattern resulting from mutations in the pendrin protein encoded by part of SLC26A4 (OMIM 605646) on chromosome 7q22.3. All patients present with bilateral hearing loss either from birth (usually detected on newborn screening tests of otoacoustic emissions and brainstem auditory evoked responses) or early childhood. Most patients develop goiter, and some of those are hypothyroid. A minority of patients complain of disequilibrium (but this has not been well characterized) and suffer from acid–base disturbances. Audiometry shows bilateral approximately symmetrical sensorineural hearing loss. On vestibular testing, cervical vestibular evoked myogenic potentials may show large amplitudes and low thresholds, probably compatible with a large vestibular aqueduct. Endocrinological testing may show hypothyroidism. MRI of the brain and internal auditory canals shows a large vestibular aqueduct and cystic, incompletely formed cochlea. Genetic testing (sequencing of SLC26A4) is confirmatory. Treatment of the hearing loss is usually with cochlear implantation. Treatment for disequilibrium has not been studied. Thyroid function should be monitored, and thyroid hormone supplemented where appropriate. If Pendred syndrome is suspected or confirmed, then referral to otolaryngology and genetic counseling is appropriate.

Fig. 67.1 Vaughan Pendred (1869–1946). (From https://litfl.com/vaughan-pendred/ (accessed 10/31/24). Reproduced with written permission from Dr. Mike Cadogan)

Introduction

Dr. Vaughan Pendred (1869–1946) was English physician. In 1896, he published a report of two sisters, aged 28 and 38 years, who were deaf and had goiters [1] (Fig. 67.1).

M. Cherchi, *Otoneurology and Vestibular Medicine*, https://doi.org/10.1007/978-3-031-94841-1_67

Genetics

Pendred syndrome is caused by biallelic mutations in the SLC26A4/PDS gene which encodes pendrin, a multifunctional anion exchanger that is found in multiple organs, including the inner ear, thyroid, and kidney. The normal pendrin protein "has affinity for chloride, iodide, and bicarbonate, among other anions" [2].

Physiology

Pendred syndrome has effects in several organ systems.

- "In the **inner ear**, pendrin functions as a chloride/bicarbonate exchanger that is essential for maintaining the composition and the [electrical] potential of the endolymph" [2].
- "In the **thyroid**, pendrin is expressed at the apical membrane of thyroid cells facing the follicular lumen. Functional studies have demonstrated that pendrin can mediate iodide efflux in heterologous cells. This, together with the [pathological] thyroid phenotype observed in humans (goiter, impaired iodine organification) suggests that pendrin could be involved in iodide efflux into the lumen, one of the steps required for thyroid hormone synthesis" [2].
- "In the **kidney**, pendrin is involved in bicarbonate secretion and chloride resorption… [It] is abundantly expressed at the apical membrane of type B, and in non-A/non-B intercalated cells in the cortical binding duct. Functionally, it is an apical chloride/bicarbonate exchanger mediating bicarbonate secretion, and it contributes to the regulation of blood pressure by modulating renal chloride absorption" [2].

Pathophysiology

For patients in whom genetic mutation has rendered pendrin dysfunctional, deficits in inner ear function, thyroid function, and renal function may develop.

In Pendred syndrome, the inner ear develops abnormally, and patients usually have a large vestibular aqueduct and a cystic, incompletely formed cochlea.

Clinical History

Patients usually present with a combination of sensorineural hearing loss and goiter, sometimes with hypothyroidism. In certain conditions, these patients may also develop metabolic alkalosis.

Patients always have significant *sensorineural hearing loss* that is bilateral (though sometimes modestly asymmetrical) and usually congenital; less commonly, it develops later in childhood.

Goiter is present in 83% of cases and usually develops after the age of 10 years; of all patients, 56% remain euthyroid, while 44% become *hypothyroid*, and of the latter, all have goiter [3]. In other words, most patients develop goiter, and of the patients who develop goiter, some (but not all) are hypothyroid.

As far as renal function is concerned, "at least some patients with Pendred syndrome may develop severe acid-base disturbances under conditions of an increased alkalai load" [3].

A minority of Pendred syndrome patients experience *disequilibrium*, and this has not been well characterized.

Testing: Audiologic

The hearing loss can be detected on newborn screening with otoacoustic emissions and auditory evoked brainstem responses.

Testing: Vestibular

Unfortunately, most vestibular literature pertains to findings in patients with large vestibular aqueduct. There is almost no vestibular literature on patients specifically with Pendred syndrome.

West and colleagues [4] studied video head impulse testing and cervical vestibular evoked myogenic potentials in 26 patients (52 ears) with Pendred syndrome who had not yet undergone cochlear implantation.

They reported that out of 52 ears, only 4 (8%) exhibited abnormal results on video head impulse testing; 2 ears showed low gain and 2 ears showed compensatory saccades. No patients had complete loss of the vestibulo-ocular reflex.

They reported that cervical vestibular evoked myogenic potentials (cVEMPs) responses were present in 76% of ears and absent in 24%. They calculated that the mean cVEMP amplitude was 192 microvolts, and that thresholds were <90 dB nHL.

West and colleagues [4] provided data from a case of Pendred syndrome in which cervical vestibular evoked myogenic potentials demonstrated abnormally large responses.

They interpreted these cVEMP findings as showing, "saccular hypersensitivity in our Pendred cohort, as the mean cVEMP thresholds were below 90 dB nHL and amplitudes were abnormally large." Extrapolating from literature on large cochlear aqueducts, they stated that, "The cause for this hypersensitivity phenomenon has previously been suggested to be attributed to a 'third window' effect due to shunting of the sound-induced energy away from the cochlea."

They further reported that, "absence of cVEMP response was associated with the presence of VHIT saccades ($p = 0.038$)," suggesting some parallel between semicircular canal dysfunction and otolith dysfunction.

To summarize the few vestibular data on patients with Pendred syndrome: Most (92%) have normal video head impulse testing. A minority (24%) have absent cervical vestibular evoked myogenic potentials (cVEMPs); of the remainder, some have abnormally large cVEMP amplitudes and low thresholds. These cVEMP findings probably reflect the presence of a large vestibular aqueduct, rather than a feature specific to Pendred syndrome.

Testing: Metabolic

Checking thyroid stimulating hormone and free thyroxine can provide support for the diagnosis [2].

Testing: Genetic

Definitive diagnosis can be secured by sequencing the coding region of the SLC26A4 gene [2].

Imaging

Temporal bone CT (or at least a head CT) can provide initial radiographic support for the diagnosis, though MRI of the internal auditory canals is considered the imaging modality of choice [5, 6].

Wemeau and colleagues [2] provide examples of patients with Pendred syndrome, one with a large vestibular aqueduct, and one with a Mondini malformation (Figs. 67.2 and 67.3).

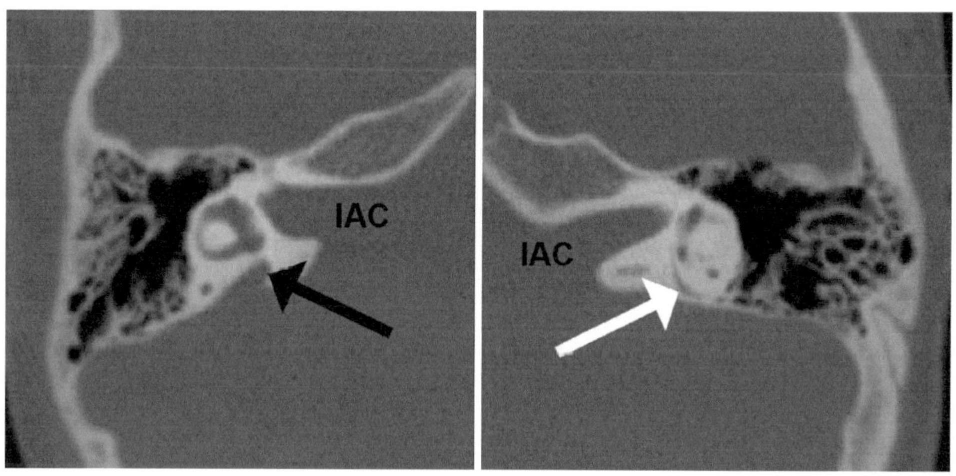

Fig. 67.2 The left panel shows an axial CT image from the right temporal of patient with Pendred syndrome demonstrating a large vestibular aqueduct (black arrow). For comparison, the right panel shows an axial CT image from the left temporal bone of a normal subject. Abbreviation: IAC internal auditory canal. (From Wemeau and Copp [2]. Reproduced from Elsevier with permission under STM signatory guidelines)

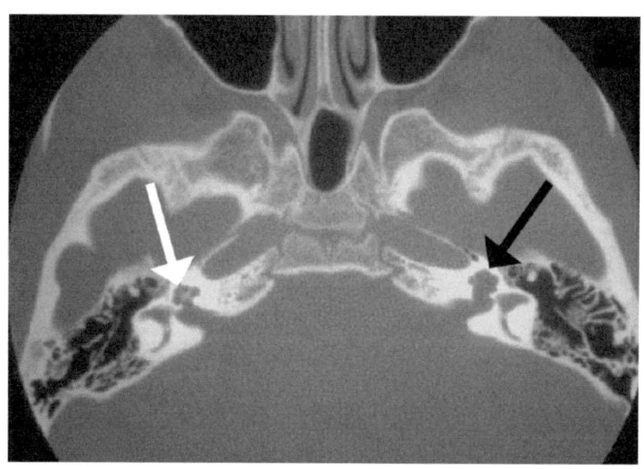

Fig. 67.3 Axial image from a temporal bone CT in a patient with Pendred syndrome. The left cochlea (black arrow) shows a Mondini malformation. The right cochlea (white arrow) shows a normal cochlea. (From Wemeau and Copp [2]. Reproduced from Elsevier with permission under STM signatory guidelines)

Treatment

Audiology consultation is appropriate. A hearing aid can be considered in patients with aidable hearing, though the majority of patients (who usually have severe to profound sensorineural hearing loss) may consider cochlear implantation.

Thyroid function should be monitored, and hypothyroidism should be treated with thyroid supplementation. Management by an endocrinologist is reasonable.

Pendred syndrome is transmitted in an autosomal recessive fashion. Referral to a genetic counselor is appropriate.

References

1. Pendred V. Deaf-mutism and goitre. Lancet. 1896;148(3808):532.
2. Wemeau JL, Kopp P. Pendred syndrome. Best Pract Res Clin Endocrinol Metab. 2017;31(2):213–24.
3. Reardon W, Coffey R, Chowdhury T, Grossman A, Jan H, Britton K, et al. Prevalence, age of onset, and natural history of thyroid disease in Pendred syndrome. J Med Genet. 1999;36(8):595–8.
4. West NC, Ryberg AC, Caye-Thomasen P. Vestibular function in Pendred syndrome: intact high frequency VOR and saccular hypersensitivity. Otol Neurotol. 2021;42(9):e1327–e32.
5. Phelps PD, Coffey RA, Trembath RC, Luxon LM, Grossman AB, Britton KE, et al. Radiological malformations of the ear in Pendred syndrome. Clin Radiol. 1998;53(4):268–73.
6. Fugazzola L, Mannavola D, Cerutti N, Maghnie M, Pagella F, Bianchi P, et al. Molecular analysis of the Pendred's syndrome gene and magnetic resonance imaging studies of the inner ear are essential for the diagnosis of true Pendred's syndrome. J Clin Endocrinol Metab. 2000;85(7):2469–75.

68 Usher Syndrome

Overview

Usher syndrome is an autosomal recessive syndrome that comprises the most common genetic cause of combined deafness and blindness, and the second most common cause of inherited hearing loss. The prevalence ranges from 1 to 10 per 30,000 individuals. There are three broad types of Usher syndrome, with multiple subtypes. Type 1 (second most common) presents with congenital, usually severe hearing loss, early visual loss, and vestibular impairment. Type 2 (most common) presents with less severe hearing loss and visual loss (though these deficits may be progressive), and normal or nearly normal vestibular function. Type 3 is variable, can be progressive, and can resemble type 1 or 2. The hearing loss is sensorineural and approximately symmetrical. Tests of vestibular function may show caloric weakness, low gain on rotatory chair testing, and impaired performance on computerized dynamic posturography. Suspected cases should be offered genetic counseling and testing. The hearing loss in Usher syndrome type 1 is often treated with cochlear implantation, while in types 2 and 3, conventional hearing aids may be attempted first. The visual loss has no known treatment, though visual accommodations can be put into place. Vestibular rehabilitation therapy can be offered to patients complaining of disequilibrium.

Introduction

Usher syndrome is the most common inherited cause of combined deafness and blindness, accounting for 50% of deaf-blind individuals younger than 65 years [1]. It is the second most common cause of inherited hearing loss after Pendred syndrome [1].

History

German ophthalmologist Friedrich Wilhelm Ernst Albrecht von Graefe (1828–1870) was the first to publish a description of this disease [2].

Scottish ophthalmologist Charles Howard Usher (1865–1942) published a large series (69 cases) and identified the pattern of inheritance [3], since which time the disease has borne his name.

Epidemiology

The prevalence of Usher syndrome has been difficult to determine. Conservative estimates are in the range of 1–10 per 30,000 individuals [1].

Among the population of deaf and hard-of-hearing individuals, Usher syndrome accounts for 3–6% of cases [4].

Usher syndrome type 1 is the second most common subtype; type 2 is the most common; type 3 is the least common [1], around 1% overall, but is higher in specific populations, accounting for 40% of cases in Finland and 40% of cases among Ashkenazi Jews [5].

Genetics

Usher syndrome is inherited in an autosomal recessive pattern.

So far three main categories of Usher syndrome have been identified, with 14 subtypes [1].

Table 68.1, adapted from Cohen and colleagues [5], is organized by type of Usher syndrome.

Table 68.2, adapted from Cohen and colleagues [5], is organized by molecular classification.

M. Cherchi, *Otoneurology and Vestibular Medicine*, https://doi.org/10.1007/978-3-031-94841-1_68

Table 68.1 The traditional clinical classification of Usher syndrome and atypical phenotypes

Clinical type	Classification	Gene locus	Auditory manifestations	Vestibular manifestations	Ocular manifestations
Type 1	Traditional	USH1A-G	Profound congenital hearing loss	Absent	Before puberty
	Atypical	USH1B	Moderate hearing loss, may be progressive	Varying from normal to total absence, intra-familial variability	Mild retinal degeneration
	Atypical	USH1D	Moderate to profound, may be progressive	Varies from normal to absent, inter and intrafamilial variability	Early-mid teens
Type 2	Traditional	USH2A-C	Prelingual onset of moderate to severe high frequency sloping hearing loss.	Normal	Second decade
	Atypical	USH2A (including 2299delG and C759F)	Progressive hearing loss, 5–10 dB per decade, particularly in third and fourth decades. USH2A have worse hearing loss than other type 2, particularly at the low frequencies. Late onset hearing loss preceded by RP caused by C759F and an unknown mutation.	Normal or mildly impaired	Second decade
Type 3	Traditional	USH3A	Variable onset of progressive hearing loss	Variable	Variable
	Additional clinical descriptions		Onset first to fifth decade of moderate to severe hearing loss. Hearing loss progresses rapidly (/20 dB per decade) especially in first and second decades.	Usually normal or mildly impaired. Rarely absent	Variable. Visual field scores significantly poorer than in USH2A patients of the same age. May precede onset of HI
Nonsyndromic RP		USH2A (C759F)	Normal	Normal	

From Cohen et al. [5]

Table 68.2 Molecular classification of Usher syndrome

Locus name	Locus	Gene	Prevalence	Gene function
USH1A	14q32	Unknown	Existence in question	
USH1B	11q13.5	MYO7A (myosin7a)	29–82% of type 1	Atypical myosin. Gating of the tip links with the deflection of the stereocilia. Important in stereocilia bundle organization and cohesion. In the eye, transporting rhodopsin between inner and outer photoreceptor segments. Expressed in ribbon synapses of photoreceptor cells
USH1C	11p15.1	USH1C (harmonin)	7–12.5% of type 1 100% of Acadian	PDZ domain protein. Fulfills a scaffolding function. Important in stereocilia bundle organization and cohesion. Expressed in outer segments, inner segments and ribbon synapses of photoreceptor cells
USH1D	10q	CDH23 (cadherin23)	10–35% of type 1	Extracellular filament protein. Important in stereocilia bundle organization and cohesion. Possibly tip link protein. Expressed in inner segments and ribbon synapses of photoreceptor cells.

(continued)

Table 68.2 (continued)

Locus name	Locus	Gene	Prevalence	Gene function
USH1E	21q	Unknown	Rare	
USH1F	10q2122	PCDH15 (protocadherin15)	11% of type 1. /20% in Ashkenazi Jews	Cellular adhesion Important in stereocilia bundle organization and cohesion
USH1G	17q2425	SANS (sans)	7% of type 1 (rare or absent in UK cohort)	Scaffold protein Forms a functional network with USH1B, C, D, and G proteins. Important in stereocilia bundle organization and cohesion. Expressed in cuticular plate and kinocilium of hair cells as well as ribbon synapses.
USH2A	1q41	USH2A (usherin)	≤/85% of type 2	Basement membrane protein?
USH2B	3p2324.2.	NBC3?	Rare	Unknown
USH2C	5q14.321.3	VLGR1 (Vlgr1/Mass1)	Rare, so far detected in females only. Causes audiogenic seizures in mice.	G protein coupled receptor. Signaling protein?
USH3	3q21q25	Clarin-1	40% of Usher in Finland and in	May be involved in hair cell and photoreceptor cell synapses.

From Cohen et al. [5]

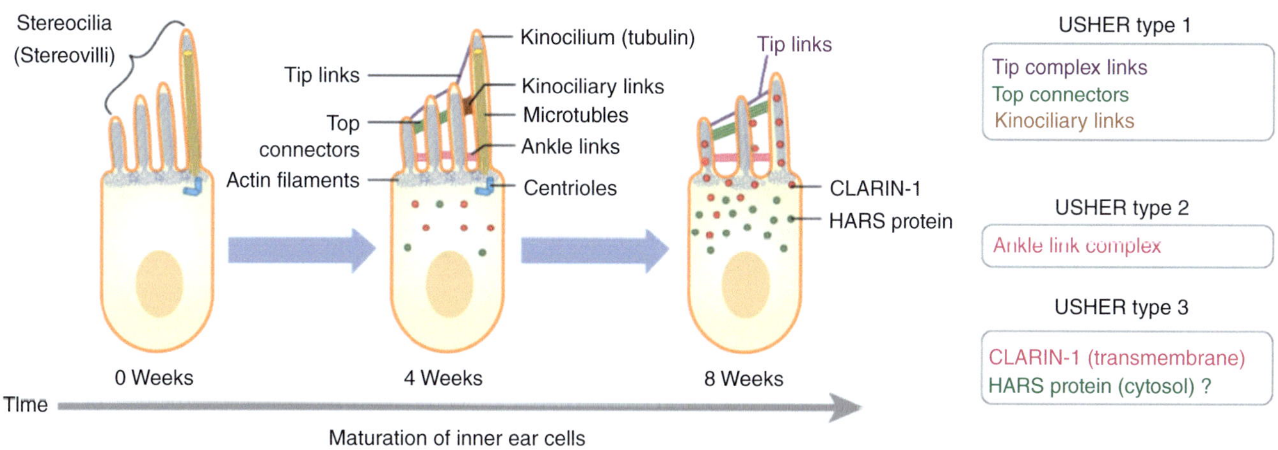

Fig. 68.1 Schematic illustration of the localization in inner ear hair cells of affected proteins in the three main types of Usher syndrome, as well as the maturation process that involves sensorineural cells in the inner ear. Only one kinocilium and many different stereocilia (stereovilli) for each cell are presented. Kinocilia are intracellular, transmembrane, and extracellular proteins that drive maturation and orientation of stereocilia in the sensorineural epithelium. Inner and outer hair cells, and vestibular type II hair cells lose their kinocilium during maturation; the kinocilium disappears in the final form. Transient links degenerate, and tip links, top connectors, and ankle links assume their final functional form. This reorganization of tip links, transient lateral links, top connectors, kinociliary links, and ankle links refers to the normal functions of actin filaments and several Usher proteins. Centrioles are also very important in leading a vital process during cell life. Hearing loss and equilibrium disorders may be derived from mutations in different genes that express proteins involved in different levels and in different times of maturation of the sensorineural cells. Clarin-1 protein (in red) is a four-transmembrane protein that is synthesized and collected in the cytosol and then transported to cellular membrane during maturation processes. The role of HARS1 gene in USH type 3 must be confirmed. (From Castiglione and Moller [1]. Reproduced from MDPI with permission under CC-BY 4.0)

Pathophysiology

Usher syndrome is sometimes classified as a sensory (nonmotile) ciliopathy because of its disruptive effects on genes that code for ciliary cell functions, but this is an oversimplification as it causes disruptions in other cellular functions as well [1]. Functionally, Usher syndrome interferes with sensory cells that express actin filaments which serve as structural proteins in hair cells and photoreceptors [1]. Other sensory nonmotile ciliopathy syndromes include Bardet–Biedl syndrome, Alström syndrome, Meckel syndrome, Joubert syndrome, and Leber congenital amaurosis [1]. Most investigators believe the auditory and vestibular deficits are due to end-organ (inner ear) dysfunction rather than to central (brain) dysfunction.

Figure 68.1, from Castiglionie and Möller [1], schematically illustrates the localization in inner ear hair cells of affected proteins in the three main types of Usher syndrome.

Clinical Presentation

The various types of Usher syndrome present with varying degrees of deficit in hearing, vision (from retinitis pigmentosa), and disequilibrium (from vestibular areflexia).

Table 68.3, from Castiglione and Möller [1], outlines the broad patterns of hearing loss, vestibular function and visual loss, in each of the subtypes of Usher syndrome.

Figure 68.2, from Castiglioni and Möller [1], shows general patterns of hearing loss across the three main types of Usher syndrome.

The *hearing loss* in all types of Usher syndrome is sensorineural and approximately symmetrical.

The *visual loss* in Usher syndrome is due to retinal degeneration that begins by affecting peripheral rod cells, and gradually progresses centrally to involve cone cells [1]. A common initial complaint is night blindness, or generally difficulty seeing in conditions of low illumination.

The *disequilibrium* may manifest in early childhood as delayed motor milestones.

Table 68.3 The broad patterns of hearing loss, vestibular function and visual loss, in each of the subtypes of Usher syndrome

Type	Hearing loss	Vestibular function	Retinitis pigmentosa (RP)	Hearing rehabilitation and communication
USH1	Congenital, severe-to-profound sensorineural hearing loss; however, exceptions with moderate hearing loss are reported.	Absent or abnormal vestibular function; however, exceptions are widely reported, from moderate to normal function	Early onset	Bilateral. Good outcomes with early cochlear implantations. Some exceptions have been reported for patients with *PCDH15* mutations. However, further studies are required to confirm this statement. Tactile signing and sign language may be necessary in selected cases.
USH2	Congenital/prelingual sloping down, moderate-to-severe sensorineural hearing loss.	Normal	Late onset, but usually in young/adult age.	Outcomes can be related to the hearing loss onset. Patients with stable hearing loss may benefit from hearing aids. Cochlear implantations have also shown good outcomes. Bilateral. Bimodal rehabilitation should be considered. They may benefit from sign language.
USH3	Variable onset of moderate-to-severe hearing loss	Variable. Potentially fast progressive.	Variable onset, but essentially during early adulthood.	Bilateral rehabilitation. Due to progression, if indicated, bilateral cochlear implantation is a valid option, but bimodal stimulation can also be effective. Efficacy of sign language in selected cases depends on onset and progression.

From Castiglione and Moller [1]

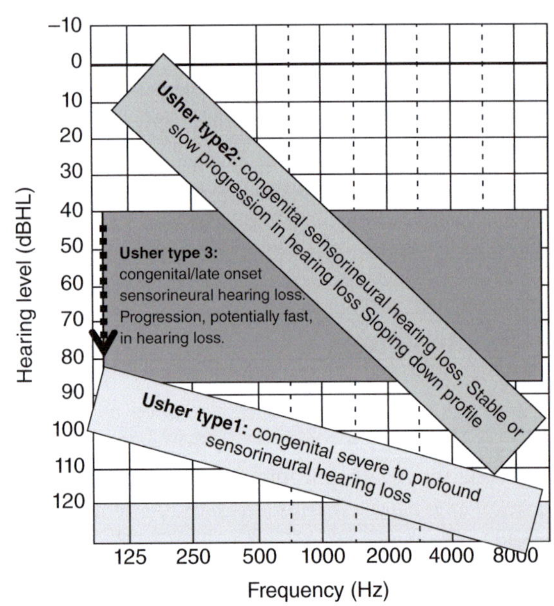

Fig. 68.2 General patterns of hearing loss across the three main types of Usher syndrome. Hearing loss is bilateral, symmetric, and sensorineural for all types. (From Castiglione and Moller [1]. Reproduced from MDPI with permission under CC-BY 4.0)

Workup: Hearing Loss

Standard audiometry shows approximately symmetrical sensorineural hearing loss that, depending on the Usher syndrome type, is variable in severity and progression.

Workup: Visual Loss

Retinitis pigmentosa at an early age can be confirmed by electroretinography [1].

Workup: Vestibular Loss

Vestibular function in Usher syndrome has been studied less than hearing and vision.

Wafa and colleagues [6] studied 90 patients (age 10.9–75.5 years) from all three major types, including eight genetic subtypes, with caloric testing, rotatory chair testing, cervical vestibular evoked myogenic potentials, and computerized dynamic posturography.

For *caloric testing*, Wafa and colleagues used air caloric stimulation, which seems like an odd choice since it is an even weaker stimulus than water caloric stimulation. They categorized responses as "normal, unilateral hypofunction, bilateral hypofunction, or absent." Of 24 patients with Usher syndrome type 1, 21 (87.5%) showed no caloric response; 3 (12.5%) had some response, reflecting the overall more severe impairment in this type. Of 51 patients with Usher syndrome type 2, 45 (88%) had normal caloric responses. Of 3 patients with Usher syndrome type 3, one had bilateral caloric weakness, one had unilateral caloric weakness, and one had normal caloric responses. The overall results of caloric testing by type are displayed in Fig. 68.3.

For *rotatory chair testing*, Wafa and colleagues checked slow harmonic acceleration at octave frequencies from 0.01 to 0.64 Hz and "interpreted [the responses] as absent, present with normal gain, or present with reduced gain." Of 24 patients with Usher syndrome type 1, 14 (58.3%) had no responses. Of 49 patients with Usher syndrome type 2, 48 (96%) had normal responses. Of 6 patients with Usher syndrome type 3, responses varied from normal to absent. The overall results of slow harmonic acceleration are displayed in Fig. 68.4.

For *cervical vestibular evoked myogenic potentials*, Wafa and colleagues "interpreted [the results] based on presence or absence of the P1-N1 response and interaural symmetry of the P1-N1 amplitude." Of 20 patients with Usher syndrome type 1, 19 (95%) had bilaterally absent responses. Of 39 patients with Usher syndrome type 2, 33 (85%) had present and symmetric responses and 6 (17%) had absent responses.

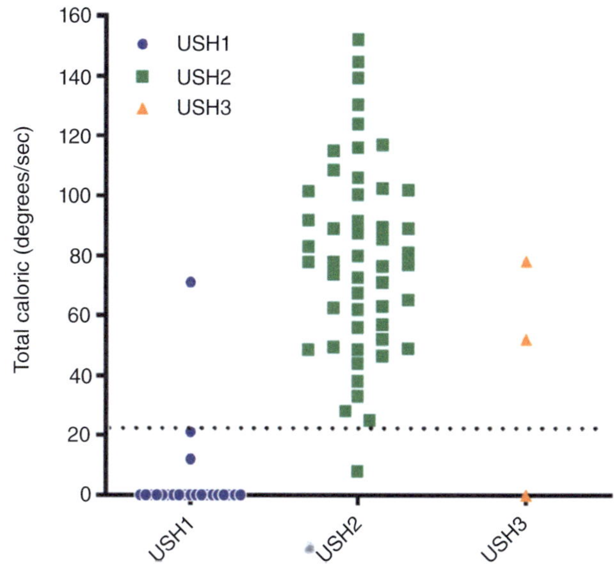

Fig. 68.3 Vestibulo-ocular reflex quantification from caloric testing in patients with Usher syndrome type 1 (USH1), Usher syndrome type 2 (USH2), and Usher syndrome type 3 (USH3). (From Wafa et al. [6]. Reproduced with written permission from Blackwell Publishing)

Of 6 patients with Usher syndrome type 3, 3 (50%) had absent responses, and 3 (50%) had present and symmetric responses.

For *computerized dynamic posturography*, Wafa and colleagues tracked the score in each of the six conditions of sensory organization testing, as well as the composite score. Of 26 patients with Usher syndrome type 1, 21 (80%) exhibited excessive postural sway resulting in falls. Of 53 patients with Usher syndrome type 2, 48 (90%) performed normally. Of 7 patients with Usher syndrome type 3, performance was variable (Fig. 68.5).

The vestibular assessments performed by Wafa and colleagues are a good starting point but are somewhat incomplete; for example, they did not include video head impulse testing or ocular vestibular evoked myogenic potentials. Nevertheless, the overall pattern is that Usher syndrome patients of type 1 tend to have severe vestibular deficits, type 2 have modest deficits, and type 3 are variable.

As far as rotatory chair testing is concerned, other studies have noted that Usher syndrome type 1 patients have significant deficits, while type 2 patients are normal [7].

As far as computerized dynamic posturography is concerned, other studies have noted that Usher syndrome type 1 patients on average perform more poorly than type 2 [7, 8].

The variability of vestibular impairment in Usher syndrome type 3 has been noted by other investigators [9, 10], and in this respect, type 3 can mimic types 1 and 2 [11].

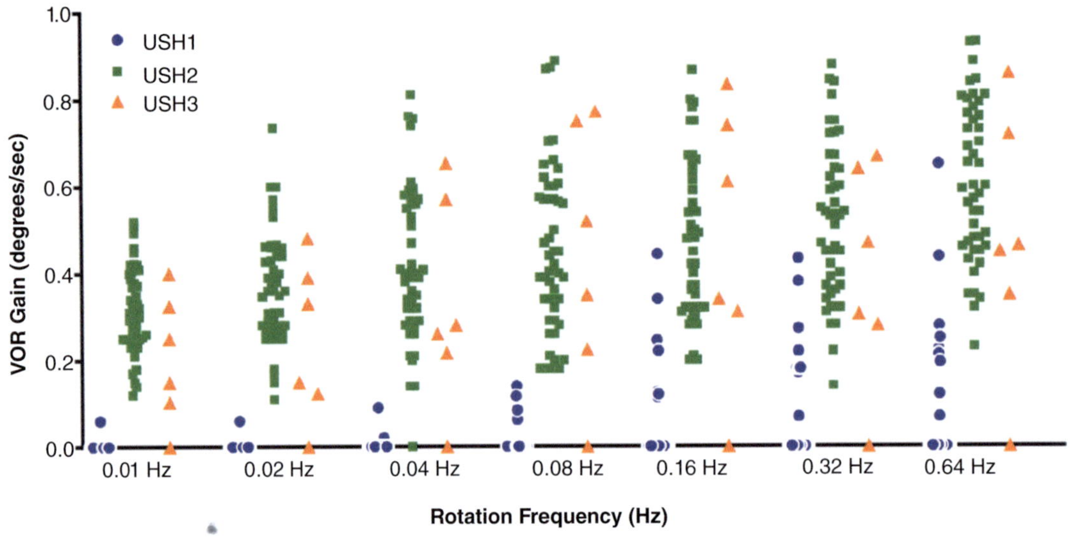

Fig. 68.4 Vestibulo-ocular reflex (VOR) gain quantification at various frequencies from rotatory chair testing in patients with Usher syndrome type 1 (USH1), Usher syndrome type 2 (USH2), and Usher syndrome type 3 (USH3). (From Wafa et al. [6]. Reproduced with written permission from Blackwell Publishing)

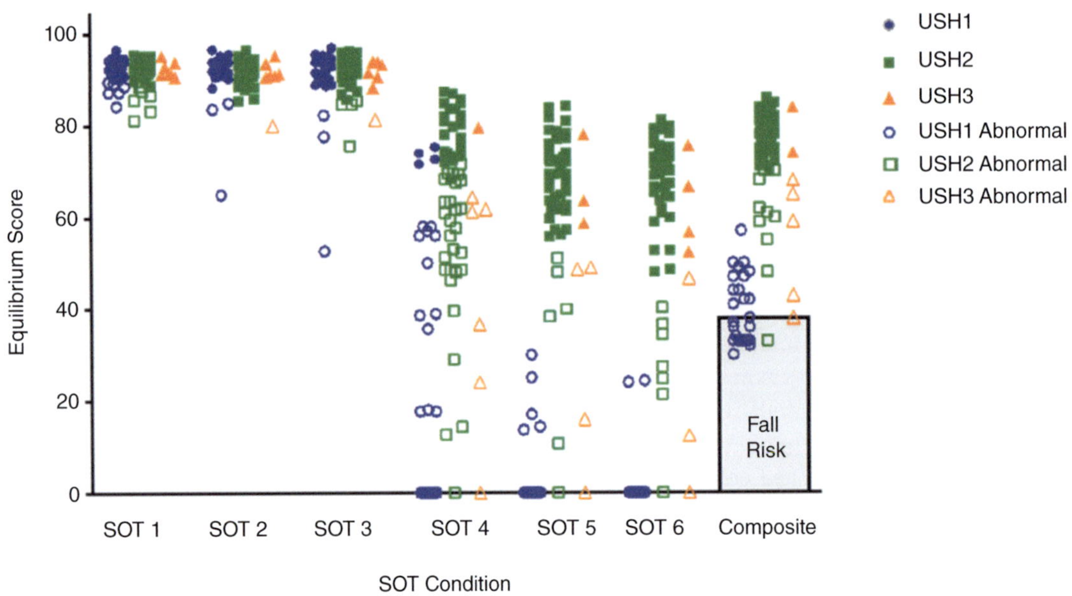

Fig. 68.5 Sensory organization testing (computerized dynamic posturography) results in patients with Usher syndrome type 1 (USH1), Usher syndrome type 2 (USH2), and Usher syndrome type 3 (USH3). SOT 1–6 = Sensory Organization Testing conditions 1–6 of computerized dynamic posturography. (From Wafa et al. [6]. Reproduced with written permission from Blackwell Publishing)

Workup: Genetic Testing

Patients with clinical features of hearing loss and visual loss from a young age should consider consultation with a medical geneticist regarding genetic testing for Usher (and other) syndromes.

Treatment

Usher syndrome type 1 usually presents with profound hearing loss. Since there is no serviceable hearing, cochlear implantation is often suggested. Usher syndrome types 2 and 3 may have less severe hearing loss, and these patients may benefit from a conventional hearing aid.

When the vision loss is recognized, accommodations can be made in terms of lighting and contrast conditions.

Rehabilitation for disequilibrium may attempt sensory substitution with vision (if there remains serviceable vision) and somatosensory input.

References

1. Castiglione A, Moller C. Usher syndrome. Audiol Res. 2022;12(1):42–65.
2. von Graefe A. Vereinzelte Beobachtungen und Bemerkungen: Exeptionelles Verhalten des Gesichtfeldes bei Pigmentartung der Netzhaut [Unique observations and remarks: Exceptional behavior of the field of vision in retinal pigmentation]. Archiv für klinische und experimentelle Ophthalmologie [Arch Clin Exp Ophthalmol]. 1858;4:250–3.
3. Usher CH. On the inheritance of retinitis pigmentosa, with notes of cases. R Lond Ophthal Hosp Rep. 1914;19:130–256.
4. Boughman JA, Vernon M, Shaver KA. Usher syndrome: definition and estimate of prevalence from two high-risk populations. J Chronic Dis. 1983;36(8):595–603.
5. Cohen M, Bitner-Glindzicz M, Luxon L. The changing face of Usher syndrome: clinical implications. Int J Audiol. 2007;46(2):82–93.
6. Wafa TT, Faridi R, King KA, Zalewski C, Yousaf R, Schultz JM, et al. Vestibular phenotype-genotype correlation in a cohort of 90 patients with Usher syndrome. Clin Genet. 2021;99(2):226–35.
7. Moller CG, Kimberling WJ, Davenport SL, Priluck I, White V, Biscone-Halterman K, et al. Usher syndrome: an otoneurologic study. Laryngoscope. 1989;99(1):73–9.
8. Caldani S, Bucci MP, Tisne M, Audo I, Van Den Abbeele T, Wiener-Vacher S. Postural instability in subjects with Usher syndrome. Front Neurol. 2019;10:830.
9. Marouf A, Johnson B, Alagramam KN. Usher syndrome IIIA: a review of the disorder and preclinical research advances in therapeutic approaches. Hum Genet. 2022;141(3–4):759–83.
10. Sadeghi M, Cohn ES, Kimberling WJ, Tranebjaerg L, Moller C. Audiological and vestibular features in affected subjects with USH3: a genotype/phenotype correlation. Int J Audiol. 2005;44(5):307–16.
11. Pennings RJ, Fields RR, Huygen PL, Deutman AF, Kimberling WJ, Cremers CW. Usher syndrome type III can mimic other types of Usher syndrome. Ann Otol Rhinol Laryngol. 2003;112(6):525–30.

Part IX

Diagnoses—Diseases Involving the Brain, Spinal Cord and Peripheral Nerves

69 Migraine-Associated Vertigo

Overview

Migraine-associated vertigo (MAV) is thought to be a version of migraine that manifests with the sensation of disequilibrium, and may or may not involve headaches meeting criteria for migraines. The disequilibrium itself can have a variety of presentations in terms of quality, frequency, duration and triggers. MAV is thought to affect nearly 1% of the population. The underlying cause of MAV is unknown; research has not yet identified specific genes responsible for MAV. Patients with MAV usually have normal physical examinations, though some may have subtle and variable nystagmus on infrared video oculography. Like migraines in general, MAV is a diagnosis of exclusion, and the main value of paraclinical testing (otovestibular testing and sometimes imaging) is to exclude competing pathologies from the differential diagnosis. There is very little literature on treatment specifically for MAV, so most practitioners extrapolate from the existing literature on regular migraine headaches. We favor starting with nonpharmacological strategies because of their relatively favorable risk profile.

Introduction

The idea that the symptom of disequilibrium can be attributed to an underlying migraine disorder goes by several terms in the literature, including "migraine-associated vertigo," "migrainous vertigo," "vestibular migraine," "vertiginous migraine," and others. We shall use the phrase "*migraine-associated vertigo*" (MAV) for convenience.

The third edition of the International Headache Society Classification of Headache Disorders [1] recognizes 15 types of migraine, 2 types of probable migraine, and two "syndromes that may be associated with migraine." Migraine can supposedly present with an impressively broad variety of symptoms [2]. The polymorphic phenotype raises the reasonable question of whether these comprise different manifestations of a single pathology, or whether they are different diseases entirely; until tools arise for confirming/disproving the diagnosis, this question will remain fodder for the perennial nosologic debate regarding "lumping" and "splitting" [3, 4].

There is a well-developed literature on MAV [5–12].

Neuhauser and Lempert [10] proposed the following criteria for "vestibular migraine:"

- *Definite vestibular migraine*
  - Episodic vestibular symptoms of at least moderate severity
  - Current or previous history of migraine according to the ICHD-II [13]
  - One of the following migrainous symptoms during two or more attacks of vertigo: migrainous headache, photophobia, phonophobia, visual aura, or other aura
  - Other causes ruled out by appropriate investigations
- *Probable vestibular migraine*
  - Episodic vestibular symptoms of at least moderate severity
  - One of the following:
    - Current or previous history of migraine according to the ICHD-II
    - Migrainous symptoms during vestibular symptoms
    - Migraine precipitants of vertigo in more than 50% of attacks (food triggers, sleep irregularities, or hormonal change)
    - Response to migraine medications in more than 50% of attacks
  - Other causes ruled out by appropriate investigations

Epidemiology

The prevalence of migraine in the general population is about 10% [14]. Approximately one-third of patients with migraines experience some degree of disequilibrium during their migraines [15]. These statistics taken together would lead one to expect a prevalence of migraine-related disequilibrium of about 3%.

M. Cherchi, *Otoneurology and Vestibular Medicine*, https://doi.org/10.1007/978-3-031-94841-1_69

However, when using the stricter diagnostic criteria of Neuhauser and Lempert [10], the prevalence of MAV turns out to be 0.98%. This leaves another 2.02% unaccounted for. One interpretation of this discrepancy is that because migraine (broadly speaking) is so common, it is statistically likely that some patients with (nonvestibular) migraine have vestibular symptoms attributable to some nonmigrainous disease. This is important to take into account when considering a differential diagnosis, since a number of "dizzy diseases" have an association with migraine that his higher than what would be expected by chance (10% of the general population, as mentioned earlier). For example:

- 33% of patients with benign paroxysmal positional vertigo have migraines [16].
- 56% of patients with Meniere's disease have migraines [17].
- 23% of patients with mal de debarquement syndrome have migraines [18].

To reiterate, this suggests that a patient with migraines and with disequilibrium may have some cause for the disequilibrium other than migraines. This is why the criteria for MAV discussed earlier explicitly state that the *symptoms must not be attributable to another disorder*. Taking this into account along with the fact that migraine is still a disease for which no confirmatory test exists (see below), it should be emphasized that *MAV is a clinical diagnosis, and a diagnosis of exclusion.*

Genetics

The study of the genetics of migraine (broadly construed) is an emerging discipline [19–24]. For MAV the evidence of heritability is still very limited [25], though several genes have been implicated [26–28]. This field is likely to advance, but as of this writing, genetic testing does not comprise a component of the standard workup.

Pathophysiological Mechanism of Disease

Despite considerable research, the underlying pathophysiology of migraines (broadly construed) remains uncertain. Research on the pain associated with migraines, and the positive response to serotonin agonists such as triptans, implicates brainstem dysfunction at the level of the serotonin pathways of the trigeminal system [29]. Research on migraine aura (in general), and specifically on the circulatory changes in the occipital cortex associated with visual auras, implicates cortical dysfunction, rooted either in vascular dysregulation, or abnormal electrical activity, or both [30–32].

The same general mechanisms have been proposed for MAV. The fact that patients with MAV may exhibit nystagmus (see below) implicates dysfunction at the level of the brainstem (particularly the vestibular nuclei). But the other obvious possibility is cerebral dysfunction at the level of the purported vestibular cortex in the temporal lobe. These proposals are logical, but remain speculative.

Clinical Presentation

Patients offer different descriptors of the disequilibrium they experience during an episode of MAV. Neuhauser and colleagues [11] cite the following descriptors and associated frequencies:

- Sensation of spinning or rotation (70%)
- Intolerance of head motion (48%)
- Disequilibrium triggered by changes in position (42%)
- Less common: motion sickness, floating, rocking, tilting, walking on an uneven surface, lightheadedness

The duration of the disequilibrium is quite variable. Neuhauser and colleagues [11] list the following:

- 5–60 minutes (33%)
- 1–24 hours (21%)
- Seconds to 5 minutes (18%)
- More than 24 hours (2%)

There are also well-documented series of MAV symptoms being chronic [33].

Migraine in general, and MAV, can present with aural symptoms. Dash and colleagues [34] report the following aural symptoms and associated frequencies:

- 66% describe sonophobia
- 63% describe tinnitus
- 32% describe hearing loss
- 11% report fluctuating hearing loss and aural fullness

These aural symptoms can, of course, can be features of nonmigrainous otovestibular diseases, and are thus another reason to keep such pathologies on the differential diagnosis.

For migraine in general, and probably also for MAV, some patients report symptom triggers such as:

- Weather changes: onset of storm fronts, changes of season
- Activities and states: physical exertion, dehydration, sleep deprivation, menses, exposure to bright light
- Dietary factors: caffeine, chocolate, alcohol, aged cheeses, monosodium glutamate, nitrites, artificial sweeteners

Patients with migraines in general, and probably also MAV, often give a history of susceptibility to motion sickness. This is reported in 45% of pediatric patients [35] and 50% of adult patients [36].

Physical Examination

Unless a patient with MAV has other medical problems, the physical examination is usually normal (but see below regarding the ocular motor examination).

Ocular Motor Examination

A patient with MAV may have eye movement abnormalities that are intermittent and fluctuating [36–38]; these may be sufficiently subtle that they are only detectable on infrared video oculography.

Testing: Auditory

While there are case reports of hearing loss in patients with migraine [39–42], our experience is that patients with MAV generally have normal hearing, unless MAV is comorbid with an otologic disorder such as Ménière's [43].

Testing: Vestibular

Studies of other otovestibular tests occasionally report abnormalities in patients with migraine, including otoacoustic emissions [44], brainstem auditory evoked responses [34, 45], vestibular evoked myogenic potentials [46], and caloric testing [47, 48], rotatory chair testing [49–52], and computerized dynamic posturography [47–49]. However, none has been rigorously evaluated in patients specifically with MAV.

Testing: Imaging

It has been reported that approximately 23% of migraine patients accumulate white matter abnormalities not attributable to other disorders. Furthermore, migraine patients are at approximately a four-fold higher risk for developing white matter abnormalities than patients without migraine, even after controlling for vascular risk factors [53]. The imaging findings typically include scattered punctate foci of T2 and FLAIR signal hyperintensity in the deep cerebral white matter. These imaging findings, which can resemble those of multiple sclerosis, are generally not accompanied by any clinical correlates. In our experience, these common MRI findings do not appear to be associated with an increased risk of migrainous infarction.

Testing: Overall

None of the abnormal findings in the paraclinical tests mentioned above (auditory, vestibular, imaging) comprise proof-positive evidence for MAV. The main benefit of undertaking these studies is to assess for competing diagnoses—such as otologic disease that could account for auditory and/or vestibular symptoms.

Differential Diagnosis

Since patients with migraines can have causes of disequilibrium unrelated to their migraines, and since MAV is a diagnosis of exclusion, it is medically reasonable to undertake a screening otovestibular workup in patients for whom a diagnosis of MAV is being considered.

For a potential MAV patient who only has vestibular symptoms (without any auditory symptoms), it is reasonable to check cervical and ocular vestibular evoked myogenic potentials, video head impulse testing and possibly videonystagmography and computerized dynamic posturography. If the patient has any auditory symptoms whatsoever, then it is reasonable to check otoacoustic emissions (if under the age of 60 years), audiometry, cervical and ocular vestibular evoked myogenic potentials, and possibly electrocochleography.

Reasons for checking neuroimaging of the brain include:

- If the headache is new, or if it has developed new associated features, or if there are focal abnormalities on neurological examination, then imaging is recommended by the American Academy of Neurology [54, 55].
- Clear lateralizing deficits on otovestibular testing compatible with specific pathologies that could be objectively (e.g., vestibular schwannoma, superior semicircular canal dehiscence).

Given the otologic diseases on the differential diagnosis (at least for MAV), MRI of the brain and internal auditory canals without and with contrast is more likely to be helpful; this images not only the brain, but also structures relevant to vestibular disease (labyrinth, vestibular nerve, brainstem, cerebellum). CT imaging is generally less helpful, except in the specific circumstance of superior semicircular canal dehiscence, in which case a temporal bone CT without contrast may be considered.

Treatment: Overview

The vast majority of research in the treatment of migraine focuses on outcomes in patients whose phenotype meets criteria for migraine headache. There is relatively sparse research on the treatment of supposed subtypes of migraine, such as migraine presenting exclusively with visual aura ("retinal migraine," "optical migraine"). Consequently, clinicians confronted with a case of a subtype of migraine will usually extrapolate from the existing literature on migraine headaches.

Management of MAV suffers from this same problem. Reviews of literature focused on the treatment of MAV note that the quantity and quality of studies is poor.

A Cochrane Review of pharmacologic prophylaxis for MAV concluded that:

> There is very limited evidence from placebo-controlled randomised trials regarding the efficacy and potential harms of pharmacological interventions for prophylaxis of vestibular migraine. We only identified evidence for two of our interventions of interest (beta-blockers and calcium channel blockers) and all evidence was of low or very low certainty. Further research is necessary to identify whether these treatments are effective at improving symptoms and whether there are any harms associated with their use [56].

A Cochrane Review of nonpharmacologic prophylaxis for MAV concluded that:

> There is a paucity of evidence for non-pharmacological interventions that may be used for prophylaxis of vestibular migraine. Only a limited number of interventions have been assessed by comparing them to no intervention or a placebo treatment, and the evidence from these studies is all of low or very low certainty. We are therefore unsure whether any of these interventions may be effective at reducing the symptoms of vestibular migraine and we are also unsure whether they have the potential to cause harm [57].

A Cochrane Review of pharmacologic abortive treatments for MAV concluded that:

> The evidence for interventions used to treat acute attacks of vestibular migraine is very sparse. We identified only two studies, both of which assessed the use of triptans. We rated all the evidence as very low-certainty, meaning that we have little confidence in the effect estimates and cannot be sure if triptans have any effect on the symptoms of vestibular migraine [58].

Treatment: Nonpharmacologic Approaches

Absent robust evidence for treatment specifically aimed at MAV, most practitioners simply extrapolate from existing literature regarding treatment of migraine in general. Since the risks of pharmacologic intervention overall probably exceed those of nonpharmacologic intervention, our usual practice is to consider nonpharmacologic strategies first.

Nutritional Supplements

The American Academy of Neurology's evidence-based guideline on nonpharmacologic treatments for migraine in general [59] found efficacy for riboflavin, coenzyme Q10 and magnesium, which are endogenous substances, and available without a prescription in the United States.

Magnesium is taken at a dose of 300–600 mg once per day. Different formulations have been studied, including magnesium citrate and magnesium oxide. The efficacy of magnesium becomes apparent within 9–12 weeks in adults [60] and within 16 weeks in children [61]. Magnesium is currently not recommended during pregnancy because of possible effects on the bone metabolism of the fetus [62].

Riboflavin (vitamin B2) is taken at a dose of 400 mg once per day. The efficacy of riboflavin becomes apparent around 12 weeks [63]. Riboflavin can be taken during pregnancy and nursing.

Vitamin D has been studied as a migraine prophylactic. Some studies [64–70] report that taking vitamin D supplementation improves migraine frequency. Vitamin D supplementation as a prophylactic treatment for migraine has been studied in both pediatric [66] and adult populations [67]. Recommended doses of vitamin D supplementation vary from 1000 to 4000 international units (I.U.) per day; one study explored 50,000 I.U. per week [68] (amounting to about 7000 mg per day). The time to onset of efficacy remains unclear; observation periods varied from 10 weeks [68] to 24 weeks [67].

Coenzyme Q10 is taken at a dose of 100 mg three times per day. The efficacy of coenzyme Q10 becomes apparent around 12 weeks [71]. Coenzyme Q10 can be taken during pregnancy and nursing.

Melatonin is taken at a dose of 3 mg once per night [72]. Melatonin has been studied in randomized controlled trials versus placebo [73, 74] and versus other medications [75]. While some reviews describe these trials as demonstrating a "positive trend" [76], most systematic reviews and meta-analyses conclude that the evidence is still emerging [77–81]. Melatonin appears to be safe in pregnancy and lactation [82]. Melatonin is currently available over-the-counter in the United States.

Herbal Supplements

The American Academy of Neurology's evidence-based guideline on nonpharmacologic treatments for migraine [59] found efficacy for several herbal supplements. These supplements do not require prescriptions.

Feverfew (also called *Tanacetum parthenium*) has been studied as a migraine prophylactic. There are several different extracts of feverfew [83]. The most commonly used extract of feverfew, called parthenolides, is taken at a dose of 100–300 mg four times per day. Another extract of feverfew, called MIG-99, is taken at a dose of 6.25 mg three times per day [84]. Several Cochrane reviews found insufficient evidence to support the use of feverfew for migraine prophylaxis [85, 86], though the most recent Cochrane review as of this writing concluded, "Since the last version of this review, one larger rigorous study has been included, reporting a difference in effect between feverfew and placebo of 0.6 attacks per month. This adds some positive evidence to the mixed and inconclusive findings of the previous review. However, this constitutes low quality evidence, which needs to be confirmed in larger rigorous trials with stable feverfew extracts and clearly defined migraine populations before firm conclusions can be drawn" [87].

A previously popular herbal supplement, *Butterbur* (also called *Petasites hybridus*), is taken at a dose of 50 to 75 mg twice per day. The efficacy of Butterbur becomes apparent around 16 weeks [88]. Some, but not all, preparations of Butterbur, contain pyrrolizidine alkaloids, which can damage the liver in humans, and can cause cancer in some animal studies [89]. Since different preparations of Butterbur vary in their content of pyrrolizidine alkaloids [90], most patients prefer to avoid this supplement.

Acupuncture

Acupuncture has been applied to an enormous number of illnesses, and in the 1980s, researchers began examining this approach systematically through randomized trials for migraines. As of this writing, there is reasonable evidence that acupuncture is effective both for migraine abortive therapy [91] and for migraine prophylaxis [92–94]. The frequency and duration of acupuncture treatment as prophylaxis for migraine remain to be clarified, though one review recommended two sessions per week for 10 weeks [95]. Acupuncture usually does not require a referral. Acupuncture is rarely covered by insurance.

Transcutaneous Electrical Neurostimulation

The technology of transcutaneous electrical neurostimulation (TENS) units has long been in use for a variety of musculoskeletal disorders. More recent studies of this therapy for migraine prophylaxis have demonstrated it to be both safe [96] and effective [97]. This therapeutic modality is currently marketed in the United States as medical devices that require a prescription.

Behavioral Therapy: Biofeedback

Biofeedback is a behavioral therapy that has been used in many medical conditions, and has been studied specifically for migraine. There is an emerging literature about the use of biofeedback for management of migraine [98–149]. Various biofeedback strategies have been studied. Patients usually learn biofeedback under the instruction of a specially trained psychologist. One can search for an appropriately trained psychologist through the website of the Association for Applied Psychophysiology and Biofeedback. Biofeedback does not require a referral and is not covered by insurance in the United States.

Behavioral Therapy: Relaxation Training

Although the biological connection between psychological stress and migraine is uncertain [150, 151], it appears that behavioral therapies such as relaxation training and stress management benefit many patients with migraine [152]. The literature regarding relaxation training is emerging [109, 125, 127, 132, 142, 148, 149, 153–161]. A variety of relaxation training strategies have been attempted. Relaxation training does not require a referral, and is not covered by insurance in the United States.

Notes on Migraine Treatment During Pregnancy and Nursing

Although more than half of women experience improvement in their migraine patterns (without any treatment) during pregnancy, some women still have migraine symptoms requiring management [162]. Unfortunately, most of the drugs commonly used in migraine are relatively contraindicated in pregnancy. Of the nonpharmacologic approaches discussed above, those applicable in pregnancy include smoking cessation, maintaining a regular sleep schedule, aerobic exercise, some nutritional supplements (riboflavin, vitamin D and coenzyme Q10), acupuncture, biofeedback, transcutaneous electrical neurostimulation and relaxation training.

Prognosis

The prognosis of MAV is probably similar to that of migraine in general, though this has not been formally assessed in large prospective studies.

References

1. Headache Classification Committee of the International Headache Society. Headache Classification Committee of the International Headache Society (IHS) the international classification of headache disorders, 3rd edition. Cephalalgia. 2018;38(1):1–211.
2. Buchholz DW, Reich SG. The menagerie of migraine. Semin Neurol. 1996;16(1):83–93.
3. Endersby J. Lumpers and splitters: Darwin, Hooker, and the search for order. Science. 2009;326(5959):1496–9.
4. Thaxton C, Goldstein J, DiStefano M, Wallace K, Witmer PD, Haendel MA, et al. Lumping versus splitting: How to approach defining a disease to enable accurate genomic curation. Cell Genom. 2022;2(5):100131.
5. Beh SC, Masrour S, Smith SV, Friedman DI. The spectrum of vestibular migraine: clinical features, triggers, and examination findings. Headache. 2019;59(5):727–40.
6. Cherchi M, Hain TC. Migraine-associated vertigo. Otolaryngol Clin N Am. 2011;44(2):367–75, viii–ix.
7. Cherchi M, Hain TC. Migraine associated vertigo. In: Dispenza F, De Stefano A, editors. Textbook of vertigo: diagnosis and management. Philadelphia: Jaypee Brothers Medical Publishers; 2014. p. 184–93.
8. Hain TC, Cherchi M. Migraine associated vertigo. In: Kesser B, Gleason T, editors. Dizziness and vertigo across the lifespan. Elsevier; 2018. p. 135–41.
9. Hain TC, Cherchi M. Migraine associated vertigo. Lea J, Pothier D, editors. Adv Otorhinolaryngol. 2019;82:119–26.
10. Neuhauser H, Lempert T. Vestibular migraine. Neurol Clin. 2009;27(2):379–91.
11. Neuhauser H, Leopold M, von Brevern M, Arnold G, Lempert T. The interrelations of migraine, vertigo, and migrainous vertigo. Neurology. 2001;56(4):436–41.
12. Neuhauser HK, Radtke A, von Brevern M, Feldmann M, Lezius F, Ziese T, et al. Migrainous vertigo: prevalence and impact on quality of life. Neurology. 2006;67(6):1028–33.
13. International Headache Society. The international classification of headache disorders: 2nd edition. Cephalalgia. 2004;24(Suppl 1):9–160.
14. Stewart W, Shechter A, Rasmussen B, Migraine prevalence. A review of population based studies. Neurology. 1994;44(suppl 4):S17–23.
15. Selby G, Lance JW. Observations on 500 cases of migraine and allied vascular headache. J Neurol Neurosurg Psychiatry. 1960;23:23–32.
16. Uneri A. Migraine and benign paroxysmal positional vertigo: an outcome study of 476 patients. Ear Nose Throat J. 2004;83(12):814–5.
17. Radtke A, Lempert T, Gresty MA, Brookes GB, Bronstein AM, Neuhauser H. Migraine and Meniere's disease: is there a link? Neurology. 2002;59(11):1700–4.
18. Hain TC, Hanna PA, Rheinberger MA. Mal de debarquement. Arch Otolaryngol Head Neck Surg. 1999;125(6):615–20.
19. Anttila V, Wessman M, Kallela M, Palotie A. Genetics of migraine. Handb Clin Neurol. 2018;148:493–503.
20. Cader MZ. The genetics of migraine and the path to precision medicine. Prog Brain Res. 2020;255:403–18.
21. de Boer I, van den Maagdenberg A, Terwindt GM. Advance in genetics of migraine. Curr Opin Neurol. 2019;32(3):413–21.
22. Ducros A. Genetics of migraine. Rev Neurol (Paris). 2021;177(7):801–8.
23. Grangeon L, Lange KS, Waliszewska-Prosół M, Onan D, Marschollek K, Wiels W, et al. Genetics of migraine: where are we now? J Headache Pain. 2023;24(1):12.
24. Sutherland HG, Albury CL, Griffiths LR. Advances in genetics of migraine. J Headache Pain. 2019;20(1):72.
25. Paz-Tamayo A, Perez-Carpena P, Lopez-Escamez JA. Systematic Review of Prevalence Studies and Familial Aggregation in Vestibular Migraine. Front Genet. 2020;11:954.
26. Oh EH, Shin JH, Cho JW, Choi SY, Choi KD, Choi JH. TRPM7 as a candidate gene for vestibular migraine. Front Neurol. 2020;11:595042.
27. Wu X, Qiu F, Wang Z, Liu B, Qi X. Correlation of 5-HTR6 gene polymorphism with vestibular migraine. J Clin Lab Anal. 2020;34(2):e23042.
28. Bahmad F Jr, DePalma SR, Merchant SN, Bezerra RL, Oliveira CA, Seidman CE, et al. Locus for familial migrainous vertigo disease maps to chromosome 5q35. Ann Otol Rhinol Laryngol. 2009;118(9):670–6.
29. Goadsby PJ. Current concepts of the pathophysiology of migraine. Neurol Clin. 1997;15(1):27–42.
30. Leão AAP. Spreading depression of activity in the cerebral cortex. J Neurophysiol. 1944;7:359–90.
31. Leão AAP, Morison RS. Propagation of spreading cortical depression. J Neurophysiol. 1945;8:33–46.
32. Olesen J, Friberg L, Olsen TS, Iversen HK, Lassen NA, Andersen AR, et al. Timing and topography of cerebral blood flow, aura, and headache during migraine attacks. Ann Neurol. 1990;28(6):791–8.
33. Waterston J. Chronic migrainous vertigo. J Clin Neurosci. 2004;11(4):384–8.
34. Dash AK, Panda N, Khandelwal G, Lal V, Mann SS. Migraine and audiovestibular dysfunction: is there a correlation? Am J Otolaryngol. 2008;29(5):295–9.
35. Barabas G, Matthews WS, Ferrari M. Childhood migraine and motion sickness. Pediatrics. 1983;72(2):188–90.
36. Kayan A, Hood JD. Neuro-otological manifestations of migraine. Brain J Neurol. 1984;107(Pt 4):1123–42.
37. von Brevern M, Zeise D, Neuhauser H, Clarke AH, Lempert T. Acute migrainous vertigo: clinical and oculographic findings. Brain J Neurol. 2005;128(Pt 2):365–74.
38. Polensek SH, Tusa RJ. Nystagmus during attacks of vestibular migraine: an aid in diagnosis. Audiol Neurootol. 2010;15(4):241–6.
39. Lee H, Lopez I, Ishiyama A, Baloh RW. Can migraine damage the inner ear? Arch Neurol. 2000;57(11):1631–4.
40. Lee H, Whitman GT, Lim JG, Yi SD, Cho YW, Ying S, et al. Hearing symptoms in migrainous infarction. Arch Neurol. 2003;60(1):113–6.
41. Viirre ES, Baloh RW. Migraine as a cause of sudden hearing loss. Headache. 1996;36(1):24–8.
42. Lipkin AF, Jenkins HA, Coker NJ. Migraine and sudden sensorineural hearing loss. Arch Otolaryngol Head Neck Surg. 1987;113(3):325–6.
43. Battista RA. Audiometric findings of patients with migraine-associated dizziness. Otol Neurotol. 2004;25(6):987–92.
44. Bolay H, Bayazit YA, Gunduz B, Ugur AK, Akcali D, Altunyay S, et al. Subclinical dysfunction of cochlea and cochlear efferents in migraine: an otoacoustic emission study. Cephalalgia. 2008;28(4):309–17.
45. Bayazit Y, Yilmaz M, Mumbuc S, Kanlikama M. Assessment of migraine-related cochleovestibular symptoms. Rev Laryngol Otol Rhinol (Bord). 2001;122(2):85–8.
46. Baier B, Stieber N, Dieterich M. Vestibular-evoked myogenic potentials in vestibular migraine. J Neurol. 2009;256(9):1447–54.
47. Celebisoy N, Gokcay F, Sirin H, Bicak N. Migrainous vertigo: clinical, oculographic and posturographic findings. Cephalalgia. 2008;28(1):72–7.
48. Teggi R, Colombo B, Bernasconi L, Bellini C, Comi G, Bussi M. Migrainous vertigo: results of caloric testing and stabilometric findings. Headache. 2009;49(3):435–44.

49. Furman JM, Sparto PJ, Soso M, Marcus D. Vestibular function in migraine-related dizziness: a pilot study. J Vestib Res Equilib Orient. 2005;15(5–6):327–32.
50. Arriaga MA, Chen DA, Hillman TA, Kunschner L, Arriaga RY. Visually enhanced vestibulo-ocular reflex: a diagnostic tool for migraine vestibulopathy. Laryngoscope. 2006;116(9):1577–9.
51. Cass SP, Furman JM, Ankerstjerne K, Balaban C, Yetiser S, Aydogan B. Migraine-related vestibulopathy. Ann Otol Rhinol Laryngol. 1997;106(3):182–9.
52. Kim EK, Sienko N, Gardi A, Krauter R, Pasquesi L, Sharon JD. Visually enhanced vestibulo-ocular reflex gain in patients with vestibular disease. Laryngoscope Investig Otolaryngol. 2023;8(4):1061–7.
53. Swartz RH, Kern RZ. Migraine is associated with magnetic resonance imaging white matter abnormalities: a meta-analysis. Arch Neurol. 2004;61(9):1366–8.
54. Practice parameter: the utility of neuroimaging in the evaluation of headache in patients with normal neurologic examinations (summary statement). Report of the Quality Standards Subcommittee of the American Academy of Neurology. Neurology. 1994;44(7):1353–4.
55. Frishberg BM. The utility of neuroimaging in the evaluation of headache in patients with normal neurologic examinations. Neurology. 1994;44(7):1191–7.
56. Webster K, Dor A, Galbraith K, Kassem LH, Harrington-Benton N, Judd O, et al. Pharmacological interventions for prophylaxis of vestibular migraine. Cochrane Database Syst Rev. 2023;2023(4):CD015187.
57. Webster KE, Dor A, Galbraith K, Haj Kassem L, Harrington-Benton NA, Judd O, et al. Non-pharmacological interventions for prophylaxis of vestibular migraine. Cochrane Database Syst Rev. 2023;4(4):CD015321.
58. Webster KE, Dor A, Galbraith K, Haj Kassem L, Harrington-Benton NA, Judd O, et al. Pharmacological interventions for acute attacks of vestibular migraine. Cochrane Database Syst Rev. 2023;4(4):CD015322.
59. Holland S, Silberstein SD, Freitag F, Dodick DW, Argoff C, Ashman E, et al. Evidence-based guideline update: NSAIDs and other complementary treatments for episodic migraine prevention in adults: report of the Quality Standards Subcommittee of the American Academy of Neurology and the American Headache Society. Neurology. 2012;78(17):1346–53.
60. Peikert A, Wilimzig C, Kohne-Volland R. Prophylaxis of migraine with oral magnesium: results from a prospective, multicenter, placebo-controlled and double-blind randomized study. Cephalalgia. 1996;16(4):257–63.
61. Wang F, Van Den Eeden SK, Ackerson LM, Salk SE, Reince RH, Elin RJ. Oral magnesium oxide prophylaxis of frequent migrainous headache in children: a randomized, double-blind, placebo-controlled trial. Headache. 2003;43(6):601–10.
62. Yokoyama K, Takahashi N, Yada Y, Koike Y, Kawamata R, Uehara R, et al. Prolonged maternal magnesium administration and bone metabolism in neonates. Early Hum Dev. 2010;86(3):187–91.
63. Schoenen J, Jacquy J, Lenaerts M. Effectiveness of high-dose riboflavin in migraine prophylaxis. A randomized controlled trial. Neurology. 1998;50(2):466–70.
64. Ghorbani Z, Togha M, Rafiee P, Ahmadi ZS, Rasekh Magham R, Haghighi S, et al. Vitamin D in migraine headache: a comprehensive review on literature. Neurol Sci. 2019;40(12):2459–77.
65. Ghorbani Z, Togha M, Rafiee P, Ahmadi ZS, Rasekh Magham R, Djalali M, et al. Vitamin D3 might improve headache characteristics and protect against inflammation in migraine: a randomized clinical trial. Neurol Sci. 2020;41:1183.
66. Kilic B, Kilic M. Evaluation of vitamin D levels and response to therapy of childhood migraine. Medicina (Kaunas). 2019;55(7):321.
67. Gazerani P, Fuglsang R, Pedersen JG, Sorensen J, Kjeldsen JL, Yassin H, et al. A randomized, double-blinded, placebo-controlled, parallel trial of vitamin D3 supplementation in adult patients with migraine. Curr Med Res Opin. 2019;35(4):715–23.
68. Mottaghi T, Askari G, Khorvash F, Maracy MR. Effect of Vitamin D supplementation on symptoms and C-reactive protein in migraine patients. J Res Med Sci. 2015;20(5):477–82.
69. Buettner C, Nir RR, Bertisch SM, Bernstein C, Schain A, Mittleman MA, et al. Simvastatin and vitamin D for migraine prevention: a randomized, controlled trial. Ann Neurol. 2015;78(6):970–81.
70. Cayir A, Turan MI, Tan H. Effect of vitamin D therapy in addition to amitriptyline on migraine attacks in pediatric patients. Braz J Med Biol Res. 2014;47(4):349–54.
71. Hershey AD, Powers SW, Vockell AL, Lecates SL, Ellinor PL, Segers A, et al. Coenzyme Q10 deficiency and response to supplementation in pediatric and adolescent migraine. Headache. 2007;47(1):73–80.
72. Peres MF, Zukerman E, da Cunha Tanuri F, Moreira FR, Cipolla-Neto J. Melatonin, 3 mg, is effective for migraine prevention. Neurology. 2004;63(4):757.
73. Alstadhaug KB, Odeh F, Salvesen R, Bekkelund SI. Prophylaxis of migraine with melatonin: a randomized controlled trial. Neurology. 2010;75(17):1527–32.
74. Ebrahimi-Monfared M, Sharafkhah M, Abdolrazaghnejad A, Mohammadbeigi A, Faraji F. Use of melatonin versus valproic acid in prophylaxis of migraine patients: a double-blind randomized clinical trial. Restor Neurol Neurosci. 2017;35(4):385–93.
75. Gonçalves AL, Martini Ferreira A, Ribeiro RT, Zukerman E, Cipolla-Neto J, Peres MF. Randomised clinical trial comparing melatonin 3 mg, amitriptyline 25 mg and placebo for migraine prevention. J Neurol Neurosurg Psychiatry. 2016;87(10):1127–32.
76. Moreno-Ajona D, Villar-Martínez MD, Goadsby PJ. Emerging targets for migraine treatment. Neurol India. 2021;69(Supplement):S98–s104.
77. Leite Pacheco R, de Oliveira Cruz Latorraca C, Adriano Leal Freitas da Costa A, Luiza Cabrera Martimbianco A, Vianna Pachito D, Riera R. Melatonin for preventing primary headache: a systematic review. Int J Clin Pract. 2018;72(7):e13203.
78. Liampas I, Siokas V, Brotis A, Vikelis M, Dardiotis E. Endogenous melatonin levels and therapeutic use of exogenous melatonin in migraine: systematic review and meta-analysis. Headache. 2020;60(7):1273–99.
79. Long R, Zhu Y, Zhou S. Therapeutic role of melatonin in migraine prophylaxis: a systematic review. Medicine (Baltimore). 2019;98(3):e14099.
80. Puliappadamb HM, Maiti R, Mishra A, Jena M, Mishra BR. Efficacy and safety of melatonin as prophylaxis for migraine in adults: a meta-analysis. J Oral Facial Pain Headache. 2022;36(3–4):207–19.
81. Yamanaka G, Kanou K, Takamatsu T, Takeshita M, Morichi S, Suzuki S, et al. Complementary and integrative medicines as prophylactic agents for pediatric migraine: a narrative literature review. J Clin Med. 2021;10(1):138.
82. Vine T, Brown GM, Frey BN. Melatonin use during pregnancy and lactation: a scoping review of human studies. Braz J Psychiatry. 2022;44(3):342–8.
83. Pareek A, Suthar M, Rathore GS, Bansal V. Feverfew (Tanacetum parthenium L.): a systematic review. Pharmacogn Rev. 2011;5(9):103–10.
84. Diener HC, Pfaffenrath V, Schnitker J, Friede M, Henneicke-von Zepelin HH. Efficacy and safety of 6.25 mg t.i.d. feverfew CO2-extract (MIG-99) in migraine prevention--a randomized, double-blind, multicentre, placebo-controlled study. Cephalalgia. 2005;25(11):1031–41.
85. Pittler MH, Ernst E. Feverfew for preventing migraine. Cochrane Database Syst Rev. 2004;(1):CD002286.

86. Pittler MH, Vogler BK, Ernst E. Feverfew for preventing migraine. Cochrane Database Syst Rev. 2000;(3):CD002286.
87. Wider B, Pittler MH, Ernst E. Feverfew for preventing migraine. Cochrane Database Syst Rev. 2015;4(4):CD002286.
88. Lipton RB, Gobel H, Einhaupl KM, Wilks K, Mauskop A. Petasites hybridus root (butterbur) is an effective preventive treatment for migraine. Neurology. 2004;63(12):2240–4.
89. Aydin AA, Zerbes V, Parlar H, Letzel T. The medical plant butterbur (Petasites): analytical and physiological (re)view. J Pharm Biomed Anal. 2013;75:220–9.
90. Avula B, Wang YH, Wang M, Smillie TJ, Khan IA. Simultaneous determination of sesquiterpenes and pyrrolizidine alkaloids from the rhizomes of Petasites hybridus (L.) G.M. et Sch. and dietary supplements using UPLC-UV and HPLC-TOF-MS methods. J Pharm Biomed Anal. 2012;70:53–63.
91. Wang LP, Zhang XZ, Guo J, Liu HL, Zhang Y, Liu CZ, et al. Efficacy of acupuncture for acute migraine attack: a multicenter single blinded, randomized controlled trial. Pain Med. 2012;13(5):623–30.
92. Linde K, Allais G, Brinkhaus B, Manheimer E, Vickers A, White AR. Acupuncture for migraine prophylaxis. Cochrane Database Syst Rev. 2009;(1):CD001218.
93. Chessman AW. Review: acupuncture reduces migraine frequency more than usual care, sham acupuncture, or prophylactic drugs. Ann Intern Med. 2016;165(8):JC44.
94. Linde K, Allais G, Brinkhaus B, Fei Y, Mehring M, Vertosick EA, et al. Acupuncture for the prevention of episodic migraine. Cochrane Database Syst Rev. 2016;(6):CD001218.
95. Zheng H, Chen M, Wu X, Li Y, Liang FR. Manage migraine with acupuncture: a review of acupuncture protocols in randomized controlled trials. Am J Chin Med. 2010;38(4):639–50.
96. Magis D, Sava S, d'Elia TS, Baschi R, Schoenen J. Safety and patients' satisfaction of transcutaneous supraorbital neurostimulation (tSNS) with the Cefaly(R) device in headache treatment: a survey of 2,313 headache sufferers in the general population. J Headache Pain. 2013;14:95.
97. Schoenen J, Vandersmissen B, Jeangette S, Herroelen L, Vandenheede M, Gerard P, et al. Migraine prevention with a supraorbital transcutaneous stimulator: a randomized controlled trial. Neurology. 2013;80(8):697–704.
98. Allen KD, Shriver MD. Enhanced performance feedback to strengthen biofeedback treatment outcome with childhood migraine. Headache. 1997;37(3):169–73.
99. Andrasik F. Biofeedback in headache: an overview of approaches and evidence. Cleve Clin J Med. 2010;77(Suppl 3):S72–6.
100. Andreychuk T, Skriver C. Hypnosis and biofeedback in the treatment of migraine headache. Int J Clin Exp Hypn. 1975;23(3):172–83.
101. Baumann RJ. Behavioral treatment of migraine in children and adolescents. Paediatr Drugs. 2002;4(9):555–61.
102. Blanchard EB, Theobald DE, Williamson DA, Silver BV, Brown DA. Temperature biofeedback in the treatment of migraine headaches: a controlled evaluation. Arch Gen Psychiatry. 1978;35(5):581–8.
103. Burke EJ, Andrasik F. Home- vs. clinic-based biofeedback treatment for pediatric migraine: results of treatment through one-year follow-up. Headache. 1989;29(7):434–40.
104. Buse DC, Andrasik F. Behavioral medicine for migraine. Neurol Clin. 2009;27(2):445–65.
105. de Tommaso M, Delussi M. Nociceptive blink reflex habituation biofeedback in migraine. Funct Neurol. 2017;32(3):123–30.
106. Drury RL, DeRisi WJ, Liberman RP. Temperature biofeedback treatment for migraine headache: a controlled multiple baseline study. Headache. 1979;19(5):278–84.
107. Ellertsen B, Nordby H, Hammerborg D, Thorlacius S. Psychophysiologic response patterns in migraine before and after temperature biofeedback. Prediction of treatment outcome. Cephalalgia. 1987;7(2):109–24.
108. Fahrion SL. Autogenic biofeedback treatment for migraine. Mayo Clin Proc. 1977;52(12):776–84.
109. Fentress DW, Masek BJ, Mehegan JE, Benson H. Biofeedback and relaxation-response training in the treatment of pediatric migraine. Dev Med Child Neurol. 1986;28(2):139–46.
110. Gamble EH, Elder ST. Multimodal biofeedback in the treatment of migraine. Biofeedback Self Regul. 1983;8(3):383–92.
111. Gauthier J, Bois R, Allaire D, Drolet M. Evaluation of skin temperature biofeedback training at two different sites for migraine. J Behav Med. 1981;4(4):407–19.
112. Gauthier J, Côté G, French D. The role of home practice in the thermal biofeedback treatment of migraine headache. J Consult Clin Psychol. 1994;62(1):180–4.
113. Gauthier J, Doyon J, Lacroix R, Drolet M. Blood volume pulse biofeedback in the treatment of migraine headache: a controlled evaluation. Biofeedback Self Regul. 1983;8(3):427–42.
114. Gauthier J, Lacroix R, Coté A, Doyon J, Drolet M. Biofeedback control of migraine headaches: a comparison of two approaches. Biofeedback Self Regul. 1985;10(2):139–59.
115. Gauthier JG, Carrier S. Long-term effects of biofeedback on migraine headache: a prospective follow-up study. Headache. 1991;31(9):605–12.
116. Gauthier JG, Fournier AL, Roberge C. The differential effects of biofeedback in the treatment of menstrual and nonmenstrual migraine. Headache. 1991;31(2):82–90.
117. Grazzi L, Bussone G. Italian experience of electromyographic-biofeedback treatment of episodic common migraine: preliminary results. Headache. 1993;33(8):439–41.
118. Hermann C, Blanchard EB, Flor H. Biofeedback treatment for pediatric migraine: prediction of treatment outcome. J Consult Clin Psychol. 1997;65(4):611–6.
119. Hoffmann E. Biofeedback training: a new therapy in the treatment of migraine and tension headache? Dan Med Bull. 1975;22(3):97–9.
120. Holmes DS, Burish TG. Effectiveness of biofeedback for treating migraine and tension headaches: a review of the evidence. J Psychosom Res. 1983;27(6):515–32.
121. Ingvaldsen SH, Tronvik E, Brenner E, Winnberg I, Olsen A, Gravdahl GB, et al. A biofeedback app for migraine: development and usability study. JMIR Form Res. 2021;5(7):e23229.
122. Johansson J, Ost LG. Self-control procedures in biofeedback: a review of temperature biofeedback in the treatment of migraine. Biofeedback Self Regul. 1982;7(4):435–42.
123. Johansson J, Ost LG. Temperature-biofeedback treatment of migraine headache. Specific effects and the effects of "generalization training". Behav Modif. 1987;11(2):182–99.
124. Kaiser RS, Primavera JP 3rd. Biofeedback and the treatment of migraine. Headache. 1987;27(3):170.
125. Kaushik R, Kaushik RM, Mahajan SK, Rajesh V. Biofeedback assisted diaphragmatic breathing and systematic relaxation versus propranolol in long term prophylaxis of migraine. Complement Ther Med. 2005;13(3):165–74.
126. Kewman D, Roberts AH. Skin temperature biofeedback and migraine headaches. A double-blind study. Biofeedback Self Regul. 1980;5(3):327–45.
127. Lacroix JM, Clarke MA, Bock JC, Doxey N, Wood A, Lavis S. Biofeedback and relaxation in the treatment of migraine headaches: comparative effectiveness and physiological correlates. J Neurol Neurosurg Psychiatry. 1983;46(6):525–32.

128. Lake A, Rainey J, Papsdorf JD. Biofeedback and rational-emotive therapy in the management of migraine headache. J Appl Behav Anal. 1979;12(1):127–40.
129. Lisspers J, Ost LG. BVP-biofeedback in the treatment of migraine. The effects of constriction and dilatation during different phases of the migraine attack. Behav Modif. 1990;14(2):200–21.
130. Martic-Biocina S, Zivoder I, Kozina G. Biofeedback and neurofeedback application in the treatment of migraine. Psychiatr Danub. 2017;29(Suppl 3):575–7.
131. Mathew RJ, Ho BT, Kralik P, Taylor D, Claghorn JL. Catecholamines and migraine: evidence based on biofeedback induced changes. Headache. 1980;20(5):247–52.
132. McGrady A, Wauquier A, McNeil A, Gerard G. Effect of biofeedback-assisted relaxation on migraine headache and changes in cerebral blood flow velocity in the middle cerebral artery. Headache. 1994;34(7):424–8.
133. Medina JL, Diamond S, Franklin MA. Biofeedback therapy for migraine. Headache. 1976;16(3):115–8.
134. Minen MT, Corner S, Berk T, Levitan V, Friedman S, Adhikari S, et al. Heartrate variability biofeedback for migraine using a smartphone application and sensor: a randomized controlled trial. Gen Hosp Psychiatry. 2021;69:41–9.
135. Mullally WJ, Hall K, Goldstein R. Efficacy of biofeedback in the treatment of migraine and tension type headaches. Pain Physician. 2009;12(6):1005–11.
136. Nestoriuc Y, Martin A. Efficacy of biofeedback for migraine: a meta-analysis. Pain. 2007;128(1–2):111–27.
137. Odawara M, Hashizume M, Yoshiuchi K, Tsuboi K. Real-time assessment of the effect of biofeedback therapy with migraine: a pilot study. Int J Behav Med. 2015;22(6):748–54.
138. Pfaller A. Efficacy of biofeedback in the treatment of migraine and tension type headaches. Pain Physician. 2010;13(1):94–6; author reply 6.
139. Prima A, Agnoli A, Tamburello A. A review of the applications of biofeedback to migraine and tension headaches. Acta Neurol (Napoli). 1979;1(6):510–21.
140. Reid GJ, McGrath PJ. Psychological treatments for migraine. Biomed Pharmacother. 1996;50(2):58–63.
141. Scharff L, Marcus DA, Masek BJ. A controlled study of minimal-contact thermal biofeedback treatment in children with migraine. J Pediatr Psychol. 2002;27(2):109–19.
142. Silver BV, Blanchard EB, Williamson DA, Theobald DE, Brown DA. Temperature biofeedback and relaxation training in the treatment of migraine headaches. One-year follow-up. Biofeedback Self Regul. 1979;4(4):359–66.
143. Sovak M, Kunzel M, Sternbach RA, Dalessio DJ. Mechanism of the biofeedback therapy of migraine: volitional manipulation of the psychophysiological background. Headache. 1981;21(3):89–92.
144. Stokes DA, Lappin MS. Neurofeedback and biofeedback with 37 migraineurs: a clinical outcome study. Behav Brain Funct. 2010;6:9.
145. Stubberud A, Tronvik E, Olsen A, Gravdahl G, Linde M. Biofeedback treatment app for pediatric migraine: development and usability study. Headache. 2020;60(5):889–901.
146. Stubberud A, Varkey E, McCrory DC, Pedersen SA, Linde M. Biofeedback as prophylaxis for pediatric migraine: a meta-analysis. Pediatrics. 2016;138(2):e20160675.
147. Turin A, Johnson WG. Biofeedback therapy for migraine headaches. Arch Gen Psychiatry. 1976;33(4):517–9.
148. Vasudeva S, Claggett AL, Tietjen GE, McGrady AV. Biofeedback-assisted relaxation in migraine headache: relationship to cerebral blood flow velocity in the middle cerebral artery. Headache. 2003;43(3):245–50.
149. Wauquier A, McGrady A, Aloe L, Klausner T, Collins B. Changes in cerebral blood flow velocity associated with biofeedback-assisted relaxation treatment of migraine headaches are specific for the middle cerebral artery. Headache. 1995;35(6):358–62.
150. Sauro KM, Becker WJ. The stress and migraine interaction. Headache. 2009;49(9):1378–86.
151. Rains JC. Epidemiology and neurobiology of stress and migraine. Headache. 2009;49(9):1391–4.
152. Becker WJ, Sauro KM. Recent studies on stress management-related treatments for migraine. Headache. 2009;49(9):1387–90.
153. Butt MN, Maryum M, Amjad I, Khan OJ, Awan L. Effects of aerobic exercise and progressive muscle relaxation on migraine. J Pak Med Assoc. 2022;72(6):1153–7.
154. D'Souza PJ, Lumley MA, Kraft CA, Dooley JA. Relaxation training and written emotional disclosure for tension or migraine headaches: a randomized, controlled trial. Ann Behav Med. 2008;36(1):21–32.
155. Fichtel A, Larsson B. Does relaxation treatment have differential effects on migraine and tension-type headache in adolescents? Headache. 2001;41(3):290–6.
156. Hay KM, Madders J. Migraine treated by relaxation therapy. J R Coll Gen Pract. 1971;21(112):664–9.
157. McGrath PJ, Humphreys P, Goodman JT, Keene D, Firestone P, Jacob P, et al. Relaxation prophylaxis for childhood migraine: a randomized placebo-controlled trial. Dev Med Child Neurol. 1988;30(5):626 31.
158. Meyer B, Keller A, Wöhlbier HG, Overath CH, Müller B, Kropp P. Progressive muscle relaxation reduces migraine frequency and normalizes amplitudes of contingent negative variation (CNV). J Headache Pain. 2016;17:37.
159. Richter IL, McGrath PJ, Humphreys PJ, Goodman JT, Firestone P, Keene D. Cognitive and relaxation treatment of paediatric migraine. Pain. 1986;25(2):195–203.
160. Sorbi M, Tellegen B. Differential effects of training in relaxation and stress-coping in patients with migraine. Headache. 1986;26(9):473–81.
161. Warner G. Relaxation therapy in migraine and chronic tension headache. Med J Aust. 1975;1(10):298–301.
162. Airola G, Allais G, Castagnoli Gabellari I, Rolando S, Mana O, Benedetto C. Non-pharmacological management of migraine during pregnancy. Neuroll Sci. 2010;31(Suppl 1):S63–5.

Cyclic Vomiting Syndrome

70

Overview

Cyclic vomiting syndrome (CVS) manifests with recurrent episodes of severe nausea and emesis lasting hours to days, occurring at erratic intervals. A patient can be afflicted with these episodes over months to decades. CVS can occur at any age; it is slightly more common in females. Gastrointestinal workups (endoscopy, gastric emptying studies) and otovestibular workups are unremarkable. The mechanism of disease is unknown; some practitioners view it as a version of migraine and may attempt treatment as such.

Introduction

Cyclic vomiting syndrome (CVS) was first described by Samuel Gee in 1882 [1]. There have been several attempts at proposing diagnostic criteria [2, 3], but these have been neither uniform nor universally accepted.

Clinical Presentation

Cyclic vomiting syndrome (CVS) manifests with recurrent episodes of severe nausea and emesis lasting hours to days, occurring at erratic intervals. A patient can be afflicted with these episodes over months to decades [4]. The data from Fleisher and Matar [5] show that the duration of episodes follows a Gaussian distribution, ranging from less than 6 hours to 1 week, with a peak of 12–24 hours.

Most descriptions of CVS comment on its intermittent quality, but some authors have noticed patients can develop increasingly frequent and prolonged episodes, to the point that symptoms appear nearly continuous; Fleisher and colleagues refer to this as "episode coalescence" [4].

Some studies focus on the variability among patients [6], while others focus on similarities [7].

Demographics

Cyclic vomiting syndrome was originally thought to affect only children, but it has become clear that it can also affect adults [4]. The data from Fleisher and Matar [5] show significant variability in age of onset (from age 1 to age 24 years) and duration (from less than a year to 20 years). It is slightly more common in females than in males [6, 8].

Mechanism of Disease

The underlying mechanism of CVS is unknown [9], and it remains a diagnosis of exclusion [10]. Repeated gastrointestinal studies have failed to find any consistent structural abnormalities that would account for CVS, so it is generally believed to reflect a disorder of physiologic function (rather than due to a structural or anatomical cause). For decades, there has been speculation that it is a variant of migraine [11, 12], and one study [11] observed that 70% of adult patients with CVS also reported a history of migraine. Another study [11] reported that 82% of pediatric patients report a history of migraine. Some studies speculate that CVS is an autonomic disorder [13].

Differential Diagnosis

The symptoms of CVS can be distressing to the point that clinicians cast a wide net seeking etiologies. One study [11] reported patients to have undergone numerous laboratory tests, imaging studies, endoscopic procedures, and laparoscopies, and had even undergone various surgeries (cholecystectomy, appendectomy, pyloroplasty, fundoplication) in an attempt to treat the problem.

Numerous diseases can present with episodic emesis. Fleisher and Matar [5] provide the following broad differential diagnosis:

M. Cherchi, *Otoneurology and Vestibular Medicine*, https://doi.org/10.1007/978-3-031-94841-1_70

- Gastrointestinal.
  - Peptic ulcer disease.
  - Pancreatitis.
  - Pancreatic pseudocyst.
  - Intestinal parasitism.
  - Recurrent subacute appendicitis.
  - Chronic idiopathic intestinal pseudo-obstruction.
  - Bochdalek's hernia.
  - Intermittent small bowel obstruction.
- Central nervous system.
  - Brain tumors causing increased intracranial pressure.
  - Brain stem tumors, with or without increased intracranial pressure.
  - Subdural hematoma or effusions.
  - Abdominal epilepsy.
  - Hydrocephalus—slit ventricle syndrome.
- Autonomic nervous system.
  - Familial dysautonomia.
- Urinary tract.
  - Obstructive uropathies.
- Endocrine.
  - Pheochromocytoma.
  - Adrenal insufficiency.
  - Diabetes mellitus.
- Metabolic.
  - Ornithine transcarbamylase deficiency.
  - Medium chain acyl coenzyme A dehydrogenase deficiency.
  - Propionic acidemia.
  - Isovaleric acidemia (chronic intermittent form).
  - Porphyria.

Testing

One study observed that patients with CVS and migraine exhibit white matter intensities on brain MRI [14], but this is unsurprising, and patients with a history of migraine alone can have such MRI findings.

The majority of patients with CVS do not complain of a disturbance of equilibrium, so commonly their evaluation and management are undertaken by gastrointestinal specialists.

Nevertheless, occasionally, these patients are referred to an otoneurologist, often after having undergone a gastrointestinal workup, sometimes with endoscopy or a gastric motility study, with unremarkable results.

For most otovestibular conditions, gastrointestinal symptoms (nausea and vomiting) will accompany or will follow the symptom of disequilibrium. However, in some patients, the gastrointestinal symptoms may predominate the clinical picture. For this reason, we will usually consider undertaking a screening otovestibular workup (usually with vestibular evoked myogenic potentials and video head impulse testing).

Treatment

Treatment guidelines have been suggested [15], though based on low quality evidence.

Given the apparent overlap between CVS and migraine, it has become somewhat common to attempt treatment with migraine prophylactic strategies [8], such as amitriptyline [10], topiramate [10, 16, 17], valproate [18], flunarizine, or cyproheptadine [6].

Overall, the prognosis of CVS is favorable [6].

References

1. Gee S. On fitful or recurrent vomiting. St Bartholomew's Hospital Rep. 1882;18:1–6.
2. Li BU, Lefevre F, Chelimsky GG, Boles RG, Nelson SP, Lewis DW, et al. North American Society for Pediatric Gastroenterology, Hepatology, and Nutrition consensus statement on the diagnosis and management of cyclic vomiting syndrome. J Pediatr Gastroenterol Nutr. 2008;47(3):379–93.
3. Hyams JS, Di Lorenzo C, Saps M, Shulman RJ, Staiano A, van Tilburg M. Functional disorders: children and adolescents. Gastroenterology. 2016;150:1456.
4. Fleisher DR, Gornowicz B, Adams K, Burch R, Feldman EJ. Cyclic Vomiting Syndrome in 41 adults: the illness, the patients, and problems of management. BMC Med. 2005;3:20.
5. Fleisher DR, Matar M. The cyclic vomiting syndrome: a report of 71 cases and literature review. J Pediatr Gastroenterol Nutr. 1993;17(4):361–9.
6. Dipasquale V, Falsaperla R, Bongiovanni A, Ruggieri M, Romano C. Clinical features and long-term outcomes in pediatric cyclic vomiting syndrome: A 9-year experience at three tertiary academic centers. Neurogastroenterol Motil. 2022;34(3):e14224.
7. Prakash C, Staiano A, Rothbaum RJ, Clouse RE. Similarities in cyclic vomiting syndrome across age groups. Am J Gastroenterol. 2001;96(3):684–8.
8. Ailani J, Kaiser EA, Mathew PG, McAllister P, Russo AF, Velez C, et al. Role of calcitonin gene-related peptide on the gastrointestinal symptoms of migraine-clinical considerations: a narrative review. Neurology. 2022;99:841.
9. Frazier R, Venkatesan T. Current understanding of the etiology of cyclic vomiting syndrome and therapeutic strategies in its management. Expert Rev Clin Pharmacol. 2022;15(11):1305–16.
10. Goldman RD. Cyclic vomiting syndrome in children. Can Fam Physician. 2021;67(11):837–8.
11. Li BU, Murray RD, Heitlinger LA, Robbins JL, Hayes JR. Is cyclic vomiting syndrome related to migraine? J Pediatr. 1999;134(5):567–72.
12. Yu ES, Priyadharsini SSY, Venkatesan T. Migraine, cyclic vomiting syndrome, and other gastrointestinal disorders. Curr Treat Options Gastroenterol. 2018;16(4):511–27.

13. Rashed H, Abell TL, Familoni BO, Cardoso S. Autonomic function in cyclic vomiting syndrome and classic migraine. Dig Dis Sci. 1999;44(8 Suppl):74S–8S.
14. Rashid S, Weaver S, Al-Robaidi K, Dure L, Singh S. Brain Magnetic Resonance Imaging (MRI) white matter hyperintensities in cyclic vomiting syndrome with or without migraine. J Child Neurol. 2022;37(3):218–21.
15. Venkatesan T, Levinthal DJ, Tarbell SE, Jaradeh SS, Hasler WL, Issenman RM, et al. Guidelines on management of cyclic vomiting syndrome in adults by the American Neurogastroenterology and Motility Society and the Cyclic Vomiting Syndrome Association. Neurogastroenterol Motil. 2019;31 Suppl 2(Suppl 2):e13604.
16. Mooers H, Srivastava S, Garacci E, Venkatesan T. Retrospective review of patients treated for cyclic vomiting syndrome with topiramate. Aliment Pharmacol Ther. 2021;54(2):153–9.
17. Sezer OB, Sezer T. A new approach to the prophylaxis of cyclic vomiting: topiramate. J Neurogastroenterol Motil. 2016;22(4):656–60.
18. Hikita T, Kodama H, Nakamoto N, Kaga F, Amakata K, Ogita K, et al. Effective prophylactic therapy for cyclic vomiting syndrome in children using valproate. Brain Dev. 2009;31(6):411–3.

71 Motion Sickness

Overview

Motion sickness, sometimes referred to in medical literature as "*constitutional motion sensitivity*," is a common human experience. The most commonly cited (though not universally accepted) theory is that motion sickness arises from a failure of sensory integration when confronted with discrepant sensory inputs (vestibular, visual, proprioceptive), which most commonly occurs when an individual is undergoing passive motion (such as in a vehicle). Heightened susceptibility to motion sickness may occur in patients with other diseases, such as migraine or vestibular lesions. Management options for motion sickness include trigger avoidance, habituation/desensitization, medication, migraine prophylaxis, motion sickness wristbands, and others.

Terminology

Medical literature often uses the phrase "*constitutional motion sensitivity*" to refer to what is more colloquially known as "*motion sickness*."

Theory of Sensory Conflict

Probably, the most widely (though not universally) accepted theory of motion sickness maintains that the underlying mechanism of disease is *sensory conflict* [1].

The brain calculates the body's orientation in, and movement through, space by taking into account multiple sensory inputs (vestibular, visual, proprioceptive, tactile, and to a small degree auditory), as well as the internal-model-predicted consequences of planned motor output (engrams). The brain integrates the various sensory inputs and tries to match them with predictions from its internal models. When the sensory inputs are concordant and match predictions from the internal model, the system functions properly. In contrast, if sensory inputs are discrepant (fail to integrate), or if they do not match predictions from the internal model, a person may experience discomfort in the form of motion sensitivity.

Motion sensitivity can be experienced when a person is in a vehicle, elevator, escalator, conveyor belt, etc. These are all examples of *passive motion*—a person's body is being moved by some external force, not by the person's voluntary activity. In this circumstance, there is likely to be a discrepancy between visual, vestibular, and somatosensory inputs.

Cybersickness [2–4] can provoke sensations similar to motion sickness, but the term "cybersickness" usually refers to a circumstance in which the predominantly variable sensory input is vision, such as when playing immersive virtual reality videogames, or scrolling on a computer screen. Similar to more mundane motion sickness, in cybersickness, there is a sensory conflict between visual cues (which suggest motion) and other inputs (vestibular, somatosensory) that suggest stasis.

Demographics

Constitutional motion sensitivity is generally thought to be "hard-wired," though can vary over the course of a person's life. It is common for a person to experience motion sensitivity more during childhood and less during adulthood, though the reverse can also occur.

Predisposing Factors

It has been frequency observed that some diseases, such as migraine [5–10], appear to be associated with higher rates of motion sensitivity.

If a person with motion sensitivity becomes afflicted with a vestibular disorder, then these may have a synergistic effect [11, 12].

M. Cherchi, *Otoneurology and Vestibular Medicine*, https://doi.org/10.1007/978-3-031-94841-1_71

Workup

Individuals with constitutional motion sensitivity who are otherwise healthy usually have normal physical examinations and no detectable abnormalities on otovestibular testing.

When an individual experiences an abrupt increase in their baseline motion sensitivity, it is medically reasonable to undertake a screening otovestibular workup to explore whether an independently occurring vestibular disease is superimposed [11, 12].

Treatment

Trigger Avoidance

Trigger avoidance is a common strategy for managing constitutional motion sensitivity. For example, an individual who becomes easily "seasick" will often simply avoid traveling by boat.

This can be successful, though in some circumstances, it may be impractical. For example, if a person is motion sensitive in buses, yet must take the bus to and from work every day, then trigger avoidance may simply be infeasible.

Habituation/Desensitization

Desensitization exercises essentially involve repeatedly exposing an individual to stimuli/situations that provoke motion sensitivity, in the expectation that such repeated exposure will eventually reduce the motion sensitivity [13, 14]. One such protocol is the Puma method, developed by a flight surgeon (Dr. Puma) for the treatment of combat pilots who were experiencing motion sickness who were not permitted to be on centrally acting medications [15]. The advantage of this approach is that it does not require medication. The disadvantage is that it entails considerable discomfort, at least initially.

Medication

A number of medications from different classes have been used in managing motion sickness. Broadly these include antihistamines, anticholinergic medications, dopamine blockers, and benzodiazepines. While sometimes effective, the disadvantage of this approach is that it involves medications and their potential adverse effects.

Treatment for Migraine

Since migraine can be construed as a state of potential hypersensitivity, treatment for migraine can be considered [16]. This is very reasonable in patients already diagnosed with migraine. What has been less explored is whether this approach (treating for migraine) is applicable for managing motion sickness in patients with no history of migraine.

Motion Sickness Wristbands and Acupressure

There is a modest literature regarding motion sickness wristbands [17, 18]. Some studies find this to be efficacious [19, 20], while others do not [18, 21]. The advantage of this approach is that it incurs no medical risk.

The supposed mechanism of motion sickness wristbands is that they act on acupressure points. Here, too, studies are mixed with respect to the efficacy of acupressure, with some studies documenting efficacy [22] and others not [23].

Horizon Glasses

Hubert Jeannin invented and patented "boarding glasses" (USOO67832.37B1) that claim to "cure motion sickness in 94% of cases in 10 minutes." These glasses consist of frames with liquid levels immediately anterior to the eyes (in the coronal plane), and also lateral to the eyes (in the sagittal plane), that are intended to "create an artificial horizon in the field of view which helps to resynchronize the eyes with the balance system." As of this writing, we had not seen any peer-reviewed literature regarding the efficacy of this device.

References

1. Zhang LL, Wang JQ, Qi RR, Pan LL, Li M, Cai YL. Motion sickness: current knowledge and recent advance. CNS Neurosci Ther. 2016;22(1):15–24.
2. Gallagher M, Ferre ER. Cybersickness: a multisensory integration perspective. Multisens Res. 2018;31(7):645–74.
3. Eftekharifar S, Thaler A, Bebko AO, Troje NF. The role of binocular disparity and active motion parallax in cybersickness. Exp Brain Res. 2021;239(8):2649–60.
4. Laessoe U, Abrahamsen S, Zepernick S, Raunsbaek A, Stensen C. Motion sickness and cybersickness – sensory mismatch. Physiol Behav. 2022;258:114015.
5. Kuritzky A, Ziegler DK, Hassanein R. Vertigo, motion sickness and migraine. Headache. 1981;21(5):227–31.
6. Cuomo-Granston A, Drummond PD. Migraine and motion sickness: what is the link? Prog Neurobiol. 2010;91(4):300–12.
7. Furman JM, Marcus DA. Migraine and motion sensitivity. Continuum (Minneap Minn). 2012;18(5 Neuro-otology):1102–17.

8. Wang J, Lewis RF. Contribution of intravestibular sensory conflict to motion sickness and dizziness in migraine disorders. J Neurophysiol. 2016;116(4):1586–91.
9. Abouzari M, Cheung D, Pham T, Goshtasbi K, Sarna B, Tajran S, et al. The relationship between vestibular migraine and motion sickness susceptibility. Otol Neurotol. 2020;41(8):1116–21.
10. Wurthmann S, Naegel S, Roesner M, Nsaka M, Scheffler A, Kleinschnitz C, et al. Sensitized rotatory motion perception and increased susceptibility to motion sickness in vestibular migraine: a cross-sectional study. Eur J Neurol. 2021;28(7):2357–66.
11. Golding JF, Patel M. Meniere's, migraine, and motion sickness. Acta Otolaryngol. 2017;137(5):495–502.
12. Strupp M, Brandt T, Huppert D, Grill E. Prevalence of motion sickness in various vestibular disorders: a study on 749 patients. J Neurol. 2018;265:95.
13. Bles W, Bos JE, Kruit H. Motion sickness. Curr Opin Neurol. 2000;13(1):19–25.
14. Cheung B, Hofer K. Desensitization to strong vestibular stimuli improves tolerance to simulated aircraft motion. Aviat Space Environ Med. 2005;76(12):1099–104.
15. Puma SC, Puma SW, inventorsMethod for conditioning to prevent motion sickness. United States2007 October 30, 2007.
16. Lee SH, Jeong SH, Kim JS, Kim HJ, Choi KD, Choi JH, et al. Effect of prophylactic medication on associated dizziness and motion sickness in migraine. Otol Neurotol. 2018;39(1):e45–51.
17. Atkinson HG. Are acupressure wrist bands helpful for motion sickness? Health News. 2003;9(1):12.
18. Miller KE, Muth ER. Efficacy of acupressure and acustimulation bands for the prevention of motion sickness. Aviat Space Environ Med. 2004;75(3):227–34.
19. Cox DJ, Singh H, Cox DM. Effectiveness of acupressure and acustimulation in minimizing driving simulation adaptation syndrome. Mil Med. 2011;176(12):1440–3.
20. Stern RM, Jokerst MD, Muth ER, Hollis C. Acupressure relieves the symptoms of motion sickness and reduces abnormal gastric activity. Altern Ther Health Med. 2001;7(4):91–4.
21. Bruce DG, Golding JF, Hockenhull N, Pethybridge RJ. Acupressure and motion sickness. Aviat Space Environ Med. 1990;61(4):361–5.
22. Hu S, Stritzel R, Chandler A, Stern RM. P6 acupressure reduces symptoms of vection-induced motion sickness. Aviat Space Environ Med. 1995;66(7):631–4.
23. Warwick-Evans LA, Masters IJ, Redstone SB. A double-blind placebo controlled evaluation of acupressure in the treatment of motion sickness. Aviat Space Environ Med. 1991;62(8):776–8.

Cybersickness

72

Overview

Cybersickness can be provoked when visual stimuli are discrepant with other sensory inputs (vestibular and proprioceptive). Common triggers include using immersive virtual reality goggles, or simply looking at a scrolling computer screen. Cybersickness is a relatively recently recognized problem (since the technology producing such visual inputs is relatively recent), and consequently, it has not been as thoroughly studied as motion sickness, and there are few data supporting efficacy of any particular treatment. Practically, most patients practice trigger avoidance; for those who cannot completely avoid the triggering visual stimuli, "breaking up" the exposure into more manageable chunks may be helpful.

Introduction

In simple terms, the word "**cybersickness**" has been used to refer to "visually induced motion sickness" (Gallagher and Ferre. Multisens Res 31(7):645–674, 2018). Much of the literature on this topic discusses it in the context of VR (virtual reality) environments (such as in video gaming), but in clinical practice, the more commonly encountered situation is that of patients who experience uncomfortable sensations (such as dizziness and nausea) associated with "screen time" (movies or television, working on a computer, etc.).

There is relatively little literature on this topic. Much of the literature about cybersickness extrapolates from previous research on constitutional motion sensitivity ("motion sickness"). There is no widely accepted method for studying cybersickness, and proposals regarding the underlying mechanism of the disease have remained conjectural.

Practically, most patients either practice trigger avoidance or "break up" the offending visual exposure into smaller, more manageable chunks. We have also seen patients counseled by neuro-optometrists to attempt strategies such as limiting the field of view (in other words, avoid very wide-screen monitors), reduce screen motion where possible, or try a variety of visual filters or lens tints. In patients with very prominent gastrointestinal symptoms, antinausea medication may be helpful, though does not eliminate the problem.

Discussion

One proposed definition of *cybersickness* is, "an unpleasant sensation… [consisting] of symptoms of disorientation, drowsiness, eyestrain, and nausea arising from exposure to immersive VR [virtual reality] environments" [1].

Whereas constitutional motion sensitivity (referred to colloquially as "motion sickness") is usually induced by actual motion (usually passive motion) of a patient, cybersickness is usually induced by artificial visual stimuli (such as by screens) designed to produce an illusion of movement (such as scrolling). Although constitutional motion sensitivity and cybersickness tend to occur in different circumstances, the common feature is thought to be sensory discrepancies—which is to say mismatches between the sensory modalities (labyrinthine, visual, proprioceptive) that the brain attempts to process and reconcile through sensory integration [2].

It appears that the longer the exposure to the offending visual stimuli, the more likely it is that an individual will experience cybersickness. The discomfort of cybersickness may persist beyond the visual exposure, and in this respect may parallel the pattern exhibited in motion sickness (where the uncomfortable sensations may persist beyond the actual exposure to motion).

There is not yet any universally agreed-upon framework for studying cybersickness. Research strategies have included assessment of symptoms (subjective, using questionnaires) and measuring physiologic parameters (objective, such as gastric motility, ocular motility or eye blinks, and electroencephalography).

Whereas constitutional motion sensitivity has been recognized for centuries, cybersickness has only been described since the advent of technologies that can induce it, so there is

M. Cherchi, *Otoneurology and Vestibular Medicine*, https://doi.org/10.1007/978-3-031-94841-1_72

much less literature about cybersickness. There is not yet any widely accepted theory about the mechanisms underlying cybersickness, and most of the literature on this topic extrapolates from that on constitutional motion sensitivity.

Since there is no consensus on how to study cybersickness, and no agreement on the theoretical underpinnings of the diagnosis, it is understandable that there has not been much progress in treatment.

Practically, most patients either practice trigger avoidance or "break up" the offending visual exposure into smaller, more manageable chunks. We have also seen patients counseled by neuro-optometrists to attempt strategies such as limiting the field of view (in other words, avoid very wide-screen monitors), reduce screen motion where possible, or try a variety of visual filters or lens tints. In patients with very prominent gastrointestinal symptoms, antinausea medication may be helpful, though does not eliminate the problem.

References

1. Gallagher M, Ferre ER. Cybersickness: a multisensory integration perspective. Multisens Res. 2018;31(7):645–74.
2. Whiley RE, Day RM. Multisensory integration of echolocation and vision in mammals. Western Undergrad Psychol J. 2019;7(1):1–14.

73 Motorist Disorientation Syndrome

Overview

Motorist disorientation syndrome (MDS) is a condition that tends to occur to drivers (less commonly to passengers) in automobiles. The incidence and prevalence are unknown. Its pathophysiology is unknown, though at some level, it is probably the result of the sensory mismatches that occur in this situation of passive motion. People appear more vulnerable to this problem when they are driving a vehicle at a higher speed in an environment with fewer (or with distant) visual cues, such as during highway travel. Other predisposing situations include driving along curves or over hills. Vision therapy exercises (usually administered under the supervision of neuro-optometry) seem more successful than vestibular therapy. Pharmacologic strategies have not been well studied, but case series report some success with agents otherwise used for migraine prophylaxis. Prognosis is uncertain.

Introduction

Ainsworth and colleagues comment that, "The term 'motorist's vestibular disorientation syndrome' was coined for… situation specific dizziness" of "patients who primarily experienced symptoms of dizziness/disorientation while in a motor car" [1]. Bronstein and Golding [2] suggest that the term *"motorist's disorientation syndrome" (MDS)* is more appropriate.

Some literature classifies motorist's disorientation syndrome (MDS) as a type of visual vertigo [3].

Epidemiology

The incidence and prevalence of MDS is unknown. Bronstein and Golding comment that:

> The number of disoriented motorists reported is too small to draw firm conclusions about epidemiology. Approximately 1% of patients referred to a well-established, London clinic specializing in vestibular and balance disorders [2].

Gresty and Ohlman state, "Each year in the neuro-otological clinic with which one of the authors (MG) is associated an average of 10 patients out of 200 seen as outpatients present with primary MDS" [4], suggesting **5%** of vestibular outpatients in that practice.

A case series [1] of 18 patients (9 female, 9 male) reported that the **mean age of onset was 41.7 years** (standard deviation of 10 years, range 20–56 years). In men, the mean duration was 3.89 years (standard deviation 1.97 years, range 1–7 years). In women, the mean duration was 8.89 (standard deviation 5.26, range 2–17 years).

Pathophysiological Mechanism of Disease

In a general sense, motorist disorientation syndrome is related to several other disorders (such as motion sickness and cybersickness) in which discrepancies (mismatches) between different sensory inputs results in the failure of multisensory integration.

Some investigators postulate that MDS results from over-reliance on visual cues (also called visual dependence). Such over-reliance on the visual system is partly a result of the fact that "the force-motion environment is potentially ambiguous" [4]. As an example of this, consider that:

- For an earth-stationary individual oriented in a backwards-leaning position, the vector of linear acceleration (from the perspective of that individual) points inferoposteriorly and is the result of gravity alone.
- For an individual oriented upright (earth vertical) in a human centrifuge facing the center of the centrifuge while being rotated by the centrifuge, the vector of linear acceleration (from the perspective of that individual) points inferoposteriorly and is the result of a combination of gravity and centrifugal force.

M. Cherchi, *Otoneurology and Vestibular Medicine*, https://doi.org/10.1007/978-3-031-94841-1_73

In both of these circumstances, there is a vector of linear acceleration pointing inferoposteriorly (from the perspective of that individual) and will stimulate the vestibular system identically. If visual cues are available, then the visual information can help disambiguate these situations. The two situations will be difficult to disambiguate if vision is distorted and impossible to disambiguate if vision is entirely unavailable.

In their original description of MDS, Page and Gresty comment that:

> False perception of the motions of the self and/or external objects in the physical world may be provoked by… unusual conditions of motion and environmental structure… [in which] the sensory inputs which are normally combined to provide an accurate model of the physical world provide false information because either the stimuli are unphysiological or important sensory information is lacking or distorted [5].

They observe that:

> Motorway driving conditions are particularly conducive to the development of these symptoms; visual background detail is at a minimum and with the eyes looking straight ahead the 'optic flow field' which moves in a series of parallel straight lines towards the driver is a weak stimulus for visual self-stabilisation… Conversely in town, buildings, trees and traffic provide a rich visual background with plentiful references to the true vertical, and there are continual angular and linear changes in the motion of the traffic [5].

They suggest:

> A theoretical explanation can be proposed to account for the illusions of turning, tilt or the feeling of being pushed sideways. Of first consideration, driving a car involves unusual visual and motion stimuli which do not occur in normal activities. Much of the vestibular stimulation involved in driving is of low frequency content and out of the normal physiological range of the labyrinth. Hence vision, both peripheral and central, is of overriding importance, not only for steering but also for assessing changes in speed, and this dominance of vision obviates problems which arise from a disordered vestibular system. The vestibular system becomes more important in circumstances when vision is reduced as when driving in an empty landscape. Hence, spatial disorientation in an empty visual field is likely to be a consequence of abnormal vestibular signals, or erroneous central interpretation of vestibular information [5].

Bronstein and Golding [2] speculate on the maladaptive sensory integration that might lead to MDS:

> Motorists learn to interpret sensory stimuli in the context of the car stabilized by its suspension and guided by steering. However, the sensory stimulation during driving is potentially ambiguous: The forces of cornering may be interpreted as tilt rather than as lateral acceleration and visual flow of the road and traffic can be interpreted to indicate veering, a form of visual vertigo. In motorists' disorientation, certain individuals appear to develop a heightened awareness of these false perceptions of car orientation, readily experiencing stereotypical symptoms of threatened rolling over on corners and veering on open highways or in streaming traffic [2].

Clinical Presentation

Gresty and Ohlmann [4] cite the following as among the most common experiences described by patients with MDS:

- "The car feels as if it veers on wide open roads such as motorways."
- "The car feels that is turning into vehicles being overtaking."
- "The car feels that it is about to turn over when descending and rounding a bend."

The case series of 18 patients described by Ainsworth and colleagues [1] reported that:

- 13 patients (72%) experienced a "unidirectional sense of pulling."
- 13 patients (72%) "only experienced symptoms on roads with high speed limits."
- 12 patients (67%) described symptom onset as acute.
- 11 patients (61%) "also reported symptoms whilst being a passenger."
- 8 patients (44%) "also experienced these symptoms when not in a car."
- 7 patients (39%) "reported feelings of acute anxiety or panic associated with their [vestibular] symptoms."
- 6 patients (33%) "reported that their symptoms were worse on bumps/hills."

Furthermore:

- 17 patients (94%) "change[d] their driving habits."
- 11 patients (61%) "stopped driving on provoking roads."
- 6 patients (33%) "stopped driving completely."

Physical Examination

In a patient with MDS who is otherwise healthy, physical examination should be normal.

Ocular Motor Examination

Pawar and colleagues [6] reported the following ocular motor findings in 17 patients (Table 73.1):

Table 73.1 Ocular motor findings in 17 patients with motorist disorientation syndrome

	Normal	Abnormal
Horizontal saccades	Normal in 17 (100%)	–
Vertical saccades	Normal in 15 (88%)	Hypometric in 2 (12%)
Horizontal smooth pursuit	Normal in 13 (77%)	Saccadic in 4 (24%)
Vertical smooth pursuit	Normal in 1 (6%)	Saccadic in 16 (94%)
Spontaneous nystagmus	Normal (absent) in 14 (82%)	Up beat nystagmus in 3 (18%)
Head-shaking nystagmus	Normal (absent) in 14 (82%)	Present in 3 (18%)
Hyperventilation-induced nystagmus	Normal (absent) in 14 (82%)	Present in 3 (18%)

From Pawar et al. [6]

Testing: Vestibular

In the case series of Ainsworth and colleagues [1], of the 15 patients who had "complete vestibular testing" (which included ocular motor testing, caloric testing, slow harmonic acceleration, step velocity testing, visual vestibulo-ocular reflex suppression, optokinetic nystagmus, optokinetic after-nystagmus), nine (60%) had at least one abnormal test reflecting vestibular asymmetry.

Differential Diagnosis

Migrainous hypersensitivity to motion and persistent postural perceptual dizziness are on the differential diagnosis.

Treatment

In the case series of 18 patients studied by Ainsworth and colleagues [1]:

- 8 patients (44%) received vestibular rehabilitation. Of those 8 patients, "5 were treated with traditional vestibular rehabilitation gaze stabilisation alone, 4 of whom found that these didn't help, and 1 found them subjectively useful."
- 3 patients (17%) "were treated with a combination of visual vertigo and gaze stabilisation exercises. All of these found the gaze stabilisation exercises not useful but the visual vertigo exercises useful."

Their series also "showed that there was a significant difference between the age of the patient and whether the treatment helped… with treatment more likely to help those who are younger" [1].

There are no controlled trials of pharmacotherapy for MDS. Pawar and colleagues [6] studied a series of 24 patients and reported some (albeit inconsistent) response to agents that are also used as migraine prophylactics, such as:

- Beta blockers (bisoprolol).
- GABA-ergic medications (gabapentin, pregabalin).
- Tricyclic compounds (amitriptyline).
- Serotonin-norepinephrine reuptake inhibitors (venlafaxine, desvenlafaxine).
- Magnesium.

Prognosis

The literature offers little regarding prognosis, which appears quite variable. Pawar and colleagues comment that "The natural history of motorist vestibular disorientation syndrome ranges from complete recovery with or without treatment to chronic suffering without relief" [6].

References

1. Ainsworth C, Davies R, Colvin I, Murdin L. Motorist disorientation syndrome; clinical features and vestibular findings. J Vestib Res. 2023;33(5):339–48.
2. Bronstein AM, Golding JF, Gresty MA. Vertigo and dizziness from environmental motion: visual vertigo, motion sickness, and drivers' disorientation. Semin Neurol. 2013;33(3):219–30.
3. Bronstein AM. The visual vertigo syndrome. Acta Otolaryngol Suppl. 1995;520(Pt 1):45–8.
4. Gresty MA, Ohlmann T, editors. Motorist Vestibular Disorientation Syndrome Revisited. RTO-HFM [Research and Technology Organization – Human Factors and Medicine Panel] Symposium on "Spatial Disorientation in Military Vehicles: Causes, consequences and Cures"; 2002; La Coruña, Spain: RTO-MP-086.
5. Page NG, Gresty MA. Motorist's vestibular disorientation syndrome. J Neurol Neurosurg Psychiatry. 1985;48(8):729–35.
6. Pawar V, Ashraf H, Dorsala S, Mary P, Hameed N, H DN, et al. Motorist's vestibular disorientation syndrome (MVDS)-proposed diagnostic criteria. J Pers Med 2023;13(5).

Mal de debarquement Syndrome

74

Overview

Mal de debarquement syndrome (MdDs) is a disorder usually triggered by prolonged exposure to passive motion (typically vehicular travel for at least several hours) whose symptoms begin at some point after that exposure to motion and usually include rocking disequilibrium that, in the majority of cases, transiently improves when the patient gets back into a moving vehicle. The mechanism of disease is unclear. The overall duration of an episode can last weeks to years. The typical age at a first episode is in the fourth or fifth decade, and the majority of patients are women. Usually, physical examination and workup (otovestibular testing, brain imaging) are normal. The main differential diagnosis is migraine-associated vertigo. Some patients notice symptomatic improvement with a low-dose benzodiazepine, SSRI, or SNRI. Nonpharmacologic interventions include vestibular rehabilitation therapy and a strategy developed by Dr. Mingjia Dai.

Introduction

Mal de debarquement syndrome (MdDs) is a condition in which an individual is exposed to motion (usually vehicular travel of some sort) for a prolonged period, and shortly after returning to solid ground, the individual continues to feel as if in motion. A striking feature of this form of disequilibrium is that the sensation *improves* (often completely disappears) when the patient gets back into a moving vehicle. Almost no other vestibular disorder behaves in this fashion.

Nomenclature

The earlier definitions of MdDs required a trigger of prolonged exposure to motion.

However, subsequent literature recognized a subset of patients who did not have any such motion exposure, yet resembled MdDs in most other respects (such as symptomatic improvement when in a moving vehicle). Classifying such patients as a "subtype of MdDs" seems peculiar since there is no "disembarkation" from any moving vehicle. However, in recognition of the other shared characteristics, this is sometimes called "nonmotion-triggered MdDs," although such terminology is rather internally contradictory.

History

Erasmus Darwin (1731–1802) is sometimes credited with having offered the first description of MdDs in volume 1 of *Zoonomia* [1]:

> Those, who have been upon the water in a boat or ship so long, that they have acquired the necessary habits of motion upon that unstable element, at their return on land frequently think in their reveries, or between sleeping and waking, that they observe the room, they sit in, or some of its furniture, to librate like the motion of the vessel. This I have experienced myself, and have been told, that after long voyages, it is some time before these ideas entirely vanish. The same is observable in a less degree after having travelled some days in a stage coach, and particularly when we lie down in bed, and compose ourselves to sleep; in this case it is observable, that the rattling noise of the coach, as well as the undulatory motion, haunts us.

Eighty-five years later, Dr. Irwin wrote [2]:

> The ocean habit teaches the [semicircular] canals to adapt themselves to the new condition of things, and to pass over unheeded erroneous impressions which were noticed at first. In fact, the new habit may become so strong that a disturbance of it, by a return to the land, will be marked by similar phenomena; hence the unsteady gait sometimes observable in a not-drunken sailor during his first few hours on shore after a long and stormy voyage.

Similar descriptions appear in various nautical literature. Jack London (1876–1916) wrote in *Sea-wolf* [3]:

> The boat's stern touched the hard shingle. I sprang out, extending my hand to Maud. The next moment she was beside me. As my fingers released hers, she clutched for my arm hastily. At the same moment I swayed, as about to fall to the sand. This was the

M. Cherchi, *Otoneurology and Vestibular Medicine*, https://doi.org/10.1007/978-3-031-94841-1_74

startling effect of the cessation of motion. We had been so long upon the moving, rocking sea that the stable land was a shock to us. We expected the beach to lift up this way and that, and the rocky walls to swing back and forth like the sides of a ship; and when we braced ourselves, automatically, for these various expected movements, their non-occurrence quite overcame our equilibrium.

And *Cruise of the Snark* [4]:

So we went ashore with them across a level flashing sea to the wonderful green land. We landed on a tiny wharf, and the dream became more insistent; for know that for twenty-seven days we had been rocking across the ocean on the tiny Snark. Not once in all those twenty-seven days had we known a moment's rest, a moment's cessation from movement. This ceaseless movement had become ingrained. Body and brain we had rocked and rolled so long that when we climbed out on the tiny wharf kept on rocking and rolling. This, naturally, we attributed to the wharf. It was projected psychology. I spraddled along the wharf and nearly fell into the water. I glanced at Charmian, and the way she walked made me sad. The wharf had all the seeming of a ship's deck. It lifted, tilted, heaved and sank; and since there were no handrails on it, it kept Charmian and me busy avoiding falling in. I never saw such a preposterous little wharf. Whenever I watched it closely, it refused to roll; but as soon as I took my attention off from it, away it went, just like the Snark. Once, I caught it in the act, just as it upended, and I looked down the length of it for two hundred feet, and for all the world it was like the deck of a ship ducking into a huge head-sea.

Epidemiology

The majority of patients are women, but the reported sex ratio varies across studies. The population of 27 patients studied by Hain and colleagues [5] consisted of 26 women (96%) and 1 man (4%). The population of 64 patients studied by Cha and colleagues [6] consisted of 48 women (75%) and 16 men (25%).

The peak age of onset varies among studies from the fourth to the fifth decades. Hain and colleagues [5] report the age of onset as 49.3 ± 10.3 years. Cha and colleagues [6] report a peak age of onset of 39 ± 13 years.

Pathophysiological Mechanism of Disease

The underlying pathophysiology of MdDs is unknown. Proposed mechanisms have included:

- Vestibular dysfunction secondary to overstimulation [7].
- "Vestibular hallucination" due to inappropriate "release of stored vestibular information from the hippocampus" [8].
- "Aberrant adaptation" [9]. Lewis elaborated that, "During passive motion there may be a mismatch between the vestibular inputs that encode head motion and other sensory and motor cues that do not correlate with the labyrinthine signals. Central adaptation may minimize the symptoms provoked by this mismatch, but when the [passive] movement ends, the brain must readapt to the stationary environment. It has been suggested that patients with MDD [mal de debarquement syndrome] undergo a physiologic adaptative process during passive motion, but for unknown reasons they do not readily adapt back to the stable environment. This impaired return to baseline results in a perception of motion when the subject is stationary."
- Dynamic multi-sensorimotor neural plasticity that adapts during exposure to motion but then maladaptively fails to revert when motion ceases [10].

Further developing the last two suggestions, some investigators reasonably hypothesize that at some level MdDs must be an example of a mis-applied internal predictive model—specifically, a predictive model that had been developed and adopted during exposure to passive motion, which then is mistakenly maintained even after return to stable ground [11]. Cha [12] describes this as "MdDs may be a disorder of oversynchronization of brain networks caused by entrainment to the background low-amplitude-oscillating environments typical of the triggers of MdDs."

For the so-called "nonmotion-triggered" type of MdDs, it is unclear how such entrainment would occur, or how the internal predictive modeling hypothesis would apply. Cha and colleagues [6] note that of 18 patients with so-called "mixed MdDs" (meaning subjects who experienced both motion-triggered and nonmotion-triggered MdDs), the first-ever episode in all of them was a motion-triggered episode. This observation suggests that the initial motion-triggered MdDs episode did trigger development of an internal predictive model, and although that initial episode resolved, it later reappeared without any exposure to motion; in other words, a latent (previously learned) internal predictive model was "unleashed." This hypothesis begs the question of what triggered the "unleashing" of the (previously) quiescent internal predictive model.

Clinical Presentation

Cha and colleagues [6] retrospectively reviewed 64 patients with MdDs. The investigators distinguished patients with "pure" MdDs (those whose episodes were only triggered by motion) from those with "mixed" MdDs (patients in whom some episodes were triggered by motion and other episodes were spontaneous). Triggers were as shown in Table 74.1.

In the series studied by Cha and colleagues [6], the 64 patients experienced a total of 206 episodes. Of these:

- 51% lasted >1 month
- 18% lasted >1 year

- 15% lasted >2 years
- 12% lasted >4 years
- 11% lasted >5 years.

Figure 74.1, from Cha and colleagues [6], shows a Kaplan–Meier curve for duration of 206 episodes of MdDs in 64 patients.

In the series studied by Hain and colleagues [5], the reported mean duration of symptoms at the time of the survey was 3.5 ± 2.5 years.

During an "episode" of MdDs, the symptoms are usually constant (except when relieved by getting into a vehicle). In the series studied by Hain and colleagues [5], 85% of patients described symptoms as constant during an episode.

In the series of 64 patients studied by Cha and colleagues [6], 41 (64%) had more than one episode.

Characteristic of classical (motion-triggered) MdDs is that symptoms usually improve when the patient gets into a moving vehicle. The study by Cha and colleagues [6] reported that 17/22 (77%) of motion-triggered episodes improved with exposure to motion.

Table 74.1 Motion triggers for patients with mal de debarquement syndrome (From Cha et al. (2008))

Triggers	All MdD total n = 64 (%)	Pure MdD total n = 46 (%)	Mixed MdD total n = 18 (%)
Boat/ Cruise	52 (81)	34 (74)	18 (100)
Airplane	26 (41)	22 (48)	4 (22)
Car/Train	10 (16)	6 (13)	4 (22)

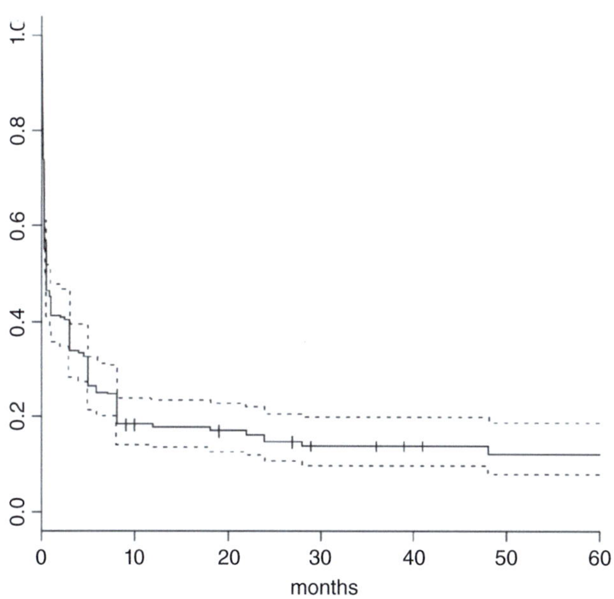

Fig. 74.1 Kaplan–Meier curve for duration of 206 episodes of MdDs in 64 patients. (From Cha and colleagues [6]. Reproduced with permission from Springer Nature)

A smaller proportion of "nonmotion-triggered" episodes of MdDs improve when a patient gets into a moving vehicle. The study by Cha and colleagues [6] reported that 4/10 (40%) of "mixed MdDs" episodes (meaning some that were motion-triggered and some that were not) improved with exposure to motion.

Cha and colleagues [6] commented that:

> The overall prevalence of migraine headache in our MdD patient group (38%) was higher than the population baseline (17.1% in women, 5.6% in men) but separating this group into the pure MdD group and the mixed MdD group showed that most of this risk came from the mixed MdD group (mixed = 73% vs pure = 22%).

The sensation of disequilibrium and other associated symptoms are variable. Tables 74.2 and 74.3 review symptoms reported by 27 MdDs patient studied by Hain and col-

Table 74.2 Disequilibrium and associated symptoms in 27 patients with mal de debarquement syndrome reported by Hain et al. [5]

Symptoms	Number (%)
Rocking	25 (93)
Swaying	22 (81)
Imbalance	20 (74)
Ear symptoms	15 (56)
Tilting	14 (52)
Nausea	10 (37)
Headache	9 (33)
Jumping vision	6 (22)
Blurred vision	5 (19)
Perioral tingling	4 (15)
Spinning (vertigo)	3 (11)
Diplopia	1 (4)
Vomiting	1 (4)
Other (eye twitch, fuzzy-headed/woozy, and pulling/ numbness in foot or lower leg	3 (11)

Table 74.3 Disequilibrium and associated symptoms in 149 patients with mal de debarquement syndrome reported by Dai et al. [13]

Symptoms	Percentage
Rocking	80
Swaying	73
Bobbing	46
Gravity pulling	54
Trampoline walking	29
Fatigue	62
Unstable in crowds	51
Brain fog	53
Sensitivity to fluorescent lights	51
Sensitivity to computer screens	54
Anxiety	49
Depression	38
Fullness of the ears	39
Tinnitus	33
Fussy vision	37
Clumsiness	37
Head pressure	28

Table 74.4 Aural symptoms reported by patients with mal de debarquement syndrome in the series studied by Hain et al. [5]

Symptom	Number (%) of patients
Fullness	19 (73)
Tinnitus	18 (69)
Hyperacusis	16 (61)
Otalgia	11 (42)
Decreased hearing	10 (39)

leagues [5], and symptoms reported by 149 MdDs patient studied by Dai and colleagues [13].

In both series [5, 13], "rocking" and "swaying" are the most common vestibular symptoms. The next most common vestibular symptoms in the series reported by Hain and colleagues [5] were "imbalance" (74%) and "tilting" (52%), while in the series reported by Dai and colleagues [13] were "bobbing" (46%) and "gravity pulling" (54%).

Hain and colleagues [5] further noted that, "Although only 15 (56%) of the [27] subjects checked ear symptoms as part of their present illness, 24 (89%) described having experienced otologic symptoms at some time since the onset of MDD [mal de debarquement syndrome] when asked about specific ear problems. Otologic symptoms reported by 27 survey subjects" are shown in Table 74.4.

Physical Examination

Patients with MdDs who are otherwise healthy will have normal physical examinations.

Ocular Motor Examination

Patients with MdDs who are otherwise healthy will usually have normal ocular motor examinations.

The study by Cha and colleagues [6] reported that 6/64 (9%) of all MdDs patients exhibited "positional torsional/downbeat nystagmus."

Testing: Audiologic and Vestibular

The study by Cha and colleagues [6] reported that, "Peripheral vestibular function testing is normal" in patients with MdDs.

Lewis [9] studied an MdDs patient in detail with audiometry, vestibular evoked myogenic potentials, subjective visual vertical, videonystagmography (including caloric testing), rotatory chair testing, and computerized dynamic posturography, and reported no abnormalities.

Nachum and colleagues [14] prospectively studied male sailors aged 18–22, including 17 patients who were "susceptible" to MdDs and 17 subjects who were not susceptible, testing them with computerized dynamic posturography before and after sailing. They found, "Abnormal SOT [sensory organization testing] results, defined as a composite and at least one equilibrium score below the normal values achieved by 95% of an age-matched, healthy population, were found in two and nine of the subjects in the SMD [susceptible to MdDs] group for the pre- and postsailing tests respectively," whereas, "All SOT results for the NSMD [nonsusceptible to MdDs] subjects were within normal limits."

The main value of otovestibular testing in patients with MdDs is to exclude competing diagnoses. Our experience has been that some patients appropriately referred for MdDs sometimes have an additional, more common disorder (such as benign paroxysmal positional vertigo) which, if unrecognized, will interfere with identification and management of MdDs.

Imaging

The study by Cha and colleagues [6] reported that "there are no consistent structural brain imaging abnormalities in patients" with MdDs. Lewis [9] also reported no abnormalities on regular brain MRI.

Some studies of MRI voxel-based morphometry report distinct features, "in (a) visual-vestibular processing areas (IPL, SPL, V3, V5/MT), (b) default mode network structures (cerebellar IX, IPL, ACC), (c) salience network structures (ACC and IFG/AI) (d) somatosensory network structures (postcentral gyrus, MCC, anterior cerebellum, cerebellar lobule VIII), and (e) a structure within the central executive network (DLPFC)" [15].

Differential Diagnosis

The Barany Society published diagnostic criteria for MdDs [16].

Gordon and colleagues [10] studied "116 members of seagoing vessels" and reported that 72% of subjects experienced "transient feeling of swinging, swaying, unsteadiness and disequilibrium" following the voyage; this resolved in ≤6 hours in 88% of subjects and in ≤48 hours in all subjects. When this phenomenon resolves within 2 days following prolonged exposure to motion, it is sometimes referred to as "land sickness," and in such patients, the sensation is usually *not* relieved by getting into a moving vehicle, so this is probably distinct from MdDs.

Cha and colleagues [6] studied a series of patients with "pure" MdDs (whose episodes were triggered by motion) and "mixed" MdDs (in whom some episodes were triggered by motion, and others were not). Among patients with "mixed" episodes, they observed that:

- Some episodes of "MdDs" in this group were not triggered by motion.
- The episodes of "MdDs" in this group were less likely to improve when the patient was re-exposed to motion.
- There was a significantly higher prevalence of migraine among this group.

Dai and colleagues [13] report follow-up of the nonpharmacologic treatment protocol that they developed. The treatment group included patients with "classic" MdDs (episodes triggered by prolonged exposure to motion) and "spontaneous" MdDs (episodes not triggered by motion). They observed that in comparison to patients with "classic" MdDs, those patients with "spontaneous" MdDs overall were less responsive to treatment, treatment effect was less sustained, and fewer achieved remission.

In our view, this constellation of characteristics raises the question of whether "nonmotion-triggered MdDs" is truly mal de debarquement syndrome. The differential diagnosis should probably include migraine-associated vertigo (MAV) [17, 18].

Treatment

Before the diagnosis of MdDs is realized, patients often attempt self-treatment with (or are prescribed) over-the-counter motion sickness medications such as dimenhydrinate or meclizine, and these are always ineffective. Prescription motion-sickness medications, such as transdermal scopolamine, are also ineffective [5].

Some patients notice short-term pharmacologic improvement with a low-dose benzodiazepine [6], but the use of such medications is limited by their significant addictive potential. In the patients studied by Hain and colleagues [5], 3/6 (50%) reported benefit from clonazepam, 1/4 (25%) from alprazolam, and 2/10 (20%) from diazepam.

Some patients notice longer-term pharmacologic improvement from an SSRI or SNRI. Our experience with venlafaxine 25–50 mg once per day has been favorable.

Some practitioners advocate simply treating MdDs as if it were migraine-associated vertigo. Ghavami and colleagues [19] treated 15 MdDs patients with a "migraine protocol" that included some combination of nortriptyline, verapamil, and topiramate and reported that, "73% of patients with persisted MdDS had improvement in their QOL [quality of life] and symptom intensity when treated with migraine lifestyle changes and prophylactic therapy."

Vestibular rehabilitation therapy is also commonly attempted for MdDs. Results are often difficult to interpret due to concomitant pharmacotherapy; for example, Hain and colleagues [5] reported that, "Fifteen subjects had balance rehabilitation (9 on these took medication during the therapy), and 10 of them reported benefit." Not all practitioners advocate a trial of vestibular rehabilitation therapy; Ghavami and colleagues [19] report that, "In our experience, MdDS patients who were treated with physical therapy or vestibular rehabilitation did not have a significant improvement in their symptomatology or QOL [quality of life]." Nevertheless, since vestibular rehabilitation therapy incurs no medical risk, we are willing to offer it to patients who express an interest in vestibular therapy, or are averse to a trial of medication.

Some patients experience longer-term nonpharmacologic improvement with the treatment protocol devised by Dr. Mingjia Dai [13, 20]. The efficacy of this protocol has been replicated in other laboratories [21]. Research is exploring whether a virtual reality version of this approach is effective [22].

Ongoing research is exploring the possible role of transcranial magnetic stimulation in the treatment of MdDs [23–26]. Transcranial alternating current stimulation has also been explored [27].

Some research suggests that different facets (symptoms) of MdDs may respond differently to different interventions. For instance, Yakushin and colleagues [28] studied whether a treatment protocol involving optokinetic stimulation with vertically moving horizontal stripes would be more helpful specifically for the "gravitational pulling sensation" (as opposed to optokinetic stimulation with horizontally moving vertical stripes for the more common "rocking" sensation of MdDs).

As with other "invisible" diseases (illnesses for which patients "look normal" to an untrained outside observer), MdDs patients often appear stigmatized, leading to additional psychosocial problems beyond the disease's original symptoms, including anxiety and depression [29]. Seeking psychiatric help for this aspect of the disease is sensible [30].

Prognosis

Prognosis is guarded because the overall duration of episodes is variable, and the response to treatments is variable.

In the majority of cases, an episode of MdDs will eventually end. The clinical question at that point then becomes: What is the risk of re-triggering an episode with another prolonged exposure to motion? There is not yet any proven algorithm to predict this, so some patients will simply practice avoidance of such exposure. Some patients will elect to travel again and anecdotally report feeling "protected" by taking a benzodiazepine prophylactically, but this has never been studied systematically.

References

1. Darwin E, Mitchill SL, Joseph Meredith Toner Collection (Library of Congress). Zoonomia, or, The laws of organic life. Vol. I. New York: Printed by T. & J. Swords, printers to the Faculty of Physic of Columbia College, No. 99 Pearl-Street; 1796. xxxii, 10, 2, 1–2, 2, 3–4, 2, 5–402, 2, 3–14, 2, 15–16, 2, 17–33, 3 p.
2. Irwin JA, Cantab MA. The pathology of sea-sickness. Lancet. 1881;118(3039):907–9.
3. London J. The Sea-wolf. New York and London; 1904.
4. London J. The cruise of the Snark. New York: The Macmillan company; 1911. xiv, 340 p.
5. Hain TC, Hanna PA, Rheinberger MA. Mal de debarquement. Arch Otolaryngol Head Neck Surg. 1999;125(6):615–20.
6. Cha YH, Brodsky J, Ishiyama G, Sabatti C, Baloh RW. Clinical features and associated syndromes of mal de debarquement. J Neurol. 2008;255(7):1038–44.
7. Murphy TP. Mal de debarquement syndrome: a forgotten entity? Otolaryngol Head Neck Surg. 1993;109(1):10–3.
8. Moeller L, Lempert T. Mal de debarquement: pseudo-hallucinations from vestibular memory? J Neurol. 2007;254(6):813–5.
9. Lewis RF. Frequency-specific mal de debarquement. Neurology. 2004;63(10):1983–4.
10. Gordon CR, Spitzer O, Doweck I, Melamed Y, Shupak A. Clinical features of mal de debarquement: adaptation and habituation to sea conditions. J Vestib Res. 1995;5(5):363–9.
11. Hain TC, Helminski JO. Therapy for mal de debarquement syndrome. In: Herdman SJ, editor. Vestibular rehabilitation. 3rd ed. Philadelphia: E.A. Davis Company; 2007. p. 202–4.
12. Cha YH. Mal de debarquement syndrome: new insights. Ann N Y Acad Sci. 2015;1343:63–8.
13. Dai M, Cohen B, Cho C, Shin S, Yakushin SB. Treatment of the mal de debarquement syndrome: a 1-year follow-up. Front Neurol. 2017;8:175.
14. Nachum Z, Shupak A, Letichevsky V, Ben-David J, Tal D, Tamir A, et al. Mal de debarquement and posture: reduced reliance on vestibular and visual cues. Laryngoscope. 2004;114(3):581–6.
15. Cha YH, Chakrapani S. Voxel based morphometry alterations in Mal de Debarquement Syndrome. PLoS One. 2015;10(8):e0135021.
16. Cha YH, Baloh RW, Cho C, Magnusson M, Song JJ, Strupp M, et al. Mal de debarquement syndrome: Diagnostic criteria consensus document of the classification committee of the Barany society. J Vestib Res. 2020;30:285.
17. Cha YH, Cui Y. Rocking dizziness and headache: a two-way street. Cephalalgia. 2013;33(14):1160–9.
18. Beh SC, Chiang HS, Sanderson C. The interconnections of Mal de Debarquement syndrome and vestibular migraine. Laryngoscope. 2021;131(5):E1653–E61.
19. Ghavami Y, Haidar YM, Ziai KN, Moshtaghi O, Bhatt J, Lin HW, et al. Management of mal de debarquement syndrome as vestibular migraines. Laryngoscope. 2017;127(7):1670–5.
20. Cohen B. Dedication to Mingjia Dai, Ph.D. for Discovery of the First Successful Treatment of the Mal de Debarquement Syndrome. Front Neurol. 2019;10:1196.
21. Mucci V, Perkisas T, Jillings SD, Van Rompaey V, Van Ombergen A, Fransen E, et al. Sham-controlled study of optokinetic stimuli as treatment for Mal de Debarquement Syndrome. Front Neurol. 2018;9:887.
22. Yakushin SB, Zink R, Clark BC, Liu C. Readaptation treatment of Mal de Debarquement syndrome with a virtual reality app: a pilot study. Front Neurol. 2020;11:814.
23. Cha YH, Cui Y, Baloh RW. Repetitive transcranial magnetic stimulation for mal de debarquement syndrome. Otol Neurotol. 2013;34(1):175–9.
24. Guofa S, Han Y, Urbano D, Yoon-Hee C, Lei D. Changes of symptom and EEG in mal de debarquement syndrome patients after repetitive transcranial magnetic stimulation over bilateral prefrontal cortex: a pilot study. Annu Int Conf IEEE Eng Med Biol Soc. 2014;2014:4294–7.
25. Cha YH, Deblieck C, Wu AD. Double-Blind Sham-Controlled Crossover Trial of Repetitive Transcranial Magnetic Stimulation for Mal de Debarquement Syndrome. Otol Neurotol. 2016;37(6):805–12.
26. Yuan H, Shou G, Gleghorn D, Ding L, Cha YH. Resting State Functional Connectivity Signature of Treatment Effects of Repetitive Transcranial Magnetic Stimulation in Mal de Debarquement Syndrome. Brain Connect. 2017;7(9):617–26.
27. Cha YH, Gleghorn D, Doudican BC. Double-blind randomized N-of-1 trial of transcranial alternating current stimulation for mal de debarquement syndrome. PLoS One. 2022;17(2):e0263558.
28. Yakushin SB, Raphan T, Cho C. Treatment of gravitational pulling sensation in patients with Mal de Debarquement Syndrome (MdDS): a model-based approach. Front Integr Neurosci. 2022;16:801817.
29. Mucci V, Canceri JM, Brown R, Dai M, Yakushin S, Watson S, et al. Mal de Debarquement Syndrome: a survey on subtypes, misdiagnoses, onset and associated psychological features. J Neurol. 2018;265(3):486–99.
30. Arroll MA, Attree EA, Cha YH, Dancey CP. The relationship between symptom severity, stigma, illness intrusiveness and depression in Mal de Debarquement Syndrome. J Health Psychol. 2016;21(7):1339–50.

Persistent Postural Perceptual Dizziness (PPPD)

75

Overview

Persistent postural perceptual dizziness (PPPD) generally follows on the heels of another "dizzy disease" and is thought to reflect failure of the vestibular system to recalibrate to the premorbid state, though the underlying pathophysiology is unknown. Current diagnostic criteria require that symptoms (nonspinning disequilibrium) be present >15 days per month for ≥3 months, are exacerbated by several factors (upright posture, active/passive motion, complex/moving visual stimuli), that the symptoms cause distress or functional impairment, and that the symptoms are not better explained by another disorder. There is no confirmatory diagnostic test for PPPD, though research studies have observed nonspecific abnormalities on volumetric MRI and posturography. There is a fairly extensive differential diagnosis which includes a number of objectively verifiable diseases, so we generally undertake a screening otovestibular workup with cervical and ocular vestibular evoked myogenic potentials, video head impulse testing, videonystagmography, rotatory chair testing and computerized dynamic posturography, and (if ear symptoms are present) audiometry. Treatments studies are generally hampered by the lack of placebo arms, so we generally proceed in order of escalating risk, starting with vestibular rehabilitation therapy, then cognitive behavioral therapy. If those fail, then a trial of sertraline is reasonable.

Introduction

Persistent postural perceptual dizziness (PPPD) is currently believed to be a disorder of equilibrium precipitated by another discrete vestibular illness. PPPD may emerge while the other vestibular illness is present, or resolving. PPPD is thought to reflect a maladaptive failure to recalibrate the vestibular system to the premorbid state. The symptoms consist of nonspinning disequilibrium, are present more than half the time, tend to be most obvious when a patient is upright, can be exacerbated by visual stimuli and spatial cues, and are frequently comorbid with psychiatric disorders (particularly anxiety). There is currently no confirmatory diagnostic test, so at present PPPD is a diagnosis of exclusion.

With the caveat that "as a rule, only poorly understood disorders require classification by a committee" [1], the diagnostic criteria for PPPD were based on a committee consensus document of the Barany Society [2] and drew on the work of several previously described entities, including phobic postural vertigo (PPV), space motion discomfort (SMD), chronic subjective dizziness (CSD), and visual vertigo (VV). PPPD has some features of each of these other entities but is not synonymous with any of them.

Precipitating Disorders

In principle, PPPD can follow nearly any other disorder of equilibrium. In descending order of frequency, the precipitating disorders are [2]:

- 25–30%: Peripheral or central vestibular disorders
- 15–20%: Vestibular migraine
- 15%: Panic attacks manifesting with dizziness
- 15%: Anxiety manifesting with dizziness
- 10–15%: Brain concussion and whiplash injury
- 7%: Autonomic disorders
- 3%: Cardiac dysrhythmias, adverse drug reactions

This distribution is fairly close to the prevalence of each category, so it is unclear whether this distribution reflects a general predisposition/vulnerability to simply any precipitant (and thus is just more commonly found following more common diseases), or whether specific diseases are actually more likely than others to trigger PPPD.

M. Cherchi, *Otoneurology and Vestibular Medicine*, https://doi.org/10.1007/978-3-031-94841-1_75

Epidemiology

The Barany consensus document states that:

> No epidemiologic studies are available for PPPD, but its prevalence and incidence may be estimated from research done on patients with PPV, VV, CST and chronic dizziness following acute vestibular syndromes. Clinical epidemiologic data from tertiary care centers with special interest in PPV and CSD showed their prevalence to be 15 – 20% among all patients presenting for evaluation of vestibular symptoms, making them the most common diagnoses among young adults and the second most common among all adults [2].

The Barany consensus document further states that symptoms, "may last for months to years," and that "the average duration of illness at the time of tertiary consultation was 4.5 years."

Diagnostic Criteria

As articulated in the Barany Society consensus document [2], the diagnostic criteria for PPPD are:

A. [***Symptoms and chronology***] One or more symptoms of dizziness, unsteadiness, or non-spinning vertigo are present on most days for 3 months or more.
   1. Symptoms last for prolonged (hours-long) periods of time but may wax and wane in severity.
   2. Symptoms need not be present continuously throughout the entire day.

B. [***Exacerbators***] Persistent symptoms occur without specific provocation but are exacerbated by three factors:
   1. Upright posture,
   2. Active or passive motion without regard to direction or position.
   3. Exposure to moving visual stimuli or complex visual patterns.

C. [***Precipitants***] The disorder is precipitated by conditions that cause vertigo, unsteadiness, dizziness, or problems with balance including acute, episodic, or chronic vestibular syndromes, other neurologic or medical illnesses, or psychological distress.
   1. When the precipitant is an acute or episodic condition, symptoms settle into the pattern of criterion A as the precipitant resolves, but they may occur intermittently at first, and then consolidate into a persistent course.
   2. When the precipitant is a chronic syndrome, symptoms may develop slowly at first and worsen gradually.

D. [***Emotional/functional sequelae***] Symptoms cause significant distress or functional impairment.

E. [***Diagnosis of exclusion***] Symptoms are not better accounted for by another disease or disorder.

Differential Diagnosis

The Barany Society consensus document [2] offers the following structured differential diagnosis:

A. "Chronic sequelae of acute precipitants." This refers to symptoms attributable to whatever antecedent "dizzy disease" triggered PPPD.

B. "Recurrent attacks of episodic precipitants," such as vestibular migraine, Ménière's disease, benign paroxysmal positional vertigo and others.

C. "Ongoing manifestations of chronic precipitants," such as:
   1. "Chronic anxiety and depressive disorders"
   2. "Postconcussive syndrome"
   3. "Autonomic disorders"

D. "Other chronic vestibular syndromes," such as:
   1. Bilateral vestibular weakness.
   2. Chronic neurologic disorders.
   3. Mal de debarquement syndrome.

E. "Adverse effects of medications."

F. "Other functional forms of vestibular disorders."

G. "Gait disorders, falls, and near falls."

Diagnostic Testing

One study [3] compared results of computerized dynamic posturography in patients diagnosed with PPPD versus patients with "acute vestibular syndromes" and normal controls. The study noted that, "Patients with PPPD had the greatest likelihood of abnormal sensory analyses," and "Patients with PPPD had significantly lower mean scores than normal individuals on conditions 2-6 and the composite." These patterns are very nonspecific.

Several studies have commented on changes in brain MRIs in patients with PPPD [4]. For example, a volumetric MRI study of PPPD patients [5] found that, "PPPD patients showed gray matter volume decrease in the temporal cortex, cingulate cortex, precentral gyrus, hippocampus, dorsolateral prefrontal cortex, caudate nucleus and the cerebellum. A negative correlation of disease duration and gray matter volume was observed in the visual cortex, supplementary motor area and somatosensory processing structures," which the authors summarized as, "In patients with PPPD areas involved in multisensory vestibular processing show gray matter volume decrease," noting that, "Longer duration of disease leads to a more pronounced gray matter alteration," and commented that, "These brain regions resemble those previously described for other vestibular disorders." In other words, these volumetric brain MRI changes do not distinguish PPPD from other vestibular disorders.

One study explored posturographic changes induced by galvanic vestibular stimulation (GVS) in patients with PPPD [6]. The study found that, "The sway increase in the baseline of PPPD may be related to an earlier transition from open- to closed-loop mode of postural control. In contrast, the smaller sway of PPPD in the foam condition under visual deprivation is in line with the known balance improvement under more demanding postural challenges in PPPD," concluding that, "GVS discloses differences between simple and complex balance tasks in PPPD."

None of these studies identified a specific pattern (of MRI volumetric changes, or posturographic changes) that was diagnostic of PPPD. Given the differential diagnosis reviewed earlier, and the fact that PPPD is a diagnosis of exclusion, when a diagnosis of PPPD is being entertained, it is medically reasonable to undertake a screening otovestibular workup with cervical and ocular vestibular evoked myogenic potentials, video head impulse testing, rotatory chair testing, and computerized dynamic posturography. If a patient complains of lateralizing hearing symptoms, then audiometry is also reasonable.

Pathophysiology

The underlying pathophysiology of PPPD is not understood. Staab and colleagues [2] proposed the following framework:

> PPPD is thought to develop via a dynamic process. In about 70% of patients, a structural vestibular syndrome (e.g., vestibular neuritis, BPPV) or other medical condition precipitates PPPD. Individuals who respond to the precipitating event with a high level of anxiety and body vigilance appear likely to progress to PPPD. Anxiety-related personality traits or pre-existing anxiety disorders appear to increase the risk of developing PPPD. In about 30% of patients, PPPD begins with acute psychological distress and then progresses to the functional disorder. PPPD may co-exist with structural or psychological illnesses, placing patients in the intersections of the functional and psychological or structural ellipses. Anxiety-related personality traits and psychological symptoms are incorporated into PPV [phobic postural vertigo], whereas they are considered predisposing factors and comorbid symptoms, respectively, in PPPD [2].

Treatments

Despite the poor understanding of the underlying pathophysiology of PPPD, a number of treatment modalities have been explored in multiple studies, including physical therapy, cognitive behavioral therapy, and pharmacotherapy [7]. Single studies of other modalities, such as noninvasive vagus nerve stimulation, have also been undertaken.

Treatment: Physical Therapy

Several studies have explored whether physical therapy is beneficial in PPPD.

One study [8] divided PPPD into two treatment groups, both of which underwent vestibular rehabilitation therapy, and one of which received additional "placebo" treatment. The study reported that, "There was a significant decrease in functional, physical, and total scores on the DHI in both groups after VRT. Adding the placebo did not have supplementary outcomes." The structure of this study strikes us as odd, as it is not truly placebo-controlled (there was no comparison of "treatment alone" versus "placebo alone").

A pilot study [9] offered vestibular rehabilitation therapy (VRT) to 30 patients and assessed outcomes using several validated scoring systems including the Dizziness Handicap Inventory (DHI), Depression, Anxiety and Stress Scale (DASS-21), and the EuroQuol quality of life scale (EQ5D). The investigators concluded, "VRT is an effective modality in significantly improving quality of life, dizziness handicap, depression, and anxiety levels within 3 months in PPPD."

These studies are encouraging but hindered by the absence of true placebo arms. However, since vestibular rehabilitation therapy incurs no significant medical risk, it is still reasonable.

Treatment: Cognitive Behavioral Therapy

Cognitive behavioral therapy (CBT) has been attempted in the treatment of PPPD. These studies are somewhat difficult to interpret because they generally combine CBT with some other treatment modality, such as vestibular rehabilitation therapy [10] or pharmacotherapy [11]. Nevertheless, since CBT incurs no medical risk, we feel it is reasonable to attempt.

Treatment: Pharmacotherapy

Several studies have explored the utility of pharmacotherapy in treating PPPD; the agents investigated are generally SSRIs (typically sertraline) or SNRIs. The majority of these publications are case–control studies, case series, and retrospective reviews [7]. There has been a single double-blind randomized controlled trial of sertraline versus verapamil [12]. As of this writing, there have been no randomized, prospective, double-blinded, placebo-controlled trials.

Treatment: Vagus Nerve Stimulation

One study [13] took the fact that many PPPD patients complain of symptoms that are potentially autonomically mediated (nausea, tachycardia, diaphoresis) as a rationale for attempting non-invasive vagus nerve stimulation (nVNS) in a protocol similar to what had been explored for cluster headaches. The investigators reported that, "Results showed that non-invasive vagus nerve stimulation significantly improved quality of life, as measured by the EQ-5D-3L ($p = 0.04$), and depression, as measured by the HADS-D ($p = 0.002$), in the nVNS group," further observing that, "in the pooled analysis (additional 4 weeks of stimulation…), less severe vertigo attacks/exacerbations ($p = 0.04$), a decrease in total postural sway path as measured by posturography ($p = 0.02$), as well as tendentious less anxiety ($p = 0.08$), occurred after stimulation," and concluded that, "These data imply that short term nVNS is a safe and promising treatment option in patients with otherwise refractory PPPD."

References

1. Fenichel G. Clinical pediatric neurology: a signs and symptoms approach. 5th ed. Philadelphia: W.B. Saunders; 2005.
2. Staab JP, Eckhardt-Henn A, Horii A, Jacob R, Strupp M, Brandt T, et al. Diagnostic criteria for persistent postural-perceptual dizziness (PPPD): consensus document of the committee for the Classification of Vestibular Disorders of the Barany Society. J Vestib Res. 2017;27(4):191–208.
3. Sohsten E, Bittar RS, Staab JP. Posturographic profile of patients with persistent postural-perceptual dizziness on the sensory organization test. J Vestib Res. 2016;26(3):319–26.
4. Sun L, Xiang K. A review on the alterations in the brain of persistent postural-perceptual dizziness patients and non-pharmacological interventions for its management. Rev Neurosci. 2020;31(6):675–80.
5. Wurthmann S, Naegel S, Schulte Steinberg B, Theysohn N, Diener HC, Kleinschnitz C, et al. Cerebral gray matter changes in persistent postural perceptual dizziness. J Psychosom Res. 2017;103:95–101.
6. Woll J, Sprenger A, Helmchen C. Postural control during galvanic vestibular stimulation in patients with persistent perceptual-postural dizziness. J Neurol. 2019;266(5):1236–49.
7. Trinidade A, Goebel JA. Persistent postural-perceptual dizziness-a systematic review of the literature for the balance specialist. Otol Neurotol. 2018;39(10):1291–303.
8. Nada EH, Ibraheem OA, Hassaan MR. Vestibular rehabilitation therapy outcomes in patients with persistent postural-perceptual dizziness. Ann Otol Rhinol Laryngol. 2019;128(4):323–9.
9. Teh CS, Abdullah NA, Kamaruddin NR, Mohd Judi KB, Fadzilah I, Zainun Z, et al. Home-based vestibular rehabilitation: a feasible and effective therapy for persistent postural perceptual dizziness (a pilot study). Ann Otol Rhinol Laryngol. 2022;132:566.
10. Popkirov S, Stone J, Holle-Lee D. Treatment of persistent postural-perceptual dizziness (PPPD) and related disorders. Curr Treat Options Neurol. 2018;20(12):50.
11. Yu YC, Xue H, Zhang YX, Zhou J. Cognitive behavior therapy as augmentation for sertraline in treating patients with persistent postural-perceptual dizziness. Biomed Res Int. 2018;2018:8518631.
12. Staab J, Eggers S, Neff B, Shepard N, editors. Vestibular migraine and persistent postural-perceptual dizziness: results of a double blind, parallel group, pharmacologic dissection trial using verapamil and sertraline. International Headache Congress of the International Headache Society; 2015.
13. Eren OE, Filippopulos F, Sonmez K, Mohwald K, Straube A, Schoberl F. Non-invasive vagus nerve stimulation significantly improves quality of life in patients with persistent postural-perceptual dizziness. J Neurol. 2018;265(Suppl 1):63–9.

Concussion and Postconcussive Disequilibrium

76

Overview

Concussion can cause a range of lesions. Lesions relevant to the audiovestibular system include otologic, neuro-ophthalmological, neurovascular, neurologic, and others. Depending on the specific lesion(s), the clinical presentation and clinical examination will vary. Depending on the symptoms and physical examination findings, appropriate testing may include audiometry, vestibular evoked myogenic potentials, video head impulse testing, videonystagmography, and rotatory chair testing. Imaging with MRI (brain and internal auditory canals) and temporal bone CT may be reasonable. Identification of specific deficits will guide treatment. Prognosis is as variable as the lesions themselves. What we have presented above pertains to the audiologic and vestibular consequences of concussion of interest to the otoneurologist; but concussion is a complex and incompletely understood phenomenon and generally requires a multidisciplinary approach to diagnosis and management. In practice, this usually requires referral to a center specializing in concussion/traumatic brain injury, which can coordinate the various evaluations and services needed to care for these patients.

Introduction

Concussion, generally classified as a subtype of traumatic brain injury, is an immense topic. Head trauma is unfortunately common and thus comprises a significant public health concern. Although concussion has attracted significant research interest, it remains incompletely understood; it lacks biomarkers for verification; convincing studies of effective treatments are lacking; and outcomes are sometimes poor.

Concussion can have a broad range of effects on the central nervous system, and we do not intend to present a comprehensive review here. Rather, we shall outline the otoneurologist's role within what is usually a multidisciplinary approach to this problem.

Mechanisms of Disequilibrium in the Context of Concussion

Head trauma can cause a variety of neurological lesions in the head and neck. Disequilibrium is fairly common in concussion, and it is on this symptom that we will focus, though this is not to suggest that other symptoms (headache, cognitive deficits) are unimportant. There are multiple mechanisms by which concussion can cause the symptom of disequilibrium [1].

- Otologic
  - Benign paroxysmal positional vertigo
  - Superior semicircular canal dehiscence
  - Labyrinthine concussion, traction injury of the vestibulocochlear nerve
  - Posttraumatic hydrops
- Neuro-ophthalmologic, particularly ocular motor disorders
  - Trochlear nerve palsy or lateral rectus palsy
  - Convergence insufficiency
  - Accommodative insufficiency
  - Saccadic deficits
- Neurovascular, e.g., vertebral artery dissection with embolization of the labyrinth
- Neurologic
  - Posttraumatic migraine with migraine associated vertigo
  - Cervicogenic vertigo
  - Brainstem lesions
  - Vestibular cortical lesions from shearing forces (at least in theory)
- Longer-term sequelae
  - Persistent postural perceptual dizziness
  - Visual vertigo

M. Cherchi, *Otoneurology and Vestibular Medicine*, https://doi.org/10.1007/978-3-031-94841-1_76

Otologic Mechanisms

The most common cause of disequilibrium over the lifespan is *benign paroxysmal positional vertigo* (BPPV). The literature is mixed with respect to whether head trauma is a risk for BPPV. Several large studies conclude that head trauma is not a risk factor for BPPV [2, 3]. Other studies conclude that head trauma is a risk factor for BPPV [4], or that it increases the risk of specific variants of BPPV, such as anterior canal [5, 6] or bilateral involvement [7]. Some studies conclude that head trauma appears to be associated with a higher risk of treatment failure in BPPV [8, 9].

Head trauma can transmit sufficient kinetic energy to the labyrinth as to cause a *labyrinthine concussion* or *traction injury of the vestibulocochlear nerve* [10–12].

Head trauma can cause *semicircular canal dehiscence*, typically of the superior canal [13].

Head trauma appears in some cases to result in *endolymphatic hydrops* [14]. This is sometimes referred to as "post-traumatic Ménière's" [15–17], but we prefer to reserve the term "Ménière's" for the idiopathic syndrome. The mechanism of posttraumatic endolymphatic hydrops is unknown. Logically, one would think that trauma causes disordered endolymphatic fluid flow, either from increased production, or decreased resorption, or both; this could in turn be due to alteration in the function of cells that produce these fluids, or in the hydraulic system through which they pass (cochlear duct, endolymphatic duct, and vestibular aqueduct).

Neuro-Ophthalmologic Mechanisms

Ocular motor dysfunction is common after head trauma. This can result from extraocular muscle palsies or from vergence deficits.

Table 76.1, from Ciuffreda and colleagues [18], shows the relative frequency of ocular motor abnormalities in a series of 160 patients who had suffered traumatic brain injury.

Table 76.1 Relative frequency of ocular motor abnormalities in a series of 160 patients who had suffered traumatic brain injury (TBI)

Ocular motor dysfunction	TBI (%)	Most common anomaly (TBI)
Accommodation	41.1	Accommodative insufficiency
Versional	51.3	Deficits of saccades
Vergence	56.3	Convergence insufficiency
Strabismus	25.6	Strabismus at near
Cranial nerve palsy	6.9	Cranial nerve III

From Ciuffreda et al. [18]

Neurovascular Mechanisms

Head trauma can result in cervical arterial dissections. *Dissection of the vertebral artery*, in particular, can predispose to embolization of the labyrinthine artery (usually via the anterior inferior cerebellar artery), resulting in *labyrinthine ischemia/infarction* [19, 20]. Embolization to any of the cerebellar arteries can also result in *cerebellar infarction* [21–23].

Neurologic Mechanisms

Migraines can develop (or recur) following concussion, and this can include migraine-associated vertigo.

Concussion is often associated with neck trauma, which in turn can predispose to cervicogenic vertigo [24].

The actual "brain damage" in concussion probably has multiple mechanisms, such as shearing forces exerted on gray and white matter, resulting in a variety of cellular dysfunctions. Magnetic resonance tractography is an evolving imaging modality for this disease process. In patients with postconcussive disequilibrium, research has identified damage to tracts relevant to vestibular function [25].

Vestibular cortex in the temporal lobe [26, 27] can be damaged in concussion. At least in theory, such lesions could give rise to disequilibrium. This has not been addressed in published studies.

Longer-Term Sequelae

Nearly any cause of disequilibrium can secondarily give rise to *persistent postural perceptual dizziness* (PPPD).

Similarly, *visual vertigo* (VV) can superimpose itself on other, pre-existing sources of disequilibrium, including post-concussive disequilibrium.

Presentation

The presentation of postconcussive disequilibrium depends on the consequent lesions.

McCrary and colleagues [13] described a case series of 14 patients and reported the following relative frequency of symptoms in posttraumatic superior semicircular canal dehiscence:

- 93% Pulsatile tinnitus
- 79% Autophony, somatic sounds
- 64% Hearing loss
- 57% Vertigo
- 36% Nonpulsatile tinnitus
- 14% Aural fullness

Physical Examination

In patients who have suffered head trauma, some physical examination findings may be suggestive of temporal bone fracture, such as ecchymosis over the mastoid process—sometimes called Battle's sign [28]—periorbital ecchymosis, or hemotympanum.

In patients who have suffered concussion and present with auditory and/or vestibular symptoms, physical examination findings will depend on the particular lesions.

Of otologic mechanisms, benign paroxysmal positional vertigo may be detectable on ocular motor examination. Unilateral vestibular weakness (such as from traction injury of the vestibulocochlear nerve, or labyrinthine infarction) may manifest with spontaneous horizontal nystagmus. Hearing loss may be evident on examination.

Of neuro-ophthalmologic mechanisms, ocular motor dysfunction may be evident on physical examination, ranging from saccadic deficiencies, vergence abnormalities, or strabismus.

Testing

Audiologic testing may show sensorineural hearing loss compatible with labyrinthine concussion, or conductive hyperacusis compatible with superior semicircular canal dehiscence.

Cervical and/or ocular vestibular evoked myogenic potentials may show findings compatible with superior semicircular canal dehiscence [29].

Unilateral vestibular weakness may be detected on vestibular evoked myogenic potentials, video head impulse testing, or caloric testing.

Bilateral vestibular weakness may be detected on rotatory chair testing.

More detailed ocular motor studies are being researched for concussion, but remain investigational. For example, Wetzel and colleagues [30] demonstrate disordered smooth pursuit in a patient who suffered concussion.

Imaging

Brain imaging in a patient with concussion may in some instances identify a discrete lesion that could account for auditory and/or vestibular symptoms. For example, MRI of the brain and internal auditory canals may identify labyrinthine hemorrhage or cerebellar infarction. Temporal bone CT may identify a temporal bone fracture or superior semicircular canal dehiscence.

But brain imaging in a patient with concussion is also necessary to evaluate for other findings that may reflect more severe traumatic brain injury, such as a parenchymal hematoma or subarachnoid hemorrhage.

Treatment

Treatment of postconcussive disequilibrium depends on the specific lesion, if identifiable.

If an identified vestibular deficit is one that is usually treated with vestibular rehabilitation, then the same principles can apply in the context of postconcussion disequilibrium [31].

Benign paroxysmal positional vertigo occurring in the context of concussion is treated the same as idiopathic BPPV [32]. Some studies conclude that head trauma appears to be associated with a higher risk of treatment failure in BPPV [8, 9].

Hearing loss, such as from labyrinthine concussion, may be aidable with amplification.

Prognosis

Postconcussive disequilibrium "may resolve in days to weeks, or persist for months to years" [1]. This variability is partly due to the range of mechanisms discussed above.

References

1. Hac NEF, Gold DR. Neuro-visual and vestibular manifestations of concussion and mild TBI. Curr Neurol Neurosci Rep. 2022;22:219.
2. Aron M, Lea J, Nakku D, Westerberg BD. Symptom resolution rates of posttraumatic versus nontraumatic benign paroxysmal positional vertigo: a systematic review. Otolaryngol Head Neck Surg. 2015;153(5):721–30.
3. Luryi AL, LaRouere M, Babu S, Bojrab DI, Zappia J, Sargent EW, et al. Traumatic versus idiopathic benign positional vertigo: analysis of disease, treatment, and outcome characteristics. Otolaryngol Head Neck Surg. 2019;160(1):131–6.
4. Pisani V, Mazzone S, Di Mauro R, Giacomini PG, Di Girolamo S. A survey of the nature of trauma of post-traumatic benign paroxysmal positional vertigo. Int J Audiol. 2015;54(5):329–33.
5. Dlugaiczyk J, Siebert S, Hecker DJ, Brase C, Schick B. Involvement of the anterior semicircular canal in posttraumatic benign paroxysmal positioning vertigo. Otol Neurotol. 2011;32(8):1285–90.
6. Jackson LE, Morgan B, Fletcher JC Jr, Krueger WW. Anterior canal benign paroxysmal positional vertigo: an underappreciated entity. Otol Neurotol. 2007;28(2):218–22.
7. Katsarkas A. Benign paroxysmal positional vertigo (BPPV): idiopathic versus post-traumatic. Acta Otolaryngol. 1999;119(7):745–9.

8. Babac S, Djeric D, Petrovic-Lazic M, Arsovic N, Mikic A. Why do treatment failure and recurrences of benign paroxysmal positional vertigo occur? Otol Neurotol. 2014;35(6):1105–10.
9. Balatsouras DG, Koukoutsis G, Aspris A, Fassolis A, Moukos A, Economou NC, et al. Benign paroxysmal positional vertigo secondary to mild head trauma. Ann Otol Rhinol Laryngol. 2017;126(1):54–60.
10. Bartholomew RA, Lubner RJ, Knoll RM, Ghanad I, Jung D, Nadol JB Jr, et al. Labyrinthine concussion: historic otopathologic antecedents of a challenging diagnosis. Laryngoscope Investig Otolaryngol. 2020;5(2):267–77.
11. Chiaramonte R, Bonfiglio M, D'Amore A, Viglianesi A, Cavallaro T, Chiaramonte I. Traumatic labyrinthine concussion in a patient with sensorineural hearing loss. Neuroradiol J. 2013;26(1):52–5.
12. Ishai R, Knoll RM, Chen JX, Wong K, Reinshagen KL, Nadol JB Jr, et al. Otopathologic changes in the cochlea following head injury without temporal bone fracture. Otolaryngol Head Neck Surg. 2018;159(3):526–34.
13. McCrary HC, Babajanian E, Patel N, Yang S, Kircher M, Carlson ML, et al. Superior semicircular canal dehiscence syndrome following head trauma: a multi-institutional review. Laryngoscope. 2021;131(11):E2810–E8.
14. DiBiase P, Arriaga MA. Post-traumatic hydrops. Otolaryngol Clin N Am. 1997;30(6):1117–22.
15. Gundrum LK. Post-traumatic Meniere's symptom-complex. Eye Ear Nose Throat Mon. 1959;38(5):385–8.
16. Lehrer JF, Poole DC. Post-traumatic Meniere's syndrome. Laryngoscope. 1984;94(1):129.
17. Paparella MM, Mancini F. Trauma and Meniere's syndrome. Laryngoscope. 1983;93(8):1004–12.
18. Ciuffreda KJ, Kapoor N, Rutner D, Suchoff IB, Han ME, Craig S. Occurrence of oculomotor dysfunctions in acquired brain injury: a retrospective analysis. Optometry. 2007;78(4):155–61.
19. Choi KD, Chun JU, Han MG, Park SH, Kim JS. Embolic internal auditory artery infarction from vertebral artery dissection. J Neurol Sci. 2006;246(1–2):169–72.
20. Eliezer M, Verillaud B, Guichard JP, Kania R, Toupet M, Herman P, et al. Labyrinthine infarction caused by vertebral artery dissection: consideration based on MRI. J Neurol. 2019;266(10):2575–7.
21. Ilker Oz I, Bozay Oz E, Serifoglu I, Kaya N, Erdem O. Cerebellar infarction in childhood: delayed-onset complication of mild head trauma. Iran J Child Neurol. 2016;10(3):82–5.
22. Kim SH, Kosnik E, Madden C, Rusin J, Wack D, Bartkowski H. Cerebellar infarction from a traumatic vertebral artery dissection in a child. Pediatr Neurosurg. 1997;27(2):71–7.
23. Tabuchi S, Nakayasu H. Traumatic vertebral artery dissection and cerebral infarction following head and neck injury with a lucid interval. Acute Med Surg. 2015;2(2):127–30.
24. Cherchi M, DiLiberto FE, Yacovino DA, Das S. The enduring controversy of cervicogenic vertigo, and its place among positional vertigo syndromes. Audiol Res. 2021;11(4):491–507.
25. Jang SH, Bae CH, Kim JW, Kwon HG. Relationship between dizziness and the core vestibular projection injury in patients with mild traumatic brain injury. Diagnostics (Basel). 2021;11(11):2070.
26. de Waele C, Baudonniere PM, Lepecq JC, Tran Ba Huy P, Vidal PP. Vestibular projections in the human cortex. Exp Brain Res. 2001;141(4):541–51.
27. Vidal PP, De Waele C, Baudonniere PM, Lepecq JC, Ba Huy PT. Vestibular projections in the human cortex. Ann N Y Acad Sci. 1999;871:455–7.
28. Battle WH. Three lectures on some points relating to injuries to the head. Br Med J. 1890;2(1542):141–7.
29. Fife TD, Colebatch JG, Kerber KA, Brantberg K, Strupp M, Lee H, et al. Practice guideline: cervical and ocular vestibular evoked myogenic potential testing: report of the guideline development, dissemination, and implementation Subcommittee of the American Academy of Neurology. Neurology. 2017;89(22):2288–96.
30. Wetzel PA, Lindblad AS, Raizada H, James N, Mulatya C, Kannan MA, et al. Eye tracking results in postconcussive syndrome versus normative participants. Invest Ophthalmol Vis Sci. 2018;59(10):4011–9.
31. Gurley JM, Hujsak BD, Kelly JL. Vestibular rehabilitation following mild traumatic brain injury. NeuroRehabilitation. 2013;32(3):519–28.
32. Ouchterlony D, Masanic C, Michalak A, Topolovec-Vranic J, Rutka JA. Treating benign paroxysmal positional vertigo in the patient with traumatic brain injury: effectiveness of the canalith repositioning procedure. J Neurosci Nurs. 2016;48(2):90–9. quiz E1.

Stroke and Disequilibrium

77

Overview

The acute vestibular syndrome (AVS) is abrupt-onset disequilibrium often associated with nausea and vomiting. In the acute setting (typically an emergency room), the main differential diagnosis is between more common "peripheral" causes (usually vestibular neuritis) and less common "central" causes (usually posterior circulation *stroke*). Posterior circulation stroke is estimated to account for 2.1–3.6% of AVS cases evaluated in emergency departments in the United States. An ocular motor approach to this scenario involves "HINTS," which is an acronym for ***h***ead ***i***mpulse test, ***n***ystagmus, and ***t***est of ***s***kew deviation; in the hands of a skilled examiner, specific combinations of these findings have been shown to distinguish between vestibular neuritis and posterior circulation stroke more accurately than MRI (which is often falsely negative in the first 48 hours).

Introduction

Stroke can affect any part of the central nervous system, and thus, its presentation is quite variegated. Vascular territories usually supply neuroanatomical areas that subserve different functions, so it is common for stroke to present with multiple symptoms and physical examination findings. Thus, although some patients with stroke may complain of disequilibrium, it is usually accompanied by other symptoms (weakness, somatosensory disturbances, language deficits, visual deficits, etc.).

However, in some cases, the lesion resulting from a stroke can be sufficiently focal that it presents with predominantly or exclusively one symptom. In a minority of cases, disequilibrium can be the presenting symptom of stroke. While such presentation is not especially common, it is important to recognize since its management and prognosis are quite different than, for example, most otologic causes of disequilibrium.

Epidemiology

The phrase "*acute vestibular syndrome*" *(AVS)* refers to disequilibrium presenting apoplectically [1].

Abrupt-onset symptoms often prompt emergency department visits. Kerber and colleagues surveyed nearly 26 million emergency department visits in the United States for the 10-year period from 1995 to 2004, and of the 285,622 sampled visits, 7160 were for "dizziness" or "vertigo," which amounts to about ***2.5%*** of all emergency department encounters [2]. Other estimates identify a slightly broader range of emergency department encounters for these symptoms, from ***2.1% to 3.6%*** [3].

In the acute setting (typically an emergency department), the main clinical concern when assessing a patient with acute vestibular syndrome (AVS) is distinguishing stroke from otologic causes. Tarnutzer and colleagues "estimate that there are about 70,000 to 100,000 strokes per year in the United States with dizziness as a prominent or presenting symptom" [4]. Given that the pathway for management of stroke is very different from most otologic causes of AVS, and given that missing a diagnosis of stroke can adversely impact outcomes, emergency department physicians have a low threshold for suspecting stroke, and this concern drives the use of imaging in the United States, which accounts for a substantial proportion of the $10 billion annually spent in the acute workup of these patients [3], even though the usual imaging modality in the emergency department—CT scanning—has relatively little diagnostic value [5]. Many papers have been published regarding missed stroke diagnoses in the evaluation of AVS patients [6]. Newman-Toker and colleagues estimate that of all AVS patients evaluated in the United States each year, 15,000–165,000 may have strokes that are missed [7].

For a patient presenting with AVS, estimates of the likelihood of stroke vary by an order of magnitude. Kerber and colleagues studied 1666 patients presenting to the emergency department or admitted to the hospital with "dizziness" and concluded that 53 (***3.2%***) were ultimately diagnosed with

M. Cherchi, *Otoneurology and Vestibular Medicine*, https://doi.org/10.1007/978-3-031-94841-1_77

stroke or transient ischemic attack [8]. Newman-Toker and colleagues report that about ***4%*** of AVS patients evaluated in the emergency department are ultimately diagnosed with stroke [9]. Doijiri and colleagues evaluated 221 consecutive AVS patients and concluded that ***11%*** with "isolated vertigo or dizziness attack" were ultimately diagnosed with stroke [10]. Tarnutzer and colleagues "estimate the true proportion of acute vestibular syndrome due to stroke to be about ***25%*** ± 15%" [4].

Pathophysiological Mechanism of Disease

In the United States, about 18% of all strokes involve the posterior circulation; however, for AVS patients in whom stroke is ultimately identified as the cause, about 83% are in the posterior circulation [4], and the usual etiologies are cerebellar stroke [11–13], often in the territory of the posterior inferior cerebellar artery [10]. Relatively few cases involve labyrinthine infarction [14, 15], which can mimic labyrinthitis [16]; such cases involve the labyrinthine artery, which usually branches off the anterior inferior cerebellar artery. Occasional cases are reported in which infarction is so focal as to involve exclusively the vestibular nucleus [17, 18].

Clinical Presentation

By definition, patients with acute vestibular syndrome present with apoplectic onset of disequilibrium, often associated with nausea and vomiting.

Physical Examination: Components of HINTS

Literature is accumulating that when evaluating an AVS patient, the combination of several physical examination findings—and specifically ocular motor findings—has greater sensitivity and specificity for stroke than MRI within the first 48 h [9, 19]. The ocular motor findings in question comprise what has come to be known by the acronym "HINTS," which stands for ***h***ead ***i***mpulse test, ***n***ystagmus, and ***t***est of ***s***kew deviation. This combination of ocular motor tests will tend to reveal different findings for "peripheral vertigo" (which essentially refers to vestibular neuritis) and "central vertigo" (which essentially refers to posterior fossa stroke).

Physical Examination: Head Impulse

The horizontal head impulse test relies on integrity/failure of the horizontal rotational vestibulo-ocular reflex (VOR) during rapid head movements. The head impulse test has been shown to distinguish posterior fossa stroke from vestibular neuritis [20].

In peripheral vestibular weakness (such as vestibular neuritis), the horizontal vestibulo-ocular reflex is inadequate, so when the patient's head is moved passively and rapidly, this will "drag" the eyes off target (this is the "low gain" of the VOR), and the patient will execute a corrective saccade to keep the eyes on the intended target—this constitutes a "positive" head impulse test.

In contrast, cerebellar lesions usually do not affect the vestibulo-ocular reflex, which is why head impulse testing tends to be "negative" (normal) in cerebellar stroke. However, exceptions to this general rule have been reported [21], and it has been estimated that about 9% of patients with posterior circulation stroke have a positive (abnormal) head impulse test [20].

Physical Examination: Nystagmus

In a unilateral peripheral vestibular lesion (such as vestibular neuritis), the "nystagmus" component of the physical examination refers to a repetitive eye movement that:

- Is spontaneous.
- The rapid component is unidirectional (usually just horizontal, but there may be an ipsitorsional component).
- May increase (in amplitude and frequency) when the patient looks in the direction of the fast phase, and decrease when the patient looks in the direction of the slow phase. This pattern is said to "obey Alexander's law."

In contrast, in a posterior fossa lesion (typically a stroke in the territory of PICA), the "nystagmus" component of the physical examination refers to a repetitive eye movement that:

- Is gaze-evoked (i.e., absent on primary position of gaze, but present on eccentric gaze) [22].
- Is direction changing. Typically, the direction of the fast phase of the nystagmus will be identical to the direction of gaze.
- The rapid component is therefore multidirectional.
- These characteristics are believed to be attributable to a failure of the "gaze holding mechanism" [23] which usually helps maintain eccentric gaze.

Physical Examination: Test of Skew Deviation

In the context of a HINTS examination, "skew deviation" refers to vertical strabismus. This finding generally reflects brainstem dysfunction and can occur in posterior fossa stroke. Skew deviation rarely occurs in vestibular neuritis [24–26].

Table 77.1 Comparison of ocular motor features of the HINTS in "peripheral" versus "central" causes of the acute vestibular syndrome

	"Peripheral"—i.e., a unilateral peripheral deficit (usually vestibular neuritis)	"Central"—i.e., posterior fossa pathology (usually an infarct in the territory of the posterior inferior cerebellar artery)
Head impulse test	"Positive," i.e., a rapid horizontal head movement elicits a refixation saccade.	"Negative," i.e., the eyes stay on target after a rapid head movement.
Nystagmus	Nystagmus is spontaneous, unidirectional, and obeys Alexander's law.	Nystagmus is gaze-evoked, direction-changing and multidirectional.
Test of skew	"Negative"—vertical skew deviation is absent.	"Positive"—vertical skew deviation is present.

Summary of HINTS

The components of HINTS, and the findings for vestibular neuritis and posterior fossa stroke, respectively, are summarized in Table 77.1.

Newman-Toker and colleagues [19] report that when the three ocular motor examination findings are combined, they exhibit a 96.5% sensitivity and an 84.4% specificity for posterior fossa stroke.

Although HINTS has demonstrated success in discriminating posterior fossa stroke from peripheral vestibular disease in AVS patients, an important caveat is that this is in the hands of an expert user, or "appropriately trained clinician" [27].

HINTS: Bedside HIT Versus Video HIT

The bedside horizontal head impulse test (bHIT) requires some skill in execution and interpretation. It turns out that highly reliable interpretation of the horizontal head impulse test can be achieved through computer algorithms applied to videonystagmography—in other words, the video head impulse test (vHIT).

Despite the greater utility of vHIT in general, there is debate regarding whether bHIT and vHIT operate differently in the context of HINTS. Some investigators have observed vHIT to be superior to bHIT in this regard [28], while others conclude that bHIT and vHIT are equivalent for HINTS [29].

Testing: Vestibular

Modest literature has examined the combination of vHIT and vestibular evoked myogenic potentials for distinguishing vestibular neuritis from posterior fossa stroke [30], or those tests plus caloric testing [31]. While such studies are interesting, it is probably not realistic to expect that vestibular evoked myogenic potentials or caloric testing could be performed in the acute setting (such as an emergency department), so this is more of academic interest rather than for practical application.

Imaging

Although MRI is a sensitive imaging modality for stroke, small strokes in the posterior fossa may not be detectable by MRI in the first 48 h [13, 32–36].

Differential Diagnosis

AVS has a broad differential diagnosis, but the main concern in the acute setting (typically an emergency department) is to distinguish "peripheral" causes (usually vestibular neuritis) from "central" causes (usually posterior circulation stroke).

Treatment and Prognosis

For the differential diagnosis above, the distinction between vestibular neuritis and a posterior circulation stroke is important because of the implications for treatment and prognosis.

- Vestibular neuritis: Immediate treatment is supportive, longer-term treatment is vestibular rehabilitation, and prognosis is generally quite good, with recurrence being uncommon.
- Posterior fossa stroke: Immediate treatment may involve tPA; long-term management depends on mechanism of stroke (e.g., anticoagulation for atrial fibrillation versus treatment of vascular risk factors for atherosclerotic disease); prognosis is variable.

References

1. Hotson JR, Baloh RW. Acute vestibular syndrome. N Engl J Med. 1998;339(10):680–5.
2. Kerber KA, Meurer WJ, West BT, Fendrick AM. Dizziness presentations in U.S. emergency departments, 1995–2004. Acad Emerg Med. 2008;15(8):744–50.
3. Edlow JA, Carpenter C, Akhter M, Khoujah D, Marcolini E, Meurer WJ, et al. Guidelines for reasonable and appropriate care in the emergency department 3 (GRACE-3): acute dizziness and vertigo in the emergency department. Acad Emerg Med Off J Soc Acad Emerg Med. 2023;30(5):442–86.

4. Tarnutzer AA, Berkowitz AL, Robinson KA, Hsieh YH, Newman-Toker DE. Does my dizzy patient have a stroke? A systematic review of bedside diagnosis in acute vestibular syndrome. CMAJ. 2011;183(9):E571–92.
5. Kerber KA, Schweigler L, West BT, Fendrick AM, Morgenstern LB. Value of computed tomography scans in ED dizziness visits: analysis from a nationally representative sample. Am J Emerg Med. 2010;28(9):1030–6.
6. Arch AE, Weisman DC, Coca S, Nystrom KV, Wira CR 3rd, Schindler JL. Missed ischemic stroke diagnosis in the emergency department by emergency medicine and neurology services. Stroke. 2016;47(3):668–73.
7. Newman-Toker DE, Moy E, Valente E, Coffey R, Hines AL. Missed diagnosis of stroke in the emergency department: a cross-sectional analysis of a large population-based sample. Diagnosis (Berl). 2014;1(2):155–66.
8. Kerber KA, Brown DL, Lisabeth LD, Smith MA, Morgenstern LB. Stroke among patients with dizziness, vertigo, and imbalance in the emergency department: a population-based study. Stroke. 2006;37(10):2484–7.
9. Newman-Toker DE, Saber Tehrani AS, Mantokoudis G, Pula JH, Guede CI, Kerber KA, et al. Quantitative video-oculography to help diagnose stroke in acute vertigo and dizziness: toward an ECG for the eyes. Stroke. 2013;44(4):1158–61.
10. Doijiri R, Uno H, Miyashita K, Ihara M, Nagatsuka K. How commonly is stroke found in patients with isolated vertigo or dizziness attack? J Stroke Cerebrovasc Dis. 2016;25(10):2549–52.
11. Perloff MD, Patel NS, Kase CS, Oza AU, Voetsch B, Romero JR. Cerebellar stroke presenting with isolated dizziness: brain MRI in 136 patients. Am J Emerg Med. 2017;35:1724.
12. Armato E, Ferri E, Pinzani A, Ulmer E. Cerebellar haemorrhage mimicking acute peripheral vestibulopathy: the role of the video head impulse test in differential diagnosis. Acta Otorhinolaryngol Ital. 2014;34(4):288–91.
13. Morita S, Suzuki M, Iizuka K. False-negative diffusion-weighted MRI in acute cerebellar stroke. Auris Nasus Larynx. 2011;38(5):577–82.
14. Eliezer M, Verillaud B, Guichard JP, Kania R, Toupet M, Herman P, et al. Labyrinthine infarction caused by vertebral artery dissection: consideration based on MRI. J Neurol. 2019;266:2575.
15. Liqun Z, Park KH, Kim HJ, Lee SU, Choi JY, Kim JS. Acute unilateral audiovestibulopathy due to embolic labyrinthine infarction. Front Neurol. 2018;9:311.
16. Nam HW, Yoo D, Lee SU, Choi JY, Yu S, Kim JS. Pearls & oy-sters: labyrinthine infarction mimicking vestibular neuritis. Neurology. 2021;97(16):787–90.
17. Kim HA, Lee H. Isolated vestibular nucleus infarction mimicking acute peripheral vestibulopathy. Stroke. 2010;41(7):1558–60.
18. Kim HJ, Lee SH, Park JH, Choi JY, Kim JS. Isolated vestibular nuclear infarction: report of two cases and review of the literature. J Neurol. 2014;261(1):121–9.
19. Newman-Toker DE, Kerber KA, Hsieh YH, Pula JH, Omron R, Saber Tehrani AS, et al. HINTS outperforms ABCD2 to screen for stroke in acute continuous vertigo and dizziness. Acad Emerg Med Off J Soc Acad Emerg Med. 2013;20(10):986–96.
20. Newman-Toker DE, Kattah JC, Alvernia JE, Wang DZ. Normal head impulse test differentiates acute cerebellar strokes from vestibular neuritis. Neurology. 2008;70(24 Pt 2):2378–85.
21. Witsch J, Ferrer M, Navaratnam D. Teaching video neuroimages: vestibulo-ocular reflex defect in cerebellar stroke. Neurology. 2018;91(9):e888–e9.
22. Strupp M, Kremmyda O, Adamczyk C, Bottcher N, Muth C, Yip CW, Bremova T. Central ocular motor disorders, including gaze palsy and nystagmus. J Neurol. 2014;261(Suppl 2):S542–58.
23. Baier B, Dieterich M. Incidence and anatomy of gaze-evoked nystagmus in patients with cerebellar lesions. Neurology. 2011;76(4):361–5.
24. Park JY, Kwon JH, Kim WJ, Kim SY, Lee EM, Yang HJ, et al. Vertical diplopia as a major manifestation of acute peripheral vestibulopathy. J Clin Neurol. 2019;15(3):401–3.
25. Eggers SDZ, Kattah JC. Approaching acute vertigo with diplopia: a rare skew deviation in vestibular neuritis. Mayo Clin Proc Innov Qual Outcomes. 2020;4(2):216–22.
26. Vibert D, Hausler R, Safran AB, Koerner F. Diplopia from skew deviation in unilateral peripheral vestibular lesions. Acta Otolaryngol. 1996;116(2):170–6.
27. Tarnutzer AA, Gold D, Wang Z, Robinson KA, Kattah JC, Mantokoudis G, et al. Impact of clinician training background and stroke location on bedside diagnostic accuracy in the acute vestibular syndrome -a meta-analysis. Ann Neurol. 2023;94:295.
28. Choi JY, Kim HJ, Kim JS. Recent advances in head impulse test findings in central vestibular disorders. Neurology. 2018;90(13):602–12.
29. Guler A, Karbek Akarca F, Eraslan C, Tarhan C, Bilgen C, Kirazli T, Celebisoy N. Clinical and video head impulse test in the diagnosis of posterior circulation stroke presenting as acute vestibular syndrome in the emergency department. J Vestib Res. 2017;27(4):233–42.
30. Calic Z, Nham B, Bradshaw AP, Young AS, Bhaskar S, D'Souza M, et al. Separating posterior-circulation stroke from vestibular neuritis with quantitative vestibular testing. Clin Neurophysiol. 2020;131(8):2047–55.
31. Nham B, Wang C, Reid N, Calic Z, Kwok BYC, Black DA, et al. Modern vestibular tests can accurately separate stroke and vestibular neuritis. J Neurol. 2023;270(4):2031–41.
32. Choi JH, Oh EH, Park MG, Baik SK, Cho HJ, Choi SY, et al. Early MRI-negative posterior circulation stroke presenting as acute dizziness. J Neurol. 2018;265(12):2993–3000.
33. Choi KD, Kim JS. Vascular vertigo: updates. J Neurol. 2018;266:1835.
34. Edlow BL, Hurwitz S, Edlow JA. Diagnosis of DWI-negative acute ischemic stroke: a meta-analysis. Neurology. 2017;89(3):256–62.
35. Frey LC, Sung GY, Tanabe J. Early false-negative diffusion-weighted imaging in brainstem infarction. J Stroke Cerebrovasc Dis. 2002;11(1):51–3.
36. Saber Tehrani AS, Kattah JC, Mantokoudis G, Pula JH, Nair D, Blitz A, et al. Small strokes causing severe vertigo: frequency of false-negative MRIs and nonlacunar mechanisms. Neurology. 2014;83(2):169–73.

White Matter Disease and Disequilibrium

78

Overview

In clinical parlance, the phrase "*white matter disease*" *(WMD)* refers to the radiographic findings of nonenhancing areas of T2 hyperintensity (on MRI) or hypointensity/hyperlucency (on CT) usually in the cerebral periventricular and deep white matter. This corresponds to several different histopathological findings and may be the result of different pathobiological mechanisms. The prevalence of WMD increases with age. In some cases, WMD seems to be associated with the symptom of disequilibrium, though the mechanism by which this occurs is unclear, and not all individuals with WMD experience disequilibrium. In a patient with WMD complaining of disequilibrium, it is sensible to screen for other more common causes. WMD is associated with poorer recovery from benign paroxysmal positional vertigo and vestibular neuritis, though whether this relationship is synergistic or merely coincidental is uncertain.

Introduction

In day-to-day clinical practice, the phrase "*white matter disease*" (WMD) usually alludes to radiographic findings, typically in the periventricular and deep white matter areas of the cerebrum, of T2 hyperintensity on MRI, or hypointensity/hyperlucency on CT.

There are various metrics for white matter disease [1], such as the ARWMC (age-related white matter changes) scale [2], the Fazekas scale [3, 4], and the Scheltens scale [5]. There are also various metrics for progression of white matter disease, such as the Rotterdam scale [6] and the Schmidt scale [7]. None of these metrics has been universally adopted.

Epidemiology

The prevalence of WMD increases with age [8].

Pathophysiological Mechanism of Disease

White matter disease goes by other names, such as "leukoaraiosis," "chronic microvascular ischemic changes," or "ischemic demyelination." These phrases may carry unintended implications for different readers. For example, since the phrase "chronic microvascular ischemic changes" contains the term "ischemia," it may seem logical to view this as a pathology resulting from discrete strokes, but this does not appear to be the case [9]. The phrase "ischemic demyelination" contains the term "demyelination," which may imply an autoimmune process, yet this does not appear to be the case either [10].

Even when WMD is radiographically well delimited, it is unclear why this would impair equilibrium. Theoretical explanations of this usually cite disruption of multiple networks as the most plausible mechanism [11, 12].

Clinical Presentation

If other causes of disequilibrium have been excluded, and WMD appears (by exclusion) to be the only pathology implicated, then the symptom of disequilibrium generally is insidious in onset and gradually progressive.

Physical Examination

There are no physical examination findings that are specific or sensitive for disequilibrium from WMD.

Ocular Motor Examination

There are no ocular motor findings that are specific or sensitive for disequilibrium from WMD.

M. Cherchi, *Otoneurology and Vestibular Medicine*, https://doi.org/10.1007/978-3-031-94841-1_78

Testing: Vestibular

There are no otovestibular test results that are specific or sensitive for disequilibrium from WMD.

Imaging

WMD typically involves the cerebral periventricular and deep white matter and usually exhibits no contrast enhancement.

Figure 78.1, from Yu and colleagues [13], shows different degrees of WMD as rated on the Fazekas scale [3, 4].

Histopathology

Autopsy studies report a variety of pathologies corresponding to the areas of radiographic abnormality, including microcystic infarcts, subependymal gliosis, and others [4], suggesting that WMD is not a monolithic process.

Differential Diagnosis: Clinical

WMD may be the "final common pathway" of several age-associated microvascular pathologies, such as hypertension and dyslipidemia. But since the prevalence of WMD

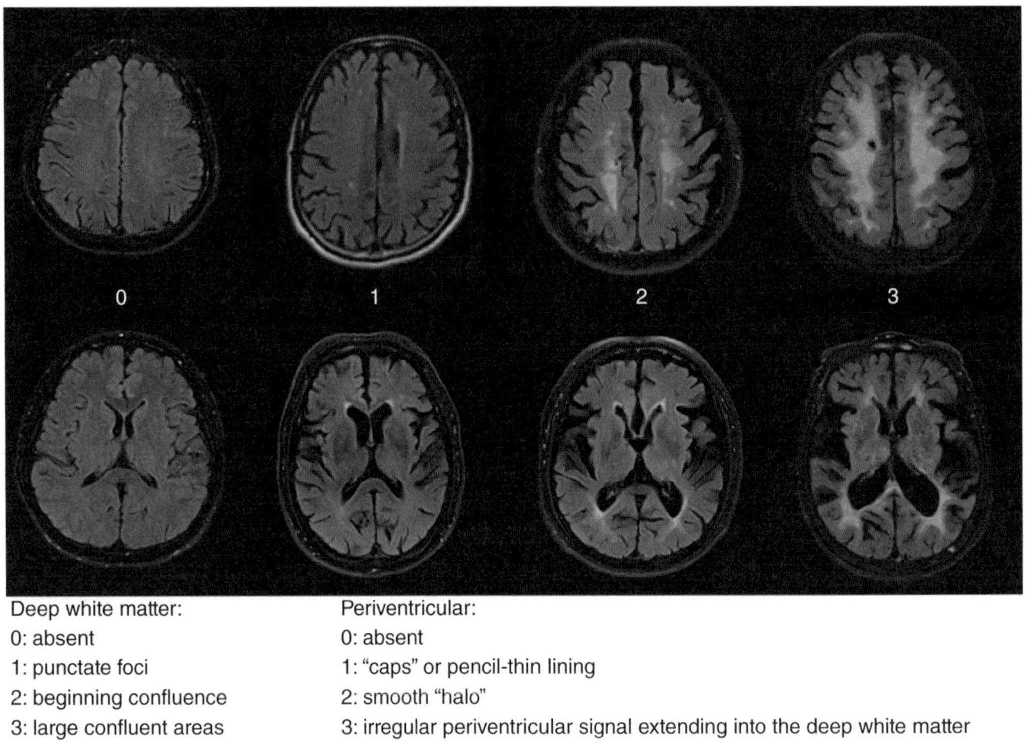

Fig. 78.1 Different degrees of white matter disease on brain MRI, graded on the Fazekas scale. (From Yu et al. [13]. Reproduced from Elsevier with permission under STM Signatory agreement)

increases with age [8], the clinician must keep in mind that there are other causes of disequilibrium whose incidence also increases with advancing age.

It is well established that the incidence of benign paroxysmal positional vertigo (BPPV) increases with age [14]. There is modest evidence that patients with WMD are at a higher risk for developing BPPV [13]. Interestingly, in patients with BPPV who also happen to have WMD, the symptom of disequilibrium tends to persist for longer than in patients without WMD [15]. The nature of this relationship is unclear; for example, it is unknown whether the factors predisposing to WMD also predispose to BPPV, or whether the greater symptomatology is simply the synergistic effect of two diseases that happen to cause the symptom of disequilibrium through independent mechanisms.

Patients with WMD who happen to suffer a bout of vestibular neuritis (VN) may have greater difficulty recovering from VN. Again, the nature of this relationship is unclear; for example, it is unknown whether the factors underlying WMD impair the compensatory mechanisms that would otherwise promote a patient's recovery, or whether mere co-occurrence of independent diseases has a synergistic effect.

It is well known that patients with a history of migraine accumulate white matter hyperintensities at a greater rate than matched controls without migraine [16–18]. Some literatures suggest that a clinical diagnosis of vestibular migraine increases the risk of WMD [19]. However, this literature only documents associations; it does not establish causality.

Since WMD tends to occur in older individuals, in some instances, it may be simply one contributor to the multifactorial disequilibrium resulting from age-related attrition of afferent and efferent functions required for properly functioning equilibrium.

Practically, if a patient with WMD is referred for a complaint of disequilibrium, it is sensible to screen for other more common causes of that symptom before concluding that WMD is the sole mechanism. A reasonable screening otovestibular workup may include cervical vestibular evoked myogenic potentials, ocular vestibular evoked myogenic potentials, video head impulse testing, videonystagmography, and rotatory chair testing.

Differential Diagnosis: Radiographic

If white matter disease exhibits enhancement with radiographic contrast dye, then diseases such as autoimmune demyelinating processes and vasculitides should be considered.

If white matter disease also exhibits areas of hemorrhage of various ages, then cerebral amyloid angiopathy should be considered.

White matter disease occurring in specific distributions may suggest particular pathologies, such as CADASIL, CARASIL, or Susac syndrome.

Treatment and Prognosis

In a patient with WMD, management of vascular risk factors should be optimized. While this will not reverse the damage that is already present, it may retard the progression.

For the symptom of disequilibrium itself, vestibular rehabilitation therapy is always reasonable to consider, though controlled studies of this intervention are lacking.

References

1. Gouw AA, van der Flier WM, van Straaten EC, Pantoni L, Bastos-Leite AJ, Inzitari D, et al. Reliability and sensitivity of visual scales versus volumetry for evaluating white matter hyperintensity progression. Cerebrovasc Dis. 2008;25(3):247–53.
2. Wahlund LO, Barkhof F, Fazekas F, Bronge L, Augustin M, Sjogren M, et al. A new rating scale for age-related white matter changes applicable to MRI and CT. Stroke. 2001;32(6):1318–22.
3. Fazekas F, Chawluk JB, Alavi A, Hurtig HI, Zimmerman RA. MR signal abnormalities at 1.5 T in Alzheimer's dementia and normal aging. AJR Am J Roentgenol. 1987;149(2):351–6.
4. Fazekas F, Kleinert R, Offenbacher H, Schmidt R, Kleinert G, Payer F, et al. Pathologic correlates of incidental MRI white matter signal hyperintensities. Neurology. 1993;43(9):1683–9.
5. Scheltens P, Barkhof F, Leys D, Pruvo JP, Nauta JJ, Vermersch P, et al. A semiquantative rating scale for the assessment of signal hyperintensities on magnetic resonance imaging. J Neurol Sci. 1993;114(1):7–12.
6. Prins ND, van Straaten EC, van Dijk EJ, Simoni M, van Schijndel RA, Vrooman HA, et al. Measuring progression of cerebral white matter lesions on MRI: visual rating and volumetrics. Neurology. 2004;62(9):1533–9.
7. Schmidt R, Fazekas F, Kapeller P, Schmidt H, Hartung HP. MRI white matter hyperintensities: three-year follow-up of the Austrian Stroke Prevention Study. Neurology. 1999;53(1):132–9.
8. Zhuang FJ, Chen Y, He WB, Cai ZY. Prevalence of white matter hyperintensities increases with age. Neural Regen Res. 2018;13(12):2141–6.
9. Fujishima M, Yao H, Terashi A, Tagawa K, Matsumoto M, Hara H, et al. Deep white matter lesions on MRI, and not silent brain infarcts are related to headache and dizziness of non-specific cause in non-stroke Japanese subjects. Intern Med. 2000;39(9):727–31.
10. Li J, Zhao YM. Clinical manifestations and imaging features of white matter demyelination in older patients. J Int Med Res. 2020;48(11):300060520966806.
11. Ibitoye RT, Castro P, Cooke J, Allum J, Arshad Q, Murdin L, et al. A link between frontal white matter integrity and dizziness in cerebral small vessel disease. Neuroimage Clin. 2022;35:103098.
12. Kaski D, Rust HM, Ibitoye R, Arshad Q, Allum JHJ, Bronstein AM. Theoretical framework for "unexplained" dizziness in the elderly: the role of small vessel disease. Prog Brain Res. 2019;248:225–40.
13. Yu T, Zhang H, Yan YM, Liu YN, Huang XF, Qiao S, et al. Correlation of idiopathic benign paroxysmal positional vertigo with cerebral small vessel disease. Am J Emerg Med. 2023;74:140–5.

14. von Brevern M, Radtke A, Lezius F, Feldmann M, Ziese T, Lempert T, Neuhauser H. Epidemiology of benign paroxysmal positional vertigo: a population based study. J Neurol Neurosurg Psychiatry. 2007;78(7):710–5.
15. Zang J, Jiang X, Feng S, Zhang H. The influence of cerebral small vessel diseases on the efficacy of repositioning therapy and prognosis of benign paroxysmal positional vertigo. Int J Med Sci. 2022;19(8):1227–34.
16. Hamedani AG, Rose KM, Peterlin BL, Mosley TH, Coker LH, Jack CR, et al. Migraine and white matter hyperintensities: the ARIC MRI study. Neurology. 2013;81(15):1308–13.
17. Swartz RH, Kern RZ. Migraine is associated with magnetic resonance imaging white matter abnormalities: a meta-analysis. Arch Neurol. 2004;61(9):1366–8.
18. Zhang W, Cheng Z, Fu F, Zhan Z. Prevalence and clinical characteristics of white matter hyperintensities in migraine: a meta-analysis. Neuroimage Clin. 2023;37:103312.
19. Shen Y, Li D, Cao Q, Hu M, Hou Z, Xu L, et al. Risk factors of vestibular migraine-related brain white matter lesions. Acta Neurol Belg. 2023;123(5):1833–9.

Encephalitis

79

Overview

Encephalitis can be caused by numerous infectious agents and can be associated with a broad variety of antibodies. Since encephalitis can affect any part of the brain, the symptoms can be variable. Ocular motor abnormalities occur in encephalitis but are not sensitive or specific for a particular infectious agent or antibody.

While one might expect there to be a one-to-one correlation between a particular infectious agent and a specific ocular motor abnormality, or between a particular antibody and a specific ocular motor abnormality, unfortunately no firm correlations have been identified. Consequently, there is no ocular motor finding that can diagnose a specific infectious agent, or particular antibody, in encephalitis.

Introduction

Encephalitis is an immense topic. Here, we will not provide a comprehensive review of its diagnosis and management. Rather, our goal is to give a sense of the breadth of the pathogens, antibodies, and ocular motor findings in encephalitis.

Pathophysiological Mechanism of Disease

In principle, any area of the central nervous system can be affected by inflammation; involvement of the brain is termed encephalitis, and involvement of the spinal cord is termed myelitis. Encephalitis can occur separately from meningitis, or concomitantly (in which case it is termed meningoencephalitis). Inflammation itself can result from a variety of etiologies, such as infectious and autoimmune; in many cases, the etiology of encephalitis remains undetermined. Neurons, glial cells, and myelin can all be damaged by inflammation.

Some nervous system tissues exhibit selective vulnerability to certain inflammatory diseases. In other cases, the source of inflammation can damage neural tissue indiscriminately.

Etiologies: Infectious

A large number of infectious agents have been detected or suspected as causes of encephalitis. Many infectious encephalitides have been reported in association with ocular motor abnormalities, including:

- Arbovirus [1]
- Coronavirus [2]
- Epstein–Barr virus [3–5]
- Flavivirus [6, 7]
- Herpes simplex virus 1 [8, 9]
- Herpes simplex virus 6 [10], *Orientia tsutsugamushi* [11]
- West Nile virus [12].

Etiologies: Immune-Mediated

In cases of encephalitis associated with ocular motor abnormalities (to be discussed below), many different autoantibodies have been detected, including:

- AMPA [13]
- CASPR2 [14]
- DPPX [15, 16]
- GAD65 [17–22]
- GD1b [23]
- GM1 [23]
- GQ1b [23]
- Hu [24]
- KLHL11 [25]
- Ma2 [26]
- mGluR65 [27]
- MOG [28]
- NMDA [29]
- Ri [30]

M. Cherchi, *Otoneurology and Vestibular Medicine*, https://doi.org/10.1007/978-3-031-94841-1_79

Clinical Presentation

The clinical presentation of encephalitis depends on which areas of the brain are affected. Common presentations include cognitive changes, seizures, and coma. Many descriptions of encephalitis do not mention whether any eye movement abnormalities are present, but it is likely that such abnormalities are underreported because the clinical picture is dominated by more dramatic features.

Ocular Motor Examination

Many reports of encephalitis comment on the presence of ocular motor abnormalities, but often they are not described [1, 4, 10, 14, 18].

Fortunately, many examiners have described in some detail the ocular motor abnormalities that can occur in association with encephalitides, such as:

- Central positional nystagmus [23]
- Spontaneous down beat nystagmus [5, 8, 9, 12, 17, 20, 21, 24, 29, 31, 32]
- Gaze evoked nystagmus [11, 19, 23, 28, 33]
- Horizontal nystagmus (not further specified) [2, 7, 22, 27]
- Ocular bobbing [34]
- Opsoclonus [6, 15]
- Optokinetic abnormalities [35]
- Periodic alternating nystagmus [23, 25, 36]
- Rebound nystagmus [37]
- Reduced caloric responses (vestibular weakness) [38]
- Saccadic hypometria [15]
- Saccadic intrusions [17]
- Skew deviation [30]
- Smooth pursuit abnormalities [35]
- Torsional nystagmus [15]
- Spontaneous up beat nystagmus [15]
- Vertical nystagmus (not further characterized) [3, 13, 23, 24]

This range of ocular motor abnormalities is not only remarkable but also unsurprising given that encephalitis can affect nearly any part of the brain.

Down beat nystagmus is probably the most commonly reported ocular motor abnormality in encephalitis, and this finding localizes to the cerebellum. The cerebellum exhibits selective vulnerability to some disease processes, such as paraneoplastic syndromes.

References

1. Suardi LR, Di Lauria N, Pozzi M, Rogasi PG, Barilaro A, Azzolini F, et al. Acute cerebellar ataxia: a rare Toscana Virus (TOSV) meningoencephalitis complication. Int J Neurosci. 2020;130(3):276–8.
2. Luis ME, Hipolito-Fernandes D, Mota C, Maleita D, Xavier C, Maio T, et al. A review of neuro-ophthalmological manifestations of human coronavirus infection. Eye Brain. 2020;12:129–37.
3. Demey HE, Martin JJ, Leus RM, Moeremans CJ, Bossaert LL. Coma as a presenting sign of Epstein-Barr encephalitis. Arch Intern Med. 1988;148(6):1459–61.
4. Hussain RS, Hussain NA. Ataxia and encephalitis in a young adult with EBV mononucleosis: a case report. Case Rep Neurol Med. 2013;2013:516325.
5. Wade CA, Toupin DN, Darpel K, Jones K, Lightner D. Downbeat nystagmus in a 7-year-old girl with Epstein-Barr virus-associated meningitis and cerebellitis. Child Neurol Open. 2021;8:2329048X211000463.
6. Pradhan S, Pandey N, Shashank S, Gupta RK, Mathur A. Parkinsonism due to predominant involvement of substantia nigra in Japanese encephalitis. Neurology. 1999;53(8):1781–6.
7. Withana M, Rodrigo C, Chang T, Karunanayake P, Rajapakse S. Dengue fever presenting with acute cerebellitis: a case report. BMC Res Notes. 2014;7:125.
8. Hirst LW, Clark AW, Wolinsky JS, Zee DS, Kaizer H, Miller NR, et al. Downbeat nystagmus. A case report of herpetic brain stem encephalitis. J Clin Neuroophthalmol. 1983;3(4):245–9.
9. Shaikh AG, Termsarasab P, Riley DE, Katirji B. The floccular syndrome in herpes simplex type 1 encephalitis. J Neurol Sci. 2013;325(1–2):154–5.
10. Carvajal E, Verdeguer A, Fernandez JM, Canete A, Castel V. Herpesvirus-6 encephalitis complicated by Wernicke-Korsakoff syndrome in a pediatric recipient of unrelated cord blood transplantation. J Pediatr Hematol Oncol. 2001;23(9):626–8.
11. Karanth SS, Gupta A, Prabhu M. Pure cerebellitis due to scrub typhus: a unique case report. Trop Doct. 2013;43(1):41–2.
12. Malhotra K, Ramanathan RS, Synowiec A, Rana S. Rare ocular manifestation in a case of West Nile virus meningoencephalitis. Ann Indian Acad Neurol. 2014;17(1):95–6.
13. Matthews E, Schmitt B, Passeri M, Mizenko C, Orjuela K, Piquet A. AMPA receptor encephalitis in a patient with metastatic breast cancer receiving palbociclib: a case report. Neurol Neuroimmunol Neuroinflamm. 2022;9(5):e200012.
14. Joubert B, Gobert F, Thomas L, Saint-Martin M, Desestret V, Convers P, et al. Autoimmune episodic ataxia in patients with anti-CASPR2 antibody-associated encephalitis. Neurol Neuroimmunol Neuroinflamm. 2017;4(4):e371.
15. Blackburn K, Vernino S, Sguigna P. Utilization of video oculography to quantify oculomotor dysfunction in anti-DPPX encephalitis: a case study. Mult Scler Relat Disord. 2020;46:102460.
16. Doherty L, Gold D, Solnes L, Probasco J, Venkatesan A. Anti-DPPX encephalitis: prominent nystagmus reflected by extraocular muscle FDG-PET avidity. Neurol Neuroimmunol Neuroinflamm. 2017;4(4):e361.
17. Baizabal-Carvallo JF. The neurological syndromes associated with glutamic acid decarboxylase antibodies. J Autoimmun. 2019;101:35–47.
18. Mirabelli-Badenier M, Morana G, Pinto F, Uccelli A, Veneselli E, Battaglia FM, et al. Anti-glutamic acid decarboxylase limbic encephalitis without epilepsy evolving into dementia with cerebellar ataxia. Arch Neurol. 2012;69(8):1064–6.
19. Muniz-Castrillo S, Vogrig A, Joubert B, Pinto AL, Goncalves D, Chaumont H, et al. Transient neurological symptoms preceding cerebellar ataxia with glutamic acid decarboxylase antibodies. Cerebellum. 2020;19(5):715–21.
20. Tsiortou P, Alexopoulos H, Dalakas MC. GAD antibody-spectrum disorders: progress in clinical phenotypes, immunopathogenesis and therapeutic interventions. Ther Adv Neurol Disord. 2021;14:17562864211003486.
21. Vale TC, Pedroso JL, Alqueres RA, Dutra LA, Barsottini OG. Spontaneous downbeat nystagmus as a clue for the diagno-

sis of ataxia associated with anti-GAD antibodies. J Neurol Sci. 2015;359(1–2):21–3.
22. Zhou H, Xie X, Zhang T, Yang M, Zhou D, Yang T. Anti-GAD65 antibody-associated autoimmune encephalitis with predominant cerebellar involvement following toripalimab treatment: a case report of a novel irAE of toripalimab. Front Immunol. 2022;13:850540.
23. Jeong SH, Nam J, Kwon MJ, Kim JK, Kim JS. Nystagmus and ataxia associated with antiganglioside antibodies. J Neuro-ophthalmol. 2011;31(4):326–30.
24. Saiz A, Bruna J, Stourac P, Vigliani MC, Giometto B, Grisold W, et al. Anti-Hu-associated brainstem encephalitis. J Neurol Neurosurg Psychiatry. 2009;80(4):404–7.
25. Bohm P, Eggenberger ER, Dubey D, Kim HW, Lopez Chiriboga AS. Teaching video neuroimages: periodic alternating nystagmus in paraneoplastic KLHL11 rhomboencephalitis. Neurology. 2021;96(21):e2668–e9.
26. Garcia-Reitboeck P, Thompson G, Johns P, Al Wahab Y, Omer S, Griffin C. Upbeat nystagmus in anti-Ma2 encephalitis. Pract Neurol. 2014;14(1):36–8.
27. Zhang Y, Lian B, Yang S, Huang X, Zhou Y, Cao L. Metabotropic glutamate receptor 5-related autoimmune encephalitis with reversible splenial lesion syndrome following SARS-CoV-2 vaccination. Medicine (Baltimore). 2023;102(7):e32971.
28. Olbert E, Brunner C, Alhani N, Nasel C, Struhal W. MOG antibody associated disease (MOGAD) presenting with extensive brain stem encephalitis: a case report. eNeurologicalSci. 2022;29:100432.
29. Tsuyusaki Y, Sakakibara R, Kishi M, Tateno F, Yoshida T. Downbeat nystagmus as the initial manifestation of anti-NMDAR encephalitis. Neurol Sci. 2014;35(1):125–6.
30. Mitchell AN, Bakhos CT, Zimmerman EA. Anti-Ri-associated paraneoplastic brainstem cerebellar syndrome with coexisting limbic encephalitis in a patient with mixed large cell neuroendocrine lung carcinoma. J Clin Neurosci. 2015;22(2):421–3.
31. Prasad S, Brown MJ, Galetta SL. Transient downbeat nystagmus from West Nile virus encephalomyelitis. Neurology. 2006;66(10):1599–600.
32. Shimizu N, Weinberger J, Yahr MD. Downbeat nystagmus as a sign of brainstem involvement in acute meningoencephalitis. Neurology. 1975;25(3):267–70.
33. Alroughani R, Thussu A, Guindi RT. Yet another atypical presentation of anti-GQ1b antibody syndrome. Neurol Int. 2015;7(2):5770.
34. Rudick R, Satran R, Eskin TA. Ocular bobbing in encephalitis. J Neurol Neurosurg Psychiatry. 1981;44(5):441–3.
35. Furman JM, Brownstone PK, Baloh RW. Atypical brainstem encephalitis: magnetic resonance imaging and oculographic features. Neurology. 1985;35(3):438–40.
36. Davis DG, Smith JL. Periodic alternating nystagmus. A report of eight cases. Am J Ophthalmol. 1971;72(4):757–62.
37. Yoshimoto Y, Koyama S. A case of acquired nystagmus alternans associated with acute cerebellitis. Acta Otolaryngol Suppl. 1991;481:371–3.
38. Ito T, Sakakibara R, Komatsuzaki A, Nakata M, Uchiyama T, Hiruma K, Hattori T. Aseptic meningoencephalitis presenting with bilateral vestibular ataxia: a case report. Intern Med. 2006;45(8):551–2.

80 Epileptic Vertigo

Overview

Epileptic vertigo refers to symptoms of disequilibrium attributable exclusively to seizure activity in vestibular cortical areas that roughly are mostly situated in the temporal and parietal areas. Some patients with seizures may have disequilibrium as an aura, but this is quickly overshadowed by more dramatic symptoms. It is unusual for seizures to remain restricted to vestibular cortex (and thus only cause disequilibrium), though this does appear to be a discrete clinical entity. For patients whose seizures truly only manifest with disequilibrium, the episodes usually start abruptly, last seconds, and end abruptly. The disequilibrium itself can have a variety of qualities, including rotational movement, translational movement, rocking, floating, and unsteadiness. Otovestibular studies such as video head impulse testing and cervical vestibular evoked myogenic potentials are sometimes abnormal in patients with epileptic vertigo, but not in any patterns that are sensitive or specific for this diagnosis. Electroencephalography often shows abnormal activity in the temporoparietal area. Brain MRI should be checked to exclude structural etiologies of seizures. The differential diagnosis includes vestibular paroxysmia. Treatment is with antiseizure medications; those reported include carbamazepine and phenytoin. If a patient's symptoms evolve to include symptoms beyond disequilibrium, then they should be under the care of a general neurologist or epileptologist.

Introduction

The phrase "*epileptic vertigo*" refers to symptoms of disequilibrium attributable exclusively to seizure activity in vestibular cortical areas. Synonymous designations in the literature include "vestibular epilepsy," "vestibular seizures," etc.

A modest proportion of seizure patients report disequilibrium as part of their seizure prodrome, but their episodes usually evolve to include much more dramatic symptoms (such as generalized tonic–clonic seizures), so the symptom of disequilibrium is likely underreported. This sequence (prodrome including disequilibrium, followed by more dramatic seizure symptoms) is likely due to the fact that it is unusual for seizure activity to remain restricted to such a small focus (in this case, just so-called vestibular cortex); it is more common for it to spread and cause other sensory, motor, and cognitive symptoms that overshadow the symptom of disequilibrium.

For example, Kim and colleagues [1] retrospectively reviewed 831 epilepsy cases and found that 40 patients (4.8%) with a mean age of 32.8 ± 11.8 years "experienced vertigo or dizziness as aura, all of whom had partial seizures."

Nevertheless, there is some evidence that epileptic vertigo (in which the disequilibrium is an isolated symptom) is a discrete clinical entity, albeit rare.

Anatomy

The areas of the cerebral cortex that mediate vestibular sensations have been imprecisely delineated. Reasonable data on this subject come from studies of intraoperative cortical stimulation, though such studies are limited by the fact that the study subjects were patients who had illnesses that warranted craniotomies, rather than healthy individuals.

Such a study by Kahane and colleagues [2] studied epilepsy patients (prior to surgical treatment of the epilepsy) by implanting surface electrodes and stimulating various areas of the cortex. They stimulated 44 sites in 28 patients. Of the 44 stimulated sites, 19 (43%) were in the temporal lobe, 14 (32%) were in the parietal lobe, 5 (11%) were in the frontal lobe, 2 (5%) were in the occipital lobe, and 1 (2%) was in the insular lobe. Some patients had multiple susceptible areas. They designated the area with the highest density of susceptible sites as the "temporo-peri-Sylvian vestibular cortex (TPSVC)."

Figure 80.1 shows the distribution of vestibularly sensitive sites in the study by Kahane and colleagues [2].

M. Cherchi, *Otoneurology and Vestibular Medicine*, https://doi.org/10.1007/978-3-031-94841-1_80

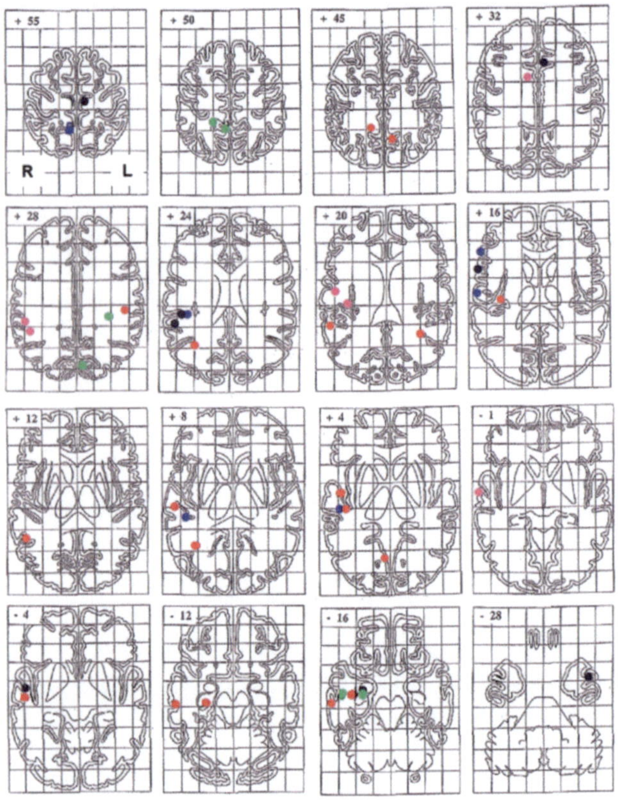

Fig. 80.1 Anatomical location of the vestibular electrical stimulation sites (44 sites in 28 patients). The vestibular sites are represented on the horizontal brain cross-sections of Talairach and Tournoux's atlas that most closely corresponds to their z millimetric coordinate. Abbreviations: R right, L left. Locations that elicit illusions of motion are indicated by red dots (yaw plan), pink dots (pitch plane), blue dots (roll plane), green dots (translations), and black dots (indefinable sensations of body motion). (From Kahane et al. [2]. Reproduced with written permission from John Wiley and Sons)

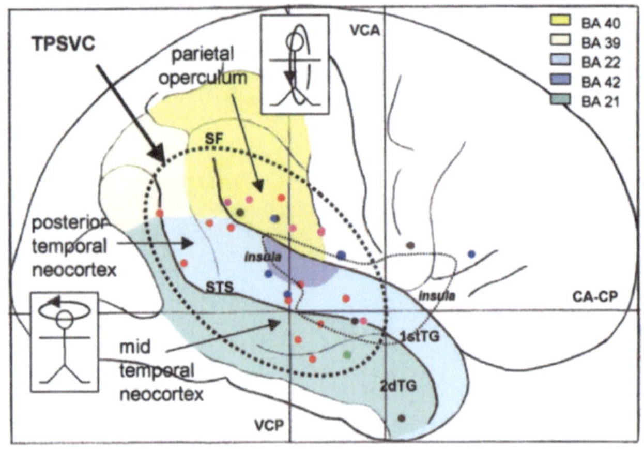

Fig. 80.2 Schematic representation of the temporo-peri-Sylvian vestibular cortex (TPSVC) . The vestibular sites located at the lateral aspect of the right or left hemispheres are projected on a lateral view of the right hemisphere normalized in the proportional stereotactic grid system of Talairach and Tournoux. Abbreviations: BA Brodmann area, CA-CP anterior commissure–posterior commissure plane, VCA vertical plane through CA, VCP vertical plane through CP, SF Sylvian fissure, STS superior temporal sulcus, 1stTG first (superior) temporal gyrus, 2dTG second (mid) temporal gyrus. Locations that elicit illusions of motion are indicated by red dots (yaw plan), pink dots (pitch plane), blue dots (roll plane), green dots (translations), and black dots (indefinable sensations of body motion). (From Kahane et al. [2]. Reproduced with written permission from John Wiley and Sons)

Figure 80.2 shows a single schematic with a composite of the distribution of vestibularly sensitive sites in the study by Kahane and colleagues [2].

Hewitt and colleagues [3] review multiple localization studies and point out that, "Partial seizures could also elicit vestibular symptoms by interfering with areas associated with representation of the body in space."

Pathophysiological Mechanism of Disease

Vestibular cortex mediates conscious perception of equilibrium. Seizure in this area should provoke symptoms of disequilibrium.

Clinical Presentation

Most case series on epileptic vertigo describe the most common *chronology* of the symptom of disequilibrium to be one of abrupt onset, brief duration (seconds), and abrupt offset. For example:

- Tarnutzer and colleagues [4] conducted a systematic literature review of 84 studies (including 11,354 patients) of what they term "epileptic vertigo and dizziness" (EVD). They drew a distinction between patients with "isolated EVD" (meaning disequilibrium and no other symptoms) and "nonisolated EVD" (meaning additional symptoms were present). They reported that the duration of episodic vestibular symptoms was "*very brief (<30 seconds)* in 69.6% of isolated EVD and 6.9% of nonisolated EVD."

- Similarly, Hewett and colleagues [3] report that, "The duration of [vestibular] seizures can be variable but is *usually brief lasting a few seconds*, though there are some patients reporting seizures lasting many minutes. The symptoms are typically paroxysmal *beginning suddenly* and unless the seizure secondarily generalizes they *usually discontinue abruptly*."
- Kogeorgos and colleagues [5] studied 30 consecutive patients age 15–65 years (mean 35 years), of whom 18 (60%) were women, and reported that, "In all 30 patients dizziness occurred in brief episodes, each *lasting no more than a few seconds*… The symptom *usually began abruptly* as dizziness or as a feeling of instability that evolved to dizziness or rotational vertigo during several seconds. In the recovery phase nausea was often experienced."

In the systematic literature review by Tarnutzer and colleagues [4], they collated data from 171 "epileptic vertigo and dizziness (EVD)" patients, of whom 166 had "nonisolated EVD" and 5 had "isolated EVD," though since "isolated EVD" was supposed to refer to patients whose seizure symptoms were limited to disequilibrium, it is unclear why those 5 patients would have been included in an analysis of symptoms other than disequilibrium. In any case, of these 171 patients, Tarnutzer and colleagues review 411 reported seizure symptoms and summarized them in Table 80.1.

The reported *quality* of the disequilibrium is variable. Figure 80.3, which Hewett and colleagues [3] adapted from Smith [6], illustrates the relative frequency of different sensations of disequilibrium described by 120 patients with epileptic vertigo.

Table 80.1 Breakdown of 411 symptoms (other than disequilibrium) in 171 patients with vestibular epilepsy studied by Tarnutzer et al. [4]

Symptoms	Number of patients (out of 411)
Sensory phenomena	133
Cephalic sensations (pressure, heat, pain)	61
Visual hallucinations	32
Auditory hallucinations	23
Simple (tinnitus, sounds, dysacusis)	19
Complex (music, voices)	4
Somatosensory, excluding cephalic sensations	16
Olfactory hallucinations	1
Motor symptoms	120
Focal motor seizures	28
Generalized tonic–clonic seizures	25
Arrest (speech, activity)	22
Falls	14
Automatisms	12
Head deviation	8
Nystagmus	8
Verbal (screaming, vocalization)	3
Disturbances of consciousness	104
Transient loss of consciousness	45
Emotional (fear, anxiety, agitation)	16
Emergence of an idea	15
Loss of contact	11
Dysmnestic (e.g., déjà vu)	6
Conscious confusion	5
Aphasia	4
Disturbance of body image	2
Autonomic/visceral phenomena	57
Vomiting	18
Nausea	12
Malaise/weakness	8
Epigastric (rising sensation)	6
Pallor	5
Thermoregulatory	5
Cardiovascular	2
Pulmonary	1

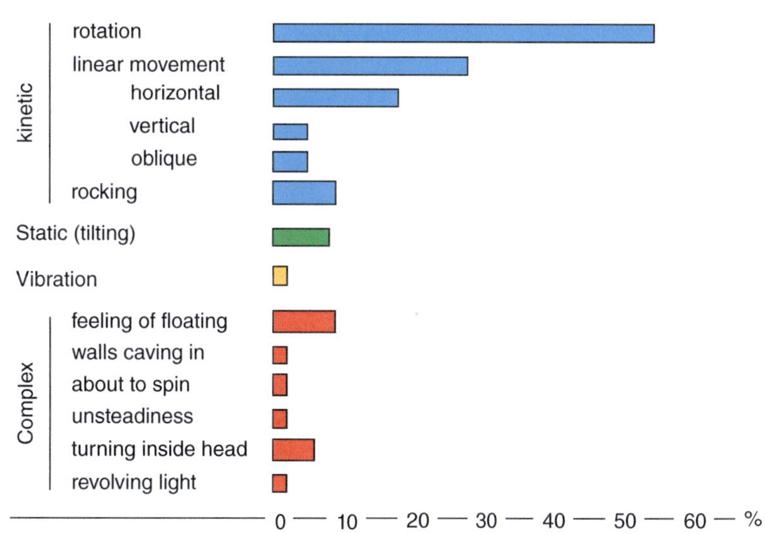

Fig. 80.3 Relative frequency of sensations of disequilibrium in 120 patients with epileptic vertigo. Heweitt et al. [3] base this figure on data originally reported in Smith [6]. (Reproduced from Frontiers Media S.A. with permission under CC-BY 4.0)

Ocular Motor Examination

Seizures often cause eye movement abnormalities, such as the classic deviation of the eyes contralateral to a unilateral frontal lobe seizure (driven by aberrant activity in the frontal eye field).

It is less common for seizures to cause nystagmus, but this can occur.

Some cortical stimulation studies have found that it is possible to induce actual nystagmus (not simply tonic deviation of the eyes). For example, Blanke and colleagues [7] demonstrated that stimulation of an area in the right middle temporal gyrus (MTG) elicited left beat nystagmus, and stimulation of an area in the superior temporal gyrus (STG) elicited rightward, downward, and "clockwise" nystagmus. The investigators concluded that, "the finding of a direction-specific nystagmus during electrical stimulation of the STG supports the view that the STG takes part in the processing of vestibular and ocular motor information."

Testing: Vestibular

Very few studies report results of any otovestibular tests in patients diagnosed with epileptic vertigo.

Surmeli and colleagues [8] studied 9 epileptic vertigo patients with electroencephalography (EEG), video head impulse testing (vHIT), and cervical vestibular evoked myogenic potentials (cVEMP). They reported that, "On vHIT examination, abnormal responses were recorded in five (55.5%), bilateral in three cases and unilateral in two. Similarly, cVEMP revealed no response in five (55.5%), bilateral in two and unilateral in three. In three patients, the side of no response to cVEMP corresponded to the side of epileptiform pathology on EEG. Two patients with bilateral abnormalities on EEG, located in temporal and temporoparietal regions, showed bilateral abnormalities on either cVEMP, or vHIT, or both."

While these observations are interesting, their value in confirming or refuting a diagnosis of epileptic vertigo is unclear.

Testing: Electroencephalography (EEG)

Tarnutzer and colleagues [4] conducted a systematic literature review of 84 studies (including 11,354 patients) of what they term "epileptic vertigo and dizziness" (EVD). They drew a distinction between patients with "isolated EVD" (meaning disequilibrium and no other symptoms) and "non-isolated EVD" (meaning additional symptoms were present). They reported that localized electroencephalography (EEG) abnormalities were mostly temporal (79.8%) rather than parietal (11.8%).

Imaging

Brain MRI without and with contrast should be undertaken to exclude structural etiologies of seizures.

Differential Diagnosis

Most episodes of epileptic vertigo start abruptly, last a few seconds, and end abruptly. That clinical history is difficult to differentiate from vestibular paroxysmia. There are two additional factors that make this distinction difficult:

- First, some patients with epileptic vertigo are reported to have otovestibular test result abnormalities, such as on video head impulse testing and cervical vestibular evoked myogenic potentials [8]. Some (though not all) patients with vestibular paroxysmia may have similar findings.
- Second, the treatments (various anti-seizure medications) of these two diseases are similar.

In other words, the history, the results of otovestibular testing, and the response to treatments do not help distinguish epileptic vertigo from vestibular paroxysmia.

At least in theory, electroencephalography (see above) should help distinguish between these two diseases. While the results of EEG may secure the diagnosis, since those results will not change management (i.e., the treatment will be the same), a case can be made for forgoing that study.

Treatment

Treatment with antiseizure medication is appropriate.

Prognosis

Kogeorgos and colleagues [5] studied 30 patients with epileptic vertigo and reported that through treatment with phenytoin or carbamazepine:

- In 23/30 (77%) patients without generalized seizures:
  - 10 achieved complete remission (mean follow-up 18 months)
  - 13 achieved "considerable reduction in the frequency and severity" (mean follow-up 2 years).
- In 7/30 (33%) of patients whose seizures secondarily generalized:
  - 6 "were completely free of both dizziness and other manifestations of epilepsy during one and half to five years of follow-up."
  - 1 patient "with persisting seizures and dizzy episodes did not comply with treatment."

References

1. Kim DW, Sunwoo JS, Lee SK. Incidence and localizing value of vertigo and dizziness in patients with epilepsy: video-EEG monitoring study. Epilepsy Res. 2016;126:102–5.
2. Kahane P, Hoffmann D, Minotti L, Berthoz A. Reappraisal of the human vestibular cortex by cortical electrical stimulation study. Ann Neurol. 2003;54(5):615–24.
3. Hewett R, Bartolomei F. Epilepsy and the cortical vestibular system: tales of dizziness and recent concepts. Front Integr Neurosci. 2013;7:73.
4. Tarnutzer AA, Lee SH, Robinson KA, Kaplan PW, Newman-Toker DE. Clinical and electrographic findings in epileptic vertigo and dizziness: a systematic review. Neurology. 2015;84(15):1595–604.
5. Kogeorgos J, Scott DF, Swash M. Epileptic dizziness. Br Med J (Clin Res Ed). 1981;282(6265):687–9.
6. Smith BH. Vestibular disturbances in epilepsy. Neurology. 1960;10(5):465–9.
7. Blanke O, Perrig S, Thut G, Landis T, Seeck M. Simple and complex vestibular responses induced by electrical cortical stimulation of the parietal cortex in humans. J Neurol Neurosurg Psychiatry. 2000;69(4):553–6.
8. Surmeli R, Yalcin AD, Surmeli M, Gunay G. Vertiginous epilepsy: documentation of clinical and electrophysiological findings of nine patients. Epileptic Disord. 2020;22(6):775–81.

81 Superficial Siderosis

Overview

In *superficial siderosis* (SS), chronic slow bleeding into cerebrospinal fluid leads to iron deposition along the surfaces of the brain, brainstem, and cranial nerves. This results in neuronal damage to which the vestibulocochlear nerve and cerebellum are selectively vulnerable. The peak age of symptom onset in SS is the fifth through seventh decades, and men are affected more than women by a ratio of 3:1. Common presenting symptoms include bilateral hearing loss and impaired gait. Less common symptoms include cognitive deficits, anosmia, extraocular muscle palsies, sensory abnormalities in the lower extremities, and urinary incontinence. Physical examination shows cerebellar findings of appendicular and gait ataxia. Ocular motor examination may show positive bedside head impulse test bilaterally. Audiologic examination shows bilateral sensorineural hearing loss, usually more pronounced in the high frequencies, which may be asymmetrical. On instrumented vestibular testing, videonystagmography shows caloric weakness (which may be asymmetrical); video head impulse testing shows low gain with compensatory saccades; rotatory chair testing may show poor visual suppression of the vestibulo-ocular reflex, the presence of saccadic breakdown of smooth pursuit, and poor optokinetic responses. Cerebrospinal fluid may show xanthochromia, red blood cells, and elevated protein. Brain MRI is the imaging modality of choice; T2 imaging shows dark outlines along the surfaces of neural structures such as the vestibulocochlear nerve, brainstem, and cerebellar folia. If a source of bleeding can be identified, it is medically reasonable to repair, but it is unknown whether this changes the symptom trajectory of SS.

Introduction

Superficial siderosis, sometimes also called superficial hemosiderosis, is a condition in which chronic, slow bleeding into the cerebrospinal fluid of the subarachnoid space results in deposition of iron in the leptomeninges and subjacent gray matter, with a predilection for the cerebellum (particularly the vermis) and certain cranial nerves (cochleovestibular and olfactory).

Dr. Ralph Hamill (1877–1961) was the first to describe the gross pathological pigmentary characteristics of what eventually became known as superficial siderosis [1]. The German pathologist Dr. H. Hoetzel [2] was the first to connect these pigmentary changes to subarachnoid hemorrhage [3].

Epidemiology

Symptomatic superficial siderosis is very uncommon [3], though the precise incidence and prevalence are unknown. Incidental discovery of superficial siderosis (by imaging or at autopsy) in patients without corresponding symptoms suggests that superficial siderosis itself may not be rare.

Figure 81.1, from Fearnley and colleagues [3], shows a histogram of the age distribution of symptom onset in patients with superficial siderosis. Symptom onset peaks in the fifth to seventh decades.

Fearnley and colleagues [3] cite a male-to-female predominance at a ratio of 3:1.

M. Cherchi, *Otoneurology and Vestibular Medicine*, https://doi.org/10.1007/978-3-031-94841-1_81

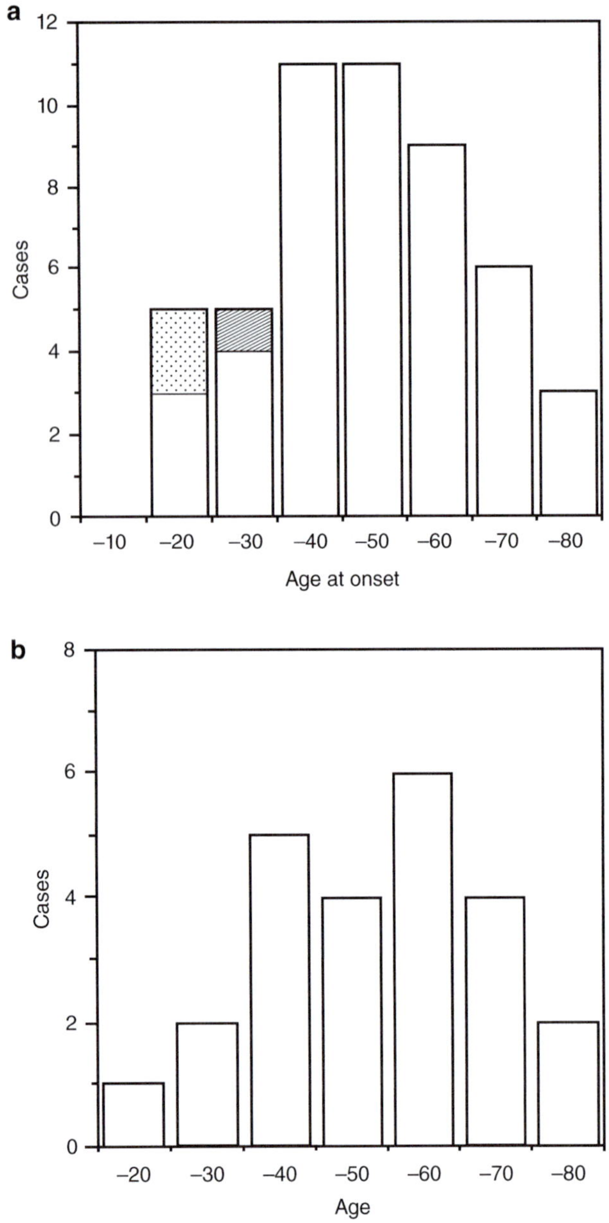

Fig. 81.1 Histograms showing age of onset of symptoms in patients with superficial siderosis of the central nervous system. Graph **a** (top): All cases, including idiopathic, in which there was sufficient information. Shaded bar areas = hemispherectomies. Graph **b** (bottom): Idiopathic cases alone in which no source of bleeding was identified. (From Fearnley et al. [3]. Reproduced with written permission from Oxford University Press)

Pathophysiological Mechanism of Disease

When extravasated blood enters the subarachnoid space, its deoxyhemoglobin gets metabolized to iron (which the brain cannot conjugate with ferritin), and thus, it gets deposited as insoluble hemosiderin in tissues with which it comes into contact [4]. The vulnerable tissues are those in close proximity to cerebrospinal fluid, which includes those in the subpial and subependymal spaces [3]. This unbound iron induces free radical formation and lipid peroxidation, resulting in tissue damage [4].

Fearnley and colleagues [3] comment that:

> There have been several cases in which superficial siderosis was found at autopsy in the absence of symptoms during life and it seems likely that there is a pre-symptomatic phase to the illness during which superficial siderosis is present, but not sufficient to cause symptoms [3].

How long does this presymptomatic phase last? Fearnley and colleagues add:

> An attempt to calculate the duration of this phase in symptomatic cases can be made by taking the interval from the onset of bleeding to the onset of symptoms referable to superficial siderosis… The pre-symptomatic phase in these cases ranged from 4 months to 30 years with a mean of 15 years [3].

Koeppen and colleagues state that, "Only CNS [central nervous system] tissues will convert heme in CSF to hemosiderin, and some tissues, such as the cerebellar cortex and the eight cranial nerves, are especially vulnerable" [5]. Why this selective vulnerability?

With regard to the apparent selective vulnerability of the vestibulocochlear nerve in superficial siderosis, Feanley and colleagues state:

> The fact that the VIIIth nerve is particularly vulnerable is due to its having a long glial segment. This results in greater length of haemosiderin deposition and, thereby, a greater chance of axonal damage. Another important factor is that the VIIIth nerve runs through the pontine cistern, which not only contains a large pool of CSF [cerebrospinal fluid] but also has a greater flow of CSF. This potentially delivers a larger amount of iron and haem causing a correspondingly greater degree of neuronal damage [3].

With regard to the apparent selective vulnerability of the cerebellum, Koeppen and colleagues state that, "The susceptibility of the cerebellar cortex is likely due to Bergmann glia that serve as conduits for heme; and the abundance of microglia" [5]. This is apparently based on, "Animal studies [that] suggest that the ferritin repressor protein-immunoreactive Bergmann glia and ferritin-containing microglia in the molecular layer of the cerebellum are involved in conversion of heme to ferritin and finally hemosiderin" [6].

Clinical Presentation

Superficial siderosis affects multiple areas of the central nervous system, corresponding with multiple dysfunctions and symptoms.

In the 87 cases reviewed by Fearnley and colleagues [3], they reported that:

- *Bilateral sensorineural hearing loss* is the most common feature, occurring in (95%) of cases [3]. The hearing loss begins insidiously and is gradually progressive.
- *Cerebellar symptoms* are almost as common as sensorineural hearing loss, occurring in 88% of cases [3].
- Other cranial nerves are less commonly affected, but the next most common is the olfactory nerve, manifesting with anosmia in as many as 17% of patients [3]. Extraocular muscle palsies are relatively infrequent, occurring in 5–10% of cases [3].
- Cognitive impairment occurred in 24% of patients in one study [3], and in 40% of patients in another series [6].
- Bilateral lower extremity paresthesias [7] or somatosensory deficits may occur in 13% of patients [3].
- Urinary incontinence is less common, occurring in 7 (8%) of patients [3].

Physical Examination

Most cases (88%) exhibit cerebellar findings of appendicular ataxia and gait ataxia [3].

A smaller proportion of cases (76%) exhibit pyramidal tract findings, and if present, these predominantly involve the lower extremities [3].

Ocular Motor Examination

Ocular motor testing occasionally shows nystagmus. This may consist of horizontal gaze evoked nystagmus or spontaneous down beat nystagmus. Even if present, in some cases, the nystagmus may only be detectable on instrumented oculography [3].

Bedside head impulse testing [8] may show a deficient vestibulo-ocular reflex gain and the presence of compensatory saccades.

Testing: Auditory

When Fearnley and colleagues reviewed the literature of 87 published cases of superficial siderosis, they commented that, "In all cases with sufficient clinical details there was a sensori-neural [hearing] loss with the high tones being more severely affected" [3].

The sensorineural hearing loss on audiometry can be predominantly high frequency, or "across the board," and may be modestly asymmetrical, as shown in Figs. 81.2, 81.3, and 81.4 from Vibert and colleagues [9].

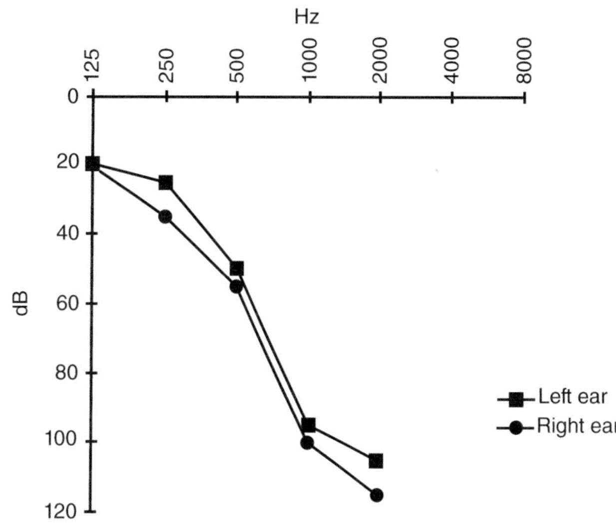

Fig. 81.2 Audiogram from a patient with superficial siderosis showing approximately symmetrical middle and high frequency hearing loss. (From Vibert et al. [9]. Reproduced with written permission from Elsevier)

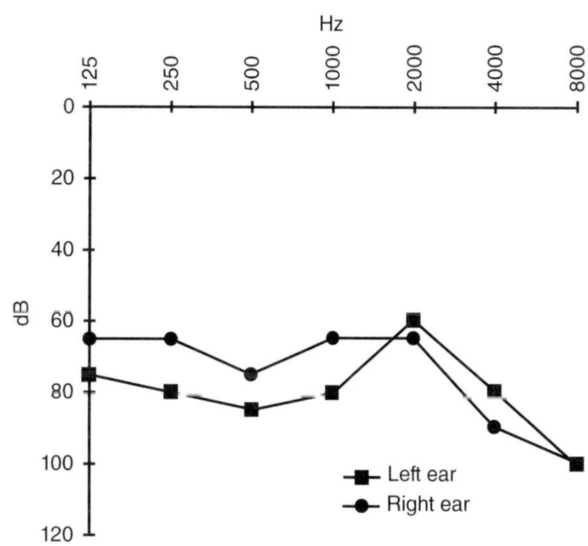

Fig. 81.3 Audiogram from a patient with superficial siderosis showing an "across-the-board" hearing loss, with modest asymmetry in the low and middle frequencies. (From Vibert et al. [9]. Reproduced with written permission from Elsevier)

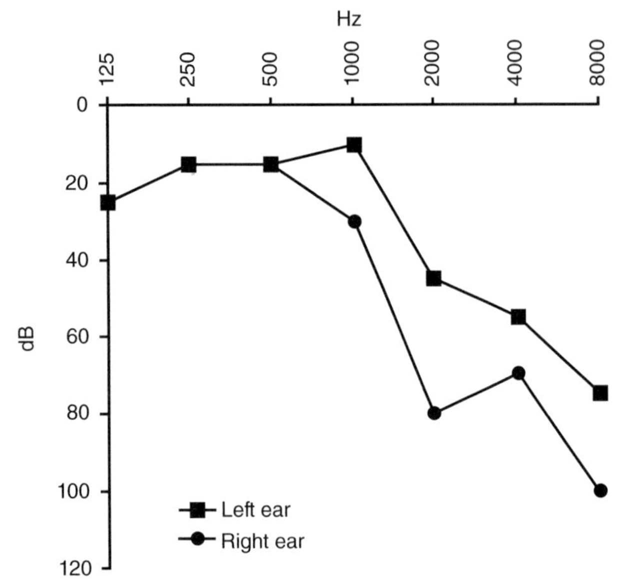

Fig. 81.4 Audiogram from a patient with superficial siderosis showing asymmetrical middle and high frequency hearing loss. (From Vibert et al. [9]. Reproduced with written permission from Elsevier)

Testing: Vestibular

When Fearnley and colleagues reviewed the literature of 87 published cases of superficial siderosis, they lamented that, "In only four papers are vestibular tests reported" [3]. This deficiency has unfortunately persisted in subsequent literature, with a few exceptions.

Vestibular studies in patients with superficial siderosis show evidence of bilateral vestibular weakness.

Several studies of videonystagmography in patients with superficial siderosis report caloric responses that are either reduced or nearly absent [7, 8].

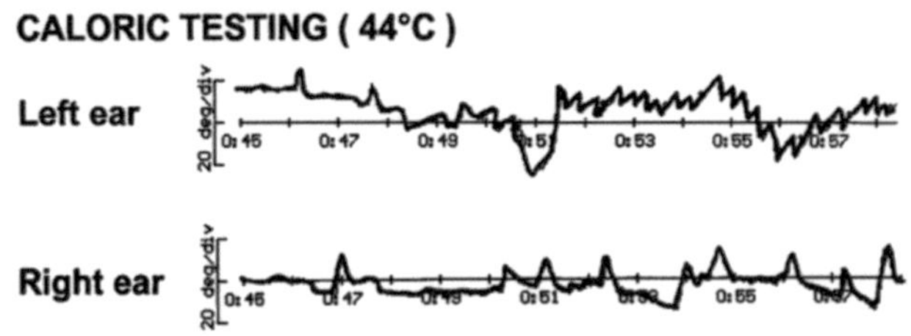

Fig. 81.5 Asymmetrical caloric responses in a patient with superficial siderosis. (From Vibert et al. [9]. Reproduced with written permission from Elsevier)

In some cases, the vestibular weakness may be asymmetrical. Figure 81.5, from Vibert and colleagues [9], demonstrates asymmetrical caloric responses, with a weaker response on the right.

Several published studies of patients with superficial siderosis have documented bilateral vestibular weakness on video head impulse testing [4, 8, 10]. Aran-Yoo and Kattah clearly document low gain and robust overt compensatory saccades in all canals [4].

Instrumented oculography may also reveal abnormalities, suggesting localization to the cerebellum, such as poor visual suppression of the vestibulo-ocular reflex, saccadic smooth pursuit, and poor optokinetic responses.

Figure 81.6, from Vibert and colleagues [9], shows incomplete visual suppression of the vestibulo-ocular reflex on rotatory chair testing, which suggests cerebellar dysfunction.

Figure 81.7, from Vibert and colleagues [9], shows saccadic pursuit, which suggests cerebellar dysfunction.

Figure 81.8, from Vibert and colleagues [9], shows poor optokinetic responses (except to the rightward 75 °/s rightward stimulus), suggesting cerebellar dysfunction.

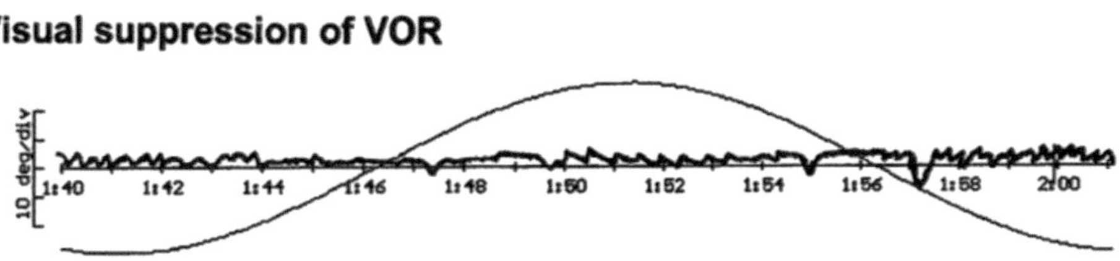

Fig. 81.6 Incomplete visual suppression of the vestibulo-ocular reflex on rotatory chair testing. (From Vibert et al. [9]. Reproduced with written permission from Elsevier)

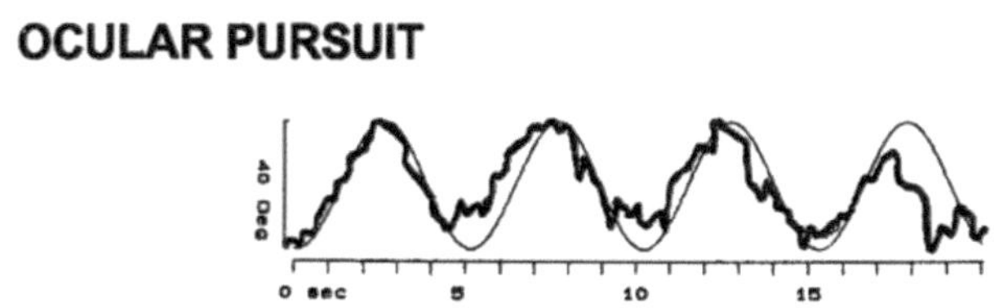

Fig. 81.7 Saccadic smooth pursuit movements in a patient with superficial siderosis. (From Vibert et al. [9]. Reproduced with written permission from Elsevier)

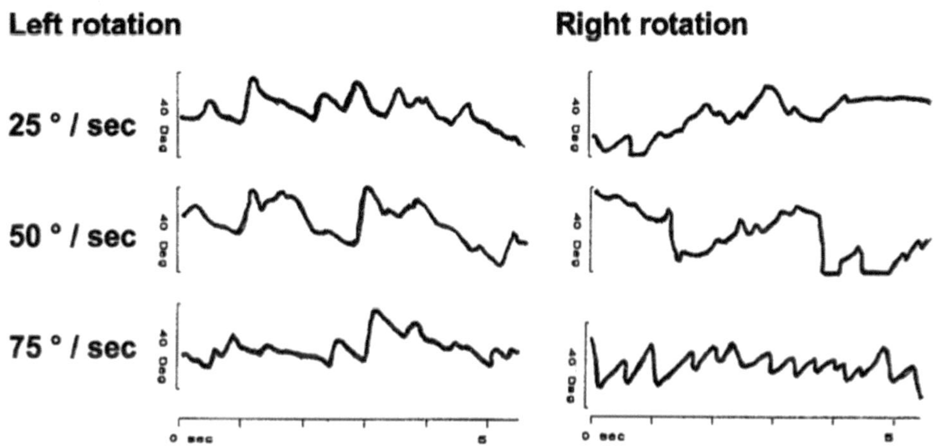

Fig. 81.8 Optokinetic responses are poor in a patient with superficial siderosis. (From Vibert et al. [9]. Reproduced with written permission from Elsevier)

Testing: Other

Cerebrospinal fluid shows several abnormalities in patients with superficial siderosis [6]. Feanley and colleagues summarize these as follows:

> Irrespective of headache, the cardinal CSF signs in superficial siderosis are haemorrhage and xanthochromia. Other features include the presence of erythrophages and siderophages and raised iron and ferritin levels. The protein may be elevated as a result of arachnoiditis and very occasionally a monoclonal or oligoclonal band may be present [3].

Practically, given the high sensitivity of brain MRI for superficial siderosis (see below), lumbar puncture no longer plays a significant role in diagnosing this disease.

Imaging

Brain MRI is the diagnostic imaging modality of choice [3] because of its very high sensitivity to hemosiderin [6]. "MRI reveals a rim of hypointensity on T2-weighted images, enveloping the surface of the brainstem, cerebellum, and sometimes the cortical fissures." However, "Even with MRI, the diagnosis may still be missed because the imaging abnormalities follow the contours of the brain and can be overlooked" [6].

Figure 81.9, from Kumar and colleagues [6], shows such abnormalities involving the cerebrum, brainstem, and spinal cord.

Figure 81.10, from Vibert and colleagues [9], demonstrates to advantage involvement of the cerebellum by superficial siderosis.

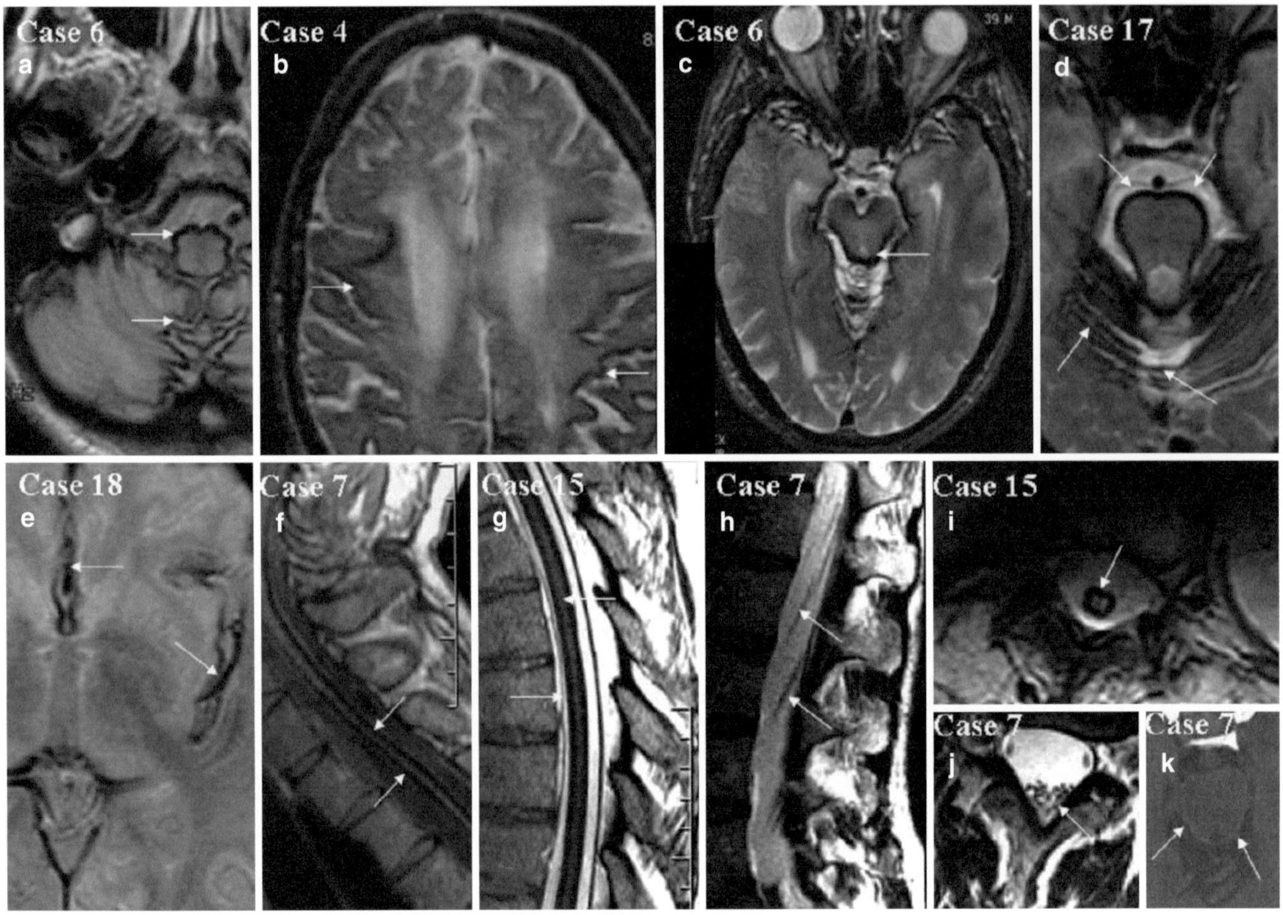

Fig. 81.9 Selected MRI images from 30 cases of superficial siderosis. Axial T2-weighted (**a** through **d**) and proton density (**e**) brain MRI in Cases 6 (**a**, **c**), 4 (**b**), 17 (**d**), and 18 (**e**), showing signal hypointensity due to hemosiderin deposition around the medulla and cerebellar vermis (**a**), along the cerebral convexities (**b**), around the midbrain (**c**), around the pons and along the cerebellar folia (**d**), and along the inter-hemispheric and Sylvian fissures (**e**). Sagittal T1-weighted (**f**) and sagittal T2-weighted (**g**, **h**) spine MRI in Cases 7 (**f**, **h**) and 15 (**g**), showing hemosiderin deposition along the cervical cord (**f**), thoracic cord (**g**), and the roots of the cauda equina (**h**). Axial T2 (**i**, **j**) spinal MRI in Cases 15 (**i**) and 7 (**j**), showing hemosiderin deposition around the cord (**i**) and along the cauda, resulting in nerve root clumping suggestive of arachnoiditis (**j**). Also presented is a CT showing a rim of hyperdensity due to hemosiderin around the brainstem in Case 7 (**k**). (From Kumar et al. [6]. Reproduced with written permission from Wolters Kluwer Health, Ltd.)

Fig. 81.10 Axial T2- and T2*-weighted images showing linear signal losses prominent primarily in the posterior fossa, particularly pronounced in the cerebellar folia at the level of the (panel **a**) hemispheres and (panel **b**) cerebellar vermis; there were also signal losses, which may be seen as a fine line along the surface of the brainstem, notably along the (**a**) mesencephalon as well as in the sulcus cinguli in the frontal interhemispheric region. (From Vibert et al. [9]. Reproduced with written permission from Elsevier)

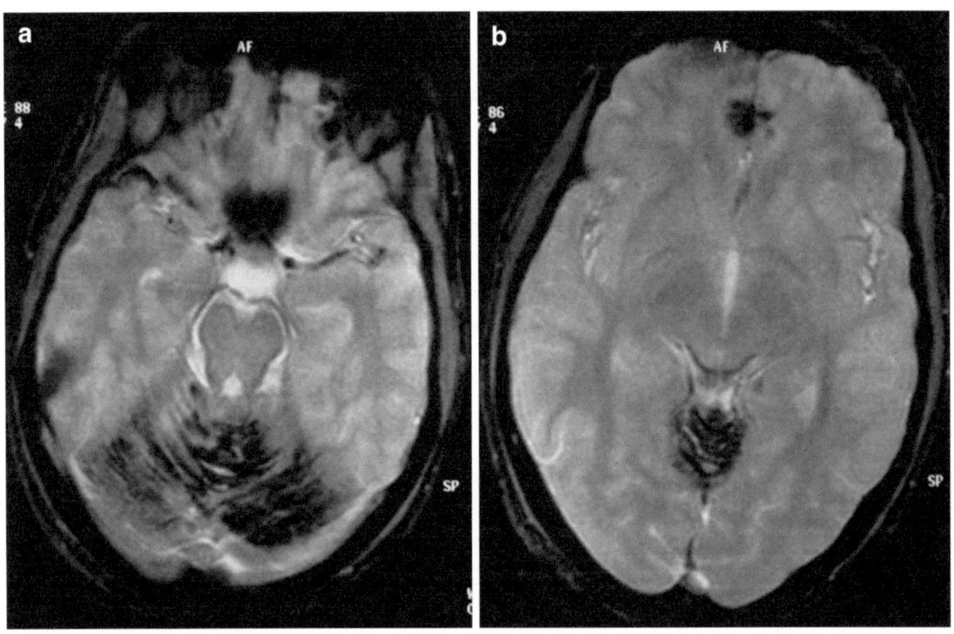

Gross Pathology and Histopathology

Fearnley and colleagues summarize the gross pathology of superficial siderosis as follows:

> The pathology of superficial siderosis is characterized by marginal siderosis, granular ependymitis and in some cases an obstructive hydrocephalus. Macroscopically, there is a brownish discoloration of the leptomeninges, usually sparing the dura mater, and of the adjacent CNS parenchyma up to a depth of 3 mm (Rosenthal, 1958). Particularly pigmented regions were the cerebellum, especially the superior vermis and crests of the folia, basal frontal lobe, olfactory bulbs, temporal cortex, brainstem and cranial nerves (I, II, V, X and especially VIII), spinal cord and spinal nerve roots [3].

With regard to histopathology, they state, "Microscopically, there is haemosiderin deposition, neuronal loss, reactive gliosis, intra-cellular ovoid bodies and demyelination" [3]. Koeppen and colleagues add, "In addition to perivascular collections of hemosiderin-laden macrophages, affected tissues displayed iron-positive anuclear foamy structures in the neuropil that resembled axonal spheroids. They were especially abundant in eighth cranial nerves and spinal cord" [5].

Differential Diagnosis

Choi and colleagues state that, "When hearing impairment was associated, superficial hemosiderosis was the most common disorder in chronic bilateral combined vestibulopathy" [8].

Nevertheless, since each of the manifestations of superficial siderosis can occur in other diseases as well, the differential diagnosis includes:

- The associated hearing loss differentiates superficial siderosis from, for example, CANVAS and normal pressure hydrocephalus.
- The slowly progressive nature differentiates superficial siderosis from, for example, ototoxicity and autoimmune inner ear disease.
- The bilaterality of the hearing loss differentiates superficial siderosis from, for example, vestibular schwannoma or labyrinthitis.

Treatment

If a source of bleeding can be identified, this should be treated, though the long-term outcome of this is unknown. It is noteworthy that in a substantial minority of cases no bleeding source can be found; Koeppen and colleagues [5] review several studies documenting that in 30–46% of cases of superficial siderosis, no source of bleeding remains undetermined.

Chelation therapy has been attempted in several cases, without success [3].

Practically, it is medically reasonable to refer patients to audiology to be evaluated for hearing amplification, and to vestibular rehabilitation therapy to be treated for bilateral vestibular weakness.

Prognosis

By the time a patient has developed symptoms, neuronal death has already occurred, and there are currently no therapies that can reverse this process.

Logically, one would think that stopping any further bleeding into the subarachnoid space should retard progression, but this remains to be demonstrated and would require longitudinal studies. Kumar and colleagues comment that, "It is unclear that elimination of the source of chronic bleeding slows or halts the clinical progression of this disorder… Judging clinical outcomes is difficult, given the slow progression of SS [superficial siderosis] and the possible episodic nature of the bleeding. It is unclear if the process continues despite eradication of the bleeding" [6].

References

1. Hamill RC. Report of a case of melanosis of the brain, cord and meninges. J Nerv Ment Dis. 1908;35(9):594.
2. Noetzel H. Diffusion von Blutfarbstoff in der inneren Randzone und äußeren Oberfläche des Zentralnervensystems bei subarachnoidaler Blutung. Arch Psychiatr Nervenkr. 1940;111(1):129–38.
3. Fearnley JM, Stevens JM, Rudge P. Superficial siderosis of the central nervous system. Brain J Neurol. 1995;118(Pt 4):1051–66.
4. Aran Yoo BS, Kattah JC. Superficial siderosis syndrome with progressive hearing loss and bilateral vestibular failure, 51 years after a neurosurgical procedure: diagnostic value of combined MRI and video head impulse test. J Neurol. 2017;264(2):391–3.
5. Koeppen AH, Michael SC, Li D, Chen Z, Cusack MJ, Gibson WM, et al. The pathology of superficial siderosis of the central nervous system. Acta Neuropathol. 2008;116(4):371–82.
6. Kumar N, Cohen-Gadol AA, Wright RA, Miller GM, Piepgras DG, Ahlskog JE. Superficial siderosis. Neurology. 2006;66(8):1144–52.
7. Revesz T, Earl CJ, Barnard RO. Superficial siderosis of the central nervous system presenting with longstanding deafness. J R Soc Med. 1988;81(8):479–81.
8. Choi SY, Kee HJ, Park JH, Kim HJ, Kim JS. Combined peripheral and central vestibulopathy. J Vestib Res Equilib Orient. 2014;24(5–6):443–51.
9. Vibert D, Hausler R, Lovblad KO, Schroth G. Hearing loss and vertigo in superficial siderosis of the central nervous system. Am J Otolaryngol. 2004;25(2):142–9.
10. Kang KW, Lee C, Kim SH, Cho HH, Lee SH. Bilateral vestibulopathy documented by video head impulse tests in superficial siderosis. Otol Neurotol. 2015;36(10):1683–6.

Cervicogenic Vertigo

82

Practical Summary

Vestibular clinicians often encounter patients with neck symptoms and vertigo in whom it may seem logical to postulate a causal relationship between the two symptoms. The possible relationship between these has been given various labels in the literature; here, we shall use the term cervicogenic vertigo (CV), but other terms include various permutations of "cervical"/"cervicogenic" and "vertigo"/"dizziness."

The most popular theory is that cervicogenic vertigo is due to an abnormality in cervical proprioception. Etiologies such as vascular compromise, even if sometimes correct, probably account for only a modest proportion of cases.

There have been numerous attempts at developing a test for cervicogenic vertigo; none appears sufficiently sensitive or specific, and none has gained wide acceptance. It seems likely that this failure is at least in part due to the fact that the system in question involves multimodal sensory integration, and in practice, it is difficult to manipulate selectively a single sensory modality while leaving the other modalities unchanged.

In the absence of a confirmatory test for CV, it remains a diagnosis of exclusion. Depending on the clinical scenario and findings on physical examination, additional workup may include otovestibular testing and imaging.

Physical therapy for the neck has been studied more than other treatment modalities and is usually described as having favorable outcomes. Given our ignorance of the underlying mechanism of disease, it is difficult to understand why physical therapy should be helpful, but one possibility is that such therapy normalizes cervical mechanics and thereby interrupts the maladaptive positive feedback loop in which each symptom (vertigo, neck pain) exacerbates the other.

Cervicogenic vertigo more commonly provokes symptoms during positional changes and other movements. Such movements can also provoke other forms of vertigo, so when formulating a differential diagnosis for CV, it is prudent to keep in mind common causes as well, such as benign paroxysmal positional vertigo.

Introduction

Vestibular clinicians frequently encounter a patient with disequilibrium and neck symptoms, in whom a thorough workup fails to identify any clear etiology for the vestibular disturbance. Although co-occurrence does not prove causality, some clinicians will consider the idea of *"cervicogenic vertigo" (CV)*.

Much of the literature regards the diagnosis of CV as "controversial" [1–4], particularly since there is no confirmatory diagnostic test [3, 5–7].

To understand *why* CV is controversial, we will discuss the proposed pathophysiological mechanisms, how those mechanisms could be assessed, why those assessments have failed, and what treatments have been attempted.

The present chapter draws heavily on our previous work with Dr. Frank DiLiberto, Dr. Dario Yacovino, and Dr. Sunit Das [4].

Assumptions, Definitions, and Conditions

For most of this discussion we use the simplifying discussion that CV is the only source of symptoms. Later we will discuss why this simplifying assumption is probably incorrect.

We use the term "vertigo" in its technical sense of a kinetic illusion, which is to say a discrepancy between perceived versus actual motion/stasis. Laypeople and many physicians may use the term "vertigo" to refer more narrowly to a spinning sensation, but many authors note that such a perception is quite uncommon in CV [7, 8]. In any case, the term "vertigo" in the phrase "cervicogenic vertigo" has been entrenched in the literature since Ryan and Cope's 1955 paper on the subject [9].

Discussions of CV explicitly or implicitly require that (1) neck symptoms are present (cervicalgia; head-on-neck or neck-on-trunk motion that is limited, excessive, unintentional, etc.) and at least partially overlap (in a temporal sense) with the symptom of disequilibrium; or (2) a neck injury occurred prior to the onset of symptoms; or (3) both [2, 8, 10].

M. Cherchi, *Otoneurology and Vestibular Medicine*, https://doi.org/10.1007/978-3-031-94841-1_82

Immediate Problems

Those assumptions, definitions, and conditions seem reasonable but immediately pose several problems.

The *first problem* concerns to the relationship between cervical disease and vertigo. The simplifying assumption mentioned earlier requires that cervical disease be the cause of vertigo—but of course that is not the only possible relationship.

One possibility is that the temporal overlap of symptoms is coincidental rather than causal [11]. The rationale for this possibility is that neck pain and disequilibrium are both common experiences, so even when each symptom is due to a separate disease, the likelihood of their coinciding (by chance) is not small. Neck symptoms are common and increasing; the prevalence among adults aged 25–84 in the United States was 14.8% in 2002 and 17.2% in 2018 [12]; similar demographics are reported in other countries [13]. In the United States, disequilibrium accounted for 20.6 million ambulatory care visits per year in 2013–2015 [14], and for four million emergency department visits in 2011 [15]. As Thompson-Harvey and Hain point out, "The main clinical problem in diagnosing cervical vertigo is that symptoms of subjects who have both neck disorders and dizziness may overlap… In other words it is difficult to differentiate between the chance coincidence of arthritis of the neck and dizziness, from the situation where arthritis of the neck causes dizziness" [10].

Another possibility is that the temporal overlap of symptoms may hold, but the disequilibrium causes the neck symptoms (rather than the other way around). On this view, the phenomenon might more correctly be termed "vertigogenic cervicalgia" (rather than "cervicogenic vertigo").

Patients experiencing (or anticipating) the symptom of disequilibrium often make unconscious compensatory postural adjustments, such as turning *en bloc,* "splinting" or "bracing" themselves, and an unintended consequence of such postural changes may be neck symptoms [7, 16]. In other words, the neck symptoms may be an effect (rather than the cause) of the vertigo.

The *second problem* concerns the relationship between CV and neck injury. Citing Mallinson and Longridge [17], Hain points out that "inner ear disorders are rare after neck trauma" [6]. And yet, traumatic injuries are rarely so focal as to affect only the neck. Even whiplash—the most common neck injury in industrialized societies—is generally not a "pure neck injury." Yacovino and Hain point out that, "Postwhiplash vertigo can combine several mechanisms. At the ear level, the otolith system is prone to suffer inertial damage" [7], and several investigators comment that acceleration-deceleration injuries can cause benign paroxysmal positional vertigo [13, 18, 19]. In general, "Dizziness following neck injury may be due to vestibular system pathologies, brain injury, or cervicogenic dizziness" [8], including "the ear (labyrinth contusion), the brainstem, the cortical and subcortical structures, and the vertebral arteries (traumatic artery dissection)" [7]. Thus, the simplifying assumption (that CV is the only source of symptoms) mentioned above is probably overly-simplifying for real-world clinical cases of neck injuries. Since whiplash injuries usually occur in the context of vehicular collision, litigation (or the prospect of litigation) and the possibility of secondary gain add layers of nonphysiologic complexity [20] whose psychological dimensions [21] and influence on subjective symptoms are difficult to assess.

Underlying Physiology

Most discussions of CV are grounded in a physiologic framework based on integration of multiple sensory inputs, planning of output, and multiple motor outputs. For example:

> The results of several studies suggest that the control of posture, perception of the orientation of the body, and the location of objects in extrapersonal space requires an integration of proprioceptive, visual, and vestibular signals as well as internally generated signals related to intended head and body movements [22].

These "intended head and body movements" are sometimes referred to as "efference copies" or "corollary discharges." In this context, efference copies feed into cortical constructs that try to predict head-on-body positions and movements in order to orchestrate appropriate somatomotor and ocular motor outputs. These cortical constructs are effectively internal models that are iteratively recalibrated based on the experience of one's planning and executing motor output, and receiving sensory feedback about the effects (out in the world) of the actual motor outputs.

Figure 82.1, from Kristjansson and Treleaven [3], depicts the multisensory integration and output relevant for CV.

In order for a person to perceive his orientation in and movement through space correctly, at some level the brain must calculate a coordinate transformation—mapping from one reference frame's coordinates on to another reference frame's coordinates [23, 24]. As Brandt describes:

> It is necessary for the sensorimotor control system to know the attitude of the head relative to the body, since the vestibular system signals only head motion relative to space and head position relative to the gravity vector. The head-mounted sensory systems must transform the rotations and accelerations they sense and correctly relate their direction to the motion and attitude of the body and the center of gravity. Neck afferents provide information about head position, and make an important contribution to the control of body and sensory spatial orientation. The perception of head or trunk rotations in space would be erroneous if only vestibular stimulation or only neck stimulation was involved. However, if the two stimuli are combined (head rotations relative to the trunk), the perception of both trunk and head rotation in space reflects the true position [1].

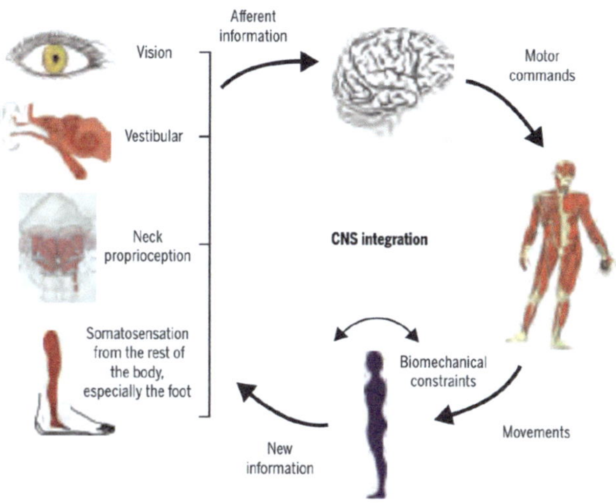

Fig. 82.1 Schematic of multisensory integration and motor output relevant for cervicogenic vertigo. CNS = central nervous system. (From Kristjansson and Treleaven [3]. Reproduced with written permission from Journal of Orthopaedic and Sports Physical Therapy Inc.)

This is a "computationally difficult adjustment between the two coordinate systems of the head and body" [7], but there is reasonable evidence that such a coordinate transformation does take place. Studies from primates "conclude that sensory vestibular signals are transformed from head-in-space coordinates to trunk-in-space coordinates on many secondary vestibular neurons in the vestibular nuclei by the addition of inputs related to head rotation on the trunk. This coordinate transformation is presumably important for controlling postural reflexes and constructing a central percept of body orientation and movement in space" [22].

Probably the most frequently proposed mechanism of CV is that it reflects a failure of multisensory integration—which is the process of reweighting and combining multiple sensory inputs to produce a coherent perception. According to this theory, the "failure" is ascribed to what are presumed to be erroneous proprioception of cervical mechanics. As Brandt states:

> Somatosensory signals from musculotendinous receptors in the neck and joints provide an accurate kinesthetic feedback of the extent of head and limb movements. These signals contribute to the perception of self-motion during active locomotion by converging with vestibular and visual input on multimodal neurons in the vestibular nuclei and thalamus, which project to cortical multisensory areas in the parietal lobe [1].

We shall review this theory and other proposed explanations for disequilibrium arising from neck pathology.

Candidate Pathophysiological Mechanisms

Several mechanisms have been suggested as the pathophysiology underlying CV. These include:

- *Hypoperfusion*. Neck movements can physically compromise cervical arterial vasculature by torque or compression, resulting in reduced blood flow to the brainstem. A related possibility is that neck movements may induce autonomic responses (e.g., through stimulation of carotid baroreceptors), also resulting in hypoperfusion.
- *Anatomic*. Neck movements may provoke brainstem compression in a person with craniocervical instability.
- *Ocular motor*. Through a variety of mechanisms, neck movements may abnormally induce or influence ocular motor responses.
- *Proprioception*. Neck movements may generate erroneous proprioceptive signals.
- *Motoric*. Neck movements may erroneously modulate efferent motor signals.
- *Migraine*. There are two possibilities here: (1) a sequence in which neck problems trigger migraine, and migraine in turn causes vertigo; (2) a "common cause" scenario in which migraine causes both neck symptoms and vertigo.

We shall review each of these mechanisms.

Hypoperfusion

Hypoperfusion secondary to vascular compromise is frequently cited as the pathophysiologic mechanism of CV. Rotational vertebral artery syndrome (RVAS), also called Bow Hunter syndrome, is a condition in which neck rotation brings about transient extrinsic compression of a vertebral artery [25–37]. This scenario appears more likely to provoke symptoms if the contralateral vertebral artery is already stenotic (e.g., from atherosclerotic disease, etc.). Such mechanics (torque and compression) may in some cases result in vascular damage, such as vertebral artery dissection [38]. There are scattered reports of RVAS manifesting with down beat nystagmus [39, 40], though such nystagmus is not specific for RVAS. The luminal reduction or obstruction from RVAS is transient, but if prolonged, it can cause infarction [41]. Despite the apparent enthusiasm in the literature, RVAS probably constitutes small proportion of cases of CV, and most discussions draw on case reports or small case series. Dynamic vascular imaging can provide corroborative evidence of RVAS, but there is little consensus regarding the mechanical effects of neck rotation on cervical vasculature [42]. Citing Mitchell

[29], Hain comments that, "Vertebral artery blood flow is compromised with full contralateral rotation in healthy individuals" [6], therefore, vascular imaging is not specific for this condition. In addition, data from animal cadaver models demonstrate that under the conditions of normal neck movements, mechanical strain on vertebral artery tissue is much lower than the forces required to induce arterial dissection [43]. Chiari malformations, in which there is anatomical "crowding" at the foramen magnum, are probably an even rarer cause of vascular compromise in CV [44].

Hypoperfusion secondary to an erroneously triggered autonomic response has been proposed as a mechanism for CV, and several etiologies have been considered. The first proposal came from Jean-Alexander Barré [45] and Young-Choen Lieou [46], who suggested that neck rotation could stimulate paravertebral sympathetic ganglia—an idea that came to be known as Barré–Lieou syndrome. This proposal was met with some initial enthusiasm [47], but subsequent investigations failed to find any reproducible evidence in support of the idea; consequently, "No sympathetic or vascular changes were subsequently identified that could account for these symptoms and this theory lost favor" [8]. Most researchers [48] adopted the view that "Positional nystagmus cannot be attributed to a disturbance of the cervical sympathetic chain as suggested by Barré" [1]. The second, and more plausible, mechanism of autonomically mediated hypoperfusion is "head turning-induced hypotension" [49] triggered when cervical rotation stimulates overly sensitive carotid sinus baroreceptors. Physical examination should be able to identify this by observing a change in blood pressure while a patient maintains the head in a position maximally rotated off the midline.

Anatomic

Abnormalities of anatomy (e.g., Chiari malformation) or mechanics (e.g., craniocervical instability) at the craniocervical junction may produce brainstem compression during cervical rotation [44], and this has been proposed as an etiology for CV. Generally, such abnormalities should be detectable on imaging. Identification of abnormal mechanics in this area may require dynamic imaging, or at least several static images with the head in different degrees of rotation with respect to the shoulders. In view of the neuroanatomical region that is involved, brainstem compression would usually cause multiple symptoms (beyond simply vertigo) and physical examination findings.

Ocular Motor Abnormalities

Several ocular motor abnormalities have been documented in association with neck rotation and neck pain, including abnormalities of the cervico-ocular reflex [50–52] and other "Deficits in oculomotor control, such as decreased smooth pursuit velocity gain, altered velocity and latency of saccadic eye movements" [3]. Some authors suggest that "The smooth pursuit neck torsion test developed by Tjell et al. [53] is considered to be specific for detecting eye movement disturbances due to altered cervical afferent input" [3], a method we shall discuss below.

Proprioception

Probably the most widely held view of CV is that it results from some abnormality of cervical proprioception, though such "Proprioception is not a function of the superficial neck muscles but of the deep short intervertebral neck muscles, which are extensively supplied with muscle spindles" [1].

Interestingly, "Of all the muscles in the body, it is the deep neck muscles that have the highest concentration of muscle spindles" [13], and consequently, "The proprioceptive system of the cervical spine… is extremely well developed, as reflected by an abundance of mechanoreceptors, especially from the gamma-muscle spindles in the deep segmental upper cervical muscles" [48].

This "dense network of mechanoreceptors in the soft tissues in this region… gives the CNS information about the orientation of the head with respect to the rest of the body via direct neurophysiological connections to the vestibular and visual systems" [3], so it is unsurprising that, "Strong connections have been demonstrated between the cervical dorsal roots and the vestibular nuclei with the neck receptors (such as proprioceptors and joint receptors) playing a role in eye-hand coordination, perception of balance, and postural adjustments" [8].

Brandt and Bronstein [16] point out that studies of various kinds of neck stimulation can alter perception. In particular, "Unilateral electrical stimulation of the neck [54] causes deviation of the subjective vertical" and "Vibration of neck muscles, which stimulates the primary endings of the muscle spindles as if the muscle were being stretched [55] elicits an illusion of head tilt and apparent movement of a visual target" [56]. Such results, as well as data from experiments involving injection of anesthetics into the neck, have led to "the current theory that cervicogenic dizziness results from abnormal input into the vestibular nuclei from the proprioceptors of the upper cervical region" [8].

These data may all provide convincing evidence of erroneous cervical proprioception, but do not in themselves explain why that abnormal sensory input should provoke the symptom of vertigo. There are two plausible mechanisms.

The first mechanism by which erroneous cervical proprioception could induce the symptom of vertigo is through "sensory mismatch"—a discrepancy between (1) the incorrect cervical proprioceptive input, and (2) the correct visual and labyrinthine input [3, 6, 10, 48].

The second mechanism by which erroneous cervical proprioception could induce the symptom of vertigo is from a discrepancy between (1) the intended motion/position (the "efference copy" or "corollary discharge" discussed above), and (2) the misperception of actual motion/position. Brandt points out that, "a multisensory mismatch would be expected to result in CV… the resultant mismatch would be maximal during active head movements (when expected and actual reafferent input do not match)" [1]. This possibility has been examined through "cervical repositioning tests" [3], and "joint position error" tests [57, 58].

Motor Mechanism

Most of the literature on CV focuses on erroneous sensory input as the main or exclusive mechanism. In contrast, very little literature considers impaired motor output as a potential mechanism. This theory postulates that CV occurs when neck pathology causes incorrect modulation of efferent (motor) pathways, and the abnormal motoric output results in true unsteadiness that the patient is correctly perceiving. If this theory is correct, then the abnormal motoric responses are probably below the threshold of observability on physical examination, because these patients generally appear normal on bedside examination. Our group explored this "erroneous motor output" theory further by testing the hypothesis that CV results in impaired vestibulospinal reflexes; we elected to compare triceps vestibular evoked myogenic potentials (whose efferent pathways pass through the cervical spinal cord) in healthy controls versus patients whose histories were compatible with CV but found no differences [59–61].

Migrainous Mechanism

The idea of "migraine-associated cervicogenic vertigo" as a mechanism for CV was advanced by Yacovino and Hain [7]. Other investigators have taken up this idea [2], and further support came from a study by Thompson-Harvey and Hain [10] who observed that a meticulously designed questionnaire could not distinguish CV from migraine-associated vertigo. The theory of "migraine-associated cervicogenic vertigo" lends itself to two potential mechanisms.

The first possible mechanism is a sequence of events in which (1) neck problems cause migraine [62–66], then (2) migraine causes vertigo [67–69].

The second possible mechanism is that neck pain and vertigo are just separate symptomatic manifestations of one underlying disease, namely migraine. In favor of this idea are copious documentation that migraine can cause both neck pain [65, 70–77] and vertigo [67–69].

The idea of "migraine-associated cervicogenic vertigo" is attractive as a unifying diagnosis, but has the drawback of exchanging one currently unverifiable diagnosis (cervicogenic vertigo) with another currently unverifiable diagnosis (migraine-associated vertigo). Nevertheless, this theory has implications for clinical practice insofar as it suggests that treatment for migraine may be a therapeutic option for managing CV.

Attempts at Developing Objective Diagnostic Tests and Their Failures

The variety of hypothetical mechanisms for CV has inspired commensurate attempts for testing those mechanisms. These include:

- *Imaging*. Imaging generally aims to identify vascular compromise triggered by head-on-neck rotation, and thus include various types of angiography (CTA, MRA, transcranial Doppler ultrasound, dedicated catheter angiography).
- *Posturography*. Computerized dynamic posturography was originally designed to assess vestibular, visual, and proprioceptive contributions to an individual's unsteadiness. Its utility in CV has also been explored.
- *Ocular motor studies*. A range of ocular motor findings have been reported in CV, such as abnormalities in spontaneous nystagmus, smooth pursuit, caloric responses, optokinetic after-nystagmus, and cervico-ocular reflex responses.

We shall review each of these attempts.

Imaging

Dynamic imaging, or static imaging of several different positions of head-on-neck rotation, may identify vascular compromise, as can occur in rotational vertebral artery syndrome, but since normal subjects can exhibit similar imaging, the finding is not specific for RVAS [6, 29].

Posturography

Computerized dynamic posturography has been explored in several studies [78–85] that report increased postural sway in patients whose histories suggest CV, but "increased postural sway is a nonspecific finding that is also evident in patients with vestibular injury" [8]. Furthermore, "postural instability can be simulated" [6].

Ocular Motor Studies

Multiple relevant sensory inputs converge at the vestibular nuclei [1, 8, 22, 48] which in turn sends projections to many locations, including the ocular motor nuclei. It is therefore logical to investigate whether particular ocular motor abnormalities might serve as biomarkers for CV. Several ocular motor findings have been reported in patients whose clinical histories are compatible with CV [86], including abnormalities in smooth pursuit, abnormal caloric responses, spontaneous and positional nystagmus [8], latent nystagmus, and abnormalities on rotatory chair testing [87]. However, such findings are neither sensitive nor specific for CV. We shall review several ocular motor tests that have been explored, including smooth pursuit, optokinetic after-nystagmus, and the cervico-ocular reflex.

Smooth Pursuit

Several researchers report smooth pursuit abnormalities in whiplash patients complaining of disequilibrium [53, 88]. However, smooth pursuit is problematic as a biomarker because, "Smooth pursuit is a complex multiple input system that is vulnerable to cognitive variables, age, and sedation. Neck pain and secondary gain, both disruptive of cognition, would also seem highly likely to influence performance of smooth pursuit. For these reasons, due to an intrinsic issue with specificity, it seems unlikely that any smooth-pursuit test could be of general utility for the diagnosis of cervical vertigo" [7].

Optokinetic After-Nystagmus

Optokinetic after-nystagmus may be abnormal in some patients whose histories are compatible with CV [89]. However, "Optokinetic after-nystagmus is difficult to obtain in humans and is generally of small velocity even in normal subjects. This makes it unlikely that this test could be sufficiently sensitive to be useful in cervical vertigo" [7].

Cervico-ocular Reflex in the Head-Still Trunk-Rotates Protocol

The "head-still trunk-rotates" protocol to assess the cervico-ocular reflex warrants discussion because it is a very logical test for CV. This test consists of stabilizing the head in space (thereby avoiding stimulation of the labyrinths and vision) while rotating the trunk in a sinusoidal fashion [47, 50–52], which should selectively isolate proprioceptive cervical input. The dependent variable is ocular motor responses. Two versions of this test have been studied. In the first version the trunk is passively oscillated by a rotatory chair while the subject's head is fixed by a mechanical stabilizer [90]. In the second version the trunk is passively oscillated by a rotatory chair, and the subject's head is free to move, but the subject is tasked with actively maintaining a laser (affixed to the forehead) on an earth-fixed target directly ahead of them [91].

Figure 82.2, from Jongkees [47], shows the experimental setup in which the patient's head is maintained in a fixed position (in space) while the body rotates underneath.

Despite the logic of this test, studies showed that it is neither sensitive nor specific for patients whose histories are compatible with CV [92]. Yacovino and Hain comment that this:

> Test of cervical rotation, actually a test of the cervico-ocular reflex, consists of rotating the body about the earth's vertical axis, while keeping the head still in space, and evaluating for nystagmus. This procedure has not been widely accepted… The cervico-ocular reflex also appears in other conditions, such as bilateral vestibular loss thus even if it were sufficiently sensitive, the finding of a cervico-ocular reflex could not be… a specific test for cervical vertigo [7].

Citing Norre [93], Brandt [1] additionally commented that "Cervical nystagmus also occurs in healthy subjects." It further turns out that even in normal subjects it is possible to induce asymmetry in the vestibulo-ocular reflex by passive sustained turning of the head-on-trunk [94], reflecting that this test is not specific for CV.

A more detailed explanation of this from Wrisley et al. [8] is that:

> The neck torsion nystagmus test, or head-fixed, body-turned maneuver is considered by some to identify cervicogenic dizziness [95]. This test requires the head of the patient to be stabilized while the body is rotated underneath [93, 96]. Theoretically, the neck proprioceptors are stimulated while the inner ear structures remain at their resting state [93]. Nystagmus is elicited in a positive test. However, this test has not been demonstrated to be specific for cervicogenic dizziness. Oosterveld et al. [97] reported that 64% of 262 patients with neck pain who presented to an otolaryngology department post-whiplash had nystagmus elicited with the head-fixed, body-turned maneuver. On the other hand, it has been demonstrated that up to 50% of subjects without cervical spine pathology have also demonstrated nystagmus with the head-fixed, body-turned maneuver [93, 95, 98]. A positive response (nystagmus) may not indicate pathology, but may instead be a manifestation of the cervical ocular reflex [93].

Why Has Testing for CV Failed?

On the surface, it appears that it should be experimentally feasible to manipulate the various sensory inputs selectively—in the case of CV, keeping labyrinthine and visual inputs constant while varying cervical proprioceptive input—and yet the failure to develop a test for CV suggests that this task is far from simple. Several investigators have noted this difficulty.

> The neck not only modulates body posture, but it also stabilizes the head in space by cervicocollic reflexes, which are similarly integrated with vestibulocollic reflexes. In healthy human

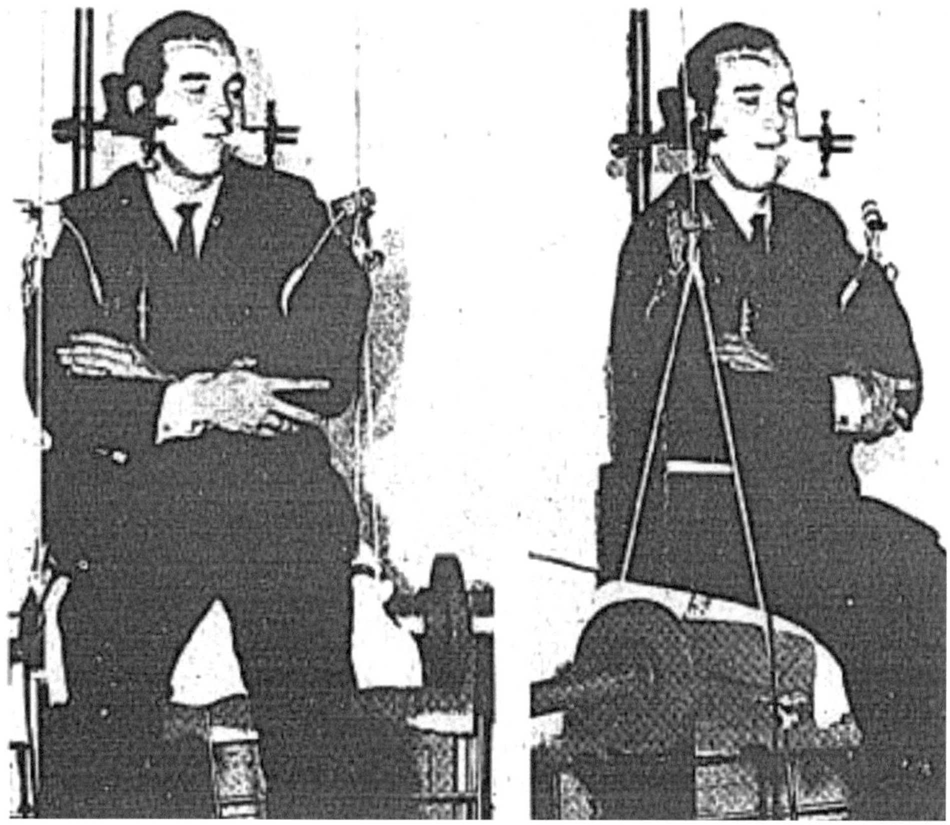

Fig. 82.2 Experimental setup in which the patient's head is maintained in a fixed position (in space) while the body rotates underneath. (From Jongkees [47]. Reproduced with written permission from John Wiley and Sons)

> beings, neck reflexes form part of the multisensory postural control mechanism, thus making it impossible for the clinician to carry out a selective test of neck function by simple postural maneuvers [1].

> Postural control is achieved through a multisensory control mechanism involving visual, vestibular and somatosensory information. These inputs are all interconnected allowing compensation of dysfunctions but making it very challenging to study the cues of one particular system without interference of another [99].

Thus, the fundamental problem is that it is difficult to vary one sensory input in complete isolation of the others.

The vestibular system is built on multimodal inputs, processing, and multimodal outputs, such that there is redundancy. This redundancy has the advantage of making the system more resilient against disease. But the redundancy can be disadvantageous for a patient when intersensory conflicts cannot be effectively reconciled, and it is also problematic for an investigator who is trying to isolate its components from each other (hence the difficulty in developing a test for CV).

If one were to design a test that was sensitive and specific for patients whose histories are compatible with CV, what would that test look like?

A general principle of sensory input is that biological sensors are better at detecting change (dynamic stimulation) than stasis (constant stimulation) [100]. This principle underlies Brandt's idea that "a multisensory mismatch would be expected to result in CV… the resultant mismatch would be maximal during active head movements (when expected and actual reafferent input do not match)" [1], and his suggestion that "Since assessment of all these measures under static conditions has so far proven inconclusive, further investigations should focus on dynamic studies" [1].

Therefore, a test for CV would probably involve dynamic input, and likely require some change in position. This is reinforced by reports (from patients whose histories are compatible with CV) that symptoms are generally more salient during movement.

Dynamic Testing May Be More Sensitive Than Static Testing, but…

Despite the argument just outlined, the fact remains that a test involving positional changes may still stimulate multiple inputs. Indeed, changes in position can trigger vertiginous symptoms in other diseases besides CV, and it is worth spending a moment on how easily one can be misled by this.

Yacovino and Hain [7] pointed out that in the very paper whose influence established the term "cervical vertigo" [9], four of the five cases described have histories that are far more compelling for benign paroxysmal positional vertigo (BPPV). We shall review the cases (#1, #2, #4, #5) from Ryan and Cope [9].

Case 1: "A man, aged 57, had been in excellent health... While he was building a wall a brick had struck him in the head; his neck was flexed at that time and the blow jarred his neck to one side... On lying down that night he had a sudden and fairly severe attack of vertigo—'everything started going round and I felt sick.' He did not vomit, and after a minute or two the vertigo subsided; but the neck pain did not allow him much sleep... He was... placed flat on a couch, and his head and trunk lowered over the end... At once he complained of vertigo, and we noticed that nystagmus appeared in association with this symptom. The vertigo and nystagmus occurred irrespective of whether the head was to the left, central, or to the right, and disappeared if the patient maintained his position for a little time."

Case 2: "A man, aged 60, was involved in an explosion... and he subsequently complained of pain and paresthesiae in the arms... He was subjected to manipulation, and thereafter slow neck traction was applied... Three days later on getting up in the morning he had an acute attack of vertigo and fell to the left. There was no vomiting, sweating, or syncope. On trying to move to the right or the left he experienced vertigo and was unable to walk straight. He stayed in bed for a week and then got up cautiously, finding that the vertigo was brought on only by rapid movement of the head. The symptoms slowly diminished and had disappeared within two weeks."

Case 4: "A married woman, aged 45... fell on to the base of her spine and also injured the back of her head... She... had slight pain in the occipital region. About three weeks after this injury she was wakened in the night by a throbbing sensation in the right side of her neck; and in the morning she sat up in bed, felt her head swimming, and fell to the left. This sensation, which was momentary, was accompanied by slight nausea. About ten minutes later she 'felt fine,' and got up to do her housework. Just before luncheon she looked up at the warming-rack and then bent down to attend to the stove, and immediately the kitchen 'seemed to spin round' to the left, and she fell on her left side."

Case 5: "A married woman, aged 40, slipped and fell on the back of her head... The next day she had a severe pain in the occipital region... Some time later she had a severe attack of vertigo, with a feeling that her environment was rotating in a clockwise direction. Since then the attacks had occurred twice a week, and on two occasions she had fallen to the ground and bruised herself."

In these cases, Ryan and Cope [9] were drawing attention to the mechanism of injury (which features quite prominently in the histories), yet the descriptions of positionally triggered spinning disequilibrium are more compatible with benign paroxysmal positional vertigo (BPPV). We suspect that these cases in which BPPV followed head/neck trauma. Since BPPV is the most common cause of vertigo over the lifespan [101], its occurrence will coincide by chance with the occurrence of other medical problems (head trauma, neck pain, etc.).

When Ryan and Cope published their article in 1995 [9], BPPV had not yet been thoroughly described and was not widely recognized. In 1920, Robert Bárány [102] reported a case of a young woman who had been suffering from vertiginous episodes triggered by lying on her right side. Numerous other reports were subsequently published. In 1952, Margaret Ruth Dix and Charles Skinner Hallpike [103, 104] described their own cases of "positional nystagmus," specifically commenting that symptoms and nystagmus were triggered "by a critical position of the head in space." Ryan and Cope [9] cited one of the 1952 papers by Dix and Hallpike, and therefore must have been aware of their work, but in retrospect it seems that the historical features of head/neck trauma (in the cases reported by Ryan and Cope) were probably red herrings.

Other authors have also noted that "Benign paroxysmal positional vertigo (BPPV) is often misdiagnosed as cervical vertigo" [2]. More broadly, "Lesions of the vestibular organs, particularly the otolithic organs, after whiplash injuries are probably underestimated by attributing dizziness and vertigo symptoms mainly to cervical damage and lesions of the central nervous system" [19]; see also Dispenza et al. [105].

A lesson to be drawn from this is that in a patient whose history of positionally triggered disequilibrium suggests CV, it is still prudent to evaluate for more common diseases, such as benign paroxysmal positional vertigo.

Many Patients Preliminarily Diagnosed with CV Are Found to Have Other Disorders

Mistaking BPPV for CV emphasizes that, when entertaining a diagnosis of CV, one should maintain a broad differential.

Yacovino and Hain comment that, "Many patients preliminarily diagnosed with such a disorder are ultimately found to have other pathologies" [7]. Similarly, Brandt observes, "Reliable and well-established signs and tests can support a convincing alternative diagnosis in almost all patients presenting with vertigo" [1].

These comments serve as a reminder that until a proof-positive diagnostic test for CV is developed, CV will remain a diagnosis of exclusion.

A Diagnosis of Exclusion

Absent a "proof positive" diagnostic test for cervicogenic vertigo, and given the frequent reports that the majority of cases provisionally diagnosed with CV are subsequently found to have a different and objectively verifiable diagnosis, it is unsurprising that reviews of the topic generally conclude that cervicogenic vertigo is a diagnosis of exclusion [2–4, 7, 8].

Early in this discussion, we set forth definitions, conditions, and assumptions for approaching CV. Taking those in conjunction with the "diagnosis of exclusion" feature brings us full circle to Wrisley's description that "the diagnosis of cervicogenic dizziness is suggested by (1) a close temporal relationship between neck discomfort and symptoms of dizziness, including time of onset and occurrence of episodes, (2) previous neck injury or pathology, and (3) elimination of other causes of dizziness" [8].

What Constitutes an Adequate Workup?

If one takes the position that CV is a diagnosis of exclusion, then the next logical question is: which other diagnoses should be excluded? Most authors agree that a thorough history and physical examination are necessary, but the literature provides consensus neither on the other diagnoses that should be excluded, nor on what sort of workup should be undertaken to exclude them.

Absent any consensus, we would suggest that (1) several objectively verifiable otovestibular diseases are very common and sometimes provoke symptoms suggestive of CV, and therefore (2) it is medically reasonable to consider a screening otovestibular workup. Depending on the features of the specific case, such a workup may include:

- Ocular vestibular evoked myogenic potentials and videonystagmography.
- Rotatory chair testing and computerized dynamic posturography may also be appropriate.
- If the patient's neck can tolerate it, cervical vestibular-evoked myogenic potentials and video head impulse testing are reasonable.
- If neck rotation provokes symptoms other than vertigo, or reveals physical examination signs of brainstem dysfunction, then imaging is sensible. At the discretion of the clinician, this could include:
  - Vascular imaging, preferably dynamic, such as CTA, MRA, or dedicated catheter angiography. There is some evidence that transcranial Doppler ultrasound may also have some role [106].
  - Imaging of the bony structures (usually by CT) and soft tissue structures (usually by MRI) of the cervical spine.

Treatment

Despite the controversy regarding CV, clinicians do encounter patients whose histories are compatible with that diagnosis, and thus face decisions on how to manage such cases. Given our relatively poor understanding of CV's pathophysiology, we do not have a solid foundation on which to base treatment recommendations.

Physical Therapy

Physical therapy (PT) for the neck has been studied in the management of CV [79, 107–114]. However, Yacovino and Hain comment that, "Regarding the treatment of proprioceptive cervical vertigo, in which pain and imbalance or vertigo are the limiting symptoms, the quality of published studies in the current literature is poor" [7].

Some literature suggests a multimodal treatment approach that includes physical therapy. For example:

> A combined approach is likely to best address the perpetuation of a vicious cycle of events where secondary adaptive changes in the sensorimotor control system could lead to altered cervical muscle function and joint mechanics further altering cervical afferent input... Physical therapy interventions such as pain management, manipulative therapy, active range-of-motion exercises, and exercises to improve neuromuscular control will all be important in reducing possible causes of altered afferent cervical input and subsequent disturbances to sensorimotor control [3].

Although the symptom of vertigo is necessary to entertain a diagnosis of CV, "Vestibular physical therapy is not a substitute for physical therapy for the neck" [6].

Other potentially useful adjunctive strategies include vision therapy.

Why Does Neck PT Work at All?

Since the underlying pathophysiology of CV is inadequately understood, one could argue that the use of neck PT is treating (partially) blindly. Nevertheless, published studies generally report favorable results of neck PT, so perhaps such an outcome is not merely by chance. But if neck PT is helpful, what is it really doing?

Whether neck pain or vertigo is the initial symptom, either one can provoke and exacerbate the other (as discussed above), leading to a mutually reinforcing cycle. Vertigo from any cause frequently provokes unintentional postural adjustments that predispose to developing neck stiffness [7, 16]. In this circumstance it is easy for a positive feedback loop to ensue. As Wrisley comments, "interconnections between the cervical proprioceptors and the vestibular nuclei may contribute to a cyclic pattern, such that cervical muscle spasms contribute to dizziness and dizziness contributes to muscle spasm" [8].

Irrespective of which symptom (neck pain, vertigo) initiated this vicious cycle, any treatment that reduces neck pain, normalizes cervical muscle tone and improves joint mobility—basically any treatments that normalize cervical mechanics [86, 115]—will interrupt this positive feedback loop, thereby increasing the opportunity for recovery. This underlies Brandt's argument that, "If CV exists, appropriate management is the same as that for the cervical pain syndrome" [1].

Medication

Several authors recommend treating CV with muscle relaxants, though supportive data from randomized controlled trials is lacking. On the theory of "migraine-associated cervicogenic vertigo" discussed above, it is reasonable to consider a trial of migraine prophylaxis. There is modest literature regarding onabotulinum toxin [116] and moxibustion [117].

Surgery and Other Invasive Procedures

Several invasive procedures have been explored as treatment for CV, including surgery for cervical spondylosis or disc herniation [99, 118, 119], percutaneous laser disc decompression [120], radiofrequency ablation nucleoplasty [121], percutaneous cervical nucleoplasty [122], and intervertebral disc replacement [123]. Reported outcomes appear variable. In view of the risks of invasive procedures, and the uncertainty of diagnosis in the first place, we regard invasive procedures as a treatment of last resort.

Alternative Therapies

Dry needling [124] and acupotomy [125] have been explored, with limited data.

References

1. Brandt T. Cervical vertigo–reality or fiction? Audiol Neurootol. 1996;1(4):187–96.
2. Li Y, Peng B. Pathogenesis, diagnosis, and treatment of cervical vertigo. Pain Physician. 2015;18(4):E583–95.
3. Kristjansson E, Treleaven J. Sensorimotor function and dizziness in neck pain: implications for assessment and management. J Orthop Sports Phys Ther. 2009;39(5):364–77.
4. Cherchi M, DiLiberto FE, Yacovino DA, Das S. The enduring controversy of cervicogenic vertigo, and its place among positional vertigo syndromes. Audiol Res. 2021;11(4):491–507.
5. Peng B. Cervical vertigo: historical reviews and advances. World Neurosurg. 2018;109:347–50.
6. Hain TC. Cervicogenic causes of vertigo. Curr Opin Neurol. 2015;28(1):69–73.
7. Yacovino DA, Hain TC. Clinical characteristics of cervicogenic-related dizziness and vertigo. Semin Neurol. 2013;33(3):244–55.
8. Wrisley DM, Sparto PJ, Whitney SL, Furman JM. Cervicogenic dizziness: a review of diagnosis and treatment. J Orthop Sports Phys Ther. 2000;30(12):755–66.
9. Ryan GM, Cope S. Cervical vertigo. Lancet. 1955;269(6905):1355–8.
10. Thompson-Harvey A, Hain TC. Symptoms in cervical vertigo. Laryngoscope Investig Otolaryngol. 2019;4(1):109–15.
11. van Leeuwen RB, van der Zaag-Loonen H. Dizziness and neck pain: a correct diagnosis is required before consulting a physiotherapist. Acta Neurol Belg. 2017;117(1):241–4.
12. Zajacova A, Grol-Prokopczyk H, Zimmer Z. Pain trends among American adults, 2002–2018: patterns, disparities, and correlates. Demography. 2021;58(2):711–38.
13. Rydevik B, Szpalski M, Aebi M, Gunzburg R. Whiplash injuries and associated disorders: new insights into an old problem. Eur Spine J. 2008;17(S3):359–416.
14. Dunlap PM, Khoja SS, Whitney SL, Freburger JK. Assessment of health care utilization for dizziness in ambulatory care settings in the United States. Otol Neurotol. 2019;40(9):e918–e24.
15. Saber Tehrani AS, Coughlan D, Hsieh YH, Mantokoudis G, Korley FK, Kerber KA, et al. Rising annual costs of dizziness presentations to U.S. emergency departments. Acad Emerg Med. 2013;20(7):689–96.
16. Brandt T, Bronstein AM. Cervical vertigo. J Neurol Neurosurg Psychiatry. 2001;71(1):8–12.
17. Mallinson AI, Longridge NS. Dizziness from whiplash and head injury: differences between whiplash and head injury. Am J Otol. 1998;19(6):814–8.
18. Lee JD, Park MK, Lee BD, Park JY, Lee TK, Sung KB. Otolith function in patients with head trauma. Eur Arch Otorrinolaringol. 2011;268(10):1427–30.
19. Vibert D, Hausler R. Acute peripheral vestibular deficits after whiplash injuries. Ann Otol Rhinol Laryngol. 2003;112(3):246–51.
20. Pearce JM. Whiplash injury: a reappraisal. J Neurol Neurosurg Psychiatry. 1989;52(12):1329–31.
21. Walton DM, Elliott JM. An integrated model of chronic whiplash-associated disorder. J Orthop Sports Phys Ther. 2017;47(7):462–71.
22. Gdowski GT, McCrea RA. Integration of vestibular and head movement signals in the vestibular nuclei during whole-body rotation. J Neurophysiol. 1999;82(1):436–49.
23. Mergner T, Huber W, Becker W. Vestibular-neck interaction and transformation of sensory coordinates. J Vestib Res Equilib Orient. 1997;7(4):347–67.
24. Mergner T, Siebold C, Schweigart G, Becker W. Human perception of horizontal trunk and head rotation in space during vestibular and neck stimulation. Exp Brain Res. 1991;85(2):389–404.
25. Kamouchi M, Kishikawa K, Matsuo R, Yasumori K, Inoue T, Okada Y, et al. Ultrasonographic detection of extracranial vertebral artery compression in bow Hunter's brain ischemia caused by neck rotation. Cerebrovasc Dis. 2003;16(3):303–5.
26. Sakaguchi M, Kitagawa K, Hougaku H, Hashimoto H, Nagai Y, Yamagami H, et al. Mechanical compression of the extracranial vertebral artery during neck rotation. Neurology. 2003;61(6):845–7.
27. Weintraub MI, Khoury A. Mechanical compression of the extracranial vertebral artery during neck rotation. Neurology. 2004;62(11):2143; author reply 4.
28. Helton T, Bavry A. Images in cardiovascular medicine. don't turn your head! Circulation. 2009;120:e162.
29. Mitchell J. Vertebral artery blood flow velocity changes associated with cervical spine rotation: a meta-analysis of the evidence with implications for professional practice. J Man Manip Ther. 2009;17(1):46–57.
30. Lu DC, Zador Z, Mummaneni PV, Lawton MT. Rotational vertebral artery occlusion-series of 9 cases. Neurosurgery. 2010;67(4):1066–72; discussion 72.
31. Cornelius JF, George B, N'Dri Oka D, Spiriev T, Steiger HJ, Hanggi D. Bow-Hunter's syndrome caused by dynamic vertebral artery stenosis at the cranio-cervical junction–a management algo-

rithm based on a systematic review and a clinical series. Neurosurg Rev. 2012;35(1):127–35; discussion 35.
32. Go G, Hwang SH, Park IS, Park H. Rotational vertebral artery compression: Bow Hunter's syndrome. J Korean Neurosurg Soc. 2013;54(3):243–5.
33. Sarkar J, Wolfe SQ, Ching BH, Kellicut DC. Bow Hunter's syndrome causing vertebrobasilar insufficiency in a young man with neck muscle hypertrophy. Ann Vasc Surg. 2014;28(4):1032 e1–10.
34. Zaidi HA, Albuquerque FC, Chowdhry SA, Zabramski JM, Ducruet AF, Spetzler RF. Diagnosis and management of Bow Hunter's syndrome: 15-year experience at barrow neurological institute. World Neurosurg. 2014;82(5):733–8.
35. Duan G, Xu J, Shi J, Cao Y. Advances in the pathogenesis, diagnosis and treatment of Bow Hunter's syndrome: a comprehensive review of the literature. Interv Neurol. 2016;5(1–2):29–38.
36. Araz Server E, Edizer DT, Yigit O, Yasak AG, Erdim C. Relationship between vertebral artery blood flow in different head positions and vertigo. Acta Otolaryngol. 2018;138(1):1–5.
37. Li Q, Xie P, Yang WS, Yan B, Davis S, Caplan LR. Vertebral artery compression syndrome. Front Neurol. 2019;10:1075.
38. Vibert D, Rohr-Le Floch J, Gauthier G. Vertigo as manifestation of vertebral artery dissection after chiropractic neck manipulations. ORL J Otorhinolaryngol Relat Spec. 1993;55(3):140–2.
39. Choi KD, Shin HY, Kim JS, Kim SH, Kwon OK, Koo JW, et al. Rotational vertebral artery syndrome: oculographic analysis of nystagmus. Neurology. 2005;65(8):1287–90.
40. Ogawa Y, Itani S, Otsuka K, Inagaki T, Shimizu S, Kondo T, et al. Intermittent positional downbeat nystagmus of cervical origin. Auris Nasus Larynx. 2014;41(2):234–7.
41. Sorensen BF. Bow Hunter's stroke. Neurosurgery. 1978;2(3):259–61.
42. Mitchell J. Doppler insonation of vertebral artery blood flow changes associated with cervical spine rotation: implications for manual therapists. Physiother Theory Pract. 2007;23(6):303–13.
43. Symons BP, Leonard T, Herzog W. Internal forces sustained by the vertebral artery during spinal manipulative therapy. J Manip Physiol Ther. 2002;25(8):504–10.
44. Rousseaux M, Salomez JL, Petit H. Syncopes and transitory neurologic manifestations revealing cervico-occipital joint malformations. Evidence of a vascular mechanism. Sem Hop. 1983;59(11):729–32.
45. Barré J-A. Sur un syndrome sympathique cervical postérieur et sa cause frequente, l'arthrite cervicale. Rev Neurol. 1926;1:1246–8.
46. Lieou Y-C, Barré J-A. Syndrome sympathique cervical postérieur et arthrite chronique de la colonne vertébrale cervicale: (étude clinique et radiologique). Strasbourg: Université de Strasbourg; 1928.
47. Jongkees LB. Cervical vertigo. Laryngoscope. 1969;79(8):1473–84.
48. Foster CA, Jabbour P. Barre-Lieou syndrome and the problem of the obsolete eponym. J Laryngol Otol. 2007;121(7):680–3.
49. Schoon Y, Olde Rikkert MG, Rongen S, Lagro J, Schalk B, Claassen JA. Head turning-induced hypotension in elderly people. PLoS One. 2013;8(8):e72837.
50. Huygen PL, Verhagen WI, Nicolasen MG. Cervico-ocular reflex enhancement in labyrinthine-defective and normal subjects. Exp Brain Res. 1991;87(2):457–64.
51. Kelders WP, Kleinrensink GJ, van der Geest JN, Schipper IB, Feenstra L, De Zeeuw CI, et al. The cervico-ocular reflex is increased in whiplash injury patients. J Neurotrauma. 2005;22(1):133–7.
52. Takemori S, Suzuki JI. Eye deviations from neck torsion in humans. Ann Otol Rhinol Laryngol. 1971;80(3):439–44.
53. Tjell C, Rosenhall U. Smooth pursuit neck torsion test: a specific test for cervical dizziness. Am J Otol. 1998;19(1):76–81.
54. Wapner S, Werner H, Chandler KA. Experiments on sensory-tonic field theory of perception. I. Effect of extraneous stimulation on the visual perception of verticality. J Exp Psychol. 1951;42(5):341–4.
55. Matthews PB. The reflex excitation of the soleus muscle of the decerebrate cat caused by vibbration applied to its tendon. J Physiol. 1966;184(2):450–72.
56. Biguer B, Donaldson IM, Hein A, Jeannerod M. Neck muscle vibration modifies the representation of visual motion and direction in man. Brain J Neurol. 1988;111(Pt 6):1405–24.
57. Treleaven J, Jull G, LowChoy N. The relationship of cervical joint position error to balance and eye movement disturbances in persistent whiplash. Man Ther. 2006;11(2):99–106.
58. L'Heureux-Lebeau B, Godbout A, Berbiche D, Saliba I. Evaluation of paraclinical tests in the diagnosis of cervicogenic dizziness. Otol Neurotol. 2014;35(10):1858–65.
59. Cherchi M, Bellinaso NP, Card K, Covington A, Krumpe A, Pfeifer MS, et al. Sound evoked triceps myogenic potentials. Otol Neurotol. 2009;30(4):545–50.
60. Cherchi M, El-Kholy W, Shirley D, Hain TC. Amplitude of sound evoked triceps myogenic potential scales with force. J Neurol Neurophysiol. 2015;6(2):286.
61. Shirley D, Cherchi M, Hain TC. Triceps acoustically evoked myogenic potentials in patients with spinal cord lesions. J Neurol Neurophysiol. 2015;6(2):288.
62. Bartsch T. Migraine and the neck: new insights from basic data. Curr Pain Headache Rep. 2005;9(3):191–6.
63. Bartsch T, Goadsby PJ. The trigeminocervical complex and migraine: current concepts and synthesis. Curr Pain Headache Rep. 2003;7(5):371–6.
64. Piovesan EJ, Kowacs PA, Oshinsky ML. Convergence of cervical and trigeminal sensory afferents. Curr Pain Headache Rep. 2003;7(5):377–83.
65. Pradhan S, Choudhury SS. Clinical characterization of neck pain in migraine. Neurol India. 2018;66(2):377–84.
66. Wober C, Brannath W, Schmidt K, Kapitan M, Rudel E, Wessely P, et al. Prospective analysis of factors related to migraine attacks: the PAMINA study. Cephalalgia. 2007;27(4):304–14.
67. Neuhauser H, Leopold M, von Brevern M, Arnold G, Lempert T. The interrelations of migraine, vertigo, and migrainous vertigo. Neurology. 2001;56(4):436–41.
68. Beh SC, Masrour S, Smith SV, Friedman DI. The spectrum of vestibular migraine: clinical features, triggers, and examination findings. Headache. 2019;59(5):727–40.
69. Hain TC, Cherchi M. Migraine associated vertigo. In: Lea J, Pothier D, editors. Advances in otorhinolaryngology, vol. 82. 2019th ed; 2019. p. 119–26.
70. Blaschek A, Milde-Busch A, Straube A, Schankin C, Langhagen T, Jahn K, et al. Self-reported muscle pain in adolescents with migraine and tension-type headache. Cephalalgia. 2012;32(3):241–9.
71. Blau JN, MacGregor EA. Migraine and the neck. Headache. 1994;34(2):88–90.
72. Calhoun AH, Ford S, Millen C, Finkel AG, Truong Y, Nie Y. The prevalence of neck pain in migraine. Headache. 2010;50(8):1273–7.
73. Dornhoffer JR, Liu YF, Donaldson L, Rizk HG. Factors implicated in response to treatment/prognosis of vestibular migraine. Eur Arch Otorrinolaringol. 2021;278(1):57–66.
74. Kelman L. Migraine pain location: a tertiary care study of 1283 migraineurs. Headache. 2005;45(8):1038–47.
75. Lampl C, Rudolph M, Deligianni CI, Mitsikostas DD. Neck pain in episodic migraine: premonitory symptom or part of the attack? J Headache Pain. 2015;16:566.
76. Ozer G, Benlier N. Neck pain: is it part of a migraine attack or a trigger before a migraine attack? Acta Neurol Belg. 2020;120(2):289–93.

77. Schoonman GG, Evers DJ, Terwindt GM, van Dijk JG, Ferrari MD. The prevalence of premonitory symptoms in migraine: a questionnaire study in 461 patients. Cephalalgia. 2006;26(10):1209–13.
78. Karlberg M, Johansson R, Magnusson M, Fransson P-A. Dizziness of suspected cervical origin distinguished by posturographic assessment of human postural dynamics. J Vestib Res Equilib Orient. 1996;6:37–47.
79. Karlberg M, Magnusson M, Malmstrom EM, Melander A, Moritz U. Postural and symptomatic improvement after physiotherapy in patients with dizziness of suspected cervical origin. Arch Phys Med Rehabil. 1996;77(9):874–82.
80. Kogler A, Lindfors J, Odkvist LM, Ledin T. Postural stability using different neck positions in normal subjects and patients with neck trauma. Acta Otolaryngol. 2000;120(2):151–5.
81. Koskimies K, Sutinen P, Aalto H, Starck J, Toppila E, Hirvonen T, et al. Postural stability, neck proprioception and tension neck. Acta Otolaryngol Suppl. 1997;529:95–7.
82. Micarelli A, Viziano A, Augimeri I, Micarelli B, Capoccia D, Alessandrini M. Diagnostic route of cervicogenic dizziness: usefulness of posturography, objective and subjective testing implementation and their correlation. Disabil Rehabil. 2021;43(12):1730–7.
83. Norre ME, Forrez G, Stevens A, Beckers A. Cervical vertigo diagnosed by posturography? Preliminary report. Acta Otorhinolaryngol Belg. 1987;41(4):574–81.
84. Rubin AM, Woolley SM, Dailey VM, Goebel JA. Postural stability following mild head or whiplash injuries. Am J Otol. 1995;16(2):216–21.
85. Williams K, Tarmizi A, Treleaven J. Use of neck torsion as a specific test of neck related postural instability. Musculoskelet Sci Pract. 2017;29:115–9.
86. Johnston JL, Daye PM, Thomson GT. Inaccurate saccades and enhanced vestibulo-ocular reflex suppression during combined eye-head movements in patients with chronic neck pain: possible implications for cervical vertigo. Front Neurol. 2017;8:23.
87. Toglia JU. Acute flexion-extension injury of the neck. Electronystagmographic study of 309 patients. Neurology. 1976;26(9):808–14.
88. Gimse R, Tjell C, Bjorgen IA, Saunte C. Disturbed eye movements after whiplash due to injuries to the posture control system. J Clin Exp Neuropsychol. 1996;18(2):178–86.
89. Karlberg M, Magnusson M. Asymmetric optokinetic afternystagmus induced by active or passive sustained head rotations. Acta Otolaryngol. 1996;116(5):647–51.
90. de Jong JM, Bles W, Bovenkerk G. Nystagmus, gaze shift, and self-motion perception during sinusoidal head and neck rotation. Ann N Y Acad Sci. 1981;374:590–9.
91. Guitton D, Kearney RE, Wereley N, Peterson BW. Visual, vestibular and voluntary contributions to human head stabilization. Exp Brain Res. 1986;64(1):59–69.
92. Treleaven J, Joloud V, Nevo Y, Radcliffe C, Ryder M. Normative responses to clinical tests for cervicogenic dizziness: clinical cervical torsion test and head-neck differentiation test. Phys Ther. 2020;100(1):192–200.
93. Norre ME. Cervical vertigo. Diagnostic and semiological problem with special emphasis upon “cervical nystagmus”. Acta Otorhinolaryngol Belg. 1987;41(3):436–52.
94. Padoan S, Karlberg M, Fransson PA, Magnusson M. Passive sustained turning of the head induces asymmetric gain of the vestibulo-ocular reflex in healthy subjects. Acta Otolaryngol. 1998;118(6):778–82.
95. Philipszoon AJ, Bos JH. Neck torsion nystagmus. Pract Otorhinolaryngol (Basel). 1963;25:339–44.
96. Fitz-Ritson D. Assessment of cervicogenic vertigo. J Manip Physiol Ther. 1991;14(3):193–8.
97. Oosterveld WJ, Kortschot HW, Kingma GG, de Jong HA, Saatci MR. Electronystagmographic findings following cervical whiplash injuries. Acta Otolaryngol. 1991;111(2):201–5.
98. Calseyde P, Ampe W, Depondt M. E.N.G. and the cervical syndrome neck torsion nystagmus. Adv Otorhinolaryngol. 1977;22:119–24.
99. Freppel S, Bisdorff A, Colnat-Coulbois S, Ceyte H, Cian C, Gauchard G, et al. Visuo-proprioceptive interactions in degenerative cervical spine diseases requiring surgery. Neuroscience. 2013;255:226–32.
100. Gray JA, Malcolm JL. The initiation of nerve impulses by mesenteric Pacinian corpuscles. Proc R Soc Lond B Biol Sci. 1950;137(886):96–114.
101. von Brevern M, Radtke A, Lezius F, Feldmann M, Ziese T, Lempert T, et al. Epidemiology of benign paroxysmal positional vertigo: a population based study. J Neurol Neurosurg Psychiatry. 2007;78(7):710–5.
102. Bárány E. Diagnose von Krankheitserscheinungen im Bereiche des Otolithenapparates [Diagnosis of symptoms in the area of the otolith apparatus]. Acta Otolaryngol. 1920;2(3):434–7.
103. Dix MR, Hallpike CS. The pathology symptomatology and diagnosis of certain common disorders of the vestibular system. Proc R Soc Med. 1952;45(6):341–54.
104. Dix MR, Hallpike CS. The pathology, symptomatology and diagnosis of certain common disorders of the vestibular system. Ann Otol Rhinol Laryngol. 1952;61(4):987–1016.
105. Dispenza F, De Stefano A, Mathur N, Croce A, Gallina S. Benign paroxysmal positional vertigo following whiplash injury: a myth or a reality? Am J Otolaryngol. 2011;32(5):376–80.
106. Simon H, Niederkorn K, Horner S, Duft M, Schrockenfuchs M. Effect of head rotation on the vertebrobasilar system. A transcranial Doppler ultrasound contribution to the physiology. HNO. 1994;42(10):614–8.
107. Alqahtani M, Kashoo F. Physical therapy in cervicogenic dizziness. Saudi J Health Sci. 2020;9:1–6.
108. Borg-Stein J, Rauch S, Krabak B. Evaluation and management of cervicogenic dizziness. Crit Rev Phys Rehabil Med. 2001;13:10.
109. Lystad RP, Bell G, Bonnevie-Svendsen M, Carter CV. Manual therapy with and without vestibular rehabilitation for cervicogenic dizziness: a systematic review. Chiropr Man Therap. 2011;19(1):21.
110. Reid SA, Callister R, Katekar MG, Rivett DA. Effects of cervical spine manual therapy on range of motion, head repositioning, and balance in participants with cervicogenic dizziness: a randomized controlled trial. Arch Phys Med Rehabil. 2014;95(9):1603–12.
111. Reid SA, Rivett DA. Manual therapy treatment of cervicogenic dizziness: a systematic review. Man Ther. 2005;10(1):4–13.
112. Yaseen K, Hendrick P, Ismail A, Felemban M, Alshehri MA. The effectiveness of manual therapy in treating cervicogenic dizziness: a systematic review. J Phys Ther Sci. 2018;30(1):96–102.
113. Yao M, Tang ZY, Cui XJ, Sun YL, Ye XL, Wang P, et al. Shi-style cervical mobilizations versus massage for cervical vertigo: a multicenter, randomized, controlled clinical trial. J Altern Complement Med. 2020;26(1):58–66.
114. Hoppes CW, Romanello AJ, Gaudette KE, Herron WK, McCarthy AE, McHale CJ, et al. Physical therapy interventions for cervicogenic dizziness in a military-aged population: protocol for a systematic review. Syst Rev. 2020;9(1):62.
115. Williams G, Sarig-Bahat H, Williams K, Tyrrell R, Treleaven J. Cervical kinematics in patients with vestibular pathology vs. patients with neck pain: a pilot study. J Vestib Res Equilib Orient. 2017;27(2–3):137–45.
116. Odderson IR. A new treatment for cervical vertigo with botulinum toxin. J Clin Neurosci. 2020;73:316–7.

117. Li H, Yu T, Cheng P, Qin S, Jiao L, Chen R. Moxibustion for cervical vertigo: a protocol for a systematic review and meta-analysis. Medicine (Baltimore). 2020;99(31):e21405.
118. Hahn T, Halatsch ME, Wirtz C, Klessinger S. Response to cervical medial branch blocks in patients with cervicogenic vertigo. Pain Physician. 2018;21(3):285–94.
119. Zhu X, Grover MJ. Cervicogenic dizziness successfully treated with upper cervical medial branch nerve radiofrequency ablation: a case report. A A Pract. 2018;10(6):150–3.
120. Ren L, Guo B, Zhang J, Han Z, Zhang T, Bai Q, et al. Mid-term efficacy of percutaneous laser disc decompression for treatment of cervical vertigo. Eur J Orthop Surg Traumatol. 2014;24(Suppl 1):S153–8.
121. Yin HD, Zhang XM, Huang MG, Chen W, Song Y, Du QJ, et al. Curative effect and mechanism of radiofrequency ablation nucleoplasty in the treatment of cervical vertigo. Br J Radiol. 2017;90(1072):20150772.
122. Li C, Qi Y, Liu G, Yin X, Jin Y, Jiang Z, et al. Long-term clinical outcomes of percutaneous cervical nucleoplasty for cervical degenerative diseases with neck pain and cervical vertigo. World Neurosurg. 2020;133:e205–e10.
123. Zheng S, Muheremu A, Sun Y, Tian W, Wu CA. Preoperative imaging differences of patients with cervical spondylosis with cervical vertigo indicate the prognosis after cervical total disc replacement. J Int Med Res. 2020;48(2):300060519877033.
124. Escaloni J, Butts R, Dunning J. The use of dry needling as a diagnostic tool and clinical treatment for cervicogenic dizziness: a narrative review & case series. J Bodyw Mov Ther. 2018;22(4):947–55.
125. Xie R, You J, Liu L, Huang C, Liang Y. Acupotomy therapy for cervical vertigo: a protocol for a systematic review and meta-analysis. Medicine (Baltimore). 2020;99(29):e20662.

83 Spinal Cord Disorders

Overview

Somatosensory input and motor output are crucial in maintaining an upright posture and locomotion. When spinal cord damage involves afferent tracts (somatosensory input) and efferent tracts (motor output), it is unsurprising that a patient may experience disequilibrium.

That being said, it is very uncommon for a patient affected exclusively by a spinal cord problem to present to an otoneurology practice. There are probably several reasons for this. First, many of the acute pathologies are so severe (e.g., spinal cord section from penetrating injury) that symptoms other than disequilibrium dominate the clinical picture, and such patients are usually under the acute care of other subspecialties (e.g., neurosurgery). Second, the more chronic diseases (e.g., infection, vitamin deficiency) are usually detected and treated by a general neurologist.

Generally, there are two circumstances in which a patient (ultimately found to have exclusively a spinal cord problem) is referred to an otoneurologist.

The first circumstance is when a patient has already been diagnosed with a spinal cord problem. Usually, the practitioners who arrived at the diagnosis are perfectly aware that a spinal cord problem can interfere with a person's equilibrium, and in this circumstance, referral to otoneurology is usually motivated by the question, "Is there anything else going on *besides* the already known spinal cord problem that could interfere with this patient's balance?" In this situation, doing screening testing for otovestibular disease (vestibular evoked myogenic potentials, video head impulse testing, videonystagmography, rotatory chair testing) is reasonable, since having a spinal cord problem does not preclude a patient's getting more common causes of disequilibrium (such as vestibular neuritis).

The second circumstance, which is rare, is when an undiagnosed patient with disequilibrium presents to an otoneurologist and has no detectable pathology above the foramen magnum. In this circumstance, it is reasonable to keep a spinal cord disorder on the differential diagnosis. In some cases, physical examination (e.g., dissociated hemisensory loss below a cord level) or history (e.g., radiation portal for cancer that includes the spinal cord) may narrow the differential diagnosis significantly. In other cases, these are still insufficient, and a reasonable workup may include:

- MRI without and with contrast of the entire spinal column. This is to catch disease such as a syrinx, demyelination (multiple sclerosis), tumor (spinal column meningioma, intramedullary tumor), compression (e.g., intervertebral disc, aneurysm), and so forth.
- Serological workup as appropriate, which may include tests for neurosyphilis (RPR, FTA), vitamin B12, copper, HTLV1, HIV, poliovirus, and West Nile virus.

Introduction

In a patient whose sole complaint is disequilibrium, it is unusual for the etiology to be an exclusively spinal cord problem.

For purposes of the discussion below, we exclude cervicogenic vertigo, which is addressed elsewhere.

Anatomy

Cho [1] offers a good summary of spinal cord anatomy.

Pathophysiology

Although any part of the spinal cord can be damaged, there are some general patterns that may be helpful in diagnosis. A good review of these syndromes, along with anatomical figures, can be found in Cho [1] (Table 83.1).

M. Cherchi, *Otoneurology and Vestibular Medicine*, https://doi.org/10.1007/978-3-031-94841-1_83

Table 83.1 Spinal cord syndromes, neuroanatomical pathways affected, resulting deficits, and etiologies

Syndrome	Neuroanatomical pathway(s) affected	Resulting deficit	Etiologies
Hemicord (Brown–Sequard) syndrome	1. Deficit in lateral corticospinal tract and rubrospinal tract. 2. Deficit in posterior column system. 3. Deficit in spinothalamic tract.	1. Ipsilateral upper motor neuron pattern weakness at and below affected level. 2. Ipsilateral vibration and position sensation loss below level. 3. Contralateral pain and temperature sensation loss below level.	Multiple sclerosis, varicella-zoster virus, penetrating trauma, asymmetric compression.
Central cord syndrome, small lesion	Deficit in anterior commissure.	Suspended pain and temperature sensation sensory level.	Syringomyelia, intramedullary tumor, hyperextension injury, neuromyelitis optica.
Central cord syndrome, small lesion	1. Deficit in lateral corticospinal tract and rubrospinal tract. 2. Deficit in lateral corticospinal tract. 3. Deficit in anterolateral system.	1. Segmental lower motor neuron pattern weakness at level. 2. Bilateral upper motor neuron pattern weakness below level (arms worse than legs). 3. Bilateral pain and temperature sensation loss below level (sacral sparing).	Syringomyelia, intramedullary tumor, hyperextension injury, neuromyelitis optica.
Posterior column syndrome	1. Deficit in posterior column system. 2. Dorsal roots.	1. Bilateral vibration and position sensation loss below level. 2. Absent reflexes (knees, ankles) when dorsal roots are involved, sensory ataxia, Lhermitte sign (if cervical).	Tabes dorsalis, cervical spondylotic myelopathy, radiation-induced myelopathy, early delayed radiation-induced myelopathy, posterior spinal artery infarction.
Posterolateral column syndrome	1. Deficit of posterior column system. 2. Deficit of lateral corticospinal tract. 3. Deficit of descending autonomic fibers.	1. Bilateral vibration and position sensation loss below level. 2. Bilateral upper motor neuron pattern of weakness below level. 3. Spastic bladder (if lesion proceeds more anteriorly). 4. Sensory ataxia, spastic gait.	Vitamin B12 deficiency (subacute combined degeneration), copper deficiency, cervical spondylotic myelopathy, HTLV1-associated myelopathy/tropical spastic paraparesis, HIV, paraneoplastic myelitis, hereditary spastic paraplegia.
Anterior horn syndrome	Deficit of lower motor neurons in the anterior horn.	Diffuse lower motor neuron pattern weakness.	Poliovirus, West Nile virus, spinal muscular atrophy, spinobulbar muscular atrophy, paraneoplastic subacute motor neuronopathy.
Combined anterior horn and corticospinal tract syndrome	1. Deficit in corticospinal tract. 2. Deficit in anterior horn.	1. Diffuse upper motor neuron pattern weakness. 2. Diffuse lower motor neuron pattern weakness. 3. Bulbar weakness. 4. Pure motor syndrome (because no afferent pathways are affected).	Cervical radiculomyelopathy, amyotrophic lateral sclerosis.
Anterior cord syndrome	1. Deficit of lower motor neurons (in gray matter). 2. Deficit of lateral corticospinal tract and rubrospinal tract. 3. Deficit of anterolateral system. 4. Deficit of descending autonomic fibers. 5. Preservation of posterior column system.	1. Bilateral lower motor neuron pattern weakness at all levels affected. 2. Bilateral upper motor neuron pattern weakness below. 3. Bilateral pain and temperature sensation loss below level. 4. Autonomic function loss below level. 5. Bilateral vibration and position sensation spared.	Poliovirus, West Nile virus, Anterior spinal artery infarction.

Treatment and Prognosis

Treatment and prognosis depend on the underlying spinal cord disorder.

Reference

1. Cho TA. Spinal cord functional anatomy. Continuum (Minneap Minn). 2015;21:13–35. https://doi.org/10.1212/01.CON.0000461082.25876.4a.

84 Peripheral Neuropathy

Overview

Somatosensory input is one of the three main afferent systems (the other two being vision and labyrinthine input) on which the brain depends for vestibular function. One would think that isolated peripheral neuropathy (which impairs somatosensory input) could cause disequilibrium, but this is difficult to study because peripheral neuropathy commonly occurs in the broader context of systemic diseases, including diseases that can interfere with vestibular function through other mechanisms (besides peripheral neuropathy); thus, it is difficult to isolate the effects of peripheral neuropathy alone.

Introduction

Somatosensory input is one of the three main afferent systems (the other two being vision and labyrinthine input) on which the brain depends for calculating one's position in, and movement through, space. Reducing somatosensory input deprives the brain of information it needs to make these calculations, so intuitively one would expect that equilibrium should deteriorate. *Peripheral neuropathy* is one mechanism of reduced somatosensory input.

The Problem

While this seems intuitive, it is somewhat difficult to disentangle *peripheral neuropathy* from other diseases because peripheral somatosensory neuropathy does not commonly occur in complete isolation from other diseases. There may be several reasons for this. The first reason is that although the vestibular nerve is a cranial nerve, cranial nerves 3–12 share many physiologic characteristics with peripheral nerves [7] and are often regarded as such; thus, one could reasonably expect that disease processes that cause peripheral neuropathy could also cause vestibular neuropathy. The second reason is that peripheral neuropathies often (not always) occur in the context of some other disease (typically a systemic disease), a disease which may interfere with equilibrium in other ways; for example, diabetes can cause peripheral somatosensory neuropathy (resulting in sensory ataxia), diabetic autonomic neuropathy (resulting in orthostatic hypotension), and diabetic retinopathy (interfering with visual input), all of which are deficits that can interfere with a person's equilibrium.

As an example of the difficulty of separating peripheral neuropathy from other diseases that can interfere with equilibrium, consider that:

- Palla and colleagues [5] showed that in patients with electrophysiologically proven axonal and demyelinating peripheral neuropathy, there was a statistically significant reduction of the vestibulo-ocular reflex gain (as assessed through head impulse testing with scleral search coils) compared to healthy controls.
- Conversely, investigators have also observed [1] that in patients with vestibular weakness, particularly when bilateral [8], there is a greater than chance association with peripheral neuropathy.

In short, patients with diagnosed peripheral neuropathy seem to have a higher rate of vestibular weakness, and patients with vestibular weakness seem to have a higher rate of peripheral neuropathy.

There are two consequences of the bidirectional association between peripheral neuropathy and vestibular weakness.

First, in a patient with a confirmed diagnosis of peripheral neuropathy whose disequilibrium seems out of proportion to the sequelae of the neuropathy alone, it is reasonable to do screening vestibular testing for vestibular weakness, because if discovered, then intervention with vestibular rehabilitation therapy is appropriate.

Second, in a patient with confirmed vestibular weakness—especially if bilateral [8]—who is not responding to appropriate vestibular rehabilitation therapy, or who has clinical

M. Cherchi, *Otoneurology and Vestibular Medicine*, https://doi.org/10.1007/978-3-031-94841-1_84

evidence of peripheral neuropathy (e.g., distal sensory deficits; muscle atrophy; reduced or absent myotatic reflexes; reduced sweating), it is reasonable to undertake a screening workup for peripheral neuropathy, or refer to a peripheral neuropathy clinic. The American Academy of Neurology's practice parameter for evaluation of distal symmetric polyneuropathy [2], which was reaffirmed on 1/22/2022, suggests that it is reasonable to check serum glucose, vitamin B12 (with methylmalonic acid with or without homocysteine), and serum protein immunofixation electrophoresis.

We have been discussing the afferent (sensory) component of neuropathy, but of course, neuropathies can also affect motor function. Hamel and Logigian [3] studied 40 patients with acute nutritional neuropathy (deficiencies of vitamin B1, vitamin B6, folate, vitamin E, copper, and vitamin B12); these patients had either pure sensory neuropathy (14/40 = 35%), pure motor neuropathy (3/40 = 7.5%), or mixed sensorimotor neuropathy (23/40 = 56%); of the 37 patients who followed-up, only 13 (35%) were walking independently; while the patients with sensory deficits likely had some component of sensory ataxia, those with only motor deficits still had impaired ambulation.

What Studies Suggest

Despite the difficulty of disentangling peripheral neuropathy from other diseases, there have been studies claiming to evaluate the sequelae of isolated peripheral neuropathy with respect to balance. The usual theme of these studies is fall risk, which is a reasonable, though imperfect, proxy for disequilibrium. For example:

- Homann and colleagues [4] studied fall risk in an older population of 228 patients (age 74.5 ± 7.8 years). Of patients with peripheral neuropathy (age 71.0 ± 8.1 years), the odds ratio of falling compared to healthy controls was 6.4 (confidence interval 3.4 to 12.3, $p = 0.000$), and 43% of these patients suffered multiple falls.
- Richardson and colleagues [6] studied 25 patients with axonal peripheral polyneuropathy compared to age- and sex-matched healthy controls and found that peripheral neuropathy was significantly associated with falling (14/25 = 56%) compared to controls (2/25 = 8%), with an odds ratio of 23.0. They also found that the neuropathy was significantly associated with repetitive falling (10/25 = 40%) compared to controls (1/25 = 4%), with an odds ratio of 6.3.

In other words, these studies claim to show that the presence of peripheral neuropathy alone is associated with a statistically significant increase in fall risk, even though the quantification of that risk varies among studies.

Taking such associations into account, Buetti and colleagues [1] note that:

> Vestibular involvement is a common and consistent finding in many peripheral neuropathies of different etiologies and should be specifically sought when there is any suggestion that unsteadiness/ataxia is out of proportion to objective signs of somatosensory loss or muscle weakness [1].

They further state that:

> In terms of clinical management, balance depends on visual, vestibular, and proprioceptive inputs and, in cases of peripheral neuropathy, imbalance is commonly attributed to proprioceptive/somatosensory/cerebellar loss. However, in the patient with peripheral neuropathy and imbalance, all three systems should be investigated. This is particularly important if the severity of imbalance appears to be out of proportion to the degree of neuropathy, as vestibular involvement may be the main cause of the unsteadiness, although complicated by the loss of somatosensory inputs and/ or associated cerebellar dysfunction [1].

They conclude that:

> From a patient perspective, imbalance may be improved with appropriate physiotherapy, which facilitates symptomatic compensation and adaptive plasticity [1].

Diagnosis

When a patient with an already diagnosed peripheral neuropathy is referred to a vestibular medicine clinic, it is medically reasonable to undertake a screening otovestibular workup to evaluate for primary vestibular disease (such as vestibular weakness) and, if identified, treat accordingly.

If an undiagnosed patient is referred to vestibular medicine and the otovestibular workup is unremarkable, or if the workup is suggestive of peripheral neuropathy (such as delayed latencies on the motor control testing component of computerized dynamic posturography), or if physical examination is suggestive of peripheral neuropathy (e.g., diminished sensation or reflexes in the extremities; high pedal arches, etc.), then referral to a specialist in peripheral neuropathies or neuromuscular medicine is reasonable.

Prognosis

The prognosis depends on the underlying diagnosis or diagnoses.

References

1. Buetti B, Luxon LM. Vestibular involvement in peripheral neuropathy: a review. Int J Audiol. 2014;53:353–9. https://doi.org/10.3109/14992027.2014.885121.
2. England JD, Gronseth GS, Franklin G, Carter GT, Kinsella LJ, Cohen JA, Asbury AK, Szigeti K, Lupski JR, Latov N, Lewis RA,

Low PA, Fisher MA, Herrmann DN, Howard JF Jr, Lauria G, Miller RG, Polydefkis M, Sumner AJ, American Academy of Neurology. Practice Parameter: evaluation of distal symmetric polyneuropathy: role of autonomic testing, nerve biopsy, and skin biopsy (an evidence-based review). Report of the American Academy of Neurology, American Association of Neuromuscular and Electrodiagnostic Medicine, and American Academy of Physical Medicine and Rehabilitation. Neurology. 2009;72:177–84. https://doi.org/10.1212/01.wnl.0000336345.70511.0f.

3. Hamel JI, Logigian EL. Clinical spectrum and prognosis in patients with acute nutritional axonal neuropathy. Neurology. 2023;100:e2134–40. https://doi.org/10.1212/WNL.0000000000207215.
4. Homann B, Plaschg A, Grundner M, Haubenhofer A, Griedl T, Ivanic G, Hofer E, Fazekas F, Homann CN. The impact of neurological disorders on the risk for falls in the community dwelling elderly: a case-controlled study. BMJ Open. 2013;3:e003367. https://doi.org/10.1136/bmjopen-2013-003367.
5. Palla A, Schmid-Priscoveanu A, Studer A, Hess K, Straumann D. Deficient high-acceleration vestibular function in patients with polyneuropathy. Neurology. 2009;72:2009–13. https://doi.org/10.1212/WNL.0b013e3181a92b7e.
6. Richardson JK, Ching C, Hurvitz EA. The relationship between electromyographically documented peripheral neuropathy and falls. J Am Geriatr Soc. 1992;40:1008–12. https://doi.org/10.1111/j.1532-5415.1992.tb04477.x.
7. Rueda-Lopes F. The cranial nerves: extensions of the central nervous system or components of the peripheral nervous system – how should we evaluate them? Radiol Bras. 2021;54:V–VI. https://doi.org/10.1590/0100-3984.2021.54.3e1.
8. Zingler VC, Cnyrim C, Jahn K, Weintz E, Fernbacher J, Frenzel C, Brandt T, Strupp M. Causative factors and epidemiology of bilateral vestibulopathy in 255 patients. Ann Neurol. 2007;61:524–32. https://doi.org/10.1002/ana.21105.

Charcot-Marie-Tooth Disease

85

Introduction

Charcot-Marie-Tooth (CMT) disease is a heterogeneous group of genetically mediated peripheral sensorimotor neuropathies.

This taxonomy of neuropathies is eponymously designated in recognition of the work of Jean-Martin Charcot and Pierre Marie [1], and Howard Henry Tooth [2], who described it nearly simultaneously. However, as various historians of medicine have commented [3, 4], these were preceded by numerous earlier descriptions [5–7].

Epidemiology

Taking all of the subtypes together, CMT has a reported prevalence of 1 in 2500 people [8]. CMT comprises the most common hereditary neuropathy [8].

Genetics

Several dozen neuropathies are classified under taxonomy of CMT. Different subtypes of CMT have different modes of inheritance. The mutated genes encode proteins expressed in axonal structures, myelin, and gap junctions.

Pathophysiological Mechanism of Disease

The genetic mutations interfere with proteins whose structural and functional abnormalities result in various combinations of axonal and demyelinating features.

CMT type 2J has been reported to cause autonomic (specifically parasympathetic) neuropathy [9].

CMT type 4C has been reported to cause cerebellar dysfunction [10].

There is evidence that some subtypes of CMT can involve cranial nerves [11]. Perhaps unsurprisingly, this can also affect the vestibulocochlear nerve.

Vestibular deficits have been reported in several types of CMT [12–14].

These features suggest that CMT can cause disequilibrium through several mechanisms.

- Perhaps, the most obvious mechanism is the peripheral neuropathy. The sensory neuropathy may manifest as a sensory ataxia. The motor neuropathy can cause weakness.
- Vestibular weakness (from affectation of the vestibular nerves) can also contribute significantly to the symptom of disequilibrium.
- Cerebellar dysfunction (in CMT type 4C) may contribute to ataxia.
- Autonomic dysfunction (in CMT type 2J) may provoke orthostatic intolerance.

Sensorineural hearing loss has been reported in association with numerous CMT subtypes [15–33].

Clinical Presentation

Sensorineural hearing loss can occur in many subtypes of CMT. This can be an early clinical feature, appearing in childhood [19, 22, 30, 31], sometimes in infancy [18, 29].

In most cases of CMT with sensorineural hearing loss, the deficit appears insidiously, is slowly progressive, and is approximately symmetrical. Rarely, the hearing loss is sudden [25] or asymmetrical [29].

M. Cherchi, *Otoneurology and Vestibular Medicine*, https://doi.org/10.1007/978-3-031-94841-1_85

Physical Examination

Patients with CMT eventually develop muscular atrophy in the legs. Station often exhibits a wide base, and gait may have a "slapping" quality.

Ocular Motor Examination

There is no literature describing bedside ocular motor findings that are sensitive or specific for CMT.

Testing: Auditory

In patients with CMT who have hearing loss, the majority appear to have bilateral sensorineural hearing loss on standard audiometry.

In some cases, auditory brainstem evoked responses may not be obtainable [23].

There is rarely reported evidence of central auditory deficits in CMT [24].

Testing: Vestibular

Various CMT subtypes have been reported to exhibit vestibular weakness on video head impulse testing [12], caloric testing [13], and/or cervical vestibular evoked myogenic potentials [14].

Testing: Other

Electromyography and nerve conduction velocities in CMT patients show various combinations of axonal and demyelinating neuropathic features.

Imaging

Imaging plays little role in CMT except to help exclude competing diagnoses.

Histopathology

A temporal bone study of a patient with CMT type 1B reported pathologic myelin of the vestibulocochlear nerves [34].

Differential Diagnosis

The usual scenario in clinical practice is that a patient already diagnosed with CMT is referred to an otoneurology or neurotology clinic for a complaint of disequilibrium, and the implicit question is whether the CMT alone is responsible for that symptom.

Although CMT in isolation can cause disequilibrium, these patients are just as susceptible to more common vestibular disorders, such as benign paroxysmal positional vertigo, vestibular neuritis, etc., so it is reasonable to undertake a screening workup, seeking treatable causes of this symptom.

It is uncommon for an otoneurologist or neurotologist to diagnose CMT. However, if an undiagnosed patient presents to an otoneurology or neurotology clinic and physical examination is compatible with peripheral sensorimotor neuropathy, then referral to a neuromuscular specialist is appropriate, particularly if there is a family history suggestive of peripheral neuropathy.

Treatment

There is not yet any curative or arrestive therapy for CMT.

In patients with CMT found to have vestibular weakness, a trial of vestibular rehabilitation therapy is reasonable.

Prognosis

The different subtypes of CMT have different tempos of progression.

References

1. Charcot J-M, Marie P. Sur une forme particulière d'atrophie musculaire progressive, souvent familiale débutant par les pieds et les jambes et atteignant plus tard les mains. Rev Méd. 1886;6:97–138.
2. Tooth HH. The peroneal type of progressive muscular atrophy. London: H.K. Lewis; 1886.
3. Brody LA, Wilkins RH. Charcot-Marie-Tooth disease. Arch Neurol. 1967;17(5):552–7.
4. Kazamel M, Boes CJ. Charcot Marie Tooth disease (CMT): historical perspectives and evolution. J Neurol. 2015;262(4):801–5.
5. Virchow R. Ein Fall von progressiver Muskelatrophie. Arch Pathol Anat Physiol Klin Med. 1855;8(4):537–40.
6. Eichhorst H. Ueber Heredität der progressiven mukelatrophie. Berl Klin Wochenschr. 1873;10:497–9.
7. Schulze. Ueber eine Eigenthümliche Progressive Atrophische Paralyse bei Mehreren Kindern derselben Familie. Berlin Klin Wschr. 1884;21:649–51.

8. Barreto LC, Oliveira FS, Nunes PS, de Franca Costa IM, Garcez CA, Goes GM, et al. Epidemiologic study of Charcot-Marie-Tooth disease: a systematic review. Neuroepidemiology. 2016;46(3):157–65.
9. Tokuda N, Noto Y, Kitani-Morii F, Hamano A, Kasai T, Shiga K, et al. Parasympathetic dominant autonomic dysfunction in Charcot-Marie-Tooth disease type 2J with the MPZ Thr124Met mutation. Intern Med. 2015;54(15):1919–22.
10. Skott H, Muntean-Firanescu C, Samuelsson K, Verrecchia L, Svenningsson P, Malmgren H, et al. The cerebellar phenotype of Charcot-Marie-Tooth neuropathy type 4C. Cerebellum Ataxias. 2019;6:9.
11. Pareyson D, Taroni F, Botti S, Morbin M, Baratta S, Lauria G, et al. Cranial nerve involvement in CMT disease type 1 due to early growth response 2 gene mutation. Neurology. 2000;54(8):1696–8.
12. Akdal G, Koçoğlu K, Tanrıverdizade T, Bora E, Bademkıran F, Yüceyar AN, et al. Vestibular impairment in Charcot-Marie-Tooth disease. J Neurol. 2021;268(2):526–31.
13. Pérez-Garrigues H, Sivera R, Vílchez JJ, Espinós C, Palau F, Sevilla T. Vestibular impairment in Charcot-Marie-Tooth disease type 4C. J Neurol Neurosurg Psychiatry. 2014;85(7):824–7.
14. Poretti A, Palla A, Tarnutzer AA, Petersen JA, Weber KP, Straumann D, et al. Vestibular impairment in patients with Charcot-Marie-tooth disease. Neurology. 2013;80(23):2099–105.
15. Alcin B, Vatovec J, Zargi M. Pure tone audiogram and speech audiometry in patients with hereditary motor and sensory neuropathy. Pflugers Arch. 2000;439(3 Suppl):R202–3.
16. Cornell J, Sellars S, Beighton P. Autosomal recessive inheritance of Charcot-Marie-Tooth disease associated with sensorineural deafness. Clin Genet. 1984;25(2):163–5.
17. Duan X, Gu W, Hao Y, Wang R, Wen H, Sun S, et al. A novel Asp121Asn mutation of myelin protein zero is associated with late-onset axonal Charcot-Marie-Tooth disease, hearing loss and pupil abnormalities. Front Aging Neurosci. 2016;8:222.
18. Hamiel OP, Raas-Rothschild A, Upadhyaya M, Frydman M, Sarova-Pinhas I, Brand N, et al. Hereditary motor-sensory neuropathy (Charcot-Marie-Tooth disease) with nerve deafness: a new variant. J Pediatr. 1993;123(3):431–4.
19. Hamsho S, Almohamed A, Haydar H, Alsaffaf Y, Sultan E, Sukkar Y, et al. Bilateral hearing impairment as an early symptom in a patient with Charcot-Marie-Tooth Type 1: the first case report from Syria. Ann Med Surg (Lond). 2024;86(5):3139–45.
20. Joo IS, Ki CS, Joo SY, Huh K, Kim JW. A novel point mutation in PMP22 gene associated with a familial case of Charcot-Marie-Tooth disease type 1A with sensorineural deafness. Neuromuscul Disord. 2004;14(5):325–8.
21. Kousseff BG, Hadro TA, Treiber DL, Wollner T, Morris C. Charcot-Marie-Tooth disease with sensorineural hearing loss--an autosomal dominant trait. Birth Defects Orig Artic Ser. 1982;18(3b):223–8.
22. Maeda Y, Kataoka Y, Sugaya A, Kariya S, Kobayashi K, Nishizaki K. Steroid-dependent sensorineural hearing loss in a patient with Charcot-Marie-Tooth disease showing auditory neuropathy. Auris Nasus Larynx. 2015;42(3):249–53.
23. Meng L, Wang K, Lv H, Wang Z, Zhang W, Yuan Y. A novel mutation in PRPS1 causes X-linked Charcot-Marie-Tooth disease-5. Neuropathology. 2019;39(5):342–7.
24. Musiek FE, Weider DJ, Mueller RJ. Audiologic findings in Charcot-Marie-Tooth disease. Arch Otolaryngol. 1982;108(9):595–9.
25. Papadakis CE, Hajiioannou JK, Kyrmizakis DE, Bizakis JG. Bilateral sudden sensorineural hearing loss caused by Charcot-Marie-Tooth disease. J Laryngol Otol. 2003;117(5):399–401.
26. Sambuughin N, de Bantel A, McWilliams S, Sivakumar K. Deafness and CMT disease associated with a novel four amino acid deletion in the PMP22 gene. Neurology. 2003;60(3):506–8.
27. Seeman P, Mazanec R, Huehne K, Suslíková P, Keller O, Rautenstrauss B. Hearing loss as the first feature of late-onset axonal CMT disease due to a novel P0 mutation. Neurology. 2004;63(4):733–5.
28. Sharma G, Zaman M, Sabouny R, Joel M, Martens K, Martino D, et al. Characterization of a novel variant in the HR1 domain of MFN2 in a patient with ataxia, optic atrophy and sensorineural hearing loss. F1000Res. 2021;10:606.
29. Sivera R, Cavalle L, Vílchez JJ, Espinós C, Pérez Garrigues H, Sevilla T. Audiological findings in Charcot-Marie-Tooth disease type 4C. J Int Adv Otol. 2017;13(1):93–9.
30. Spaans F, Faber CG, Smeets HJ, Hofman PA, Braida C, Monckton DG, et al. Encephalopathic attacks in a family co-segregating myotonic dystrophy type 1, an intermediate Charcot-Marie-Tooth neuropathy and early hearing loss. J Neurol Neurosurg Psychiatry. 2009;80(9):1029–35.
31. Stojkovic T, Latour P, Vandenberghe A, Hurtevent JF, Vermersch P. Sensorineural deafness in X-linked Charcot-Marie-Tooth disease with connexin 32 mutation (R142Q). Neurology. 1999;52(5):1010–4.
32. Synofzik M, Müller vom Hagen J, Haack TB, Wilhelm C, Lindig T, Beck-Wödl S, et al. X-linked Charcot-Marie-Tooth disease, Arts syndrome, and prelingual non-syndromic deafness form a disease continuum: evidence from a family with a novel PRPS1 mutation. Orphanet J Rare Dis. 2014;9:24.
33. Verhagen WI, Huygen PL, Gabreëls-Festen AA, Engelhart M, van Mierlo PJ, van Engelen BG. Sensorineural hearing impairment in patients with Pmp22 duplication, deletion, and frameshift mutations. Otol Neurotol. 2005;26(3):405–14.
34. Nadol JB Jr, Hedley-Whyte ET, Amr SS, O'Malley JT, Kamakura T. Histopathology of the inner ear in Charcot-Marie-Tooth syndrome caused by a missense variant (p.Thr65Ala) in the MPZ gene. Audiol Neurootol. 2018;23(6):326–34.

CANVAS

86

Overview

CANVAS (*c*erebellar *a*taxia, *n*europathy, *v*estibular *a*reflexia *s*yndrome) is a rare progressive neurodegenerative disorder that causes multifactorial disequilibrium through vestibular weakness (due to a vestibular ganglionopathy), sensory neuropathy (due to sensory ganglionopathy), and cerebellar ataxia. Most cases appear to be due to a mutation in the RFC1 gene; familial cases exhibit an autosomal recessive pattern more than an autosomal dominant pattern (though this latter may be pseudodominance due to compound heterozygosity). It affects men and women approximately equally. The mean age of onset is in the sixth decade (range 20s–70s). Symptoms start insidiously and progress slowly; chronic cough may precede the other symptoms by years to decades. Physical examination findings include cerebellar ataxia and diminished small and large fiber sensation in all limbs. Ocular motor examination should show deficiencies in the visual-vestibulo-ocular reflex (VVOR) and may also show deficits in smooth pursuit, gaze evoked nystagmus or spontaneous down beat nystagmus. Vestibular testing (rotatory chair testing, video head impulse testing) shows clear evidence of vestibular weakness. Electromyography with nerve conduction velocities shows nonlength-dependent small- and large-fiber sensory neuropathy which can have either axonal or demyelinating features or both, usually in all limbs. There may be evidence of dysautonomia. MRI shows cerebellar atrophy, particularly in the dorsal vermis. Genetic testing can confirm the diagnosis by identifying mutations in RFC1 and exclude competing diagnoses such as spinocerebellar ataxias and Friedreich's ataxia. There is no effective treatment, though a trial of vestibular rehabilitation therapy is reasonable.

Introduction

CANVAS is an acronym for *c*e*rebellar* a*taxia, n*e*uropathy, and* v*estibular* a*reflexia* s*yndrome*.

In 2004, Migliaccio and colleagues [1] reported findings in four patients with a syndrome of bilateral vestibular weakness and cerebellar ataxia. Three of the patients also had electrophysiological evidence of a peripheral sensory neuropathy, and in one of those patients, biopsy revealed a severe axonal neuropathy. Subsequent cases suggested that the associated sensory neuropathy was a fairly consistent feature [2].

Additional cases were subsequently reported in the literature, and eventually, the constellation of findings (cerebellar ataxia, bilateral vestibular weakness, and peripheral neuropathy) came to be described as a syndrome, CANVAS.

These patients come to the attention of otoneurology by way of referral for the symptom of disequilibrium. In CANVAS, the disequilibrium is multifactorial, resulting from a combination two afferent defects (sensory neuropathy, vestibular weakness) and one coordination defect (cerebellar ataxia).

Epidemiology

CANVAS is rare. However, emerging research suggests that the underlying genetic mutation (in the RFC1 gene) may be more common than previously suspected [3]. Furthermore, as we will discuss below, the "RFC1 spectrum of diseases" can manifest with various combinations of features (cerebellar ataxia, neuropathy, vestibular weakness). Consequently, it may be that the incidence/prevalence of CANVAS is underestimated because those cases presenting with "incomplete" forms will, only with the passage of time, become "full-blown" CANVAS.

In the 27 cases reported by Szmulewicz and colleagues [2], they found an approximately equal distribution by sex (13 males, 14 females), and reported the mean age of onset of reported symptoms was 60 years (range 33–71 years).

In the 105 cases reported by Cortese and colleagues [4], the mean age of symptom onset (excluding cough) was 52 years (range 19–76 years).

M. Cherchi, *Otoneurology and Vestibular Medicine*, https://doi.org/10.1007/978-3-031-94841-1_86

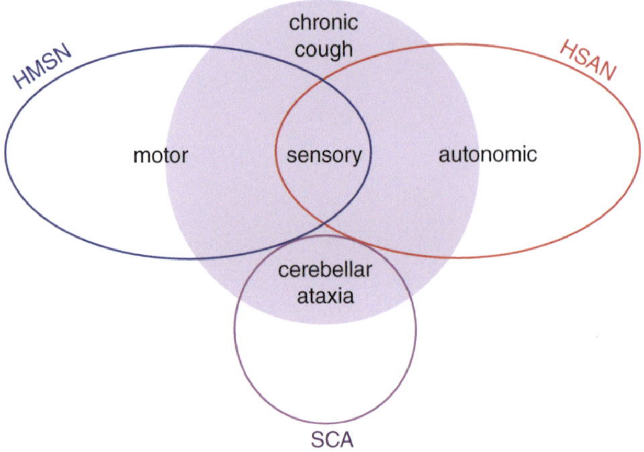

Fig. 86.1 Spectrum of RFC1 diseases. HMSN hereditary motor and sensory neuropathy, HSAN hereditary sensory and autonomic neuropathy; SCA spinocerebellar ataxia. (From Beijer et al. [10]. Reproduced with written permission from Blackwell Publishing)

Genetics

CANVAS (OMIM 614575) results from a mutation resulting in a repeat expansion in the gene encoding for RFC1 (replication factor C subunit 1) (OMIM 102579) on chromosome 4p14, a pentameric DNA polymerase accessory protein involved in DNA replication and repair [5, 6].

Mutations in RFC1 are identified in 100% of familial cases of CANVAS and in 92% of sporadic cases [6].

In a series of 44 patients with genetically confirmed CANVAS reported by Huin and colleagues [7], 19 (43%) exhibited a pattern of autosomal recessive transmission, 15 (34%) were sporadic, and 10 (23%) appeared to exhibit an autosomal dominant pattern, though this latter may be a pseudodominant manifestation [8] of compound heterozygosity [9].

Emerging literature suggests that CANVAS may be one manifestation of a broader spectrum of diseases resulting from RFC1 mutations. For example, Beijer and colleagues [10] depict RFC1 diseases as a group of partly overlapping neuropathies, as depicted in Fig. 86.1.

Emerging evidence suggests that RFC1 mutations may also account for isolated sensory neuropathy [11].

Table 86.1 Stimulus/task and reflex systems mediating ocular motor responses relevant in CANVAS

Stimulus or task	Reflex system(s)
Fixation on moving small target (vision available)	Smooth pursuit system (SP) using visual fixation
Movement of entire visual scene (vision available)	Optokinetic system (OPK)
Movement of head (vision deprived)	Labyrinthine-to-ocular motor system: vestibulo-ocular reflex (VOR)
Movement of head while vision is available	Both visual systems (smooth pursuit, optokinetic) plus vestibulo-ocular system

Pathophysiological Mechanism of Disease

In order to understand the key ocular motor finding in CANVAS, it is helpful to review the reflex systems that drive eye movements during visual and vestibular stimulation, as shown in Table 86.1.

In their patients (whose diagnosis would later be designated CANVAS), Migliaccio and colleagues [12] observed, "bursts of compensatory saccades when the patient's head was slowly and smoothly oscillated from side to side while fixating an earth-fixed target straight ahead." They interpreted this as reflecting, "low VVOR gain," and hypothesized that, "This abnormality is due to combined impairment of VOR [vestibulo-ocular reflex] and SP [smooth pursuit]." They concluded that this was a "characteristic sign on clinical examination: impairment of the VVOR [visual vestibulo-ocular reflex]."

Clinical Presentation

In the 27 cases of CANVAS described by Szmulewicz and colleagues [2], they state that the presenting features were gait imbalance in 20 (74%), dysesthesia in 8 (30%), oscillopsia in 5 (19%), "dizziness" in 2 (7%), and falls in 1 (4%); some patients had more than one presenting feature.

In some cases, chronic cough precedes the other symptoms by years or even decades [6, 13].

Cortese and colleagues [14] graph the percentage of patients that have each symptom at onset, and overall.

Figure 86.2, from Cortese and colleagues [4], shows the age of onset for each symptom.

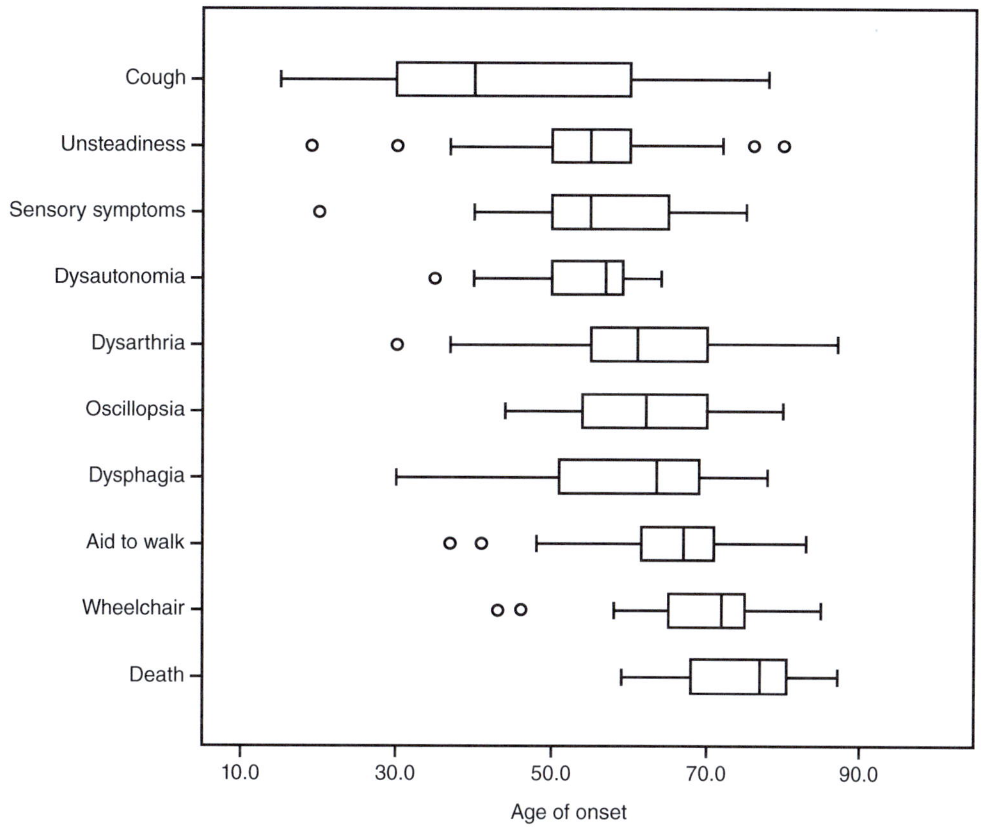

Fig. 86.2 Age of onset for each symptom of 105 patients with CANVAS. (From Cortese at al [4]. Reproduced from Oxford University Press with permission under CC-BY 4.0)

Physical Examination

Patients usually have signs of cerebellar appendicular ataxia.

Patients usually have signs of sensory neuropathy and sensory ataxia. Regarding neuropathic signs, in the 27 cases of CANVAS described by Szmulewicz and colleagues [2], they reported that "15/27 [56%] had absent ankle jerks, 21/27 [78%] had impaired vibration sensation (with abnormal joint position sense in a further one), and 21/27 [78%] had a deficit of pin-prick sensation," while "only two [7%] lacked any clinical evidence of a non-length-dependent multimodality sensory deficit."

Patients may also have evidence of dysautonomia, such as orthostatic hypotension [15, 16].

Cortese and colleagues [14] graph the percentage of patients exhibiting each physical examination finding.

Rarely, patients with RFC1 mutations exhibit fasciculations [7, 17].

Ocular Motor Examination

Evidence of bilateral vestibular weakness should be apparent on ophthalmoscopy and dynamic visual acuity testing.

Szmulewicz and colleagues [2] state that, "An impaired VVOR can be demonstrated clinically by turning a patient's head slowly (at about 0.5 Hz) from side-to-side while the patient starts at an earth-fixed target and observing that the compensatory eye movements are saccadic rather than smooth." They commented that other clinical features may include, "saccadic smooth pursuit, gaze-evoked nystagmus, gait ataxia, dysarthria and appendicular ataxia." They additionally commented that "Gaze-evoked horizontal or down beat nystagmus was found in all but two patients" of the 27 they analyzed.

Terryn and colleagues [18] provide a clear video of an abnormal visual vestibulo-ocular reflex in a CANVAS patient.

Huin and colleagues [7] reported the following ocular motor findings for a series of 34 patients with RFC1 mutations:

- Saccadic pursuit in 22 (65%)
- Saccadic dysmetria in 15 (44%)
- Down beat nystagmus in 13 (38%)
- Gaze evoked nystagmus in 11 (32%)
- Fixation instability in 9 (26%)
- Slow saccades in 8 (24%)
- Down beat nystagmus plus gaze evoked nystagmus in 3 (9%)

Table 86.2 Otovestibular test results in patients with CANVAS

Subject	Impairment of smooth pursuit (electronystagmography)	Reduced vestibulo-ocular reflex gain (rotational chair)	Reduced vestibulo-ocular gain (calorics)	Absent vestibular evoked myogenic potentials
1	Not recorded	Not recorded	Not recorded	Not recorded
2	√	√	Not recorded	Not recorded
3	√	√	Not recorded	Not recorded
4	Not recorded	Not recorded	Not recorded	Not recorded
5	√	√	Not recorded	Not recorded
6	√	√	Not recorded	Not recorded
7	√	√	Not recorded	Not recorded
8	Not recorded	Not recorded	Not recorded	Not recorded
9	√	Not recorded	√	√
10	√	Not recorded	√	Not recorded
11	√	Not recorded	Not recorded	Not recorded
12	√	Not recorded	√	√
13	Not recorded	√	√	√
14	Not recorded	√	Not recorded	Not recorded
15	√	√	√	Not recorded
16	√	√	√	Not recorded
17	√	√	√	Not recorded
18	Not recorded	Not recorded	Not recorded	Not recorded

From Szmulewicz et al. [2]

Testing: Vestibular

Tests of the vestibulo-ocular reflex should show vestibular weakness. Szmulewicz and colleagues [2] studied a CANVAS patient with video head impulse testing and clearly demonstrated low gain and compensatory saccades in all canals. Halmagyi and Szmulewicz [19] report similar video head impulse testing findings.

In another paper of the same year, Szmulewicz and colleagues [2] reported on 18 patients who had some combination of rotatory chair testing, caloric testing, and vestibular evoked myogenic potentials performed. In every case where one of the tests was performed, there was evidence of vestibular weakness, either in the form of reduced VOR gain on rotatory chair testing, reduced caloric responses, or absent vestibular evoked myogenic potentials. Table 86.2 summarizes their results.

Other investigators reported only inconsistent abnormalities on vestibular evoked myogenic potentials [15].

Testing: Electromyography and Nerve Conduction Velocities

Szmulewicz and colleagues [2] reported that, "All 20 patients in whom a four-limb nerve conduction study was performed had a loss of upper and lower limb SNAPs" (sensory nerve action potentials). They also stated that the sensory neuropathy can be either axonal or demyelinating, noting that, "One sural nerve was biopsied in six of the patents and revealed severe axonal loss in all, with evidence of coexistent demyelination in one."

The neuropathy affects both small and large diameter fiber types [2].

Szmulewicz and colleagues [2] reported further that the neuropathy is "non-length-dependent" and affects both upper and lower limbs, and concluded that "the total body involvement in some suggested a *neuron*opathy" (as opposed to merely a *neur*opathy).

Such observations, coupled with the histopathological findings reported in the vestibular nerves (see below), have led to speculation that the vestibular deficit may be due to a vestibular ganglionopathy (as opposed to merely vestibular neuropathy) [20], which was subsequently confirmed on temporal bone studies [21].

Testing: Genetic

Mutations in RFC1 are identified in 100% of familial cases of CANVAS and in 92% of sporadic cases [6].

Genetic testing, or at least referral to a genetic counselor, is reasonable. This may help confirm a diagnosis of CANVAS by identifying a mutation in RFC1 [5, 22], as well as exclude differential diagnoses such as spinocerebellar ataxia (particularly SCA3) and Friedreich's ataxia.

Imaging

Szmulewicz and colleagues [2] reported that, "MRI was done in all patients, and cerebellar atrophy was evident in 22/27 [81%] patients," adding that, "While the severity varied, a consistent pattern of anterior and dorsal vermis atrophy (vermal lobules VI, VIIa, and VIIb), and laterally, atrophy predominantly affecting crus 1 (corresponding to vermal lobule VII), was observed." Szmulewicz and colleagues [2] provide good examples of sagittal MRIs demonstrating clear cerebellar atrophy.

Histopathology

In a case studied by Szmulewicz and colleagues [23], they observed "severe atrophy of both vestibular nerves," "severe loss of Scarpa's ganglion cells," and "macroscopic cerebellar cortical atrophy of the anterior vermis and the superomedial aspect of the cerebellar hemispheres."

A temporal bone series of 5 patients with CANVAS showed that the vestibular deficit is due to a vestibular ganglionopathy [21].

There is emerging evidence that some CANVAS patients (with genetically confirmed RFC1 mutations) have a component of motor neuronopathy [7].

Differential Diagnosis

It is important to bear in mind that, although the majority of CANVAS patients eventually develop symptoms/signs/findings of all three components to which its name alludes (cerebellar, neuropathic, vestibular), not all patients with RFC1 mutations exhibit the full clinical picture.

Figure 86.3, from Cortese and colleagues [4], shows Venn diagrams of the incomplete overlap among the three components.

From this figure, the frequency of findings, in descending order, is as follows:

- All 3 components are present (cerebellar, neuropathic, vestibular). This is the most common manifestation.
- Cerebellar plus neuropathic is the second most common.
- Neuropathic in isolation is the third most common.
- Neuropathic plus vestibular.
- The rarest combinations are isolated cerebellar, combined cerebellar and vestibular, and isolated vestibular.

Because of this variety, the differential diagnosis is somewhat broad.

Szmulewicz and colleagues [2] suggest that, "The major differential diagnoses of CANVAS are SCA3 (Machado–Joseph disease), Friedreich's ataxia (FRDA), multiple system atrophy of cerebellar type (MSA-C), and Wernicke's encephalopathy." Szmulewicz and colleagues [16] state that they usually also screen for SCA1, 2, 6, and 7.

Cortese and colleagues [14] offer the more extensive differential diagnosis in Table 86.3.

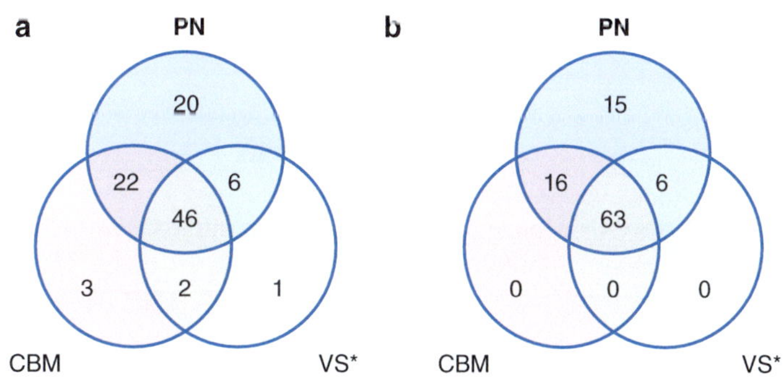

Fig. 86.3 Partial overlap of cerebellar, neuropathic and vestibular findings in 105 patients with RFC1 mutations. Involvement of peripheral nerve, cerebellum and vestibular system in RFC1 biallelic expansion. Panel (**a**): Symptoms and signs. Clinical involvement of peripheral nerve (PN), vestibular system (VS) and cerebellum (CBM) is defined as follows: (i) peripheral nerve: sensory neuropathy: sensory symptoms (loss of feeling, paresthesia, dysesthesia, neuropathic pain) and/or abnormal sensory examination including sensory ataxia; (ii) vestibular system: bilateral vestibulopathy: presence of head movement-induced oscillopsia and/or abnormal head impulse test; (iii) cerebellum: cerebellar dysfunction: cerebellar dysarthria and/or dysphagia, abnormal eye movements (nystagmus, dysmetric saccades, broken pursuits). Panel (**b**) Symptoms, signs, and investigations. In addition to the clinical involvement as defined in A, investigational involvement of the three main system affected in CANVAS was further identified by (i) peripheral nerve: sensory neuropathy/neuronopathy on nerve conduction studies; (ii) vestibular system: bilaterally abnormal video head impulse test, bithermal caloric or motorized rotational tests; and (iii) cerebellum: cerebellar atrophy on brain MRI. Abnormal in 55 of 59 (93%) with clinical information available. Abnormal in 69 of 74 (93%) with clinical and/or investigational information available. (From Cortese at al [4]. Reproduced from Oxford University Press with permission under CC-BY 4.0)

Table 86.3 Differential diagnosis of CANVAS suggested by Cortese at al [4]

	Causes	Key features
Sensory neuropathy/neuronopathy		
Acquired causes	Paraneoplastic	History of neoplasms, subacute onset, antineuronal antibodies, cerebrospinal fluid analysis
	Immune-mediated	Autoantibody screen, systemic involvement
	Toxic/metabolic (diabetes mellitus type 2, B12 deficiency, copper deficiency)	Blood Platinum-derived treatment, B6 intoxication
Hereditary causes	Hereditary sensory neuropathy RNF170	Early onset, insensitivity to pain, painless ulcerations Dominant family history, normal cerebellar function
Late-onset cerebellar ataxia		
Acquired causes	Multiple system atrophy	Early and severe dysautonomia, parkinsonism, rapid eye movement sleep behavior disorders, MR scan of brain findings (hot-cross bun sign)
	Paraneoplastic	Subacute onset, antineuronal antibodies, CSF analysis
	Toxic/metabolic	Thiamine deficiency/alcoholic abuse
	Vascular	Acute onset, cerebrovascular risk factors
	Inflammatory	Subacute onset, MR scan of brain findings, CSF analysis
	Idiopathic late onset cerebellar ataxia	Progressive late-onset ataxia without known cause
Hereditary causes	Friedreich's ataxia	Onset <25 years, pyramidal signs, scoliosis, pes cavus, systemic involvement (cardiac, endocrine, ocular)
	SCA3 (late-onset Machado–Joseph disease)	Autosomal dominant, extrapyramidal features, 5-M neuropathy, progressive external ophthalmoplegia
	Gerstmann–Straussler–Scheinker syndrome	Cognitive impairment, more rapid progression
Bilateral vestibular areflexia		
Acquired causes	Ototoxic drugs, Ménière's syndrome, vestibular neuritis, tumors, infections, inflammation	More commonly unilateral, cochlear symptoms
Hereditary causes	SCA (SCA types 1, 2, 3, 6)	Autosomal dominant, mild vestibular impairment, additional signs
	Charcot-Marie-tooth disease	Sensorimotor neuropathy
	Usher syndrome type I and II	Deafness, retinitis pigmentosa
Complex neurological diseases		
Mitochondrial diseases	Polymerase gamma	Ophthalmoplegia, multisystem involvement

Treatment

There is no established treatment for CANVAS. Management of disequilibrium is difficult since there is multimodal impairment.

Szmulewicz and colleagues [16] state that, "Anecdotally, we find significant benefit in referring our patients for individualised combination neurological and vestibular rehabilitation, which are performed regularly by the patient and reviewed periodically by a specialist physiotherapist."

Practically, referral for vestibular rehabilitation therapy is reasonable, on the logic that strategies aimed at bilateral vestibular weakness can be applied—recognizing that sensory substitution strategies will be limited due to the neuropathy, and motor compensatory strategies will be limited by the incoordination from the cerebellar ataxia.

Prognosis

There is no curative or arrestive therapy. Cortese and colleagues comment that, "The disease shows a slowly progressive course. Half of the patients needed a stick when walking after 10 years, and a quarter were wheelchair-dependent 5 years later. However, life expectancy was not affected" [14].

References

1. Waterston JA, Barnes GR, Grealy MA, Luxon LM. Coordination of eye and head movements during smooth pursuit in patients with vestibular failure. J Neurol Neurosurg Psychiatry. 1992;55(12):1125–31.
2. Szmulewicz DJ, Waterston JA, Halmagyi GM, Mossman S, Chancellor AM, McLean CA, et al. Sensory neuropathy as part of the cerebellar ataxia neuropathy vestibular areflexia syndrome. Neurology. 2011;76(22):1903–10.

3. Traschutz A, Heindl F, Bilal M, Hartmann AM, Dufke C, Riess O, et al. Frequency and phenotype of RFC1 repeat expansions in bilateral vestibulopathy. Neurology. 2023;101(10):e1001–e13.
4. Cortese A, Tozza S, Yau WY, Rossi S, Beecroft SJ, Jaunmuktane Z, et al. Cerebellar ataxia, neuropathy, vestibular areflexia syndrome due to RFC1 repeat expansion. Brain. 2020;143(2):480–90.
5. Davies K, Szmulewicz DJ, Corben LA, Delatycki M, Lockhart PJ. RFC1-related disease: molecular and clinical insights. Neurol Genet. 2022;8(5):e200016.
6. Cortese A, Simone R, Sullivan R, Vandrovcova J, Tariq H, Yau WY, et al. Biallelic expansion of an intronic repeat in RFC1 is a common cause of late-onset ataxia. Nat Genet. 2019;51(4):649–58.
7. Huin V, Coarelli G, Guemy C, Boluda S, Debs R, Mochel F, et al. Motor neuron pathology in CANVAS due to RFC1 expansions. Brain. 2022;145(6):2121–32.
8. Falcone GMI, Tessa A, Arena IG, Barghigiani M, Migliorato A, Incensi A, et al. Pseudodominance in RFC1-spectrum disorder. Cerebellum. 2024;23(6):2622–8.
9. Ronco R, Perini C, Curro R, Dominik N, Facchini S, Gennari A, et al. Truncating variants in RFC1 in cerebellar ataxia, neuropathy, and vestibular Areflexia syndrome. Neurology. 2023;100(5):e543–e54.
10. Beijer D, Dohrn MF, De Winter J, Fazal S, Cortese A, Stojkovic T, et al. RFC1 repeat expansions: a recurrent cause of sensory and autonomic neuropathy with cough and ataxia. Eur J Neurol. 2022;29(7):2156–61.
11. Curro R, Salvalaggio A, Tozza S, Gemelli C, Dominik N, Galassi Deforie V, et al. RFC1 expansions are a common cause of idiopathic sensory neuropathy. Brain. 2021;144(5):1542–50.
12. Migliaccio AA, Halmagyi GM, McGarvie LA, Cremer PD. Cerebellar ataxia with bilateral vestibulopathy: description of a syndrome and its characteristic clinical sign. Brain. 2004;127(Pt 2):280–93.
13. Dupre M, Hermann R, Froment TC. Update on cerebellar ataxia with neuropathy and bilateral vestibular Areflexia syndrome (CANVAS). Cerebellum. 2021;20(5):687–700.
14. Cortese A, Curro R, Vegezzi E, Yau WY, Houlden H, Reilly MM. Cerebellar ataxia, neuropathy and vestibular areflexia syndrome (CANVAS): genetic and clinical aspects. Pract Neurol. 2022;22(1):14–8.
15. Moreno-Ajona D, Alvarez-Gomez L, Manrique-Huarte R, Rivas E, Martinez-Vila E, Perez-Fernandez N. VEMPs and dysautonomia assessment in definite cerebellar ataxia, neuropathy, vestibular Areflexia syndrome (CANVAS): a case series study. Cerebellum. 2019;20:717.
16. Szmulewicz DJ, McLean CA, MacDougall HG, Roberts L, Storey E, Halmagyi GM. CANVAS an update: clinical presentation, investigation and management. J Vestib Res. 2014;24(5–6):465–74.
17. Scriba CK, Beecroft SJ, Clayton JS, Cortese A, Sullivan R, Yau WY, et al. A novel RFC1 repeat motif (ACAGG) in two Asia-Pacific CANVAS families. Brain. 2020;143(10):2904–10.
18. Terryn J, Van Eesbeeck A, Vermeer S, Vandenberghe W. The characteristic eye movement disorder of RFC1-linked CANVAS. Mov Disord Clin Pract. 2020;7(2):230–1.
19. Halmagyi GM, Szmulewicz DJ. Vestibular function testing in patients with RFC1 mutations. J Neurol. 2021;268(12):4894–6.
20. Yacovino DA, Zanotti E, Hain TC. Is cerebellar ataxia, neuropathy, and vestibular Areflexia syndrome (CANVAS) a vestibular ganglionopathy? J Int Adv Otol. 2019;15(2):304–8.
21. Ishai R, Seyyedi M, Chancellor AM, McLean CA, Rodriguez ML, Halmagyi GM, et al. The pathology of the vestibular system in CANVAS. Otol Neurotol. 2021;42(3):e332–e40.
22. Traschutz A, Cortese A, Reich S, Dominik N, Faber J, Jacobi H, et al. Natural history, phenotypic spectrum, and discriminative features of multisystemic RFC1 disease. Neurology. 2021;96(9):e1369–e82.
23. Szmulewicz DJ, Merchant SN, Halmagyi GM. Cerebellar ataxia with neuropathy and bilateral vestibular areflexia syndrome: a histopathologic case report. Otol Neurotol. 2011;32(8):e63–5.

87 Susac Syndrome

Overview

Susac syndrome (retinocochleocerebral vasculopathy) is believed to be an autoimmune disorder (perhaps mediated by CD8+ T cells) preferentially affecting the precapillary arterioles of the brain, retina, and labyrinth, manifesting with some combination of visual deficits, auditory symptoms (hearing loss, tinnitus), encephalopathy, and headache. The disease is rare and affects women more than men, with a peak age of onset of around 30 years (range 14–63 years). The hearing loss often affects low and middle more than high frequencies and can be unilateral or bilateral and fluctuating. Vestibular workup may reveal unilateral weakness on vestibular evoked myogenic potentials (which may recover), caloric testing, or video head impulse testing. Fluorescein angiography shows branch retinal artery occlusions. Brain MRI shows T2/FLAIR lesions in the corpus callosum, periventricular, thalamic, and internal capsular regions. There are no randomized treatment trials, but treatment is sometimes attempted with corticosteroids and other immunosuppressants or immunomodulatory agents.

Introduction

In 1979, neuro-ophthalmologist John O. Susac (1940–2012), pathologist John O. Hardman (1933–2006), and neurologist John B. Selhorst reported two cases of young women with neurological symptoms (headache, behavioral disturbances) and visual deficits who were found to have microangiopathy of the brain and retina [1]. Subsequent reports also noted hearing loss.

This clinical triad (visual abnormalities, sensorineural hearing loss, and cognitive dysfunction) resulting from *retinocochleocerebral vasculopathy* became known as *Susac syndrome*. This is currently believed to be an immune-mediated vascular disorder preferentially affecting blood vessels of the eye, ear, and brain (Fig. 87.1).

Fig. 87.1 Dr. John O. Susac (1940–2012). (From Egan [2]. Reproduced with written permission from Wolters Kluwer Health)

Epidemiology

Susac syndrome is rare. Bose and colleagues state that from the original description in 1979 until 2021, there were only 450 cases reported in the medical literature [3].

Bose and colleagues [3] studied a series of 16 patients and reported a female-to-male ratio of 2.2 to 1, with a mean onset of 30.5 years (±9.6 years).

M. Cherchi, *Otoneurology and Vestibular Medicine*, https://doi.org/10.1007/978-3-031-94841-1_87

Hardy and colleagues [4] studied a series of 30 patients and reported a female-to-male ratio of 17 to 13 (1.3 to 1) with a mean age of onset of 36.8 (range 14–63).

Genetics

There has been little study of the genetics of Susac syndrome. David and colleagues [5] performed whole-exome sequencing on 40 patients with Susac syndrome and identified a de novo missense variant in the ZNF76 (zinc finger protein 76) at chromosome 6p21.31 (OMIM 194549).

Pathophysiological Mechanism of Disease

Susac syndrome is a "microangiopathy affecting the precapillary arterioles of the brain, retina and inner ear (cochlea and semicircular canals)" manifesting with the "triad of encephalopathy, branch retinal artery occlusions… and sensorineural hearing loss" [3]. Immunological studies suggest that the vasculopathy results from malfunctioning CD8+ T cells [4].

Clinical Presentation

Of the three cardinal clinical features of Susac syndrome (visual deficits, hearing loss, encephalopathy), usually only a subset is present initially. It has been reported that of patients eventually diagnosed with Susac syndrome, only 13% present with all three symptoms at onset [3].

Hardy and colleagues [4] described a series of 30 Susac patients and reported:

> Neuro-otological involvement is relatively uncommon at the onset of SuS [Susac syndrome] but 93% of patients reported experiencing hearing loss, 54% tinnitus, and 67% vertigo or disequilibrium at some point during their disease course over an average of 4.5 years of follow-up. Moreover, 97% of patients experienced at least one of these symptoms during their disease course and 87% were left with residual neuro-otological symptoms [4].

Mean time to diagnosis has been reported to range from 3 weeks to 3 months [3].

Bose and colleagues reported a series of 16 Susac patients [3]. Table 87.1 shows presenting complaints and results of preliminary workup.

Notably, headache (75%) is the most common presenting feature. Encephalopathy occurs at some stage of the disease in 69% of cases.

Hardy and colleagues reported a series of 30 Susac patients [4]. Table 87.2 shows the distribution of symptoms.

The encephalopathic features can range "from minor cognitive symptoms to frank confusion" [4].

Table 87.1 Presenting complaints and results of workup in 16 patients with Susac syndrome

Feature	Number of patients	Percentage (%)
History of travel prior to onset of symptoms	4/16	25
Presenting complaints		
Headache	12/16	75
Eye symptoms	8/16	50
Ear symptoms	9/16	56.25
Confusion/forgetfulness	6/16	37.5
Encephalopathy (at any stage)	11/16	68.75
Psychiatric comorbidities	5/16	31.25
CSF—raised protein	12/24	85.7
MRI		
Callosal lesions	16/16	100.0
Pericallosal lesions	14/16	87.5
Infratentorial lesions	9/16	56.25
Enhancement	6/16	37.5
DWI	6/16	37.5
Hearing loss (at any stage)	12/16	75
Audiometry abnormalities (at any stage)	11/14	78.6
Objective evidence of retinal ischemia, arterial occlusions on FFA/Fundus (at any stage)	11/16	68.75
Monophasic course	11/16	68.75

From Bose et al. [3]

Table 87.2 Symptoms and other features of 30 patients with Susac syndrome

Neuro-otological characteristic	Patients (n = 30)
Hearing impairment	93% (28/30)
Bilateral	57% (17/30)
Unilateral	37% (11/30)
At onset of Susac syndrome	17% (5/29)
At final follow-up	67% (20/30)
Tinnitus	**54% (13/24)**
Bilateral	4
Unilateral	5
Not stated if bilateral or unilateral	4
At onset of Susac syndrome	10% (3/29)
At final follow-up	50% (12/24)
Vertigo or disequilibrium	**67% (18/27)**
At onset	21% (6/29)
At final follow-up	0% (0/27)
Hearing loss, tinnitus, vertigo, or disequilibrium	
During disease course	97% (29/30)
At final follow-up	87% (26/30)
Hearing aid	**14% (4/28)**
Cochlear implant	**7% (2/28)**

From Hardy et al. [4]

The visual symptoms may manifest abruptly (with visual loss, visual field changes, scintillating scotoma, and other features), but visual loss may also develop insidiously and thus not be volunteered by the patient [3].

The hearing loss may be abrupt or insidious in onset [3], can progress in a step-wise fashion, and can be unilateral or bilateral [4].

In the series reported by Hardy and colleagues [4], disequilibrium was reported by 18 (67%) of 27 patients, was the initial symptom (or one of several presenting symptoms) in 6 (21%) of 29 patients, and was the only otoneurological symptom in 2 (7%) of patients. It is unusual for vestibular symptoms to be the exclusive initial presentation of Susac syndrome [6].

Testing (Auditory): Audiometry and Otoacoustic Emissions

When Susac syndrome involves hearing loss, pure tone audiometry "often shows low to mid-frequency sensorineural hearing loss in the early stages, but the pantonal loss can also occur in severe cases" [4]. The initial involvement of low and middle frequencies, particularly if unilateral, may resemble Ménière's disease.

Hardy and colleagues [4] performed serial audiometry in a patient with Susac syndrome demonstrating fluctuating sensorineural hearing loss (more pronounced in the high frequencies).

In the series of Susac patients studied by Hardy and colleagues [4], some patients had the paradoxical finding of hearing loss on pure tone audiometry yet preserved otoacoustic emissions, suggesting dysfunction of only the type 1 (inner) cochlear hair cells or retrocochlear pathways (such as an auditory neuropathy). The remaining patients with hearing loss on pure tone audiometry had commensurate loss on otoacoustic emissions.

Testing (Vestibular): Caloric Testing

In the series of Susac patients studied by Hardy and colleagues [4], 10 had caloric testing, of which one showed bilateral weakness, and two showed asymmetrical weakness. The case studied by River and colleagues [7] reported a 56% right unilateral caloric weakness.

Testing (Vestibular): Vestibular Evoked Myogenic Potentials

Oron and colleagues [8] reported unilaterally absent cervical vestibular evoked myogenic potentials in a patient with Susac syndrome (Fig. 87.2).

Hardy and colleagues [4] reported that patients initially found to have unilateral weakness on ocular vestibular evoked myogenic potentials or cervical vestibular evoked myogenic potentials subsequently appeared to recover and exhibit normal responses on these tests.

Suman and colleagues [9] reported a case of Susac syndrome in which serial cervical and ocular vestibular evoked myogenic potentials were performed. Cervical vestibular evoked myogenic potentials were unilaterally absent (left side) on the initial study but present on the second and third studies. Ocular vestibular evoked myogenic potentials were normal in all three studies.

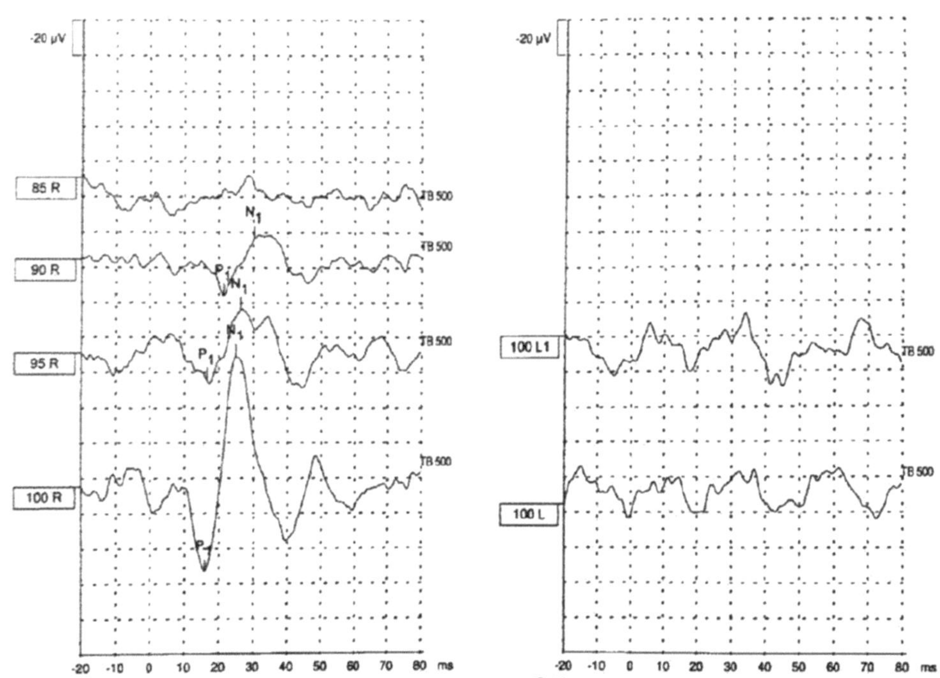

Fig. 87.2 Cervical vestibular evoked myogenic potentials are absent on the patient's left side (right panel) in this individual with Susac syndrome. (From Oron et al. [8]. Reproduced with written permission from IOS Press)

Testing (Vestibular): Video Head Impulse Testing

Marrodan and colleagues [10] reported the case of a 19-year-old lady diagnosed with Susac syndrome who underwent video head impulse testing showing that the right posterior semicircular canal exhibited low gain and over compensatory saccades. Of note, the patient had no vestibular complaints.

Testing: Other

Fluorescein angiography shows branch retinal artery occlusions.

Imaging

Brain MRI usually shows lesions on FLAIR sequences involving the corpus callosum, periventricular, and thalamic regions. The MRI lesions have been described with colorful terms such as "snowballs" (referring to the round lesions in the corpus callosum), "spokes and icicles" (for linear lesions in the corpus callosum), and "string of beads" pattern in the internal capsule [10], and may also involve the periventricular and thalamic regions [3].

Histopathology

Bose and colleagues [3] reported on the brain biopsy of one patient, and the autopsy of another patient, which together showed "multiple microinfarcts involving grey/white matter, deep grey nuclei, brain stem and corpus callosum with endothelial cell necrosis and perivascular lymphocytic infiltration."

Francis and colleagues studied the temporal bone of a 51-year-old woman with Susac syndrome and reported, "atrophy and degeneration involving the apical halves of the cochlear duct without inflammation or infection. The findings were consistent with capillary occlusion as being responsible for the atrophy" [11].

Differential Diagnosis

The cognitive and visual symptoms of Susac syndrome may resemble demyelinating disease (such as multiple sclerosis, neuromyelitis optica, acute demyelinating encephalomyelitis), but imaging usually helps differentiate these.

The headache and encephalopathy of Susac syndrome may resemble CADASIL, but the visual and auditory symptoms may differentiate these.

Cases of Susac syndrome manifesting with abrupt onset cochlear symptoms and found to have ipsilateral vestibular deficits [7] may resemble labyrinthitis, but the cognitive symptoms will usually differentiate these.

Susac syndrome can also be mistaken for a CNS vasculitis.

Treatment and Prognosis

There are no prospective, randomized, double-blinded, placebo-controlled treatment trials for Susac syndrome, so treatment decisions usually draw from case reports, case series, and expert opinion. Attempted treatments have included corticosteroids, immunosuppressants (rituximab, cyclophosphamide, mycophenolate mofetil, azathioprine), and immunomodulatory therapies (intravenous immunoglobulin).

Hearing loss and tinnitus, once established, are usually permanent [4].

References

1. Susac JO, Hardman JM, Selhorst JB. Microangiopathy of the brain and retina. Neurology. 1979;29(3):313–6.
2. Egan R. John O. Susac, MD (1940–2012). J Neuroophthalmol. 2012;32:292.
3. Bose S, Papathanasiou A, Karkhanis S, Appleton JP, King D, Batra R, et al. Susac syndrome: neurological update (clinical features, long-term observational follow-up and management of sixteen patients). J Neurol. 2023;270(12):6193–206.
4. Hardy TA, Taylor RL, Qiu J, O'Brien B, Gopinath S, Trewin B, et al. The neuro-otology of Susac syndrome. J Neurol. 2020;267(12):3711–22.
5. David C, Papo T, Ba I, Ollivier E, Boileau C, Dieude P, et al. Hunting for the genetic basis of Susac syndrome. Eur J Neurol. 2021;28(7):e57–e9.
6. Magliulo G, Al-Ansi W, Parrotto D, Gagliardi S, Stasolla A. Susac syndrome and vestibular-evoked myogenic potentials. Otolaryngol Head Neck Surg. 2008;138(4):542–3.
7. River Y, Shupak A, Tiosano B, Danilov V, Braverman I. Susac's syndrome: a case with unusual cardiac vestibular and imaging manifestations. Case Rep Neurol Med. 2015;2015:419408.
8. Oron Y, Handzel O, Habot-Wilner Z, Regev K, Karni A, Zur D, et al. Vestibular function assessment of Susac syndrome patients by the video head impulse test and cervical vestibular-evoked myogenic potentials. J Vestib Res. 2020;30(6):393–9.
9. Suman NS, Poovaiah PP, Rangarajan A, Tiwari R, Nashi S, Yuvraj P, et al. Cervical and ocular vestibular evoked myogenic potential recovery in susac syndrome: a case report. Am J Audiol. 2022;31(4):1059–66.
10. Marrodan M, Laffue A, Fiol MP, Correale J, Gualtieri F. Video head impulse test contributes to Susac syndrome diagnosis. J Neurol. 2019;266(8):2070–2.
11. Francis HW, Makary C, Halpin C, Crane BT, Merchant SN. Temporal bone findings in a case of Susac's syndrome. Otol Neurotol. 2011;32(8):1198–204.

Part X

Diagnoses—Ataxias

88 Spinocerebellar and Other Genetic Ataxias

Overview

Spinocerebellar ataxias and a few other ataxias (e.g., dentato-rubro-pallido-luysian atrophy, fragile X tremor ataxia syndrome, and others) are genetically mediated ataxias, nearly all of which are inherited in an autosomal dominant fashion. The various ataxias have different features (age of onset, ocular motor findings, dysmorphism, association with cognitive and other neurological deficits) but many resemble each other, such that a confident diagnosis is usually not achievable based on history and physical examination alone; genetic testing is the diagnostic modality of choice. None of these ataxias yet has specific treatment; management is symptomatic. The main value in identifying them lies in planning one's future and advising offspring.

Discussion

Spinocerebellar ataxias are genetically mediated ataxias. Nearly all of them are inherited in an autosomal dominant fashion. Many of the initially described diseases had symptoms or signs that potentially localized to spinal tracts; thus, the group became referred to as spinocerebellar ataxias (SCAs). Almost (but not quite) all of these are inherited in an autosomal dominant fashion. There are other ataxias not classified as "spinocerebellar," but are still reasonable to group with the SCAs because of certain shared clinical features.

The number of genetically identified ataxias is long and growing. The most commonly encountered ones were the first to be identified (SCA types 1–8). The remainder are rare.

Table 88.1 is a summary of these ataxias. It draws on information in UpToDate, though we have additionally incorporated information pertaining to ocular motor findings

Out of the 50 ataxias listed in the table above, over half [27] have been documented as associated with various eye movement abnormalities compatible with cerebellar or brainstem dysfunction [34], and although these are of special interest to otoneurologists, the eye movements are neither sensitive nor specific for these diseases.

In other respects as well, many of these ataxias can resemble each other fairly closely, and even an extremely astute clinician would be reluctant to make a diagnosis based on history and physical examination alone. Practically, when a genetic ataxia is suspected, it is reasonable to refer the patient to an ataxia specialist, neurogeneticist, or genetic counselor; these clinical practitioners will usually order a genetic panel which tests for all (or for a large subset) of the known ataxias.

Some of these genetic ataxias may be referred to an otoneurology clinic because of bilateral vestibular weakness (or at least reduced vestibulo-ocular reflexes), which has been reported in the following:

- SCA1 [35].
- SCA3 [35–38].
- SCA6 [36, 39].
- SCA7 [36].
- SCA27b [40–44].

Practically, what features should raise the suspicion of a genetic ataxia?

- Slowly progressive ataxia, beginning at any age.
- Family history of ataxia or balance problems, especially if it occurs at progressively younger ages with each passing generation, since some of these diseases exhibit anticipation (progressively earlier onset due to trinucleotide repeat expansion with each successive generation).
- Syndromic presentations (e.g., spasticity, dysmorphism, mental retardation, dementia, seizures).

None of these ataxias yet has specific treatment; management is symptomatic. The main value in identifying them lies in planning one's future and advising offspring.

M. Cherchi, *Otoneurology and Vestibular Medicine*, https://doi.org/10.1007/978-3-031-94841-1_88

Table 88.1 Spinocerebellar ataxias and other genetic ataxias

Disorder	Distinguishing features	Gene or locus	Protein and mutation	OMIM#
SCA1	Epidemiology: Accounts for 3–16% of autosomal dominant cerebellar ataxias. Broad age range, but peak onset in 3rd–4th decades. Clinical features: Pyramidal signs, peripheral neuropathy. ***Ocular motor: Ophthalmoplegia.*** Pathology: Degeneration of cerebellar Purkinje cells, brainstem cranial nerve nuclei, inferior olivary nuclei and spinocerebellar tracts.	*ATXN1* (6p22.3)	Ataxin 1; CAG repeat	*601556*
SCA2	Epidemiology: Accounts for 6–18% of spinocerebellar ataxia kindreds. Infancy to late adulthood; age of onset inversely proportional to number of trinucleotide repeats. Clinical features: Occasionally, parkinsonism, myoclonus, areflexia, dementia. ***Ocular motor: Slow saccades.***	*ATXN2* (12q24)	Ataxin 2; CAG repeat	*601517*
SCA3 (MJD)	Epidemiology: This is the most common autosomal dominant spinocerebellar ataxia, accounting for 21–23% of cases in the United States, 12% in Australia, and 48% in China. Clinical features: Extrapyramidal signs, muscle cramps and fasciculations, peripheral neuropathy, autonomic dysfunction, cognitive dysfunction, brainstem dysfunction (persistent stare from lid retraction dysarthria, dysphagia, tongue fasciculations). ***Ocular motor: Slow saccades, saccadic pursuit.***	*ATXN3* (14q32)	Ataxin 3 (MJD1); CAG repeat	*607047*
SCA4	Clinical features: Sensory axonal neuropathy.	16q22.1		*600223*
SCA5	Epidemiology: Early onset (3rd decade) but slow progression. Clinical features: Almost purely cerebellar syndrome. Imaging: Cerebellar atrophy.	*SPTBN2*	Spectrin beta, nonerythrocytic 2	*604985*
SCA6	Epidemiology: 15–17% of dominant cerebellar ataxias. Onset between 20 and 60 years. May lack family history. Clinical features: Similar to SCA5. ***Ocular motor: "Horizontal and vertical nystagmus," abnormal vestibulo-ocular reflex.*** Imaging: Global cerebellar degeneration.	*CACNA1A* (19p13)	Calcium voltage-gated channel subunit alpha1 A; CAG repeat	*601011*
SCA7	Epidemiology: 2–5% of autosomal dominant ataxias. Infancy to adulthood; age of onset inversely proportional to number of trinucleotide repeats. Clinical features: Ataxia, retinal degeneration. When onset is in childhood, may include seizures, myoclonus and cardiac involvement.	*ATXN7* (3p12)	Ataxin 7; CAG repeat	*607640*
SCA8	Epidemiology: Accounts for 3–5% of spinocerebellar ataxias worldwide. Clinical features: Mild disease. ***Ocular motor: Nystagmus at extreme lateral gaze*** [1]. Imaging: Marked cerebellar atrophy. Pathology: Neurodegeneration of Purkinje, inferior olivary and nigral cells, and periaqueductal gliosis, described in one family.	*ATXN8OS*	ATXN8 opposite strand lncRNA; CTG*CAG repeat	*603680*
SCA9	Not assigned			
SCA10	Clinical features: Variable; can have purely cerebellar manifestations, or include generalized seizures or extrapyramidal findings.	*ATXN10* (22q13)	Ataxin 10; ATTCT repeat	*611150*
SCA11	Clinical features: Mild disease, purely cerebellar.	*TTBK2* (15q14-q21.3	Tau tubulin kinase 2	*611695*
SCA12	Clinical features: Ataxia, often accompanied by tremor and dementia.	*PPP2R2B* (5q31-q33)	Protein phosphatase 2 regulatory subunit Bbeta; CAG repeat in 5′ region	*604325*

(continued)

Table 88.1 (continued)

Disorder	Distinguishing features	Gene or locus	Protein and mutation	OMIM#
SCA13	Clinical features: Cerebellar ataxia and intellectual disability.	*KCNC3* (19q13)	Potassium voltage-gated channel subfamily C member 3	*176264*
SCA14	Epidemiology: Broad age range of onset. Clinical features: Earlier onset can present with myoclonus followed by ataxia. Later onset (≥39 years) can be purely cerebellar. ***Ocular motor: May show saccadic pursuit*** [2].	*PRKCG* (19q13)	Protein kinase C gamma	*176980*
SCA15/16	Note: SCA15 and SCA16 were originally thought to be separate disorders. Epidemiology: Onset from mid-childhood to middle age. Clinical features: Slowly progressive; sometimes with myoclonus, dystonia, postural/action tremor. Imaging: Atrophy of cerebellum, particularly in the vermis.	*ITPR1* (3p26.2pter)	Inositol 1,4,5-triphosphate receptor type 1	*147265*
SCA17	Epidemiology: 19–48 years. Age of onset inversely proportional to number of trinucleotide repeats. Described in four Japanese families. Clinical features: Gait ataxia, dementia; can present with predominant chorea; may develop bradykinesia, dysmetria, dysdiadochokinesia, hyperreflexia.	*TBP*	TATA-box binding protein; CAG repeat	*600075*
SCA18	Epidemiology: Described in an American family of Irish ancestry. Clinical features: Pyramidal signs, weakness, sensory axonal neuropathy	7q22-q32		*607458*
SCA19/22	Note: SCA19 and SCA22 were originally thought to be separate disorders. Epidemiology: SCA19 was originally described in a Dutch family, and SCA22 in a Taiwanese family. Clinical features: Predominantly cerebellar syndrome. More severe cases may include cognitive impairment, myoclonus, hand tremor, neuropathy. ***Ocular motor: May have saccadic abnormalities and down beat nystagmus*** [3].	*KCND3*	Potassium voltage-gated channel subfamily D member 3	*605411*
SCA20	Epidemiology: 19–64 years (mean 46.5 years). Described in one family of Anglo-Celtic ancestry and 6 Portuguese families. Clinical features: Dysarthria is the most common initial symptom, and is usually followed by gait or upper limb ataxia. Unusual features include palatal tremor and dysphonia. ***Ocular motor: May show hypermetric saccades on downward gaze or lateral gaze; saccadic intrusions into smooth pursuit; square wave jerks on primary position of gaze; impaired visual suppression of the vestibulo-ocular reflex*** [4], and ***down beat nystagmus*** [5]. Imaging: Dentate nucleus calcification on CT.	11q12	Contiguous gene duplication syndrome	*608687*
SCA21	Epidemiology: Early onset. First identified in four generations of a French family. Clinical features: Slow progression. Mild to severe cognitive impairment; motor clumsiness.	*TMEM240* (1p36.33)	Transmembrane protein 240	*616101*
SCA23	Epidemiology: Late onset (age >40 years). Described in a Dutch family. Clinical: Slowly progressive. Distal sensory deficits. Variable dysarthria. ***Ocular motor: Some have ocular dysmetria and slow saccades*** [6].	*PDYN* (20p13)	Prodynorphin	*131340*

(continued)

Table 88.1 (continued)

Disorder	Distinguishing features	Gene or locus	Protein and mutation	OMIM#
SCA24	Notes: Autosomal *recessive* ataxia, so has been redesignated as SCA*R*4. Epidemiology: Described in a Slovenian family. Clinical features: Progressive ataxia. May also have corticospinal signs and axonal sensorimotor neuropathy. ***Ocular motor: Large horizontal saccadic intrusions disrupt visual fixation*** [7, 8], ***horizontal saccadic hypermetria, microsaccadic oscillations, and increased velocity of larger saccades*** [9].	*VPS13D* (1p36)	Vacuolar protein sorting 13 homolog D	*607317*
SCA25	Epidemiology: Variable age of onset. Described in a large French family and 3 generations of an Australian family. Clinical features: Variably severe symptoms of ataxia, sensory neuropathy, facial tics, gastrointestinal symptoms. ***Ocular motor: May exhibit "horizontal" or "vertical" nystagmus, instability of central fixation*** [10], ***square wave jerks, hypermetric saccades, down beat nystagmus, slow saccades, impaired vestibulo-ocular reflex, "mild torsional/ see-saw nystagmus," "very mild up beat nystagmus on upgaze," "restriction of upper and lateral gaze," "oculomotor apraxia"*** [11].	*PNPT1*	Polyribonucleotide nucleotidyltransferase 1	*608703*
SCA26	Epidemiology: Described in a Norwegian family. Clinical features: Slowly progressive pure cerebellar ataxia.	*EEF2* (19p13.3)	Eukaryotic translation elongation factor 2	*130610*
SCA27	Epidemiology: Early onset. Described in a large Dutch family and in two generations of a family of French Canadian origin. Clinical features: Cognitive impairment; early-onset hand tremor, orofacial dyskinesia. Ocular motor: Members of the French Canadian family exhibited ***"sustained horizontal nystagmus," upbeat and downbeat nystagmus*** [12]. Another report of a multigenerational family described ***multidirectional nystagmus, "horizontal nystagmus," "vertical nystagmus," abnormal smooth pursuit and vertical oscillopsia (!)*** [13]. Note: A second variant (GAA repeat expansion in the first intron of FGF14) is recognized as "SCA27B." this variant has a median onset of 55 years, and about half of patients have episodic ataxia and ***ocular motor finding of downbeat nystagmus***.	*FGF14** (13q34)	Fibroblast growth factor 14	*601515*
SCA28	Epidemiology: Juvenile onset (mean age 19.5 years). Described in an Italian family. Clinical features: Ptosis. In some cases pyramidal tract signs. ***Ocular motor: Ophthalmoparesis***. One report merely described ***"minor abnormalities in ocular movements"*** [14]. Another report described ***ocular motor apraxia*** [15].	*AFG3L2* (18p)	AFG3-like matrix AAA peptidase subunit 2	*604581*
SCA29	Epidemiology: Early-onset. Clinical features: Nonprogressive ataxia; may be an allelic variant of SCA15. ***Ocular motor: May have up beat nystagmus*** [16, 17] ***on primary position of gaze*** [18], ***upgaze evoked up beat nystagmus*** [19], ***gaze evoked horizontal nystagmus*** [20–22]. More detailed descriptions also identified ***intermittent pendular oscillations, rebound nystagmus, defective horizontal pursuit, absent horizontal optokinetic nystagmus, defective visual suppression of rotationally induced nystagmus*** [23], ***rotatory nystagmus, and impaired vestibulo-ocular reflex gain*** [24]. Imaging: Variable atrophy of cerebellar vermis.	3p26		*117360*
SCA30	Epidemiology: Mid- to late-life onset. Described in an Australian family of Anglo-Celtic ancestry. Clinical features: Slowly progressive, relatively pure ataxia. May have minor pyramidal tract signs. ***Ocular motor: All have hypermetric saccades, some may have slight gaze evoked nystagmus*** [25]. Imaging: Atrophy of cerebellar vermis and hemispheres.	4q34.3-q35.1		*613371*

(continued)

Table 88.1 (continued)

Disorder	Distinguishing features	Gene or locus	Protein and mutation	OMIM#
SCA31	Epidemiology: Late onset. Clinical features: Ataxia, decreased muscle tone. Variable ***hearing loss, vertigo***.	*BEAN1* (16q22.1)	Brain expressed associated with NEDD4 1; (TGGAA)n repeat	*612051*
SCA32	Epidemiology: Described in a single Chinese family. Clinical features: Cognitive impairment, affected males with azoospermia and testicular atrophy.	7q32-q33		*613909*
SCA33	Not assigned			
SCA34	Epidemiology: Described in a French Canadian kindred and two Japanese families. Clinical features: Ataxia and pyramidal signs. French Canadian kindred also has skin lesions consisting of papulosquamous erythematous ichthyosiform plaques. Two Japanese families lack the cutaneous findings.	*ELOVL4*	ELOVL fatty acid elongase 4	*605512*
SCA35	Epidemiology: Several families of Han Chinese descent. Late onset, with mean age of onset 44 years. Clinical features: Slowly progressive gait and limb ataxia; cervical dystonia; mild dysarthria.	*TGM6* (20p13)	Transglutaminase 6	*613900*
SCA36	Epidemiology: Late onset. Initially reported in Japanese and Spanish families, but subsequently found in other populations. Clinical features: Truncal ataxia, dysarthria. Japanese families show variable motor neuron disease. Spanish families have ***sensorineural hearing loss***.	*NOP56*	NOP56 ribonucleoprotein; GGCCTG repeat	*614154*
SCA37	Epidemiology: Late onset (mean 48 years, range 38–64 years). Described in a Spanish family. Clinical features: Falls, dysarthria, clumsiness. ***Ocular motor: Early on patients have abnormal vertical eye movements; later they may develop horizontal eye movement abnormalities as well; the abnormalities consist of hypometric saccades and cogwheel pursuit*** [26]. Imaging: Cerebellar atrophy.	*DAB1*	DAB adaptor protein 1	*603448*
SCA38	Epidemiology: Late onset (34–51 years). Described in several European families. Clinical features: Slowly progressive relatively pure cerebellar phenotype; truncal ataxia, limb ataxia; neuropathy. ***Ocular motor: Almost all patients have gaze evoked nystagmus; some also have slow saccades*** [27].	*ELOVL5*	ELOVL fatty acid elongase 5	*611805*
SCA39	Not assigned			
SCA40	Epidemiology: Late onset (5th decade). Described in a family from China. Clinical features: Truncal and gait ataxia, dysarthria, pyramidal signs (hyperreflexia and spasticity), tremor, parkinsonism. ***Ocular motor: Ocular dysmetria.***	*CCDC88C*	Coiled-coil domain-containing 88C	*611204*
SCA41	Epidemiology: Described in a single 40-year-old male. Clinical features: Gait ataxia.	*TRPC3*	Transient receptor potential cation channel subfamily C member 3	*602345*
SCA42	Epidemiology: Variable age of onset. Reported in multiple French and Japanese families. Clinical features: Early motor delay, hypotonia, speech delay, severe intellectual disability, ataxia, facial myokymia. ***Ocular motor: Saccadic eye movements, nystagmus.***	*CACNA1G*	Calcium voltage-gated channel subunit alpha 1G	*604065*
SCA43	Epidemiology: Identified in 5 generations of a Belgian family. Clinical features: Gait ataxia, neuropathy, hyporeflexia, tremor.	*MME*	Membrane metalloendopeptidase	*120520*
SCA44	Epidemiology: Three generations of an English family. Early onset has also been described. Clinical features: Ataxia, dysarthria, dysmetria, dysphagia.	*GRM1*	Glutamate metabotropic receptor 1	*604473*
SCA45	Epidemiology: Two generations of a family. Clinical features: Limb and gait ataxia, dysarthria. ***Ocular motor: Downbeat nystagmus.***	*FAT2* (5q)	FAT atypical cadherin 2	*604269*

(continued)

Table 88.1 (continued)

Disorder	Distinguishing features	Gene or locus	Protein and mutation	OMIM#
SCA46	Epidemiology: Described in a Dutch family. Clinical features: Neuropathy and sensory ataxia affecting lower limbs more than upper limbs. ***Ocular motor: Saccadic interruption of smooth pursuit, square wave jerks unsuppressed by visual fixation, slow pursuit, hypometric saccades*** [28]. Imaging: Cerebellar atrophy.	*PLD3*	Phospholipase D family member 3	*615698*
SCA47	Epidemiology: Onset usually in adulthood, but early onset form has also been described. Clinical features: Early onset variant has developmental disability, ataxia, chorea, seizures. Later onset variant has ataxia, dysarthria, dysmetria, and diplopia.	*PUM1*	Pumilio RNA binding family member 1	*607204*
SCA48	Epidemiology: Reported in multiple European families. Clinical features: Gait ataxia, dysarthria, dysphagia, cognitive dysfunction (executive dysfunction) in adulthood, psychiatric manifestations (depression, anxiety, apathy). ***Ocular motor: May have horizontal nystagmus and dysmetric saccades*** [29].	*STUB1*	STIP1 homology and U-box containing protein 1	*607207*
ATX-ATN1 (DRPLA)	Notes: Dentato-rubro-pallido-luysian atrophy. Epidemiology: Wide variability in age of onset. More common in Japan (mean age of onset 47 years), less common in European and North American populations. A variant called "Haw River syndrome" (which lacks the myoclonic epilepsy) has been reported in several families of African American descent in North Carolina. Clinical features: Choreoathetosis, seizures (myoclonic epilepsy), myoclonus, rigidity, hyperreflexia, dementia. ***Ocular motor: Slow saccades.*** Imaging: Atrophy of cerebellum and brainstem, calcification of basal ganglia, leukodystrophy.	*ATN1* (12p)	Atrophin 1; CAG repeat	*607462*
FXTAS	Note: Fragile X tremor ataxia syndrome. Epidemiology: Adult-onset. Mostly, though not exclusively, in males. Clinical features: Progressive ataxia and postural tremor. Additional features may include short-term memory loss, executive dysfunction, cognitive decline, parkinsonism, peripheral neuropathy, proximal weakness of lower limbs and autonomic dysfunction. ***Ocular motor: May have square wave jerks and horizontal nystagmus*** [30]. Imaging: MRI shows brain atrophy, T2/FLAIR signal abnormalities in the middle cerebellar peduncle and splenium of the corpus callosum. Pathology: White matter disease with astrocytic pathology, intranuclear inclusions in neurons and astrocytes in brain and spinal cord, amyloid beta within capillaries, and cerebral microbleeds.	*FMR1*	Fragile X messenger ribonucleoprotein 1 (FMR1).	300623
ATX-DAB1	Epidemiology: Adult-onset. Clinical features: Slowly progressive.	*DAB1*	DAB adaptor protein 1	*615945*
ATX-DNMT1	Clinical features: ***Sensorineural deafness***, narcolepsy, dementia.	*DNMT1*	DNA methyltransferase 1	*126375*
ATX-EBF3	Clinical features: Hypotonia, ataxia, and delayed development syndrome (HADDS).	*EBF3*	EBF transcription factor 3	*617330*
ATX-LMNB1	Epidemiology: Adult-onset. Clinical features: Autosomal dominant, demyelinating leukodystrophy (ADLD)	*LMNB1*	Lamin B1	*169500*
ATX-SAMD9L	Ataxia-pancytopenia syndrome (ATXPC). ***Ocular motor: Nystagmus*** [31], ***particularly horizontal and vertical nystagmus and dysmetria*** [32].	*SAMD9L*	Sterile alpha motif domain containing 9 like	*159550*
ATX-SNAP25b	Epidemiology: Early onset. Clinical features: Myasthenia, developmental delay, intellectual disability, seizures, craniofacial dysmorphism, rarely tremor.	*SNAP25b*	Synaptosome-associated protein 25b	*616330*

(continued)

Table 88.1 (continued)

Disorder	Distinguishing features	Gene or locus	Protein and mutation	OMIM#
ATX-TUBB2A	Clinical features: Spasticity, developmental delay, seizures, distal amyotrophy, rarely optic atrophy.	*TUBB2A*	Tubulin beta 2A class IIa	*615101*
ATX/ HSP-VAMP1(SPAX1)	Clinical features: Spastic ataxia. ***Ocular motor: "Initially there are problems with saccadic gaze greater than pursuit for vertical movements. Downward gaze is often more impaired than upward and eventually horizontal gaze also becomes affected. However, lateral eyeball movements are clearly much less affected than vertical"*** [33].	*VAMP1*	Vesicle-associated membrane protein 1	*108600*

References

1. Tazon B, Badenas C, Jimenez L, Munoz E, Mila M. SCA8 in the Spanish population including one homozygous patient. Clin Genet. 2002;62(5):404–9.
2. Morita H, Yoshida K, Suzuki K, Ikeda SI. A Japanese case of SCA14 with the Gly128Asp mutation. J Hum Genet. 2006;51(12):1118–21.
3. Duarri A, Jezierska J, Fokkens M, Meijer M, Schelhaas HJ, den Dunnen WF, et al. Mutations in potassium channel kcnd3 cause spinocerebellar ataxia type 19. Ann Neurol. 2012;72(6):870–80.
4. Knight MA, Gardner RJ, Bahlo M, Matsuura T, Dixon JA, Forrest SM, et al. Dominantly inherited ataxia and dysphonia with dentate calcification: spinocerebellar ataxia type 20. Brain. 2004;127(Pt 5):1172–81.
5. Coutinho P, Cruz VT, Tuna A, Silva SE, Guimaraes J. Cerebellar ataxia with spasmodic cough: a new form of dominant ataxia. Arch Neurol. 2006;63(4):553–5.
6. Bakalkin G, Watanabe H, Jezierska J, Depoorter C, Verschuuren-Bemelmans C, Bazov I, et al. Prodynorphin mutations cause the neurodegenerative disorder spinocerebellar ataxia type 23. Am J Hum Genet. 2010;87(5):593–603.
7. Seong E, Insolera R, Dulovic M, Kamsteeg EJ, Trinh J, Bruggemann N, et al. Mutations in VPS13D lead to a new recessive ataxia with spasticity and mitochondrial defects. Ann Neurol. 2018;83(6):1075–88.
8. Swartz BE, Burmeister M, Somers JT, Rottach KG, Bespalova IN, Leigh RJ. A form of inherited cerebellar ataxia with saccadic intrusions, increased saccadic speed, sensory neuropathy, and myoclonus. Ann N Y Acad Sci. 2002;956:441–4.
9. Swartz BE, Li S, Bespalova I, Burmeister M, Dulaney E, Robinson FR, et al. Pathogenesis of clinical signs in recessive ataxia with saccadic intrusions. Ann Neurol. 2003;54(6):824–8.
10. Stevanin G, Bouslam N, Thobois S, Azzedine H, Ravaux L, Boland A, et al. Spinocerebellar ataxia with sensory neuropathy (SCA25) maps to chromosome 2p. Ann Neurol. 2004;55(1):97–104.
11. Barbier M, Bahlo M, Pennisi A, Jacoupy M, Tankard RM, Ewenczyk C, et al. Heterozygous PNPT1 variants cause spinocerebellar ataxia type 25. Ann Neurol. 2022;92(1):122–37.
12. Choquet K, La Piana R, Brais B. A novel frameshift mutation in FGF14 causes an autosomal dominant episodic ataxia. Neurogenetics. 2015;16(3):233–6.
13. Piarroux J, Riant F, Humbertclaude V, Remerand G, Hadjadj J, Rejou F, et al. FGF14-related episodic ataxia: delineating the phenotype of episodic ataxia type 9. Ann Clin Transl Neurol. 2020;7(4):565–72.
14. Di Bella D, Lazzaro F, Brusco A, Plumari M, Battaglia G, Pastore A, et al. Mutations in the mitochondrial protease gene AFG3L2 cause dominant hereditary ataxia SCA28. Nat Genet. 2010;42(4):313–21.
15. Pierson TM, Adams D, Bonn F, Martinelli P, Cherukuri PF, Teer JK, et al. Whole-exome sequencing identifies homozygous AFG3L2 mutations in a spastic ataxia-neuropathy syndrome linked to mitochondrial m-AAA proteases. PLoS Genet. 2011;7(10):e1002325.
16. Dudding TE, Friend K, Schofield PW, Lee S, Wilkinson IA, Richards RI. Autosomal dominant congenital non-progressive ataxia overlaps with the SCA15 locus. Neurology. 2004;63(12):2288–92.
17. Tomiwa K, Baraitser M, Wilson J. Dominantly inherited congenital cerebellar ataxia with atrophy of the vermis. Pediatr Neurol. 1987;3(6):360–2.
18. Jen JC, Lee H, Cha YH, Nelson SF, Baloh RW. Genetic heterogeneity of autosomal dominant nonprogressive congenital ataxia. Neurology. 2006;67(9):1704–6.
19. Furman JM, Baloh RW, Chugani H, Waluch V, Bradley WG. Infantile cerebellar atrophy. Ann Neurol. 1985;17(4):399–402.
20. Fenichel GM, Phillips JA. Familial aplasia of the cerebellar vermis. Possible X-linked dominant inheritance. Arch Neurol. 1989;46(5):582–3.
21. Huang L, Chardon JW, Carter MT, Friend KL, Dudding TE, Schwartzentruber J, et al. Missense mutations in ITPR1 cause autosomal dominant congenital nonprogressive spinocerebellar ataxia. Orphanet J Rare Dis. 2012;7:67.
22. Imamura S, Tachi N, Oya K. Dominantly inherited early-onset non-progressive cerebellar ataxia syndrome. Brain Dev. 1993;15(5):372–6.
23. Kattah JC, Kolsky MP, Guy J, O'Doherty D. Primary position vertical nystagmus and cerebellar ataxia. Arch Neurol. 1983;40(5):310–4.
24. Parolin Schnekenberg R, Perkins EM, Miller JW, Davies WI, D'Adamo MC, Pessia M, et al. De novo point mutations in patients diagnosed with ataxic cerebral palsy. Brain. 2015;138(Pt 7):1817–32.
25. Storey E, Bahlo M, Fahey M, Sisson O, Lueck CJ, Gardner RJ. A new dominantly inherited pure cerebellar ataxia, SCA 30. J Neurol Neurosurg Psychiatry. 2009;80(4):408–11.
26. Serrano-Munuera C, Corral-Juan M, Stevanin G, San Nicolas H, Roig C, Corral J, et al. New subtype of spinocerebellar ataxia with altered vertical eye movements mapping to chromosome 1p32. JAMA Neurol. 2013;70(6):764–71.
27. Di Gregorio E, Borroni B, Giorgio E, Lacerenza D, Ferrero M, Lo Buono N, et al. ELOVL5 mutations cause spinocerebellar ataxia 38. Am J Hum Genet. 2014;95(2):209–17.
28. van Dijk GW, Wokke JH, Oey PL, Franssen H, Ippel PF, Veldman H. A new variant of sensory ataxic neuropathy with autosomal dominant inheritance. Brain. 1995;118(Pt 6):1557–63.
29. Pakdaman Y, Berland S, Bustad HJ, Erdal S, Thompson BA, James PA, et al. Genetic dominant variants in STUB1, segregating in families with SCA48, display in vitro functional impairments indistinctive from recessive variants associated with SCAR16. Int J Mol Sci. 2021;22(11):5870.
30. Biancalana V, Toft M, Le Ber I, Tison F, Scherrer E, Thibodeau S, et al. FMR1 premutations associated with fragile X-associated tremor/ataxia syndrome in multiple system atrophy. Arch Neurol. 2005;62(6):962–6.
31. Tesi B, Davidsson J, Voss M, Rahikkala E, Holmes TD, Chiang SCC, et al. Gain-of-function SAMD9L mutations cause a syn-

drome of cytopenia, immunodeficiency, MDS, and neurological symptoms. Blood. 2017;129(16):2266–79.
32. Chen DH, Below JE, Shimamura A, Keel SB, Matsushita M, Wolff J, et al. Ataxia-pancytopenia syndrome is caused by missense mutations in SAMD9L. Am J Hum Genet. 2016;98(6):1146–58.
33. Grewal KK, Stefanelli MG, Meijer IA, Hand CK, Rouleau GA, Ives EJ. A founder effect in three large Newfoundland families with a novel clinically variable spastic ataxia and supranuclear gaze palsy. Am J Med Genet A. 2004;131(3):249–54.
34. Salari M, Etemadifar M, Rashedi R, Mardani S. A review of ocular movement abnormalities in hereditary cerebellar ataxias. Cerebellum. 2024;23(2):702–21.
35. Burk K, Fetter M, Abele M, Laccone F, Brice A, Dichgans J, et al. Autosomal dominant cerebellar ataxia type I: oculomotor abnormalities in families with SCA1, SCA2, and SCA3. J Neurol. 1999;246(9):789–97.
36. Kim JM, Nam TS, Choi SM, Kim BC, Lee SH. Clinical value of vestibulo-ocular reflex in the differentiation of spinocerebellar ataxias. Sci Rep. 2023;13(1):14783.
37. Gordon CR, Joffe V, Vainstein G, Gadoth N. Vestibulo-ocular arreflexia in families with spinocerebellar ataxia type 3 (Machado-Joseph disease). J Neurol Neurosurg Psychiatry. 2003;74(10):1403–6.
38. Buttner N, Geschwind D, Jen JC, Perlman S, Pulst SM, Baloh RW. Oculomotor phenotypes in autosomal dominant ataxias. Arch Neurol. 1998;55(10):1353–7.
39. Lee SU, Kim JS, Kim HJ, Choi JY, Park JY, Kim JM, et al. Evolution of the vestibular function during head impulses in spinocerebellar ataxia type 6. J Neurol. 2020;267(6):1672–8.
40. Abou Chaar W, Eranki AN, Stevens HA, Watson SL, Wong DY, Avila VS, et al. Clinical, radiological and pathological features of a large American Cohort of Spinocerebellar Ataxia (SCA27B). Ann Neurol. 2024;96:1092–103.
41. Pellerin D, Wilke C, Traschutz A, Nagy S, Curro R, Dicaire MJ, et al. Intronic FGF14 GAA repeat expansions are a common cause of ataxia syndromes with neuropathy and bilateral vestibulopathy. J Neurol Neurosurg Psychiatry. 2024;95(2):175–9.
42. Rafehi H, Read J, Szmulewicz DJ, Davies KC, Snell P, Fearnley LG, et al. An intronic GAA repeat expansion in FGF14 causes the autosomal-dominant adult-onset ataxia SCA27B/ATX-FGF14. Am J Hum Genet. 2023;110(6):1018.
43. Pellerin D, Heindl F, Wilke C, Danzi MC, Traschutz A, Ashton C, et al. GAA-FGF14 disease: defining its frequency, molecular basis, and 4-aminopyridine response in a large downbeat nystagmus cohort. EBioMedicine. 2024;102:105076.
44. Nuzhnyi E, Abramycheva N, Protsenko A, Belyakova-Bodina A, Larina E, Fedotova E, et al. Clinical and video-oculographic characteristics of spinocerebellar ataxia type 27B (GAA-FGF14 ataxia): a single-center retrospective study. Clin Transl Neurosci [Internet]. 2024;8(4):29.

89 Episodic Ataxias and Acetazolamide-Responsive Ataxias

Overview

Episodic ataxias (EA), and more broadly the so-called *acetazolamide-responsive ataxias*, are due to genetic defects in ion channels. Most of these diseases are familial, inherited in an autosomal dominant fashion. Symptoms often, though not always, begin in childhood. EAs are rare, but among them, EA1 and EA2 were the first described and are the most common. EA3–6 are much rarer. In many cases, episodes are triggered by "stress" (emotional duress), physical exertion, or dietary factors (e.g., caffeine). The frequency of the episodes is extremely variable. The duration is also variable, though generally, the episodes in EA1 last minutes, while those in EA2 last hours. Typical symptoms include disequilibrium, ataxia/incoordination, and dysarthria; less common symptoms include visual symptoms, weakness, tremor, headache, and nausea. EA1 and EA2 have been connected to a broad range (dozens each) of genetic mutations, though the mechanism by which these mutations provoke neurological dysfunction is unclear. During an episode, patients with EA1 and EA2 may exhibit a variety of eye movement abnormalities. Interictally, some patients exhibit subtle eye movement abnormalities as well. Most of these eye movement abnormalities broadly localize to the cerebellum, but none is specific or sensitive for these disorders. For most of these disorders, many patients will exhibit some improvement with acetazolamide; EA2 patients tend to be most responsive. Several other medications have been tried in the treatment of episodic ataxias, though none has been studied as well as acetazolamide.

In a patient complaining of episodes disequilibrium and incoordination lasting minutes to hours, sometimes triggered by stress, by exertion, or by dietary factors, who feels normal (or nearly so) interictally, whose examination is unremarkable, and who has a family history of similar symptoms, EA is a reasonable consideration. Because EAs are rare, the clinician should keep a broader differential diagnosis that includes much more common entities, such as migraine associated vertigo. Imaging (usually an MRI of the brain) is reasonable to exclude structural lesions. A trial of acetazolamide (absent any contraindications) is reasonable. If that is unsuccessful, then referral to a specialist in ataxia, or a genetics counselor, is reasonable.

Introduction

Some sources [1, 2] credit Harry Lee Parker (1894–1959) with the original description of a cohort of patients with what eventually became known as familial episodic ataxia [3] (Fig. 89.1).

Fig. 89.1 Harry Lee Parker (1894–1959) is often credited with having provided the first description of a cohort of patients with what eventually became known as episodic ataxia. (From Klaas et al. [2]. Reproduced with written permission from Mayo Foundation for Medical Education and Research)

M. Cherchi, *Otoneurology and Vestibular Medicine*, https://doi.org/10.1007/978-3-031-94841-1_89

Familial episodic ataxias (EA) are channelopathies manifesting with episodic ataxia and disequilibrium [4]. Episodic ataxia type 1 (EA1) is caused by missense mutations in the potassium channel gene KCNA1 on chromosome 12p13. Episodic ataxia type 2 (EA2) is caused by missense and nonsense mutations in the calcium channel gene CACNA1A on chromosome 19p.

Episodic ataxias types 1 and 2 are quite rare but have been relatively well described. More recently reported episodic ataxias (types 3, 4, 5, and 6) are even more rare and are still being studied and debated.

These diseases are often grouped together because they share some features, yet there is probably no single feature that unifies them all. They are probably all due to genetic mutations, though for some variants, the culprit gene has not yet been identified. They are mostly hereditary, although EA6 has only been identified in a single individual (perhaps as a de novo mutation), so the term "familial" is probably inappropriate. Patients experience episodic symptoms, but interictally may experience milder chronic symptoms at baseline. Most patients exhibit ataxia, but not all. Most types have at least some patients whose symptoms improve with acetazolamide, but not all (such as EA4).

Episodic Ataxia Type 1

Introduction

In 1975, VanDyke and colleagues described a kindred in which 11 people over 3 generations exhibited episodic ataxia and continuous myokymia [5]. Subsequent reports [6] identified similar findings in other families. In 1994, linkage analysis identified mutations at a locus on chromosome 12p13 [7] that was identified as a gene that encodes the KCNA1 potassium channel [8]. This eventually led to this being designated as a familial episodic ataxia, and it was ultimately called "episodic ataxia type 1," when other types were discovered.

Episodic Ataxia Type 1, Various Genetic Mutations

EA1 is inherited in an autosomal dominant fashion.

A broad range of genetic mutations has been described in EA1 [9–22].

For each mutation, there have been various proposals for the mechanism by which it causes neurological dysfunction [15, 23–34].

Episodic Ataxia Type 1, Clinical Presentation

Episodes of EA1 can begin in early childhood or adolescence [35].

Some episodes of EA1 have no apparent trigger. In some patients, the triggers are relatively consistent. Graves and colleagues studied 39 patients (51% male), age 15–65 years (median 37 years), with genetically confirmed EA1, and reported the triggers listed in Table 89.1 from Graves and colleagues [36].

EA1 attacks usually last minutes [4, 37], less commonly hours [36]. In some patients, the episodes can occur with impressive frequency, up to dozens of times per day [35]. Symptoms usually include midline cerebellar dysfunction (truncal ataxia), limb ataxia, dysarthria, dystonia [37], and sometimes visual symptoms such as oscillopsia and visual blurring [38]. Table 89.2 lists the symptoms reported by the patients studied in the series by Graves and colleagues [36].

Table 89.1 Triggers of episodic ataxia type 1

Trigger	Number (n = 33)	Percent
Exertion/exercise	29	87.9
Stress/emotional upset	28	84.8
Environmental temperature	18	54.5
Fever	10	30.3
Caffeine	9	27.3
Alcohol	9	27.3
Sudden movement	9	27.3
Diet	9	27.3
Rest after exertion/exercise	9	27.3
Prolonged rest	7	21.2
Pregnancy	6 (out of 19 females)	31.6
Menstruation	5 (of 19 females)	26.3
Tiredness	2	6.1
Strong smells	1	3.0
Anxiety	1	3.0
Bending over/looking down	1	3.0

From Graves et al. [36]

Table 89.2 Symptoms of episodic ataxia type 1

Symptom	Number (n = 33)	Percent
Imbalance	31	93.9
Slurred speech	31	93.9
Incoordination of hands	28	84.8
Weakness	28	84.8
Tremors	27	81.8
Muscle twitching	26	78.8
Muscle stiffness	21	63.6
Spinning sensation	14	42.4
Visual disturbance	13	39.4
Nausea	6	18.2
Headache	4	12.1
Vomiting	1	3.0

From Graves et al. [36]

Although the disease is genetically mediated, it appears that nongenetic factors also influence the clinical manifestations [39, 40].

Episodic Ataxia Type 1, Ocular Motor Manifestations

Some of these patients exhibit *down beat nystagmus* [41].

Episodic Ataxia Type 1, Interictal Symptoms

In between episodes, patients with EA1 often exhibit myokymia, but may also experience other symptoms, such as headache or pseudodystonia [42].

Episodic Ataxia Type 1, Treatments

Acetazolamide was fortuitously found to have some effect on symptoms of EA1 [43], and subsequently, it was reported that approximately ***50%*** of patients improve with acetazolamide [4].

There are a few reports of EA1 responding to *phenytoin* [37].

Episodic Ataxia Type 2

Episodic Ataxia Type 2, Epidemiology

The prevalence of EA2 has been estimated at 1 in 100,000 [35]. One report of several generations of a Korean family found decreasing age of onset with increasing number of trinucleotide repeats, suggesting anticipation [44].

Episodic Ataxia Type 2, Clinical Presentation

Episodes of EA2 usually begin in infancy or childhood. Some mutations appear to have much later onset; Cuenca-Leon and colleagues reported a patient in whom symptom onset was in the sixth decade [45].

Episodes of EA2 can be triggered by stress, exercise, and caffeine [4]. The mechanism for these triggers is unclear, but animal models suggest that emotional and physiologic duress activate the locus coeruleus and trigger release of norepinephrine in the cerebellum, which has downstream effects of interfering with Purkinje cell function. This postulated mechanism is illustrated in Fig. 89.2 from Benarroch [46].

The frequency of the episodes in EA2 is extremely variable, ranging from a few per year to a few per week; they can last for hours [35].

Ictal symptoms of EA2 include a sensation of vertigo, dysarthria, and truncal ataxia [35], and may be accompanied by headache [37]. Jen and colleagues [47] state that the most common neurologic symptoms are ataxia, vertigo, and fluctuating weakness. Additional features are listed in Table 89.3.

Some patients describe experiencing vertigo *without* ataxia [48].

The clinical presentation cannot be predicted from the specific mutation (discussed below) [47]. Multiple patients in the same family with the same genetic mutation can have different clinical manifestations [49].

Episodic Ataxia Type 2, Genetics

EA2 is inherited in an autosomal dominant fashion.

Linkage analysis identified a locus on chromosome 19p [50, 51] that was found to encode the CACNA1A calcium channel gene [52]. Most of the CACNA1A mutations in EA2 are nonsense mutations leading to a truncated protein with loss of function [35]. However, over 30 mutations in this gene have been found to manifest as EA2, and these mutations are of various types, including nonsense mutations, missense mutations, deletions, an insertion, and a donor splice site mutation [47].

A large number of mutations has been described in patients diagnosed with EA2 [45, 47, 49–104].

The CACNA1A-encoded calcium channels are heavily expressed in Purkinje and granule cells in the cerebellum.

Other mutations in the CACNA1A gene manifest as other neurological disorders: familial hemiplegic migraine type 1, and spinocerebellar ataxia type 6 [35, 58, 87, 105].

Episodic Ataxia Type 2, Pathophysiology

The exact mechanism by which EA2 mutations provoke dysfunction of calcium channels is unclear. Possibilities include "reduced current density, increased rate of inactivation, and a shift in the voltage dependence of activation to more positive values" [94].

MR spectroscopy on patients with EA2 reported abnormal pH (alkalosis) in the *cerebellum* [1, 106]. Harno and colleagues [107] speculate that acetazolamide may "stabilize the dysfunctioning P/Q-type calcium channels, which express highly in the cerebellum." Identification of abnormal metabolism in the cerebellar cortex seems logical given the clinical manifestations of EA2.

However, there is also some evidence of abnormalities in *cerebral cortical function*. Helmich and colleagues [108] studied six EA2 patients and 13 healthy controls with transcranial magnetic stimulation and reported that, "patients with episodic ataxia type 2 have an excessive increase in

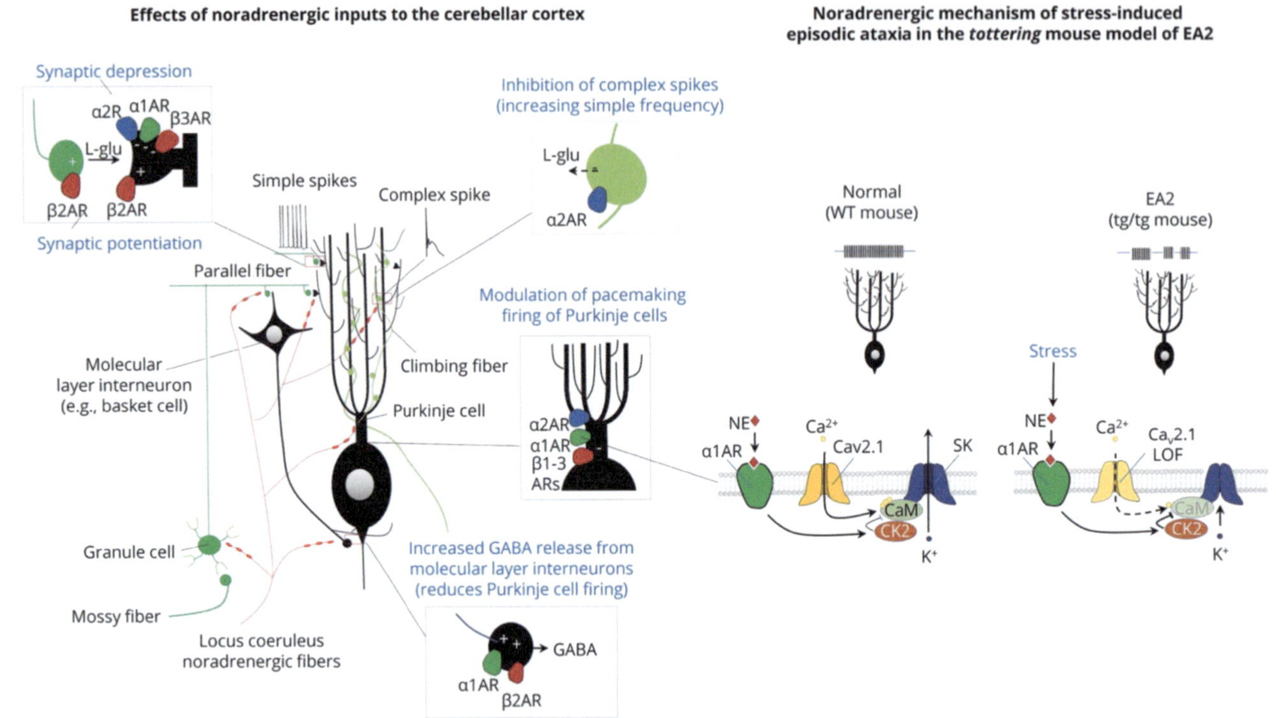

Fig. 89.2 Effects of noradrenergic inputs to the cerebellar cortex and mechanisms of stress-induced episodic ataxia type 2 (EA2). The cerebellar cortex receives noradrenergic input from the locus coeruleus (LC). Norepinephrine (NE), acting through α1-, α2-, and β-adrenergic receptors (adrenoceptors [ARs]), has a complex modulatory effect on Purkinje cell activity through its effect on their intrinsic electrophysiologic activity, their parallel fiber input resulting in simple spikes, their climbing fiber input producing complex spikes, and the influence of molecular layer GABAergic inhibitory neurons. Norepinephrine, acting through α1ARs, α2ARs, and β3ARs in Purkinje cells, produces depression at the parallel fiber–Purkinje cell synapse. By contrast, activation of β2ARs facilitates long-term potentiation at this synapse through both presynaptic and postsynaptic mechanisms. Activation of presynaptic α2Rs inhibits glutamate release from climbing fibers and reduces complex spike activity in Purkinje cells. This would promote an increased frequency of single-spike firing. However, this effect is overshadowed by the inhibitory effects through molecular layer interneurons, as NE acting through presynaptic α1Rs and β2ARs promotes the release of GABA at their inhibitory synapses on Purkinje cells. Norepinephrine may also affect the intrinsic firing of Purkinje cells by directly affecting ion channel function. The normal spontaneous activity of Purkinje cells requires the functional coupling between P/Q (Cav2.1) channels and small conductance Ca2+ -activated potassium channels (SK) responsible for slow after hyperpolarization. The Ca2+ sensor calmodulin (CaM) forms a complex with the SK channels and with casein kinase 2 (CK2), which phosphorylates CaM and accelerates SK deactivation. The tottering mice harbors a genetic variant affecting Cav2.1 channels in Purkinje cells and is a model of episodic ataxia type 2 (EA2). In this model, stress induces LC activation and NE release in the cerebellum leading to the activation of α1ARs, which promotes CK2-dependent phosphorylation of CaM resulting in reduced probability of the SK channels in Purkinje cells, thus impairing precise Purkinje cell firing. (From Benarroch [46]. Reproduced with written permission from Wolters Kluwer Health Limited)

Table 89.3 Clinical features associated with episodic ataxia type 2

Onset	Before age 20 [years]
Duration	Hours to days
Trigger	Emotional stress Physical exertion
Associated symptoms	Progressive ataxia Migraine, hemiplegic migraine Fluctuating generalized weakness Vertigo Dystonia Seizures
Signs	Gaze evoked and rebound nystagmus Slurring of speech Truncal ataxia

From Jen et al. [47]

motor cortex excitability following a strong facilitatory input." Indelicato and colleagues [109] studied 19 genetically confirmed EA2 patients with electroencephalography and found 12 (63%) were abnormal at baseline (outside of any attack), usually showing "lateralized intermittent slowly, mainly affecting the temporal region," and 7 (37%) showed "interictal epileptic discharges."

There is also some evidence of *peripheral nerve dysfunction* in EA2. Krishnan and colleagues [110] reported on 3 members of a family with EA2 and found that, "Nerve excitability testing demonstrated significant abnormalities, with all patients outside the normal 95 % confidence limits in having a high rheobase and reduced early hyperpolarizing threshold electrotonus. On average there were also significant reductions in refractoriness, late sub excitability and early depolarizing threshold electrotonus." Tomlinson and colleagues [97] studied 8 patients from 2 families with EA2 and reported that, "All patients had similar excitability abnormalities, including a high electrical threshold and increased responses to hyperpolarizing and depolarizing currents in threshold electrotonus," and 2 patients, showed "increased jitter in single-fibre EMG studies indicated unstable neuromuscular transmission."

Episodic Ataxia Type 2, Physical Examination

If an EA2 patient is examined during an attack, one may identify *spontaneous nystagmus*. Early in the disease, the patient's examination may be normal interictally, but later there may be *spontaneous down beat nystagmus* even between attacks [35] and *gaze evoked nystagmus* [111].

Kim and colleagues [112] reported the case of an EA2 patient with *rebound up beat nystagmus after lateral gaze holding*—in effect, perverted rebound nystagmus.

Shervin Badv and Niksirat [113] reported *downward gaze palsy* during attacks of a patient with EA2.

Episodic Ataxia Type 2, Vestibular Testing

Harno and colleagues [114] studied three patients with genetically confirmed EA2 with nystagmography while the patients were off, and then back on, treatment with acetazolamide. The first patient while off treatment exhibited *hypermetric saccades*; this improved on treatment. The second patient, while off treatment, exhibited *spontaneous* left beat *nystagmus* of 5 deg/s that increased with head turns; the nystagmus improved (but did not cease) on treatment. The third patient while off treatment exhibited *spontaneous up beat nystagmus* of 3 deg/s and *horizontal rebound nystagmus* of 2 deg/s.

Harno and colleagues [114] also studied computerized dynamic posturography on the second patient, and reported that, "the *sway velocities in posturography were pathologically increased* without medication."

Choi and colleagues [114] studied 16 patients with EA2 interictally and reported that more than half of the patients exhibited abnormalities in at least one of each vestibular test of video head impulse testing (vHIT), subjective visual vertical (SVV), ocular vestibular evoked myogenic potentials (oVEMP) or cervical vestibular evoked myogenic potentials (cVEMP). Specifically, "Patients with EA2 commonly showed *abnormal VOR [vestibulo-ocular reflex] responses at least for one SCC [semicircular canal] with high-acceleration, high-frequency head impulses* (14/16, 88%), and impaired visual-vestibular interaction (7/12, 58%). In response to low acceleration and frequency stimuli, the VOR gains were generally normal. The majority of EA2 patients had impairments in at least one of the otolith function tests (13/16, 81%): SVV [*subjective visual vertical*] tilt or variability (7/14, 50%), oVEMP [*ocular vestibular evoked myogenic potentials*] (8/15, 53%), and cVEMP [*cervical vestibular evoked myogenic potentials*] (4/16, 25%). Vestibular impairments are common in EA2 even during the interictal periods" [115].

Another report of four patients observed, "*deficits in tracking* behavior with common features. *Ocular tracking tended to result in hypermetric saccades at longer than normal latencies. Smooth pursuit tracking was absent* in 1 patient and had lower than normal gain in the others" [116].

Gordon and colleagues [117] studied four EA2 patients using scleral search coils. They reported that 2 (50%) exhibited "abnormal VOR" (vestibulo-ocular reflex) and "*showed essentially no SP*" (*smooth pursuit*).

Kipfer and colleagues [118] reported on a family of three EA2 patients who demonstrated *slow horizontal saccades*.

Kim and colleagues [119] reported the case of an EA2 patient with *slow vertical saccades* (both upward and downward).

Rucker and colleagues [120] reported on two patients with EA2 who exhibited *slowing of adducting saccades*. Shimmura and colleagues [121] described *slow adducting movements during smooth pursuit* in a patient with EA2.

Some investigators point out that there are some similarities in eye movement abnormalities between EA2 and spinocerebellar ataxia type 6 [122].

Episodic Ataxia Type 2, Imaging

MR spectroscopy on patients with EA2 reported abnormal pH (alkalosis) in the cerebellum [1, 106]. Harno and colleagues [107] speculate that acetazolamide may "stabilize the dysfunctioning P/Q-type calcium channels, which express highly in the cerebellum."

Episodic Ataxia Type 2, Treatments

Acetazolamide remains the best studied treatment for EA2, with response rates ranging from ***70%*** [123] to ***90%*** [4]. This remains the best studied treatment.

One report described symptomatic improvement in EA2 through a combination of acetazolamide and *valproate* [124].

There is also some literature regarding 4-AP (*4-aminopyridine*, also called dalfampridine) [125–132]. An extended release of 4-AP called fampridine has also been studied [133].

One report advocated the combination of 4-AP plus *topiramate* in patients unresponsive to acetazolamide [134].

Other proposed treatments have included *chlorzoxazone* [135], *levetiracetam* [136, 137], and flunarizine [138].

Episodic Ataxia Type 3

In 2001, Steckley and colleagues [139] reported on a large Canadian family of Mennonite ancestry. In this kindred, there was a broad age range of symptom onset (1–42 years). Episodes were sometimes provoked by movement (kinesigenic); involved symptoms of vertigo, nausea, and tinnitus; tended to last minutes, rarely hours; and responded to acetazolamide [139]. Linkage analysis identified a locus on chromosome 1q42 [140], but the specific gene has not yet been identified.

Episodic Ataxia Type 4

Episodic ataxia type 4, also called periodic vestibulocerebellar ataxia, is another autosomal dominant ataxia originally described in two families from North Carolina [141]. Its symptoms are similar to those of EA2 [35], though the age of onset is later (3rd–6th decade), and the attacks do not respond to acetazolamide.

Episodic Ataxia Type 5

Episodic ataxia type 5 was described by Escayg and colleagues [142] who reported on a French Canadian kindred whose members suffered from attacks very similar to EA2, although the age of onset was somewhat later (3rd–4th decade); symptoms included vertigo, ataxia, and dysarthria, lasting hours at a time, improved with acetazolamide [35]. The disorder is linked to a mutation in a gene for a different calcium channel (CACNB4).

Episodic Ataxia Type 6

Episodic ataxia type 6 was described in a single child in a report by Jen and colleagues [143]. Aside from symptoms of ataxia, the episodes are different in that they involve hemiplegia and seizures. The disease has been attributed to a mutation in an astrocyte glutamate transporter [35].

References

1. Bain PG, O'Brien MD, Keevil SF, Porter DA. Familial periodic cerebellar ataxia: a problem of cerebellar intracellular pH homeostasis. Ann Neurol. 1992;31(2):147–54.
2. Klaas JP, Burkholder DB, Singer W, Boes CJ. Harry Lee Parker and paroxysmal dysarthria and ataxia. Neurology. 2013;80(3):311–4.
3. Parker HL. Periodic ataxia. Coll Papers Mayo Clin. 1946;38:642–5.
4. Baloh RW, Jen JC. Genetics of familial episodic vertigo and ataxia. Ann N Y Acad Sci. 2002;956:338–45.
5. VanDyke DH, Griggs RC, Murphy MJ, Goldstein MN. Hereditary myokymia and periodic ataxia. J Neurol Sci. 1975;25(1):109–18.
6. Hanson PA, Martinez LB, Cassidy R. Contractures, continuous muscle discharges, and titubation. Ann Neurol. 1977;1(2):120–4.
7. Litt M, Kramer P, Browne D, Gancher S, Brunt ER, Root D, et al. A gene for episodic ataxia/myokymia maps to chromosome 12p13. Am J Hum Genet. 1994;55(4):702–9.
8. Browne DL, Gancher ST, Nutt JG, Brunt ER, Smith EA, Kramer P, et al. Episodic ataxia/myokymia syndrome is associated with point mutations in the human potassium channel gene, KCNA1. Nat Genet. 1994;8(2):136–40.
9. Bretschneider F, Wrisch A, Lehmann-Horn F, Grissmer S. Expression in mammalian cells and electrophysiological characterization of two mutant Kv1.1 channels causing episodic ataxia type 1 (EA-1). Eur J Neurosci. 1999;11(7):2403–12.
10. Chen SH, Fu SJ, Huang JJ, Tang CY. The episodic ataxia type 1 mutation I262T alters voltage-dependent gating and disrupts protein biosynthesis of human Kv1.1 potassium channels. Sci Rep. 2016;6:19378.
11. D'Adamo MC, Gallenmuller C, Servettini I, Hartl E, Tucker SJ, Arning L, et al. Novel phenotype associated with a mutation in the KCNA1(Kv1.1) gene. Front Physiol. 2014;5:525.
12. Eunson LH, Rea R, Zuberi SM, Youroukos S, Panayiotopoulos CP, Liguori R, et al. Clinical, genetic, and expression studies of mutations in the potassium channel gene KCNA1 reveal new phenotypic variability. Ann Neurol. 2000;48(4):647–56.
13. Imbrici P, Gualandi F, D'Adamo MC, Masieri MT, Cudia P, De Grandis D, et al. A novel KCNA1 mutation identified in an Italian family affected by episodic ataxia type 1. Neuroscience. 2008;157(3):577–87.
14. Karalok ZS, Megaro A, Cenciarini M, Guven A, Hasan SM, Taskin BD, et al. Identification of a New de Novo mutation underlying regressive episodic ataxia type I. Front Neurol. 2018;9:587.
15. Klein A, Boltshauser E, Jen J, Baloh RW. Episodic ataxia type 1 with distal weakness: a novel manifestation of a potassium channelopathy. Neuropediatrics. 2004;35(2):147–9.
16. Knight MA, Storey E, McKinlay Gardner RJ, Hand P, Forrest SM. Identification of a novel missense mutation L329I in the episodic ataxia type 1 gene KCNA1–a challenging problem. Hum Mutat. 2000;16(4):374.
17. Lassche S, Lainez S, Bloem BR, van de Warrenburg BP, Hofmeijer J, Lemmink HH, et al. A novel KCNA1 mutation causing episodic ataxia type I. Muscle Nerve. 2014;50(2):289–91.
18. Lee GB, Kim GY, Jeong IH, Kim N, Kim JW. A novel KCNA1 mutation in an episodic ataxia type 1 patient with asterixis and falls. J Clin Neurol. 2021;17(2):333–5.
19. Mestre TA, Manole A, MacDonald H, Riazi S, Kraeva N, Hanna MG, et al. A novel KCNA1 mutation in a family with episodic ataxia and malignant hyperthermia. Neurogenetics. 2016;17(4):245–9.
20. Petitjean D, Kalstrup T, Zhao J, Blunck R. A disease mutation causing episodic ataxia type I in the S1 links directly to the voltage sensor and the selectivity filter in Kv channels. J Neurosci. 2015;35(35):12198–206.
21. Shook SJ, Mamsa H, Jen JC, Baloh RW, Zhou L. Novel mutation in KCNA1 causes episodic ataxia with paroxysmal dyspnea. Muscle Nerve. 2008;37(3):399–402.
22. Yuan H, Yuan H, Wang Q, Ye W, Yao R, Xu W, et al. Two novel KCNA1 variants identified in two unrelated Chinese families affected by episodic ataxia type 1 and neurodevelopmental disorders. Mol Genet Genomic Med. 2020;8(10):e1434.
23. D'Adamo MC, Imbrici P, Sponcichetti F, Pessia M. Mutations in the KCNA1 gene associated with episodic ataxia type-1 syndrome

impair heteromeric voltage-gated K(+) channel function. FASEB J. 1999;13(11):1335–45.
24. D'Adamo MC, Liu Z, Adelman JP, Maylie J, Pessia M. Episodic ataxia type-1 mutations in the hKv1.1 cytoplasmic pore region alter the gating properties of the channel. EMBO J. 1998;17(5):1200–7.
25. Ferrick-Kiddie EA, Rosenthal JJ, Ayers GD, Emeson RB. Mutations underlying Episodic Ataxia type-1 antagonize Kv1.1 RNA editing. Sci Rep. 2017;7:41095.
26. Heeroma JH, Henneberger C, Rajakulendran S, Hanna MG, Schorge S, Kullmann DM. Episodic ataxia type 1 mutations differentially affect neuronal excitability and transmitter release. Dis Model Mech. 2009;2(11–12):612–9.
27. Imbrici P, D'Adamo MC, Grottesi A, Biscarini A, Pessia M. Episodic ataxia type 1 mutations affect fast inactivation of K+ channels by a reduction in either subunit surface expression or affinity for inactivation domain. Am J Physiol Cell Physiol. 2011;300(6):C1314–22.
28. Imbrici P, D'Adamo MC, Kullmann DM, Pessia M. Episodic ataxia type 1 mutations in the KCNA1 gene impair the fast inactivation properties of the human potassium channels Kv1.4-1.1/Kvbeta1.1 and Kv1.4-1.1/Kvbeta1.2. Eur J Neurosci. 2006;24(11):3073–83.
29. Manganas LN, Akhtar S, Antonucci DE, Campomanes CR, Dolly JO, Trimmer JS. Episodic ataxia type-1 mutations in the Kv1.1 potassium channel display distinct folding and intracellular trafficking properties. J Biol Chem. 2001;276(52):49427–34.
30. Maylie B, Bissonnette E, Virk M, Adelman JP, Maylie JG. Episodic ataxia type 1 mutations in the human Kv1.1 potassium channel alter hKvbeta 1-induced N-type inactivation. J Neurosci. 2002;22(12):4786–93.
31. Peters CJ, Werry D, Gill HS, Accili EA, Fedida D. Mechanism of accelerated current decay caused by an episodic ataxia type-1-associated mutant in a potassium channel pore. J Neurosci. 2011;31(48):17449–59.
32. Tomlinson SE, Tan SV, Kullmann DM, Griggs RC, Burke D, Hanna MG, et al. Nerve excitability studies characterize Kv1.1 fast potassium channel dysfunction in patients with episodic ataxia type 1. Brain. 2010;133(Pt 12):3530–40.
33. Zhao J, Petitjean D, Haddad GA, Batulan Z, Blunck R. A common kinetic property of mutations linked to episodic ataxia type 1 studied in the shaker Kv channel. Int J Mol Sci. 2020;21(20):7602.
34. Zhu J, Alsaber R, Zhao J, Ribeiro-Hurley E, Thornhill WB. Characterization of the Kv1.1 I262T and S342I mutations associated with episodic ataxia 1 with distinct phenotypes. Arch Biochem Biophys. 2012;524(2):99–105.
35. Kotagal V. Acetazolamide-responsive ataxia. Semin Neurol. 2012;32(5):533–7.
36. Graves TD, Cha YH, Hahn AF, Barohn R, Salajegheh MK, Griggs RC, et al. Episodic ataxia type 1: clinical characterization, quality of life and genotype-phenotype correlation. Brain. 2014;137(Pt 4):1009–18.
37. Dressler D, Benecke R. Diagnosis and management of acute movement disorders. J Neurol. 2005;252(11):1299–306.
38. Rajakulendran S, Schorge S, Kullmann DM, Hanna MG. Episodic ataxia type 1: a neuronal potassium channelopathy. Neurotherapeutics. 2007;4(2):258–66.
39. Gilbert GJ, Graves TD, Kullmann DM. Nongenetic factors influence severity of episodic ataxia type 1 in monozygotic twins. Neurology. 2011;76(5):490.
40. Graves TD, Rajakulendran S, Zuberi SM, Morris HR, Schorge S, Hanna MG, et al. Nongenetic factors influence severity of episodic ataxia type 1 in monozygotic twins. Neurology. 2010;75(4):367–72.
41. Jorge A, Melancia D, Figueiredo C, Galego O, Oliveira J, Martins AI, et al. Downbeat nystagmus in episodic ataxia type 1 associated with a novel KCNA1 mutation. Mov Disord. 2022;37(2):430–2.
42. Bhattacharjee S, Deenadayalu A, Paramanandam V. Interictal headache, pseudodystonia, and persistent ataxia in episodic ataxia type 1 due to a novel KCNA1 gene mutation. Mov Disord Clin Pract. 2022;9(2):272–4.
43. Griggs RC, Moxley RT 3rd, Lafrance RA, McQuillen J. Hereditary paroxysmal ataxia: response to acetazolamide. Neurology. 1978;28(12):1259–64.
44. Choi KD, Yook JW, Kim MJ, Kim HS, Park YE, Kim JS, et al. Possible anticipation associated with a novel splice site mutation in episodic ataxia type 2. Neurol Sci. 2013;34(9):1629–32.
45. Cuenca-Leon E, Banchs I, Serra SA, Latorre P, Fernandez-Castillo N, Corominas R, et al. Late-onset episodic ataxia type 2 associated with a novel loss-of-function mutation in the CACNA1A gene. J Neurol Sci. 2009;280(1–2):10–4.
46. Benarroch E. What is the role of norepinephrine in cerebellar modulation and stress-induced episodic ataxia? Neurology. 2023;100(8):383–6.
47. Jen J, Kim GW, Baloh RW. Clinical spectrum of episodic ataxia type 2. Neurology. 2004;62(1):17–22.
48. Ling X, Zhao DH, Zhao J, Shen B, Yang X. Episodic ataxia type 2 characterised by recurrent dizziness/vertigo: a report of four cases. Int J Neurosci. 2019;129(2):103–9.
49. Jung J, Testard H, Tournier-Lasserve E, Riant F, Vallet AE, Berroir S, et al. Phenotypic variability of episodic ataxia type 2 mutations: a family study. Eur Neurol. 2010;64(2):114–6.
50. Kramer PL, Yue Q, Gancher ST, Nutt JG, Baloh R, Smith E, et al. A locus for the nystagmus-associated form of episodic ataxia maps to an 11-cM region on chromosome 19p. Am J Hum Genet. 1995;57(1):182–5.
51. von Brederlow B, Hahn AF, Koopman WJ, Ebers GC, Bulman DE. Mapping the gene for acetazolamide responsive hereditary paryoxysmal cerebellar ataxia to chromosome 19p. Hum Mol Genet. 1995;4(2):279–84.
52. Ophoff RA, Terwindt GM, Vergouwe MN, van Eijk R, Oefner PJ, Hoffman SM, et al. Familial hemiplegic migraine and episodic ataxia type-2 are caused by mutations in the Ca2+ channel gene CACNL1A4. Cell. 1996;87(3):543–52.
53. Ahuja AS, Rozen TD, Atwal PS. A sleep modulated Channelopathy: a novel CACNA1A pathogenic variant identified in episodic Ataxia type 2 and a potential link to sleep alleviated migraine. BMC Neurol. 2019;19(1):246.
54. Algahtani H, Shirah B, Algahtani R, Al-Qahtani MH, Abdulkareem AA, Naseer MI. A novel mutation in CACNA1A gene in a Saudi female with episodic ataxia type 2 with no response to acetazolamide or 4-aminopyridine. Intractable Rare Dis Res. 2019;8(1):67–71.
55. Balck A, Tunc S, Schmitz J, Hollstein R, Kaiser FJ, Bruggemann N. A novel frameshift CACNA1A mutation causing episodic ataxia type 2. Cerebellum. 2018;17(4):504–6.
56. Dahimene S, Page KM, Nieto-Rostro M, Pratt WS, D'Arco M, Dolphin AC. A CaV2.1 N-terminal fragment relieves the dominant-negative inhibition by an Episodic ataxia 2 mutant. Neurobiol Dis. 2016;93:243–56.
57. Denier C, Ducros A, Durr A, Eymard B, Chassande B, Tournier-Lasserve E. Missense CACNA1A mutation causing episodic ataxia type 2. Arch Neurol. 2001;58(2):292–5.
58. Denier C, Ducros A, Vahedi K, Joutel A, Thierry P, Ritz A, et al. High prevalence of CACNA1A truncations and broader clinical spectrum in episodic ataxia type 2. Neurology. 1999;52(9):1816–21.
59. Dorgans K, Salvi J, Bertaso F, Bernard L, Lory P, Doussau F, et al. Characterization of the dominant inheritance mechanism of episodic ataxia type 2. Neurobiol Dis. 2017;106:110–23.
60. Eunson LH, Graves TD, Hanna MG. New calcium channel mutations predict aberrant RNA splicing in episodic ataxia. Neurology. 2005;65(2):308–10.

61. Graves TD, Imbrici P, Kors EE, Terwindt GM, Eunson LH, Frants RR, et al. Premature stop codons in a facilitating EF-hand splice variant of CaV2.1 cause episodic ataxia type 2. Neurobiol Dis. 2008;32(1):10–5.
62. Guida S, Trettel F, Pagnutti S, Mantuano E, Tottene A, Veneziano L, et al. Complete loss of P/Q calcium channel activity caused by a CACNA1A missense mutation carried by patients with episodic ataxia type 2. Am J Hum Genet. 2001;68(3):759–64.
63. Hirose H, Arayama T, Takita J, Igarashi T, Hayashi Y, Nagao Y. A family of episodic ataxia type 2: no evidence of genetic linkage to the CACNA1A gene. Int J Mol Med. 2003;11(2):187–9.
64. Hommersom MP, van Prooije TH, Pennings M, Schouten MI, van Bokhoven H, Kamsteeg EJ, et al. The complexities of CACNA1A in clinical neurogenetics. J Neurol. 2022;269(6):3094–108.
65. Hu Y, Jiang H, Wang Q, Xie Z, Pan S. Identification of a novel nonsense mutation p.Tyr1957Ter of CACNA1A in a Chinese family with episodic ataxia 2. PLoS One. 2013;8(2):e56362.
66. Imbrici P, Eunson LH, Graves TD, Bhatia KP, Wadia NH, Kullmann DM, et al. Late-onset episodic ataxia type 2 due to an in-frame insertion in CACNA1A. Neurology. 2005;65(6):944–6.
67. Isaacs DA, Bradshaw MJ, Brown K, Hedera P. Case report of novel CACNA1A gene mutation causing episodic ataxia type 2. SAGE Open Med Case Rep. 2017;5:2050313X17706044.
68. Jen J, Wan J, Graves M, Yu H, Mock AF, Coulin CJ, et al. Loss-of-function EA2 mutations are associated with impaired neuromuscular transmission. Neurology. 2001;57(10):1843–8.
69. Jen J, Yue Q, Nelson SF, Yu H, Litt M, Nutt J, et al. A novel nonsense mutation in CACNA1A causes episodic ataxia and hemiplegia. Neurology. 1999;53(1):34–7.
70. Jeng CJ, Chen YT, Chen YW, Tang CY. Dominant-negative effects of human P/Q-type Ca2+ channel mutations associated with episodic ataxia type 2. Am J Physiol Cell Physiol. 2006;290(4):C1209–20.
71. Jeng CJ, Sun MC, Chen YW, Tang CY. Dominant-negative effects of episodic ataxia type 2 mutations involve disruption of membrane trafficking of human P/Q-type Ca2+ channels. J Cell Physiol. 2008;214(2):422–33.
72. Kaja S, Van De Ven RC, Frants RR, Ferrari MD, Van Den Maagdenberg AM, Plomp JJ. Reduced ACh release at neuromuscular synapses of heterozygous leaner Ca(v)2.1-mutant mice. Synapse. 2008;62(5):337–44.
73. Kaunisto MA, Harno H, Kallela M, Somer H, Sallinen R, Hamalainen E, et al. Novel splice site CACNA1A mutation causing episodic ataxia type 2. Neurogenetics. 2004;5(1):69–73.
74. Kim JM, Kim JS, Ki CS, Jeon BS. Episodic ataxia type 2 due to a deletion mutation in the CACNA1A gene in a Korean family. J Clin Neurol. 2006;2(4):268–71.
75. Kinder S, Ossig C, Wienecke M, Beyer A, von der Hagen M, Storch A, et al. Novel frameshift mutation in the CACNA1A gene causing a mixed phenotype of episodic ataxia and familiar hemiplegic migraine. Eur J Paediatr Neurol. 2015;19(1):72–4.
76. Labrum RW, Rajakulendran S, Graves TD, Eunson LH, Bevan R, Sweeney MG, et al. Large scale calcium channel gene rearrangements in episodic ataxia and hemiplegic migraine: implications for diagnostic testing. J Med Genet. 2009;46(11):786–91.
77. Lance S, Mossman S, Poke G. A novel CACNA1A nonsense variant [c.4054C>T (p.Arg1352($_*$))] causing episodic ataxia type 2. Case Rep Neurol Med. 2018;2018:5802650.
78. Lipman AR, Fan X, Shen Y, Chung WK. Clinical and genetic characterization of CACNA1A-related disease. Clin Genet. 2022;102(4):288–95.
79. Maksemous N, Roy B, Smith RA, Griffiths LR. Next-generation sequencing identifies novel CACNA1A gene mutations in episodic ataxia type 2. Mol Genet Genomic Med. 2016;4(2):211–22.
80. Maksemous N, Smith RA, Sutherland HG, Sampaio H, Griffiths LR. Whole-exome sequencing implicates SCN2A in episodic ataxia, but multiple ion channel variants may contribute to phenotypic complexity. Int J Mol Sci. 2018;19(10):3113.
81. Mantuano E, Romano S, Veneziano L, Gellera C, Castellotti B, Caimi S, et al. Identification of novel and recurrent CACNA1A gene mutations in fifteen patients with episodic ataxia type 2. J Neurol Sci. 2010;291(1–2):30–6.
82. Nachbauer W, Nocker M, Karner E, Stankovic I, Unterberger I, Eigentler A, et al. Episodic ataxia type 2: phenotype characteristics of a novel CACNA1A mutation and review of the literature. J Neurol. 2014;261(5):983–91.
83. Nardello R, Plicato G, Mangano GD, Gennaro E, Mangano S, Brighina F, et al. Two distinct phenotypes, hemiplegic migraine and episodic Ataxia type 2, caused by a novel common CACNA1A variant. BMC Neurol. 2020;20(1):155.
84. Nielsen EN, Asbjornsdottir B, Moller LB, Nielsen JE, Lindquist SG. Episodic ataxia type 2 (EA2) with interictal myokymia and focal dystonia. Cold Spring Harb Mol Case Stud. 2022;8(6):a006236.
85. Nikaido K, Tachi N, Ohya K, Wada T, Tsutsumi H. New mutation of CACNA1A gene in episodic ataxia type 2. Pediatr Int. 2011;53(3):415–6.
86. Petrovicova A, Brozman M, Kurca E, Gobo T, Dluha J, Kalmarova K, et al. Novel missense variant of CACNA1A gene in a Slovak family with episodic ataxia type 2. Biomed Pap Med Fac Univ Palacky Olomouc Czech Repub. 2017;161(1):107–10.
87. Pradotto L, Mencarelli M, Bigoni M, Milesi A, Di Blasio A, Mauro A. Episodic ataxia and SCA6 within the same family due to the D302N CACNA1A gene mutation. J Neurol Sci. 2016;371:81–4.
88. Riant F, Lescoat C, Vahedi K, Kaphan E, Toutain A, Soisson T, et al. Identification of CACNA1A large deletions in four patients with episodic ataxia. Neurogenetics. 2010;11(1):101–6.
89. Riant F, Mourtada R, Saugier-Veber P, Tournier-Lasserve E. Large CACNA1A deletion in a family with episodic ataxia type 2. Arch Neurol. 2008;65(6):817–20.
90. Robbins MS, Lipton RB, Laureta EC, Grosberg BM. CACNA1A nonsense mutation is associated with basilar-type migraine and episodic ataxia type 2. Headache. 2009;49(7):1042–6.
91. Scoggan KA, Chandra T, Nelson R, Hahn AF, Bulman DE. Identification of two novel mutations in the CACNA1A gene responsible for episodic ataxia type 2. J Med Genet. 2001;38(4):249–53.
92. Sintas C, Carreno O, Fernandez-Castillo N, Corominas R, Vila-Pueyo M, Toma C, et al. Mutation spectrum in the CACNA1A gene in 49 patients with episodic ataxia. Sci Rep. 2017;7(1):2514.
93. Sivak S, Kurca E, Krajciova A, Hikkelova M, Simko J, Misovicova N, et al. Novel missense variant of CACNA1A gene: a case report of a family with episodic ataxia type 2. J Neurol Sci. 2017;376:119–20.
94. Spacey SD, Hildebrand ME, Materek LA, Bird TD, Snutch TP. Functional implications of a novel EA2 mutation in the P/Q--type calcium channel. Ann Neurol. 2004;56(2):213–20.
95. Spacey SD, Materek LA, Szczygielski BI, Bird TD. Two novel CACNA1A gene mutations associated with episodic ataxia type 2 and interictal dystonia. Arch Neurol. 2005;62(2):314–6.
96. Subramony SH, Schott K, Raike RS, Callahan J, Langford LR, Christova PS, et al. Novel CACNA1A mutation causes febrile episodic ataxia with interictal cerebellar deficits. Ann Neurol. 2003;54(6):725–31.
97. Tomlinson SE, Tan SV, Burke D, Labrum RW, Haworth A, Gibbons VS, et al. In vivo impact of presynaptic calcium channel dysfunction on motor axons in episodic ataxia type 2. Brain. 2016;139(Pt 2):380–91.
98. van den Maagdenberg AM, Kors EE, Brunt ER, van Paesschen W, Pascual J, Ravine D, et al. Episodic ataxia type 2. Three novel truncating mutations and one novel missense mutation in the CACNA1A gene. J Neurol. 2002;249(11):1515–9.

99. Verriello L, Carrera P, Pauletto G, Bernardini A, Valente M, Gigli GL. Case report and ten-year follow-up of episodic ataxia type 2 due to a novel variant in CACNA1A. eNeurologicalSci. 2021;23:100334.
100. Verriello L, Pauletto G, Nilo A, Lonigro I, Betto E, Valente M, et al. Epilepsy and episodic ataxia type 2: family study and review of the literature. J Neurol. 2021;268(11):4296–302.
101. Wan J, Carr JR, Baloh RW, Jen JC. Nonconsensus intronic mutations cause episodic ataxia. Ann Neurol. 2005;57(1):131–5.
102. Wan J, Mamsa H, Johnston JL, Spriggs EL, Singer HS, Zee DS, et al. Large genomic deletions in CACNA1A cause episodic ataxia type 2. Front Neurol. 2011;2:51.
103. Xu Y, Wang Z, Sun Q, Zhou L, Xu H, Hu Y. Clinical features and CACNA1A gene mutation in a family with episodic ataxia type 2. Zhong Nan Da Xue Xue Bao Yi Xue Ban. 2022;47(6):801–8.
104. Zafeiriou DI, Lehmann-Horn F, Vargiami E, Teflioudi E, Ververi A, Jurkat-Rott K. Episodic ataxia type 2 showing ictal hyperhidrosis with hypothermia and interictal chronic diarrhea due to a novel CACNA1A mutation. Eur J Paediatr Neurol. 2009;13(2):191–3.
105. Jen J. Calcium channelopathies in the central nervous system. Curr Opin Neurobiol. 1999;9(3):274–80.
106. Sappey-Marinier D, Vighetto A, Peyron R, Broussolle E, Bonmartin A. Phosphorus and proton magnetic resonance spectroscopy in episodic ataxia type 2. Ann Neurol. 1999;46(2):256–9.
107. Harno H, Heikkinen S, Kaunisto MA, Kallela M, Hakkinen AM, Wessman M, et al. Decreased cerebellar total creatine in episodic ataxia type 2: a 1H MRS study. Neurology. 2005;64(3):542–4.
108. Helmich RC, Siebner HR, Giffin N, Bestmann S, Rothwell JC, Bloem BR. The dynamic regulation of cortical excitability is altered in episodic ataxia type 2. Brain. 2010;133(Pt 12):3519–29.
109. Indelicato E, Unterberger I, Nachbauer W, Eigentler A, Amprosi M, Zeiner F, et al. The electrophysiological footprint of CACNA1A disorders. J Neurol. 2021;268(7):2493–505.
110. Krishnan AV, Bostock H, Ip J, Hayes M, Watson S, Kiernan MC. Axonal function in a family with episodic ataxia type 2 due to a novel mutation. J Neurol. 2008;255(5):750–5.
111. Bertholon P, Chabrier S, Riant F, Tournier-Lasserve E, Peyron R. Episodic ataxia type 2: unusual aspects in clinical and genetic presentation. Special emphasis in childhood. J Neurol Neurosurg Psychiatry. 2009;80(11):1289–92.
112. Kim HJ, Kim JS, Choi JH, Shin JH, Choi KD, Zee DS. Rebound upbeat nystagmus after lateral gaze in episodic ataxia type 2. Cerebellum. 2014;13(3):411–3.
113. Shervin Badv R, Niksirat A. Downward vertical gaze palsy as a prominent manifestation of episodic ataxia type 2: a case report. Iran J Child Neurol. 2013;7(4):58–60.
114. Harno H, Hirvonen T, Kaunisto MA, Aalto H, Levo H, Isotalo E, et al. Acetazolamide improves neurotological abnormalities in a family with episodic ataxia type 2 (EA-2). J Neurol. 2004;251(2):232–4.
115. Choi JH, Oh EH, Choi SY, Kim HJ, Lee SK, Choi JY, et al. Vestibular impairments in episodic ataxia type 2. J Neurol. 2022;269(5):2687–95.
116. Engel KC, Anderson JH, Gomez CM, Soechting JF. Deficits in ocular and manual tracking due to episodic ataxia type 2. Mov Disord. 2004;19(7):778–87.
117. Gordon CR, Caspi A, Levite R, Zivotofsky AZ. Mechanisms of vestibulo-ocular reflex (VOR) cancellation in spinocerebellar ataxia type 3 (SCA-3) and episodic ataxia type 2 (EA-2). Prog Brain Res. 2008;171:519–25.
118. Kipfer S, Jung S, Lemke JR, Kipfer-Kauer A, Howell JP, Kaelin-Lang A, et al. Novel CACNA1A mutation(s) associated with slow saccade velocities. J Neurol. 2013;260(12):3010–4.
119. Kim S, Kim S, Lee S, Kim HJ. Vertical saccadic slowing in episodic ataxia type 2. J Clin Neurol. 2022;18(6):726–8.
120. Rucker JC, Jen J, Stahl JS, Natesan N, Baloh RW, Leigh RJ. Internuclear ophthalmoparesis in episodic ataxia type 2. Ann N Y Acad Sci. 2005;1039:571–4.
121. Shimmura M, Uehara T, Yamashita K, Shigeto H, Yamasaki R, Ishikawa K, et al. Slowed abduction during smooth pursuit eye movement in episodic ataxia type 2 with a novel CACNA1A mutation. J Neurol Sci. 2017;381:4–6.
122. Ying SH, Jen JC, Baloh RW. Similar oculomotor phenotypes in episodic ataxia type 2 and spinocerebellar atrophy type 6. Ann N Y Acad Sci. 2001;942:508–9.
123. Strupp M, Zwergal A, Brandt T. Episodic ataxia type 2. Neurotherapeutics. 2007;4(2):267–73.
124. Scoggan KA, Friedman JH, Bulman DE. CACNA1A mutation in a EA-2 patient responsive to acetazolamide and valproic acid. Can J Neurol Sci. 2006;33(1):68–72.
125. Alvina K, Khodakhah K. The therapeutic mode of action of 4-aminopyridine in cerebellar ataxia. J Neurosci. 2010;30(21):7258–68.
126. Claassen J, Teufel J, Kalla R, Spiegel R, Strupp M. Effects of dalfampridine on attacks in patients with episodic ataxia type 2: an observational study. J Neurol. 2013;260(2):668–9.
127. Lohle M, Schrempf W, Wolz M, Reichmann H, Storch A. Potassium channel blocker 4-aminopyridine is effective in interictal cerebellar symptoms in episodic ataxia type 2–a video case report. Mov Disord. 2008;23(9):1314–6.
128. Malamud E, Otallah SI. Use of dalfampridine in a young child with episodic ataxia type 2. Child Neurol Open. 2022;9:2329048X221075447.
129. Strupp M, Kalla R, Claassen J, Adrion C, Mansmann U, Klopstock T, et al. A randomized trial of 4-aminopyridine in EA2 and related familial episodic ataxias. Neurology. 2011;77(3):269–75.
130. Strupp M, Kalla R, Dichgans M, Freilinger T, Glasauer S, Brandt T. Treatment of episodic ataxia type 2 with the potassium channel blocker 4-aminopyridine. Neurology. 2004;62(9):1623–5.
131. Strupp M, Kalla R, Glasauer S, Wagner J, Hufner K, Jahn K, et al. Aminopyridines for the treatment of cerebellar and ocular motor disorders. Prog Brain Res. 2008;171:535–41.
132. Strupp M, Teufel J, Zwergal A, Schniepp R, Khodakhah K, Feil K. Aminopyridines for the treatment of neurologic disorders. Neurol Clin Pract. 2017;7(1):65–76.
133. Muth C, Teufel J, Schols L, Synofzik M, Franke C, Timmann D, et al. Fampridine and acetazolamide in EA2 and related familial EA: a prospective randomized placebo-controlled trial. Neurol Clin Pract. 2021;11(4):e438–e46.
134. Gonzalez-Mingot C, Lopez-Ortega R, Brieva-Ruiz L. The efficacy of combining topiramate and 4-aminopyridine to reduce relapses and interictal progression in two cases of episodic ataxia type 2. Neurol Sci. 2022;43(8):5099–101.
135. Alvina K, Khodakhah K. KCa channels as therapeutic targets in episodic ataxia type-2. J Neurosci. 2010;30(21):7249–57.
136. Lee H, Jang DH, Jang JH, Kim T. Effectiveness of levetiracetam in an acetazolamide-unresponsive patient with episodic ataxia type 2 by a novel CACNA1A nonsense mutation. Eur J Neurol. 2017;24(7):e43–e4.
137. Na S, Kim T. Efficacy of levetiracetam in patients with episodic ataxia type 2 caused by CACNA1A mutation: three case reports. Neurol Sci. 2021;42(9):3897–9.
138. Yuan X, Zheng Y, Gao F, Sun W, Wang Z, Zhao G. Case report: a novel CACNA1A mutation caused flunarizine-responsive type 2 episodic ataxia and hemiplegic migraine with abnormal MRI of cerebral white matter. Front Neurol. 2022;13:899813.
139. Steckley JL, Ebers GC, Cader MZ, McLachlan RS. An autosomal dominant disorder with episodic ataxia, vertigo, and tinnitus. Neurology. 2001;57(8):1499–502.

140. Cader MZ, Steckley JL, Dyment DA, McLachlan RS, Ebers GC. A genome-wide screen and linkage mapping for a large pedigree with episodic ataxia. Neurology. 2005;65(1):156–8.
141. Damji KF, Allingham RR, Pollock SC, Small K, Lewis KE, Stajich JM, et al. Periodic vestibulocerebellar ataxia, an autosomal dominant ataxia with defective smooth pursuit, is genetically distinct from other autosomal dominant ataxias. Arch Neurol. 1996;53(4):338–44.
142. Escayg A, De Waard M, Lee DD, Bichet D, Wolf P, Mayer T, et al. Coding and noncoding variation of the human calcium-channel beta4-subunit gene CACNB4 in patients with idiopathic generalized epilepsy and episodic ataxia. Am J Hum Genet. 2000;66(5):1531–9.
143. Jen JC, Wan J, Palos TP, Howard BD, Baloh RW. Mutation in the glutamate transporter EAAT1 causes episodic ataxia, hemiplegia, and seizures. Neurology. 2005;65(4):529–34.

90 Gluten Ataxia

Overview

Gluten ataxia (GA) is an autoimmune-mediated cerebellar ataxia believed to be provoked by exposure to gluten (a protein contained in wheat and several other grains). GA can occur in the context of other gluten-sensitive diseases (such as celiac disease) or in isolation. The age of onset is a broad range (from the second to the ninth decades), though most studies report a mean age of onset in the fifth or sixth decade. GA affects males and females approximately equally. Some studies suggest that GA may account for 25–36% of sporadically occurring cerebellar ataxias. Symptoms usually start insidiously and progress slowly, though occasional cases have been reported of abrupt onset or rapid progression. Some literature suggests that alcohol intake may increase the risk of GA. The most common presenting symptoms include gait ataxia, limb ataxia, and dysarthria. Physical examination may show midline and hemispheric cerebellar signs, as well as ocular motor findings compatible with cerebellar dysfunction. The serum antibodies detected are usually anti-gliadin IgG antibodies (more common) and anti-transglutaminase-6 IgA antibodies (less common). These antibodies are thought to attack Purkinje cells, though autopsy studies also show damage of cerebellar granular neurons, cerebellar nuclei, and inferior olivary neurons. MR imaging often shows cerebellar atrophy, but MR spectroscopy can show abnormal cerebellar metabolism even in the absence of atrophy. The mainstay of treatment is a strict gluten-free diet, though this should not be taken lightly as it is not risk-free. Less studied (and thus second-line) therapies include intravenous immunoglobulin, plasmapheresis, corticosteroids, or other immunosuppressants. Earlier treatment (including dietary treatment) retards progression in some (but not all) patients.

The main value of identifying (or suspecting) gluten ataxia is that it is one of the few cerebellar ataxias for which treatment (usually gluten-free diet) has the potential to reverse symptoms or retard their progression [1].

In a patient with biopsy-proven celiac disease who has ataxia, it is likely that the ataxia is also gluten-mediated, though it is still reasonable for the clinician to keep in mind a differential diagnosis that includes other causes of ataxia (such as alcoholism). A patient with celiac disease will likely be instructed by their gastroenterologist to initiate and maintain a gluten-free diet, and if in so doing the ataxia also improves, then *ex juvantibus* the ataxia was likely gluten-related.

In a patient without celiac disease and without gastrointestinal symptoms who presents in adulthood with insidious onset, gradually progressive ataxia, and no family history suggestive of any genetically mediated ataxia, it is reasonable to suspect gluten ataxia. Obtaining a brain MRI is reasonable as this may (1) show cerebellar atrophy and (2) help exclude other diseases. Checking for anti-gliadin antibodies and anti-transglutaminase-6 antibodies is also reasonable. If there is reasonable evidence and suspicion for GA, then initiating and maintaining a strict gluten-free diet is reasonable, though this dietary change should ideally be supervised by a medical dietician or nutritionist.

Introduction

The topic of *gluten ataxia* is a "contentious concept" [2], with opposing views expressed in the same journal issue, such as "Gluten ataxia: an overestimated condition?" [3] and "Gluten ataxia: an underdiagnosed condition" [4].

Mitoma and colleagues [5] credit Jean-Martin Charcot with having published the first description of a cerebellar ataxia as an immune-mediated illness in 1868 [6]. Some investigators [7] credit Cooke and Smith with having provided the "first comprehensive report of neurological manifestations in the context of histologically confirmed coeliac disease" in 1966 [8].

Hadjivassiliou and colleagues began publishing on this topic in 1998 [9] and has contributed substantially to the literature [1, 4, 7, 9–24].

Celiac disease is an autoimmune gastrointestinal illness whose diagnosis can be secured from duodenal biopsy. Dicke

M. Cherchi, *Otoneurology and Vestibular Medicine*, https://doi.org/10.1007/978-3-031-94841-1_90

and colleagues are credited with having identified the wheat protein *gluten* as the antigen [25]. When an individual with biopsy-proven celiac disease develops other symptoms, it is reasonable to speculate that the other symptoms may also be mediated by an autoimmune mechanism. In the case of cerebellar ataxias, this seems especially reasonable since a range of autoimmune-mediated cerebellar ataxias have already been well-described.

The classification outlined below, from Mitoma and colleagues [26], lists the main cerebellar ataxias that have been attributed to an autoimmune mechanism.

1. Autoimmunity that mainly targets the cerebellum or its related structures.
   a. Cerebellar autoimmunity not triggered by another disease:
      i. Anti-GAD antibody-associated cerebellar ataxia
      ii. Cerebellar type of Hashimoto's encephalopathy
      iii. Primary autoimmune cerebellar ataxia
      iv. Others
   b. Cerebellar autoimmunity triggered by another disease or condition:
      i. Gluten ataxia (gluten sensitivity)
      ii. Acute cerebellitis (infection)
      iii. Miller Fisher syndrome (infection)
      iv. Paraneoplastic cerebellar degenerations (neoplasm)
2. Autoimmunity that simultaneously targets various parts of the central nervous system:
   a. Multiple sclerosis
   b. Ataxia in the context of connective tissue diseases such as systemic lupus erythematosus

In some cases, an individual with cerebellar ataxia but no gastrointestinal symptoms may be found on duodenal biopsy to have celiac disease, and in such circumstances, the patient is often described as having "subclinical celiac disease" [27, 28].

The scenario becomes more complex when an individual has neither gastrointestinal symptoms nor histopathologic evidence of celiac disease [15, 29, 30]. This situation is sometimes referred to as "nonceliac *gluten sensitivity*" (NCGS)—but the broader concept is that an *immunological response to gluten is somehow causing an ataxia.*

Epidemiology of Gluten Ataxia

In the series of 100 gluten ataxia patients studied by Hadjivassiliou and colleagues, the mean age at onset of ataxia was 53 years [10], though the age range is quite broad. Hadjivassiliou and colleagues studied 68 patients with gluten ataxia and reported the age of onset to range from 14 to 81 years, with a mean of 48 years [15]. Although this is predominantly a disease of older patients, it can (rarely) also occur in the pediatric population [31].

In a series of 404 patients with sporadically occurring progressive ataxia, Hadjivassiliou reported that 101 (25%) had gluten ataxia [10] which, if correct, makes gluten ataxia among the most common of all sporadically occurring ataxias.

A case series of sporadically occurring cerebellar ataxias in Japan reported that 5/14 (36%) of patients were found to nave anti-gliadin antibodies [32].

In the series analyzed by Hadjivassiliou, the disease is nearly equally distributed between males (49%) and females (51%).

Risk Factors

There is modest evidence that alcohol intake may increase susceptibility to gluten ataxia [33]. The mechanism for this is not known; it is not clear whether cerebellar toxicity of alcohol simply adds to what would otherwise have been an isolated gluten-mediated autoimmune cerebellar insult, or whether alcohol somehow potentiates the autoimmune process itself.

Clinical Presentation

Table 90.1, from Hadjivassiliou [7], lists other clinical features of patients with gluten ataxia.

Symptoms less commonly reported in association with gluten ataxia include palatal tremor [34], cognitive impairment [35], and myoclonus [36]. One might speculate that these are "gluten ataxia plus" syndromes—in other words,

Table 90.1 Clinical and immunological characteristics of 100 patients with gluten ataxia. (From Hadjivassiliou et al. [7])

Clinical and immunological characteristics of 100 patients with gluten ataxia	
Male:female ratio	49:51
Mean age at onset of ataxia	53 years
Ocular signs	80%
Upper limb ataxia	70%
Lower limb ataxia	90%
Gait ataxia	100%
Peripheral neuropathy	60%
Enteropathy on biopsy	28%
Anti-gliadin antibody positive	100%
Antiendomysium antibody positive	22%
Transglutaminase antibody positive	56%
HLA DQ2	70%
Presence of oligoclonal bands	50%

cases in which gluten ataxia is only one facet of a broader gluten-triggered autoimmune disease.

Physical Examination

Hadjivassiliou and colleagues reported that in a series of 68 patients with gluten ataxia, 100% exhibited gait ataxia, 90% exhibited lower limb ataxia, and 70% exhibited upper limb ataxia [15].

Burk and colleagues [37] studied 104 patients with sporadic cerebellar ataxia associated with gluten sensitivity and reported that 100% exhibited ataxia of stance and gait, 100% exhibited dysarthria, and 97% exhibited limb ataxia.

Ocular Motor Examination

Hadjivassiliou and colleagues reported that in a series of 68 patients with gluten ataxia, 84% exhibited "ocular signs," which they qualify as "spontaneous and gaze-evoked nystagmus and/or abnormal saccades" [15].

Burk and colleagues [37] studied 104 patients with sporadic cerebellar ataxia and reported that 66.7% exhibited gaze evoked nystagmus, 33.3% exhibited spontaneous nystagmus, 25% exhibited slowing of saccades, and 16.7% exhibited an upward gaze palsy.

Pathophysiological Mechanism of Disease

Similar to other exogenously triggered autoimmune diseases, in gluten ataxia, it has been suspected that antibody cross-reactivity (molecular mimicry) plays a significant mechanistic role [38]. Indeed, Hadjivassiliou and colleagues report experimental evidence suggesting such cross-reactivity between antigenic epitopes on Purkinje cells and gluten peptides [13]. The specific target of these autoantibodies is not known, but some research suggests that it may be synapsin I [39].

Antibodies

A number of antibodies have been reported in the serum of patients with gluten ataxia. The best studied include *anti-gliadin antibodies* [37, 40] and *anti-transglutaminase-6 antibodies* [11, 12, 41].

Antibodies: Anti-gliadin Antibodies

Lock and colleagues [42] studied a series of 20 patients with "idiopathic ataxia" and reported that *anti-gliadin antibodies* (either IgG or IgA) were found in 40% of patients.

Liu and colleagues [43] reported on a series of 361 patients with sporadic ataxia. They found that 9% of patients had "gluten sensitivity." In these gluten-sensitive patients, they found anti-gliadin IgG and IgA antibodies, and *t-transglutaminase IgA antibodies*. They reported that *anti-gliadin IgG antibodies* were "more specific to the disease of sporadic ataxia."

Rashtak and colleagues [44] report that levels of anti-gliadin antibodies are higher in gluten-sensitive patients with both ataxia and enteropathy (which is to say, patients with celiac disease and ataxia) compared to patients with only ataxia but no enteropathy.

Not all investigators find a strong correlation between anti-gliadin antibodies and gluten ataxia [45].

Antibodies: Anti-gliadin Antibodies Occur in the Healthy Population

Anti-gliadin antibodies are present in about 10% of asymptomatic individuals [46].

Antibodies: Anti-gliadin Antibodies Can Also Occur in Patients with Hereditary Ataxias

One series of 50 patients reported the presence of anti-gliadin antibodies in 7/26 (27%) patients with sporadic ataxias, and 9/24 (37%) of patients with genetically confirmed autosomal dominant ataxias [47].

Antibodies: Anti-transglutaminase-6 Antibodies

Hadjivassiliou and colleagues reported that, "The prevalence of *TG6 [transglutaminase-6] antibodies* was 21 of 65 (32%) in idiopathic sporadic ataxia, 35 of 48 (73%) in GA, 16 of 50 (32%) in CD, 4 of 82 (5%) in neurology controls, and 2 of 57 (4%) in healthy controls" [11].

Notably, in gluten ataxia patients with anti-transglutaminase-6 antibodies, the ataxia can be present in the absence of any gastrointestinal symptoms [12].

Antibodies: Anti-transglutaminase-6 Antibodies Occur in Other Diseases

Anti-transglutaminase-6 antibodies occur in other diseases as well [48]. For example, anti-transglutaminase-6 antibodies have been reported in gluten-sensitive patients with neuropathy [49].

Antibodies: Antiendomysial Antibodies

Hadjivassiliou and colleagues report that antiendomysial antibodies may be found in 22% of patients with gluten ataxia [10].

Antibodies: Antiganglioside Antibodies

Several investigators have identified antiganglioside antibodies in patients with gluten ataxia [50, 51].

Imaging

Hadjivassiliou and colleagues reported that in a series of 68 patients with gluten ataxia, "MRI revealed atrophy of the cerebellum in 79% and white matter hyperintensities in 19%" [15].

It was recognized that even in the absence of cerebellar atrophy, patients with gluten ataxia could still exhibit metabolic abnormalities in the cerebellum detectable on magnetic resonance spectroscopy (MRS) [52]. Hadjivassiliou and colleagues reported that "up to 60% of patients with gluten ataxia have evidence of cerebellar atrophy on MR imaging" [7], but on MRS "all patients have spectroscopic abnormalities primarily affecting the vermis" [22], suggesting that, "MRS abnormalities and atrophy do not necessarily occur concurrently" [24].

These metabolic abnormalities on MR spectroscopy can improve in patients who maintain a strict gluten-free diet [17, 18].

Histopathology

Clark [53] reviewed multiple cases and autopsy series of patients with gluten ataxia (in some cases with other neurological symptoms as well). He concluded that, "Nearly all patients studied at autopsy had significant loss of Purkinje neurons with Bergmann gliosis. The other commonly affected targets were cerebellar granular neurons, inferior olivary neurons, and cerebellar nuclei."

Prognosis

Similar to other untreated autoimmune cerebellar ataxias, gluten ataxia tends to appear insidiously and progress relatively slowly [54], though there are also reports of more rapid progression [55].

Treatment

The mainstay of therapy is initiating and maintaining a gluten-free diet. Second-line therapies include intravenous immunoglobulin, plasmapheresis, corticosteroids, or other immunosuppressants. Since autoimmune disease brings about neuronal damage and neuronal death, some researchers reasonably maintain the position that "time is brain" or "time is cerebellum" [56] and therefore suggest that treatment be initiated as early as possible [57].

Treatment: Dietary Restriction

For patients with gluten sensitivity, including gluten ataxia, several studies report improvement in at least some symptoms [1, 58].

In gluten-sensitive patients with neurological symptoms, eliminating gluten from the diet may bring about improvement in some of the neurological symptoms, but not others. For example, in patients with gluten sensitivity who exhibit both ataxia and neuropathy, maintaining a strict gluten-free diet may improve both symptoms [59], or only the neuropathy but not the ataxia [60].

There has been considerable media attention directed toward the putative ills of gluten and the putative benefits of a gluten-free diet.

A gluten-free diet has achieved some popularity among health faddists [61], which in the United States has been promoted through the media by commercial interests, and this has given rise to spurious claims of health benefits based on little or no evidence. It is important to keep in mind that a gluten-free diet is not without risk. For example, maintaining a strict gluten-free diet increases the risk of certain nutritional deficiencies [58]. Marketing of gluten-free products often downplays their disadvantages [62]. Thus, the "intervention" of initiating and maintaining a gluten-free diet should not be taken lightly; if there is reasonable evidence that a patient's symptoms can be confidently connected to gluten, then dietary modification is sensible, but invoking a gluten-free diet in an individual with no evidence of gluten-related disease for purposes of prophylaxis or on the basis of poorly articulated "health benefits" should be discouraged [63].

Gluten is found primarily in wheat products, but also in other grains. In the United States, food labeling regulations from the Food and Drug Administration have made it somewhat easier to distinguish gluten-containing foods from gluten-free foods in a supermarket, and some restaurants will also clarify the gluten content of foods on their menus. Given the discussion above regarding the nutritional risks of maintaining a gluten-free diet, if a patient is considering instituting a gluten-free diet, it is sensible to consult a medical dietician or nutritionist to make sure that the patient can maintain a diet that is gluten-free yet still provides adequate nutrition in other respects.

Treatment: Other Methods

Less well-studied interventions of gluten ataxia include intravenous immunoglobulin [64, 65], plasmapheresis [57], corticosteroids, and other immunosuppressants [66]. These are generally regarded as second-line therapies, and there are no clear guidelines regarding when to consider them. Hadjivassiliou and colleagues [10] suggest, "Treatment with immunosuppressants and intravenous immunoglobulins could be considered if a strict gluten-free diet for at least a year has not resulted in any improvement of the ataxia or if the ataxia is rapidly progressive."

References

1. Hadjivassiliou M, Davies-Jones GA, Sanders DS, Grunewald RA. Dietary treatment of gluten ataxia. J Neurol Neurosurg Psychiatry. 2003;74(9):1221–4.
2. Serratrice J, Attarian S, Disdier P, Weiller PJ, Serratrice G. Neuromuscular diseases associated with antigliadin antibodies. A contentious concept. Acta Myol. 2004;23(3):146–50.
3. Meira AT, Franklin GL, Ferreira MG, Martinez ARM, Franca MC Jr, Pedroso JL, et al. Gluten Ataxia: an overestimated condition? Cerebellum. 2022;21(4):617–9.
4. Hadjivassiliou M, Grupsilonnewald RA. Gluten Ataxia: an underdiagnosed condition. Cerebellum. 2022;21(4):620–2.
5. Mitoma H, Manto M, Hadjivassiliou M. Immune-mediated cerebellar ataxias: clinical diagnosis and treatment based on immunological and physiological mechanisms. J Mov Disord. 2021;14(1):10–28.
6. Charcot J-M. Séance du 14 mars. CR Soc Biol (Paris). 1868;20:13.
7. Hadjivassiliou M, Sanders DS, Grunewald RA, Woodroofe N, Boscolo S, Aeschlimann D. Gluten sensitivity: from gut to brain. Lancet Neurol. 2010;9(3):318–30.
8. Cooke WT, Smith WT. Neurological disorders associated with adult coeliac disease. Brain: a journal of neurology. 1966;89(4):683–722.
9. Hadjivassiliou M, Grunewald RA, Chattopadhyay AK, Davies-Jones GA, Gibson A, Jarratt JA, et al. Clinical, radiological, neurophysiological, and neuropathological characteristics of gluten ataxia. Lancet. 1998;352(9140):1582–5.
10. Hadjivassiliou M. Immune-mediated acquired ataxias. Handb Clin Neurol. 2012;103:189–99.
11. Hadjivassiliou M, Aeschlimann P, Sanders DS, Maki M, Kaukinen K, Grunewald RA, et al. Transglutaminase 6 antibodies in the diagnosis of gluten ataxia. Neurology. 2013;80(19):1740–5.
12. Hadjivassiliou M, Aeschlimann P, Strigun A, Sanders DS, Woodroofe N, Aeschlimann D. Autoantibodies in gluten ataxia recognize a novel neuronal transglutaminase. Ann Neurol. 2008;64(3):332–43.
13. Hadjivassiliou M, Boscolo S, Davies-Jones GA, Grunewald RA, Not T, Sanders DS, et al. The humoral response in the pathogenesis of gluten ataxia. Neurology. 2002;58(8):1221–6.
14. Hadjivassiliou M, Duker AP, Sanders DS. Gluten-related neurologic dysfunction. Handb Clin Neurol. 2014;120:607–19.
15. Hadjivassiliou M, Grunewald R, Sharrack B, Sanders D, Lobo A, Williamson C, et al. Gluten ataxia in perspective: epidemiology, genetic susceptibility and clinical characteristics. Brain. 2003;126(Pt 3):685–91.
16. Hadjivassiliou M, Grunewald RA, Lawden M, Davies-Jones GA, Powell T, Smith CM. Headache and CNS white matter abnormalities associated with gluten sensitivity. Neurology. 2001;56(3):385–8.
17. Hadjivassiliou M, Grunewald RA, Sanders DS, Shanmugarajah P, Hoggard N. Effect of gluten-free diet on cerebellar MR spectroscopy in gluten ataxia. Neurology. 2017;89(7):705–9.
18. Hadjivassiliou M, Grunewald RA, Sanders DS, Zis P, Croal I, Shanmugarajah PD, et al. The significance of low Titre Antigliadin antibodies in the diagnosis of Gluten Ataxia. Nutrients. 2018;10(10):1444.
19. Hadjivassiliou M, Maki M, Sanders DS, Williamson CA, Grunewald RA, Woodroofe NM, et al. Autoantibody targeting of brain and intestinal transglutaminase in gluten ataxia. Neurology. 2006;66(3):373–7.
20. Hadjivassiliou M, Manto M, Mitoma H. Rare etiologies in immune-mediated cerebellar ataxias: diagnostic challenges. Brain Sci. 2022;12(9):1165.
21. Hadjivassiliou M, Martindale J, Shanmugarajah P, Grunewald RA, Sarrigiannis PG, Beauchamp N, et al. Causes of progressive cerebellar ataxia: prospective evaluation of 1500 patients. J Neurol Neurosurg Psychiatry. 2017;88(4):301–9.
22. Hadjivassiliou M, Sanders DD, Aeschlimann DP. Gluten-related disorders: gluten ataxia. Dig Dis. 2015;33(2):264–8.
23. Hadjivassiliou M, Sanders DS, Woodroofe N, Williamson C, Grunewald RA. Gluten ataxia. Cerebellum. 2008;7(3):494–8.
24. Hadjivassiliou M, Wallis LI, Hoggard N, Grunewald RA, Griffiths PD, Wilkinson ID. MR spectroscopy and atrophy in Gluten, Friedreich's and SCA6 ataxias. Acta Neurol Scand. 2012;126(2):138–43.
25. Dicke WK, Weijers HA, Van De Kamer JH. Coeliac disease. II. The presence in wheat of a factor having a deleterious effect in cases of coeliac disease. Acta Paediatr (Stockh). 1953;42(1):34–42.
26. Mitoma H, Adhikari K, Aeschlimann D, Chattopadhyay P, Hadjivassiliou M, Hampe CS, et al. Consensus paper: neuroimmune mechanisms of cerebellar ataxias. Cerebellum. 2016;15(2):213–32.
27. Jain D, Bhaduri G, Giri K, Balania V, Jain P. Underlying coeliac disease in a case of cerebellar ataxia with unknown etiology. Arch Clin Cases. 2019;6(1):22–5.
28. Pellecchia MT, Scala R, Perretti A, De Michele G, Santoro L, Filla A, et al. Cerebellar ataxia associated with subclinical celiac disease responding to gluten-free diet. Neurology. 1999;53(7):1606–8.
29. Guan WJ, Liu XJ, Tang BS, Liu YT, Zhou Y, Jiang H, et al. Gluten ataxia of sporadic and hereditary cerebellar ataxia in patients from mainland China. Neurol India. 2013;61(3):226–30.
30. Rodrigo L, Hernandez-Lahoz C, Lauret E, Rodriguez-Pelaez M, Soucek M, Ciccocioppo R, et al. Gluten ataxia is better classified as non-celiac gluten sensitivity than as celiac disease: a comparative clinical study. Immunol Res. 2016;64(2):558–64.

31. Gordon N. Cerebellar ataxia and gluten sensitivity: a rare but possible cause of ataxia, even in childhood. Dev Med Child Neurol. 2000;42(4):283–6.
32. Ihara M, Makino F, Sawada H, Mezaki T, Mizutani K, Nakase H, et al. Gluten sensitivity in Japanese patients with adult-onset cerebellar ataxia. Intern Med. 2006;45(3):135–40.
33. Currie S, Hoggard N, Clark MJ, Sanders DS, Wilkinson ID, Griffiths PD, et al. Alcohol induces sensitization to gluten in genetically susceptible individuals: a case control study. PLoS One. 2013;8(10):e77638.
34. Kheder A, Currie S, Romanowski C, Hadjivassiliou M. Progressive ataxia with palatal tremor due to gluten sensitivity. Mov Disord. 2012;27(1):62–3.
35. Makhlouf S, Messelmani M, Zaouali J, Mrissa R. Cognitive impairment in celiac disease and non-celiac gluten sensitivity: review of literature on the main cognitive impairments, the imaging and the effect of gluten free diet. Acta Neurol Belg. 2018;118(1):21–7.
36. Sarrigiannis PG, Hoggard N, Aeschlimann D, Sanders DS, Grunewald RA, Unwin ZC, et al. Myoclonus ataxia and refractory coeliac disease. Cerebellum Ataxias. 2014;1:11.
37. Burk K, Bosch S, Muller CA, Melms A, Zuhlke C, Stern M, et al. Sporadic cerebellar ataxia associated with gluten sensitivity. Brain. 2001;124(Pt 5):1013–9.
38. Green PH, Alaedini A, Sander HW, Brannagan TH 3rd, Latov N, Chin RL. Mechanisms underlying celiac disease and its neurologic manifestations. Cell Mol Life Sci. 2005;62(7-8):791–9.
39. Alaedini A, Okamoto H, Briani C, Wollenberg K, Shill HA, Bushara KO, et al. Immune cross-reactivity in celiac disease: anti-gliadin antibodies bind to neuronal synapsin I. J Immunol. 2007;178(10):6590–5.
40. Abele M, Schols L, Schwartz S, Klockgether T. Prevalence of antigliadin antibodies in ataxia patients. Neurology. 2003;60(10):1674–5.
41. Osman D, Umar S, Muhammad H, Nikfekr E, Rostami K, Ishaq S. Neurological manifestation of coeliac disease with particular emphasis on gluten ataxia and immunological injury: a review article. Gastroenterol Hepatol Bed Bench. 2021;14(1):1–7.
42. Lock RJ, Tengah DP, Williams AJ, Ward JJ, Bingley PJ, Wills AJ, et al. Cerebellar ataxia, peripheral neuropathy, "gluten sensitivity" and anti-neuronal autoantibodies. Clin Lab. 2006;52(11-12):589–92.
43. Liu CS, Soong BW, Lee YC, Chen WL, Kuo CL, Cheng WL, et al. Gluten sensitivity: associated sporadic cerebellar ataxia in Taiwan. Acta Neurol Taiwan. 2010;19(4):263–9.
44. Rashtak S, Rashtak S, Snyder MR, Pittock SJ, Wu TT, Gandhi MJ, et al. Serology of celiac disease in gluten-sensitive ataxia or neuropathy: role of deamidated gliadin antibody. J Neuroimmunol. 2011;230(1-2):130–4.
45. Wong D, Dwinnel M, Schulzer M, Nimmo M, Leavitt BR, Spacey SD. Ataxia and the role of antigliadin antibodies. Can J Neurol Sci. 2007;34(2):193–6.
46. Lock RJ, Pengiran Tengah DS, Unsworth DJ, Ward JJ, Wills AJ. Ataxia, peripheral neuropathy, and anti-gliadin antibody. Guilt by association? J Neurol Neurosurg Psychiatry. 2005;76(11):1601–3.
47. Bushara KO, Goebel SU, Shill H, Goldfarb LG, Hallett M. Gluten sensitivity in sporadic and hereditary cerebellar ataxia. Ann Neurol. 2001;49(4):540–3.
48. Ferlazzo E, Polidoro S, Gobbi G, Gasparini S, Sueri C, Cianci V, et al. Epilepsy, cerebral calcifications, and gluten-related disorders: are anti-transglutaminase 6 antibodies the missing link? Seizure. 2019;73:17–20.
49. Zis P, Rao DG, Sarrigiannis PG, Aeschlimann P, Aeschlimann DP, Sanders D, et al. Transglutaminase 6 antibodies in gluten neuropathy. Dig Liver Dis. 2017;49(11):1196–200.
50. Shill HA, Alaedini A, Bushara KO, Latov N, Hallett M. Anti-ganglioside antibodies in idiopathic and hereditary cerebellar degeneration. Neurology. 2003;60(10):1672–3.
51. Volta U, De Giorgio R, Granito A, Stanghellini V, Barbara G, Avoni P, et al. Anti-ganglioside antibodies in coeliac disease with neurological disorders. Dig Liver Dis. 2006;38(3):183–7.
52. Wilkinson ID, Hadjivassiliou M, Dickson JM, Wallis L, Grunewald RA, Coley SC, et al. Cerebellar abnormalities on proton MR spectroscopy in gluten ataxia. J Neurol Neurosurg Psychiatry. 2005;76(7):1011–3.
53. Clark HB. The neuropathology of Autoimmune Ataxias. Brain Sci. 2022;12(2):257.
54. Newrick L, Hoggard N, Hadjivassiliou M. Recognition and management of rapid-onset gluten ataxias: case series. Cerebellum Ataxias. 2021;8(1):16.
55. Hermaszewski RA, Rigby S, Dalgleish AG. Coeliac disease presenting with cerebellar degeneration. Postgrad Med J. 1991;67(793):1023–4.
56. Mitoma H, Manto M, Hampe CS. Immune-mediated cerebellar ataxias: from bench to bedside. Cerebellum Ataxias. 2017;4:16.
57. Mitoma H, Manto M, Hampe CS. Immune-mediated cerebellar ataxias: practical guidelines and therapeutic challenges. Curr Neuropharmacol. 2019;17(1):33–58.
58. Aljada B, Zohni A, El-Matary W. The gluten-free diet for celiac disease and beyond. Nutrients. 2021;13(11):3993.
59. Mearns ES, Taylor A, Thomas Craig KJ, Puglielli S, Leffler DA, Sanders DS, et al. Neurological manifestations of neuropathy and ataxia in celiac disease: a systematic review. Nutrients. 2019;11(2):380.
60. Anandacoomaraswamy D, Ullal J, Vinik AI. A 70-year-old male with peripheral neuropathy, ataxia and antigliadin antibodies shows improvement in neuropathy, but not ataxia, after intravenous immunoglobulin and gluten-free diet. J Multidiscip Healthc. 2008;1:93–6.
61. Palmieri B, Vadala M, Laurino C. Gluten-free diet in non-celiac patients: beliefs, truths, advantages and disadvantages. Minerva Gastroenterol Dietol. 2019;65(2):153–62.
62. Khairuddin MAN, Lasekan O. Gluten-free cereal products and beverages: a review of their health benefits in the last five years. Foods. 2021;10(11):2523.
63. Niland B, Cash BD. Health benefits and adverse effects of a gluten-free diet in non-celiac disease patients. Gastroenterol Hepatol (N Y). 2018;14(2):82–91.
64. Nanri K, Okita M, Takeguchi M, Taguchi T, Ishiko T, Saito H, et al. Intravenous immunoglobulin therapy for autoantibody-positive cerebellar ataxia. Intern Med. 2009;48(10):783–90.
65. Burk K, Melms A, Schulz JB, Dichgans J. Effectiveness of intravenous immunoglobin therapy in cerebellar ataxia associated with gluten sensitivity. Ann Neurol. 2001;50(6):827–8.
66. Mitoma H, Hadjivassiliou M, Honnorat J. Guidelines for treatment of immune-mediated cerebellar ataxias. Cerebellum Ataxias. 2015;2:14.

Friedreich Ataxia

91

Overview

Friedreich ataxia is the most common hereditary ataxia, transmitted in an autosomal recessive fashion. The genetic abnormality is a trinucleotide repeat expansion in the frataxin gene on chromosome 9q13 that encodes for an iron trafficking protein in mitochondria, malfunction of which leads to the production of free radicals resulting in oxidative stress and cell death. The greater the number of trinucleotide repeats, the more severe and more rapidly progressive the disease. The disease usually begins insidiously in childhood or young adulthood, and the initial symptoms usually include gait ataxia and dysarthria. Aside from the prominent abnormalities in the nervous system, other organ systems are also affected, including cardiovascular (hypertrophic cardiomyopathy), endocrine (diabetes), and skeletal (pes cavus, scoliosis). Physical examination usually shows appendicular ataxia and weakness, cerebellar dysarthria, reduced proprioception and vibratory sensation, and the skeletal abnormalities mentioned before. Ocular motor examination may show eye movements compatible with cerebellar dysfunction, including spontaneous nystagmus, saccadic dysmetria, and poor smooth pursuit. Audiometry may often show hearing loss and abnormal brainstem auditory evoked responses. Instrumented vestibular testing usually shows a variety of abnormalities as well. Brain MRI may show atrophy of the superior cerebellar vermis. No pharmacotherapy has yet been shown to alter the course of the disease, so management is symptomatic and usually involves multiple specialists (neurology, cardiology, endocrinology, speech pathology, occupational and physical therapy, and medical genetics).

Introduction

Friedreich ataxia (FA) is the most common hereditary ataxia and is transmitted in an autosomal recessive fashion.

The disease was first described by the German neuropathologist Nikolaus Friedreich (1825–1922) in 1863 [1] (Fig. 91.1).

M. Cherchi, *Otoneurology and Vestibular Medicine*, https://doi.org/10.1007/978-3-031-94841-1_91

Fig. 91.1 Nikolaus Friedreich (1825–1922) (From https://en.wikipedia.org/wiki/Nikolaus_Friedreich, public domain)

Epidemiology

Friedreich ataxia is the most common hereditary ataxia in the Caucasian population [2]. The estimated prevalence of FA is 1 in 50,000, with a carrier frequency in the Caucasian population of 1 in 90 [3].

Genetics

Friedreich ataxia (OMIM 229300) results from a trinucleotide (GAA) repeat expansion in the frataxin gene located at chromosome 9q21.11 (OMIM 606829). This abnormality is found in the first intron of the gene in 98% of individuals [3]. There is an inverse relationship between the length of the trinucleotide repeat expansion and the onset/severity of symptoms.

Pathophysiological Mechanism of Disease

The genetic mutation of this disease affects frataxin, a nuclear-encoded protein expressed at the inner mitochondrial membrane that is involved in the influx/efflux of iron into and out of mitochondria. It is suspected that this faulty iron metabolism results in elevated levels of free radicals, leading to mitochondrial damage and eventually cell death [3].

Clinical Presentation

The usual age of onset is between 5 and 15 years of age [3], though late-onset cases have been reported. Durr and colleagues [4] studied 140 FA patients and reported the age of onset to range from 2 to 51 years (mean 15.5 ± 8 years).

The most common symptoms include ataxia, clumsiness, and dysarthria [3]. In the series of 140 FA patients studied by Durr and colleagues [4], the authors state, "The presenting symptom was gait ataxia in all patients except for several in whom scoliosis was diagnosed before the ataxia." The meta-analysis of 650 genetically confirmed FA patients by Reetz and colleagues [5] reported that the most common presenting symptoms were instability (77.2%), scoliosis (23.1%), and falls (19.8%). Further details are displayed in Fig. 91.2.

Table 91.1 outlines the clinical features in 140 FA patients studied by Durr and colleagues [4].

There is variability in clinical presentation, some of which appears to be correlated with the specific pattern of genetic mutation [3]. The later-onset cases appear to be related to a more limited number of trinucleotide repeats in the abnormal frataxin gene.

Table 91.2, from Durr and colleagues [4], shows that a higher number of trinucleotide repeats correlates with earlier mean age of onset, shorter time to becoming wheelchair-bound, greater prominence of physical examination abnormalities, greater frequency of diabetes, and cardiomyopathy.

Reetz and colleagues [5] present a meta-analysis of studies including 650 genetically confirmed FA patients. They describe in detail the nonataxia symptoms in FA and report that, "The most frequent clinical features beyond afferent ataxia were abnormal eye movements (90.5%), scoliosis (73.5%), deformities of the feet (58.8%), urinary dysfunction (42.8%), cardiomyopathy and cardiac hypertrophy (40.3%), followed by decreased visual acuity (36.8%); less frequent features were, among others, depression (14.1%) and diabetes (7.1%)."

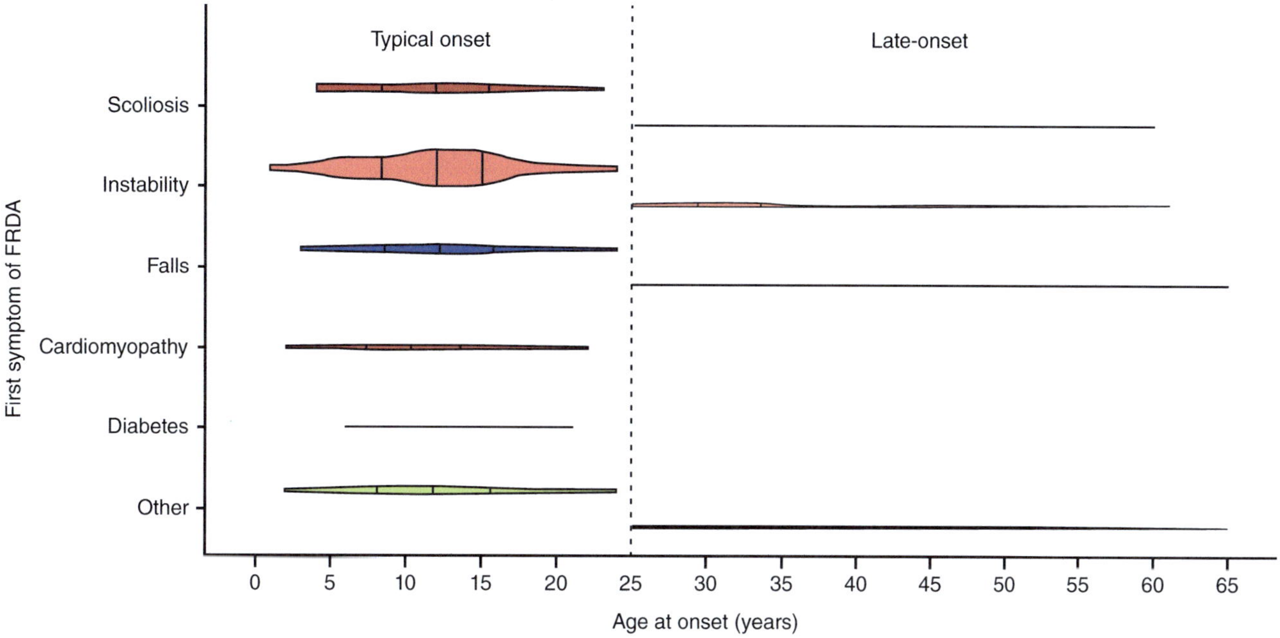

Fig. 91.2 Distribution of first symptoms at initial presentation (age of onset) of Friedreich ataxia for typical- and late-onset groups: the thicker the shape at a given point, the more patients presented this symptom at the respective age at onset; vertical lines within each shape represent quartiles. FRDA = Friedreich ataxia (From Reetz et al. [5]. Reproduced with written permission from Wolters Kluwer Health Limited)

Table 91.1 Frequency of clinical features in 140 patients with Friedreich ataxia (From Durr et al. [4])

Clinical sign	This study (Durr et al.), percent of patients	Harding study, percent of patients
Gait ataxia	100	–
Limb ataxia	99	99
Dysarthria	91	97
Lower-limb areflexia	87	99
Knee jerks present	12	0.9
Loss of vibration sense	78	73
Extensor plantar reflexes	79	89
Muscle weakness in lower limbs	67	88
Scoliosis	60	79
Pes cavus	55	55
Horizontal nystagmus	40	20
Wasting of lower limbs	39	39
Saccadic-pursuit eye movements	30	12
Swallowing difficulties	27	–
Wasting of upper limbs	25	49
Sphincter disturbances	23	–
Reduced visual acuity	13	18
Hearing loss	13	8
Axonal neuropathy (abnormal motor-nerve conduction velocity in the upper limbs with small or absent sensory action potentials), total of 63 patients tested in the study by Durr et al.	98	96
Cardiomyopathy on echocardiography, total of 75 patients in the study by Durr et al.	63	–
Abnormal brain-stem evoked potentials, total of 29 patients in the study by Durr et al.	61	–
Abnormal visual evoked potentials, total of 34 patients in the study by Durr et al.	34	–
Diabetes or impaired glucose tolerance, total of 61 patients in the study by Durr et al.	32	10

Table 91.2 In Friedreich ataxia, a greater number of trinucleotide repeats correlates with earlier age of onset and more rapid progression. (From Durr et al. [4])

Clinical feature	Number of GAA repeats			p value
	<520 repeats	520–780 repeats	>780 repeats	
Number of patients	48	54	38	
Mean age at onset in years	22.5 ± 9	13.3 ± 5	9.7 ± 5	<0.001
Mean time until wheelchair use in years	13.8 ± 6	11.5 ± 5	7.9 ± 5	0.01
Mean duration of disease in years	15.8 ± 6	14.9 ± 9	16.9 ± 11	Not significant
Areflexia, upper limbs	66	90	90	<0.005
Areflexia, lower limbs	79	91	92	Not significant
Extensor plantar response (%)	66	78	97	<0.005
Pes cavus (%)	38	54	78	<0.005
Scoliosis (%)	41	61	81	<0.005
Muscle weakness in lower limbs (%)	56	64	67	0.01
Wasting in lower limbs (%)	33	31	57	<0.05
Diabetes (%, number/number tested)	23 (3/13)	33 (9/27)	33 (7/21)	Not significant
Cardiomyopathy on echocardiography (%, number, number/number tested)	44 (8/18)	67 (22/33)	83 (20/24)	<0.05

Physical Examination

Physical examination usually shows appendicular ataxia, muscle weakness, reduced or absent myotatic reflexes, present Babinski signs, reduced or absent vibratory sensation in the lower limbs, and dysarthria in a cerebellar pattern [3].

Patients also develop skeletal abnormalities including scoliosis and pes cavus [6].

Ocular Motor Examination

Ocular motor abnormalities are common in FA. On face-to-face examination, or with infrared video Frenzel goggles, one may find spontaneous nystagmus, saccadic dysmetria, and poor smooth pursuit (all of which suggest cerebellar localization).

Table 91.3 summarizes the ocular motor findings and some vestibular test result abnormalities of 30 genetically confirmed FA patients reported by Ziegelboim and colleagues [7].

Table 91.3 Oculomotor findings and vestibular test result abnormalities of 30 genetically confirmed FA patients reported by Zeigelboim et al. [7]

Vestibular and ocular motor findings	Number	Percent
Bilateral labyrinthine hyporeflexia	16	53.3
Multiple semispontaneous nystagmus	12	40.0
Absent rotary nystagmus	11	36.7
Asymmetrical optokinetic nystagmus	10	33.4
Spontaneous nystagmus	6	20.0
Pendulum tracking alterations	5	16.7
Bilateral labyrinthine hyperreflexia	5	16.7
Bidirectional semispontaneous nystagmus	2	6.7
Unidirectional semispontaneous nystagmus	1	3.4
Unilateral labyrinthine hyperreflexia	1	3.4

Testing: Auditory

FA patients generally do not complain of hearing loss. Nevertheless, the majority have abnormalities on hearing tests.

Ell and colleagues [8] studied 10 patients with genetically confirmed FA. All 10 patients underwent audiometry, of which 9 (***90%***) were abnormal.

We find it striking that the hearing loss is more pronounced in the low frequencies; this suggests greater pathology at the cochlear apex or its afferents.

Ell and colleagues [8] studied auditory evoked brainstem responses (ABRs) in 7 genetically confirmed FA patients and reported that 6 (***86%***) were abnormal.

Testing: Vestibular

The majority of FA patients exhibit ocular motor abnormalities. Table 91.4 compiles the statistics reported in several studies.

Some of the outliers in Table 91.4 (such as the 12.2% of patients with abnormal smooth pursuit reported by Harding and colleagues) were probably due to methodological differences.

Some studies report results in a continuous rather than dichotomous fashion. For example, Spieker and colleagues [9] noted the following statistically significant differences of FA patients compared to controls: reduced saccade velocity, reduced optokinetic gain, reduced vestibulo-ocular reflex gain, reduced time constant, reduced vestibulo-ocular reflex suppression, and reduced total caloric response.

There are also reports of FA patients exhibiting periodic alternating nystagmus [10, 11].

Table 91.4 Comparison of ocular motor abnormalities reported in Friedreich ataxia in several studies

	Fahey et al. 2009	Kirkham et al. 1979	Ell et al. 1984	Furman et al. 1983	Baloh et al. 1975	Harding et al. 1981	Luis et al. 2016	Ribal et al. 2007
Saccades: dysmetria		71%		88%				
Saccades: hypermetria	37%							
Saccades: hypometria	9%							
Saccades: latency prolonged	100%							
Smooth pursuit: abnormal	100%	86%	90%	100%		12.2%		
Nystagmus: spontaneous down beat	45%							
Nystagmus: horizontal		80%						
Nystagmus: rebound		57%	50%					
Nystagmus: gaze evoked					80%			
Nystagmus: gaze paretic		64%						
Nystagmus: gaze evoked paretic, horizontal	60%		80%					
Nystagmus: gaze evoked paretic, vertical			20%					
"Nystagmus" (not further specified)						20%		
Square wave jerks		43%	70%					94%
"Fixation instability"				100%				
Flutter			50%					
Optokinetic responses: abnormal		43%	20%					
Caloric responses reduced or absent		100%			100%			
vHIT: reduced VOR gain	100%						100%	
vHIT: overt saccades present							89%	
VOR suppression: abnormal		86%	60%					
RCT: reduced VOR gain				31%				
RCT: reduced VVOR gain				100%				
RCT: increased VFX gain				100%				

In summary, the most common oculomotor abnormalities observed in FA include saccadic dysmetria, prolonged saccadic latencies, abnormal smooth pursuit, abnormal nystagmus (spontaneous, rebound), and square wave jerks. Instrumented vestibular testing shows abnormalities on caloric testing (reduced responses), video head impulse testing (reduced vestibulo-ocular reflex gain, presence of overt compensatory saccades), and rotatory chair testing (reduced visual vestibulo-ocular reflex gain, increased visual fixation suppression gain).

Testing: Other

Patients invariably develop cardiac abnormalities including electrocardiographic or echocardiographic signs of hypertrophic cardiomyopathy.

About 30% of patients develop diabetes [3].

Imaging

Brain MRI often shows cerebellar atrophy, particularly in the superior vermis [12].

Histopathology

Autopsy studies of patients with FA [13] typically show central nervous system findings that include degeneration of the posterior columns, ventral and dorsal spinocerebellar tracts [14], loss of dorsal root ganglion cells, loss of neurons in the nucleus dorsalis of the spinal cord, the spinal and principal trigeminal nuclei, and the mesencephalic trigeminal nucleus. Peripheral nervous system findings include distal axonal peripheral neuropathy (tending to affect larger myelinated fibers that transmit proprioception and vibratory sensation).

Differential Diagnosis

Other diseases that can (at least initially) resemble FA include ataxia with vitamin E deficiency [4], ataxia-telangiectasia, Charcot-Marie-Tooth disease, abetalipoproteinemia, and Refsum disease.

Treatment

Given the suspicion that the mechanism of disease involves free radical production due to abnormal iron trafficking in mitochondria [3], it is logical to explore interventions such as antioxidant therapies [15].

Idebenone, a synthetic analogue of ubiquinone (coenzyme Q10), is an antioxidant that may improve performance of the electron transport chain in mitochondria.

Despite this logical therapeutic target, neurological outcomes of treatment with idebenone have been disappointing. A phase 3 prospective, randomized, double-blinded, placebo-controlled trial of idebenone over 6 months showed no different in neurological outcome in the drug versus placebo [16]. A placebo-controlled study of idebenone found no statistically significant difference in patient perception of symptomatic benefit [17].

Cardiac outcomes of treatment with idebenone may be different. For example, a prospective, randomized, placebo-controlled trial of idebenone reported "significant reductions of interventricular septal wall thickness and left ventricular mass in the idebenone group versus the placebo group" [18].

One prospective, open-label, single-arm trial combined idebenone with the iron chelator, deferiprone and reported that the combination brought about a "stabilizing effect in neurologic dysfunctions due to an improvement in the kinetic functions, [but] with a worsening of gait and posture scores. Heart hypertrophy parameters and iron deposits in dentate nucleus improved significantly" [19]. Given that this was an open-label study, we can draw no confident conclusions from these results.

Some experimental work suggested that interferon gamma 1b may increase frataxin levels, but a placebo-controlled trial of this drug found no difference in outcomes [20].

Since there is no therapy to retard, arrest, or reverse the underlying pathology, management usually focuses on symptomatic management. Given the multifaceted nature of FA, management usually involves multiple subspecialties, including a neurologist (if possible, an ataxia specialist), cardiology, endocrinology (to monitor for and treat glucose intolerance or diabetes), speech pathology (for swallowing evaluations), occupational and physical therapy, and medical genetics.

Prognosis

The disease is relentlessly progressive, eventually leading to inability to ambulate, leaving a patient wheelchair-bound.

Survival in FA is usually limited by cardiac complications, the degree of which appears dependent on the extent of the trinucleotide repeat expansion [21]. In the series of 133 genetically confirmed FA patients studied by Pousset and colleagues [21], the cause of death in 80% was cardiac in origin, usually from heart failure.

References

1. Friedreich N. Ueber degenerative Atrophie der spinalen Hinterstränge. Archiv für pathologische Anatomie und Physiologie und für klinische Medicin. 1863;26(5):433–59.
2. Wallace SE, Bird TD. Molecular genetic testing for hereditary ataxia: what every neurologist should know. Neurol Clin Pract. 2018;8(1):27–32.
3. Cossee M, Durr A, Schmitt M, Dahl N, Trouillas P, Allinson P, et al. Friedreich's ataxia: point mutations and clinical presentation of compound heterozygotes. Ann Neurol. 1999;45(2):200–6.
4. Durr A, Cossee M, Agid Y, Campuzano V, Mignard C, Penet C, et al. Clinical and genetic abnormalities in patients with Friedreich's ataxia. N Engl J Med. 1996;335(16):1169–75.
5. Reetz K, Dogan I, Hohenfeld C, Didszun C, Giunti P, Mariotti C, et al. Nonataxia symptoms in Friedreich Ataxia: report from the registry of the European Friedreich's Ataxia Consortium for translational studies (EFACTS). Neurology. 2018;91(10):e917–e30.
6. Abrahao A, Pedroso JL, Braga-Neto P, Bor-Seng-Shu E, de Carvalho AP, Barsottini OG. Milestones in Friedreich ataxia: more than a century and still learning. Neurogenetics. 2015;16(3):151–60.
7. Zeigelboim BS, Mesti JC, Fonseca VR, Faryniuk JH, Marques JM, Cardoso RC, et al. Otoneurological abnormalities in patients with Friedreich's Ataxia. Int Arch Otorhinolaryngol. 2017;21(1):79–85.
8. Ell J, Prasher D, Rudge P. Neuro-otological abnormalities in Friedreich's ataxia. J Neurol Neurosurg Psychiatry. 1984;47(1):26–32.
9. Spieker S, Schulz JB, Petersen D, Fetter M, Klockgether T, Dichgans J. Fixation instability and oculomotor abnormalities in Friedreich's ataxia. J Neurol. 1995;242(8):517–21.
10. Gorman WF, Brock S. Periodic alternating nystagmus in Friedreich's ataxia. Am J Ophthalmol. 1950;33(6):860–4.
11. Gorman WF, Brock S, Kestenbaum A. Periodic alternating nystagmus in Friedreich's ataxia. J Nerv Ment Dis. 1950;112(5):437–9.
12. Hadjivassiliou M, Wallis LI, Hoggard N, Grunewald RA, Griffiths PD, Wilkinson ID. MR spectroscopy and atrophy in Gluten, Friedreich's and SCA6 ataxias. Acta Neurol Scand. 2012;126(2):138–43.
13. Jitpimolmard S, Small J, King RH, Geddes J, Misra P, McLaughlin J, et al. The sensory neuropathy of Friedreich's ataxia: an autopsy study of a case with prolonged survival. Acta Neuropathol. 1993;86(1):29–35.
14. Lubozynski MF, Roelofs RI. Friedreich's ataxia. South Med J. 1975;68(6):757–63.
15. Barca E, Emmanuele V, DiMauro S, Toscano A, Quinzii CM. Anti-oxidant drugs: novelties and clinical implications in cerebellar ataxias. Curr Neuropharmacol. 2019;17(1):21–32.

16. Lynch DR, Perlman SL, Meier T. A phase 3, double-blind, placebo-controlled trial of idebenone in friedreich ataxia. Arch Neurol. 2010;67(8):941–7.
17. Cook A, Boesch S, Heck S, Brunt E, Klockgether T, Schols L, et al. Patient-reported outcomes in Friedreich's ataxia after withdrawal from idebenone. Acta Neurol Scand. 2019;139(6):533–9.
18. Mariotti C, Solari A, Torta D, Marano L, Fiorentini C, Di Donato S. Idebenone treatment in Friedreich patients: one-year-long randomized placebo-controlled trial. Neurology. 2003;60(10):1676–9.
19. Velasco-Sanchez D, Aracil A, Montero R, Mas A, Jimenez L, O'Callaghan M, et al. Combined therapy with idebenone and deferiprone in patients with Friedreich's ataxia. Cerebellum. 2011;10(1):1–8.
20. Lynch DR, Hauser L, McCormick A, Wells M, Dong YN, McCormack S, et al. Randomized, double-blind, placebo-controlled study of interferon-gamma 1b in Friedreich Ataxia. Ann Clin Transl Neurol. 2019;6(3):546–53.
21. Pousset F, Legrand L, Monin ML, Ewenczyk C, Charles P, Komajda M, et al. A 22-year follow-up study of long-term cardiac outcome and predictors of survival in Friedreich Ataxia. JAMA Neurol. 2015;72(11):1334–41.

Part XI

Diagnoses—Movement Disorders

92 Parkinsonism

Overview

Distinguishing among the various *parkinsonian disorders* can be difficult, especially in the early stages of the diseases. When an otoneurologist is consulted to help make this distinction, the relevant expertise is careful instrumented examination of eye movements with videonystagmography and, if possible, with rotatory chair testing. Other otovestibular tests (vestibular evoked myogenic potentials, video head impulse testing, computerized dynamic posturography) may also show abnormalities in parkinsonian disorders, but the main value of these additional tests is to exclude competing diagnoses. For example, identification of unilateral or bilateral vestibular weakness may provide a vestibular therapist a narrower therapeutic target.

Introduction

The term "*parkinsonism*" refers to a constellation of physical examination findings including bradykinesia (paucity of movement), resting tremor, cogwheel rigidity, and postural instability. Multiple diseases can present with some subset of these findings.

Idiopathic Parkinson's disease (IPD) was the first to be recognized and is the archetype of parkinsonism, usually including most or all of the clinical features described earlier. This was first described by Dr. James Parkinson in "An essay on the shaking palsy," published in 1817. It is unique among parkinsonian disorders in that most patients exhibit a favorable response to levodopa.

All other types of parkinsonism tend to differ in one or more features from idiopathic Parkinson's disease.

- *Drug-related parkinsonism.* Dopamine antagonists (such as some antipsychotic agents and several antiemetic agents) have the potential to induce parkinsonism. Withdrawal of the offending drug may prevent further progression; in some cases, the symptoms may improve.
- *Toxin-related parkinsonism.* Etiologies include carbon monoxide, manganese, MPTP (1-methyl-4-phenyl-1,2,3,6-tetrahydropyridine), organic solvents, and carbon disulfide.
- *Vascular parkinsonism.* This refers to infarctions in the basal ganglia. This can exhibit many of the features of IPD. It does not respond to dopamine.
- Neurodegenerative etiologies:
  - *Progressive supranuclear palsy* (PSP), previously eponymously designated Steele–Richardson–Olszewsky syndrome. It includes some parkinsonian features (bradykinesia, sometimes rigidity, and postural instability) but lacks others (tremor). Additional features include prominent and early falls; dysphagia; rigidity of the neck; and supranuclear palsy of vertical (more than horizontal) eye movements.
  - *Corticobasal ganglionic degeneration* (CBGD), sometimes also referred to as just "corticobasal degeneration." This disorder usually has very asymmetrical motor findings (sometimes affecting just one limb) that may include akinesia, dystonia, rigidity, and myoclonus, as well as the unusual features of ideomotor apraxia and alien limb phenomenon. There are usually accompanying cognitive deficits, such as executive dysfunction, aphasia, apraxia, visuospatial dysfunction, and behavioral changes.
  - *Lewy body disease*, sometimes also referred to as "dementia with Lewy bodies." Aside from parkinsonism, the other features of Lewy body disease include dementia with visual hallucinations, fluctuating cognition, and REM sleep behavior disorder.
  - *Bilateral striopallidodentate calcinosis*, also eponymously designated Fahr disease. The sporadically occurring form is called *idiopathic basal ganglia calcification* (IBGC). It can also occur in a familial pat-

M. Cherchi, *Otoneurology and Vestibular Medicine*, https://doi.org/10.1007/978-3-031-94841-1_92

tern, usually in an autosomal dominant fashion, in which case it is called *primary familial brain calcification* (PFBC). Clinical features are variable; besides parkinsonism, these may include chorea, dystonia, ataxia, and cognitive impairment.

- *Multiple systems atrophy* (MSA) is an umbrella term *for the following three neurodegenerative parkinsonian* syndromes:
  - *Olivopontocerebellar atrophy* (OPCA), sometimes known as "MSA-C," meaning "MSA with prominent cerebellar features."
  - *Striatonigral degeneration*, sometimes referred to as "MSA-P," meaning "MSA with predominant parkinsonian features."
  - *Shy-Drager syndrome*, sometimes referred to as "MSA-A," meaning, "MSA with predominant autonomic features."

Because these diseases often share parkinsonian features, they can be difficult to distinguish from one another, particularly early in the course of the disease. If there is uncertainty about the diagnosis, it is medically sensible to refer a patient to a movement disorders specialist.

Occasionally, an otoneurologist may be called upon to help make this distinction because some of these diseases are more likely to have specific eye movement abnormalities. For example, progressive supranuclear palsy (PSP) characteristically has square wave jerks, supranuclear gaze palsy (with preserved reflex vestibulo-ocular reflex eye movements), slowing of vertical (more than horizontal) saccades, and "hang-up" at end-trajectory on optokinetic testing. These ocular motor findings may emerge gradually on serial oto-vestibular testing, and if they do, then a diagnosis of PSP is significantly more likely.

Parkinson's Disease

93

For Clinicians

Overview

Idiopathic Parkinson's disease (IPD) is the most common movement disorder. It results from cell death of dopamine-producing neurons in the substantia nigra. Symptoms start insidiously and progress gradually. The stereotypical presentation on physical examination includes bradykinesia, tremor, rigidity, and postural instability. Ocular motor examination may identify convergence insufficiency. The differential diagnosis includes other parkinsonian disorders. The most common pharmacotherapy is carbidopa-levodopa; less common approaches include amantadine and COMT inhibitors. Invasive therapies include deep brain stimulation and ablative procedures (thalamotomy, pallidotomy). With treatment, many patients can maintain an acceptable level of function with nearly normal life expectancy. The management of Parkinson's disease is often initiated by a general neurologist, and if it proves difficult to control, then referral to a movement disorders specialist is appropriate.

The role of otoneurology is limited in idiopathic Parkinson's disease (IPD). If a patient originally suspected of having IPD is failing to respond to therapy, then they are sometimes referred to otoneurology. In this situation, the role of the otoneurologist is usually to help distinguish IPD from other parkinsonian syndromes. The usual example in this circumstance is when the appearance of specific ocular motor findings begins to increase the suspicion for progressive supranuclear palsy and lower the suspicion for IPD.

Introduction

In 1817, the English physician, Dr. James Parkinson (1755–1824), published, "An Essay on the Shaking Palsy" [1], and this is usually credited as the first clinical description of this entity. During life, this is a clinical diagnosis, based on the history (gradually progressive symptoms), cardinal physical examination findings (bradykinesia, tremor, rigidity, and postural instability), and the absence of evidence for an alternative cause of parkinsonism. This is called *idiopathic Parkinson's disease (IPD)* .

Epidemiology

IPD is the most common movement disorder. The prevalence of IPD is 1–2 per 1000 people [2]. The onset of IPD younger than age 50 years is uncommon. IPD affects 1% of the population age 60 years and up [3] and up to 4% in the highest age groups [4].

Etiology

The majority of cases of Parkinson's disease have no identifiable underlying cause, and these cases are classified as idiopathic.

Genetics

An increasing number of studies report possible genetic sources of what otherwise appeared to be IPD, perhaps up to 14.8% [5]. Mutations in the LRRK2 (leucine-rich repeat kinase) account for the bulk of genetic cases [6].

Pathophysiological Mechanism of Disease

IPD is due to cell death of dopamine-producing neurons in the substantia nigra pars compacta.

M. Cherchi, *Otoneurology and Vestibular Medicine*, https://doi.org/10.1007/978-3-031-94841-1_93

Clinical Presentation

Symptoms start insidiously and progress gradually.

Some familial forms of parkinsonism may present at an earlier age.

Some nonmotor symptoms of IPD may occur years prior to the onset of motor symptoms. The main prodromal symptoms include hyposmia [7] and REM sleep behavior disorder [7].

Physical Examination

Classically patients with IPD exhibit the following:

- *Bradykinesia*, meaning that their movements are reduced in velocity and amplitude; this applies to appendicular movements (e.g., small handwriting), pharyngeal movements (e.g., reduced volume of voice), facial movements (masked facies), stance (narrow base), and ambulation (e.g., reduced stride length, reduced arm swing, shuffling).
- *Tremor*. This is present to varying degrees. It often presents asymmetrically. The appendicular tremor is present at rest (e.g., when a patient is sitting quietly, or when walking with arms at the sides), and is often described as a "pill rolling tremor," referring to the regular oscillatory pronation–supination movement of the forearm. Vocal tremor and head tremor may also be present.
- *Rigidity*. This refers to resistance to passive movement detected on physical examination. The so-called cogwheel rigidity can be detected as rhythmic intermittent resistance to passive rotation at the wrists and is due to underlying tremor.
- *Postural instability*. This is due in part to bradykinesia, and in part to reduced reflex postural adjustments in response to unexpected perturbations. It can be detected on the "pull test," in which an examiner gently but quickly pulls the patient backward at the shoulders; in response to this, a healthy person will usually place one heel further backward for support, whereas a patient with IPD will fail to make this postural adjustment and will simply fall backward.

Posture is usually stooped, and the base of support is narrow.

The diagnostic criteria for IPD have evolved. At one point, postural instability was considered one of the cardinal symptoms [8], but this was subsequently relegated to a "supportive" finding [9].

Ocular Motor Examination

Patients with IPD often exhibit convergence insufficiency to variable degrees.

Testing: Ocular Motor

Instrumented ocular motor testing, such as *videonystagmography* (VNG), may show findings not detectable on face-to-face examination. Compared to controls, IPD patients have prolonged saccadic latency, decreased saccadic accuracy, and reduced gain on smooth pursuit [10]. Some research has also explored whether the degree of square wave jerks can help distinguish IPD from other parkinsonian disorders [11].

Testing: Video Head Impulse Testing (vHIT)

Research on *video head impulse testing* (vHIT) in IPD has provided conflicting results. Some investigators find no difference in the vestibulo-ocular reflex (VOR) gain on vHIT [12, 13], while other researchers report reduced VOR gain [14, 15]. Other vHIT protocols, such as the SHIMP (suppression head impulse) paradigm, are reportedly abnormal in IPD patients [13].

Testing: Vestibular Evoked Myogenic Potentials (VEMPs)

Vestibular evoked myogenic potentials (VEMPs) may be abnormal in IPD.

Scarpa and colleagues [16] studied 15 IPD patients and found cVEMPs responses were absent bilaterally in 47% and absent unilaterally in 13%. Berkiten and colleagues [14] studied 40 IPD patients and found cVEMPs responses generally had lower amplitudes; responses were absent bilaterally in 20% and unilaterally in 15%.

IPD patients may have abnormal cVEMP latencies, but not in a consistent pattern. For example, Scarpa and colleagues [16] reported that the p13 latency was significantly *increased,* whereas Berkiten and colleagues [14] reported that the p13 latency was *decreased.*

Berkiten and colleagues [14] studied 40 IPD patients and reported that oVEMP responses generally had lower amplitudes; responses were oVEMPs were absent bilaterally in 15% and unilaterally in 13%.

Testing: Computerized Dynamic Posturography (CDP)

Computerized dynamic posturography (CDP) often quantifies postural instability in IPD patients [17–23], but the pattern is not specific for IPD.

Testing: Other

The sense of smell is often reduced in IPD [7].

Autonomic testing may show some degree of dysfunction. For example, tilt table test may show orthostatic hypotension in patients with IPD.

Imaging

A standard brain MRI will not confirm IPD, but may help exclude competing diagnoses; for example, infarctions involving the basal ganglia may suggest vascular parkinsonism; midbrain atrophy (sometimes called the "hummingbird sign") suggests progressive supranuclear palsy.

In 2011, the FDA approved the radiopharmaceutical agent ioflupane I 123 for SPECT (single-photon emission computed tomography) scans; this agent contains a radioactive isotope of iodine (I-123) which has a high binding affinity for presynaptic dopamine transporters, and its decay can be detected by a gamma camera. This nuclear medicine procedure can visualize striatal dopamine transporter uptake, and thus, it is known as a *DaT scan (dopamine transporter scan)*. It can help distinguish some parkinsonian syndromes (idiopathic Parkinson's disease, progressive supranuclear palsy, and multiple systems atrophy) from others.

Histopathology

Histopathology shows Lewy bodies (accumulation of alpha synuclein) and loss of dopaminergic cells in the substantia nigra. With disease progression, the Lewy bodies spread outside of the substantia nigra toward the cerebral cortex.

Differential Diagnosis

The differential diagnosis of IPD is that of other parkinsonian disorders. Early in these diseases, they can be difficult to distinguish.

A favorable (and sustained) response to levodopa is strongly suggestive of IPD *ex juvantibus,* as other parkinsonian disorders do not respond to that medication.

Exclusion criteria [24] for IPD are the presence of prominent cerebellar abnormalities (which suggests olivopontocerebellar atrophy), the presence of supranuclear gaze palsy (which suggests progressive supranuclear palsy), the presence of frontotemporal cognitive changes (which suggests Lewy body disease), and other cognitive findings such as apraxia (which suggests corticobasal ganglionic degeneration), absence of any response to levodopa therapy, and a normal DaT scan.

"Red flag" criteria which lower the suspicion for IPD [24] are early gait impairment, absence of progression, early bulbar dysfunction, inspiratory respiratory dysfunction, severe autonomic failure during the first year of the disease, recurrent falls due to reduced balance, early antecollis, pyramidal tract signs, bilateral symmetric parkinsonism, and the absence of any of the common nonmotor features of PD (such as sleep disturbances, autonomic dysfunction, or hyposmia).

Treatment

The most widely used treatment for IPD is carbidopa-levodopa. Other pharmacologic agents used include amantadine and inhibitors of COMT (catechol-O-methyltransferase). Problems with pharmacologic treatment of IPD include that (1) the efficacy of these drugs tends to decrease as the disease progresses; (2) the drugs are associated with symptom fluctuation; (3) the drugs often have side effects.

If pharmacotherapy fails, then invasive treatments may be considered in selective patients. These include deep brain stimulation and ablative procedures (thalamotomy, pallidotomy).

Unfortunately, the postural instability is the feature that tends to be least responsive to treatment.

There is evidence that physical activity may help preserve the level of function for a patient with IPD.

Prognosis

The treatment of IPD has improved over the years, and most patients can preserve a reasonable level of function with nearly normal life expectancy.

References

1. Parkinson J. An essay on the shaking palsy. London: Sherwood, Neely and Jones; 1817.
2. von Campenhausen S, Bornschein B, Wick R, Botzel K, Sampaio C, Poewe W, et al. Prevalence and incidence of Parkinson's disease in Europe. Eur Neuropsychopharmacol. 2005;15(4):473–90.

3. de Lau LM, Breteler MM. Epidemiology of Parkinson's disease. Lancet Neurol. 2006;5(6):525–35.
4. de Rijk MC, Launer LJ, Berger K, Breteler MM, Dartigues JF, Baldereschi M, et al. Prevalence of Parkinson's disease in Europe: a collaborative study of population-based cohorts. Neurologic diseases in the elderly research group. Neurology. 2000;54(11 Suppl 5):S21–3.
5. Westenberger A, Skrahina V, Usnich T, Beetz C, Vollstedt EJ, Laabs BH, et al. Relevance of genetic testing in the gene-targeted trial era: the Rostock Parkinson's disease study. Brain. 2024;147(8):2652–67.
6. Gilks WP, Abou-Sleiman PM, Gandhi S, Jain S, Singleton A, Lees AJ, et al. A common LRRK2 mutation in idiopathic Parkinson's disease. Lancet. 2005;365(9457):415–6.
7. Haehner A, Hummel T, Hummel C, Sommer U, Junghanns S, Reichmann H. Olfactory loss may be a first sign of idiopathic Parkinson's disease. Mov Disord. 2007;22(6):839–42.
8. Gelb DJ, Oliver E, Gilman S. Diagnostic criteria for Parkinson disease. Arch Neurol. 1999;56(1):33–9.
9. Postuma RB, Berg D, Stern M, Poewe W, Olanow CW, Oertel W, et al. MDS clinical diagnostic criteria for Parkinson's disease. Mov Disord. 2015;30(12):1591–601.
10. Zhou MX, Wang Q, Lin Y, Xu Q, Wu L, Chen YJ, et al. Oculomotor impairments in de novo Parkinson's disease. Front Aging Neurosci. 2022;14:985679.
11. Rascol O, Sabatini U, Simonetta-Moreau M, Montastruc JL, Rascol A, Clanet M. Square wave jerks in parkinsonian syndromes. J Neurol Neurosurg Psychiatry. 1991;54(7):599–602.
12. Hawkins KE, Chiarovano E, Paul SS, Burgess AM, MacDougall HG, Curthoys IS. Vestibular semicircular canal function as detected by video head impulse test (vHIT) is essentially unchanged in people with Parkinson's disease compared to healthy controls. J Vestib Res. 2022;32(3):261–9.
13. Hawkins KE, Rey-Martinez J, Chiarovano E, Paul SS, Valldeperes A, MacDougall HG, et al. Suppression head impulse test paradigm (SHIMP) characteristics in people with Parkinson's disease compared to healthy controls. Exp Brain Res. 2021;239(6):1853–62.
14. Berkiten G, Tutar B, Atar S, Kumral TL, Salturk Z, Akan O, et al. Assessment of the clinical use of vestibular evoked myogenic potentials and the video head impulse test in the diagnosis of early-stage Parkinson's disease. Ann Otol Rhinol Laryngol. 2023;132(1):41–9.
15. Lv W, Guan Q, Hu X, Chen J, Jiang H, Zhang L, et al. Vestibulo-ocular reflex abnormality in Parkinson's disease detected by video head impulse test. Neurosci Lett. 2017;657:211–4.
16. Scarpa A, Cassandro C, Vitale C, Ralli M, Policastro A, Barone P, et al. A comparison of auditory and vestibular dysfunction in Parkinson's disease and multiple system atrophy. Parkinsonism Relat Disord. 2020;71:51–7.
17. Ebersbach G, Gunkel M. Posturography reflects clinical imbalance in Parkinson's disease. Mov Disord. 2011;26(2):241–6.
18. Johnson L, James I, Rodrigues J, Stell R, Thickbroom G, Mastaglia F. Clinical and posturographic correlates of falling in Parkinson's disease. Mov Disord. 2013;28(9):1250–6.
19. Lee JM, Koh SB, Chae SW, Seo WK, Kwon DY, Kim JH, et al. Postural instability and cognitive dysfunction in early Parkinson's disease. Can J Neurol Sci. 2012;39(4):473–82.
20. Nonnekes J, de Kam D, Geurts AC, Weerdesteyn V, Bloem BR. Unraveling the mechanisms underlying postural instability in Parkinson's disease using dynamic posturography. Expert Rev Neurother. 2013;13(12):1303–8.
21. Rossi-Izquierdo M, Basta D, Rubio-Rodriguez JP, Santos-Perez S, Ernst A, Sesar-Ignacio A, et al. Is posturography able to identify fallers in patients with Parkinson's disease? Gait Posture. 2014;40(1):53–7.
22. Souza CO, Voos MC, Barbosa AF, Chen J, Francato DCV, Milosevic M, et al. Relationship between posturography, clinical balance and executive function in Parkinson s disease. J Mot Behav. 2019;51(2):212–21.
23. Visser JE, Oude Nijhuis LB, Janssen L, Bastiaanse CM, Borm GF, Duysens J, et al. Dynamic posturography in Parkinson's disease: diagnostic utility of the "first trial effect". Neuroscience. 2010;168(2):387–94.
24. Tysnes OB, Storstein A. Epidemiology of Parkinson's disease. J Neural Transm (Vienna). 2017;124(8):901–5.

Progressive Supranuclear Palsy

94

Overview

Progressive supranuclear palsy (PSP) is a relatively rare progressive neurodegenerative tauopathy. The mean age of onset is 63 years. The disease appears to occur sporadically, though there is suggestion of genetic mutation. The only confirmed risk factor for PSP is lower educational attainment. Patients usually present with gait problems, and eventually additional symptoms such as dysarthria, dysphagia, and urinary incontinence. Several subtypes of PSP have been described. On physical examination, patients usually have neck hypertonia, a "frightened" appearance of facies, convergence insufficiency, and reduced amplitude and velocity of voluntary vertical (more than horizontal) saccades, with preservation of the vestibulo-ocular reflex. Instrumented ocular motor examination may reveal square wave jerks, reduced gain on smooth pursuit, and "hang up" at end trajectory of optokinetic responses. Other otovestibular tests (vestibular evoked myogenic potentials, computerized dynamic posturography) may also be abnormal, but not in a pattern that is sensitive or specific for PSP. Brain MRI often shows midbrain atrophy. FDG-SPECT scans may aid in the diagnosis. Tau-binding radioligand PET studies remain investigational. There is no proven cure or arrestive therapy, and the disease is uniformly fatal from 5.3 to 9.7 years (mean 7.4 years) after symptom onset.

The role of otoneurology is limited in PSP. If a patient originally suspected of having idiopathic Parkinson's disease is failing to respond to therapy, then they are sometimes referred to otoneurology. Otovestibular studies (particularly videonystagmography and rotatory chair testing) may identify some of the ocular motor abnormalities that increase suspicion for PSP. This may help the patient avoid needless trials of Parkinson's drugs, or invasive treatments.

Usually, PSP patients are managed by a movement disorders neurologist.

Introduction

Progressive supranuclear palsy (PSP) is a neurodegenerative disease often classified under the broader umbrella of "Parkinson's plus" syndromes, as it can exhibit some parkinsonian features.

At the University of Toronto, the Canadian neurologist Dr. John Clifford Richardson (1909–1986) observed a series of patients with the clinical findings of this disease. He worked with two colleagues at the same institution: the Polish neuropathologist Dr. Jerzy Olszewski (1913–1964) and another Canadian neurologist Dr. John C. Steele (1934–2022). In 1963, they published the first description of this clinical entity [1, 2], which for some time was referred to by the eponymous designation, *Steele–Richardson–Olszewski syndrome*, but eventually came to be known as *progressive supranuclear palsy (PSP)* (Figs. 94.1, 94.2, and 94.3).

PSP is relatively rare, though it has attracted disproportionate attention because it has afflicted several high-profile celebrities over the years, such as Dudley Moore, Linda Ronstadt, Patricia Richardson, Laurence Richardson, and Kathryn Leigh Scott.

M. Cherchi, *Otoneurology and Vestibular Medicine*, https://doi.org/10.1007/978-3-031-94841-1_94

Fig. 94.1 Dr. John Clifford Richardson (1909–1986). (From Williams et al. [3]. Reproduced with written permission from the University of Toronto)

Fig. 94.2 Dr. Jerzy Olszewski (1913–1964). (From Williams et al. [3]. Reproduced with written permission from the University of Toronto)

Fig. 94.3 John C. Steele (1934–2022). (From Morris and Lees [4]. Reproduced with written permission from John Wiley & Sons)

Epidemiology

PSP has a prevalence of 5–6 cases per 100,000 people, with a mean age of onset of 63 years [5].

The only confirmed risk factor for PSP is lower educational attainment [6, 7].

Genetics

PSP generally does not exhibit obvious familial patterns. However, there is evidence that certain polymorphisms of the MAPT (microtubule associated protein tau) gene on chromosome 17q21 (which encodes the tau protein) is overrepresented in PSP patients [8]. The specific mutations involve inversion, deletions, and partial reinversions [5].

The evidence for other genetic abnormalities is weaker but includes mutations in EIF2AK3 (the gene that encodes protein kinase R-like endoplasmic reticulum kinase) which regulates protein transcription; syntaxin-6 (STX6) which is involved in vesicular exocytosis; and MOBP (myelin oligodendrocyte basic protein) which is a component of myelin [5].

Pathophysiological Mechanism of Disease

PSP is a tauopathy [9] and shares this molecular feature with Alzheimer's disease and Pick disease. The mechanism by which accumulation of tau protein causes leads to cell dysfunction and death is unknown, though there is some suggestion that mitochondrial failure plays a role [10].

Clinical Presentation

Gait difficulty is the presenting motor symptom in about 60% of cases [5].

Dysarthria is present in 41% of patients within 2 years of initial symptom onset [5].

Dysphagia is present in 18% of patients within 2 years of initial symptom onset, and in 46% of patients within 5 years of symptom onset [5]. Dysphagia is important to recognize because it can lead to aspiration pneumonia.

Urinary incontinence is present in 42% of patients by 3.5 years from initial symptom onset [5].

Several subtypes of PSP have been proposed, including the following:

- Richardson's syndrome.
- PSP-parkinsonism.
- PSP with progressive gait freezing.
- PSP-corticobasal syndrome.
- PSP-speech language.
- PSP with frontal presentation.
- PSP with predominant cerebellar ataxia.
- PSP with mixed pathology.

It is unclear whether these are distinct diseases or variants of a single underlying disease. The latter possibility is suggested by what has been noted by many investigators, namely that, "The regional distribution of tau pathology and neuronal loss is a source of pathological and, consequentially, clinical heterogeneity in PSP" [11].

Physical Examination

On physical examination, a PSP patient often has *neck hypertonia* that may be more pronounced with distraction (such as alternately opening and closing the fists). Facies are often described as having a "frightened" appearance. Face-to-face ocular motor examination may show clear *convergence insufficiency*; vertical (more than horizontal) *saccades* may be slow and of limited excursion, but reflex vestibulo-ocular reflex movements are normal.

There have been attempts at systematizing bedside ocular motor examination with the aim of distinguishing PSP from, for example, idiopathic Parkinson's disease [12], but none has been widely accepted or applied.

Ocular Motor Examination

On video Frenzel oculography, one often can observe *square wave jerks* (SWJ). Figure 94.4, from Chen and colleagues [13], compares the very low amplitude refixation saccades in a healthy individual (top panel) with the square wave jerks in a PSP patient (bottom panel).

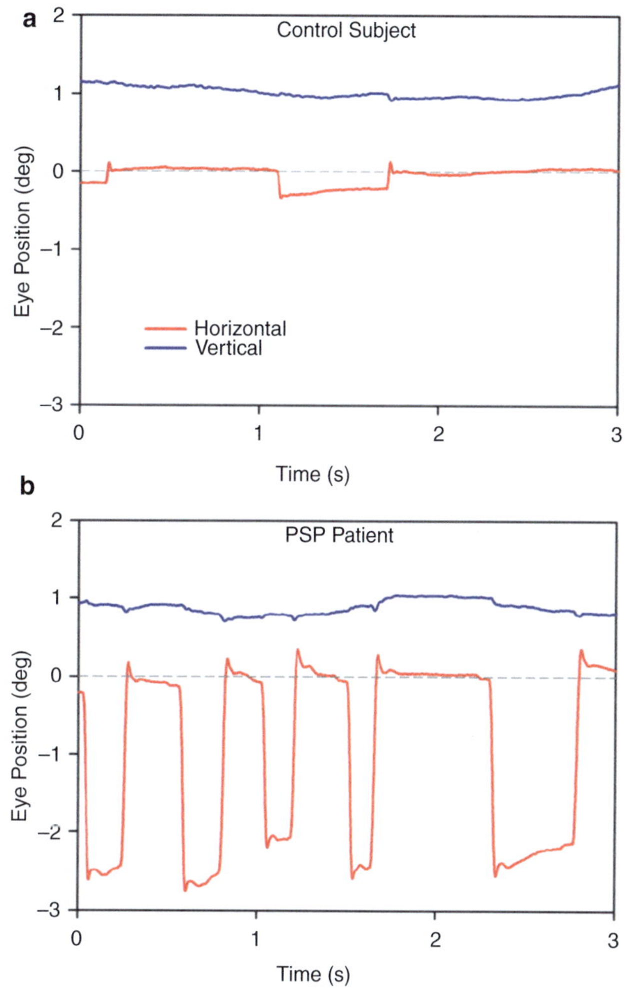

Fig. 94.4 Comparison of fixation behavior in a normal elderly subject (panel A) and in a patient with PSP (progressive supranuclear palsy) (panel B). Both show saccadic intrusions (square-wave jerks), but they are larger and more frequent in the record from the PSP patient. The gray-dashed line corresponds to the desired horizontal position of fixation. The vertical channel has been offset in both records to aid clarity. Positive values indicate rightward and upward movements (From Chen et al. [13]. Reproduced from Frontiers Media S.A. with permission under CC-BY 4.0)

Videonystagmography (VNG) shows that *vertical saccades exhibit reduced velocity and amplitude*. Figure 94.5, from Chen and colleagues [13], compares vertical saccadic amplitude and velocity in a healthy individual (top panels) with those of a PSP patient (bottom panel).

Videonystagmography (VNG) shows that *vertical smooth pursuit exhibits low gain*. Figure 94.6, from Chen and colleagues [13], compares vertical smooth pursuit in a healthy individual (top panels) with that of a PSP patient (bottom panel).

Vestibular physiologists sometimes refer to PSP as a "fast phase disorder" given the dysfunctional saccades. Since the abnormalities of saccades (reduced amplitude and velocity) are more pronounced than the gain in pursuit (reduced gain), optokinetic stimulation will tend to drag the eyes in the direction of stimulation, because the slow movement (in the direction of the optokinetic stimulus) will overwhelm the contraversive saccade (in the direction opposite that of the optokinetic stimulus), and the eyes will tend to "hang up" and the end of the trajectory. In PSP, this can happen both with vertical [14] and horizontal optokinetic testing.

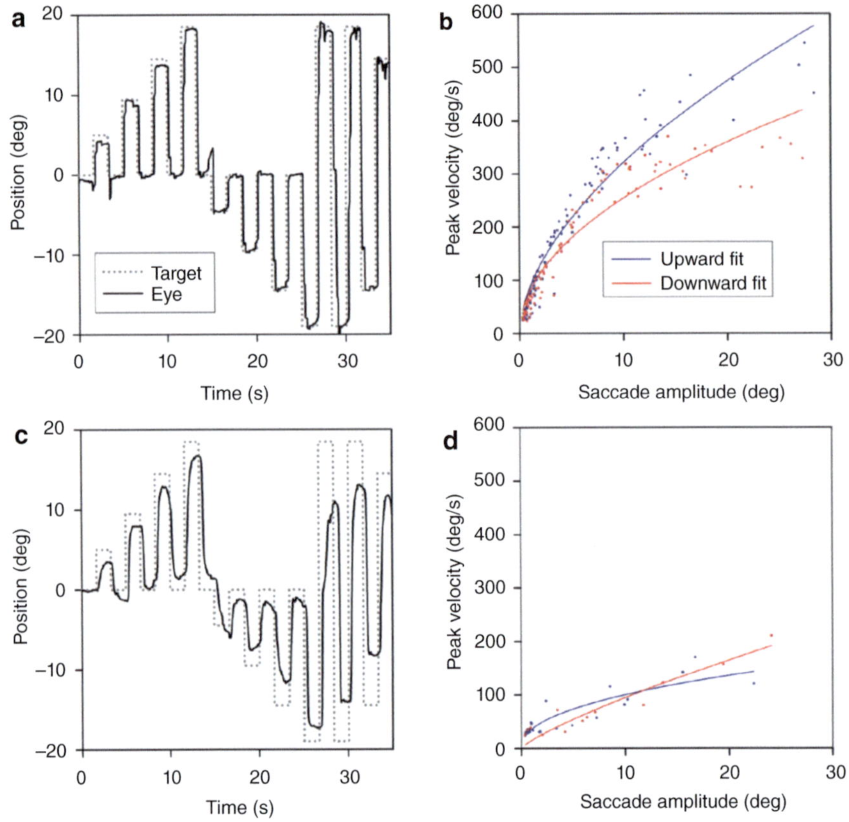

Fig. 94.5 Vertical saccades made by a normal subject and a patient with PSP (progressive supranuclear palsy). Panel A: Representative healthy elderly subject's target-directed vertical saccades. Panel B: Peak velocity–amplitude relationship of these saccades, as modeled by the equation (peak velocity = K × amplitude^{L}). Both fit curves have $R^2 > 0.9$. This particular subject made faster upward saccades than downward saccades. Panel C: PSP patient's saccades to similar target jumps as for the control in Panel A. Panel D: PSP patient's peak velocity–amplitude fit curves; both have $R^2 > 0.75$. The curve fits for this PSP patient's upward and downward saccades show overlap (From Chen et al. [13]. Reproduced from Frontiers Media S.A. with permission under CC-BY 4.0)

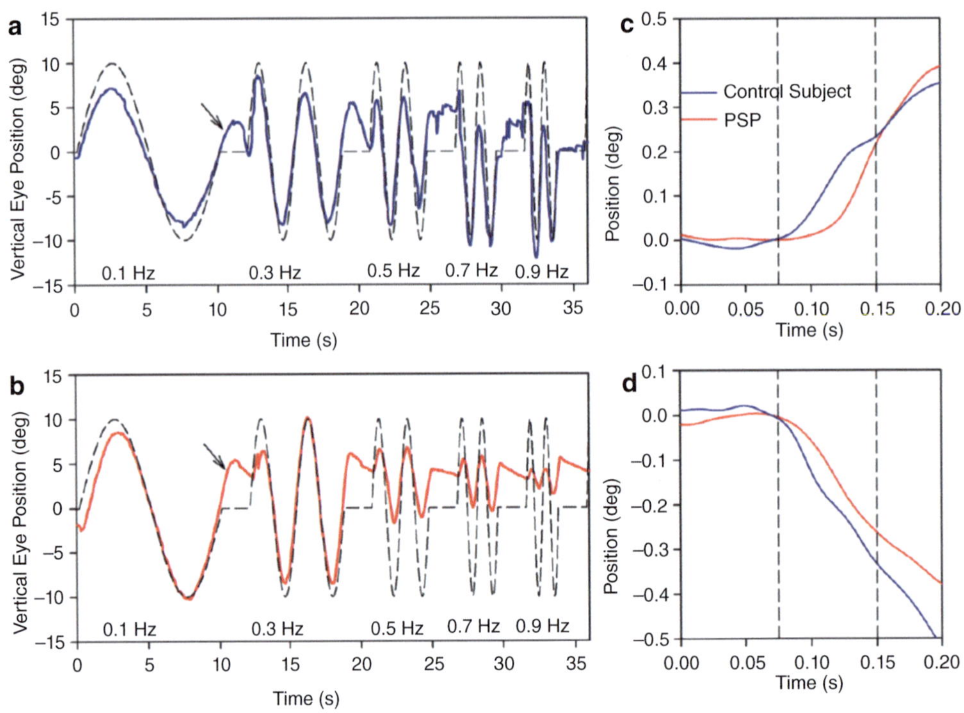

Fig. 94.6 Comparison of vertical smooth pursuit (panels A and B) and ocular following responses (panels C and D) of a normal subject and a PSP (progressive supranuclear palsy) patient. Both sets of responses were evoked as subjects viewed vertically-moving sine-wave gratings on a monitor subtending 50° × 37.5°. For pursuit, the stimulus was sinusoidal motion of a 0.27 cycles per degree grating over the range of temporal frequencies shown. For OFR (ocular following responses), the stimulus was ramp motion for 200 ms of a 0.17 cycles per degree grating. At lower frequencies of motion, both the normal subject (panel A) and the PSP patient (panel B) showed smooth tracking with some predictive properties (arrows); at high frequencies, the PSP (panel B) showed substantial decrease in the size of the pursuit movements compared with the control subject. Both upward (panel C) and downward (panel D) grating motion induced similar-sized OFR from the control subject (blue line) and PSP patient (red line), although the latency to onset was larger in the patient (From Chen et al. [13]. Reproduced from Frontiers Media S.A. with permission under CC-BY 4.0)

Testing: Vestibular Evoked Myogenic Potentials (VEMPs))

Different studies report conflicting results on vestibular evoked myogenic potentials in PSP patients.

Carpinelli and colleagues [15] reported that cervical vestibular evoked myogenic potentials (cVEMPs) and ocular vestibular evoked myogenic potentials (oVEMPs) tend to be absent more than in patients with idiopathic Parkinson's disease, and more than in healthy controls.

In contrast, Goldschagg and colleagues [16] report normal cVEMPs and oVEMPs in 16 PSP patients.

Testing: Caloric Testing

Goldschagg and colleagues [16] studied 16 PSP patients and reported normal responses on caloric testing.

Testing: Video Head Impulse Testing (vHIT))

Goldschagg and colleagues [16] studied 16 PSP patients and reported normal vestibulo-ocular reflex responses on video head impulse testing (vHIT).

Testing: Computerized Dynamic Posturography (CDP))

Dale and colleagues [17] studied computerized dynamic posturography (CDP) in 12 PSP patients and reported that perception of toes up surface tilt was reduced in comparison to patients with idiopathic Parkinson's disease and control subjects.

Goldschagg and colleagues [16] studied CDP in 16 PSP patients and reported that the "total root-mean-square body sway was significantly increased in patients with PSP" compared to healthy control subjects.

Ondo and colleagues [18] studied CDP in 20 PSP patients and reported that the total limits of stability time were significantly prolonged, total path sway was significantly larger, and total sensory organization testing scores were significantly worse, when compared to patients with idiopathic Parkinson's disease.

Pasha and colleagues [19] studied CDP in 29 PSP patients and reported "significantly poor DP [dynamic posturography] scores" compared to healthy controls.

In summary, different CDP studies report several abnormalities in CDP, including reduced perception of toes up surface tilt, increased body sway, impaired limits of stability, and poorer sensory organization testing scores.

Imaging: MRI

It has been observed for some time that conventional MRI often shows disproportionate midbrain atrophy in PSP patients, but this finding alone is insufficiently sensitive and specific.

Josephs and colleagues [20] studied 13 PSP patients; they reported brainstem atrophy and involvement of the cortex and subjacent white matter.

Massey and colleagues [21] studied 22 PSP patients and compared them to 13 patients with multiple systems atrophy, 7 patients with idiopathic Parkinson's disease, 6 patients with corticobasal ganglionic degeneration, and 9 healthy controls. They compared conventional MRI with histopathology of each patient and concluded that PSP patients could be distinguished from the others 73% of the time based on radiographic criteria.

Quattrone and colleagues [22] developed what they termed the "magnetic resonance parkinsonism index (MRPI)," which consists of [pons area / midbrain area] * [middle cerebellar peduncle width / superior cerebellar peduncle width]. Their original paper attempted to distinguish PSP from idiopathic Parkinson's disease (IPD) and from the parkinsonian variant of multiple systems atrophy (MSA-P). The study took 33 PSP patients, 108 IPD patients, 19 MSA-P patients, and 50 healthy controls. They found that:

- In PSP patients, the midbrain area and SCP width were significantly smaller than in patients with PD, MSA-P, and controls, though with some overlap.
- In PSP patients, the [pons area/midbrain area] and the [middle cerebellar peduncle width/superior cerebellar peduncle width] ratios were significantly larger than in patients with PD, MSA-P, and controls, though with some overlap.
- In PSP patients, the MRPI (which is the [pons area/ midbrain area] * [middle cerebellar peduncle width/ superior cerebellar peduncle width]) was significantly larger than in patients with PD, MSA-P, and controls. They noted that the MRPI correctly identified all patients with PSP.

Hussl and colleagues [23] also found that the MRPI was reasonably effective at distinguishing PSP from other groups (IPD, MSA-P, and healthy controls).

Morelli and colleagues [24] similarly found that the MRPI correctly distinguished PSP from IPD and from healthy controls. Morelli and colleagues also found the MRPI to be more accurate than clinical criteria [25] in predicting the development of PSP from patients with "clinically unclassifiable parkinsonism (CUP)."

Imaging: FDG-SPECT

Several studies have examined the application of fluorine-18-labeled fluorodeoxyglucose single-photon emission computed tomography (FDG-SPECT) to patients with PSP.

Tang and colleagues [26] studied 167 patients diagnosed with PSP, idiopathic Parkinson's disease, and multiple systems atrophy. They found that FDG-SPECT showed 88% sensitivity, 94% specificity, 91% positive predictive value, and 92% negative predictive value for PSP.

Imaging: Tau Binders

Since PSP is a tauopathy, it is reasonable to consider whether tau-binding radiotracers could identify PSP patients. The ligand ^{18}F-flortaucipir (AV-1451) binds to paired helical filaments of tau, and several investigations have explored the utility of this radiotracer in positron emission tomography for diagnosing PSP.

A study by Whitwell and colleagues [27] reported that the distribution of ^{18}F-flortaucipir (AV-1451) is different in PSP than in Alzheimer's disease.

However, studies by Cho and colleagues [28], and by Smith and colleagues [29], did not find ^{18}F-flortaucipir (AV-1451) to be able to distinguish PSP from healthy controls adequately.

Currently, PET scanning remains investigational for PSP.

Histopathology

The histopathologic criteria for PSP [30] require findings of either neurofibrillary tangles, or tau protein neuropil threads, or both, in the basal ganglia and brainstem [11]. Other supportive features include neuronal loss, gliosis, tufted astrocytes, and oligodendroglial coiled bodies [31].

Differential Diagnosis

Early in the disease, it can be difficult to distinguish PSP from other parkinsonian disorders. The following table, from Boxer and colleagues [11], reviews some of the imaging, CSF, serum and physiological markers that have been explored as possible biomarkers (Table 94.1).

Table 94.1 Features that have been explored as potential biomarkers for progressive supranuclear palsy. (From Boxer et al. [11])

Class	Measure	Potential uses						Supported by autopsy confirmed cases	Key insights
		Bedside	Animal model	Predictive of PSP pathology	Differential diagnosis	Surrogate outcome for clinical trial	Pharmacodynamic for clinical trial		
Imaging									
MRI									
	Visual read on conventional MRI	X			X			Yes	Hummingbird and morning glory signs specific for PSP, but don't differentiate from CBD pathology
	Volumetric (VBM, BSI, etc.)		X					Yes	Brain atrophy responds to anti-tau therapy in tau transgenic mouse models
				X		X		Yes	PSP-like atrophy detectable in nfvPPA and PPAOS
					X			Yes	Midbrain atrophy not specific to PSP; also found in CBD pathology
	Midbrain/pons ratio(s) such as MPRI			X	X			Yes	Ratio differentiates path confirmed PSP from PD, MSA; predicts gaze palsy and other PSP-RS signs (clinically diagnosed cases)
						X		Yes	Distinct atrophy rate measurable in PSP
	Diffusion tensor imaging (DTI)					X		No	DTI difference in PSP phenotypes; DTI sensitive to disease progression
	Free water imaging				X			No	Improved differential diagnosis: PSP from PD
	Resting-state fMRI						X	No	Reduced connectivity in midbrain-anchored network also present in normals
	Task fMRI			X			X	No	Differential diagnosis: PSP from PD
PET									
	FDG	X		X	X			Yes	Medial frontal, caudate, thalamic hypometabolism; network changes in metabolism identify PSP
	Tau (T807/AV1451, PBB3)			X	X		X	Yes	Anecdotal reports of in vivo binding for most tracers; T807/AV1451 low affinity for 4R tau in post mortem brain
MEG				X			X	No	Measurable evoked responses in PSP

(continued)

Table 94.1 (continued)

Class	Measure	Potential uses						Supported by autopsy confirmed cases	Key insights
		Bedside	Animal model	Predictive of PSP pathology	Differential diagnosis	Surrogate outcome for clinical trial	Pharmacodynamic for clinical trial		
Fluid									
CSF									
	Tau							No	Total tau comparable to controls
	p181 tau							No	pTau comparable to or lower than controls
	Neurofilament light chain (NfL)	X	X	X	X	X	X	No	NfL elevated in PSP, increases over time. Changes correlated with clinical and imaging. Elevated in tau transgenic mice
	YKL-40				X			No	Elevated in PSP
Blood	NfL		X	X	X	X	?	No	Plasma NfL elevated in PSP and predictive of clinical/imaging changes
Physiological									
Eye movements	Infrared oculography	X			X	X		Yes	Decreased saccade velocity and gain in PSP
Eyelid function	Spontaneous and evoked blink rate				X	X		No	Blink rate parameters associated with disease
Retinal thickness	Optical coherence tomography			X	X			No	Decreased retinal thickness in PSP
Activity	Wrist actigraphy		X			X		No	Disrupted circadian rhythms and sleep
Finger taps	Quantitative					X		No	Distinct abnormalities: PSP-RS vs. PD

Treatment

There is currently no curative or arrestive therapy for PSP. Some patients exhibit a modest and transient response to levodopa therapy. Trials of coenzyme Q10, deep brain stimulation, riluzole [32], tideglusib [33], and davunetide [34] failed to show any benefit.

Prognosis

PSP is a uniformly fatal disease. From the time of symptom onset, survival ranges from 5.3 to 9.7 years, with a mean survival of 7.4 years [5].

References

1. Richardson JC, Steele J, Olszewski J. Supranuclear ophthalmoplegia, pseudobulbar palsy, nuchal dystonia and dementia. A clinical report on eight cases of "Heterogenous system degeneration". Trans Am Neurol Assoc. 1963;88:25–9.
2. Steele JC, Richardson JC, Olszewski J. Progressive Supranuclear Palsy. A heterogeneous degeneration involving the brain stem, basal ganglia and cerebellum with vertical gaze and pseudobulbar palsy, nuchal dystonia and dementia. Arch Neurol. 1964;10:333–59.
3. Williams DR, Lees AJ, Wherrett JR, Steele JC. J. Clifford Richardson and 50 years of progressive supranuclear palsy. Neurology. 2008;70(7):566–73.
4. Morris HR, Lees AJ. Obituary for Dr. John C. Steele, 1934–2022. Mov Disord Clin Pract. 2023;10(1):7–8.
5. Golbe LI. Progressive supranuclear palsy. Semin Neurol. 2014;34(2):151–9.
6. Golbe LI, Rubin RS, Cody RP, Belsh JM, Duvoisin RC, Grosmann C, et al. Follow-up study of risk factors in progressive supranuclear palsy. Neurology. 1996;47(1):148–54.
7. Vidal JS, Vidailhet M, Derkinderen P, de Gaillarbois TD, Tzourio C, Alperovitch A. Risk factors for progressive supranuclear palsy: a case-control study in France. J Neurol Neurosurg Psychiatry. 2009;80(11):1271–4.
8. Baker M, Litvan I, Houlden H, Adamson J, Dickson D, Perez-Tur J, et al. Association of an extended haplotype in the tau gene with progressive supranuclear palsy. Hum Mol Genet. 1999;8(4):711–5.
9. Pollock NJ, Mirra SS, Binder LI, Hansen LA, Wood JG. Filamentous aggregates in pick's disease, progressive supranuclear palsy, and Alzheimer's disease share antigenic determinants with microtubule-associated protein, tau. Lancet. 1986;2(8517):1211.
10. Albers DS, Augood SJ, Park LC, Browne SE, Martin DM, Adamson J, et al. Frontal lobe dysfunction in progressive supranuclear palsy: evidence for oxidative stress and mitochondrial impairment. J Neurochem. 2000;74(2):878–81.
11. Boxer AL, Yu JT, Golbe LI, Litvan I, Lang AE, Hoglinger GU. Advances in progressive supranuclear palsy: new diagnostic criteria, biomarkers, and therapeutic approaches. Lancet Neurol. 2017;16(7):552–63.
12. Pagonabarraga J, Horta-Barba A, Busteed L, Bejr-Kasem H, Illan-Gala I, Aracil-Bolanos I, et al. Quantitative evaluation of oculomotor disturbances in progressive supranuclear palsy. Parkinsonism Relat Disord. 2021;85:63–8.
13. Chen AL, Riley DE, King SA, Joshi AC, Serra A, Liao K, et al. The disturbance of gaze in progressive supranuclear palsy: implications for pathogenesis. Front Neurol. 2010;1:147.
14. Garbutt S, Riley DE, Kumar AN, Han Y, Harwood MR, Leigh RJ. Abnormalities of optokinetic nystagmus in progressive supranuclear palsy. J Neurol Neurosurg Psychiatry. 2004;75(10):1386–94.
15. Carpinelli S, Valko PO, Waldvogel D, Buffone E, Baumann CR, Straumann D, et al. Distinct vestibular evoked myogenic potentials in patients with Parkinson disease and progressive Supranuclear Palsy. Front Neurol. 2020;11:598763.
16. Goldschagg N, Bremova-Ertl T, Bardins S, Dinca N, Feil K, Krafczyk S, et al. No evidence of a contribution of the vestibular system to frequent falls in progressive Supranuclear Palsy. J Clin Neurol. 2019;15(3):339–46.
17. Dale ML, Horak FB, Wright WG, Schoneburg BM, Nutt JG, Mancini M. Impaired perception of surface tilt in progressive supranuclear palsy. PLoS One. 2017;12(3):e0173351.
18. Ondo W, Warrior D, Overby A, Calmes J, Hendersen N, Olson S, et al. Computerized posturography analysis of progressive supranuclear palsy: a case-control comparison with Parkinson's disease and healthy controls. Arch Neurol. 2000;57(10):1464–9.
19. Pasha SA, Yadav R, Ganeshan M, Saini J, Gupta A, Sandhya M, et al. Correlation between qualitative balance indices, dynamic posturography and structural brain imaging in patients with progressive supranuclear palsy and its subtypes. Neurol India. 2016;64(4):633–9.
20. Josephs KA, Whitwell JL, Dickson DW, Boeve BF, Knopman DS, Petersen RC, et al. Voxel-based morphometry in autopsy proven PSP and CBD. Neurobiol Aging. 2008;29(2):280–9.
21. Massey LA, Micallef C, Paviour DC, O'Sullivan SS, Ling H, Williams DR, et al. Conventional magnetic resonance imaging in confirmed progressive supranuclear palsy and multiple system atrophy. Mov Disord. 2012;27(14):1754–62.
22. Quattrone A, Nicoletti G, Messina D, Fera F, Condino F, Pugliese P, et al. MR imaging index for differentiation of progressive supranuclear palsy from Parkinson disease and the Parkinson variant of multiple system atrophy. Radiology. 2008;246(1):214–21.
23. Hussl A, Mahlknecht P, Scherfler C, Esterhammer R, Schocke M, Poewe W, et al. Diagnostic accuracy of the magnetic resonance parkinsonism index and the midbrain-to-pontine area ratio to differentiate progressive supranuclear palsy from Parkinson's disease and the Parkinson variant of multiple system atrophy. Mov Disord. 2010;25(14):2444–9.
24. Morelli M, Arabia G, Salsone M, Novellino F, Giofre L, Paletta R, et al. Accuracy of magnetic resonance parkinsonism index for differentiation of progressive supranuclear palsy from probable or possible Parkinson disease. Mov Disord. 2011;26(3):527–33.
25. Morelli M, Arabia G, Novellino F, Salsone M, Giofre L, Condino F, et al. MRI measurements predict PSP in unclassifiable parkinsonisms: a cohort study. Neurology. 2011;77(11):1042–7.
26. Tang CC, Poston KL, Eckert T, Feigin A, Frucht S, Gudesblatt M, et al. Differential diagnosis of parkinsonism: a metabolic imaging study using pattern analysis. Lancet Neurol. 2010;9(2):149–58.
27. Whitwell JL, Lowe VJ, Tosakulwong N, Weigand SD, Senjem ML, Schwarz CG, et al. [(18) F]AV-1451 tau positron emission tomography in progressive supranuclear palsy. Mov Disord. 2017;32(1):124–33.
28. Cho H, Choi JY, Hwang MS, Lee SH, Ryu YH, Lee MS, et al. Subcortical (18) F-AV-1451 binding patterns in progressive supranuclear palsy. Mov Disord. 2017;32(1):134–40.
29. Smith R, Schain M, Nilsson C, Strandberg O, Olsson T, Hagerstrom D, et al. Increased basal ganglia binding of (18) F-AV-1451 in patients with progressive supranuclear palsy. Mov Disord. 2017;32(1):108–14.
30. Litvan I, Hauw JJ, Bartko JJ, Lantos PL, Daniel SE, Horoupian DS, et al. Validity and reliability of the preliminary NINDS

neuropathologic criteria for progressive supranuclear palsy and related disorders. J Neuropathol Exp Neurol. 1996;55(1):97–105.
31. Dickson DW, Ahmed Z, Algom AA, Tsuboi Y, Josephs KA. Neuropathology of variants of progressive supranuclear palsy. Curr Opin Neurol. 2010;23(4):394–400.
32. Bensimon G, Ludolph A, Agid Y, Vidailhet M, Payan C, Leigh PN, et al. Riluzole treatment, survival and diagnostic criteria in Parkinson plus disorders: the NNIPPS study. Brain. 2009;132(Pt 1):156–71.
33. Tolosa E, Litvan I, Hoglinger GU, Burn D, Lees A, Andres MV, et al. A phase 2 trial of the GSK-3 inhibitor tideglusib in progressive supranuclear palsy. Mov Disord. 2014;29(4):470–8.
34. Boxer AL, Lang AE, Grossman M, Knopman DS, Miller BL, Schneider LS, et al. Davunetide in patients with progressive supranuclear palsy: a randomised, double-blind, placebo-controlled phase 2/3 trial. Lancet Neurol. 2014;13(7):676–85.

Multiple System Atrophy

95

Overview

Multiple system atrophy (MSA) is a sporadically occurring progressive neurodegenerative alpha synucleinopathy that causes various combinations of parkinsonism, ataxia, and autonomic dysfunction. Several subtypes have been reported according to which feature is most prominent. The disease is rare (0.6–0.7 cases per 100,000 person years), with a peak age of onset in the sixth decade. Common presenting symptoms include falls, dysphagia, and REM sleep behavior disorder. Depending on the subtype, physical examination may show findings compatible with parkinsonism or ataxia. Face-to-face ocular motor examination may reveal spontaneous nystagmus, gaze evoked nystagmus, and choppy smooth pursuit. Instrumented ocular motor testing may additionally show square wave jerks, fixation abnormalities, saccadic dysmetria, and reduced vestibulo-ocular reflex. Vestibular evoked myogenic potentials may be reduced or absent. Autonomic testing may reveal orthostatic hypotension and abnormalities on thermoregulatory sweat testing (less commonly on quantitative sudomotor axon reflex test). Brain MRI may show the so-called "hot cross buns" sign or "putaminal slit" sign. FDG-PET may show hypometabolism in the putamina. The differential diagnosis is the other parkinsonian disorders. There is no curative or arrestive treatment, so management is symptomatic and often involves multiple subspecialties (neurology, psychiatry, cardiology, gastroenterology, urology, physical, and occupational therapy). Prognosis is poor, with a mean survival from symptom onset of 9.8 years.

Introduction

Multiple system atrophy (MSA) is a sporadically occurring alpha synucleinopathy that manifests as a relentlessly progressive neurodegenerative disorder, and whose subtypes exhibit varying degrees of parkinsonism, ataxia, and autonomic dysfunction [1, 2]. Three main forms have been described.

- Dr. Joseph Jules Dejerine and Dr. Andre Thomas reported on a patient with what they termed *olivopontocerebellar atrophy* (OPCA) [3].
- Dr. Raymond Delacy Adams (1911–2008), Dr. Ludo Van Bogaert (1897–1989), and Dr. Henri Marie Hector Vander Eecken (born 1920) described what eventually became known as *striatonigral degeneration* (SND) [4].
- Dr. George Milton Shy and Dr. Glenn Albert Drager reported a series of patients with autonomic failure [5]. For a time, this was known as *Shy–Drager syndrome* (SDS).

Dr. J.G. Graham and Dr. D.R. Oppenheimer first used the term *multiple system atrophy* (MSA) as an umbrella term for these various syndromes [6].

Since then, a number of other "atypical variants" of MSA have been described [7], including the following:

- MSAs with greater than expected cognitive impairment.
  - *MSA with Lewy body disease*. Since MSA and Lewy body disease are both alpha synucleinopathies, this clinical entity seems more likely to be one in which the neuropathology from one disease extends beyond its usual neuroanatomical area, provoking additional symptoms suggestive of the other disease.
  - *MSA with cognitive impairment/dementia*. This sounds similar to the "MSA with Lewy body disease" mentioned earlier.
  - *MSA with severe hippocampal atrophy*.
- MSAs with additional motor symptoms.
  - *MSA with dystonia*.
  - *MSA with spinal myoclonus*.
- *MSA with tau pathology*. We find it difficult to understand how this would be classified as an "atypical MSA" since MSA is an alpha synucleinopathy rather than a tauopathy. We think this entity is more likely to be an "atypical

M. Cherchi, *Otoneurology and Vestibular Medicine*, https://doi.org/10.1007/978-3-031-94841-1_95

tauopathy" whose pathology unusually happens to be in the neuroanatomical distribution that causes an MSA-like clinical picture.

The different presentations and deficits in MSA are likely due to variability in the distribution and degree of neuropathological involvement [2]. Multiple organ systems can be affected in MSA [1], including the following:

- Neurological
  - Parkinsonism
  - Cerebellar dysfunction
  - Pyramidal signs
  - Frontal lobe executive dysfunction
- Sleep
  - REM sleep behavior disorder
  - Sleep apnea
  - Excessive daytime sleepiness
  - Restless leg syndrome
- Psychiatric
  - Depression
  - Anxiety
- Otolaryngological
  - Stridor
  - Dysarthria
  - Dysphonia
- Cardiologic
  - Syncope
  - Orthostatic hypotension
  - Postprandial hypotension
  - Nocturnal hypertension
  - Leg edema
- Respiratory
  - Aspiration pneumonia
- Gastroenterological
  - Dysphagia
  - Constipation
  - Diarrhea
- Genitourinary
  - Sexual dysfunction
  - Urinary urgency and frequency
  - Nocturia
  - Urinary urge incontinence
  - Urinary retention
  - Recurrent urinary tract infections
- Dermatological
  - Hypohidrosis or anhidrosis
  - Vasomotor abnormalities

Epidemiology

The mean incidence of MSA is 0.6 to 0.7 cases (range 0.1–2.4 cases) per 100,000 person-years. The point prevalence is 3.4–4.9 cases per 100,000 people, increasing to 7.8 per 1000,000 people older than 40 years [2].

The peak age of onset of MSA is in the sixth decade, though earlier (fourth–fifth decades) and later (eighth decade) onsets have been reported [1]. So-called "young onset MSA" is rare; Kim and colleagues [8] studied 455 MSA patients and reported that 4 (0.9%) had symptom onset prior to the age of 40 years. Late-onset MSA is similarly unusual; Lee and colleagues [9] studied 1425 MSA patients and reported that 39 (2.7%) developed first symptoms at or after 76.8 years.

Most cases of MSA appear to be sporadic. Several pedigrees have been reported in which MSA appears to occur in an autosomal dominant or autosomal recessive fashion [1, 7].

Pathophysiological Mechanism of Disease

The underlying pathobiology of MSA remains incompletely understood [1]. The normal alpha synuclein protein appears to be involved in synaptic vesicle transport and neurotransmitter release [10]. The abnormal alpha synuclein protein appears to have a predilection for oligodendrocytes [11, 12].

Clinical Presentation

Synucleinopathies are often associated with REM sleep behavior disorder (RBD). MSA is no exception, and often, the RBD precedes other symptoms by years. Iranzo and colleagues [13] studied 26 MSA patients and found that 54% had RBD prior to developing any waking motor symptoms.

O'Sullivan and colleagues [14] reviewed the records of 83 cases of MSA and reported symptom onset (differentiated by presence/absence of early autonomic symptoms) as follows (Table 95.1):

In the MSA-A variant, early autonomic dysfunction occurs in 60% of patients and can involve erectile dysfunction, lower urinary tract symptoms, anhidrosis, and cardiovascular autonomic symptoms such as recurrent syncope [1].

The MSA-C and MSA-P variants are distinguished by whether the predominant motor features are cerebellar (MSA-C) or parkinsonian (MSA-P) [15].

Table 95.1 Reported symptom onset, differentiated by presence/absence of early autonomic symptoms, in 83 cases of multiple systems atrophy. (From O'Sullivan et al. [14])

	MSA with early autonomic symptoms		MSA without early autonomic symptoms	
	Number (percent)	Years since initial symptom onset	Number (percent)	Years since initial symptom onset
Frequent falls	22 (55%)	5.7 ± 2.2	23 (64%)	5.4 ± 2.4
Wheelchair dependent	23 (55%)	6.6 ± 2.5	19 (56%)	6.9 ± 1.8
Unintelligible speech	14 (33%)	6.7 ± 2.0	19 (53%)	7.6 ± 2.0
Severe dysphagia	13 (32%)	7.1 ± 2.4	12 (33%)	7.6 ± 2.3
Urinary catheter	29 (71%)	5.8 ± 2.7	18 (50%)	6.9 ± 2.3
Cognitive impairment	5 (13%)	6.6 ± 2.5	5 (14%)	6.6 ± 2.0

Physical Examination

On physical examination, patients have some combination of parkinsonism and ataxia. Bedside examination of orthostatics may reveal orthostatic hypotension.

Ocular Motor Examination

MSA patients may exhibit spontaneous nystagmus or gaze evoked nystagmus [1].

Anderson and colleagues [16] studied 30 patients with probable MSA (22 with MSA-P and 8 with MSA-C) by clinical examination. They reported "excessive square wave jerks" in 70% (21/30), "mild supranuclear gaze palsy" in 27% (8/30), gaze evoked nystagmus in 40% (12/30), "positioning down beat nystagmus" in 40% (10/25), mild or moderate saccadic hypometria in 73% (22/30), "broken up" smooth pursuit in 93% (28/30), and reduced vestibulo-ocular reflex suppression in 67% (16/24). They further reported that formal videonystagmography with caloric testing revealed no additional findings.

Testing: Videonystagmography (VNG)

Pinkhardt and colleagues [17] studied instrumented ocular motor testing in 19 MSA patients (11 with MSA-P and 8 with MSA-C) and compared them with 27 patients with idiopathic Parkinson's disease (IPD) and 23 healthy controls. They reported that smooth pursuit eye movements were similar between the MSA-P and MSA-C patients, but there was significant overlap with the IPD and healthy controls.

Pinnock and colleagues [18] studied instrumented ocular motor testing in 44 patients with various parkinsonian disorders (MSA, progressive supranuclear palsy, and idiopathic Parkinson's disease) compared to 50 age-matched healthy controls. They reported that the patients "exhibited larger and more frequent SI [saccadic intrusions] as well as greater displacement from the fixation target," adding that, "Patients with multiple system atrophy show increased frequency of SI both with and without a visible target."

Klotz and Klockgether [19] reported a case of an MSA patient with very prominent macro square wave jerks.

Zhou and colleagues [20] studied instrumented ocular motor testing in 45 patients with MSA (24 with MSA-P and 21 with MSA-C) and 40 age-matched healthy controls. They reported that overall 93% of MSA patients exhibited ocular motor abnormalities, broken down as follows: 38% had abnormalities on fixation and gaze holding; 51% had abnormalities without fixation; 73% had abnormalities on saccades; 71% had abnormalities on smooth pursuit; and 38% had abnormalities on optokinetic testing.

Zhou and colleagues [21] studied instrumented ocular motor testing in 33 MSA patients (18 with MSA-P and 15 with MSA-C), 96 patients with idiopathic Parkinson's disease, and 40 healthy controls. They reported that 64% of MSA patients exhibited saccadic dysmetria (both hypometria and hypermetria), and that this was statistically significantly greater than the other groups.

Brooks and colleagues [22] studied prosaccade and antisaccade tasks in 11 MSA patients and compared them to 21 patients with idiopathic Parkinson's disease and 20 healthy controls. They reported that MSA patients had longer prosaccade latencies than the other groups.

Venhovens and colleagues [23] studied 6 patients with MSA. They reported that 6 (100%) had saccadic abnormalities, 6 (100%) had abnormal smooth pursuit, 5 (83%) had optokinetic abnormalities, 4 (67%) had abnormalities of fixation, 2 (33%) had abnormalities on rotatory chair testing, and 1 (17%) had spontaneous nystagmus.

In Table 95.2, we have compiled the various ocular motor abnormalities from both instrumented and face-to-face studies.

In summary, in studies that report frequency of occurrence, the ocular motor abnormalities found in ≥50% of patients with MSA include square wave jerks and other fixation abnormalities, gaze holding abnormalities, saccadic abnormalities (usually dysmetria), choppy smooth pursuit, and reduced vestibulo-ocular reflex.

Table 95.2 Ocular motor abnormalities reported in several studies of multiple system atrophy

	Anderson et al. 2008	Zhou et al. 2022	Zhou et al. 2021	Venhovens et al. 2020	Klotz et al. 2005	Brooks et al. 2017	Pinkhardt et al. 2009	Pinnock et al. 2010
Supranuclear gaze palsy	27%							
Nystagmus, gaze evoked	40%							
Nystagmus, positioning down beat	40%							
Nystagmus, spontaneous				17%				
Saccades, "abnormal"		73%		100%				
Saccadic dysmetria (hyper- and hypo-)			64%					
Saccadic hypometria	73%							
Saccades, prolonged prosaccadic latencies						√		
Smooth pursuit, "broken up"	93%	71%		100%			√	
Vestibulo-ocular reflex, reduced	67%							
Square wave jerks	70%				√			
Saccadic intrusions								√
Fixation and gaze holding abnormal		38%		67%				
Gaze abnormal without fixation (perhaps spontaneous nystagmus?)		51%						
Optokinetic nystagmus abnormal		38%			83%			
Abnormalities on rotatory chair testing				33%				

Testing: Vestibular Evoked Myogenic Potentials (VEMPs)

Venhovens and colleagues [23] studied 6 patients (12 ears) with MSA. On *cervical vestibular evoked myogenic potentials*, responses were delayed unilaterally in 2 (33%) patients and bilaterally in 1 (17%) patient; responses were absent in 1 patient (17%). On *ocular vestibular evoked myogenic potentials*, responses were absent unilaterally in 3 (50%) patient and bilaterally in 1 (17%) patient.

Scarpa and colleagues [24] studied 16 MSA patients with cervical vestibular evoked myogenic potentials and reported them to be absent bilaterally in 27% and unilaterally in 13% of patients.

Testing: Subjective Visual Vertical (SVV)

Venhovens and colleagues [23] studied 6 patients with MSA and reported that 1 (17%) had an abnormal response on subjective visual vertical testing.

Testing: Other

Given the pronounced autonomic dysfunction in many cases of MSA, orthostatic hypotension is often detectable at the bedside. If bedside testing does not identify this yet clinical suspicion for MSA is high, then checking a tilt table test is reasonable.

Up to 80% of patients with MSA have decreased sweating, and up to 45% have frank anhidrosis [25]. The autonomic dysfunction is usually attributed to the disease's preganglionic deficits [26]. However, mounting evidence suggests that MSA involves postganglionic autonomic fibers as well [27–29]. Thus, thermoregulatory sweat testing (TST) (which assesses both pre- and postganglionic autonomic function) is abnormal in most cases of MSA, whereas quantitative sudomotor axon reflex test (QSART) (which only assesses postganglionic autonomic function) is only abnormal in about 30% of cases [30].

Imaging

The usual diagnostic dilemma is in distinguishing MSA from other parkinsonian syndromes such as idiopathic Parkinson's disease, progressive supranuclear palsy, and corticobasal ganglionic degeneration. Radiographic findings are sometimes helpful in this regard, but imperfect [31].

Some brain MRI findings have reasonably good predictive values for a diagnosis of MSA [1]. These include the following:

- The "hot cross buns" sign, which is a cruciform T2 signal hyperintensity in the pons [32].

- The "putaminal slit" sign, which is a linear T2 signal hyperintensity at the dorsolateral margin of the putamen [32].

Positron emission tomography (PET) with the radionuclide tracer ^{18}F-fluorodeoxyglucose can reportedly differentiate MSA-P from idiopathic Parkinson's disease (IPD) with 95% sensitivity and 100% specificity. On this study, MSA-P patients exhibit hypometabolism in the putamen, whereas IPD patients exhibit hypermetabolism [33].

Histopathology

A diagnosis of MSA may be suspected antemortem, but currently a definitive diagnosis relies on postmortem examination of the brain [15].

The predominant histopathological feature of MSA is the presence of glial cytoplasmic inclusions [34], the main component of which is alpha synuclein [35]. Other alpha synucleinopathies include idiopathic Parkinson's disease and Lewy body disease. MSA's glial cytoplasmic inclusions can involve smaller amounts of other abnormal proteins as well, such as ubiquitin, tau, LRRK2, DJ-1, p25α, GFAP, and MBP [1].

Lesions involve neuronal loss in the striatonigral and olivopontocerebellar systems, but can involve other areas in the central nervous system, peripheral nervous system, and autonomic nervous system [36]. The variability in distribution and severity of this neuropathology likely account for the variable clinical manifestations [2].

Differential Diagnosis

The differential diagnosis includes other parkinsonian syndromes (idiopathic Parkinson's disease, progressive supranuclear palsy, corticobasal ganglionic degeneration, Lewy body disease), and ataxias.

Gilman and colleagues [15] list the following criteria for *probable MSA*:

A sporadic, progressive, adult (>30 year old)-onset disease characterized by:

- Autonomic failure involving urinary incontinence (inability to control the release of urine from the bladder, with erectile dysfunction in males) or an orthostatic decrease of blood pressure within 3 min of standing by at least 30 mm Hg systolic or 15 mm Hg diastolic, *and.*
- Either:
  - Poorly levodopa-responsive parkinsonism (bradykinesia with rigidity, tremor, or postural instability), *or.*
  - A cerebellar syndrome (gait ataxia with cerebellar dysarthria, limb ataxia, or cerebellar oculomotor dysfunction).

Gilman and colleagues [15] list the following criteria for *possible MSA*:

A sporadic, progressive, adult (>30 year old)-onset disease characterized by:

- Parkinsonism (bradykinesia with rigidity, tremor, or postural instability), *or.*
- A cerebellar syndrome (gait ataxia with cerebellar dysarthria, limb ataxia, or cerebellar oculomotor dysfunction) *and.*
  - At least one feature suggesting autonomic dysfunction (otherwise unexplained urinary urgency, frequency or incomplete bladder emptying, erectile dysfunction in males, or significant orthostatic blood pressure decline that does not meet the level required in probable MSA) *and.*
  - At least one of the following features:
    - Possible MSA-P or MSA-C:
      - Babinski sign with hyperreflexia.
      - Stridor.
    - Possible MSA-P:
      - Rapidly progressive parkinsonism.
      - Poor response to levodopa.
      - Postural instability within 3 years of motor onset.
      - Gait ataxia, cerebellar dysarthria, limb ataxia, or cerebellar oculomotor dysfunction.
      - Dysphagia within 5 years of motor onset.
      - Atrophy on MRI of putamen, middle cerebellar peduncle, pons, or cerebellum.
    - Possible MSA-C:
      - Parkinsonism (bradykinesia and rigidity).
      - Atrophy on MRI of putamen, middle cerebellar peduncle, or pons.
      - Hypometabolism on FDG-PET in putamen.
      - Presynaptic nigrostriatal dopaminergic denervation on SPECT or PET.

Fanciulli and colleagues [15] reviewed a broad range of possible biomarkers studied for MSA. The majority are of "suboptimal diagnostic accuracy" or show "insufficient evidence" of diagnostic power.

Treatment

Some MSA patients exhibit an initial, modest, and transient response to levodopa treatment, which is often attempted when parkinsonian features are more prominent. This response is never sustained.

Multiple clinical trials of different interventions have been shown not to alter the disease course [1, 37]. These have included trials of agents with neuroprotective potential, such as human growth hormone [38], minocycline [39], riluzole [40], rifampicin [41], rasagiline [42], and epigallocatechin-gallate [43].

Autologous mesenchymal stem cell administration was studied in a randomized, placebo-controlled trial [44] and an open-label trial [45], and seemed modestly to attenuate clinical progression, but remains experimental.

Since there is no curative or arrestive treatment for MSA, management strategies are symptomatic. Patients with dystonia are sometimes offered treatment with onabotulinum toxin. Physical therapy may play some role in managing parkinsonian, cerebellar, and dystonic features. Autonomic features sometimes respond to pharmacotherapy (such as anticholinergics for urinary incontinence; sildenafil for erectile dysfunction; orthostatic hypotension may respond to midodrine).

Given that MSA by definition involves multiple body systems, management generally entails multiple subspecialists, which depending on the patient's needs may be some combination of neurology, psychiatry (for mood disorders), otolaryngology (for stridor), pulmonology, sleep medicine (for REM sleep behavior disorder), cardiology (for orthostatic hypotension), gastroenterology (for constipation), urology (for neurogenic bladder), and physical and occupational therapy.

Prognosis

MSA is a uniformly fatal disease [1, 2]. The mean survival from symptom onset is 9.8 years [46], with a range of 6–10 years [2, 37]. Few patients survive beyond 15 years; Petrovic and colleagues [47] studied 135 MSA cases and found that 4 (3%) survived ≥15 years beyond symptom onset.

References

1. Fanciulli A, Stankovic I, Krismer F, Seppi K, Levin J, Wenning GK. Multiple system atrophy. Int Rev Neurobiol. 2019;149:137–92.
2. Fanciulli A, Wenning GK. Multiple-system atrophy. N Engl J Med. 2015;372(3):249–63.
3. Dejerine JJ, Thomas A. L'atrophie olivo-ponto-cérébelleuse. Nouvelle iconographie de la Salpêtriere. 1900;13:330–70.
4. Adams RD, Vanbogaert L, Vandereecken H. Striato-nigral degeneration. J Neuropathol Exp Neurol. 1964;23:584–608.
5. Shy GM, Drager GA. A neurological syndrome associated with orthostatic hypotension: a clinical-pathologic study. Arch Neurol. 1960;2:511–27.
6. Graham JG, Oppenheimer DR. Orthostatic hypotension and nicotine sensitivity in a case of multiple system atrophy. J Neurol Neurosurg Psychiatry. 1969;32(1):28–34.
7. Jellinger KA. Heterogeneity of multiple system atrophy: an update. Biomedicine. 2022;10(3):599.
8. Kim HJ, Jeon BS, Lee JY, Yun JY, Kim YE, Paek SH. Young-onset multiple system atrophy. J Neurol Sci. 2012;319(1–2):168–70.
9. Lee YH, Ando T, Lee JJ, Baek MS, Lyoo CH, Kim SJ, et al. Later-onset multiple system atrophy: a multicenter Asian study. Mov Disord. 2020;35(9):1692–3.
10. Monzio Compagnoni G, Di Fonzo A. Understanding the pathogenesis of multiple system atrophy: state of the art and future perspectives. Acta Neuropathol Commun. 2019;7(1):113.
11. Ahmed Z, Asi YT, Sailer A, Lees AJ, Houlden H, Revesz T, et al. The neuropathology, pathophysiology and genetics of multiple system atrophy. Neuropathol Appl Neurobiol. 2012;38(1):4–24.
12. Jellinger KA. Multiple System Atrophy: An Oligodendroglioneural Synucleinopathy. J Alzheimer's Dis. 2018;62:1141–79.
13. Iranzo A, Santamaria J, Rye DB, Valldeoriola F, Marti MJ, Munoz E, et al. Characteristics of idiopathic REM sleep behavior disorder and that associated with MSA and PD. Neurology. 2005;65(2):247–52.
14. O'Sullivan SS, Massey LA, Williams DR, Silveira-Moriyama L, Kempster PA, Holton JL, et al. Clinical outcomes of progressive supranuclear palsy and multiple system atrophy. Brain. 2008;131(Pt 5):1362–72.
15. Gilman S, Wenning GK, Low PA, Brooks DJ, Mathias CJ, Trojanowski JQ, et al. Second consensus statement on the diagnosis of multiple system atrophy. Neurology. 2008;71(9):670–6.
16. Anderson T, Luxon L, Quinn N, Daniel S, David Marsden C, Bronstein A. Oculomotor function in multiple system atrophy: clinical and laboratory features in 30 patients. Mov Disord. 2008;23(7):977–84.
17. Pinkhardt EH, Kassubek J, Sussmuth S, Ludolph AC, Becker W, Jurgens R. Comparison of smooth pursuit eye movement deficits in multiple system atrophy and Parkinson's disease. J Neurol. 2009;256(9):1438–46.
18. Pinnock RA, McGivern RC, Forbes R, Gibson JM. An exploration of ocular fixation in Parkinson's disease, multiple system atrophy and progressive supranuclear palsy. J Neurol. 2010;257(4):533–9.
19. Klotz L, Klockgether T. Multiple system atrophy with macrosquare-wave jerks. Mov Disord. 2005;20(2):253–4.
20. Zhou H, Sun Y, Wei L, Wang X, Jiang Y, Li F, et al. Quantitative assessment of oculomotor function by videonystagmography in multiple system atrophy. Clin Neurophysiol. 2022;141:15–23.
21. Zhou H, Wang X, Ma D, Jiang Y, Li F, Sun Y, et al. The differential diagnostic value of a battery of oculomotor evaluation in Parkinson's disease and multiple system atrophy. Brain Behav. 2021;11(7):e02184.
22. Brooks SH, Klier EM, Red SD, Mehta ND, Patel SS, Chuang AZ, et al. Slowed prosaccades and increased antisaccade errors as a potential behavioral biomarker of multiple system atrophy. Front Neurol. 2017;8:261.
23. Venhovens J, Meulstee J, Bloem BR, Verhagen WIM. Neurovestibular dysfunction and falls in Parkinson's disease and atypical parkinsonism: a prospective 1 year follow-up study. Front Neurol. 2020;11:580285.

24. Scarpa A, Cassandro C, Vitale C, Ralli M, Policastro A, Barone P, et al. A comparison of auditory and vestibular dysfunction in Parkinson's disease and multiple system atrophy. Parkinsonism Relat Disord. 2020;71:51–7.
25. Iodice V, Lipp A, Ahlskog JE, Sandroni P, Fealey RD, Parisi JE, et al. Autopsy confirmed multiple system atrophy cases: mayo experience and role of autonomic function tests. J Neurol Neurosurg Psychiatry. 2012;83(4):453–9.
26. Donadio V, Nolano M, Elam M, Montagna P, Provitera V, Bugiardini E, et al. Anhidrosis in multiple system atrophy: a preganglionic sudomotor dysfunction? Mov Disord. 2008;23(6):885–8.
27. Coon EA, Fealey RD, Sletten DM, Mandrekar JN, Benarroch EE, Sandroni P, et al. Anhidrosis in multiple system atrophy involves pre- and postganglionic sudomotor dysfunction. Mov Disord. 2017;32(3):397–404.
28. Doppler K, Weis J, Karl K, Ebert S, Ebentheuer J, Trenkwalder C, et al. Distinctive distribution of phospho-alpha-synuclein in dermal nerves in multiple system atrophy. Mov Disord. 2015;30(12):1688–92.
29. Provitera V, Nolano M, Caporaso G, Stancanelli A, Manganelli F, Iodice R, et al. Postganglionic sudomotor denervation in patients with multiple system atrophy. Neurology. 2014;82(24):2223–9.
30. Palma JA, Norcliffe-Kaufmann L, Kaufmann H. Diagnosis of multiple system atrophy. Auton Neurosci. 2018;211:15–25.
31. Schrag A, Good CD, Miszkiel K, Morris HR, Mathias CJ, Lees AJ, et al. Differentiation of atypical parkinsonian syndromes with routine MRI. Neurology. 2000;54(3):697–702.
32. Horimoto Y, Aiba I, Yasuda T, Ohkawa Y, Katayama T, Yokokawa Y, et al. Longitudinal MRI study of multiple system atrophy – when do the findings appear, and what is the course? J Neurol. 2002;249(7):847–54.
33. Arbizu J, Luquin MR, Abella J, de la Fuente-Fernández R, Fernandez-Torrón R, García-Solís D, et al. Functional neuroimaging in the diagnosis of patients with parkinsonism: Update and recommendations for clinical use. Rev Esp de Med Nucl e Imagen Mol. 2014;33(4):215–26.
34. Papp MI, Kahn JE, Lantos PL. Glial cytoplasmic inclusions in the CNS of patients with multiple system atrophy (striatonigral degeneration, olivopontocerebellar atrophy and Shy-Drager syndrome). J Neurol Sci. 1989;94(1–3):79–100.
35. Spillantini MG, Crowther RA, Jakes R, Cairns NJ, Lantos PL, Goedert M. Filamentous alpha-synuclein inclusions link multiple system atrophy with Parkinson's disease and dementia with Lewy bodies. Neurosci Lett. 1998;251(3):205–8.
36. Ubhi K, Low P, Masliah E. Multiple system atrophy: a clinical and neuropathological perspective. Trends Neurosci. 2011;34(11):581–90.
37. Meissner WG, Fernagut PO, Dehay B, Peran P, Traon AP, Foubert-Samier A, et al. Multiple system atrophy: recent developments and future perspectives. Mov Disord. 2019;34(11):1629–42.
38. Holmberg B, Johansson JO, Poewe W, Wenning G, Quinn NP, Mathias C, et al. Safety and tolerability of growth hormone therapy in multiple system atrophy: a double-blind, placebo-controlled study. Mov Disord. 2007;22(8):1138–44.
39. Dodel R, Spottke A, Gerhard A, Reuss A, Reinecker S, Schimke N, et al. Minocycline 1-year therapy in multiple-system-atrophy: effect on clinical symptoms and [(11)C] (R)-PK11195 PET (MEMSA-trial). Mov Disord. 2010;25(1):97–107.
40. Bensimon G, Ludolph A, Agid Y, Vidailhet M, Payan C, Leigh PN, et al. Riluzole treatment, survival and diagnostic criteria in Parkinson plus disorders: the NNIPPS study. Brain. 2009;132(Pt 1):156–71.
41. Low PA, Robertson D, Gilman S, Kaufmann H, Singer W, Biaggioni I, et al. Efficacy and safety of rifampicin for multiple system atrophy: a randomised, double-blind, placebo-controlled trial. Lancet Neurol. 2014;13(3):268–75.
42. Poewe W, Seppi K, Fitzer-Attas CJ, Wenning GK, Gilman S, Low PA, et al. Efficacy of rasagiline in patients with the parkinsonian variant of multiple system atrophy: a randomised, placebo-controlled trial. Lancet Neurol. 2015;14(2):145–52.
43. Levin J, Maass S, Schuberth M, Giese A, Oertel WH, Poewe W, et al. Safety and efficacy of epigallocatechin gallate in multiple system atrophy (PROMESA): a randomised, double-blind, placebo-controlled trial. Lancet Neurol. 2019;18(8):724–35.
44. Lee PH, Lee JE, Kim HS, Song SK, Lee HS, Nam HS, et al. A randomized trial of mesenchymal stem cells in multiple system atrophy. Ann Neurol. 2012;72(1):32–40.
45. Singer W, Berini SE, Sandroni P, Fealey RD, Coon EA, Suarez MD, et al. Pure autonomic failure: predictors of conversion to clinical CNS involvement. Neurology. 2017;88(12):1129–36.
46. Low PA, Reich SG, Jankovic J, Shults CW, Stern MB, Novak P, et al. Natural history of multiple system atrophy in the USA: a prospective cohort study. Lancet Neurol. 2015;14(7):710–9.
47. Petrovic IN, Ling H, Asi Y, Ahmed Z, Kukkle PL, Hazrati LN, et al. Multiple system atrophy-parkinsonism with slow progression and prolonged survival: a diagnostic catch. Mov Disord. 2012;27(9):1186–90.

96 Huntington's Disease

Practical Summary

Huntington's disease (HD) is an autosomal dominant inherited movement disorder that manifests primarily with chorea, rigidity, and dementia. The majority of patients also have a variety of ocular motor disorders that may be most easily detected on instrumented testing; there is debate regarding whether the ocular motor abnormalities can effectively diagnose HD prior to the onset of motor symptoms, or be used to monitor progress of the disease. HD patients also exhibit abnormal performance on computerized dynamic posturography. Structural imaging and functional imaging show abnormalities in the head of the caudate and striatum. HD is uniformly fatal; treatment is symptomatic; tetrabenazine and deutetrabenazine may modestly and transiently diminish chorea.

Introduction

Huntington's disease (HD), also known as Huntington's chorea, is a progressive neurodegenerative disorder affecting the basal ganglia and other structures in the central nervous system (Fig. 96.1).

Vale and Cardoso [1] state that Charles Oscar Waters published the first description of what was probably HD in 1841, but George Huntington provided a more complete description [2], and the disease is eponymously designated after him. Huntington's original article was republished in 2003 [3]. Wechsler [4] states that the phrase "Huntington's chorea" first appeared in 1887 [5].

Fig. 96.1 George Huntington (1850–1916). (From https://en.wikipedia.org/wiki/George_Huntington, public domain)

Epidemiology

A systematic review and meta-analysis reported that the prevalence of HD is 2.71 per 100,000 people, and the incidence is 0.38 per 100,000 person-years [6].

Genetics

Huntington's disease (OMIM 143100) is a heritable disorder transmitted in an autosomal dominant fashion. It results from a trinucleotide repeat (CAG) expansion in the Huntingtin gene, HTT (OMIM 613004), located on 4p16.3. Like other trinucleotide repeat diseases, the number of repeats usually expands with each generation, leading to anticipation (presentation of the disease earlier in life with each successive generation).

M. Cherchi, *Otoneurology and Vestibular Medicine*, https://doi.org/10.1007/978-3-031-94841-1_96

Pathophysiological Mechanism of Disease

Aggregates of abnormal huntingtin protein accumulate in the cytoplasm and nucleus of affected cells throughout the body, but predominantly in the central nervous system, and most prominently in the caudate and putamen. It is suspected that these protein aggregates are toxic to cellular function.

An autopsy study of the brains of four HD patients who had slow vertical saccades during life revealed abnormalities (reduced number of larger neurons in one patient, nonspecific gliosis in three patients) in the rostral interstitial nucleus of the median longitudinal fasciculus, suggesting that this pathology is one underlying substrate for slowed vertical saccades [7].

Clinical Presentation

The initial manifestation of HD may be psychiatric or behavioral symptoms, which may precede other symptoms by 10 years. The other manifestations include motor (progressive chorea and rigidity) and cognitive decline (dementia), whose peak onset is in the fourth decade.

Physical Examination

In a Huntington patient with no other diseases, physical examination shows chorea and rigidity.

Ocular Motor Examination

Face-to-face ocular motor examination may appear normal.

Testing: Instrumented Ocular Motor Testing

The majority of HD patients exhibit a variety of ocular motor abnormalities detectable on instrumented testing.

Table 96.1, from Leigh and colleagues [8], provides a good overview of ocular motor abnormalities that can be observed in HD, and their relative frequency.

Although ocular motor abnormalities are common in HD, Beenen and colleagues noted that "75% of the HD patients had pathologically slow saccades. This number rose to 87% if other ocular motor deficits were included. Thus *a small, but definite number of HD patients had normal eye movements. This included patients with disease durations of more than 14 years*" [9].

A number of studies have explored ocular motor function in "presymptomatic" HD patients, which in this context generally refers to genetically confirmed HD individuals who have not yet developed motor symptoms. Many researchers have observed ocular motor abnormalities in these presymptomatic HD patients [10–17]. However, not all investigators have found this to be reliable; Collewijn and colleagues conducted scleral search coil studies in HD patients and their offspring compared to controls, and concluded that "The material as a whole suggests that ocular motor dysfunction does not develop prior to, but concurrently with and as a part of generalized, progressive deterioration of motor control. The implication is that ocular motor screening of clinically healthy at-risk subjects does not reliably contribute to an earlier diagnosis of future HD" [18].

Some investigators have studied the ocular motor abnormalities in HD as a "biomarker" not just for diagnosis, but also for disease progression [10, 14–17, 19–26]. While this may be merely of academic interest now, if therapies for HD become available in the future, then such ocular motor "biomarkers" would have the potential to quantify treatment efficacy.

While the ocular motor findings in HD are fascinating and have implications for the underlying neuroanatomical sub-

Table 96.1 Ocular motor abnormalities (and their frequency) in 50 patients with Huntington's disease (From Leigh et al. [8])

Eye movement	Nature of abnormality	Percentage of 50 patients affected
Fixation	Steady fixation disrupted by inappropriate saccades	73
Saccades	Impaired initiation reflected by:	
	Increased latency.	89
	Obligatory head movement in vertical plane.	89
	Obligatory head movement in horizontal plane.	92
	Obligatory blink.	35
	Slow saccades	
	In vertical plane.	62
	In horizontal plane.	50
	Hypometria and reduced range of saccadic movement	
	In vertical plane.	56
	In horizontal plane.	46
Smooth pursuit	Disrupted by inappropriate saccades; reduced gain in more advanced cases	
	In vertical plane.	56
	In horizontal plane.	60
Vergence	Convergence movements usually impaired because of extraneous saccades that took the eye off target	33
Eccentric gaze holding	Eyes brought back to primary position by inappropriate saccades	11
Vestibulo-ocular reflex	Decreased gain (eye velocity / head velocity)	0

strates [14, 17, 24, 25, 27–32], clinicians would not rely on such findings to secure the diagnosis.

Testing: Other Vestibular

Several abnormalities have been described in computerized dynamic posturography in HD patients [33–41], including the following:

- Impaired postural control at the limits of stability [33].
- Greater postural sway [39], resulting in poor performance on all conditions of sensory organization testing [38].
- Increased latencies on motor control testing [40].

Testing: Other

A diagnosis of HD is confirmed by genetic testing which reveals more than 36 trinucleotide repeats in the HTT gene.

Imaging

Brain imaging has been studied extensively in HD [42], both in structural [43] and functional [44] modalities. Broadly, structural imaging shows atrophy of the caudate head; functional imaging shows reduced metabolic activity in the striatum.

Differential Diagnosis

The differential diagnosis of HD includes other basal ganglionic disorders.

Treatment

If a diagnosis of HD is being entertained, a patient should strongly consider consultation with a genetic counselor.

There is no curative or arrestive therapy for HD. Treatment is symptomatic. Tetrabenazine and deutetrabenazine may modestly and transiently diminish chorea.

Tetrabenazine has been reported to improve performance on computerized dynamic posturography in sensory organization testing in which a single modality (either vision or proprioception) is sway-referenced [35].

Deep brain stimulation (DBS) has been studied for a variety of movement disorders, including HD. Fawcett and colleagues reported that pallidal DBS improved some ocular motor abnormalities; specifically, "Pallidal DBS decreased pro-saccade latency, total movement time, and the number of correctly executed trials, as well as increasing saccade gain" [45].

Prognosis

HD is uniformly fatal. From the time that motor symptoms become manifest, death occurs on average within 17 years.

References

1. Vale TC, Cardoso F. Chorea: a journey through history. Tremor Other Hyperkinetic Mov. 2015;5:tre-5-296.
2. Huntington G. On chorea. Med Surg Rep Philadelphia. 1872;26(15):317–21.
3. Huntington G. On chorea. George Huntington, M.D. J Neuropsychiatry Clin Neurosci. 2003;15(1):109–12.
4. Wexler A. A brief prehistory of Huntington's disease. J Huntingtons Dis. 2013;2(3):231–7.
5. Huber A. Chorea hereditaria er Erwachsenen. Virchow's Archiv für pathologische Anatomie. 1887;108:267–85.
6. Medina A, Mahjoub Y, Shaver L, Pringsheim T. Prevalence and incidence of Huntington's disease: an updated systematic review and meta-analysis. Mov Disord. 2022;37(12):2327–35.
7. Leigh RJ, Parhad IM, Clark AW, Buettner-Ennever JA, Folstein SE. Brainstem findings in Huntington's disease. Possible mechanisms for slow vertical saccades. J Neurol Sci. 1985;71(2–3):247–56.
8. Leigh RJ, Newman SA, Folstein SE, Lasker AG, Jensen BA. Abnormal ocular motor control in Huntington's disease. Neurology. 1983;33(10):1268–75.
9. Beenen N, Büttner U, Lange HW. The diagnostic value of eye movement recordings in patients with Huntington's disease and their offspring. Electroencephalogr Clin Neurophysiol. 1986;63(2):119–27.
10. Ali FR, Michell AW, Barker RA, Carpenter RH. The use of quan titative oculometry in the assessment of Huntington's disease. Exp Brain Res. 2006;169(2):237–45.
11. Antoniades CA, Altham PM, Mason SL, Barker RA, Carpenter R. Saccadometry: a new tool for evaluating presymptomatic Huntington patients. Neuroreport. 2007;18(11):1133–6.
12. Blekher T, Johnson SA, Marshall J, White K, Hui S, Weaver M, et al. Saccades in presymptomatic and early stages of Huntington disease. Neurology. 2006;67(3):394–9.
13. Blekher TM, Yee RD, Kirkwood SC, Hake AM, Stout JC, Weaver MR, et al. Oculomotor control in asymptomatic and recently diagnosed individuals with the genetic marker for Huntington's disease. Vis Res. 2004;44(23):2729–36.
14. Hicks SL, Robert MP, Golding CV, Tabrizi SJ, Kennard C. Oculomotor deficits indicate the progression of Huntington's disease. Prog Brain Res. 2008;171:555–8.
15. Robert MP, Nachev PC, Hicks SL, Golding CV, Tabrizi SJ, Kennard C. Saccadometry of conditional rules in presymptomatic Huntington's disease. Ann N Y Acad Sci. 2009;1164:444–50.
16. Rupp J, Blekher T, Jackson J, Beristain X, Marshall J, Hui S, et al. Progression in prediagnostic Huntington disease. J Neurol Neurosurg Psychiatry. 2010;81(4):379–84.
17. Vaca-Palomares I, Coe BC, Brien DC, Campos-Romo A, Munoz DP, Fernandez-Ruiz J. Voluntary saccade inhibition deficits

correlate with extended white-matter cortico-basal atrophy in Huntington's disease. Neuroimage Clin. 2017;15:502–12.
18. Collewijn H, Went LN, Tamminga EP, Vegter-Van der Vlis M. Oculomotor defects in patients with Huntington's disease and their offspring. J Neurol Sci. 1988;86(2–3):307–20.
19. Antoniades CA, Xu Z, Mason SL, Carpenter RH, Barker RA. Huntington's disease: changes in saccades and hand-tapping over 3 years. J Neurol. 2010;257(11):1890–8.
20. Blekher T, Weaver MR, Cai X, Hui S, Marshall J, Jackson JG, et al. Test-retest reliability of saccadic measures in subjects at risk for Huntington disease. Invest Ophthalmol Vis Sci. 2009;50(12):5707–11.
21. Grabska N, Rudzińska M, Wójcik-Pędziwiatr M, Michalski M, Sławek J, Szczudlik A. Saccadic eye movements in juvenile variant of Huntington disease. Neurol Neurochir Pol. 2014;48(4):236–41.
22. Patel SS, Jankovic J, Hood AJ, Jeter CB, Sereno AB. Reflexive and volitional saccades: biomarkers of Huntington disease severity and progression. J Neurol Sci. 2012;313(1–2):35–41.
23. Peltsch A, Hoffman A, Armstrong I, Pari G, Munoz DP. Saccadic impairments in Huntington's disease. Exp Brain Res. 2008;186(3):457–69.
24. Rupp J, Dzemidzic M, Blekher T, West J, Hui S, Wojcieszek J, et al. Comparison of vertical and horizontal saccade measures and their relation to gray matter changes in premanifest and manifest Huntington disease. J Neurol. 2012;259(2):267–76.
25. Vaca-Palomares I, Brien DC, Coe BC, Ochoa-Morales A, Martínez-Ruano L, Munoz DP, et al. Implicit learning impairment identified via predictive saccades in Huntington's disease correlates with extended cortico-striatal atrophy. Cortex. 2019;121:89–103.
26. Winder JY, Roos RAC. Premanifest Huntington's disease: examination of oculomotor abnormalities in clinical practice. PLoS One. 2018;13(3):e0193866.
27. Fielding J, Georgiou-Karistianis N, Millist L, Fahey M, White O. Saccadic trajectory in Huntington's disease. J Int Neuropsychol Soc. 2006;12(4):455–64.
28. Lasker AG, Zee DS. Ocular motor abnormalities in Huntington's disease. Vis Res. 1997;37(24):3639–45.
29. Lasker AG, Zee DS, Hain TC, Folstein SE, Singer HS. Saccades in Huntington's disease: initiation defects and distractibility. Neurology. 1987;37(3):364–70.
30. Lasker AG, Zee DS, Hain TC, Folstein SE, Singer HS. Saccades in Huntington's disease: slowing and dysmetria. Neurology. 1988;38(3):427–31.
31. Tian JR, Zee DS, Lasker AG, Folstein SE. Saccades in Huntington's disease: predictive tracking and interaction between release of fixation and initiation of saccades. Neurology. 1991;41(6):875–81.
32. Winograd-Gurvich CT, Georgiou-Karistianis N, Evans A, Millist L, Bradshaw JL, Churchyard A, et al. Hypometric primary saccades and increased variability in visually-guided saccades in Huntington's disease. Neuropsychologia. 2003;41(12):1683–92.
33. Medina LD, Pirogovsky E, Salomonczyk D, Goldstein J, Panzera R, Gluhm S, et al. Postural limits of stability in premanifest and manifest Huntington's disease. J Huntingtons Dis. 2013;2(2):177–84.
34. Beckmann H, Bohlen S, Saft C, Hoffmann R, Gerss J, Muratori L, et al. Objective assessment of gait and posture in premanifest and manifest Huntington disease – a multi-center study. Gait Posture. 2018;62:451–7.
35. Fekete R, Davidson A, Ondo WG, Cohen HS. Effect of tetrabenazine on computerized dynamic posturography in Huntington disease patients. Parkinsonism Relat Disord. 2012;18(7):896–8.
36. Nanetti L, Contarino VE, Castaldo A, Sarro L, Bachoud-Levi AC, Giavazzi M, et al. Cortical thickness, stance control, and arithmetic skill: an exploratory study in premanifest Huntington disease. Parkinsonism Relat Disord. 2018;51:17–23.
37. Reilmann R, Rumpf S, Beckmann H, Koch R, Ringelstein EB, Lange HW. Huntington's disease: objective assessment of posture–a link between motor and functional deficits. Mov Disord. 2012;27(4):555–9.
38. Reyes A, Salomonczyk D, Teo WP, Medina LD, Bartlett D, Pirogovsky-Turk E, et al. Computerised dynamic posturography in premanifest and manifest individuals with Huntington's disease. Sci Rep. 2018;8(1):14615.
39. Salomonczyk D, Panzera R, Pirogovsky E, Goldstein J, Corey-Bloom J, Simmons R, et al. Impaired postural stability as a marker of premanifest Huntington's disease. Mov Disord. 2010;25(14):2428–33.
40. Tian JR, Herdman SJ, Zee DS, Folstein SE. Postural control in Huntington's disease (HD). Acta Otolaryngol Suppl. 1991;481:333–6.
41. Valade D, Davous P, Rondot P. Comparative study of posturography and electrooculography in at-risk subjects for Huntington's disease. Eur Neurol. 1984;23(4):252–64.
42. Johnson EB, Gregory S. Huntington's disease: brain imaging in Huntington's disease. Prog Mol Biol Transl Sci. 2019;165:321–69.
43. Wilson H, Dervenoulas G, Politis M. Structural magnetic resonance imaging in Huntington's disease. Int Rev Neurobiol. 2018;142:335–80.
44. Gregory S, Scahill RI. Functional magnetic resonance imaging in Huntington's disease. Int Rev Neurobiol. 2018;142:381–408.
45. Fawcett AP, Moro E, Lang AE, Lozano AM, Hutchison WD. Pallidal deep brain stimulation influences both reflexive and voluntary saccades in Huntington's disease. Mov Disord. 2005;20(3):371–7.

97 Stiff Person Syndrome

Overview

Stiff person syndrome (SPS) is a rare movement disorder that usually presents with muscle stiffness/rigidity of the trunk and limbs, that often has a chronic (baseline) component punctuated by episodic exacerbations. Less common features include ataxia, ocular motor abnormalities, autonomic dysfunction, and psychiatric features. The majority of cases are associated with autoantibodies (usually anti-GAD65 antibodies, but numerous others have been reported), and a small percentage are paraneoplastic (with a variety of reported malignancies). The ocular motor findings (mainly reported in cases with anti-GAD65 antibodies) localize predominantly to the cerebellum. In cases with ataxia, brain MRI may show cerebellar atrophy. The muscular symptoms sometimes respond to GABA-ergic medications (such as benzodiazepines). In cases with autoantibodies, immunosuppression or immunomodulatory therapies may be considered.

Introduction

In 1956, Frederick Moersch and Henry Woltman reported a case series of 14 patients with progressive muscular rigidity of the trunk and thighs that they termed "stiff man syndrome" [1]. This was subsequently renamed *stiff person syndrome (SPS)*, which is now usually classified as a movement disorder characterized by progressive, recurrent episodes of muscle stiffness variously involving the trunk, neck, and extremities. Most cases have been associated with autoantibodies; a minority of cases appear to be idiopathic (i.e., unassociated with autoantibodies). Of the cases with an autoimmune etiology, some exhibit additional features such as ataxia, ocular motor abnormalities, or psychiatric symptoms.

Epidemiology

Stiff person syndrome is rare, estimated to affect 1 to 2 per million people; women are affected almost twice as often as men [2]. About 5% of cases are found to be of paraneoplastic etiology [3].

Pathophysiological Mechanism of Disease

The hypothesis of an autoimmune mechanism of disease arose because of SPS occurring in the context of other autoimmune pathologies such as type 1 diabetes mellitus [4], myasthenia gravis [5], autoimmune thyroiditis [6], vitiligo, pernicious anemia, and others.

Several different autoantibodies have been identified in SPS.

- The most common is antibodies to *glutamic acid decarboxylase (GAD65)*, which is reported to account for 40% [7] to 80% [3] of cases. Glutamic acid decarboxylase is an intracellular enzyme that catalyzes the conversion of glutamate to the inhibitory neurotransmitter gamma-aminobutyric acid (GABA), so interference with this enzyme would effectively reduce the amount of available GABA.
- Antibodies have also been detected against the *postsynaptic GABA$_A$ receptor* [8], thereby reducing the efficacy of existing GABA.
- Antibodies against *amphiphysin*, an intracellular protein that promotes cleavage of intracellular vesicles and would thus interfere with exocytosis of neurotransmitters. Antiamphiphysin antibodies have been associated with breast cancer [9].
- Antibodies against the alpha-1 subunit of the *glycine receptor* has been reported in some patients whose SPS is of the "progressive encephalomyelitis with rigidity and myoclonus (PERM)" subtype [10, 11]. Glycine is an inhibitory neurotransmitter.

M. Cherchi, *Otoneurology and Vestibular Medicine*, https://doi.org/10.1007/978-3-031-94841-1_97

- Antibodies against *thyroid peroxidase* have been reported [6].
- Antibodies against *acetylcholine receptors* have been reported [12].
- Rare cases have been reported of patients positive for more than one autoantibody. Thomas and colleagues reported a patient who tested positive for anti-GAD antibodies and antiacetylcholine receptor antibodies [5]. Nguyen and colleagues reported a patient who tested positive for anti-GAD and anti-SOX1 antibodies [3].

Clinical Presentation

Muscle stiffness is the common feature of this syndrome, produced by the involuntary cocontraction of agonist and antagonist muscles. This can have chronic and episodic components. These muscular symptoms are the presenting feature in 95% of patients, often beginning with leg stiffness, followed by paraspinal muscle rigidity and painful spasms [13].

Some patients with SPS additionally exhibit what appears to be a cerebellar ataxia [14, 15].

Patients with autoimmune-mediated SPS, particularly with anti-GAD65 antibodies, may exhibit a variety of ocular motor abnormalities.

Some patients exhibit neuropsychiatric features such as specific phobias [16–20].

Some cases of sudden death have been reported in SPS. Various mechanisms for this have been considered [21], such as respiratory failure from chest wall muscle rigidity [22, 23] or autonomic failure (in cases with antiacetylcholine receptor antibodies) [24].

The reported variations in SPS presentation have proliferated, even when considering cases associated with a particular autoantibody. Because of this, some researchers have begun to view this as a "spectrum" of disorders. For example, SPS associated with anti-GAD65 antibodies has been reported to manifest variously with cerebellar ataxia, autoimmune epilepsy, limbic encephalitis, progressive encephalomyelitis with rigidity and myoclonus (PERM), and ocular motor abnormalities. As such, some investigators have proposed classifying these as "GAD antibody spectrum disorders" [25, 26].

Physical Examination

Muscle stiffness and rigidity in the trunk, and sometimes limbs and neck, is characteristic. This can fluctuate in intensity.

Ocular Motor Examination

Ocular motor abnormalities are sometimes observable on face-to-face examination, though in some cases, this may only be discernible on instrumented testing. Thomas and colleagues described an SPS patient with anti-GAD antibodies (and eventually antiacetylcholine receptor antibodies) who on face-to-face examination exhibited "variable alternating esotropia," "bilateral abduction deficits," a tendency towards slow horizontal saccades, and "slight end-point nystagmus" [5].

Testing: Instrumented Ocular Motor Studies

It is uncommon for a patient with stiff person syndrome to be referred for otoneurological evaluation; in such cases, the usual clinical query is for ataxia or unsteadiness. A variety of ocular motor abnormalities have been described in association with SPS associated with anti-GAD65 antibodies; these are interesting, but not essential for the diagnosis.

- Ances and colleagues described spontaneous down beat nystagmus in an SPS patient with anti-GAD antibodies [27]. They additionally reported appendicular ataxia.
- Economides and Horton described an SPS patient with anti-GAD antibodies who exhibited ***asymmetric low gain on horizontal smooth pursuit; hypometric horizontal saccades***; and abnormal gaze-evoked nystagmus in all directions [28].
- Hac and colleagues described an SPS patient with anti-GAD antibodies who exhibited ***spontaneous down beat nystagmus, gaze-evoked nystagmus in all directions;*** rebound nystagmus; centripetal nystagmus; and ***apogeotropic direction-changing positional nystagmus*** on Dix–Hallpike testing [29].
- Oskarsson and colleagues described an SPS patient with anti-GAD antibodies who exhibited ***progressively lengthening latencies of horizontal and vertical saccades; restricted range of horizontal pursuit*** (with normal gain), that fatigued to cessation; ***fatiguing optokinetic responses***; and ***fatiguing vestibulo-ocular reflexes*** [30].
- Tilikete and colleagues described an SPS patient with anti-GAD antibodies who exhibited periodic alternating nystagmus [31].
- Wang and colleagues reviewed 22 cases of SPS with anti-GAD antibodies and described ***abnormalities in vergence*** (divergence insufficiency), skew (***alternating skew deviation***), spontaneous down beat nystagmus (with or without fixation, sometimes increased by hyperventilation), ***saccadic*** abnormalities (both ***hypermetria and hypometria***),

pursuit abnormalities (*largely saccadic*), ***direction-changing positional nystagmus on Dix–Hallpike testing***, and sometimes ***abnormalities on head impulse testing*** [32].
- Zivotofsky and colleagues reported an SPS patient with anti-GAD antibodies whose ocular motor abnormalities included ***abnormal vertical smooth pursuit (reduced gain***, much more so in the downward direction), ***abnormal horizontal vertical saccades*** (amounting to a series of small saccades in a ***staircase***-like fashion), spontaneous down beat nystagmus, and square wave jerks [15]. Additional features in this patient included cerebellar ataxia.

These ocular motor findings predominantly localize to the cerebellum (spontaneous down beat nystagmus; dysmetric saccades; abnormalities of saccadic latency; gaze-evoked nystagmus; abnormal gain on smooth pursuit; abnormal optokinetic responses; periodic alternating nystagmus; rebound nystagmus).

Other findings suggest brainstem involvement (e.g., vergence abnormalities). The decline of vestibulo-ocular reflexes [30] is compatible with an impaired (or fatigable) velocity storage mechanism, which may result from interference with the function of commissural inhibition between the vestibular nuclei due to reduced GABA activity [33].

Testing: Vestibular

Teggi and colleagues [33] reported on an SPS patient with anti-GAD antibodies who experienced recurrent vertiginous episodes during which there was spontaneous horizontal nystagmus. They identified a 58% right unilateral caloric weakness on caloric testing.

Testing: Other

Cerebrospinal fluid analysis of SPS patients may identify oligoclonal bands for anti-GAD antibodies [34].

Electrodiagnostic studies in patients with SPS usually show coactivation of agonist–antagonist muscles [35], which may be continuous and diminish (though not disappear) during sleep [36].

Imaging

Some studies of SPS patients report cerebellar atrophy on brain MRI [28, 31].

Histopathology

There are very few published autopsy studies of patients with SPS.

Mutsomoto and colleagues reported two cases of patients with SPS who had sudden death attributed to autonomic instability [24]. In one patient, there was perivascular gliosis of the spinal cord and brain stem. In the other patient, there was lymphocytic perivascular infiltration in the spinal cord, brain stem, and basal ganglia.

Meinck and colleagues [37] reported one case in which there were "scattered lymphocyte cuffs around leptomeningeal, intracerebral, and particularly intraspinal vessels, suggesting a mild inflammatory process."

Witherick and colleagues [38] reported a case in which microscopic examination of the spinal cord showed vacuolation of motor neurons in the caudal segments; electron microscopy suggested that these vacuoles might have been "hyperswollen lysosomes." Histology of dorsal root ganglion showed neuronal loss with macrophage infiltration and attack of ganglion cells. On immunohistochemical examination, CD68 showed microglial reaction in gray matter, and GFAP showed marked gray matter gliosis.

Differential Diagnosis

The differential diagnosis of SPS is broad and depends on the presentation. When the syndrome presents exclusively with muscle stiffness/rigidity, then the differential may include myelopathies, muscular dystrophies, and basal ganglionic disorders. When the syndrome includes ataxia, cerebellar disorders should be considered.

Treatment

Treatment of SPS is usually overseen by a specialist in neuromuscular disorders (less commonly movement disorders), sometimes in conjunction with immunology and (if paraneoplastic) oncology.

Muscle spasticity is often treated with benzodiazepines (often diazepam or clonazepam, less commonly lorazepam), though this has not been studied in randomized controlled trials. In cases that are unresponsive to benzodiazepines, baclofen may be considered.

In cases with identified autoantibodies, immunosuppression (often with steroids) or immunomodulatory therapies (intravenous immunoglobulin) may be considered.

For cases that are believed to be paraneoplastic, management should include treatment of the underlying malignancy.

Prognosis

A longitudinal study that followed patients for up to 8 years reported that SPS is a progressive disease leading to physical disability over time [13].

References

1. Moersch FP, Woltman HW. Progressive fluctuating muscular rigidity and spasm ("stiff-man" syndrome); report of a case and some observations in 13 other cases. Proc Staff Meet Mayo Clin. 1956;31(15):421–7.
2. Hadavi S, Noyce AJ, Leslie RD, Giovannoni G. Stiff person syndrome. Pract Neurol. 2011;11(5):272–82.
3. Nguyen PM, Vu DD, Vu KD, Nguyen HT, Nguyen DV. Stiff person syndrome: a case report with sudden onset and coexistence of seropositive antibodies to glutamic acid decarboxylase and anti-SOX1 antibodies. Case Rep Neurol. 2022;14(2):237–44.
4. Solimena M, Folli F, Aparisi R, Pozza G, De Camilli P. Autoantibodies to GABA-ergic neurons and pancreatic beta cells in stiff-man syndrome. N Engl J Med. 1990;322(22):1555–60.
5. Thomas S, Critchley P, Lawden M, Farooq S, Thomas A, Proudlock FA, et al. Stiff person syndrome with eye movement abnormality, myasthenia gravis, and thymoma. J Neurol Neurosurg Psychiatry. 2005;76(1):141–2.
6. Nie DA, Abud A, Serrano-Gonzalez M, Harappanahally G, Patil R. Pediatric stiff limb syndrome with polyautoimmunity of anti-GAD-65, anti-islet cell, and thyroid peroxidase antibodies: a case report and review of literature. J Neuroimmunol. 2022;367:577865.
7. Vincent A, Grimaldi LM, Martino G, Davenport C, Todd I. Antibodies to 125I-glutamic acid decarboxylase in patients with stiff man syndrome. J Neurol Neurosurg Psychiatry. 1997;62(4):395–7.
8. Raju R, Rakocevic G, Chen Z, Hoehn G, Semino-Mora C, Shi W, et al. Autoimmunity to GABAA-receptor-associated protein in stiff-person syndrome. Brain. 2006;129(Pt 12):3270–6.
9. Folli F, Solimena M, Cofiell R, Austoni M, Tallini G, Fassetta G, et al. Autoantibodies to a 128-kd synaptic protein in three women with the stiff-man syndrome and breast cancer. N Engl J Med. 1993;328(8):546–51.
10. Martinez-Hernandez E, Arino H, McKeon A, Iizuka T, Titulaer MJ, Simabukuro MM, et al. Clinical and immunologic investigations in patients with stiff-person spectrum disorder. JAMA Neurol. 2016;73(6):714–20.
11. McKeon A, Martinez-Hernandez E, Lancaster E, Matsumoto JY, Harvey RJ, McEvoy KM, et al. Glycine receptor autoimmune spectrum with stiff-man syndrome phenotype. JAMA Neurol. 2013;70(1):44–50.
12. Lester J, Cojab J, Kleriga E. Stiff person syndrome and acetylcholine receptor ganglionic neuronal antibodies. Case Rep Neurol. 2020;12(1):24–6.
13. Rakocevic G, Alexopoulos H, Dalakas MC. Quantitative clinical and autoimmune assessments in stiff person syndrome: evidence for a progressive disorder. BMC Neurol. 2019;19(1):1.
14. Kono S, Miyajima H, Sugimoto M, Suzuki Y, Takahashi Y, Hishida A. Stiff-person syndrome associated with cerebellar ataxia and high glutamic acid decarboxylase antibody titer. Internal medicine. 2001;40(9):968–71.
15. Zivotofsky AZ, Siman-Tov T, Gadoth N, Gordon CR. A rare saccade velocity profile in stiff-person syndrome with cerebellar degeneration. Brain Res. 2006;1093(1):135–40.
16. Ameli R, Snow J, Rakocevic G, Dalakas MC. A neuropsychological assessment of phobias in patients with stiff person syndrome. Neurology. 2005;64(11):1961–3.
17. Henningsen P, Meinck HM. Specific phobia is a frequent non-motor feature in stiff man syndrome. J Neurol Neurosurg Psychiatry. 2003;74(4):462–5.
18. Marano M, Motolese F, Lanzone J, Di Santo A, Rossi M, Bevacqua MG, et al. Pseudoagoraphobia, a diagnostic clue in stiff-limb syndrome. Mov Disord Clin Pract. 2020;7(3):313–7.
19. Nasri A, Gharbi A, Ouali U, Mrabet S, Souissi A, Jomli R, et al. Psychiatric symptoms in stiff-person syndrome: a systematic review and a report of two cases. J Acad Consult Liaison Psychiatry. 2023;64(2):183–91.
20. Souissi MA, Bellakhal S, Gharbi E, Abdelkefi I, Douggui MH. A stiff person syndrome misdiagnosed as a psychiatric illness: a case report. Curr Rheumatol Rev. 2020;16(4):343–5.
21. Goetz CG, Klawans HL. On the mechanism of sudden death in Moersch-Woltman syndrome. Neurology. 1983;33(7):930–2.
22. Bheemavarapu B, Singh A, Ranjini NJ, Abhilash Meda VS, Patil D. Recurrent acute on chronic respiratory failure in stiff person syndrome. Case Rep Neurol. 2023;15(1):187–91.
23. Jachiet V, Laine L, Gendre T, Henry C, Da Silva D, de Montmollin E. Acute respiratory failure in a patient with stiff-person syndrome. Neurocrit Care. 2016;25(3):455–7.
24. Mitsumoto H, Schwartzman MJ, Estes ML, Chou SM, La Franchise EF, De Camilli P, Solimena M. Sudden death and paroxysmal autonomic dysfunction in stiff-man syndrome. J Neurol. 1991;238(2):91–6.
25. Dalakas MC. Stiff-person syndrome and GAD antibody-spectrum disorders: GABAergic neuronal excitability, immunopathogenesis and update on antibody therapies. Neurotherapeutics. 2022;19(3):832–47.
26. Bai L, Ren H, Liang M, Lu Q, Lin N, Liu M, et al. Neurological disorders associated with glutamic acid decarboxylase 65 antibodies: clinical spectrum and prognosis of a cohort from China. Front Neurol. 2022;13:990553.
27. Ances BM, Dalmau JO, Tsai J, Hasbani MJ, Galetta SL. Downbeating nystagmus and muscle spasms in a patient with glutamic-acid decarboxylase antibodies. Am J Ophthalmol. 2005;140(1):142–4.
28. Economides JR, Horton JC. Eye movement abnormalities in stiff person syndrome. Neurology. 2005;65(9):1462–4.
29. Hac NEF, Murphy OC, Butala AA, Newsome SD, Gold DR. Centripetal nystagmus, slow saccades, cerebellar ataxia, and parkinsonism in a patient with anti-GAD65-associated stiff person syndrome spectrum disorder. J Neuro-Ophthalmol. 2023;43(2):273–6.
30. Oskarsson B, Pelak V, Quan D, Hall D, Foster C, Galetta S. Stiff eyes in stiff-person syndrome. Neurology. 2008;71(5):378–80.
31. Tilikete C, Vighetto A, Trouillas P, Honnorat J. Anti-GAD antibodies and periodic alternating nystagmus. Arch Neurol. 2005;62(8):1300–3.
32. Wang Y, Tourkevich R, Bosley J, Gold DR, Newsome SD. Ocular motor and vestibular characteristics of antiglutamic acid decarboxylase-associated neurologic disorders. J Neuroophthalmol. 2021;41(4):e665–e71.

33. Teggi R, Piccioni LO, Martino G, Bellini C, Bussi M. Stiff-person syndrome with acute recurrent peripheral vertigo: possible evidence of gamma aminobutyric acid as a neurotransmitter in the vestibular periphery. J Laryngol Otol. 2008;122(6):636–8.
34. Jarius S, Stich O, Speck J, Rasiah C, Wildemann B, Meinck HM, Rauer S. Qualitative and quantitative evidence of anti-glutamic acid decarboxylase-specific intrathecal antibody synthesis in patients with stiff person syndrome. J Neuroimmunol. 2010;229(1–2):219–24.
35. Li Y, Thakore N. An appraisal of electrodiagnostic studies in stiff person syndrome. J Clin Neuromuscul Dis. 2020;22(2):84–9.
36. Armon C, McEvoy KM, Westmoreland BF, McManis PG. Clinical neurophysiologic studies in stiff-man syndrome: use of simultaneous video-electroencephalographic-surface electromyographic recording. Mayo Clin Proc. 1990;65(7):960–7.
37. Meinck HM, Ricker K, Hulser PJ, Schmid E, Peiffer J, Solimena M. Stiff man syndrome: clinical and laboratory findings in eight patients. J Neurol. 1994;241(3):157–66.
38. Witherick J, Highley JR, Hadjivassiliou M. Pathological findings in a case of stiff person syndrome with anti-GAD antibodies. Mov Disord. 2011;26(11):2138–9.

Fahr Disease

98

Overview

Fahr disease (FD) is a rare neurodegenerative disorder in which symmetrical calcification occurs in various brain regions, particularly the basal ganglia and cerebellum. Some cases are idiopathic, and others occur in a familial pattern (autosomal dominant or recessive). Clinical presentation includes movement disorders, cognitive and psychiatric disorders, and ataxia. Some patients exhibit imbalance, but this has not been well studied from the otovestibular perspective. The intracranial calcifications can be identified on head CT. Treatment is supportive.

Introduction

In 1930, the German pathologist Karl Theodor Fahr (Pirmasens, Germany 1877–Hamburg, Germany 1945) published a case report of a patient suffering from what was eventually designated *Fahr disease (FD)* or *syndrome* [1], though Manyam and colleagues [2] credit Delacour [3] with having published the first description of calcification of the basal ganglia.

FD is a neurodegenerative disorder in which calcium deposits accumulate, usually symmetrically, in various areas of the brain, including the basal ganglia, striatum, internal capsule, cerebral cortex and subcortical regions, thalamus, hippocampus, and cerebellum (usually the dentate nucleus).

Some cases of FD occur sporadically, and these are sometimes termed *idiopathic basal ganglia calcification (IBGC).*

Other cases of FD occur in familial patterns, and these are sometimes called *primary familial brain calcification (PFBC*).

Epidemiology

FD is rare. Age at symptom onset ranges from the third to the seventh decades [2, 4].

Genetics

Several genes have been implicated in the familial form of FD and exhibit autosomal dominant and autosomal recessive patterns.

Autosomal dominant variants:

- IBGC1 (OMIM 213600) is caused by a mutation in the solute carrier family 20 member 2 (SLC20A2) on chromosome 8p11.2 (OMIM 158378) [5].
- IBGC4 (OMIM 615007) is caused by a mutation in the platelet-derived growth factor receptor beta (PDGFRB) gene on chromosome 5q32 (OMIM 173410) [6].
- IBGC6 (OMIM 616413) is caused by a mutation in the xenotropic and polytropic retrovirus receptor 1 (XPR1) gene on chromosome 1q25.3 (OMIM 605237) [7].
- IBGC5 (OMIM 615483) is caused by a mutation in the platelet-derived growth factor subunit B (PDGFB) gene on chromosome 22q13.1 (OMIM 190040) [8, 9].

Autosomal recessive variants:

- IBGC7 (OMIM 618317) is caused by a mutation in the myogenesis regulating glycosidase (MYORG) gene on chromosome 9p13.3 (OMIM 618255) and is transmitted in an autosomal recessive pattern [10].
- IBGC8 (OMIM 618824) is caused by a mutation in the junctional adhesion molecule 2 (JAM2) gene on chromosome 21q21.3 (OMIM 606870) and is transmitted in an autosomal recessive pattern [11, 12].

Clinical Presentation

In view of the diverse neuroanatomical areas that can be involved by FD, it is unsurprising that the clinical presentation is variable.

M. Cherchi, *Otoneurology and Vestibular Medicine*, https://doi.org/10.1007/978-3-031-94841-1_98

Manyam and colleagues [2] conducted a literature and registry review of patients with FD. Of 67 symptomatic patients, they documented:

- 37 (55%) exhibited movement disorders
- 26 (39%) exhibited cognitive deficits
- 24 (36%) exhibited speech deficits
- 24 (36%) exhibited cerebellar dysfunction
- 21 (31%) exhibited psychiatric disorders
- 15 (22%) exhibited pyramidal disorders
- 12 (18%) exhibited gait disorders
- 11 (16%) complained of pain and other sensory disorders
- 9 (13%) exhibited genitourinary dysfunction
- 8 (12%) exhibited gastrointestinal dysfunction
- 6 (9%) exhibited seizures.

Of the 37 patients with movement disorders, they documented:

- 21 (57%) had parkinsonism
- 7 (19%) had chorea
- 3 (8%) had tremor
- 3 (8%) had dystonia
- 2 (5%) had athetosis
- 1 (3%) had orofacial dyskinesias.

There are several case reports of FD patients suffering from imbalance [13–16].

Testing: Vestibular

As of this writing, there were no published reports of instrumented vestibular testing in FD patients.

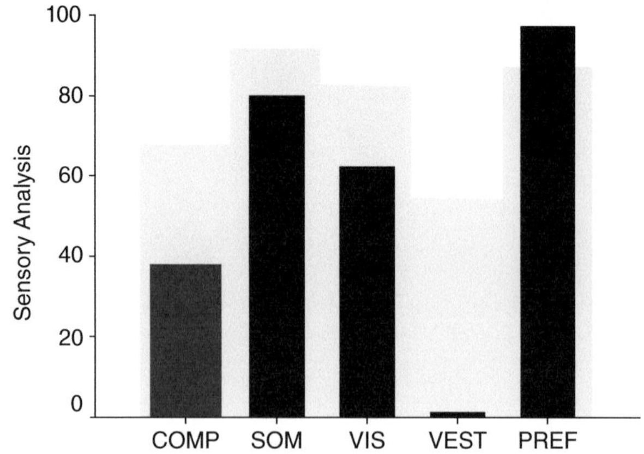

Fig. 98.1 Sensory organization testing in a patient with Fahr disease. (From Scarano et al. [17]. Reproduced from Frontiers Media S.A. with permission under CC-BY 4.0)

Testing: Other

Scarano and colleagues studied computerized dynamic posturography in a 69-year-old male with FD [17]. Sensory organization testing showed primarily a vestibular pattern, as shown in Fig. 98.1. This pattern is also found in other disorders with basal ganglionic involvement (even though they are not primary otovestibular diseases).

Imaging

The calcifications in Fahr disease are revealed on CT scans. Figure 98.2, from Scarano and colleagues [17], shows approximately symmetrical calcification in the basal ganglia and cerebellum.

Note that abnormal calcium deposits, including in patterns compatible with FD, are found incidentally on any-

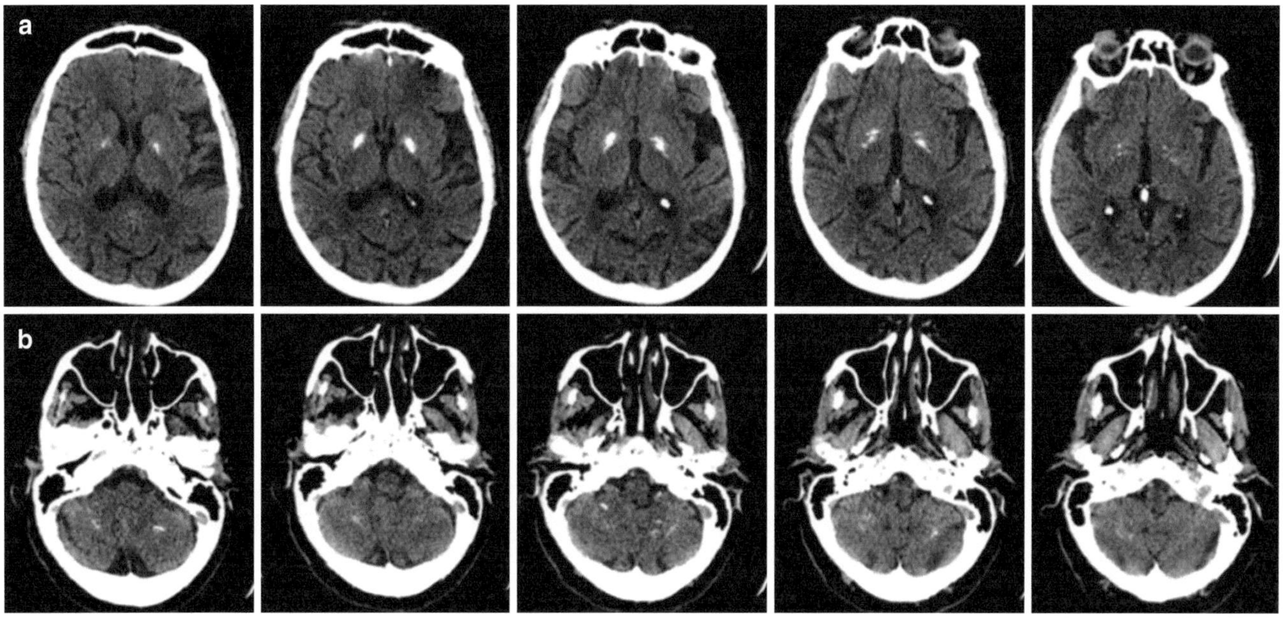

Fig. 98.2 Axial head CT sections of a patient with Fahr disease. The upper row shows (**a**) the bilateral calcifications within the globus pallidus and cerebral white matter, and the lower row shows (**b**) the calcifications within the cerebellar hemispheres. (From Scarano et al. [17]. Reproduced from Frontiers Media S.A. with permission under CC-BY 4.0)

where from 0.3% [18] to 1.2% [19] of head CTs of patients exhibiting no clinical features of FD.

Differential Diagnosis

The differential diagnosis of FD includes basal ganglionic and cerebellar disorders.

Treatment and Prognosis

There is no direct therapy for FD. Treatment is supportive.

References

1. Fahr KT. Idiopathische Verkalkung der Hirngefässe Zentralblatt für allgemeine Pathologie und pathologische. Anatomie. 1930;50:129–33.
2. Manyam BV, Walters AS, Narla KR. Bilateral striopallidodentate calcinosis: clinical characteristics of patients seen in a registry. Mov Disord. 2001;16(2):258–64.
3. Delacour A. Ossification des capillaires du cerveau. Ann Med Psychol (Paris). 1850;2:458–61.
4. Ellie E, Julien J, Ferrer X. Familial idiopathic striopallidodentate calcifications. Neurology. 1989;39(3):381–5.
5. Wang C, Li Y, Shi L, Ren J, Patti M, Wang T, et al. Mutations in SLC20A2 link familial idiopathic basal ganglia calcification with phosphate homeostasis. Nat Genet. 2012;44(3):254–6.
6. Nicolas G, Pottier C, Maltete D, Coutant S, Rovelet-Lecrux A, Legallic S, et al. Mutation of the PDGFRB gene as a cause of idiopathic basal ganglia calcification. Neurology. 2013;80(2):181–7.
7. Legati A, Giovannini D, Nicolas G, Lopez-Sanchez U, Quintans B, Oliveira JR, et al. Mutations in XPR1 cause primary familial brain calcification associated with altered phosphate export. Nat Genet. 2015;47(6):579–81.
8. Keogh MJ, Pyle A, Daud D, Griffin H, Douroudis K, Eglon G, et al. Clinical heterogeneity of primary familial brain calcification due to a novel mutation in PDGFB. Neurology. 2015;84(17):1818–20.
9. Keller A, Westenberger A, Sobrido MJ, Garcia-Murias M, Domingo A, Sears RL, et al. Mutations in the gene encoding PDGF-B cause brain calcifications in humans and mice. Nat Genet. 2013;45(9):1077–82.
10. Yao XP, Cheng X, Wang C, Zhao M, Guo XX, Su HZ, et al. Biallelic mutations in MYORG cause autosomal recessive primary familial brain calcification. Neuron. 2018;98(6):1116–23 e5.
11. Schottlaender LV, Abeti R, Jaunmuktane Z, Macmillan C, Chelban V, O'Callaghan B, et al. Bi-allelic JAM2 variants Lead to early-onset recessive primary familial brain calcification. Am J Hum Genet. 2020;106(3):412–21.
12. Cen Z, Chen Y, Chen S, Wang H, Yang D, Zhang H, et al. Biallelic loss-of-function mutations in JAM2 cause primary familial brain calcification. Brain. 2020;143(2):491–502.
13. Castello E, Pallestrini EA. [Otoneurological manifestations in Fhar's disease. A case report]. (0392-100X (Print)).
14. Parasram M, Levi Chazen J, Sarva H. Primary familial brain calcification. J Am Osteopath Assoc. 2020;120
15. Wang H, Shao B, Wang L, Ye Q. Fahr's disease in two siblings in a family: a case report. Exp Ther Med. 2015;9(5):1931–3.
16. Salli M, D'Arienzo A, Bonanno M, Morello S, Sanfilippo A, Mauro GL, et al. Femoral neck's fracture in Fahr's syndrome: case report. Clin Cases Miner Bone Metab. 2016;13(3):265–7.
17. Scarano S, Rota V, Tesio L, Perucca L, Robecchi Majnardi A, Caronni A. Balance impairment in Fahr's disease: mixed signs of parkinsonism and cerebellar disorder. A case study. Front Hum Neurosci. 2022;16:832170.
18. Koller WC, Cochran JW, Klawans HL. Calcification of the basal ganglia: computerized tomography and clinical correlation. Neurology. 1979;29(3):328–33.
19. Forstl H, Krumm B, Eden S, Kohlmeyer K. Neurological disorders in 166 patients with basal ganglia calcification: a statistical evaluation. J Neurol. 1992;239(1):36–8.

Part XII

Diagnoses—Posterior Fossa Lesions

99 Chiari Malformation

Overview

Chiari malformations (CMs) are developmental anomalies that neuroanatomically involve structures in the posterior fossa and craniocervical junction, but can also involve other organ systems. The more common types are CM1 (58.7% of cases), CM2 (22.9% of cases), and CM4 (15.3% of cases). The remaining types (0, 0.5, 1.5, 3.5, 4, 5) are very rare, and would not usually present to an otoneurologist.

In an otoneurology practice, a patient may be referred (usually for a complaint of disequilibrium, or for findings of nystagmus noted on physical examination) who has had brain imaging performed that identified cerebellar tonsillar descent. Whether or not the tonsillar descent is sufficient to meet criteria for Chiari 1 malformation, the usual clinical query is whether the position of the cerebellar tonsils accounts for the patient's disequilibrium, and if so, whether the patient should be referred to neurosurgery (usually suboccipital decompression, sometimes with high cervical laminectomy).

From an otoneurological perspective, the unifying features of CM are oculomotor and vestibular abnormalities related to cerebellar dysfunction, as the cerebellum is either being physically impinged upon at the craniocervical junction (due to herniation into the spinal canal), or is hypoplastic. Oculomotor findings have been well described in CM1, fairly described in CM2, and not described well (or at all) for the remaining types. The most common vestibular oculomotor findings include gaze evoked nystagmus (horizontal more than vertical), spontaneous vertical nystagmus (down beat more than up beat), deficits in smooth pursuit and optokinetic responses, all of which localize to the cerebellum, and are not specific for CM. Less common vestibular oculomotor findings include periodic alternating nystagmus and see-saw nystagmus. Other (nonvestibular) oculomotor findings include strabismus, specifically acquired esotropia (usually comitant), and rarely exotropia.

Some aspects of the nystagmus (at least of CM1) may respond to pharmacologic treatment, though data supporting efficacy are weak.

Surgical decompression is sometimes attempted in management of CM1. Reports of surgical outcomes are variable. Moreover, since having a rare disease (such as CM) does not preclude getting a more common disease, even in a patient with radiologically confirmed CM we feel it is still medically reasonable to undertake a screening workup (we usually check vestibular evoked myogenic potentials, video head impulse testing, videonystagmography and rotatory chair testing) for more common causes of disequilibrium, such as vestibular neuritis. While it is true that unilateral vestibular weakness has been reported in CM1, such a deficit is much more commonly caused by vestibular neuritis, so a trial of vestibular rehabilitation therapy is reasonable. If workup for alternative diagnoses is unrevealing, and if a patient's symptoms interfere with level of function, and if vestibular and oculomotor findings are compatible with what is known about CM, then referral to neurosurgery is reasonable.

Audiologically, some CM patients exhibit sensorineural hearing loss, which may be asymmetrical. There are case reports of this hearing loss improving after surgical decompression.

Introduction

Carmel and Markesbery [1] point out that John Cleland (1835–1925) was probably the first to publish [2] a description of posterior fossa anomalies that would later come to be known as *Chiari malformations*, but Dr. Cleland has not received appropriate credit for the discovery.

Dr. Hans Chiari (1851–1916) described three types of malformations in 1891 [3] and a fourth type in 1895 [4]. Over the years, investigators have described additional types (0, 0.5, 1.5, 3.5, 5) (Fig. 99.1).

According to Carmel and Markesbery [1], the 1894 paper by Dr. Julius Arnold (1835–1915) [6] "described multiple anomalies in a newborn infant. Much of this paper is devoted to details of visceral and lung abnormalities, and only scant attention was given to the brain stem." Despite the limited description in Arnold's work, Carmel and Markesbery [1]

M. Cherchi, *Otoneurology and Vestibular Medicine*, https://doi.org/10.1007/978-3-031-94841-1_99

Fig. 99.1 Dr. Hans Chiari (1851–1916). (From Tubbs et al. [5]. Reproduced with written permission from Springer Nature B.V.)

note that a subsequent publication by Schwalbe and Gredig [7] referred to the second type of malformation as the "malformation of Arnold and Chiari," so for a time this was known as an "Arnold-Chiari malformation."

In the present discussion, we will refer to these as "Chiari malformations," abbreviated CM.

Table 99.1, from Russel and Wick [8], provides a very brief overview of the four main types.

Horn and colleagues [9] analyzed a nationwide database of 520,561 reported cases of CM. Of these, they identified that CM1 accounted for 305,726 (58.7%), CM2 for 119,632 (22.9%), CM3 for 15,540 (3%), and CM4 for 79,663 (15.3%). Thus, the relative prevalence, in descending order, is CM1 > CM2 > CM4 > CM3.

Horn and colleagues also reported that 44.3% of cases were associated with at least one concurrent anomaly (other than the Chiari malformation itself), which can be in almost any other organ system.

Table 99.1 Four main types of Chiari malformations as summarized by Russel and Wick [8]

Type	Signs
Chiari type I	Cerebellar tonsils, floccule, and medial aspects of the caudal medulla in the upper cervical canal. May not appear until early adulthood.
Chiari type II	Same as type I, but caudal portion of the fourth ventricle positioned in the upper cervical canal. May not appear until early adulthood.
Chiari type III	Smaller cerebellum from physical appearance. Cerebellum, lower brainstem, and fourth ventricle pushed into the upper cervical canal. Present at birth.
Chiari type IV	Cerebellum and brainstem located within the posterior fossa. Present at birth.

Radiologic Landmarks

The anatomical landmark by which cerebellar tonsillar descent is judged is the McRae line [10] which, on a midsagittal image, connects the basion (anterior margin of the foramen magnum) to the opisthion (posterior margin of the foramen magnum).

Type 0 Chiari Malformation (CM0)

Bogdanov and colleagues [11] comment that, "Some investigators have used CM0 [Chiari malformation type 0] to refer to patients with and without syringomyelia with minimal or absent TH [cerebellar tonsillar herniation], radiological features of SPCF [small posterior cranial fossa], and typical clinical signs and symptoms of CM1, including occipital headache, posterior cervical pain, and cerebellomedullary dysfunction," adding that, "investigations support classifying CM0 without syringomyelia as a classical CM1 [Chiari malformation type 1] variant because CM0 and CM1 share findings of an SPCF and caudally located obex."

In other words, CM0 is very similar to CM1 except that the tonsillar descent does not meet criteria for CM1.

Figure 99.2, from Bogdanov and colleagues [11], shows several examples of sagittal and axial images of CM0 cases.

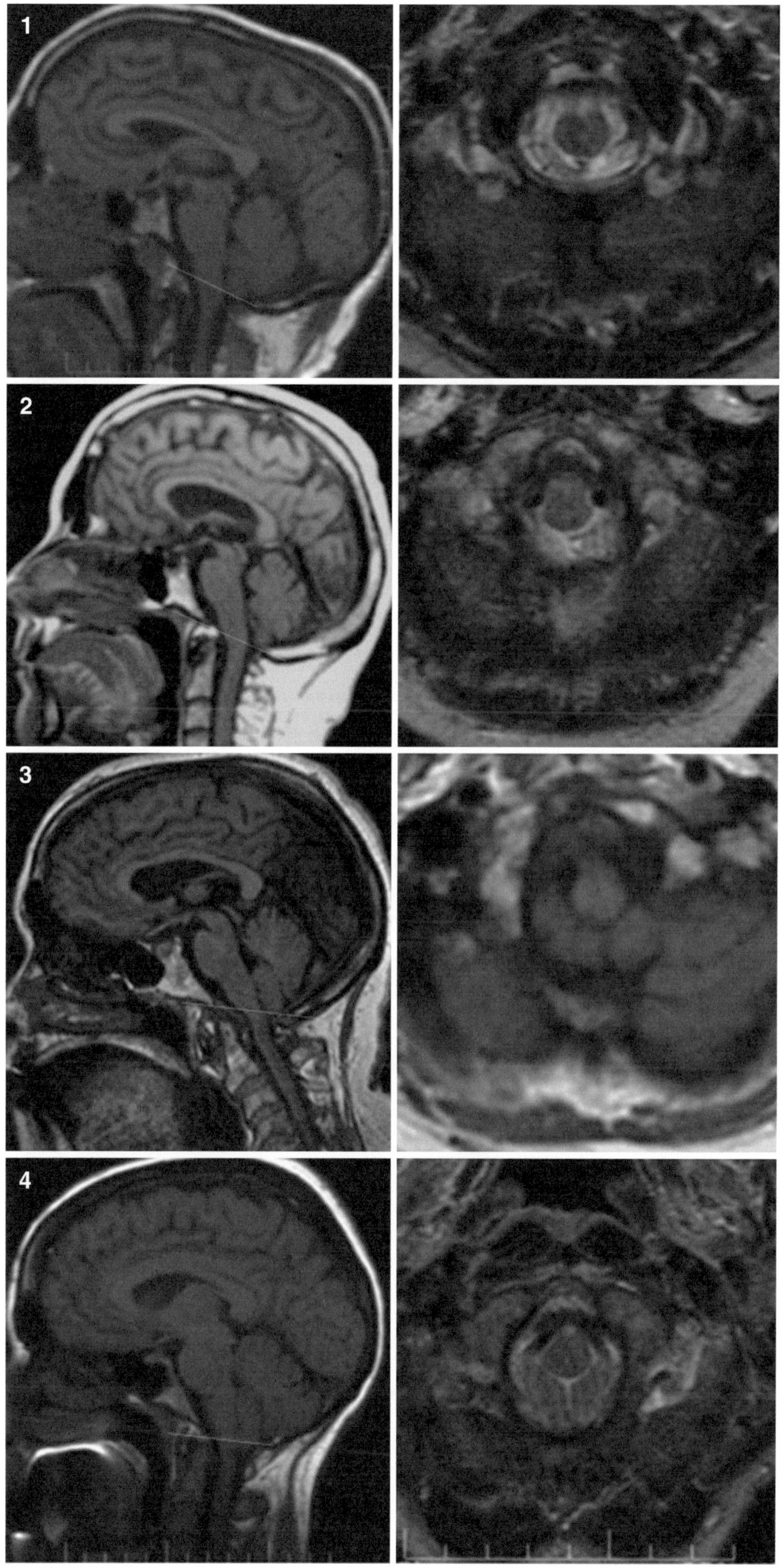

Fig. 99.2 Types of CM0 without syringomyelia. The McRae line (white line) is shown on the T1-weighted sagittal images (left panels). Axial images at McRae's line are shown (right panels) on T1-weighted (**c**) and T2-weighted (**a**, **b**, **d**) images. Adult symptomatic CM0-only patients with tonsillar herniation ≤2 mm (panels (**a–c**)) and borderline CM1-only patients with tonsillar herniation 3 mm (**d**). All patients demonstrate short bones (CL and SO <40 mm), crowdedness of the PCF, tight foramen magnum, and CM1-like clinical manifestations: transient localized suboccipital cough-related headaches (**a**) or constant and transient suboccipital headaches (**b–d**), truncal ataxia, vertigo, dizziness, and upper motor neuron signs. The sagittal images in (**a**, **c**) (left panels) also show spinal canal narrowing at the level of the odontoid. (From Bogdanov et al. [11]. Reproduced from MDPI with permission under CC-BY 4.0)

Type 0.5 Chiari Malformation (CM0.5)

Morgenstern and colleagues [12] identified a group of patients in whom the cerebellar tonsils were not merely downwardly displaced, but also ventrally herniated, which they define as, "the tonsils crossing a line that bisects the caudal medulla at the level of the foramen magnum."

Figure 99.3, from Morgenstern and colleagues [12], demonstrates axial images in which the cerebellar tonsils wrap ventrally around the medulla.

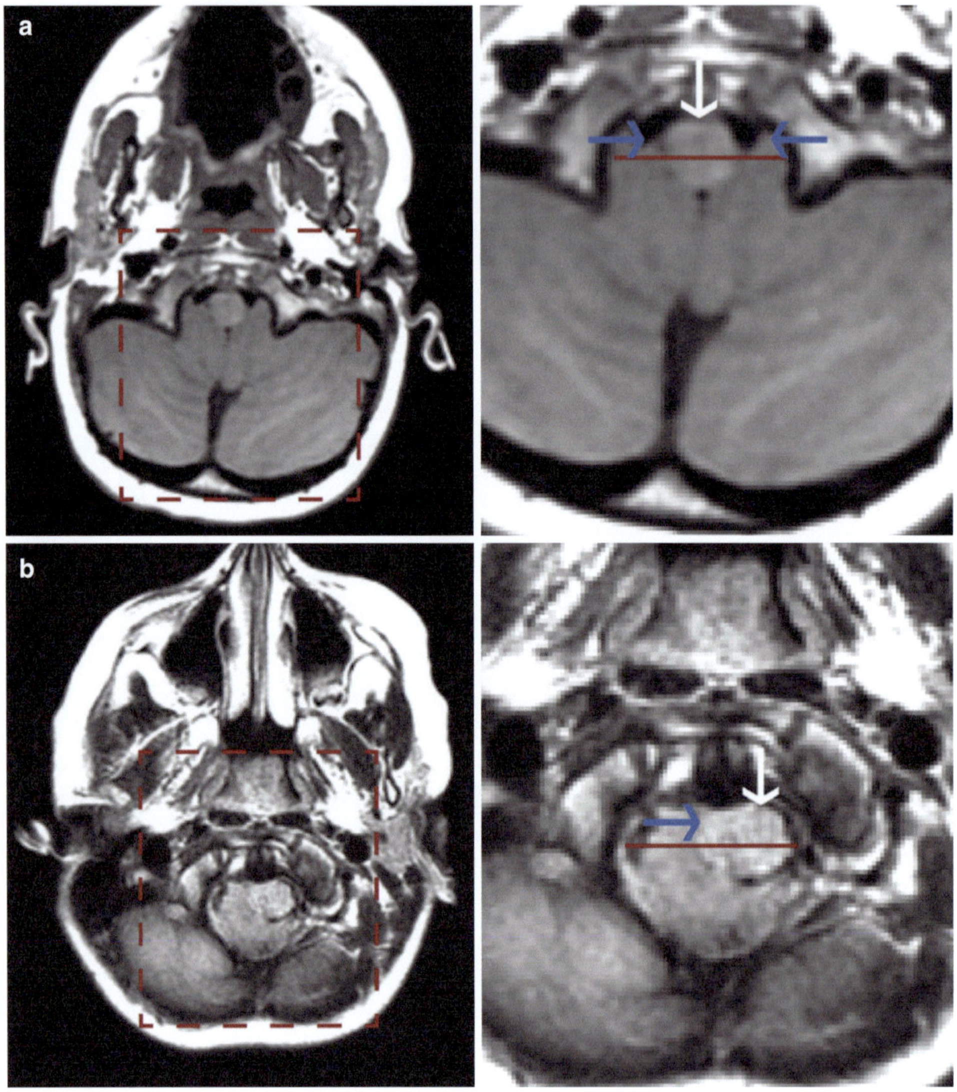

Fig. 99.3 Axial magnetic resonance imaging findings from 2 patients with ventral tonsillar herniation, as defined by 1 or both tonsils crossing the line bisecting the caudal medulla. Panel (**a**): Bilateral ventral herniation. Panel (**b**): Unilateral ventral herniation. Blue arrows indicate the ventrally located cerebellar tonsils. Areas in left panels outlined in red are magnified in right panels. White arrow indicates the medulla showing right to left compression from the unilateral ventral compression. (From Morgenstern et al. [12]. Reproduced with permission from Elsevier under STM Signatory agreement)

Type 1 Chiari Malformation (CM1)

Chiari's original description [3] was:

> The first type is the lengthening of the tonsils and the medial parts of the inferior lobes of the cerebellum to form cone-shaped processes which accompany the medulla oblongata into the vertebral canal.

Cesmebasi and colleagues [13] provide a more detailed list of the anatomical features associated with CM1:

Brain: Usually normal except for the cerebellar tonsils.
Ventricles/cisterns: In some cases, the fourth ventricle is elongated.
Cerebellar tonsils: Caudal displacement of cerebellar tonsils more than 3–5 mm inferior to the plane of the foramen magnum.
Skull: Basilar skull and craniocervical junction abnormalities are found in about 50% of cases and can include underdevelopment of the occipital bone; increased size of the foramen magnum; shortened clivus.
Meninges: Slope of the tentorium cerebelli is often elevated; arachnoid mater at the level of the foramen magnum is often thickened.
Spinal canal: Commonly associated findings include a Klippel–Feil deformity (maldeveloped vertebra and/or fusion between two or more cervical vertebrae); atlanto-axial assimilation; retroflexion of the odontoid process; thickening of the ligamentum flavum; scoliosis.
Spinal cord: Syrinx is present in 50–75% of cases.

Barkovich and colleagues [14] reviewed MRIs from 25 patients with CM1 and compared them to 200 normal controls. They reported that a cutoff point of 3 or 4 mm of cerebellar tonsillar descent had 96% sensitivity and 99.5% specificity for identifying CM1, and that a cutoff point of 5 mm of cerebellar tonsillar descent had a sensitivity of 92% and specificity of 100% for identifying CM1. Based on this study, many investigators use cerebellar tonsillar descent of ≥5 mm as a criterion for CM1.

Epidemiology of CM1

The prevalence of CM1 is estimated at 96 cases per 100,000 people, with a median age at diagnosis of 10.5 years [12].

The age of symptom onset is variable. Some investigators report that most patients with CM1 develop symptoms in the second or third decade of life [15], while others report the third or fourth decade of life [16].

Clinical Presentation of CM1

Piper and colleagues [17] took information from Milhorat and colleagues [18] regarding the categories and frequencies of symptoms in CM1 patients and reorganized them by category of anatomical localization and by frequency (Table 99.2).

Table 99.2 Symptoms of Chiari 1 malformation organized by category of neuroanatomical localization and by frequency

General symptoms	Ocular symptoms	Otoneurological symptoms	Hindbrain symptoms	Spinal cord symptoms
81%: Headache	78%: Any ocular symptom	74%: Any otoneurological symptom		84%: Any spinal cord complaint
58%: Chronic fatigue	65%: Retro-orbital pain	57%: Dizziness		59%: Paresthesia or hyperesthesia
	55%: Visual phenomena	52%: Disequilibrium	52%: Any hindbrain symptom	57%: Muscle weakness
	48%: Blurred vision	46%: Pressure in ears	43%: Dysphagia	44%: Nonradicular pain
39%: Memory disturbance		36%: Decreased hearing	38%: Sleep apnea	40%: Anesthesia
34%: Neck pain			31%: Hoarseness	30%: Spasticity
			26%: Tremors 26%: Palpitations	29%: Burning dysesthesia
24%: Back pain				24%: Impotence in males
			23%: Poor coordination	23%: Poor position sense
				22%: Impaired temperature sensation
	20%: Diplopia	20%: Vertigo		20%: Trophic phenomena
18%: Intermittent nausea and/or vomiting			16%: Facial pain or tenderness 16%: Throat pain	17%: Urinary incontinence
		13%: Oscillopsia	13%: Syncope	
			9%: Dyspnea	9%: Muscular atrophy
	7%: Field cuts	7%: Hyperacusis		2%: Fecal incontinence

From Piper et al. [17] and Milhorat et al. [18]

Audiologic Findings in CM1

Kumar and colleagues [19] reported on 77 patients with CM1. Of these patients, they noted 10 had bilateral sensorineural hearing loss, and 22 had unilateral sensorineural hearing loss.

The sensorineural hearing loss can be symmetrical or asymmetrical, and the patterns variable, as shown in the examples studied by Rydell and colleagues [20].

Oculomotor Findings in CM1

Regrettably, many reports may list "nystagmus" as a physical examination finding in CM1 but do not characterize it further, which limits their utility. For instance, Hida and colleagues [21] commented that 2/16 (13%) of their CM1 patients exhibited "nystagmus," but provided no details.

Shaikh and Ghasia [22] comment that overall, nystagmus (of all types) is the most common oculomotor abnormality in CM1. They state that in CM1, *gaze-evoked nystagmus* (sometimes with *rebound nystagmus*) is present in 30% of patients, and *down beat nystagmus* is present in 4–6% of patients.

Kumar and colleagues [19] reported on 77 patients with CM1. Of these patients, they noted that 27 (35%) had *horizontal nystagmus*, 3 (4%) had *spontaneous up beat nystagmus*, 4 (5%) had *spontaneous down beat nystagmus*, 4 (5%) had *saccadic dysmetria*, 3 (4%) had impaired *optokinetic nystagmus*, 12 (16%) had impaired *smooth pursuit*, and 9 (12%) patients had "*positional nystagmus*."

Liebenberg and colleagues [23] reviewed 40 CM1 patients. They reported that 12/40 (30%) had oculomotor and vestibulo-ocular abnormalities. Of these twelve, 4/12 (33%) had "*horizontal nystagmus*," 3/12 (25%) had *down beat nystagmus*, and 1/12 (8%) had "*jerky pursuit*."

Numerous reports describe spontaneous *down beat* nystagmus in CM1 [15, 24–26]. This spontaneous down beat nystagmus may augment on lateral gaze [15, 27]. Sometimes this down beat nystagmus is intermittent [28, 29].

Rarely, CM1 may produce spontaneous *up beat nystagmus* [30].

A number of cases of CM1 have been reported presenting primarily or exclusively with *acquired esotropia* [16, 31–38], usually comitant.

Ghasia and colleagues [39] reported intermittent *exotropia* and gaze-evoked nystagmus depending on viewing position.

There are several reports of *periodic alternating nystagmus* (PAN) in CM1 [40, 41].

Rarely, CM1 can cause *see-saw nystagmus* [42].

Instrumented Vestibular Testing in CM1: Videonystagmography (VNG) or Electronystagmography (ENG)

Goldschagg and colleagues [43] treated ten adult CM1 patients with suboccipital decompression and unilateral tonsillectomy. They reported that 9 of these patients underwent videonystagmography (VNG) preoperatively, and that 3/9 (33%) exhibited pathological caloric asymmetry.

Yee and colleagues [29] reported a case of CM1 in which electronystagmography showed geotropic direction changing positional nystagmus.

Mossman and colleagues [44] studied electronystagmography on one CM1 patient and observed convergence nystagmus, which is a very unusual finding.

Instrumented Vestibular Testing in CM1: Scleral Search Coils

Lewis and colleagues [35] studied five CM1 patients with scleral search coils and reported that, "All patients reported onset of horizontal diplopia due to acquired esotropia as an initial manifestation of the Arnold-Chiari I malformation." The authors concluded that, "Acquired esotropia, often in association with other eye movement abnormalities, may be an early sign of Arnold-Chiari I malformation."

Pieh and Gottlob [27] studied two CM1 patients with scleral search coils and reported "nystagmus of skew," meaning they observed synchronous down beat nystagmus of one eye and up beat nystagmus of the other eye.

Instrumented Vestibular Testing in CM1: Video Head Impulse Testing (vHIT)

Kim and colleagues [45] studied a CM1 patient and reported that horizontal head impulses elicited upward ocular deviation during the impulse, followed by a corrective downward saccade—in other words, perverted nystagmus.

Instrumented Vestibular Testing in CM1: Rotatory Chair Testing (RCT)

Baloh and colleagues [46] studied a CM1 patient with serial videonystagmography and rotatory chair testing over 3 years. They reported that on rotatory chair testing, the step velocity component showed progressive decline in the duration of postrotatory nystagmus (in other words, progressive decrease in the time constant); slow harmonic acceleration showed no responses at 0.2 Hz (responses were present at higher frequencies); and disappearance of optokinetic nystagmus. These findings are similar to what would be seen in bilateral vestibular weakness.

Yee and colleagues [29] reported a case of CM1 in which caloric responses were absent bilaterally, and slow harmonic acceleration showed reduced vestibulo-ocular reflex responses at low frequencies. Again, these findings are similar to what would be seen in bilateral vestibular weakness.

Summary of Oculomotor Findings

Russel and Wick [8] summarize possible oculomotor findings in CM1 and CM2 as follows:

> Patients typically have down-beat nystagmus caused by herniation of the cerebellar tonsils downward through the foramen magnum. The imbalance of the otolith-ocular reflexes, brought on by faulty communication between the vestibular neurons, the vertical subnuclei of the oculomotor nucleus, and the interstitial nucleus, results in vertical pendular eye movements. Loss of the flocculus inhibition due to compression of the cerebellum frequently prevents fixation suppression, causing constant shifts in fixation as the patient's eye tries to find a fixation point. Clinical evaluation of eye movements generally reveals that the saccade system is intact, but smooth pursuits, optokinetic nystagmus (OKN), and fixation suppression are all significantly affected. It is possible, although less likely, to have down beat nystagmus associated with gaze deviation internuclear ophthalmoplegia [and] see-saw nystagmus.

Instrumented Vestibular Testing in CM1: Vestibular Evoked Myogenic Potentials (VEMPs)

Goldschagg and colleagues [43] treated ten adult CM1 patients with suboccipital decompression and unilateral tonsillectomy. They reported that two of the patients had abnormal cervical vestibular evoked myogenic potentials, and that these both improved postoperatively.

Instrumented Vestibular Testing in CM1: Computerized Dynamic Posturography (CDP)

Goldschagg and colleagues [43] treated ten adult CM1 patients with suboccipital decompression and unilateral tonsillectomy. They reported no changes in static posturography from preoperative to postoperative performance.

Imaging in CM1

Figure 99.4, from Piper and colleagues [17], provides clear MRI findings from an infant with CM1.

Rarely, serial imaging of a CM1 patient may show slow, progressive descent of the cerebellar tonsils. Such a case was captured by Pokharel and Siatkowski [16] in Fig. 99.5.

Differential Diagnosis of CM1

The main differential diagnosis of CM1 includes other cerebellar disorders.

Because CM1 can produce intermittent symptoms of extraocular muscle abnormalities, bulbar dysfunction and proximal limb weakness, it can be mistaken for myasthenia gravis [47].

Uncommonly CM1 can present with only esotropia [16, 31–38], usually comitant, and thus be mistaken for primary strabismus [36].

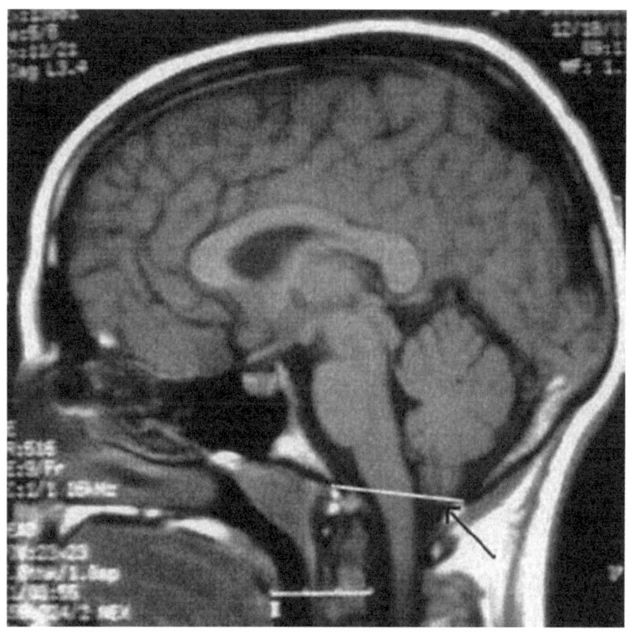

Fig. 99.4 Sagittal MRI showing Chiari 1 malformation with 1–2 mm of cerebellar tonsillar descent below McRae's line. (From Pokharel and Siatkowski [16]. Reproduced with permission from Elsevier under STM Signatory guidelines)

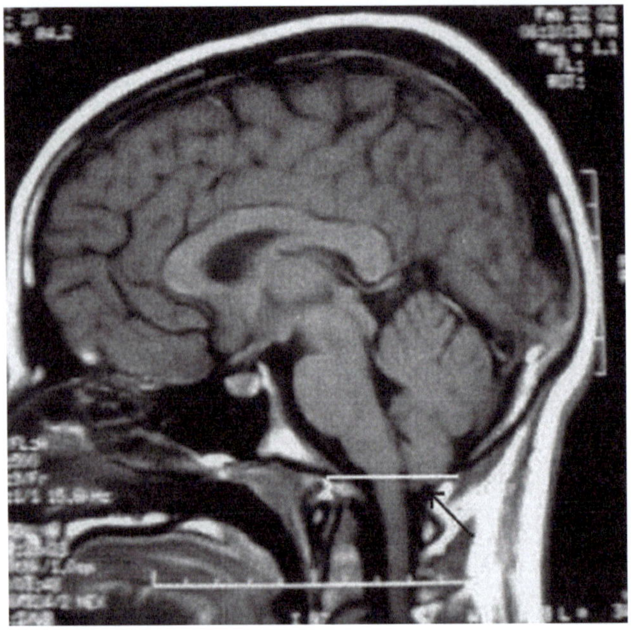

Fig. 99.5 Sagittal MRI showing the same Chiari 1 malformation 2 years later with 3–4 mm of cerebellar tonsillar descent below McRae's line. (From Pokharel and Siatkowski [16]. Reproduced with permission from Elsevier under STM Signatory guidelines)

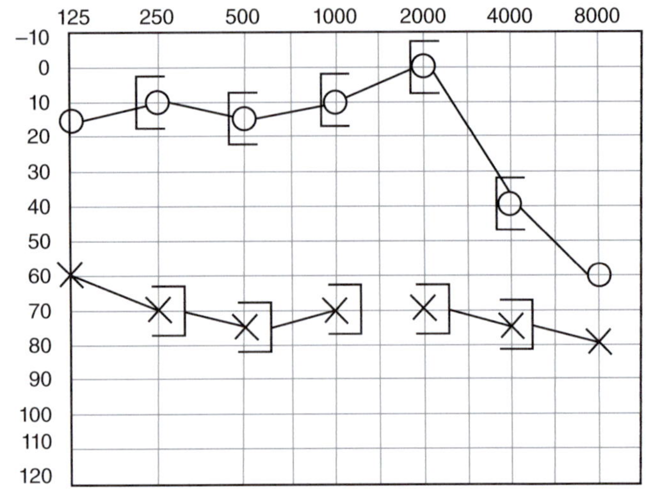

Fig. 99.6 Audiogram of a patient with Chiari 1 malformation showing left-sided moderate to severe sensorineural hearing loss. (From Albers and Engels [57]. Reproduced with written permission from Cambridge University Press)

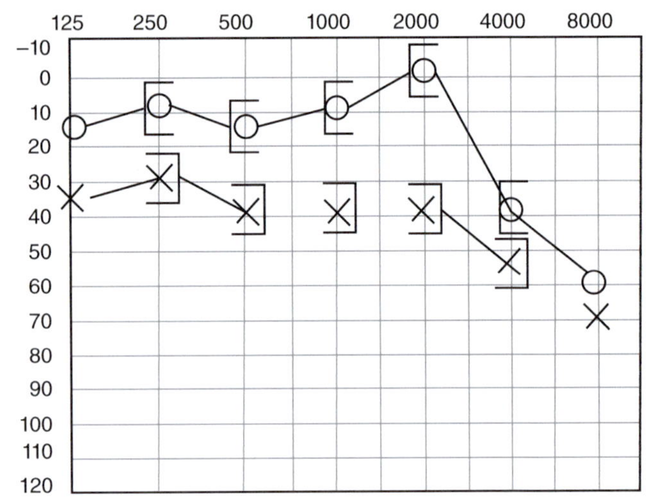

Fig. 99.7 Audiogram of the same patient with Chiari 1 malformation 1 year after surgical decompression, showing some improvement in the left-sided sensorineural hearing loss. (From Albers and Engels [57]. Reproduced with written permission from Cambridge University Press)

Treatment of CM1: Pharmacologic

In some patients, the oculomotor abnormalities in CM1 may be ameliorated pharmacologically. Options are reviewed by Shaikh and Ghasia [22].

Several agents have been studied for the treatment of down beat nystagmus in CM1, including clonazepam [48, 49], baclofen [50], gabapentin [51], 3,4-diaminopyridine [52] and 4-aminopyridine [53]. These studies were either small, or had unsatisfactory results.

For periodic alternating nystagmus, baclofen is the first-line agent [54, 55]; memantine has also been studied [56].

Treatment of CM1: Surgical

Albers and Ingels [57] reported some improvement in asymmetrical sensorineural hearing loss in CM1 after decompression surgery.

Figures 99.6 and 99.7, from Albers and Ingels [57], show a preoperative audiogram from a patient with CM1 who underwent posterior fossa decompression and cervical laminectomy, and a postoperative audiogram performed 1 year later. There was modest improvement in the left-sided hearing loss.

Numerous reports describe improvement in nystagmus after decompression surgery for CM1 [15], but often these observations are simply from case reports. For example:

- Pederson and colleagues [28] studied a CM1 patient with intermittent down beat nystagmus and oscillopsia and reported that after suboccipital decompression surgery, "There was a dramatic diminution of the visual complaints and downbeat nystagmus."
- Gingold and Winfield [58] reported on one CM1 patient whose oscillopsia completely resolved after suboccipital decompression and upper cervical laminectomy.
- Pieh and Gottlob [27] reported on one CM1 patient with "nystagmus of skew" that improved within 3 months of

having undergone posterior fossa decompression and upper cervical laminectomy.

- Yee and colleagues [29] reported on one CM1 patient with episodic vertical oscillopsia and spontaneous down beat nystagmus whose symptoms resolved following sub-occipital decompression.
- Zimmerman and colleagues [42] reported a case of CM1 with oscillopsia and see-saw nystagmus whose symptoms and nystagmus diminished (but did not resolve) after foramen magnum decompression and C1 laminectomy.

For trends in surgical outcomes, we find case series more informative.

Liebenberg and colleagues [23] studied 12 CM1 patients who underwent posterior fossa decompression. They reported that oculomotor and vestibulo-ocular abnormalities resolved in 8/12 (66%) of patients and improved in 1/12 (8%). Of the 8 patients whose abnormalities resolved, the mean time to complete resolution was 15.5 months (range 3–71 months).

Some reports of surgical intervention for CM1 are more guarded. For example, Dones colleagues [59] retrospectively reviewed records of 27 CM1 patients who had undergone surgery. They reported that the distribution depended on the absence/presence of a syrinx (Table 99.3).

Goldschagg and colleagues [43] studied ten CM1 patients immediately before surgery, immediately after surgery, and 3 months postoperatively. The surgery was *suboccipital decompression with unilateral cerebellar tonsillectomy*. Table 99.4 summarizes their outcomes.

Table 99.3 Preoperative and postoperative nystagmus and vertigo in Chiari 1 malformation without and with syrinx

	Preoperative	Postoperative	Change
Nystagmus, total	8/27 (30%)	2/27 (7%)	From 8 to 2 = 6/8 (75%)
Nystagmus, with syrinx	3/16 (19%)	0/16 (0%)	From 3 to 0 = 0/3 (100%)
Nystagmus, without syrinx	5/11 (45%)	2/11 (18%)	From 5 to 2 = 3/5 (60%)
Vertigo, total	6/27 (22%)	3/27 (11%)	From 6 to 3 = 3/6 (50%)
Vertigo, with syrinx	0/16 (0%)	0/16 (0%)	–
Vertigo, without syrinx	6/11 (55%)	3/11 (27%)	From 6 to 3 = 3/6 (50%)

From Dones et al. [59]

Our interpretation of these results is as follows:

- *Improvement*:
  - Patient 6 improved: Preoperatively there was an abnormal ocular tilt reaction, abnormal subjective visual vertical, and abnormal visual fixation suppression. At 3 months postoperatively, these had all improved.
- *No change*:
  - Patients 2, 3, and 5 were unchanged at 3 months postoperatively.
- *Change* (some abnormalities improved, but new ones appeared):
  - Patient 1 changed: Preoperatively had gaze evoked nystagmus, head shaking nystagmus and abnormal video head impulse testing on the right. At 3 months postoperatively, there was deficient horizontal and vertical smooth pursuit.
  - Patient 4 changed: Preoperatively had head shaking nystagmus. At 3 months postoperatively, there was spontaneous down beat nystagmus.
  - Patient 8 changed: Preoperatively there was bilaterally abnormal video head impulse testing, head shaking nystagmus and spontaneous down beat nystagmus. At 3 months postoperatively, those features had improved, but subjective visual vertical was abnormal.
- *Worsening*:
  - Patient 7 worsened: Preoperatively was normal. At 3 months postoperatively, there was abnormal smooth pursuit.
  - Patient 9 worsened: Preoperatively was normal. At 3 months postoperatively, there was gaze evoked nystagmus, rebound nystagmus and abnormal smooth pursuit.
  - Patient 10 worsened: Preoperatively was normal. At 3 months postoperatively, there was abnormal smooth pursuit.

In short, for this series of CM1 patients studied by Goldschagg and colleagues, 1/10 (10%) of patients improved, 3/10 (30%) had no change, 3/10 (30%) developed new abnormalities, and 3/10 (30%) worsened. However, note that these patients underwent surgery that involved not only sub-occipital decompression, but also *unilateral cerebellar tonsillectomy*.

Table 99.4 Preoperative and postoperative outcomes in ten Chart 1 malformation patients

P	Pathological ocular motor findings			Pathological neurological findings		
	Before	After 1 week	After 3 months	Before	After 1 week	After 3 months
1	GEN r > l, PN l, HIT r	SP v and h	SP v and h	Finger-chase dysmetria b, Heel-shin slide dysmetria, dysesthesia in both hands	Finger-chase dysmetria l	Finger-chase dysmetria l
2	Strabismus sursoadductorius l	md	Unchanged	None	md	None
3	n	md	n	Finger-chase dysmetria l, Hypoesthesia l hand and arm	md	None
4	PN l	n	DBN	Atrophy and paresis of the interosseal muscles, finger-nose dysmetria l	Gait ataxia, atrophy and paresis unchanged, finger-nose dysmetria r > l	Improvement of paresis and atrophy of hand muscles, finger-chase dysmetria l
5	n	SVV r	n	Unsteady gait, hypesthesia right arm and leg	Unsteady stance, finger-chase dysmetria and diadochokinesis r	Finger-chase dysmetria
6	OTR r (head tilt, SVV), VFS v	n	n	Intention tremor l	Finger-chase dysmetria l	Hearing impairment l, finger-chase dysmetria l
7	n	n	SP v	Dysdiadochokinesis r, hypesthesia in both hands, pallhypesthesia	none	Lateral canal BPPV r
8	GEN h, RN, SP v and h, hypermetric saccades v and h, VFS h > v, HIT l and r, PN l, DBN	GEN h, RN, SP v and h, hypermetric saccades v and h, VFS h > v, PN l	GEN h, RN, SP h and v, hypermetric saccades h, VFS h > v, SVV l	Isolated ocular motor findings	md	Unchanged
9	n	md	GEN h, RN, SP v	None	md	4/5 paresis and hyperesthesia of the right upper and lower limb with reduced reflexes, dysdiadochokinesis r
10	n	SVV r	SP h	Paresis of triceps and interosseal muscles, finger-nose dysmetria l > r, hypesthesia in the fingers	md	Improvement of the paresis and hypesthesia

From Goldschagg et al. [43]

Abbreviations: *P* patient number, *r* right, *l* left, *b* both, *h* horizontal, *v* vertical, *n* none, *md* missing data, *GEN* Gaze-evoked nystagmus, *PN* head-shaking nystagmus, *HIT* head impulse test, *SP* smooth pursuit, *DBN* downbeat nystagmus, *SVV* deviation of the subjective visual vertical >2.5°, *OTR* ocular tilt reaction, *VFS* visual fixation suppression of the vestibulo-ocular reflex, *RN* rebound nystagmus

In terms of instrumented vestibular testing, Goldschagg and colleagues reported that 2/10 (20%) had abnormal asymmetry on cervical vestibular evoked myogenic potentials, and that these both improved postoperatively. They reported that 3/9 (33%) of patients had a pathological caloric asymmetry, and that this improved in 1/3 (33%) postoperatively.

Giammattei and colleagues [60] performed a retrospective review of 38 adult CM1 patients followed for ≥5 years. They reported that gait impairment improved in 83% of patients.

Type 1.5 Chiari Malformation (CM1.5)

Cesmebasi and colleagues [13] state that this category is intended to "address patients with Chiari type I malformation with an added component of brainstem and fourth ventricle elongation/caudal descent. Chiari type 1.5 patients have symptoms similar to Chiari type I patients and may have a higher incidence of developing syringomyelia."

Tubbs and colleagues [61] provided good sagittal MR images of what they classify as CM1.5.

Type 2 Chiari Malformation (CM2)

Chiari's original description was:

> A second type is formed by the elongation of parts of the cerebellum into the dilated vertebral canal within the elongated 4th ventricle, extending into the vertebral canal.

Cesmebasi and colleagues [13] list the following anatomical features of CM2:

Brain: Multiple anomalies at all levels (telencephalon, diencephalon, mesencephalon, metencephalon, myelencephalon).

- *Telencephalon*: There may be partial or complete agenesis of the corpus callosum, and partial or complete absence of the septum pellucidum. When the corpus callosum is absent, there may be interdigitation of the cerebral hemispheres. The cerebral cortex may have sulcal deformities. The olfactory bulb may be partially or completely absent. The cingulate gyrus may be absent. The internal aspect of the lateral ventricles may have a granular appearance due to heterotopic gray matter bulging into them.
- *Diencephalon*: 75–90% of patients exhibit an enlarged massa intermedia that is often anteriorly displaced. The hypothalamus may be elevated. The habenular commissure and the pineal gland may be elongated.
- *Mesencephalon*: The midbrain is usually elongated. The quadrigeminal plate is usually shortened. Tectal beaking may be present. The cerebral aqueduct may be deformed (stretched, laterally compressed, forked, stenotic, posteriorly kinked). Cranial nerve nuclei may be hypoplastic.
- *Metencephalon*: The cerebellum is smaller and may upwardly herniate over the tentorium cerebelli. The cerebellum is often laterally displaced. The vermis is herniated through the foramen magnum. Cerebellar folia may be absent or dysplastic. The lateral cerebellar edges may extent anteriorly toward the brainstem and basilar artery ("cerebellar inversion"). The pons is flattened and elongated. The ventral pontine surface may be indented by a displayed vertebral artery or by the anterior rim of the foramen magnum. There may be dysplasia of the basal pontine nuclei, tegmental nuclei and cranial nerve nuclei.
- *Myelencephalon*: The medulla may be flattened and elongated, giving it a "trumpet-like" appearance. About 70% of patients have a protrusion caudal to the gracile and cuneate tubercles at C2–C4 levels, sometimes termed a cervicomedullary kink, buckle, hump or spur.
- *Other*: The malformations tend to affect cranial and upper cervical nerves that take a vertical course. The vertebrobasilar system and its branches may be caudally displaced. The great vein of Galen may be elongated.
- *Ventricles/cisterns*: About 90% of patients have hydrocephalus. The lateral ventricles are often asymmetric. The frontal horns are more pointed.

Cerebellar tonsils: Herniation of the cerebellar vermis with caudal descent of the brainstem and fourth ventricle.

Skull: Craniolacunia/lückenschädel ("beaten copper" appearance of the bone). There may be scalloping of the anterior portion of the frontal bone, posteromedial aspect of the petrous part of the temporal bones and jugular tubercles. The scalloping of the petrous part of the petrous part of the temporal bone may shorten the internal auditory canal. The foramen magnum is usually enlarged. The posterior cranial fossa is smaller and flattened. The opisthion (mid-point on the posterior margin of the foramen magnum) may be notched; cranioschisis (congenital failure of the skull to completely close) is sometimes seen.

Meninges: Tentorium cerebelli is usually heart- or V-shaped, widened, hypoplastic and low-lying.

Spinal canal: There may be scalloping of the odontoid process, enlargement of the cervical spinal canal, incomplete formation of the posterior arch of the atlas, and Klippel–Feil anomaly.

Spinal cord: Most patients have a myelomeningocele. About 6% of these patients also have an associated split cord malformation (diastematomyelia).

Oculomotor Findings in CM2

Arnold and colleagues [62] studied 3 cases of CM2. All 3 patients had bilateral internuclear ophthalmoplegia, deficits in smooth pursuit, deficits in optokinetic responses, and vestibulo-ocular responses. Two had abduction paresis and deficits in smooth pursuit.

Longridge and Mallinson [63] studied a 29-year-old CM2 patient. On face-to-face oculomotor examination, there was subtle spontaneous right beat nystagmus; there was also right greater than left-sided gaze evoked nystagmus; up beat nystagmus on up gaze and down beat nystagmus on down gaze (in other words, both horizontal and vertical gaze evoked nystagmus), saccadic hypermetria and saccadic smooth pursuit. On dynamic visual acuity testing, she went from 20/20 to worse than 20/100.

Tubbs and colleagues [64] studied children with CM2 and concluded that "there is a correlation between the presence and severity of nystagmus and the severity of tectal beaking."

Instrumented Oculomotor Testing in CM2: Videonystagmography (VNG)

Salman and colleagues [65] studied saccades using videonystagmography (VNG) in 21 patients (age 8–19 years) with CM2 compared to 39 healthy controls. Surprisingly, they reported no differences in saccades between the CM2 and the control groups. In a subsequent publication [66], these investigators concluded that this normality was due to saccadic adaptation.

Salman and colleagues [67] studied smooth pursuit using videonystagmography in the same group of 21 patients compared to 38 healthy controls and reported that in the 8 patients with nystagmus, smooth pursuit gain was abnormal.

Instrumented Oculomotor Testing in CM2: Electronystagmography (ENG)

Longridge and Mallinson [63] studied a 29-year-old CM2 patient with electronystagmography (ENG). They reported bilateral horizontal gaze evoked nystagmus, subtle rebound nystagmus, saccadic hypermetria, saccadic smooth pursuit, abnormal optokinetic responses and abnormal visual fixation suppression.

Imaging in CM2

Piper and colleagues [17] provide a good sagittal MRI of a patient with CM2.

Type 3 Chiari malformation (CM3)

Chiari's original description, as translated by Ivashchuk and colleagues [68], was:

> Of the third type of serial changes of the cerebellum caused by chronic congenital hydrocephalus, there was only one case available to me. It demonstrated the greatest degree of displacement of the cerebellum, out of the cranial cavity through the foramen magnum into the vertebral canal, involving the displacement of nearly the entire cerebellum, which was itself hydrocephalic, into a cervical spina bifida.

Ivashchuk and colleagues [68] identified the following as the most salient anatomical features of the Chiari 3 malformation:

- Encephalocele/high cervical or occipital
- Cervical/occipital bony defect
- Small posterior cranial fossa
- Caudal displacement of hindbrain
- Hydrocephalus
- Aberrant venous drainage
- Syringomyelia

Cesmebasi and colleagues [13] state that, "Chiari III malformation are defined as findings of a Chiari II malformation with the addition of a posterior cranial fossa encephalocele that contains cerebellar and brainstem tissue." They list the following anatomical abnormalities associated with CM3:

Brain: In all cases, the cerebellum is herniated into a low occipital or high cervical encephalocele. Some patients also have herniation of the brainstem, occipital lobes, dural venous sinuses, and subarachnoid or ventricular cerebrospinal fluid spaces. Additionally, partial or complete dysplasia of the corpus callosum is seen.

Skull: Scalloping of the dorsal clivus and petrous ridge. Occasionally the parietal bones are hypoplastic. There is usually an enlarged foramen magnum.

Spine: Generally CM3 patients have osseous deformities of the upper cervical spine. Some have posterior cervical agenesis.

Oculomotor Findings in CM3

Gorgoglione and colleagues [69] reported the case of an infant with CM3 who exhibited "nystagmus," which they did not describe in any further detail.

Imaging in CM3

Figure 99.8, from Ivashchuk and colleagues [68], shows sagittal and coronal views of a patient with CM3.

Type 3.5 Chiari Malformation (CM3.5)

In 1894, Dr. Giuseppe Muscatello [70] published a case that he described as a variant of CM3. Fisahn and colleagues [71] viewed this case as unique ("the only reported case") and proposed calling it "Chiari 3.5."

Figure 99.9, from Fisahn and colleagues [71], is Muscatello's [70] original illustration.

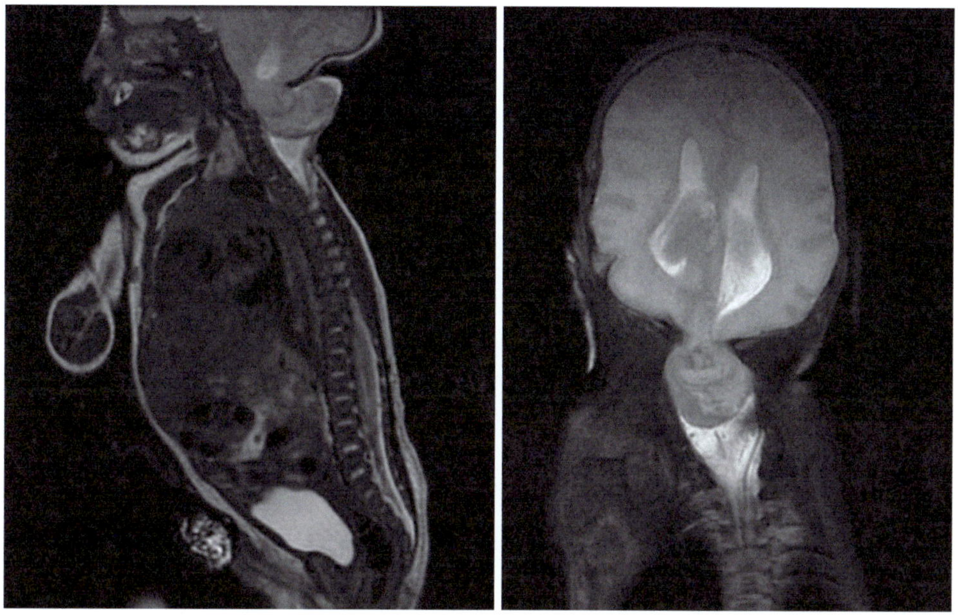

Fig. 99.8 Sagittal MRI (left panel) and coronal MRI (right panel) views of a patient with a Chiari malformation type 3. (From From Ivashchuk et al. [68]. Reproduced with written permission from Springer Nature B.V.)

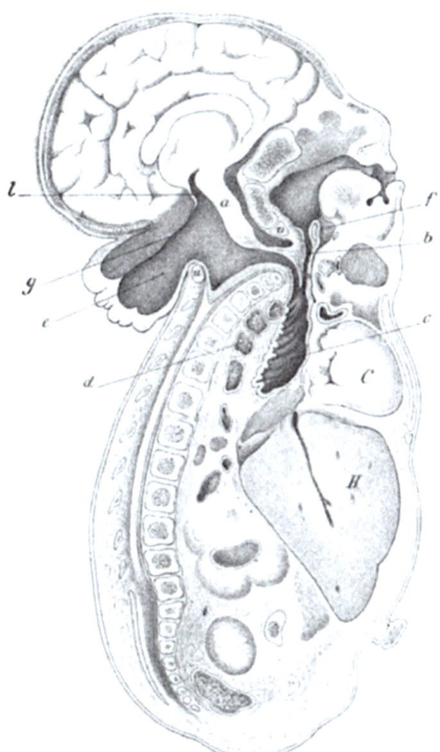

Fig. 99.9 Giuseppe Muscatello's (1894) illustration of a variant of a Chiari malformation type 3. Sagittal section through the head, neck, and torso of the case reported by Muscatello. a = Pons and medulla oblongata; b = lumen of the esophagus; c = stomach; d = intestines; e = cave of the encephalomyelocystocele with connection to the esophagus; f = glottis; g = anterior medullary velum; l = corpora quadrigemina; G = heart; H = liver. (Fisahn et al. [71] propose calling this "Chiari 3.5." Reproduced with written permission from Springer Nature)

Type 4 Chiari Malformation (CM4)

Chiari's original description [4], as translated by Tubbs and colleagues [5], was:

Medical history: The encephalocele consists of a cylindrical, broad based, 6 cm long, and 3 cm wide skin sac. The sac has an ulcerated opening measuring about 8 cm. Through this opening, another, 4 cm wide inner sac, similarly ulcerated at the end, protrudes. That second, inner sac is made up of meninges with longitudinally oriented folds, almost of the size of a little finger. The inner sac communicates with the intracranial compartment through a 1 cm^2 wide opening located just below the posterior fontanel. The outer sac is contingent with the scalp. In surgery, the inner sac was disconnected and resected at the intracranial opening and closed with the excess skin overlying it. The patient passed 36 h after surgery.

Pathological-anatomical diagnosis: Meningitis cerebralis suppurativa et suppuratio in ventriculis cerebri ex ulceratione hydrencephaloceles occipitalis (horas XXXVI ante mortem amputatae). Dilatatio congenita ventriculorium cerebri. Pneumonia lobularis bilateralis. Sclerema textus adipose subcutanei.

Dissection of the central nervous system: The child's head is disproportionally large compared to its height (47 cm), just a little more narrow. Hair shaved. Inferior to the posterior fontanel there is a skin bulge measuring 6 × 3 × 2 cm with a 6 cm long closed incision at the top. The edges are edematous and erythematous.

The soft skull is ecchymotic, pale, in the area of the bulge edematous, serous, and purulent. Circumferentially the skull measures 27 cm. The sutures are split. The sagittal suture is 0.5 cm wide. The anterior fontanel is larger measuring 4 cm in length and 2.5 cm in width. The posterior fontanel is normal in size. The skull bones are fairly rigid. In the squama ossis occiptialis, 2 cm inferior to the posterior fontanel, there is an opening of the size of the nail of a small finger. The opening is surrounded by the split of the processus sagittalis major of the pachymenix. Through that opening a hemorrhagic and necrotic mass of inner meninges and brain parenchyma are extruding. In the sinus durae matris, there is dark, liquid, and also freshly coagulated blood. The sinus sagittalis major is split into two sinuses from the area of the posterior fontanel that diverge to form the sinus transversi. The tentorium cerebelli is narrow, the processus falciformis minor is absent, the posterior fossa is small. The inner surface of the skull base has deep impressiones digitatae. The foramen occipital magnum is remarkably narrow. The inner meninges are pale and infiltrated by serous and purulent liquid.

The gyri and sulci of the cerebrum appear normal. The lateral ventricles and third ventricle are enlarged and contain serous and purulent liquid. The brain matter is pale, moist, and soft. Originating from the precuneus are two processes of brain matter that contain the occipital horns and plexus choroidei of the lateral ventricles and protrude through the previously mentioned opening.

The cerebellum is only half the size of a normal cerebellum. It is 3.5 cm wide and 2 cm long. The largest diameter from the upper to the lower surface is 3 cm. It is asymmetric with the left hemisphere significantly smaller than the right. The lobi superiores, posteriores et inferiores are easily distinguished, the tonsils and flocculi are rudimentary. The inferior vermis is more prominent and the incisura marginalis posterior is absent. The pons and medulla oblongata are 14 and 11 mm long, respectively. Both are thin.

A median sagittal section of the cerebellum, pons, medulla oblongata, and upper cervical spinal cord shows unusual rolling of the lower portion of the arbor vitae resulting in a deep location of the pyramis (a), high location of the uvula (b), and location of the nodulus (c) of the inferior vermis on the upper end of the rolling of the arbor vitae. The dorsal wall of the fourth ventricle (d) is bowed due to the rolling and the velum medullare posterius (e) is very long. The entry of the tela choroidea ventriculi IV is well demarcated (f) and at the usual location. The ponticuli (g) are longer and thicker, as is the velum medullare posterius. The corpora dentata cerebelli are smaller, otherwise normally shaped. The pons and medulla are normal.

The spinal cord is of dense consistency. The inner meninges of medulla oblongata, pons, and cerebellum are not purulent.

Microscopic examination of the encephalocele occipitalis, vermis, pons, medulla oblongata, corpus dentatum of the right cerebellar hemisphere, the fifth cervical segment, fifth and ninth thoracic segments, and third segment of the spinal cord is performed after solidification with liquor mülleri.

The inner sac of the encephalocele has two layers. The outer layer is highly vascularized and hemorrhagic as are the inner meninges while the inner layer is of sclerotic brain matter. The latter makes up the previously mentioned longitudinally oriented folds. The inner surface of the sac is lined with cylindrical epithelium.

The vermis, pons, and medulla oblongata are examined using the right half of the section. The vermis has normal texture. Its cortex has an external granular layer, a molecular layer, and an inner granular layer. There are numerous Purkinje cells. The medulla is mostly black on copper hematoxylin stain. The pons and medulla oblongata are normal. Only the pyramids are devoid of medullary substance. The fourth ventricle is lined with normal epithelium. The elongated velum medullare posterius and taenia plexus chorioidei ventriculi IV are sclerotic.

The corpus dentatum cerbelli has prominent folding and numerous multipolar ganglion cells. The medulla spinalis has a mild dilatation of the central canal lined with epithelium and the pyramidal tract is devoid of medullary substance.

Tubbs and colleagues [5] translated sections of this paper and identified the following as the most salient anatomical features of the Chiari 4 malformation:

- Occipital encephalocele (through the posterior fontanelle) with occipital lobe and horn as contents
- Splitting of the superior sagittal sinus around the occipital encephalocele
- Hydrocephalus
- Cerebellar hypoplasia but *not* aplasia

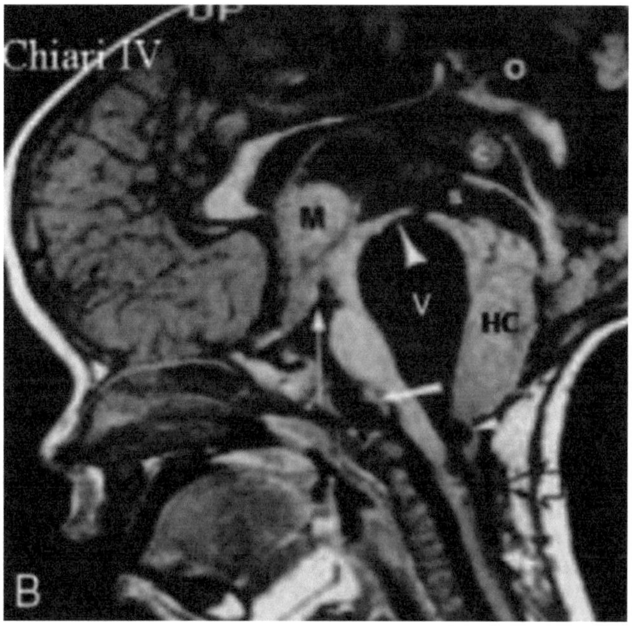

Fig. 99.10 Sagittal MRI of Chiari malformation type 4. Abbreviations: C cerebellum, HC hypoplastic cerebellum; V fourth ventricle. (From Singh et al. [72]. Reproduced with written permission from Wolters Kluwer Health Limited)

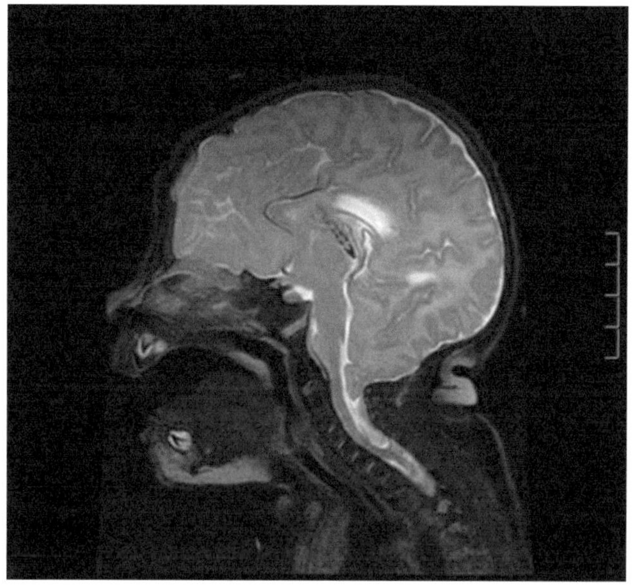

Fig. 99.11 Sagittal MRI showing herniation of the occipital lobe through the foramen magnum, and absence of the cerebellum. Tubbs et al. [73] proposed designating this a Chiari malformation type 5. (Reproduced with written permission from Springer Nature B.V.)

- Absence of falx cerebelli
- Narrowed tentorium cerebelli
- Small posterior fossa
- Narrow foramen magnum
- Elongated inferior medullary velum "Rolled" inferior vermis
- Indentation of the posterior wall of the fourth ventricle

Figure 99.10, from Singh and colleagues [72], shows a CM4.

Type 5 Chiari Malformation (CM5)

Tubbs and colleagues [73] reported the case of an infant with cerebellar agenesis and herniation of the occipital lobe through the foramen magnum. They proposed designating this "Chiari 5 malformation."

Figure 99.11, from Tubbs and colleagues [73], shows herniation of the occipital lobes through the foramen magnum, and absence of the cerebellum.

References

1. Carmel PW, Markesbery WR. Early descriptions of the Arnold-Chiari malformation. The contribution of John Cleland. J Neurosurg. 1972;37(5):543–7.
2. Cleland J. Contribution to the study of spina bifida, encephalocele, and anencephalus. J Anat Physiol. 1883;17(Pt 3):257–92.
3. Chiari H. Ueber Veränderungen des Kleinhirns infolge von Hydrocephalie des Grosshirns [Regarding changes of the cerebellum from cerebral hydrocephalus]. Dtsch Med Wochenschr. 1891;17(42):1172–5.
4. Chiari H. Über Veränderungen des Kleinhirns, des Pons und der Medulla oblongata in Folge von congenitaler Hydrcephalie des Grosshirns. Denkschr Akad Wiss Wien. 1895;63:71–116.
5. Tubbs RS, Demerdash A, Vahedi P, Griessenauer CJ, Oakes WJ. Chiari IV malformation: correcting an over one century long historical error. Childs Nerv Syst. 2016;32(7):1175–9.
6. Arnold J. Myelocyste, transposition von gewebskeimen und sympodie. Beitr Path Anat. 1894;16:1–28.
7. Schwalbe E, Gredig M. Ueber Entwicklungsstörungen des Kleinhirns, Hirnstamms und Halsmarks bei Spina bifida (Arnold'sche und Chiari'sche Missbildung) [On developmental disorders of the cerebellum, brainstem and cervical spinal cord in spina bifida (Arnold's and Chiari's malformations)]. Beitr Path Anat. 1907;40:132–94.

8. Russell GE, Wick B, Tang RA. Arnold-Chiari malformation. Optom Vis Sci. 1992;69(3):242–7.
9. Horn SR, Shepard N, Vasquez-Montes D, Bortz CA, Segreto FA, De La Garza RR, et al. Chiari malformation clusters describe differing presence of concurrent anomalies based on Chiari type. J Clin Neurosci. 2018;58:165–71.
10. McRae DL. Bony abnormalities in the region of the foramen magnum: correlation of the anatomic and neurologic findings. Acta Radiol. 1953;40(2–3):335–54.
11. Bogdanov EI, Faizutdinova AT, Heiss JD. The small posterior cranial fossa syndrome and Chiari malformation type 0. J Clin Med. 2022;11(18):5472.
12. Morgenstern PF, Tosi U, Uribe-Cardenas R, Greenfield JP. Ventrolateral tonsillar position defines novel Chiari 0.5 classification. World Neurosurg. 2020;136:444–53.
13. Cesmebasi A, Loukas M, Hogan E, Kralovic S, Tubbs RS, Cohen-Gadol AA. The Chiari malformations: a review with emphasis on anatomical traits. Clin Anat. 2015;28(2):184–94.
14. Barkovich AJ, Wippold FJ, Sherman JL, Citrin CM. Significance of cerebellar tonsillar position on MR. AJNR Am J Neuroradiol. 1986;7(5):795–9.
15. Goodwin D, Halvorson AR. Chiari I malformation presenting as downbeat nystagmus: clinical presentation, diagnosis, and management. Optometry. 2012;83(2):80–6.
16. Pokharel D, Siatkowski RM. Progressive cerebellar tonsillar herniation with recurrent divergence insufficiency esotropia. J AAPOS. 2004;8(3):286–7.
17. Piper RJ, Pike M, Harrington R, Magdum SA. Chiari malformations: principles of diagnosis and management. BMJ. 2019;365:l1159.
18. Milhorat TH, Chou MW, Trinidad EM, Kula RW, Mandell M, Wolpert C, et al. Chiari I malformation redefined: clinical and radiographic findings for 364 symptomatic patients. Neurosurgery. 1999;44(5):1005–17.
19. Kumar A, Patni AH, Charbel F. The Chiari I malformation and the neurotologist. Otol Neurotol. 2002;23(5):727–35.
20. Rydell RE, Pulec JL. Arnold-Chiari malformation. Neuro-otologic symptoms. Arch Otolaryngol. 1971;94(1):8–12.
21. Hida K, Iwasaki Y, Koyanagi I, Abe H. Pediatric syringomyelia with chiari malformation: its clinical characteristics and surgical outcomes. Surg Neurol. 1999;51(4):383–90; discussion 90–1.
22. Shaikh AG, Ghasia FF. Neuro-ophthalmology of type 1 Chiari malformation. Expert Rev Ophthalmol. 2015;10(4):351–7.
23. Liebenberg WA, Georges H, Demetriades AK, Hardwidge C. Does posterior fossa decompression improve oculomotor and vestibulo-ocular manifestations in Chiari 1 malformation? Acta Neurochir. 2005;147(12):1239–40; discussion 40.
24. Chait GE, Barber HO. Arnold-Chiari malformation--some otoneurological features. J Otolaryngol. 1979;8(1):65–70.
25. Faria MA Jr, Spector RH, Tindall GT. Downbeat nystagmus as the salient manifestation of the Arnold-Chiari malformation. Surg Neurol. 1980;13(5):333–6.
26. Hosford DA, Spector RH. Arnold-Chiari malformation in a geriatric patient. South Med J. 1985;78(5):618–21.
27. Pieh C, Gottlob I. Arnold-Chiari malformation and nystagmus of skew. J Neurol Neurosurg Psychiatry. 2000;69(1):124–6.
28. Pedersen RA, Troost BT, Abel LA, Zorub D. Intermittent downbeat nystagmus and oscillopsia reversed by suboccipital craniectomy. Neurology. 1980;30(11):1239–42.
29. Yee RD, Baloh RW, Honrubia V. Episodic vertical oscillopsia and downbeat nystagmus in a Chiari malformation. Arch Ophthalmol. 1984;102(5):723–5.
30. Hatashita S, Shimoji T, Shimizu A, Ishizawa A, Tanaka M. A case of up-beat nystagmus associated with Chiari malformation type I (author's transl). No Shinkei Geka. 1980;8(10):989–94.
31. Biousse V, Newman NJ, Petermann SH, Lambert SR. Isolated comitant esotropia and Chiari I malformation. Am J Ophthalmol. 2000;130(2):216–20.
32. Defoort-Dhellemmes S, Denion E, Arndt CF, Bouvet-Drumare I, Hache JC, Dhellemmes P. Resolution of acute acquired comitant esotropia after suboccipital decompression for Chiari I malformation. Am J Ophthalmol. 2002;133(5):723–5.
33. Imes RK, Quinn TA. Acute comitant esotropia in Chiari 1 malformation. Ophthalmology. 2001;108(5):834.
34. Kowal L, Yahalom C, Shuey NH. Chiari 1 malformation presenting as strabismus. Binocul Vis Strabismus Q. 2006;21(1):18–26.
35. Lewis AR, Kline LB, Sharpe JA. Acquired esotropia due to Arnold-Chiari I malformation. J Neuro-ophthalmol. 1996;16(1):49–54.
36. Passo M, Shults WT, Talbot T, Palmer EA, Acquired esotropia. A manifestation of Chiari I malformation. J Clin Neuroophthalmol. 1984;4(3):151–4.
37. Weeks CL, Hamed LM. Treatment of acute comitant esotropia in Chiari I malformation. Ophthalmology. 1999;106(12):2368–71.
38. Hentschel SJ, Yen KG, Lang FF. Chiari I malformation and acute acquired comitant esotropia: case report and review of the literature. J Neurosurg. 2005;102(4 Suppl):407–12.
39. Ghasia FF, Gulati D, Westbrook EL, Shaikh AG. Viewing condition dependence of the gaze-evoked nystagmus in Arnold Chiari type 1 malformation. J Neurol Sci. 2014;339(1–2):134–9.
40. Korres S, Balatsouras DG, Zournas C, Economou C, Gatsonis SD, Adamopoulos G. Periodic alternating nystagmus associated with Arnold-Chiari malformation. J Laryngol Otol. 2001;115(12):1001–4.
41. Al-Awami A, Flanders ME, Andermann F, Polomeno RC. Resolution of periodic alternating nystagmus after decompression for Chiari malformation. Can J Ophthalmol. 2005;40(6):778–80.
42. Zimmerman CF, Roach ES, Troost BT. See-saw nystagmus associated with Chiari malformation. Arch Neurol. 1986;43(3):299–300.
43. Goldschagg N, Feil K, Ihl F, Krafczyk S, Kunz M, Tonn JC, et al. Decompression in Chiari malformation: clinical, ocular motor, cerebellar, and vestibular outcome. Front Neurol. 2017;8:292.
44. Mossman SS, Bronstein AM, Gresty MA, Kendall B, Rudge P. Convergence nystagmus associated with Arnold-Chiari malformation. Arch Neurol. 1990;47(3):357–9.
45. Kim SH, Kim HJ, Kim JS. Perverted downward corrective saccades during horizontal head impulses in Chiari malformation. Cerebellum. 2019;18(3):333–9.
46. Baloh RW, Beykirch K, Tauchi P, Yee RD, Honrubia V. Ultralow vestibuloocular reflex time constants. Ann Neurol. 1988;23(1):32–7.
47. Muthusamy P, Matte G, Kosmorsky G, Chemali KR. Chiari type I malformation: a mimicker of myasthenia gravis. Neurologist. 2011;17(2):86–8.
48. Currie JN, Matsuo V. The use of clonazepam in the treatment of nystagmus-induced oscillopsia. Ophthalmology. 1986;93(7):924–32.
49. Young YH, Huang TW. Role of clonazepam in the treatment of idiopathic downbeat nystagmus. Laryngoscope. 2001;111(8):1490–3.
50. Dieterich M, Straube A, Brandt T, Paulus W, Buttner U. The effects of baclofen and cholinergic drugs on upbeat and downbeat nystagmus. J Neurol Neurosurg Psychiatry. 1991;54(7):627–32.
51. Averbuch-Heller L, Tusa RJ, Fuhry L, Rottach KG, Ganser GL, Heide W, et al. A double-blind controlled study of gabapentin and baclofen as treatment for acquired nystagmus. Ann Neurol. 1997;41(6):818–25.
52. Strupp M, Schuler O, Krafczyk S, Jahn K, Schautzer F, Buttner U, et al. Treatment of downbeat nystagmus with 3,4-diaminopyridine: a placebo-controlled study. Neurology. 2003;61(2):165–70.

53. Claassen J, Spiegel R, Kalla R, Faldon M, Kennard C, Danchaivijitr C, et al. A randomised double-blind, cross-over trial of 4-aminopyridine for downbeat nystagmus--effects on slowphase eye velocity, postural stability, locomotion and symptoms. J Neurol Neurosurg Psychiatry. 2013;84(12):1392–9.
54. Stahl JS, Plant GT, Leigh RJ. Medical treatment of nystagmus and its visual consequences. J R Soc Med. 2002;95(5):235–7.
55. Straube A, Leigh RJ, Bronstein A, Heide W, Riordan-Eva P, Tijssen CC, et al. EFNS task force--therapy of nystagmus and oscillopsia. Eur J Neurol. 2004;11(2):83–9.
56. Kumar A, Thomas S, McLean R, Proudlock FA, Roberts E, Boggild M, et al. Treatment of acquired periodic alternating nystagmus with memantine: a case report. Clin Neuropharmacol. 2009;32(2):109–10.
57. Albers FW, Ingels KJ. Otoneurological manifestations in Chiari-I malformation. J Laryngol Otol. 1993;107(5):441–3.
58. Gingold SI, Winfield JA. Oscillopsia and primary cerebellar ectopia: case report and review of the literature. Neurosurgery. 1991;29(6):932–6.
59. Dones J, De Jesus O, Colen CB, Toledo MM, Delgado M. Clinical outcomes in patients with Chiari I malformation: a review of 27 cases. Surg Neurol. 2003;60(2):142–7; discussion 7–8.
60. Giammattei L, Messerer M, Daniel RT, Aghakhani N, Parker F. Long-term outcome of surgical treatment of Chiari malformation without syringomyelia. J Neurosurg Sci. 2020;64(4):364–8.
61. Tubbs RS, Iskandar BJ, Bartolucci AA, Oakes WJ. A critical analysis of the Chiari 1.5 malformation. J Neurosurg. 2004;101(2 Suppl):179–83.
62. Arnold AC, Baloh RW, Yee RD, Hepler RS. Internuclear ophthalmoplegia in the Chiari type II malformation. Neurology. 1990;40(12):1850–4.
63. Longridge NS, Mallinson AI. Arnold-Chiari malformation and the otolaryngologist: place of magnetic resonance imaging and electronystagmography. Laryngoscope. 1985;95(3):335–9.
64. Tubbs RS, Soleau S, Custis J, Wellons JC, Blount JP, Oakes WJ. Degree of tectal beaking correlates to the presence of nystagmus in children with Chiari II malformation. Childs Nerv Syst. 2004;20(7):459–61.
65. Salman MS, Sharpe JA, Eizenman M, Lillakas L, To T, Westall C, et al. Saccades in children with spina bifida and Chiari type II malformation. Neurology. 2005;64(12):2098–101.
66. Salman MS, Sharpe JA, Eizenman M, Lillakas L, To T, Westall C, et al. Saccadic adaptation in Chiari type II malformation. Can J Neurol Sci. 2006;33(4):372–8.
67. Salman MS, Sharpe JA, Lillakas L, Steinbach MJ, Dennis M. Smooth ocular pursuit in Chiari type II malformation. Dev Med Child Neurol. 2007;49(4):289–93.
68. Ivashchuk G, Loukas M, Blount JP, Tubbs RS, Oakes WJ. Chiari III malformation: a comprehensive review of this enigmatic anomaly. Childs Nerv Syst. 2015;31(11):2035–40.
69. Gorgoglione ML, Laera R, Curcio A, Caruso G, Delia G, Germano AF. Treatment of Chiari III malformation in infant with 4K 3D ORBEYE exoscope. World Neurosurg. 2023;171:144.
70. Muscatello G. Ueber die angeborenen Spalten des Schädels und der Wirbelsäule [About the hereditary clefts of the skull and the spine]. Arch Klin Chir. 1894;47:162–301.
71. Fisahn C, Shoja MM, Turgut M, Oskouian RJ, Oakes WJ, Tubbs RS. The Chiari 3.5 malformation: a review of the only reported case. Childs Nerv Syst. 2016;32(12):2317–9.
72. Singh R, Arora R, Kumar R. Clinical notes on Chiari malformation. J Craniofac Surg. 2018;29(4):e417–e21.
73. Tubbs RS, Muhleman M, Loukas M, Oakes WJ. A new form of herniation: the Chiari V malformation. Childs Nerv Syst. 2012;28(2):305–7.

Basilar Invagination

100

Overview

The phrase "*basilar invagination*" usually refers to a craniocervical junction abnormality in which the odontoid process ascends excessively. This anatomical configuration can impinge on the medulla and cause a variety of symptoms that localize to the brainstem (usually the medulla) or craniocervical junction. Basilar invagination sometimes presents with nystagmus [1, 2], and specifically down beat nystagmus [3, 4]. Management is surgical decompression.

There is very little literature regarding the ocular motor findings in patients with basilar invagination, so we shall illustrate this with a case.

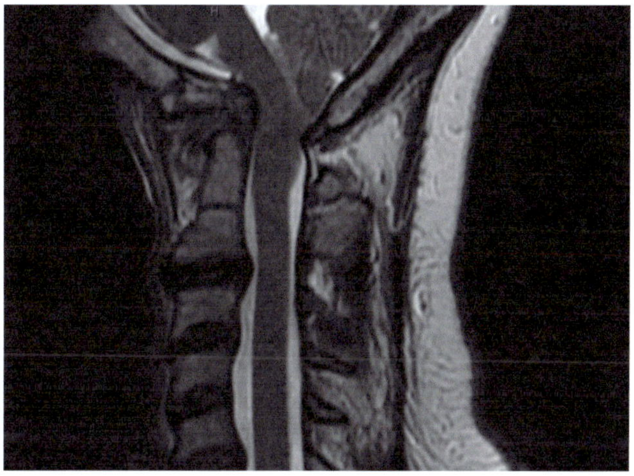

Fig. 100.1 Mid-sagittal T2 cervical spine MRI demonstrating that the inferior clivus protrudes posteriorly and impinges on the medulla

Case

A patient who carried a diagnosis of basilar invagination was referred to us to be evaluated for chronic, gradually progressive unsteadiness when upright.

In general, neurological examination station was characterized by a wide base; gait appeared mildly ataxic; myotatic reflexes were diffusely brisk, with positive Hoffman's and Troemner's signs bilaterally, though plantar responses were flexor.

Cervical spine MRI is shown in Figs. 100.1 and 100.2.

Videonystagmography showed spontaneous down beat nystagmus that dramatically augmented on lateral gaze, as displayed in Figs. 100.3, 100.4, and 100.5.

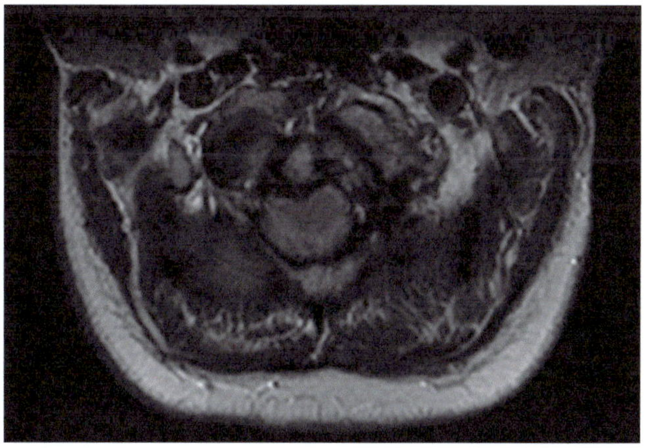

Fig. 100.2 Axial T2 cervical MRI demonstrating that the cerebellar tonsils descend through the foramen magnum, resulting in crowding and no visible cerebrospinal fluid around the cord

M. Cherchi, *Otoneurology and Vestibular Medicine*, https://doi.org/10.1007/978-3-031-94841-1_100

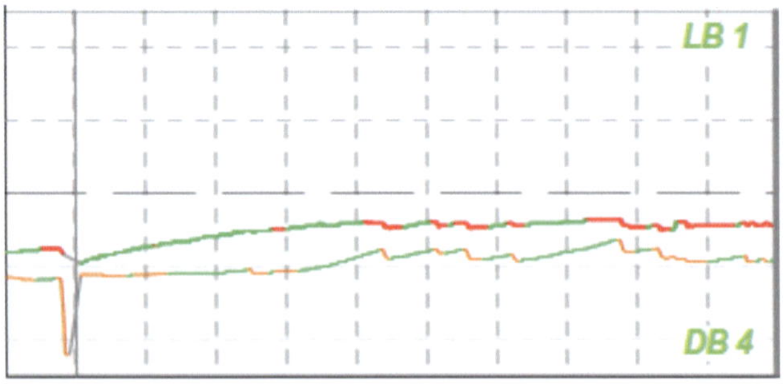

Fig. 100.3 Videonystagmogram tracing in a case of basilar invagination in which there is spontaneous down beat nystagmus on primary position of gaze. Abbreviation: DB down beat nystagmus

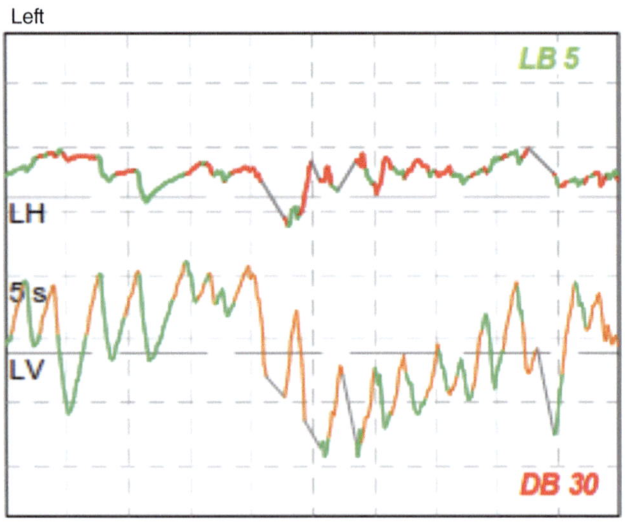

Fig. 100.4 Videonystagmogram tracing in a case of basilar invagination in which left gaze significantly augmented the baseline down beat nystagmus. Abbreviations: DB down beat nystagmus, LH left eye horizontal tracing, LV left eye vertical tracing

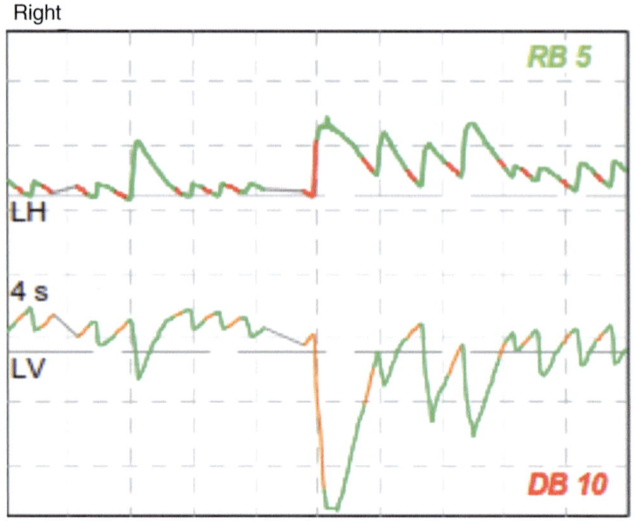

Fig. 100.5 Videonystagmogram tracing in a case of basilar invagination in which right gaze significantly augmented the baseline down beat nystagmus. Abbreviations: DB down beat nystagmus, LH left eye horizontal tracing, LV left eye vertical tracing

References

1. de Oliveira Filho IT, Romero PC, Fontoura EAF, Botelho RV. Chiari malformation and types of basilar invagination with/without syringomyelia. Surg Neurol Int. 2019;10:206.
2. Ridder T, Anderson RC, Hankinson TC. Ventral decompression in Chiari malformation, basilar invagination, and related disorders. Neurosurg Clin N Am. 2015;26(4):571–8.
3. Pratiparnawatr P, Tiamkao S, Tanapaisal C, Kanpittaya J, Jitpimolmard S. Downbeating nystagmus and postural hypotension due to basilar invagination. J Med Assoc Thail. 2000;83(12):1530–4.
4. Yeow YK, Tjia TL. The localizing value of downbeat nystagmus. Singapore Med J. 1989;30(3):273–6.

Paraneoplastic Syndrome

101

Overview

Paraneoplastic syndromes arise when the immune system's recognition of a malignancy results in misdirected immune-mediated damage of healthy tissues. Although any part of the body, and any part of the nervous system, can be affected, what usually comes to the attention of an otoneurologist is a paraneoplastic cerebellar disorder (which usually presents with ataxia), sometimes with a cochleovestibular component (which can present with vertigo, hearing loss, and tinnitus). These symptoms can begin rapidly, or evolve gradually, and they may precede recognition of the cancer by months or even years. Paraneoplastic syndromes have been described with a broad variety of malignancies and antibodies. Physical examination usually shows signs of appendicular and/or midline cerebellar dysfunction; ocular motor abnormalities usually include spontaneous down beat nystagmus, but broad array of ocular motor abnormalities have been reported. These ocular motor abnormalities may be more easily detected and characterized with instrumented testing. Other vestibular tests will not confirm a paraneoplastic syndrome, but may help exclude alternative diagnoses. Brain MRI sometimes will show cerebellar atrophy, less commonly enhancement. The differential diagnosis includes other cerebellar diseases. In a patient with stigmata of malignancy (e.g., weight loss, blood in stool, hemoptysis) in whom no cancer has yet been identified, it is reasonable to consider a paraneoplastic etiology of symptoms. If an apparently paraneoplastic syndrome occurs, then a search for malignancy should be undertaken, usually with CT of the chest/abdomen/pelvis without and with oral and intravenous contrast. If that is unrevealing, then a full-body FDG-PET scan is reasonable. The main treatment is removal of the cancer. If that fails, then there are no clear guidelines about what to try next, though a variety of immunosuppressive/immunomodulatory strategies have been attempted. Prognosis is unfavorable.

Introduction

When the immune system identifies "non-self" antigens, it may mount an immune response. "Non-self" antigens may be present on infectious agents (viruses, bacteria), and the immune system response serves the purpose of destroying the invading pathogen. Cancerous cells, which are the body's own cells whose genes have mutated, may express surface antigens that to the immune system appear "non-self," and the immune system response serves the purpose of destroying the cancer cells (sometimes referred to as "immune surveillance").

In some cases, the "non-self" antigens (in the infectious agent or on the cancer cell) are similar to antigens of the body's own healthy cells—sometimes called "molecular mimicry." This is problematic because the immune system response to the infectious agent or cancer cell will mistakenly also attack the body's own healthy cells. When this misdirected immune attack is triggered by an infectious agent, it is called a "para-infectious" phenomenon. When this misdirected immune attack is triggered by a cancer, it is called a "paraneoplastic" phenomenon.

These immune-mediated phenomena can attack any tissue in the body, including the nervous system. They can attack any part of the peripheral or central nervous system [1]. Within the central nervous system, the cerebellum is often affected; when triggered by an infectious agent, this is called "para-infectious cerebellitis"; when triggered by a malignancy, this is called "*paraneoplastic cerebellar disease*."

In this section, we will focus on paraneoplastic syndromes primarily or exclusively involving the cerebellum and audiovestibular system.

M. Cherchi, *Otoneurology and Vestibular Medicine*, https://doi.org/10.1007/978-3-031-94841-1_101

Epidemiology

The prevalence of paraneoplastic syndromes is unknown [1], and estimates vary dramatically, from 1 per 100,000 cancer patients (0.01%) to 553 per 60,000 cancer patients (0.9%)—a difference of nearly two orders of magnitude. Paraneoplastic syndromes have been reported with many different types of cancers.

Pathophysiological Mechanism of Disease

In a paraneoplastic syndrome, an immune response, whether humoral or cell-mediated [1], erroneously attacks the body's own healthy cells, presumably because of similarity between the epitopes of the cancer and the epitopes of healthy cells. A variety of antibodies have been identified.

Shams'ili and colleagues [2] studied 50 patients with paraneoplastic cerebellar degeneration, and identified antibodies listed in Table 101.1.

Clinical Presentation

Paraneoplastic syndromes have a wide range of tempo in symptom onset, ranging from hours to months.

Hammami and colleagues [3] studied 26 patients with paraneoplastic syndromes and reported the chronology, in months after initial presentation, at which any of the following began: hearing impairment, acute vertigo, rhombencephalitis, encephalomyelitis, wheelchair dependence, tumor diagnosis, and death. We have extracted and reanalyzed their data, and present the following observations:

- Onset of acute vertigo is relatively "front-loaded" in that it is a common presenting symptom (13 patients, 50%), but otherwise occurs within about 5 months after initial onset of some other symptom.
- Onset of hearing impairment is a common presenting symptom (9 patients, 35%), but otherwise occurs within about 20 months of initial onset after some other symptom.

Table 101.1 Paraneoplastic antibodies in paraneoplastic cerebellar degeneration

Antibody	Clinical syndromes	Immunohistochemistry	Western blot cerebellar extract	Gene	Associated cancer
Anti-Yo	Cerebellar ataxia	Cytoplasm of Purkinje cells and large brainstem neurons	34, 52 and 62 kDa	cdr34, cdr62-1, cdr62-2	Ovarian, breast
Anti-Hu	Cerebellar ataxia, PEM/SN	Nuclei of all neurons, nucleolar sparing	35 ± 40 kDa	HuD, HuC, Hel-N1	SCLC
Anti-Ri	Cerebellar ataxia, OM	Nuclei of all central neurons, with nucleolar sparing	55 and 80 kDa	NOVA-1, NOVA-2	Breast, gynecological, SCLC
Anti-Tr	Cerebellar ataxia	Cytoplasm and dendrites of Purkinje cells	–	Unknown	Hodgkin's lymphoma
Anti-VGCC	Cerebellar ataxia, LEMS	–	–	CACNA1A	SCLC (60%)
Anti-Ma	Cerebellar ataxia, brainstem dysfunction	Nuclei and cytoplasm of neurons	37 and 40 kDa	Ma1-5	Many
Anti-Ta/Ma2	Limbic encephalopathy, cerebellar ataxia	Nuclei and cytoplasm of neurons	40 kDa	Ma2	Testis
Anti-CRMP5/CV2	PEM/SN, cerebellar ataxia	Cytoplasm of oligodendrocytes	66 kDa	CRMP5	SCLC, thymoma, gynecological
Anti-mGluR1	Cerebellar ataxia	Cytoplasm of Purkinje cells and brush cells, climbing fibers	–	MGluR1	Hodgkin's lymphoma

From Shams-ilii et al. [2]
Abbreviations: *OM* opsoclonus/myoclonus, *VGCC* voltage-gated calcium channels, *LEMS* Lambert–Eaton myasthenic syndrome

- Rhombencephalitis is a somewhat common presenting symptom (6 patients, 23%), but otherwise occurs within about 15 months after initial onset of some other symptom.
- Encephalomyelitis is a somewhat uncommon presenting symptom (4 patients, 15%), but otherwise occurs within about 6 months after initial onset of some other symptom.
- Wheelchair dependence is never the presenting symptom. Its onset is fairly evenly distributed over about 32 months after initial symptom onset, but can occur as late as 48 months.
- Tumor diagnosis is fairly evenly distributed over about 29 months after initial onset.

Physical Examination

The neurological symptoms of patients with paraneoplastic syndromes often manifests with cerebellar symptoms such as ataxia, dysarthric speech, and abnormal eye movements.

Audiologic Evaluation

While not common, paraneoplastic syndromes can cause hearing loss, and on audiometry, this can be unilateral, bilateral and symmetric, or bilateral and asymmetrical.

Hammami and colleagues [3] studied 26 patients with "cochleovestibular" paraneoplastic manifestations, and stated that 16/26 (62%) of these patients presented with both cochlear and vestibular symptoms, 8/26 (31%) presented only with hearing loss, and 2/26 (8%) presented only with tinnitus. Of the patients with hearing loss, 13/26 (50%) had asymmetrical progression.

Kattah and colleagues [4] studied a 45-year-old man with a testicular seminoma whose paraneoplastic syndrome included bilateral, modestly asymmetrical, high greater than low frequency sensorineural hearing loss.

The hearing loss in paraneoplastic syndromes depends on the presence of malignancy and thus can occur at any age. Since cancer tends to be more a disease of adults, it is unsurprising that a larger proportion of these cases occur in older age [5], though it can occur in middle age [4] or even early childhood. For example, paraneoplastic hearing loss has been identified in a 24-month-old child (by brainstem auditory evoked responses) with an anti-Hu neuroblastoma [6].

In paraneoplastic syndrome patients with cochlear and/or vestibular symptoms, the variety of antibodies and associated cancers are broad.

Hammami and colleagues [3] studied 26 patients with paraneoplastic syndromes who had cochlear and/or vestibular symptoms, and reported the following proportions of antibodies and associated cancers:

- Antibody associations
  - 46%: KLHL11+
  - 31%: KLHL11+/LUZP4+
  - 11%: ANNA1
  - 8%: Amphiphysin
  - 4%: LUZP4+
- Cancer associations
  - 54%: Seminoma
  - 25%: Regressed germ cell tumor
  - 9%: Ductal breast carcinoma
  - 4%: Non-small cell lung cancer
  - 4%: Lung adenocarcinoma
  - 4%: Squamous cell carcinoma

Ocular Motor Examination

An impressive variety of ocular motor abnormalities has been reported in association with paraneoplastic syndromes, including the following:

- Centripetal nystagmus [7, 8]
- Spontaneous down beat nystagmus [4, 5, 9–20]
- Elliptical nystagmus [21]
- Abnormal visual fixation suppression [4, 5, 14, 18]
- Gaze evoked nystagmus, both horizontal and vertical [10, 22]
- Ocular flutter [5, 19, 23, 24]
- Opsoclonus, usually in the context of opsoclonus-myoclonus [6, 25]
- Periodic alternating nystagmus [18, 26]
- Perverted nystagmus [27]
- Positional nystagmus [5, 10]
- Abnormalities on rotatory chair testing, such as low gain and phase lead [4], suggestive of bilateral vestibular weakness
- Rebound nystagmus [7]
- Saccadic dysmetria [28, 29]
- Saccadic intrusions [23, 29]
- See-saw nystagmus [30]
- Skew deviation [15, 17, 22]
- Smooth pursuit abnormalities, usually with saccadic intrusions [4, 5, 14, 15, 28, 29]
- Spontaneous up beat nystagmus [5, 28, 29, 31, 32]
- Windmill nystagmus [33]

The majority of these ocular motor abnormalities localize to the cerebellum; some localize to the brainstem.

Testing: Vestibular

Instrumented ocular motor testing (such as with videonystagmography) may characterize the eye movement abnormalities better than a face-to-face examination.

Computerized dynamic posturography may show increased postural sway [34], but this is neither sensitive nor specific for paraneoplastic cerebellar degeneration.

Other tests (such as vestibular evoked myogenic potentials, video head impulse testing) have not been studied in paraneoplastic cerebellar degeneration. Their main value is to exclude competing diagnoses.

Imaging

Brain MRI may show *cerebellar atrophy* in paraneoplastic cerebellar degeneration [35–41]. Less commonly, there may be *enhancement*, particularly of the *cerebellar folia* [1, 42].

Figure 101.1, from Dalmau and colleagues [1], shows contrast enhancement of the sulci of the cerebellar vermis.

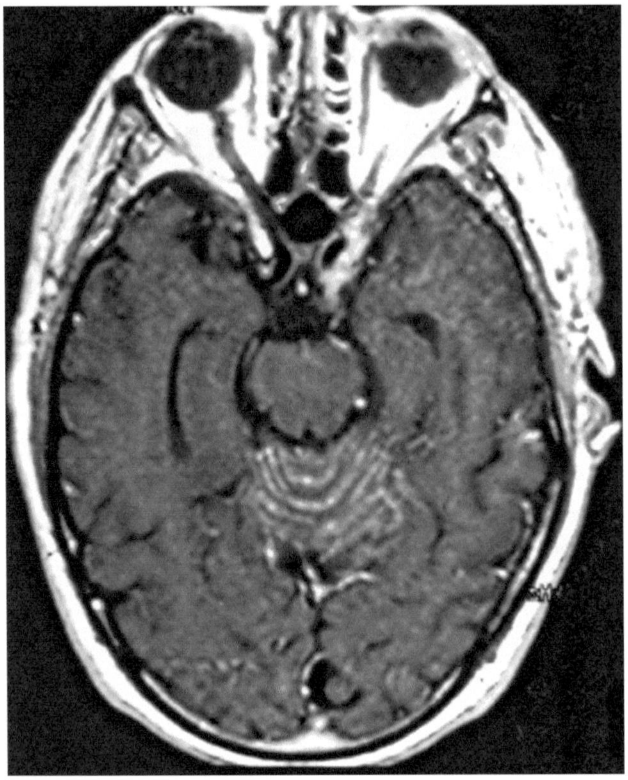

Fig. 101.1 Contrast enhancement in the sulci of the cerebellar vermis in a patient with a paraneoplastic cerebellar syndrome from non-Hodgkin's lymphoma with anti-Tr antibodies. (From Dalmau et al. [1]. Reproduced with written permission from Elsevier under STM Signatory guidelines)

Histopathology

Paraneoplastic cerebellar degeneration characteristically shows extensive loss of Purkinje cells, as well as degeneration in the deep cerebellar nuclei and inferior olivary nuclei [1].

Differential Diagnosis

In a patient with neurological symptoms compatible with a paraneoplastic syndrome, in whom a cancer has already been identified, and paraneoplastic antibodies have already been detected, the diagnosis of a paraneoplastic syndrome is secure.

However, the clinical scenario is often not so straightforward; either the symptom presentation may not be "classic" for a paraneoplastic syndrome, or no malignancy is yet been identified, or no antibodies have been detected [1]. For this reason, diagnostic criteria have been suggested.

Graus and colleagues [1] review the diagnostic criteria for definite and for possible paraneoplastic neurological syndromes as follows:

Definite paraneoplastic syndrome

1. Classical syndrome and cancer that develops within 5 years of the diagnosis of the neurological disorder.
2. A nonclassical syndrome that resolves or significantly improves after cancer treatment without concomitant immunotherapy, provided that the syndrome is not susceptible to spontaneous remission.
3. A nonclassical syndrome with onconeural antibodies (well characterized or not) and cancer that develops within 5 years of the diagnosis of the neurological disorder.
4. A neurological syndrome (classical or not) with well-characterized onconeural antibodies (Hu, Yo, CV2, Ri, Ma2, or amphiphysin), and no cancer.

Possible paraneoplastic syndrome

1. A classical syndrome, no onconeural antibodies, no cancer but at high risk to have an underlying tumor.
2. A neurological syndrome (classical or not) with partially characterized onconeural antibodies and no cancer.
3. A nonclassical syndrome, no onconeural antibodies, and cancer present within 2 years of diagnosis.

The differential diagnosis of a paraneoplastic cerebellar syndrome includes other cerebellar insults. Since some cases of paraneoplastic cerebellar syndromes evolve over many months, they may be mistaken for a primary neurodegenerative process, such as multiple systems atrophy or a late-onset cerebellar ataxia.

Treatment

The standard of care in paraneoplastic syndromes with an identified malignancy is to remove the cancer. After that, if symptoms persist or worsen, there are no clear guidelines on how to proceed [1, 42]. A variety of immune system manipulation strategies have been attempted, including steroids such as *methylprednisolone* [43], other immunosuppressants—*cyclophosphamide* [43], *tacrolimus* [44], and *rituximab* [45]—*plasmapheresis* [46], *intravenous immunoglobulin* [43, 47–49], and others. While there are occasional reports of dramatic responses to such interventions [50], in most cases, the outcome is not satisfactory.

For a patient with suspected paraneoplastic syndrome, a search should be undertaken for malignancy. This often involves a CT of the chest, abdomen, and pelvis, with and without intravenous and oral contrast. If that is unrevealing, then a full-body FDG-PET scan is reasonable [51]. Some investigators advocate doing CT and FDG-PET simultaneously [52, 53]. If no malignancy is identified, then close follow-up is appropriate, with the idea that an occult malignancy will eventually declare itself.

Prognosis

Even when malignancy is treated and immunomodulatory therapies are attempted, prognosis is unfavorable.

References

1. Dalmau J, Rosenfeld MR. Paraneoplastic syndromes of the CNS. Lancet Neurol. 2008;7(4):327–40.
2. Shams'ili S, Grefkens J, de Leeuw B, van den Bent M, Hooijkaas H, van der Holt B, et al. Paraneoplastic cerebellar degeneration associated with antineuronal antibodies: analysis of 50 patients. Brain. 2003;126(Pt 6):1409–18.
3. Hammami MB, Eggers SDZ, Madhavan A, Montalvo MJ, Pittock SJ, Dubey D. Paraneoplastic cochleovestibulopathy: clinical presentations, oncological and serological associations. J Neurol Neurosurg Psychiatry. 2021;92(11):1181–5.
4. Kattah JC, Eggers SD, Bach SE, Dubey D, McKeon AB. Paraneoplastic progressive downbeat nystagmus, ataxia and sensorineural hearing loss due to the ANTI-Kelch-11 protein antibody. J Neuroophthalmol. 2021;41(2):261–5.
5. Kearsley JH, Johnson P, Halmagyi GM. Paraneoplastic cerebellar disease. Remission with excision of the primary tumor. Arch Neurol. 1985;42(12):1208–10.
6. Fisher PG, Wechsler DS, Singer HS. Anti-Hu antibody in a neuroblastoma-associated paraneoplastic syndrome. Pediatr Neurol. 1994;10(4):309–12.
7. Hac NEF, Murphy OC, Butala AA, Newsome SD, Gold DR. Centripetal nystagmus, slow saccades, cerebellar Ataxia, and Parkinsonism in a patient with anti-gad65-associated stiff person syndrome spectrum disorder. J Neuroophthalmol. 2023;1–5:273.
8. Cherchi M, Hac NEF. Paraneoplastic gaze-evoked centripetal nystagmus in ovarian teratoma. J Neuroophthalmol. 2024;44(2):e274–e5.
9. Bussiere M, Al-Khotani A, Steckley JL, Nicolle M, Nicolle D. Paraneoplastic downbeat nystagmus. Can J Ophthalmol. 2008;43(2):243–5.
10. Choi SY, Park SH, Kim HJ, Kim JS. Paraneoplastic downbeat nystagmus associated with cerebellar hypermetabolism especially in the nodulus. J Neurol Sci. 2014;343(1–2):187–91.
11. Guy JR, Schatz NJ. Paraneoplastic downbeating nystagmus. A sign of occult malignancy. J Clin Neuroophthalmol. 1988;8(4):269–72.
12. Hammack J, Kotanides H, Rosenblum MK, Posner JB. Paraneoplastic cerebellar degeneration. II. Clinical and immunologic findings in 21 patients with Hodgkin's disease. Neurology. 1992;42(10):1938–43.
13. Khatib L, Do LD, Benaiteau M, Villagran-Garcia M, Scharf M, Meyer P, et al. Autoimmune cerebellar ataxia associated with anti-glutamate receptor delta2 antibodies: a rare but treatable entity. Cerebellum. 2023;23:260–6.
14. Kherallah B, Samaha E, Bach SE, Guede C, Kattah JC. Case report: acute vestibular syndrome and cerebellitis in anti-Yo paraneoplastic syndrome. Front Neurol. 2022;13:960584.
15. Martin AN, Dillon PM, Jones DE, Brenin DR, Lapides DA. Anti-Yo mediated paraneoplastic cerebellar degeneration associated with pseudobulbar affect in a patient with breast cancer. Case Rep Oncol Med. 2017;2017:8120689.
16. McCrystal M, Anderson NE, Jones RW, Evans BD. Paraneoplastic cerebellar degeneration in a patient with chemotherapy-responsive ovarian cancer. Int J Gynecol Cancer. 1995;5(5):396–9.
17. McLellan R, Currie JL, Royal W, Rosenshein NB. Ovarian carcinoma and paraneoplastic cerebellar degeneration. Obstet Gynecol. 1988;72(6):922–5.
18. Ogawa E, Sakakibara R, Kawashima K, Yoshida T, Kishi M, Tateno F, et al. VGCC antibody-positive paraneoplastic cerebellar degeneration presenting with positioning vertigo. Neurol Sci. 2011;32(6):1209–12.
19. Shivaram S, Tallapalli AVR, Gupta M, Nashi S, Kulkarni GB, Alladi S. Spontaneous downbeat nystagmus in anti-GAD-antibody-associated paraneoplastic syndrome. J Neurosci Rural Pract. 2022;13(3):546–9.
20. Tsuyusaki Y, Sakakibara R, Kishi M, Tateno F, Yoshida T. Downbeat nystagmus as the initial manifestation of anti-NMDAR encephalitis. Neurol Sci. 2014;35(1):125–6.
21. Mistry EA, Lee AG, Lai EC. Teaching video neuroImages: P/Q--type voltage-gated calcium channel-associated paraneoplastic elliptical nystagmus. Neurology. 2016;87(10):e109.
22. Taylor RB, Mason W, Kong K, Wennberg R. Reversible paraneoplastic encephalomyelitis associated with a benign ovarian teratoma. Can J Neurol Sci. 1999;26(4):317–20.
23. Furman JM, Eidelman BH, Fromm GH. Spontaneous remission of paraneoplastic ocular flutter and saccadic intrusions. Neurology. 1988;38(3):499–501.
24. Kruger JM, Yonekawa Y, Skidd P, Cestari DM. Ocular flutter as the presenting sign of lung adenocarcinoma. Digit J Ophthalmol. 2014;20(1):4–6.
25. Desai J, Mitchell WG. Acute cerebellar ataxia, acute cerebellitis, and opsoclonus-myoclonus syndrome. J Child Neurol. 2012;27(11):1482–8.
26. Eggers SD, Pittock SJ, Shepard NT, Habermann TM, Neff BA, Klebig RR. Positional periodic alternating vertical nystagmus with PCA-Tr antibodies in Hodgkin lymphoma. Neurology. 2012;78(22):1800–2.

27. Inui R, Saito K, Shimomura Y, Yamashita D, Kawamoto M, Ishikawa T. Anti-Ma-associated paraneoplastic cerebellar degeneration in a patient with nodular lymphocyte-predominant Hodgkin lymphoma: a case report. BMC Neurol. 2020;20(1):355.
28. Wray SH, Dalmau J, Chen A, King S, Leigh RJ. Paraneoplastic disorders of eye movements. Ann N Y Acad Sci. 2011;1233(1):279–84.
29. Wray SH, Martinez-Hernandez E, Dalmau J, Maheshwari A, Chen A, King S, et al. Paraneoplastic upbeat nystagmus. Neurology. 2011;77(7):691–3.
30. Rizvi MT, Cameron L, Kilbane C, Shaikh AG. Paraneoplastic see-saw nystagmus and opsoclonus provides evidence for hyperexcitable reciprocally innervating mesencephalic network. J Neurol Sci. 2018;390:239–45.
31. Cherchi M. Paraneoplastic upbeat nystagmus in renal cell carcinoma. J Neurol Neurophysiol. 2015;6(287):2.
32. Garcia-Reitboeck P, Thompson G, Johns P, Al Wahab Y, Omer S, Griffin C. Upbeat nystagmus in anti-Ma2 encephalitis. Pract Neurol. 2014;14(1):36–8.
33. Lee SU, Kim HJ, Oh SW, Song EY, Choi JY, Kim JS. Pearls & Oy-sters: windmill nystagmus in paraneoplastic cerebellar degeneration. Neurology. 2018;91(19):e1831–e3.
34. Wessel K, Diener HC, Dichgans J, Thron A. Cerebellar dysfunction in patients with bronchogenic carcinoma: clinical and posturographic findings. J Neurol. 1988;235(5):290–6.
35. Venkatraman A, Opal P. Paraneoplastic cerebellar degeneration with anti-Yo antibodies – a review. Ann Clin Transl Neurol. 2016;3(8):655–63.
36. Fancellu R, Corsini E, Bernardi G, Buzzo P, Ferrari ML, Lamantea E, et al. Paraneoplastic cerebellar ataxia associated with anti-Hu antibodies and benign ganglioneuroma. Funct Neurol. 2014;29(4):277–80.
37. Akpinar S, Berk O, Karaca L. Paraneoplastic cerebellar degeneration associated with multiple myeloma–3.4 years follow up. J Neuro-Oncol. 1990;9(1):63–8.
38. Scheid R, Voltz R, Briest S, Kluge R, von Cramon DY. Clinical insights into paraneoplastic cerebellar degeneration. J Neurol Neurosurg Psychiatry. 2006;77(4):529–30.
39. Escudero-Fernandez JM, Garcia-Carpintero AS, Delgado-Alvarez I, Castellote-Alonso A, Vazquez-Mendez EJ. Brain paraneoplastic syndromes in a patient with mediastinal ganglioneuroma. Radiol Case Rep. 2020;15(11):2059–62.
40. Schlake HP, Husstedt IW, Grotemeyer KH, Potter R. Paraneoplastic subacute cerebellar degeneration in Hodgkin's disease. Report of three cases and review of the literature. Clin Neurol Neurosurg. 1989;91(4):329–35.
41. Emir S, Kutluk MT, Gogus S, Buyukpamukcu M. Paraneoplastic cerebellar degeneration and Horner syndrome: association of two uncommon findings in a child with Hodgkin disease. J Pediatr Hematol Oncol. 2000;22(2):158–61.
42. Darnell RB, Posner JB. Paraneoplastic syndromes involving the nervous system. N Engl J Med. 2003;349(16):1543–54.
43. Keime-Guibert F, Graus F, Fleury A, Rene R, Honnorat J, Broet P, et al. Treatment of paraneoplastic neurological syndromes with antineuronal antibodies (Anti-Hu, anti-Yo) with a combination of immunoglobulins, cyclophosphamide, and methylprednisolone. J Neurol Neurosurg Psychiatry. 2000;68(4):479–82.
44. Albert ML, Austin LM, Darnell RB. Detection and treatment of activated T cells in the cerebrospinal fluid of patients with paraneoplastic cerebellar degeneration. Ann Neurol. 2000;47(1):9–17.
45. Esposito M, Penza P, Orefice G, Pagano A, Parente E, Abbadessa A, et al. Successful treatment of paraneoplastic cerebellar degeneration with Rituximab. J Neuro-Oncol. 2008;86(3):363–4.
46. Cocconi G, Ceci G, Juvarra G, Minopoli MR, Cocchi T, Fiaccadori F, et al. Successful treatment of subacute cerebellar degeneration in ovarian carcinoma with plasmapheresis. A case report. Cancer. 1985;56(9):2318–20.
47. Phuphanich S, Brock C. Neurologic improvement after high-dose intravenous immunoglobulin therapy in patients with paraneoplastic cerebellar degeneration associated with anti-Purkinje cell antibody. J Neuro-Oncol. 2007;81(1):67–9.
48. Uchuya M, Graus F, Vega F, Rene R, Delattre JY. Intravenous immunoglobulin treatment in paraneoplastic neurological syndromes with antineuronal autoantibodies. J Neurol Neurosurg Psychiatry. 1996;60(4):388–92.
49. Widdess-Walsh P, Tavee JO, Schuele S, Stevens GH. Response to intravenous immunoglobulin in anti-Yo associated paraneoplastic cerebellar degeneration: case report and review of the literature. J Neuro-Oncol. 2003;63(2):187–90.
50. Croteau D, Owainati A, Dalmau J, Rogers LR. Response to cancer therapy in a patient with a paraneoplastic choreiform disorder. Neurology. 2001;57(4):719–22.
51. Marchand V, Graveleau J, Lanctin-Garcia C, Bourbouloux E, Bridji B, Resche I, et al. A rare gynecological case of paraneoplastic cerebellar degeneration discovered by FDG-PET. Gynecol Oncol. 2007;105(2):545–7.
52. Frings M, Antoch G, Knorn P, Freudenberg L, Bier U, Timmann D, et al. Strategies in detection of the primary tumour in anti-Yo associated paraneoplastic cerebellar degeneration. J Neurol. 2005;252(2):197–201.
53. Linke R, Schroeder M, Helmberger T, Voltz R. Antibody-positive paraneoplastic neurologic syndromes: value of CT and PET for tumor diagnosis. Neurology. 2004;63(2):282–6.

Part XIII

Diagnoses—Intracranial Pressure Disorders

Hydrocephalus

102

Overview

Hydrocephalus is the presence of excessive cerebrospinal fluid (CSF). Hydrocephalus is usually, though not always, associated with increased intracranial pressure and dilation of CSF ventricles. Hydrocephalus is usually classified as either obstructive or nonobstructive, and as congenital or acquired. The causes in children and adults tend to be different. Common presenting symptoms include headache, blurry vision, double vision, and sometimes cognitive changes. Ophthalmoscopic examination may show optic nerve head edema. Ocular motor examination may show a variety of abnormalities, the most common of which include trochlear nerve palsy. When hydrocephalus is clinically suspected, brain imaging is appropriate. Treatment depends on the underlying etiology.

Introduction

The term *hydrocephalus* refers to the presence of excessive cerebrospinal fluid (CSF) . In most cases, this is associated with an increase in intracerebral pressure (an exception being normal pressure hydrocephalus) and dilation of the CSF ventricles.

Cerebrospinal fluid is synthesized by the choroid plexus in the lateral ventricles, third and fourth ventricles. From the most proximal site of synthesis in the lateral ventricles, CSF flows through the interventricular foramina of Monro into the third ventricle, then through the cerebral aqueduct into the fourth ventricle. From there, CSF passes through the medial aperture (foramen of Magendie) and lateral apertures (foramina of Luschka) of the fourth ventricle into the subarachnoid space at the skull base. CSF then flows through the subarachnoid space over the cerebral convexities and into the spinal canal. Finally, CSF is resorbed by the arachnoid villi (located in the superior sagittal sinus, intracranial venous sinuses, and around spinal nerve roots) into the blood stream. The movement of CSF is driven by the pressure waves from pulsation of the cerebral arteries (which expands the vascular component of intracerebral cerebral space and drives an equal volume of CSF to exit), propelling CSF along the pathway just described.

Figure 102.1 is a schematic of CSF production (from serum), through the central nervous system, back into the bloodstream.

M. Cherchi, *Otoneurology and Vestibular Medicine*, https://doi.org/10.1007/978-3-031-94841-1_102

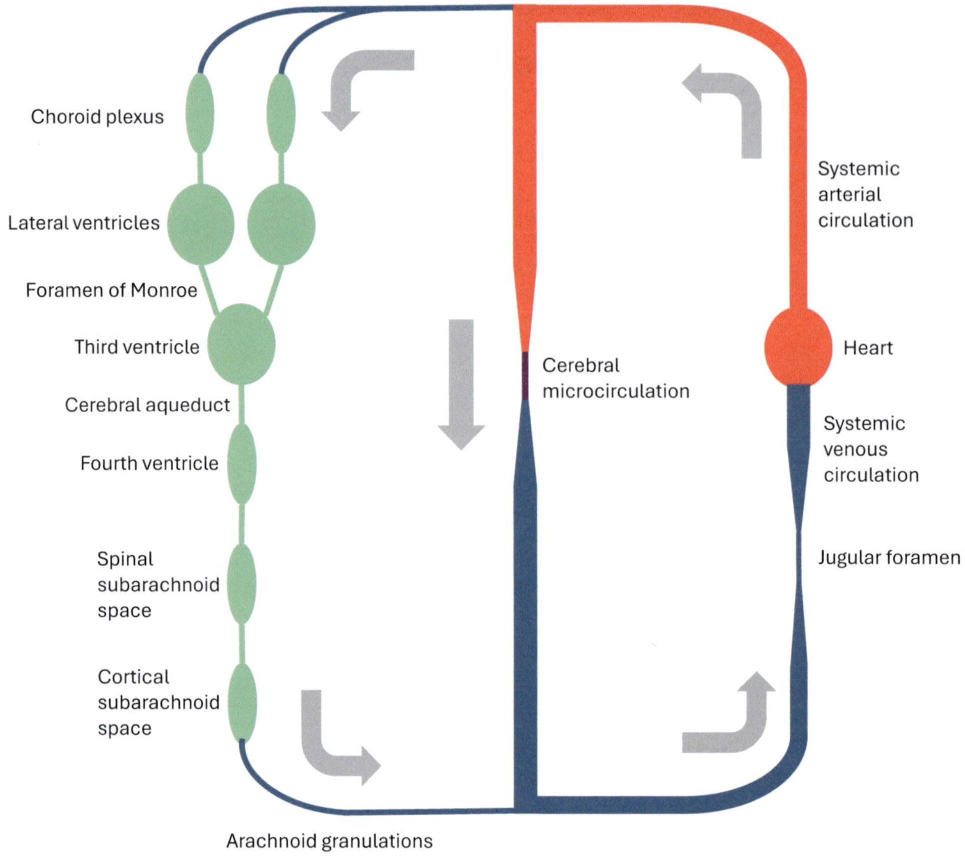

Fig. 102.1 Schematic of cerebrospinal fluid flow (By Marcello Cherchi)

Pathophysiological Mechanism of Disease

A traditional classification of hydrocephalus distinguishes:

- *Obstructive hydrocephalus* (also called *noncommunicating hydrocephalus*). This refers to a structural blockage somewhere between the site(s) of CSF production (at the choroid plexus in the lateral, third, and fourth ventricles) and resorption (arachnoid granulations in the superior sagittal sinus, intracranial venous sinuses, and around spinal nerve roots). For obstructive hydrocephalus, the usual etiologies include tumors, intraventricular hemorrhage, vascular malformation, and compression from edema.
- *Nonobstructive hydrocephalus* (also called *communicating hydrocephalus*). This refers to hydrocephalus from something other than obstructive (structural) causes, usually due to overproduction of CSF or impaired resorption of CSF, or both. In the "overproduction of CSF," category is usually cited choroid plexus papilloma. In the "underresorption of CSF," category the differential diagnosis includes damage to the arachnoid villi (such as from meningitis) or venous outflow obstruction (such as central venous sinus thrombosis).

A second dimension along which hydrocephalus is classified is whether it is *congenital* or *acquired* [1]. Etiologies of hydrocephalus in children and adults tend to be different.

Epidemiology

Hydrocephalus can occur at any age but is more common in infancy. Consequently, the incidence depends on the age distribution of the population.

Figure 102.2, from Dewan and colleagues [2], shows the global annual incidence of hydrocephalus, which varies from 500 per 100,000 in low-income regions to 120 per 100,000 in middle- and high-income regions. There is a direct relationship between birth rate and incidence of hydrocephalus.

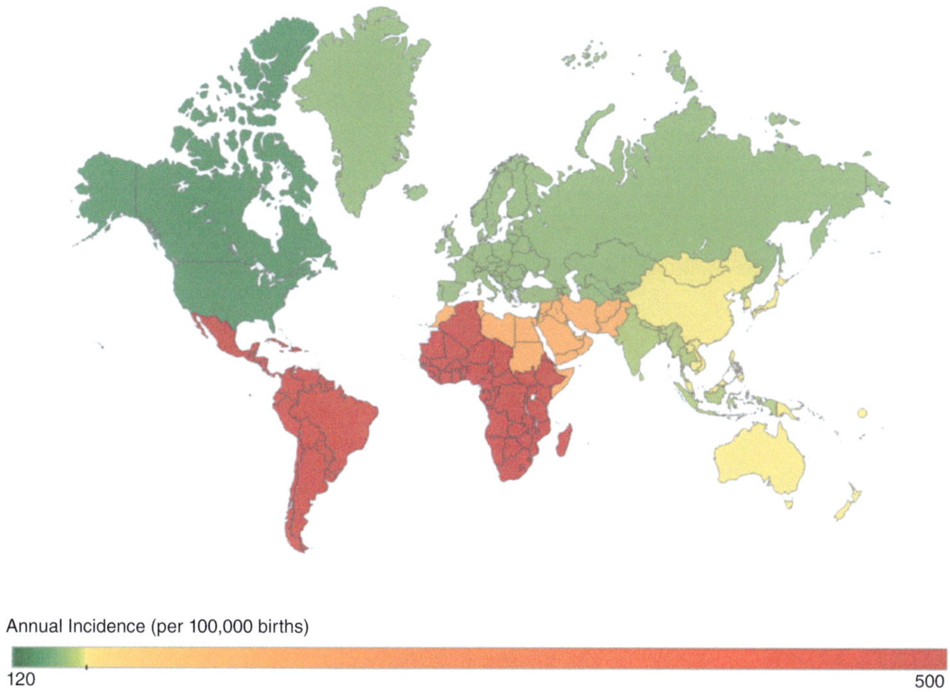

Fig. 102.2 Global incidence of hydrocephalus. (From Dewan et al. [2]. Reproduced from OpenStreetMap with permission under Fair Use)

Genetics

A number of genes have been implicated in hydrocephalus (particularly congenital hydrocephalus) [1, 3], but the mechanism by which these genes bring about hydrocephalus remains unknown.

Clinical Presentation

The clinical presentation of a patient with hydrocephalus will depend in part on the etiology of the hydrocephalus. However, in hydrocephalus from any cause, common symptoms include headache (presumably from traction on the innervated structures in the subarachnoid space), blurry vision (from optic nerve edema), diplopia (from various extraocular muscle palsies), and cognitive changes which can (if severe) include depressed level of consciousness (from herniation).

Physical Examination

Physical examination of a patient with hydrocephalus will depend in part on the etiology of the hydrocephalus. From an otoneurological perspective, we will omit from this discussion etiologies that would independently cause eye movement abnormalities, such as cerebellar hemorrhage [4].

Hydrocephalus often (though not always) causes optic disc edema which may be discernible on funduscopic examination.

Ocular Motor Examination

Many cases of hydrocephalus cause diplopia from strabismus. This usually happens secondary to trochlear nerve palsy, [5, 6], but can also occur with palsies of the abducens nerve [7] or ocular motor nerve [8].

Some cases of hydrocephalus exert pressure on the quadrigeminal plate and thereby induce an upgaze palsy.

A variety of other ocular motor abnormalities have been reported in association with hydrocephalus [9, 10], including the following:

- Convergence nystagmus, or convergence-retraction nystagmus [11–15], in isolation or as part of pretectal syndrome [16].
- Spontaneous down beat nystagmus [17].
- Opsoclonus [18].
- Abnormal optokinetic responses [19].
- Positional nystagmus [20, 21].
- Pseudobobbing [13].
- Skew deviation [22] not due to extraocular muscle palsy.
- Spontaneous up beat nystagmus [23].
- Wall-eyed bilateral internuclear ophthalmoplegia (WEBINO) [24].

Many patients with hydrocephalus, especially in the early stages, have normal eye movements. Consequently, the diagnosis of hydrocephalus does not depend on the presence of ocular motor abnormalities.

Imaging

When hydrocephalus is clinically suspected, brain imaging is appropriate.

Treatment

The treatment of hydrocephalus depends on the underlying etiology.

Prognosis

Hydrocephalus, if presenting acutely and progressing rapidly, runs the risk of various types of brain herniation (subfalcine, central transtentorial, uncal transtentorial, upward cerebellar, downward cerebellar tonsillar) depending on the etiology of hydrocephalus. Brain herniation is life-threatening; these patients require emergent neurosurgical evaluation and are usually observed in a neurointensive care unit.

References

1. Zhang J, Williams MA, Rigamonti D. Genetics of human hydrocephalus. J Neurol. 2006;253(10):1255–66.
2. Dewan MC, Rattani A, Mekary R, Glancz LJ, Yunusa I, Baticulon RE, et al. Global hydrocephalus epidemiology and incidence: systematic review and meta-analysis. J Neurosurg. 2018;130:1065–79.
3. Garcia-Bonilla M, McAllister JP, Limbrick DD. Genetics and molecular pathogenesis of human hydrocephalus. Neurol India. 2021;69(Supplement):S268–S74.
4. Pannu AK, Saroch A, Sharma N. Intracerebellar hemorrhage in a young adult. Indian J Crit Care Med. 2020;24(1):69–70.
5. Mantopoulos D, Hunter DG, Cestari DM. Isolated bilateral fourth cranial nerve palsies as the presenting sign of hydrocephalus. Case Rep Ophthalmol. 2011;2(2):211–4.
6. Guy JR, Friedman WF, Mickle JP. Bilateral trochlear nerve paresis in hydrocephalus. J Clin Neuroophthalmol. 1989;9(2):105–11.
7. Alver I, Abuzayed B, Kafadar AM, Muhammedrezai S, Sanus GZ, Akar Z. Primary fourth ventricular meningioma: case report and review of the literature. Turk Neurosurg. 2011;21(2):249–53.
8. Cultrera F, D'Andrea M, Battaglia R, Chieregato A. Unilateral oculomotor nerve palsy: unusual sign of hydrocephalus. J Neurosurg Sci. 2009;53(2):67–70.
9. Gaston H. Ophthalmic complications of spina bifida and hydrocephalus. Eye (Lond). 1991;5(Pt 3):279–90.
10. Lopponen H, Sorri M, Serlo W, von Wendt L. ENG findings of shunt-treated hydrocephalus in children. Int J Pediatr Otorhinolaryngol. 1992;23(1):35–44.
11. Chattha AS, Delong GR. Sylvian aqueduct syndrome as a sign of acute obstructive hydrocephalus in children. J Neurol Neurosurg Psychiatry. 1975;38(3):288–96.
12. Hashizume A, Watanabe H, Matsuo K, Katsuno M, Tanaka F, Nagatani T, et al. Endoscopic third ventriculotomy improves parkinsonism following a ventriculo-peritoneal shunt in a patient with non-communicating hydrocephalus secondary to idiopathic aqueduct stenosis. J Neurol Sci. 2011;309(1–2):148–50.
13. Keane JR. Pretectal pseudobobbing. Five patients with 'V'-pattern convergence nystagmus. Arch Neurol. 1985;42(6):592–4.
14. Maroulis H, Halmagyi GM, Heard R, Cook RJ. Sylvian aqueduct syndrome with slit ventricles in shunted hydrocephalus due to adult aqueduct stenosis. J Neurosurg. 2008;109(5):939–43.
15. Sato Y, Maehara F, Utsunomiya H, Hayashi T, Moritaka K, Kaji M. Convergence and retraction nystagmus in hydrocephalus due to shunt malfunction. Kurume Med J. 1983;30(4):125–8.
16. Keane JR. The pretectal syndrome: 206 patients. Neurology. 1990;40(4):684–90.
17. Phadke JG, Hern JE, Blaiklock CT. Downbeat nystagmus–a false localising sign due to communicating hydrocephalus. J Neurol Neurosurg Psychiatry. 1981;44(5):459.
18. Shetty T, Rosman NP. Opsoclonus in hydrocephalus. Arch Ophthalmol. 1972;88(6):585–9.
19. Suzuki M, Harada Y, Ishida M, Wada H, Ohta M, Sakoda K, et al. Aqueductal stenosis–results of vestibular function tests. J Laryngol Otol. 1985;99(2):151–61.
20. Power L, Murray K, Bullus K, Drummond KJ, Trost N, Szmulewicz DJ. Central conditions mimicking benign paroxysmal positional vertigo: a case series. J Neurol Phys Ther. 2019;43(3):186–91.
21. Sweeney PJ, Hahn JF, McHenry MC, Mitsumoto H. Mucormycosis presenting as positional nystagmus and hydrocephalus: case report. J Neurosurg. 1980;52(2):270–2.
22. Moguel-Ancheita S, Castellanos-Perez Bolde CG, Orozco-Gomez LP, Skew deviation. Strabismological diagnosis and treatment alternatives. Cir Cir. 2009;77(4):267–73. 49–55.
23. Kim JS, Yoon B, Choi KD, Oh SY, Park SH, Kim BK. Upbeat nystagmus: clinicoanatomical correlations in 15 patients. J Clin Neurol. 2006;2(1):58–65.
24. Keereman V, Platteau E, Crevits L, Algoed L. Wall-eyed bilateral internuclear ophthalmoplegia (WEBINO) in a patient with idiopathic intracranial hypertension. Neurologist. 2018;23(5):157–9.

Normal-Pressure Hydrocephalus

103

Overview

Idiopathic *normal-pressure hydrocephalus (NPH)* is, as the name suggests, hydrocephalus with normal intracranial pressure. The mean age of onset is in the mid-70s. Symptoms begin insidiously and consist of some combination of (1) cognitive impairment, (2) gait impairment, and (3) urinary incontinence. Most cases occur without any obvious trigger, while some appear to follow a specific neurological insult (such as head trauma or infection). The mechanism of the disease is unclear. It is also unclear whether any genetic factors are contributory. The tempo of symptoms is variable; in some patients, it is progressive, while in others, it is relatively static. Several radiographic diagnostic criteria have been proposed, but none widely accepted. Treatment is placement of a shunt (usually ventriculoperitoneal, less commonly lumboperitoneal), but response rates are variable, and response is often unsustained; thus, prognosis is guarded. Nevertheless, NPH is important to recognize because it remains one of the few potentially treatable causes of gait impairment and cognitive decline.

Fig. 103.1 Dr. Salomon Hakim (1922–2011). (From Pedro et al. [1])

Fig. 103.2 Dr. Raymond Delacy Adams (1911–2008). (Reproduced with written permission from Harvard University Archives)

Introduction

Neurosurgeon Dr. *Salomon Hakim* (1922–2011) [1], neurologist Dr. *Raymond Delacy Adams* (1911–2008) [2], and colleagues published several papers in 1965 [3, 4] regarding a constellation of findings associated with radiographic hydrocephalus in patients whose intracranial pressure appeared normal (Figs. 103.1 and 103.2).

In the first paper [4], they reported on three patients, each of whom exhibited three characteristics (sometimes referred to as "Hakim's triad"), which they described as follows:

- "*Impairment of consciousness*" which "varied from mild inattentiveness with failure of daily events to be registered in mind and later retained, to a state of virtual coma. In this state the patient often was manifestly unreceptive of stimuli as well as unresponsive to them. The latter was at times more prominent than the former. The psychomotor retardation also varied in degree; when mild, a lack of impulsivity, expressed as apathy, disinterest, and lack of spontaneity, was associated with reduced levels of verbal

M. Cherchi, *Otoneurology and Vestibular Medicine*, https://doi.org/10.1007/978-3-031-94841-1_103

and motorial function. When severe, it amounted to total immobility and absence of consecutive psychic activity, a state akin to akinetic mutism."

- "*Incontinence of sphincters*" which "was essentially of cerebral type, the patient being unaware of the contents of bladder and bowel and incapable of making any arrangement for the somewhat precipitate action of these organs."
- "*Alteration of gait*," which "varied from a mild disorder in which locomotion appeared careless, the steps slow and somewhat shortened and the balance slightly uncertain, to one in which there was total incapacity to stand and walk. With the latter the plantar reflexes were usually extensor and the tendon reflexes brisk, particularly in the legs. An element of ataxia seemed to have been present but was always difficult to evaluate, and clearcut intention tremor and dysmetria were not demonstrable."

They also specifically noted that "the CSF pressure was indeed at all times within normal limits."

Epidemiology

"The precise incidence and prevalence of INPH [idiopathic normal pressure hydrocephalus] are not known" [5].

Mendes and colleagues [6] prospectively studied 30 patients who underwent placement of a Strata programmable valve ventriculoperitoneal shunt at a mean age of 77.1 years (77 years and 1 month). They reported that, "Symptoms of dementia had been present for 35 months before surgery on average, whereas the average onset of gait-related symptoms and urinary incontinence was 25 months and 22 months before surgery, respectively." Thus, in this cohort, the average age of symptom onset of dementia was 74 years and 2 months, of gait problems was 75 years, and urinary incontinence was 75 years and 3 months.

Genetics

Familial occurrence of NPH has been reported [7–11], but the extent to which genetic factors play a role in NPH remains unclear.

- Yang and colleagues reported that 15% of the unrelated NPH patients they analyzed had mutations in CWH43, which "modifies the lipid anchor of glycosylphosphatidylinositol-anchored proteins" [12].
- Morimoto and colleagues reported a loss-of-function variant in CFAP43 in a Japanese family with several members who had NPH [13].
- Korhonen and colleagues reported an expansion in C9ORF72 in 8/487 (1.6%) of patients with "possible idiopathic normal pressure hydrocephalus" [14]. Mutations in the C9ORF72 gene have also been implicated in some cases of the frontotemporal dementia—amyotrophic lateral sclerosis complex.
- Sato and colleagues reported that of 50 patients with shunt-responsive idiopathic NPH, 26% had a "copy number loss in intron 2 of the SFMBT1 gene" [15].

Pathophysiological Mechanism of Disease

Much of the literature draws a distinction between idiopathic NPH (arising without any clear precipitant) and secondary NPH (occurring possibly as a consequence of some preceding insult such as trauma or infection).

Regarding the disorder of gait, in their original paper, Hakim and Adams [4] stated that, "One can only vaguely speculate in general terms that the dilatation of the ventricles subjects the major long tracts in the cerebral white matter and corpus callosum to compression and stretching, with a resulting physiological block of function which is to a considerable extent reversible when the pressure is reduced. This hypothesis would be in harmony with the observation that it is the white mater of the cerebrum which appears to yield maximally to great expansion of the ventricular system."

Subsequent research suggested that:

> Electromyographic evidence reveals contraction of antagonistic muscle groups and abnormally increased activity in the antigravity muscles acting on hip and knee joints [2, 9]. This suggests that the gait disorder of INPH is a disturbance in the phased activation of muscle groups, as would be seen in a disorder of subcortical motor control rather than a primary pyramidal tract disturbance. With progression of INPH and/or the evolution of extensive subcortical white matter changes, pyramidal tract involvement may become more likely [5]

Clinical Presentation

The tempo of symptom evolution is variable:

> There is considerable variation in the nature, severity, and course of progression of the symptoms of INPH. The natural history of untreated INPH has not been well characterized, and it is not clear whether all cases ultimately progress, nor is the tempo of progression established for the majority of cases. Progression of symptoms is generally expected but may not be uniform over the course of the disease. Symptoms of INPH in the early stage and late stages of the disease may differ dramatically, as may symptoms in previously treated versus untreated patients [5].

The combination of symptoms, and the intensity of any given symptom, is also variable:

All three components of the so-called classic triad of INPH need not be present for the diagnosis of INPH to be made. Gait disturbance tends to be the most readily recognized feature of INPH. Cognitive disturbances do not occur in all patients, and the severity of cognitive and motor symptoms does not necessarily correlate at baseline or progress in parallel. Incontinence in INPH is difficult to distinguish from urinary symptoms associated with other common disorders. The signs and symptoms of INPH are typically bilateral but may appear lateralized when superimposed on coexisting conditions, such as stroke, radiculopathy, and peripheral neuropathy [5].

Regarding the quality of gait impairment:

The pattern of gait seen in INPH patients has been variably described as 'apractic,' 'bradykinetic,' 'glue-footed,' 'magnetic,' 'parkinsonian,' 'short-stepped,' and 'shuffling.' Gait problems may emerge as difficulty in ascending or descending stairs and curbs and in walking at the expected pace. Patients may complain of difficulty rising from a chair, 'give-way' weakness of the lower extremities, and fatigue brought on by walking. As the disease progresses, turning in place becomes tenuous and typically requires multiple steps (*en bloc*) [5].

Physical Examination

Regarding further specific characteristics of gait:

Computerized analysis of the gait of INPH patients has shown 1) reduced cadence, 2) decreased step height, and 3) reduced counterrotation of the shoulders relative to the pelvis during walking. The upward angular movement of the foot during stepping is inadequate in INPH, and stride length is diminished. The step width and foot rotation angles are reportedly increased, and there is less step-to-step variability in these parameters than in normal controls [5].

Regarding further characteristics of stance:

The stance in INPH may be more forward leaning than in healthy normal individuals, and INPH patients tend to show a wider sway and imbalance that may be accentuated by eye closure [5].

Regarding cognition, Relkin and colleagues [5] state that, "The principal cognitive symptoms seen in INPH are suggestive of a subcortical dementing process, including slowing of thought, inattentiveness, and apathy, as well as encoding and recall problems," and they offer Table 103.1 comparing the cognitive deficits of NPH and Alzheimer's.

Table 103.1 Comparison of cognitive deficits in Alzheimer's disease and idiopathic normal-pressure hydrocephalus

Cognitive skills	Alzheimer's disease	Idiopathic normal-pressure hydrocephalus
Impaired	Memory Learning Orientation Attention concentration Executive functions Writing	Psychomotor slowing Fine motor speed Fine motor accuracy
Borderline impairment	Motor and psychomotor skills Visuospatial skills Language Reading	Auditory memory (immediate and delayed) Attention concentration Executive function Behavioral or personality changes

From Relkin et al. [5]

Testing: Vestibular

Vestibular testing does not play a significant role in confirming a diagnosis of NPH. Interestingly, Inui and colleagues [16] report that 25% of NPH patients exhibit a shift in the tuning curve of cervical vestibular evoked myogenic potentials, which they interpret reflecting that in hydrocephalus the "excessive CSF accumulation propagated to the endolymph and perilymph."

The main value of vestibular testing lies in excluding competing causes of chronic disequilibrium (such as vestibular weakness).

Imaging

Several radiologic criteria have been studied for the diagnosis of NPH [17, 18], but none has gained wide acceptance.

Differential Diagnosis

Various diagnostic criteria have been proposed for NPH [6, 19–22], but none has been widely accepted.

Relkin and colleagues [5] state that idiopathic NPH "should be considered in the differential diagnosis of any unexplained disturbance of gait, continence, and cognition that begins insidiously in an adult. This is most commonly a gait disturbance, followed in frequency by cognitive impairments and least often, urinary incontinence."

Relkin and colleagues [5] provide the following extensive differential diagnosis of NPH:

- Neurodegenerative disorders
  - Alzheimer's disease
  - Parkinson's disease
  - Lewy body disease
  - Huntington's disease
  - Frontotemporal dementia
  - Corticobasal degeneration
  - Progressive supranuclear palsy
  - Amyotrophic lateral sclerosis
  - Multisystem atrophy
  - Spongiform encephalopathy
- Vascular dementia
  - Cerebrovascular disease
  - Stroke
  - Multi-infarct state
  - Binswanger's disease
  - Cerebral autosomal dominant arteriopathy, subcortical infarcts and leukoencephalopathy (CADASIL)
  - Vertebrobasilar insufficiency
- Other hydrocephalus disorders
  - Aqueductal stenosis
  - Arrested hydrocephalus
  - Long-standing overt ventriculomegaly syndrome
  - Noncommunicating hydrocephalus
- Infectious diseases
  - Lyme
  - Human immunodeficiency virus
  - Syphilis
- Urological disorders
  - Urinary tract infection
  - Bladder or prostate cancer
  - Benign prostatic enlargement
- Miscellaneous
  - Vitamin B12 deficiency
  - Collagen vascular disorders
  - Epilepsy
  - Depression
  - Traumatic brain injury
  - Spinal stenosis
  - Chiari malformation
  - Arrested hydrocephalus
  - Wernicke's encephalopathy
  - Carcinomatous meningitis
  - Spinal cord tumor

Treatment

The usual treatment of NPH is placement of a ventriculoperitoneal shunt—less commonly a lumboperitoneal shunt [23]—but these procedures come with risk, and response rates are often unsatisfactory (see below). Consequently, the question becomes one of trying to predict which patients are likely to respond to shunting, and guidance on patient selection is weak [22].

Boon and colleagues [20] concluded that CSF outflow resistance was the best predictor of shunt responsiveness.

Thavarajasingam and colleagues [18] conducted a systematic literature review and meta-analysis and concluded that no radiologic criterion can serve as a sole predictor of responsiveness to shunting.

Vanneste and colleagues [19] recommended using a combination of radiologic and clinical criteria for predicting responsiveness to shunting.

Prognosis

The lack of widely accepted diagnostic criteria has been an impediment to interpreting the results of treatment trials. Response rates are variable, but difficult to compare across studies given the variability in diagnostic criteria they employ.

Relkin and colleagues [5] comment that, "On the basis of clinical presentation alone, evidence shows that favorable response to shunting will vary from 46% to 63%."

Kahlon and colleagues [24] studied 54 NPH patients who underwent shunting and reported that, "At the 6-month follow-up examination, 83% of the operated patients improved in gait, 65% improved in reaction time, 46% improved in memory, and 31% improved in identical forms tests; 96% found themselves subjectively improved."

Trungu and colleagues [25] studied 181 NPH patients who underwent ventriculoperitoneal shunting and reported that an impressive 91.2% of patients exhibited improvement after 12 months.

Even in those patients who do appear to exhibit an initial response to shunting, the improvement is usually not sustained. Kahlon and colleagues [24] comment that, "only 20% of the operated patients demonstrated continued functional improvements after 5 years."

References

1. Pedro MKF, Silva J, Rocha S, Germiniani FMB, Ramina R, Teive HAG, et al. Salomon Hakim: the man behind normal pressure hydrocephalus. Arq Neuropsiquiatr. 2019;77(10):746–8.
2. Fisher CM. Dr. Raymond Delacy Adams (1911-2008): an appreciation. Neurologist. 2010;16(3):141–2.
3. Adams RD, Fisher CM, Hakim S, Ojemann RG, Sweet WH. Symptomatic occult hydrocephalus with "Normal" cerebrospinal-fluid pressure. A treatable syndrome. N Engl J Med. 1965;273:117–26.
4. Hakim S, Adams RD. The special clinical problem of symptomatic hydrocephalus with normal cerebrospinal fluid pressure. Observations on cerebrospinal fluid hydrodynamics. J Neurol Sci. 1965;2(4):307–27.
5. Relkin N, Marmarou A, Klinge P, Bergsneider M, Black PM. Diagnosing idiopathic normal-pressure hydrocephalus. Neurosurgery. 2005;57(3 Suppl):S4–16. discussion ii–v.
6. Mendes GAS, de Oliveira MF, Pinto FCG. The timed up and go test as a diagnostic criterion in normal pressure hydrocephalus. World Neurosurg. 2017;105:456–61.
7. Huovinen J, Kastinen S, Komulainen S, Oinas M, Avellan C, Frantzen J, et al. Familial idiopathic normal pressure hydrocephalus. J Neurol Sci. 2016;368:11–8.
8. Cusimano MD, Rewilak D, Stuss DT, Barrera-Martinez JC, Salehi F, Freedman M. Normal-pressure hydrocephalus: is there a genetic predisposition? Can J Neurol Sci. 2011;38(2):274–81.
9. Takahashi Y, Kawanami T, Nagasawa H, Iseki C, Hanyu H, Kato T. Familial normal pressure hydrocephalus (NPH) with an autosomal-dominant inheritance: a novel subgroup of NPH. J Neurol Sci. 2011;308(1–2):149–51.
10. Portenoy RK, Berger A, Gross E. Familial occurrence of idiopathic normal-pressure hydrocephalus. Arch Neurol. 1984;41(3):335–7.
11. McGirr A, Cusimano MD. Familial aggregation of idiopathic normal pressure hydrocephalus: novel familial case and a family study of the NPH triad in an iNPH patient cohort. J Neurol Sci. 2012;321(1–2):82–8.
12. Yang HW, Lee S, Yang D, Dai H, Zhang Y, Han L, et al. Deletions in CWH43 cause idiopathic normal pressure hydrocephalus. EMBO Mol Med. 2021;13(3):e13249.
13. Morimoto Y, Yoshida S, Kinoshita A, Satoh C, Mishima H, Yamaguchi N, et al. Nonsense mutation in CFAP43 causes normal-pressure hydrocephalus with ciliary abnormalities. Neurology. 2019;92(20):e2364–e74.
14. Korhonen VE, Remes AM, Helisalmi S, Rauramaa T, Sutela A, Vanninen R, et al. Prevalence of C9ORF72 expansion in a large series of patients with idiopathic normal-pressure hydrocephalus. Dement Geriatr Cogn Disord. 2019;47(1–2):91–103.
15. Sato H, Takahashi Y, Kimihira L, Iseki C, Kato H, Suzuki Y, et al. A segmental copy number loss of the SFMBT1 gene is a genetic risk for shunt-responsive, idiopathic normal pressure hydrocephalus (iNPH): a case-control study. PLoS One. 2016;11(11):e0166615.
16. Inui T, Haginomori SI, Kajimoto Y, Kuriyama T, Shirai T, Kinoshita I, et al. Asymmetry and tuning shift of the cervical vestibular evoked myogenic potential indicate saccular dysfunction in idiopathic normal pressure hydrocephalus. Clin Neurophysiol. 2022;134:43–9.
17. Ryska P, Slezak O, Eklund A, Malm J, Salzer J, Zizka J. Radiological markers of idiopathic normal pressure hydrocephalus: relative comparison of their diagnostic performance. J Neurol Sci. 2020;408:116581.
18. Thavarajasingam SG, El-Khatib M, Vemulapalli K, Iradukunda HAS, Sajeenth Vishnu K, Borchert R, et al. Radiological predictors of shunt response in the diagnosis and treatment of idiopathic normal pressure hydrocephalus: a systematic review and meta-analysis. Acta Neurochir. 2023;165(2):369–419.
19. Vanneste J, Augustijn P, Tan WF, Dirven C. Shunting normal pressure hydrocephalus: the predictive value of combined clinical and CT data. J Neurol Neurosurg Psychiatry. 1993;56(3):251–6.
20. Boon AJ, Tans JT, Delwel EJ, Egeler-Peerdeman SM, Hanlo PW, Wurzer HA, et al. The Dutch normal-pressure hydrocephalus study. How to select patients for shunting? An analysis of four diagnostic criteria. Surg Neurol. 2000;53(3):201–7.
21. Krauss JK, Halve B. Normal pressure hydrocephalus: survey on contemporary diagnostic algorithms and therapeutic decision-making in clinical practice. Acta Neurochir. 2004;146(4):379–88. discussion 88.
22. Halperin JJ, Kurlan R, Schwalb JM, Cusimano MD, Gronseth G, Gloss D. Practice guideline: idiopathic normal pressure hydrocephalus: response to shunting and predictors of response: report of the guideline development, dissemination, and implementation Subcommittee of the American Academy of Neurology. Neurology. 2015;85(23):2063–71.
23. Liu JT, Su PH. The efficacy and limitation of lumboperitoneal shunt in normal pressure hydrocephalus. Clin Neurol Neurosurg. 2020;193:105748.
24. Kahlon B, Sjunnesson J, Rehncrona S. Long-term outcome in patients with suspected normal pressure hydrocephalus. Neurosurgery. 2007;60(2):327–32. discussion 32.
25. Trungu S, Scollato A, Ricciardi L, Forcato S, Polli FM, Miscusi M, et al. Clinical outcomes of shunting in normal pressure hydrocephalus: a multicenter prospective observational study. J Clin Med. 2022;11(5):1286.

Intracranial Pressure Fluctuations and Hearing Loss

104

Overview

A common model (probably an oversimplification) of labyrinthine fluid dynamics is that the intracranial compartment communicates through the cochlear aqueduct to the perilymphatic space, and the endolymphatic space communicates via the endolymphatic duct to the endolymphatic sac. Changes in intracranial pressure can be transmitted to the labyrinth, and such changes (whether increase or decrease) appear to affect cochlear function adversely. The resulting cochlear dysfunction is probably a combination of mechanical disruption and alteration in chemical composition of the perilymphatic fluid and endolymphatic fluid.

When the cochlear dysfunction (manifesting as hearing loss) is the result of increased intracranial pressure (such as from hydrocephalus), normalization of intracranial pressure can bring about improvement in hearing.

When cochlear dysfunction (manifesting as hearing loss) is the result of decreased intracranial pressure (such as after a lumbar puncture, or overshunting of normal pressure hydrocephalus), normalization of intracranial pressure can bring about improvement in hearing.

Hydraulics of the Labyrinth

Outlining the "hydraulics" of the labyrinth can aid in understanding the mechanism by which intracranial pressure fluctuations may affect hearing.

The bony labyrinth is a convoluted cavity within the temporal bone that is filled with perilymphatic fluid. Within that perilymph-filled space is also located the membranous labyrinth, which is similarly shaped as the bony labyrinth (though slightly smaller), and the membranous labyrinth is in turn filled with endolymphatic fluid.

The perilymphatic fluid space communicates with the intracranial space via the cochlear aqueduct.

This relationship between the perilymph and the cerebrospinal fluid is schematically depicted in Fig. 104.1 from Stover and colleagues [1].

Although the cochlear aqueduct permits passage of fluid between the perilymphatic space and the intracranial space, a study of 250 human cadavers found that the patency of the cochlear aqueduct gradually decreases over the course of life [2]. Figure 104.2 from Marchbanks and Reid [3] redepicts data originally from Wlodyka [2].

The production and flow of perilymphatic fluid and endolymphatic fluid is complex [4]. Perilymphatic fluid has a composition fairly similar to that of cerebrospinal fluid. Endolymphatic fluid has a very different ionic composition from that of perilymphatic fluid [5], yet there is some evidence that endolymph is derived from perilymph [6].

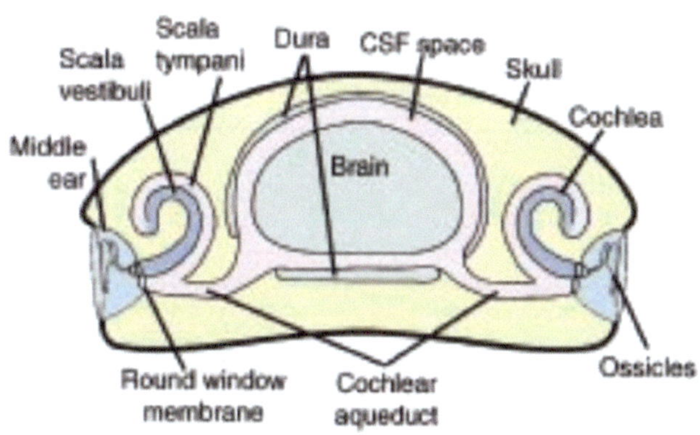

Fig. 104.1 Schematic of ear anatomy, emphasizing the connection between the perilymphatic space of the cochlea and the cerebrospinal fluid space. (From Stover et al. [1]. Reproduced with written permission from Springer Nature B.V.)

M. Cherchi, *Otoneurology and Vestibular Medicine*, https://doi.org/10.1007/978-3-031-94841-1_104

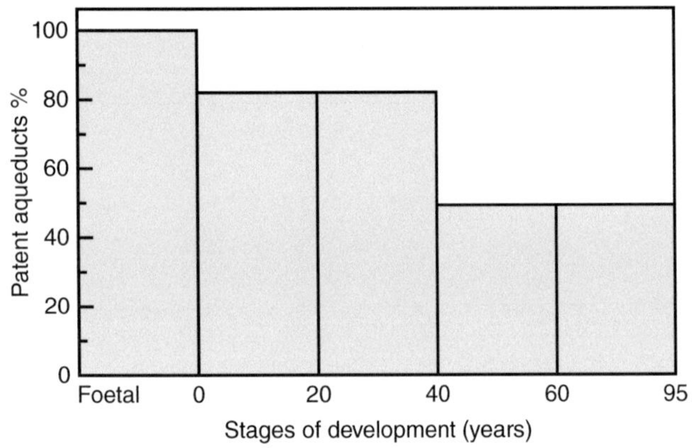

Fig. 104.2 Percent of patent cochlear aqueducts by age. (From Marchbanks and Reid [3], adapted from Wlodyka [2]. Reproduced with written permission from Informa UK Limited)

Pressure Fluctuations, Intracranial, and Intralabyrinthine

The relationships between endolymphatic fluid, perilymphatic fluid, and cerebrospinal fluid is inadequately understood. A commonly proposed model regarding, "perilymphatic and endolymphatic pressure balance is generally linked to the theory that the endolymphatic sac transmits cerebrospinal fluid (CSF) pressure changes to the endolymph to equalize CSF pressure changes transmitted to the perilymph via the cochlear aqueduct" [7]. This model's simplicity is appealing, but animal experiments suggest that the relationship is not so straightforward [3, 7–10].

Endolymph "is believed to be reabsorbed along the scala media by the stria vascularis, Reissner's membrane, and the planum semilunatum, and at the saccus endolymphaticus which lies intracranially among the dural sheaths" [11].

It is thought that when intracranial pressure decreases, this results in perilymphatic fluid exiting the perilymphatic space and entering the intracranial compartment. This effectively permits the endolymphatic space to expand and "inflate" with more endolymphatic fluid, which results in "endolymphatic hydrops" (excessive endolymphatic fluid) . This state of endolymphatic hydrops is similar to what occurs in Ménière's disease, and is also associated with a similar pattern of hearing loss (predominantly low frequency sensorineural).

In contrast, when intracranial pressure increases, equilibration of that increased pressure will entail an increase in pressure in the perilymphatic space, a state of affairs sometimes referred to as perilymphatic hypertension [12, 13]. Since the bony labyrinth is incompressible, perilymphatic hypertension will exert pressure on the membranous labyrinth, which may decrease the volume of endolymphatic fluid.

Both increased endolymphatic fluid ("endolymphatic hydrops") and decreased endolymphatic fluid can have deleterious effects on cochlear function.

Increased Intracranial Pressure

Numerous case reports have documented hearing loss associated with ***increased intracranial pressure*** [14], and correction of intracranial pressure results in improvement (if not complete restoration) of hearing [11, 15–17].

Dixon and colleagues [15] performed serial audiometry in a 14-year-old girl with hydrocephalus who underwent ventriculostomy without shunt placement and demonstrated clear improvement in sensorineural hearing postoperatively compared to preoperatively (Figs. 104.3 and 104.4).

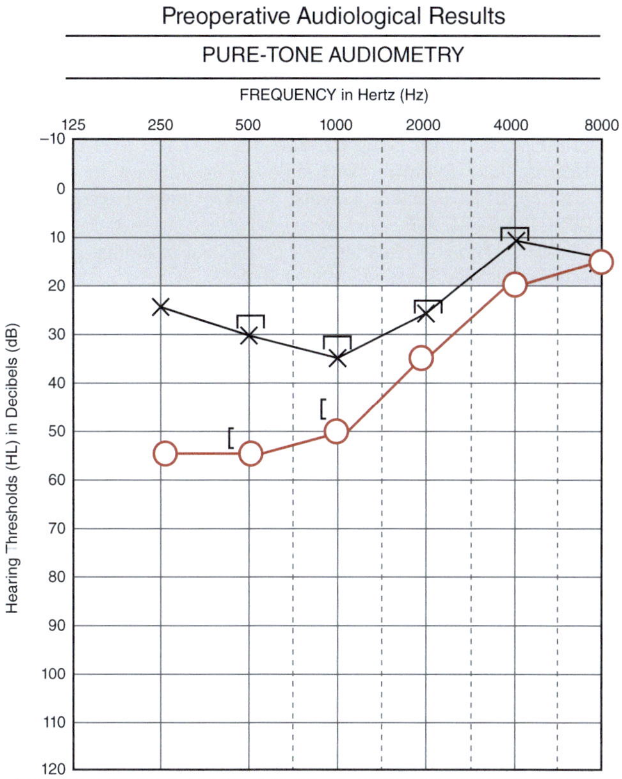

SPEECH AUDIOMETRY

Ear	PTA	SRT	Word Rec (%)
Right	47 dB HL	50 dB HL	88%
Left	30 dB HL	25 dB HL	100%

Fig. 104.3 Audiogram prior to ventriculostomy exhibiting mild left and moderate right up-sloping sensorineural hearing loss. (From Dixon et al. [15]. Reproduced with written permission from John Wiley and Sons Limited)

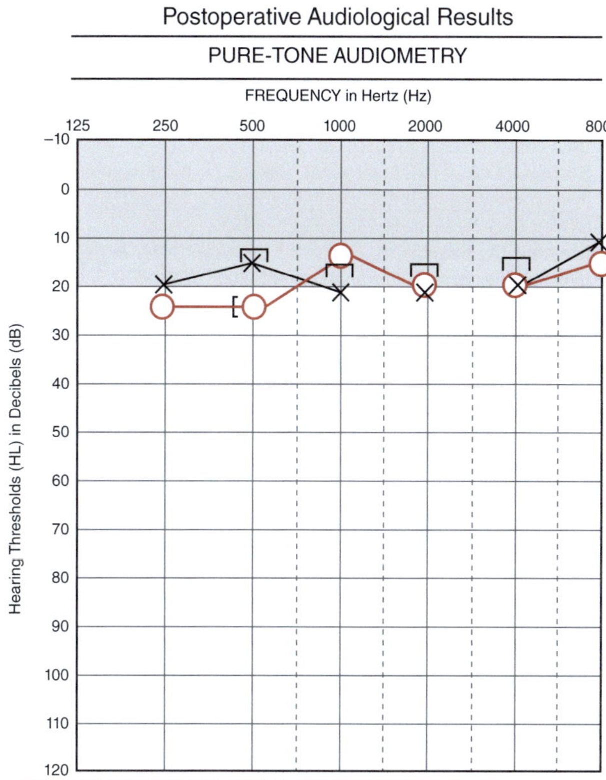

SPEECH AUDIOMETRY

Ear	PTA	SRT	Word Rec (%)
Right	20 dB HL	10 dB HL	96%
Left	17 dB HL	5 dB HL	100%

Fig. 104.4 Audiogram 2 months after ventriculostomy revealing resolution of sensorineural hearing loss. (Reproduced with written permission from John Wiley and Sons Limited)

Decreased Intracranial Pressure

Numerous case reports have documented that ***decreased intracranial pressure*** (such as following neurosurgical procedures in which cerebrospinal fluid is temporarily depleted) is also associated with hearing loss [18–27]. Walsted [25] studied 126 patients who underwent "different neurosurgical procedures that involved puncture or incision of the dura" and observed that, "The size of the [resulting] hearing loss, the number of affected patients, and the number of frequencies involved increased with increasing amount of CSF [cerebrospinal fluid] lost." Walsted and colleagues [26] also conducted a prospective study of 32 neurosurgical patients and found that the postsurgical hearing loss recovered in 1 week.

Note that this hearing loss can occur after shunt placement even in patients who (presurgically) had ***normal pressure*** hydrocephalus [28].

References

1. Stover T, Yagi M, Raphael Y. Transduction of the contralateral ear after adenovirus-mediated cochlear gene transfer. Gene Ther. 2000;7(5):377–83.
2. Włodyka J. Studies on cochlear aqueduct patency. Ann Otol Rhinol Laryngol. 1978;87(1 Pt 1):22–8.
3. Marchbanks RJ, Reid A. Cochlear and cerebrospinal fluid pressure: their inter-relationship and control mechanisms. Br J Audiol. 1990;24(3):179–87.
4. Sterkers O, Ferrary E, Amiel C. Production of inner ear fluids. Physiol Rev. 1988;68(4):1083–128.
5. Gagov H, Chichova M, Mladenov M. Endolymph composition: paradigm or inevitability? Physiol Res. 2018;67(2):175–9.
6. Thalmann R, Thalmann I. Source and role of endolymph macromolecules. Acta Otolaryngol. 1999;119(3):293–6.
7. Carlborg BI, Farmer JC Jr. Transmission of cerebrospinal fluid pressure via the cochlear aqueduct and endolymphatic sac. Am J Otolaryngol. 1983;4(4):273–82.
8. Kishimoto S, Nagahara K, Fisch U, Dillier N. Inner ear pressure measurements. Effects of obstruction of the cochlear aqueduct and endolymphatic duct on the perilymphatic pressure. Otolaryngol Clin N Am. 1983;16(1):21–35.

9. Carlborg B, Densert B, Densert O. Functional patency of the cochlear aqueduct. Ann Otol Rhinol Laryngol. 1982;91(2 Pt 1):209–15.
10. Carlborg BI, Konradsson KS, Carlborg AH, Farmer JC Jr, Densert O. Pressure transfer between the perilymph and the cerebrospinal fluid compartments in cats. Am J Otol. 1992;13(1):41–8.
11. Barlas O, Gokay H, Turantan MI, Baserer N. Adult aqueductal stenosis presenting with fluctuating hearing loss and vertigo: report of two cases. J Neurosurg. 1983;59(4):703–5.
12. Miranda JA, Suzuki FA, de Carvalho Borges MH. Perilymphatic hyperension. Braz J Otorhinolaryngol. 2006;72(3):430.
13. Paparella MM, Schachern PA, Goycoolea MV. Perilymphatic hypertension. Otolaryngol Head Neck Surg. 1988;99(4):408–13.
14. Tandon PN, Sinha A, Kacker SK, Saxena RK, Singh K. Auditory function in raised intracranial pressure. J Neurol Sci. 1973;18(4):455–67.
15. Dixon JF, Jones RO. Hydrocephalus-associated hearing loss and resolution after ventriculostomy. Otolaryngol Head Neck Surg. 2012;146(6):1037–9.
16. Jamshidi A, Glidewell C, Murnick J, Magge S, Reilly BK. Resolution of bilateral sensorineural hearing loss following ventriculoperitoneal shunt and literature review. Int J Pediatr Otorhinolaryngol. 2017;100:141–4.
17. Sammons VJ, Jacobson E, Lawson J. Resolution of hydrocephalus-associated sensorineural hearing loss after insertion of ventriculoperitoneal shunt. J Neurosurg Pediatr. 2009;4(4):394–6.
18. Girardi FP, Cammisa FP Jr, Sangani PK, Parvataneni HK, Khan SN, Grewal H, et al. Sudden sensorineural hearing loss after spinal surgery under general anesthesia. J Spinal Disord. 2001;14(2):180–3.
19. Guillaume DJ, Knight K, Marquez C, Kraemer DF, Bardo DM, Neuwelt EA. Cerebrospinal fluid shunting and hearing loss in patients treated for medulloblastoma. J Neurosurg Pediatr. 2012;9(4):421–7.
20. Hussain SS, Heard CM, Bembridge JL. Hearing loss following spinal anaesthesia with bupivacaine. Clin Otolaryngol Allied Sci. 1996;21(5):449–54.
21. Lim HW, Shim BS, Yang CJ, Kim JH, Cho YH, Cho YS, et al. Hearing loss following ventriculoperitoneal shunt in communicating hydrocephalus patients: a pilot study. Laryngoscope. 2014;124(8):1923–7.
22. Michel O, Brusis T. Hearing loss as a sequel of lumbar puncture. Ann Otol Rhinol Laryngol. 1992;101(5):390–4.
23. Pogodzinski MS, Shallop JK, Sprung J, Weingarten TN, Wong GY, McDonald TJ. Hearing loss and cerebrospinal fluid pressure: case report and review of the literature. Ear Nose Throat J. 2008;87(3):144–7.
24. Spirakis SE, Hurley RM. Unilateral hearing loss in children with shunt-treated hydrocephalus. J Am Acad Audiol. 2003;14(9):510–7.
25. Walsted A. Effects of cerebrospinal fluid loss on hearing. Acta Otolaryngol Suppl. 2000;543:95–8.
26. Walsted A, Nielsen OA, Borum P. Hearing loss after neurosurgery. The influence of low cerebrospinal fluid pressure. J Laryngol Otol. 1994;108(8):637–41.
27. Walsted A, Salomon G, Thomsen J, Tos M. Hearing decrease after loss of cerebrospinal fluid. A new hydrops model? Acta Otolaryngol. 1991;111(3):468–76.
28. van Veelen-Vincent ML, Delwel EJ, Teeuw R, Kurt E, de Jong DA, Brocaar MP, et al. Analysis of hearing loss after shunt placement in patients with normal-pressure hydrocephalus. J Neurosurg. 2001;95(3):432–4.

Part XIV

Diagnoses—Neurologically-Mediated Visual Disorders

Visual Snow Syndrome

105

Practical Summary

Visual snow (VS) refers to the experience of perceiving the entire visual scene as distorted by television static. In most patients, VS occurs in the context of several other visual and nonvisual symptoms, the constellation of which is referred to as *visual snow syndrome* (VSS). This is uncommon but probably underrecognized. In most patients, VSS begins spontaneously (primary VSS), while in others, it appears to occur following some other insult or event (secondary VSS), such as head trauma, use of medication, or recreational drugs. There is some clinical overlap with migrainous visual aura, and there is debate in the literature about whether VSS lies on a continuum with migraine, or represents a discrete clinical entity. The underlying pathophysiology is unknown, and investigational studies report imaging or electrophysiologic abnormalities at multiple levels in the visual system. The peak age of onset is the third decade, with a slight (53%) female preponderance. Symptoms may begin as intermittent, but in most patients eventually become continuous. Physical examination, bedside ocular motor examination, and standard brain MRI are normal. Patients should be examined by a general ophthalmologist or neuro-ophthalmologist to exclude primary ophthalmic disease. Treatment has been attempted with a broad variety of medications, with largely unsatisfactory results. There is emerging evidence of benefit from vision therapy or treatment with visual filters, and these interventions incur no medical risk. Prognosis is guarded; symptoms can last for years.

Introduction

Although patients with visual *snow syndrome (VSS)* do not usually describe any symptoms of disequilibrium, they are nevertheless sometimes referred to otoneurology. Outside of ophthalmology and neuro-optometry, VSS is not known by most medical practitioners, and absent any vestibular symptoms, the rationale for referral to otoneurology is unclear.

In 1995, Liu and colleagues [1] reported on ten migraine patients with "persistent positive visual phenomena lasting months to years." As additional cases were reported in the literature, this phenomenon eventually became known as visual snow (VS).

Visual snow is currently taken to the experience of a "continuous visual disturbance resembling a badly tuned analogue television" [2], "seeing numerous tiny snow-like dots throughout the visual field" [3], "innumerable flickering dots throughout the visual field, similar to 'TV static'" [4].

Although VS can occur in isolation, it more commonly occurs in conjunction with several other visual and sometimes nonvisual symptoms, in which case the constellation of symptoms is referred to as visual snow syndrome (VSS).

Schankin and colleagues [5] studied 78 patients with VSS and reported the additional symptoms listed in Table 105.1.

Table 105.1 Visual and nonvisual symptoms reported in 78 patients with visual snow syndrome

	All patients with visual snow $n = 78$	
Palinopsia ("after images")	67	86%
Palinopsia ("trailing")	47	60%
Floaters	63	81%
Blue field entoptic phenomenon	62	79%
Spontaneous photopsia	49	63%
Self-light of the eye	41	53%
Photophobia	58	74%
Light is too bright	43	74%
Light is painful	15	26%
Nyctalopia	53	68%
Due to visual snow	25	47%
Due to palinopsia	3	6%
Due to halos	3	6%
Due to a combination	22	41%
Tinnitus	48	62%
Concentration problems	47	60%
Lethargy	43	55%
Irritability	29	37%

From Schankin et al. [5]

M. Cherchi, *Otoneurology and Vestibular Medicine*, https://doi.org/10.1007/978-3-031-94841-1_105

Some investigators draw a distinction between primary and secondary VSS [4, 6]. Primary VSS occurs spontaneously, whereas secondary VSS appears to occur following some insult or event, such as concussion [6–8], medication [9], infection or use of hallucinogenic and other recreational drugs including cannabis, ecstasy, cocaine, or amphetamines [5, 6].

There are no universally accepted criteria for diagnosing VSS, but Metzler and Robinson [4], building on work by Schankin and colleagues [5], proposed the following diagnostic criteria for visual snow syndrome:

1. Visual Snow: dynamic, continuous, tiny dots in the entire visual field lasting longer than 3 months
2. Presence of at least two additional visual symptoms from the following categories:
   (a) Palinopsia: afterimages or trailing of moving objects
   (b) Photophobia
   (c) Nyctalopia (impaired night vision)
   (d) Other persistent positive visual phenomena including (but not limited to): enhanced entoptic phenomena (excessive floaters or blue field entoptic phenomenon), kaleidoscope-type colors with eyes open or closed, spontaneous photopsias
3. Symptoms are not consistent with typical migraine visual aura
4. Symptoms are not better explained by another disorder

Some investigators maintain that visual snow is distinct from migrainous visual aura [5], but not all investigators agree that visual snow syndrome comprises a discrete clinical entity [10].

Table 105.2 Demographics of patients with visual snow syndrome

	All patients with visual snow $n = 78$	
Age	30 ± 10	
Females (female:Male)	41	1.1: 1
History of headache	68	87%
Worsening with visual snow	29 of 68	37%
Chronic headache	17 of 68	22%
History of migraine	46	59%
History of typical migraine aura	21	27%
Only visual	18 of 21	86%
Visual and others	3 of 21	14%
Family history of headache	44	56%
Family history of headache in patients without history of headache	7 of 10	70%
History of anxiety	12 of 53	23%
History of depression	11 of 53	21%
Visual snow present since childhood	19	24%
Later onset of visual snow	*59*	*76%*
Start age (years)	21 ± 9	
Duration (years)	8 ± 10	
Continuous from beginning	54 of 59	92%
Initial episodic	5 of 59	8%
Constant from start	*32*	*41%*
Duration of constant symptoms (years)	14 ± 18	
Progressive worsening	*33*	*42%*
Duration of progressive worsening (years)	6 ± 8	
Now constant symptoms	16 of 33	48%
Duration of constant symptoms (years)	3 ± 3	
Stepwise worsening	*10*	*13%*
Now constant	10 out of 10	100%
Duration of constant symptoms (years)	3 ± 3	
Both progressive and stepwise	*3*	*4%*

From Schankin et al. [5]

Epidemiology

Visual snow syndrome is uncommon. In 2018, Metzler and Robertson [4] stated that only 200 reports had been published. Referral patterns since then suggest that the diagnosis is being increasingly recognized, or at least increasingly suspected.

Schankin and colleagues [5] studied 78 patients with VSS and reported the demographics listed in Table 105.2.

Notable features include the age of onset was 21 ± 9 years, symptom duration was 8 ± 10 years, and there was a slight female preponderance at 41/78 (53%) of patients.

Genetics

As of this writing, VSS has not been described as occurring in familial patterns, nor has any genetic pattern been published.

Pathophysiological Mechanism of Disease

The underlying pathobiology of VSS is unknown. Data from imaging and electrophysiology studies suggest several possibilities at multiple levels of the visual system.

Some studies suggest dysfunction as early in the visual pathway as the retina:

- Alissa and colleagues [11] studied 7 patients with visual snow (VS) and 9 healthy controls by exposing subjects to various colored stimuli and measuring pupillary responses. They observed that, "The pupil responses in three of the VS patients lacked the rapid recovery phase following the initial constriction to stimulus onset," and concluded that, "The absence of recovery following the initial constriction of the pupil in three of the VS patients deviates from normal responses and suggests the presence of a more sustained retinal afferent signal that drives the pupil response." In other words, they are suggesting that at least part of the pathophysiology localizes to the *retina*.

Some studies identify dysfunction in primary visual cortex:

- McKendrick and colleagues [12] compared 16 visual snow patients with 18 healthy controls on a series of visual perceptual tasks. They reported that, "Our study demonstrates that visual perceptual measures involving the suprathreshold processing of contrast and luminance are abnormal in a group of individuals with visual snow," and concluded that "Our data are consistent with elevated excitability in *primary visual cortex*; however, further research is required to provide more direct evidence for this proposed mechanism."
- White and colleagues [13] performed a literature review on visual snow syndrome and concluded that "VSS is likely associated with either *hyperactive visual cortices* or, alternatively, impaired processing of simultaneous afferent information projecting to cortex."
- Yildiz and colleagues [14] compared repetitive pattern reversal visual evoked potentials from transcranial magnetic stimulation-induced phosphene in 17 VS patients (10 with migraine, 7 without migraine) and 12 healthy controls. They reported that "The loss of habituation and lower threshold for occipital cortex excitability were demonstrated electrophysiologically in VS patients."

Some studies suggest that the dysfunction lies in secondary visual cortex involved in higher-level visual processing:

- Schankin and Goadsby [15] review fMRI data in patients with visual snow (VS) and note "hypermetabolism in the *supplementary visual cortex*," adding that, "the primary visual cortex (cuneus) did not show hypermetabolism in VS patients arguing for VS being a *disorder of visual post-processing outside of the retino-geniculo-cortical pathway*."
- Eren and colleagues [16] compared visual evoked potentials in 18 visual snow patients with age-matched migraineurs and healthy controls. They reported that "Visual evoked potentials from patients with VS demonstrated increased N145 latency and reduced N75-P100 amplitudes," and concluded that "These findings are in agreement with the idea that the primary disturbance in VS is a dysfunction of the *visual association cortex*."
- Aldusary and colleagues [17] compared fMRI in 19 visual snow patients and 16 age- and sex-matched healthy control subjects. They reported that "Patients with VS showed hyperconnectivity between extrastriate visual and inferior temporal brain regions and also between prefrontal and parietal (angular cortex) regions," and interpreted this as showing "VS to be associated with both *functional and structural changes in the early and higher visual cortex, as well as the temporal cortex*."

Some studies characterize VSS as a dysfunction at a broader "network" level:

- Klein and Schankin [18] performed a literature review of 53 studies on VSS and concluded that "The studies that have been included in this review demonstrate structural, functional, and metabolic alterations in the *primary and/or secondary visual areas* of the brain. Beyond that, results indicate a disruption in the *pre-cortical visual pathways and large-scale networks* including the default mode network and the salience network."
- Fraser [19] reviewed literature suggesting that VSS "is due to a widespread cortical dysfunction. *Cortical hyper-excitability* coupled with changes in *thalamocortical pathways* and *higher level salience network* controls have all shown differences in patients with visual snow syndrome compared to controls."
- Strik and colleagues [20] compared resting-state MRI in 40 VSS patients and 60 healthy controls. They reported that "This study revealed reduced dynamic variation in modularity and local efficiency strength in the VSS brain, suggesting that *brain network dynamics* are less variable in terms of segregation and local clustering."

Some investigators suggest that VSS is one presentation along a continuum of disease that may include migraine [21] or other "perceptual disorders" such as persistent postural perceptual dizziness, fibromyalgia, and chronic tinnitus [22].

Clinical Presentation

Symptoms of visual snow syndrome (VSS) include not just visual snow itself, but also other visual symptoms, and often nonvisual symptoms as well.

Schankin and colleagues [5] studied 78 patients with VSS and reported that, besides the symptom of visual snow itself, other visual and nonvisual symptoms included the following:

- Visual symptoms:
  - Palinopsia. This refers to the perception that, after an object moves, it leaves after-images (in 86% of patients), visual trailing (in 60%), or visual ghosting.
  - Excessive floaters (81%).
  - "Blue field entopic phenomenon" (79%) which refers to "little bright dots traveling on a wiggly path when looking at a blue background."
  - Photophobia, manifesting as light being perceived as too bright (74%) or painful (26%).
  - Spontaneous photopsia (63%) (flashes of light).
  - "Self-light of the eye" (53%) which refers to "clouds, swirls, or waves with eyes closed."
  - Nyctalopia (68%) (impaired night vision).
- Cognitive symptoms:
  - Concentration problems (60%).
  - Lethargy (55%).
  - Irritability (37%).
- Tinnitus (62%).

Lauschke and colleagues [23] studied a cohort of 32 patients with VS and reported that 22% complained of tremor.

Solly and colleagues [24] add that VSS patients endorse considerable psychiatric comorbidity.

We find it noteworthy that, besides perhaps nyctalopia, all of the visual abnormalities are positive phenomena. Notably absent are negative visual phenomena (such as scotomata) or visual distortions (tunnel vision, macropsia) that occur in migraine with visual aura.

Figure 105.1, from Metzler and Robertson [4], illustrates some of the visual phenomena experienced in visual snow syndrome.

Fig. 105.1 Examples of some of the visual phenomena experienced in visual snow syndrome. Panel (**a**): Normal vision. Panel (**b**): Visual snow. Panel (**c**): Palinopsia (both visual trailing behind the car and a ghost-like afterimage next to the sign). Panel (**d**): Other frequent or persistent positive phenomena (including increased floaters, blue field entopic phenomena, photopsia, and colored blobs). Panel (**e**): Nyctalopia (impaired night vision). Panel (**f**): Composite illustration showing the possible level of visual disability these symptoms may cause when combined. (From Metzler and Robertson [4]. Reproduced with written permission from Mayo Foundation for Medical Education and Research)

With respect to chronology, Schankin and colleagues [5] studied 78 patients with VSS and reported that:

- Of 78 patients, 19 (24%) described symptoms since childhood ("as long as the patient can recall"), and 59 (76%) at some point thereafter ("patient recalls period of life without visual snow").
- Of 78 patients, 32 (41%) described symptoms as being continuous since onset, 33 (42%) as gradually progressive worsening, 10 (13%) as stepwise worsening, and 3 (4%) as both progressive and stepwise worsening.

Physical Examination

In patients with VSS who are otherwise healthy, physical examination should be normal [25].

Ocular Motor Examination

As of this writing, there were no published studies of bedside ocular motor findings in VSS.

Testing: Vestibular

As of this writing, there were no published studies of standard otovestibular test results in patients with VSS.

However, several studies describe various abnormalities on specific saccade tasks in VSS patients compared to controls.

- Solly and colleagues [26] compared 64 VSS patients (32 with migraine and 32 without migraine) and 23 healthy controls on several saccade tasks: prosaccades (PS), antisaccades (AS), and interleaved AS-PS tasks. They observed that "compared to controls, the VSS group generated significantly shortened PS latencies ($p = 0.029$) and an increased rate of AS errors ($p = 0.001$), irrespective of the demands placed on visual processing (i.e., task context)," and concluded that "The absence of any additional switch cost on the AS-PS task in VSS suggests that the PS latency and AS error differences are attributable to a speeded PS response rather than to impaired executive processes more commonly implicated in poorer AS performance."
- Foletta and colleagues [27] compared 40 VSS patients (20 with migraine, 20 without migraine) and 20 healthy controls on the "inhibition of return" (IOR) phenomenon that "describes the relative slowing of a response to a validly cued location compared to invalidly cued location with longer cue/target intervals." They concluded that "Our results are consistent with the proposal that in VSS, a shift of attention elicits a stronger increase in saccade-related activity than healthy controls."

These tests are interesting, though not (yet) broadly applied in clinical practice because of the specialized equipment needed. Both of the cited studies used EyeLink® 1000 video oculography hardware that has a frame rate acquisition of 1000 Hz (which is not available on most commercial videonystagmography software).

Imaging

Several MRI studies report morphological abnormalities in VSS patients.

- Aldusary and colleagues [17] compared fMRI in 19 visual snow patients and 16 age- and sex-matched healthy control subjects. They reported that VSS patients "showed increased GMV [gray matter volume] in the right lingual gyrus."
- Puledda and colleagues [28] compared MRI-based voxel morphometry in 24 VSS patients and 24 age- and sex-matched healthy controls. They reported that "Compared to healthy controls, patients with visual snow syndrome had increased gray matter volume in the left primary and secondary visual cortices, the left visual motion area V5, and the left cerebellar crus I/lobule VI area."
- Strik and colleagues [29] used a 7-Tesla MRI to compare 40 VSS patients with 43 healthy controls. They reported that "widespread changes in grey matter microstructure, which followed a caudal-rostral pattern and affected the occipital cortices most profoundly."

The studies cited above lie in the domain of research. In clinical practice, regular MRI is unlikely to reveal findings that confirm a diagnosis of VSS; the main value of MRI is to exclude competing diagnoses.

Histopathology

As of this writing, there were no published autopsy reports of patients with VSS.

Differential Diagnosis

Each individual component of VSS can occur in other diseases as well. Since the majority of the symptoms in VSS involve visual perception, it is important for patients to have a general ophthalmological, or neuro-ophthalmological evaluation, in order to exclude primary ophthalmological disease; patients with VSS who are otherwise healthy should have normal ophthalmological examinations [30].

Patel and colleagues [31] report a case of a patient diagnosed with VSS who in fact had birdshot chorioretinopathy.

Treatment: Visual Rehabilitation

There is emerging evidence that vision therapy may play a role in the management of VSS [32]. We favor a trial of vision therapy as it incurs no medical risk.

Treatment: Visual Filters

Lauschke and colleagues [23] studied a cohort of 32 patients with visual snow (VS) and reported that "ninety-two percent of patients felt that symptoms improved during the testing with a particular coloured filter," "particularly in the yellow-blue color spectrum." This finding still has to be replicated, but we support a trial of this intervention given that it incurs no medical risk. Eren and Schankin [25] note that this visual intervention can be applied for everyday use.

Treatment: Medication

Bou Ghannam and Pelak [3] reviewed treatment trials medications used for treatment "persistent visual phenomena and/or visual snow." The findings are summarized by results and class of evidence in Table 105.3.

In short, there was only one Class I study, which concluded that sumatriptan was ineffective. There was one Class II study which concluded that ketamine was effective. There were two Class II studies that nimodipine and metoprolol were ineffective. The remaining studies (showing efficacy, inefficacy, or mixed results) were all Class III or Class IV.

Despite these weak data, Bou Ghannam and colleagues [3] recommended the following treatments:

- Lamotrigine, starting at 25 mg per day, tapering up to 200–300 mg per day in divided doses.
- Acetazolamide, starting at 250 mg per day, tapering up to 1000 mg per day in divided doses.
- Verapamil ER, starting at 120 mg per day, tapering up to 240 mg per day.

Eren and Schankin [25] reviewed studies of other medications as well, including fluoxetine, esketamine, magnesium, amitriptyline, sertraline, baclofen, gabapentin, carbamazepine, coenzyme Q10, methylphenidate, flubiprofen, duloxetine, paroxetine, pizotifen, and dothiepin. They concluded that "Currently, there is no treatment available that would effectively improve VSS," but added that "The best data is available for lamotrigine. It has been used in 36 patients, was effective in 22.2% including one total response, and did not result in worsening. The second most often used medication was topiramate. It is both, an anti-epileptic drug and usually highly effective for migraine. Only two out of 13 patients with VSS (15.4%) had some effect, and one had worsening."

Puledda and colleagues [33] concluded that "medications are generally ineffective in VSS."

Table 105.3 Summary of studies of medications that have been tried in management of visual snow syndrome

	Effective	Ineffective	Mixed results
Class I evidence	–	Sumatriptan (one RCT)	–
Class II evidence	Ketamine (one RCT)	Metoprolol (one RCT) Nimodipine (one RCT)	–
Class III evidence	Lamotrigine (one study)	Nifedipine (one RCT)	–
Class IV evidence	Acetazolamide (one case report) Aspirin (one case report) Furosemide (two case series) Ketamine (one study, one case report) Lamotrigine (four studies) Naproxen (one case report) Nifedipine (one case report) Picotamide (one study)	Acetazolamide (two series) Divalproex (one case series) Flunarizine (one case report, one case series) Ketamine (one case series) Lamotrigine (one case series) Nortriptyline (one case series) Propranolol (one case series) Topiramate (one study) Verapamil (one case report)	Lamotrigine (one case series) Propranolol (one case series)

Data are from Bou Ghannam and Pelak [3]

Prognosis

With or without treatment, symptoms of VSS can last years [2].

References

1. Liu GT, Schatz NJ, Galetta SL, Volpe NJ, Skobieranda F, Kosmorsky GS. Persistent positive visual phenomena in migraine. Neurology. 1995;45(4):664–8.
2. Graber M, Scutelnic A, Klein A, Puledda F, Goadsby PJ, Schankin CJ. Natural course of visual snow syndrome: a long-term follow-up study. Brain Commun. 2022;4(5):fcac230.

3. Bou Ghannam A, Pelak VS. Visual snow: a potential cortical hyperexcitability syndrome. Curr Treat Options Neurol. 2017;19(3):9.
4. Metzler AI, Robertson CE. Visual snow syndrome: proposed criteria, clinical implications, and pathophysiology. Curr Neurol Neurosci Rep. 2018;18(8):52.
5. Schankin CJ, Maniyar FH, Digre KB, Goadsby PJ. 'Visual snow' – a disorder distinct from persistent migraine aura. Brain. 2014;137(Pt 5):1419–28.
6. Mehta DG, Garza I, Robertson CE. Two hundred and forty-eight cases of visual snow: a review of potential inciting events and contributing comorbidities. Cephalalgia. 2021;41(9):1015–26.
7. Ciuffreda KJ, Han ME, Tannen B, Rutner D. Visual snow syndrome: evolving neuro-optometric considerations in concussion/mild traumatic brain injury. Concussion. 2021;6(2):CNC89.
8. Werner RN, Gustafson JA. Case report: visual snow syndrome after repetitive mild traumatic brain injury. Optom Vis Sci. 2022;99(4):413–6.
9. Naguy A, Naguy C, Singh A. Probable methylphenidate-related reversible "visual snow" in a child with ADHD. Clin Neuropharmacol. 2022;45(4):105–6.
10. Sampatakakis SN, Lymperopoulos L, Mavridis T, Karagiorgis G, Papadopoulos C, Deligianni CI, et al. Visual snow: a systematic review and a case series. Cephalalgia. 2022;42(13):1409–19.
11. Alissa R, Bi W, Bessero A, Plant G, Barbur JL. Vision in subjects with hyperawareness of afterimages and "visual snow". Acta Ophthalmol. 2012;90(s249). https://doi.org/10.1111/j.1755-3768.2012.F066.x.
12. McKendrick AM, Chan YM, Tien M, Millist L, Clough M, Mack H, et al. Behavioral measures of cortical hyperexcitability assessed in people who experience visual snow. Neurology. 2017;88(13):1243–9.
13. White OB, Clough M, McKendrick AM, Fielding J. Visual snow: visual misperception. J Neuroophthalmol. 2018;38(4):514–21.
14. Yildiz FG, Turkyilmaz U, Unal-Cevik I. The clinical characteristics and neurophysiological assessments of the occipital cortex in visual snow syndrome with or without migraine. Headache. 2019;59(4):484–94.
15. Schankin CJ, Goadsby PJ. Visual snow–persistent positive visual phenomenon distinct from migraine aura. Curr Pain Headache Rep. 2015;19(6):23.
16. Eren O, Rauschel V, Ruscheweyh R, Straube A, Schankin CJ. Evidence of dysfunction in the visual association cortex in visual snow syndrome. Ann Neurol. 2018;84(6):946–9.
17. Aldusary N, Traber GL, Freund P, Fierz FC, Weber KP, Baeshen A, et al. Abnormal connectivity and brain structure in patients with visual snow. Front Hum Neurosci. 2020;14:476.
18. Klein A, Schankin CJ. Visual snow syndrome as a network disorder: a systematic review. Front Neurol. 2021;12:724072.
19. Fraser CL. Visual snow: updates on pathology. Curr Neurol Neurosci Rep. 2022;22(3):209–17.
20. Strik M, Clough M, Solly EJ, Glarin R, White OB, Kolbe SC, et al. Brain network dynamics in people with visual snow syndrome. Hum Brain Mapp. 2023;44(5):1868–75.
21. Puledda F, Schankin C, Goadsby PJ. Visual snow syndrome: a clinical and phenotypical description of 1,100 cases. Neurology. 2020;94(6):e564–e74.
22. Klein A, Schankin CJ. Visual snow syndrome, the spectrum of perceptual disorders, and migraine as a common risk factor: a narrative review. Headache. 2021;61(9):1306–13.
23. Lauschke JL, Plant GT, Fraser CL. Visual snow: a thalamocortical dysrhythmia of the visual pathway? J Clin Neurosci. 2016;28:123–7.
24. Solly EJ, Clough M, Foletta P, White OB, Fielding J. The psychiatric symptomology of visual snow syndrome. Front Neurol. 2021;12:703006.
25. Eren O, Schankin CJ. Insights into pathophysiology and treatment of visual snow syndrome: a systematic review. Prog Brain Res. 2020;255:311–26.
26. Solly EJ, Clough M, McKendrick AM, Foletta P, White OB, Fielding J. Ocular motor measures of visual processing changes in visual snow syndrome. Neurology. 2020;95(13):e1784–e91.
27. Foletta PJ, Clough M, McKendrick AM, Solly EJ, White OB, Fielding J. Delayed onset of inhibition of return in visual snow syndrome. Front Neurol. 2021;12:738599.
28. Puledda F, Bruchhage M, O'Daly O, Ffytche D, Williams SCR, Goadsby PJ. Occipital cortex and cerebellum gray matter changes in visual snow syndrome. Neurology. 2020;95(13):e1792–e9.
29. Strik M, Clough M, Solly EJ, Glarin R, White OB, Kolbe SC, et al. Microstructure in patients with visual snow syndrome: an ultra-high field morphological and quantitative MRI study. Brain Commun. 2022;4(4):fcac164.
30. Yoo YJ, Yang HK, Choi JY, Kim JS, Hwang JM. Neuro-ophthalmologic findings in visual snow syndrome. J Clin Neurol. 2020;16(4):646–52.
31. Patel RC, Vitale AT, Creel DJ, Digre KB. Not all that flickers is snow. J Neuroophthalmol. 2021;41(1):e97–e9.
32. Tsang T, Shidlofsky C, Mora V. The efficacy of neuro-optometric visual rehabilitation therapy in patients with visual snow syndrome. Front Neurol. 2022;13:999336.
33. Puledda F, Vandenbussche N, Moreno-Ajona D, Eren O, Schankin C, Goadsby PJ. Evaluation of treatment response and symptom progression in 400 patients with visual snow syndrome. Br J Ophthalmol. 2022;106(9):1318–24.

106 Misokinesia

Overview

Misokinesia ("hatred of movement") has been described as "a psychological phenomenon that is defined as a strong negative affective or emotional response to the sight of someone else's small and repetitive movements, such as seeing someone mindlessly fidgeting with a hand or foot" [2]. This has been recognized relatively recently (2021), and there is almost no literature on the topic. In some respects, it resembles misophonia ("hatred of sounds"), about which there is a better developed literature [1, 3–5, 7].

As of this writing, one case report [6] of a patient with both misophonia and misokinesia described symptomatic improvement on propranolol 60 mg.

We have observed misokinesia in a subpopulation of oto-neurology patients—specifically, patients with migraine and migraine-associated vertigo (MAV). At present, we view this as a manifestation of migrainous visual hypersensitivity, but research on the topic is needed.

References

1. Ferrer-Torres A, Gimenez-Llort L. Misophonia: a systematic review of current and future trends in this emerging clinical field. Int J Environ Res Public Health. 2022;19 https://doi.org/10.3390/ijerph19116790.
2. Jaswal SM, De Bleser AKF, Handy TC. Misokinesia is a sensitivity to seeing others fidget that is prevalent in the general population. Sci Rep. 2021;11:17204. https://doi.org/10.1038/s41598-021-96430-4.
3. Neacsiu AD, Szymkiewicz V, Galla JT, Li B, Kulkarni Y, Spector CW. The neurobiology of misophonia and implications for novel, neuroscience-driven interventions. Front Neurosci. 2022;16:893903. https://doi.org/10.3389/fnins.2022.893903.
4. Potgieter I, MacDonald C, Partridge L, Cima R, Sheldrake J, Hoare DJ. Misophonia: a scoping review of research. J Clin Psychol. 2019;75:1203–18. https://doi.org/10.1002/jclp.22771.
5. Siepsiak M, Dragan W. Misophonia – a review of research results and theoretical concepts. Psychiatr Pol. 2019;53:447–58. https://doi.org/10.12740/PP/92023.
6. Webb J. Beta-blockers for the treatment of misophonia and misokinesia. Clin Neuropharmacol. 2022;45:13–4. https://doi.org/10.1097/WNF.0000000000000492.
7. Zai G, Dembo J, Levitsky N, Richter MA. Misophonia: a detailed case series and literature review. Prim Care Companion CNS Disord. 2022;24 https://doi.org/10.4088/PCC.21cr03124.

M. Cherchi, *Otoneurology and Vestibular Medicine*, https://doi.org/10.1007/978-3-031-94841-1_106

Part XV

Diagnoses—Neurologically-Mediated Auditory Disorders

Cortical Deafness

107

Overview

The afferent auditory system decussates early in its course (beginning at the dorsal and ventral cochlear nuclei), and multiple times during its ascent to the auditory cortex located in the superior temporal gyrus of each temporal lobe. Consequently, a unilateral lesion of the auditory cortex does not cause hearing loss. Bilateral lesions of the auditory cortex are very unusual, but such cases can result in *cortical deafness*. Of these rare cases, the most common cause is stroke, and in such cases, the symptom onset is abrupt. Most patients will complain of hearing loss. Physical examination and audiometry will show sensorineural hearing loss. Middle latency responses and late latency responses may be abnormal or absent. Imaging (preferably MRI) should be performed to characterize the lesions. Treatment depends on the etiology. Prognosis is poor; recovery is usually minimal.

Introduction: Review of Anatomy

Anatomically, the cochlea exhibits tonotopic mapping, with type 1 hair cells mediating the highest frequencies at the base of the cochlea, and those mediating the lowest frequencies at its apex [1], as shown in Fig. 107.1, from Li and colleagues [2].

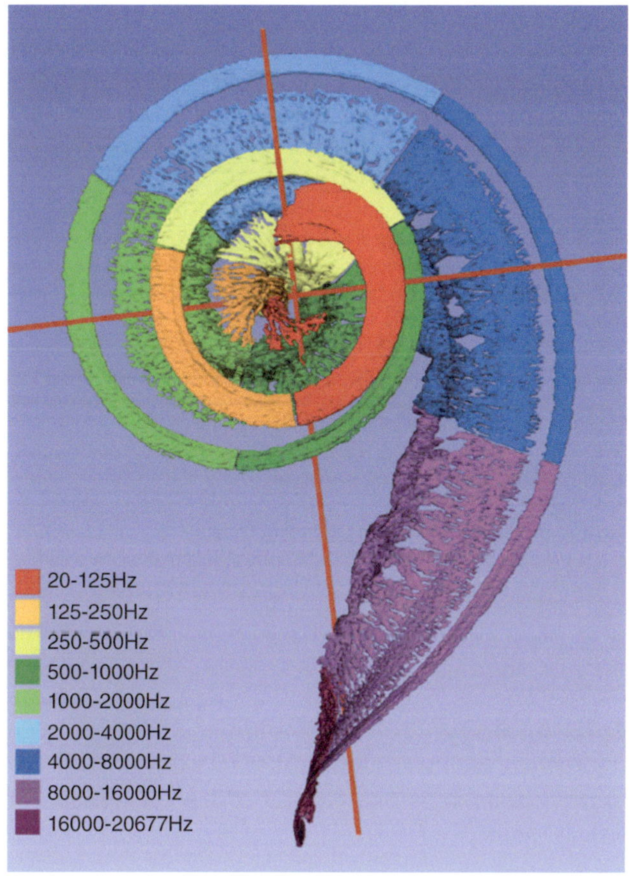

Fig. 107.1 Tonotopic mapping of the cochlea. (From Li et al. [2]. Reproduced with written permission from Springer Nature B.V.)

This tonotopic mapping is preserved throughout the entire course of afferent auditory signals, from the spiral ganglion all the way up to the auditory cortex, as illustrated in Fig. 107.2, from Saenz and Langers [3]. Primary auditory cortex is located in the superior temporal gyrus of each temporal lobe, which is irrigated by the middle cerebral artery.

Auditory signals travel from the spiral ganglion through the auditory component of the cochleovestibular nerve and synapse in the dorsal and ventral cochlear nuclei in the brainstem. From there, some fibers decussate immediately via commissural pathways through the trapezoid body, while other fibers ascend. There are multiple other decussations during this ascent through the afferent auditory system, including at the levels of the nuclei of the lateral lemniscus (via the commissure of Probst), the inferior colliculus (via the commissure of the inferior colliculus), and the auditory cortex (via the corpus callosum). Thus, by the time auditory signals have reached primary auditory cortex, there is significant bilaterality of encoding. For this reason, a lesion of one primary auditory cortex does not cause deafness, though it can cause a variety of other acoustic perceptual abnormalities.

M. Cherchi, *Otoneurology and Vestibular Medicine*, https://doi.org/10.1007/978-3-031-94841-1_107

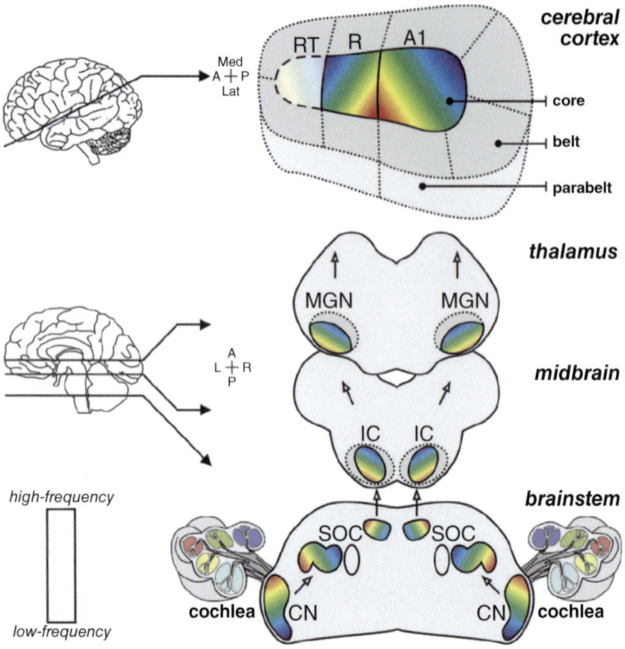

Fig. 107.2 Tonotopic mapping in the afferent auditory system from the cochlea through the auditory cortex. All nuclei that form part of the classical lemniscal auditory pathway are tonotopically organized. These include various subdivisions of the cochlear nucleus (CN), superior olivary complex (SOC), inferior colliculus (IC), and medial geniculate nucleus (MGN). In the auditory cerebral cortex in the superior part of the temporal lobe, expected divisions of core, belt, and parabelt are based on the non-human primate model of auditory cortical organization. Human neuroimaging consistently shows at least two primary tonotopic gradients ("high-to-low-to-high") in the auditory cortex, homologous to primary fields A1 and rostral field R in the monkey cortex. In some primate studies, a third rostrotemporal (RT) field is delineated, but neuroimaging evidence for a similar field in humans is sparse. (From Saenz and Langers [3]. Reproduced from Elsevier with permission under the STM Signatory agreement)

Epidemiology

Cortical deafness is rare because of the combination of lesions required.

Pathophysiological Mechanism of Disease

Like any area of the central nervous system, primary auditory cortex is vulnerable to all manner of lesions, including infarction, demyelination, and tumor.

Because of the multiple decussations mentioned earlier, a unilateral lesion of the auditory cortex does not cause deafness. However, bilateral lesions can, and do, cause cortical deafness. It is unusual to suffer lesions in bilateral auditory cortices; consequently, cortical deafness is quite rare [4]. In my first 15 years of subspecialty practice, I encountered only one convincing case of cortical deafness.

Although cortical deafness is rare, the most common cause is bilateral temporal lobe infarctions, which is documented by numerous case reports in the literature [4–13]. In some cases, the infarctions may be subcortical [14, 15] rather than cortical.

The location of such infarctions can be variable, as is evident from the 46 cases of infarct-related cortical deafness studied by Silva and colleagues [16].

Less common causes of cortical deafness include meningitis [17] and multiple sclerosis [18]. There is one report of Creutzfeldt–Jakob disease presenting with cortical deafness [19]. One report describes hearing loss following head trauma [20], but in this case, the trauma caused a subdural hematoma and other significant intracranial lesions. .

Clinical Presentation

Most patients present with a complaint of hearing loss. Some patients will deny hearing loss, suggesting a type of auditory agnosia analogous to the visual agnosia experienced in Anton syndrome (cortically blind patients who deny vision loss).

Physical Examination

Patients have sensorineural hearing loss on examination.

Ocular Motor Examination

Unless the patient has suffered additional lesions outside of the auditory cortices, ocular motor examination should be normal.

Testing: Auditory

In most cases of cortical deafness, audiometry shows profound bilateral sensorineural hearing loss, as demonstrated in the audiogram in Fig. 107.3, from Narayanan and colleagues [12].

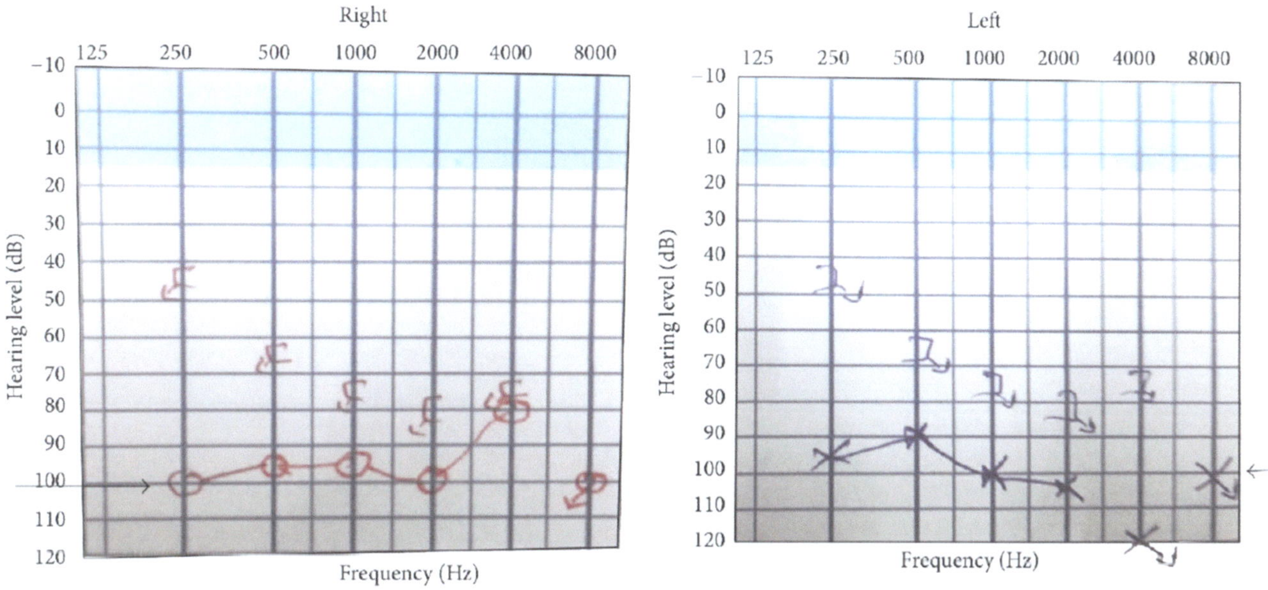

Fig. 107.3 Audiogram from a patient with cortical deafness showing bilateral sensorineural hearing loss. (From Narayanan et al. [12]. Reproduced from Hidawi with permission under CC-BY 4.0)

Testing: Brainstem Auditory Evoked Responses (BAERs)

Some reported cases in which electrophysiological testing was performed report absence of middle latency evoked responses [21, 22] or late latency responses [17].

Imaging

Figures 107.4 and 107.5, from Narayanan and colleagues [12], show bilateral temporal lobe infarctions with left-sided hemorrhagic transformations.

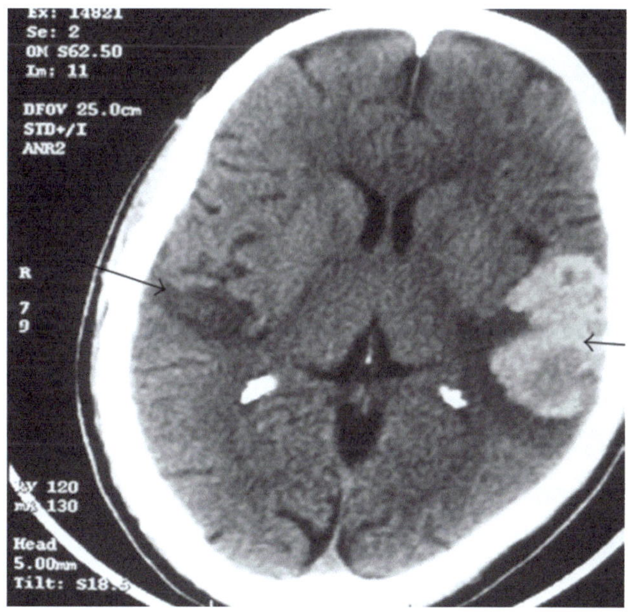

Fig. 107.4 Axial CT head showing infarcts in bilateral temporal lobes with hemorrhagic transformation on left. (From Narayanan et al. [12]. Reproduced from Hidawi with permission under CC-BY 4.0)

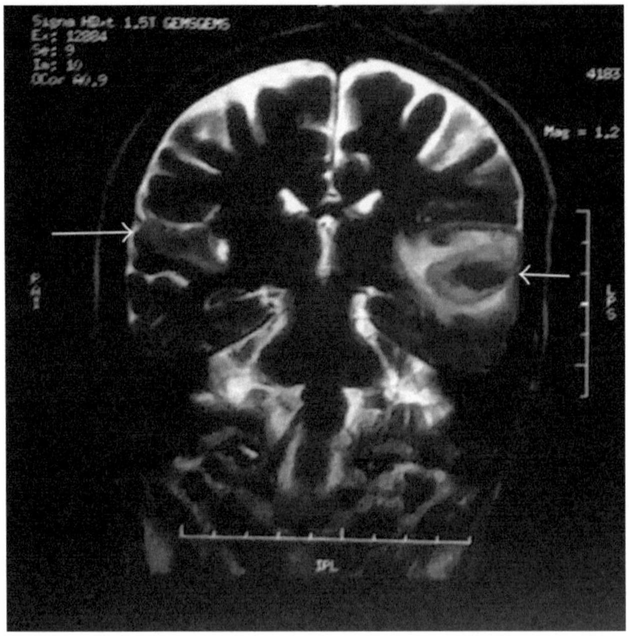

Fig. 107.5 Coronal MRI brain showing infarcts in bilateral Heschl gyri. (From Narayanan et al. [12]. Reproduced from Hidawi with permission under CC-BY 4.0)

Differential Diagnosis

In a patient with abrupt-onset total bilateral hearing loss, cortical deafness (from infarction) is a consideration. If the underlying etiology is a slower process, such as multiple sclerosis or tumor, then the onset of hearing loss may occur over a longer period, and could perhaps resemble diseases such as autoimmune inner ear disease.

Treatment

Treatment involves management of the underlying disease, if possible. Recovery is usually poor.

Prognosis

Prognosis for recovery of any hearing is poor. Brody and colleagues [4] state that "Presentation and recovery of hearing are dependent on the extent of the initial lesions. The majority of patients can expect improvements in pure-tone auditory thresholds over time; however patients should be counseled that recovery of the ability to understand speech is unlikely."

References

1. Pickles JO. Auditory pathways: anatomy and physiology. Handb Clin Neurol. 2015;129:3–25.
2. Li H, Helpard L, Ekeroot J, Rohani SA, Zhu N, Rask-Andersen H, et al. Three-dimensional tonotopic mapping of the human cochlea based on synchrotron radiation phase-contrast imaging. Sci Rep. 2021;11(1):4437.
3. Saenz M, Langers DR. Tonotopic mapping of human auditory cortex. Hear Res. 2014;307:42–52.
4. Brody RM, Nicholas BD, Wolf MJ, Marcinkevich PB, Artz GJ. Cortical deafness: a case report and review of the literature. Otol Neurotol. 2013;34(7):1226–9.
5. Adams AE, Rosenberger K, Winter H, Zollner C. A case of cortical deafness. Arch Psychiatr Nervenkr. 1970;224(3):213–20.
6. Bahls FH, Chatrian GE, Mesher RA, Sumi SM, Ruff RL. A case of persistent cortical deafness: clinical, neurophysiologic, and neuropathologic observations. Neurology. 1988;38(9):1490–3.
7. Graham J, Greenwood R, Lecky B. Cortical deafness--a case report and review of the literature. J Neurol Sci. 1980;48(1):35–49.
8. Kaga K, Shinjo Y, Enomoto C, Shindo M. A case of cortical deafness and loss of vestibular and somatosensory sensations caused by cerebrovascular lesions in bilateral primary auditory cortices, auditory radiations, and postcentral gyruses – complete loss of hearing despite normal DPOAE and ABR. Acta Otolaryngol. 2015;135(4):389–94.
9. Kneebone CS, Burns RJ. A case of cortical deafness. Clin Exp Neurol. 1981;18:91–7.
10. Lachowska M, Pastuszka A, Sokolowsk J, Szczudlik P, Niemczyk K. Cortical deafness due to ischaemic strokes in both temporal lobes. J Audiol Otol. 2021;25(3):163–70.
11. Michel F, Peronnet F, Schott B. A case of cortical deafness: clinical and electrophysiological data. Brain Lang. 1980;10(2):367–77.
12. Narayanan S, Majeed KA, Subramaniam G, Narayanan A, Navaf KM. A case of cortical deafness due to bilateral heschl gyrus infarct. Case Rep Med. 2017;2017:6816748.
13. Silva J, Sousa M, Mestre S, Nzwalo I, Nzwalo H. Cortical deafness of following bilateral temporal lobe stroke. J Stroke Cerebrovasc Dis. 2020;29(7):104827.
14. Earnest MP, Monroe PA, Yarnell PR. Cortical deafness: demonstration of the pathologic anatomy by CT scan. Neurology. 1977;27(12):1172–5.
15. Tanaka Y, Kamo T, Yoshida M, Yamadori A. 'So-called' cortical deafness. Clinical, neurophysiological and radiological observations. Brain J Neurol. 1991;114(Pt 6):2385–401.
16. Silva G, Goncalves R, Taveira I, Mouzinho M, Osorio R, Nzwalo H. Stroke-associated cortical deafness: a systematic review of clinical and radiological characteristics. Brain Sci. 2021;11(11)
17. Sasidharan M, Itty JE, Hinduja G, Hasna S, Priya D. Cortical deafness as a sequela to meningitis: a single case study. Intractable Rare Dis Res. 2020;9(4):247–50.

18. Tabira T, Tsuji S, Nagashima T, Nakajima T, Kuroiwa Y. Cortical deafness in multiple sclerosis. J Neurol Neurosurg Psychiatry. 1981;44(5):433–6.
19. Tobias E, Mann C, Bone I, de Silva R, Ironside J. A case of Creutzfeldt-Jakob disease presenting with cortical deafness. J Neurol Neurosurg Psychiatry. 1994;57(7):872–3.
20. Singh M, Ahluwalia H, Lal KK, Chauhan SP. Cortical deafness following head injury – a case report. Indian J Otolaryngol Head Neck Surg. 1997;49(Suppl 1):81–4.
21. Ozdamar O, Kraus N, Curry F. Auditory brain stem and middle latency responses in a patient with cortical deafness. Electroencephalogr Clin Neurophysiol. 1982;53(2):224–30.
22. Vedder JS, Barrs DM, Fifer RC. The use of middle latency response in the diagnosis of cortical deafness. Otolaryngol Head Neck Surg. 1988;98(4):333–7.

108 Misophonia

Overview

Misophonia ("dislike of sound") is an aversive behavioral, emotional, and autonomic response to specific sounds that are usually (though not always) human-generated, often by orofacial movements (e.g., chewing, lip smacking, swallowing). Misophonia can occur in isolation, though risk factors include various psychiatric disorders (particularly obsessive-compulsive disorder) and auditory disorders (hyperacusis, tinnitus). The underlying pathophysiology remains unknown, and no set of diagnostic criteria has been widely accepted. Workup may include tests for related conditions such as various etiologies of hyperacusis. There have been no replicated randomized, prospective, double-blinded, placebo-controlled treatment trials, so management may include trigger avoidance, cognitive-behavioral therapy, and psychiatric approaches. Prognosis is uncertain.

Introduction

The term "*misophonia*" (meaning, "dislike of sound") was introduced by Jastreboff and Jastreboff [1]. Synonymous designations in the medical literature include "selective sound sensitivity syndrome" [2].

To clarify this nomenclature, it may be helpful to put it in the context of related terms, including "decreased sound tolerance," "hyperacusis," and "phonophobia."

Figure 108.1 depicts the hierarchical relationships among these terms in the form of a Venn diagram.

Decreased sound tolerance is a broad category, with respect to which they state that "decreased sound tolerance consists of more than one problem. It is not necessarily loud sounds, but even quiet sounds, which can cause discomfort. Decreased sound tolerance might reflect a physical discomfort, or can be related to a dislike or a fear of sound" [1]. This supercategory of decreased sound tolerance subsumes hyperacusis and misophonia.

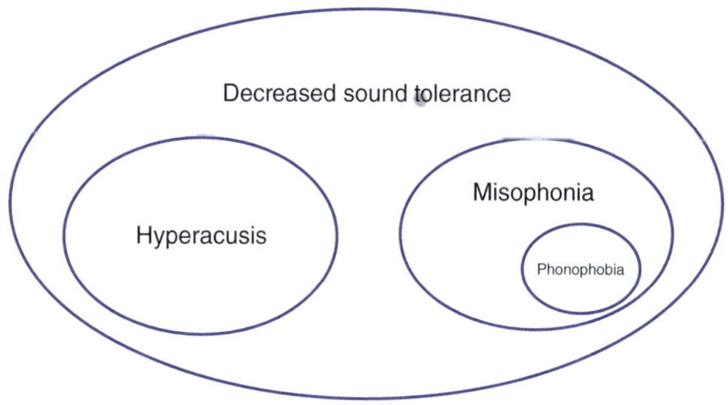

Fig. 108.1 Venn diagram depicting the hierarchical relationships among decreased sound tolerance, hyperacusis, misophonia, and photophobia. (By Marcello Cherchi)

Hyperacusis is a subcategory of decreased sound tolerance that "can be defined as an abnormally strong reaction to sound occurring within the auditory pathways. At the behavioral level, it is manifested by a patient experiencing physical discomfort as a result of exposure to sound… The same sound would not evoke a similar reaction in the average listener. The strength of the reaction is controlled by the physical characteristics of the sound, e.g., its spectrum and intensity" [1].

Misophonia is a subcategory of decreased sound tolerance, and phonophobia is a subcategory of misophonia. Jastreboff and Jastreboff state:

> Misophonia and phonophobia can be defined as abnormally strong reactions of the autonomic and limbic systems resulting from enhanced connections between the auditory and limbic systems. Importantly, misophonia and phonophobia do not involve a significant activation of the auditory system. At the behavioral level, patients have a negative attitude to sound (misophonia), or are afraid of sound (phonophobia). In cases of misophonia and phonophobia, the strength of the patient's reaction is only partially determined by the physical characteristics of the upsetting sound. It is also dependent on the patient's previous evaluation and recollection of the sound (e.g., sound as a potential threat, and/or the belief that the sound can be harmful), the patient's psychological profile and the context in which the sound is presented. [1]

M. Cherchi, *Otoneurology and Vestibular Medicine*, https://doi.org/10.1007/978-3-031-94841-1_108

Misophonia is a subcategory of decreased sound tolerance. "Misophonic patients are simply disliking these sounds without necessarily fearing them, which are then creating negative emotional responses" [1]. Furthermore, "Misophonia [is] a separate disorder since, unlike hyperacusis… the symptomatology was associated exclusively with one type of sound; and it [is] distinct from phonophobia (fear of sounds) because the primary emotional response differed" [3].

Phonophobia is a subcategory of misophonia. The term "describes a specific type of misophonia, when fear is the dominant emotion involved in the dislike of the sound. The majority of patients with decreased sound tolerance have misophonia, but only some of them are phonophobic. A common reason for phonophobia is the fear that sounds, frequently 'normal' environmental sound, may damage the ear, or make symptoms worse" [1].

They further state that:

> Clinical observations reveal that in many cases, decreased sound tolerance consists of more than one problem. It is not necessarily loud sounds, but even quiet sounds, which can cause discomfort. Decreased sound tolerance might reflect a physical discomfort, or can be related to a dislike or a fear of sound. [1]

It is important to recognize that the auditory stimuli in question do not necessarily need to be of large amplitude ("loud"); sometimes, they can be sounds which healthy people would perceive as rather quiet. As Jastreboff and Jastreboff state, "Please note that neither hyperacusis, nor misophonia nor phonophobia have any relation to hearing thresholds. Patients with hyperacusis, misophonia or phonophobia may have normal hearing, or they may be hearing impaired" [1].

Thus, these definitions by Jastreboff and Jastreboff depend more on *behavioral* characteristics (emotions elicited by sounds) and on *physiological* characteristics (such as autonomic responses) of the misophonic response, rather than on particular features of the misophonic stimulus. Ferrer-Torres and Gimenez-Llort [3] view the misophonic response as consisting of several components:

- An emotional component
  - Anger
  - Anxiety
  - Disgust
- A physiological response
  - Muscle contraction
  - Tachycardia
  - Difficulty breathing
  - Chest pressure
- A behavioral response
  - Escape
  - Avoidance
  - Violent impulses

Ferrer-Torres and Gimenez-Llort [3] summarize misophonia as "a complex neurophysiological and behavioral disorder of multifactorial origin [that] is characterized by an increased physiological and emotional response produced by intolerance to specific auditory stimuli."

Numerous self-administered questionnaires have been explored for diagnosing misophonia, but none has been universally accepted; these are reviewed by Ferrer-Torres and Gimenez-Llort [3]. There have been several proposals for diagnostic criteria of misophonia [4–6], but none has been universally accepted. As Ferrer-Torres and Gimenez-Llort bluntly state, "there is currently no consensus on the criteria to establish the diagnosis" [3].

Nevertheless, it is instructive to review some of the proposals. Building on the work of Schroder and colleagues [4], Dozier and colleagues [5] propose the following criteria, which focus on characteristics of the misophonic stimulus and of the response, and their contextual dependency:

1. A misophonic stimulus can be produced by any source (i.e., human, animal, electronic, and equipment, and not limited to stimuli from humans).
2. A misophonic stimulus can be virtually any sensory modality, with auditory and visual stimuli being the most common and not limited to auditory stimuli.
3. The eliciting stimulus is a conditioned stimulus and so excludes responses to innate sensitivities, such as those with sensory processing disorder or sensory overresponsiveness.
4. The strength of the response is controlled by the context and experience with the stimulus and not the physical characteristics of the stimulus. Therefore, a minimal intensity stimulus will elicit the response.
5. The response is elicited by a single or small number of minimal intensity stimulus instances, and is not the response from persistent, annoying stimuli.
6. The stimulus elicits an immediate physical response, which is often a skeletal muscle response, but can be any physical response.
7. A moderate duration of the stimulus (e.g., 15 s) elicits physiological arousal.

A partially overlapping list of criteria is suggested by Aazh [6], who additionally requires that the symptoms not be better explained by another diagnosis:

1. *General description.* Misophonia is characterized by decreased tolerance to specific sounds or stimuli associated with such sounds (known as triggers). Exposure to triggers can evoke disproportionate emotional, physiological, and behavioral reactions leading to distress and/or impairment in social, occupational, or aca-

demic functioning. Triggers often, but not exclusively, are stimuli generated by another human being's body.

2. *Misophonic triggers*. Most common triggers are auditory, but some may react to visual triggers too. Common triggers include but are not limited to sounds associated with oral functions (e.g., chewing, eating, smacking lips, slurping, coughing, throat clearing, and swallowing), nasal sounds (e.g., breathing and sniffing), nonoral/nasal sounds produced by people (e.g., pen clicking, keyboard typing, finger or foot tapping, and shuffling footsteps), as well as sounds produced by objects (e.g., clock ticking) or sounds generated by animals. Examples of visual triggers are cracking knuckles and jiggling or swinging legs or watching someone eat.
3. *Reactions to misophonic triggers*. Emotional: Anger, irritation, disgust, and anxiety. Physiological: Increased muscular tension, increased heart rate, and sweating. Behavioral reactions: Agitation, aggression, avoidance, seeking to discontinue the triggering stimuli, and mimicking.
4. *Influences on reactions*. The strength of the reaction can be influenced by [1] the context, [2] the individual's perceived degree of control, and [3] the relationship with individual who is the source of the trigger.
5. *Functional impairments*. Impaired occupational and/or academic functioning, concentration difficulties, impaired social functioning, strained social relationships, and social isolation.
6. *Relationship to other conditions/disorders*. The symptoms of misophonia should not be better explained by any co-occurring disorders including but not limited to hearing impairment, tinnitus, hyperacusis, anxiety disorders, mood disorders, personality disorders, obsessive-compulsive-related disorders, posttraumatic stress disorder, autism spectrum disorder, and attention-deficit hyperactivity disorder.

We find some of these criteria to be broader than may be clinically warranted given the current state of knowledge of this condition. For instance, the idea that misophonia is "not limited to auditory stimuli" suggests that misophonia may belong to a broader sensory processing disorder that can involve other sensory modalities—but taking that step conflates problems such as misokinesia with misophonia, even though these may be distinct clinical entities.

Several brief reviews [7, 8] and more extensive reviews [3, 9, 10] summarize well the emerging literature on this topic.

Epidemiology

The range of reported prevalence of misophonia is quite broad depending on the population studied, varying by over an order of magnitude, ranging from 3.2% [11] to 19.9% of undergraduates [12] to 49.3% of medical students [13]. Even if one takes the low end of these estimates, it suggests that misophonia is quite common.

One study reported the mean age of symptom onset to be 13 years [4], though the mean age at the time of diagnosis may be 37 years [4] to 39 years [14].

There appear to be psychiatric and auditory risk factors for misophonia. Aazh and colleagues [15] retrospectively reviewed records of 257 patients consulting at an audiology clinic for symptoms of hyperacusis and/or tinnitus; the patients' mean age was 53 ± 16 years (range 17–97 years). They noted that 23% of patients reported symptoms compatible with what the authors judged to be misophonia, and that the risk of misophonia increased with increasing impact of tinnitus, with increasing impact of hyperacusis, and with symptoms of anxiety and depression.

Genetics

Familial clusters of misophonia have been reported [16]—sometimes with different family members experiencing aversion to the same misophonic trigger—implicating a genetic contribution, but so far, the genetics of misophonia remain unknown [17].

Pathophysiological Mechanism of Disease

Research has not yet clarified the pathophysiological mechanism(s) underlying misophonia.

Psychiatric etiologies of misophonia have been considered [18] on the basis of the emotional responses that are provoked in misophonia [19–27], and the greater than chance comorbidity with several psychiatric disorders [28–39], including obsessive-compulsive disorders. However, most investigators conclude that there is insufficient evidence to warrant classifying misophonia as a purely psychiatric disorder.

Given the core clinical feature of an auditory trigger, another logical candidate etiology would be that misophonia is a primary auditory disorder, but this probably cannot be the full picture. Ferrer-Torres and Gimenez-Llort comment that:

> Misophonia cannot be classified as an auditory disorder since no relationship has been found between it and hearing thresholds, as the disorder can occur in people with normal hearing, with hearing loss, or with some auditory pathology… Additionally, the specificity of the triggering stimuli suggests that the symptoms are unlikely to be caused by an alteration of the auditory system. [3]

Brout and colleagues [9] observed similarities between misophonia and misokinesia, and suggested that it may be fruitful for future research to consider these disorders in the broader context of sensory overresponsivity. Additional support for such an approach comes from observations that patients with misophonia often exhibit overresponsivity in other sensory modalities as well [40].

After reviewing evidence from multiple studies of individual candidate biomarkers, Brout and colleagues [9] comment that seeking a single diagnostic biomarker is likely to fail, and that misophonia will probably be more productively studied as a network-level disorder.

Clinical Presentation

Despite the absence of any consensus on diagnostic criteria, most clinical descriptions of misophonia are fairly similar. The following is an accessible example:

> Misophonia is a set of symptoms which some people experience when exposed to certain sounds. Strong emotions such as anger, irritation, disgust, or anxiety are evoked immediately when people with misophonia hear particular sounds. Somatic responses are also present—pressure in the chest, arms, head, or the whole body, as well as increased heart rate, increased body temperature, physical pain, or difficulties with breathing. [41]

Physical Examination

Physical examination should be normal in an individual with misophonia who is otherwise healthy.

Testing: Auditory

At minimum, given the differential diagnosis (see below), when a diagnosis of misophonia is being considered, it is medically reasonable to check:

- *Audiometry* (with air and bone conduction, and loudness discomfort levels) and *otoacoustic emissions* (OAEs).
  - Hearing loss (detected on audiometry and/or otoacoustic emissions) can be associated with sound sensitivity through the mechanism of loudness recruitment.
  - Identification of conductive hyperacusis (on audiometry) may increase suspicion for superior semicircular canal dehiscence.
- *Cervical and ocular vestibular evoked myogenic potentials* (VEMPs). This helps evaluate for labyrinthine causes of sound sensitivity such as superior semicircular canal dehiscence. Some patients with misophonia may be unable to tolerate VEMP testing.

Testing: Other

Emerging evidence reviewed elsewhere [3, 9] reports that patients with misophonia may exhibit abnormalities on various tests such as electroencephalography, structural findings on MRI [42], activation findings on functional magnetic resonance imaging [43–45], galvanic (conductive) skin responses, and autonomic function. These are interesting candidate biomarkers for misophonia, but have not yet been incorporated into any widely accepted diagnostic criteria.

Differential Diagnosis

Each characteristic of misophonia can be found in other diseases as well, so the differential diagnosis is somewhat broad. Table 108.1, from Ferrer-Torres and colleagues [3], provides a differential diagnosis of misophonia.

Table 108.1 Differential diagnosis of misophonia

	Disorder	Similarities with misophonia	Differences with misophonia
Psychiatric disorder	Specific phobia	A triggering stimulus may evoke a negative response, as well as avoidance behaviors	In specific phobia, one experiences mostly anxiety and fear, whereas with misophonia a high degree of anger and aggression is perceived
	Phonophobia	Fear of a specific sound	The main symptom is not anxiety or fear as in phonophobia, but the feeling of irritation, disgust, or anger
	Social phobia	Habitual avoidance of social situations, due to experiencing anxiety and stress	In social phobia, the reason is a hypersensitivity towards negative social evaluation; whereas in misophonia, the social situation is avoided to prevent an encounter with the misophonic sound
	Posttraumatic stress disorder (PTSD)	Aversive reaction to a stimulus and avoidance behaviors are shared	The person with PTSD had to experience a traumatic event. In the case of misophonia, no such association has been demonstrated
	Obsessive-compulsive disorder (OCD)	They share an excessive preoccupation towards a specific stimulus, as well as the feeling of anxiety	People with OCD often perform compulsive behaviors to reduce anxiety. In their case, there are no behaviors such as aggression and anger that may occur in misophonia
	Intermittent explosive disorder	Here, the shared factor is anger	In misophonia, the triggering stimulus of anger is always a sound, for explosive disorder it can be any stimulus. Loss of control does not usually occur in misophonia
	Eating behavior disorders	The most frequent emotional trigger in misophonia, as in the eating disorder, is food	For the person with misophonia, the trigger is the sound of food; for an eating disorder, it is the ingestion of food
	Obsessive-compulsive personality disorder (OCPD)	The shared factors are anger or aggression	People with misophonia respond to the same auditory stimulus, people with OCPD do not relate to triggering sounds
	Autism spectrum disorders (ASD)	Auditory hyper-reactivity is observed	People with ASD show intolerance to unexpected and loud noises. People with misophonia can react to any type of auditory stimulus
	Sensory processing disorder (SPD)	Auditory hyper-reactivity is observed	The person with TPS reacts to unexpected and loud noises, there is also hyperreactivity to other stimuli. The person with misophonia reacts to any auditory stimulus
	Personality disorders with impulsive aggression	Difficulty in controlling anger and impulsivity occurs	The reactions are not necessarily related to a specific sound, as is the case with misophonia
Auditive disorder	Tinnitus	Can provoke negative emotions; anxiety	It is perceived in one or both ears in the absence of acoustic source
	Hyperacusis	Negative reaction to any auditory stimulus with physical characteristics (loudness and frequency)	The stimulus characteristics are neutral or of very low frequency and intensity and context-independent

Adapted from Ferrer-Torres and Gimenez-Llort [3]

Treatment

There are not yet any randomized, prospective, double-blinded, placebo-controlled treatment trials for misophonia [3]. As Palumbo and colleagues bluntly note, "Currently there is no cure or pharmaceutical agent for misophonia" [46]. Before such trials can be designed and their results scrutinized, considerable work remains to be done in terms of coming to some reasonably widely accepted set of diagnostic criteria, and some agreement on whether (and which) diagnostic testing should be part of the workup.

Until such research emerges, most patients with misophonia end up practicing trigger avoidance, though this may not always be practical. For patients who are receptive to the idea, consultation with psychiatry is appropriate to help manage the emotional burdens imposed by misophonia. There is modest evidence that cognitive-behavioral therapy may be helpful [47–54]. Consultation with an audiologist experienced in the management of hyperacusis is also sensible if such a specialist is available. Some audiologists may elect to employ strategies similar to those used in tinnitus retraining therapy [3, 55].

Novel proposals regarding the possible application of neurostimulation [56] are being explored.

We remain cautious about apparently favorable responses to pharmaceutical agents cited in the literature, as these are all case reports, such as for risperidone [57, 58], methylphenidate [59], fluoxetine [60], beta blockers [61], 3,4-methelyn edeoxy-methamphetamine [62], and sertraline [63].

Prognosis

Given the lack of uniformity in diagnostic criteria for misophonia, very few longitudinal studies are available [64] to give a clear sense of the natural history of this disorder, so prognosis is uncertain [65].

References

1. Jastreboff MM, Jastreboff PJ. Components of decreased sound tolerance: hyperacusis, misophonia, phonophobia. Inst Transl Health Sci Newslett. 2001;2(5-7):1–5.
2. Cavanna AE. What is misophonia and how can we treat it? Expert Rev Neurother. 2014;14(4):357–9.
3. Ferrer-Torres A, Gimenez-Llort L. Misophonia: a systematic review of current and future trends in this emerging clinical field. Int J Environ Res Public Health. 2022;19(11)
4. Schroder A, Vulink N, Denys D. Misophonia: diagnostic criteria for a new psychiatric disorder. PLoS One. 2013;8(1):e54706.
5. Dozier TH, Lopez M, Pearson C. Proposed diagnostic criteria for misophonia: a multisensory conditioned aversive reflex disorder. Front Psychol. 2017;8:1975.
6. Aazh H. Commentary: consensus definition of misophonia. Front Neurosci. 2022;16:1077097.
7. Cavanna AE, Seri S. Misophonia: current perspectives. Neuropsychiatr Dis Treat. 2015;11:2117–23.
8. Zai G, Dembo J, Levitsky N, Richter MA. Misophonia: a detailed case series and literature review. Prim Care Companion CNS Disord. 2022;24(5)
9. Brout JJ, Edelstein M, Erfanian M, Mannino M, Miller LJ, Rouw R, et al. Investigating misophonia: a review of the empirical literature, clinical implications, and a research agenda. Front Neurosci. 2018;12:36.
10. Potgieter I, MacDonald C, Partridge L, Cima R, Sheldrake J, Hoare DJ. Misophonia: a scoping review of research. J Clin Psychol. 2019;75(7):1203–18.
11. Jastreboff PJ, Jastreboff MM. Treatments for decreased sound tolerance (hyperacusis and misophonia). Semin Hear. 2014;35(02):105–20.
12. Wu MS, Lewin AB, Murphy TK, Storch EA. Misophonia: incidence, phenomenology, and clinical correlates in an undergraduate student sample. J Clin Psychol. 2014;70(10):994–1007.
13. Naylor J, Caimino C, Scutt P, Hoare DJ, Baguley DM. The prevalence and severity of misophonia in a UK undergraduate medical student population and validation of the Amsterdam Misophonia Scale. Psychiatr Q. 2021;92(2):609–19.
14. Quek TC, Ho CS, Choo CC, Nguyen LH, Tran BX, Ho RC. Misophonia in Singaporean psychiatric patients: a cross-sectional study. Int J Environ Res Public Health. 2018;15(7)
15. Aazh H, Erfanian M, Danesh AA, Moore BCJ. Audiological and other factors predicting the presence of misophonia symptoms among a clinical population seeking help for tinnitus and/or hyperacusis. Front Neurosci. 2022;16:900065.
16. Sanchez TG, Silva FED. Familial misophonia or selective sound sensitivity syndrome: evidence for autosomal dominant inheritance? Braz J Otorhinolaryngol. 2018;84(5):553–9.
17. Smit DJA, Bakker M, Abdellaoui A, Hoetink AE, Vulink N, Denys D. A genome-wide association study of a rage-related misophonia symptom and the genetic link with audiological traits, psychiatric disorders, and personality. Front Neurosci. 2022;16:971752.
18. Taylor S. Misophonia: a new mental disorder? Med Hypotheses. 2017;103:109–17.
19. Bagrowska P, Pionke-Ubych R, Gaweda L. Do they make these sounds to hurt me? The mediating role of emotion regulation, anxiety and hostile attributions in the relationship between misophonia and paranoia-like thoughts. Schizophr Res. 2022;250:137–42.
20. Cassiello-Robbins C, Anand D, McMahon K, Guetta R, Trumbull J, Kelley L, et al. The mediating role of emotion regulation within the relationship between neuroticism and misophonia: a preliminary investigation. Front Psychiatry. 2020;11:847.
21. Daniels EC, Rodriguez A, Zabelina DL. Severity of misophonia symptoms is associated with worse cognitive control when exposed to misophonia trigger sounds. PLoS One. 2020;15(1):e0227118.
22. Ferrer-Torres A, Gimenez-Llort L. Sounds of silence in times of COVID-19: distress and loss of cardiac coherence in people with misophonia caused by real, imagined or evoked triggering sounds. Front Psychiatry. 2021;12:638949.
23. Ferrer-Torres A, Gimenez-Llort L. Confinement and the hatred of sound in times of COVID-19: a molotov cocktail for people with misophonia. Front Psychiatry. 2021;12:627044.
24. Guetta RE, Cassiello-Robbins C, Trumbull J, Anand D, Rosenthal MZ. Examining emotional functioning in misophonia: the role of affective instability and difficulties with emotion regulation. PLoS One. 2022;17(2):e0263230.
25. Rinaldi LJ, Simner J, Koursarou S, Ward J. Autistic traits, emotion regulation, and sensory sensitivities in children and adults with misophonia. J Autism Dev Disord. 2023;53(3):1162–74.
26. Savard MA, Sares AG, Coffey EBJ, Deroche MLD. Specificity of affective responses in misophonia depends on trigger identification. Front Neurosci. 2022;16:879583.
27. Wang Q, Vitoratou S, Uglik-Marucha N, Gregory J. Emotion processes predicting outbursts and functional impact in misophonia. Front Psychol. 2022;13:903142.
28. Siepsiak M, Rosenthal MZ, Raj-Koziak D, Dragan W. Psychiatric and audiologic features of misophonia: use of a clinical control group with auditory over-responsivity. J Psychosomat Res. 2022;156:110777.
29. Banker SM, Na S, Beltran J, Koenigsberg HW, Foss-Feig JH, Gu X, et al. Disrupted computations of social control in individuals with obsessive-compulsive and misophonia symptoms. iScience. 2022;25(7):104617.
30. Cassiello-Robbins C, Anand D, McMahon K, Brout J, Kelley L, Rosenthal MZ. A preliminary investigation of the association between misophonia and symptoms of psychopathology and personality disorders. Front Psychol. 2020;11:519681.
31. Erfanian M, Kartsonaki C, Keshavarz A. Misophonia and comorbid psychiatric symptoms: a preliminary study of clinical findings. Nord J Psychiatry. 2019;73(4-5):219–28.
32. Guzick AG, Cervin M, Smith EEA, Clinger J, Draper I, Goodman WK, et al. Clinical characteristics, impairment, and psychiatric morbidity in 102 youth with misophonia. J Affect Disord. 2023;324:395–402.
33. McKay D, Kim SK, Mancusi L, Storch EA, Spankovich C. Profile analysis of psychological symptoms associated with misophonia: a community sample. Behav Ther. 2018;49(2):286–94.

34. Robinson S, Hedderly T, Conte G, Malik O, Cardona F. Misophonia in children with tic disorders: a case series. J Dev Behav Pediatr. 2018;39(6):516–22.
35. Rosenthal MZ, McMahon K, Greenleaf AS, Cassiello-Robbins C, Guetta R, Trumbull J, et al. Phenotyping misophonia: psychiatric disorders and medical health correlates. Front Psychol. 2022;13:941898.
36. Sharan R, Sharma V. A case of bipolar disorder and misophonia. Prim Care Companion CNS Disord. 2020;22(3)
37. Siepsiak M, Sobczak AM, Bohaterewicz B, Cichocki L, Dragan WL. Prevalence of misophonia and correlates of its symptoms among inpatients with depression. Int J Environ Res Public Health. 2020;17(15)
38. Webber TA, Johnson PL, Storch EA. Pediatric misophonia with comorbid obsessive-compulsive spectrum disorders. Gen Hosp Psychiatry. 2014;36(2):231 e1–2.
39. Yektatalab S, Mohammadi A, Zarshenas L. The prevalence of misophonia and its relationship with obsessive-compulsive disorder, anxiety, and depression in undergraduate students of Shiraz University of medical sciences: a cross-sectional study. Int J Commun Based Nurs Midwifery. 2022;10(4):259–68.
40. Efraim Kaufman A, Weissman-Fogel I, Rosenthal MZ, Kaplan Neeman R, Bar-Shalita T. Opening a window into the riddle of misophonia, sensory over-responsiveness, and pain. Front Neurosci. 2022;16:907585.
41. Siepsiak M, Dragan W. Misophonia – a review of research results and theoretical concepts. Psychiatr Pol. 2019;53(2):447–58.
42. Eijsker N, Schroder A, Liebrand LC, Smit DJA, van Wingen G, Denys D. White matter abnormalities in misophonia. Neuroimage Clin. 2021;32:102787.
43. Eijsker N, Schroder A, Smit DJA, van Wingen G, Denys D. Structural and functional brain abnormalities in misophonia. Eur Neuropsychopharmacol. 2021;52:62–71.
44. Schroder A, van Wingen G, Eijsker N, San Giorgi R, Vulink NC, Turbyne C, et al. Misophonia is associated with altered brain activity in the auditory cortex and salience network. Sci Rep. 2019;9(1):7542.
45. Grossini E, Stecco A, Gramaglia C, De Zanet D, Cantello R, Gori B, et al. Misophonia: analysis of the neuroanatomic patterns at the basis of psychiatric symptoms and changes of the orthosympathetic/parasympathetic balance. Front Neurosci. 2022;16:827998.
46. Palumbo DB, Alsalman O, De Ridder D, Song JJ, Vanneste S. Misophonia and potential underlying mechanisms: a perspective. Front Psychol. 2018;9:953.
47. Aazh H, Landgrebe M, Danesh AA, Moore BC. Cognitive behavioral therapy for alleviating the distress caused by tinnitus, hyperacusis and misophonia: current perspectives. Psychol Res Behav Manag. 2019;12:991–1002.
48. Jager I, Vulink N, van Loon A, van der Pol M, Schroder A, Slaghekke S, et al. Synopsis and qualitative evaluation of a treatment protocol to guide systemic group-cognitive behavioral therapy for misophonia. Front Psychiatry. 2022;13:794343.
49. Jager IJ, Vulink NCC, Bergfeld IO, van Loon A, Denys D. Cognitive behavioral therapy for misophonia: a randomized clinical trial. Depress Anxiety. 2020;38(7):708–18.
50. Kamody RC, Del Conte GS. Using dialectical behavior therapy to treat misophonia in adolescence. Prim Care Companion CNS Disord. 2017;19(5)
51. McGuire JF, Wu MS, Storch EA. Cognitive-behavioral therapy for 2 youths with misophonia. J Clin Psychiatry. 2015;76(5):573–4.
52. Rappoldt LR, van der Pol MM, de Wit C, Slaghekke S, Houben C, Sondaar T, et al. Effectiveness of an innovative treatment protocol for misophonia in children and adolescents: design of a randomized controlled trial. Contemp Clin Trials Commun. 2023;33:101105.
53. Roushani K, Mehrabizadeh HM. The effectiveness of cognitive-behavioral therapy on anger in female students with misophonia: a single-case study. Iran J Med Sci. 2021;46(1):61–7.
54. Schroder AE, Vulink NC, van Loon AJ, Denys DA. Cognitive behavioral therapy is effective in misophonia: an open trial. J Affect Disord. 2017;217:289–94.
55. Jastreboff PJ, Jastreboff MM. Decreased sound tolerance: hyperacusis, misophonia, diplacousis, and polyacousis. Handb Clin Neurol. 2015;129:375–87.
56. Neacsiu AD, Szymkiewicz V, Galla JT, Li B, Kulkarni Y, Spector CW. The neurobiology of misophonia and implications for novel, neuroscience-driven interventions. Front Neurosci. 2022;16:893903.
57. Naguy A, Al-Humoud AM, Pridmore S, Abuzeid MY, Singh A, Elsori D. Low-dose risperidone for an autistic child with comorbid ARFID and misophonia. Psychopharmacol Bull. 2022;52(1):91–4.
58. Pan EJ, Weleff J, Anand A, Barnett BS. Treatment of misophonia with risperidone in a patient with autism spectrum disorder. Case Rep Psychiatry. 2022;2022:3169834.
59. Osuagwu FC, Osuagwu VC, Machoka AM. Methylphenidate ameliorates worsening distractibility symptoms of misophonia in an adolescent male. Prim Care Companion CNS Disord. 2020;22(5)
60. Sarigedik E, Yurteri N. Misophonia successfully treated of with fluoxetine: a case report. Clin Neuropharmacol. 2021;44(5):191–2.
61. Webb J. Beta-blockers for the treatment of misophonia and misokinesia. Clin Neuropharmacol. 2022;45(1):13–4.
62. Webb J, Keane S. MDMA for the treatment of misophonia, a proposal. Front Psychiatry. 2022;13:983285.
63. Zuschlag ZD, Leventhal KC. Rapid and sustained resolution of misophonia-type hyperacusis with the selective serotonin reuptake inhibitor sertraline. Prim Care Companion CNS Disord. 2021;23(3)
64. Dibb B, Golding SE. A longitudinal investigation of quality of life and negative emotions in misophonia. Front Neurosci. 2022;16:900474.
65. Rinaldi LJ, Smees R, Ward J, Simner J. Poorer well-being in children with misophonia: evidence from the sussex misophonia scale for adolescents. Front Psychol. 2022;13:808379.

109 Musical Hallucinations

Overview

Musical hallucinations (MH) are well-formed auditory perceptions in the absence of any auditory stimulus. Hearing loss appears to be the most significant risk factor for MH, but it has also been reported in association with several psychiatric disorders (usually obsessive-compulsive disorder or schizophrenia) and neurological disorders. The mean age of onset is in the seventh decade, and it is more common in females. The underlying mechanism is unknown. Management strategies include reassurance, expectant management (since some cases remit spontaneously), various medications (including antiepileptics, antidepressants, antipsychotics, and anticholinergics) and psychological treatments, though none of these interventions has been studied in a randomized controlled trial.

Introduction

The medical definition of a hallucination is a perception in the absence of any stimulus. *Musical hallucinations (MH)*, as the name suggests, are perceptions of music in the absence of any musical stimulus. While sometimes classified under the broader category of tinnitus, musical hallucinations are complex, well-formed perceptions, not merely tinnitus with a hissing, ringing, or buzzing quality. There is debate regarding whether MH and simple tinnitus are distinct phenomena, or different manifestations of a single underlying disease process [1].

Epidemiology

There are numerous case reports and series of musical hallucinations (MH) occurring in association with hearing loss [2–8], and this appears to be the most significant risk factor [9]. This emergence of well-formed auditory hallucinations during auditory deprivation (from hearing loss) is somewhat analogous to the emergence of well-formed visual hallucinations during visual deprivation (from vision loss), which has given rise to the idea that musical hallucinations are the *auditory analogue of Charles Bonnet syndrome* [10–12].

Some literature cites a female preponderance of MH, ranging from 70% [13, 14] to 80% [15] of cases.

Evers and Ellger conducted a literature review of cases and reported that the mean age of onset of MH was 61.5 years [14]. Warner and Aziz report that the incidence of MH in people 65 and older is 1 per 10,000 [16]. From statistics such as these, age greater than 60 years has been cited as a risk factor for MH [13, 15], though it is unclear whether this risk is distinct from age-related hearing loss.

Pathophysiological Mechanism of Disease

The mechanism of MH is unknown [13]. Since many cases occur in the context of hearing loss, "the most accepted physiopathology of musical hallucination associated to hearing loss (caused by cochlear lesion, cochlear nerve lesion or by interruption of mesencephalon or pontine auditory information) is the disinhibition of auditory memory circuits due to sensory deprivation" [7].

There are reports of patients exhibiting both palinacusis (perseverant perception of an auditory stimulus after the actual stimulus has ended) and musical hallucinations, and such overlap raises questions about whether these phenomena share a common mechanism [17].

Some investigators maintain that a component of otologic disease must be present [18, 19]. This strikes us as untenable since MH has been reported in patients without hearing loss or any other evidence of otologic disease.

M. Cherchi, *Otoneurology and Vestibular Medicine*, https://doi.org/10.1007/978-3-031-94841-1_109

Clinical Presentation

With gradually progressive hearing loss (such as age-related hearing loss), the emergence of musical hallucinations can be similarly insidious. Abrupt-onset MH has been reported in cases of sudden hearing loss [8].

Testing: Auditory

Given that hearing loss appears to be a risk factor for MH, it is medically reasonable to check audiometry.

Imaging

No uniform imaging features have been identified in MH.

Differential Diagnosis

MH can occur in patients after cochlear implantation [20]. Low and colleagues [21] studied a series of 82 cochlear implantation patient and reported that 11 patients (13%) developed MH after implantation.

MH has been reported in association with a variety of medical conditions, including the following:

- Psychiatric disorders
  - Schizophrenia [22–24]
  - Hermesh and colleagues [25] studied 190 consecutive psychiatric patients diagnosed with anxiety, affective disorders, and schizophrenic disorders, and reported that "Musical hallucinations occurred in more than one fifth of all diagnoses. The prevalence of musical hallucinations was highest in OCD [obsessive-compulsive disorder] patients (41%)" [25].
- Stroke [26]
- Parkinson's disease [27]
- Epilepsy [28, 29]
- Lyme disease [30]

There are case reports of MH occurring after various medical procedures, including the following:

- Resection of brain tumors [31, 32] and epileptogenic temporal lobe tissue [33]
- Electroconvulsive therapy [34]
- Hemodialysis [35]
- Transplant [36]
- Various surgical procedures [37, 38]

There are case reports of MH occurring in association with various medications, including the following:

- Amantadine [39]
- Amiodarone [40]
- Betamethasone [41]
- Ceftazidime [42]
- Clomipramine [43]
- Imipramine [44]
- Lorazepam [45]
- Mirtazapine [46]
- Morphine [47]
- Oxycodone [48]
- Paroxetine [49]
- Pramipexole [50]
- Propranolol [51]
- Tramadol [52]
- Voriconazole [53]

We would interpret such case reports with caution, since these medications are generally being deployed to treat a pre-existing disease. For example, there are case reports of MH being "triggered" when amantadine [39] or pramipexole [50] were used to treat Parkinson's disease, but MH can occur even in untreated Parkinson's disease patients

Treatment and Prognosis

MH can remit spontaneously [54], so in many cases, reassurance and expectant management are sufficient.

In cases where the underlying etiology of MH appears to be hearing loss, amplification may ameliorate symptoms; masking may also play a role in management [9].

For cases in which MH arises following initiation of medication, it is reasonable to try discontinuing that medication (if medically feasible).

Literature regarding pharmacologic treatment of MH is largely limited to case reports, including the following:

- Carbamazepine [55, 56]
- Donepezil [57, 58]
- Escitalopram [59]
- Gabapentin [60]
- Lamotrigine [61]
- Olanzapine [62, 63]

- Quetiapine [62, 64, 65]
- Risperidone [63]
- Rivastigmine [66]

Other treatment modalities reported include repetitive transcranial magnetic stimulation [67, 68] and various cognitive approaches [69].

References

1. Gandhi R, Narang P, Lippmann S. Are musical and auditory hallucinations related? Psychosomatics. 2016;57(4):442–3.
2. Ali JA. Musical hallucinations and deafness: a case report and review of the literature. Neuropsychiatry Neuropsychol Behav Neurol. 2002;15(1):66–70.
3. Brunner JP, Amedee RG. Musical hallucinations in a patient with presbycusis: a case report. Ochsner J. 2015;15(1):89–91.
4. Fenton GW, McRae DA. Musical hallucinations in a deaf elderly woman. Br J Psychiatry. 1989;155:401–3.
5. Marschall TM, van Dijk P, Kluk K, Koops S, Linszen MMJ, Griffiths TD, et al. Hallucinations in hearing impairment: how informed are clinicians? Schizophr Bull. 2023;49(12 Suppl 2):S33–s40.
6. Niranjan V, Rastogi P, Razdan RG. "Musical ear syndrome"-musical hallucinations in a patient with severe hearing loss – a report. Asian J Psychiatr. 2017;29:101–2.
7. Sanchez TG, Rocha SC, Knobel KA, Kii MA, Santos RM, Pereira CB. Musical hallucination associated with hearing loss. Arquivos de neuro-psiquiatria. 2011;69(2b):395–400.
8. Tanriverdi N, Sayilgan MA, Ozçürümez G. Musical hallucinations associated with abruptly developed bilateral loss of hearing. Acta Psychiatr Scand. 2001;103(2):153–5.
9. Colon-Rivera HA, Oldham MA. The mind with a radio of its own: a case report and review of the literature on the treatment of musical hallucinations. Gen Hosp Psychiatry. 2014;36(2):220–4.
10. Aziz VM, Nessim M, Warner NJ. Charles-Bonnet syndrome and musical hallucination. Int Psychogeriatr. 2004;16(4):489–91.
11. Serrador García M, Santos-Bueso E, Sáenz-Francés F, Díaz-Valle D, Martínez-De-La-Casa-Borrelia JM, García-Feijóo J. Charles Bonnet plus syndrome: apropos of a case. Eur J Ophthalmol. 2012;22(5):836–9.
12. Singh A, Karasin J, Madhusoodanan S. The sound of music: a rare case of auditory Charles Bonnet syndrome in an elderly male. Ann Clin Psychiatry. 2019;31(2):107–10.
13. Evers S. Musical hallucinations. Curr Psychiatry Rep. 2006;8(3):205–10.
14. Evers S, Ellger T. The clinical spectrum of musical hallucinations. J Neurol Sci. 2004;227(1):55–65.
15. Alvarez Perez P, Garcia-Antelo MJ, Rubio-Nazabal E. "Doctor, I hear music": a brief review about musical hallucinations. Open Neurol J. 2017;11:11–4.
16. Warner N, Aziz V. Hymns and arias: musical hallucinations in older people in Wales. Int J Geriatr Psychiatry. 2005;20(7):658–60.
17. Terao T, Matsunaga K. Musical hallucinations and palinacousis. Psychopathology. 1999;32(2):57–9.
18. Cope TE, Baguley DM. Is musical hallucination an otological phenomenon? A review of the literature. Clin Otolaryngol. 2009;34(5):423–30.
19. Gordon AG. Do musical hallucinations always arise from the inner ear? Med Hypotheses. 1997;49(2):111–22.
20. Auffarth IS, Kropp S. Musical hallucination in a patient after cochlear implantation. J Neuropsychiatry Clin Neurosci. 2009;21(2):230–1.
21. Low WK, Tham CA, D'Souza VD, Teng SW. Musical ear syndrome in adult cochlear implant patients. J Laryngol Otol. 2013;127(9):854–8.
22. Baba A, Hamada H. Musical hallucinations in schizophrenia. Psychopathology. 1999;32(5):242–51.
23. Baba A, Hamada H, Kocha H. Musical hallucinations in schizophrenia. 2. Relations with verbal hallucinations. Psychopathology. 2003;36(2):104–10.
24. Saba PR, Keshavan MS. Musical hallucinations and musical imagery: prevalence and phenomenology in schizophrenic inpatients. Psychopathology. 1997;30(4):185–90.
25. Hermesh H, Konas S, Shiloh R, Dar R, Marom S, Weizman A, Gross-Isseroff R. Musical hallucinations: prevalence in psychotic and nonpsychotic outpatients. J Clin Psychiatry. 2004;65(2):191–7.
26. Serby MJ, Hagiwara M, O'Connor L, Lalwani AK. Musical hallucinations associated with pontine lacunar lesions. J Neuropsychiatry Clin Neurosci. 2013;25(2):153–6.
27. Mittal M, Giron LT Jr. Elvis is back: musical hallucinations in a Parkinson disease patient. South Med J. 2010;103(8):837–41.
28. Borelli P, Vedovello M, Braga M, Pederzoli M, Beretta S. Persistent interictal musical hallucination in a patient with mesial temporal sclerosis-related epilepsy: first case report and etiopathological hypothesis. Cogn Behav Neurol. 2016;29(4):217–21.
29. Coebergh JAF, Lauw RF, Sommer IEC, Blom JD. Musical hallucinations and their relation with epilepsy. J Neurol. 2019;266(6):1501–15.
30. Stricker RB, Winger EE. Musical hallucinations in patients with Lyme disease. South Med J. 2003;96(7):711–5.
31. Isolan GR, Bianchin MM, Bragatti JA, Torres C, Schwartsmann G. Musical hallucinations following insular glioma resection. Neurosurg Focus. 2010;28(2):E9.
32. Keshavan MS, Kahn EM, Brar JS. Musical hallucinations following removal of a right frontal meningioma. J Neurol Neurosurg Psychiatry. 1988;51(9):1235–6.
33. Williams VG, Tremont G, Blum AS. Musical hallucinations after left temporal lobectomy. Cogn Behav Neurol. 2008;21(1):38–40.
34. Janakiraman R, Wildgoose K, Seelam K. ECT associated musical hallucinations in an elderly patient: a case report. Ann Gen Psychiatry. 2006;5:10.
35. Fukunishi I, Kitaoka T, Shirai T, Watanabe S. Musical hallucinations after childbirth in a female patient on hemodialysis. Nephron. 1998;79(1):105.
36. Fukunishi I, Kita Y, Harihara Y, Kubota K, Takayama T, Kawarasaki H, Makuuchi M. Musical hallucinations after living-donor liver transplantation. Psychosomatics. 1999;40(6):530–1.
37. Iijima M, Maeda T, Tsubouchi K, Iwamoto J. Musical hallucinations with post-pharyngolaryngoesophagectomy aphonia. Gen Hosp Psychiatry. 2000;22(2):135–6.
38. Wong CL, Bhalerao S. Post-surgery musical hallucinations of a patriotic Canadian song. Can J Neurol Sci. 2010;37(6):901–3.
39. Gondim Fde A, Costa HA, Taunay TC, de Oliveira GR, Ferreira JM, Rola FH. Transient amantadine-induced musical hallucinations in a patient with Parkinson's disease. Mov Disord. 2010;25(10):1505–6.
40. Molinaro J, DeVries P, Ha J, Knight JM. New-onset hallucinations with amiodarone: a case report. Ann Gen Psychiatry. 2022;21(1):34.
41. Kanemura S, Tanimukai H, Tsuneto S. Can "steroid switching" improve steroid-induced musical hallucinations in a patient with terminal cancer? J Palliat Med. 2010;13(12):1495–8.
42. Song CI, Jung YE. Musical hallucination caused by ceftazidime in a woman with a hearing impairment. Clin Psychopharmacol Neurosci. 2019;17(2):326–8.
43. Vallada HP, Gentil V. Musical hallucinations triggered by clomipramine? Br J Psychiatry. 1991;159:888–9.
44. Terao T. Tricyclic-induced musical hallucinations and states of relative sensory deprivation. Biol Psychiatry. 1995;38(3):192–3.

45. Curtin F, Redmund C. Musical hallucinations during a treatment with benzodiazepine. Can J Psychiatry. 2002;47(8):789–90.
46. Lee GH, Stewart JT. A case of musical hallucinations related to mirtazapine. Clin Neuropharmacol. 2018;41(6):222–3.
47. Davies AN, Quinn T. Opioid-related musical hallucinations. J Pain Symptom Manage. 2005;29(4):327–8.
48. Moore TA. Musical hallucinations induced by oxycodone. Am J Geriatr Psychiatry. 2003;11(4):470.
49. Muraosa H, Suzuki A, Noto K, Otani K. Musical Hallucinations Induced by Conventional Doses of Paroxetine. Am J Case Rep. 2020;21:e926735.
50. Kataoka H, Ueno S. Auditory musical hallucinations associated with extended-release pramipexole in an elderly patient with Parkinson's disease. Medicine (Baltimore). 2014;93(27):e251.
51. Fernandez A, Crowther TR, Vieweg WV. Musical hallucinations induced by propranolol. J Nerv Ment Dis. 1998;186(3):192–4.
52. Keeley PW, Foster G, Whitelaw L. Hear my song: auditory hallucinations with tramadol hydrochloride. BMJ. 2000;321(7276):1608.
53. Agrawal AK, Sherman LK. Voriconazole-induced musical hallucinations. Infection. 2004;32(5):293–5.
54. Coebergh JA, Lauw RF, Bots R, Sommer IE, Blom JD. Musical hallucinations: review of treatment effects. Front Psychol. 2015;6:814.
55. Aizawa S, Terao T, Hatano K, Ishii N. Musical hallucinations responding to a further increase of carbamazepine. BMJ Case Rep. 2014;
56. Terao T, Tani Y. Carbamazepine treatment in a case of musical hallucinations with temporal lobe abnormalities. Aust N Z J Psychiatry. 1998;32(3):454–6.
57. Ukai S, Yamamoto M, Tanaka M, Shinosaki K, Takeda M. Donepezil in the treatment of musical hallucinations. Psychiatry Clin Neurosci. 2007;61(2):190–2.
58. Zilles D, Zerr I, Wedekind D. Successful treatment of musical hallucinations with the acetylcholinesterase inhibitor donepezil. J Clin Psychopharmacol. 2012;32(3):422–4.
59. Bergman J, Pashinian A, Weizman A, Poyurovsky M. The beneficial effect of escitalopram on obsessive-compulsive-related musical hallucinations in elderly patients with hearing impairment: a case series. Int Clin Psychopharmacol. 2014;29(5):263–5.
60. Holroyd S, Sabeen S. Successful treatment of hallucinations associated with sensory impairment using gabapentin. J Neuropsychiatry Clin Neurosci. 2008;20(3):364–6.
61. Huntley JD, Sandall A, Philpot M. "In the midnight hour": a case report of musical hallucinations with multiple etiological factors treated with lamotrigine. Int Psychogeriatr. 2011;23(2):322–4.
62. Mansoor D, Ganzini L. Musical hallucinations successfully treated with antipsychotic medications: three case reports. Psychosomatics. 2014;55(2):191–3.
63. Peritogiannis V, Menti N, Mavreas V. Atypical antipsychotics for the treatment of musical hallucinations in an elderly patient without a psychiatric disorder. J Neurosci Rural Pract. 2016;7(4):606–8.
64. David RR, Fernandez HH. Quetiapine for hypnogogic musical release hallucinations. J Geriatr Psychiatry Neurol. 2000;13(4):210–1.
65. Kitamura H, Saito M, Someya T. Two cases of musical hallucination successfully treated with quetiapine. Psychiatry Clin Neurosci. 2017;71(11):789.
66. Blom JD, Coebergh JA, Lauw R, Sommer IE. Musical hallucinations treated with acetylcholinesterase inhibitors. Front Psychiatry. 2015;6:46.
67. Cosentino G, Giglia G, Palermo A, Panetta ML, Lo Baido R, Brighina F, Fierro B. A case of post-traumatic complex auditory hallucinosis treated with rTMS. Neurocase. 2010;16(3):267–72.
68. Nordberg J, Taiminen T, Virtanen L, Jääskeläinen SK, Scheinin NM. Successful suppression of musical hallucinations with low-frequency rTMS of the left temporo-parietal junction: a case report. Brain Stimul. 2021;14(6):1467–9.
69. Shergill SS, Murray RM, McGuire PK. Auditory hallucinations: a review of psychological treatments. Schizophr Res. 1998;32(3):137–50.

Part XVI

Diagnoses—Specific Ocular Motor Syndromes

110 Centripetal Nystagmus

Overview

Centripetal nystagmus is an ocular motor pattern in which at extreme lateral gaze there is jerk nystagmus whose fast phase beats toward primary position of gaze. This rare pattern has been observed in several neurodegenerative processes probably involving the cerebellum.

Introduction

Centripetal nystagmus is an ocular motor pattern in which at extreme lateral gaze there is jerk nystagmus whose fast phase beats toward primary position of gaze—in other words, at extreme right gaze, there is left beat nystagmus, and at extreme left gaze, there is right beat nystagmus.

Epidemiology

Centripetal nystagmus is a very rare ocular motor finding, reported with cerebellar disease [1], Creutzfeldt–Jakob disease [2], anti-GAD65-associated stiff person syndrome [3], and as part of an anti-NMDA paraneoplastic syndrome [4].

Pathophysiological Mechanism of Disease

Control systems analysis models predict that centripetal nystagmus can happen when gain of the neural integrator for horizontal eye movements in the nucleus prepositus hypoglossi (NPH) is greater than 1 [5]. In principle, this should be possible with abnormal excitation. However, the diseases in which centripetal nystagmus occurs are destructive processes, so we believe it is more likely that inhibitory input to the NPH is impaired, leading to disinhibition [4]. Candidate inhibitory afferents to the NPH include projections from the cerebellar flocculus, paraflocculus, and fastigial nuclei [6].

Clinical Presentation

The ocular motor findings of centripetal nystagmus are very striking, though the clinical picture is usually dominated by other features of the various neurodegenerative processes in which it occurs.

Ocular Motor Examination

At extreme lateral gaze, there is jerk nystagmus whose fast phase beats toward primary position of gaze. In several of the reported cases, centripetal nystagmus may occur in conjunction with other ocular motor abnormalities.

Instrumented Ocular Motor Testing

Figure 110.1 shows part of the ocular motor tracings of a patient with centripetal nystagmus occurring as part of an NMDA paraneoplastic syndrome secondary to a mature ovarian teratoma [4].

M. Cherchi, *Otoneurology and Vestibular Medicine*, https://doi.org/10.1007/978-3-031-94841-1_110

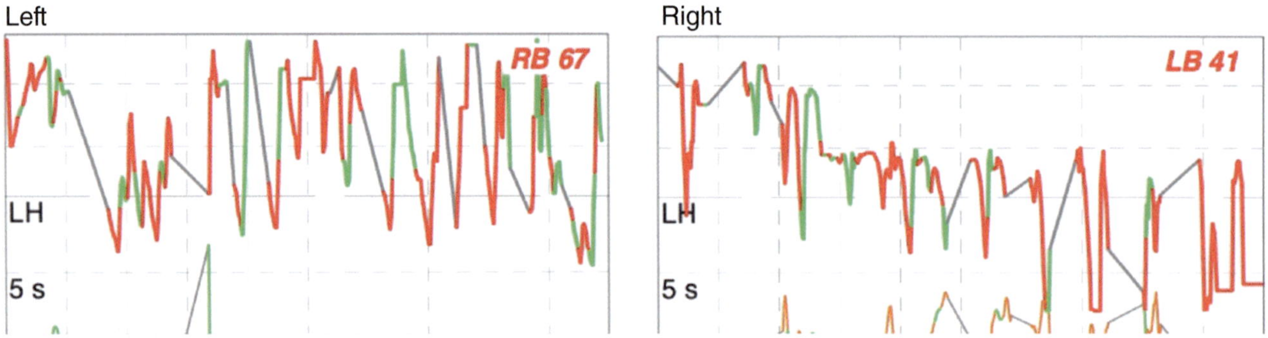

Fig. 110.1 Centripetal nystagmus in anti-NMDA paraneoplastic syndrome from a mature ovarian teratoma. Each panel shows 5 seconds ("5 s") of the ocular motor tracing of the left eye horizontal channel ("LH"). By convention, on horizontal tracings, an upward deflection represents rightward eye movement and a downward deflection represents leftward eye movement. The left panel shows right beat nystagmus of 67° per second ("RB 67") at extreme left gaze. The right panel shows left beat nystagmus of 41° per second ("LB 41") at extreme right gaze. In each tracing, the nearly vertically oriented segments represent the fast phase of horizontal nystagmus while the oblique segments represent the slow phase of horizontal nystagmus. (From Cherchi and Hac [4]. Reproduced with written permission from Wolters Kluwer Health)

Differential Diagnosis

The differential diagnosis of centripetal nystagmus primarily consists of cerebellar lesions.

Treatment and Prognosis

The treatment (if any) and prognosis depend on the underlying etiology.

References

1. Leech J, Gresty M, Hess K, Rudge P. Gaze failure, drifting eye movements, and centripetal nystagmus in cerebellar disease. Br J Ophthalmol. 1977;61(12):774–81.
2. Helmchen C, Büttner U. Centripetal nystagmus in a case of Creutzfeldt-Jacob disease. Neuro-Ophthalmology. 1995;15(4):187–92.
3. Hac NEF, Murphy OC, Butala AA, Newsome SD, Gold DR. Centripetal nystagmus, slow saccades, cerebellar ataxia, and parkinsonism in a patient with anti-gad65-associated stiff person syndrome spectrum disorder. J Neuro-Ophthalmol. 2023;1–5
4. Cherchi M, Hac NEF. Paraneoplastic Gaze-Evoked Centripetal Nystagmus in Ovarian Teratoma. Journal of Neuro-Ophthalmology. 2024;44(2):e274–e5.
5. Optican LM, Zee DS. A hypothetical explanation of congenital nystagmus. Biol Cybern. 1984;50(2):119–34.
6. McCrea RA, Baker R, Delgado-Garcia J. Afferent and efferent organization of the prepositus hypoglossi nucleus. Prog Brain Res. 1979;50:653–65.

111 Congenital Nystagmus

Overview

In otoneurology, the phrase, *"congenital nystagmus" (CN)*, refers to an eye movement abnormality that becomes evident around 3 months of age, consists of involuntary cyclical jerk nystagmus or pendular nystagmus that is usually conjugate and usually predominantly horizontal, which worsens on attempted fixation, and may improve when the eyes are in a certain orientation within the orbits (called a "null point"). CN is rare. CN can occur in isolation, or in conjunction with other diseases (such as albinism), including ophthalmologic disease (such as achromatopsia). The most common presenting symptom is poor vision, which results from the fact that the involuntary eye movements interfere with foveation of a target. The underlying pathophysiologic mechanism of CN is unknown. A number of genetic mutations have been reported in association with CN. CN is usually evident on face-to-face examination, though accurate characterization requires instrumented oculographic testing. Some imaging studies report increased gray matter volume in certain temporo-occipital regions and parts of the cerebellum. Patients with CN can also have vestibular disease, and it is reasonable to screen for these. CN patients can also have unrelated ophthalmic disease, so evaluation by ophthalmology or neuro-ophthalmology is appropriate. Small case series report improvement of nystagmus with gabapentin or memantine. Surgery is sometimes attempted. CN is a lifelong condition.

Introduction

The unqualified phrase "congenital nystagmus" (CN) refers broadly to any abnormal pattern of eye movements identified at or shortly after birth. However, in otoneurology, the term is often used to refer more narrowly to a set of oscillatory eye movement disorders (recognized at birth or during infancy) of variable rhythmicity that result from a *deficit of the visual fixation and smooth pursuit systems*. Clinically, what distinguishes these abnormal eye movements is that they *worsen (become more pronounced) when the patient is attempting to fixate* [1, 2]. This contrasts with most forms of nystagmus arising from peripheral vestibular disease, in which fixation, tends to reduce or suppress the nystagmus.

Congenital nystagmus can appear in a variety of forms and is often difficult to characterize. Dell'Osso and Daroff [2] offer the following classification of congenital nystagmus.

1. Pendular nystagmus.
   - (a) *Pure pendular nystagmus* (PP). As the name implies, the nystagmus is pendular, meaning there is an approximately sinusoidal (and therefore symmetrical) movement.
   - (b) *Asymmetric pendular nystagmus* (AP). While still pendular, the movement in one direction is rapid enough that it is sometimes mistaken for a saccade. "This form usually occurs as the subject looks laterally, but is occasionally present in primary position."
   - (c) *Pendular nystagmus with foveating saccades* (Pfs). This form exhibits "small 'braking' saccades [that] stop the SEM [slow eye movement] after it has bypassed the target and, in addition, the saccades achieve refoveation of the target."
2. Jerk nystagmus.
   - (a) Unidirectional jerk nystagmus.
     - (i) Saccadic foveation. "In both [forms], the saccades are of sufficient amplitude to fully refoveate the target."
       1. *Pure jerk nystagmus* (PJ). This has a "classical saw-tooth appearance."
       2. *Jerk nystagmus with extended foveation* (Jef). In this form, "There are long periods of foveation following each saccade during which eye is motionless and good vision possible."
     - (ii) SEM (slow eye movement) foveation. "In the other two types of unidirectional jerk nystagmus, the saccades, although corrective in nature, are

M. Cherchi, *Otoneurology and Vestibular Medicine*, https://doi.org/10.1007/978-3-031-94841-1_111

of insufficient amplitude to fully refoveate the target."

1. *Pseudo-cycloid nystagmus* (PC). This form is "often misidentified clinically as pendular [but actually] consists of an accelerating… drift off target which is terminated by a small… braking saccade and following by a SEM [slow eye movement] which refoveates the target."
2. *Pseudo-jerk nystagmus* (PJ). In this form, "the actual direction of the 'fast' phase is opposite to the apparent direction."

(b) Bidirectional jerk nystagmus. These forms "occur at gaze angles in the transition (neutral) zone in which the direction of the jerk nystagmus is reversed. As such, they are transient, quite variable, and usually not conducive to good vision."

(i) Pseudo-pendular nystagmus.

1. *Pure pseudo-pendular nystagmus* (PP). This form consists of "alternating accelerating SEM [slow eye movement] and braking saccades."
2. *Pseudo-pendular nystagmus with foveating saccades* (PPfs). In this form, "the waveform is biased (as is Pfs) such that the braking saccades that actually foveate the target are larger than those at the other peaks. They are also of variable amplitudes depending upon the actual overshoot of the target."

(ii) *Triangular nystagmus* (T). "When bidirectional jerk waveforms occur at the termination of SEM [slow eye movement] that are not rapidly accelerating, a triangular waveform (T) results."

(iii) *Bidirectional jerk nystagmus* (BDJ). In this form, "one of the two saccades actually foveates the target."

3. *Dual jerk nystagmus* (DJ). This form "consists of the simultaneous admixture of jerk and pendular nystagmus with the superimposition of a rapid small amplitude sinusoidal oscillation upon the larger amplitude jerk nystagmus."

The classification described by Dell'Osso and Daroff [2] has been adopted by other investigators [1].

The nystagmus is nearly always conjugate, though monocular CN has been described [3]. The eye movements are predominantly horizontal, though modest torsional [1, 4] or vertical [5] components may also be present.

Some forms of CN can be modulated by vergence [1]—typically the spontaneous nystagmus improves with convergence.

Epidemiology

Huang and colleagues [6] identified 50 patients with congenital nystagmus out of a clinic population of 32,050 patients who had presented for "vertigo/dizziness, tinnitus, hearing loss or pathological eye movements," amounting to 0.16% of that population, and they described this as comparable to the prevalence rate of 1 in 1000–1500, citing Self and Lotery [7]. In their 50 patients (27 male, 23 female), the mean age at presentation was 36 years, with a range of 3 to 76 years.

Genetics

A range of genetic defects has been reported in association with CN [7–28]. The mechanism by which these mutations cause or contribute to CN is unknown.

Pathophysiological Mechanism of Disease

The underlying pathophysiological mechanism of CN is unknown, though Optican and Zee [29] have proposed a model based on control systems analysis.

Clinical Presentation

"The commonest symptom of congenital nystagmus is reduced visual acuity which can range from mild to severe" [5], but this symptom in isolation is not a common reason for referral to otoneurology.

Common reasons for CN patients to be referred to otoneurology include observation (by a family member or another clinician) of abnormal eye movements. It is uncommon for CN patients to be referred for the symptom of disequilibrium ("dizziness," "vertigo"), but insofar as poor vision can interfere with equilibrium, this is sometimes the basis for the referral.

CN may be apparent at birth, but typically becomes evident around the age of 3 months when infants begin visually to fixate [6], and it is a lifelong condition.

It has been well documented that CN reduces visual acuity [30], and in fact, "For all cases of idiopathic congenital nystagmus visual acuity is reduced" [31]. This is because "normal visual acuity requires a stationary retinal image on the fovea. If fixation instabilities cause movement of the retinal image across the fovea for a few degrees, visual acuity is diminished" [32].

Since untreated congenital nystagmus is constant and lifelong, some investigators have suggested that the chronic interference with foveation may additionally introduce a component of amblyopia, further interfering with vision [33].

The intrusive involuntary eye movements make it difficult to execute visual tasks requiring precise control of eye movements, such as reading [34].

Despite the spontaneous oscillatory nature of eye movements in CN, oscillopsia is not a common complaint. Gresty and colleagues comment that "when viewing a stationary target under normal illumination oscillopsia is generally not a symptom of congenital nystagmus regardless of eye position" [5]. However, patients whose CN manifests as periodic alternating nystagmus (PAN) may intermittently complain of oscillopsia [1].

Physical Examination

In a patient with CN who is otherwise healthy, aside from the ocular motor abnormalities and possible head tilt (see below), the remainder of the physical examination should be normal.

However, note that CN is often comorbid with other pathologies (such as albinism) and other ophthalmic diseases (such as achromatopsia).

Ocular Motor Examination

CN can often be identified on face-to-face examination.

CN often has a "null point," meaning a position of the eyes within the orbits at which the nystagmus waveform is smallest. "If the null point is eccentric this leads to a head turn in the opposite direction" [5], which is why many CN patients assume some type of head tilt or rotation when looking "straight ahead."

Testing: Oculographic

Detailed oculographic recordings are often helpful in characterizing CN and distinguishing it from alternative diagnoses.

CN patients often have impaired optokinetic responses, with decreased gain of optokinetic nystagmus and reduced or absent optokinetic afternystagmus [35]. These findings are sensitive but not specific for CN.

Some studies of patients with CN report the vestibulo-ocular reflex (VOR) to be normal [36], while others report it to be consistently abnormal [37]. There have been very few studies of the high frequency end of the vestibular spectrum [38].

Gresty and colleagues [5] report that "Abnormalities of responses to rotational testing occur in about 50% of asymptomatic patients with congenital nystagmus" [5].

Curiously, CN patients have reduced motion aftereffects [39]. Presumably, this is because consistent foveation (which CN patients lack) is required to build the habituation that generates motion aftereffect illusions when the visual stimulus stops moving.

Imaging

Hufner and colleagues [40] compared MRIs of 14 CN patients (7 males, mean age 28.3 ± 9.38 years) and 18 healthy controls (11 males, mean age 28.56 ± 8.58 years) using voxel-based morphometry. Compared to controls, the CN patients exhibited "significant volume increases in [the gray matter of] the human motion sensitive complex V5/MT+, the fusiform gyrus, and the middle occipital gyrus bilaterally," and speculated that "these volume increases may be associated with excess visual motion stimulation due to involuntary retinal slip of the visual scene" (Fig. 111.1).

They further observed a "positive correlation (linear model) of nystagmus sway path with cerebellar GM [gray matter] volume… in the following areas: vermal parts VIII-X as well as hemisphere lobule II, hemisphere VI, crus I, crus II, and lobule VII-IX bilaterally," and speculated that this "might be related to the subjects' attempt to maintain fixation, rather than be due to the generation of nystagmus" (Fig. 111.2).

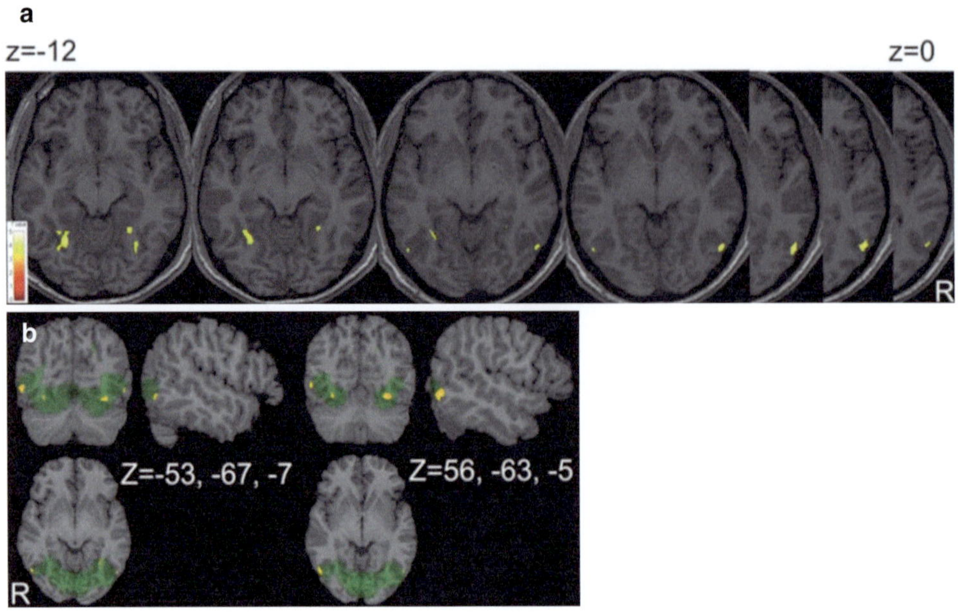

Fig. 111.1 Results of the contrast congenital nystagmus (CN) subjects vs. controls. Panel (A): The areas of volume increase in CN subjects were projected onto an individual anatomical brain chosen at random. The slices were chosen from $z = -12$ to $z = 0$ (distance 2 mm). They show volume changes in the fusiform and lingual gyri as well as V5/MT+ bilaterally. Panel (B): The functional ROI (green) is superimposed; x, y, z, views at the two cluster maxima of V5/MT+ are shown. (From Hufner et al. [40]. Reproduced from Frontiers Media S.A. with permission under CC-BY 4.0)

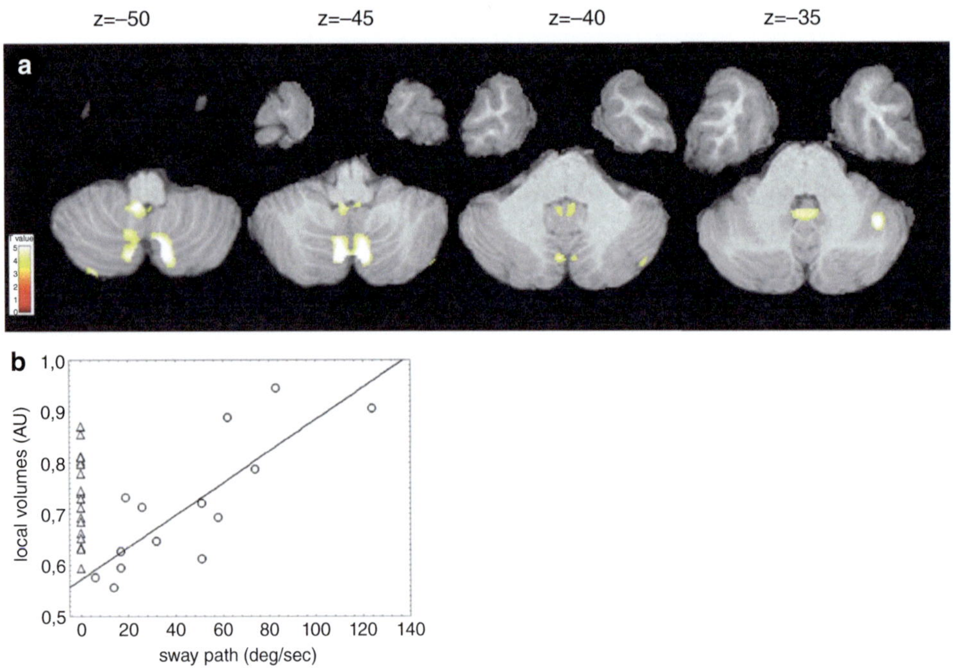

Fig. 111.2 Results of correlation analysis of cerebellar gray matter volume with nystagmus severity in congenital nystagmus (CN) subjects. Panel (A): A positive correlation of cerebellar gray matter volume with nystagmus sway path was found in CN subjects (linear model), while no overall changes in gray matter compared to controls were detected. Panel (B): The normalized local cerebellar volume values were plotted against nystagmus sway path for display purposes. Normalized local cerebellar volume values were extracted at $x = 6$ $y = -72$ $z = -48$ (local cluster maximum) within a specified volume of interest (VOI) to obtain an indirect measure of gray matter thickness (arbitrary units, AU). The triangles in panel (B) depict the wide variability of cerebellar volume in controls. In the CN subjects the linear correlation between the cerebellar volume and the nystagmus severity is demasked only by adding nystagmus sway path as a variable [circles in panel (B)]. The correlation statistics were calculated for the CN subjects at $r = 0.85$, $p < 0.001$. The r values must be interpreted with caution, since such values tend to be overestimated in this kind of analysis. (From Hufner et al. [40]. Reproduced from Frontiers Media S.A. with permission under CC-BY 4.0)

Differential Diagnosis

CN is rare. It can occur in isolation, though is sometimes associated with other pathologies (such as albinism) and other ophthalmic diseases (such as achromatopsia).

It should be kept in mind that CN patients are just as vulnerable to common otovestibular diseases as the general population. Of the 50 CN patients studied by Huang and Hsu [6], they report that:

> Audiometry revealed normal mean hearing level (MHL) in 66 ears and abnormal MHL in 34 ears (34%). The caloric test showed normal responses in 75 ears, canal paresis in 6 ears and caloric areflexia in 19 ears, representing 25% (25/100) abnormality. The oVEMP test was performed in 8 patients (16 ears) and showed normal oVEMPs in 11 ears, reduced oVEMPs in 1 ear and absent oVEMP in 4 ears, accounting for 31% (5/16) abnormality. The cVEMP test was performed in 16 patients (32 ears) and identified normal, delayed and absent responses in 22, 2 and 8 ears, respectively, representing 31% (10/32) abnormality. In sum, approximately one-third of the congenital nystagmus patients (n = 17) had comorbid inner ear disorders including labyrinthine insufficiency 4, endolymphatic hydrops 4, sequela of otitis media 2, congenital deafness 2, acoustic trauma 2 and others 3. The remaining 33 patients, about two-thirds of the congenital nystagmus patient, lacked inner ear disorders" [6].

This emphasizes the point that screening for other otovestibular disorders may be worthwhile, as some may be treatable; for example, vestibular weakness may be managed with vestibular rehabilitation therapy.

Since poor vision is a common reason for referral, it should be kept in mind that the differential diagnosis of poor vision is very broad and includes (non-otoneurological) primary ophthalmologic diseases as well. Holmstrom and colleagues [41] studied 62 patients referred for "nystagmus" and found that only 2 (3%) were ultimately diagnosed with CN; the remaining diagnoses included foveal hypoplasia, albinism, achromatopsia, rod-cone dystrophy, and degenerative high myopia—all primary ophthalmological diseases. This emphasizes the point that even in a patient with CN, evaluation by ophthalmology or neuro-ophthalmology is important.

Treatment

In a retrospective review, Sherry and colleagues [42] observed improvement in visual acuity and nystagmus in two patients with CN treated with gabapentin and memantine. Consequently, McLean and colleagues [43] performed a prospective randomized, double-blinded study of 48 CN patients treated with memantine (n = 16), gabapentin (n = 16), and placebo (n = 15). They reported that at 56 days, outcomes in the treatment groups (memantine and gabapentin) were similar to each other and were statistically significantly superior to placebo in terms of improvement in visual acuity, nystagmus intensity, and foveation.

Periodic alternating nystagmus (PAN) can be acquired, in some cases appearing after cerebellar lesions, and in some such cases, the nystagmus may improve with baclofen [44]. Case reports [45] and case series [46] have also observed improvement in patients with congenital periodic alternating nystagmus (cPAN) treated with baclofen.

A variety of nonpharmacologic approaches have been attempted in CN, including contact lenses [47], acupuncture [48], biofeedback [49], auditory feedback [50], extraocular muscle injection with onabotulinum toxin [51], and cannabis [52].

A variety of surgical approaches to CN have been proposed [53–61].

Prognosis

CN is a lifelong condition.

References

1. Abadi RV, Dickinson CM. Waveform characteristics in congenital nystagmus. Doc Ophthalmol. 1986;64(2):153–67.
2. Dell'Osso LF, Daroff RB. Congenital nystagmus waveforms and foveation strategy. Doc Ophthalmol. 1975;39(1):155–82.
3. Kamel KS, Congenital nystagmus. Report of a rare case of uniocular congenital nystagmus. Bull Ophthalmol Soc Egypt. 1974;67:349–52.
4. Averbuch-Heller L, Dell'Osso LF, Leigh RJ, Jacobs JB, Stahl JS. The torsional component of "horizontal" congenital nystagmus. J Neuro-ophthalmol. 2002;22(1):22–32.
5. Gresty M, Page N, Barratt H. The differential diagnosis of congenital nystagmus. J Neurol Neurosurg Psychiatry. 1984;47(9):936–42.
6. Huang YC, Hsu SY, Young YH. Comorbid inner ear disorders in 50 patients with congenital nystagmus. Clin Otolaryngol. 2018;43(4):1171–7.
7. Self J, Lotery A. The molecular genetics of congenital idiopathic nystagmus. Semin Ophthalmol. 2006;21(2):87–90.
8. Arshad MW, Shabbir MI, Asif S, Shahzad M, Leydier L, Rai SK. FRMD7 gene alterations in a pakistani family associated with congenital idiopathic nystagmus. Genes (Basel). 2023;14(2)
9. Du W, Bu J, Dong J, Jia Y, Li J, Liang C, et al. A novel frame-shift mutation in FRMD7 causes X-linked idiopathic congenital nystagmus in a Chinese family. Mol Vis. 2011;17:2765–8.
10. Gupta S, Pathak E, Chaudhry VN, Chaudhry P, Mishra R, Chandra A, et al. A novel mutation in FRMD7 causes X-linked idiopathic congenital nystagmus in a North Indian family. Neurosci Lett. 2015;597:170–5.
11. Gutmann DH, Brooks ML, Emanuel BS, McDonald-McGinn DM, Zackai EH. Congenital nystagmus in a (46,XX/45,X) mosaic woman from a family with X-linked congenital nystagmus. Am J Med Genet. 1991;39(2):167–9.

12. He X, Gu F, Wang Z, Wang C, Tong Y, Wang Y, et al. A novel frameshift mutation in FRMD7 causing X-linked idiopathic congenital nystagmus. Genet Test. 2008;12(4):607–13.
13. Hoffmann S, Becker A, Hoerle S, Metz A, Oertel WH, Sommer N, et al. Autosomal dominant congenital nystagmus is not linked to 6p12, 7p11, and 15q11 in a German family. Am J Ophthalmol. 2004;138(3):439–43.
14. Hu J, Liang D, Xue J, Liu J, Wu L. A novel GPR143 splicing mutation in a Chinese family with X-linked congenital nystagmus. Mol Vis. 2011;17:715–22.
15. Hu Y, Shen J, Zhang S, Yang T, Huang S, Yuan H. A novel splicing mutation of the FRMD7 gene in a Chinese family with X-linked congenital nystagmus. Mol Vis. 2012;18:87–91.
16. Kerrison JB, Arnould VJ, Barmada MM, Koenekoop RK, Schmeckpeper BJ, Maumenee IH. A gene for autosomal dominant congenital nystagmus localizes to 6p12. Genomics. 1996;33(3):523–6.
17. Kerrison JB, Koenekoop RK, Arnould VJ, Zee DS, Maumenee IH. Clinical features of autosomal dominant congenital nystagmus linked to chromosome 6p12. Am J Ophthalmol. 1998;125(1):64–70.
18. Kim US, Cho E, Kim HJ. A novel nonsense mutation of GPR143 gene in a Korean kindred with X-linked congenital nystagmus. Int J Ophthalmol. 2016;9(9):1367–70.
19. Kohmoto T, Okamoto N, Satomura S, Naruto T, Komori T, Hashimoto T, et al. A FRMD7 variant in a Japanese family causes congenital nystagmus. Hum Genome Var. 2015;2:15002.
20. Li Y, Jiang L, Wang L, Wang C, Liu C, Guo A, et al. p.His16Arg of STXBP1 (MUNC18-1) associated with syntaxin 3B causes autosomal dominant congenital nystagmus. Front Cell Dev Biol. 2020;8:591781.
21. Liu JY, Ren X, Yang X, Guo T, Yao Q, Li L, et al. Identification of a novel GPR143 mutation in a large Chinese family with congenital nystagmus as the most prominent and consistent manifestation. J Hum Genet. 2007;52(6):565–70.
22. Patton MA, Jeffery S, Lee N, Hogg C. Congenital nystagmus cosegregating with a balanced 7;15 translocation. J Med Genet. 1993;30(6):526–8.
23. Pearce WG. Congenital nystagmus – genetic and environmental causes. Can J Ophthalmol. 1978;13(1):1–9.
24. Peng Y, Meng Y, Wang Z, Qin M, Li X, Dian Y, et al. A novel GPR143 duplication mutation in a Chinese family with X-linked congenital nystagmus. Mol Vis. 2009;15:810–4.
25. Schorderet DF, Tiab L, Gaillard MC, Lorenz B, Klainguti G, Kerrison JB, et al. Novel mutations in FRMD7 in X-linked congenital nystagmus. Mutation in brief #963. Online. Hum Mutat. 2007;28(5):525.
26. Self J, Lotery A. A review of the molecular genetics of congenital Idiopathic Nystagmus (CIN). Ophthalmic Genet. 2007;28(4):187–91.
27. Wang XF, Chen H, Huang PJ, Feng ZK, Hua ZQ, Feng X, et al. Genotype-phenotype analysis and mutation spectrum in a cohort of chinese patients with congenital nystagmus. Front Cell Dev Biol. 2021;9:627295.
28. Xiu Y, Yao Y, Yang T, Pan M, Yang H, Fang W, et al. Identification of a novel idiopathic congenital nystagmus-causing missense mutation, p.G296C, in the FRMD7 gene. Mol Med Rep. 2018;18(3):2816–22.
29. Optican LM, Zee DS. A hypothetical explanation of congenital nystagmus. Biol Cybern. 1984;50(2):119–34.
30. Simmers AJ, Gray LS, Winn B. The effect of abnormal fixational eye movements upon visual acuity in congenital nystagmus. Curr Eye Res. 1999;18(3):194–202.
31. Abadi RV, Sandikcioglu M. Visual resolution in congenital pendular nystagmus. Am J Optom Physiol Opt. 1975;52(9):573–81.
32. Dorn L, Curkovic T, Dorn V. Foveation period and waveforms of congenital ocular nystagmus. Collegium antropologicum. 2005;29(Suppl 1):99–101.
33. Bedell HE. Visual and perceptual consequences of congenital nystagmus. Semin Ophthalmol. 2006;21(2):91–5.
34. Kudo Y, Kaga K, Ito K, Nakamura M. Ocular movement during reading in patients with congenital nystagmus. Acta oto-laryngologica Supplementum. 1995;520(Pt 2):282–4.
35. Yee RD, Baloh RW, Honrubia V. Study of congenital nystagmus: optokinetic nystagmus. Br J Ophthalmol. 1980;64(12):926–32.
36. Yee RD, Baloh RW, Honrubia V, Kim YS. A study of congenital nystagmus: vestibular nystagmus. J Otolaryngol. 1981;10(2):89–98.
37. Gresty MA, Barratt HJ, Page NG, Ell JJ. Assessment of vestibulo-ocular reflexes in congenital nystagmus. Ann Neurol. 1985;17(2):129–36.
38. Denia-Lafuente A, Lombardero B. Vestibular function measured using the video head impulse test in congenital nystagmus and vertigo: a case report. Front Neurol. 2021;12:690402.
39. Shallo-Hoffmann J, Wolsley CJ, Acheson JF, Bronstein AM. Reduced duration of a visual motion aftereffect in congenital nystagmus. Doc Ophthalmol. 1998;95(3-4):301–14.
40. Hufner K, Stephan T, Flanagin VL, Deutschlander A, Dera T, Karch C, et al. Cerebellar and visual gray matter brain volume increases in congenital nystagmus. Front Neurol. 2011;2:60.
41. Holmstrom G, Bondeson ML, Eriksson U, Akerblom H, Larsson E. 'Congenital' nystagmus may hide various ophthalmic diagnoses. Acta Ophthalmol. 2014;92(5):412–6.
42. Shery T, Proudlock FA, Sarvananthan N, McLean RJ, Gottlob I. The effects of gabapentin and memantine in acquired and congenital nystagmus: a retrospective study. Br J Ophthalmol. 2006;90(7):839–43.
43. McLean R, Proudlock F, Thomas S, Degg C, Gottlob I. Congenital nystagmus: randomized, controlled, double-masked trial of memantine/gabapentin. Ann Neurol. 2007;61(2):130–8.
44. Halmagyi GM, Rudge P, Gresty MA, Leigh RJ, Zee DS. Treatment of periodic alternating nystagmus. Ann Neurol. 1980;8(6):609–11.
45. Solomon D, Shepard N, Mishra A. Congenital periodic alternating nystagmus: response to baclofen. Ann N Y Acad Sci. 2002;956:611–5.
46. Comer RM, Dawson EL, Lee JP. Baclofen for patients with congenital periodic alternating nystagmus. Strabismus. 2006;14(4):205–9.
47. Allen ED, Davies PD. Role of contact lenses in the management of congenital nystagmus. Br J Ophthalmol. 1983;67(12):834–6.
48. Blekher T, Yamada T, Yee RD, Abel LA. Effects of acupuncture on foveation characteristics in congenital nystagmus. Br J Ophthalmol. 1998;82(2):115–20.
49. Mezawa M, Ishikawa S, Ukai K. Changes in waveform of congenital nystagmus associated with biofeedback treatment. Br J Ophthalmol. 1990;74(8):472–6.
50. Sharma P, Tandon R, Kumar S, Anand S. Reduction of congenital nystagmus amplitude with auditory biofeedback. J AAPOS. 2000;4(5):287–90.
51. Carruthers J. The treatment of congenital nystagmus with Botox. J Pediatr Ophthalmol Strabismus. 1995;32(5):306–8.
52. Pradeep A, Thomas S, Roberts EO, Proudlock FA, Gottlob I. Reduction of congenital nystagmus in a patient after smoking cannabis. Strabismus. 2008;16(1):29–32.
53. Bagheri A, Ale-Taha M, Abrishami M, Salour H. Effect of horizontal rectus surgery on clinical and paraclinical indices in congenital nystagmus. J Ophthalmic Vis Res. 2008;3(1):6–15.

54. Dell'Osso LF, Flynn JT. Congenital nystagmus surgery. A quantitative evaluation of the effects. Arch Ophthalmol. 1979;97(3):462–9.
55. Erbagci I, Gungor K, Bekir NA. Effectiveness of retroequatorial recession surgery in congenital nystagmus. Strabismus. 2004;12(1):35–40.
56. Flynn JT, Dell'Osso LF. The effects of congenital nystagmus surgery. Ophthalmology. 1979;86(8):1414–27.
57. Nelson LB, Wagner RS, Harley RD. Congenital nystagmus surgery. Int Ophthalmol Clin. 1985;25(4):133–8.
58. Spielmann A. Clinical rationale for manifest congenital nystagmus surgery. J AAPOS. 2000;4(2):67–74.
59. Taylor JN, Jesse K. Surgical management of congenital nystagmus. Aust N Z J Ophthalmol. 1987;15(1):25–34.
60. Wagdy FM, Ismael ME, Sarhan AE. Evaluation of the role of displacement surgery in the management of congenital nystagmus. Electron Physician. 2017;9(1):3672–7.
61. Whittle J. The effectiveness of surgery for congenital nystagmus. Eye (Lond). 1999;13(Pt 1):1–2.

Convergence-Retraction Nystagmus

112

Overview

Convergence-retraction nystagmus (CRN) is a relatively rare, acquired pattern of usually intermittent abnormal eye movements in which an attempt at upward gaze instead elicits either convergence (bilateral adduction) or retraction (simultaneous cocontraction of the medial and lateral recti of each eye, pulling the globe posteriorly into the orbit) or both. This ocular motor finding localizes to the midbrain. It can occur in isolation but often is found in conjunction with other ocular motor abnormalities, depending on the etiology and extent of the lesion. Presenting symptoms may include difficulty reading, diplopia, and disequilibrium. Etiologies include compressive lesions (such as a pineal mass), ischemic or hemorrhagic infarcts, hydrocephalus and seizures; consequently, imaging (preferably brain MRI) is appropriate. Treatment and prognosis depend on the underlying etiology.

Introduction

Convergence-retraction nystagmus (CRN) is a relatively rare, acquired pattern of usually intermittent abnormal eye movements in which an attempt at upward gaze instead elicits either convergence (bilateral adduction) or retraction (simultaneous cocontraction of the medial and lateral recti of each eye, pulling the globe posteriorly into the orbit) or both. This ocular motor finding localizes to the midbrain. It can occur in isolation but often is found in conjunction with other ocular motor abnormalities, depending on the etiology and extent of the lesion.

Epidemiology

Convergence-retraction nystagmus (CRN) is relatively rare. It occurs in a subset of cases of destructive lesions (tumor, stroke) that affect the midbrain, and less commonly in seizures with periodic lateralizing epileptiform discharges.

Pathophysiological Mechanism of Disease

The mechanism of convergence-retraction nystagmus (CRN) is incompletely understood, in part because of the different ocular motor abnormalities that comprise this phenomenon [1, 2]. Some investigators maintain that CRN is fundamentally a disorder of saccades [3], while others maintain that it is a disorder of vergence movements [4].

Rambold and colleagues state that, "lesions associated with CRN are thought to be localized in the *midbrain, lateral and anterior to the oculomotor nucleus*" [4].

Figure 112.1, from a discussion of Parinaud's syndrome, Ortiz and colleagues [2], shows several pathologies, but as far as convergence-retraction nystagmus is concerned, they state that "CRN arises from a dysfunction in the riMLF [rostral interstitial nucleus of the median longitudinal fasciculus] and PC [posterior commissure]."

M. Cherchi, *Otoneurology and Vestibular Medicine*, https://doi.org/10.1007/978-3-031-94841-1_112

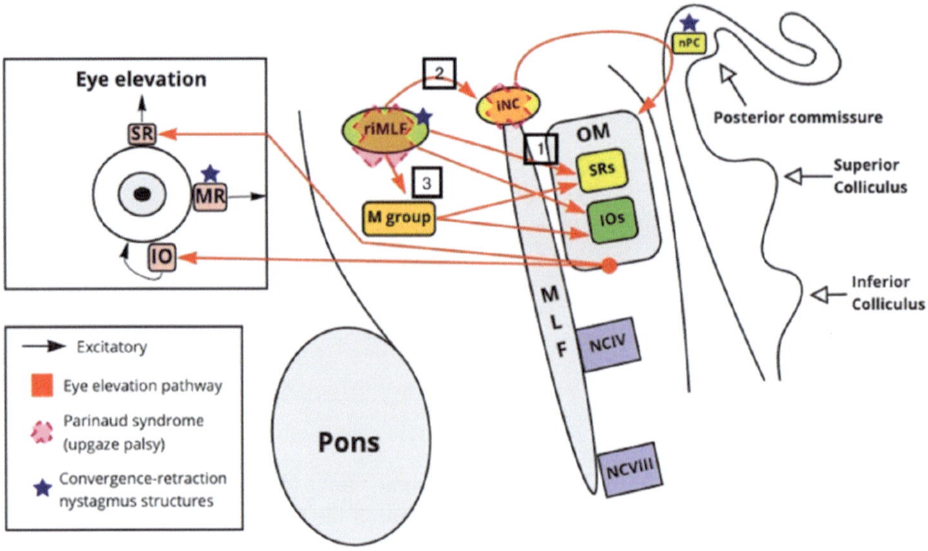

Fig. 112.1 Diagram of the neural structures involved in Parinaud's syndrome, including those for convergence-retraction nystagmus (CRN) and upward gaze palsy. The superior rectus (SR) and inferior oblique muscle (iOM) are key for the vertical elevation of the gaze and the oculomotor (OM) nucleus. The riMLF (rostral interstitial nucleus of the medial longitudinal fasciculus) sends information to three destinations: (1) directly to the superior rectus (SR) and inferior oblique muscle (IOs), (2) to the interstitial nucleus of Cajal (iNC), which send their axons to the nPC (nucleus of the posterior commissure), (3) to the M group, and CCN (central caudal nuclear group) which also innervate the RM and iOM nuclei, assisting with upward gaze. CRN is caused by a dysfunction in the riMLF and the nPC, the riMLF receives afferent projections for the center of convergence and divergence. The center of convergence and divergence is located dorsolateral to the oculomotor nerve in the dorsal midbrain. (From Ortiz et al. [2]. Reproduced from MDPI with permission under CC-BY 4.0)

Clinical Presentation

Although speaking of the broader pretectal syndrome (and not only of convergence-retraction nystagmus), Keane's review of 206 patients [5] concluded that "Symptoms varied with the etiology, which in many patients had a rapid or catastrophic onset." Most published cases describe abrupt onset of symptoms, typically from infarction [6–8].

Gradual onset over months is less common and usually occurs in the context of a degenerative etiology [9], or a gradually developing compressive etiology such as a tumor [10] or hydrocephalus [3, 11].

Presenting symptoms can include the following:

- Difficulty reading [3]
- Diplopia [3, 4, 9–12]
- "Dizziness" [4]
- "Droopy eyelids" [9]
- Convulsions if occurring in the context of seizures [6]
- Coma [13]

Physical Examination

In a patient with isolated convergence-retraction nystagmus (CRN), general physical examination may be normal [12]. Depending on the extent and location of the underlying lesion, other physical examination findings may be present [5].

Ocular Motor Examination

Convergence-retraction nystagmus (CRN) can occur in isolation [12], but often it occurs in conjunction with other ocular motor disorders, such as in the context of the pretectal syndrome [5]. It occurs in about 34% of cases of pretectal syndrome [14].

When the patient tries to execute an upward saccade (attempted voluntarily or in response to downward-moving optokinetic stimuli; see below), there is either retraction or convergence or some combination thereof. The convergence is, of course, bilateral adduction, brought about by simultaneous contraction of the medial recti and relaxation of the lateral recti. The retraction—posterior movement of the globes within the orbits—is probably about by simultaneous cocontraction of the medial and lateral recti, though some investigators state that there is cocontraction of all extraocular muscles [7].

Convergence-retraction nystagmus can be elicited by instructing a patient to execute an upward saccade. Another method is to use an optokinetic drum with stripes oriented horizontally and moving downward; this elicits expected slow phase downward (pursuit) eye movements, but instead of these being followed by a fast phase upward "resetting" saccade, there is a rapid (saccadic) movement that is either convergence, or retraction, or some combination thereof [1].

In those patients with only convergence nystagmus (and no retraction nystagmus), the convergence is brought about by a fast phase (saccadic) movement, while the divergence back to primary position of gaze is brought about by a slow phase movement [2]. Pullicino and colleagues comment that "convergence-retraction nystagmus (CNR) is not a true form of nystagmus but consists of closely spaced, alternating, adducting saccades" [14].

Material in various video case reports demonstrates the appearance of CRN on face-to-face examination [10].

Imaging

Given the potential etiologies and differential diagnosis (see below), brain imaging (usually an MRI) is appropriate for patients with convergence-retraction nystagmus.

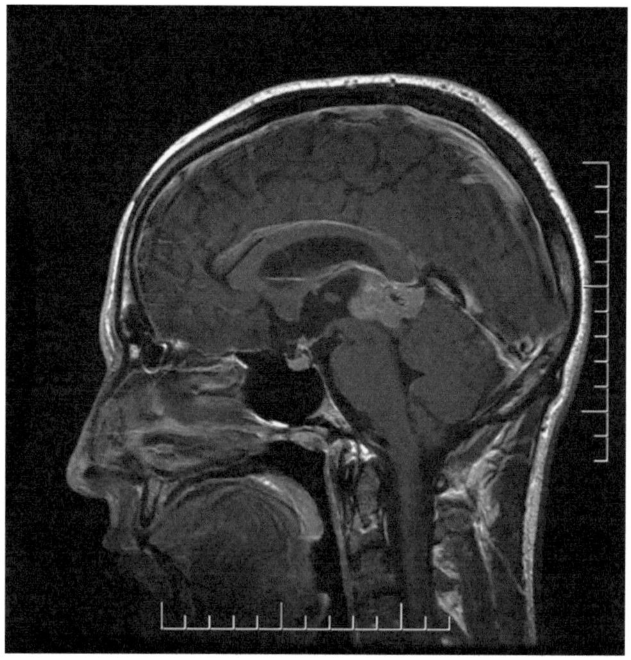

Fig. 112.2 Mid-sagittal MRI showing a pineal mass compressing the quadrigeminal plate in a patient with convergence-retraction nystagmus. (From Margolin and Emami [10]. Reproduced with written permission from Wolters Kluwer Health Limited)

Figure 112.2, from Margolin and Emami [10], is a mid-sagittal MRI demonstrating a pineal mass (primary germinoma) compressing the quadrigeminal plate.

Figure 112.3, from Scheitler and colleagues [13], is a mid-sagittal MRI demonstrating intraparenchymal hemorrhage involving the right thalamus extending into the dorsal mesencephalon.

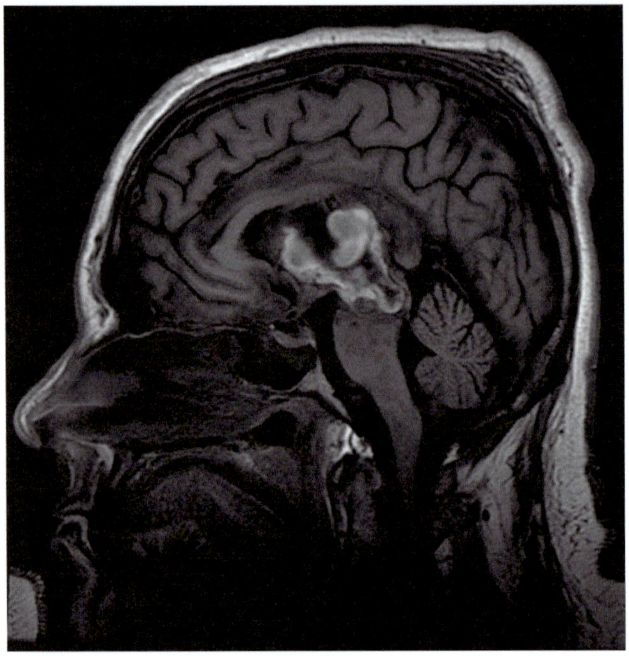

Fig. 112.3 Mid-sagittal MRI demonstrating intraparenchymal hemorrhage involving the right thalamus extending into the dorsal mesencephalon. (From Scheitler et al. [13]. Reproduced with written permission from Springer Nature B.V.)

Differential Diagnosis

Convergence-retraction nystagmus is one of several possible ocular motor findings in the pretectal syndrome [5], variously referred to as Parinaud's syndrome, Sylvian aqueduct syndrome, and dorsal midbrain syndrome.

Convergence-retraction nystagmus has been reported in the following:

- Pineal mass [12].
- Lesions that damage the posterior commissure, such as pineal tumors [3, 14, 15].
- Leigh syndrome (subacute necrotizing encephalopathy), reported in a 6-year-old boy [9].
- Infarctions (ischemic or hemorrhagic) involving the pretectal area, posterior commissure, dorsomedial thalamic nuclei [7], and thalamo mesencephalic region [8, 13].
- Chiari malformation.
- Hydrocephalus [11].
- Seizures. Examples of reported cases include a 65-year-old man [6] and a 57-year-old woman [16]. In these cases, the patients had complex focal seizures that initially presented with lateralizing motor symptoms and impaired consciousness, but those ceased and were replaced by periodic lateralizing epileptiform discharges synchronous with convergence-retraction nystagmus.

Treatment

Treatment of convergence-retraction nystagmus depends on the underlying etiology.

Prognosis

Prognosis of convergence-retraction nystagmus depends on the underlying etiology.

References

1. Leigh JR, Rucker JC. Nystagmus and related ocular motility disorders. In: Miller NR, Newman NJ, editors. Walsh & Hoyt's clinical neuro-ophthalmology, vol. 1. 6th ed. Baltimore: Lippincott Williams & Wilkins; 2005. p. 1133–73.
2. Ortiz JF, Eissa-Garces A, Ruxmohan S, Cuenca V, Kaur M, Fabara SP, et al. Understanding Parinaud's syndrome. Brain Sci. 2021;11:11.
3. Ochs AL, Stark L, Hoyt WF, D'Amico D. Opposed adducting saccades in convergence-retraction nystagmus: a patient with sylvian aqueduct syndrome. Brain J Neurol. 1979;102(3):497–508.
4. Rambold H, Kompf D, Helmchen C. Convergence retraction nystagmus: a disorder of vergence? Ann Neurol. 2001;50(5):677–81.
5. Keane JR. The pretectal syndrome: 206 patients. Neurology. 1990;40(4):684–90.
6. Brenner RP, Carlow TJ. PLEDs and nystagmus retractorius. Ann Neurol. 1979;5(4):403.
7. Luda E, Sicuro L, Albera R, Roberto C, De Lucchi R. Retraction-convergence nystagmus: clinical and radiological study of 4 cases of vascular origin. Ital J Neurol Sci. 1992;13(2):119–24.
8. Man BL, Fu YP. See-saw nystagmus, convergence-retraction nystagmus and contraversive ocular tilt reaction from a paramedian thalamomesencephalic infarct. BMJ Case Rep. 2014;
9. Chen W, Feny C, Chu S, Zhao G, Sun X, Li Z, et al. Convergence-retraction nystagmus and ophthalmoplegia as the presenting sign of Leigh syndrome in a young boy. J Neuro-ophthalmol. 2022;
10. Margolin E, Emami S. Teaching video neuroimages: convergence retraction nystagmus due to compressive lesion of the dorsal midbrain. Neurology. 2020;94(21):e2285–e6.
11. Sato Y, Maehara F, Utsunomiya H, Hayashi T, Moritaka K, Kaji M. Convergence and retraction nystagmus in hydrocephalus due to shunt malfunction. Kurume Med J. 1983;30(4):125–8.
12. Woodward K, Sitaram A, Peters S. Isolated convergence-retraction nystagmus secondary to intralesional haemorrhage of a pineal cyst: an easily missed neurological finding with potentially life-threatening consequences. BMJ Case Rep. 2020;13(9)
13. Scheitler KM, Mustafa R, Wijdicks EFM. Lid and convergence retraction nystagmus in thalamic-midbrain hematoma. Neurocrit Care. 2021;34(2):691–2.
14. Pullicino P, Lincoff N, Truax BT. Abnormal vergence with upper brainstem infarcts: pseudoabducens palsy. Neurology. 2000;55(3):352–8.
15. Daroff RB, Hoyt WF. Supranuclear disorders of ocular control systems in man: clinical, anatomical and physiological correlations. In: Bach-y-Rita P, Collins CC, Hyde JE, editors. The control of eye movements. New York: Academic; 1971. p. 175–235.
16. Young GB, Brown JD, Bolton CF, Sibbald WM. Periodic lateralized epileptiform discharges (PLED's) and nystagmus retractorius. Ann Neurol. 1977;2(1):61–2.

Internuclear Ophthalmoplegia

113

Overview

Internuclear ophthalmoplegia (INO) refers to the ocular motor disorder resulting from a lesion in the medial longitudinal fasciculus (interfering with excitatory projections to the ipsilateral oculomotor nucleus from the contralateral abducens nucleus) in which, for saccades directed contraversive to that lesion, the eye ipsilateral to the lesion fails to adduct fully, whereas the eye contralateral to the lesion abducts fully but then may exhibit abducting nystagmus. (This deficit does not affect saccades directed ipsiversive to the lesion and does not affect vergence movements.) This disconjugate eye movement can manifest clinically with intermittent diplopia, and some patients experience this visual deficit as interfering with their equilibrium. INO can be unilateral or bilateral. The most common causes of INO in younger patients include demyelination, and in older patients include infarction. MRI is the preferred imaging modality for identifying lesions of the medial longitudinal fasciculus.

Introduction

Internuclear ophthalmoplegia (INO) refers to the ocular motor disorder resulting from a lesion in the medial longitudinal fasciculus in which, for saccades directed contraversive to that lesion, the eye ipsilateral to the lesion fails to adduct fully, whereas the eye contralateral to the lesion abducts fully but then may exhibit abducting nystagmus.

Pathophysiological Mechanism of Disease

In the phrase "internuclear ophthalmoplegia," the term "internuclear" refers to the fact that one of the pathways in the medial longitudinal fasciculus lies between two nuclei relevant to ocular motor function; specifically, the pathway begins in the abducens (CN VI) nucleus, decussates and ascends to the oculomotor (CN III) nucleus.

Figure 113.1, from Frohman and colleagues [1], depicts the relevant neuroanatomy.

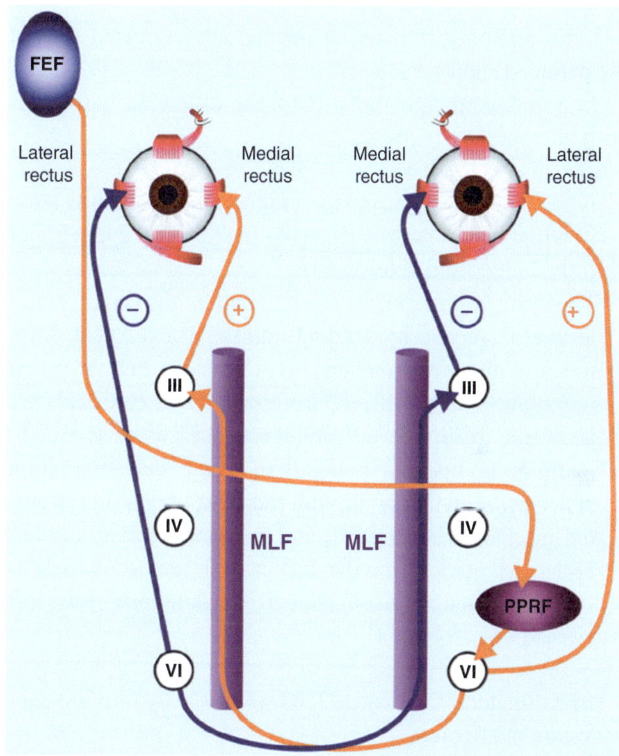

Fig. 113.1 Neuroanatomical substrate for a leftward-directed saccade. Excitatory pathways are shown in orange and the reciprocal inhibitory pathways are shown in blue. The particular pathway shown emanates from the frontal eye field (FEF), which projects through the anterior limb of the internal capsule, decussates to the opposite side at the midbrain–pontine junction, and then innervates the paramedian pontine reticular formation (PPRF). From there, projections directly innervate the lateral rectus (ipsilateral to the PPRF). A second decussation, back to the side of origin of FEF activation, via the MLF, innervates the medial rectus subnucleus of cranial nerve III and then neurons here project to innervate the medial rectus muscle. The right FEF command to trigger a saccade culminates in conjugate eye movements to the left. According to Herring's law, the horizontal yoke pair, the medial and lateral recti, are activated in synchrony. (From Frohman et al. [1]. Reproduced with written permission from Wolters Kluwer Health Limited)

M. Cherchi, *Otoneurology and Vestibular Medicine*, https://doi.org/10.1007/978-3-031-94841-1_113

When executing a voluntary saccade directed toward the patient's *left*, the normal physiology is as follows:

1. The **right** frontal eye field projects to the *left* abducens nucleus.
2. The *left* abducens nucleus sends an excitatory projection to the *left* lateral rectus muscle, resulting in abduction of the *left* eye. (There is a separate inhibitory pathway for the *left* medial rectus.)
3. The *left* abducens nucleus also sends a projection that decussates, ascends through the **right** medial longitudinal fasciculus, and synapses in **right** the oculomotor nucleus.
4. The **right** oculomotor nucleus projects to the **right** medial rectus, resulting in adduction of the **right** eye. (There is a separate inhibitory pathway for the **right** lateral rectus.)

(This anatomy is reversed for saccades directed toward the patient's right.)

In a unilateral right-sided INO, the following will occur with attempted leftward saccades:

- A lesion in the right medial longitudinal fasciculus (step 3) will result in failure of step 4, and the right eye will not fully adduct (move leftward).
- The brain keeps attempting to complete the (inadequate) leftward saccade by sending additional excitation from the left abducens nucleus. By Hering's law of equal innervation [2–4], these additional excitatory signals are projected from the left abducens nucleus to the right oculomotor nucleus (but fail to reach that destination because the pathway through the right medial longitudinal fasciculus is lesioned) and also project from the left abducens nucleus to the left lateral rectus—which is successful, but results in abduction nystagmus of the left eye [5].

In a unilateral right-sided INO, rightwardly directed saccades are unaffected.

(In a unilateral left-sided INO, the above descriptions would be reversed.)

INO usually leaves convergence movements (which are mediated by different pathways) intact [5].

Clinical Presentation

While internuclear ophthalmoplegia might not be considered a "vestibular" disorder, its ocular motor effects interfere with vision, resulting in intermittent diplopia, and can be experienced by some patients as a form of disequilibrium.

Ocular Motor Examination

Unilateral or bilateral INO can often be identified on face-to-face ocular motor examination, though may evade detection if subtle.

Testing: Instrumented Ocular Motor Testing

Figure 113.2, from Omary and colleagues [6], shows ocular motor plots of eye position and velocity, and an MRI demonstrating a lesion involving the right medial longitudinal fasciculus.

Figure 113.3, from our own clinical material, shows binocular oculography of horizontal saccades in the patient with bilateral INO.

- The leftmost panel shows that latencies for horizontal saccades are normal in both eyes.
- The middle panels show plots of horizontal saccade velocity; on rightward saccades, the velocities for the abducting **right** eye (red) are normal, while the velocities for the adducting *left* eye (blue) are diminished; on leftward saccades, the velocities for the abducting *left* eye are normal, while the velocities for the adducting **right** eye are diminished.
- The rightmost panel shows plots of horizontal saccadic accuracy; on rightward saccades, the abducting **right** eye (red) is eumetric, while the adducting *left* eye (blue) is hypometric; on leftward saccades, the abducting *left* eye is hypermetric, and the adducting **right** eye is hypometric.
- The bottom panel shows a binocular tracing. On rightward saccades (represented by an upward deflection), the **right** eye (red tracing) is hypermetric; on leftward saccades, (represented by a downward deflection) the *left* eye (blue tracing) is hypermetric.

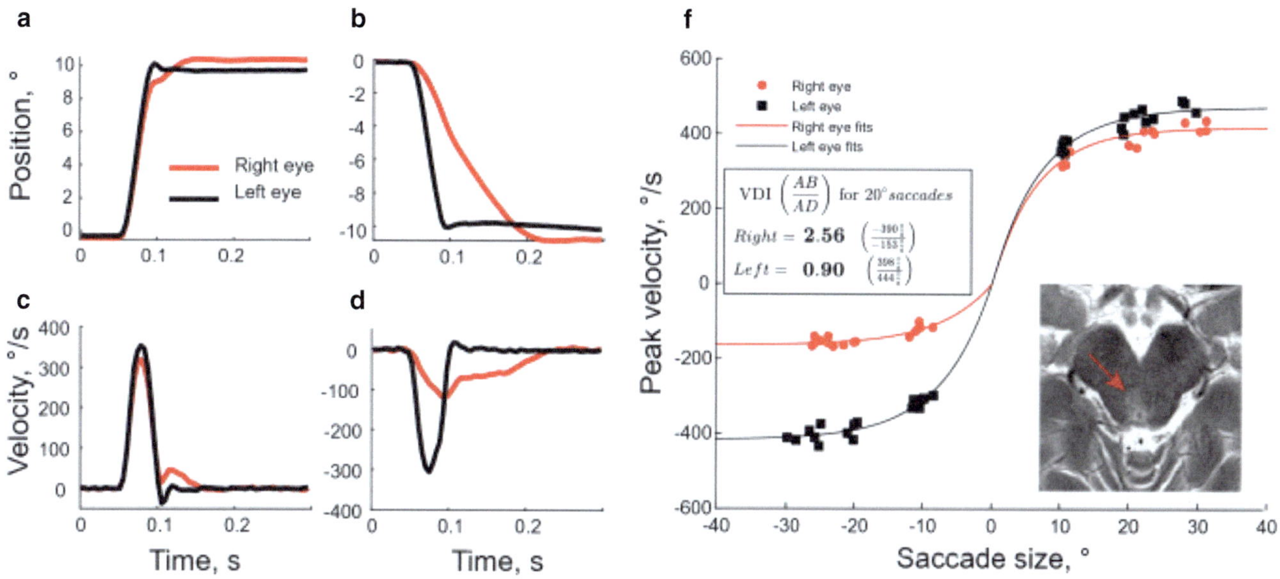

Fig. 113.2 Ocular motor plots of eye position and velocity, and an MRI demonstrating a lesion involving the right medial longitudinal fasciculus [6]. Saccadic eye movements from a patient with right INO. Saccadic eye movements and summary results of a patient with right INO. (A) 10° rightward saccade, with peak velocities around 350°/s. (C) Both eyes show a similar velocity profile for this saccade. (B) 10° leftward saccade, with velocity shown in (D). The left abducting eye (black) moves faster and reaches the target earlier than the right adducting eye (red), indicating a right INO. (E) Peak eye velocity is plotted against saccade size for all trials. For rightward (positively directed) saccades, both eyes have approximately the same peak velocity, but for leftward saccades, the right adducting eye (red) moves with a peak velocity of less than half that of the left abducting eye (black). We defined the versional disconjugacy index (VDI) as the ratio of the peak velocity of abducting to adducting eyes, for 20° saccades based on exponential fits, resulting in an increased VDI of 2.56 for saccades to the left (indicating a right INO) and a normal VDI of 0.9 to the right. The inset shows the corresponding MR image from this patient, where all three of our raters diagnosed a lesion of the right MLF in the brainstem. (From Omary et al. [6]. Reproduced with written permission from Springer Nature B.V.)

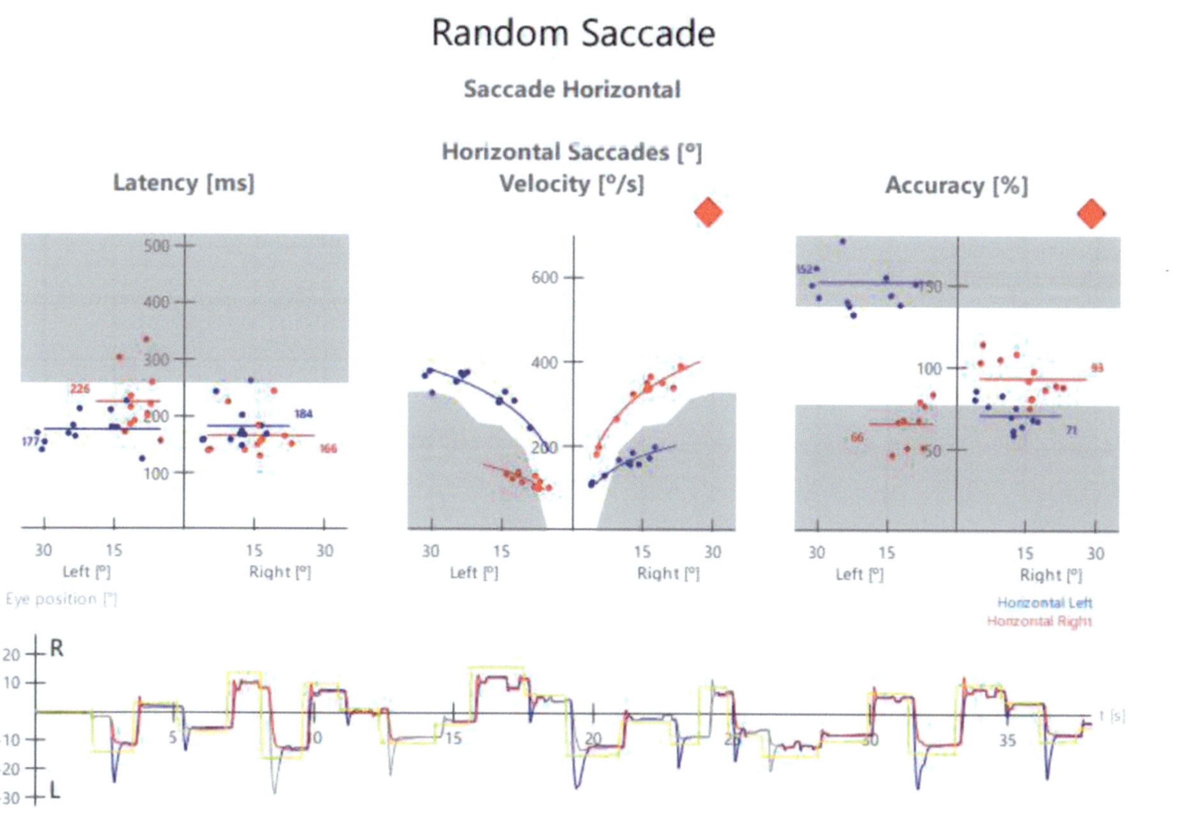

Fig. 113.3 Analysis of horizontal saccades in a case of bilateral internuclear ophthalmoplegia. See text for details. (Courtesy of Marcello Cherchi)

Testing: Vestibular

The ocular motor consequences of INO can interfere with vestibular tests that rely on ocular motor function, such as video head impulse testing.

INO can also interfere with results of ocular vestibular evoked myogenic potentials [7].

Imaging

MRI is the imaging modality of choice to detect lesions involving the medial longitudinal fasciculus from etiologies such as infarction, demyelination, hemorrhage, and neoplasm.

Figure 113.4, from Deleau and colleagues [8], shows a focal area of high T2 signal in the left medial longitudinal fasciculus, attributed to ischemic infarction.

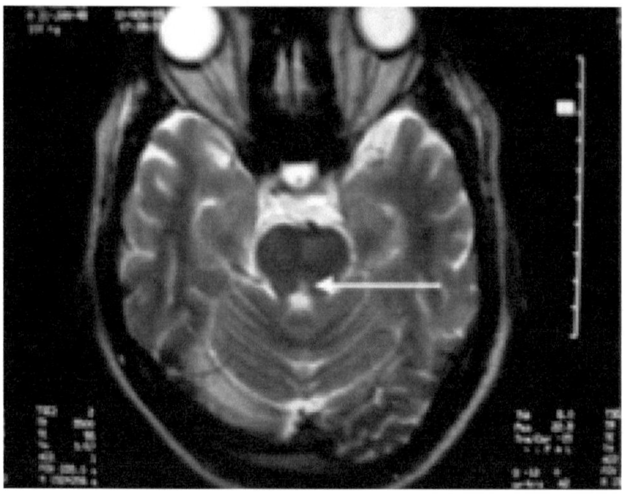

Fig. 113.4 Brain MRI showing focal area of high T2 signal in the left medial longitudinal fasciculus in a patient with left-sided internuclear ophthalmoplegia. (From Deleu et al. [8]. Reproduced with written permission from Acta Medica Belgica)

Differential Diagnosis

The differential diagnosis of INO includes any disease process that can lesion the medial longitudinal fasciculus. In younger patients, the most common etiology is demyelination, typically from multiple sclerosis. In older patients, the most common etiology is infarction [9]. Less common etiologies include other autoimmune diseases (such as sarcoidosis, systemic lupus erythematosus), infections, neoplasm, and head trauma. It can also occur in the context of Wernicke encephalopathy.

Treatment

Treatment of the underlying etiology of INO is appropriate. For example, if the underlying etiology is an ischemic infarct, then secondary stroke prophylaxis is appropriate; if the underlying etiology is from multiple sclerosis, then treatment for that diagnosis may be advanced.

In addition, if (despite such treatments) the visual deficit persists, then consultation with neuro-ophthalmology or neuro-optometry may be appropriate.

Prognosis

The prognosis of INO depends on the underlying etiology.

Eggenberger and colleagues [10] state that "most ischemic-based INO become asymptomatic in primary position over 2 to 3 months."

Kanhai and colleagues report a prospective, randomized, placebo-controlled trial of patients with internuclear ophthalmoplegia from multiple sclerosis and reported improvement in certain INO parameters (such peak velocity) in the treatment group [11].

References

1. Frohman TC, Galetta S, Fox R, Solomon D, Straumann D, Filippi M, et al. Pearls & Oy-sters: The medial longitudinal fasciculus in ocular motor physiology. Neurology. 2008;70(17):e57–67.
2. Hering KEK. Die lehre vom binocularen sehen. Leipzig: Engelmann; 1868.
3. Hering KEK. Der Raumsinn und die Bewegung des Auges. In: Hermann L, editor. Handbuch der Physiologie, vol. III, part 1. Leipzig; 1879. p. 343–601.
4. Hering KEK. Spatial sense and movements of the eye. Baltimore: American Academy of Optometry; 1942.
5. Zee DS, Hain TC, Carl JR. Abduction nystagmus in internuclear ophthalmoplegia. Ann Neurol. 1987;21(4):383–8.
6. Omary R, Bockisch CJ, De Vere-Tyndall A, Pazahr S, Barath K, Weber KP. Lesion follows function: video-oculography compared with MRI to diagnose internuclear ophthalmoplegia in patients with multiple sclerosis. J Neurol. 2023;270(2):917–24.
7. Rosengren SM, Colebatch JG. Ocular vestibular evoked myogenic potentials are abnormal in internuclear ophthalmoplegia. Clin Neurophysiol. 2011;122(6):1264–7.
8. Deleu D, Sokrab T, Salim K, El Siddig A, Hamad AA. Pure isolated unilateral internuclear ophthalmoplegia from ischemic origin: report of a case and literature review. Acta Neurol Belg. 2005;105(4):214–7.
9. Kim JS. Internuclear ophthalmoplegia as an isolated or predominant symptom of brainstem infarction. Neurology. 2004;62(9):1491–6.
10. Eggenberger E, Golnik K, Lee A, Santos R, Suntay A, Satana B, et al. Prognosis of ischemic internuclear ophthalmoplegia. Ophthalmology. 2002;109(9):1676–8.
11. Kanhai KMS, Nij Bijvank JA, Wagenaar YL, Klaassen ES, Lim K, Bergheanu SC, et al. Treatment of internuclear ophthalmoparesis in multiple sclerosis with fampridine: a randomized double-blind, placebo-controlled cross-over trial. CNS Neurosci Ther. 2019;25(6):697–703.

Latent Nystagmus

114

Overview

Latent nystagmus (LN) is disorder of ocular motility thought to result from developmental failure of binocular vision early in life, the most common cause of which is infantile strabismus or amblyopia. The characteristic findings include that (1) when one eye is occluded there will be nystagmus whose fast phase beats toward the viewing eye; (2) optokinetic testing exhibits low gain in the temporalward direction of the viewing eye. The mechanism of LN is not known, but several lines of evidence implicate binocular disparity (due to failure to develop stereopsis early in life). Since most individuals use binocular vision for most activities, most patients with LN can function normally and may come to attention if the abnormal eye movement is observed by an examiner. Since most patients with LN can function normally, treatment is usually unnecessary. In cases where LN is believed to be the cause of symptoms, a variety of cycloplegic procedures have been attempted.

Introduction

Latent nystagmus (LN), sometimes also called "infantile nystagmus," is an abnormal eye movement that appears (manifests) under monocular viewing conditions and disappears (returns to its latent state) when binocular vision is available. The nystagmus is conjugate, horizontal, and beats toward the viewing eye, and thus, the direction changes depending upon which eye is viewing. In other words, when the patient is viewing out of the right eye, there will be conjugate right beat nystagmus; when the patient is viewing out of the left eye, there will be conjugate left beat nystagmus. In some, though not all patients, the nystagmus obeys Alexander's law [1].

Latent nystagmus (LN) was initially described by Fromaget and Fromaget [2] in 1912. They wrote:

> [When] the patient, [with] both eyes open, looks at the eye chart, both eyes are motionless. We place a screen in front of the right eye which is still open. Instantly an oscillatory, horizontal, very rapid nystagmus appears, of 18 to 20 beats every quarter of a minute. The vision [visual acuity] also drops to 0.4. If you uncover the right eye, immediately the nystagmus disappears. If we repeat the experiment by closing the left eye then a horizontal, slow and irregular nystagmus appears of 5 to 6 beats per quarter of a minute and the visual acuity does not reach 0.1. [2]

This description is incomplete, in the sense that Fromaget and Fromaget merely note that the nystagmus appears when one eye is occluded, but they do not specify the direction of the nystagmus that they observed. Fromaget and Fromaget reported on several other cases of latent nystagmus [3, 4]. The cases reported by Fromaget and Fromaget are reviewed by Sorsby [5].

Abadi and Scallan [6] provide a good case of LN with clear ocular motor tracings.

As described above, LN becomes evidence when one eye is occluded. However, in some circumstances, it can become apparent even when vision is available to both eyes, and in this situation, it is called *manifest latent nystagmus* (MLN) [7], first described by Faucon [8] in 1872.

Epidemiology

Latent nystagmus (LN) is "strongly linked to binocular maldevelopment in infancy, from either strabismus or deprivation of monocular spatial vision (amblyopia)" [7].

Latent nystagmus can occur in patients with uncorrected childhood strabismus [9]. Gresty and colleagues [10] studied 30 patients with LN and found that 29 reported a history of strabismus. Many, though not all, of these patients have amblyopia [10].

Genetics

Latent nystagmus (LN) has been described in familial patterns, such as in identical twins [11], but it is unknown whether there is a genetic susceptibility to developing LN.

M. Cherchi, *Otoneurology and Vestibular Medicine*, https://doi.org/10.1007/978-3-031-94841-1_114

Pathophysiological Mechanism of Disease

The mechanism of latent nystagmus is unknown [12]. At some level, both the vestibular system and the optokinetic system are probably involved. As Brodsky and Tusa note:

> Visual stabilization mechanisms act in concert with labyrinthine reflexes. In normal life, optokinetic responses are elicited mainly by head movements, which also stimulate the vestibular system. Because vestibular neurons receive such prominent visual and vestibular inputs, disrupting either input reduces the tonic activation of these neurons, with the effect of disturbing the responses to the other sensory modality… According to Miles, each sensory modality 'has played such a major role in the evolution of the other that it is impossible to understand the operation of either one in isolation.' [13]

Brodsky and Tusa further state:

> Various theories have been advanced to explain latent nystagmus. These include a primitive tonus imbalance, an egocentric disorder, a disorder of the subcortical optokinetic system, a subcortical maldevelopment of retinal slip control, abnormal cortical motion processing, a disorder of proprioception, and an evolutionary preponderance of the nasal half of the retina. [13]

Several proposals rely on the idea that in strabismic patients there has been a developmental failure of visual fusion, such that no synaptic connections provide one neuron with binocular information.

This synaptic architecture is illustrated in Figs. 114.1 and 114.2, from Tychsen and colleagues [7], which show the normal synaptic configuration in the visual cortex for binocular vision, and abnormal synaptic configuration in failure of binocular vision.

Tychsen then provides Fig. 114.3, with the following accompanying explanation:

> Neural network diagrams showing visual signal flow for pursuit and gaze holding in strabismic and normal primates. A paucity of mature binocular connections explains behavioral asymmetries evident as asymmetric pursuit/optokinetic nystagmus and latent fixation nystagmus. In all primates, pursuit area neurons in each hemisphere encode ipsilaterally directed pursuit. Signal flow is initiated by a moving stimulus in the monocular visual field (VF), which evokes a response in visual area neurons V1 and MT. Each eye at birth has access, through innate monocular connections, to the pursuit area neurons in MSTd of the contralateral hemisphere. Access to pursuit neurons of the ipsilateral hemisphere requires mature binocular connections. In fusion maldevelopment (right column), retinal ganglion cell fibers from the nasal and temporal hemiretina (eye) decussate at the optic chiasm (chi), synapse at the lateral geniculate nucleus (LGN), and project to alternating rows of ODCs in V1 (visual area rectangles). In each V1, ODCs representing the nasal hemiretina (temporal visual hemifield) occupy slightly more cortical territory than those representing the temporal hemiretina (nasal hemifield), but each ODC contains neurons sensitive to nasally directed and temporally directed motion (half circles shaped like the matching hemifield; arrows indicate directional preference). Visual area neurons, including those beyond V1 in area MT, encoding nasally directed motion are wired innately—through

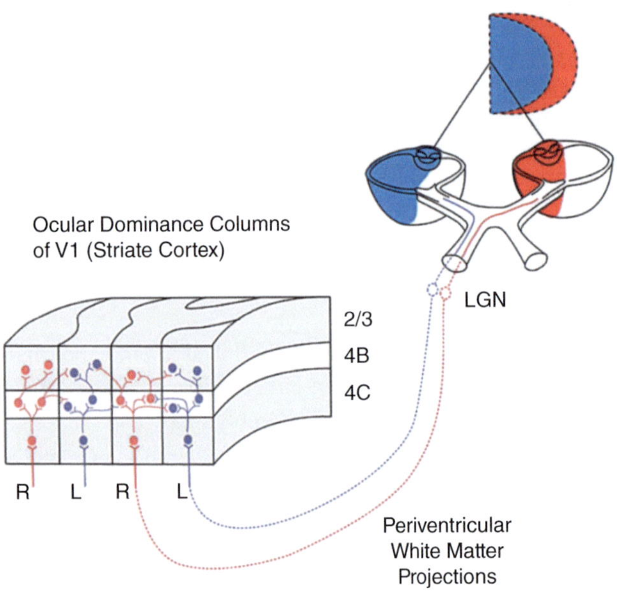

Fig. 114.1 Neuroanatomic basis for binocular vision. Monocular retinogeniculate projections from left eye (temporal retina, nasal visual hemifield) and right eye (nasal retina, temporal hemifield) remain segregated up to and within the input layer of ocular dominance columns (ODCs) in V1, layer 4C (striate visual cortex). Binocular vision is made possible by horizontal connections between ODCs of opposite ocularity in upper layers 4B and 2/3 (as well as lower layers 5/6, not shown). Right eye inputs are shown in red; left eye inputs are shown in blue. (From Tychsen et al. [7]. Reproduced with written permission from Wolters Kluwer Health Limited)

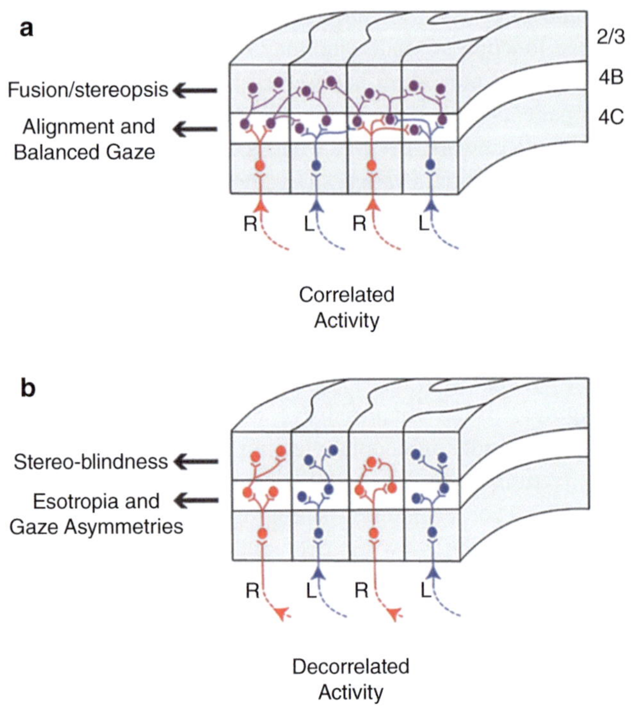

Fig. 114.2 Horizontal connections for binocular vision to layer 2–4B of V1 in normal (correlated activity) primates (A) and strabismic (decorrelated) primates (B). V1 of normal primates is characterized by equal numbers of monocular and binocular connections. In strabismic primates, the connections are predominantly monocular. Right eye inputs are shown in red; left eye inputs are shown in blue; binocular connections are shown in violet. (From Tychsen et al. [7]. Reproduced with written permission from Wolters Kluwer Health Limited)

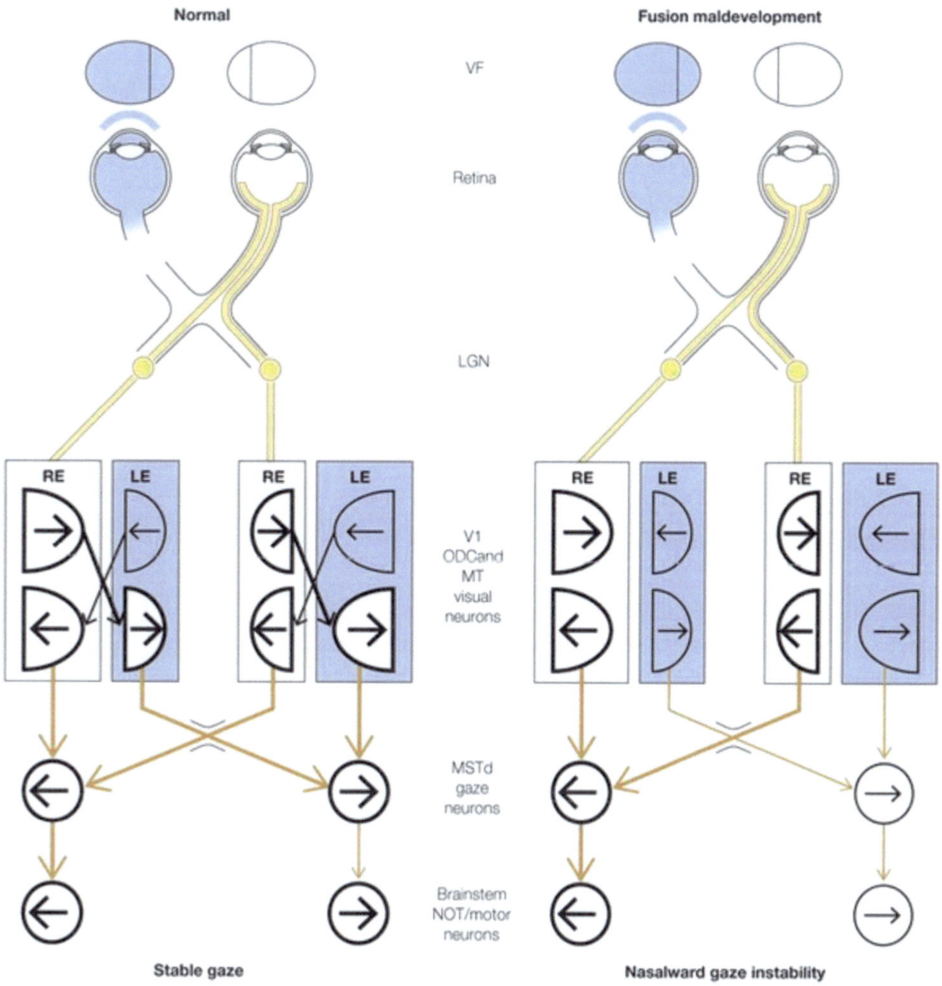

Fig. 114.3 Comparison of visual input and ocular motor output in a normal patient versus a patient with latent nystagmus. (From Tychsen et al. [7]. Reproduced with written permission from Wolters Kluwer Health Limited)

monocular connections—to the pursuit area. In normal primates (left column), binocular connections are present, linking neurons with similar orientation/directional preferences within ODCs of opposite ocularity (diagonal lines between columns). When the subject views with the right eye, visual neurons preferring nasally directed motion project to the left hemisphere pursuit area and visual neurons preferring temporally directed motion project to the right hemisphere pursuit area. Temporally directed visual area neurons gain access to pursuit area neurons only through binocular connections. Call = corpus callosum. Bold lines = active neurons and neuronal projections. [7]

Schor [14] alludes to this idea when stating:

> Latent nystagmus and fixation drifts of amblyopic patients could be described in terms of functional suppression of cortical projections to the pretectal OKN [optokinetic nystagmus] system. Bilateral reductions of cortical projections to the pretectum would, during attempted steady monocular fixation, result in subcortical stimulation of the contralateral NOT [nucleus of the optic tract] without a counterbalancing stimulation from the cortical projection to the ipsilateral NOT. Thus, stimulation of either eye alone would cause a nasal drift bias to the side of the closed eye. [14]

Similarly, Kommerell and Zee [12] reported on two patients who could voluntarily elicit LN. They hypothesized that "this phenomenon could be the result of voluntary control of the visual input contributed by the amblyopic eye and/or a direct influence of will on the slow eye movement and fixation systems." In other words, these two patients could "choose" to pay attention to the visual input of only one eye and ignore that of the fellow eye, bringing about the same effect as "occluding" the ignored eye. Abadi [15] comments that van Vliet [16, 17] also reported that some LN patients could induce nystagmus "by the 'intention' of looking with one eye.."

Brodsky and Tusa [13] describe this as follows:

> The vestibular nucleus presides over motion input from the eyes and labyrinths. Latent nystagmus corresponds to the optokinetic component of ocular rotation that is driven monocularly by nasal optic flow during a turning movement of the body in lateral-eyed animals. Congenital esotropia alters visual pathway development from the visual cortex to subcortical centers that project to the vestibular nucleus, allowing this primitive subcortical motion detection system to generate latent nystagmus under conditions of monocular fixation. [13]

A different view is expressed by Gresty and colleagues [10], who studied the combinations of LN, strabismus, stereopsis, amblyopia, and optokinetic abnormalities in 29 patients and concluded that:

> Dissociations found between the presence and severity of strabismus, stereopsis, amblyopia and optokinetic abnormalities point to these features being relatively independent although associated in typical clusterings. This is evidence against the theory that strabismus and LN are directly caused by nasotemporal optokinetic imbalance which persists because of failure to develop binocular vision. The variability of findings favours the view that LN and CN [congenital nystagmus] arise from a genetic or acquired embryological disorder with various degrees and directions of expression. [10]

Brodsky and Tusa comment that "the magnitude of the resulting manifest latent nystagmus is proportional to the degree of the interocular visual disparity" [13].

Clinical Presentation

Most, though not all, patients with latent nystagmus lack binocular vision. Gresty and colleagues [10] studied 30 LN patients and reported that 24 of them lacked binocular vision, while the remaining 6 had "varying degrees of stereopsis."

LN patients do not usually complain of disequilibrium, but the lack of stereopsis can interfere with depth perception, which may cause difficulties that the patient reports as unsteadiness, difficulty navigating stairs, etc.

Physical Examination

In a patient with latent nystagmus who is otherwise healthy, general physical examination should be normal.

Ocular Motor Examination

In a patient with latent nystagmus (LN), at baseline, when one eye is occluded, on attempted visual fixation on primary position of gaze, relative to the viewing eye, there should be conjugate nasalward slow drift and temporalward corrective saccades (in other words, nystagmus beating toward the side of the viewing eye).

Aside from the baseline pattern of LN evident with monocular viewing, LN patients also have difficulty with smooth pursuit during monocular viewing that Brodsky and Tusa describe as, "nasotemporal asymmetry of the horizontal optokinetic response during monocular viewing" [13], which they say was first reported by Roelofs [18]. During modular viewing, the baseline eye movement consists of slow nasalward drift (relative to the viewing eye) with a corrective temporalward saccade (relative to the viewing eye); thus, nasalward smooth pursuit appears normal, whereas temporalward smooth pursuit exhibits low gain and thus corrective temporalward saccades are triggered.

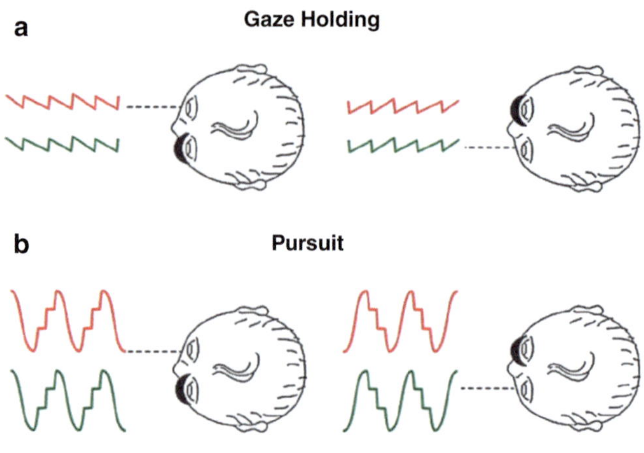

Fig. 114.4 Nasalward gaze asymmetries in strabismic human and monkey. Panel (A): Latent nystagmus in gaze holding. When the subject views with the right eye, both eyes have a nasalward slow-phase drift, followed by temporalward refoveating fast-phase microsaccades. The direction of the nystagmus reverses instantaneously when the left eye fixates so that the slow phase is nasalward with respect to the fixating eye. Panel (B): Pursuit. Horizontal smooth pursuit is asymmetric during monocular viewing. Pursuit is smooth (normal) when target motion is nasalward in the visual field (VF). Pursuit is cogwheel (low gain) when the target moves temporalward. The movements of the 2 eyes are conjugate, and the direction of the asymmetry reverses instantaneously with a change of fixating eye so that the direction of robust pursuit is always for nasalward motion in the visual field (likewise for optokinetic nystagmus). (From Tychsen et al. [7]. Reproduced with written permission from Wolters Kluwer Health Limited)

These two patterns (attempted fixed gaze during monocular viewing, and attempted smooth pursuit during monocular viewing) are demonstrated in Fig. 114.4, from Tychsen and colleagues [7].

There are several studies of LN involving vertical and torsional components [19–21].

Testing: Vestibular

Vestibular testing is usually unrevealing in latent nystagmus (LN). However, note that ocular motor studies can be misleading if they are performed with the camera always over the same eye, because this will give the mistaken impression of a unidirectional nystagmus at baseline—in other words, it may miss the alternating nature of LN.

Differential Diagnosis

Latent nystagmus (LN) is usually quite distinctive and most easily detected on face-to-face examination.

LN can co-occur with other disorders, including ophthalmic and other ocular motor disorders, so it is reasonable to consult ophthalmology or neuro-ophthalmology.

Treatment

In most patients, latent nystagmus (LN) is present only with monocular viewing. Since most people use binocular vision for most daily activities, LN usually remains latent and patients have no symptoms, or minimal symptoms, and thus do not require treatment [22].

For patients in whom LN provokes symptoms, various treatments have been attempted, including injection of extraocular muscles with onabotulinum toxin [22] and cyclopentolate eyedrops [23, 24]. Occlusion therapy (blocking vision in one eye) has also been reported as sometimes effective [25], which is unexpected given that monocular viewing generally worsens the nystagmus of LN.

Prognosis

Latent nystagmus (LN) is usually a lifelong condition. In the majority of patients, LN in itself does not interfere with daily functioning.

References

1. Dell'Osso LF, Schmidt D, Daroff RB. Latent, manifest latent, and congenital nystagmus. Arch Ophthalmol. 1979;97(10):1877–85.
2. Fromaget C, Fromaget H. Nystagmus latent (nystagmus et strabisme). Annales d'Oculistique. 1912;147:344–52.
3. Fromaget H. Nystagmus latent. Annales d'Oculistique. 1913;149:241–50.
4. Fromaget C, Fromaget H. Nystagmus latent. Annales d'Oculistique. 1916;153:465–72.
5. Sorsby A. Latent nystagmus. Br J Ophthalmol. 1931;15(1):1–18.
6. Abadi RV, Scallan CJ. Waveform characteristics of manifest latent nystagmus. Invest Ophthalmol Vis Sci. 2000;41(12):3805–17.
7. Tychsen L, Richards M, Wong A, Foeller P, Bradley D, Burkhalter A. The neural mechanism for Latent (fusion maldevelopment) nystagmus. J Neuro-ophthalmol. 2010;30(3):276–83.
8. Faucon A. Nystagmus par insuffisance des droits externes. J Ophthal (Paris). 1872;1:223–9.
9. Dell'Osso L, Traccis S, Abel L. Strabismus – a necessary condition for latent and manifest latent nystagmus. Neuro-Ophthalmology. 2009;3:247–57.
10. Gresty MA, Metcalfe T, Timms C, Elston J, Lee J, Liu C. Neurology of latent nystagmus. Brain J Neurol. 1992;115(Pt 5):1303–21.
11. Abramson DH. Latent nystagmus in identical twins. Arch Ophthalmol. 1973;89(3):256–7.
12. Kommerell G, Zee DS, Latent nystagmus. Release and suppression at will. Invest Ophthalmol Vis Sci. 1993;34(5):1785–92.
13. Brodsky MC, Tusa RJ. Latent nystagmus: vestibular nystagmus with a twist. Arch Ophthalmol. 2004;122(2):202–9.
14. Schor CM. Subcortical binocular suppression affects the development of latent and optokinetic nystagmus. Am J Optom Physiol Opt. 1983;60(6):481–502.
15. Abadi RV. Pattern contrast thresholds in latent nystagmus. Acta Ophthalmol (Copenh). 1980;58(2):210–20.
16. van Vliet AG. On the central mechanism of latent nystagmus. Monogr Hum Genet. 1972;6:204.
17. van Vliet AG. On the central mechanism of latent nystagmus. Acta Ophthalmol (Copenh). 1973;51(6):772–81.
18. Roelofs CO. Nystagmus latens. Archiv für Augenheilkunde. 1928;98:401–47.
19. Guyton DL, Cheeseman EW Jr, Ellis FJ, Straumann D, Zee DS. Dissociated vertical deviation: an exaggerated normal eye movement used to damp cyclovertical latent nystagmus. Trans Am Ophthalmol Soc. 1998;96:389–424. discussion -9
20. Hertle RW, Dell'Osso LF. Vertical latent nystagmus component and vertical saccadic asymmetries in subjects with dissociated vertical deviation. J AAPOS. 1999;3(4):257–8.
21. Irving EL, Goltz HC, Steinbach MJ, Kraft SP. Vertical latent nystagmus component and vertical saccadic asymmetries in subjects with dissociated vertical deviation. J AAPOS. 1998;2(6):344–50.
22. Liu C, Gresty M, Lee J. Management of symptomatic latent nystagmus. Eye (Lond). 1993;7(Pt 4):550–3.
23. Windsor CE. Modification of latent nystagmus. 2. Arch Ophthalmol. 1968;80(3):352–3.
24. Windsor CE, Burian HM. Modification of latent nystagmus. I. Arch Ophthalmol. 1968;80(5):657–63.
25. von Noorden GK, Avilla C, Sidikaro Y, LaRoche R. Latent nystagmus and strabismic amblyopia. Am J Ophthalmol. 1987;103(1):87–9.

Opsoclonus

115

Overview

Opsoclonus is an eye movement abnormality consisting of bursts of involuntary, seemingly random, high-frequency, conjugate saccades that are multidirectional (with horizontal, vertical, and torsional components), occurring in sequence without any intersaccadic interval delay. It often occurs in conjunction with myoclonus involving any combination of limbs, trunk, and craniocervical region. In the pediatric population, this is almost always a paraneoplastic syndrome related to neuroblastoma. In adults, the differential diagnosis includes paraneoplastic (with various cancers), parainfectious (with various infections), and other rare causes. The underlying pathophysiology is unclear, but likely involves dysfunction of omnipause neurons or their upstream regulators. The clinical presentation is distinct; onset can be over hours to days and may consist of the ocular motor findings, myoclonus, and encephalopathy or other behavioral disturbances. Workup should include a neoplastic workup and (if that is normal) a search for paraneoplastic onconeuronal antibodies and infection. If neoplasm or infection is discovered, then those should be treated. Whether or not neoplasm and infection are found, many investigators also advocate intervention with immunosuppressive or immunomodulatory therapies (steroids, plasma exchange, intravenous immunoglobulin). Treatment of myoclonus is sometimes attempted with clonazepam and/or antiseizure medications (gabapentin, levetiracetam, valproate).

Introduction

Opsoclonus, sometimes referred to as saccadomania [1], is an unusual eye movement abnormality consisting of bursts of involuntary, seemingly random, high-frequency, conjugate saccades that are multidirectional (with horizontal, vertical, and torsional components), occurring in sequence without any intersaccadic interval delay [2].

The Polish neurologist Dr. Kazimierz Orzechowski (1878–1942) [3] was the first to suggest the term "opsoclonus" in a 1927 paper [4] as a description for this pattern of eye movements.

The term opsoclonus is probably more generally recognized by clinicians as one part of the broader *opsoclonus-myoclonus syndrome* (OMS), later described by the Austrian pediatric neurologist, Dr. Marcel Kinsbourne [5].

In children, OMS usually occurs as a paraneoplastic manifestation of neuroblastoma [6].

Though even rarer, opsoclonus (with or without myoclonus) can also occur in adults [7, 8]. In many cases, the underlying etiology is unknown. Among cases with an identified etiology, most appear due to a paraneoplastic syndrome, less to a parainfectious process.

Hsu and Young [9] reported a case and provided a video showing opsoclonus in a 39-year-old woman that was attributed to a parainfectious phenomenon related to herpes simplex virus infection.

Epidemiology

Adult-onset OMS is rare. It occurs either as a paraneoplastic syndrome (more common) or parainfectious syndrome (less common). Armangue and colleagues [10] reviewed records of 114 adult patients with OMS and determined that 69 (61%) were "idiopathic," meaning that no underlying cause could be discerned.

Klaas and colleagues [9] reviewed 21 adult cases of OMS from their own institution (Mayo Clinic) and 116 previously reported adult cases in the literature. Of their own 21 cases, the associated malignancies included primary breast adenocarcinoma and small cell lung cancer. Of the 116 previous cases they reviewed, malignancies were identified in 55 (47.4%), and these included lung carcinoma (small cell and non-small cell), breast adenocarcinoma, melanoma, ovarian carcinoma, ovarian teratoma, renal cell carcinoma, thyroid

M. Cherchi, *Otoneurology and Vestibular Medicine*, https://doi.org/10.1007/978-3-031-94841-1_115

medullary carcinoma, esthesioneuroblastoma, B-cell lymphoma, cervical carcinoma, bladder carcinoma, and gastric adenocarcinoma.

Armangue and colleagues [11] reviewed 114 adult-onset cases of OMS and reported that 45 (39%) were paraneoplastic. The most common neoplasms were lung cancer in 19 (42%), breast cancer in 10 (22%), other cancers in 5 (11%), and ovarian teratoma in 8 (18%).

Of the 116 cases of adult-onset OMS reviewed by Klaas and colleagues [9], infectious causes were identified in 15 (12.9%), and these included HIV in 7 (47%), *Borrelia burgdorferi* in 2 (13%), Streptococcus in 1 (7%), Salmonella in 1 (7%), cytomegalovirus in 1 (7%), Epstein–Barr virus in 1 (7%), Coxsackie virus in 1 (7%), and West Nile Virus in 1 (7%). Reports are also emerging of OMS occurring in the context of COVID-19 infection [12].

Pathophysiological Mechanism of Disease

Whether paraneoplastic or parainfectious, opsoclonus (with or without myoclonus) is thought to result from a misdirected immune response that mistakenly targets healthy "self" cells, rather than only "non-self" cancerous cells or infectious pathogens, presumably due to epitope similarity (molecular mimicry).

Klaas and colleagues [8] note very rare reports of adult-onset OMS occurring apparently due to nonparaneoplastic, nonparainfectious causes, such as in association with anti-GAD65 antibody [13], and during pregnancy [14] or the postpartum period [15]. There is also a reported case of OMS occurring after the use of intranasal cocaine [16].

Oh and colleagues [2] reviewed three theories regarding the pathophysiologic mechanism of opsoclonus.

- The "omnipause neuron" hypothesis is that opsoclonus results from damage of omnipause neurons in the nucleus raphe interpositus of the pons. In the healthy state, these neurons suppress unwanted saccades by inhibiting burst neurons in the paramedian pontine reticular formation and the rostral interstitial nucleus of Cajal.
- The "brainstem theory" suggests that saccadic oscillations result from either (1) alterations in the membrane properties of saccadic burst neurons that make them prone to excessive postinhibitory rebound excitation after sustained inhibition from the omnipause neurons, or (2) reduction in the efficacy of omnipause neuron inhibition. The second possibility sounds similar to the "omnipause neuron" hypothesis.
- The "cerebellar theory" hypothesizes that dysfunctional cerebellar Purkinje cells fail to inhibit the cerebellar fastigial nuclei, which in turn reinforces inhibition of omnipause neurons, thereby releasing (disinhibiting) burst neurons.

While these hypotheses have merit, they remain speculative.

Clinical Presentation

Symptoms usually develop over hours to days.

Klaas and colleagues [8] reviewed 21 cases of adult-onset OMS from their own institution (Mayo Clinic) and stated that "a flulike prodrome before neurologic symptom onset was reported by 6 patients" (28%). They reported that the initial neurologic symptoms were dizziness in 14 (67%), imbalance from myoclonus in 14 (67%), nausea/vomiting in 10 (48%), visual abnormalities from opsoclonus in 6 (28%), tremor in 4 (19%), and altered speech due to myoclonus in 2 (10%).

Klaas and colleagues [8] also reviewed 116 previously reported cases of adult-onset OMS. Of 93 cases for which presenting symptoms were documented, these included dizziness in 44 (47%), imbalance in 35/93 (38%), nausea in 32 (34%), tremor in 20 (22%), visual disturbance in 15 (16%), seizure in 3 (3%), fever in 2/93 (2%), headache in 2 (2%), hearing loss in 1 (1%), and anxiety in 1 (1%); some patients had multiple presenting symptoms.

Armangue and colleagues [11] retrospectively reviewed the records of 45 patients with adult-onset OMS and noted that 6 (13%) had an infectious prodrome, 22(49%) presented with "acute vertigo," and 11 (24%) presented with "subacute ataxia or myoclonus." Other features at some point in the course of the disease included encephalopathy in 13 (29%) and "severe behavioral changes" in 6 (13%).

Physical Examination

Opsoclonus is usually evident on face-to-face examination and is not subtle.

In patients who also have myoclonus, there is considerable variability in the anatomical area of involvement [8]; this can include any combination of upper extremities, lower extremities, craniocervical muscles, and trunk.

Testing: Serological

Klaas and colleagues [8] also reviewed 116 previously reported cases of adult-onset OMS. Of the 60 cases thought to be paraneoplastic in nature, the associated antibodies included ANNA-2 (anti-Ri), NMDA-R, calcium channel P/Q-type or N-type, ANNA-1 (anti-Hu), and amphiphysin IgG. Among these 116 cases they reviewed, and in the 21 cases from their own institution (Mayo Clinic), they stated that "lung and breast carcinomas were the most common cancers identified, and ANNA-2 [anti-Ri] was the most commonly reported paraneoplastic antibody."

Armangue and colleagues [11] similarly reported that in paraneoplastic cases of OMS, 7/10 (70%) of breast cancer patients had anti-ANNA2 (anti-Ri) antibodies.

Differential Diagnosis

Opsoclonus (with or without myoclonus) is usually such a distinctive presentation that there is a relatively limited number of categories of differential diagnoses. In the pediatric population, the main diagnostic consideration is neuroblastoma. In adults, the differential is somewhat broader (multiple neoplasms, multiple infections).

Oh and colleagues [11] suggest the following diagnostic algorithm when assessing opsoclonus myoclonus:

- Paraneoplastic OMS (opsoclonus-myoclonus syndrome):
  - Intracellular antibodies
    - (i) Check antibodies against Ri, amphiphysin, Hu and CV2/CRMPS. Check CT chest/abdomen/pelvis, FDG-PET to assess for lung and breast cancer.
    - (ii) Check antibodies against Yo. Check CT chest/abdomen/pelvis, FDG-PET. Consider oophorectomy.
    - (iii) Check antibodies against Ma1 and Ma2. Check CT chest/abdomen/pelvis to assess for testicular and colorectal cancer.
  - Membrane antibodies
    - (i) Check antibodies against NMDAr. Check CT chest/abdomen/pelvis and pelvic MRI to assess for ovarian teratoma.
  - Seronegative
    - (i) Check CT chest/abdomen/pelvis
      If CT identifies cancer, then treat accordingly.
      If CT does not identify cancer, then consider whole-body 18F-FDG PET.
- Idiopathic OMS:
  - Para-infectious: Check for HIV, Mycoplasma pneumonia, Salmonella, rotavirus, cytomegalovirus, human herpesvirus, hepatitis C, *Rickettsia conorii*, and others.
  - Others:
    - (i) Consider checking for paraneoplastic antibodies.
    - (ii) Consider checking CT chest/abdomen/pelvis.

Treatment

For adult-onset OMS cases in which malignancy is identified, treatment should include removal of the cancer. For cases in which infection is identified, treatment should include antimicrobial agents (where available).

There is otherwise no standardized protocol for management. Given the presumed immunologically mediated mechanism of disease, a variety of immunosuppressive interventions have been attempted, including plasma exchange, corticosteroids (prednisone, methylprednisolone), and intravenous immunoglobulin. In the 116 cases of adult-onset OMS reviewed by Klaas and colleagues [8], they calculated that 79% of patients who received some kind of immunotherapy achieved remission or improvement.

Acute (symptomatic) management of myoclonus is often attempted with clonazepam [17], gabapentin [18], levetiracetam [19], or valproate [7, 16, 19].

Prognosis

Of the 21 cases of adult-onset OMS studied by Klaas and colleagues [8], 12 (57%) achieved "neurologic remission," 3 (14%) died, 1 (5%) had "mild symptoms," 1 (5%) was "improved," and in 4 (19%), the outcome was unknown.

References

1. Espinosa PS, Berger JR. Volitional opsoclonus. Neurology. 2005;65(5):E11.
2. Oh SY, Kim JS, Dieterich M. Update on opsoclonus-myoclonus syndrome in adults. J Neurol. 2019;266(6):1541–8.
3. Magowska AM, Owecki MK, Bogusz H. Kazimierz Orzechowski (1878–1942). J Neurol. 2018;265(8):1953–5.
4. Orzechowski K. De l'ataxie des yeux dite myoclonique (opsoclonic, opsochorie). J Psychol Neurol(Lpz). 1927;35:1.
5. Kinsbourne M. Myoclonic encephalopathy of infants. J Neurol Neurosurg Psychiatry. 1962;25(3):271–6.
6. Rossor T, Yeh EA, Khakoo Y, Angelini P, Hemingway C, Irani SR, et al. Diagnosis and management of opsoclonus-myoclonus-ataxia syndrome in children: an international perspective. Neurol Neuroimmunol Neuroinflamm. 2022;9(3)
7. Digre KB. Opsoclonus in adults. Report of three cases and review of the literature. Arch Neurol. 1986;43(11):1165–75.
8. Klaas JP, Ahlskog JE, Pittock SJ, Matsumoto JY, Aksamit AJ, Bartleson JD, et al. Adult-onset opsoclonus-myoclonus syndrome. Arch Neurol. 2012;69(12):1598–607.
9. Hsu SY, Young YH. Opsoclonus recorded by a smartphone. JAMA Otolaryngol Head Neck Surg. 2017;143(8):847–8.

10. Pike M. Opsoclonus-myoclonus syndrome. Handb Clin Neurol. 2013;112:1209–11.
11. Armangue T, Sabater L, Torres-Vega E, Martinez-Hernandez E, Arino H, Petit-Pedrol M, et al. Clinical and immunological features of opsoclonus-myoclonus syndrome in the era of neuronal cell surface antibodies. JAMA Neurol. 2016;73(4):417–24.
12. Sanguinetti SY, Ramdhani RA. Opsoclonus-myoclonus-ataxia syndrome related to the novel coronavirus (COVID-19). J Neuro-ophthalmol. 2021;41(3):e288–e9.
13. Markakis I, Alexiou E, Xifaras M, Gekas G, Rombos A. Opsoclonus-myoclonus-ataxia syndrome with autoantibodies to glutamic acid decarboxylase. Clin Neurol Neurosurg. 2008;110(6):619–21.
14. Koide R, Sakamoto M, Tanaka K, Hayashi H. Opsoclonus-myoclonus syndrome during pregnancy. J Neuro-ophthalmol. 2004;24(3):273.
15. Adhikari S, Thuringer A, Maali L, Jassam Y. Opsoclonus myoclonus syndrome in a postpartum period. Mult Scler Relat Disord. 2021;50:102862.
16. Scharf D. Opsoclonus-myoclonus following the intranasal usage of cocaine. J Neurol Neurosurg Psychiatry. 1989;52(12):1447–8.
17. Bartos A. Effective high-dose clonazepam treatment in two patients with opsoclonus and myoclonus: GABAergic hypothesis. Eur Neurol. 2006;56(4):240–2.
18. Moretti R, Torre P, Antonello RM, Nasuelli D, Cazzato G. Opsoclonus-myoclonus syndrome: gabapentin as a new therapeutic proposal. Eur J Neurol. 2000;7(4):455–6.
19. Lou E, Hensley ML, Lassman AB, Aghajanian C. Paraneoplastic opsoclonus-myoclonus syndrome secondary to immature ovarian teratoma. Gynecol Oncol. 2010;117(2):382–4.

Oculopalatal Tremor

116

Overview

Oculopalatal tremor (OPT) usually begins several months after suffering an acute insult (typically an ischemic or hemorrhagic stroke) in the brainstem or cerebellum interrupting a circuit known as the triangle of Guillain–Mollaret. Symptoms include chronic rhythmic clicking tinnitus (from the palatal tremor) and a disturbance of vision or equilibrium (from the pendular nystagmus). Examination shows a 2–3 Hz palatal tremor with synchronous pendular nystagmus (with some combination of vertical, torsional, and horizontal components) that can be conjugate or dissociated. Brain MRI shows T2/FLAIR signal abnormality in the inferior olives that reflects hypertrophic degeneration. Small trials of several drugs find some efficacy of gabapentin and memantine, with gabapentin being the better tolerated medication.

Introduction

"Palatal myoclonus" was the term originally given to a 2–3 Hz oscillatory rising and lowering of the soft palate that persists during sleep. Since the movement has a sinusoidal pattern rather than a "spike" pattern, this phenomenon was eventually instead designated "*palatal tremor*." Dr. Adam Politzer is usually credited with having published the first description [1].

The movement of the soft palate is often synchronous with an oculomotor abnormality, and in such cases, it is designated "*oculopalatal tremor*" (OPT). Spencer [2] is usually credited with having published the first description of this. OPT is an "acquired pendular nystagmus synchronous with a tremor of the soft palate/pharynx" [3], with the abnormal pharyngeal muscular activity "affecting muscles derived from branchial arches including muscles of the palate, pharynx," and sometimes "head, neck and diaphragm" [4].

Demographics

Oculopalatal tremor is rare, but the precise prevalence and incidence are unknown. A study of 82 patients by Jang and Borruat [5] reported that the average age of onset was 54 years with a male predominance.

Clinical Presentation

Most cases of OPT occur after an acute lesion of the brainstem or cerebellum, typically an ischemic or hemorrhagic stroke. A patient's original symptoms may improve, but within several months, they develop rhythmic clicking tinnitus (from the palatal tremor) and a disturbance of vision or equilibrium (from the pendular nystagmus). These symptoms are chronic.

Pathophysiology

OPT "result[s] from the interruption of the *dentato–rubro–olivary pathway*" [3] which runs from (1) the red nucleus in the midbrain, via the central tegmental tract to (2) the inferior olivary nucleus in the medulla, via the inferior cerebellar

M. Cherchi, *Otoneurology and Vestibular Medicine*, https://doi.org/10.1007/978-3-031-94841-1_116

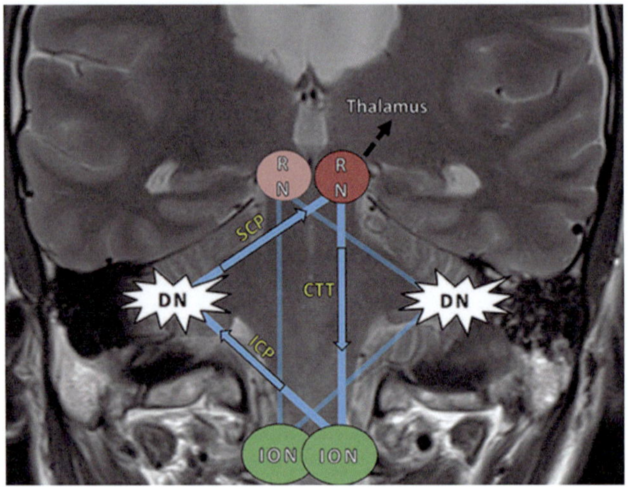

Fig. 116.1 Coronal MRI demonstrating the Guillain–Mollaret triangle (GMT) (blue) and its unidirectional course of inhibitory fibers, which ascend from the dentate nucleus (DN) to the contralateral red nucleus (RN) and then descend ipsilaterally to the inferior olivary nucleus (ION). The bilateral overlapping Guillain–Mollaret triangles together form a "tilted star of David configuration." Fibers of the GMT, which is also called the dentato–rubro–olivary-pathway, overlap with dentato–thalamo–cortical pathways proceeding into both hemispheres (black arrow). Abbreviations: HOD = hypertrophic olivary degeneration; SCP = superior cerebellar peduncle; ICP = inferior cerebellar peduncle; RN = red nucleus; CTT = central tegmental tract; DN = dentate nucleus; ION = inferior olivary nucleus. (From Schaller-Paule et al. [7]. Reproduced from Frontiers Media S.A. with permission under CC-BY 4.0)

peduncle to (3) the dentate nucleus in the cerebellum, via the dentatorubral tract back up to the red nucleus. This pathway was originally described by Dr. George Charles Guillain and Dr. Pierre Mollaret [6] and is thus usually referred to as the *triangle of Guillain–Mollaret* (Fig. 116.1).

Anatomically, OPT is associated with hypertrophy of the inferior olive [3], originally described by Oppenheim [2]. The histopathological change is one of "trans-synaptic degeneration" in the inferior olive and is "attributed to increased water content and gliosis. Pathologically, hypertrophy of the olive corresponds to enlargement of both neurons and astrocytes and the neurons become vacuolated" [3] .

Control systems analysis has led to models of specific cellular mechanisms likely responsible for the oscillatory activity [8], though some studies suggest that more than one mechanism may be at play [9]. Oddly, an FDG-PET scan study of six patients with unilateral OPT "had hypometabolism in ipsilesional pontine tegmentum and hypermetabolism in contralesional thalamus" [10].

The inferior olivary hypertrophic degeneration is generally unilateral, and it is "ipsilateral to the [causative] lesion if the lesion is in the brain stem or contralateral to the lesion if the lesion is in the cerebellum" [3].

Rare cases of bilateral involvement have been documented [11].

Etiology

OPT is the delayed result of a discrete lesion anywhere along the pathways of the Guillain–Mollaret triangle, typically in the brainstem or cerebellum [4]. Most cases of OPT are the delayed result of an acute lesion; it has been estimated that about 80% of cases are due to a brainstem vascular lesion [5], such as ischemic or hemorrhagic infarction [12].

Less commonly, OPT occurs in the context of neoplasm. Spencer's original description was a case of a neoplasm in a young girl [2]. Occurrence with medulloblastoma has also been described [11].

OPT can occur in multiple sclerosis, though its behavior appears distinct from that of other brainstem lesions in some respects [13]. In cases of OPT occurring in multiple sclerosis (MS), it is believed that an MS lesion has caused "instability of the neural integrator, which normally ensures steady gaze" [14].

Even less common are cases that appear to occur spontaneously [15], without any detectable discrete event or lesion.

Physical Examination

In a patient with OPT, examination of the oropharynx should show a fairly regular 2–3 Hz oscillation of the soft palate. Oculomotor examination in about 90% of cases shows prominent pendular nystagmus with vertical and torsional components [5] synchronous with the palatal tremor. The oropharyngeal and oculomotor components may be present to a greater or lesser degree.

Jang and Borruat [16] reported a case with accompanying video that demonstrates these findings to advantage.

The nystagmus can be conjugate or dissociated. Jang and Borruat [5] studied 12 cases and reported that dissociated nystagmus tended to occur in unilateral lesions (with the oculomotor finding contralateral to the lesion), whereas conjugate nystagmus tended to occur with bilateral lesions. Similar patterns of conjugate versus dissociated nystagmus were described the case reported by Tilikete and colleagues [13, 17], and in the series of 22 patients reported by Kim and Moon [18].

Figure 116.2 from East and Hazell [19] shows impedance tympanometry in a patient with palatal myoclonus, with spikes occurring at a frequency of about 80 per minute.

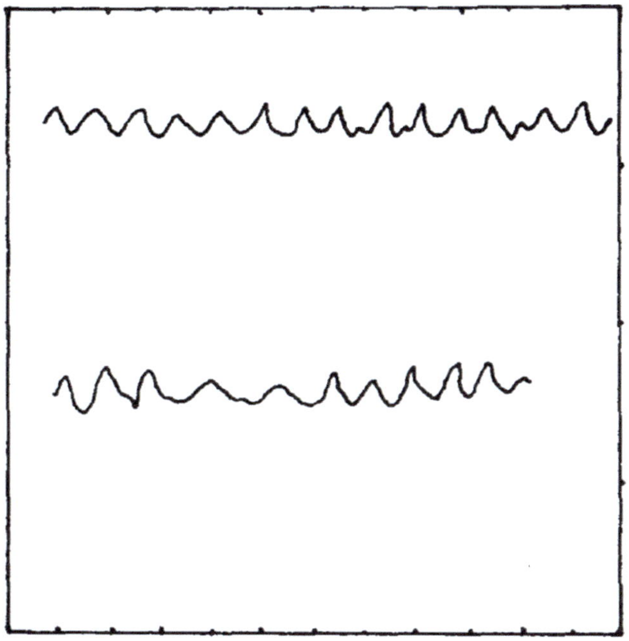

seconds

Fig. 116.2 Immittance tympanometry in a patient with palatal myoclonus showing spikes occurring at a frequency of about 80 per minute. (From East and Hazell 1987). Reproduced with written permission from Cambridge University Press)

Testing

It is desirable to perform binocular oculographic studies in order to document that the eye movements are compatible with OPT. Taking a video of, or at least documenting, palatal tremor is also important.

Imaging

The inferior olivary hypertrophic degeneration of OPT is reflected radiographically on MRI by an enlarged inferior olive with abnormal T2 and FLAIR signal hyperintensity. This imaging finding "appears as early as 3 weeks after a causal event but usually 6–8 months and disappears by 4 years" [3]. Good cases with clear imaging findings are reported by Jang and Borruat [16] and by Behzadi et al. [11].

Treatment

A study comparing the use of gabapentin and baclofen in the treatment of 21 patients with "acquired nystagmus" (including 2 patients with OPM) reported better results with gabapentin [20].

One study of comparing the use of gabapentin, baclofen, and memantine reported that the magnitude of the pendular nystagmus diminished with gabapentin and memantine but not with baclofen [14].

A crossover trial of gabapentin and memantine of 12 patients with acquired pendular nystagmus, including four patients with OPT following brainstem stroke, reported improvement in nystagmus with each drug, though gabapentin was better tolerated than memantine [21].

Another trial of gabapentin and memantine of 16 patients with acquired pendular nystagmus, including six patients with OPT, found that both gabapentin and memantine improved visual outcomes, but that gabapentin was better tolerated than memantine [22].

References

1. Politzer A. Lehrbuch der Ohrenheilkunde für praktische Ärtze und Studeierende. Stuttgart: Verlag von Ferdinand Enke; 1878.
2. Oppenheim H. Über Olivendegeneration bei Atheromatose der basalen Hirnarterien. Berl Klin Wochenschr. 1887;34:638–9.
3. Jun B. The development of hypertrophic inferior olivary nucleus in oculopalatal tremor. Neuroophthalmology. 2016;40(6):297–9.
4. Borruat FX. Oculopalatal tremor: current concepts and new observations. Curr Opin Neurol. 2013;26(1):67–73.
5. Jang L, Borruat FX. Oculopalatal tremor: variations on a theme by Guillain and Mollaret. Eur Neurol. 2014;72(3-4):144–9.
6. Guillain GC, Mollaret P. Deux cas de myoclonies synchrones et rythmees velo-pharyngo-laryngo-oculo-diaphragmatiques: Le probleme anatomique et phsiologique [Two cases of synchronous and rhythmical velo-pharyngo-laryngo-oculo-diaphragmatic myoclonus]. Rev Neurol. 1931;2:545–66.
7. Schaller-Paule MA, Steidl E, Shrestha M, Deichmann R, Steinmetz H, Seiler A, et al. Multicenter prospective analysis of hypertrophic olivary degeneration following infratentorial stroke (HOD-IS): evaluation of disease epidemiology, clinical presentation, and MR-imaging aspects. Front Neurol. 2021;12:675123.
8. Hong S, Leigh RJ, Zee DS, Optican LM. Inferior olive hypertrophy and cerebellar learning are both needed to explain ocular oscillations in oculopalatal tremor. Prog Brain Res. 2008;171:219–26.
9. Shaikh AG, Hong S, Liao K, Tian J, Solomon D, Zee DS, et al. Oculopalatal tremor explained by a model of inferior olivary hypertrophy and cerebellar plasticity. Brain J Neurol. 2010;133(Pt 3):923–40.
10. Moon SY, Cho SS, Kim YK, Kim SE, Kim JH, Kim JS. Cerebral glucose metabolism in oculopalatal tremor. Eur J Neurol. 2008;15(1):42–9.
11. Behzadi F, Fiester PJ, Rao D. Bilateral Hypertrophic Olivary Degeneration Following Brainstem Insult: A Retrospective Review and Examination of Causative Pathology. Neurosci Insights. 2021;16:26331055211007445.
12. Alstadhaug KB. Oculopalatal and cerebellar limb tremor due to hypertrophic olivary degeneration. Eur J Neurol. 2007;14(5):e6–7.
13. Tilikete C, Jasse L, Pelisson D, Vukusic S, Durand-Dubief F, Urquizar C, et al. Acquired pendular nystagmus in multiple sclerosis and oculopalatal tremor. Neurology. 2011;76(19):1650–7.
14. Shaikh AG, Thurtell MJ, Optican LM, Leigh RJ. Pharmacological tests of hypotheses for acquired pendular nystagmus. Ann N Y Acad Sci. 2011;1233:320–6.

15. Oliveira CA, Negreiros Junior J, Cavalcante IC, Bahmad F Jr, Venosa AR. Palatal and middle-ear myoclonus: a cause for objective tinnitus. Int Tinnitus J. 2003;9(1):37–41.
16. Jang L, Borruat FX. Micronystagmus of oculopalatal tremor. Neurology. 2013;80(3):e27.
17. Tilikete C, Desestret V. Hypertrophic olivary degeneration and palatal or oculopalatal tremor. Front Neurol. 2017;8:302.
18. Kim JS, Moon SY, Choi KD, Kim JH, Sharpe JA. Patterns of ocular oscillation in oculopalatal tremor: imaging correlations. Neurology. 2007;68(14):1128–35.
19. East CA, Hazell JW. The suppression of palatal (or intra-tympanic) myoclonus by tinnitus masking devices. A preliminary report. J Laryngol Otol. 1987;101(12):1230–4.
20. Averbuch-Heller L, Tusa RJ, Fuhry L, Rottach KG, Ganser GL, Heide W, et al. A double-blind controlled study of gabapentin and baclofen as treatment for acquired nystagmus. Ann Neurol. 1997;41(6):818–25.
21. Thurtell MJ, Joshi AC, Leone AC, Tomsak RL, Kosmorsky GS, Stahl JS, et al. Crossover trial of gabapentin and memantine as treatment for acquired nystagmus. Ann Neurol. 2010;67(5):676–80.
22. Nerrant E, Abouaf L, Pollet-Villard F, Vie AL, Vukusic S, Berthiller J, et al. Gabapentin and memantine for treatment of acquired pendular nystagmus: effects on visual outcomes. J Neuro-ophthalmol. 2020;40(2):198–206.

117 Periodic Alternating Nystagmus

Overview

Periodic alternating nystagmus (PAN) is spontaneous horizontal nystagmus that alternates in direction, whose cycles are usually 2–4 min. Most cases of PAN are acquired; some are congenital. Most cases are spontaneous; rare familial cases have been reported. The underlying pathophysiology is unknown, though the idea of a slowly oscillating null point has been proposed. Patients with PAN often complain of oscillopsia. PAN can occur in isolation, though some cases occur in the context of other pathologies, such as encephalitis, diseases affecting the cerebellum (neoplasm, infarct, demyelination, Chiari malformation) or posterior fossa (arachnoid cyst, posterior reversible encephalopathy syndrome), albinism, or ataxias (Friedreich's ataxia, spinocerebellar ataxia). Rare cases appear as adverse effects of medications (lithium, primidone, phenobarbital). Oculography is useful for documenting and characterizing the ocular motor abnormality. Rotatory chair testing may show elevated gain on slow harmonic acceleration. In a patient with PAN, it is sensible to check a brain MRI to identify structural causes (e.g., Chiari malformation, neoplasm) and acute causes (e.g., infarction, demyelination). In cases with irreversible underlying causes, an attempt at treatment with baclofen is reasonable.

Introduction

Periodic alternating nystagmus (PAN) is spontaneous horizontal nystagmus that alternates in direction. Baloh and colleagues [1] provide tracings demonstrating this clearly.

Epidemiology

PAN is often an acquired disorder, though it can also occur as a form of congenital nystagmus [2, 3].

Genetics

Familial patterns of PAN have been reported [4], but the genetic contribution to PAN is unknown.

Pathophysiological Mechanism of Disease

The underlying pathophysiology of PAN is unknown. Several mechanisms have been proposed.

Several investigators have suggested a slowly oscillating shift of the null point as the likely mechanism [5, 6].

Based on analysis of abnormalities observed in saccades and smooth pursuit during the nystagmus-free intervals of PAN, Meienberg and Hoyt postulated cerebellar disinhibition as the mechanism of PAN [7]. Rudge and Leech further suggested that PAN may be due to "alternating activity of the vestibular nuclei on either side of the brain-stem as a result of a removal of cerebellar inhibition, and that this caused periodic shifts in the null region of Daroff and Dell'Osso" [8].

Leigh and colleagues offer a control systems analysis model of PAN and propose that "PAN arises from (1) an instability in the brain-stem neural networks that generate slow phases of vestibular and optokinetic nystagmus, (2) the action of an adaptive network that normally acts to null prolonged, inappropriate nystagmus, and (3) an inability to use retinal-error-velocity information" [9].

Furman and colleagues [10] propose a control systems analysis model of PAN. In a subsequent piece, Furman and colleagues [11] concluded that "PAN is caused by an instability in the velocity storage element, a hypothetical neural circuit that perseverates the eye movement response to both vestibular and optokinetic stimulation. Further, we postulate that PAN may be caused by lesions of the cerebellar uvula and nodulus or their connections with the brainstem vestibular nuclei."

M. Cherchi, *Otoneurology and Vestibular Medicine*, https://doi.org/10.1007/978-3-031-94841-1_117

Although most investigators postulate a neurological mechanism for PAN, there have also been reports of PAN occurring with peripheral (labyrinthine) disease. For example, Murofushi and colleagues described PAN in a patient with Ménière's disease [12]; Kim and colleagues described PAN in patients with Ménière's disease, acute labyrinthitis, and vestibular schwannoma [13]; Taki and colleagues described PAN during caloric stimulation [14]. The mechanism by which peripheral labyrinthine disease or stimulation could provoke PAN is unknown.

Clinical Presentation

Patients with PAN commonly complain of oscillopsia.

Physical Examination

Aside from the ocular motor abnormalities (see below), an individual with PAN who is otherwise healthy should have a normal physical examination

Ocular Motor Examination

PAN can occur as an isolated ocular motor manifestation, though it can also occur in association with other ocular motor abnormalities, such as spontaneous down beat nystagmus [15] and pendular see-saw nystagmus. [16]

The periodicity of PAN—meaning the time from the start of the phase of nystagmus beating in one direction, through the null period, through the phase of nystagmus beating in the other direction—is variable. Baloh and colleagues [1] state that the periodicity "cycle length has varied from 1 to 6 min[utes] with null periods varying from 2 to 20 s[econds]."

Testing: Vestibular

Furman and colleagues studied four PAN patients with rotatory chair testing and observed that on slow harmonic acceleration:

> As in normal subjects, the gain of the VOR [vestibulo-ocular reflex] of the patients with PAN increases with increasing frequency. However, in general, the patients' gain was greater than normal. [11]

In addition, they reported that:

> The phase of patients' horizontal VOR shows a surprising result in that their phase lead at 0.05 Hz was less than the average for normal (in 3 out of 4 cases), while at 0.0025 Hz, each patient's phase lead was much greater than normal. This pattern of abnormality has not been reported previously for any vestibular disorder. Phase leads at 0.0025 Hz have not been reported previously for any human subjects. [11]

Imaging

Imaging in patients with isolated PAN is usually normal, though it helps to exclude competing diagnoses.

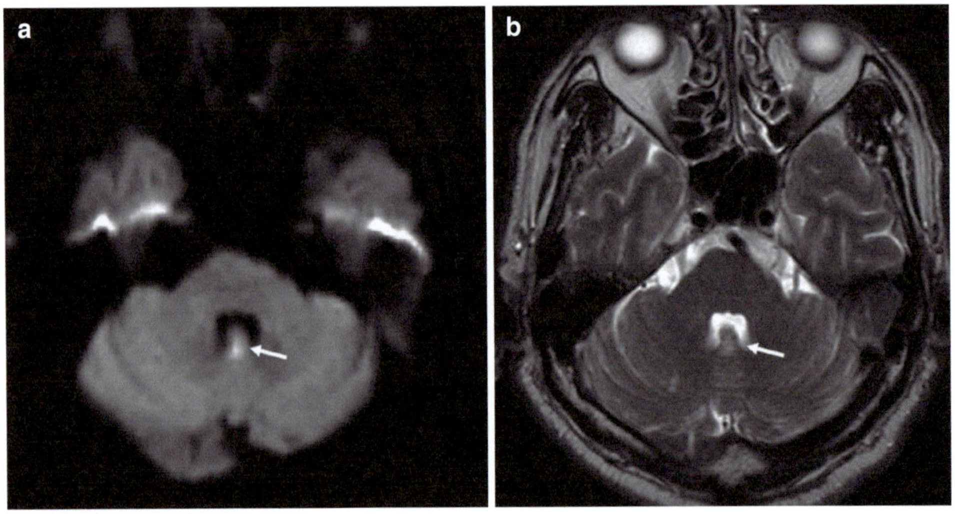

Fig. 117.1 Axial MRI showing infarction of the cerebellar nodulus in a patient with acute onset periodic alternating nystagmus. Panel A shows a diffusion-weighted image. Panel B shows a T2-weighted image. (From Jeong et al. [17]. Reproduced with written permission from Wolters Kluwer Health Limited)

Acute acquired PAN usually warrants obtaining a brain MRI as it can result from infarction of the cerebellar nodulus, as shown in Fig. 117.1 from Jeong and colleagues. [17]

Histopathology

Baloh and colleagues [1] reviewed autopsies of three patients with PAN. The first case involved a cyst of the fourth ventricle; the second case involved a Chiari malformation; the third case involved multiple sclerosis [15]. Baloh and colleagues state that "in each [case], multiple regions of the brain-stem have been involved." Furman and colleagues [11] cited several additional autopsy studies [18, 19]; the study by Karp and Rorke reported the case of an 11-year-old girl with an astrocytoma involving the vestibular nuclei at the pontomedullary junction on one side [18].

Differential Diagnosis

While PAN is an interesting ocular motor finding, in some cases, it is associated with a diagnosable disease. Baloh and colleagues [1] list a variety of conditions with which PAN has occurred, including encephalitis, brainstem ischemia, demyelinating disease, syringobulbia, syphilis, and trauma. Furman and colleagues [11] list additional associated diagnoses including spinocerebellar degeneration, posterior fossa neoplasm, posterior fossa arachnoid cyst, and cryptococcal infection. There are additionally case reports of PAN occurring in the context of acute lesions of the cerebellar nodulus [17, 20] albinism, [21] posterior reversible encephalopathy syndrome [22], Chiari malformation [23], Friedreich's ataxia [24, 25] spinocerebellar ataxia type 6 [26], multiple sclerosis [27], and epilepsy [28]. It also appears that PAN can be induced by certain medications such as lithium [29, 30] and primidone/phenobarbital [31].

Treatment

Several papers have observed that baclofen can improve PAN, [32] whether it is acquired [33, 34] or congenital [35, 36].

Other treatments are based largely on case reports.

Lee and colleagues [37] reported improvement of PAN with amantadine.

Argente-Escrig and colleagues [38] reported improvement of PAN in a patient treated with intravenous immunoglobulin and postulated *ex juvantibus* an autoimmune mechanism.

Kumar and colleagues [39] reported improvement of PAN with memantine.

Surgery on extraocular muscles has also been attempted as treatment for PAN [40, 41].

Prognosis

The prognosis of PAN depends on the underlying etiology. There are very few reversible causes of PAN (e.g., medication toxicity). For the remainder, prognosis is guarded, and treatment will depend on the etiology; for example, disease-modifying therapy for multiple sclerosis or resection for posterior fossa tumor.

References

1. Baloh RW, Honrubia V, Konrad HR. Periodic alternating nystagmus. Brain J Neurol. 1976;99(1):11–26.
2. Shallo-Hoffmann J, Riordan-Eva P. Recognizing periodic alternating nystagmus. Strabismus. 2001;9(4):203–15.
3. Shallo-Hoffmann J, Faldon M, Tusa RJ. The incidence and waveform characteristics of periodic alternating nystagmus in congenital nystagmus. Invest Ophthalmol Vis Sci. 1999;40(11):2546–53.
4. Hertle RW, Yang D, Kelly K, Hill VM, Atkin J, Seward A. X-linked infantile periodic alternating nystagmus. Ophthalmic Genet. 2005;26(2):77–84.
5. Daroff RB, Dell'Osso LF. Periodic alternating nystagmus and the shifting null. Can J Otolaryngol. 1974;3:367–71.
6. Robb RM. Periodic alteration of null point in congenital nystagmus. Association with alternating gaze deviation and esotropia. Arch Ophthalmol. 1972;87(2):169–73.
7. Meienberg O, Hoyt WF. Ocular motor control disorder during the neutral phase of periodic alternating nystagmus. J Neurol. 1980;223(4):309–12.
8. Rudge P, Leech J. Analysis of a case of periodic alternating nystagmus. J Neurol Neurosurg Psychiatry. 1976;39(4):314–9.
9. Leigh RJ, Robinson DA, Zee DS. A hypothetical explanation for periodic alternating nystagmus: instability in the optokinetic-vestibular system. Ann N Y Acad Sci. 1981;374:619–35.
10. Furman JM, Hain TC, Paige GD. Central adaptation models of the vestibulo-ocular and optokinetic systems. Biol Cybern. 1989;61(4):255–64.
11. Furman JM, Wall C 3rd, Pang DL. Vestibular function in periodic alternating nystagmus. Brain J Neurol. 1990;113(Pt 5):1425–39.
12. Murofushi T, Chihara Y, Ushio M, Iwasaki S. Periodic alternating nystagmus in Meniere's disease: the peripheral type? Acta Otolaryngol. 2008;128(7):824–7.
13. Kim SH, Chung WK, Kim BG, Hwang CS, Kim MJ, Lee WS. Periodic alternating nystagmus of peripheral vestibular origin. Laryngoscope. 2014;124(4):980–3.
14. Taki M, Hasegawa T, Adachi N, Fujita T, Sakaguchi H, Hisa Y. Periodic alternating nystagmus during caloric stimulation. Auris Nasus Larynx. 2014;41(2):211–4.
15. Keane JR. Periodic alternating nystagmus with downward beating nystagmus. A clinicoanatomical case study of multiple sclerosis. Arch Neurol. 1974;30(5):399–402.

16. Jeong SH, Hwang JM, Kim JS. Co-occurrence of periodic alternating and pendular seesaw nystagmus in blindness. J Neurol Sci. 2009;285(1-2):257–8.
17. Jeong HS, Oh JY, Kim JS, Kim J, Lee AY, Oh SY. Periodic alternating nystagmus in isolated nodular infarction. Neurology. 2007;68(12):956–7.
18. Karp JS, Rorke LB. Letter: perioditic alternating nystagmus. Arch Neurol. 1975;32(6):422–3.
19. Susac JO, Henry JM, editors. Periodic alternating nystagmus: clinicopathologic correlation. Neuro-ophthalmology, symposium of the University of Miami and the Bascom Palmer Eye Institute. St. Louis: Mosby Comp; 1975.
20. Oh YM, Choi KD, Oh SY, Kim JS. Periodic alternating nystagmus with circumscribed nodular lesion. Neurology. 2006;67(3):399.
21. Guyer DR, Lessell S. Periodic alternating nystagmus associated with albinism. J Clin Neuroophthalmol. 1986;6(2):82–5.
22. Mackay DD, Zepeda Garcia R, Galetta SL, Prasad S. Periodic alternating gaze deviation and nystagmus in posterior reversible encephalopathy syndrome. Neurol Clin Pract. 2014;4(6):482–5.
23. Korres S, Balatsouras DG, Zournas C, Economou C, Gatsonis SD, Adamopoulos G. Periodic alternating nystagmus associated with Arnold-Chiari malformation. J Laryngol Otol. 2001;115(12):1001–4.
24. Gorman WF, Brock S. Periodic alternating nystagmus in Friedreich's ataxia. Am J Ophthalmol. 1950;33(6):860–4.
25. Gorman WF, Brock S, Kestenbaum A. Periodic alternating nystagmus in Friedreich's ataxia. J Nerv Ment Dis. 1950;112(5):437–9.
26. Hashimoto T, Sasaki O, Yoshida K, Takei Y, Ikeda S. Periodic alternating nystagmus and rebound nystagmus in spinocerebellar ataxia type 6. Mov Disord. 2003;18(10):1201–4.
27. Matsumoto S, Ohyagi Y, Inoue I, Oishi A, Goto H, Nakagawa T, et al. Periodic alternating nystagmus in a patient with MS. Neurology. 2001;56(2):276–7.
28. Moster ML, Schnayder E. Epileptic periodic alternating nystagmus. J Neuro-Ophthalmol. 1998;18(4):292–3.
29. Lee MS, Lessell S. Lithium-induced periodic alternating nystagmus. Neurology. 2003;60(2):344.
30. Rust H, Lutz N, Honegger F, Fischer-Barnicol D, Welge-Luessen A, Kappos L, et al. Periodic alternating nystagmus in a patient on long-term lithium medication. J Neurol Sci. 2016;369:252–3.
31. Schwankhaus JD, Kattah JC, Lux WE, Masucci EF, Kurtzke JF. Primidone/phenobarbital-induced periodic alternating nystagmus. Ann Ophthalmol. 1989;21(6):230–2.
32. DiBartolomeo JR, Yee RD. Periodic alternating nystagmus. Otolaryngol Head Neck Surg. 1988;99(6):552–7.
33. Halmagyi GM, Rudge P, Gresty MA, Leigh RJ, Zee DS. Treatment of periodic alternating nystagmus. Ann Neurol. 1980;8(6):609–11.
34. Isago H, Tsuboya R, Kataura A. A case of periodic alternating nystagmus: with a special reference to the efficacy of baclofen treatment. Auris Nasus Larynx. 1985;12(1):15–21.
35. Comer RM, Dawson EL, Lee JP. Baclofen for patients with congenital periodic alternating nystagmus. Strabismus. 2006;14(4):205–9.
36. Solomon D, Shepard N, Mishra A. Congenital periodic alternating nystagmus: response to baclofen. Ann N Y Acad Sci. 2002;956:611–5.
37. Lee SH, Lee SY, Choi SM, Kim BC, Kim MK, Kim JS. Resolution of periodic alternating nystagmus with amantadine. J Neurol Sci. 2016;364:65–7.
38. Argente-Escrig H, Bataller L, Krstulovic Roa C, Perez Guillen V, Perez Garrigues H, Casanova EB. Atypical periodic alternating nystagmus responding to high-dose intravenous immunoglobulins: a case report. J Neuroinflammation. 2017;14(1):71.
39. Kumar A, Thomas S, McLean R, Proudlock FA, Roberts E, Boggild M, et al. Treatment of acquired periodic alternating nystagmus with memantine: a case report. Clin Neuropharmacol. 2009;32(2):109–10.
40. Mimura O, Ishikawa H, Kimura N, Kimura A, Borlongan CV. Retrospective study of recession of four horizontal rectus muscle in periodic alternating nystagmus. Clin Ophthalmol. 2014;8:2523–8.
41. Thomas R, Mathai A, Braganza A, Billson F. Periodic alternating nystagmus treated with retrobulbar botulinum toxin and large horizontal muscle recession. Indian J Ophthalmol. 1996;44(3):170–2.

Rebound Nystagmus

118

Overview

Rebound nystagmus (RN) is an ocular motor finding in which, after a patient has maintained prolonged horizontal eccentric gaze and then returns to primary position of gaze, there is transient jerk nystagmus whose slow phase is in the direction of the previous horizontal eccentric gaze, and the fast phase is back toward primary position. RN often occurs in association with horizontal gaze-evoked nystagmus. The mechanism of RN is unknown, but it is thought to reflect a deficit in the neural integrator required for gaze holding. RN can occur in the context of many diseases affecting the cerebellum and brainstem. Treatment and prognosis depend on the underlying cause.

Introduction

Rebound nystagmus (RN) is "a transient nystagmus that occurs when the eyes return to straight ahead position following a period of prolonged eccentric gaze holding" [1] in which "Quick phases of the nystagmus are away from the direction of a previously attempted eccentric gaze" [2], and thus, the "slow phases of rebound nystagmus are in the direction of prior eccentric gaze" [1].

Epidemiology

Over the course of 7 years (1991–1998), Lin and Young [2] studied 9156 patients who presented to a single hospital "with vertigo or tinnitus" and reported that 10 (0.1%) "showed rebound nystagmus by ENG" (electronystagmography). This sample is somewhat odd in that patients with "vertigo *or tinnitus*" were included—so presumably some patients had only auditory symptoms and no vestibular symptoms. Nevertheless, Lin and Young are suggesting that rebound nystagmus is uncommon.

Pathophysiological Mechanism of Disease

Rebound nystagmus "is thought to be related to the function of the oculomotor neural integrator—the circuit that ensures accurate gaze holding after any eye movement—but the exact mechanism of rebound nystagmus is unknown" [1].

Shallo-Hoffman and colleagues [3] comment that:

> End-point nystagmus (EPN) [their term for horizontal gaze-evoked nystagmus] is evoked by far lateral gaze… Rebound nystagmus (RN) consists of a reversal of nystagmus that is evoked by a gaze shift to the primary position (PP) after a period of lateral gaze. The direction of RN is always the reverse of the EPN [3].

In other words, horizontal gaze-evoked nystagmus and rebound nystagmus co-occur.

Bogli and colleagues [4] hypothesize that both gaze-evoked nystagmus and rebound nystagmus reflect similar underlying "cerebellar loss of function." Specifically:

> A brain stem/cerebellar neural integrator enables stable eccentric gaze. Cerebellar loss-of-function can cause an inability to maintain gaze eccentrically (gaze-evoked nystagmus). Moreover, after returning gaze to straight ahead, the eyes may drift toward the prior eye position (rebound nystagmus). Typically, gaze-evoked nystagmus decays during continuously held eccentric gaze. We hypothesized this adaptive behavior to be prerequisite for rebound nystagmus and thus predicted a correlation between the velocity decay of gaze-evoked nystagmus and the initial velocity of rebound nystagmus [4].

Bogli and colleagues studied 11 patients with cerebellar degeneration and reported a correlation, concluding that "rebound nystagmus can be explained as gaze-evoked nystagmus relative to a set point (position with least eye drift) away from straight-ahead eye position" [4].

Shallo-Hoffmann and colleagues state that "the cause of RN [rebound nystagmus] is a velocity bias of the neural integrator evolved during long eccentric gaze" [3].

Note that not all rebound nystagmus is pathological; "after eccentric gaze, healthy subjects also show rebound nystagmus when fixation is removed" [1].

M. Cherchi, *Otoneurology and Vestibular Medicine*, https://doi.org/10.1007/978-3-031-94841-1_118

Clinical Presentation

The clinical presentation depends on the underlying disease with which rebound nystagmus is associated.

Physical Examination

Rebound nystagmus can occur in association with diseases of the cerebellum and brainstem. In some cases, there will be corresponding signs on physical examination (such as appendicular cerebellar dysmetria), while in other cases, the general neurological examination may be unrevealing.

Ocular Motor Examination

Rebound nystagmus is usually difficult to observe on face-to-face ocular motor examination.

Testing: Instrumented Ocular Motor Examination

Rebound nystagmus can be subtle, so examination under infrared video oculography is often revealing.

Figure 118.1, from Otero-Millan and colleagues [1], illustrates rebound nystagmus.

Imaging

If the clinical scenario merits imaging, the area in question (posterior fossa) is usually best visualized by MRI.

Differential Diagnosis

Lin and Young suggest that "rebound nystagmus implies a lesion in the **cerebellum** or **brainstem** for which MRI examination is 100% sensitive" [2].

Rebound nystagmus has been reported in association with a range of posterior fossa lesions.

Hood and colleagues [5] studied 25 patients exhibiting rebound nystagmus and reported the following pathologies:

- Cerebellar degeneration or atrophy.
- Cerebellar lesion, unspecified.
- Demyelinating disease (location unspecified).
- Foramen magnum lesions.
- Posterior fossa tumors.
- Vascular lesions of the posterior fossa.
- Drug intoxication.
- Undetermined posterior fossa lesion with familial history.

Lin and Young [2] studied 10 patients exhibiting rebound nystagmus and reported the following pathologies:

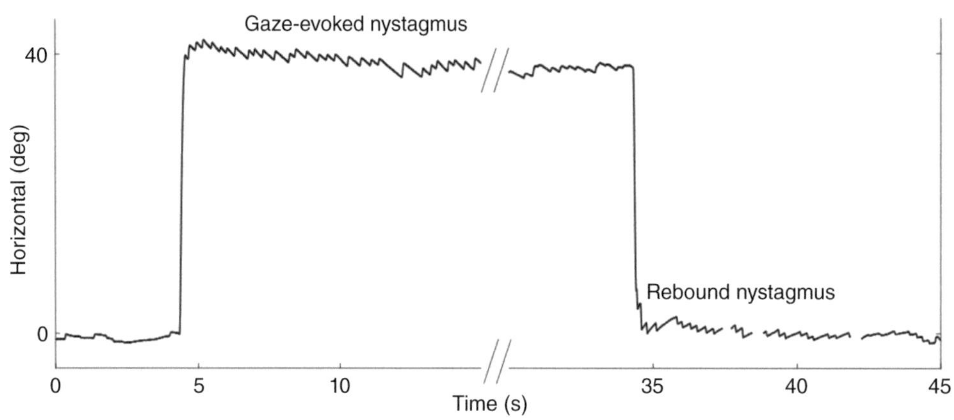

Fig. 118.1 Example of rebound nystagmus. A healthy subject looks at a flashing target, first at center (i.e., straight ahead), then eccentrically at 40 degrees to the right for 30 seconds, and then back to the center. During eccentric fixation at 40 degrees, there is a gaze-evoked nystagmus with the slow phase drifting toward the center, which gradually decreases in intensity. During the rebound period, the slow phase drifts toward the previously held, eccentric position. (From Otero-Millan et al. [1]. Reproduced with written permission from Elsevier B.V.)

- Cerebellopontine angle schwannoma (three cases).
- Cerebellopontine angle meningioma.
- Cerebellar anaplastic astrocytoma.
- Brainstem astrocytoma.
- Brainstem lymphoma.
- Cerebellar hemorrhage.
- Dissecting aneurysm of vertebral artery.
- Dolichoectasia of vertebrobasilar artery.

Scattered case reports also describe rebound nystagmus occurring in various additional diseases, often involving cerebellar pathways. For example, Bondar and colleagues report rebound nystagmus in a patient with olivopontocerebellar atrophy [6].

Lin and Young state that rebound nystagmus "has a certain lateralizing value, with its direction away from the lesioned side" [2]. Similarly, Morales-Garcia and colleagues report that "in three out of four patients with unilateral lesions rebound nystagmus was ipsilateral with respect to the side of the lesion" [7]. This differs from our own experience, in which rebound nystagmus does not reliably lateralize the lesion.

Treatment and Prognosis

Treatment and prognosis of a patient exhibiting rebound nystagmus will depend on the underlying disease.

References

1. Otero-Millan J, Colpak AI, Kheradmand A, Zee DS. Rebound nystagmus, a window into the oculomotor integrator. Prog Brain Res. 2019;249:197–209.
2. Lin CY, Young YH. Clinical significance of rebound nystagmus. Laryngoscope. 1999;109(11):1803–5.
3. Shallo-Hoffmann J, Schwarze H, Simonsz HJ, Muhlendyck H. A reexamination of end-point and rebound nystagmus in normals. Invest Ophthalmol Vis Sci. 1990;31(2):388–92.
4. Bogli SY, Straumann D, Schuknecht B, Bertolini G, Tarnutzer AA. Cerebellar rebound nystagmus explained as gaze-evoked nystagmus relative to an eccentric set point: implications for the clinical examination. Cerebellum. 2021;20(5):751–9.
5. Hood JD, Kayan A, Leech J. Rebound nystagmus. Brain. 1973;96(3):507–26.
6. Bondar RL, Sharpe JA, Lewis AJ. Rebound nystagmus in olivocerebellar atrophy: a clinicopathological correlation. Ann Neurol. 1984;15(5):474–7.
7. Morales-Garcia C, Cardenas JL, Arriagada C, Otte J. Clinical significance of rebound nystagmus in neuro-otological diagnosis. Ann Otol Rhinol Laryngol. 1978;87(2 Pt 1):238–42.

119 See-Saw Nystagmus

Overview

See-saw nystagmus (SSN) is a rare eye movement disorder in which the eyes move in a see-saw pattern, meaning that there is a cycle of pendular nystagmus in which (1) one eye lowers and extorts while the other eye simultaneously rises and intorts, then (2) the first eye rises and intorts while the other eye simultaneously lowers and extorts. The term "hemi-see-saw" nystagmus refers to a pattern that is similar except that part of the cycle is slow (pendular nystagmus) and part of the cycle is fast (jerk nystagmus). SSN can be congenital or acquired. The mechanism is unknown. Usually, it occurs in the context of visual deficits; less commonly, it occurs in individuals whose vision is intact. Many (though not all) patients are found to have a disorder at the optic chiasm (typically associated with bitemporal hemianopsia). A smaller proportion of patients have pathology involving the brainstem (such as infarction). In a patient with SSN, it is appropriate to check imaging (brain MRI) and refer for an ophthalmologic and/or neuro-ophthalmologic evaluation. Treatment and prognosis depend on the underlying etiology; for example, in a patient with a tumor compressing the optic chiasm, resection of the tumor may bring about some improvement. There are a few reported cases of SSN improving with clonazepam.

Introduction

The British ophthalmologist Dr. Ernest Edmund Maddox (1863–1933) [1] was the first to describe see-saw nystagmus [2]. Since then, dozens of cases have been reported in the literature.

The term *see-saw nystagmus (SSN)* refers to a disconjugate eye movement abnormality in which (1) the first eye rises and intorts, while the fellow eye lowers and extorts; (2) the first eye lowers and extorts, while the fellow eye rises and intorts; (3) the cycle repeats. This movement is said to resemble that of a see-saw with its fulcrum at the nasal bridge.

The nomenclature for see-saw nystagmus is confusing. The unqualified term "*see-saw nystagmus*" refers to a situation in which the described eye movements are slow (pendular) throughout their entire cycle. In contrast, the term "*hemi-see-saw nystagmus*" refers to the same eye movement when the initial eccentric movement (vertical and torsional) is slow (similar to the pendular variety), but the return to primary position of gaze is fast (jerk nystagmus); in other words, only half of the cycle ("hemi") is pendular nystagmus, while the other half is jerk nystagmus. Contrary to what one might expect, note that the term "hemi-see-saw nystagmus" does *not* refer to a situation in which only one eye exhibits this pattern.

Epidemiology

See-saw nystagmus (SSN) is rare, though its incidence and prevalence are unknown.

Several articles include summary reviews of previously published cases [3–6].

Most cases of see-saw nystagmus (SSN) are acquired. Occasional congenital cases have been reported, usually in patients with congenital optic chiasm abnormalities [7, 8] or congenital retinitis pigmentosa [9].

Some cases of SSN occur in the context of blindness [10] or significant visual loss from retinitis pigmentosa (RP), whether the onset of visual loss is in childhood [9] or adulthood [11].

Some cases of SSN occur in the context of congenital optic chiasm defects, usually achiasma [8], hemichiasma [12], or chiasmic dysplasia [13] such as optic fiber nondecussation [7, 13]. In other cases, the optic chiasm has been damaged, typically by compression from a space-occupying lesion [14] such as craniopharyngioma [4, 15–18], pituitary adenoma [18], or suprasellar epidermoid cyst [19]. Generally,

M. Cherchi, *Otoneurology and Vestibular Medicine*, https://doi.org/10.1007/978-3-031-94841-1_119

these patients with optic chiasm defects have bitemporal hemianopsia.

Some cases of acquired SSN result from infarction [14]. Halmagyi and colleagues [20] described two patients with see-saw nystagmus who had suffered left thalamic hemorrhages. Kanter and colleagues [21] reported a case of see-saw nystagmus in a patient who had suffered a "presumed brainstem vascular event" in the paramedian ventral midbrain with "involvement of the right interstitial nucleus of Cajal." Lee and Kim [22] describe see-saw nystagmus in a patient who had suffered an infarction involving the left anterior rostral cerebellar hemisphere and rostral vermis and superior cerebellar peduncle. Man and Fu [23] report on a patient with a paramedian thalamomesencephalic infarct who developed a combination of see-saw nystagmus, convergence-retraction nystagmus, and contraversive ocular tilt.

Rarely, acquired SSN occurs in the context of multiple sclerosis [24, 25].

One case of see-saw nystagmus has been reported in the context of a paraneoplastic syndrome [26] in a patient with breast cancer and antineuronal antibody type 2.

Hemi-see-saw nystagmus is less common than "plain" see-saw nystagmus. Choi and colleagues [27] reported a case of pontine infarction presenting with hemi-see-saw nystagmus and internuclear ophthalmoplegia. Lee and colleagues [28] reported on a patient with a medial medullary infarction who exhibited up beat torsional nystagmus that evolved into hemi-see-saw nystagmus.

Pathophysiological Mechanism of Disease

The mechanism of see-saw nystagmus is unknown.

Because many patients with SSN have significant visual defects, several investigators have proposed that the reduced visual input is key to explaining this phenomenon [29]. For example:

> Based on analysis of the ocular oscillation characteristics of the cases of seesaw nystagmus reported in the literature… we postulate that seesaw nystagmus is another type of ocular oscillation brought about by an unstable visuovestibular interaction control system. Nonavailability of retinal error signals to the inferior olivary nucleus essential for vestibuloocular reflex adaptation due to complete chiasmal dissection makes the system less stable. This system instability is further accentuated by the pursuit feedback element. The intact inferior olivary nucleus-nodulus connections in seesaw nystagmus would explain the 180 degree phase difference that distinguishes it from the midline form of oculopalatal myoclonus, where these connections are likely disrupted [3].

And:

> Therefore, in the presence of normal labyrinthine signals, insufficient visual input via accessory optic system can adaptively increase the excitability of the INC neurons. Such maladaptive increase in excitability leads to oscillations comprising pendular seesaw nystagmus [26].

Whether this proposal is correct, or at least whether it is the only mechanism of see-saw nystagmus, is brought into question by the fact that some patients with SSN have no detectable visual impairment. Rizvi and colleagues [26] studied a case of acquired see-saw nystagmus in the context of a paraneoplastic syndrome in a patient with breast cancer and antineuronal antibody type 2 who had no visual deficits and no cerebellar findings on examination. Using a control systems analysis, they proposed:

> The hypothesis that abnormal increases in the excitability of the INC [interstitial nucleus of Cajal] excitatory burst neurons (due to cross immune etiology) that comprise reciprocally excitatory circuit can cause reverberations leading to pendular or pendular seesaw nystagmus [26].

Some cases of see-saw nystagmus have been reported following head trauma [3, 30, 31] and may be delayed [32, 33]. Usually, such cases also involve bitemporal hemianopsia [31, 32], which may be complete or partial [3] or optic atrophy [33]. In these cases, it is unclear whether the SSN results from the visual deficit, or from some other damage consequent to the head trauma.

Clinical Presentation

The most common complaint in patients with see-saw nystagmus is visual disturbance, such as oscillopsia, and this is the usual reason for referral to otoneurology. In cases that develop in the context of slowly progressive visual loss, the patient usually is initially seen by ophthalmology.

Physical Examination

Patients with isolated see-saw nystagmus usually have otherwise normal general physical and neurological examinations. However, given the frequent association with visual loss, an examiner should do a beside check of visual fields.

See-saw nystagmus can also present as part of a broader problem (see the section on differential diagnosis below), and such patients often have other abnormalities on physical examination.

Ocular Motor Examination

In some cases, see-saw nystagmus (SSN) is subtle and may only be detectable on instrumented oculography. However, in most cases, the SSN can be observed on face-to-face examination. It may be modulated by various voluntary eye movements; for instance, it may diminish on convergence [33], or increase on eccentric gaze.

In many cases, the eye movement is slow and thus would be classified as pendular nystagmus [4, 5, 14, 25, 34], while in other cases, it has more the quality of a jerk nystagmus [20, 35]. See the discussion above regarding the distinction between "see-saw nystagmus" and "hemi-see-saw nystagmus."

Rudich and Lesser [36] report a case with accompanying video that shows see-saw nystagmus evident on face-to-face examination.

Some cases of SSN may be more difficult to recognize when the SSN occurs in conjunction with another ocular motor abnormality, such as congenital nystagmus [7], bowtie nystagmus [37], internuclear ophthalmoplegia [27], latent nystagmus ("infantile nystagmus") [38], periodic alternating nystagmus [10], or convergence-retraction nystagmus [23]. In some of these cases, the SSN may develop after some other form of nystagmus [28, 37].

Testing: Instrumented Oculography

Figure 119.1, from Nguyen and colleagues [8], shows oculographic tracings from a patient with see-saw nystagmus. The salient characteristic is that the tracings of horizontal eye movements (blue = right eye, orange = left eye) are conjugate, while the tracings of vertical eye movements (yellow = right eye, purple = left eye) are 180 degrees out of phase. Unfortunately, there are no tracings for the torsional eye movements (which one would expect to be conjugate).

Differential Diagnosis

See-saw nystagmus can occur in association with a number of ophthalmologic and neurologic conditions. The following list combines differential diagnoses listed in Eggenberger [32] and Leigh and Rucker [39].

- Achiasma.
- Albinism.
- Brainstem stroke, particularly upper brainstem infarction.
- Chiari malformation.
- Congenital, including agenesis of the optic chiasm.
- Head trauma.
- Hemichiasma.
- Meso-diencephalic disease.
- Midbrain tumor.
- Multiple sclerosis.
- Parasellar masses, sometimes with extension.
- Septo-optic dysplasia.
- Syringobulbia.
- Vision loss from retinitis pigmentosa or rod cone dystrophy.

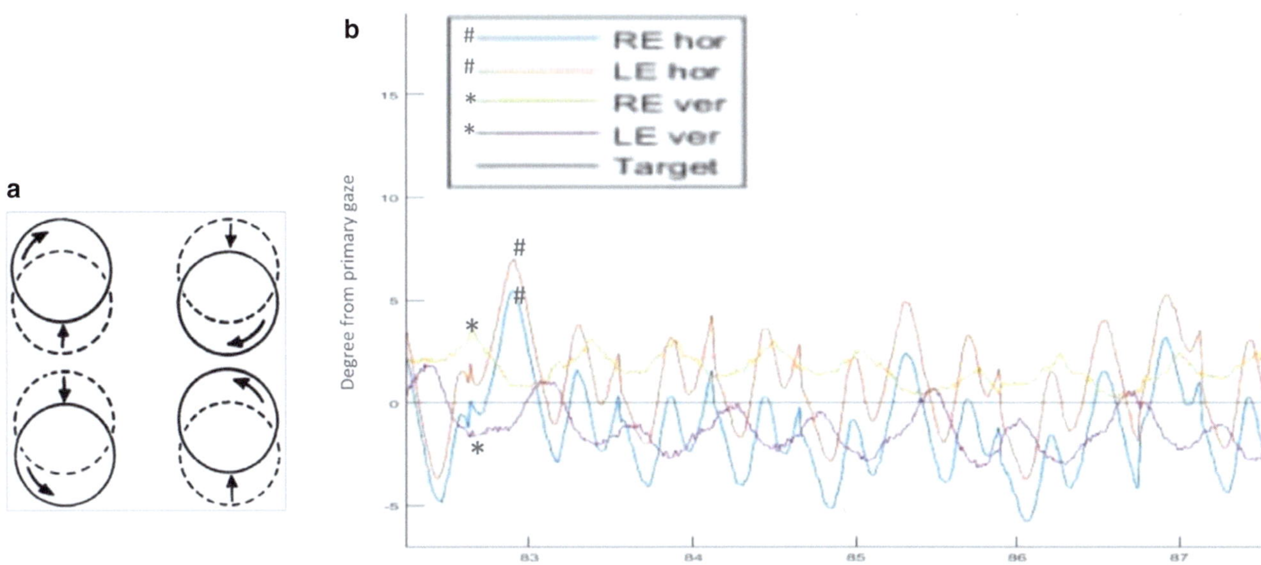

Fig. 119.1 Schematic diagram (panel **a**) and oculography trace (panel **b**) showing see-saw nystagmus consisting of conjugate torsional ocular movement, with superimposed dissociated vertical movements, such that the intorting eye rises, while the other eye, which is extorting, falls to give a see-saw pattern. The (*) indicates right and left eye vertical trace; the (#) indicates horizontal trace. (From Nguyen et al. [8]. Reproduced with written permission from Elsevier B.V.)

When visual symptoms develop abruptly and see-saw nystagmus is present on examination, vascular causes should be suspected and imaging (brain MRI) is appropriate.

When see-saw nystagmus is observed in a patient with a history of gradually progressive visual loss, especially if physical examination demonstrates bitemporal hemianopsia, it is appropriate to check imaging (brain MRI) in order to assess for a structural lesion compressing the optic chiasm.

Consultation with ophthalmology and/or neuro-ophthalmology is appropriate for all cases of see-saw nystagmus.

Treatment

Treatment of reversible causes of see-saw nystagmus (such as resection of a tumor compressing the optic chiasm) is appropriate. Surgical removal of a causative tumor sometimes also brings about improvement in see-saw nystagmus [15, 16, 19]. In some cases, the postsurgical improvement is transient, and the see-saw nystagmus eventually returns [18].

There are several case reports of see-saw nystagmus improving with clonazepam [32, 40], sometimes in combination with baclofen [32].

Some reports notice transient diminution of see-saw nystagmus after ingestion of alcohol [31].

Prognosis

Prognosis of see-saw nystagmus depends on its etiology.

References

1. Cooper JB. Dr. Ernest E. Maddox. Br J Ophthalmol. 1934;18(1):55.
2. Maddox EE. See-saw nystagmus with bitemporal hemianopia. Proc R Soc Med. 1914;7:12–3.
3. Nakada T, Kwee IL. Seesaw nystagmus. Role of visuovestibular interaction in its pathogenesis. J Clin Neuroophthalmol. 1988;8(3):171–7.
4. Druckman R, Ellis P, Kleinfeld J, Waldman M. Seesaw nystagmus. Arch Ophthalmol. 1966;76(5):668–75.
5. Drachman DA. See-saw nystagmus. J Neurol Neurosurg Psychiatry. 1966;29(4):356–61.
6. Fein JM, Williams DB. See-saw nystagmus. J Neurol Neurosurg Psychiatry. 1969;32(3):202–7.
7. Apkarian P, Bour LJ. See-saw nystagmus and congenital nystagmus identified in the non-decussating retinal-fugal fiber syndrome. Strabismus. 2001;9(3):143–63.
8. Nguyen CT, Goh C, Desmond P, Abel LA, Lim CHL, Andrew Symons RC, et al. Congenital achiasma and see-saw nystagmus in VATER syndrome association with hydrocephalus. J Clin Neurosci. 2018;51:63–5.
9. Bergin DJ, Halpern J. Congenital see-saw nystagmus associated with retinitis pigmentosa. Ann Ophthalmol. 1986;18(12):346–9.
10. Jeong SH, Hwang JM, Kim JS. Co-occurrence of periodic alternating and pendular seesaw nystagmus in blindness. J Neurol Sci. 2009;285(1–2):257–8.
11. Chiew YR. See saw Nystagmus associated with retinitis Pigmentosa. QJM. 2023;116:572.
12. Dell'Osso LF, Daroff RB. Two additional scenarios for see-saw nystagmus: achiasma and hemichiasma. J Neuro-ophthalmol. 1998;18(2):112–3.
13. Davis GV, Shock JP. Septo-optic dysplasia associated with see-saw nystagmus. Arch Ophthalmol. 1975;93(2):137–9.
14. Daroff RB. See-Saw Nystagmus. Neurology. 1965;15:874–7.
15. Pal R, Dutta P, Chatterjee D, Ahuja C, Singh A. Pendular see-saw Nystagmus: a rare presenting manifestation of Craniopharyngioma. Neuroophthalmology. 2020;45(2):126–9.
16. Kinder RS, Howard GM. See-Saw Nystagmus. An unusual sign of lesions near the third ventricle. Am J Dis Child. 1963;106:331–2.
17. Saluja G, Samdani A, Bhatia P. See-saw nystagmus in giant craniopharyngioma. BMJ Case Rep. 2020;13(6)
18. Schurr PH. See-Saw Nystagmus. Proc R Soc Med. 1963;56(9):808–10.
19. Smith JL, Mark VH. See-saw nystagmus with suprasellar epidermoid tumor. AMA Arch Ophthalmol. 1959;62(2):280–3.
20. Halmagyi GM, Aw ST, Dehaene I, Curthoys IS, Todd MJ. Jerk-waveform see-saw nystagmus due to unilateral meso-diencephalic lesion. Brain. 1994;117(Pt 4):789–803.
21. Kanter DS, Ruff RL, Leigh RJ, Modic M. See-saw nystagmus and brainstem infarction: MRI findings. Neuroophthalmology. 1987;7(5):279–83.
22. Lee H, Kim HA. Hemi-seesaw nystagmus in a patient with acute infarction in the superior cerebellar artery territory. J Neurol. 2018;265(5):1219–21.
23. Man BL, Fu YP. See-saw nystagmus, convergence-retraction nystagmus and contraversive ocular tilt reaction from a paramedian thalamomesencephalic infarct. BMJ Case Rep. 2014;2014
24. Samkoff LM, Smith CR. See-saw nystagmus in a patient with clinically definite MS. Eur Neurol. 1994;34(4):228–9.
25. Sandramouli S, Benamer HT, Mantle M, Chavan R. See-saw nystagmus as the presenting sign in multiple sclerosis. J Neuro-ophthalmol. 2005;25(1):56–7.
26. Rizvi MT, Cameron L, Kilbane C, Shaikh AG. Paraneoplastic see-saw nystagmus and opsoclonus provides evidence for hyperexcitable reciprocally innervating mesencephalic network. J Neurol Sci. 2018;390:239–45.
27. Choi SY, Kim DH, Lee JH, Kim J. Jerky hemi-seesaw nystagmus and head tilt reaction combined with internuclear ophthalmoplegia from a pontine infarction. J Clin Neurosci. 2009;16(3):456–8.
28. Lee SU, Park SH, Jeong SH, Kim HJ, Kim JS. Evolution of torsional-upbeat into hemi-seesaw nystagmus in medial medullary infarction. Clin Neurol Neurosurg. 2014;118:80–2.
29. Arnott EJ. Vertical see-saw nystagmus. Trans Ophthalmol Soc U K (1962). 1964(84):251–7.
30. Eggenberger ER. Mystery case: Pendular see-saw nystagmus as a delayed complication of traumatic brain injury. Neurology. 2015;84(5):547.
31. Frisen L, Wikkelso C. Posttraumatic seesaw nystagmus abolished by ethanol ingestion. Neurology. 1986;36(6):841–4.
32. Eggenberger ER. Delayed-onset seesaw nystagmus posttraumatic brain injury with bitemporal hemianopia. Ann N Y Acad Sci. 2002;956:588–91.
33. Yunusov F, Park JH, Huh YE, Kim HJ, Kim JS. Mystery case: pendular see-saw nystagmus as a delayed complication of traumatic brain injury. Neurology. 2014;82(17):e147–8.
34. Porta-Etessam J, Casanova I, Pajuelo B, Di Capua D, del Val J, Garcia ME, et al. See-saw nystagmus in a patient with Wallenberg syndrome. J Neuro-ophthalmol. 2009;29(1):73–4.

35. Oh K, Chang JH, Park KW, Lee DH, Choi KD, Kim JS. Jerky see-saw nystagmus in isolated internuclear ophthalmoplegia from focal pontine lesion. Neurology. 2005;64(7):1313–4.
36. Rudich DS, Lesser RL. See-saw nystagmus in a patient with septo-optic dysplasia. J Neuro-ophthalmol. 2009;29(3):251–2.
37. Choi KD, Jung DS, Park KP, Jo JW, Kim JS. Bowtie and upbeat nystagmus evolving into hemi-seesaw nystagmus in medial medullary infarction: possible anatomic mechanisms. Neurology. 2004;62(4):663–5.
38. Dell'Osso LF, Jacobs JB, Serra A. The sub-clinical see-saw nystagmus embedded in infantile nystagmus. Vis Res. 2007;47(3):393–401.
39. Leigh JR, Rucker JC. Nystagmus and related ocular motility disorders. In: Miller NR, Newman NJ, editors. Walsh & Hoyt's clinical neuro-ophthalmology. 1. 6th edition ed. 530 Walnut Street, Philadelphia, Pennsylvania 19106 USA; 351 West Camden Street, Baltimore, Maryland 21201–2436 USA: Lippincott Williams & Wilkins; 2005. p. 1133–73.
40. Cochin JP, Hannequin D, Do Marcolino C, Didier T, Augustin P. Intermittent Sea-saw nystagmus successfully treated with clonazepam. Rev Neurol (Paris). 1995;151(1):60–2.

Spontaneous Down Beat Nystagmus

120

Overview

Spontaneous down beat jerk nystagmus (SDBJN) on primary position of gaze nearly always reflects underlying pathology and can occur in a variety of diseases, some of which are reversible, most of which are not. There are several potential mechanisms (discussed below) that can cause SDBJN. The presentation depends on the underlying cause, though in some cases, the nystagmus itself can be experienced as oscillopsia. SDBJN is often evident on face-to-face ocular motor examination; in some cases, it may be enhanced in amplitude and/or frequency at the extremes of lateral gaze. In some cases, the nystagmus is subtle and can be suppressed by visual fixation, but will be discernible on infrared video oculography. Treatment of SDBJN depends on the underlying cause. In cases due to an irreversible lesion, the prognosis is poor, but treatment is sometimes attempted with 4-aminopyridine (dalfampridine) or 3,4-diaminopyridine.

Introduction

Spontaneous down beat jerk nystagmus (SDBJN) is an abnormal eye movement in which on primary position of gaze, the patient's eyes slowly drift up, and then snap back down, and the cycle repeats. If this nystagmus is subtle, it may be suppressed by visual fixation.

Pathophysiological Mechanism of Disease

As with many forms of jerk nystagmus, the pathological eye movement is the slow phase (drift), whereas the fast phase (saccade) is a corrective movement for refixation. This means that in spontaneous down beat jerk nystagmus, the pathological eye movement is actually the upward drift. There are several mechanisms by which this can occur.

The *first mechanism* by which spontaneous upward drift of the eyes can occur is by failure of the inferior rectus to contract (leading to unopposed activity of the superior rectus). As Leigh and Rucker explain:

> Excitatory projections for the vertical vestibulo-ocular reflex from the posterior semicircular canals, which mediate downward eye movements, synapse in the medial vestibular nucleus and then cross dorsally in the medulla beneath the nucleus prepositus hypoglossi to reach the contralateral medial longitudinal fasciculus (MLF). Experimental lesions that damage this pathway cause upward eye drifts and downbeat nystagmus [1].

The reason the "experimental lesions" cause this eye movement is that they interfere with this oligoneuronal circuit:

1. Afferent signal from posterior semicircular canal sends an *excitatory* projection that synapses on a neuron in the medial vestibular nucleus.
2. The neuron in medial vestibular nucleus sends ascending *excitatory* projection that decussates, passes through the contralateral medial longitudinal fasciculus, and synapses in the contralateral oculomotor nucleus.
3. Neuron in contralateral oculomotor nucleus projects to the inferior rectus muscle, causes contraction, and leads to a slow downward eye movement.

Interruption anywhere along this pathway will result in reduced contraction of the inferior rectus and, given the unopposed normal tonic stimulation of the superior rectus, the eye will drift upward; this will be followed by a downward corrective saccade. As this cycle repeats, it will manifest as spontaneous down beat jerk nystagmus.

M. Cherchi, *Otoneurology and Vestibular Medicine*, https://doi.org/10.1007/978-3-031-94841-1_120

The *second mechanism* by which spontaneous upward drift of the eyes can occur is by inappropriate contraction of the superior rectus. As Leigh and Rucker explain:

> On the other hand, it appears that excitatory connections from the anterior semicircular canals, which mediate upward eye movements, take different routes; more than one pathway may contribute… The case for the cerebellar flocculus being an important structure in the production of downbeat nystagmus rests on the finding that Purkinje cells send inhibitory projections to the central connections of the anterior semicircular canal but not to the posterior canal. This asymmetry of inhibitory projections accounts for the finding that experimental flocculectomy causes downbeat nystagmus. This lesion disinhibits the projections to the anterior canal but not to the posterior canal, causing the eyes to drift up and producing downbeat nystagmus [1].

The oligoneuronal circuit in question is this:

1. Neurons in the superior vestibular nucleus receive several inputs, including:
   (a) An excitatory signal from the ipsilateral anterior semicircular canal.
   (b) An *inhibitory* signal from the ipsilateral cerebellar flocculus.
2. The neuron in the superior vestibular nucleus sends an *excitatory* projection through the brachium conjunctivum to synapse on a neuron in the contralateral oculomotor nucleus.
3. The neuron in the oculomotor nucleus sends a projection that decussates (to the eye ipsilateral to the original anterior semicircular canal) and synapses on the ipsilateral superior rectus, leading to a slow upward eye movement.

A lesion involving the cerebellar flocculus will reduce or eliminate signaling from its inhibitory projection to the ipsilateral superior vestibular nucleus, with the downstream result of excessive contraction of the superior rectus that will overcome the normal tonic opposing contraction of the inferior rectus, and result in slow upward drift of the eyes, which will then be followed by a downward corrective saccade. As this cycle repeats, it will manifest as spontaneous down beat jerk nystagmus.

A schematic of these first two neuroanatomical mechanisms is provided by Baloh and Spooner [2].

The *third mechanism* by which spontaneous down beat jerk nystagmus can theoretically occur is with lesions that affect the smooth pursuit system. This mechanism was proposed by Zee and colleagues [3] based on a control systems analysis.

Clinical Presentation

The clinical presentation will depend on the overall disease context. In general, if spontaneous down beat jerk nystagmus is not suppressed by visual fixation, then the patient may experience oscillopsia.

Physical Examination

Findings on the general physical and neurological examination will depend on the underlying etiology of the spontaneous down beat jerk nystagmus.

Ocular Motor Examination

Spontaneous down beat jerk nystagmus may be evident on face-to-face examination. If the nystagmus is mild, then it may be suppressed by visual fixation, and not be evident on face-to-face examination. In some cases, the down beat jerk nystagmus may enhance (in amplitude and/or frequency) at the extremes of lateral gaze. In other cases, the down beat jerk nystagmus may not be spontaneous (i.e., not present on primary position of gaze), but may only appear at the extremes of lateral gaze (in which case it is no longer classified as spontaneous).

Testing: Vestibular

Instrumented oculography, such as infrared videonystagmography, will document spontaneous down beat jerk nystagmus. This nystagmus may augment at the extremes of lateral gaze, as shown in Fig. 120.1 from Leigh and Zee [4].

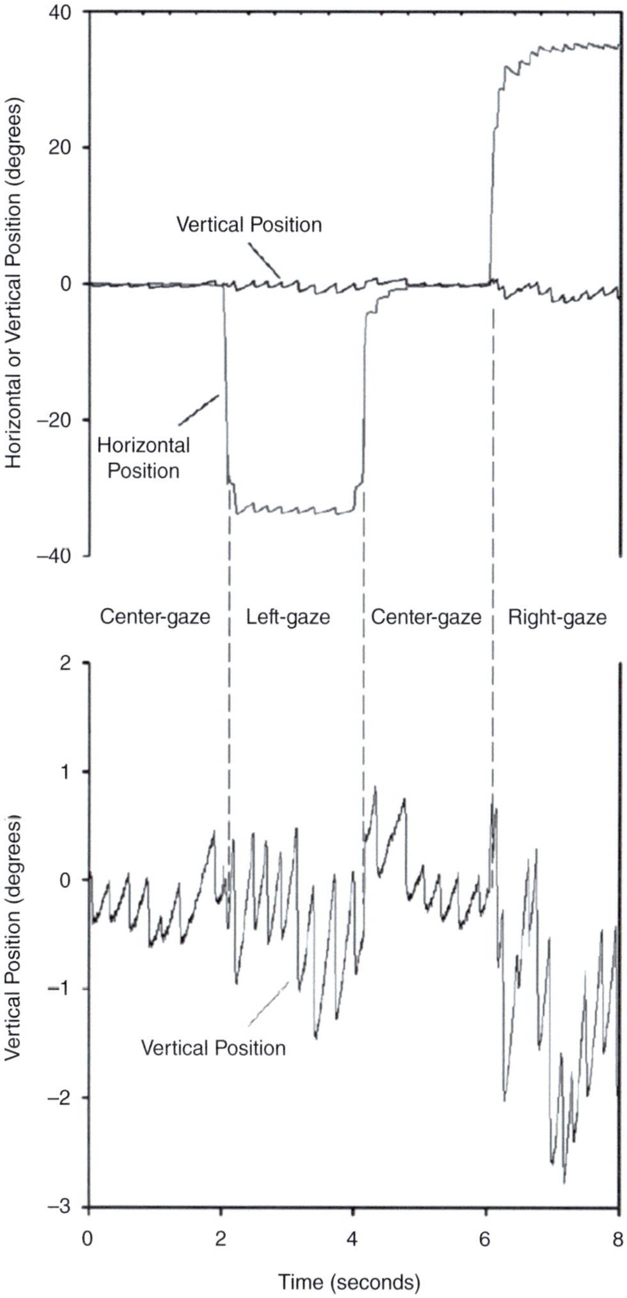

Fig. 120.1 Example of down beat nystagmus that was brought out during lateral gaze, when it was accompanied by a horizontal gaze evoked nystagmus that causes a diagonal (down and lateral) beating nystagmus. The top panel shows horizontal and vertical eye position plotted on the same scale. The bottom panel plots vertical eye position on a more sensitivity scale, and it is evident that the nystagmus increases in either left gaze or right gaze compared with center gaze. Positive values indicate rightward or upward rotations from the viewpoint of the patient. The single-position traces are offset to aid clarity of display. (From Leigh and Zee [4]. Reproduced with written permission from Oxford Publishing Limited)

Table 120.1 Etiologies of spontaneous down beat jerk nystagmus

Cerebellar ataxias, including the following:
Familial episodic ataxia.
Ataxia telangiectasia.
Celiac disease.
Paraneoplastic degeneration.
Autoimmune states (such as antibodies against glutamic acid decarboxylase).
Multiple systems atrophy.
Craniocervical anomalies, including the following:
Chiari malformation.
Paget's disease.
Basilar invagination.
Syringobulbia.
Cerebellar infarction.
Brainstem strokes.
Dolichoectasia of the vertebrobasilar artery or compression of the vertebral artery.
Amyotrophic lateral sclerosis.
Multiple sclerosis.
Encephalitis (such as from HTLV1).
Head trauma.
Heat stroke.
Increased intracranial pressure and hydrocephalus.
Toxic-metabolic etiologies:
Anticonvulsant medications.
Lithium intoxication.
Acute alcohol intoxication.
Cerebellar degeneration associated with alcohol.
Wernicke's encephalopathy.
Magnesium depletion.
Amiodarone.
Opioids.
Deficiency of vitamin B12.
Toluene abuse.
Ciguatera-fish poisoning.
Tetanus.
Mitochondrial encephalopathy with lactic acidosis and stroke-like episodes (MELAS).
Migraine, including vestibular and hemiplegic variants.
Transient manifestation of anterior canal benign paroxysmal positional vertigo.
Transient finding in infants.
Transient finding in children and adults with episodic ataxia type 2.
"Idiopathic" form.

Differential Diagnosis

Spontaneous down beat jerk nystagmus can arise in a variety of disease states. A differential diagnosis of spontaneous down beat jerk nystagmus is listed in Table 120.1, adapted from Leigh and Zee [4].

Treatment

Treatment of spontaneous down beat jerk nystagmus depends on the underlying etiology. For example, if it is due to medication toxicity, then one may consider discontinuing or reducing the offending medication.

For cases of spontaneous down beat jerk nystagmus resulting from irreversible lesions, treatment options are poor. In cases due to cerebellar lesions, Leigh and Rucker offer the following rationale for pharmacologic intervention:

> The cerebellum inhibits vestibular circuits mediating upward, but not downward, eye movements. Consequently, impaired cerebellar inhibition could cause uninhibited upward drifts of the eyes, which evoke corrective rapid downward movements: downbeat nystagmus. Potassium channels are abundant on cerebellar Purkinje cells—the output neurons from cerebellar cortex—and a related agent, 4-aminopyridine, is reported to increase the discharge of these neurons by affecting the slowly depolarizing potential. Enhancement of Purkinje cell activity due to 3,4-diaminopyridine could restore to normal levels the inhibitory influence of the cerebellar cortex upon vertical vestibular eye movements. Studies are under way to determine the long-term effects of these drugs on nystagmus and its visual consequences, and to compare 3,4-diaminopyridine and 4-aminopyridine; the latter penetrates the blood-brain barrier better, has a longer half-life, and is generally better tolerated [1].

The types of studies to which they are alluding include a placebo-controlled study of 3,4-diaminopyridine [5] and a crossover study of 4-aminopyridine and 3,4-diaminopyridine [6].

In the United States, the drug 4-aminopyridine, also called dalfampridine, is FDA approved, whereas 3,4-diaminopyridine is not and must be obtained through compounding pharmacies.

Prognosis

The prognosis of spontaneous down beat jerk nystagmus due to irreversible lesions is poor.

References

1. Leigh JR, Rucker JC. Nystagmus and related ocular motility disorders. In: Miller NR, Newman NJ, editors. Walsh & Hoyt's clinical neuro-ophthalmology. 6th ed. Philadelphia/Baltimore: Lippincott Williams & Wilkins; 2005. p. 1133–73.
2. Baloh RW, Spooner JW. Downbeat nystagmus: a type of central vestibular nystagmus. Neurology. 1981;31(3):304–10.
3. Zee DS, Friendlich AR, Robinson DA. The mechanism of downbeat nystagmus. Arch Neurol. 1974;30(3):227–37.
4. Leigh RJ, Zee DS. The neurology of eye movements. 5th ed. Oxford/New York: Oxford University Press; 2015. xx, 1109 pages p
5. Strupp M, Schuler O, Krafczyk S, Jahn K, Schautzer F, Buttner U, et al. Treatment of downbeat nystagmus with 3,4-diaminopyridine: a placebo-controlled study. Neurology. 2003;61(2):165–70.
6. Polman CH, Bertelsmann FW, de Waal R, van Diemen HA, Uitdehaag BM, van Loenen AC, et al. 4-Aminopyridine is superior to 3,4-diaminopyridine in the treatment of patients with multiple sclerosis. Arch Neurol. 1994;51(11):1136–9.

121 Spontaneous Upbeat Nystagmus

Overview

Spontaneous upbeat jerk nystagmus (SUBJN) on primary position of gaze often reflects pathology, but can also occur in some otherwise apparently normal individuals. The mechanism of SUBJN is unclear and may vary depending on the underlying pathology. SUBJN is sometimes suppressed by visual fixation and thus may not be evident on face-to-face examination, only becoming apparent under infrared video oculography.

Introduction

Spontaneous upbeat jerk nystagmus (SUBJN) is an abnormal eye movement in which, on primary position of gaze, the patient's eyes slowly drift down and then snap back up, and the cycle repeats.

Pathophysiological Mechanism of Disease

The mechanism of spontaneous upbeat nystagmus is unclear and may vary depending on the disease in question. For example, some investigators suspect that the spontaneous upbeat nystagmus found with nicotine is due to an influence on the vestibular ocular reflex [1, 2].

Clinical Presentation

The clinical presentation depends on the underlying disease (see below on differential diagnosis).

Oculomotor Examination

Leigh and Zee [3] make the following general observations about spontaneous upbeat nystagmus:

- Unlike spontaneous downbeat nystagmus, upbeat nystagmus does not usually increase on lateral gaze.
- Upbeat nystagmus that is present in central position usually follows Alexander's law, becoming greatest in up gaze. Sometimes, however, the nystagmus is accentuated on looking down, and then the slow phase is more likely to be increasing-velocity rather than linear.
- Prevention of visual fixation (e.g., by using infrared goggles) may alter the frequency of quick phases but usually does not influence slow-phase velocity.
- Convergence enhances the nystagmus in some patients, suppresses it in others, and converts it to downbeat nystagmus in others.

Testing: Vestibular

Spontaneous upbeat nystagmus may be suppressed by visual fixation and only become apparent when a patient is examined under infrared video oculography.

Differential Diagnosis

Spontaneous upbeat nystagmus is uncommon, and its differential diagnosis is somewhat broad. The following list is adopted from Leigh and Zee [3].

- Infarction of medulla, pontomesencephalic junction, or superior cerebellar peduncle.
- Wernicke's encephalopathy.
- Multiple sclerosis.
- Tumors of the medulla.
- Ataxia telangiectasia.
- Cerebellar degeneration.
- Brainstem encephalitis.
- Creutzfeldt-Jacob disease.
- Cerebellar vermis tumor.

M. Cherchi, *Otoneurology and Vestibular Medicine*, https://doi.org/10.1007/978-3-031-94841-1_121

- Paraneoplastic phenomenon.
- Associated with anti-Ma2 antibodies.
- Meningitis.
- Associated with anti-GAD antibodies, including stiff-person syndrome.
- Leber's congenital amaurosis and other congenital disorders of the anterior visual pathways.
- Tobacco and nicotine.
- Organophosphate poisoning and organoarsenic poisoning.
- Transient finding in infants.
- Can be found in some normal adults.

Our clinical experience is that spontaneous upbeat nystagmus is most commonly encountered in association with nicotine [2, 4, 5].

Treatment and Prognosis

Treatment and prognosis of spontaneous upbeat nystagmus depends on the underlying cause. Glasauer and colleagues report that 4-aminopyridine may improve fixation suppression of spontaneous upbeat nystagmus [6].

References

1. Pereira CB, Strupp M, Eggert T, Straube A, Brandt T. Nicotine-induced nystagmus: three-dimensional analysis and dependence on head position. Neurology. 2000;55(10):1563–6.
2. Pereira CB, Strupp M, Holzleitner T, Brandt T. Smoking and balance: correlation of nicotine-induced nystagmus and postural body sway. Neuroreport. 2001;12(6):1223–6.
3. Leigh RJ, Zee DS. The neurology of eye movements. 5th ed. Oxford/New York: Oxford University Press; 2015. xx, 1109 pages
4. Kim JI, Somers JT, Stahl JS, Bhidayasiri R, Leigh RJ. Vertical nystagmus in normal subjects: effects of head position, nicotine and scopolamine. J Vestib Res. 2000;10(6):291–300.
5. Sibony PA, Evinger C, Manning KA. Tobacco-induced primary-position upbeat nystagmus. Ann Neurol. 1987;21(1):53–8.
6. Glasauer S, Strupp M, Kalla R, Buttner U, Brandt T. Effect of 4-aminopyridine on upbeat and downbeat nystagmus elucidates the mechanism of downbeat nystagmus. Ann N Y Acad Sci. 2005;1039:528–31.

Square Wave Jerks

122

Overview

Square wave jerks (SWJ) are small, involuntary, and (usually) horizontal eye movements (≤0.5 degrees) that take the eyes off target, then after a brief intersaccadic interval, return the eyes to the target. Some investigators suggest that SWJ are "enhanced microsaccades and are closely related to the instability of the fixation eye movements" [1]. The neuroanatomical substrate of SWJ is not well understood, though a control systems analysis proposed a model involving a leaky neural integrator [2].

Shallo-Hoffmann and colleagues [3] report that SWJ are more prominent in darkness and with the eyes closed; they are less prominent when the subject is fixating on a target.

Some investigators report that SWJ are more common in children [4], while other investigators report no clear correlation with age [5].

SWJ can be more pronounced in specific diseases, such as:

- Cerebellar disease [6].
- Cerebral lesions [7].
- Friedreich's ataxia [8].
- Multiple sclerosis with brainstem involvement [9].
- Multiple system atrophy [10].
- Parkinson's disease [11, 12].
- Progressive supranuclear palsy [13, 14].
- Spinocerebellar degenerations [15].
- Tobacco smoking [16, 17].

There are rare reports of *vertical* square wave jerks occurring in neurodegenerative disorders such as Gaucher disease [18], Huntington's chorea [19], and Friedreich's ataxia [20]. The mechanism of vertical square wave jerks is unknown, but experimental studies in monkeys suggest that it can result from lesions in the nucleus reticularis tegmenti pontis (NRTP) [21].

References

1. Ohtsuka K, Mukuno K, Ukai K, Ishikawa S. The origin of square wave jerks: conditions of fixation and microsaccades. Jpn J Ophthalmol. 1986;30(2):209–15.
2. Suzuki Y, Kase M, Hashimoto M, Ohtsuka K. Leaky neural integration observed in square-wave jerks. Jpn J Ophthalmol. 2003;47(6):535–6.
3. Shallo-Hoffmann J, Petersen J, Muhlendyck H. How normal are "normal" square wave jerks? Invest Ophthalmol Vis Sci. 1989;30(5):1009–11.
4. Salman MS, Sharpe JA, Lillakas L, Steinbach MJ. Square wave jerks in children and adolescents. Pediatr Neurol. 2008;38(1):16–9.
5. Shallo-Hoffmann J, Sendler B, Muhlendyck H. Normal square wave jerks in differing age groups. Invest Ophthalmol Vis Sci. 1990;31(8):1649–52.
6. Daroff RB. Ocular oscillations. Ann Otol Rhinol Laryngol. 1977;86(1 Pt 1):102–7.
7. Sharpe JA, Herishanu YO, White OB. Cerebral square wave jerks. Neurology. 1982;32(1):57–62.
8. Dale RT, Kirby AW, Jampel RS. Square wave jerks in Friedreich's ataxia. Am J Ophthalmol. 1978;85(3):400–6.
9. Dell'Osso LF, Troost BT, Daroff RB. Macro square wave jerks. Neurology. 1975;25(10):975–9.
10. Fukazawa T, Tashiro K, Hamada T, Kase M. Multisystem degeneration: drugs and square wave jerks. Neurology. 1986;36(9):1230–3.
11. Rascol O, Sabatini U, Simonetta-Moreau M, Montastruc JL, Rascol A, Clanet M. Square wave jerks in parkinsonian syndromes. J Neurol Neurosurg Psychiatry. 1991;54(7):599–602.
12. White OB, Saint-Cyr JA, Tomlinson RD, Sharpe JA. Ocular motor deficits in Parkinson's disease. II. Control of the saccadic and smooth pursuit systems. Brain. 1983;106(Pt 3):571–87.

M. Cherchi, *Otoneurology and Vestibular Medicine*, https://doi.org/10.1007/978-3-031-94841-1_122

13. Fisk JD, Goodale MA, Burkhart G, Barnett HJ. Progressive supranuclear palsy: the relationship between ocular motor dysfunction and psychological test performance. Neurology. 1982;32(7):698–705.
14. Troost BT, Daroff RB. The ocular motor defects in progressive supranuclear palsy. Ann Neurol. 1977;2(5):397–403.
15. Yamamoto H, Saito S, Sobue I. Bedside and electro-oculographic analysis of abnormal ocular movements in spinocerebellar degenerations: effects of thyrotropin-releasing hormone. Neurology. 1988;38(1):110–4.
16. Sibony PA, Evinger C, Manning KA. The effects of tobacco smoking on smooth pursuit eye movements. Ann Neurol. 1988;23(3):238–41.
17. Thaker GK, Ellsberry R, Moran M, Lahti A, Tamminga C. Tobacco smoking increases square-wave jerks during pursuit eye movements. Biol Psychiatry. 1991;29(1):82–8.
18. Biegstraaten M, van Schaik IN, Aerts JM, Langeveld M, Mannens MM, Bour LJ, et al. A monozygotic twin pair with highly discordant Gaucher phenotypes. Blood Cells Mol Dis. 2011;46(1):39–41.
19. Bollen E, Reulen JP, Den Heyer JC, Van der Kamp W, Roos RA, Buruma OJ. Horizontal and vertical saccadic eye movement abnormalities in Huntington's chorea. J Neurol Sci. 1986;74(1):11–22.
20. Fahey MC, Cremer PD, Aw ST, Millist L, Todd MJ, White OB, et al. Vestibular, saccadic and fixation abnormalities in genetically confirmed Friedreich ataxia. Brain. 2008;131(Pt 4):1035–45.
21. Averbuch-Heller L, Kori AA, Rottach KG, Dell'Osso LF, Remler BF, Leigh RJ. Dysfunction of pontine omnipause neurons causes impaired fixation: macrosaccadic oscillations with a unilateral pontine lesion. Neuroophthalmology. 1996;16(2):99–106.

Staircase Saccades

123

Overview

Staircase saccades comprise a series of small consecutive saccades in a single direction. These have been associated with lesions of the superior colliculi and frontal eye fields and have also been described in some genetically mediated ataxias (ataxia with oculomotor apraxia types 1 and 2), patients with hemispatial neglect, Parkinson's disease, and Whipple's disease.

Introduction

The phrase "*staircase saccades*" refers to a series of small consecutive saccades in a single direction. The phrase describes the appearance of this saccadic pattern on oculographic tracings, as shown in Tilikete [1].

Genetics

Staircase saccades in themselves are not genetically mediated, though there are several genetic ataxias in which staircase saccades occur.

Staircase saccades can be found in ataxia with oculomotor apraxia type 1 (OMIM 208920), an autosomal recessive cerebellar ataxia with hypoalbuminemia resulting from a mutation in the APTX gene on chromosome 9p21.1 which encodes aprataxin (OMIM 606350) [2].

Staircase saccades can be found in ataxia with oculomotor apraxia type 2 (OMIM 606002), an autosomal recessive spinocerebellar ataxia with axonal neuropathy resulting from a mutation in the SETX gene on chromosome 9q34.13 which encodes senataxin (OMIM 608465) [3].

Pathophysiological Mechanism of Disease

Studies in monkeys have shown that staircase saccades can be generated by stimulation in a specific layer of the *superior colliculus* [4, 5]. Studies in monkeys have shown that staircase saccades can be generated by stimulation of the *frontal eye field* [6], and studies in humans have shown that staircase saccades can result from lesions in the frontal eye fields [7].

A control systems analysis model proposed by Dell'Osso [8] postulates that the generation of staircase saccades requires two discrete types of dysfunction: (1) [1] transient loss of retinal position information and (2) a transient change in sampled, reconstructed retinal error. A study in healthy humans emulating these conditions was able to elicit staircase saccades [9]. Other proposed models are reviewed by Gancarz and Grossberg [10].

Ocular Motor Examination

Staircase saccades are sometimes visible on face-to-face examination.

Testing: Vestibular

On tracings of instrumented infrared video oculography, staircase saccades exhibit the appearance described earlier.

M. Cherchi, *Otoneurology and Vestibular Medicine*, https://doi.org/10.1007/978-3-031-94841-1_123

Differential Diagnosis

Given the pathophysiological mechanisms described earlier, the finding of staircase saccades should raise suspicion for a lesion in the superior colliculi and/or frontal eye fields, though the differential diagnosis is somewhat broader.

- Lesions of the superior colliculi.
- Lesions in the frontal eye fields [7].
- Shaikh and colleagues [11] reported staircase saccades in Parkinson's disease.
- Girotti and colleagues [12] report that in patients with unilateral neglect, saccades in the neglected hemifield either exhibited lengthened reaction time or a staircase saccadic pattern.
- Shaikh and Ghasia [13] reported staircase saccades in a patient with Whipple's disease.
- Ataxia with oculomotor apraxia type 1 [2].
- Ataxia with oculomotor apraxia type 2 [3].

Treatment and Prognosis

There is no known treatment for staircase saccades resulting from static lesions (e.g., strokes or demyelination in the superior colliculi or frontal eye fields) or genetically mediated ataxias. In patients with Parkinson's disease or Whipple's disease exhibiting staircase saccades, it is unknown whether treatment for those diseases alters the course of this ocular motor finding.

References

1. Froment Tilikete C. How to assess eye movements clinically. Neurol Sci. 2022;43(5):2969–81.
2. Le Ber I, Moreira MC, Rivaud-Pechoux S, Chamayou C, Ochsner F, Kuntzer T, et al. Cerebellar ataxia with oculomotor apraxia type 1: clinical and genetic studies. Brain. 2003;126(Pt 12):2761–72.
3. Panouillères M, Frismand S, Sillan O, Urquizar C, Vighetto A, Pélisson D, Tilikete C. Saccades and eye-head coordination in ataxia with oculomotor apraxia type 2. Cerebellum. 2013;12(4):557–67.
4. Schiller PH, Stryker M. Single-unit recording and stimulation in superior colliculus of the alert rhesus monkey. J Neurophysiol. 1972;35(6):915–24.
5. Paul K, Gnadt JW. Activity of omnipause neurons during "staircase saccades" elicited by persistent microstimulation of the superior colliculus. Vis Res. 2006;46(20):3430–42.
6. Robinson DA, Fuchs AF. Eye movements evoked by stimulation of frontal eye fields. J Neurophysiol. 1969;32(5):637–48.
7. Rivaud S, Müri RM, Gaymard B, Vermersch AI, Pierrot-Deseilligny C. Eye movement disorders after frontal eye field lesions in humans. Exp Brain Res. 1994;102(1):110–20.
8. Dell'Osso LF. Ocular motor system control models and the cerebellum: hypothetical mechanisms. Cerebellum. 2019;18(3):605–14.
9. Rucker JC, Dell'Osso LF, Garbutt S, Jacobs JB. "Staircase" saccadic intrusions plus transient yoking and neural integrator failure associated with cerebellar hypoplasia: a model simulation. Semin Ophthalmol. 2006;21(4):229–43.
10. Gancarz G, Grossberg S. A neural model of the saccade generator in the reticular formation. Neural Netw. 1998;11(7–8):1159–74.
11. Shaikh AG, Xu-Wilson M, Grill S, Zee DS. 'Staircase' square-wave jerks in early Parkinson's disease. Br J Ophthalmol. 2011;95(5):705–9.
12. Girotti F, Casazza M, Musicco M, Avanzini G. Oculomotor disorders in cortical lesions in man: the role of unilateral neglect. Neuropsychologia. 1983;21(5):543–53.
13. Shaikh AG, Ghasia FF. Novel eye movement disorders in Whipple's disease-staircase horizontal saccades, gaze-evoked nystagmus, and Esotropia. Front Neurol. 2017;8:321.

Windmill Nystagmus

124

Overview

Windmill nystagmus is rare. This is an acquired disorder of ocular motility in which there is spontaneous nystagmus whose direction rotates. Windmill nystagmus usually results from significant bilateral visual loss that has usually been gradually progressive (e.g., macular degeneration, retinitis pigmentosa), though it can also be subacute. One case of paraneoplastic windmill nystagmus has been reported in which the patient's vision was normal and the nystagmus developed over 3 weeks. When windmill nystagmus occurs in the context of visual loss, treatment involves managing the visual loss, and the causes of visual loss usually have a poor prognosis. When windmill nystagmus occurs as a paraneoplastic syndrome, the treatment and prognosis depend on the underlying malignancy.

Introduction

Windmill nystagmus is an abnormal eye movement in which there is spontaneous nystagmus, of which the fast phase direction varies in its horizontal and vertical components, usually in a regular pattern. The vector of the fast phase rotates, and the direction of the rotation (clockwise or counterclockwise) depends on the amplitude and phase relationship between the horizontal and vertical components.

Epidemiology

Windmill nystagmus is rare.

Most described cases have been reported in patients with significant visual limitations. Usually the visual loss is the result of slowly progressive degenerative processes (e.g., retinitis pigmentosa, macular degeneration, etc.) [1], though it has also been reported as soon as 3 months after subacute visual loss [2].

Lee and colleagues [3] described windmill nystagmus occurring in paraneoplastic cerebellar degeneration in a patient with small cell lung carcinoma and positive antibodies for anti-SOX1 and anti-Hu. In this case, the patient complained of unsteadiness that developed over 3 weeks.

Pathophysiological Mechanism of Disease

The underlying pathophysiological mechanism of windmill nystagmus is unknown. Choi and colleagues emphasize the relevance of low vision:

> The underlying mechanisms [of windmill nystagmus] remain uncertain. An attractive hypothesis is that a malfunction within the visual-motor 'adaptation' pathways, and specifically in the mechanisms that normally sense image motion on the retina, leads to the unwanted oscillations. Without such [visual] information, central gaze-holding networks, which are normally finely tuned to the exact 'gains' needed to hold gaze stable,

M. Cherchi, *Otoneurology and Vestibular Medicine*, https://doi.org/10.1007/978-3-031-94841-1_124

> become susceptible to the 'noise' and drift that is inherent in all neural networks [1].

Choi and colleagues [1] comment on the relationship between windmill nystagmus and (horizontal) period alternating nystagmus:

> Windmill nystagmus is characterized by a clock-like rotation of the beating direction. It resembles periodic alternating nystagmus (PAN) , which is restricted to the horizontal plane, in that the direction of nystagmus changes in a predictable pattern. In contrast to PAN, however, windmill nystagmus alternates direction in both the horizontal and vertical planes. Thus, windmill nystagmus has been considered a periodic alternating horizontal nystagmus with a superimposed periodic alternating vertical nystagmus, with the periods of the two oscillations being 90 degrees out of phase [1].

They also comment on the relationship between windmill nystagmus and elliptical pendular nystagmus:

> Windmill (jerk) nystagmus is in some respects analogous to circular or elliptical (pendular) nystagmus, though the period of the oscillation of pendular nystagmus is much lower than that of windmill nystagmus. Both represent a summation of simultaneous horizontal and vertical oscillations. The trajectory of the eyes is determined by the amplitude and phase relationships of the horizontal and vertical components. When the two pendular oscillations are either in phase or 180 degrees out of phase, the nystagmus is diagonal. In contrast, the nystagmus is circular when the two components are equal and 90 degrees out of phase, and elliptical when the amplitudes are unequal but still 90 degrees out of phase [1].

Clinical Presentation

Most patients present with gradually progressive visual deterioration that has resulted in near blindness.

Physical Examination

The most common underlying cause of windmill nystagmus is severe, gradually progressive visual loss, which should be apparent on physical examination.

In the case of windmill nystagmus occurring as a paraneoplastic syndrome [3], the authors reported that the patient was unsteady to the point of being unable to walk unaided.

Ocular Motor Examination

Lee and colleagues [3] report a case with an accompanying video that demonstrates windmill nystagmus well.

Testing: Vestibular

Instrumented oculography is the best way to capture the windmill pattern of nystagmus.

Figure 124.1 below, from Choi and colleagues [1], shows plots of the horizontal and vertical components of windmill nystagmus, as well as tracings of the nystagmus.

Figure 124.2 below, from Lee and colleagues [2], shows plots of the horizontal and vertical components of windmill nystagmus, as well as tracings of the nystagmus.

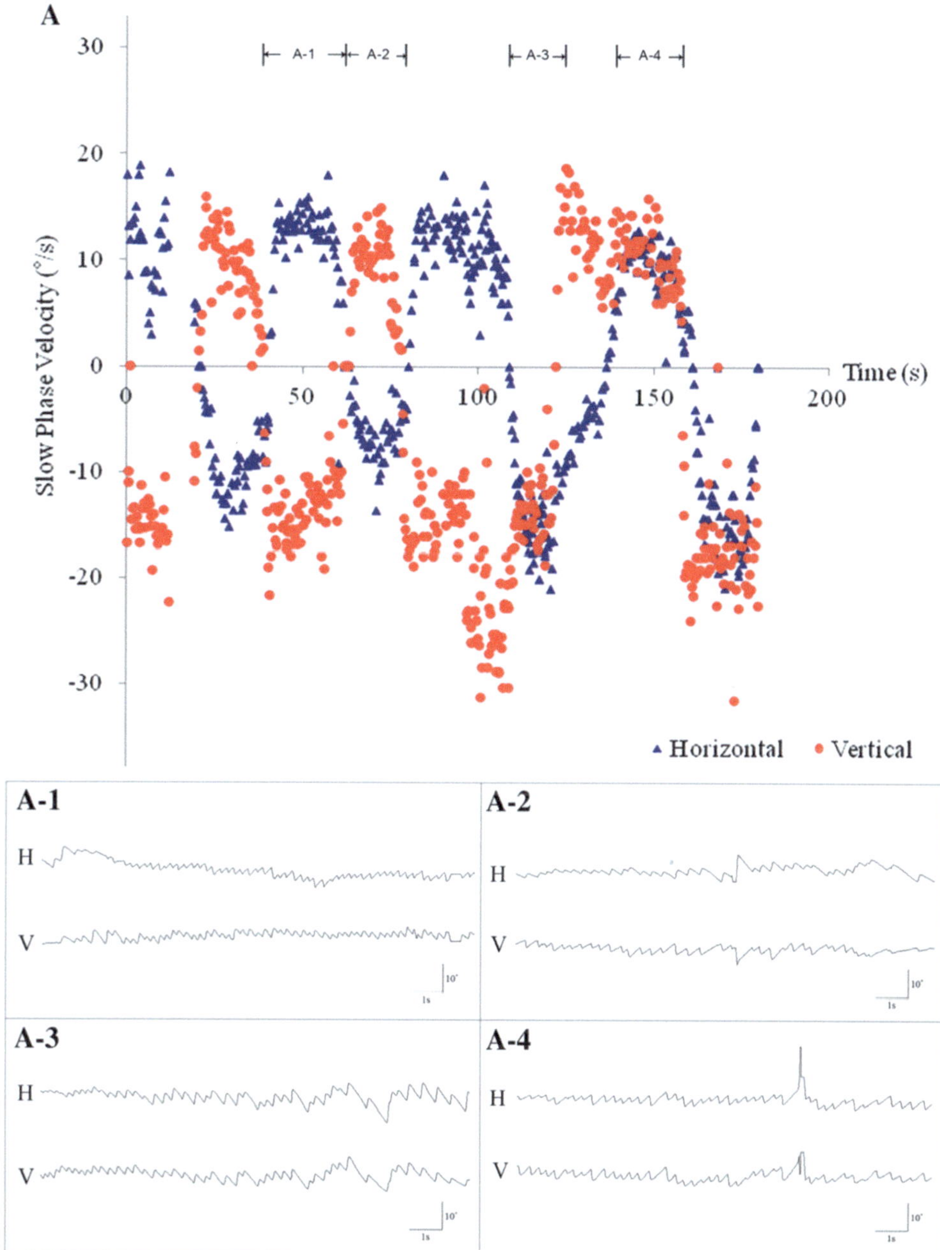

Fig. 124.1 Windmill nystagmus. Panel A: plot of the slow-phase velocities of a patient with windmill nystagmus. The top panel shows a plot of the alternating direction and amplitude of the slow-phase velocity of horizontal (blue triangles) and vertical (red circles) components of nystagmus. The nystagmus is mostly diagonal and reverses its direction before evolving into the other diagonal plane. A complete cycle of reversing diagonal nystagmus in both oblique planes lasts about 270 s. Panels A1-A4: recordings of the nystagmus show left-upbeating (A-1), right-downbeating (A-2), right-upbeating (A-3), and then left-downbeating nystagmus (A-4). Upward deflection indicates rightward and upward eye motion. (From Choi et al. [1]. Reproduced with written permission from Springer Nature B.V)

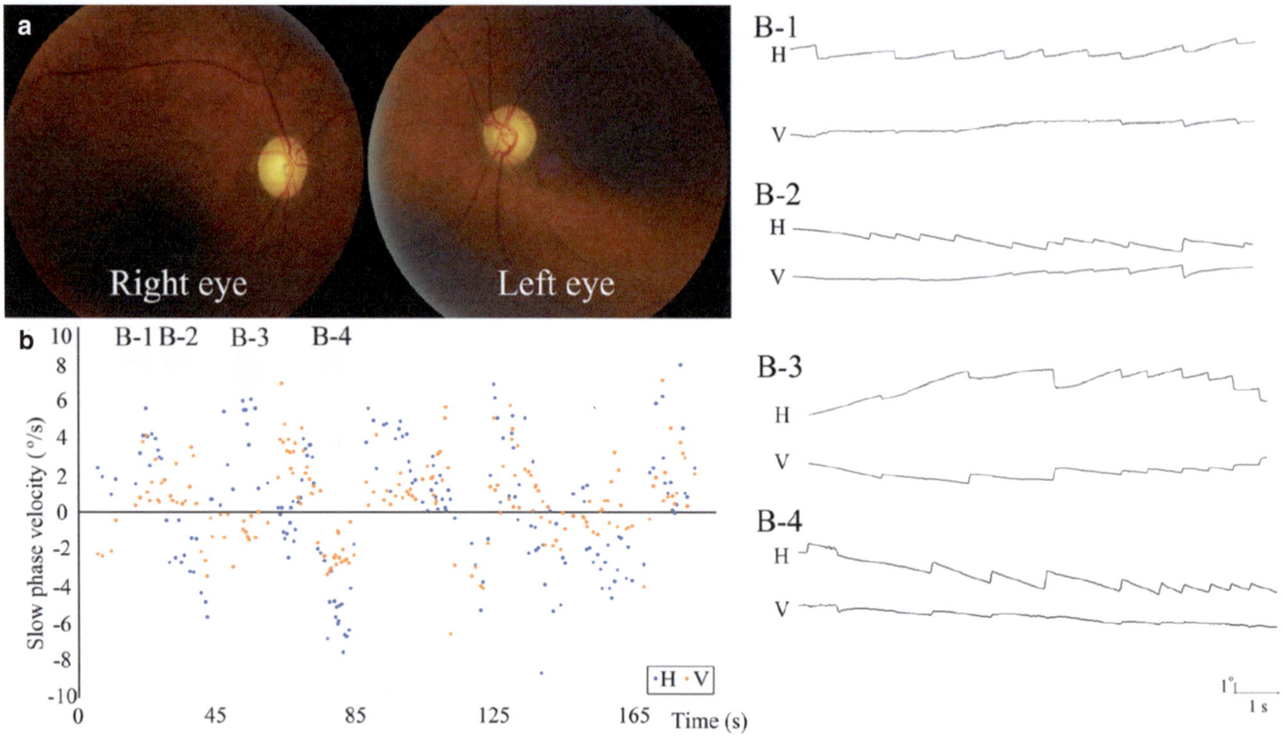

Fig. 124.2 Windmill nystagmus. The left panel shows a plot of the alternating direction and amplitude of the slow-phase velocity of horizontal (blue dots) and vertical (red dots) components of nystagmus. The right panels show tracings of the nystagmus. (From Lee et al. [2]. Reproduced with written permission from Springer Nature B.V)

Differential Diagnosis

Most cases of windmill nystagmus appear in patients with gradual but relentlessly progressive bilateral visual loss (e.g., macular degeneration, retinitis pigmentosa). In such cases, an ophthalmologist should be involved in the patient's care.

Windmill nystagmus of subacute onset (weeks) has also been reported as a paraneoplastic syndrome in a patient with normal vision. Therefore, if windmill nystagmus develops subacutely in a patient with otherwise normal vision, a neoplastic workup should be undertaken.

Treatment

When windmill nystagmus is the result of visual loss, management is the treatment (where possible) of the visual loss.

When windmill nystagmus occurs as a paraneoplastic syndrome, management is the treatment of the underlying cancer.

Prognosis

In cases where windmill nystagmus is the result of visual loss, the prognosis of the visual loss itself is poor because of the underlying causes (macular degeneration, retinitis pigmentosa).

In cases where windmill nystagmus is the result of a paraneoplastic syndrome, the prognosis depends on the underlying malignancy. In the case reported by Lee and colleagues [3], the windmill nystagmus improved after chemotherapy and radiotherapy.

References

1. Choi KD, Shin HK, Kim JS, Kim SH, Choi JH, Kim HJ, et al. Variants of windmill nystagmus. J Neurol. 2016;263(7):1375–81.
2. Lee SU, Jo SY, Kim HJ, Choi JY, Kim JS. Windmill nystagmus in a patient with subacute visual loss. J Neurol. 2018;265(11):2737–9.
3. Lee SU, Kim HJ, Oh SW, Song EY, Choi JY, Kim JS. Pearls & Oy-sters: windmill nystagmus in paraneoplastic cerebellar degeneration. Neurology. 2018;91(19):e1831–e3.

Part XVII

Diagnoses—Eye Movements Elicited by Specific Provocative Maneuvers

Head-Shaking Nystagmus

125

Overview

Head-shaking nystagmus (HSN) was first described in 1907 by Robert Bárány [1]. Subsequent studies [2–5] described different techniques of eliciting the nystagmus and proposed various hypotheses regarding its pathophysiology [6–11].

Head-shaking nystagmus (HSN) may follow a prolonged sinusoidal head oscillation. A common technique for eliciting HSN is as follows: The patient is seated in an upright position and instrumented so that fixation is removed but horizontal and vertical eye movements can be observed; this is best performed using ENG, VNG, or VFO. Optical Frenzel goggles can be used, but are probably less sensitive because they allow some vision. Performing the test without eliminating fixation is not diagnostic.

Eye movements are observed in darkness for 10 seconds to obtain a baseline. Next, the examiner grasps the patient's head firmly and moves it briskly back and forth in the yaw plane (around the vertical axis), aiming for a frequency of about 2 Hz and a displacement of the head of approximately 30 degrees to either side. If possible, the head should be tilted approximately 20 degrees anteriorly (with respect to vertical) so that the axis of rotation is close to being parallel to the axes of the lateral semicircular canals, though small deviations from this are probably not clinically significant. The headshaking is continued for 20 cycles and then abruptly stopped.

In normal subjects, or patients with symmetrical vestibular loss (such as bilateral vestibular loss), no nystagmus is expected; one study [12] found that not a single individual out of 60 healthy subjects exhibited HSN; in contrast, other studies report HSN in healthy controls ranging from 9.6% (7 out of 73) [13] of individuals to 14.3% (4 out of 28) [14] of individuals. In persons with a dynamic imbalance between the ears (such as due to unilateral vestibular neuritis or an acoustic neuroma), a nystagmus is often seen (usually with the fast phase beating toward the "better" ear [8, 15]) which decays over about 30 seconds. This is often referred to as the first phase of nystagmus, because in some cases, it is followed by a second phase of nystagmus that is weaker, decays more slowly, and is in the opposite direction (i.e., with the fast phase directed toward the "bad" ear). Rarely, horizontal headshaking produces a nonhorizontal nystagmus, such as vertical nystagmus [16] or torsional nystagmus [17]; this is called a "perverted" head-shaking nystagmus.

Variations in the technique of the headshaking test mainly involve the use of a different axis of rotation. These include oscillation of the head vertically, about the naso-occipital axis (by alternating lateral flexion of the neck), or using a combination of horizontal and vertical movement so that the nose traces out a circular trajectory [18]. Some studies have also explored headshaking with the body in positions other than upright [19, 20], but these techniques are not in wide clinical use. Generally, headshaking using these other axes of rotation or other body positions does not have a well-established clinical indication.

Hain and Spindler [6] review several possible mechanisms of HSN and point out that it likely involves more than one mechanism since HSN can have central or peripheral etiologies.

The first hypothesis postulates the existence of a tone asymmetry that predisposes to an underlying potential for nystagmus that is usually absent due to compensatory mechanisms, but becomes manifest when the compensatory mechanisms are disrupted (such as by headshaking). (We avoid here the term "latent nystagmus," which refers to a specific pathological process.)

The second hypothesis postulates asymmetries in vestibular gain due to a variety of factors, such as loss of vestibular hair cells, lesions of the vestibular nerve or its root entry zone, asymmetrical central gain, or asymmetrical cervical afferent input. Asymmetry in the peripheral vestibular input leads to nystagmus by virtue of Ewald's second law, which in its general form states that excitation is a relatively better vestibular stimulus than is inhibition [21], and which in its specific form states that ampullopetal endolymph flow in the horizontal canal causes a greater response than ampullofugal endolymph flow [22, 23]. Ewald's second law is thought to

M. Cherchi, *Otoneurology and Vestibular Medicine*, https://doi.org/10.1007/978-3-031-94841-1_125

be due to the inability of inhibitory stimuli to decrease vestibular nerve firing rates to less than zero [6, 24].

The third hypothesis postulates asymmetries in timing. Hain and Spindler [6] conceptualized this class of mechanisms as deriving from errors in central head movement response storage. Storage mechanisms (often termed "leaky integrators") play a role in many central nervous system circuits by providing low-pass filtering and averaging. Responses are stored in three places in the vestibular system (the semicircular canals, the central velocity storage mechanism, and adaptation circuitry); consequently, there are three corresponding potential sources of asymmetry in timing. First, response storage asymmetry in the semicircular canals could result, for example, from asymmetrical endolymph viscosity between the ears, leading to a timing mismatch and filtering across the two ears. Second, there can be asymmetry in the central velocity storage mechanism. The central velocity storage mechanism prolongs the raw vestibular signal [21]; its neuroanatomical substrate is thought to include the medial vestibular nuclei of Schwalbe and their commissural connections [21, 22]. Third, there can be asymmetry in the compensatory changes involved in central and/or peripheral adaptation [21].

The fourth hypothesis postulates simple mechanical causes, such as the improper technique of eliciting HSN. When head movement is not confined to a single plane, such as in a "circular trajectory" [18], there is an effective rotation about an axis of the head that induces a per-rotatory or post-rotatory nystagmus. Thus, one must be careful to maintain a constant and consistent axis when doing headshaking to avoid extraneous nystagmus.

Since there are several possible mechanisms for HSN, it probably has a relatively high sensitivity and low specificity.

There are few studies on the normative values of incidence of HSN in healthy individuals, and the available studies describe a rather broad range. Li [12] states that of 60 healthy individuals, none had head-shaking nystagmus. Vicini et al. [13] studied 73 normal subjects and observed HSN in 7 (9.6%). Levo and colleagues [25] studied 20 healthy subjects using videonystagmography; they reported that head-shaking nystagmus was observed in 35% of individuals.

References

1. Bárány R. Untersuchungen über das Verhalten des Vestibularapparates bei Kopftraumen und ihre praktische Bedeutng. Verhandungen der Deutschen otol Gessellschaft. 1907:252–6.
2. Borries GV. Klinische Untersuchungen über die durch Kopfbewegungen und Kopfstellungen ausgelösten Nystagmusantälle. Monatschr Ohrenheilk. 1923;57:644–83.
3. Klestadt W. In: Bumke O, Foerster O, editors. Handbuch der Neurologie. Berlin: J. Springer Verlag; 1936.
4. Moritz W. Auswertungen des Kopfschüttel-nystagmus. Arch Otolrhinolaryngol. 1951;209:59–67.
5. Vogel K. Differentialdiagnostische Anhaltspunkte für die Erkennung von Schädigungen des Gleichgewichsapparates nach Schädelverletzungen. Deutsche Med Wochenschr. 1929;7:268–70.
6. Hain TC, Spindler J. Head-shaking nystagmus. In: Sharpe JA, Barber HO, editors. The vestibulo-ocular reflex and vertigo. New York: Raven Press; 1993. p. 217–28.
7. Perez P, Llorente JL, Gomez JR, Del Campo A, Lopez A, Suarez C. Functional significance of peripheral head-shaking nystagmus. Laryngoscope. 2004;114(6):1078–84.
8. Katsarkas A, Smith H, Galiana H. Head-shaking nystagmus (HSN): the theoretical explanation and the experimental proof. Acta Otolaryngol. 2000;120(2):177–81.
9. Minagar A, Sheremata WA, Tusa RJ. Perverted head-shaking nystagmus: a possible mechanism. Neurology. 2001;57(5):887–9.
10. Halmagyi GM, Curthoys IS. A clinical sign of canal paresis. Arch Neurol. 1988;45(7):737–9.
11. Fetter M, Zee DS, Koenig E, Dichgans J. Head-shaking nystagmus during vestibular compensation in humans and rhesus monkeys. Acta Otolaryngol. 1990;110(3–4):175–81.
12. Li Z. Head shaking nystagmus. Zhonghua Er Bi Yan Hou Ke Za Zhi. 1991;26(6):356–7. 82–3
13. Vicini C, Casani A, Ghilardi P. Assessment of head shaking test in neuro-otological practice. ORL J Otorhinolaryngol Relat Spec. 1989;51(1):8–13.
14. Choi KD, Oh SY, Park SH, Kim JH, Koo JW, Kim JS. Head-shaking nystagmus in lateral medullary infarction: patterns and possible mechanisms. Neurology. 2007;68(17):1337–44.
15. Hain TC, Fetter M, Zee DS. Head-shaking nystagmus in patients with unilateral peripheral vestibular lesions. Am J Otolaryngol. 1987;8(1):36–47.
16. Wu ZM, Zhang SZ, Zhou N, Xiang SJ, Yang WY, Han DY. Cross-check of caloric test and head shaking nystagmus. Zhonghua Er Bi Yan Hou Tou Jing Wai Ke Za Zhi. 2005;40(8):575–8.
17. Califano L, Capparuccia PG, Di Maria D, Melillo MG, Villari G. Sign-report: torsional nystagmus from head-shaking in patients with labyrinth lithiasis of the vertical canal. Acta Otorhinolaryngol Ital. 2001;21(6):337–40.
18. Haslwanter T, Minor LB. Nystagmus induced by circular head shaking in normal human subjects. Exp Brain Res. 1999;124(1):25–32.
19. Palla A, Marti S, Straumann D. Head-shaking nystagmus depends on gravity. J Assoc Res Otolaryngol. 2005;6(1):1–8.
20. Kamei T, Iizuka T, Matsuzaki M. Influence of lateral head tilting on head-shaking nystagmus. Eur Arch Otorrinolaringol. 1997;254(Suppl 1):S58–60.
21. Leigh RJ, Zee DS. The neurology of eye movements. 4th ed. Oxford University Press; 2006.
22. Baloh RW, Honrubia V. Clinical neurophysiology of the vestibular system. 3rd ed. Oxford University Press; 2001.
23. Ewald JR. Physiologische Untersuchungen über das Endorgan des Nervus octavus. Wiesbaden: Bergmann; 1892.
24. Baloh RW, Honrubia V, Konrad HR. Ewald's second law re-evaluated. Acta Otolaryngol. 1977;83(5–6):475–9.
25. Levo H, Aalto H, Petteri Hirvonen T. Nystagmus measured with video-oculography: methodological aspects and normative data. ORL J Otorhinolaryngol Relat Spec. 2004;66(3):101–4.

Hyperventilation-Induced Nystagmus

126

Overview

In 1970, Tsutsui and Fukai [1] described the effect of hyperventilation on spontaneous nystagmus. In 1972, Drachman and Hart [2] documented "a syndrome, not previously described, of positional vertigo occurring only after HV [hyperventilation]."

Hyperventilation by rapid, large breaths through the mouth can induce nystagmus, termed *hyperventilation-induced nystagmus (HVIN)*. Instrumentation that eliminates fixation (ENG, VNG, VFO) is sensitive for HVIN. Methods that limit but do not eliminate fixation (optical Fresnel lenses) are less sensitive.

Most clinicians have the patient hyperventilate for 30 seconds, though some recommend 90 seconds [3–5] or longer [6]. The nystagmus induced by hyperventilation is generally weak. The fast phase of HVIN is usually directed contraversive to (away from) the diseased side—suggesting that the lesion is paretic (inhibitory); this behavior is rarely useful. However, in some disease processes, the fast phase is directed ipsiversive (toward) the diseased side—suggesting that the lesion is irritative (excitatory) [7, 8]; in this circumstance, it can be directed oppositely to the nystagmus associated with the head-shaking test or vibration test. HVIN is usually horizontal, less commonly vertical, and rarely torsional [9]. It can be delayed and follow hyperventilation rather than appearing during the hyperventilation itself.

Several possible mechanisms [10] have been proposed for hyperventilation-induced nystagmus. Hyperventilation primarily affects blood oxygenation and pH, but has other metabolic effects as well. Hyperventilating alert patients during testing can increase spontaneous nystagmus and may also affect central compensatory mechanisms [11]. Ideally, a hypothesis regarding the underlying mechanism of HVIN would account for why its fast phase is sometimes directed toward the lesioned side, while in other instances is directed away from the lesioned side.

The most plausible explanation for hyperventilation-induced nystagmus is the direct effect of hyperventilation, which induces hypocapnia, raising the serum pH, and thereby improving neural conduction [7, 12]. Transient improvement of vestibular function on the lesioned side (physiologically, probably an increase in the tonic firing rate along the vestibular nerve) would resemble rotating the head toward the lesioned side, resulting in the slow phase of nystagmus directed away from the lesion and the fast phase toward the lesion; in other words, this mechanism would account for HVIN whose *fast phase is directed ipsiversive to a subtotal paretic lesion,* and thus explain the pattern observed in diseases that limit (but do not completely destroy) the function of the vestibular nerve (e.g., subtotal vestibular neuritis, a small vestibular schwannoma). Minor and colleagues [7] performed electronystagmography on six patients with vestibular schwannomas and concluded that the vector of hyperventilation-induced nystagmus corresponds to the plane of the semicircular canal whose partially demyelinated afferent fibers exhibit a transient increase in activity—in other words, the fast phase of the nystagmus was directed ipsiversive to the lesion. Hyperventilation can also induce a nystagmus whose *fast phase is directed contraversive to the side of a complete vestibular lesion* [11]. This can be seen, for example, in persons who are postlabyrinthectomy or vestibular neurectomy, and who thus have no remaining peripheral vestibular system on that side; thus, there is no opportunity for that residual vestibular function to "improve" with hyperventilation. Other potential explanations for this pattern include augmentation of vestibular nystagmus through alerting or other mechanisms [13, 14] and disruption of central compensation [11].

References

1. Tsutsui J, Fukai S. Effect of hyperventilation and apnea on spontaneous nystagmus. Am J Ophthalmol. 1970;69(1):130–2.
2. Drachman DA, Hart CW. An approach to the dizzy patient. Neurology. 1972;22(4):323–34.

M. Cherchi, *Otoneurology and Vestibular Medicine*, https://doi.org/10.1007/978-3-031-94841-1_126

3. Bance ML, O'Driscoll M, Patel N, Ramsden RT. Vestibular disease unmasked by hyperventilation. Laryngoscope 1998;108(4, part 1):610-4.
4. Baloh RW, Honrubia V. Clinical neurophysiology of the vestibular system. 3rd ed. Oxford University Press; 2001.
5. Robichaud J, DesRoches H, Bance ML. Is hyperventilation-induced nystagmus more common in retrocochlear vestibular disease than in end-organ vestibular disease? J Otolaryngol. 2002;31(3):140–3.
6. Wilson WR, Kim JW. Study of ventilation testing with electronystagmography. Ann Otol Rhinol Laryngol. 1981;90(1, part 1):56–9.
7. Minor LB, Haslwanter T, Straumann D, Zee DS. Hyperventilation-induced nystagmus in patients with vestibular schwannoma. Neurology. 1999;53(9):2158–68.
8. Choi KD, Cho HJ, Koo JW, Park SH, Kim JS. Hyperventilation-induced nystagmus in vestibular schwannoma. Neurology. 2005;64(12):2062.
9. Rosignoli M, Almadori G, Maurizi M. Effect of voluntary hyperventilation on rotatory nystagmus. Acta Otorhinolaryngol Ital. 1984;4(6):667–77.
10. Cherchi M, Hain TC. Provocative maneuvers for vestibular disorders. In: Eggers S, Zee D, editors. Vertigo and imbalance: clinical neurophysiology of the vestibular system. Handbook of clinical neurophysiology. Amsterdam: Elsevier; 2009. p. 111–34.
11. Sakellari V, Bronstein AM, Corna S, Hammon CA, Jones S, Wolsley CJ. The effects of hyperventilation on postural control mechanisms. Brain. 1997;120(Pt 9):1659–73.
12. Somjen GG, Allen BW, Balestrino M, Aitken PG. Pathophysiology of pH and Ca2+ in bloodstream and brain. Can J Physiol Pharmacol. 1987;65(5):1078–85.
13. Monday LA, Tetreault L. Hyperventilation and vertigo. Laryngoscope. 1980;90(6 Pt 1):1003–10.
14. Theunissen EJ, Huygen PL, Folgering HT. Vestibular hyperreactivity and hyperventilation. Clin Otolaryngol Allied Sci. 1986;11(3):161–9.

Valsalva-Induced Nystagmus

127

Overview

The Valsalva maneuver was first described by Antonio Maria Valsalva in 1704 [1]. The maneuver originally described by Valsalva was a method of inflating the middle ear by forced nasal expiration with the nose and lips closed. For this technique, there is increased intrathoracic pressure as well as increased upper airway pressure. Subsequently, the method of doing the "Valsalva maneuver" was modified so that it was performed against a closed glottis; this technique avoids changing middle ear pressure (since the pressure differential ends at the glottis and does not get transmitted to the pharyngeal orifice of the Eustachian tube) but retains the effect on intrathoracic pressure. Here we will be discussing the form with the glottis closed, but the reader should be aware that the term "Valsalva maneuver" is used for two distinct techniques.

Performing the Valsalva maneuver against a closed glottis results in a nearly simultaneous increase in intrathoracic pressure, which is transmitted intraspinally and thence intracranially [2, 3]. In patients with superior semicircular canal dehiscence (SSCD) [4], the increased intracranial pressure can be transmitted from there through the dehiscence into the superior semicircular canal and (since the labyrinth is a continuous space) the remaining vestibular and cochlear structures. That pressure differential can stimulate the superior semicircular canal and provoke nystagmus in the plane of that canal [5]. Sometimes, the naturally occurring intracranial pressure oscillations resulting from the cardiac pulse are sufficient to be transmitted to the superior semicircular canal and provoke nystagmus (even without performing the Valsalva maneuver) [6, 7].

The Valsalva maneuver normally provokes no nystagmus in healthy individuals, thus there is no normative range associated with this examination of the ocular motor examination.

References

1. Valsalva AM. De Aure Humana Tractus. Bologna 1704.
2. Hamilton WF, Woodbury RA, Harper HT Jr. Physiologic relationships between intrathoracic, intraspinal and arterial pressures. JAMA J Am Med Assoc. 1936;107(11):853–6.
3. Williams B. Simultaneous cerebral and spinal fluid pressure recordings. Acta Neurochir. 1981;58(3–4):167–85.
4. Minor LB, Solomon D, Zinreich JS, Zee DS. Sound- and/or pressure-induced vertigo due to bone dehiscence of the superior semicircular canal. Arch Otolaryngol Head Neck Surg. 1998;124(3):249–58.
5. Cremer PD, Minor LB, Carey JP, Della Santina CC. Eye movements in patients with superior canal dehiscence syndrome align with the abnormal canal. Neurology. 2000;55(12):1833–41.
6. Tilikete C, Krolak-Salmon P, Truy E, Vighetto A. Pulse-synchronous eye oscillations revealing bone superior canal dehiscence. Ann Neurol. 2004;56(4):556–60.
7. Younge BR, Khabie N, Brey RH, Driscoll CL. Rotatory nystagmus synchronous with heartbeat: a treatable form of nystagmus. Trans Am Ophthalmol Soc. 2003;101:113–7. discussion 7–8

M. Cherchi, *Otoneurology and Vestibular Medicine*, https://doi.org/10.1007/978-3-031-94841-1_127

Vibration-Induced Nystagmus

128

Overview

Georg von Békésy's observation in 1935 [1] that applying vibration to various locations on the skull could alter vestibular perception has been replicated [2] many times, and is sometimes referred to as the "propriogyral illusion" [3]. But that actual nystagmus that could be induced by applying vibration to the neck was first demonstrated by Lücke in 1973 [4]. Subsequent studies have investigated its occurrence, clinical significance, and mechanisms [5–14]. We describe here the technique involving a lateralized application of vibration to the neck or mastoid. Stimulation at the midline (midfrontal, vertex, etc.) has also been studied but is not used in most clinical laboratories.

Vibration-induced nystagmus (VIN) can be elicited by applying low-frequency vibration to the neck or mastoid area. A common technique is as follows: The patient is seated upright and instrumented with electronystagmography (ENG), videonystagmography (VNG), or video Frenzel oculography (VFO), so that eye movements can be observed in the dark, avoiding fixation suppression. Direct visualization, or using optical Fresnel lenses, is not sufficient. Spontaneous nystagmus is recorded for at least 10 seconds in complete darkness to obtain a baseline. Next, a source of low-frequency vibration is firmly applied to the neck over the belly of the sternocleidomastoid muscle. A recent review [15] notes that vibration frequencies from 40 Hz to 150 Hz have been studied; occasional studies use higher frequencies (e.g., 500 Hz [16] to 800 Hz [17], though for skull vibration); many laboratories use 100 Hz, but this has not been standardized. The location of the vibration is not critical as long as there is a firm connection between the vibrator and the muscle. During the vibration, the eyes are observed for nystagmus. Ten seconds of vibration is adequate. After eye movements are recorded with vibration applied to one side, the source of vibration is moved to the other side.

Vibration-induced nystagmus is commonly observed in patients with vestibular asymmetry (such as a person who has had vestibular neuritis, removal of a vestibular schwannoma, or treatment with intratympanic gentamicin) [7]. In these contexts, vibration on either side of the neck provokes a modest (about 5 deg/sec) horizontal nystagmus whose fast phase beats toward the intact side. People with symmetrical vestibular function—whether normal subjects (with symmetrical and intact vestibulo-ocular reflexes), or patients with bilateral weakness (in whom the deficiency is symmetrical)—often have no VIN. Some normal subjects have a direction-changing VIN that can be either ipsiversive (fast phase beating toward the side of vibration) or contraversive (fast phase beating away from the side of vibration) [18]. One study [12] found VIN in a surprising 81.6% (31 out of 38) normal subjects, but added that the magnitude of the nystagmus did not exceed 2.8 deg/sec.

In some patients with unilateral vestibular weakness, the VIN may additionally have a torsional component or be exclusively torsional. Presumably, this is due to the involvement of utricular afferents.

Variations in the technique of the vibration test mainly involve the location of the vibration. Vibration may be applied to the bony prominences around the ear with the mastoid process being the most favored (but this is loud and can therefore introduce the confounding effect of additionally eliciting Tullio's phenomenon in patients with perilymphatic fistulae or third window syndromes), or may be applied to the posterior neck rather than the anterior neck muscles [19].

Hypothetical mechanisms [20] for VIN fall into three general categories: (1) direct generation by the neck ("cervical nystagmus"), perhaps through proprioceptors, (2) generation from the inner ear itself, and (3) through an interaction between the neck and central vestibular processing ("neck fixation").

The first hypothesis, namely that vibration-induced nystagmus is generated from the neck, relies on the observation that vibration stimulates neck proprioceptors [8, 21, 22]. This hypothesis does not account for the reliable directionality of VIN.

M. Cherchi, *Otoneurology and Vestibular Medicine*, https://doi.org/10.1007/978-3-031-94841-1_128

The second hypothesis, namely that VIN is generated from the inner ear, is based on the idea that vibration stimulates labyrinthine structures [8], including the semicircular canals and the otolith organs [13, 23–26]. Karlberg and colleagues [8], citing work from Hudspeth [25], describe this logic as follows: "Most [vestibular hair] cells show saturation in the negative stimulus direction but no abrupt or complete saturation for positive stimuli. Thus, movements of the hair bundle toward the kinocilium will depolarize the cell more than movements of the same amplitude in the opposite direction will hyperpolarize it. The net effect of an oscillating mechanical stimulus to a hair bundle will be excitatory." If this hypothesis were true (and if it were the only mechanism at play), then since vibration acts more powerfully on the side on which the vibration is applied, one would expect that this would temporarily induce a vestibular asymmetry in normal subjects and thereby elicit nystagmus beating toward the ear being stimulated, as well as greater nystagmus in patients with unilateral loss for stimulation on the intact side. VIN is usually generated equally well from vibration on either sternocleidomastoid, but the patterns predicted above are seen in some subjects. Accordingly, there is some plausibility to this hypothesis, and it may account for VIN in some patients.

The third hypothesis, namely that VIN results from an interaction between neck and central vestibular processing (specifically from the release of neck fixation), is based on the idea that the vestibular system is merely one of multiple sensory inputs to central circuits that estimate one's orientation in space and generate compensatory movements. Vision and ankle proprioception are sensory inputs that are most commonly considered as part of the somatosensory integration process, but some suggest that in order for input from the vestibular system (situated in the head) to be properly used to stabilize the body, the position of the neck on the trunk is critical as it defines a coordinate transformation [27]. By the same reasoning, neck input may be helpful in the brain's determination of whether or not the head is moving; if the neck muscles' inputs are stable, then the head is probably stationary (with respect to the trunk). Thus, the neck fixation hypothesis conjectures that the central vestibular system can conclude that the head is not moving in space when the neck is unstimulated, and therefore can suppress spontaneous nystagmus. Additional evidence in support of this hypothesis comes from data suggesting that cervical afferents influence the velocity storage mechanism [28] and interposed nuclei of the cerebellum [29]. Overall, we find this hypothesis more plausibly explains the findings observed in VIN.

References

1. von Békésy G. Über akustische Reizung des Vestibularapparates. Arch Ges Physiol. 1935;236:59–72.
2. Betts GA, Barone M, Karlberg M, MacDougall H, Curthoys IS. Neck muscle vibration alters visually-perceived roll after unilateral vestibular loss. Neuroreport. 2000;11(12):2659–62.
3. Lackner JR, Levine MS. Changes in apparent body orientation and sensory localization induced by vibration of postural muscles: vibratory myesthetic illusions. Aviat Space Environ Med. 1979;50(4):346–54.
4. Lücke K. A vibratory stimulus of 100 Hz for provoking pathological nystagmus. Z Laryngol Rhinol Otol. 1973;52(10):716–20.
5. Choi KD, Oh SY, Kim HJ, Koo JW, Cho BM, Kim JS. Recovery of vestibular imbalances after vestibular neuritis. Laryngoscope. 2007;117(7):1307–12.
6. Dumas G, De Waele C, Hamann KF, Cohen B, Negrevergne M, Ulmer E, Schmerber S. Skull vibration induced nystagmus test. Ann Otolaryngol Chir Cervicofac. 2007;
7. Hamann KF, Schuster EM. Vibration-induced nystagmus – a sign of unilateral vestibular deficit. ORL J Otorhinolaryngol Relat Spec. 1999;61(2):74–9.
8. Karlberg M, Aw ST, Black RA, Todd MJ, MacDougall HG, Halmagyi GM. Vibration-induced ocular torsion and nystagmus after unilateral vestibular deafferentation. Brain. 2003;126(Pt 4):956–64.
9. Michel J, Dumas G, Lavieille JP, Charachon R. Diagnostic value of vibration-induced nystagmus obtained by combined vibratory stimulation applied to the neck muscles and skull of 300 vertiginous patients. Rev Laryngol Otol Rhinol (Bord). 2001;122(2):89–94.
10. Ohki M, Murofushi T, Nakahara H, Sugasawa K. Vibration-induced nystagmus in patients with vestibular disorders. Otolaryngol Head Neck Surg. 2003;129(3):255–8.
11. Park H, Shin J, Shim D. Mechanisms of vibration-induced nystagmus in normal subjects and patients with vestibular neuritis. Audiol Neurootol. 2007;12(3):189–97.
12. Perez N. Vibration induced nystagmus in normal subjects and in patients with dizziness. A videonystagmography study. Rev Laryngol Otol Rhinol (Bord). 2003;124(2):85–90.
13. Ulmer E, Chays A, Bremond G. Vibration-induced nystagmus: mechanism and clinical interest. Ann Otolaryngol Chir Cervicofac. 2004;121(2):95–103.
14. Xie SJ, Yang WY, Zhang SZ, Wu ZM, Chen YS, Jia HB, et al. Clinical significance of vibration-induced nystagmus in patients with unilateral peripheral vestibular disorders. Zhonghua Er Bi Yan Hou Tou Jing Wai Ke Za Zhi. 2006;41(10):736–9.
15. Dumas G, Curthoys IS, Lion A, Perrin P, Schmerber S. The skull vibration-induced nystagmus test of vestibular function-a review. Front Neurol. 2017;8:41.
16. Aw ST, Aw GE, Todd MJ, Bradshaw AP, Halmagyi GM. Three-dimensional vibration-induced vestibulo-ocular reflex identifies vertical semicircular canal dehiscence. J Assoc Res Otolaryngol. 2011;12(5):549–58.
17. Dumas G, Tan H, Dumas L, Perrin P, Lion A, Schmerber S. Skull vibration induced nystagmus in patients with superior semicircular canal dehiscence. Eur Ann Otorhinolaryngol Head Neck Dis. 2019;136:263.
18. Popov KE, Lekhel H, Faldon M, Bronstein AM, Gresty MA. Visual and oculomotor responses induced by neck vibration in normal subjects and labyrinthine-defective patients. Exp Brain Res. 1999;128(3):343–52.
19. Dumas G, Perrin P, Schmerber S, Lavieille JP. Nystagmus and vibration test research of mechanisms, theoretical methods: on 52 cases of unilateral vestibular lesions. Rev Laryngol Otol Rhinol (Bord). 2003;124(2):75–83.

20. Cherchi M, Hain TC. Provocative maneuvers for vestibular disorders. In: Eggers S, Zee D, editors. Vertigo and imbalance: clinical neurophysiology of the vestibular system. Handbook of clinical neurophysiology. Amsterdam: Elsevier; 2009. p. 111–34.
21. Burke D, Hagbarth KE, Lofstedt L, Wallin BG. The responses of human muscle spindle endings to vibration of non-contracting muscles. J Physiol. 1976;261(3):673–93.
22. Roll JP, Vedel JP, Ribot E. Alteration of proprioceptive messages induced by tendon vibration in man: a microneurographic study. Exp Brain Res. 1989;76(1):213–22.
23. Young ED, Fernandez C, Goldberg JM. Responses of squirrel monkey vestibular neurons to audio-frequency sound and head vibration. Acta Otolaryngol. 1977;84(5–6):352–60.
24. Wit HP, Bleeker JD, Mulder HH. Responses of pigeon vestibular nerve fibers to sound and vibration with audiofrequencies. J Acoust Soc Am. 1984;75(1):202–8.
25. Hudspeth AJ. Mechanoelectrical transduction by hair cells of the bullfrog's sacculus. Prog Brain Res. 1989;80:129–35. discussion 7–8
26. Christensen-Dalsgaard J, Narins PM. Sound and vibration sensitivity of VIIIth nerve fibers in the frogs Leptodactylus albilabris and Rana pipiens pipiens. J Comp Physiol A. 1993;172(6):653–62.
27. Nashner LM, Wolfson P. Influence of head position and proprioceptive cues on short latency postural reflexes evoked by galvanic stimulation of the human labyrinth. Brain Res. 1974;67(2):255–68.
28. Karlberg M, Magnusson M. Asymmetric optokinetic after-nystagmus induced by active or passive sustained head rotations. Acta Otolaryngol. 1996;116(5):647–51.
29. Luan H, Gdowski MJ, Newlands SD, Gdowski GT. Convergence of vestibular and neck proprioceptive sensory signals in the cerebellar interpositus. J Neurosci. 2013;33(3):1198–210a.

Part XVIII

Diagnoses—Other Central Nervous System Pathologies

129 Diseases That Can Affect Any Part of the Central Nervous System

Overview

There are several general pathologic processes that can cause a discrete lesion in any part of the central nervous system (CNS). The main ones are:

- *Infarction*: Hemorrhagic and ischemic strokes can affect any part of the brain or spinal cord.
- *Demyelination*: Any part of the central nervous system is vulnerable to demyelination. Some demyelinating diseases have a predilection for specific areas of the central nervous system (e.g., neuromyelitis optica spectrum disorders).
- *Space-Occupying Lesions*: Any part of the brain and spinal cord can be affected by tumors (malignant and nonmalignant), vascular anomalies, and other processes that exert mass effect on neural structures.

Other disease processes that can, at least in principle, affect any part of the CNS include infection and sarcoidosis.

As with any subdiscipline of neurology, diagnostic thinking proceeds through the steps of:

1. *Anatomy*—localizing the lesion.
2. *Pathophysiology*—what physiologic dysfunction is occurring.
3. *Etiology*—what is the underlying cause of the pathophysiology?

A good example of this thought process in general neurology might be a young healthy patient brought to the emergency room after having been found with abrupt-onset left-sided weakness (face, arm, and leg) that is now rapidly improving, and on examination, she has left-sided Hoffman's, Tromner's, and Babinski signs.

1. Anatomy—Long tract signs from the left upper motor neuron to the right side of body.
2. Pathophysiology—Transient upper motor neuron dysfunction, rapidly improving. Processes that can behave in this fashion include transient ischemia and a postseizure state.
3. Etiology—Transient ischemic attack (such as from embolization) would be somewhat unlikely in a young patient with no vascular risk factors. If this weakness is postictal (i.e., following a seizure), then cortical irritation, such as from a tumor, is a consideration.

Diagnosis in otoneurology would follow a similar thought process. Take the example of a 70-year-old patient with atrial fibrillation who abruptly developed disequilibrium and total left-sided hearing loss that have remained constant since onset several days ago, and who, on examination has left-sided sensorineural hearing loss and spontaneous right-beat nystagmus that increases on rightward gaze and decreases on leftward gaze.

1. Anatomy—The left-sided sensorineural hearing loss and spontaneous right-beat nystagmus both implicate a left-sided auditory and vestibular deficit.
2. Pathophysiology—The labyrinth could be dysfunctional, or the vestibulocochlear nerve could be dysfunctional.
3. Etiology—Left-sided labyrinthitis is a consideration, as it is common. Left-sided labyrinthine infarction is uncommon, but still a consideration in an older patient with vascular risk factors. A vestibular schwannoma, while possible, would be unlikely to present so abruptly.

Thus, from the perspective of arriving at a diagnosis in otoneurology, it is not usually productive to speak in general terms about infarction, demyelination, or tumors. Those etiologies may be worth considering depending on the anatomy and pathophysiology suggested by the history, physical examination, and perhaps ancillary test results.

M. Cherchi, *Otoneurology and Vestibular Medicine*, https://doi.org/10.1007/978-3-031-94841-1_129

Selected Vascular Phenomena in Otoneurology

130

Introduction

Several audiologic and vestibular problems have been reported in association with a variety of vascular abnormalities. Overall, vascular causes of such symptoms are rare. We discuss a selection of these here, including rotational vertebral artery syndrome, vascular impingement on the vestibulocochlear nerve, fibromuscular dysplasia, and glomus tympanicum.

Rotational Vertebral Artery Syndrome (RVAS)

Cerebral hypoperfusion secondary to vascular compromise is often discussed as a mechanism of cervicogenic vertigo (CV). *Rotational vertebral artery syndrome (RVAS)*, also called bowhunter syndrome, is a condition in which one of the vertebral arteries is transiently extrinsically compressed during neck rotation [1–13]. This appears more likely to provoke symptoms if the contralateral vertebral artery is already narrowed, such as by atherosclerotic disease. In some cases, this compression, perhaps combined with torquing of the artery, results in damage to the artery itself, such as a dissection [14]. There are a few reports of RVAS manifesting with predominantly downbeat nystagmus [15, 16], though such a pattern of nystagmus is not specific for the condition. If the vascular compromise is prolonged, it may culminate in infarction [17]. The discussions of RVAS are based primarily on case reports or small case series. Despite the apparent enthusiasm in the literature, this condition probably comprises only a small proportion of cases of cervicogenic vertigo. There is no consensus on the range of effects that neck turning can have on the vertebral arteries [18], and while dynamic vascular imaging can provide corroborative evidence of RVAS, Hain [19] pointed out that "Vertebral artery blood flow is compromised with full contralateral rotation in healthy individuals," citing Mitchell [5]; thus, vascular imaging is not specific for this condition. Further, cadaveric animal model evidence suggests typical physiologic motions at the neck produce vertebral artery strains substantially lower than the failure point (dissection) [20].

Subclavian Steal Syndrome (SSS)

Subclavian steal syndrome (SSS) is a vascular phenomenon in which occlusion or stenosis (usually secondary to atherosclerotic disease) of either the subclavian or innominate artery (proximal to the origin of the vertebral artery) alters hemodynamics such that blood supply to the ipsilateral upper extremity is maintained at the expense of inducing retrograde flow in the ipsilateral vertebral artery, potentially compromising the function of brainstem structures supplied by the vertebral artery. In some cases, the retrograde flow in the vertebral artery is insufficient to cause symptoms at rest, but exercising the ipsilateral arm (thereby increasing circulatory demands in that extremity) augments the retrograde vertebral artery flow and can bring on symptoms, including disequilibrium.

The prevalence of SSS has been estimated at 1.3%, has a 2:1 female:male ratio, and is more common over the age of 55 years [21]. Suspicion for subclavian steal syndrome is raised when blood pressure in one arm is lower than the other by 20 mm Hg, but the diagnosis usually requires confirmation by Doppler ultrasound.

Although SSS can cause a presyncopal sensation [21, 22], curiously it has been reported as the cause of inversion illusions as well [23].

Vascular Impingement on the Vestibulocochlear Nerve: As a Cause of Vestibular Paroxysmia

The original descriptions by Dr. Peter Jannetta [24] attributed vestibular paroxysmia to "*vascular cross compression*" of the vestibular nerve by a *loop of the anterior inferior cer-*

M. Cherchi, *Otoneurology and Vestibular Medicine*, https://doi.org/10.1007/978-3-031-94841-1_130

ebellar artery, though subsequent literature described vascular cross compression by other blood vessels as well, such as the posterior inferior cerebellar artery, a vertebral artery or a vein [25], or the superior cerebellar artery [26]. This was thought to be analogous to some cases of trigeminal neuralgia, another condition in which a cranial nerve (the trigeminal nerve) is sometimes found to have vascular impingement [27].

Although some radiologic literature concluded that there was a correlation between vascular cross compression and audiovestibular symptoms [28–36], other literature found no such correlation [37–42].

Since some patients whose clinical history is compatible with vestibular paroxysmia have no evidence of vascular impingement, and since some asymptomatic individuals have vascular impingement, this brought into question the hypothesis of a purely vascular etiology of this disease, and in fact, the Barany Society consensus document states, "The role of imaging in the diagnosis and identification of the affected side is not clear because of the high rate of blood vessels close to the 8th cranial nerve in healthy subjects" [43].

Although there is some logic to the hypothesis of vascular cross compression, there are certainly other neural pathologies that can damage axons and/or Schwann cells (from demyelination, infection, ischemia, and others) that presumably could provoke episodic irritability of the nerve.

Vascular Impingement on the Vestibulocochlear Nerve: As a Cause of Sensorineural Hearing Loss

Vascular cross-compression of the vestibulocochlear nerve has also been suggested as a cause of sensorineural hearing loss. There have been several attempts at developing radiographic/anatomical classifications of vascular impingement on the vestibulocochlear nerve [44–47]. A minority of literature concludes that there is some correlation between such vascular impingement on the vestibulocochlear nerve and sensorineural hearing loss [33], but the majority of the literature and systematic reviews conclude that there is no such correlation [37, 39, 42, 48, 49]. Furthermore, literature regarding surgical intervention in cases of vascular impingement on the vestibulocochlear nerve reports no improvement in sensorineural hearing loss [28, 50].

Fibromuscular Dysplasia (FMD)

Fibromuscular dysplasia (FMD) is a noninflammatory, nonatherosclerotic arteriopathy manifesting with morphological abnormalities such as tortuosity, dissection, aneurysm, stenosis, and other occlusions. The etiology is unclear, but research increasingly suggests genetic predisposition [51–54]. It has been reported in nearly all vasculature, but most commonly involves the renal, internal carotid, and vertebral arteries. Olin and colleagues [55] studied 447 patients with FMD and reported that affected arteries included the renal (79.7%), extracranial carotid (74.3%), and vertebral (36.6%); presenting symptoms included pulsatile tinnitus (27.5%), dizziness (26.0%) "tinnitus" (presumably nonpulsatile) (18.8%), and stroke (6.9%). The literature reports various kinds of tinnitus [55–73] much more commonly than vestibular symptoms [74–77], with the mechanism for the latter usually being cited as stroke [77, 78].

Glomus Tympanicum

A *glomus tympanicum (GT)* tumor is a paraganglioma arising at various locations in the middle ear. It is highly vascular, and the classic appearance under otomicroscopy/otoscopy is a red pulsatile mass behind the eardrum. It is usually a disease of adulthood, and is more common in women. A number of diseases can resemble GTs, and GTs can mimic other lesions. About 15% of GTs secrete catecholamines. The most common presenting symptoms are pulsatile tinnitus and hearing loss. Rarely, GTs can cause disequilibrium and epistaxis. GTs are usually unilateral; they can occur with other paragangliomas anywhere along the sympathetic chain. Audiometry usually shows mixed conductive and sensorineural hearing loss. CT is usually the imaging modality of choice, but MRI and various forms of angiography may also play a role. Treatment is usually surgical, though radiation therapy has become more common. Recurrence is uncommon and occurs more frequently in surgically treated cases. If a clinician observes a red pulsatile mass behind the eardrum, referral to a neuro-otologic otolaryngologist is appropriate.

Vascular Causes of Pulsatile Tinnitus

Pulsatile tinnitus has been reported in association with a number of arterial and venous abnormalities, including jugular bulb wall dehiscence, transverse sinus stenosis, sigmoid sinus diverticulum, sigmoid sinus wall dehiscence, enlarged oblique occipital sinus, stenoses of various dural venous sinuses from arachnoid granulations, superficial middle cerebral vein, dilated mastoid emissary vein, carotid artery stenosis, fibromuscular dysplasia, arteriovenous fistula, and others.

References

1. Kamouchi M, Kishikawa K, Matsuo R, Yasumori K, Inoue T, Okada Y, et al. Ultrasonographic detection of extracranial vertebral artery compression in bow hunter's brain ischemia caused by neck rotation. Cerebrovasc Dis. 2003;16(3):303–5.
2. Sakaguchi M, Kitagawa K, Hougaku H, Hashimoto H, Nagai Y, Yamagami H, et al. Mechanical compression of the extracranial vertebral artery during neck rotation. Neurology. 2003;61(6):845–7.
3. Weintraub MI, Khoury A. Mechanical compression of the extracranial vertebral artery during neck rotation. Neurology. 2004;62(11):2143.
4. Helton T, Bavry A. Images in cardiovascular medicine. Don't turn your head! Circulation. 2009;120:e162.
5. Mitchell J. Vertebral artery blood flow velocity changes associated with cervical spine rotation: a meta-analysis of the evidence with implications for professional practice. J Man Manip Ther. 2009;17(1):46–57.
6. Lu DC, Zador Z, Mummaneni PV, Lawton MT. Rotational vertebral artery occlusion-series of 9 cases. Neurosurgery. 2010;67(4):1066–72. discussion 72
7. Cornelius JF, George B, N'Dri Oka D, Spiriev T, Steiger HJ, Hanggi D. Bow-hunter's syndrome caused by dynamic vertebral artery stenosis at the cranio-cervical junction--a management algorithm based on a systematic review and a clinical series. Neurosurg Rev. 2012;35(1):127–35. discussion 35
8. Go G, Hwang SH, Park IS, Park H. Rotational vertebral artery compression: bow hunter's syndrome. J Korean Neurosurg Soc. 2013;54(3):243–5.
9. Sarkar J, Wolfe SQ, Ching BH, Kellicut DC. Bow hunter's syndrome causing vertebrobasilar insufficiency in a young man with neck muscle hypertrophy. Ann Vasc Surg. 2014;28(4):1032.e1–e10.
10. Zaidi HA, Albuquerque FC, Chowdhry SA, Zabramski JM, Ducruet AF, Spetzler RF. Diagnosis and management of bow hunter's syndrome: 15-year experience at barrow neurological institute. World Neurosurg. 2014;82(5):733–8.
11. Duan G, Xu J, Shi J, Cao Y. Advances in the pathogenesis, diagnosis and treatment of Bow hunter's syndrome: a comprehensive review of the literature. Interv Neurol. 2016;5(1–2):29–38.
12. Araz Server E, Edizer DT, Yigit O, Yasak AG, Erdim C. Relationship between vertebral artery blood flow in different head positions and vertigo. Acta Otolaryngol. 2018;138(1):1–5.
13. Li Q, Xie P, Yang WS, Yan B, Davis S, Caplan LR. Vertebral artery compression syndrome. Front Neurol. 2019;10:1075.
14. Vibert D, Rohr-Le Floch J, Gauthier G. Vertigo as manifestation of vertebral artery dissection after chiropractic neck manipulations. ORL J Otorhinolaryngol Relat Spec. 1993;55(3):140–2.
15. Choi KD, Shin HY, Kim JS, Kim SH, Kwon OK, Koo JW, et al. Rotational vertebral artery syndrome: oculographic analysis of nystagmus. Neurology. 2005;65(8):1287–90.
16. Ogawa Y, Itani S, Otsuka K, Inagaki T, Shimizu S, Kondo T, et al. Intermittent positional downbeat nystagmus of cervical origin. Auris Nasus Larynx. 2014;41(2):234–7.
17. Sorensen BF. Bow hunter's stroke. Neurosurgery. 1978;2(3):259–61.
18. Mitchell J. Doppler insonation of vertebral artery blood flow changes associated with cervical spine rotation: implications for manual therapists. Physiother Theory Pract. 2007;23(6):303–13.
19. Hain TC. Cervicogenic causes of vertigo. Curr Opin Neurol. 2015;28(1):69–73.
20. Symons BP, Leonard T, Herzog W. Internal forces sustained by the vertebral artery during spinal manipulative therapy. J Manip Physiol Ther. 2002;25(8):504–10.
21. Psillas G, Kekes G, Constantinidis J, Triaridis S, Vital V. Subclavian steal syndrome: neurotological manifestations. Acta Otorhinolaryngol Ital. 2007;27(1):33–7.
22. Porwal V, Verma A. Otoneurological manifestation in subclavian steal syndrome. J Evol Med Dent Sci. 2013;2:1634–7.
23. Arntzen K, Alstadhaug KB. Room tilt illusion and subclavian steal - a case report. BMC Neurol. 2020;20(1):369.
24. Jannetta PJ. Neurovascular cross-compression in patients with hyperactive dysfunction symptoms of the eighth cranial nerve. Surg Forum. 1975;26:467–9.
25. Best C, Gawehn J, Kramer HH, Thomke F, Ibis T, Muller-Forell W, et al. MRI and neurophysiology in vestibular paroxysmia: contradiction and correlation. J Neurol Neurosurg Psychiatry. 2013;84(12):1349–56.
26. Ryu H, Yamamoto S, Sugiyama K, Nozue M. Neurovascular compression syndrome of the eighth cranial nerve. What are the most reliable diagnostic signs? Acta Neurochir. 1998;140(12):1279–86.
27. Gronseth G, Cruccu G, Alksne J, Argoff C, Brainin M, Burchiel K, et al. Practice parameter: the diagnostic evaluation and treatment of trigeminal neuralgia (an evidence-based review): report of the quality standards Subcommittee of the American Academy of neurology and the European Federation of Neurological Societies. Neurology. 2008;71(15):1183–90.
28. Moosa S, Fezeu F, Kesser BW, Ramesh A, Sheehan JP. Sudden unilateral hearing loss and vascular loop in the internal auditory canal: case report and review of literature. J Radiosurg SBRT. 2015;3(3):247–55.
29. De Ridder D, De Ridder L, Nowe V, Thierens H, Van de Heyning P, Moller A. Pulsatile tinnitus and the intrameatal vascular loop: why do we not hear our carotids? Neurosurgery. 2005;57(6):1213–7. discussion –7
30. Herzog JA, Bailey S, Meyer J. Vascular loops of the internal auditory canal: a diagnostic dilemma. Am J Otol. 1997;18(1):26–31.
31. Jannetta PJ, Moller MB, Moller AR. Disabling positional vertigo. N Engl J Med. 1984;310(26):1700–5.
32. McCabe BF, Harker LA. Vascular loop as a cause of vertigo. Ann Otol Rhinol Laryngol. 1983;92(6 Pt 1):542–3.
33. McDermott AL, Dutt SN, Irving RM, Pahor AL, Chavda SV. Anterior inferior cerebellar artery syndrome: fact or fiction. Clin Otolaryngol Allied Sci. 2003;28(2):75–80.
34. Nowe V, De Ridder D, Van de Heyning PH, Wang XL, Gielen J, Van Goethem J, et al. Does the location of a vascular loop in the cerebellopontine angle explain pulsatile and non-pulsatile tinnitus? Eur Radiol. 2004;14(12):2282–9.
35. Shinn JB, Bush ML, Jones RO. Correlation of central auditory processing deficits and vascular loop syndrome. Ear Nose Throat J. 2009;88(10):E34–7.
36. Wuertenberger CJ, Rosahl SK. Vertigo and tinnitus caused by vascular compression of the vestibulocochlear nerve, not intracanalicular vestibular schwannoma: review and case presentation. Skull Base. 2009;19(6):417–24.
37. Gorrie A, Warren FM 3rd, de la Garza AN, Shelton C, Wiggins RH 3rd. Is there a correlation between vascular loops in the cerebellopontine angle and unexplained unilateral hearing loss? Otol Neurotol. 2010;31(1):48–52.
38. Hoekstra CE, Prijs VF, van Zanten GA. Diagnostic yield of a routine magnetic resonance imaging in tinnitus and clinical relevance of the anterior inferior cerebellar artery loops. Otol Neurotol. 2015;36(2):359–65.
39. Makins AE, Nikolopoulos TP, Ludman C, O'Donoghue GM. Is there a correlation between vascular loops and unilateral auditory symptoms? Laryngoscope. 1998;108(11 Pt 1):1739–42.
40. Parnes LS, Shimotakahara SG, Pelz D, Lee D, Fox AJ. Vascular relationships of the vestibulocochlear nerve on magnetic resonance imaging. Am J Otol. 1990;11(4):278–81.
41. Reisser C, Schuknecht HF. The anterior inferior cerebellar artery in the internal auditory canal. Laryngoscope. 1991;101(7 Pt 1):761–6.

42. van der Steenstraten F, de Ru JA, Witkamp TD. Is microvascular compression of the vestibulocochlear nerve a cause of unilateral hearing loss? Ann Otol Rhinol Laryngol. 2007;116(4):248–52.
43. Strupp M, Lopez-Escamez JA, Kim JS, Straumann D, Jen JC, Carey J, et al. Vestibular paroxysmia: Diagnostic criteria. J Vestib Res. 2016;26(5–6):409–15.
44. Sirikci A, Bayazit Y, Ozer E, Ozkur A, Adaletli I, Cüce MA, et al. Magnetic resonance imaging based classification of anatomic relationship between the cochleovestibular nerve and anterior inferior cerebellar artery in patients with non-specific neuro-otologic symptoms. Surg Radiol Anat. 2005;27(6):531–5.
45. Kazawa N, Togashi K, Ito J. The anatomical classification of AICA/PICA branching and configurations in the cerebellopontine angle area on 3D-drive thin slice T2WI MRI. Clin Imaging. 2013;37(5):865–70.
46. Kim SH, Ju YR, Choi JE, Jung JY, Kim SY, Lee MY. Anatomical location of AICA loop in CPA as a prognostic factor for ISSNHL. PeerJ. 2019;7:e6582.
47. Di Stadio A, Dipietro L, Ralli M, Faralli M, Della Volpe A, Ricci G, et al. Loop characteristics and audio-vestibular symptoms or hemifacial spasm: is there a correlation? A multiplanar MRI study. Eur Radiol. 2020;30(1):99–109.
48. Papadopoulou AM, Bakogiannis N, Sofokleous V, Skrapari I, Bakoyiannis C. The impact of vascular loops in the cerebellopontine angle on audio-vestibular symptoms: a systematic review. Audiol Neurootol. 2022;27(3):200–7.
49. Leng Y, Lei P, Liu Y, Chen C, Xia K, Liu B. Vascular loops in cerebellopontine angle in patients with unilateral idiopathic sudden sensorineural hearing loss: evaluations by three radiological grading systems. Laryngoscope Investig Otolaryngol. 2022;7(5):1532–40.
50. Esposito G, Messina R, Carai A, Colafati GS, Savioli A, Randi F, et al. Cochleovestibular nerve compression syndrome caused by Intrameatal anterior inferior cerebellar artery loop: synthesis of Best evidence for clinical decisions. World Neurosurg. 2016;96:556–61.
51. Ganesh SK, Morissette R, Xu Z, Schoenhoff F, Griswold BF, Yang J, et al. Clinical and biochemical profiles suggest fibromuscular dysplasia is a systemic disease with altered TGF-beta expression and connective tissue features. FASEB J. 2014;28(8):3313–24.
52. Gornik HL, Persu A, Adlam D, Aparicio LS, Azizi M, Boulanger M, et al. First international consensus on the diagnosis and management of fibromuscular dysplasia. Vasc Med. 2019;24(2):164–89.
53. Perdu J, Boutouyrie P, Bourgain C, Stern N, Laloux B, Bozec E, et al. Inheritance of arterial lesions in renal fibromuscular dysplasia. J Hum Hypertens. 2007;21(5):393–400.
54. Persu A, Dobrowolski P, Gornik HL, Olin JW, Adlam D, Azizi M, et al. Current progress in clinical, molecular, and genetic aspects of adult fibromuscular dysplasia. Cardiovasc Res. 2022;118(1):65–83.
55. Olin JW, Froehlich J, Gu X, Bacharach JM, Eagle K, Gray BH, et al. The United States registry for fibromuscular dysplasia: results in the first 447 patients. Circulation. 2012;125(25):3182–90.
56. Arnold M, De Marchis GM, Stapf C, Baumgartner RW, Nedeltchev K, Buffon F, et al. Triple and quadruple spontaneous cervical artery dissection: presenting characteristics and long-term outcome. J Neurol Neurosurg Psychiatry. 2009;80(2):171–4.
57. Bagh I, Olin JW, Froehlich JB, Kline-Rogers E, Gray B, Kim ESH, et al. Association of Multifocal Fibromuscular Dysplasia in elderly patients with a more benign clinical phenotype: data from the US registry for fibromuscular dysplasia. JAMA Cardiol. 2018;3(8):756–60.
58. Dicks AB, Gornik HL, Gu X, Bacharach JM, Fendrikova Mahlay N, Froehlich JB, et al. Association of Fibromuscular Dysplasia and Pulsatile Tinnitus: a report of the US registry for fibromuscular dysplasia. J Am Heart Assoc. 2021;10(17):e021962.
59. Dufour JJ, Lavigne F, Plante R, Caouette H. Pulsatile tinnitus and fibromuscular dysplasia of the internal carotid. J Otolaryngol. 1985;14(5):293–5.
60. Foyt D, Carfrae MJ, Rapoport R. Fibromuscular dysplasia of the internal carotid artery causing pulsatile tinnitus. Otolaryngol Head Neck Surg. 2006;134(4):701–2.
61. Gruber B, Hemmati M. Fibromuscular dysplasia of the vertebral artery: an unusual cause of pulsatile tinnitus. Otolaryngol Head Neck Surg. 1991;105(1):113–4.
62. Henrard C, Belge H, Fastré S, Di Monaco S, Revencu N, Hammer F, et al. Cervical artery dissection: fibromuscular dysplasia versus vascular Ehlers-Danlos syndrome. Blood Press. 2019;28(2):139–43.
63. Hofmann E, Behr R, Neumann-Haefelin T, Schwager K. Pulsatile tinnitus: imaging and differential diagnosis. Dtsch Arztebl Int. 2013;110(26):451–8.
64. Kesav P, Manesh Raj D, John S. Cerebrovascular Fibromuscular Dysplasia – a practical review. Vasc Health Risk Manag. 2023;19:543–56.
65. Kim ES, Serhal M. The diagnosis and treatment of fibromuscular dysplasia: an update for cardiologists. Curr Treat Options Cardiovasc Med. 2016;18(6):37.
66. Kythreotou A, Weerakkody RA, Koysombat K, Marzouqa N, Baker DM. A retrospective cohort study of cerebrovascular fibromuscular dysplasia. Ann Vasc Surg. 2023;92:104–10.
67. Narula N, Kadian-Dodov D, Olin JW. Fibromuscular dysplasia: contemporary concepts and future directions. Prog Cardiovasc Dis. 2018;60(6):580–5.
68. Raj RK, Gandhi RT, Katzen BT. Fibromuscular dysplasia-related carotid pseudoaneurysm and pulsatile tinnitus. J Vasc Interv Radiol. 2012;23(12):1657.
69. Topal H, Lemmens R, Fourneau I. Possible familial presentation in two siblings with carotid fibromuscular dysplasia. Acta Chir Belg. 2015;115(1):83–6.
70. Vanzin JR, Martio AE, Manzato LB, Abud DG. Casper stent in the treatment of pulsatile tinnitus in fibromuscular dysplasia: therapeutic review and case report. Brain Circ. 2021;7(4):281–4.
71. Waldvogel D, Mattle HP, Sturzenegger M, Schroth G. Pulsatile tinnitus–a review of 84 patients. J Neurol. 1998;245(3):137–42.
72. Weissman JL, Hirsch BE. Imaging of tinnitus: a review. Radiology. 2000;216(2):342–9.
73. Zhang Y, Zheng H, Zhou M, He L. Teaching neuroimages: carotid-cavernous fistula caused by fibromuscular dysplasia. Neurology. 2014;82(15):e134–5.
74. De Groote M, Van der Niepen P, Hemelsoet D, Callewaert B, Vermassen F, Billiouw JM, et al. Fibromuscular dysplasia – results of a multicentre study in Flanders. Vasa. 2017;46(3):211–8.
75. Krittanawong C, Kumar A, Johnson KW, Kaplin S, Virk HUH, Wang Z, et al. Prevalence, presentation, and associated conditions of patients with fibromuscular dysplasia. Am J Cardiol. 2019;123(7):1169–72.
76. Netsch C, Gross AJ, Brüning R. Symptomatic hydronephrosis from renal artery aneurysm associated with fibromuscular dysplasia: management with transarterial embolization. J Endourol. 2011;25(4):569–72.
77. Tsien MZ, Brorson JR, Lukas RV, Dawson EC. A 16-year-old male with dizziness, parasthesias, and ataxia. Pediatr Ann. 2014;43(5):e116–9.
78. Chen YL, Hsu YD, Lee JT. Fibromuscular dysplasia of the vertebral artery presenting with lateral medullary syndrome: a case report. Acta Neurol Taiwanica. 2004;13(4):198–202.

Part XIX

Diagnoses—Intersections Between Otovestibular Disease, Psychology and Psychiatry

131 Psychologic Disequilibrium

Overview

Disequilibrium can occur in conjunction with psychiatric illness. We refer to this as "*psychologic disequilibrium*" in order to remain neutral with respect to the direction of causality. Several factors make this topic difficult to study, including the variability of presentation, but there is some evidence that primary otovestibular disorders can provoke psychological symptoms, and evidence that psychiatric disease can cause symptoms of disequilibrium. It is plausible that many cases involve variable overlap and variable directionality of causation. We find explanatory frameworks based on internal modeling compelling. Vestibular testing of patients with anxiety and panic disorder reports variable abnormalities, but none is sensitive or specific for these psychiatric diagnoses, and in our view, such findings should prompt a search for primary otovestibular disorders. Practically, since psychiatric disease and the symptom of disequilibrium can be mutually reinforcing, we encourage psychiatric consultation to interrupt this maladaptive positive feedback loop.

Introduction

Although Sigmund Freud (1856–1939) is generally known as the founder of psychoanalysis, his training was as a neurologist at the University of Vienna. In 1895, he published a paper entitled, "On the grounds for detaching a particular syndrome from neurasthenia under the description 'anxiety neurosis'" [1], in which he stated:

> 'Vertigo' occupies a prominent place in the group of symptoms of anxiety neurosis. In its mildest form it is best described as 'giddiness'; in its severer manifestations, as 'attacks of vertigo' (with or without anxiety), it must be classed among the gravest symptoms of the neurosis. The vertigo of anxiety neurosis is not rotatory nor does it especially affect certain planes or directions, like Meniere's vertigo. It belongs to the class of locomotor or co-ordinatory vertigo, as does the vertigo in oculomotor paralysis. It consists in a specific state of discomfort, accompanied by sensations of the ground rocking, of the legs giving way and of its being impossible to stand up any more; while the legs feel as heavy as lead and tremble or the knees bend. This vertigo never leads to a fall.

Freud was probably not the first to propose the idea of a relationship between psychological factors and the somatic symptom of "vertigo." In the past few decades, there have been numerous other attempts to characterize this relationship, with a range of nomenclatures referencing a variety of symptoms, putative causes, and putative effects.

What Framework Can We Use to Approach This Issue?

A Dichotomous Nosology Is Probably Not Helpful

As we discuss elsewhere, the various terms subsumed under the concept of disequilibrium have, at best, marginal diagnostic value in themselves. Conceptually, we find it helpful to regard the perception of disequilibrium as an "output," and the underlying etiology of that symptom as the "input." There is an extraordinary array of disorders ("inputs") that can lead to the symptom ("output") of disequilibrium.

During the process of evaluating a patient, we prefer to use the relatively neutral phrase "*psychologic disequilibrium*" to reflect that there may be a relationship between psychological factors and the symptom of disequilibrium, without claiming that the relationship is causal (in either direction). We prefer to avoid the term "psychogenic disequilibrium" because this implies (even before a diagnosis has been established) that psychological factors are the *cause* ("-genic") of the symptom of disequilibrium, when they may ultimately turn out not to be so. We also prefer to avoid the term "psychiatric disequilibrium" [2], because this implies that the underlying psychological experience is pathological or requires treatment ("-iatric"), which it may ultimately turn out not to be.

M. Cherchi, *Otoneurology and Vestibular Medicine*, https://doi.org/10.1007/978-3-031-94841-1_131

Furthermore, we would concur with authors who caution against imposing a dichotomous nosology on this problem. For example:

- "A categorical classification of dizziness does not fully appreciate how these conditions occur together and interact" [3].
- Jacob and Furman believe that it is flawed to regard these symptoms as "either organic [meaning with a detectable physiologic source] or psychogenic but not both" [4].
- "The traditional dichotomous classification of vestibular symptoms as 'organic' or 'psychogenic' is unworkable" [5].

Finally, Dieterich and colleagues [5] comment that:

> Psychiatric disorders may be the primary causes or secondary complications of vestibular syndromes. Primary psychiatric disorders occur without other preceding vestibular syndromes. Secondary psychiatric disorders develop after the onset of primary vestibular disorders and often outlast their remission. In some cases, psychiatric disorders occur first, but do not manifest vertigo, unsteadiness, or dizziness. Then, with the onset of a vestibular disorder, they increase in severity and contribute to overall morbidity in an interactive manner. Psychiatric disorders also may trigger functional vestibular syndromes such as PPV [phobic postural vertigo] or CSD [chronic subjective dizziness] [5].

There are compelling data to suggest that psychological factors (in isolation) can cause the symptom of disequilibrium, and equally compelling data to suggest that the perception of disequilibrium can cause psychological symptoms. A study by Staab and Ruckenstein [6] concluded:

> These data support the hypothesis that physical neurotologic conditions may trigger psychopathy as often as primary anxiety disorders cause dizziness. A third pattern appears to be equally common wherein physical neurotologic conditions exacerbate preexisting psychiatric illnesses… There is a bidirectional relationship between neurotologic conditions and psychiatric disorders [6].

In other words, it is likely that many cases have a mix of directionality.

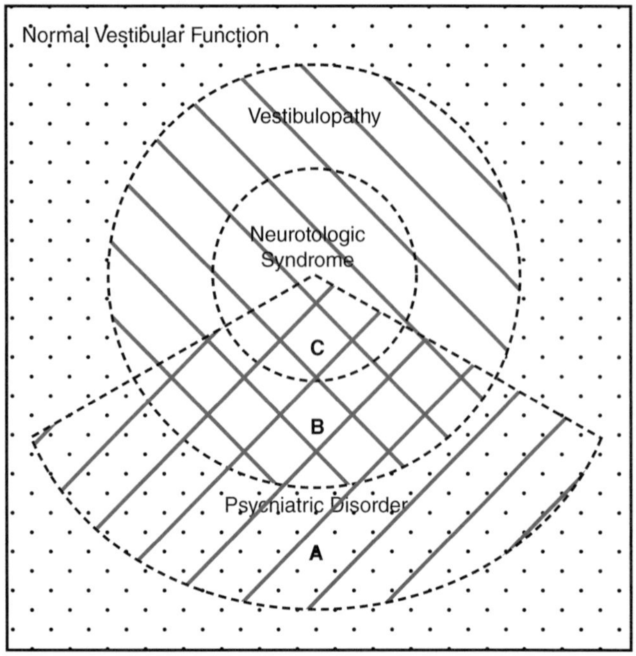

Fig. 131.1 A taxonomy of dizziness, focusing on the intersection between otovestibular disorders and psychiatric disorders. From Furman and Jacob [2]. Reproduced with written permission from Wolters Kluwer Health Limited.

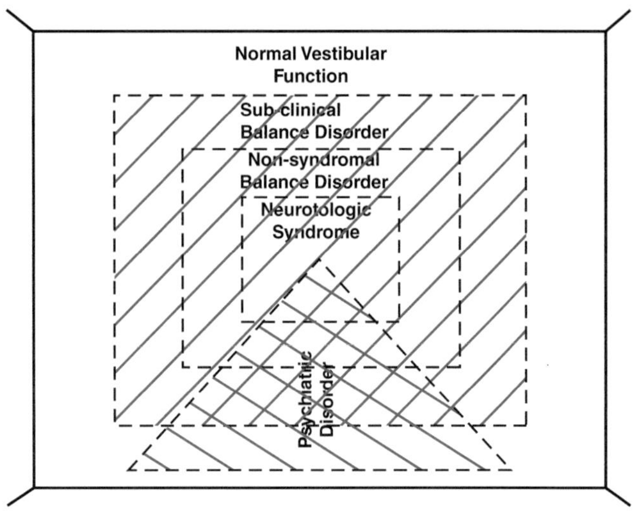

Fig. 131.2 A taxonomy of dizziness, focusing on the intersection between otovestibular disorders and psychiatric disorders. From Furman and Jacob [7]. Reproduced with written permission from Elsevier B.V.

A More Nuanced Nosology Is Appropriate

Furman and Jacob published and revised a "taxonomy" of disequilibrium and psychiatric factors in the context of otoneurology [2, 4, 7], selections of which are illustrated in Figs. 131.1 and 131.2 below.

The progression in their thinking was essentially one of increasing granularity of detectable otoneurological conditions. The organization of concentric rectangles in the final iteration [7] is articulated elsewhere [4] as follows, progressing from innermost to outermost:

1. "Syndromal vestibular disorders, e.g., Ménière's disease and benign paroxysmal positional vertigo."
2. "Nonsyndromal vestibular disorders, i.e., abnormal laboratory vestibular testing, dizziness symptoms, and ancillary testing, e.g., magnetic resonance imaging, which do not fit into a pattern of a specific syndrome."

3. "Subclinical vestibular disorders, i.e., symptoms suggestive of vestibular dysfunction without vestibular laboratory abnormalities."
4. "No vestibular disorder, i.e., medical disorders such as orthostatic hypotension and certain psychiatric disorders."

Furman and Jacob [7] offer an alternative graphic representation of these relationships which may be more instructive, as illustrated in Fig. 131.3 below. They describe this as depicting a "Framework for interface between balance disorders and psychiatric disorders. The interface between balance disorders and psychiatric disorders is depicted as several levels of interaction ranging from noncoexistent disorders to a complete overlap in pathophysiology" [7].

Ultimately, these levels are intended to convey a gradation of overlap between symptoms (disequilibrium, psychological symptoms) and pathologies (otovestibular disorders, psychiatric disease) without making claims pertaining to causality. Other investigators have proposed classifications that are similar in spirit [3].

The culmination of these gradations of overlap is that by the final point on the scale (corresponding to the outermost space in the *Figures* presented earlier), there is no detectable otovestibular or other medical pathology that would manifest with a perceived disturbance of disequilibrium, and thus by exclusion, the symptom of disequilibrium must be associated exclusively with a psychological process. Is this the point at which terms such as "psychogenic disequilibrium" or "psychiatric disequilibrium" would be appropriate? We would argue that such terms are still inappropriate for two reasons.

First, Furman and Jacob [2] add the following humbling reminders of the limitations that experience and technology impose on our diagnostic capabilities:

> The diagnosis of a defined neurotologic syndrome depends on the knowledge and expertise of the treating physician and the availability of a vestibular laboratory. The boundary between vestibulopathy and normal vestibular function is similarly fluid, because the probability of uncovering objective findings depends on the composition of the battery of vestibular tests included in the assessment. In particular, negative findings on physical examination and laboratory testing do not guarantee normality. Clinical vestibular tests leave large areas of the vestibular system unexamined [2].

In other words, as knowledge and technology advance [8], it is likely that previously undetectable vestibular pathology will become identifiable. Therefore, we would hesitate to commit to the term "psychogenic" disequilibrium.

Second, not all psychological experiences should be classified as pathological. There are life circumstances in which it is unsurprising to experience emotions compatible with formal diagnoses of depression, anxiety, etc. Therefore, we would hesitate to employ the term "psychiatric" disequilibrium, because that would suggest (perhaps erroneously) that the emotional experience is pathological and therefore requires treatment.

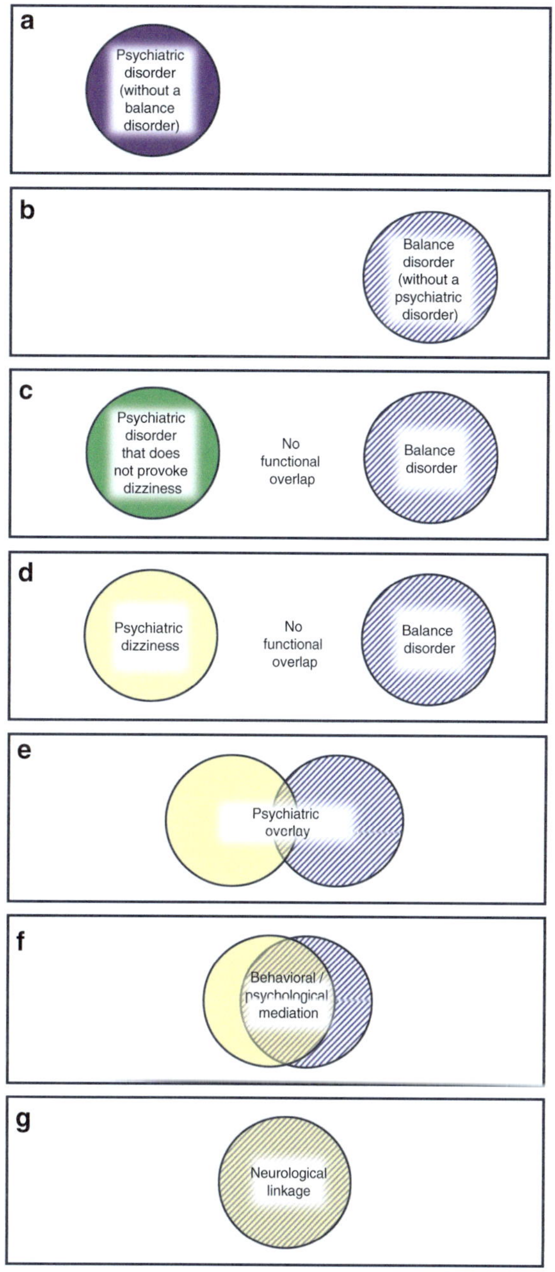

Fig. 131.3 Framework for interface between balance disorders and psychiatric disorders. Panel A: psychiatric disorder without a balance disorder. Panel B: balance disorder without a psychiatric disorder. Panel C: cooccurrence of a psychiatric disorder (that does not provoke dizziness) and a balance disorder, with no functional overlap. Panel D: cooccurrence of psychiatric dizziness and a balance disorder, with no functional overlap. Panel E: cooccurrence of psychiatric dizziness and a balance disorder, with psychiatric overlay. Panel F: cooccurrence of psychiatric dizziness and a balance disorder, with behavioral/psychological mediation. Panel G: neurological linkage of psychiatric dizziness and a balance disorder. Based on Furman and Jacob [7]

Pathophysiological Mechanism of Disease: Consider Internal Models

Despite everything we have just discussed, there remains the reasonable possibility that there are instances in which the symptom of disequilibrium is due exclusively to psychological factors and nothing else.

Even though we may not yet be in a position to identify such cases with complete confidence, it is certainly plausible that such instances exist, analogous to somatization disorders. So, if this is the case, what is a plausible mechanism?

At a general level, "Vestibular and anxiety disorders are related via both somatopsychic and psychosomatic mechanisms, and also are linked via overlapping neural circuits" [4], but this concept is so broad that it affords little possibility of any deeper understanding.

Figure 131.4 below, originally from Brandt [9] and later modified in Dieterich and Staab [5], shows an early version of how "phobic postural vertigo" could arise from a mismatch between planned motion ("efference copy") and perception of actual motion ("afference").

Broadly speaking, this idea relies on a movement *illusion* (in the medical sense), meaning misperception of an actual stimulus.

But what about the perception of motion in the absence of any such stimulus? In medical terms, this would be classified as a movement *hallucination,* meaning perception in the absence of any stimulus. Reuten and colleagues [10] construct a persuasive argument for an explanatory model of this. They begin by reviewing data from other research demonstrating that cognitive factors influence vestibular perceptions [11–13], and specifically that perception of actual motion can be modulated by a variety of factors including mental imagery [14, 15], expectations [16], and contextual information [17–19]. They reasoned that:

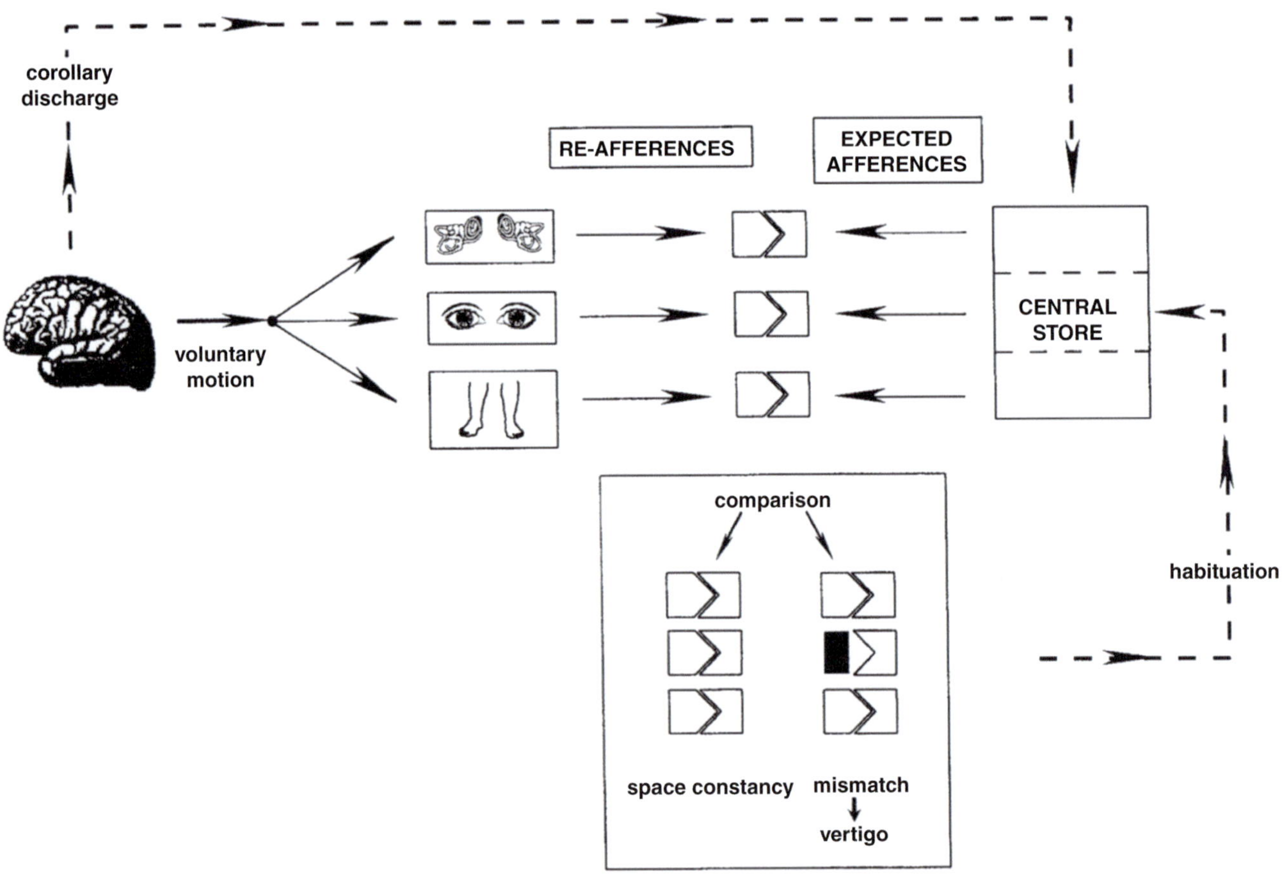

Fig. 131.4 Model of how phobic postural vertigo might arise from discrepancies between planned motion ("efference copy") and perception of actual motion ("afference"). This schematic diagram depicts the sensorimotor derangement, or the neural mismatch concept, of phobic postural vertigo as caused by decoupling of the efference copy signal. An active movement leads to sensory stimulation, the messages of which are compared with a multisensory pattern of expectation calibrated by early experience of motions (central store). The pattern of expectation is either prepared by the efference copy signal, which is emitted parallel to and simultaneously with the motion impulse, or by vestibular excitation during passive vehicular transportation. If concurrent sensory stimulation and the pattern of expectation are in agreement, self-motion is perceived while space constancy is maintained. If the efference copy signal (corollary discharge) is inappropriate, there is a sensorimotor mismatch, and a self-generated head motion or body sway may be erroneously perceived as external perturbations, causing the subjective postural vertigo. From Brandt [9]. Reproduced with written permission from Wolters Kluwer Health Limited

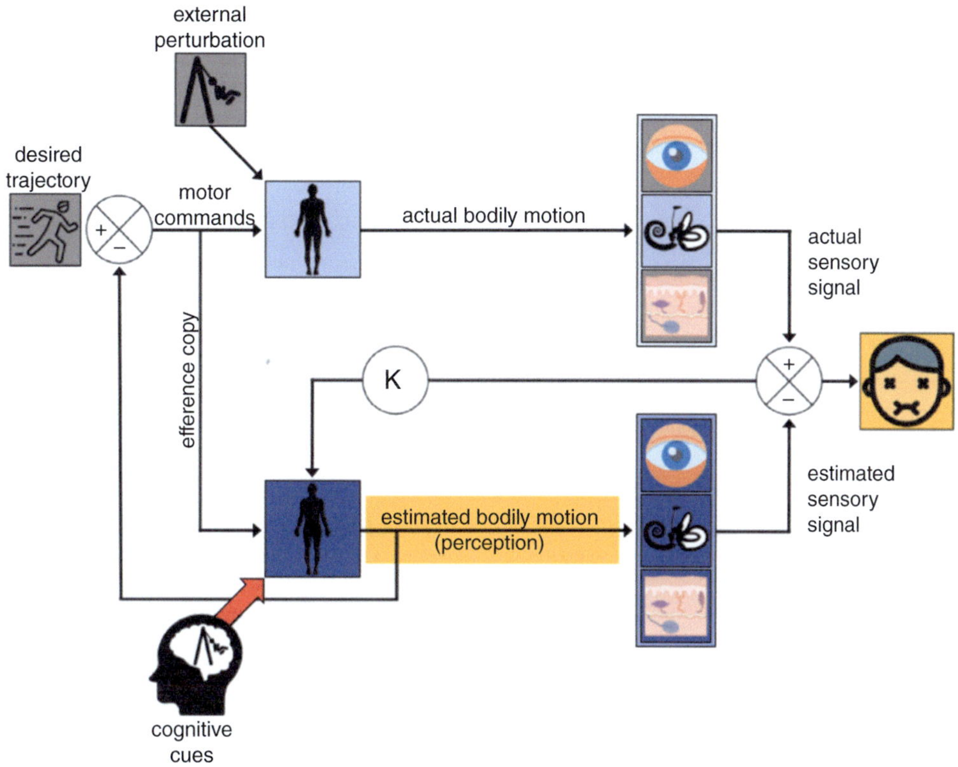

Fig. 131.5 A proposal of how cognitive factors can influence internal models of motion perception, from Reuten et al. [10]. This depicts a simplified model of sensorimotor control, including the perception of motion and the origin of motion sickness. Light blue boxes represent the actual motor and sensory systems; the dark blue boxes represent the internal models of these systems. In the study by Reuten et al., the investigators minimized physical (active and passive), visual, somatosensory, and auditory cues about self-motion (in gray). The remaining inputs were cognitive and vestibular cues on (the lack of) motion. The measures of interest for Reuten et al. were perceived self-motion and motion sickness (yellow elements). Reproduced with written permission from Springer Nature B.V.

> These studies all concerned experiments with motion stimuli and thus reflect modulations of a percept of self-motion that is elicited by sensory stimulation. Because we are not aware of any study on self-motion perception without a motion stimulus, we performed a study investigating whether cognitive cues can elicit a percept of self-motion in the absence of sensory motion. In specific, we minimized physical (inertial), visual, somatosensory and auditory cues about self-motion [10].

Their experiments demonstrated that, indeed, it is possible to elicit (using various cognitive cues) perception of motion even in the absence of any true movement (and thus without stimulating any vestibular input). In order to unify findings from previous research (showing that cognitive factors could modulate perception of actual movement) with their own research (showing that cognitive factors could provoke perception of movement in the absence of any such movement), they proposed that the underlying mechanism has to do with internal modeling [20]. Reuten and colleagues [10] provide a schematic summary of this idea in Fig. 131.5 below.

They explain their model as follows:

> The internal model of the bodily dynamics (dark blue box) estimates the bodily motion that would result from the motor commands. This estimation controls our perception of self-motion. Under optimal conditions, it equals the actual self-motion produced by the real body (light blue box). As this prediction lacks the delay and other peculiarities of the sensorimotor system, it is the best input for feedback control of self-motion. To ensure that this estimate is indeed accurate, this signal is fed to the internal model predicting the sensory signals (visual, vestibular, and somatosensory; dark blue). If there is a difference between the resulting integrated estimated and actual sensory signal, their discrepancy will be used to update the internal model and hence the estimated bodily motion… The updating of the internal model is weighted relative to the noise of the actual sensory systems by a gain K. A high uncertainty about those signals is then accounted for by a low gain and, vice versa, a low uncertainty by a high gain. In the current study, we are interested whether not only efference copies but also motion expectations generated by cognitive cues influence the estimated bodily motion. Given that the updating of the internal model is based on a Kalman gain, motion will only be reported when there is a low signal to noise ratio of the senses. If so, cognitive cues could result in the perception of self-motion in the actual absence of motion [10].

We find this model compelling, as it dovetails with other research regarding internal modeling [21] as applied to sensorimotor control [22].

Epidemiology

The epidemiology of psychological disequilibrium has been difficult to estimate. Ardic and Atesci comment on the broad range of estimates: "The ratio of the patients with psychogenic dizziness reported in different studies varies between 9 and 55%" [23]. This range is likely due to differences in methodology and to the lack of uniformity in definitions between studies.

Additional factors that complicate such estimations include the relative timing of the symptoms and different assumptions regarding causality. Dieterich and Staab [5] comment:

> Retrospective, cross-sectional, and prospective studies conducted over three decades suggest that 8–10% of patients with vestibular symptoms have an anxiety or depressive disorder as the primary cause of their illness. One in eight will develop a de novo anxiety or depressive disorder triggered by an acute vestibular disease, and a total of 30–50% will manifest anxiety or depressive morbidity over the course of their vestibular illness. Another 15–20% will primarily manifest the functional vestibular syndromes of PPV [phobic postural vertigo] or CSD [chronic subjective dizziness]. Thus, functional and psychiatric disorders may be primary, secondary, or comorbid problems in many patients presenting for evaluation of vestibular symptoms. In other words, these conditions are common causes, consequences, and complications of vestibular symptoms [5].

Clinical Presentation

Some researchers conclude that a patient's description of symptoms alone can successfully discriminate psychological disequilibrium from "otologic" disequilibrium [3].

Other authors are more cautious in this regard. For example, Dieterich and colleagues [5] comment that:

> The clinical history is paramount because functional and psychiatric vestibular disorders are diagnosed by history. There are no single pathognomonic symptoms for these illnesses, but key elements of clinical history are more common in structural versus functional versus psychiatric syndromes. Combinations of symptoms and their temporal associations are keys to properly making (i.e., ruling in) diagnoses. Importantly, however, the existence of features in one category does not exclude disorders in other categories from consideration. High rates of comorbidity make it impossible to use the presence of one set of features to eliminate (i.e., rule out) other illnesses [5].

To this end, Dieterich and colleagues [5] offer the following *Table* 131.1 comparing symptom tendencies in "structural disorders," "functional disorders," and "psychiatric disorders."

Finally, the clinician should bear in mind that these symptoms are protean:

> Patients' dominant symptoms may change over time as one disorder begets another, so it is important to follow the temporal evolution of their illnesses. Misattribution of all symptoms to the initial illnesses may incorrectly keep the focus of diagnostic evaluations and treatment on those conditions, even if they have remitted [5].

Table 131.1 Symptom tendencies in "structural disorders," "functional disorders," and "psychiatric disorders."From Dieterich et al. [5]

		Structural disorders	Functional disorders	Psychiatric disorders
Acute or episodic syndromes	Vestibular symptoms	Vertigo attacks		Dizziness Unsteadiness
	Motion sensitivity	Vertigo in specific positions Vertigo with rapid head tilts or turns		Fear of provocative motion
	Aural symptoms	Fluctuating hearing loss Unilateral tinnitus		
	Autonomic and vegetative symptoms	Nausea Emesis	Mild queasiness without emesis Mild diaphoresis, tremulousness	Chest pain, palpitations, dyspnea, tremulousness, paresthesias
	Posture and gait abnormalities	Direction-specific pulsion or falls Drop attacks Ataxia	Variable gait and stance deficits Excessive upper-body motion	Excessive caution
	Emotional symptoms			Fear of falling or heights Fear of becoming incapacitated or accidentally harming others

(continued)

Table 131.1 (continued)

		Structural disorders	Functional disorders	Psychiatric disorders
Chronic syndromes	Vestibular symptoms	Unsteadiness Dizziness	Unsteadiness Dizziness	Unsteadiness Dizziness
	Motion sensitivity	Vertigo, unsteadiness, or oscillopsia with rapid head movements	Unsteadiness or dizziness with motion of the self in any direction or movement in the environment	Avoidance of provocative motion environments
	Aural symptoms	Progressive hearing loss Unilateral tinnitus		Variable tinnitus
	Autonomic and vegetative symptoms		Chronic fatigue	Tension, restlessness, insomnia, fatigue, weight change Cognitive complaints
	Postural and gait abnormalities	Progressive decline in gait speed or fluidity Gradually increasing frequency of falls	Postural unsteadiness or dizziness Variable sensations of impending falls, toppling over, or floating off the ground	Excessive caution Unnecessary use of gait aids
	Emotional symptoms			Catastrophic worries about consequences of symptoms Demoralization or sadness Pessimism or hopelessness

Testing: Vestibular

Several studies have exposed patients with *anxiety* to various optic stimuli and identified increased sway on static posturography compared to healthy controls [24, 25].

Several studies have explored vestibular tests in patients with *panic disorder*. Hoffman and colleagues studied vestibular autorotation tests in 19 patients diagnosed with panic disorder and reported asymmetries in vestibulo-ocular reflex gain in 100% of patients [26]. Jacob and colleagues [27] studied patients with variable combinations of panic disorder, anxiety, depression, and agoraphobia and reported that, "Vestibular abnormalities were common in all the groups but most prevalent in the patients with panic disorder with moderate to severe agoraphobia." Jacob and colleagues [28] studied computerized dynamic posturography in patients with variable combinations of panic disorder, anxiety, depression, and agoraphobia and reported that "Agoraphobics rely on proprioceptive cues for maintenance of upright balance." Jacob and colleagues [29] studied several vestibular tests in 21 patients (8 with panic disorder, 13 with agoraphobia and panic attacks) and reported that 67% exhibited positional or spontaneous nystagmus, 56% exhibited abnormal calorics, 35% exhibited abnormalities on rotational testing, and 32% exhibited abnormal results on posturography. In a subsequent study, Jacob and colleagues [30] studied several vestibular tests in eight patients with panic disorder and reported that six (75%) had abnormal results on computerized dynamic posturography, four (50%) had unilateral caloric weakness, and two (25%) showed directional preponderance on rotational testing. Sklare and colleagues [31] studied 17 patients diagnosed with panic disorder (with and without agoraphobia) and reported that 71% exhibited abnormalities on electronystagmography, though other studies did not find such abnormalities [32]. Some of the vestibular studies in patients with panic disorder are surprising and difficult to interpret. For example, it does not seem plausible that purely psychological factors could cause significant caloric asymmetry; common causes of unilateral vestibular weakness (such as vestibular neuritis) should probably be entertained, irrespective of any psychiatric diagnoses.

In any case, the various vestibular test result abnormalities found in patients with anxiety and panic disorder (with and without agoraphobia) are neither sensitive nor specific to these conditions, and in our view, such findings should prompt consideration of primary otovestibular disorders.

Differential Diagnosis

Psychological dizziness as we have defined it here is different from diagnoses such as phobic postural vertigo [9, 33], space and motion discomfort [34], visually induced dizziness [35], chronic subjective dizziness [36], and persistent postural perceptual dizziness [37].

Although the otoneurologist and neuro-otologist are tasked with diagnosing and managing disequilibrium, in psychological disequilibrium, there may also be a diagnosable psychiatric condition that merits management.

Dieterich and colleagues offer the following list of "Psychiatric disorders that may cause or contribute to vestibular symptoms, ranked by likelihood of doing so" [5]:

1. Anxiety and phobic disorders.
2. Traumatic stress and obsessive-compulsive disorders (via associated panic attacks and chronic anxiety).
3. Depressive disorders.
4. Somatic symptom disorder/bodily distress disorder.
5. Dissociative disorders (including depersonalization/derealization syndromes).

In Table 131.2 below, Brandt [9] provides a more detailed, though unprioritized, list of psychiatric diagnoses in association with which symptoms of disequilibrium ("vertigo and dizziness") can occur.

Table 131.2 Psychiatric diagnoses in association with which disequilibrium has been described. From Brandt [9]

DSM-IV classification	Description
Vertigo as an associated symptom	
(295.xxx)	Schizophrenia
(297.1)	Delusional (paranoid) disorder (somatic type)
(296.2)	Major depressive disorder (300.4) dysthymic disorder (or depressive neurosis)
(300.4)	Dysthymic disorder (or depressive neurosis)
(300.02)	Generalized anxiety disorder
(300.11)	Conversion disorder (or hysterical neurosis conversion type)
(300.7)	Hypochondriasis (or hypochondriacal neurosis)
(300.81)	Somatoform disorder
(300.6)	Depersonalization disorder (or depersonalization neurosis)
(300.19)	Factitious disorder with physical symptoms
(309.9)	Adjustment disorders unspecified (with physical complaints)
Vertigo as a defined syndrome	
(300.21)	Panic disorder with agoraphobia
(300.22)	Agoraphobia without a history of panic disorder
(300.29)	Acrophobia (specific phobia) Phobic postural vertigo Space phobia
Psychogenic overlay of organic vertigo syndromes	
	Predisposed personalities Manifest psychiatric disorders Deliberate aggravation of symptoms
(V 65.2)	Malingering

Treatment

It is generally reasonable to consider psychiatric consultation. The rationale for this comes from the potential for mutual reinforcement that psychological symptoms and symptoms of disequilibrium can have on each other. As Jacob and Furman [4] comment:

> We have seen that psychological factors can exacerbate vestibular symptoms, and that vestibular dysfunction may result in anxiety. In fact, it is likely that vestibular symptoms and anxiety are mutually enhancing, leading to chronicity of both the vestibular and psychiatric symptoms [4].

Thus, whether psychological factors are the exclusive cause of the symptom of disequilibrium, or psychological responses occur during the disequilibrium or following the disequilibrium, psychiatric consultation may help to "break" the maladaptive positive feedback loop ("vicious cycle") of mutually amplifying symptoms.

In our experience, the main barrier to this intervention is patients' reluctance to view any component of their symptoms as "psychological." We find this regrettable, as many patients probably miss an opportunity for improvement.

Prognosis

Prognosis depends on the success of psychological and/or psychiatric intervention.

References

1. Freud S. On the grounds for detaching a particular syndrome from neurasthenia under the description 'anxiety neurosis'. The standard edition of the complete psychological works of Sigmund Freud, vol. 3. London: Hogarth Press; 1964. p. 85–115.
2. Furman JM, Jacob RG. Psychiatric dizziness. Neurology. 1997;48(5):1161–6.
3. Clark MR, Sullivan MD, Fischl M, Katon WJ, Russo JE, Dobie RA, et al. Symptoms as a clue to otologic and psychiatric diagnosis in patients with dizziness. J Psychosom Res. 1994;38(5):461–70.
4. Jacob RG, Furman JM. Psychiatric consequences of vestibular dysfunction. Curr Opin Neurol. 2001;14(1):41–6.
5. Dieterich M, Staab JP, Brandt T. Functional (psychogenic) dizziness. Handb Clin Neurol. 2016;139:447–68.
6. Staab JP, Ruckenstein MJ. Which comes first? Psychogenic dizziness versus otogenic anxiety. Laryngoscope. 2003;113(10):1714–8.

7. Furman JM, Jacob RG. A clinical taxonomy of dizziness and anxiety in the otoneurological setting. J Anxiety Disord. 2001;15(1–2):9–26.
8. Cherchi M, Yacovino DA. Dysfunction along the continuum of vestibulocochlear anatomy, and the corresponding spectrum of clinical presentation: how little we know, and what else we need to learn. Hear Balance Commun. 2021;19:246–57.
9. Brandt T. Phobic postural vertigo. Neurology. 1996;46(6):1515–9.
10. Reuten AJC, Smeets JBJ, Martens MH, Bos JE. Self-motion perception without sensory motion. Exp Brain Res. 2022;240(10):2677–85.
11. Ferre ER, Haggard P. Vestibular cognition: state-of-the-art and future directions. Cogn Neuropsychol. 2020;37(7–8):413–20.
12. Ferre ER, Harris LR. Introduction to vestibular cognition special issue: Progress in vestibular cognition. Multisens Res. 2015;28(5–6):393–6.
13. Mast FW, Ellis AW. Internal models, vestibular cognition, and mental imagery: conceptual considerations. Multisens Res. 2015;28(5–6):443–60.
14. Mertz S, Belkhenchir S, Lepecq JC. Evidence of imagined passive self-motion through imagery-perception interaction. Acta Psychol. 2000;105(1):57–78.
15. Nigmatullina Y, Arshad Q, Wu K, Seemungal BM, Bronstein AM, Soto D. How imagery changes self-motion perception. Neuroscience. 2015;291:46–52.
16. Ellis AW, Klaus MP, Mast FW. Vestibular cognition: the effect of prior belief on vestibular perceptual decision making. J Neurol. 2017;264(Suppl 1):74–80.
17. D'Amour S, Harris LR, Berti S, Keshavarz B. The role of cognitive factors and personality traits in the perception of illusory self-motion (vection). Atten Percept Psychophys. 2021;83(4):1804–17.
18. Riecke BE. Cognitive and higher-level contributions to illusory self-motion perception ("vection"): Does the possibility of actual motion affect vection?(Spatio-temporal integration of multimodal sensations,Symposium 2 at the 27th Annual Meeting). Jpn J Psycho Sci. 2009;28(1):135–9.
19. Wertheim AH, Mesland BS, Bles W. Cognitive suppression of tilt sensations during linear horizontal self-motion in the dark. Perception. 2001;30(6):733–41.
20. Peterka RJ. Sensory integration for human balance control. Handb Clin Neurol. 2018;159:27–42.
21. Francis BA, Wonham WM. The internal model principle of control theory. Automatica. 1976;12(5):457–65.
22. Wolpert DM, Ghahramani Z, Jordan MI. An internal model for sensorimotor integration. Science. 1995;269(5232):1880–2.
23. Ardic FN, Atesci FC. Is psychogenic dizziness the exact diagnosis? Eur Arch Otorrinolaringol. 2006;263(6):578–81.
24. Jacob RG, Redfern MS, Furman JM. Optic flow-induced sway in anxiety disorders associated with space and motion discomfort. J Anxiety Disord. 1995;9(5):411–25.
25. Redfern MS, Furman JM, Jacob RG. Visually induced postural sway in anxiety disorders. J Anxiety Disord. 2007;21(5):704–16.
26. Hoffman DL, O'Leary DP, Munjack DJ. Autorotation test abnormalities of the horizontal and vertical vestibulo-ocular reflexes in panic disorder. Otolaryngol Head Neck Surg. 1994;110(3):259–69.
27. Jacob RG, Furman JM, Durrant JD, Turner SM. Panic, agoraphobia, and vestibular dysfunction. Am J Psychiatry. 1996;153(4):503–12.
28. Jacob RG, Furman JM, Durrant JD, Turner SM. Surface dependence: a balance control strategy in panic disorder with agoraphobia. Psychosom Med. 1997;59(3):323–30.
29. Jacob RG, Moller MB, Turner SM, Wall C 3rd. Otoneurological examination in panic disorder and agoraphobia with panic attacks: a pilot study. Am J Psychiatry. 1985;142(6):715–20.
30. Jacob RG, Lilienfeld SO, Furman JMR, Durrant JD, Turner SM. Panic disorder with vestibular dysfunction: further clinical observations and description of space and motion phobic stimuli. J Anxiety Disord. 1989;3(2):117–30.
31. Sklare DA, Stein MB, Pikus AM, Uhde TW. Dysequilibrium and audiovestibular function in panic disorder: symptom profiles and test findings. Am J Otol. 1990;11(5):338–41.
32. Swinson RP, Cox BJ, Rutka J, Mai M, Kerr S, Kuch K. Otoneurological functioning in panic disorder patients with prominent dizziness. Compr Psychiatry. 1993;34(2):127–9.
33. Brandt T, Huppert D, Dieterich M. Phobic postural vertigo: a first follow-up. J Neurol. 1994;241(4):191–5.
34. Jacob RG, Woody SR, Clark DB, Lilienfeld SO, Hirsch BE, Kucera GD, et al. Discomfort with space and motion: a possible marker of vestibular dysfunction assessed by the situational characteristics questionnaire. J Psychopathol Behav Assess. 1993;15(4):299–324.
35. Steenerson KK, Hoskin J, Fife TD. Visually induced dizziness. Curr Opin Neurol. 2022;35(1):113–7.
36. Staab JP. Chronic subjective dizziness. Continuum (Minneap Minn). 2012,18(5 Neuro-otology):1118–41.
37. Staab JP, Eckhardt-Henn A, Horii A, Jacob R, Strupp M, Brandt T, et al. Diagnostic criteria for persistent postural-perceptual dizziness (PPPD): consensus document of the committee for the classification of vestibular disorders of the Barany Society. J Vestib Res. 2017;27(4):191–208.

Dissociative Experiences in Vestibular Disorders

132

Overview

Dissociative experiences such as depersonalization and derealization are those in which there is a mismatch between an expected perception and an experienced perception, whether the discrepancy is in point of view (self vs. outside of self), in self-perception (customary or baseline body image vs. a distorted body image) or in reality (expected vs. actual perception of the environment). There is an emerging literature regarding dissociative experiences accompanying peripheral and central vestibular disorders. From the perspective of internal modeling, it is plausible that these symptoms reflect multisensory integration's failure to reconcile the mismatch between expected versus experienced vestibular input in patients with vestibular disease.

Introduction

Dissociative experiences such as depersonalization and derealization are those in which there is a mismatch between an expected perception and an experienced perception, whether the discrepancy is in point of view (self vs. outside of self), in self-perception (customary or baseline body image vs. a distorted body image) or in reality (expected vs. actual perception of the environment).

The terminology in this area is not uniform and includes:

- *Out-of-Body Experiences* (OBE) [1] and *Autoscopic* (AS) Hallucinations [2]: "During an out-of-body experience (OBE), the experient seems to be awake and to see his body and the world from a location outside the physical body. A closely related experience is autoscopy (AS), which is characterized by the experience of seeing one's body in extrapersonal space" [3].
- *Depersonalization* [4, 5].
- *Derealization* [4, 5].
- *Distorted body image (body dysmorphism)* [2].

Blanke and colleagues note that these abnormal perceptions "are frequently associated with pathological sensations of position, movement, and perceived completeness of one's own body. These include vestibular sensations (such as floating, flying, elevation, and rotation), visual body-part illusions (such as the illusory shortening, transformation, or movement of an extremity), and the experience of seeing one's body only partially" [3].

Epidemiology

Of the few studies that comment on the epidemiology of these symptoms, some describe a higher prevalence in women and in patients with visual/vestibular hypersensitivity, migraine, and motion sickness [6]. Other studies report a higher prevalence and frequency of these symptoms in patients with anxiety [7].

Pathophysiological Mechanism of Disease

The underlying mechanism of these symptoms is unknown, but a series of observations has led to reasonable hypotheses. For example, several studies have noted that patients with peripheral otologic vestibular disorders report symptoms of depersonalization and derealization [8–10] and out-of-body experiences [1]. Similar findings have been reported in a series of patients with peripheral and central vestibular disorders [6, 11].

Several studies have proposed that these experiences result from failed multisensory integration in the temporoparietal cortex [2, 3], including failure to reconcile vestibular inputs with other sensory inputs [12, 13].

Experimental support for this idea comes from the observation that subjects undergoing caloric testing (which by its nature elicits a nonphysiological vestibular perception that is discrepant with other sensory inputs) sometimes report

M. Cherchi, *Otoneurology and Vestibular Medicine*, https://doi.org/10.1007/978-3-031-94841-1_132

depersonalization/derealization [4, 9] or distorted self-body perceptions [5]. Similar findings are reported in patients undergoing unilateral utricular stimulation during unilateral utricular centrifugation [14].

Blake and colleagues speculate that:

> The complex experiences of OBE and AS represent paroxysmal disorders of body perception and cognition (or body schema). The processes of body perception and cognition, and the unconscious creation of central representation(s) of one's own body based on proprioceptive, tactile, visual, and vestibular information—as well as their integration with sensory information of extrapersonal space—are a prerequisite for rapid and effective action with our surroundings. Based on our findings, we speculate that ambiguous input from these different sensory systems is an important mechanism of OBE and AS, and thus the intriguing experience of seeing one's body in a position that does not coincide with its felt position. We suggest that OBE and AS are related to a failure to integrate proprioceptive, tactile, and visual information with respect to one's own body (disintegration in personal space) and by a vestibular dysfunction leading to an additional disintegration between personal (vestibular) space and extrapersonal (visual) space. We argue that both disintegrations (personal; personal-extrapersonal) are necessary for the occurrence of OBE and AS [3].

From the perspective of control systems analysis, these experiences reflect a discrepancy between predictions of internal models (of self and environment) [15, 16] and experienced perceptions.

Final Comments

In our clinical experience, patients presenting to an otoneurology clinic frequently have difficulty articulating their symptoms of disequilibrium, and often use phrases such as "I just feel weird," "Something's just not right," "I feel a disconnect," and so forth. We suspect that these reflect the dissociative experiences that can accompany vestibular disorders irrespective of etiology.

References

1. Lopez C, Elzière M. Out-of-body experience in vestibular disorders – a prospective study of 210 patients with dizziness. Cortex. 2018;104:193–206.
2. Mohr C, Blanke O. The demystification of autoscopic phenomena: experimental propositions. Curr Psychiatry Rep. 2005;7(3):189–95.
3. Blanke O, Landis T, Spinelli L, Seeck M. Out-of-body experience and autoscopy of neurological origin. Brain. 2004;127(Pt 2):243–58.
4. Jáuregui-Renaud K, Sang FY, Gresty MA, Green DA, Bronstein AM. Depersonalisation/derealisation symptoms and updating orientation in patients with vestibular disease. J Neurol Neurosurg Psychiatry. 2008;79(3):276–83.
5. Lopez C, Nakul E, Preuss N, Elzière M, Mast FW. Distorted own-body representations in patients with dizziness and during caloric vestibular stimulation. J Neurol. 2018;265(Suppl 1):86–94.
6. Toupet M, Van Nechel C, Hautefort C, Heuschen S, Duquesne U, Cassoulet A, Bozorg GA. Influence of Visual and vestibular hypersensitivity on derealization and depersonalization in chronic dizziness. Front Neurol. 2019;10:69.
7. Kolev OI, Georgieva-Zhostova SO, Berthoz A. Anxiety changes depersonalization and derealization symptoms in vestibular patients. Behav Neurol. 2014;2014:847054.
8. Grigsby JP, Johnston CL. Depersonalization, vertigo and Ménière's disease. Psychol Rep. 1989;64(2):527–34.
9. Sang FY, Jáuregui-Renaud K, Green DA, Bronstein AM, Gresty MA. Depersonalisation/derealisation symptoms in vestibular disease. J Neurol Neurosurg Psychiatry. 2006;77(6):760–6.
10. Sun Y, Coltisor A, Jacobson GP, Roberts RA. Case study: depersonalization and vestibular impairment. J Am Acad Audiol. 2021;32(5):324–30.
11. Elyoseph Z, Geisinger D, Zaltzman R, Gordon CR, Mintz M. How vestibular dysfunction transforms into symptoms of depersonalization and derealization? J Neurol Sci. 2023;444:120530.
12. Cheyne JA, Girard TA. The body unbound: vestibular-motor hallucinations and out-of-body experiences. Cortex. 2009;45(2):201–15.
13. Jáuregui RK. Vestibular function and depersonalization/derealization symptoms. Multisens Res. 2015;28(5–6):637–51.
14. Aranda-Moreno C, Jáuregui-Renaud K. Derealization during utricular stimulation. J Vestib Res. 2016;26(5–6):425–31.
15. Vogeley K, Kurthen M, Falkai P, Maier W. Essential functions of the human self model are implemented in the prefrontal cortex. Conscious Cogn. 1999;8(3):343–63.
16. Saini F, Ponzo S, Silvestrin F, Fotopoulou A, David AS. Depersonalization disorder as a systematic downregulation of interoceptive signals. Sci Rep. 2022;12(1):22123.

Psychiatric Disease and Vestibular Dysfunction

133

Overview

There is a modest literature regarding ocular motor abnormalities in psychiatric disease, and most (though not all) of this literature pertains to schizophrenia. This is a difficult area to study because it can be challenging to determine whether a given ocular motor finding is attributable [1] to psychiatric disease, or [2] to the centrally acting medication used to treat that psychiatric disease, or [3] to an independently occurring vestibular problem. Practically, a clinician needs to keep these different possibilities in mind when interpreting ocular motor findings in patients with psychiatric disease.

Introduction

Ocular motor abnormalities have been reported in patients with various psychiatric diseases who do not complain of vestibular symptoms [1, 2]. Some studies report similar ocular motor findings in schizophrenics and patients with other psychiatric pathology [3, 4], and therefore conclude that such findings cannot distinguish among these psychopathologies. Some studies conclude that ocular motor abnormalities cannot even distinguish patients with psychiatric disease from those without [5].

Most literature regarding ocular motor findings in psychiatric disease pertains to schizophrenia. Less commonly studied psychiatric diseases include psychosis [6] and "manic-depression" (now renamed bipolar disorder) [4, 7].

Epidemiology

As far as schizophrenia is concerned, some studies observe that up to 80% of patients have detectable ocular motor abnormalities [8].

Genetics

A number of studies of smooth-pursuit abnormalities in schizophrenics and their (nonschizophrenic) relatives suggest that there is likely a genetic mechanism underlying these ocular motor findings [1, 8–10]. For example:

- Holzman and colleagues note that, "Disordered smooth-pursuit eye movements occur in a high percentage of schizophrenic patients and their first-degree relatives. A test of the hypothesis that these disorders represent a genetic indicator of schizophrenia was undertaken by testing pursuit eye movements in a sample of monozygotic and dizygotic twins discordant for clinical schizophrenia. Deviant eye tracking is significantly concordant within monozygotic twin pairs, and less so within dizygotic twin pairs discordant for schizophrenia. A genetic interpretation is consistent with these results" [9].
- As far as schizophrenia is concerned, Latham and colleagues note that, "Between 65% and 80% of schizophrenics and about 45% of their first-degree relatives have disordered horizontal pursuit eye movements and that vertical pursuit is similarly impaired," and since such ocular motor abnormalities appear "in a significant proportion of unaffected first-degree relatives of schizophrenic patients… it seems to have a strong genetic component" [8].

While these conclusions are suggestive, no specific genes have yet been identified to explain the ocular motor findings.

Pathophysiological Mechanism of Disease

The mechanism by which psychiatric disease may be connected with ocular motor abnormalities is not known. Latham and colleagues [8] suggest that cortical dysfunction underlies abnormal optokinetic and pursuit eye movements in schizophrenics.

M. Cherchi, *Otoneurology and Vestibular Medicine*, https://doi.org/10.1007/978-3-031-94841-1_133

Testing: Instrumented Ocular Motor Examination

Latham and colleagues [8] review literature concluding that, "Between 65% and 80% of schizophrenics… have disordered horizontal-pursuit eye movements and that vertical pursuit is similarly impaired."

Latham and colleagues studied full-field optokinetic responses and partial-field optokinetic responses by electronystagmography in 12 schizophrenics and five normal subjects and reported that, "The large differences in dysrhythmia between the nystagmus types in all 12 subjects are consistent with two separate nystagmus mechanisms. The reflex form of OKN (full-field OKN), which involves a simple, direct brain mechanism, is without exception more regular and stable than partial-field OKN, although there was more dysrhythmia in the responses of the seven schizophrenics than in the five controls" and conclude that this may be "due to a cortical dysfunction" [8].

Several electronystagmographic studies of optokinetic responses report consistent differences between schizophrenics and normal control subjects [8, 11].

Testing: Vestibular

There is disagreement in the literature regarding whether schizophrenic patients exhibit consistently different caloric responses from normal controls [11–13].

Differential Diagnosis

A significant difficulty in studying ocular motor function in psychiatric disease is that many of these patients are being treated with one or more centrally acting medications that can in themselves provoke ocular motor abnormalities. Thus, in a patient being treated for psychiatric disease who has ocular motor abnormalities, it can be difficult to determine whether the findings are attributable to the psychiatric disease itself, or to its treatment, or whether the patient has independently occurring vestibular pathology.

Treatment

Although a number of studies report ocular motor abnormalities in patients with various psychiatric diseases, few or no studies explore whether treatment of the psychiatric disease influences the ocular motor findings.

Prognosis

As far as schizophrenia is concerned, Latham and colleagues report that the ocular motor abnormalities tend to remain stable over time [8].

Final Comments

The ocular motor abnormalities discussed here are not used to confirm or refute the associated psychiatric diagnoses. Rather, we bring this to readers' attention so that when one is evaluating a patient with diagnosed psychiatric disease, one will bear in mind that some ocular motor findings may be associated with the psychiatric disease itself, rather than reflect primary vestibular dysfunction.

References

1. Lipton RB, Levy DL, Holzman PS, Levin S. Eye movement dysfunctions in psychiatric patients: a review. Schizophr Bull. 1983;9(1):13–32.
2. Kiyomizu K, Matsuda K, Torihara K, Nakayama M, Komaki S, Tono T, et al. Nystagmus using video-oculography in psychiatric patients. Eur Arch Otorhinolaryngol. 2009;266(8):1167–74.
3. Yee RD, Baloh RW, Marder SR, Levy DL, Sakala SM, Honrubia V. Eye movements in schizophrenia. Invest Ophthalmol Vis Sci. 1987;28(2):366–74.
4. Lipton RB, Levin S, Holzman PS. Horizontal and vertical pursuit eye movements, the oculocephalic reflex, and the functional psychoses. Psychiatry Res. 1980;3(2):193–203.
5. Levy DL, Holzman PS, Proctor LR. Vestibular dysfunction and psychopathology. Schizophr Bull. 1983;9(3):383–438.
6. Holzman PS, Kringlen E, Levy DL, Haberman SJ. Deviant eye tracking in twins discordant for psychosis. A replication. Arch Gen Psychiatry. 1980;37(6):627–31.
7. Jones AM, Pivik RT. Abnormal visual—vestibular interactions in psychosis. Biol Psychiatry. 1983;18(1):45–61.
8. Latham C, Holzman PS, Manschreck TC, Tole J. Optokinetic nystagmus and pursuit eye movements in schizophrenia. Arch Gen Psychiatry. 1981;38(9):997–1003.
9. Holzman PS, Kringlen E, Levy DL, Proctor LR, Haberman SJ, Yasillo NJ. Abnormal-pursuit eye movements in schizophrenia. Evidence for a genetic indicator. Arch Gen Psychiatry. 1977;34(7):802–5.
10. Holzman PS, Levy DL, Proctor LR. Smooth pursuit eye movements, attention, and schizophrenia. Arch Gen Psychiatry. 1976;33(12):1415–20.
11. Pawlak-Osińska K, Kaźmierczak H, Osiński P, Michorzewski A. Electronystagmographic study in chronic schizophrenia. Int Tinnitus J. 2000;6(2):172–4.
12. Levy DL, Holzman PS, Proctor LR. Vestibular responses in schizophrenia. Arch Gen Psychiatry. 1978;35(8):972–81.
13. Kiyomizu K, Matsuda K, Torihara K, Nakayama M, Ishida Y, Yoshida K, Tono T. Neuro-otological findings in psychiatric patients with nystagmus. Eur Arch Otorhinolaryngol. 2011;268(12):1713–9.

Cognitive Impairment from Disequilibrium

134

Overview

The relationship between the symptom of disequilibrium and cognitive deficits is still poorly understood, but research is beginning to delineate potential mechanisms of this connection. A variety of visuospatial, nonspatial, and emotional deficits appear to be associated with vestibular dysfunction. While cognitive deficits may co-occur with the symptom of disequilibrium itself, it appears that some cognitive deficits may persist even between discrete episodes of disequilibrium. Some literature suggests that different vestibular diseases are associated with specific patterns of cognitive dysfunction.

How Do We Measure the Cognitive Deficits Associated with Disequilibrium?

Some investigators suspect that standard psychometric testing may not be adequate to capture the cognitive deficits associated with disequilibrium [1]. Other research concludes that standard "dizziness questionnaire" metrics, such as the Dizziness Handicap Inventory [2], do not elicit sufficient information to draw conclusions about cognitive deficits [3].

Consequently, some investigators have developed psychometric tests specific to the cognitive deficits occurring in the context of disequilibrium [4, 5].

What Sorts of Cognitive Deficits Occur with Disequilibrium?

Some investigators conclude that the predominant adverse cognitive effects of disequilibrium lie in the domain of visuospatial abilities [6, 7]; for example, "The strongest evidence exists in connecting vestibular function to the cognitive domain of visuospatial ability, which includes spatial memory, navigation, mental rotation, and mental representation of three-dimensional space" [8].

One study observed that cognitive deficits involve spatial and nonspatial tasks [9, 10] and that these deficits are more pronounced in patients with bilateral vestibular loss compared with unilateral vestibular loss [11]. Surprisingly, some investigators identified different degrees of cognitive deficits depending on the lateralization of vestibular weakness [12].

Cognitive Deficits from Disequilibrium Appear to Be Independent of Age and Emotional Distress

Although normal healthy aging is known to be associated with declines in vestibular function [13] and cognition, research suggests that vestibular disease can cause cognitive dysfunction beyond what is expected for age [14].

Disequilibrium from any cause appears to have the potential to cause emotional disorders [15, 16], disturbances of sleep and wakefulness [17], and cognitive deficits. Some investigators have concluded that the cognitive deficits are beyond what would be expected from emotional distress alone [18].

Are These Cognitive Deficits Only Present When Perceiving Disequilibrium?

Some investigators have noted that patients suffering from episodic vestibular disorders may exhibit cognitive deficits interictally—meaning deficits even during "dizzy-free" periods between discrete episodes of disequilibrium [19], suggesting that disequilibrium may cause longer-term neurological changes.

M. Cherchi, *Otoneurology and Vestibular Medicine*, https://doi.org/10.1007/978-3-031-94841-1_134

Can Different "Dizzy Diseases" Cause Different Patterns of Cognitive Deficits?

Some literature discusses cognitive impairment from disequilibrium caused by specific diseases, such as migraine-associated vertigo [20]. In a similar vein, some literature identifies different degrees of cognitive impairment in different vestibular diseases [21].

What Is the Mechanism by Which Disequilibrium Causes Cognitive Deficits?

Concerning the underlying mechanism of the reported cognitive deficits, some research identifies neuroanatomical projections from the vestibular nuclei to the cerebral cortex [22], hippocampus, and amygdala [23]. One study showed that patients with chronic bilateral vestibular loss exhibit hippocampal atrophy on MRI [24].

Some research observes significant variability in cognitive deficits associated with disequilibrium, and postulates that this variability is due to the fact that vestibular compensation progresses through different stages, each of which places different demands on attention, and thus proposes "the idea of a limited quantity of cognitive resources that can be allocated to cognitive tasks during the compensation stages. This basic mechanism of attentional limitations may lead to different compensation profiles in patients, with or without cognitive dysfunction, depending on the compensation stage" [25].

References

1. Donaldson LB, Yan F, Liu YF, Nguyen SA, Rizk HG. Does cognitive dysfunction correlate with dizziness severity in patients with vestibular migraine? Am J Otolaryngol. 2021;103124:103124.
2. Jacobson GP, Newman CW. The development of the dizziness handicap inventory. Arch Otolaryngol Head Neck Surg. 1990;116(4):424–7.
3. Dornhoffer JR, Liu YF, Zhao EE, Rizk HG. Does cognitive dysfunction correlate with dizziness severity in Meniere's disease patients. Otol Neurotol. 2021;42(3):e323–e31.
4. Lacroix E, Deggouj N, Salvaggio S, Wiener V, Debue M, Edwards MG. The development of a new questionnaire for cognitive complaints in vertigo: the Neuropsychological Vertigo Inventory (NVI). Eur Arch Otorrinolaringol. 2016;273(12):4241–9.
5. Jacobson GP, Piker EG, Hatton K, Roberts RA. A factor analytic assessment of the English translation of the neuropsychological vertigo inventory (NVI). J Otol. 2020;15(2):45–9.
6. Zhang X, Huang Y, Xia Y, Yang X, Zhang Y, Wei C, et al. Vestibular dysfunction is an important contributor to the aging of visuospatial ability in older adults-data from a computerized test system. Front Neurol. 2022;13:1049806.
7. Xie Y, Bigelow RT, Frankenthaler SF, Studenski SA, Moffat SD, Agrawal Y. Vestibular loss in older adults is associated with impaired spatial navigation: data from the triangle completion task. Front Neurol. 2017;8:173.
8. Bigelow RT, Agrawal Y. Vestibular involvement in cognition: visuospatial ability, attention, executive function, and memory. J Vestib Res. 2015;25(2):73–89.
9. Smith PF, Zheng Y, Horii A, Darlington CL. Does vestibular damage cause cognitive dysfunction in humans? J Vestib Res. 2005;15(1):1–9.
10. Risey J, Briner W. Dyscalculia in patients with vertigo. J Vestib Res. 1990;1(1):31–7.
11. Popp P, Wulff M, Finke K, Ruhl M, Brandt T, Dieterich M. Cognitive deficits in patients with a chronic vestibular failure. J Neurol. 2017;264(3):554–63.
12. Hufner K, Hamilton DA, Kalla R, Stephan T, Glasauer S, Ma J, et al. Spatial memory and hippocampal volume in humans with unilateral vestibular deafferentation. Hippocampus. 2007;17(6):471–85.
13. Agrawal Y, Van de Berg R, Wuyts F, Walther L, Magnusson M, Oh E, et al. Presbyvestibulopathy: diagnostic criteria consensus document of the classification committee of the Barany Society. J Vestib Res. 2019;29(4):161–70.
14. Rizk HG, Sharon JD, Lee JA, Thomas C, Nguyen SA, Meyer TA. Cross-sectional analysis of cognitive dysfunction in patients with vestibular disorders. Ear Hear. 2020;41(4):1020–7.
15. Beh SC. The neuropsychology of dizziness and related disorders. Otolaryngol Clin N Am. 2021;54(5):989–97.
16. Brandt T, Dieterich M. 'Excess anxiety' and 'less anxiety': both depend on vestibular function. Curr Opin Neurol. 2020;33(1):136–41.
17. Smith L, Wilkinson D, Bodani M, Bicknell R, Surenthiran SS. Short-term memory impairment in vestibular patients can arise independently of psychiatric impairment, fatigue, and sleeplessness. J Neuropsychol. 2019;13(3):417–31.
18. Xie D, Welgampola MS, Miller LA, Young AS, D'Souza M, Breen N, Rosengren SM. Subjective cognitive dysfunction in patients with dizziness and vertigo. Audiol Neurootol. 2021;1–11:122.
19. Chari DA, Liu YH, Chung JJ, Rauch SD. Subjective cognitive symptoms and dizziness handicap inventory (DHI) performance in patients with vestibular migraine and Meniere's disease. Otol Neurotol. 2021;42(6):883–9.
20. Wang N, Huang HL, Zhou H, Yu CY. Cognitive impairment and quality of life in patients with migraine-associated vertigo. Eur Rev Med Pharmacol Sci. 2016;20(23):4913–7.
21. Liu YF, Locklcar TD, Sharon JD, Lacroix E, Nguyen SA, Rizk HG. Quantification of cognitive dysfunction in dizzy patients using the neuropsychological vertigo inventory. Otol Neurotol. 2019;40(7):e723–e31.
22. Hanes DA, McCollum G. Cognitive-vestibular interactions: a review of patient difficulties and possible mechanisms. J Vestib Res. 2006;16(3):75–91.
23. Bigelow RT, Semenov YR, du Lac S, Hoffman HJ, Agrawal Y. Vestibular vertigo and comorbid cognitive and psychiatric impairment: the 2008 National Health Interview Survey. J Neurol Neurosurg Psychiatry. 2016;87(4):367–72.
24. Brandt T, Schautzer F, Hamilton DA, Bruning R, Markowitsch HJ, Kalla R, et al. Vestibular loss causes hippocampal atrophy and impaired spatial memory in humans. Brain. 2005;128(Pt 11):2732–41.
25. Lacroix E, Deggouj N, Edwards MG, Van Cutsem J, Van Puyvelde M, Pattyn N. The cognitive-vestibular compensation hypothesis: how cognitive impairments might be the cost of coping with compensation. Front Hum Neurosci. 2021;15:732974.

Cognitive Decline Associated with Hearing Loss

135

Overview

It is very well established that some degree of hearing loss comes with normal healthy aging, and insofar as such hearing loss is "expected," the term *presbycusis* refers to age-associated hearing loss. The degree of "expected" age-associated hearing loss is somewhat broad, and distinguishing presbycusis from hearing loss attributable to diagnosable disease is a nuanced task generally undertaken by an audiologist.

Some degree of cognitive decline is also "expected" with normal healthy aging; we might apply the term "*presbymentia*" to age-associated cognitive decline, but we are not aware of this term being used in the literature. Determining "age-appropriate" cognitive deficits from deficits due to diagnosable disease (such as various dementias) is a complex task, appropriate evaluation of which is generally undertaken by a cognitive neurologist.

Since presbycusis and "presbymentia" appear to occur with healthy aging, it is difficult to ascertain whether these influence each other; specifically, it is difficult to distinguish whether both declines are due to a common generalized age-associated attrition of function, or whether there is a causal link between them. However, a growing body of research suggests that hearing loss can accelerate cognitive decline beyond what would be expected for age [1].

We tend to view this observation as being in keeping with other bodies of research demonstrating that depriving an individual of a sensory modality from which they have had lifelong input appears to increase the risk of cognitive decline. This has been frequently observed in the association between visual decline and cognitive decline [2–7].

Such associations, if causal, raise the possibility that ameliorating the sensory deficit may offer an avenue for retarding or reversing the associated cognitive decline. Strikingly, this has been demonstrated for the sensory modality of hearing; specifically, it has been shown that the use of hearing restorative devices can decrease the risk of long-term cognitive decline [1].

This has potential relevance for individuals diagnosed with dementia, most forms of which are not directly treatable. If such individuals have correctable hearing loss, then this would be one of the few interventions with demonstrable efficacy.

Practically, we suggest that if an individual perceives a limitation in hearing that is interfering with their ability to function, then it is always reasonable to consult an audiologist for evaluation. If an individual with cognitive decline (of whatever cause) has never had a hearing evaluation, then the research cited above suggests that this would be medically reasonable.

References

1. Yeo BSY, Song H, Toh EMS, Ng LS, Ho CSH, Ho R, et al. Association of hearing aids and cochlear implants with cognitive decline and dementia: a systematic review and meta-analysis. JAMA Neurol. 2022;80:134.
2. Fang IM, Fang YJ, Hu HY, Weng SH. Association of visual impairment with cognitive decline among older adults in Taiwan. Sci Rep. 2021;11(1):17593.
3. Lim ZW, Chee ML, Soh ZD, Cheung N, Dai W, Sahil T, et al. Association between visual impairment and decline in cognitive function in a multiethnic Asian population. JAMA Netw Open. 2020;3(4):e203560.
4. Mohammed Z, Kee QT, Fadzil NM, Abd Rahman MH, Din NC, Mariappan V, Shahar S. Association between vision and cognitive function among community-dwelling older adults in Selangor, Malaysia. Int J Ophthalmol. 2023;16(1):115–20.
5. Ong SY, Cheung CY, Li X, Lamoureux EL, Ikram MK, Ding J, et al. Visual impairment, age-related eye diseases, and cognitive function: the Singapore Malay Eye study. Arch Ophthalmol. 2012;130(7):895–900.
6. Tai CJ, Tseng TG, Yeh CJ, Liao WC, Hsiao YH, Lee SH, et al. The association between visual trajectories and cognitive impairment: insights from the Taiwan Longitudinal Study on Aging. Aging Clin Exp Res. 2022;34(9):2129–37.
7. Vu TA, Gupta P, Leow FZY, Fenwick EK, Man REK, Tham YC, et al. The longitudinal association between cognitive impairment and incident visual impairment in a multiethnic Asian population: a prospective cohort study. Age Ageing. 2022;51(5)

M. Cherchi, *Otoneurology and Vestibular Medicine*, https://doi.org/10.1007/978-3-031-94841-1_135

Cognitive Impairment from Tinnitus

136

Overview

The relationship between tinnitus and cognitive deficits is poorly understood, and research on this topic is still in its early stages.

Introduction

It has been well documented that individuals suffering from tinnitus exhibit measurable cognitive deficits [1, 2].

Tinnitus is known to cause psychiatric disorders, and a number of psychiatric disorders can interfere with cognition. However, since some patients without psychiatric disease can also exhibit cognitive deficits, it is suspected that some degree of cognitive deficiency must be attributable to tinnitus itself, independent of any psychiatric mechanism [3].

Tinnitus often accompanies hearing loss [4], and hearing loss has been shown to predispose to cognitive deficits [5], consequently it has been difficult to disentangle the cognitive effects of hearing loss and tinnitus. Somewhat surprisingly, one study concluded that normal hearing tinnitus patients (i.e., those who have tinnitus but no measurable hearing loss) do not have cognitive deficits [6] though the authors acknowledge that this conclusion differs from that of other researchers.

What Are the Mechanisms by Which Tinnitus Might Cause Cognitive Deficits?

Some research postulates that tinnitus interferes with various facets of attention [7–9], perhaps by adversely affecting executive control [10], which in turn causes "cognitive inefficiency" [11].

Other studies conclude that, "tinnitus affects cognition to the extent that it reduces cognitive capacity needed to perform tasks that require voluntary, conscious, effortful, and strategic control" [12].

References

1. Andersson G, Eriksson J, Lundh LG, Lyttkens L. Tinnitus and cognitive interference: a stroop paradigm study. J Speech Lang Hear Res. 2000;43(5):1168–73.
2. Andersson G, McKenna L. The role of cognition in tinnitus. Acta Otolaryngol Suppl. 2006;556:39–43.
3. Jackson JG, Coyne IJ, Clough PJ. A preliminary investigation of potential cognitive performance decrements in non-help-seeking tinnitus sufferers. Int J Audiol. 2014;53(2):88–93.
4. Lockwood AH, Salvi RJ, Burkard RF. Tinnitus. N Engl J Med. 2002;347(12):904–10.
5. Jafari Z, Kolb BE, Mohajerani MH. Age-related hearing loss and cognitive decline: MRI and cellular evidence. Ann N Y Acad Sci. 2021;1500(1):17–33.
6. Waechter S, Brannstrom KJ. The impact of tinnitus on cognitive performance in normal-hearing individuals. Int J Audiol. 2015;54(11):845–51.
7. Stevens C, Walker G, Boyer M, Gallagher M. Severe tinnitus and its effect on selective and divided attention. Int J Audiol. 2007;46(5):208–16.
8. Mohamad N, Hoare DJ, Hall DA. The consequences of tinnitus and tinnitus severity on cognition: a review of the behavioural evidence. Hear Res. 2016;332:199–209.
9. Cardon E, Jacquemin L, Mertens G, Van de Heyning P, Vanderveken OM, Topsakal V, et al. Cognitive performance in chronic tinnitus patients: a cross-sectional study using the RBANS-H. Otol Neurotol. 2019;40(9):e876–e82.
10. Tegg-Quinn S, Bennett RJ, Eikelboom RH, Baguley DM. The impact of tinnitus upon cognition in adults: a systematic review. Int J Audiol. 2016;55(10):533–40.
11. Hallam RS, McKenna L, Shurlock L. Tinnitus impairs cognitive efficiency. Int J Audiol. 2004;43(4):218–26.
12. Rossiter S, Stevens C, Walker G. Tinnitus and its effect on working memory and attention. J Speech Lang Hear Res. 2006;49(1):150–60.

M. Cherchi, *Otoneurology and Vestibular Medicine*, https://doi.org/10.1007/978-3-031-94841-1_136

Part XX

Diagnoses—Ophthalmologic and Visual Factors in Disequilibrium

137 Visual Vertigo

Overview

Visual vertigo [1] refers to a perceived disturbance of equilibrium in response to visual stimuli. A broad range of visual experiences can trigger visual vertigo, including moving visual stimuli (such as movie screens), visual stimuli that are stationary but detailed or complex (such as a patterned carpet), or visual stimuli that conflict with other sensory input (such as when riding an escalator).

The brain relies heavily on visual input in order to calculate one's position in, and movement through, space. The influence of visual stimuli is such that simply showing someone videos or even stationary pictures can induce the patient to adjust his or her posture [2–5]. In some people, the brain appears to "pay more attention" to visual stimuli than is appropriate (which is sometimes called "visual dependence"), with the result being that these individuals experience sensitivity to visual stimuli that can manifest as a sensation of disturbed balance [6, 7].

In some cases, an individual with another balance disorder (such as from an inner ear problem) can additionally have a component of visual vertigo [6] due to visual dependence [8]. In such cases, it is especially important to recognize that there are multiple sources of the impaired balance. It is for this reason that a thorough assessment [9] is warranted before embarking on a treatment plan. Such assessment may include some vestibular ("inner ear") tests, balance tests (such as computerized dynamic posturography) [10], and a neuro-optometric examination—in other words, a multidisciplinary approach is often required.

A neuro-optometry examination evaluates specific ocular motor and visual functions, such as accommodation, convergence/divergence, phorias, fixation, binocularity, fusion, depth perception, and other aspects of visual processing, as well as the ways in which these phenomena interact with the vestibular system. A neuro-optometry examination may include refraction, a retinal examination, a slit lamp examination, and a Lancaster test.

Patients with neuro-optometric deficits may experience eye strain, double vision, visual fatigue, headaches, and visual vertigo. Such patients may have difficulty with specific visual tasks, such as reading, looking at computer screens, or traveling in vehicles.

There is controversy in the medical community regarding treatment for visual vertigo. Visual rehabilitation ("vision therapy") appears promising [11]. The evidence supporting this approach is modest, but since it appears to pose no medical risk, we offer it to patients for whom other treatment options are unavailable or impractical. A neuro-optometrist oversees this aspect of a patient's care. The goals of treatment may include decreasing sensitivity to visual motion, increasing visual comfort, and improving visual-vestibular integration.

In patients who have visual vertigo in addition to some other source of imbalance, a combination of treatments may be appropriate, including vision therapy, physical therapy, and medical therapy.

Discussion

Approximately 50% of all of the brain's pathways are related in some way to vision [12], so it is unsurprising that many nonvisual neurological functions interface in some way with visual functions. The maintenance of equilibrium is among these.

Visual vertigo (VV) remains a controversial topic, and one about which neuro-ophthalmologists and neuro-optometrists often disagree, both on the existence of the diagnosis and its treatment [13].

Insofar as visual vertigo is construed as a "visual processing" problem, candidate mechanisms include neurological disease such as *migraine*, which in itself can cause a variety of visual abnormalities, as well as photophobia and other types of visual hypersensitivity. Visual vertigo is probably distinct from misokinesia [14, 15].

M. Cherchi, *Otoneurology and Vestibular Medicine*, https://doi.org/10.1007/978-3-031-94841-1_137

It also seems likely that visual vertigo can arise from primary ophthalmologic processes (see ophthalmologic visual deficits and disequilibrium). Visual vertigo is often suspected in patients with unilateral (or asymmetrical) cataracts or macular degeneration, or strabismus. Common to these disease processes are *interocular disparities* which interfere with the brain's ability to fuse the images from each eye into a single coherent percept, and thus have the potential to *impair the stereopsis required for depth perception*. This type of visual impairment will limit the reliability/utility of visual input when the brain attempts to arrive at conclusions regarding a person's orientation and movement. Of particular relevance are diseases such as:

- *Anisometropia* (significantly different refractive needs between the two eyes). Although it is possible to provide each eye with a lens of different refractive power, one of the results of this is that the size of the image projected on each retina will be different (which is called *aniseikonia*), and the larger that difference, the more likely it is that the brain will fail to fuse those images, and thus depth perception will be impaired.
- *Intraocular lens implants* following cataract surgery. These implants cannot change shape (and therefore cannot adjust refraction for different viewing distances) the way that a native lens does, and thus have a fixed distance of best acuity.
- *Status post LASIK surgery*, particularly when configured for monovision (one eye used for near viewing, the other eye for far viewing).
- *Strabismus* (ocular misalignment) of any type. There are several possible consequences of ocular misalignment.
  (a) First, a patient will (maladaptively) continue to attempt using both eyes (even though they are misaligned) and thus will perceive two images (called diplopia), which is disorienting.
  (b) Second, a patient may simply suppress (ignore) the visual input from one eye (usually the nondominant eye). This strategy achieves haplopia (perception of a single object) at the expense of sacrificing stereopsis, and thus impairs depth perception.
  (c) Third, a patient may develop anomalous retinal correspondence (ARC) (more on this below).
- *Orbitopathies* in which the globe (eyeball) itself is vertically displaced, such as hypoglobus from silent sinus syndrome [16]. This situation creates a misalignment of the visual axes due to an anatomical abnormality, one that cannot be overcome by extra-ocular muscle adaptation.

Retinal correspondence merits further discussion as it is often overlooked in this context.

In normal retinal correspondence (NRC), correctly functioning ocular motility directs the eyes such that light emitted from a point source will fall on the same location of each retina, and the viewer will experience this perception as "seeing a single object" (haplopia). In other words, normal retinal correspondence is needed in order for the brain to fuse the images from the two eyes successfully into a single percept.

Anomalous retinal correspondence (ARC) was first described in 1826 by Johannes Peter Müller [17]. ARC occurs when malfunctioning ocular motility results in a situation in which light emitted from a point source falls on different (noncongruent, noncorresponding) locations in each retina. This means that when an individual tries to focus on a point of interest, that point will fall on the fovea (area of highest acuity) of one eye, but will fall on some extra-foveal location (of lower acuity) of the other eye. This attempts to fuse the images of the two eyes (in order to achieve haplopia), but one image will be of poorer acuity than the other—in other words, the attempt at fusion is only partially successful.

A neuro-optometrist or neuro-ophthalmologist may diagnose ARC using a synoptophore [18], Bagolini striated glasses [19, 20], the Worth 4-dot test [21] or the after-image test [22], although these tests sometimes give discrepant results [18, 23].

References

1. Bronstein AM. The visual vertigo syndrome. Acta Otolaryngol Suppl. 1995;520(Pt 1):45–8.
2. Guerraz M, Bronstein AM. Mechanisms underlying visually induced body sway. Neurosci Lett. 2008;443(1):12–6.
3. Guerraz M, Yardley L, Bertholon P, Pollak L, Rudge P, Gresty MA, Bronstein AM. Visual vertigo: symptom assessment, spatial orientation and postural control. Brain. 2001;124(Pt 8):1646–56.
4. Kuno S, Kawakita T, Kawakami O, Miyake Y, Watanabe S. Postural adjustment response to depth direction moving patterns produced by virtual reality graphics. Jpn J Physiol. 1999;49(5):417–24.
5. Lestienne F, Soechting J, Berthoz A. Postural readjustments induced by linear motion of visual scenes. Exp Brain Res. 1977;28(3–4):363–84.
6. Bronstein AM. Vision and vertigo: some visual aspects of vestibular disorders. J Neurol. 2004;251(4):381–7.
7. Bronstein AM, Golding JF, Gresty MA. Vertigo and dizziness from environmental motion: visual vertigo, motion sickness, and drivers' disorientation. Semin Neurol. 2013;33(3):219–30.
8. Cousins S, Cutfield NJ, Kaski D, Palla A, Seemungal BM, Golding JF, et al. Visual dependency and dizziness after vestibular neuritis. PLoS One. 2014;9(9):e105426.
9. Pavlou M, Davies RA, Bronstein AM. The assessment of increased sensitivity to visual stimuli in patients with chronic dizziness. J Vestib Res. 2006;16(4–5):223–31.
10. Bronstein AM. Visual vertigo syndrome: clinical and posturography findings. J Neurol Neurosurg Psychiatry. 1995;59(5):472–6.

11. Cohen AH. Vision rehabilitation for visual-vestibular dysfunction: the role of the neuro-optometrist. NeuroRehabilitation. 2013;32(3):483–92.
12. Patil SA, Grossman S, Kenney R, Balcer LJ, Galetta S. Where's the Vision? The Importance of Visual Outcomes in Neurologic Disorders: The 2021 H. Houston Merritt Lecture. Neurology. 2023;100(5):244–53.
13. Subramanian PS, Barton JJS, Ranalli P, Smith C, Francis CE, Frishberg B. Consensus statement on visual rehabilitation in mild traumatic brain injury. Neurol Clin Pract. 2022;12(6):422–8.
14. Jaswal SM, De Bleser AKF, Handy TC. Misokinesia is a sensitivity to seeing others fidget that is prevalent in the general population. Sci Rep. 2021;11(1):17204.
15. Webb J. Beta-blockers for the treatment of Misophonia and Misokinesia. Clin Neuropharmacol. 2022;45(1):13–4.
16. Pula JH, Mehta M. Silent sinus syndrome. Curr Opin Ophthalmol. 2014;25(6):480–4.
17. Simonsz HJ. First description of anomalous retinal correspondence by Johannes Peter Muller in 1826. Strabismus. 2010;18(3):116.
18. Rutstein RP, Daum KM, Eskridge JB. Clinical characteristics of anomalous correspondence. Optom Vis Sci. 1989;66(7):420–5.
19. Bagolini B. Sensorial anomalies in strabismus. (suppression, anomalous correspondence, amblyopia). Doc Ophthalmol. 1976;41(1):1–22.
20. Bagolini B. Scientific essay: [Bagolini's] striated GLASSES: a reappraisal. Binocul Vis Strabismus Q. 1999;14(4):266–71.
21. Worth CA. Squint: its causes, pathology, and treatment. London: Balliere, Tindall and Cox; 1903.
22. Hansen AK. After-image transfer test in anomalous retinal correspondence. AMA Arch Ophthalmol. 1954;52(3):369–74.
23. Garg MK, Jain IS, Gupta SD. Abnormal retinal correspondence, evaluation of the diagnostic procedures. J All India Ophthalmol Soc. 1969;17(6):242–4.

Ophthalmologic Visual Deficits and Disequilibrium

138

Overview

Vision is one of the three main sensory modalities involved in maintaining equilibrium (the other two are somatosensory input and vestibular input). Visual input can be abnormal in many ways (reduced acuity, reduced contrast sensitivity, impaired stereopsis, and depth perception) from a range of diseases (refractive errors, cataracts, glaucoma, retinopathy, macular degeneration). Ophthalmologic visual deficits interfere with visual input and can thereby cause disequilibrium. Studies have shown that ophthalmic visual deficits can lead to impaired postural stability and increased fall risk at any age. Most (though not all) studies show that treatment of the ophthalmologic visual deficit brings about improvement in postural stability and reduces fall risk. If treatment of the ophthalmologic visual deficit is not possible, or if it is unsuccessful, then a trial of physical therapy is medically reasonable; a basic mechanism of such therapy is to decrease reliance on visual input and increase reliance on other (intact) sensory modalities. Although ophthalmic visual deficits may be the main or only cause of disequilibrium, it is medically reasonable to check for common otovestibular diseases by undertaking a screening workup.

Introduction

It has frequently been observed that patients who fall are found to have higher rates of visual impairment. For example, Schwartz and colleagues [1], citing work by Lord and colleagues [2] comment that, "Elderly patients admitted to hospitals, especially those who have fallen, have a high prevalence of visual impairment."

This observation of this associative relationship (falling is associated with visual impairment) raises the question of whether there is also a causal relationship (visual impairment causes an increase in falls). Schwartz and colleagues comment that, "Vision is one of the three basic input channels controlling balance, together with the vestibular and somatosensory subsystems. It is not surprising, therefore, that visual dysfunction has been reported to be associated with an increased risk of falling" [1].

How would one study the relationship between visual problems and disequilibrium? Two broad considerations are relevant.

- Many factors can adversely impact *visual function*; the most commonly studied visual deficits include reduction in visual acuity, reduction in contrast sensitivity, and impaired depth perception [3–5]. The effects of each type of deficit may differ; for example, Lord [6] reported that impaired depth perception and reduced contrast sensitivity may have more of an impact than poor visual acuity. A variety of ocular diseases can cause these visual deficits, including refractive errors, cataracts, glaucoma, diabetic retinopathy, macular degeneration, and others [7, 8].
- The presence and degree of *disequilibrium* can be studied by a variety of methods. A common approach is to infer the presence of disequilibrium from an increased rate of falls—in other words, use reported falls as a proxy quantitative biomarker for disequilibrium. Another approach is to use a specific measure of equilibrium; these types of studies commonly analyze instrumented assessments of postural sway (such as static or dynamic computerized dynamic posturography). It must also be kept in mind that disequilibrium can result from a very broad range of other problems (vestibular, neurologic, orthopedic, cardiovascular, and others).

Thus, there is a many-to-many relationship between types of visual impairment and measures of disequilibrium. Consequently, it would be ideal if:

- Studies of the relationship between visual impairment and disequilibrium should, at minimum, begin by observing correlations between a particular measurable visual deficit and a particular measurable assessment of equilibrium.

M. Cherchi, *Otoneurology and Vestibular Medicine*, https://doi.org/10.1007/978-3-031-94841-1_138

- In studies of the relationship between visual impairment and disequilibrium, each group of patients should be studied against carefully matched healthy individuals to control for age and medical histories.

Most studies are not ideally designed. Nevertheless, below we review several broad categories of observations.

Visual Impairment Increases Fall Risk

Visual dysfunction (of undifferentiated etiologies) has been repeatedly reported to be associated with increased fall risk [5, 9–11].

Other studies focus on particular ocular diseases. For example:

- Reduced *visual acuity* has been reported to increase the rate of falls [12].
- *Cataracts* have been reported to increase fall risk in the elderly, even independent of visual acuity [13].
- Alexander and colleagues [14] report that age-related macular degeneration, especially in conditions of low lighting, may increase fall risk.

Visual Impairment Decreases Postural Stability

Visual dysfunction (of undifferentiated etiologies) has been shown to correlate with decreased postural stability [15–18].

Other studies focus on particular ocular diseases or specific visual functions. For example:

- Reduced *visual acuity* and reduced *contrast sensitivity* have been associated with decreased postural stability [19].
- *Cataract simulation* (via refractive blur) has been reported to have an adverse impact on postural stability [16].
- Black and colleagues [20] report that in older adults with *glaucoma*, greater visual field loss and thinner retinal nerve fiber layer thickness are associated with reduced postural stability.
- Chatard and colleagues [21] report that age-related *macular degeneration* (whether unilateral or bilateral) causes increased postural instability. Other studies report similar findings [22, 23].
- In the pediatric population, visual impairment from *strabismus* [24–26], with or without amblyopia [27], has been reported to impair postural stability.

Improved Vision Reduces Fall Risk

Some studies report that the treatment of *cataracts* reduces fall risk [28]. Other studies do not find such improvement in fall risk [29]. Some studies report that cataract surgery improved the symptoms of dizziness but did not affect fall risk [30]. Some studies even report an increased risk of falls after vision-improving interventions [31].

In view of the mixed results, Desapriya and colleagues [32] conducted a systematic review of the literature and performed a meta-analysis that included 737 participants from 22 studies. They reported that, "Pooling of data from two RCTs [randomized controlled trials] of 535 participants showed a nonsignificant reduction in the incidence of falls after expedited cataract surgery."

Improved Vision Increases Postural Stability

Some studies report improvement in postural stability following *cataract* surgery [1, 33].

Some studies of *strabismic* children report improvement in postural stability after correction [34, 35].

The Role of Stereopsis

A factor often overlooked in studies is whether the visual impairment in question is monocular or binocular. This is important because interocular disparities in visual function can impair stereopsis, and thereby adversely impact depth perception. Impaired depth perception can, in turn, cause or worsen disequilibrium.

Some studies are sufficiently astute to recognize this. For example, Felson and colleagues [9] observed that patients with good vision in one eye and poor vision in the other eye actually had a *higher fall risk* than in patients with comparably poor vision in both eyes; they stated, "This suggests that good stereoscopic vision may be necessary to prevent falls."

Epidemiology

The relationship between visual problems and disequilibrium is worth studying in all age groups. The prevalence of visual impairment, on average, increases with age [3] due to age-related changes such as cataracts and macular degeneration. Vision in the pediatric population can be adversely affected as well, although the etiologies of visual impairment may be different, such as strabismus and amblyopia [24, 27].

Pathophysiological Mechanism of Disease

At a basic level, as mentioned earlier, vision is one of the three main sensory modalities upon which the brain draws in order to maintain equilibrium, so impaired vision is likely to interfere with a person's stability and may increase fall risk.

Clinical Presentation

Many of the ophthalmological diseases we have discussed tend to develop gradually (macular degeneration, glaucoma, cataracts, diabetic retinopathy). Some may be congenital (e.g., strabismus). Often, when these patients are referred to otoneurology, their ophthalmologic disease has already been diagnosed, and the clinical query is whether some other (nonophthalmologic) factor is causing the symptom of disequilibrium.

Differential Diagnosis

In some cases, a patient's visual deficit may be the main contributor, or even the exclusive cause, of the symptom of disequilibrium. However, in order to reach that conclusion, it is medically reasonable to undertake a screening workup at least for the more common otovestibular etiologies of disequilibrium. Depending on the clinical scenario, we usually consider checking audiometry, cervical and ocular vestibular evoked myogenic potentials, video head impulse testing, videonystagmography, rotatory chair testing, and computerized dynamic posturography. If a patient has discrete visual complaints and has not yet had a thorough eye examination, then referral to ophthalmology is appropriate.

Treatment

If a visual deficit is believed to be contributing to disequilibrium, then that visual deficit (if treatable) is a reasonable therapeutic target. Some (though not all) of the studies cited earlier describe improvements in postural stability and falls after treatment aimed at improving vision.

If the visual deficit is untreatable, or treatment is unsuccessful, then other approaches can be considered.

Physical therapy is a reasonable, low-risk intervention for patients in whom visual deficits appear to be contributing to disequilibrium.

How does such an intervention work? In general, if a patient has disequilibrium due to deficient input of one sensory modality, the patient may unconsciously develop compensatory strategies that place greater reliance on other (intact) remaining sensory inputs. For example, Lions and colleagues [25] compared postural stability in strabismic children and healthy controls, and observed that strabismic children "use proprioceptive information more than control age-matched children to control their posture."

It is possible to leverage this ability—in other words, develop a treatment strategy whose purpose is to increase reliance on intact sensory inputs and reduce reliance on deficient ones. For example, Radvay and colleagues [36] studied 54 patients with age-related macular degeneration and 55 healthy control subjects. They commented that:

> Sixteen of these patients and 14 controls subsequently received balance training sessions on a postural platform (Multitest) stressing sensorimotor coordination by selectively inhibiting or disturbing either, visual, vestibular or somatosensory input. Producing a conflict between two inputs reinforces the use of the third [36].

In other words, physical therapy can be configured to increase a patient's reliance on intact sensory modalities and rely less on impaired sensory modalities.

Prognosis

Prognosis of disequilibrium from visual deficits will depend on the underlying etiology.

References

1. Schwartz S, Segal O, Barkana Y, Schwesig R, Avni I, Morad Y. The effect of cataract surgery on postural control. Invest Ophthalmol Vis Sci. 2005;46(3):920–4.
2. Lord SR, Ward JA, Williams P, Anstey KJ. Physiological factors associated with falls in older community-dwelling women. J Am Geriatr Soc. 1994;42(10):1110–7.
3. Abdelhafiz AH, Austin CA. Visual factors should be assessed in older people presenting with falls or hip fracture. Age Ageing. 2003;32(1):26–30.
4. Lord SR, Dayhew J. Visual risk factors for falls in older people. J Am Geriatr Soc. 2001;49(5):508–15.
5. Harwood RH. Visual problems and falls. Age Ageing. 2001;30(Suppl 4):13–8.
6. Lord SR. Visual risk factors for falls in older people. Age Ageing. 2006;35 Suppl 2:ii42–i5.
7. Kamel HK, Guro-Razuman S, Shareeff M. The activities of daily vision scale: a useful tool to assess fall risk in older adults with vision impairment. J Am Geriatr Soc. 2000;48(11):1474–7.
8. Kotecha A, Chopra R, Fahy RT, Rubin GS. Dual tasking and balance in those with central and peripheral vision loss. Invest Ophthalmol Vis Sci. 2013;54(8):5408–15.
9. Felson DT, Anderson JJ, Hannan MT, Milton RC, Wilson PW, Kiel DP. Impaired vision and hip fracture. The Framingham study. J Am Geriatr Soc. 1989;37(6):495–500.

10. Ivers RQ, Cumming RG, Mitchell P, Attebo K. Visual impairment and falls in older adults: the Blue Mountains eye study. J Am Geriatr Soc. 1998;46(1):58–64.
11. Klein BE, Klein R, Lee KE, Cruickshanks KJ. Performance-based and self-assessed measures of visual function as related to history of falls, hip fractures, and measured gait time. The Beaver Dam Eye Study. Ophthalmology. 1998;105(1):160–4.
12. Coleman AL, Stone K, Ewing SK, Nevitt M, Cummings S, Cauley JA, et al. Higher risk of multiple falls among elderly women who lose visual acuity. Ophthalmology. 2004;111(5):857–62.
13. McCarty CA, Fu CL, Taylor HR. Predictors of falls in the Melbourne visual impairment project. Aust N Z J Public Health. 2002;26(2):116–9.
14. Alexander MS, Lajoie K, Neima DR, Strath RA, Robinovitch SN, Marigold DS. Effects of age-related macular degeneration and ambient light on curb negotiation. Optom Vis Sci. 2014;91(8):975–89.
15. Black FO, Wall C 3rd, Rockette HE Jr, Kitch R. Normal subject postural sway during the Romberg test. Am J Otolaryngol. 1982;3(5):309–18.
16. Anand V, Buckley JG, Scally A, Elliott DB. Postural stability changes in the elderly with cataract simulation and refractive blur. Invest Ophthalmol Vis Sci. 2003;44(11):4670–5.
17. Turano K, Rubin GS, Herdman SJ, Chee E, Fried LP. Visual stabilization of posture in the elderly: fallers vs. nonfallers. Optom Vis Sci. 1994;71(12):761–9.
18. Manchester D, Woollacott M, Zederbauer-Hylton N, Marin O. Visual, vestibular and somatosensory contributions to balance control in the older adult. J Gerontol. 1989;44(4):M118–27.
19. Lord SR, Clark RD, Webster IW. Visual acuity and contrast sensitivity in relation to falls in an elderly population. Age Ageing. 1991;20(3):175–81.
20. Black AA, Wood JM, Lovie-Kitchin JE, Newman BM. Visual impairment and postural sway among older adults with glaucoma. Optom Vis Sci. 2008;85(6):489–97.
21. Chatard H, Tepenier L, Jankowski O, Aussems A, Allieta A, Beydoun T, et al. Effects of age-related macular degeneration on postural sway. Front Hum Neurosci. 2017;11:158.
22. Dev MK, Wood JM, Black AA. The effect of low light levels on postural stability in older adults with age-related macular degeneration. Ophthalmic Physiol Opt. 2021;41(4):853–63.
23. Turano KA, Dagnelie G, Herdman SJ. Visual stabilization of posture in persons with central visual field loss. Invest Ophthalmol Vis Sci. 1996;37(8):1483–91.
24. Ezane MD, Lions C, Bui Quoc E, Milleret C, Bucci MP. Spatial and temporal analyses of posture in strabismic children. Graefes Arch Clin Exp Ophthalmol. 2015;253(10):1629–39.
25. Lions C, Bui Quoc E, Wiener-Vacher S, Bucci MP. Postural control in strabismic children: importance of proprioceptive information. Front Physiol. 2014;5:156.
26. Papalia GF, Mangano G, Diaz Balzani LA, Cupo G, Giurazza G, Di Zazzo A, et al. Strabismus and postural control: a systematic review. Musculoskelet Surg. 2022;106(4):345–56.
27. Zipori AB, Colpa L, Wong AMF, Cushing SL, Gordon KA. Postural stability and visual impairment: assessing balance in children with strabismus and amblyopia. PLoS One. 2018;13(10):e0205857.
28. Brannan S, Dewar C, Sen J, Clarke D, Marshall T, Murray PI. A prospective study of the rate of falls before and after cataract surgery. Br J Ophthalmol. 2003;87(5):560–2.
29. McGwin G Jr, Gewant HD, Modjarrad K, Hall TA, Owsley C. Effect of cataract surgery on falls and mobility in independently living older adults. J Am Geriatr Soc. 2006;54(7):1089–94.
30. Supuk E, Alderson A, Davey CJ, Green C, Litvin N, Scally AJ, Elliott DB. Dizziness, but not falls rate, improves after routine cataract surgery: the role of refractive and spectacle changes. Ophthalmic Physiol Opt. 2016;36(2):183–90.
31. Cumming RG, Ivers R, Clemson L, Cullen J, Hayes MF, Tanzer M, Mitchell P. Improving vision to prevent falls in frail older people: a randomized trial. J Am Geriatr Soc. 2007;55(2):175–81.
32. Desapriya E, Subzwari S, Scime-Beltrano G, Samayawardhena LA, Pike I. Vision improvement and reduction in falls after expedited cataract surgery systematic review and metaanalysis. J Cataract Refract Surg. 2010;36(1):13–9.
33. Durmus B, Emre S, Cankaya C, Baysal O, Altay Z. Gain in visual acuity after cataract surgery improves postural stability and mobility. Bratisl Lek Listy. 2011;112(12):701–5.
34. Legrand A, Bui-Quoc E, Bucci MP. Re-alignment of the eyes, with prisms and with eye surgery, affects postural stability differently in children with strabismus. Graefes Arch Clin Exp Ophthalmol. 2012;250(6):849–55.
35. Lions C, Colleville L, Bui-Quoc E, Bucci MP. Importance of visual inputs quality for postural stability in strabismic children. Neurosci Lett. 2016;617:127–33.
36. Radvay X, Duhoux S, Koenig-Supiot F, Vital-Durand F. Balance training and visual rehabilitation of age-related macular degeneration patients. J Vestib Res. 2007;17(4):183–93.

139 Superior Oblique Myokymia

Overview

Myokymia of the *superior oblique* (less commonly inferior oblique) muscle can cause torsional saccades in the affected eye. This condition is typically unilateral and thus presents clinically with monocular oscillopsia. Case series reports the age of onset to range from 22 to 50 years. Most cases appear to be idiopathic; in some cases, microvascular compression is implicated, or traction on the trochlear nerve by a space-occupying lesion. Often, the diagnosis is based on the history alone; less commonly, the clinician may observe the unilateral torsional eye movements. Proposed treatments have included a variety of oral medications, timolol eye drops, and several surgical procedures.

Introduction

Myokymia of extraocular muscles can cause oscillopsia, which some patients can also experience as a disturbance of equilibrium. The superior oblique muscle is usually implicated, but myokymia of the inferior oblique muscle can also cause this [1].

Superior oblique myokymia was first described by Alexander Duane in 1906 [2].

Epidemiology

A series of 16 patients with superior oblique myokymia reported the age of onset ranged from 22 to 50 years, with 9 (56%) male and 7 (44%) female [3].

Pathophysiological Mechanism of Disease

Superior oblique myokymia reflects abnormal activity of the superior oblique muscle, possibly from aberrant firing of the trochlear nerve. In many cases, this appears to be idiopathic.

A number of imaging studies support the hypothesis of microvascular compression of the trochlear nerve as the etiology of superior oblique myokymia [4–8].

Some imaging studies of patients with SOM report that the affected superior oblique muscle is smaller than the contralateral (normal) side, which has led some investigators to conclude that, "These anatomical changes in the superior oblique muscle of patients with myokymia suggest that an antecedent injury to the trochlear nerve has occurred" [4].

SOM has also been reported in patients with dural arteriovenous malformations [9].

In some cases of SOM, the irritability of the trochlear nerve may be due to traction secondary to a space-occupying lesion in the posterior fossa, such as an arachnoid cyst [10] or neoplasm [11].

Clinical Presentation

Superior oblique myokymia (SOM) is generally unilateral, so the most common symptom of SOM is unilateral oscillopsia.

M. Cherchi, *Otoneurology and Vestibular Medicine*, https://doi.org/10.1007/978-3-031-94841-1_139

Physical Examination

In a patient with superior oblique myokymia who is otherwise healthy, the general neurological examination should be normal.

Ocular Motor Examination

Given the action of the superior oblique muscle, the ocular motor manifestation of SOM is bursts of torsional saccades. This is sometimes observable on face-to-face examination [12].

Testing: Vestibular

Several instrumented ocular motor studies have been conducted on patients with SOM [13–18].

Testing: Other

Rare studies have assessed electromyographic activity of the superior oblique muscle during SOM [9].

Imaging

Some imaging studies (generally high-resolution MRI of the orbits) report microvascular compression of the trochlear nerve [4–8]. Other studies report that the affected superior oblique muscle is smaller than its normal counterpart [4].

Differential Diagnosis

Myokymia of the eyelid (rather than the superior or inferior oblique muscles) can sometimes be experienced as oscillopsia [19, 20]. This is important to recognize because eyelid myokymia may be treated differently, such as with onabotulinum toxin injections.

Treatment

A number of treatments for SOM have been attempted or suggested, including:

- Oral beta blockers:
  - Betaxolol [21].
  - Levobunol [22].
  - Propranolol [23, 24].
- Topical beta blockers (delivered as eye drops):
  - Timolol [25].
- Antiepileptic medications:
  - Carbamazepine [24].
  - Gabapentin [26, 27].
  - Phenytoin [24].
  - Valproate [24].
- Other medications:
  - Memantine [28].
- Surgery.
  - Microvascular decompression [29–32].
  - Other surgeries [33–40].
- Prism lenses.

Prognosis

Superior oblique myokymia can be a chronic or recurrent condition [3]. Although medical therapy is not always successful, we nevertheless advocate starting with lower-risk, noninvasive approaches before entertaining the option of surgical intervention.

References

1. Smith TA, Cornblath WT. Alternating superior and inferior oblique myokymia. JAMA Ophthalmol. 2014;132(7):898–9.
2. Duane A. Unilateral rotary nystagmus. Trans Am Ophthalmol Soc. 1906;11(Pt 1):63–7.
3. Brazis PW, Miller NR, Henderer JD, Lee AG. The natural history and results of treatment of superior oblique myokymia. Arch Ophthalmol. 1994;112(8):1063–7.
4. Mehta AM, Demer JL. Magnetic resonance imaging of the superior oblique muscle in superior oblique myokymia. J Pediatr Ophthalmol Strabismus. 1994;31(6):378–83.
5. Yousry I, Dieterich M, Naidich TP, Schmid UD, Yousry TA. Superior oblique myokymia: magnetic resonance imaging support for the neurovascular compression hypothesis. Ann Neurol. 2002;51(3):361–8.
6. Yousry I, Moriggl B, Dieterich M, Naidich TP, Schmid UD, Yousry TA. MR anatomy of the proximal cisternal segment of the trochlear nerve: neurovascular relationships and landmarks. Radiology. 2002;223(1):31–8.
7. Hashimoto M, Ohtsuka K, Hoyt WF. Vascular compression as a cause of superior oblique myokymia disclosed by thin-slice magnetic resonance imaging. Am J Ophthalmol. 2001;131(5):676–7.
8. Kang S, Kim JS, Hwang JM, Choi BS, Kim JH. Mystery case: superior oblique myokymia due to vascular compression of the trochlear nerve. Neurology. 2013;80(13):e134–5.
9. Kommerell G, Schaubele G. Superior oblique myokymia. An electromyographical analysis. Trans Ophthalmol Soc U K (1962). 1980;100(4):504–6.
10. Donaldson L, van Adel B, Rodriguez AR. Superior Oblique Myokymia Presumed Due to Large Posterior Fossa Arteriovenous Malformation. Can J Neurol Sci. 2020;47(6):824–5.
11. Morrow MJ, Sharpe JA, Ranalli PJ. Superior oblique myokymia associated with a posterior fossa tumor: oculographic correlation with an idiopathic case. Neurology. 1990;40(2):367–70.

12. Zhao CW, Snow JP, Gheihman G, Prasad S. Teaching Video NeuroImage: Monocular Rotational Nystagmus Caused by Superior Oblique Myokymia. Neurology. 2024;102(2):e208062.
13. Kattah JC, FitzGibbon EJ. Superior oblique myokymia. Curr Neurol Neurosci Rep. 2003;3(5):395–400.
14. Ma J, Labbe S, Micieli JA. Successful Treatment of Superior Oblique Myokymia With Cannabidiol Oil. J Neuro-ophthalmol. 2021;41(2):e192–e3.
15. Suzuki Y, Washio N, Hashimoto M, Ohtsuka K. Three-dimensional eye movement analysis of superior oblique myokymia. Am J Ophthalmol. 2003;135(4):563–5.
16. Thinda S, Chen YR, Liao YJ. Cardinal features of superior oblique myokymia: an infrared oculography study. Am J Ophthalmol Case Rep. 2017;7:115–9.
17. Thurston SE, Saul RF. Superior oblique myokymia: quantitative description of the eye movement. Neurology. 1991;41(10):1679–81.
18. Hayakawa Y, Takagi M, Hasebe H, Hasegawa S, Takada R, Usui T, et al. A case of superior oblique myokymia observed by an image-analysis system. J Neuro-ophthalmol. 2000;20(3):163–5.
19. Krohel GB, Rosenberg PN. Oscillopsia associated with eyelid myokymia. Am J Ophthalmol. 1986;102(5):662–3.
20. Reinecke RD. Translated myokymia of the lower eyelid causing uniocular vertical pseudonystagmus. Am J Ophthalmol. 1973;75(1):150–1.
21. Bibby K, Deane JS, Farnworth D, Cappin J. Superior oblique myokymia--a topical solution? Br J Ophthalmol. 1994;78(11):882.
22. Zhang M, Gilbert AL, Hunter DG. Superior oblique myokymia treated with levobunolol. J AAPOS. 2018;22(1):67–9 e2.
23. Tyler TD, Ruiz RS. Propranolol in the treatment of superior oblique myokymia. Arch Ophthalmol. 1990;108(2):175–6.
24. Williams PE, Purvin VA, Kawasaki A. Superior oblique myokymia: efficacy of medical treatment. J AAPOS. 2007;11(3):254–7.
25. Borgman CJ. Topical timolol in the treatment of monocular oscillopsia secondary to superior oblique myokymia: a review. J Optom. 2014;7(2):68–74.
26. Deokule S, Burdon M, Matthews T. Superior oblique myokymia improved with gabapentin. J Neuro-ophthalmol. 2004;24(1):95–6.
27. Tomsak RL, Kosmorsky GS, Leigh RJ. Gabapentin attenuates superior oblique myokymia. Am J Ophthalmol. 2002;133(5):721–3.
28. Jain S, Farooq SJ, Gottlob I. Resolution of superior oblique myokymia with memantine. J AAPOS. 2008;12(1):87–8.
29. Fam MD, Scott C, Forster A, Kamel MH. Microvascular decompression for superior oblique myokymia: case report. Br J Neurosurg. 2014;28(4):552–5.
30. Mikami T, Minamida Y, Ohtsuka K, Houkin K. Resolution of superior oblique myokymia following microvascular decompression of trochlear nerve. Acta Neurochir 2005;147(9):1005–6; discussion 6.
31. Samii M, Rosahl SK, Carvalho GA, Krzizok T. Microvascular decompression for superior oblique myokymia: first experience. Case report. J Neurosurg. 1998;89(6):1020–4.
32. Scharwey K, Krzizok T, Samii M, Rosahl SK, Kaufmann H. Remission of superior oblique myokymia after microvascular decompression. Ophthalmologica. 2000;214(6):426–8.
33. Agarwal S, Kushner BJ. Results of extraocular muscle surgery for superior oblique myokymia. J AAPOS. 2009;13(5):472–6.
34. de Sa LC, Good WV, Hoyt CS. Surgical management of myokymia of the superior oblique muscle. Am J Ophthalmol. 1992;114(6):693–6.
35. Kawasaki T, Fujitsu K, Ichikawa T, Miyahara K, Okada T, Tanino S, et al. Superior oblique myokymia: a case report of surgical treatment, review of the literature, and consideration of surgical approach. World Neurosurg. 2019;131:197–9.
36. Law MX, Hoyt CS, Horton JC. Recurrent superior oblique myokymia treated by distal tendon extirpation. J Neuro-ophthalmol. 2019;39(3):345–7.
37. Noro S, Seo Y, Honjo K, Okuma M, Asayama B, Amano Y, et al. lateral supracerebellar infratentorial approach for superior oblique myokymia: a case series. Oper Neurosurg (Hagerstown). 2022;22(3):101–5.
38. Palmer EA, Shults WT. Superior oblique myokymia: preliminary results of surgical treatment. J Pediatr Ophthalmol Strabismus. 1984;21(3):96–101.
39. Ruttum MS, Harris GJ. Superior oblique myectomy and trochlectomy in recurrent superior oblique myokymia. Graefes Arch Clin Exp Ophthalmol. 1988;226(2):145–7.
40. Ruttum MS, Harris GJ. Superior oblique myectomy and trochlear resection for superior oblique myokymia. Am J Ophthalmol. 2009;148(4):563–5.

140 Illusions of Tilt and Inversion

Overview

An *inversion illusion* refers to the perception that the world has flipped 180 degrees in the coronal plane, but similar misperceptions of smaller angles or in different planes can also occur, so this problem can more generally be referred to as an *illusion of tilt*. Illusions of tilt are rare; they more commonly occur in males (60%–66%), with a mean age of onset in the sixth decade, though with a broad range (12–85 years). The episodes are sometimes accompanied by nausea and/or vomiting. Physical examination sometimes shows various ocular motor abnormalities, gait ataxia, truncal ataxia, or other findings. The etiology is unknown, but it is generally thought to arise either when actually conflicting peripheral inputs (visual, vestibular, somatosensory) cannot be reconciled, or a central disturbance fails to integrate inputs properly. In patients experiencing tilt illusions, a variety of anatomical localizations have been described from a range of central (neurologic) and peripheral (otologic) etiologies. An initial workup may include imaging (such as MRI of the brain and internal auditory canals without and with contrast) and otovestibular testing. Workup of an isolated (nonrecurrent) episode sometimes reveals a "central" etiology such as ischemia/infarction of the brainstem or cerebellum. Etiologies of recurrent episodes include peripheral vestibular disease, migraine, and seizure.

Introduction

Sierra-Hidalgo and colleagues [1] credit Bishopp [2] as having published the first report of an *inversion illusion* in 1805. This phenomenon has been described under various terminologies, including:

- "Illusion of tilting of the visual environment" [3]
- "Inverted vision" [4]
- "Reversal of vision metamorphopsia" [5, 6]
- "Room tilt illusion" [1, 7–9]
- "Upside-down reversal of vision" [8, 10, 11]
- "Upside-down vision" [4, 12]
- "Upside-down visual inversion" [10, 13].

The illusion is often "complete inversion" (meaning a 180 degree flip) in the coronal plane, but it can be in other angles or in other planes—sometimes referred to as "oblique" [5]. Many papers report that tilt illusions involve coronal rotations of 180 or 90 degrees, which has led some authors to conclude that, "Illusory rotations of the visual scene tend to be oriented along one of the principal orthogonals" [4]. However, broader literature reviews conclude that many papers "do not include the spectrum of manifestations of the RTI [room tilt illusion] phenomenon that may occur in a plane other than the coronal and/or in 90° steps" [1]. For these reasons, we prefer the phrase "*tilt illusion*" because it accommodates that the illusory perception of deviation from verticality can occur over a range, not only a complete inversion.

Sierra-Hidalgo and colleagues [1] provide the following Fig. 140.1 to illustrate the various misperceived orientations involved in tilt illusions.

M. Cherchi, *Otoneurology and Vestibular Medicine*, https://doi.org/10.1007/978-3-031-94841-1_140

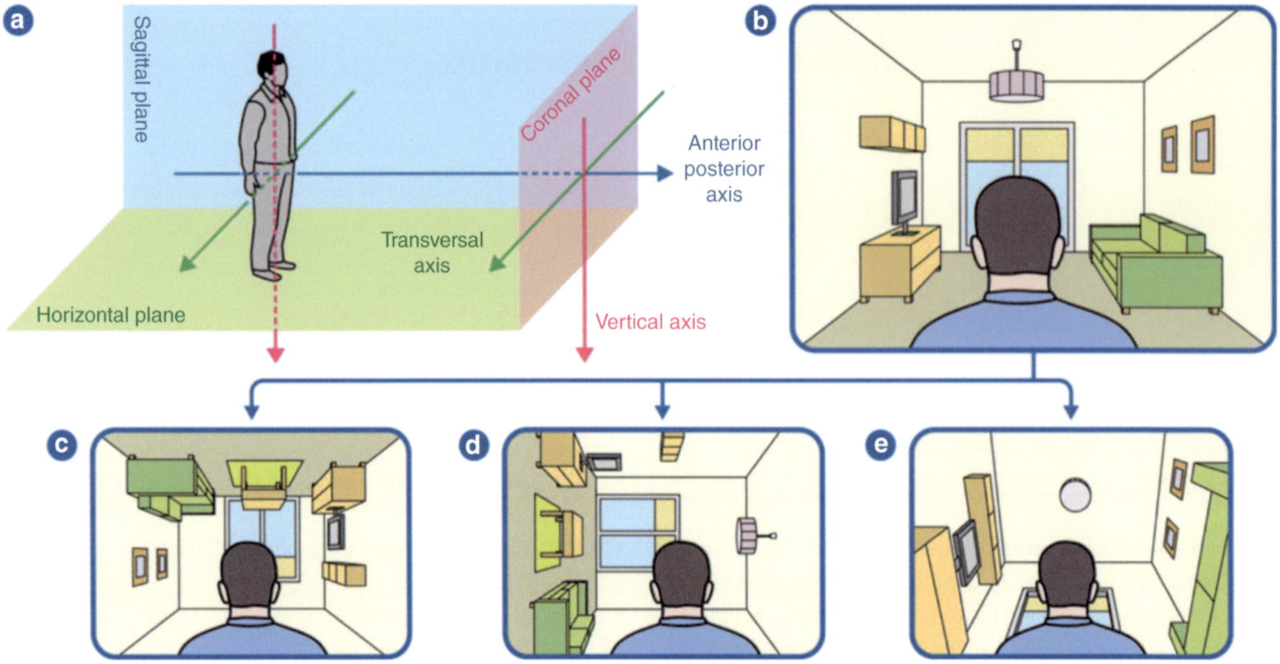

Fig. 140.1 Illustration of the various misperceived orientations involved in tilt illusions: (**a**) Three-dimensional representation of axes and planes of the space in relation to a subject; (**b**) representation of a domestic scene from a subject's view; (**c**) view of the scene during a room tilt illusion of 180° rotation following the coronal plane; (**d**) view of the scene during a 90° rotation following the coronal plane; (**e**) view of the scene during a 90° rotation following the sagittal plane. (From Sierra-Hidalgo et al. [1]. Reproduced with written permission from Springer Nature B.V)

Epidemiology

Tilt illusions are rare, and it is difficult to estimate their incidence or prevalence [6].

In case series, most authors observe a male preponderance. Sierra-Hidalgo and colleagues [1] studied 135 cases and reported that 60.2% were male. Yap conducted a literature review of 52 cases from 28 articles and reported that 66% of patients were male.

Sierra-Hidalgo and colleagues [1] reported the mean age at onset was 51.2 ± 20.3 years. Yap [6] reported the mean age of onset to be 52.5 ± 20.2 years (range 12–85 years).

Pathophysiological Mechanism of Disease

"The perception of verticality is considered to be mediated by inputs from vestibular, somatosensory, and visual symptoms and processed in the vestibular nuclei and then projected to the cortex via thalamic nuclei" [7], and tilt illusions are usually "attributed to an erroneous cortical mismatch of the visual and vestibular three-dimensional coordinate maps" [1], because "lesions involving the necessary inputs to create a verticality perception" [1] result in "disintegration between vestibular, visual, and somatosensory systems" [7].

Most literature broadly agrees that tilt illusions can arise from both central and peripheral disorders [7, 14].

Sierra-Hidalgo and colleagues [1] studied 13 cases of their own patients with tilt illusions, as well as those from a review of the literature, amounting to a total of 135 cases. They provided the localization of tilt illusions in the following Table 140.1.

Kirsch and colleagues [9] studied functional connectivity MRI in a patient with recurrent tilt illusions and reported "bilateral changes of vestibular and visual cortex activity… during RTI" (room tilt illusions).

Note that tilt illusions, including complete (180°) inversion illusions, can occur in healthy individuals during parabolic flight [15], in which case it is usually attributed to erroneous otolith signaling from zero gravity conditions [16].

Clinical Presentation

Episodes of tilt illusions are often isolated, but can be recurrent. In a series of 13 of their own patients, Sierra-Hidalgo and colleagues [1] reported that 8 (62%) "had two or more episodes," and "only one patient [8%] suffered from recurrent RTI [room tilt illusions] for 1 year."

Table 140.1 Localization in 135 cases of tilt illusions. From Sierra-Hidalgo et al. [1]

	Number	Percent
Central nervous system	97	61.4
Posterior fossa	48	30.4
"Vertebrobasilar territory"	15	9.5
Cerebellum	14	8.9
Medulla	10	6.3
Pons	5	3.2
Mesencephalon	1	0.6
"Brainstem"	3	1.9
Supratentorial	48	30.4
Parieto-occipital	29	18.4
Parietal lobe	7	4.4
Occipital lobe	3	1.9
Frontal lobe	3	1.9
Temporal lobe	3	1.9
Thalamus	3	1.9
Spinal cord	1	0.6
Vestibular system	31	19.6
VIII cranial nerve	11	7.0
Inner ear	20	12.7
Peripheral nervous system	4	2.5
Diffuse/multiple	6	3.8
Unknown/not specified	20	12.7
Total	158	100.0

The duration of tilt illusions is variable, lasting from seconds to hours. Yap reviewed 52 cases from 28 articles and reported the mean duration to be 16 minutes [6].

Some authors report that during a tilt illusion, patients will complain of feeling "dizzy" [3]. Of the 52 cases reviewed by Yap [6], they noted the following symptoms associated with the tilt illusion:

- "Vertigo" in 40%
- Vomiting in 25%.
- Nausea in 23%.
- "Dizziness" in 8%.

Physical Examination and Ocular Motor Examination

Of the 52 cases reviewed by Yap [6], 44% had a "normal clinical examination at the time of presentation." Of those with physical examination findings, these included:

- Ocular motor and ophthalmological abnormalities:
  - "Nystagmus" (not further specified) in 33%
  - "Abnormal saccade" in 2%
  - "Apraxia" in 2%
  - Internuclear ophthalmoplegia in 2%.
  - "Gaze palsy" (not further specified) in 2%
  - "Skew" in 2%
  - Hemianopia in 2%.
- Ataxia in 23%, including:
  - Gait ataxia in 8%.
  - Truncal ataxia in 8%.
- "Abnormal sensory findings" (not further specified) in 14%
- "Weakness" (not further specified) in 8%
- "Abnormal reflexes" (not further specified) in 6%
- "Hearing deficits" (not further specified) in 4%.

Testing: Auditory

Very few studies report any audiologic testing in patients with tilt illusions. In one of the patients reported by Stracciari and colleagues, "Brainstem auditory evoked potentials (BAER) revealed increased latency of V wave on right stimulation" [17].

Testing: Vestibular

Very few studies report on any vestibular testing in patients with tilt illusions.

In the patient described by Kirsch and colleagues [9], they reported normal caloric testing.

Tiliket and colleagues studied three patients "with persistent skew deviation caused by a brain stem lesion" in whom "vestibular stimulation" elicited "a sudden 90 degrees room tilt illusion (RTI)" [18]. They studied these three patients with rotatory chair testing, "including earth-vertical axis rotation for canal stimulation and off-vertical axis rotation (OVAR) for otolith stimulation. Measurement of the subjective visual vertical (SVV) was also performed." They reported that, "The otolith-ocular reflex registered by OVAR was impaired in the 3 patients with skew deviation and the SVV in 2 patients."

Imaging

Imaging findings are as variable as the etiologies. A straightforward case of an acute right medial cerebellar infarct in the territory of the right posterior inferior cerebellar artery was reported as the etiology of a tilt illusion in the case described by Horga Hernandez and colleagues [8], as shown in Fig. 140.2 below.

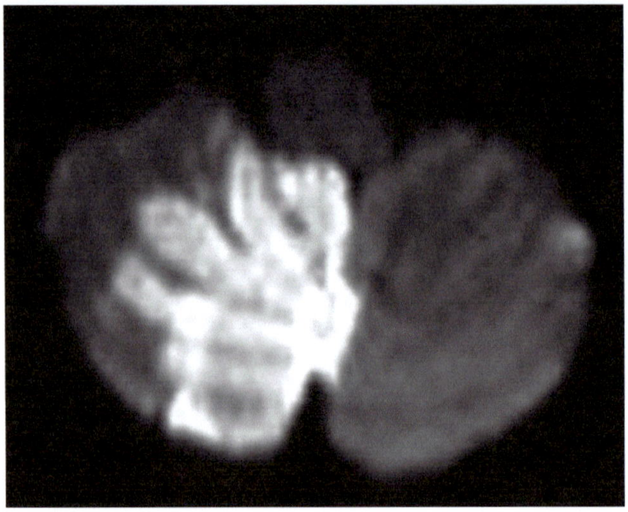

Fig. 140.2 Acute infarct of the right medial cerebellum in the territory of the right anterior inferior cerebellar artery. (From Horga Hernandez et al. [8]. Reproduced with written permission from Springer Nature B.V)

Differential Diagnosis

Of cases attributed to central etiologies, the most commonly reported include ischemia or infarction [8, 12, 17, 19] and multiple sclerosis [10, 13].

Sierra-Hidalgo and colleagues [1] studied 13 cases of their own patients with tilt illusions, as well as those from a review of the literature, amounting to a total of 135 cases. They provided the etiologies organized in the following Table 140.2.

Subclavian steal syndrome (SSS) has also been reported as a mechanism of illusions of inversion and tilt [20].

In view of this differential diagnosis, a medically reasonable workup may include:

- Auditory testing, such as audiometry.
- Vestibular testing, such as cervical vestibular evoked myogenic potentials, ocular vestibular evoked myogenic potentials, video head impulse testing, videonystagmography, and rotatory chair testing.
- MRI of the brain and internal auditory canals without and with contrast.
- If physical examination reveals a difference in blood pressure of ≥20 mm Hg between the arms, then consider checking a Doppler ultrasound of the subclavian arteries to evaluate for subclavian steal syndrome.

Table 140.2 Etiologies in 135 cases of tilt illusions

Disorder	Number	%
Central nervous system disorders	98	66.2
Focal disorders	84	56.8
Infarction	32	21.6
Traumatic brain injury	27	18.2
Transient ischemic attack	9	6.1
Epilepsy	5	3.4
Intracranial tumor	5	3.4
Hematoma	3	2.0
Surgical brain injury	2	1.4
Brain abscess	1	0.7
Multifocal/diffuse disorders	14	9.5
Migraine	5	3.4
Multiple sclerosis	3	2.0
Neurocysticercosis	1	0.7
Opioid intoxication	1	0.7
multiple system atrophy	1	0.7
Parkinson's disease	1	0.7
Susac's syndrome	1	0.7
encephalitis	1	0.7
Vestibular disorders	29	19.6
Ménière's disease	13	8.8
Vestibular neuritis	6	4.1
Benign paroxysmal positional vertigo	4	2.7
VIII cranial nerve section	3	2.0
Ramsay-hunt syndrome	1	0.7
Perilymphatic fistula	1	0.7
Labyrinthectomy	1	0.7
Peripheral nervous system disorders	4	2.7
Guillain-Barré syndrome	3	2.0
Critical illness neuropathy	1	0.7
Unknown mechanism	16	10.8
After mild cranial trauma	1	0.7
Other/not specified	15	10.1
Hysteria	1	0.7
Total	148	100.0

Form Sierra-Hidalgo et al. [1]

Treatment and Prognosis

Yip states that, "There is no current evidence that acute management has any beneficial effect on reducing RVM [reversal of vision metamorphopsia] episode duration" [6].

There is no literature to support any specific intervention for prophylaxis against recurrent episodes. However:

- If otovestibular workup provides evidence of disease of the labyrinth or vestibular nerve [1, 6, 7, 14], then it is logical to treat that underlying disease.

- Seizure appears to be a very rare cause of tilt illusions [21], but if there are grounds to suspect this (and if episodes are recurrent), then a trial of an antiepileptic (or referral to an epileptologist) is reasonable.
- If workup provides no evidence of structural abnormalities (neurologic ischemia, demyelination, tumor, etc.) and an otovestibular workup provides no evidence of an otologic source, then (by exclusion) it is medically reasonable to entertain the possibility of migraine [1, 6, 7, 22] and (if episodes are recurrent) consider a trial of migraine prophylaxis.

References

1. Sierra-Hidalgo F, de Pablo-Fernandez E, Herrero-San Martin A, Correas-Callero E, Herreros-Rodriguez J, Romero-Munoz JP, et al. Clinical and imaging features of the room tilt illusion. J Neurol. 2012;259(12):2555–64.
2. Bishopp T. Case of optical illusion from hysteria. Med Phys J. 1805;14(78):117–8.
3. Ropper AH. Illusion of tilting of the visual environment. Report of five cases. J Clin Neuroophthalmol. 1983;3(2):147–51.
4. Solms M, Kaplan-Solms K, Saling M, Miller P. Inverted vision after frontal lobe disease. Cortex. 1988;24(4):499–509.
5. River Y, Ben Hur T, Steiner I. Reversal of vision metamorphopsia: clinical and anatomical characteristics. Arch Neurol. 1998;55(10):1362–8.
6. Yap JA. Upside-down vision: a systematic review of the literature. BMJ Neurol Open. 2022;4(2):e000337.
7. Akdal G, Toydemir HE, Tanriverdizade T, Halmagyi GM. Room tilt illusion: a symptom of both peripheral and central vestibular disorders. Acta Neurol Belg. 2017;117(1):363–5.
8. Horga Hernandez A, Pujadas F, Purroy F, Delgado P, Huertas R, Alvarez-Sabin J. Upside down reversal of vision due to an isolated acute cerebellar ischemic infarction. J Neurol. 2006;253(7):953–4.
9. Kirsch V, Keeser D, Becker-Bense S, Karali T, Ertl-Wagner B, Brandt T, et al. Vestibular and visual cortex activity during room tilt illusion. J Neurol. 2017;264(Suppl 1):70–3.
10. Deniz O, Keklikoglu HD, Vural G, Temel S, Dilbaz FA. Acute "upside-down" visual inversion in a patient with multiple sclerosis. Neurol Sci. 2012;33(3):635–7.
11. Pamir MN, Ozer AF, Siva A, Ozek MM, Kansu T. "Upside down" reversal of vision after third ventriculostomy. J Clin Neuroophthalmol. 1990;10(4):271–2.
12. Charles N, Froment C, Rode G, Vighetto A, Turjman F, Trillet M, et al. Vertigo and upside down vision due to an infarct in the territory of the medial branch of the posterior inferior cerebellar artery caused by dissection of a vertebral artery. J Neurol Neurosurg Psychiatry. 1992;55(3):188–9.
13. Okuyucu EE, Okuyucu S, Karazincir S, Akoglu E, Duman T. Acute 'upside-down' visual inversion due to multiple sclerosis. Mult Scler. 2008;14(2):266–7.
14. Malis DD, Guyot JP. Room tilt illusion as a manifestation of peripheral vestibular disorders. Ann Otol Rhinol Laryngol. 2003;112(7):600–5.
15. Roberts TD. Inversion illusion in the so-called zero-gravity conditions of parabolic flight. Aerosp Med. 1973;44(5):484–7.
16. Graybiel A, Kellogg RS. Inversion illusion in parabolic flight: its probable dependence on otolith function. Aerosp Med. 1967;38(11):1099–103.
17. Stracciari A, Guarino M, Ciucci G, Pazzaglia P. Acute upside down reversal of vision in vertebrobasilar ischaemia. J Neurol Neurosurg Psychiatry. 1993;56(4):423.
18. Tiliket C, Ventre-Dominey J, Vighetto A, Grochowicki M. Room tilt illusion. A central otolith dysfunction. Arch Neurol. 1996;53(12):1259–64.
19. Steiner I, Shahin R, Melamed E. Acute "upside down" reversal of vision in transient vertebrobasilar ischemia. Neurology 1987;37(10):1685–1686.
20. Arntzen K, Alstadhaug KB. Room tilt illusion and subclavian steal – a case report. BMC Neurol. 2020;20(1):369.
21. Unal A, Cila A, Saygi S. Reversal of vision metamorphopsia: a manifestation of focal seizure due to cortical dysplasia. Epilepsy Behav. 2006;8(1):308–11.
22. Lopez Dominguez JM, Rojas-Marcos I, Sanz Fernandez G, Diaz Espejo C. Room tilt illusion: a rare symptom of migraine aura. Neurologia. 2007;22(1):58–60.

Part XXI

Diagnoses—Cardiovascular Causes of Disequilibrium

Orthostatic Hypotension and Postural Orthostatic Tachycardia

141

Overview

Of the cardiovascular etiologies of disequilibrium, orthostatic intolerance is probably the most common. Orthostatic intolerance comprises two related conditions that can occur separately or together; these are orthostatic hypotension (overly robust drop in blood pressure when going from a supine to a standing position) and postural orthostatic tachycardia (overly robust increase in heart rate when going from a supine to a standing position). Orthostatic intolerance is often just one feature of a broader disease process. If the underlying cause cannot be treated, then direct treatment of orthostatic intolerance can be considered.

Introduction

The phrase *orthostatic hypotension* refers to an inappropriately large drop in blood pressure when a person assumes an upright posture (sitting or standing). Overly robust drops in blood pressure can cause people to become lightheaded when they stand up, or even to faint. There is no universal consensus on the criteria for diagnosis, but commonly used figures include a *drop of systolic blood pressure (the "top number") by more than 20 mm Hg*, or a *drop of diastolic blood pressure (the "bottom number") by more than 10 mm Hg* within 3 min of standing [1–4]. Orthostatic hypotension in itself is not usually an independently occurring disease, but may be a manifestation of a variety of underlying medical conditions.

A related condition is *postural orthostatic tachycardia syndrome* (sometimes abbreviated "POTS"), which is an inappropriately large increase in heart rate that occurs when a person assumes an upright (sitting or standing) posture. Overly robust increases in heart rate when a person assumes an upright posture can cause dizziness. There is no universal consensus on the criteria for diagnosis [5], but a commonly used figure is an *increase in heart rate of more than 30 beats per minute* [6]. Orthostatic tachycardia in itself is not usually an independently occurring disease, but can reflect underlying medical conditions.

Orthostatic hypertension, another orthostatic phenomenon, is less common and discussed elsewhere.

Epidemiology

Ricci and colleagues state that different studies report the prevalence of orthostatic hypotension to range from 6% to 35% depending on age and associated comorbidities, "showing a strong association with elevated BP [blood pressure]" [7].

Pathophysiological Mechanism of Disease

Part of the underlying mechanism of orthostatic hypotension and postural orthostatic tachycardia is that when a person stands up, gravity pulls blood into the legs. This pooling of blood in the legs means that less blood is available in the thorax to be pumped upward and perfuse the brain adequately. A healthy individual's body quickly senses that such pooling is occurring and invokes strategies for preventing it, including a modest increase in heart rate, an increase in the amount of blood that the heart ejects each time it pumps (the ejection fraction), and constriction of peripheral blood vessels. If disease interferes with the body's ability to detect pooling, or if it impedes the compensatory strategies, then the brain does not receive the blood it needs, and a person may feel lightheaded or experience frank syncope.

Clinical Presentation

In the context of an otoneurology clinic, patients who are ultimately found to have orthostatic intolerance typically complain of disequilibrium while upright. Patients with orthostatic hypotension frequently describe lightheadedness

M. Cherchi, *Otoneurology and Vestibular Medicine*, https://doi.org/10.1007/978-3-031-94841-1_141

and, in some cases, may experience syncope. Patients with postural orthostatic tachycardia may or may not complain of palpitations, but may still notice lightheadedness.

Physical Examination

There is no firm consensus on the diagnostic criteria in terms of change in blood pressure (for orthostatic hypotension) or change in heart rate (for postural orthostatic tachycardia syndrome). It is important to keep in mind that both conditions can occur in the same patient.

Ocular Motor Examination

Some patients with orthostatic hypotension may exhibit nystagmus [8], though this is neither sensitive nor specific for the condition.

Testing: Vestibular

There is evidence that vestibular function can modulate cerebral blood flow [9], so one might expect that vestibular dysfunction could interfere with such modulation. Vestibular deficits, particularly otolith dysfunction [10], do appear to adversely impact autonomic regulation [11].

While this may be true, the implications for diagnosis and management are unclear. For example, if a patient with orthostatic hypotension is found to have otolith dysfunction (such as reduced vestibular evoked myogenic potentials), does this change the treatment strategy?

Testing: Other

Tilt table testing (TTT) is the criterion standard for diagnosing orthostatic hypotension and postural orthostatic tachycardia syndrome. While many cases can be diagnosed at the bedside, the reality is that orthostatic hypotension [12] and postural orthostatic tachycardia can also occur in a delayed fashion, and thus may not be identified on physical examination.

Differential Diagnosis

Ricci and colleagues [7] provide the following differential diagnosis of orthostatic hypotension:

- Primary
  - Parkinson's disease
  - Multiple system atrophy
  - Pure autonomic failure
  - Lewy body dementia
  - Autoimmune autonomic ganglionopathy
  - Hereditary disorders (familial dysautonomia, dopamine beta-hydroxylase deficiency)
- Secondary
  - Pharmacologic (iatrogenic)
  - Diabetes mellitus
  - Cardiovascular diseases (sick-sinus syndrome, atrioventricular block, cardiac failure, aortic stenosis, pulmonary hypertension, essential hypertension)
  - Renal failure
  - Autoimmune diseases
  - Volume depletion
  - Venous pooling
  - Alcoholic neuropathy
  - Endocrine disorders (adrenal insufficiency, thyroid diseases, diabetes insipidus)
  - Amyloidosis
  - Multiple myeloma
  - Paraneoplastic syndromes
  - Cerebrovascular disease
  - Multiple sclerosis
  - Spinal cord diseases

Robertson and Robertson [13] studied 100 consecutive patients with orthostatic hypotension and reported that 27% had primary autonomic failure, 35% had secondary autonomic failure, and 38% had no autonomic failure, broken down as follows:

- *Hypotension without autonomic failure 38*
  - Antidepressant therapy 9
  - Mitral valve prolapse 6
  - Mastocytosis 5
  - Antihypertensive therapy 4
  - Phenylpropanolamine abuse 3
  - Carotid hypersensitivity 2
  - Micturition syncope 2
  - Overdose 2
  - Addison's disease 1
  - Marijuana abuse 1
  - Undiagnosed 3
- *Secondary autonomic failure 35*
  - Diabetes mellitus 15
  - Paraneoplastic 10
  - Amyloidosis 8
  - Intoxication 2
- *Primary autonomic failure 27*
  - Shy-Drager syndrome 15
  - Bradbury-Eggleston syndrome 9
  - Baroreflex failure 2
  - Dopamine beta-hydroxylase deficiency 1

Treatment: Nonpharmacologic

Although there are several ways of approaching orthostatic intolerance, we favor starting with methods that do not involve adding medications [14]. Not all methods are applicable in all cases.

Below we discuss a range of nonpharmacologic strategies. Not every patient is able to (or wants to) use every single strategy. Some patients may require the use of several strategies simultaneously to achieve improvement in symptoms.

- ☐ We encourage patients to *review their medications with their other clinicians* to see whether any of the medications may be contributing to this problem. Pharmacology should be simplified where medically feasible, and always under the supervision of the prescribing physician.
- ☐ *Increase fluid intake*. Overall, this has been shown to be the most effective strategy for patients [15]. There are no exact criteria on how much of an increase is appropriate. We usually recommend that patients try to drink at least twice as much fluid as usual. The strategy of increasing fluid intake may not be appropriate in all patients—for instance, patients with renal failure or cardiac failure may be advised by their doctor not to increase fluid intake.
- ☐ *Increase dietary sodium intake*. This strategy may slightly raise and stabilize blood pressure. Some patients should not increase their dietary sodium for instance, many patients with cardiac disease are specifically instructed to *reduce* salt in their diet—so this strategy must first be approved by their primary care physician or cardiologist.
- ☐ *Wear Jobst stockings*. Jobst stockings are basically tights. By squeezing the legs and lower body, they reduce lower extremity pooling of blood. Ideally, the *stockings should come up to the waist,* because this prevents pooling of blood not only in the legs but also in the pelvic venous plexus. Stockings that only come up to the knee or to the hip are less effective. Jobst stockings can be obtained in medical supply stores. They do not require a prescription. There are different degrees of "tightness." Patients should choose a grade that they can tolerate, bearing in mind that if the stockings are too loose, then they will not be effective.
- ☐ *Wear an abdominal binder*. An abdominal binder is like a girdle. By modestly compressing the abdomen, it promotes the return of blood from the abdominal venous plexus up into the thorax, and can benefit patients with orthostatic hypotension [16, 17]. Abdominal binders are available at medical supply stores. They come in different sizes and degrees of tightness, and do not require a prescription.
- ☐ *Adjust your bed so that the head is higher than the pelvis, and the pelvis is higher than the feet*. This is called the "reverse Trendelenburg position." This strategy promotes modest pooling of blood in the lower extremities at night, thereby maintaining a modest activation of the renin-angiotensin-aldosterone axis, and thus, when the patient arises in the morning, they will be less likely to suffer a large drop in blood pressure. Patients should be advised that it is not adequate simply to sleep with a few pillows beneath the head; rather, the entire bed must be inclined. Usually, this can be accomplished by putting 1 or 2 bricks beneath each post at the head of the bed. An angle of 15° to 20° is adequate.
- ☐ *Avoid eating large meals*. After consuming a large meal, the body may shunt up to 20% of the blood to the gastrointestinal tract to aid in digestion [18]. This means that less blood is available for other tasks, such as circulating to the brain. When feasible, it is preferable to eat a greater number of small meals rather than a smaller number of large meals. This does not mean that one must reduce the total amount of food consumed in a day, but rather that food consumption should be more evenly distributed.
- ☐ *Avoid hot environments*. When a person perspires or feels warm, by body purposefully shunts blood to circulate just beneath the skin in an attempt to dissipate heat. This means that less blood is available for other tasks, such as circulating to the brain.
- ☐ *Avoid straining at stool*. Straining at stool, or straining during lifting heavy objects, often entails the Valsalva maneuver, which actually induces a *decrease* in blood pressure both during the maneuver itself, and after "releasing" from the straining [19]. For patients with constipation (leading to straining at stool), it may be helpful to treat the constipation with increased hydration, increased dietary fiber, other dietary modifications (e.g., prunes), or consult their internist for pharmacological intervention (such as with laxatives).
- ☐ *Get up gradually in the morning*. Many patients instinctively realize that they need to arise from bed slowly. Other patients must be explicitly instructed that when they first wake up, they should first sit at the side of the bed for a few minutes before actually standing up and walking.
- ☐ *Orthostatic training*. This consists of several relatively simple exercises that can be taught by a physical therapist familiar with the problem of orthostatic hypotension.

The main exercise is simple and can be done at home. It consists simply of *standing up against a wall* for 30 min.

Fig. 141.1 Physical maneuvers that may raise orthostatic blood pressure. (From Low and Singer [30]. Reproduced with written permission from Elsevier B.V.)

Initially, patients may find they are unable to maintain this posture for a full 30 min, but over time, they increase their tolerance and usually are able to achieve this target. One study of an orthostatic training program followed 24 patients who actually experienced syncope (fainting) due to their orthostatic hypotension [20]. The intention was for patients to stand against a wall without moving for 30 min twice per day. However, most patients only ended up doing the exercise once per day. Nevertheless, after 5–18 months, 22 patients (92%) no longer suffered syncope. Most other studies have found similar results [21–26]. However, a minority of studies found no significant difference between those patients who performed the exercises and those who did not [27–29].

For patients who have failed other therapies, we usually recommend trying orthostatic training since there is no significant medical risk. The more frequently one performs the exercises, and the greater the length of time (i.e., preferably months rather than simply weeks), the more likely the patient is to benefit from them.

Even when the abovementioned exercise is helpful, some patients may still experience occasional lightheadedness when upright. If a patient has this sensation and fears that she might faint, then there are several "*counter-maneuvers*" that reduce its likelihood. The maneuvers include toe raising, leg crossing, thigh contraction, and bending at the waist, all of which reduce venous capacity and increase total peripheral resistance [30]. Illustrations of these maneuvers are given below (Fig. 141.1).

In order to monitor progress, it is helpful to compare pre-intervention and postintervention features. Maintaining a symptom diary is useful, but it is also very helpful for a patient to acquire a home blood pressure cuff so that they can additionally track objective measurements of blood pressure and cardiac pulse.

Patients should be educated to check their blood pressure and pulse (while in a lying down position) after they have been lying flat for 3 min, then stand up and recheck their blood pressure (while still in a standing position). Measuring these at approximately the same time (or times) every day will minimize variability that could be introduced by other factors (such as activity and diet).

Treatment: Pharmacologic

Probably the most widely studied agent for the treatment of orthostatic hypotension is midodrine, either alone [31–37] or in combination with other agents [38, 39]. Midodrine may not be appropriate for all patients, particularly those with comorbid cardiac disease. Even in patients without cardiac disease, one of the risks of midodrine is supine hypertension.

Prognosis

The prognosis of orthostatic intolerance depends on the underlying cause.

References

1. Consensus statement on the definition of orthostatic hypotension, pure autonomic failure, and multiple system atrophy. J Neurol Sci. 1996;144(1–2):218–9.
2. Consensus statement on the definition of orthostatic hypotension, pure autonomic failure, and multiple system atrophy. The Consensus Committee of the American Autonomic Society and the American Academy of Neurology. Neurology. 1996;46(5):1470.
3. The definition of orthostatic hypotension, pure autonomic failure, and multiple system atrophy. J Auton Nerv Syst. 1996;58(1–2):123–4.
4. Kaufmann H. Consensus statement on the definition of orthostatic hypotension, pure autonomic failure and multiple system atrophy. Clin Auton Res. 1996;6(2):125–6.
5. Olshansky B, Cannom D, Fedorowski A, Stewart J, Gibbons C, Sutton R, et al. Postural orthostatic tachycardia syndrome (POTS): a critical assessment. Prog Cardiovasc Dis. 2020;63(3):263–70.
6. McGee S, Abernethy WB 3rd, Simel DL. The rational clinical examination. Is this patient hypovolemic? JAMA. 1999;281(11):1022–9.

7. Ricci F, De Caterina R, Fedorowski A. Orthostatic hypotension: epidemiology, prognosis, and treatment. J Am Coll Cardiol. 2015;66(7):848–60.
8. Choi JH, Seo JD, Kim MJ, Choi BY, Choi YR, Cho BM, et al. Vertigo and nystagmus in orthostatic hypotension. Eur J Neurol. 2015;22(4):648–55.
9. Serrador JM, Schlegel TT, Black FO, Wood SJ. Vestibular effects on cerebral blood flow. BMC Neurosci. 2009;10:119.
10. Aoki M, Sakaida Y, Tanaka K, Mizuta K, Ito Y. Evidence for vestibular dysfunction in orthostatic hypotension. Exp Brain Res. 2012;217(2):251–9.
11. Yates BJ. Autonomic reaction to vestibular damage. Otolaryngol Head Neck Surg. 1998;119(1):106–12.
12. Gibbons CH, Freeman R. Delayed orthostatic hypotension. Auton Neurosci. 2020;229:102724.
13. Robertson D, Robertson RM. Causes of chronic orthostatic hypotension. Arch Intern Med. 1994;154(14):1620–4.
14. Freeman R. Treatment of orthostatic hypotension. Semin Neurol. 2003;23(4):435–42.
15. Newton JL, Frith J. The efficacy of nonpharmacologic intervention for orthostatic hypotension associated with aging. Neurology. 2018;91(7):e652–e6.
16. Figueroa JJ, Singer W, Sandroni P, Sletten DM, Gehrking TL, Gehrking JA, et al. Effects of patient-controlled abdominal compression on standing systolic blood pressure in adults with orthostatic hypotension. Arch Phys Med Rehabil. 2015;96(3):505–10.
17. Smit AA, Wieling W, Fujimura J, Denq JC, Opfer-Gehrking TL, Akarriou M, et al. Use of lower abdominal compression to combat orthostatic hypotension in patients with autonomic dysfunction. Clin Auton Res. 2004;14(3):167–75.
18. Victor M, Ropper AH. Adams and Victor's principles of neurology. 7th ed. New York: McGraw-Hill; 2001.
19. Looga R. The Valsalva manoeuvre–cardiovascular effects and performance technique: a critical review. Respir Physiol Neurobiol. 2005;147(1):39–49.
20. Abe H, Kondo S, Kohshi K, Nakashima Y. Usefulness of orthostatic self-training for the prevention of neurocardiogenic syncope. Pacing Clin Electrophysiol. 2002;25(10):1454–8.
21. Reybrouck T, Heidbuchel H, Van de Werf F, Ector H. Tilt training: a treatment for malignant and recurrent neurocardiogenic syncope. Pacing Clin Electrophysiol. 2000;23(4 Pt 1):493–8.
22. Ector H, Reybrouck T, Heidbuchel H, Gewillig M, Van de Werf F. Tilt training: a new treatment for recurrent neurocardiogenic syncope and severe orthostatic intolerance. Pacing Clin Electrophysiol. 1998;21(1 Pt 2):193–6.
23. Kinay O, Yazici M, Nazli C, Acar G, Gedikli O, Altinbas A, et al. Tilt training for recurrent neurocardiogenic syncope: effectiveness, patient compliance, and scheduling the frequency of training sessions. Jpn Heart J. 2004;45(5):833–43.
24. Reybrouck T, Ector H. Tilt training: a new challenge in the treatment of neurally mediated syncope. Acta Cardiol. 2006;61(2):183–9.
25. Abe H, Kohshi K, Nakashima Y. Effects of orthostatic self-training on head-up tilt testing and autonomic balance in patients with neurocardiogenic syncope. J Cardiovasc Pharmacol. 2003;41(Suppl 1):S73–6.
26. Abe H, Sumiyoshi M, Kohshi K, Nakashima Y. Effects of orthostatic self-training on head-up tilt testing for the prevention of tilt-induced neurocardiogenic syncope: comparison of pharmacological therapy. Clin Exp Hypertens. 2003;25(3):191–8.
27. On YK, Park J, Huh J, Kim JS. Is home orthostatic self-training effective in preventing neurally mediated syncope? Pacing Clin Electrophysiol. 2007;30(5):638–43.
28. Duygu H, Zoghi M, Turk U, Akyuz S, Ozerkan F, Akilli A, et al. The role of tilt training in preventing recurrent syncope in patients with vasovagal syncope: a prospective and randomized study. Pacing Clin Electrophysiol. 2008;31(5):592–6.
29. Foglia-Manzillo G, Giada F, Gaggioli G, Bartoletti A, Lolli G, Dinelli M, et al. Efficacy of tilt training in the treatment of neurally mediated syncope. A randomized study. Europace. 2004;6(3):199–204.
30. Low PA, Singer W. Management of neurogenic orthostatic hypotension: an update. Lancet Neurol. 2008;7(5):451–8.
31. Smith W, Wan H, Much D, Robinson AG, Martin P. Clinical benefit of midodrine hydrochloride in symptomatic orthostatic hypotension: a phase 4, double-blind, placebo-controlled, randomized, tilt-table study. Clin Auton Res. 2016;26(4):269–77.
32. Low PA, Gilden JL, Freeman R, Sheng KN, McElligott MA. Efficacy of midodrine vs placebo in neurogenic orthostatic hypotension. A randomized, double-blind multicenter study. Midodrine study group. JAMA. 1997;277(13):1046–51.
33. Chen JJ, Han Y, Tang J, Portillo I, Hauser RA, Dashtipour K. Standing and supine blood pressure outcomes associated with Droxidopa and Midodrine in patients with neurogenic orthostatic hypotension: a Bayesian meta-analysis and mixed treatment comparison of randomized trials. Ann Pharmacother. 2018;52(12):1182–94.
34. Singer W, Joyner MJ, Sandroni P, Benarroch EE, Fealey RD, Mandrekar J, et al. Midodrine efficacy in orthostatic hypotension. J Gen Intern Med. 2014;29(11):1440–1.
35. Murad MH, Singh B, Parsaik AA. Midodrine efficacy in orthostatic hypotension. J Gen Intern Med. 2014;29(11):1442–3.
36. Parsaik AK, Singh B, Altayar O, Mascarenhas SS, Singh SK, Erwin PJ, et al. Midodrine for orthostatic hypotension: a systematic review and meta-analysis of clinical trials. J Gen Intern Med. 2013;28(11):1496–503.
37. Jankovic J, Gilden JL, Hiner BC, Kaufmann H, Brown DC, Coghlan CH, et al. Neurogenic orthostatic hypotension: a double-blind, placebo-controlled study with midodrine. Am J Med. 1993;95(1):38–48.
38. Shi SJ, Platts SH, Ziegler MG, Meck JV. Effects of promethazine and midodrine on orthostatic tolerance. Aviat Space Environ Med. 2011;82(1):9–12.
39. Byun JI, Moon J, Kim DY, Shin H, Sunwoo JS, Lim JA, et al. Efficacy of single or combined midodrine and pyridostigmine in orthostatic hypotension. Neurology. 2017;89(10):1078–86.

Orthostatic Hypertension

142

Introduction

Orthostatic hypertension falls under the broader category of orthostatic intolerance, but has been less well studied than other orthostatic phenomena, such as orthostatic hypotension and postural orthostatic tachycardia.

Jordan and colleagues review a dozen definitions of orthostatic hypertension and observe that "A uniform definition of orthostatic hypertension has never been established" [1]. Nevertheless, Jordan and colleagues proposed consensus statements defining orthostatic hypertension as an increase in systolic blood pressure ≥20 mm Hg when going from a supine to a standing position [2, 3].

Epidemiology

Jordan and colleagues report that, "In studies defining orthostatic hypertension as a sustained increase in systolic blood pressure ≥20 mm Hg and/ or diastolic blood pressure ≥10 mmHg within 3 min of standing, the reported prevalence of orthostatic hypertension ranged between 5% and 30%" [1].

Orthostatic hypertension can occur at any age, including in the pediatric population [4–7]. Some investigators report a similar prevalence across the lifespan; for example, Wecht and colleagues [8] studied 286 veterans and reported that the prevalence of orthostatic hypertension was similar in patients aged 20–30 years (17%) and patients over 70 years (19%). Other investigators reach different conclusions; for example, Jordan and colleagues report that, "The proportion of individuals with orthostatic hypertension increases with advancing age, greater body mass index, and chronic cardiovascular conditions such as essential hypertension" [1].

Citing Yoshinari and colleagues [9], Jordan and colleagues comment that "Patients with diabetes mellitus showed a fivefold higher orthostatic hypertension prevalence compared with a nondiabetic population" [1].

Pathophysiological Mechanism of Disease

Many diseases have been reported in association with orthostatic hypertension. Renal diseases are probably the most common, but other associated conditions include dysautonomia, diabetes mellitus, and older age.

- Citing McCann and colleagues [10], Jordan and colleagues comment that, "The term orthostatic hypertension, which is an increase in blood pressure with standing, began to appear in the medical literature as early as the 1940s and 1950s, typically in relation to *kidney diseases*" [1].
- *Chronic kidney disease* (CKD) appears to be a risk factor for orthostatic hypertension. Ghazi and colleagues [11] reported that out of 1386 adult patients with chronic kidney disease, 153 (11.0%) were found to have orthostatic hypertension. Rouabhi and colleagues [12] reported that out of 3873 patients with chronic kidney disease, 81 (2.1%) were found to have orthostatic hypertension.
- Orthostatic hypertension has been "observed in association with a variety of other clinical conditions, including *essential hypertension, dysautonomias, and type 2 diabetes mellitus*" [13].
- "Conditions associated with higher activity of the sympathetic nervous system and/or *baroreflex dysregulation, such as diabetes, essential hypertension, and aging* have been found to be often associated with [orthostatic hypertension]" [14].

Such associations are suggestive, but they do not prove causality.

Given the rapidity of the hypertensive response, most investigators speculate that such response is mediated by the autonomic nervous system:

- "*Dysfunction of the autonomous nervous system* is considered to be the primary pathophysiological disturbance" [15].

M. Cherchi, *Otoneurology and Vestibular Medicine*, https://doi.org/10.1007/978-3-031-94841-1_142

- "The underlying pathophysiology is thought to involve *activation of the sympathetic nervous system*" [13].
- "*Sympathetic overactivation*" [4].
- "Most conditions associated with orthostatic hypertension are associated with *altered* cardiovascular *adrenergic control mechanisms*" [1].

Even if these speculations are correct, they leave open the question of whether orthostatic hypertension is a primary autonomic disorder, or whether such autonomic dysfunction is secondary to some other process.

The significantly greater prevalence of orthostatic hypertension in diabetics [9] suggests (though does not prove) an autonomic neuropathy.

Another category of proposed mechanisms relates orthostatic hypertension to vascularly based phenomena.

- Reports of renal arterial abnormalities suggest that circulatory disturbances (essentially, inadequate renal perfusion) inappropriately trigger an autonomic response. For example, Tsukamoto and colleagues [16] reported a case in which they concluded that orthostatic hypertension in their patient was due to *renal artery stenosis and nephroptosis* (defined as a "condition in which the kidney descends more than two vertebral bodies (or >5 cm) during a position change from supine to upright" [17]).
- Streeten and colleagues [18] suggest that orthostatic hypertension "involves excessive orthostatic blood pooling, which results in decreased venous return, decreased cardiac output, *increased sympathetic stimulation* (presumably through low-pressure cardiopulmonary receptors), and excessive arteriolar, but not venular, constriction."

These explanations have their merits, but most investigators reach conclusions such as:

- "The actual etiology is poorly understood" and therefore "in need of further investigation" [13] and "deserves more attention" [1].
- "The pathogenetic mechanisms leading to [orthostatic hypertension] are still controversial and may vary with age" [14].

Clinical Presentation

Curiously, the presentation of orthostatic hypertension can closely resemble orthostatic hypotension, even though these appear to be diametrically opposed blood pressure phenomena. Specifically, patients with orthostatic hypertension may complain of lightheadedness upon standing.

Physical Examination

As mentioned earlier, various definitions of orthostatic hypertension have been proposed [1]. Practically, most clinicians use the consensus statement proposed by Jordan and colleagues, defining orthostatic hypertension as an increase in systolic blood pressure ≥20 mm Hg when going from a supine to a standing position [2, 3].

Somewhat similarly to orthostatic hypotension, orthostatic hypertension can be variably present, and may depend in part on the time of day during which it is measured [11]. Given its intermittent nature, if a practitioner does not find orthostatic hypertension on physical examination, yet has a high index of suspicion for it, it is medically reasonable to reexamine the patient subsequently, or refer them for tilt table testing (on the possibility that there is delayed orthostatic hypertension).

Testing

Although orthostatic hypertension can often be detected on a bedside examination, tilt table testing is the modality of choice for securing the diagnosis [19, 20], in part because the manifestation of hypertension can occur at any point after the head is upright tilted [19], including after some delay. Jordan and colleagues [9] provide a clear example of this.

Differential Diagnosis

The main differential diagnosis of orthostatic hypertension includes other disorders in which disequilibrium is triggered upon assuming an upright posture, such as orthostatic hypotension, or (less commonly) benign paroxysmal positional vertigo.

Treatment

Given the proposed mechanisms of disease, and given the other medical conditions in whose context orthostatic hypertension occurs, it is sensible for these patients to be referred to a cardiologist and/or nephrologist.

References

1. Jordan J, Ricci F, Hoffmann F, Hamrefors V, Fedorowski A. Orthostatic hypertension: critical appraisal of an overlooked condition. Hypertension. 2020;75(5):1151–8.
2. Jordan J, Biaggioni I, Kotsis V, Nilsson P, Grassi G, Fedorowski A, et al. Consensus statement on the definition of orthostatic hypertension endorsed by the American Autonomic Society

and the Japanese Society of Hypertension. Hypertens Res. 2023;46(2):291–4.
3. Jordan J, Biaggioni I, Kotsis V, Nilsson P, Grassi G, Fedorowski A, et al. Consensus statement on the definition of orthostatic hypertension endorsed by the American Autonomic Society and the Japanese Society of Hypertension. Clin Auton Res. 2023;33(1):69–73.
4. Hu Y, Wang Y, He B, Wang Y, Han Z, Tao C, et al. Sympathetic overactivation from supine to upright is associated with orthostatic hypertension in children and adolescents. Front Pediatr. 2020;8:54.
5. Zhang Q, Li J, Xie Y, Zhao J, Du J. Orthostatic hypertension in children and adolescents with postural tachycardia syndrome. J Trop Pediatr. 2014;60(6):461–6.
6. Hu Y, He B, Han Z, Wang Y, Tao C, Wang Y, et al. Risk factors for orthostatic hypertension in children. J Pediatr. 2020;227:212–7.e1.
7. Hu Y, Jin H, Du J. Orthostatic hypertension in children: an update. Front Pediatr. 2020;8:425.
8. Wecht JM, Weir JP, Martinez S, Eraifej M, Bauman WA. Orthostatic hypotension and orthostatic hypertension in American veterans. Clin Auton Res. 2016;26(1):49–58.
9. Yoshinari M, Wakisaka M, Nakamura U, Yoshioka M, Uchizono Y, Iwase M. Orthostatic hypertension in patients with type 2 diabetes. Diabetes Care. 2001;24(10):1783–6.
10. McCann WS, Romansky MJ. Orthostatic hypertension: the effect of nephroptosis on the renal blood flow. JAMA J Am Med Assoc. 1940;115(8):573–8.
11. Ghazi L, Cohen JB, Townsend RR, Drawz PE, Rahman M, Pradhan N, et al. Orthostatic hypotension, orthostatic hypertension, and ambulatory blood pressure in patients with chronic kidney disease in CRIC. J Hypertens. 2024;42(2):329–36.
12. Rouabhi M, Durieux J, Al-Kindi S, Cohen JB, Townsend RR, Rahman M. Orthostatic hypertension and hypotension and outcomes in CKD: the CRIC (chronic renal insufficiency cohort) study. Kidney Med. 2021;3(2):206–15.e1.
13. Fessel J, Robertson D. Orthostatic hypertension: when pressor reflexes overcompensate. Nat Clin Pract Nephrol. 2006;2(8):424–31.
14. Palatini P. Orthostatic hypertension: a newcomer among the hypertension phenotypes. Hypertension. 2023;80(10):1993–2002.
15. Magkas N, Tsioufis C, Thomopoulos C, Dilaveris P, Georgiopoulos G, Doumas M, et al. Orthostatic hypertension: from pathophysiology to clinical applications and therapeutic considerations. J Clin Hypertens (Greenwich). 2019;21(3):426–33.
16. Tsukamoto Y, Komuro Y, Akutsu F, Fujii K, Marumo F, Kusano S, et al. Orthostatic hypertension due to coexistence of renal fibromuscular dysplasia and nephroptosis. Jpn Circ J. 1988;52(12):1408–14.
17. Murari SB, Gadepalli T, Rao VP, Ram R. Renal scintigraphy in diagnosis and management of nephroptosis. Indian J Nucl Med. 2012;27(1):52–4.
18. Streeten DH, Auchincloss JH Jr, Anderson GH Jr, Richardson RL, Thomas FD, Miller JW. Orthostatic hypertension. Pathogenetic studies. Hypertension. 1985;7(2):196–203.
19. Petersen Á, Salas-Herrera C, Lerma C, Brown-Escobar C, Kostin A, Sierra-Beltrán M, et al. Transient orthostatic hypertension during head-up tilt test in young adults: a phenotype of blood pressure variability. J Clin Neurophysiol. 2021;38(3):242–9.
20. Lee H, Kim HA. Orthostatic hypertension: an underestimated cause of orthostatic intolerance. Clin Neurophysiol. 2016;127(4):2102–7.

143 Cardiac Arrhythmia

Overview

Cardiac arrhythmias can cause a transient decrease in cerebral perfusion. If this decrease is sufficiently severe and prolonged, it can culminate in syncope. If not, it can manifest with presyncope, usually experienced as lightheadedness [1].

However, given that patients may use terms with different intended meanings than what a clinician expects, they may describe a presyncopal sensation using words such as "dizziness" or "vertigo," which serves as a reminder that it is unwise to diagnose the etiology of disequilibrium based only on a patient's description [2].

However, while uncommon, some patients appear truly to experience a transient decrease in cerebral perfusion as a spinning sensation [3], and thus may offer terms such as "spinning" and "vertigo" in response to an interviewer's queries about the quality of the sensation. This can easily lead a clinician to consider an otologic source of symptoms.

Practically, if a patient reports episodes of disequilibrium (of whatever quality) that culminate in syncope, cardiovascular causes should lead the differential diagnosis.

In cases that do not culminate in syncope, cardiovascular causes should be lower on the differential diagnosis, though they should not be completely excluded. Reasons to consider a cardiac arrhythmia as the cause of disequilibrium include:

- A reasonable search for other etiologies, including oto-vestibular diseases, has been unrevealing.
- A patient experiences palpitations or shortness of breath preceding, or in association with, the disequilibrium.
- There is a known history, or grounds for suspecting, cardiovascular disease.

In such cases, cardiology referral is appropriate.

References

1. Susanto M. Dizziness: if not vertigo could it be cardiac disease? Aust Fam Physician. 2014;43(5):264–9.
2. Newman-Toker DE, Camargo CA Jr. 'Cardiogenic vertigo' – true vertigo as the presenting manifestation of primary cardiac disease. Nat Clin Pract Neurol. 2006;2(3):167–72. quiz 73
3. Gordon M. Occult cardiac arrhythmias associated with falls and dizziness in the elderly: detection by Holter monitoring. J Am Geriatr Soc. 1978;26(9):418–23.

M. Cherchi, *Otoneurology and Vestibular Medicine*, https://doi.org/10.1007/978-3-031-94841-1_143

144 Vestibular Syncope

Overview

The phrase "*vestibular syncope*" refers to the circumstance in which a vestibular event causes a subsequent episode of syncope. The mechanism of this is unknown, but proposals have included maladaptive cardiovascular/hemodynamic changes in response to the vestibular hallucination (induced by vestibular disease). Primary cardiovascular problems are more common causes of syncope than vestibular problems, so a cardiovascular workup (such as with tilt table testing and cardiac event monitoring) is reasonable. If no cardiovascular cause is found, then it is reasonable to undertake a workup for vestibular disease; if such is found, then treatment (if it exists) is appropriate.

Introduction

The term syncope refers to an unexpected loss of consciousness. The phrase "*vestibular syncope*" implies that vestibular dysfunction causes the syncopal event.

Epidemiology

There is only a modest literature regarding vestibular syncope, none of which details the epidemiology of this clinical entity.

Pathophysiological Mechanism of Disease

Most instances of syncope result from transient hypoperfusion of the brain, so the proximal mechanism is inadequate blood flow.

One study of recurrent "cardiogenic vertigo" found that the "vertigo" was unassociated with syncope 52% of the time, preceded syncope 37% of the time, and followed syncope 11% of the time [1].

For cases in which "vertigo" is followed by syncope, the sequence of events raises the possibility of a causal relationship. The mechanism underlying such a relationship ("vertigo" causing syncope) remains unclear, but several proposals have emerged.

Raphan, summarizing work done with Bernard Cohen, suggests, "the idea that there was a velocity-storage mechanism or integrator that not only is a focus for visual-vestibular interaction but also codes spatial orientation relative to gravity as referenced by the otoliths" [2].

Developing this idea further, Kwon and colleagues suggest that, "By integrating the velocity-storage (VS) circuit in the brainstem and cerebellum, we propose that the vestibular syncope develops as a result of dyssynergia of the vestibulosympathetic and baroreflexes in which centrally estimated downward inertial acceleration during the vertigo attacks acts as a trigger" [3]. In particular, "The dual reflex pathways incorporating the function of the velocity-storage circuit in the brainstem and cerebellum suggest that vestibular syncope is a neurally mediated reflex syncope associated with a sudden hemodynamic change during vertigo. This change can be arterial hypertension triggered by a false downward inertial cue, as suggested previously, or hypotension driven by a false upward inertial cue" [4].

The key scenarios for this are:

- In normal circumstances, when a healthy individual is subjected to *true* upward acceleration, that acceleration will cause pooling of intravascular volume in the lower extremities, which results in correspondingly less intravascular volume available in the thorax for the heart to pump upward and perfuse the brain adequately. In order to offset this, several cardiovascular mechanisms are activated (increased heart rate, increased ejection fraction, peripheral vasoconstriction). In contrast, if the individual is subjected to a *hallucination* (false perception) of upward acceleration (when in fact there is no such acceleration), then the same cardiovascular mechanisms will

M. Cherchi, *Otoneurology and Vestibular Medicine*, https://doi.org/10.1007/978-3-031-94841-1_144

(mistakenly) be activated, resulting in *hyper*perfusion of the brain.

- In normal circumstances, when a healthy individual is subjected to *true* downward acceleration, that acceleration will cause pooling of intravascular volume in the upper body and brain. In order to offset this, several cardiovascular mechanisms are activated (decreased heart rate, decreased ejection fraction, peripheral vasodilation). In contrast, if the individual is subjected to a *hallucination* (false perception) of downward acceleration (when in fact there is no such acceleration), then the same cardiovascular mechanisms will (mistakenly) be activated, resulting in *hypo*perfusion of the brain.

A reasonable "test scenario" for the latter circumstance would be a drop attack—a hallucination of abrupt downward acceleration. In otoneurology and neurotology, a paradigm example of such drop attacks can be found in the otolithic catastrophe of Tumarkin, which can occur in Meniere's disease, and this has been documented in several studies [5–7].

A drop attack is an abrupt and very pronounced hallucination of downward acceleration, but it is likely that the severity of such hallucinations can occur over a continuum, and thus it is logically plausible that less abrupt and less severe vestibular hallucinations may still provoke some degree of maladaptive (inappropriate compensatory) hemodynamic changes.

While the above logic may seem speculative, it is grounded in theoretical considerations [8–11] and broader evidence from studies regarding the relationship between the vestibular system and the autonomic nervous system, which has been reviewed elsewhere in some detail [12–19]. Briefly, there is some experimental and clinical evidence in humans that vestibular lesions can interfere with autonomic function [12, 20–23], though not all studies reach the same conclusion [24–28]. It appears that autonomic dysfunction following vestibular lesions can recover quickly in some cases [16]. There is evidence that with aging, the vestibular contribution to autonomic control declines [29–31]. A variety of neuroanatomical structures and pathways have been implicated in the relationship between vestibular function and autonomic function, including the medial vestibular nucleus, the rostral ventrolateral medulla, and caudal medullary raphe nuclei to the preganglionic neurons in the thoracic spinal cord [14], the insular cortex [32] and regions in the medial aspect of the cerebellar cortex [9]. Some of these, and other neuroanatomical pathways, are instructively diagrammed by Yates and colleagues [19].

Some of these, and other neuroanatomical pathways, are depicted in Fig. 144.1.

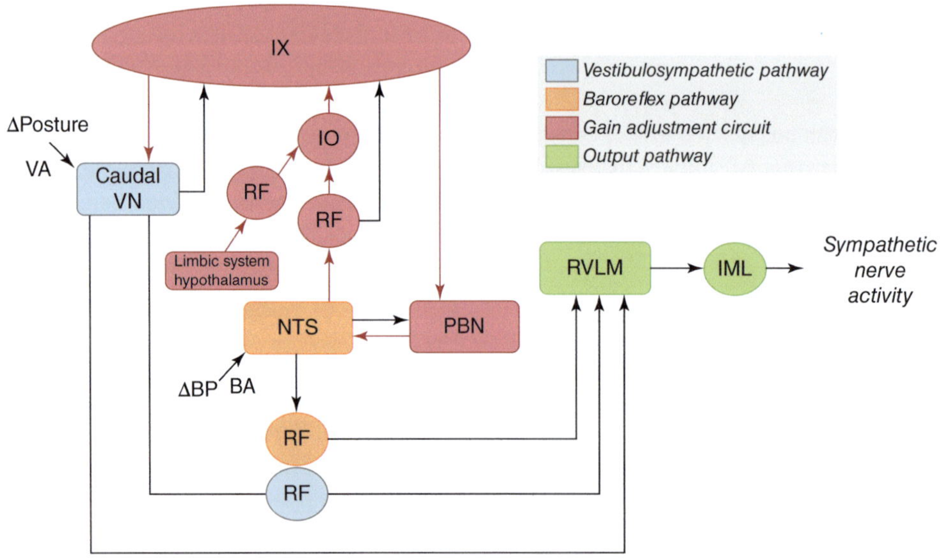

Fig. 144.1 Neural regions that generate and modify the gains of vestibulosympathetic reflexes (VSR) and baroreceptor reflexes. Abbreviations: *BA* baroreceptor afferent, *BP* blood pressure, *IML* intermediolateral cell column, *IO* inferior olivary nucleus, *IX* cerebellar lobule IX, uvula, *NTS* nucleus tractus solitarius, *PBN* parabrachial nucleus, *RF* reticular formation, *RVLM* rostral ventrolateral medulla, *VA* vestibular afferent, *VN* vestibular nucleus complex. (From Yates et al. [19]. Reproduced with written permission from John Wiley and Sons Limited.)

Clinical Presentation

In order for vestibular-mediated syncope to occur, there must be a vestibular event of some kind. Absent such a vestibular event, it is difficult to make a convincing case that the syncope is mediated by the vestibular system.

Testing: Vestibular

For patients in whom vestibular-mediated syncope is suspected, and whose cardiovascular workup has been unremarkable, it is medically reasonable to undertake a screening otovestibular workup with cervical vestibular evoked myogenic potentials, ocular vestibular evoked myogenic potentials, video head impulse testing, and videonystagmography.

Testing: Other

If a patient is referred for a clinical query of vestibular-mediated syncope but has *not* had a cardiac workup, then this should be considered, since primary cardiovascular causes of syncope are more common than vestibular ones. This aspect of the workup is usually undertaken by cardiology, and may include tilt table testing and cardiac event monitoring.

Differential Diagnosis

As mentioned earlier, in order for vestibular-mediated syncope to occur, there must be a vestibular event of some kind. If there is no such vestibular event preceding the syncopal episode, then the differential diagnosis must include primary cardiac factors (arrhythmia, orthostatic intolerance, etc.).

Treatment and Prognosis

If an adequate cardiovascular workup fails to reveal a cause for syncope, and if the history is compatible with vestibular syncope, and if an adequate otovestibular workup reveals evidence for vestibular disease, then it is medically reasonable to entertain a diagnosis of vestibular syncope.

There are no guidelines for treatment in this circumstance, but it is logical to treat the identified vestibular disease, if such treatment exists.

The prognosis depends on the vestibular disease itself.

Final Comments

Earlier, we alluded to the logically plausible view that vestibular influences on the autonomic nervous system probably operate along a continuum. Thus, although we have been discussing vestibularly mediated *syncope,* it is plausible that vestibular dysfunction could provoke more modest autonomic changes that cause *presyncope,* rather than culminating in frank loss of consciousness. We speculate that this may be part of the mechanism by which diseases that usually cause sensations of rotation (such as benign paroxysmal positional vertigo or vestibular neuritis) may additionally provoke some degree of presyncope.

To this point, our discussion has pertained to vestibular influences on the autonomic nervous system function. There is additionally some literature regarding autonomic etiologies of vestibular sensations [33, 34]. Thus, there is the potential for a maladaptive positive feedback loop.

References

1. Kim HA, Ahn J, Park HS, Lee SM, Choi SY, Oh EH, et al. Cardiogenic vertigo: characteristics and proposed diagnostic criteria. J Neurol. 2021;268(3):1070–5.
2. Raphan T. Vestibular, locomotor, and vestibulo-autonomic research: 50 years of collaboration with Bernard Cohen. J Neurophysiol. 2020;123(1):329–45.
3. Kwon E, Lee JY, Kim HJ, Choi JY, Kim JS. Can dyssynergia of vestibulosympathetic and baroreflexes cause vestibular syncope? The hypothesis based on the velocity-storage function. Cerebellum. 2022;21(2):244–52.
4. Kwon H, Kwon E, Kim HJ, Choi JY, Kim JS. Vestibular syncope: clinical characteristics and mechanism. Ann Clin Transl Neurol. 2022;9(10):1616–25.
5. Pyykkö I, Manchaiah V, Zou J, Levo H, Kentala E. Vestibular syncope: a disorder associated with drop attack in Ménière's disease. Auris Nasus Larynx. 2018;45(2):234–41.
6. Pyykkö I, Manchaiah V, Zou J, Levo H, Kentala E. Association between Syncope and Tumarkin Attacks in Ménière's disease. J Int Adv Otol. 2019;15(1):135–40.
7. Pyykkö I, Manchaiah V, Zou J, Levo H, Kentala E. Do patients with Ménière's disease have attacks of syncope? J Neurol. 2017;264(Suppl 1):48–54.
8. Balaban CD. Vestibular autonomic regulation (including motion sickness and the mechanism of vomiting). Curr Opin Neurol. 1999;12(1):29–33.
9. Balaban CD, Porter JD. Neuroanatomic substrates for vestibulo-autonomic interactions. J Vestib Res Equilib Orient. 1998;8(1):7–16.
10. Previc FH. Do the organs of the labyrinth differentially influence the sympathetic and parasympathetic systems? Neurosci Biobehav Rev. 1993;17(4):397–404.
11. Yates BJ. Autonomic reaction to vestibular damage. Otolaryngol Head Neck Surg. 1998;119(1):106–12.
12. Carter JR, Ray CA. Sympathetic responses to vestibular activation in humans. Am J Physiol Regul Integr Comp Physiol. 2008;294(3):R681–8.

13. Kerman IA, McAllen RM, Yates BJ. Patterning of sympathetic nerve activity in response to vestibular stimulation. Brain Res Bull. 2000;53(1):11–6.
14. Yates BJ. Vestibular influences on the sympathetic nervous system. Brain Res Brain Res Rev. 1992;17(1):51–9.
15. Yates BJ, Miller AD. Physiological evidence that the vestibular system participates in autonomic and respiratory control. J Vestib Res Equilib Orient. 1998;8(1):17–25.
16. Yates BJ, Bronstein AM. The effects of vestibular system lesions on autonomic regulation: observations, mechanisms, and clinical implications. J Vestib Res Equilib Orient. 2005;15(3):119–29.
17. Furman JM, Jacob RG, Redfern MS. Clinical evidence that the vestibular system participates in autonomic control. J Vestib Res Equilib Orient. 1998;8(1):27–34.
18. Bogle JM, Benarroch E, Sandroni P. Vestibular-autonomic interactions: beyond orthostatic dizziness. Curr Opin Neurol. 2022;35(1):126–34.
19. Yates BJ, Bolton PS, Macefield VG. Vestibulo-sympathetic responses. Compr Physiol. 2014;4(2):851–87.
20. Kim JG, Lee JH, Lee SU, Choi JY, Kim BJ, Kim JS. Utricular dysfunction in patients with orthostatic hypotension. Clin Auton Res. 2022;32(6):431–44.
21. Jáuregui-Renaud K, Hermosillo AG, Gómez A, Márquez MF, Cárdenas M, Bronstein AM. Vestibular function interferes in cardiovascular reflexes [corrected]. Arch Med Res. 2003;34(3):200–4.
22. Radtke A, Popov K, Bronstein AM, Gresty MA. Vestibulo-autonomic control in man: short- and long-latency vestibular effects on cardiovascular function. J Vestib Res Equilib Orient. 2003;13(1):25–37.
23. Cooke WH, Carter JR, Kuusela TA. Human cerebrovascular and autonomic rhythms during vestibular activation. Am J Physiol Regul Integr Comp Physiol. 2004;286(5):R838–43.
24. Aghababaei Ziarati M, Taziki MH, Hosseini SM. Autonomic laterality in caloric vestibular stimulation. World J Cardiol. 2020;12(4):144–54.
25. Biaggioni I, Costa F, Kaufmann H. Vestibular influences on autonomic cardiovascular control in humans. J Vestib Res Equilib Orient. 1998;8(1):35–41.
26. Heidenreich KD, Weisend S, Fouad-Tarazi FM, White JA. The incidence of coexistent autonomic and vestibular dysfunction in patients with postural dizziness. Am J Otolaryngol. 2009;30(4):225–9.
27. Kalderon L, Chaimoff M, Katz-Leurer M. The cardiac autonomic control system response to vestibular stimulation in subjects with BPPV compared with healthy controls. NeuroRehabilitation. 2023;52(2):273–8.
28. Watenpaugh DE, Cothron AV, Wasmund SL, Wasmund WL, Carter R 3rd, Muenter NK, et al. Do vestibular otolith organs participate in human orthostatic blood pressure control? Auton Neurosci. 2002;100(1-2):77–83.
29. Monahan KD, Ray CA. Vestibulosympathetic reflex during orthostatic challenge in aging humans. Am J Physiol Regul Integr Comp Physiol. 2002;283(5):R1027–32.
30. Ray CA, Monahan KD. Aging attenuates the vestibulosympathetic reflex in humans. Circulation. 2002;105(8):956–61.
31. Sauder CL, Conboy EE, Chin-Sang SA, Ray CA. Otolithic activation on visceral circulation in humans: effect of aging. Am J Physiol Renal Physiol. 2008;295(4):F1166–9.
32. Nagai M, Scheper V, Lenarz T, Förster CY. The insular cortex as a vestibular area in relation to autonomic function. Clin Auton Res. 2021;31(2):179–85.
33. Nakagawa H, Ohashi N, Kanda K, Watanabe Y. Autonomic nervous system disturbance as etiological background of vertigo and dizziness. Acta oto-laryngologica Supplementum. 1993;504:130–3.
34. Pappas DG Jr. Autonomic related vertigo. Laryngoscope. 2003;113(10):1658–71.

Part XXII

Diagnoses—Medical Causes of Disequilibrium

145 Pharmacologic Causes of Disequilibrium

For Clinicians

Medications can have unintended effects, both desirable and undesirable.

In otoneurology, the question sometimes arises whether a given medication is responsible for vestibular symptoms (less commonly for auditory symptoms). In some cases, there may be clear grounds for suspecting such a relationship. A straightforward example of this would be when a patient is started on an antihypertensive medication and within days, develops lightheadedness upon standing, and is found on examination to have orthostatic hypotension; in such a case, the likely mechanism of the symptom is orthostatic hypotension, and the likely culprit is the antihypertensive medication.

But in many cases, the relationship is not so obvious, such as when there is no clear temporal proximity between starting the drug and the onset of symptoms, or when the drug in question does not have disequilibrium listed as a common adverse effect. In such circumstances, if the history and physical examination are unrevealing, then it is medically reasonable to undertake a screening otovestibular workup; if those results do not discern a primary otovestibular disturbance, then it is also medically reasonable to suspect the medication as the cause of this symptom, and the prescriber may consider reducing the dose of the medication, discontinuing it, or substituting it.

M. Cherchi, *Otoneurology and Vestibular Medicine*, https://doi.org/10.1007/978-3-031-94841-1_145

Postprandial Disequilibrium

146

Overview

Some individuals experience *disequilibrium* (usually manifesting as lightheadedness) *postprandially*. Potential proximal causes include postprandial hypotension and postprandial hypoglycemia. The underlying mechanisms of these problems have not been fully elucidated. Postprandial hypotension may be captured on ambulatory blood pressure monitoring. Postprandial hypoglycemia may be captured on ambulatory glucose monitoring.

Introduction

Some individuals experience disequilibrium during or after eating. The main mechanisms implicated in *postprandial disequilibrium* include postprandial hypotension and postprandial hypoglycemia.

Pathophysiological Mechanism of Disease

Postprandial hypotension can occur in otherwise healthy individuals, but appears more common in the elderly [1] and in the context of various neurological diseases [2]. The mechanism of postprandial hypotension is unclear; proposals have included impaired sympathetic baroreflex function, release of vasodilatory peptides, gastric distension, and splanchnic blood pooling [1, 3–5]. Postprandial hypotension appears to be a different clinical entity from orthostatic hypotension [3]. It is probably best assessed by ambulatory blood pressure monitoring before a meal, and during or after a meal when the patient is symptomatic [6, 7].

In nondiabetic patients, there has been debate about how common *postprandial hypoglycemia* is [8]. Although the usual method for assessing postprandial hypoglycemia would be a glucose tolerance test, this appears relatively insensitive to the phenomenon [9, 10], and some researchers suggest that measurement of blood glucose *during a postprandial episode* by ambulatory glucose monitoring (rather than merely in a laboratory setting) is more likely to secure the diagnosis [11].

Testing

Practically, in a patient who experiences postprandial disequilibrium (usually manifesting as lightheadedness) fairly consistently, their internist may suggest ambulatory glucose monitoring and/or ambulatory blood pressure monitoring. Depending on those results, referral to cardiology or endocrinology may be appropriate.

Treatment

The materials distributed to patients with postprandial hypotension suggest generous hydration before a meal, eating smaller meals, reducing intake of rapidly digested carbohydrates, and simply being aware that the nadir of blood pressure tends to occur 30–60 min after a meal [12].

There are relatively few treatment trials for idiopathic postprandial hypoglycemia [13]. Some studies of idiopathic postprandial hypoglycemia suggest that treatment with acarbose (taken with each meal) may be helpful [14].

References

1. Trahair LG, Horowitz M, Jones KL. Postprandial hypotension: a systematic review. J Am Med Dir Assoc. 2014;15(6):394–409.
2. Pavelic A, Krbot Skoric M, Crnosija L, Habek M. Postprandial hypotension in neurological disorders: systematic review and meta-analysis. Clin Auton Res. 2017;27(4):263–71.

M. Cherchi, *Otoneurology and Vestibular Medicine*, https://doi.org/10.1007/978-3-031-94841-1_146

3. Jansen RW, Lipsitz LA. Postprandial hypotension: epidemiology, pathophysiology, and clinical management. Ann Intern Med. 1995;122(4):286–95.
4. O'Mara G, Lyons D. Postprandial hypotension. Clin Geriatr Med. 2002;18(2):307–21.
5. Madden KM, Feldman B, Meneilly GS. Baroreflex function and postprandial hypotension in older adults. Clin Auton Res. 2021;31(2):273–80.
6. Luciano GL, Brennan MJ, Rothberg MB. Postprandial hypotension. Am J Med. 2010;123(3):281 e1–6.
7. Nakajima S, Otsuka K, Yamanaka T, Omori K, Kubo Y, Toyoshima T, et al. Ambulatory blood pressure and postprandial hypotension. Am Heart J. 1992;124(6):1669–71.
8. Hofeldt FD, Adler RA, Herman RH. Postprandial hypoglycemia. Fact or fiction? JAMA. 1975;233(12):1309.
9. Brun JF, Fedou C, Mercier J. Postprandial reactive hypoglycemia. Diabetes Metab. 2000;26(5):337–51.
10. Lefebvre PJ, Andreani D, Marks V, Creutzfeldt W. Statement on postprandial hypoglycemia. Diabetes Care. 1988;11(5):439–40.
11. Palardy J, Havrankova J, Lepage R, Matte R, Belanger R, D'Amour P, Ste-Marie LG. Blood glucose measurements during symptomatic episodes in patients with suspected postprandial hypoglycemia. N Engl J Med. 1989;321(21):1421–5.
12. No Authors. Eating can cause low blood pressure. Postprandial hypotension can show up as dizziness or falling after a meal. Harv Heart Lett. 2010;20(11):2.
13. Galati SJ, Rayfield EJ. Approach to the patient with postprandial hypoglycemia. Endocr Pract. 2014;20(4):331–40.
14. Peter S. Acarbose and idiopathic reactive hypoglycemia. Horm Res. 2003;60(4):166–7.

Part XXIII

Diagnoses—Infectious Causes of Disequilibrium

147 COVID-19

Overview

Literature documents the occurrence of audiologic and vestibular symptoms in the context of COVID-19 (both immediate and in "long-haul COVID-19" scenarios), though it is difficult to ascertain whether this relationship is causal or associative [1]. The mechanism by which COVID-19 might cause audiovestibular symptoms remains unclear. Even when a patient has been diagnosed with COVID-19, it is prudent to bear in mind that audiologic and vestibular symptoms have a broad differential diagnosis.

Auditory and vestibular symptoms have also been reported in association with the vaccines and treatments for COVID-19, though again, it has been difficult to ascertain whether this relationship is causal or associative [1]. However, in view of the evidence of the potential for "long-haul COVID-19" symptoms, and the absence of compelling evidence for similarly lasting symptoms from vaccines and treatments for COVID-19, it appears that the risk of harm from the vaccines and treatments is lower than the risks of COVID-19 itself.

The material in this chapter draws heavily from our prior review of this topic [1].

Introduction

The most common manifestations of COVID-19 infection include respiratory (cough) and constitutional (fever, fatigue), but the virus can impact nearly every organ system. Auditory and vestibular symptoms are not the most common manifestations, but nor are they rare. We shall review what is known and unknown about auditory and vestibular symptoms that have been reported in association with COVID-19. We close with a review of auditory and vestibular symptoms reported in association with the vaccines and treatments for COVID-19.

Pathobiology of COVID-19

COVID-19 is caused by the severe acute respiratory syndrome coronavirus type 2 ("SARS-CoV-2"). It is well understood how this virus enters the body through the respiratory tract. It is less clear how the virus gains access to the central nervous system.

How SARS-CoV-2 Infects the Body

SARS-CoV-2 is a member of a family of single-stranded RNA coronaviruses. Spike glycoprotein "hooks" protrude from the virus's lipid bilayer envelope, and these facilitate entry into host cells [2], where they replicate, and whence they propagate. The spike glycoproteins bind to angiotensin-converting enzyme receptors [3, 4] that are densely expressed in cells in the pulmonary system.

How SARS-CoV-2 Infects the Nervous System

The factors that make a person more susceptible to central nervous system (CNS) involvement by SARS-CoV-2 remain unclear (CNS) [5, 6], though there is evidence [7] of cells within the CNS that express angiotensin-converting enzyme receptors—even if at a lower density than found in the lower respiratory tract. It appears that after these viruses have invaded the CNS, they can advance via axonal transport [6].

It is well known that other coronaviruses can invade the central nervous system, as they have been isolated in the brain, spinal cord tissue, and cerebrospinal fluid of patients suffering from a variety of neurological diseases such as encephalitis [8–12], acute disseminated encephalomyelitis [13], optic neuritis [14], Parkinson's disease [15], and multiple sclerosis [16–18]. For some of these diseases (e.g., multiple sclerosis, Parkinson's disease), it seems somewhat

M. Cherchi, *Otoneurology and Vestibular Medicine*, https://doi.org/10.1007/978-3-031-94841-1_147

unlikely that coronaviruses play a causative role, whereas in others (e.g., optic neuritis, encephalitis, and encephalomyelitis), it is more plausible.

Early in the COVID-19 pandemic, reports emerged of neurologic symptoms in individuals infected with SARS-CoV-2 [19, 20] who had been hospitalized [21], especially among those in intensive care units [22]. Data suggest that SARS-CoV-2 [23–25] has some similarities to other coronaviruses' ability to infect the CNS [26], though there are also several differences [4], such as the tendency of SARS-CoV-2 to cause anosmia and ageusia [27–31], suggesting involvement of cranial nerves. Thus, it is reasonable to consider whether cranial nerve involvement by SARS-CoV-2 is the mechanism of other focal neurological manifestations, including vestibular symptoms [32], auditory symptoms (such as hearing loss or tinnitus), or both [33]. Indeed, the ability of COVID-19 to cause auditory and vestibular symptoms appears to be distinct from the other SARS coronaviruses [34].

Auditory and Vestibular Symptoms Associated with COVID-19

We shall review literature regarding auditory and vestibular symptoms occurring in the context of SARS-CoV-2 infection; the difficulty of ascertaining whether this relationship is associative or causal; and several hypothesized pathophysiologic mechanisms.

COVID-19 Associated with Auditory Disorders

There is an emerging literature regarding the association between SARS-CoV-2 infection and hearing loss. While some work reported no relationship between SARS-CoV-2 infection and hearing loss [35, 36], accumulating case reports [37–43] and case series [44–48] document hearing loss occurring in the context of COVID-19. De Luca and colleagues [49] review numerous case reports and a case series of COVID-19 patients, including a broad range of ages (18–84 years), both sexes, with unilateral or bilateral, usually sudden onset sensorineural hearing loss. Fancello et al. [50] similarly reviewed case reports and two case series of COVID-19 patients, including both sexes, with a broad range of ages, with unilateral (more common) or bilateral (less common) sudden sensorineural hearing loss. The prospective single-institution study by Dusan and colleagues [44] reported a similar variety.

A more modest literature has also documented an association between SARS-CoV-2 infection and tinnitus [38, 39, 41, 46, 47, 50–55]. Some papers describe the tinnitus occurring with hearing loss [38, 39, 41]; other papers describe tinnitus occurring in the absence of hearing loss [55]; other papers document both [54].

COVID-19 Associated with Disequilibrium

A case series of 799 patients from Wuhan, China, reported dizziness as a symptom in 8% of confirmed cases COVID-19 [56]. Another study of 214 from Wuhan confirmed cases of COVID-19 that explicitly asked patients about neurological symptoms reported dizziness in 16.8% of patients [29]. A study of 509 patients from several hospitals in Chicago reported dizziness in 29.7% of cases [21]. The variability reflected in this broad range (8–29.7%) among individual studies has been noted in reviews [32], though there is increasing recognition among clinicians of an association between COVID-19 and vestibular disorders [50].

Potential Mechanisms for Auditory and Vestibular Symptoms from SARS-CoV-2

Given the evidence of vestibular symptoms (dizziness) and cochlear symptoms (hearing loss, tinnitus) occurring in the context of SARS-CoV-2 infection, many reviews conclude that the relationship between COVID-19 and audiovestibular dysfunction may be causal rather than associative [34, 50, 57–66].

Audiovestibular test results are sometimes abnormal in patients infected with COVID-19 [64], but the underlying mechanisms by which SARS-CoV-2 might cause this remain unclear. We shall discuss several potential mechanisms.

First, SARS-CoV-2 appears capable of affecting individual cranial nerves, manifesting with anosmia [27–31], optic neuritis [14], and ophthalmoparesis [67], so it is reasonable to consider the possibility that SARS-CoV-2 involves the vestibulocochlear nerve, manifesting as vestibular neuritis [68–71] and/or cochlear neuritis or labyrinthitis. Some of the literature [72] reporting "vestibular neuritis in COVID-19" appears to base the diagnosis on clinical criteria rather than on objective vestibular testing, so such reports must be interpreted with caution.

Second, there is some evidence that SARS-CoV-2 can directly infect inner ear hair cells [73], which would constitute a plausible mechanism for cases of labyrinthitis occurring in the context of COVID-19 [74]. Whether such involvement by SARS-CoV-2 in turn increases the risk for other specific otologic diseases, such as benign paroxysmal positional vertigo [75, 76] or Ménière's disease [77] remains unclear.

Third, ischemia secondary to hypoxia [33] comprises a potential mechanism for inner ear dysfunction related to SARS-CoV-2 infection. This is plausible since COVID-19 is known to deoxygenate erythrocytes and increase the risk of thrombosis, which could easily compromise the inner ear's circulation that depends exclusively on the small, tenuous labyrinthine artery.

Fourth, intralabyrinthine hemorrhage is a candidate mechanism [37, 78], though reports of this are rare.

Fifth, since COVID-19 can cause autonomic dysfunction [79–83], orthostatic intolerance (typically in the form of postural orthostatic tachycardia) is a plausible mechanism of disequilibrium [84–90], though it would be unlikely to explain auditory symptoms.

Treatment Suggestions

The reports in the literature of vestibular neuritis occurring in the context of COVID-19 [68–71] do not prove that an acute vestibular syndrome is caused by the infection. Nevertheless, it is clinically relevant to identify vestibular hypofunction in such patients because vestibular rehabilitation therapy may still benefit them, whether the "vestibular neuritis" is attributable to COVID-19 or some other etiology. For patients diagnosed with COVID-19 (or for whom there is a high index of suspicion for COVID-19), it is medically reasonable to check a screening otovestibular workup (cervical and ocular vestibular evoked myogenic potentials, video head impulse testing and videonystagmography), and if results suggest unilateral or bilateral vestibular hypofunction, then it is sensible to offer a trial of vestibular rehabilitation therapy.

Auditory and Vestibular Symptoms Associated with Vaccination for SARS-CoV-2 and Treatment for COVID-19

Concerted research and pharmaceutical efforts have been directed at developing vaccines for SARS-CoV-2 and treatments for COVID-19.

The emergence of vaccines against SARS-CoV-2 and treatments for COVID-19 have raised questions about their potential adverse effects, including the potential for vestibular and auditory symptoms. We shall review several reports and studies about these agents. We do not cover the use of interleukin modulators (anakinra, tocilizumab, and sarilumab) and other immunomodulating agents being studied for COVID-19 patients who are in (or who are at risk of entering) intensive care units [91].

Vaccines for COVID-19

In the United States, several vaccines against SARS-CoV-2 were FDA-approved [92–94]. A broader review of possible adverse effects associated with the vaccines can be found elsewhere [95]. There are still insufficient data regarding whether these agents can specifically cause auditory or vestibular symptoms, or whether they can exacerbate preexisting otovestibular diseases. Our clinical experience is that apparent adverse effects from the vaccines have been transient, lasting days to weeks; while symptoms from SARS-CoV-2 infection can be very mild, they can also be devastating or lethal. In general, the risks incurred by the vaccines against SARS-CoV-2 and treatments for COVID-19 appear to be greatly outweighed by the risks of the disease itself.

Vaccines for COVID-19 and Dizziness

The literature regarding potential adverse effects from the Pfizer vaccine (BNT162b2 mRNA) is difficult to interpret as it pertains to vestibular symptoms. The original trial [93] included 43,548 participants (of whom 21,720 received the intervention and 21,828 received a placebo). The report and its supplementary material do not list dizziness/vertigo as an adverse event. But a small, uncontrolled study [96] of 1245 vaccine recipients reported that of 803 patients with "generalized" or "neurological" symptoms, 67 (8.34%) reported "dizziness," and 20 (2.49%) reported "vertigo." There has also been a case report of postural orthostatic tachycardia following receipt of this vaccine [97]. Finally, a case of vestibular neuritis (supposedly corroborated on video head impulse testing) was reported in a patient who received the Pfizer vaccine [71], but whether this relationship was associative or causal is unknown.

The literature regarding the Moderna vaccine's (mRNA-1273) association with vestibular symptoms has also been difficult to interpret. The original trial [92] studied 30,420 participants (of whom 15,210 received the vaccine and 15,210 received a placebo). Symptoms of "dizziness and vertigo" were reported as adverse effects in zero patients in the trial's supplementary material. In contrast, a small study of 432 vaccine recipients [98] reported "vertigo-like symptoms" in 15 (3.47%) patients and "dizziness" in 63 (14.58%) patients.

The literature regarding Johnson & Johnson's Janssen vaccine (Ad26.COV2.S) is also difficult to interpret. The original trial [94] included 39,260 participants (of whom 19,630 received the vaccine and 19,630 received a placebo). The supplementary material from the trial reported that no vaccine recipients reported symptoms of "dizziness" or "vertigo." Yet, the product monograph (https://covid-vaccine.canada.ca/info/pdf/janssen-covid-19-vaccine-pm-en.pdf, accessed June 26, 2022) that reviewed data from 43,783 participants (of whom 21,895 received the vaccine and 21,888 received a placebo) stated that 13 patients in the treatment group and seven patients in the placebo group reported the symptom of "vertigo."

Vaccines for COVID-19 and Hearing Loss

Data pertaining to the two mRNA vaccines (Pfizer and Moderna) is publicly available on the Centers for Disease Control's Vaccine Adverse Effects Reporting System (CDC VAERS). Formeister and colleagues [99] analyzed these data and concluded that 40 cases of unilateral sensorineural hearing loss (confirmed on audiometric testing) were "most likely" to be attributable to the vaccines themselves. The calculated incidence was 0.3 cases per 100,000 individuals (i.e., 3 per million). They further reported that in these cases, the unilateral hearing loss occurred within 3 weeks (mean 4 days) of having received the vaccines.

Vaccines for COVID-19 and Tinnitus

Tinnitus has been less studied as a potential adverse effect of SARS-CoV-2 vaccines, but emerging literature explores this [100, 101].

Antiviral Therapies for COVID-19

For an individual already infected with SARS-CoV-2, the vaccines play no role in acute management, but treatments aiming at the virus itself are being investigated; these are reviewed elsewhere [102]. Some of these agents (e.g., ivermectin, chloroquine, and hydroxychloroquine) have documented ototoxicity.

Paxlovid received emergency use authorization in the United States in December of 2021, and full FDA approval on May 25 of 2023, for treatment of "mild-to-moderate COVID-19 in adults who are at high risk for progression to severe disease." Paxlovid is a combination antiviral agent containing ritonavir (also used to treat HIV) and nirmatrelvir (an investigational drug). As of this writing, there was very little literature about the potential ototoxicity of these agents. Ritonavir, which is also one component of Kaletra (the second component of which is lopinavir), has been reported to cause reversible bilateral hearing loss [103], but whether that property is due to ritonavir or lopinavir remains unclear.

Bamlanivimab (LY-CoV555 by Lilly) is a neutralizing IgG1 antibody for SARS-CoV-2 [104]. Dougan and colleagues [105] conducted a randomized phase 3 trial of bamlanivimab in 1035 subjects (518 subjects received bamlanivimab plus etesevimab; 517 subjects received placebo). They reported that the symptom of "dizziness" was endorsed by four patients (0.8%) in the treatment group and three patients (0.6%) in the control group. Since these rates are similar, and since the intervention arm received both bamlanivimab and etesevimab, the risk attributable to bamlanivimab alone is unclear.

Molnupiravir (MK-4882/EIDD-2801 by Merck) is a nucleotide analogue that inhibits SARS-CoV-2 replication. Jayk Bernal and colleagues [106] conducted a phase 3 trial of 1411 subjects (710 subjects received molnupiravir, 701 subjects received a placebo). They reported that the symptom of "dizziness" was endorsed by 1% of subjects in the treatment group and 0.7% of subjects in the placebo group. The similar rates leave unclear what risk is attributable to molnupiravir.

Conclusions

Literature reports auditory and vestibular symptoms associated with SARS-CoV-2 infection, with its vaccines, and with treatments for COVID-19. Whether the relationship between these symptoms and the disease or its treatments is causal or associative remains unclear. Overall, the risks of the disease itself seem to outweigh the risks of its vaccines and treatments. Bearing in mind that auditory and vestibular symptoms have a large differential diagnosis, it is medically reasonable to undertake a screening otovestibular workup in COVID-19 patients complaining of such symptoms, as this may impact management. For example, identification of vestibular weakness may prompt a trial of vestibular rehabilitation therapy; identification of hearing loss may prompt consultation with audiology.

References

1. Cherchi M. Otovestibular symptoms of COVID-19 and its vaccines/treatments. Encyclopedia. 2022;2(3):1200–9.
2. De Masi L, Argenio MA, Giordano D, Facchiano A. Molecular aspects of spike – ACE2 interaction. Encycl COVID-19. 2022;2(1):96–108.
3. Yan R, Zhang Y, Li Y, Xia L, Guo Y, Zhou Q. Structural basis for the recognition of SARS-CoV-2 by full-length human ACE2. Science. 2020;367(6485):1444–8.
4. Ceccarelli M, Berretta M, Venanzi Rullo E, Nunnari G, Cacopardo B. Differences and similarities between severe acute respiratory syndrome (SARS)-CoronaVirus (CoV) and SARS-CoV-2. Would a rose by another name smell as sweet? Eur Rev Med Pharmacol Sci. 2020;(2284-0729):24. (Electronic))
5. Dales S. Factors controlling coronavirus infections and disease of the central nervous system. A review. Adv Exp Med Biol. 1995;380:13–22.
6. Dube M, Le Coupanec A, Wong AHM, Rini JM, Desforges M, Talbot PJ. Axonal transport enables neuron-to-neuron propagation of human coronavirus OC43. J Virol. 2018;92(17)
7. Baig AM, Khaleeq A, Ali U, Syeda H. Evidence of the COVID-19 virus targeting the CNS: tissue distribution, host-virus interaction, and proposed neurotropic mechanisms. ACS Chem Neurosci. 2020;11(7):995–8.

8. Morfopoulou S, Brown JR, Davies EG, Anderson G, Virasami A, Qasim W, et al. Human coronavirus OC43 associated with fatal encephalitis. N Engl J Med. 2016;375(5):497–8.
9. Arabi YM, Harthi A, Hussein J, Bouchama A, Johani S, Hajeer AH, et al. Severe neurologic syndrome associated with Middle East respiratory syndrome corona virus (MERS-CoV). Infection. 2015;43(4):495–501.
10. Li Y, Li H, Fan R, Wen B, Zhang J, Cao X, et al. Coronavirus infections in the central nervous system and respiratory tract show distinct features in hospitalized children. Intervirology. 2016;59(3):163–9.
11. Nilsson A, Edner N, Albert J, Ternhag A. Fatal encephalitis associated with coronavirus OC43 in an immunocompromised child. Infect Dis (Lond). 2020;52(6):1–4.
12. Shnayder NA, Sirbiladze TK, Demko IV, Petrova MM, Nasyrova RF. Limbic encephalitis associated with COVID-19. Encyclopedia. 2022;2(1)
13. Yeh EA, Collins A, Cohen ME, Duffner PK, Faden H. Detection of coronavirus in the central nervous system of a child with acute disseminated encephalomyelitis. Pediatrics. 2004;113(1 Pt 1):e73–6.
14. Dessau RB, Lisby G, Frederiksen JL. Coronaviruses in spinal fluid of patients with acute monosymptomatic optic neuritis. Acta Neurol Scand. 1999;100(2):88–91.
15. Fazzini E, Fleming J, Fahn S. Cerebrospinal fluid antibodies to coronavirus in patients with Parkinson's disease. Mov Disord. 1992;7(2):153–8.
16. Burks JS, DeVald BL, Jankovsky LD, Gerdes JC. Two coronaviruses isolated from central nervous system tissue of two multiple sclerosis patients. Science. 1980;209(4459):933–4.
17. Salmi A, Ziola B, Hovi T, Reunanen M. Antibodies to coronaviruses OC43 and 229E in multiple sclerosis patients. Neurology. 1982;32(3):292–5.
18. Hovanec DL, Flanagan TD. Detection of antibodies to human coronaviruses 229E and OC43 in the sera of multiple sclerosis patients and normal subjects. Infect Immun. 1983;41(1):426–9.
19. Koralnik IJ, Tyler KL. COVID-19: a global threat to the nervous system. Ann Neurol. 2020;88(1):1–11.
20. Pinzon RT, Wijaya VO, Buana RB, Al Jody A, Nunsio PN. Neurologic characteristics in coronavirus disease 2019 (COVID-19): a systematic review and meta-analysis. Front Neurol. 2020;11:565.
21. Liotta EM, Batra A, Clark JR, Shlobin NA, Hoffman SC, Orban ZS, et al. Frequent neurologic manifestations and encephalopathy-associated morbidity in Covid-19 patients. Ann Clin Transl Neurol. 2020;7:2221.
22. Pinna P, Grewal P, Hall JP, Tavarez T, Dafer RM, Garg R, et al. Neurological manifestations and COVID-19: experiences from a tertiary care center at the frontline. J Neurol Sci. 2020;415:116969.
23. Nath A. Neurologic complications of coronavirus infections. Neurology. 2020;94:809.
24. Poyiadji N, Shahin G, Noujaim D, Stone M, Patel S, Griffith B. COVID-19-associated acute hemorrhagic necrotizing encephalopathy: CT and MRI features. Radiology. 2020;296(2):E119–E20.
25. Filatov A, Sharma P, Hindi F, Espinosa PS. Neurological complications of coronavirus disease (COVID-19): encephalopathy. Cureus. 2020;12(3):e7352.
26. Zubair AS, McAlpine LS, Gardin T, Farhadian S, Kuruvilla DE, Spudich S. Neuropathogenesis and neurologic manifestations of the coronaviruses in the age of coronavirus disease 2019: a review. JAMA Neurol. 2020;77(8):1018–27.
27. Giacomelli A, Pezzati L, Conti F, Bernacchia D, Siano M, Oreni L, et al. Self-reported olfactory and taste disorders in SARS-CoV-2 patients: a cross-sectional study. Clin Infect Dis. 2020;71:889.
28. Gautier JF, Ravussin Y. A new symptom of COVID-19: loss of taste and smell. Obesity (Silver Spring). 2020;
29. Mao L, Jin H, Wang M, Hu Y, Chen S, He Q, et al. Neurologic manifestations of hospitalized patients with coronavirus disease 2019 in Wuhan, China. JAMA Neurol. 2020;77
30. Mermelstein S. Acute anosmia from COVID-19 infection. Pract Neurol. 2020;20(4):343–4.
31. Vavougios GD. Potentially irreversible olfactory and gustatory impairments in COVID-19: indolent vs. fulminant SARS-CoV-2 neuroinfection. Brain Behav Immun. 2020;87:107–8.
32. Saniasiaya J, Kulasegarah J. Dizziness and COVID-19. Ear Nose Throat J. 2021;100(1):29–30.
33. Kaliyappan K, Chen YC, Krishnan Muthaiah VP. Vestibular Cochlear manifestations in COVID-19 cases. Front Neurol. 2022;13:850337.
34. Almufarrij I, Uus K, Munro KJ. Does coronavirus affect the audio-vestibular system? A rapid systematic review. Int J Audiol. 2020;59(7):487–91.
35. Bhatta S, Sharma S, Sharma D, Maharjan L, Bhattachan S, Sah MK, et al. Study of hearing status in COVID-19 patients: a multi-centered review. Indian. J Otolaryngol Head Neck Surg. 2021:1–7.
36. Chari DA, Parikh A, Kozin ED, Reed M, Jung DH. Impact of COVID-19 on presentation of sudden sensorineural hearing loss at a single institution. Otolaryngol Head Neck Surg. 2021;165(1):163–5.
37. Chern A, Famuyide AO, Moonis G, Lalwani AK. Bilateral sudden sensorineural hearing loss and intralabyrinthine hemorrhage in a patient with COVID-19. Otol Neurotol. 2021;42(1):e10–e4.
38. Chirakkal P, Al Hail AN, Zada N, Vijayakumar DS. COVID-19 and tinnitus. Ear Nose Throat J. 2021;100(2_suppl):160S–2S.
39. Degen C, Lenarz T, Willenborg K. Acute profound sensorineural hearing loss after COVID-19 pneumonia. Mayo Clin Proc. 2020;95(8):1801–3.
40. Edwards M, Muzaffar J, Naik P, Coulson C. Catastrophic bilateral sudden sensorineural hearing loss following COVID-19. BMJ Case Rep. 2021;14(6):e243157.
41. Gerstacker K, Speck I, Riemann S, Aschendorff A, Knopf A, Arndt S. Deafness after COVID-19? HNO. 2021;69(Suppl 2):92–5.
42. Koumpa FS, Forde CT, Manjaly JG. Sudden irreversible hearing loss post COVID-19. BMJ Case Rep. 2020;13(11):e238419.
43. Lamounier P, Franco Goncalves V, Ramos HVL, Gobbo DA, Teixeira RP, Dos Reis PC, et al. A 67-year-old woman with sudden hearing loss associated with SARS-CoV-2 infection. Am J Case Rep. 2020;21:e927519.
44. Dusan M, Milan S, Nikola D. COVID-19 caused hearing loss. Eur Arch Otorrinolaringol. 2021;279(5):2363–72.
45. Fidan V, Akin O, Koyuncu H. Rised sudden sensorineural hearing loss during COVID-19 widespread. Am J Otolaryngol. 2021;42(5):102996.
46. Gosavi S, Nagarajan S, Shah NJ, Thomas NAT, Kumar KR, Sangole V. ENT symptomology in active COVID-19 patients in our tertiary care centre. Indian J Otolaryngol Head Neck Surg. 2021:1–6.
47. Munro KJ, Uus K, Almufarrij I, Chaudhuri N, Yioe V. Persistent self-reported changes in hearing and tinnitus in post-hospitalisation COVID-19 cases. Int J Audiol. 2020;59(12):889–90.
48. Ricciardiello F, Pisani D, Viola P, Cristiano E, Scarpa A, Giannone A, et al. Sudden sensorineural hearing loss in mild COVID-19: case series and analysis of the literature. Audiol Res. 2021;11(3):313–26.
49. De Luca P, Scarpa A, Ralli M, Tassone D, Simone M, De Campora L, et al. Auditory disturbances and SARS-CoV-2 infection: brain inflammation or Cochlear affection? Systematic review and discussion of potential pathogenesis. Front Neurol. 2021;12:707207.
50. Fancello V, Hatzopoulos S, Corazzi V, Bianchini C, Skarzynska MB, Pelucchi S, et al. SARS-CoV-2 (COVID-19) and audio-vestibular disorders. Int J Immunopathol Pharmacol. 2021;35:20587384211027373.

51. Ali ST, Kang AK, Patel TR, Clark JR, Perez-Giraldo GS, Orban ZS, et al. Evolution of neurologic symptoms in non-hospitalized COVID-19 "long haulers". Ann Clin Transl Neurol. 2022;9:950.
52. Beukes E, Ulep AJ, Eubank T, Manchaiah V. The impact of COVID-19 and the pandemic on tinnitus: a systematic review. J Clin Med. 2021;10(13)
53. Beukes EW, Baguley DM, Jacquemin L, Lourenco M, Allen PM, Onozuka J, et al. Changes in tinnitus experiences during the COVID-19 pandemic. Front Public Health. 2020;8:592878.
54. Elibol E. Otolaryngological symptoms in COVID-19. Eur Arch Otorrinolaringol. 2021;278(4):1233–6.
55. Liang Y, Xu J, Chu M, Mai J, Lai N, Tang W, et al. Neurosensory dysfunction: a diagnostic marker of early COVID-19. Int J Infect Dis. 2020;98:347–52.
56. Chen T, Wu D, Chen H, Yan W, Yang D, Chen G, et al. Clinical characteristics of 113 deceased patients with coronavirus disease 2019: retrospective study. BMJ. 2020;368:m1091.
57. Karimi-Galougahi M, Naeini AS, Raad N, Mikaniki N, Ghorbani J. Vertigo and hearing loss during the COVID-19 pandemic – is there an association? Acta Otorhinolaryngol Ital. 2020;40(6):463–5.
58. Maharaj S, Bello Alvarez M, Mungul S, Hari K. Otologic dysfunction in patients with COVID-19: a systematic review. Laryngoscope Investig Otolaryngol. 2020;5(6):1192–6.
59. AlJasser A, Alkeridy W, Munro KJ, Plack CJ. Is COVID-19 associated with self-reported audio-vestibular symptoms? Int J Audiol. 2021;1-9
60. Almufarrij I, Munro KJ. One year on: an updated systematic review of SARS-CoV-2, COVID-19 and audio-vestibular symptoms. Int J Audiol. 2021;60(12):935–45.
61. Chao CH, Young YH. Evolution of incidence of audiovestibular disorders during the pandemic COVID-19 period. Eur Arch Otorrinolaringol. 2021;279(7):3341–5.
62. Gallus R, Melis A, Rizzo D, Piras A, De Luca LM, Tramaloni P, et al. Audiovestibular symptoms and sequelae in COVID-19 patients. J Vestib Res Equilib Orient. 2021;31(5):381–7.
63. Jafari Z, Kolb BE, Mohajerani MH. Hearing loss, tinnitus, and dizziness in COVID-19: a systematic review and meta-analysis. Can J Neurol Sci. 2021:1–12.
64. Tan M, Cengiz DU, Demir I, Demirel S, Colak SC, Karakas O, et al. Effects of Covid-19 on the audio-vestibular system. Am J Otolaryngol. 2021;43(1):103173.
65. Viola P, Ralli M, Pisani D, Malanga D, Sculco D, Messina L, et al. Tinnitus and equilibrium disorders in COVID-19 patients: preliminary results. Eur Arch Otorrinolaringol. 2021;278(10):3725–30.
66. Wichova H, Miller ME, Derebery MJ. Otologic manifestations after COVID-19 vaccination: the house ear clinic experience. Otol Neurotol. 2021;42(9):e1213–e8.
67. Dinkin M, Gao V, Kahan J, Bobker S, Simonetto M, Wechsler P, et al. COVID-19 presenting with ophthalmoparesis from cranial nerve palsy. Neurology. 2020;95(5):221–3.
68. Malayala SV, Raza A. A case of COVID-19-induced vestibular neuritis. Cureus. 2020;12(6):e8918.
69. Vanaparthy R, Malayala SV, Balla M. COVID-19-induced vestibular neuritis, hemi-facial spasms and Raynaud's phenomenon: a case report. Cureus. 2020;12(11):e11752.
70. Mat Q, Noel A, Loiselet L, Tainmont S, Chiesa-Estomba CM, Lechien JR, et al. Vestibular neuritis as clinical presentation of COVID-19. Ear Nose Throat J. 2021:145561321995021.
71. Jeong J. Vestibular neuritis after COVID-19 vaccination. Hum Vaccin Immunother. 2021;17(12):5126–8.
72. Malayala SV, Mohan G, Vasireddy D, Atluri P. A case series of vestibular symptoms in positive or suspected COVID-19 patients. Infez Med. 2021;29(1):117–22.
73. Jeong M, Ocwieja KE, Han D, Wackym PA, Zhang Y, Brown A, et al. Direct SARS-CoV-2 infection of the human inner ear may underlie COVID-19-associated audiovestibular dysfunction. Commun Med. 2021;1(1):44.
74. Perret M, Bernard A, Rahmani A, Manckoundia P, Putot A. Acute labyrinthitis revealing COVID-19. Diagnostics (Basel). 2021;11(3):482.
75. Picciotti PM, Passali GC, Sergi B, De Corso E. Benign paroxysmal positional vertigo (BPPV) in COVID-19. Audiol Res. 2021;11(3):418–22.
76. Maslovara S, Kosec A. Post-COVID-19 benign paroxysmal positional vertigo. Case Rep Med. 2021;2021:9967555.
77. Lovato A, Frosolini A, Marioni G, de Filippis C. Higher incidence of Meniere's disease during COVID-19 pandemic: a preliminary report. Acta Otolaryngol. 2021;141(10):921–4.
78. Narozny W, Tretiakow D, Skorek A. Intralabyrinthine hemorrhage in a patient with Covid-19 – a challenge for an otologist. Otol Neurotol. 2021;42(7):e967.
79. Shouman K, Vanichkachorn G, Cheshire WP, Suarez MD, Shelly S, Lamotte GJ, et al. Autonomic dysfunction following COVID-19 infection: an early experience. Clin Auton Res. 2021;31(3):385–94.
80. Dani M, Dirksen A, Taraborrelli P, Torocastro M, Panagopoulos D, Sutton R, et al. Autonomic dysfunction in 'long COVID': rationale, physiology and management strategies. Clin Med (Lond). 2021;21(1):e63–e7.
81. Goodman BP, Khoury JA, Blair JE, Grill MF. COVID-19 Dysautonomia. Front Neurol. 2021;12:624968.
82. Lo YL. COVID-19, fatigue, and dysautonomia. J Med Virol. 2021;93(3):1213.
83. Eshak N, Abdelnabi M, Ball S, Elgwairi E, Creed K, Test V, et al. Dysautonomia: an overlooked neurological manifestation in a critically ill COVID-19 patient. Am J Med Sci. 2020;360(4):427–9.
84. Miglis MG, Prieto T, Shaik R, Muppidi S, Sinn D-I, Jaradeh S. A case report of postural tachycardia syndrome after COVID-19. Clin Auton Res. 2020;30(5):449–51.
85. Johansson M, Ståhlberg M, Runold M, Nygren-Bonnier M, Nilsson J, Olshansky B, et al. Long-haul post-COVID-19 symptoms presenting as a variant of postural orthostatic tachycardia syndrome: the Swedish experience. JACC Case Rep. 2021;3(4):573–80.
86. Kanjwal K, Jamal S, Kichloo A, Grubb BP. New-onset postural orthostatic tachycardia syndrome following coronavirus disease 2019 infection. J Innov Card Rhythm Manag. 2020;11(11):4302–4.
87. Schofield JR. Persistent antiphospholipid antibodies, mast cell activation syndrome, postural orthostatic tachycardia syndrome and post-COVID syndrome: 1 year on. Eur J Case Rep Intern Med. 2021;8(3):002378.
88. Goldstein DS. The possible association between COVID-19 and postural tachycardia syndrome. Heart Rhythm. 2021;18(4):508–9.
89. Novak P. Post COVID-19 syndrome associated with orthostatic cerebral hypoperfusion syndrome, small fiber neuropathy and benefit of immunotherapy: a case report. eNeurologicalSci. 2020;21:100276.
90. Blitshteyn S, Whitelaw S. Postural orthostatic tachycardia syndrome (POTS) and other autonomic disorders after COVID-19 infection: a case series of 20 patients. Immunol Res. 2021;69(2):205–11.
91. Skarzynska MB, Matusiak M, Skarzynski PH. Adverse audio-vestibular effects of drugs and vaccines used in the treatment and prevention of COVID-19: a review. Audiol Res. 2022;12(3):224–48.
92. Baden LR, El Sahly HM, Essink B, Kotloff K, Frey S, Novak R, et al. Efficacy and safety of the mRNA-1273 SARS-CoV-2 vaccine. N Engl J Med. 2021;384(5):403–16.
93. Polack FP, Thomas SJ, Kitchin N, Absalon J, Gurtman A, Lockhart S, et al. Safety and efficacy of the BNT162b2 mRNA Covid-19 vaccine. N Engl J Med. 2020;383(27):2603–15.

94. Sadoff J, Gray G, Vandebosch A, Cardenas V, Shukarev G, Grinsztejn B, et al. Safety and efficacy of single-dose Ad26.COV2.S vaccine against Covid-19. N Engl J Med. 2021;384(23):2187–201.
95. Ajmera KM. Adverse events profile of COVID-19 preventative strategies. Encyclopedia. 2022;2(1):457–65.
96. Kadali RAK, Janagama R, Peruru S, Malayala SV. Side effects of BNT162b2 mRNA COVID-19 vaccine: a randomized, cross-sectional study with detailed self-reported symptoms from healthcare workers. Int J Infect Dis. 2021;106:376–81.
97. Reddy S, Reddy S, Arora M. A case of postural orthostatic tachycardia syndrome secondary to the messenger RNA COVID-19 vaccine. Cureus. 2021;13(5):e14837.
98. Kadali RAK, Janagama R, Peruru S, Gajula V, Madathala RR, Chennaiahgari N, et al. Non-life-threatening adverse effects with COVID-19 mRNA-1273 vaccine: a randomized, cross-sectional study on healthcare workers with detailed self-reported symptoms. J Med Virol. 2021;93(7):4420–9.
99. Formeister EJ, Chien W, Agrawal Y, Carey JP, Stewart CM, Sun DQ. Preliminary analysis of association between COVID-19 vaccination and sudden hearing loss using US Centers for Disease Control and Prevention Vaccine Adverse Events Reporting System Data. JAMA Otolaryngol Head Neck Surg. 2021;147(7):674–6.
100. Ahmed SH, Waseem S, Shaikh TG, Qadir NA, Siddiqui SA, Ullah I, et al. SARS-CoV-2 vaccine-associated-tinnitus: a review. Ann Med Surg (Lond). 2022;75:103293.
101. Parrino D, Frosolini A, Gallo C, De Siati RD, Spinato G, de Filippis C. Tinnitus following COVID-19 vaccination: report of three cases. Int J Audiol. 2021;61(6):526–9.
102. Coffin AB, Boney R, Hill J, Tian C, Steyger PS. Detecting novel ototoxins and potentiation of ototoxicity by disease settings. Front Neurol. 2021;12:725566.
103. Williams B. Ototoxicity may be associated with protease inhibitor therapy. Clin Infect Dis. 2001;33(12):2100–2.
104. No Authors. An EUA for Bamlanivimab-a monoclonal antibody for COVID-19. JAMA. 2021;325(9):880–1.
105. Dougan M, Nirula A, Azizad M, Mocherla B, Gottlieb RL, Chen P, et al. Bamlanivimab plus etesevimab in mild or moderate Covid-19. N Engl J Med. 2021;385(15):1382–92.
106. Jayk Bernal A, Gomes da Silva MM, Musungaie DB, Kovalchuk E, Gonzalez A, Delos Reyes V, et al. Molnupiravir for oral treatment of Covid-19 in nonhospitalized patients. N Engl J Med. 2022;386(6):509–20.

Epstein-Barr Virus

148

Overview

Some patients affected with the *Epstein-Barr virus (EBV)* develop meningitis, encephalitis, or cerebellitis. Some patients with EBV complain of disequilibrium, but the mechanism of this symptom is unclear. Since some EBV patients exhibit ocular motor abnormalities that potentially localize to the cerebellum, and since some patients with EBV are found to have cerebellitis, it is possible (though still unproven) that EBV impacts the vestibular system by adversely affecting cerebellar function.

Introduction

The *Epstein-Barr virus (EBV)* is the eponymous designation for the human herpes virus 4 (HHV4), a double-stranded DNA virus. The virus was first described by the English pathologist and virologist, Dr. Michael Anthony Epstein (b. 1921), and the English virologist, Dr. Yvonne M. Barr (1932–1916), in several publications in which they isolated virus particles from cultured Burkitt lymphoma cells [1–3] (Figs. 148.1 and 148.2).

Fig. 148.1 Michael Anthony Epstein. (From Epstein [4]. Reproduced with written permission from John Wiley and Sons Limited)

Fig. 148.2 Yvonne M. Barr. (From Epstein [4]. Reproduced with written permission from John Wiley and Sons Limited)

M. Cherchi, *Otoneurology and Vestibular Medicine*, https://doi.org/10.1007/978-3-031-94841-1_148

Epidemiology

EBV is endemic among humans. Most infected humans are asymptomatic, but some develop complications such as infectious mononucleosis, nasopharyngeal carcinomas, and hematologic malignancies (B-cell lymphomas, T-cell lymphomas, Hodgkin's lymphoma).

Pathophysiological Mechanism of Disease

EBV is capable of causing meningitis [5], encephalitis [6], and cerebellitis [5, 7, 8]. Potential mechanisms include inflammation as a direct effect of infection or inflammation as a para-infectious process.

The mechanism by which EBV infection might cause disequilibrium is unclear. There are a handful of studies that describe nystagmus occurring in the context of EBV, suggesting that EBV or its downstream effects may adversely impact the vestibular system.

We will see below that some patients with EBV develop periodic alternating nystagmus (PAN) and downbeat nystagmus. Since these eye movements can both result from cerebellar dysfunction, and since post-EBV cerebellitis has been described [5, 7, 8], it is plausible that the mechanism by which EBV provokes disequilibrium is through cerebellar affectation, though this remains conjectural.

Clinical Presentation

We have encountered patients with laboratory-confirmed EBV infection who complain of disequilibrium, which can manifest days or months after the apparent infection.

Ocular Motor Examination

D'Ambrosio and colleagues [7] report a patient with post-EBV cerebellitis, and Hussain and colleagues [6] report a patient with EBV encephalitis; these reports both document the finding of "bilateral nystagmus," without further description.

Del Giudice and colleagues [8] report a patient with post-EBV cerebellar ataxia, and document periodic alternating nystagmus (PAN).

Wade and colleagues [5] report a patient with EBV-associated meningitis and cerebellitis, and document downbeat nystagmus.

Testing: Vestibular

As of this writing, there were no peer-reviewed publications of instrumented otovestibular studies on patients infected with EBV complaining of disequilibrium.

Treatment

Treatment of EBV infection is largely symptomatic. In acute infection, acyclovir may be helpful; the use of corticosteroids is controversial. There is no clear treatment for latent infection.

References

1. Epstein MA, Achong BG, Barr YM. Virus particles in cultured lymphoblasts from Burkitt's lymphoma. Lancet. 1964;1(7335):702–3.
2. Epstein MA, Barr YM. Cultivation in vitro of human lymphoblasts from Burkitt's malignant lymphoma. Lancet. 1964;1(7327):252–3.
3. Epstein MA, Henle G, Achong BG, Barr YM. Morphological and biological studies on a virus in cultured lymphoblasts from Burkitt's lymphoma. J Exp Med. 1965;121(5):761–70.
4. Epstein A. Burkitt lymphoma and the discovery of Epstein-Barr virus. Br J Haematol. 2012;156(6):777–9.
5. Wade CA, Toupin DN, Darpel K, Jones K, Lightner D. Downbeat nystagmus in a 7-year-old girl with Epstein-Barr virus-associated meningitis and cerebellitis. Child Neurol Open. 2021;8:2329048X211000463.
6. Hussain RS, Hussain NA. Ataxia and encephalitis in a young adult with EBV mononucleosis: a case report. Case Rep Neurol Med. 2013;2013:516325.
7. D'Ambrosio E, Khalighinejad F, Ionete C. Intravenous immunoglobulins in an adult case of post-EBV cerebellitis. BMJ Case Rep. 2020;13(2):e231661.
8. Del Giudice E, Mondì F, Bazzanella GR, Marcellino A, Martucci V, Pontrelli G, et al. Post-infectious acute cerebellar ataxia treatment, a case report and review of literature. Children (Basel). 2023;10(4):668.

Human Immunodeficiency Virus and Otovestibular Symptoms

149

Overview

The *human immunodeficiency virus (HIV)*, which is responsible for acquired immunodeficiency syndrome (AIDS), compromises a patient's immune system and increases the risk of opportunistic infections, as well as noninfectious problems. The presence of such HIV-associated medical problems often makes it difficult to study which symptoms and deficits are due to HIV itself. Nevertheless, it appears that HIV may be responsible for increased rates of auditory symptoms (hearing loss, tinnitus) and vestibular symptoms (disequilibrium) in all age groups. In otoneurology, the usual clinical scenario is that a patient already diagnosed with HIV is referred for evaluation of the auditory and vestibular symptoms. Workup may reveal audiologic abnormalities (hearing loss on audiometry and otoacoustic emissions; abnormalities on brainstem auditory evoked responses) and vestibular abnormalities (vestibular weakness on caloric testing). Screening for other (non-HIV-related) causes of these symptoms is reasonable. If other causes are excluded with a reasonable degree of confidence, then one may conclude that HIV itself is responsible for the auditory and vestibular symptoms. The auditory symptoms are usually managed by an audiologist. If vestibular symptoms are present, referral to vestibular rehabilitation therapy is reasonable.

Epidemiology

Auditory and vestibular symptoms appear to be more common in HIV+ patients than in healthy individuals. Fokouo and colleagues [1] studied 90 HIV+ patients and 90 healthy controls. They reported that, "The HIV-positive patients had more otologic symptoms (hearing loss, dizziness, tinnitus, and otalgia) than HIV-negative patients (41 vs. 13, $p = 0.04$)."

Pathophysiological Mechanism of Disease

As we discuss below, most, though not all, studies of otoacoustic emissions in HIV+ patients find a greater rate of abnormalities compared to healthy controls [2–4], suggesting cochlear (end-organ) dysfunction, as opposed to dysfunction of the auditory nerve or higher auditory processing. Furthermore, temporal bone studies (also discussed below) report evidence of HIV infection in the cochlea itself [5]. Taken together, these data suggest a direct effect of HIV infection on the peripheral auditory system.

There is also evidence (discussed below) of vestibular dysfunction on clinical testing, but no corresponding histopathological evidence from temporal bone studies.

Patients with HIV+ are at increased risk for other infectious and noninfectious diseases, some of which may also affect the ear. For example, Michaels and colleagues [6] studied 49 temporal bones from 25 AIDS patients and found evidence of labyrinthine infection by cryptococcosis and cytomegalovirus, and Kaposi's sarcoma in the vestibulocochlear nerve. This often makes it difficult to isolate the effects attributable to HIV alone.

Clinical Presentation: Audiologic

Hearing loss occurring in the context of HIV generally begins insidiously and progresses gradually, though abrupt-onset cases have been reported [7].

Millar and colleagues [8] compared the case histories and pure tone audiometry in 60 HIV+ patients and 32 HIV- subjects. They found no statistically significant difference in auditory thresholds, but did find a significantly higher rate of tinnitus in the HIV+ patients compared to controls.

M. Cherchi, *Otoneurology and Vestibular Medicine*, https://doi.org/10.1007/978-3-031-94841-1_149

Soucek and Michaels [2] studied 62 AIDS patients (mean age 39 years, length of HIV positivity ranging from 3 to 10 years), and reported that, "Ten [16%] patients, all of whom had sensorineural hearing loss, complained of tinnitus. This was bilateral in six patients and unilateral in four."

Clinical Presentation: Vestibular

Soucek and Michaels [2] studied 62 AIDS patients (mean age 39 years, length of HIV positivity ranging from 3 to 10 years), and reported that, "Sixteen [26%] patients had vertigo."

Palacios and colleagues [9] studied 23 HIV+ children (median age 4.5 years, range 5 months to 16 years) and reported that, "Four children (17.4%), all over 5 years old, referred to disequilibrium disorders when questioned."

Testing: General Comments

There are case reports [9] in the literature of HIV+ patients presenting with auditory and vestibular symptoms, whose workups reveal various deficits in auditory tests and vestibular tests. While case reports are a good start, since HIV+ patients are vulnerable to a variety of other illnesses, such deficits are better explored through case series or cohort studies.

Testing: Audiologic

Many studies report no statistically significant differences in air-conducted pure tone audiometry in HIV+ patients [3, 4, 10].

In contrast, Fokouo and colleagues [1] studied pure tone audiometry in 90 HIV+ patients (divided into three subgroups of HAART-naïve patients, patients receiving first-line HAART, and patients receiving second-line HAART) versus 90 healthy control subjects, all age 15–49 years. They reported a statistically significantly higher odds ratio of hearing loss among all subgroups of HIV+ patients compared to controls.

Similarly, Soucek and Michaels [2] studied 62 adult AIDS patients (mean age 39 years, length of HIV positivity ranging from 3 to 10 years). They interpreted the findings as follows:

> This is mainly mild, but there are significant numbers with more severe losses. All frequencies are represented in the hearing losses, but, as the mean threshold audiograms also indicate… the higher and lower frequencies have greater numbers of ears with losses than the middle range. A survey of the individual audiograms showed that 39% of the patients tested had this type of audiogram, with notable losses at higher or lower frequencies but little or no loss at the middle frequency range. Accepting a threshold of 20 dB or worse at any frequency as an indication of hearing loss in this group, the *total number of subjects affected was 43 (69%)* [2].

Soucek and Michaels [2] additionally studied brainstem auditory evoked responses and found that:

> In 16 patients investigated by auditory brainstem evoked responses (ABR), five [31%] were abnormal. All patients tested had some degree of sensorineural hearing loss. In three of the five patients, there was an abnormal I–III interpeak interval suggesting a lower brainstem lesion. In one case, wave III on both sides was delayed. This patient had a cryptococcal meningitis. In the fifth patient, who had migraine and had had a head injury in the past, there was an abnormal I–V central conduction time [2].

Palacios and colleagues [9] studied 23 HIV+ children (median age 4.5 years, range 5 months to 16 years) and reported that, "Pure tone audiometry was carried out in 12 children over 4 years of age: 4 (33%) showed hearing loss, 2 were conductive."

Studies of otoacoustic emissions provide (OAEs) mixed results. Most studies report a greater rate of abnormalities in HIV+ patients compared to controls [2–4]. Some report no difference in OAEs between HIV+ and HIV- patients [11].

Some studies find evidence of central auditory problems rather than peripheral auditory dysfunction. For instance, Buckey and colleagues [11] studied 107 HIV+ individuals (age 37 ± 11.5 years) and 72 HIV- individuals (age 27 ± 17 years) using two tests of peripheral auditory function (pure tone audiometry and distortion product otoacoustic emissions) and one test of central auditory function (gaps-in-noise detection). The authors reported that compared to the HIV- subjects, in the HIV+ patients, the peripheral tests were normal, while the central tests were abnormal. They concluded that, "The gap detection results from the study show effects on central processing in HIV+ individuals, supporting the origin of HIV-related hearing complaints in the central auditory system" rather than the peripheral auditory system.

The studies cited above mostly pertain to adult HIV+ patients. Data from pediatric HIV+ patients may suggest different or additional mechanisms of auditory dysfunction. Maro and colleagues [12] studied 131 HIV+ patients and 113 HIV- subjects, all under age 18 years, and found that, "Children with HIV+ were more likely to have a history of ear drainage, and to have abnormal audiograms… [and] the HIV+ group… did have significantly lower DPOAE [distortion product otoacoustic emissions] magnitudes." They concluded, "These data suggest that (1) HIV+ children often have middle ear damage which complicates understanding the direct effects of HIV on the hearing system, and (2) even when corrected for confounders, DPOAEs were lower in the HIV+ group." Palacios and colleagues [9] similarly reported evidence of "conductive hearing loss associated with previous otitis media events" in a population of 23 HIV+ children (median age 4.5 years, range 5 months to 16 years).

Testing: Vestibular

Palacios and colleagues [9] studied 23 HIV+ children (median age 4.5 years, range 5 months to 16 years) and reported that, "ENG [electronystagmography] was carried out in six of 23 patients; alterations in smooth pursuit were detected in one [17%] patient of the six tested. Caloric tests were abnormal in the four [100%] patients tested, showing unilateral vestibular hyporeflexia. Rotatory tests were abnormal in the two [100%] patients for whom tests could be performed, with asymmetries in both vestibular and otovestibular tests."

Karaca and colleagues [13] studied video head impulse testing in 48 HIV+ patients and 24 healthy controls, and found no differences between the two groups. They commented that, "Because the video head impulse test is considered a high-frequency test of vestibulo-ocular reflex, it is possible that vestibular deficits of the human immunodeficiency virus can be confined to low frequencies."

Millar and colleagues [14] studied video head impulse testing in 60 HIV+ patients and 32 healthy controls, and reported that, "The prevalence of peripheral vestibular impairment in the HIV-positive group was not significantly different from that of the HIV-negative group."

Histopathology

There is some evidence that HIV can directly involve the cochlea. Pappas and colleagues [5] studied eight temporal bones and reported that:

> Extracellular viral-like particles with morphologic characteristics of HIV-1 were identified on the tectorial membrane in three cases. Numerous viral-like particles with cores appeared trapped within lacunae along the tectorial membrane in one specimen. Intracellular viral-like particles, appearing essentially similar to identified HIV-1 particles of infected lymphocyte cultures, were found within the cytoplasm of connective tissue cells. These same cells had gradients of enlarged viral envelopes forming cisterns that contained cores and varying amounts of cistern particulate. Some sectioned profiles of enlarged cytoplasmic cisterns revealed the attachment and inward protrusion of the core from the cistern wall. Other ultrastructural findings included cytoplasmic globular-dense particulate bodies, aggregates of smaller viral-like particles, and mitochondrial changes [5].

Differential Diagnosis

As far as HIV is concerned, the usual clinical scenario in otoneurology is that a patient already diagnosed with HIV is referred for auditory or vestibular complaints. In this circumstance, as discussed below, it is appropriate to undertake an otovestibular workup to screen for other treatable diseases. If the workup is unrevealing, then by exclusion, it is reasonable to conclude that HIV itself is the mechanism of the symptoms in question.

The reverse scenario is extraordinarily rare—that is, a patient with auditory or vestibular complaints is found to have (previously undiagnosed) HIV. Auditory and vestibular symptoms in themselves do not constitute grounds to screen for HIV; rather, the decision to monitor for HIV would be driven by the presence of other risk factors, and the patient's HIV status would usually be checked by their internist.

Treatment

Most of the studies that have compared audiologic findings in HIV+ patients who have received antiretroviral therapies to those in untreated HIV+ patients found no statistically significant differences [12], suggesting that HIV infection itself (rather than the antiretroviral therapies) is responsible. However, not all studies reach the same conclusion; for example, Maro and colleagues [3] report that HIV+ patients receiving ART (antiretroviral therapy) had "significantly higher mean gap-detection threshold [a central auditory processing test] compared with the ART-group." Despite our interest in preserving and improving auditory and vestibular function, we believe that treatment of HIV should take priority, even if there is a minority of literature suggesting potential audiovestibular risks to antiretroviral therapy.

Patients who are HIV+ and have subjective hearing loss should be evaluated and managed by an audiologist. Such management may include amplification for hearing loss and masking for tinnitus.

In patients who are HIV+ and have subjective disequilibrium, it is medically reasonable to consider undertaking a screening vestibular workup (cervical and ocular vestibular evoked myogenic potentials, video head impulse testing, videonystagmography, rotatory chair testing). If evidence of vestibular weakness is found, then a trial of vestibular rehabilitation therapy is appropriate.

Prognosis

Absent sufficient longitudinal studies, the prognosis of auditory and vestibular symptoms in HIV+ patients is unknown.

References

1. Fokouo JV, Vokwely JE, Noubiap JJ, Nouthe BE, Zafack J, Minka Ngom ES, et al. Effect of HIV infection and highly active antiretroviral therapy on hearing function: a prospective case-control study from Cameroon. JAMA Otolaryngol Head Neck Surg. 2015;141(5):436–41.

2. Soucek S, Michaels L. The ear in the acquired immunodeficiency syndrome: II. Clinical and audiologic investigation. Am J Otol. 1996;17(1):35–9.
3. Maro II, Moshi N, Clavier OH, MacKenzie TA, Kline-Schoder RJ, Wilbur JC, et al. Auditory impairments in HIV-infected individuals in Tanzania. Ear Hear. 2014;35(3):306–17.
4. Niemczak CE, White-Schwoch T, Fellows A, Magohe A, Gui J, Rieke C, et al. Peripheral auditory function in young HIV-positive adults with clinically normal hearing. Otolaryngol Head Neck Surg. 2022;167(1):155–62.
5. Pappas DG Jr, Chandra Sekhar HK, Lim J, Hillman DE. Ultrastructural findings in the cochlea of AIDS cases. Am J Otol. 1994;15(4):456–65.
6. Michaels L, Soucek S, Liang J. The ear in the acquired immunodeficiency syndrome: I. Temporal bone histopathologic study. Am J Otol. 1994;15(4):515–22.
7. Grimaldi LM, Luzi L, Martino GV, Furlan R, Nemni R, Antonelli A, et al. Bilateral eighth cranial nerve neuropathy in human immunodeficiency virus infection. J Neurol. 1993;240(6):363–6.
8. Millar A, Joubert K, Naude A. Prevalence of hearing loss and tinnitus in a group of adults with Human Immunodeficiency Virus. S Afr J Commun Disord. 2020;67(1):e1–7.
9. Palacios GC, Montalvo MS, Fraire MI, Leon E, Alvarez MT, Solorzano F. Audiologic and vestibular findings in a sample of human immunodeficiency virus type-1-infected Mexican children under highly active antiretroviral therapy. Int J Pediatr Otorhinolaryngol. 2008;72(11):1671–81.
10. Luque AE, Orlando MS, Leong UC, Allen PD, Guido JJ, Yang H, et al. Hearing function in patients living with HIV/AIDS. Ear Hear. 2014;35(6):e282–90.
11. Buckey JC, Fellows AM, Magohe A, Maro I, Gui J, Clavier O, et al. Hearing complaints in HIV infection originate in the brain not the ear. AIDS. 2019;33(9):1449–54.
12. Maro II, Fellows AM, Clavier OH, Gui J, Rieke CC, Wilbur JC, et al. Auditory impairments in HIV-infected children. Ear Hear. 2016;37(4):443–51.
13. Karaca S, Kalcioglu MT, Sargin F, Topcu MT, Uzun L, Tekin M, et al. Evaluation of effects of the human immunodeficiency virus on the vestibular system by using the video head impulse test. Eurasian J Med. 2022;54(2):138–44.
14. Millar A, Joubert K, Naude A. Prevalence of peripheral vestibular impairment in adults with human immunodeficiency virus. J Audiol Otol. 2021;25(1):36–42.

Lyme Disease

150

Overview

Lyme disease results from infection by Borrelia burgdorferi. It can present in many ways, including with multiple cranial neuropathies. From the otovestibular perspective, it can manifest with various combinations of symptoms referable to the vestibulocochlear nerve, including any combination of hearing loss, tinnitus, otalgia, and disequilibrium. Otovestibular testing may show sensorineural hearing loss and caloric weakness.

Introduction

Borrelia burgdorferi is a tick-borne spirochete bacterial infection that causes *Lyme disease* in humans. Lyme disease sometimes presents in classical fashion with a "bullseye rash," and can develop a broad range of systemic and organ-specific manifestations, including multiple cranial neuropathies. Here we limit the discussion to otovestibular manifestations of this disease.

Epidemiology

The epidemiology and geographic distribution of Lyme disease have been well characterized [1].

Pathophysiological Mechanism of Disease

Despite considerable research, there remains uncertainty about the pathophysiological mechanism by which Lyme disease causes symptoms [2].

Clinical Presentation

Lyme disease can present with multiple symptoms referable to the vestibulocochlear nerve, including hearing loss, tinnitus, disequilibrium, and otalgia [3, 4], or some subset of those symptoms [5]. For example, Lyme disease has been reported to present (initially) with only sudden hearing loss [6, 7]; in the late stage of infection, it has been reported to present solely with disequilibrium [8, 9].

One study analyzed 38 patients (11 male, 27 female) with serologically confirmed Lyme disease, all of whom complained of disequilibrium. Of these 38 patients, 11 (29%) had also experienced some form of hearing loss, 2 (5%) specifically had sudden hearing loss, and 29 (76%) complained of tinnitus.

Sowula and colleagues [10] recruited 50 patients (35 women, 15 men) with serologically confirmed Lyme disease and studied them with audiometry, auditory brainstem evoked responses, and videonystagmography. Of these 50 patients, 38 (76%) reported symptoms of "vertigo" and 29 (76%) complained of tinnitus. Of the 38 patients who complained of vertigo, 11 (29%) experienced hearing loss. Of those 11 patients with hearing loss, the hearing loss was bilateral in two, and presented with sudden onset in two. The investigators reported there to be a statistically significant association between the "degree of labyrinth damage" (as measured by caloric weakness) and the presence of high-frequency hearing loss (not low-frequency hearing loss).

Testing: Vestibular

One study compared 91 patients with Ménière's disease and 11 patients with serologically confirmed Lyme disease. Of the 11 patients with Lyme disease, 6 (55%) had symptoms of

M. Cherchi, *Otoneurology and Vestibular Medicine*, https://doi.org/10.1007/978-3-031-94841-1_150

sensorineural hearing loss, tinnitus and disequilibrium—in other words, the phenotype of Ménière's disease, and their workup also showed similar results (such as endolymphatic hydrops on electrocochleography) [11]. However, as noted earlier in this discussion, another study [10] reported that patients with Lyme disease complaining of vertigo have high-frequency hearing loss (rather than the low-frequency hearing loss that would be more typical of Ménière's disease).

Testing: Other

One study of 47 consecutive patients with sudden hearing loss (and no other symptoms) reported that 21% of the cases were found to have positive Lyme titers, though there was no difference in the clinical course of the seropositive and seronegative patients [12].

Imaging

Brain MRI in patients with Lyme disease may demonstrate enhancement of cranial nerves, including the vestibulocochlear nerve [13].

Differential Diagnosis

For each otovestibular symptom that Lyme disease causes, there is a broad differential diagnosis. It should be kept in mind that other tick-borne illnesses (besides Lyme disease) can manifest with otovestibular symptoms [14].

Treatment

Lyme disease is usually treated with doxycycline, though specific management may depend on the particular organ system involvement.

References

1. Mead PS. Epidemiology of Lyme disease. Infect Dis Clin N Am. 2015;29(2):187–210.
2. Donta ST, States LJ, Adams WA, Bankhead T, Baumgarth N, Embers ME, et al. Report of the pathogenesis and pathophysiology of lyme disease subcommittee of the HHS Tick Borne Disease Working Group. Front Med (Lausanne). 2021;8:643235.
3. Goldfarb D, Sataloff RT. Lyme disease: a review for the otolaryngologist. Ear Nose Throat J. 1994;73(11):824–9.
4. Moscatello AL, Worden DL, Nadelman RB, Wormser G, Lucente F. Otolaryngologic aspects of Lyme disease. Laryngoscope. 1991;101(6 Pt 1):592–5.
5. Vos FI, Merkus P, van Nieuwkerk EB, Hensen EF. Rare cause of bilateral sudden deafness. BMJ Case Rep. 2016;2016:bcr2016216004.
6. Espiney Amaro C, Montalvao P, Huins C, Saraiva J. Lyme disease: sudden hearing loss as the sole presentation. J Laryngol Otol. 2015;129(2):183–6.
7. Peeters N, van der Kolk BY, Thijsen SF, Colnot DR. Lyme disease associated with sudden sensorineural hearing loss: case report and literature review. Otol Neurotol. 2013;34(5):832–7.
8. Jozefowicz-Korczynska M, Zamyslowska-Szmytke E, Piekarska A, Rosiak O. Vertigo and severe balance instability as symptoms of lyme disease-literature review and case report. Front Neurol. 2019;10:1172.
9. Peltomaa M, Pyykko I, Seppala I, Viljanen M. Lyme borreliosis—an unusual cause of vertigo. Auris Nasus Larynx. 1998;25(3):233–42.
10. Sowula K, Szaleniec J, Dworak M, Przeklasa M, Maraj M, Ceranowicz P, Tomik J. Vertigo as one of the symptoms of lyme disease. J Clin Med. 2021;10(13):2814.
11. Selmani Z, Pyykko I, Ishizaki H, Ashammakhi N. Use of electrocochleography for assessing endolymphatic hydrops in patients with Lyme disease and Meniere's disease. Acta Otolaryngol. 2002;122(2):173–8.
12. Lorenzi MC, Bittar RS, Pedalini ME, Zerati F, Yoshinari NH, Bento RF. Sudden deafness and Lyme disease. Laryngoscope. 2003;113(2):312–5.
13. Piche-Renaud PP, Branson H, Yeh EA, Morris SK. Lyme disease presenting with multiple cranial neuropathies on MRI. IDCases. 2018;12:117–8.
14. Sowula K, Skladzien J, Szaleniec J, Gawlik J. Otolaryngological symptoms in patients treated for tick-borne diseases. Otolaryngol Pol. 2018;72(1):30–4.

Otosyphilis

151

Overview

Otosyphilis refers to otovestibular involvement of infection by *Treponema pallidum.* The incidence of syphilis began to increase during the AIDS epidemic. The proportion of patients with syphilis that go on to develop otosyphilis is unknown. Otosyphilis can manifest at any stage of syphilitic infection. The audiovestibular symptoms (hearing loss, tinnitus, disequilibrium) may be the initial manifestations of the disease. When characterized by audiometry, the hearing loss in otosyphilis is usually (though not always) sensorineural and can be unilateral and bilateral (symmetrical or asymmetrical) and affect any frequencies; in short, any pattern is possible. Some patients may also have abnormal brainstem auditory evoked responses. Few studies of vestibular testing in otosyphilis patients are available. Imaging may show cranial osteolytic lesions involving the labyrinth (on temporal bone CT) and soft tissue damage or space-occupying intracranial gummas (on brain MRI). The audiovestibular manifestations of otosyphilis may resemble autoimmune inner ear disease or Ménière's disease. Separately from otosyphilis, neurosyphilis can cause tabes dorsalis (degeneration of the posterior columns of the spinal cord with impairment of vibration and proprioception) associated with sensory ataxia and thus disequilibrium. Treatment of otosyphilis is the same as treatment for syphilis in general. In some patients, hearing loss improves, though this improvement may be transient. Vestibular symptoms sometimes respond to treatment. For patients in whom the audiologic and/or vestibular symptoms do not respond to treatment for syphilis, referral to audiology and vestibular physical therapy is medically reasonable.

Introduction

Treponema pallidum is the spirochetal pathogen responsible for syphilis. Syphilis can affect many organ systems and thus have a variety of manifestations. When syphilis involves the audiovestibular system, it is sometimes called "*otosyphilis*."

Epidemiology

Syphilis was previously somewhat uncommon in the United States, but its incidence began to increase during the AIDS pandemic in 1980s. Ramchandani and colleagues [1], citing data from the Centers for Disease Control, state that the rate of early syphilis infection in the united states increased 71% from 2014 to 2018. Increases of this magnitude have led to descriptions of it as a resurgence [2, 3].

It is not known what percentage of syphilis patients go on to develop otosyphilis. Dombrowski and colleagues [4] analyzed a case series of 573 patients with "early or unknown duration syphilis" and found that 22 (3.9%) of patients reported hearing loss and 15 (2.7%) reported tinnitus.

Otosyphilis can occur at any stage of syphilis infection, but its symptoms are more likely to develop in early stages—meaning primary syphilis, secondary syphilis, and early latent syphilis [5]. The otologic [5] and vestibular [6] symptoms may be the initial manifestations of the disease.

Pathophysiological Mechanism of Disease

In syphilis, cellular damage occurs both as a result of the pathogen (Treponema pallidum) itself and a series of immunological responses [7] and inflammation.

Clinical Presentation

Otosyphilis can present with any combination of hearing loss, tinnitus, or disequilibrium, and these may be the initial manifestations of the disease [5, 6]. Table 151.1 below, from Yimtae and colleagues [1], reports the prevalence of these symptoms in a case series of 85 patients.

Fayad and Linthicum [8] state that "Vestibular disturbances could be present in as many as 80% of patients with otosyphilis."

M. Cherchi, *Otoneurology and Vestibular Medicine*, https://doi.org/10.1007/978-3-031-94841-1_151

Table 151.1 Audiologic and vestibular symptoms in a series of 85 patients with otosyphilis

Characteristics	Number	Percentage
Presenting symptoms	85	100.0
Hearing loss	77	90.6
Tinnitus	62	72.9
Vertigo	45	52.9
Onset of symptoms	85	100.0
Insidious hearing loss	64	75.3
Symmetrical bilateral	36	42.4
Asymmetrical bilateral	28	32.9
Sudden hearing loss	20	23.5
Symmetrical bilateral	9	10.6
Asymmetrical bilateral	8	9.4
Unilateral	3	3.5
Acute vertigo	1	1.2

From Yimtae et al. [1]

Mishra and colleagues [9] studied eight patients with otosyphilis and concomitant HIV infection and reported that all complained of hearing loss, while seven (88%) complained of tinnitus.

Physical Examination

Patients with otosyphilis generally have sensorineural hearing loss, which may be unilateral or bilateral.

Ocular Motor Examination

There are few reports in the literature of symptomatic otosyphilis patients exhibiting nystagmus. Klemm and Wollina [10] described a patient with spontaneous and positional nystagmus.

Testing: Auditory

Ramchandani and colleagues [11] comment that:

> As a general rule, otosyphilis in adults causes SNHL [sensorineural hearing loss], although conductive hearing loss may be present in those with long-standing disease. Hearing loss can be unilateral or bilateral, often with sudden onset and rapid progression, and those patients who report unilateral impairment may have bilateral disease on audiogram. Bilateral hearing loss may be symmetric or asymmetric and can present in a variety of ways: mild or moderate and low to high frequency with poor or excellent speech discrimination scores. There is not one classic type of presentation [11].

In other words, no pattern of hearing loss is sensitive or specific for otosyphilis. Song and colleagues [12] provide examples demonstrating this variability.

Lowhagen and colleagues [13] studied 26 patients with secondary or early latent syphilis and found that seven (27%) had abnormal brainstem auditory evoked responses.

Testing: Vestibular

Very few publications report results of vestibular tests in otosyphilis.

Song and colleagues [12] reported the case of a patient with bilateral otosyphilitic involvement. They performed rotatory chair testing and interpreted the slow harmonic acceleration as showing, "bilateral total vestibular function loss."

Testing: Other

A diagnosis of syphilis is usually secured by checking treponemal tests (e.g., fluorescent treponemal antibody absorption) and non-treponemal tests (venereal disease research laboratory).

Imaging

Otosyphilis can cause cranial osteolytic lesions that involve the labyrinth. Figure 151.1 below, from Hosseini and colleagues [14], shows images from a temporal bone CT of such a case.

Similar CT findings are reported in other cases in the literature [15, 16].

Otosyphilis can also manifest as a space-occupying intracranial gumma [17] impinging on audiovestibular structures. Figure 151.2 below, from Chandrasekharan and colleagues [18], shows an axial MRI of such a case.

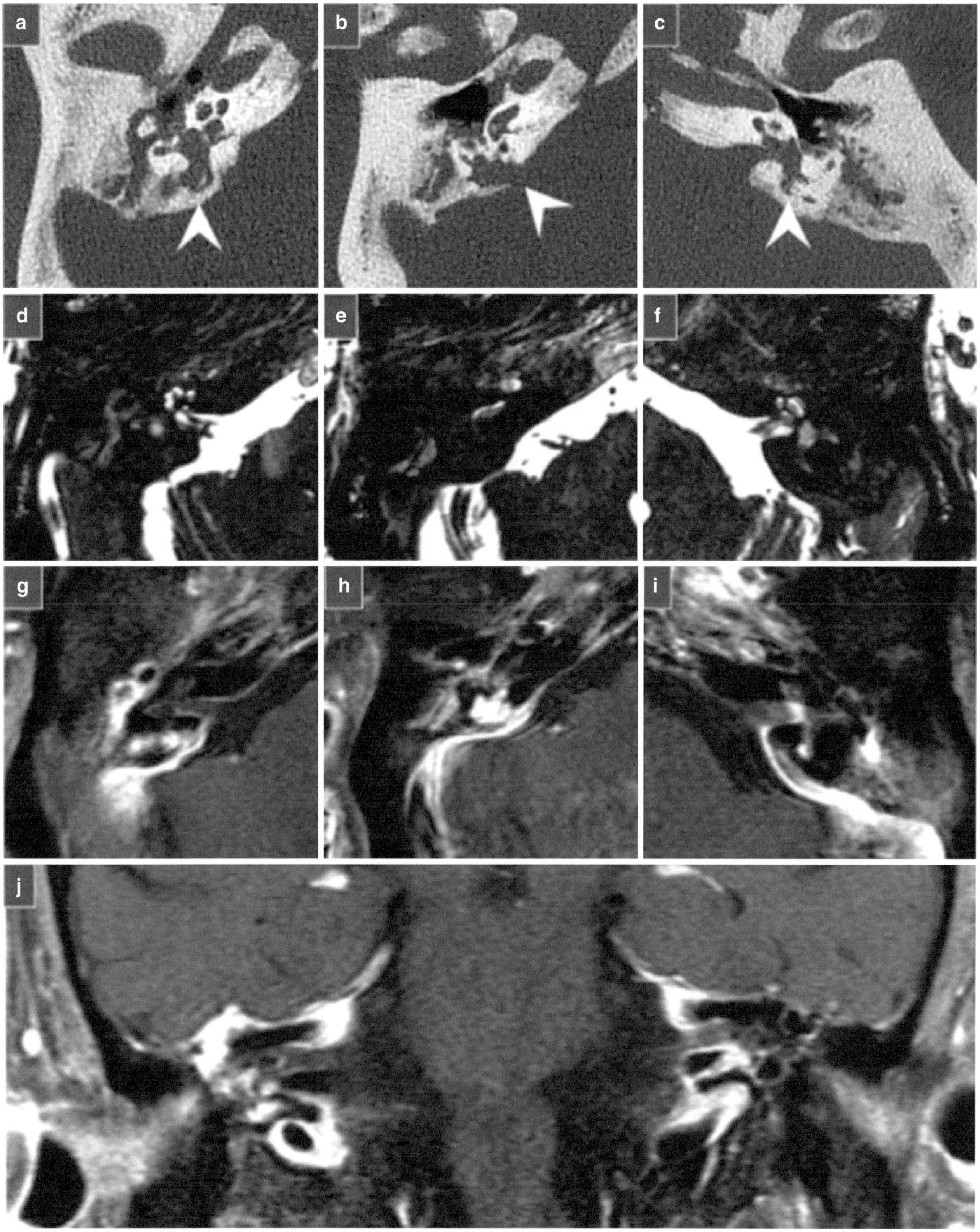

Fig. 151.1 Some imaging findings in otosyphilis. CT (panels **a–c**) showing lytic lesions involving bilateral otic capsules, including the right lateral semicircular canal, the right posterior petrous face, and left vestibule (arrowheads). Corresponding MRI, including axial T2 (panels **d–f**), axial T1 + contrast (panels **g–i**), and coronal T1 + contrast images (panel **j**). Note decreased T2 hyperintensity in the right labyrinth and scala tympani of the basal turn, and the left cochlea and labyrinth. Postcontrast images show enhancement of the posterior and middle fossa dura and of the lytic areas seen on CT. (From Hosseini et al. [14]. Reproduced with written permission from Wolters Kluwer Health Limited)

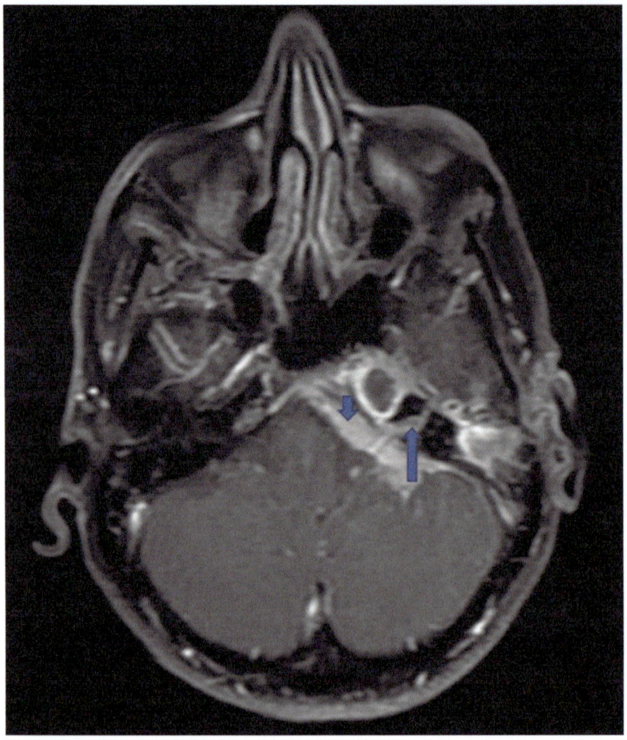

Fig. 151.2 Axial MRI through the internal auditory canals shows homogeneously enhancing dural-based soft tissue lesion (short arrow) in the premedullary cistern and cerebellopontine region extending to the ipsilateral internal acoustic meatus. Enhancing lesions (long arrow) in the petromastoid region with enhancement of the vestibulocochlear nerve. (From Chandrasekharan et al. [18]. Reproduced with written permission from Georg Thieme Verlag)

Histopathology

Very few temporal bone studies of patients with otosyphilis have been published [8].

Differential Diagnosis

Otosyphilis can present similarly to autoimmune inner ear disease [19] and Ménière's disease [8].

Note that syphilis can impair equilibrium through other mechanisms besides labyrinthine involvement. In particular, neurosyphilis can cause tabes dorsalis (degeneration of the posterior columns of the spinal cord) which impairs large fiber sensation (vibration and proprioception), leading to a sensory ataxia.

Treatment

Treatment of otosyphilis is treatment of syphilis itself. Treatment is usually supervised by an internist or infectious disease specialist and commonly involves intramuscular penicillin with oral probenecid or intravenous penicillin [11].

Prognosis

Treatment of syphilis may reverse hearing loss in some cases of otosyphilis [20–22], though this has not been studied systematically.

Traidl and colleagues [23], citing data from Yimtae and colleagues [23], state that, "in a retrospective analysis of 206 patients only 6.6% revealed an improvement of the hearing symptoms after treatment." In some cases, the improvement in hearing is not sustained [24].

Figure 151.3 below, from Etemady-Deylamy and Abate [22], shows improvement in left-sided sensorineural hearing loss on audiometry from before treatment (left panel) and after treatment (right panel).

In the case described by Etemady-Deylamy and Abate [22], the patient also presented with "vertigo," and this had resolved at 6 months after treatment.

Klemm and Wollina [10] described a patient with spontaneous and positional nystagmus that completely remitted after treatment.

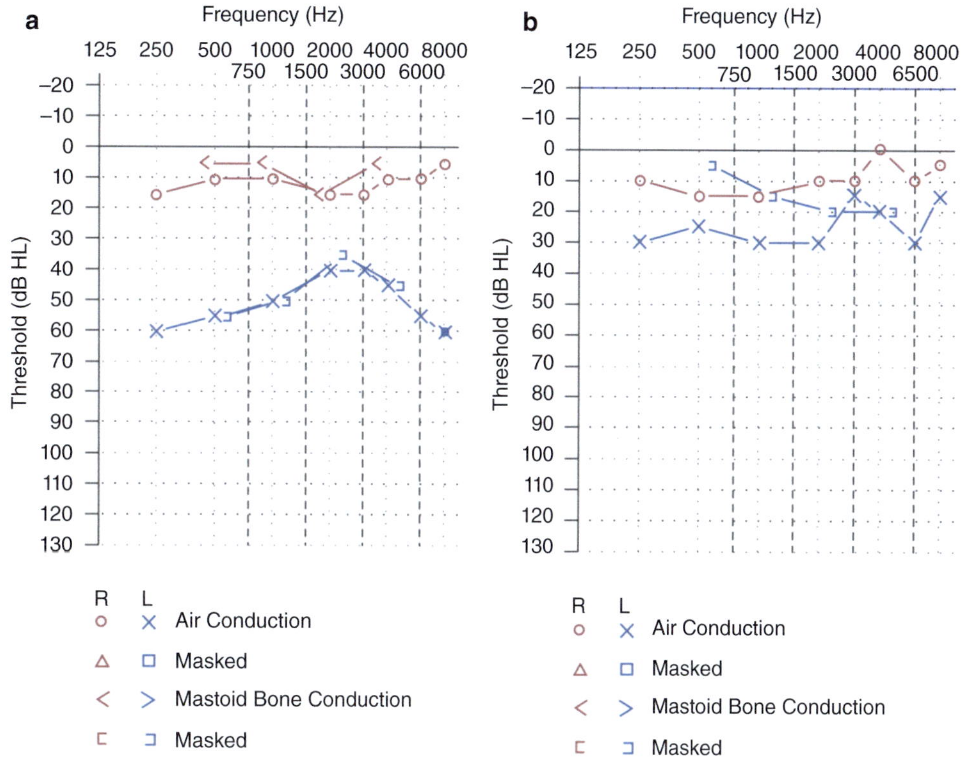

Fig. 151.3 Improvement in left-sided sensorineural hearing loss on audiometry from before treatment (left panel) and after treatment (right panel). (From Etemady-Deylamy and Abate [22]. Reproduced from Hindawi with permission under CC-BY 4.0)

References

1. Yimtae K, Srirompotong S, Lertsukprasert K. Otosyphilis: a review of 85 cases. Otolaryngol Head Neck Surg. 2007;136(1):67–71.
2. Theeuwen H, Whipple M, Litvack JR. Otosyphilis: resurgence of an old disease. Laryngoscope. 2019;129(7):1680–4.
3. Papalini C, Cagini C, Ricci G, Pasticci MB. Ocular and oto-syphilis: not a thing of the past. Infection. 2017;45(5):725–6.
4. Dombrowski JC, Pedersen R, Marra CM, Kerani RP, Golden MR. Prevalence estimates of complicated syphilis. Sex Transm Dis. 2015;42(12):702–4.
5. Jones-Vanderleest JG. Neurosyphilis, ocular syphilis, and otosyphilis: detection and treatment. Am Fam Physician. 2022;106(2):122–3.
6. Hungerbuhler JP, Regli F. Cochleovestibular involvement as the first sign of syphilis. J Neurol. 1978;219(3):199–204.
7. Carlson JA, Dabiri G, Cribier B, Sell S. The immunopathobiology of syphilis: the manifestations and course of syphilis are determined by the level of delayed-type hypersensitivity. Am J Dermatopathol. 2011;33(5):433–60.
8. Fayad JN, Linthicum FH Jr. Temporal bone histopathology case of the month: otosyphilis. Am J Otol. 1999;20(2):259–60.
9. Mishra S, Walmsley SL, Loutfy MR, Kaul R, Logue KJ, Gold WL. Otosyphilis in HIV-coinfected individuals: a case series from Toronto, Canada. AIDS Patient Care STDS. 2008;22(3):213–9.
10. Klemm E, Wollina U. Otosyphilis: report on six cases. J Eur Acad Dermatol Venereol. 2004;18(4):429–34.
11. Ramchandani MS, Litvack JR, Marra CM. Otosyphilis: a review of the literature. Sex Transm Dis. 2020;47(5):296–300.
12. Song JJ, Lee HM, Chae SW, Hwang SJ. Bilateral otosyphilis in a patient with HIV infection. Eur Arch Otorrinolaringol. 2005;262(12):972–4.
13. Lowhagen GB, Rosenhall U, Andersson M, Blomstrand C, Lindholm L, Roupe G. Central nervous system involvement in early syphilis. Part II. Correlation between auditory brain stem responses (ABR) and cerebrospinal fluid abnormalities. Acta Derm Venereol. 1983;63(6):530–5.
14. Hosseini DK, Aaron KA, Alyono JC. Hearing loss due to otosyphilis: imaging case of the month. Otol Neurotol. 2021;42(10):e1634–e5.
15. Ogungbemi A, Dudau C, Morley S, Beale T. Computed tomography features of luetic osteitis (otosyphilis) of the temporal bone. J Laryngol Otol. 2014;128(2):185–8.
16. Sonne JE, Zeifer B, Linstrom C. Manifestations of otosyphilis as visualized with computed tomography. Otol Neurotol. 2002;23(5):806–7.
17. Little JP, Gardner G, Acker JD, Land MA. Otosyphilis in a patient with human immunodeficiency virus: internal auditory canal gumma. Otolaryngol Head Neck Surg. 1995;112(3):488–92.
18. Chandrasekharan R, Kulkarni C, Pullara SK, Moorthy S. Magnetic resonance imaging in otosyphilis: a rare manifestation of neurosyphilis. Indian J Radiol Imaging. 2022;32(2):278–84.
19. Garcia-Berrocal JR, Gorriz C, Ramirez-Camacho R, Trinidad A, Ibanez A, Rodriguez Valiente A, et al. Otosyphilis mimics immune disorders of the inner ear. Acta Otolaryngol. 2006;126(7):679–84.
20. Sarsak EW, Omer WE, Al Bishawi AA, Maslamani MA, Abdulmajed AAB. Otosyphilis: a rare case of reversible hearing loss in a young man with secondary syphilis. IDCases. 2023;31:e01666.
21. He S, Messner AH, Mirani G. Otosyphilis: a rare cause of reversible hearing loss in a teenage male. Cureus. 2022;14(3):e23468.
22. Etemady-Deylamy A, Abate G. Otosyphilis with clinical and serologic responses with weekly intramuscular penicillin: a case report and literature review. Case Rep Infect Dis. 2022;2022:3152383.
23. Traidl S, Angela Y, Stender A, Kulberg A, Tronnier M, Prenzler NK, et al. A rare case of neuro- and otosyphilis in secondary syphilis. J Eur Acad Dermatol Venereol. 2021;35(11):e823–e6.
24. Dobbin JM, Perkins JH. Otosyphilis and hearing loss: response to penicillin and steroid therapy. Laryngoscope. 1983;93(12):1540–3.

Part XXIV

Diagnoses—Endocrinological, Nutritional and Metabolic Causes of Disequilibrium

152 Thyroid Disease and Disequilibrium

Overview

Thyroid disease is common [1]. A small proportion of patients with thyroid disease report disequilibrium, but the relationship between that symptom and the endocrine abnormality is unclear, and studies reach different conclusions about whether any such relationship exists. Given this uncertainty, we do not find it practical to screen for thyroid disease in patients complaining of vestibular symptoms. If a patient is diagnosed with thyroid disease, that should be managed (usually by an internist or endocrinologist) irrespective of the presence of vestibular symptoms. If a patient with thyroid disease complains of disequilibrium, it is medically reasonable to undertake a screening otovestibular workup to check for common vestibular diseases.

Introduction

Thyroid hormone plays a crucial role in energy metabolism in every cell in the body, so it is unsurprising that thyroid disorders can manifest with a broad variety of symptoms. A relatively small proportion of patients with thyroid disease report disequilibrium, but the relationship between the endocrine abnormality and the vestibular symptom is unclear. The few studies that exist reach different conclusions about whether there is any such relationship.

Studies Reporting an Association Between Thyroid Disease and Vestibular Disease

Choi and colleagues [2] studied 19,071 patients with benign paroxysmal positional vertigo and 76,284 controls and reported that goiter, hypothyroidism, thyroiditis, and hyperthyroidism were associated with BPPV (each $p < 0.05$).

Kim and colleagues [3] studied 8183 patients with Ménière's disease and 32,732 controls and reported that goiter, hypothyroidism, and hyperthyroidism were associated with Ménière's disease ($p < 0.05$).

Modugno and colleagues [4] studied 70 patients with benign paroxysmal positional vertigo and reported that, "In 34 cases (48.5%), autoimmune alterations were found: in 19 cases (27.1%), the level of antithyroid antibodies far exceeded the normal values with a significant incidence in comparison with a control group ($p < 0.01$). No other 'risk factors' were present."

Teggi and colleagues [5] retrospectively analyzed thyroid autoantibodies in 3042 patients with benign paroxysmal positional vertigo and reported that, "antithyroid autoantibodies may play a role in recurrences in subjects with initial manifestations [of BPPV] between 40 and 60 years."

Papi and colleagues [6] conducted a multivariate analysis of several serum tests (thyroid-stimulating hormone, thyroid peroxidase antibodies, thyroglobulin antibodies) in 134 benign paroxysmal positional vertigo patients against controls and reported that, "BPPV is strongly associated with both hypothyroidism (mainly, subclinical hypothyroidism) and thyroid autoimmunity."

Studies Reporting No Association Between Thyroid Disease and Vestibular Disease

Sari and colleagues [7] studied thyroid antibodies (thyroid peroxidase antibodies, thyroglobulin antibodies) in 50 patients with benign paroxysmal positional vertigo, 52 patients with non-BPPV disequilibrium, and 60 controls, and reported that they "didn't find any relation between BPPV and thyroid autoimmunity."

Miskiewicz-Orczyk and colleagues [8, 9] studied several otovestibular tests (audiometry, tympanometry, Dix-Hallpike maneuver, caloric testing, rotatory chair testing) in 28 women with Hashimoto's thyroiditis "and coexisting chronic vertigo" (19 benign paroxysmal positional vertigo, 7 Ménière's

M. Cherchi, *Otoneurology and Vestibular Medicine*, https://doi.org/10.1007/978-3-031-94841-1_152

disease, 2 vestibular neuritis) and reported that, "No correlation was found between… the results of thyroid function tests or the assessment of the vestibular end organ."

The Data from Different Studies Are Difficult to Reconcile

Studies reach different conclusions regarding the relationship between thyroid disease and vestibular disorders. We are particularly struck by the completely opposite conclusions reached by Papi and colleagues [6] and Sari and colleagues [7].

Pathophysiological Mechanism of Disease

The mechanism by which thyroid disease might affect vestibular function is unknown. Modugno and colleagues [4] studied thyroid autoantibodies in patients with benign paroxysmal positional vertigo and stated, "It can be hypothesized that the diffusion of immune-complexes in the inner ear could change the composition of the endolymphatic fluid, exerting a mechanical stimulation of the receptors and provoking the typical vertigo."

Notes

Since thyroid function does not appear to influence the results of instrumented vestibular tests [8], there are currently no grounds for conducting screening vestibular tests in patients with thyroid disease. However, if a patient with thyroid disease complains of disequilibrium, it is medically reasonable to undertake a screening otovestibular workup, not for the purpose of "verifying thyroid-related vestibular disease," but rather to check for common vestibular disorders.

References

1. Taylor PN, Albrecht D, Scholz A, Gutierrez-Buey G, Lazarus JH, Dayan CM, Okosieme OE. Global epidemiology of hyperthyroidism and hypothyroidism. Nat Rev Endocrinol. 2018;14(5):301–16.
2. Choi HG, Song YS, Wee JH, Min C, Yoo DM, Kim SY. Analyses of the relation between BPPV and thyroid diseases: a nested case-control study. Diagnostics (Basel). 2021;11(2):329.
3. Kim SY, Song YS, Wee JH, Min C, Yoo DM, Choi HG. Association between Ménière's disease and thyroid diseases: a nested case-control study. Sci Rep. 2020;10(1):18224.
4. Modugno GC, Pirodda A, Ferri GG, Montana T, Rasciti L, Ceroni AR. A relationship between autoimmune thyroiditis and benign paroxysmal positional vertigo? Med Hypotheses. 2000;54(4):614–5.
5. Teggi R, Guidetti R, Gatti O, Guidetti G. Recurrence of benign paroxysmal positional vertigo: experience in 3042 patients. Acta Otorhinolaryngol Ital. 2021;41(5):461–6.
6. Papi G, Corsello SM, Milite MT, Zanni M, Ciardullo AV, Donato CD, Pontecorvi A. Association between benign paroxysmal positional vertigo and autoimmune chronic thyroiditis. Clin Endocrinol. 2009;70(1):169–70.
7. Sari K, Yildirim T, Borekci H, Akin I, Aydin R, Ozkiris M. The relationship between benign paroxysmal positional vertigo and thyroid autoimmunity. Acta Otolaryngol. 2015;135(8):754–7.
8. Miśkiewicz-Orczyk K, Kos-Kudła B, Lisowska G. The function of the vestibular organ in Hashimoto's thyroiditis. Endokrynol Pol. 2022;73(6):935–41.
9. Miśkiewicz-Orczyk K, Vlaykov A, Lisowska G, Strzelczyk J, Kos-Kudła B. Does thyroid hormone metabolism correlate with the objective assessment of the vestibular organ in patients with vertigo? J Clin Med. 2022;11(22):6771.

153 Vitamin B1 (Thiamine) Deficiency

Overview

Deficiency of vitamin B1 (thiamine) can result from malnutrition and can occur in alcoholism. Severe cases may manifest as Wernicke-Korsakoff syndrome, originally described by Karl Wernicke (1848–1905) [1, 2] and Sergei Sergeievich Korsakoff (1853–1900) [3, 4]. Ocular motor abnormalities may be the first manifestations of vitamin B1 deficiency.

Vitamin B1, also called thiamine, is usually obtained in the diet from animal products (fish, meat) and whole grains. In industrialized countries, many foods are fortified by the addition of vitamin B1. Vitamin B1 functions as a cofactor for enzymes involved in the metabolism of carbohydrates and branched-chain amino acids.

Vitamin B1 deficiency can occur due to dietary deprivation, such as starvation and alcoholism.

In severe cases, vitamin B1 deficiency can present as Wernicke-Korsakoff syndrome, which involves amnesia (and thus confabulation), psychosis, ataxia, and oculomotor abnormalities.

The ocular motor abnormalities may be among the first manifestations of incipient vitamin B1 deficiency, and thus are important to recognize as the possible harbinger of florid Wernicke-Korsakoff syndrome.

The eye movement abnormalities described in association with vitamin B1 deficiency include the following:

- Ophthalmoparesis [5]
- Symmetric vestibular weakness, manifesting as decreased vestibulo-ocular reflex gain [5, 6]
- Horizontal gaze-evoked nystagmus [6, 7]

Treatment is parenteral administration of vitamin B1, followed by oral supplementation.

References

1. Wernicke K. Die akute, hämorrhagische Polioencephalitis superior [Acute hemorrhagic superior polioencephalitis]. Lehrbuch der Gehirnkrankheiten. 2. Kassel: Fischer; 1881. p. 229–42.
2. Thomson AD, Cook CC, Guerrini I, Sheedy D, Harper C, Marshall EJ. Wernicke's encephalopathy revisited. Translation of the case history section of the original manuscript by Carl Wernicke 'Lehrbuch der Gehirnkrankheiten fur Aerzte and Studirende' (1881) with a commentary. Alcohol Alcohol. 2008;43(2):174–9.
3. Korsakoff SS. Об Алкогольном паралич [On alcoholic paralysis]. Vestnick Klin Psychiat Neurol. 1887;4:1–102.
4. Victor M, Yakovlev PI. S.S. Korsakoff's psychic disorhic disorder in conjunction with peripheral neuritis; a translation of Korsakoff's original article with comments on the author and his contribution to clinical medicine. Neurology. 1955;5(6):394–406.
5. Kattah JC. The spectrum of vestibular and ocular motor abnormalities in thiamine deficiency. Curr Neurol Neurosci Rep. 2017;17(5):40.
6. Kattah JC, Dhanani SS, Pula JH, Mantokoudis G, Tehrani ASS, Toker DEN. Vestibular signs of thiamine deficiency during the early phase of suspected Wernicke encephalopathy. Neurol Clin Pract. 2013;3(6):460–8.
7. Kattah JC. Early signs of thiamine deficiency: a case report. Ann Intern Med. 2020;173(1):72–3.

M. Cherchi, *Otoneurology and Vestibular Medicine*, https://doi.org/10.1007/978-3-031-94841-1_153

Vitamin B12 Deficiency

154

Overview

Vitamin B12, also called *cyanocobalamin*, is usually obtained in the diet from animal products (meat, poultry, fish, eggs, dairy products). A deficiency of vitamin B12 can occur due to inadequate diet or due to a variety of malabsorption syndromes [1].

Vitamin B12 is essential for cellular energy metabolism, so a *deficiency of vitamin B12* can have a wide variety of presentations. As far as otoneurology is concerned, the main mechanisms by which vitamin B12 deficiency can interfere with a person's equilibrium include the following:

- Spinal cord subacute combined degeneration [2]
- Peripheral neuropathy [3]
- Ocular motor disturbances

Ocular motor disturbances are not commonly documented in vitamin B12 deficiency, but may also be underreported if the clinical picture is dominated by other features (such as spinal cord subacute combined degeneration).

There are several case reports and case series in the literature [4–6] regarding ocular motor abnormalities observed in vitamin B12 deficiency. Most cases involve spontaneous downbeat nystagmus in the primary position of gaze. A smaller proportion of cases present with internuclear ophthalmoplegia [4]. One case report noted spontaneous upbeat nystagmus in the primary position of gaze [5].

Treatment of significant vitamin B12 deficiency is usually initiated with intramuscular injections, eventually transitioning to oral supplementation. If this fails, then a search is usually undertaken for a vitamin B12 malabsorption syndrome.

References

1. Gueant JL, Gueant-Rodriguez RM, Alpers DH. Vitamin B12 absorption and malabsorption. Vitam Horm. 2022;119:241–74.
2. Kostick N, Chen E, Eckert T, Sirotkin I, Baldinger E, Frontera A. Clinical presentation of subacute combined degeneration in a patient with chronic B12 deficiency. Fed Pract. 2022;39(3):142–6.
3. Solomon LR. Vitamin B12-responsive neuropathies: a case series. Nutr Neurosci. 2016;19(4):162–8.
4. Akdal G, Yener GG, Ada E, Halmagyi GM. Eye movement disorders in vitamin B12 deficiency: two new cases and a review of the literature. Eur J Neurol. 2007;14(10):1170–2.
5. Ouchi H, Ishikawa K, Inoue K, Shibano K, Hara K. Primary position upbeat nystagmus with low serum vitamin B12. Intern Med. 2020;59(24):3229–33.
6. Puri V, Chaudhry N, Satyawani M. Down beat nystagmus in vitamin B 12 deficiency syndrome. Electromyogr Clin Neurophysiol. 2006;46(2):101–4.

M. Cherchi, *Otoneurology and Vestibular Medicine*, https://doi.org/10.1007/978-3-031-94841-1_154

155 Vitamin D Deficiency

Overview

There is disagreement in the literature regarding the relationship between serum vitamin D levels and the occurrence of benign paroxysmal positional vertigo (BPPV). There is some evidence that *vitamin D deficiency* has adverse neuromuscular consequences (such as loss of muscle mass and strength), can adversely affect balance (in the form of increased postural sway), and can increase fall risk.

Introduction

Vitamin D plays a crucial role in the body's metabolism of calcium and phosphorus. For purposes of otoneurology, the main question is whether vitamin D levels influence the occurrence/recurrence of benign paroxysmal positional vertigo (BPPV). Vitamin D is also relevant to the maintenance of muscle mass and strength, and these neuromuscular factors may be relevant to balance and fall risk.

As far as BPPV is concerned, some literature suggests that low serum levels of vitamin D correlate with a higher incidence (first occurrence) of BPPV [1–9], and therefrom infer that low vitamin D levels are a risk factor for BPPV. Other studies find no relationship between vitamin D levels and BPPV [10–12], and therefrom conclude that low vitamin D levels are not a risk factor for BPPV. Other studies reach more nuanced conclusions. For example, some studies conclude that low vitamin D levels are a risk factor for *recurrent* BPPV [2, 8, 10], rather than for a first episode of BPPV. Other studies note that vitamin D levels have some correlation with gender and age, and that all three factors must be taken into consideration when trying to assess the risk of BPPV [13, 14].

As far as neuromuscular factors are concerned, there is some evidence that low vitamin D levels are associated with poorer balance and with increased fall risk [15].

Epidemiology

As far as benign paroxysmal positional vertigo (BPPV) is concerned, some literature has reported seasonal patterns in the occurrence of BPPV [4, 6], and in particular, that BPPV is more common in winter months. The higher occurrence of BPPV in seasons with reduced light raised suspicion of a relationship with vitamin D.

Pathophysiological Mechanism of Disease

The mechanism by which low serum levels of vitamin D might influence the risk of BPPV is unclear [5]. Otoliths are comprised of calcium carbonate, and vitamin D plays an essential role in calcium metabolism, but these facts do not in themselves prove the mechanism.

Regarding the neuromuscular role of vitamin D, it is known that vitamin D receptors are expressed in muscle tissue, and patients with vitamin D deficiency show atrophy of type 2 muscle fibers [16]. Studies have further shown that the degree of postural sway [17, 18] and fall risk [19] are inversely correlated with vitamin D levels, though the underlying mechanisms of this inverse correlation are not clear.

Treatment

As far as benign paroxysmal positional vertigo (BPPV) is concerned, since some literature reports that lower vitamin D levels correlate with increased risk of BPPV, it is natural to wonder whether supplementing vitamin D might influence the occurrence of BPPV. There is emerging evidence that, after a first episode of BPPV, taking vitamin D and perhaps calcium may reduce the likelihood of recurrence of BPPV [20]. For instance, a randomized controlled trial [21] reported

M. Cherchi, *Otoneurology and Vestibular Medicine*, https://doi.org/10.1007/978-3-031-94841-1_155

that supplementation with vitamin D 400 I.U. and calcium carbonate 500 mg twice per day reduces the annual recurrence rate of BPPV. Note that increasing one's level of vitamin D (even without supplementing additional calcium) may increase serum calcium concentrations, which at least in theory can increase the risk of developing kidney stones [22], though it is unclear whether this increased risk is clinically meaningful; one might think that vitamin D supplementation should perhaps be avoided in persons who already have a history of kidney stone formation, but this does not seem to be borne out by the evidence [23]; a person considering supplementing their vitamin D intake (with or without calcium supplementation) should first consult their primary care physician.

As far as neuromuscular factors are concerned, there is some evidence that supplementing vitamin D in patients who are deficient in vitamin D may reduce postural sway [17] and reduce fall risk [19].

These factors (risk of BPPV, increased postural sway, increased fall risk) add to the list of medical reasons for treating vitamin D deficiency (such as reducing the risk of bone loss and fracture risk).

References

1. Chen J, Zhang S, Cui K, Liu C. Risk factors for benign paroxysmal positional vertigo recurrence: a systematic review and meta-analysis. J Neurol. 2021;268(11):4117–27.
2. Ding J, Liu L, Kong WK, Chen XB, Liu X. Serum levels of 25-hydroxy vitamin D correlate with idiopathic benign paroxysmal positional vertigo. Biosci Rep. 2019;39(4)
3. Jeong SH, Kim JS. Impaired calcium metabolism in benign paroxysmal positional vertigo: a topical review. J Neurol Phys Ther. 2019;43(Suppl 2):S37–41.
4. Meghji S, Murphy D, Nunney I, Phillips JS. The seasonal variation of benign paroxysmal positional vertigo. Otol Neurotol. 2017;38(9):1315–8.
5. Sfakianaki I, Binos P, Karkos P, Dimas GG, Psillas G. Risk factors for recurrence of benign paroxysmal positional vertigo. A clinical review. J Clin Med. 2021;10(19)
6. Shu L, Wu J, Jiang CY, Sun XH, Pan H, Fang J, et al. Seasonal variation of idiopathic benign paroxysmal positional vertigo correlates with serum 25-hydroxyvitamin D levels: a six-year registry study in Shanghai, China. Sci Rep. 2019;9(1):16230.
7. Yang B, Lu Y, Xing D, Zhong W, Tang Q, Liu J, Yang X. Association between serum vitamin D levels and benign paroxysmal positional vertigo: a systematic review and meta-analysis of observational studies. Eur Arch Otorhinolaryngol. 2020;277(1):169–77.
8. Talaat HS, Abuhadied G, Talaat AS, Abdelaal MSS. Low bone mineral density and vitamin D deficiency in patients with benign positional paroxysmal vertigo. Eur Arch Oto-Rhino-Laryngol. 2015;272(9):2249–53.
9. Rhim GI. Serum vitamin D and recurrent benign paroxysmal positional vertigo. Laryngoscope Investig Otolaryngol. 2016;1(6):150–3.
10. AlGarni MA, Mirza AA, Althobaiti AA, Al-Nemari HH, Bakhsh LS. Association of benign paroxysmal positional vertigo with vitamin D deficiency: a systematic review and meta-analysis. Eur Arch Otorhinolaryngol. 2018;275(11):2705–11.
11. Goldschagg N, Teupser D, Feil K, Strupp M. No evidence for a specific vitamin D deficit in benign paroxysmal positional vertigo. Eur J Neurol. 2021;
12. Li S, Wang Z, Liu Y, Cao J, Zheng H, Jing Y, et al. Risk factors for the recurrence of benign paroxysmal positional vertigo: a systematic review and meta-analysis. Ear Nose Throat J. 2022;101(3):NP112–NP34.
13. Song P, Zhao X, Xu Y, Zhao Z, Wang L, Liu Y, Gao Q. Correlation between benign paroxysmal positional vertigo and 25-hydroxyvitamin D. Front Neurol. 2020;11:576.
14. Yang CJ, Kim Y, Lee HS, Park HJ. Bone mineral density and serum 25-hydroxyvitamin D in patients with idiopathic benign paroxysmal positional vertigo. J Vestib Res. 2017;27(5-6):287–94.
15. Dawson-Hughes B. Vitamin D and muscle function. J Steroid Biochem Mol Biol. 2017;173:313–6.
16. Ceglia L, Niramitmahapanya S, da Silva MM, Rivas DA, Harris SS, Bischoff-Ferrari H, et al. A randomized study on the effect of vitamin D(3) supplementation on skeletal muscle morphology and vitamin D receptor concentration in older women. J Clin Endocrinol Metab. 2013;98(12):E1927–35.
17. Cangussu LM, Nahas-Neto J, Orsatti CL, Poloni PF, Schmitt EB, Almeida-Filho B, Nahas EA. Effect of isolated vitamin D supplementation on the rate of falls and postural balance in postmenopausal women fallers: a randomized, double-blind, placebo-controlled trial. Menopause. 2016;23(3):267–74.
18. Pfeifer M, Begerow B, Minne HW, Suppan K, Fahrleitner-Pammer A, Dobnig H. Effects of a long-term vitamin D and calcium supplementation on falls and parameters of muscle function in community-dwelling older individuals. Osteoporos Int. 2009;20(2):315–22.
19. Bischoff-Ferrari HA, Dawson-Hughes B, Staehelin HB, Orav JE, Stuck AE, Theiler R, et al. Fall prevention with supplemental and active forms of vitamin D: a meta-analysis of randomised controlled trials. BMJ. 2009;339:b3692.
20. Jeong SH, Lee SU, Kim JS. Prevention of recurrent benign paroxysmal positional vertigo with vitamin D supplementation: a meta-analysis. J Neurol. 2020;
21. Jeong SH, Kim JS, Kim HJ, Choi JY, Koo JW, Choi KD, et al. Prevention of benign paroxysmal positional vertigo with vitamin D supplementation: a randomized trial. Neurology. 2020;95(9):e1117–e25.
22. Letavernier E, Daudon M, Vitamin D. Hypercalciuria and kidney stones. Nutrients. 2018;10:3.
23. Ganji MR, Shafii Z, Hakemi MS. Vitamin D supplementation and risk of hypercalciuria in stone formers. Iran J Kidney Dis. 2019;13(1):27–31.

Wilson's Disease

156

Overview

Wilson's disease (*hepatolenticular degeneration*) is a genetic disorder of copper metabolism wherein copper accumulates in the liver, brain, and other organs. Initial manifestations of the liver disease include jaundice and cirrhosis. Neurological manifestations may include dysarthria, cerebellar ataxia, dystonia, tremor, Parkinsonism, choreoathetosis, cognitive impairment, and myoclonus. Most patients with neurological symptoms, and half of the patients without neurological symptoms, exhibit copper accumulation at the corneal periphery, observable as Kaiser-Fleischer rings. Most patients with neurological Wilson's disease also exhibit abnormalities in ocular motor function, including abnormalities of vertical (more than horizontal) smooth pursuit, optokinetic responses, and saccades. Clues to diagnosis come from physical examination, laboratory findings (e.g., elevated serum ceruloplasmin, elevated urinary copper excretion), brain imaging, and liver biopsy (elevated copper concentration); definitive diagnosis comes from genetic testing.

Introduction

Dr. Samuel Alexander Kinnier Wilson (Cedarville, New Jersey, USA 1878–London, England, 1937) published a detailed report and analysis of 12 cases of *hepatolenticular degeneration* [1], a disease which eventually bore his name Fig. 156.1. In his article, he summarized (page 436):

> Progressive lenticular degeneration may be defined as a disease which occurs in young people, which is often familial but not congenital or hereditary; it is essentially and chiefly a disease of the extrapyramidal motor system, and is characterised by involuntary movements, usually of the nature of tremor, dysarthria, dysphagia, muscular weakness, spasticity, and contractures with progressive emaciation; with these may be associated emotionalism and certain symptoms of a mental nature. It is progressive, and, after a longer or shorter period, fatal. Pathologically it is characterised predominantly by bilateral degeneration of the lenticular nucleus, and in addition cirrhosis of the liver is constantly found, the latter morbid condition rarely, if ever, giving rise to symptoms during the life of the patient [1].

Fig. 156.1 Samuel Alexander Kinnier Wilson (1878–1937). From https://en.wikipedia.org/wiki/Samuel_Alexander_Kinnier_Wilson, public domain

At 214 pages, Wilson's article was the longest ever published in the journal *Brain* [2].

M. Cherchi, *Otoneurology and Vestibular Medicine*, https://doi.org/10.1007/978-3-031-94841-1_156

Epidemiology

Reilly and colleagues [3] reported that the incidence of Wilson's disease in Ireland was 17 per one million. A literature review by Sandahl and colleagues [4] found the prevalence of Wilson's disease to range from 1 in 30,000 to 1 in 40,000.

Genetics

Wilson's disease (OMIM 277900) is now known to be an autosomal recessively transmitted defect in the ATP7B gene (OMIM 606882) on chromosome 13q14.3 that encodes for adenosine triphosphatase responsible for the excretion of excess copper. The defect permits accumulation of the excess copper predominantly in the liver, cornea, and brain (basal ganglia, thalamus, midbrain, cerebellum, prefrontal cortex). Less commonly affected organs include the kidney, heart, skin, endocrine system, and musculoskeletal system.

Pathophysiological Mechanism of Disease

As far as the ocular motor abnormalities in Wilson's disease are concerned, Ingster-Moati and colleagues state that, "The extrapyramidal fibers originating from the striated nucleus and the pallidum play a role of relay in the Cajal and Darkschewitsch nuclei, providing afferent fibers to the ocular motor nuclei through the medial longitudinal fasciculus" [5].

Clinical Presentation

Lesniak and colleagues [6] reported that in about 40% of patients, the initial symptoms are related to hepatic dysfunction; whereas in about 15% of patients, the initial symptoms are neuropsychiatric, including psychosis, antisocial behavior, irritability, impulsiveness, mood disorders, and frontal executive dysfunction.

The copper deposition in the deep epithelial layers at the outer border of the cornea (Descemet's membrane) can be observed as brownish, grayish, or greenish rings; these can be found in about 90% of patients with neurological symptoms, and 50% of patients without neurological symptoms. This ocular finding was described by German ophthalmologists, Bernhard Kayser [7] and Bruno Fleischer [8], and came to be known as Kayser-Fleischer rings [9]. Kaiser-Fleischer rings can sometimes be seen simply on face-to-face inspection, though are more easily observed on a slit lamp examination. Kaiser-Fleischer rings can also be seen on anterior segment optical coherence tomography [10].

For a time, it was thought that Kayser-Fleischer rings were pathognomonic for Wilson's disease, but it was eventually realized that these can also be found in neonatal cholestasis [11], other liver diseases, and conditions that result in copper overload.

Physical Examination

In Wilson's disease patients with neurological involvement, physical examination findings may include dysarthria, cerebellar ataxia, dystonia, tremor, Parkinsonism, choreoathetosis, cognitive impairment, and myoclonus [12].

Ocular Motor Examination

Although ocular motor abnormalities are quite common in Wilson's disease [5], Wilson's original description [1] only commented on this in two out of the 12 cases he described [13]. Some of these may be difficult to observe on face-to-face examination, and only become apparent on instrumented ocular motor testing.

Testing: Vestibular

A rather broad variety of ocular motor abnormalities has been reported in Wilson's disease, including the following:

- Slow vertical saccades [5, 14]
- Abnormal vertical smooth pursuit [5]
- Abnormal horizontal and vertical optokinetic responses [5]
- Gaze distractibility [15]
- Oculogyric crisis [16]
- Lid opening apraxia [17]

Ingster-Moati and colleagues [5] studied 34 Wilson's disease patients using electro-oculography. They reported that 31 (91%) of patients had one or more ocular motor abnormalities, with vertical eye movements affected more than horizontal eye movements. Of all 34 patients:

- 29/34 (85%) had abnormalities of vertical pursuit.
- 14/34 (41%) had abnormalities of vertical optokinetic nystagmus.
- 14/34 (41%) had abnormalities of horizontal pursuit.
- 10/34 (30%) had abnormalities of vertical saccades.
- 9/34 (27%) had abnormalities of horizontal optokinetic nystagmus.
- 5/34 (15%) had abnormalities of horizontal saccades.
- 3/34 (9%) had abnormalities of visual fixation.

Given this variety of abnormalities, it is understandable that patients have difficulty reading [13].

Testing: Other

Findings that increase suspicion for Wilson's disease include the presence of Kaiser-Fleischer rings, elevated serum ceruloplasmin, elevated urinary copper excretion, or elevated copper concentration on liver biopsy. Diagnosis can be confirmed with genetic testing, for which a medical genetics consultation is recommended.

Imaging

Prashanth and colleagues [18] retrospectively reviewed 56 patients with Wilson's disease and observed that:

- 75% had tectal plate hyperintensity.
- 63% had central pontine myelinolysis-like abnormalities.
- 55% had signal changes in the basal ganglia, thalamus, and brainstem.
- 14% of cases had the "face of a giant panda" sign.

They concluded that, "Besides 'Face of giant panda' sign, hyperintensities in tectal-plate and central pons (CPM-like), and simultaneous involvement of basal ganglia, thalamus, and brainstem are virtually pathognomonic of WD [Wilson's disease]" [18].

Fig. 156.2, from Prashanth and colleagues [18], shows these imaging findings.

A number of abnormalities are detectable on functional imaging in patients with Wilson's disease [19], but this is not part of the standard diagnostic workup.

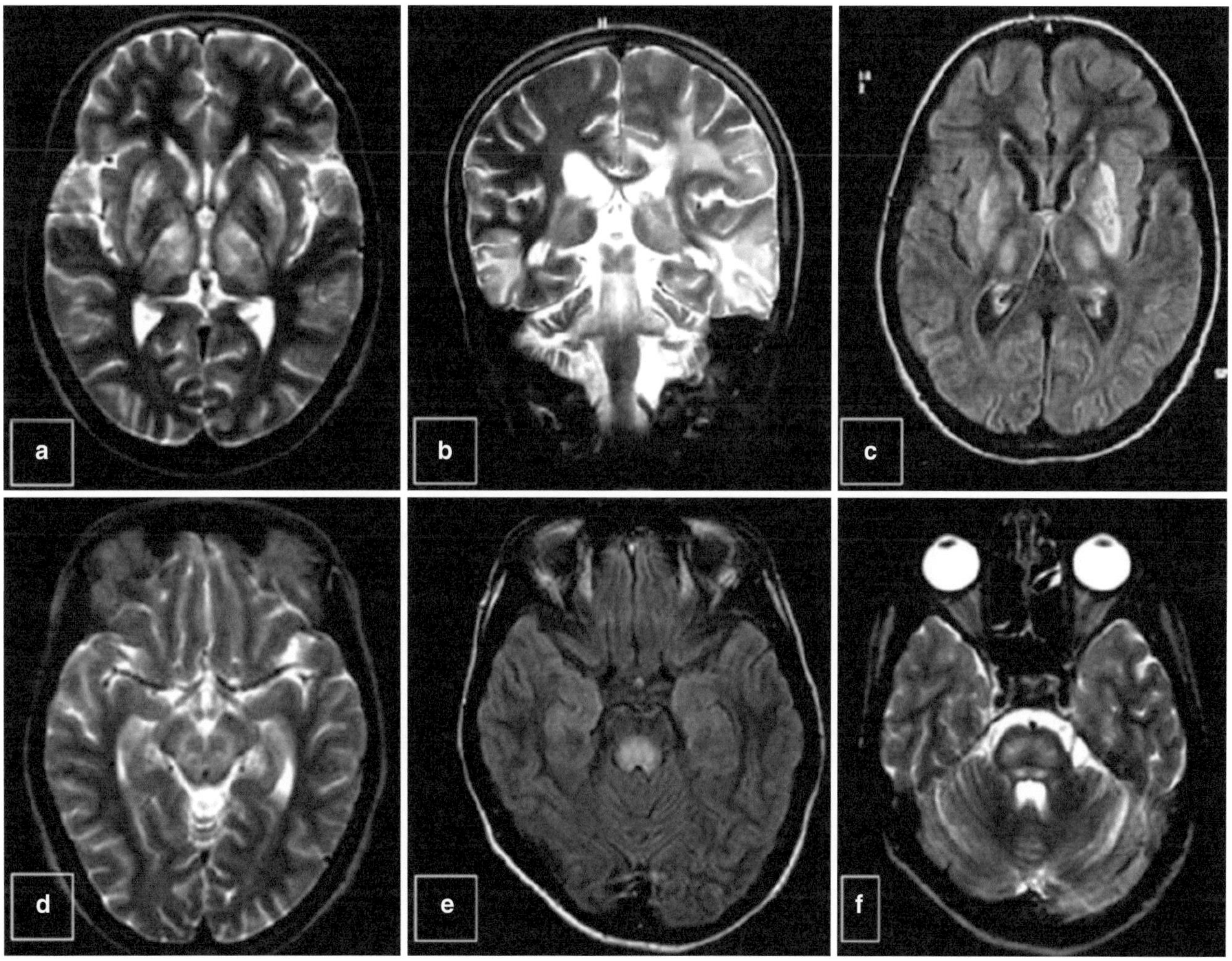

Fig. 156.2 MRI (brain) in patients with Wilson's disease (WD): Panel A: axial section of T2WI showing signal changes in bilateral putamen and thalamus in a patient with WD. Panel B: coronal section of T2WI showing signal changes involving white matter, basal ganglia, and brain stem structures in a patient with WD. Panel C: FLAIR sequence revealing signal changes in bilateral caudate, putamen, and thalamus in a patient with WD. Panel D: axial T2W image at the midbrain level showing the "Face of Giant Panda" sign. Panel E: FLAIR axial sequence through midbrain revealing hyperintense signal changes in the tectum. Panel F: axial T2WI through pons showing central pontine myelinolysis (CPM)-like signal changes. From Prashanth et al. [18]. Reproduced with written permission from John Wiley and Sons Limited

Histopathology

Liver biopsy shows cirrhosis and elevated copper concentration.

Differential Diagnosis

The differential diagnosis of Wilson's disease includes other basal ganglionic disorders.

Treatment and Prognosis

In the United States, the most common treatment for Wilson's disease is the chelating agent D-penicillamine. Less commonly used agents include trientene hydrochloride, trientene tetrahydrochloride, ammonium tetrathiomolybdate, or sodium dimercaptosuccinate. Zinc supplementation (e.g., zinc acetate) is sometimes also used for maintenance therapy because zinc competes with copper absorption. Treatment can decelerate and, in some cases, partially reverse disease progression.

References

1. Wilson SAK. Progressive lenticular degeneration: a familial nervous dysease associated with cirrhosis of the liver. Brain. 1912;34(4):295–507.
2. Compston A. Progressive lenticular degeneration: a familial nervous disease associated with cirrhosis of the liver, by S. A. Kinnier Wilson, (from the National Hospital, and the Laboratory of the National Hospital, Queen Square, London) brain 1912: 34; 295-509. Brain. 2009;132(Pt 8):1997–2001.
3. Reilly M, Daly L, Hutchinson M. An epidemiological study of Wilson's disease in the Republic of Ireland. J Neurol Neurosurg Psychiatry. 1993;56(3):298–300.
4. Sandahl TD, Laursen TL, Munk DE, Vilstrup H, Weiss KH, Ott P. The prevalence of Wilson's disease: an update. Hepatology. 2020;71(2):722–32.
5. Ingster-Moati I, Bui Quoc E, Pless M, Djomby R, Orssaud C, Guichard JP, et al. Ocular motility and Wilson's disease: a study on 34 patients. J Neurol Neurosurg Psychiatry. 2007;78(11):1199–201.
6. Lesniak M, Czlonkowska A, Seniow J. Abnormal antisaccades and smooth pursuit eye movements in patients with Wilson's disease. Mov Disord. 2008;23(14):2067–73.
7. Kayser B. Über einen fall von angeborener grünlicher Verfärbung der cornea [On a case of congenital greenish discoloration of the cornea]. Klin Monatsbl Augenheilkd. 1902;40(2):22–5.
8. Fleischer B. Zwei weitere Falle von grunlicher Verfarbung der Kornea [Two more cases of greenish discoloration of the cornea]. Klin Monatsbl Augenheilkd. 1903;41(1):489–91.
9. Pearce JMS. Historical note: the Kayser-Fleischer ring. Adv Clin Neurosci Rehabil. 2019;19(4):48–9.
10. Broniek-Kowalik K, Dziezyc K, Litwin T, Czlonkowska A, Szaflik JP. Anterior segment optical coherence tomography (AS-OCT) as a new method of detecting copper deposits forming the Kayser-Fleischer ring in patients with Wilson disease. Acta Ophthalmol. 2019;97(5):e757–e60.
11. Dunn LL, Annable WL, Kliegman RM. Pigmented corneal rings in neonates with liver disease. J Pediatr. 1987;110(5):771–6.
12. Lorincz MT. Neurologic Wilson's disease. Ann N Y Acad Sci. 2010;1184:173–87.
13. Hyman NM, Phuapradit P. Reading difficulty as a presenting symptom in Wilson's disease. J Neurol Neurosurg Psychiatry. 1979;42(5):478–80.
14. Kirkham TH, Kamin DF. Slow saccadic eye movements in Wilson's disease. J Neurol Neurosurg Psychiatry. 1974;37(2):191–4.
15. Lennox G, Jones R. Gaze distractibility in Wilson's disease. Ann Neurol. 1989;25(4):415–7.
16. Lee MS, Kim YD, Lyoo CH. Oculogyric crisis as an initial manifestation of Wilson's disease. Neurology. 1999;52(8):1714–5.
17. Keane JR. Lid-opening apraxia in Wilson's disease. J Clin Neuroophthalmol. 1988;8(1):31–3.
18. Prashanth LK, Sinha S, Taly AB, Vasudev MK. Do MRI features distinguish Wilson's disease from other early onset extrapyramidal disorders? An analysis of 100 cases. Mov Disord. 2010;25(6):672–8.
19. Tinaz S, Arora J, Nalamada K, Vives-Rodriguez A, Sezgin M, Robakis D, et al. Structural and functional brain changes in hepatic and neurological Wilson disease. Brain Imaging Behav. 2021;15(5):2269–82.

Zinc Deficiency

157

Overview

Zinc influences the activity of several neurotransmitter recep tor systems, including ones relevant to vestibular function, such as GABA, glutamate, potassium, and NMDA. There is almost no literature on the vestibular effects of zinc deficiency, but one study reported abnormalities in horizontal and vertical saccades, and horizontal smooth pursuit.

Introduction

Zinc influences the activity of several neurotransmitter receptor systems, including some that are relevant to vestibular function.

Pathophysiological Mechanism of Disease

Zinc modulates activity at numerous neurotransmitter receptors, including GABA, glutamate, potassium, and NMDA [1]. Since all of these neurotransmitter systems are found in the vestibular system [2], it is reasonable to suspect that abnormalities in zinc metabolism may have vestibular effects. Candidate neuroanatomical locations of these effects include vestibular hair cells, vestibular nuclei, and the cerebellum [3].

Testing: Vestibular

Abnormalities in zinc metabolism have been little studied from the otoneurological perspective. Ferreira and Cury [3] studied videonystagmography in nine patients with zinc deficiency and reported that they exhibited abnormalities in horizontal saccades (specifically leftward saccadic latencies and bilateral saccadic accuracies), vertical saccades (upward saccadic latencies), and reduced smooth pursuit gain. However, it should be noted that this single study involved patients who (1) had malabsorption syndrome due to small intestinal resection for treatment of mesenteric thrombosis; (2) had received perioperative aminoglycoside antibiotics; and (3) had magnesium deficiency. In other words, the observed abnormalities may not be attributable only to zinc deficiency.

Treatment

Zinc deficiency may be treated with supplementation. Supplementation may be administered orally, but should be given parenterally (by injection) in patients with malabsorption syndromes.

References

1. Smart TG, Xie X, Krishek BJ. Modulation of inhibitory and excitatory amino acid receptor ion channels by zinc. Prog Neurobiol. 1994;42(3):393–441.
2. Soto E, Vega R, Sesena E. Neuropharmacological basis of vestibular system disorder treatment. J Vestib Res. 2013;23(3):119–37.
3. Ferreira GD, Cury MC, Oliveira JA, Manfredi AK, Vannucchi H. Vestibular evaluation using videonystagmography of chronic zinc deficient patients due to short bowel syndrome. Braz J Otorhinolaryngol. 2009;75(2):290–4.

M. Cherchi, *Otoneurology and Vestibular Medicine*, https://doi.org/10.1007/978-3-031-94841-1_157

Altitude Sickness

158

Overview

"*Altitude sickness*," also called "acute mountain sickness," is a colloquial term for a nonlife-threatening type of high-altitude illness. More severe forms of high-altitude illness include pulmonary edema and cerebral edema. Higher altitudes and more rapid ascent are the main risk factors for altitude sickness. Acetazolamide and ginkgo biloba appear superior to placebo for treating altitude sickness.

Introduction

At higher altitudes, there is lower barometric pressure and lower oxygen concentration, resulting in hypobaric hypoxia. Such conditions can cause several types of high-altitude illness (HAI), including acute mountain sickness, high-altitude pulmonary edema, and high-altitude cerebral edema. The risk of HAI is greater at higher altitudes and with more rapid ascents. Younger people are more vulnerable to HAI. Some studies suggest that women are more vulnerable than men. The best treatment is prevention. Some studies suggest that acetazolamide or ginkgo biloba is superior to placebo. Some genes appear to confer greater resilience against high-altitude illness.

Epidemiology

Maggiorini and colleagues [1] studied 466 recreational climbers in the Swiss Alps and found a very clear correlation between increasing altitude and increasing risk of acute mountain sickness (AMS), specifically, "The prevalence of acute mountain sickness correlated with altitude: it was 9% at 2850 m, 13% at 3050 m, 34% at 3650 m, and 53% at 4559 m."

Burtscher and colleagues [2] reviewed 21 studies (total 11,021 patients) who ascended to altitudes of 2200–4559 meters. They found that air travel (rapid ascent) was associated with a 4.5-fold steeper increase in acute mountain sickness (AMS), compared to more gradual ascent.

Most studies report that younger age is associated with a higher risk of AMS [2].

Some [2], though not all [3] studies report that women are more vulnerable to AMS than men.

Genetics

Some human populations (e.g., inhabitants of the Tibetan Plateau or the Andes) appear better adapted to altitude [4], and the study of such populations has revealed that specific genes confer more rapid acclimatization to hypobaric hypoxia [5].

Pathophysiological Mechanism of Disease

West comments that, "The deleterious effects of high altitude are primarily caused by the low inspired P_{O2}" [4]. The main mechanism for this is that at high altitudes, there is lower barometric pressure and lower oxygen concentration, leading to hypobaric hypoxia [5].

Several physiologic mechanisms can enable a person to acclimatize to hypobaric hypoxia, and different mechanisms act in different time frames when adapting to such conditions acutely (minutes, days) and chronically (months) [5]. These mechanisms include hyperventilation, polycythemia, hypoxic pulmonary vasoconstriction, changes in oxygen affinity of hemoglobin, increases in oxidative enzymes, and increased concentration of capillaries in peripheral muscle [4].

M. Cherchi, *Otoneurology and Vestibular Medicine*, https://doi.org/10.1007/978-3-031-94841-1_158

Clinical Presentation

Potential manifestations of "high-altitude illness" (HAI) include the following [5]:

- *Acute mountain sickness (AMS)*, the most common type, usually occurs at altitudes of greater than 2500 meters. Symptoms manifest within hours or days, and consist of some combination of headache, weakness, fatigue, listlessness, nausea, and anorexia [5].
- *High-altitude pulmonary edema (HAPE)* is characterized by symptoms such as shortness of breath, coughing, and frothy or blood-stained sputum [5].
- *High-altitude cerebral edema (HACE)* is characterized by confusion, ataxia, and a decreasing level of consciousness [5].

Imaging

In a patient with the most severe manifestation of high-altitude illness, brain imaging may demonstrate cerebral edema.

Treatment

The most effective management strategy is prevention. Hackett and colleagues [3] state that, "Prevention consists in slow ascent, rapid recognition of warning signs, and prompt descent to avoid progression."

A systematic review [6] concluded there is moderate quality evidence that *acetazolamide* (250–700 mg) is superior to placebo at "reducing the proportion of people who develop altitude sickness." The same review concluded that acetazolamide is more effective than ginkgo biloba, but acetazolamide is also associated with a higher risk of adverse effects (paresthesias).

Prognosis

A systematic review [6] found "no reliable data on prognosis. It is widely held that if no further ascent is attempted, then the symptoms of acute mountain sickness tend to resolve over a few days. We found no reliable data about long-term sequelae in people whose symptoms have completely resolved."

References

1. Maggiorini M, Buhler B, Walter M, Oelz O. Prevalence of acute mountain sickness in the Swiss Alps. BMJ. 1990;301(6756):853–5.
2. Burtscher J, Swenson ER, Hackett PH, Millet GP, Burtscher M. Flying to high-altitude destinations: is the risk of acute mountain sickness greater? J Travel Med. 2023;30(4):taad011.
3. Hackett PH, Rennie D, Levine HD. The incidence, importance, and prophylaxis of acute mountain sickness. Lancet. 1976;2(7996):1149–55.
4. West JB. High-altitude medicine. Am J Respir Crit Care Med. 2012;186(12):1229–37.
5. Mallet RT, Burtscher J, Pialoux V, Pasha Q, Ahmad Y, Millet GP, et al. Molecular mechanisms of high-altitude acclimatization. Int J Mol Sci. 2023;24(2):1698.
6. Murdoch D. Altitude sickness. BMJ Clin Evid. 2010;2010:1209.

Part XXV

Diagnoses—Toxic Causes of Disequilibrium

Alcohol

159

Overview

Ethanol (the form of *alcohol* consumed by humans) has a broad range of short-term and long-term physiologic effects across multiple organ systems. From the otoneurological perspective, acute alcohol intoxication can affect a broad range ocular motor functions, including the vestibulo-ocular reflex, semicircular canal function, the velocity-storage mechanism, optokinetic nystagmus, tracking, visual fixation suppression, saccades (saccadic velocity, prosaccades and antisaccades), smooth pursuit (including its gain and velocity), and postrotatory nystagmus; it can also cause positional alcohol nystagmus, alcohol gaze nystagmus, and spontaneous down beat nystagmus; finally, it can impair visual processing tasks such as motion parallax and depth perception. Acute alcohol intoxication also impairs cerebellar function. On computerized dynamic posturography, acute alcohol intoxication increases body sway. Longer-term effects of alcohol include damage to the cerebellum, cerebral white matter, and peripheral nerves. Chronic alcoholism can cause lesions in the mammillary bodies, resulting in Wernicke-Korsakoff syndrome.

Introduction

Ethanol is the form of *alcohol* consumed by humans. Some of the effects of alcohol, particularly on cognition, have been recognized for millennia. But alcohol has a rather broad range of physiologic effects across multiple organ systems [1]. Robert Barany was one of the first vestibular physicians to study the otologic and neurologic effects of acute alcohol intoxication [2, 3]. Here we will discuss some of the vestibular, ocular motor, and neurological consequences of alcohol toxicity.

Acute Alcohol Intoxication: Ocular Motor Findings

Many articles describe the acute effects of alcohol on the ocular motor system [4–8].

Alcohol can affect *semicircular canal* function [9]. It can also affect *otolith* function [10], though perhaps this effect is greater on ocular vestibular evoked myogenic potentials than for cervical vestibular evoked myogenic potentials [11].

Acute effects of alcohol can impair a variety of eye movements, including the function of the *velocity-storage mechanism* [12], *optokinetic nystagmus* [13–15], *tracking* [16], *visual fixation suppression* of vestibularly driven nystagmus [17–20], *saccadic velocity* [15, 21], both *prosaccades and antisaccades* [21], *smooth pursuit* [17, 19], *smooth pursuit gain* [21], the *smooth pursuit velocity* [15, 18], and *postrotatory nystagmus* [22].

Numerous studies have reported that acute alcohol intoxication can impair the *vestibulo-ocular reflex* (VOR) [23, 24]. One study [12] reported that mid-frequency VOR gain (as measured by slow harmonic acceleration in rotatory chair testing) was unaffected, whereas *high-frequency VOR gain* (as measured by video head impulse testing) was reduced, and *time constants* (as measured by step velocity testing in rotatory chair testing) were reduced.

A specific ocular motor abnormality related to acute alcohol intoxication is *positional alcohol nystagmus*, sometimes abbreviated PAN [25, 26]. (The reader should be aware that, somewhat confusingly, the phenomenon of periodic alternating nystagmus is also abbreviated PAN, though it is an entirely different pathology.) The underlying mechanism of positional alcohol nystagmus (PAN) is incompletely understood [9, 27–29], but may be a consequence of the differential effect of alcohol on the specific gravity of endolymphatic fluid versus perilymphatic fluid. The manifestation of PAN

M. Cherchi, *Otoneurology and Vestibular Medicine*, https://doi.org/10.1007/978-3-031-94841-1_159

goes through different phases depending on the latency from alcohol ingestion [30, 31]. PAN can be modulated by gravity [32] and linear acceleration [33].

Acute alcohol intoxication can also cause *spontaneous downbeat nystagmus* [34, 35] and "vertical velocity offset" [36].

One paper suggested that alcohol leads to a "disturbance of the visual-oculomotor system" as a mechanism underlying poor performance on *dynamic visual acuity testing* [37].

One paper explored the effect that acute alcohol intoxication has on slow eye movements and concluded that this is the mechanism by which alcohol *impairs motion parallax and thus depth perception* [38].

It has been observed that acute alcohol intoxication can affect the nystagmus evoked at the extremes of lateral gaze, and this was the basis for the so-called "*alcohol gaze nystagmus*" test [39] which has been used by law enforcement for some years as a proxy for blood alcohol concentration [40, 41]. The pattern of alcohol gaze evoked nystagmus can resemble that of cerebellar degenerative processes [42]. However, it appears that, for example, police officers certified in this technique do not always adhere to the guidelines and requirements for administering this test [43]. The validity of this test for law enforcement purposes has also come under scrutiny [44].

The acute effects of alcohol on level of alertness are variable; some individuals find alcohol to be sedating, while others experience it as stimulating. *The ocular motor effects of acute alcohol intoxication appear to be independent of the subjective state of sedation/stimulation* [45].

It should also be pointed out that a minority of literature has concluded that, "there was no significant correlation between duration of alcohol dependence and abnormal ENG [electronystagmography]" [46].

Acute Alcohol Intoxication: Posturographic Findings

Some papers refer rather vaguely to the effects that acute alcohol intoxication can have on an individual's balance [47]. Other papers have studied this more objectively, observing that acute alcohol intoxication tends to *increase body sway*, as measured by stabilometry (a form of *posturography*) [48] and computerized dynamic posturography [12, 49, 50].

Acute Alcohol Intoxication: Cerebellar Toxicity

Alcohol can affect numerous areas of the central nervous system. The cerebellum appears especially vulnerable to the effects of alcohol [51].

Alcohol also "*delays the conduction of electric signals* from the central nervous system to the muscles controlling posture and impairs the integration of sensory inputs required for maintaining vertical stance. Consequently, alcohol intoxication delays the ability to detect postural changes and enact the appropriate response" [52].

Longer-Term Effects of Alcohol

Alcohol can also have longer-term deleterious consequences for ocular motor function [53], presumably due to permanent damage from the longer-term toxic effects of alcohol. However, there is also some evidence that several of the longer-term ocular motor sequelae of chronic alcoholism can improve over several years of abstinence [54].

The cerebellum appears to be selectively vulnerable to the effects of chronic alcohol use. The mechanism of cerebellar dysfunction appears to involve impairment of synaptic plasticity of cerebellar Purkinje cells [55], and death of Purkinje cells and granule cells [56].

Chronic alcohol use can cause damage in other areas of the central nervous system relevant for balance beyond cerebellar atrophy, including loss of cerebral white matter and peripheral nerves [57].

There is modest evidence that alcohol can impair recovery from vestibular lesions [58].

Acute on Chronic Manifestations: Wernicke-Korsakoff Syndrome

The mamillary bodies are selectively vulnerable to thiamine deficiency, and chronic alcohol use often results in thiamine deficiency. Thiamine deficiency develops gradually, but beyond a certain threshold, its clinical effects may manifest abruptly in the form of Wernicke-Korsakoff syndrome. Relevant to equilibrium, Wernicke-Korsakoff syndrome can manifest with several eye movement abnormalities, including in the vestibulo-ocular reflex, gaze-evoked nystagmus, and ocular motor paresis [59, 60].

References

1. Birkova A, Hubkova B, Cizmarova B, Bolerazska B. Current view on the mechanisms of alcohol-mediated toxicity. Int J Mol Sci. 2021;22(18):9686.
2. Bárány R. Experimentelle alkoholintoxikation (Experimental alcohol intoxication). Monatsschr Ohrenheilkd. 1911;45:959–62.
3. Bárány R, Rothfeld J. Untersuchungen des Vestibularapparates bei akuter Alkoholintoxication und bei delirium tremens (Investigations of the vestibular apparatus in acute alcohol intoxication and delirium tremens). Dtsch Zsch f Nervenhk. 1913;50:133–57.
4. Howells DE. Nystagmus and other eye signs in acute alcoholism. Br Med J. 1952;2(4789):862–4.
5. Howells DE. Nystagmus as a physical sign in alcoholic intoxication. Br Med J. 1956;1(4980):1405–6.
6. Aschan G. Different types of alcohol nystagmus. Acta Otolaryngol Suppl. 1958;140:69–78.
7. Umeda Y, Sakata E. Alcohol and the oculomotor system. Ann Otol Rhinol Laryngol. 1978;87(3 Pt 1):392–8.
8. Jozefowicz-Korczynska M, Lukomski M, Kurnatowski P. Vestibular system in alcohol and drug addicts. Mater Med Pol. 1994;26(4):123–6.
9. Money KE, Johnson WH, Corlett BM. Role of semicircular canals in positional alcohol nystagmus. Am J Phys. 1965;208:1065–70.
10. Hafstrom A, Modig F, Karlberg M, Fransson PA. Increased visual dependence and otolith dysfunction with alcohol intoxication. Neuroreport. 2007;18(4):391–4.
11. Rosengren SM, Weber KP, Hegemann SC, Roth TN. The effect of alcohol on cervical and ocular vestibular evoked myogenic potentials in healthy volunteers. Clin Neurophysiol. 2014;125(8):1700–8.
12. Choi HG, Hong SK, Park SK, Lee HJ, Chang J. Acute alcohol intake impairs the velocity storage mechanism and affects both high-frequency vestibular-ocular reflex and postural control. Int J Environ Res Public Health. 2022;19(7):3911.
13. Mizoi Y, Hishida S, Maeba Y. Diagnosis of alcohol intoxication by the optokinetic test. Q J Stud Alcohol. 1969;30(1):1–14.
14. Schroeder DJ. Some effects of alcohol on nystagmus and vertigo during caloric and optokinetic stimulation. Ann Otol Rhinol Laryngol. 1972;81(2):218–29.
15. Baloh RW, Sharma S, Moskowitz H, Griffith R. Effect of alcohol and marijuana on eye movements. Aviat Space Environ Med. 1979;50(1):18–23.
16. Schroeder DJ, Gilson RD, Guedry FE, Collins WE. Effects of alcohol on nystagmus and tracking performance during laboratory angular accelerations about the Y and Z axes. Aerosp Med. 1973;44(5):477–87.
17. Guedry FE Jr, Gilson RD, Schroeder DJ, Collins WE. Some effects of alcohol on various aspects of oculomotor control. Aviat Space Environ Med. 1975;46(8):1008–13.
18. Barnes GR. The effects of ethyl alcohol on visual pursuit and suppression of the vestibulo-ocular reflex. Acta Otolaryngol Suppl. 1984;406:161–6.
19. Takahashi M, Akiyama I, Tsujita N, Yoshida A. The effect of alcohol on the vestibulo-ocular reflex and gaze regulation. Arch Otorhinolaryngol. 1989;246(4):195–9.
20. Harder T, Reker U. Influence of low dose alcohol on fixation suppression. Acta Otolaryngol Suppl. 1995;520(Pt 1):33–6.
21. Roche DJ, King AC. Alcohol impairment of saccadic and smooth pursuit eye movements: impact of risk factors for alcohol dependence. Psychopharmacology. 2010;212(1):33–44.
22. Karlovsek MZ, Balazic J. Evaluation of the post-rotational nystagmus test (PRN) in determining alcohol intoxication. J Anal Toxicol. 2005;29(5):390–3.
23. Tianwu H, Watanabe Y, Asai M, Shimizu K, Takada S, Mizukoshi K. Effects of alcohol ingestion on vestibular function in postural control. Acta Otolaryngol Suppl. 1995;519:127–31.
24. Martellucci S, Ralli M, Attanasio G, Russo FY, Marcelli V, Greco A, et al. Alcohol binge-drinking damage on the vestibulo-oculomotor reflex. Eur Arch Otorrinolaringol. 2021;278(1):41–8.
25. Harris CS, Guedry FE, Graybiel A. Positional alcohol nystagmus in relation to labyrinthine function. Res Rep U S Nav Sch Aviat Med. 1962;76:19.
26. Bergstedt M, Kus J. Positional alcohol nystagmus in vertigo cases. Confin Neurol. 1968;30(3):145–60.
27. Oosterveld WJ. On the origin of positional alcohol nystagmus. Acta Otolaryngol. 1973;75(2):252–8.
28. Aschan G, Bergstedt M. Positional alcoholic nystagmus (PAN) in man following repeated alcohol doses. Acta Otolaryngol Suppl. 1975;330:15–29.
29. Han K, Lee JY, Shin JE, Kim CH. Positional alcohol nystagmus and serum osmolality: new insights into dizziness associated with acute alcohol intoxication. Med Hypotheses. 2020;138:109606.
30. Hill RJ, Collins WE, Schroeder DJ. Influence of alcohol on positional nystagmus over 32-hour periods. Ann Otol Rhinol Laryngol. 1973;82(1):103–10.
31. Odkvist LM. The effect of gravity of positional alcohol nystagmus phase II in man. Acta Otolaryngol. 1975;80(3–4):214–9.
32. Oosterveld WJ. Effect of gravity on positional alcohol nystagmus (PAN). Aerosp Med. 1970;41(5):557–60.
33. Oosterveld WJ, Meineri G, Paolucci G. Quantitative effect of linear acceleration on positional alcohol nystagmus. Aerosp Med. 1974;45(7):695–700.
34. Costin JA, Smith JL, Emery S, Tomsak RL. Alcoholic downbeat nystagmus. Ann Ophthalmol. 1980;12(10):1127–31.
35. Zasorin NL, Baloh RW. Downbeat nystagmus with alcoholic cerebellar degeneration. Arch Neurol. 1984;41(12):1301–2.
36. Fetter M, Haslwanter T, Bork M, Dichgans J. New insights into positional alcohol nystagmus using three-dimensional eye-movement analysis. Ann Neurol. 1999;45(2):216–23.
37. Schmal F, Thiede O, Stoll W. Effect of ethanol on visual-vestibular interactions during vertical linear body acceleration. Alcohol Clin Exp Res. 2003;27(9):1520–6.
38. Nawrot M, Nordenstrom B, Olson A. Disruption of eye movements by ethanol intoxication affects perception of depth from motion parallax. Psychol Sci. 2004;15(12):858–65.
39. Goding GS, Dobie RA. Gaze nystagmus and blood alcohol. Laryngoscope. 1986;96(7):713–7.
40. Good GW, Augsburger AR. Use of horizontal gaze nystagmus as a part of roadside sobriety testing. Am J Optom Physiol Optic. 1986;63(6):467–71.
41. Halperin E, Yolton RL. Is the driver drunk? Oculomotor sobriety testing. J Am Optom Assoc. 1986;57(9):654–7.
42. Romano F, Tarnutzer AA, Straumann D, Ramat S, Bertolini G. Gaze-evoked nystagmus induced by alcohol intoxication. J Physiol. 2017;595(6):2161–73.
43. Booker JL. End-position nystagmus as an indicator of ethanol intoxication. Sci Justice. 2001;41(2):113–6.
44. Booker JL. The horizontal gaze nystagmus test: fraudulent science in the American courts. Sci Justice. 2004;44(3):133–9.
45. Holdstock L, de Wit H. Ethanol impairs saccadic and smooth pursuit eye movements without producing self-reports of sedation. Alcohol Clin Exp Res. 1999;23(4):664–72.
46. Verma RK, Panda NK, Basu D, Raghunathan M. Audiovestibular dysfunction in alcohol dependence. Are we worried? Am J Otolaryngol. 2006;27(4):225–8.
47. Belle M, Sartori Sdo A, Rossi AG. Alcoholism: effects on the cochleo-vestibular apparatus. Braz J Otorhinolaryngol. 2007;73(1):110–6.

48. Kubo T, Sakata Y, Koshimune A, Sakai S, Ameno K, Ijiri I. Positional nystagmus and body sway after alcohol ingestion. Am J Otolaryngol. 1990;11(6):416–9.
49. Ledin T, Odkvist LM. Effect of alcohol measured by dynamic posturography. Acta Otolaryngol Suppl. 1991;481:576–81.
50. Nieschalk M, Ortmann C, West A, Schmal F, Stoll W, Fechner G. Effects of alcohol on body-sway patterns in human subjects. Int J Legal Med. 1999;112(4):253–60.
51. Alekseeva N, McGee J, Kelley RE, Maghzi AH, Gonzalez-Toledo E, Minagar A. Toxic-metabolic, nutritional, and medicinal-induced disorders of cerebellum. Neurol Clin. 2014;32(4):901–11.
52. Vassar RL, Rose J. Motor systems and postural instability. Handb Clin Neurol. 2014;125:237–51.
53. Nsamba C, al-Marashi MS. Nystagmus in alcoholics. Q J Stud Alcohol. 1972;33(3):706–13.
54. Sasa M, Takaori S, Matsuoka I, Miyazaki T, Miyazaki K. Peripheral and central vestibular disorders in alcoholics. A three-year follow-up study. Arch Otorhinolaryngol. 1981;230(1):93–101.
55. Valenzuela CF, Lindquist B, Zamudio-Bulcock PA. A review of synaptic plasticity at Purkinje neurons with a focus on ethanol-induced cerebellar dysfunction. Int Rev Neurobiol. 2010;91:339–72.
56. Luo J. Mechanisms of ethanol-induced death of cerebellar granule cells. Cerebellum. 2012;11(1):145–54.
57. de la Monte SM, Kril JJ. Human alcohol-related neuropathology. Acta Neuropathol. 2014;127(1):71–90.
58. Zingler VC, Strupp M, Krafczyk S, Karch C, Brandt T. Does alcohol cancel static vestibular compensation? Ann Neurol. 2004;55(1):144–5.
59. Kattah JC, Dhanani SS, Pula JH, Mantokoudis G, Tehrani ASS, Toker DEN. Vestibular signs of thiamine deficiency during the early phase of suspected Wernicke encephalopathy. Neurol Clin Pract. 2013;3(6):460–8.
60. Kattah JC. The Spectrum of vestibular and ocular motor abnormalities in thiamine deficiency. Curr Neurol Neurosci Rep. 2017;17(5):40.

Heavy Metals

160

Overview

Toxicity to humans from *heavy metals* is largely the consequence of industrial activity that results in accidental environmental contamination [1] and subsequent human exposure through air, water, and food [2]. Less commonly, heavy metal exposure results from intentional poisoning (such as arsenic or thallium). There is no uniformly accepted definition of heavy metals. The most medically relevant elements are usually cited as arsenic, cadmium, lead, mercury, and thallium.

Arsenic

Clinicians occasionally encounter cases of intentional arsenic poisoning. The mechanisms of injury from arsenic toxicity are thought to include damage to capillary endothelium, thiol binding, uncoupling of oxidative phosphorylation (interfering with ATP synthesis), and disruption of neurotransmitter homeostasis [2].

Reported otoneurological symptoms of arsenic toxicity can include bilateral sensorineural hearing loss, tinnitus, and disequilibrium [3]. Reported ocular motor findings in arsenic toxicity include upbeat nystagmus on upward gaze [4, 5].

Other clues to the diagnosis of arsenic toxicity include pancytopenia, dermatological abnormalities, hepatic and renal failure [2].

Cadmium

On 12/15/22, Consumer Reports published an article (https://www.consumerreports.org/health/food-safety/lead-and-cadmium-in-dark-chocolate-a8480295550/, last accessed on 3/23/25) documenting the presence of cadmium in several brands of chocolate. In response to that article, patients began asking us about the relationship between cadmium and disequilibrium.

The mechanisms of injury in cadmium toxicity are thought to include DNA damage (and thus genomic instability), generation of radical oxygen species, and thence oxidative stress [2].

As of this writing, there are very few studies on the question of otoneurological consequences of cadmium toxicity. Most of the data are from animal studies. The few human studies are inconclusive.

Other clues to the diagnosis of cadmium toxicity include bone degeneration, pulmonary injury, and hepatorenal dysfunction [2].

Animal Studies on Cadmium

The data on the relationship between cadmium and vestibular function comes mostly from animal studies. Most of these are in vitro studies that describe the effects of cadmium on calcium and potassium channel currents in vestibular neurons. A few of these are in vivo studies that report on vestibular function in live animals.

One study reported that in vestibular neurons in mice, in various calcium channel subtypes (L-, N-, P-, Q-), barium currents were mostly blocked by nifedipine, and the residual current was blocked by cadmium [6].

One study of type II vestibular hair cells in guinea pigs identified a cadmium-sensitive inward-flowing calcium current [7].

One study in guinea pigs reported that in type II vestibular hair cells from the saccule, acetylcholine-gated large conductance potassium channels were inhibited by cadmium [8]. A study of vestibular nerve afferents in rats identified a similar blocking effect of cadmium on voltage-sensitive potassium currents [9].

One study of lead and cadmium exposure in mice reported that lead caused a decline in vestibular function, whereas cadmium did not [10].

M. Cherchi, *Otoneurology and Vestibular Medicine*, https://doi.org/10.1007/978-3-031-94841-1_160

Human Studies on Cadmium

There are very few studies in humans, and their results are difficult to interpret as they do not separate the effects of cadmium and lead.

One study of balance in humans reported an inverse relationship between (1) performance on the Romberg test of standing balance on firm and compliant support surfaces, and (2) blood concentrations of lead and cadmium [11]. Note that this study did not separate the levels of cadmium and lead, so it could not ascertain whether any apparent effects were due to one element or the other.

A human study of balance in children reported no relationship to prenatal exposure to cadmium and lead [12].

Lead

Lead toxicity in the pediatric population usually comes to medical attention due to cognitive/behavioral abnormalities and developmental delay.

In adults, lead toxicity more commonly results from acute or acute-on-chronic chemical exposure, including sniffing volatile substances that include compounds such as tetraethyl lead [13].

The mechanisms of injury from lead toxicity are thought to include an increase in inflammatory cytokines (IL-1β, TNF-α, IL-6), increased serum endothelin-1, nitric oxide and eosinophil peroxidase, inactivation of δ-aminolevulinic acid dehydratase, and ferrochelatase (interfering with heme biosynthesis), and reduced levels of glutathione, superoxide dismutase, catalase, and glutathione peroxidase [2].

From the otoneurological perspective, there are reports of lead exposure manifesting with "nystagmus," but the available literature from human cases is unfortunately not more descriptive than this [14]. Animal studies report impaired postrotatory nystagmus [15].

Other clues to the diagnosis of lead toxicity include cognitive problems, cardiopulmonary dysfunction, hepatic dysfunction, and anemia [2].

Mercury

Much of the early literature on mercury toxicity arose from studying the consequences of the massive industrial accident in Minamata Bay, Japan, where a chemical factory released methylmercury wastewater into the ocean from the 1930s to the 1960s, and this organic mercury contaminated fish that entered the food chain consumed by humans.

The mechanisms of injury from mercury toxicity are thought to include thiol binding, enzyme inhibition, production of radical oxygen species, reduction of aquaporin mRNA, inhibition of glutathione peroxidase, and increased expression of c-Fos [2].

From the otoneurological perspective, mercury poisoning can manifest with spontaneous and positional nystagmus, and impaired horizontal and vertical optokinetic nystagmus [16–18].

Additional clues to the diagnosis of mercury toxicity include other neurological damage, hepatorenal dysfunction, gastrointestinal ulceration, and teratogenicity [2].

Thallium

Thallium poisoning sometimes occurs through the consumption of food grown in contaminated soil. Clinicians also occasionally encounter cases of intentional thallium poisoning.

The mechanisms of injury in thallium toxicity are thought to include disruption of sulfhydryl groups on mitochondrial membranes, thereby interfering with sodium-potassium ATP-ase [19].

There is very little literature on otoneurological effects of thallium. Existing reports only comment on the symptom of "vertigo" [20] and the presence of "nystagmus" [19] without further characterization.

Other clues to the diagnosis of thallium toxicity may include diffuse weakness, painful acral paresthesias, optic neuritis and other cranial nerve dysfunction, choreoathetosis, and alopecia. Initially, cases of thallium toxicity are sometimes misdiagnosed as Guillain-Barre syndrome [19, 21].

Final Comments

The otoneurological findings in heavy metal toxicity are nonspecific (with the possible exception of impaired vertical optokinetic nystagmus in mercury toxicity), so these diagnoses are more likely to be suspected from the history and laboratory tests (such as serum markers). If heavy metal toxicity is suspected, a medical toxicology consultation is appropriate.

References

1. Tchounwou PB, Yedjou CG, Patlolla AK, Sutton DJ. Heavy metal toxicity and the environment. Exp Suppl. 2012;101:133–64.
2. Balali-Mood M, Naseri K, Tahergorabi Z, Khazdair MR, Sadeghi M. Toxic mechanisms of five heavy metals: mercury, lead, chromium, cadmium, and arsenic. Front Pharmacol. 2021;12:643972.
3. Suo A, Wang Y, Niu H, Yang X. Arsenic poisoning-induced sensorineural hearing loss: a case report. J Int Adv Otol. 2024;20(2):182–5.
4. Nakamagoe K, Ishii K, Tamaoka A, Shoji S. Upward gaze-evoked nystagmus with organoarsenic poisoning. Neurology. 2006;66(1):131–2.

5. Nakamagoe K, Fujizuka N, Koganezawa T, Shimizu K, Takiguchi S, Horaguchi T, et al. Residual central nervous system damage due to organoarsenic poisoning. Neurotoxicol Teratol. 2013;37:33–8.
6. Desmadryl G, Chambard JM, Valmier J, Sans A. Multiple voltage-dependent calcium currents in acutely isolated mouse vestibular neurons. Neuroscience. 1997;78(2):511–22.
7. Rennie KJ, Ashmore JF. Ionic currents in isolated vestibular hair cells from the Guinea-pig crista ampullaris. Hear Res. 1991;51(2):279–91.
8. Kong WJ, Guo CK, Zhang XW, Chen X, Zhang S, Li GQ, et al. The coupling of acetylcholine-induced BK channel and calcium channel in Guinea pig saccular type II vestibular hair cells. Brain Res. 2007;1129(1):110–5.
9. Ramakrishna Y, Sadeghi SG. Activation of GABA(B) receptors results in excitatory modulation of calyx terminals in rat semicircular canal cristae. J Neurophysiol. 2020;124(3):962–72.
10. Klimpel KE, Lee MY, King WM, Raphael Y, Schacht J, Neitzel RL. Vestibular dysfunction in the adult CBA/CaJ mouse after lead and cadmium treatment. Environ Toxicol. 2017;32(3):869–76.
11. Min KB, Lee KJ, Park JB, Min JY. Lead and cadmium levels and balance and vestibular dysfunction among adult participants in the National Health and Nutrition Examination Survey (NHANES) 1999-2004. Environ Health Perspect. 2012;120(3):413–7.
12. Taylor CM, Humphriss R, Hall A, Golding J, Emond AM. Balance ability in 7- and 10-year-old children: associations with prenatal lead and cadmium exposure and with blood lead levels in childhood in a prospective birth cohort study. BMJ Open. 2015;5(12):e009635.
13. Cairney S, Maruff P, Burns C, Currie B. The neurobehavioural consequences of petrol (gasoline) sniffing. Neurosci Biobehav Rev. 2002;26(1):81–9.
14. Cairney S, Maruff P, Burns CB, Currie J, Currie BJ. Neurological and cognitive impairment associated with leaded gasoline encephalopathy. Drug Alcohol Depend. 2004;73(2):183–8.
15. Mameli O, Caria MA, Melis F, Solinas A, Tavera C, Ibba A, et al. Neurotoxic effect of lead at low concentrations. Brain Res Bull. 2001;55(2):269–75.
16. Mizukoshi K, Nagaba M, Ohno Y, Ishikawa K, Aoyagi M, Watanabe Y, et al. Neurotological studies upon intoxication by organic mercury compounds. ORL J Otorhinolaryngol Relat Spec. 1975;37(2):74–87.
17. Mizukoshi K, Watanabe Y, Kobayashi H, Nakano Y, Koide C, Inomata S, et al. Neurotological follow-up studies upon Minamata disease. Acta Otolaryngol Suppl. 1989;468:353–7.
18. Mizukoshi K, Watanabe Y, Shojaku H, Aso S, Asai M, Inukai K, et al. Neurotological long-term follow-up in Minamata disease in Niigata, Japan. Nihon Jibiinkoka Gakkai Kaiho. 2002;105(3):225–31.
19. Misra UK, Kalita J, Yadav RK, Ranjan P. Thallium poisoning: emphasis on early diagnosis and response to haemodialysis. Postgrad Med J. 2003;79(928):103–5.
20. Jimenez O, Caceres H, Gimenez L, Soto L, Montenegro M, Rueda JAA. Thallium poisoning: a case report. J Yeungnam Med Sci. 2023;40(3):311–4.
21. Cvjetko P, Cvjetko I, Pavlica M. Thallium toxicity in humans. Arh Hig Rada Toksikol. 2010;61(1):111–9.

Caffeine

161

Overview

Caffeine, the most widely used psychoactive substance worldwide, functions as a stimulant. After crossing the blood-brain barrier, caffeine's main effect is that it binds to adenosine receptors and blocks the activity of endogenous adenosine; it also influences the activity of several other neurotransmitter systems (acetylcholine, noradrenaline, dopamine). From the otoneurological perspective, caffeine has measurable effects on some ocular motor parameters (such as reducing saccadic latency, increasing saccadic velocity, increasing the peak velocity and amplitude of optokinetic nystagmus, and improving performance on certain tracking tasks). It is less clear whether caffeine has consistent effects on the auditory system. Some Ménière's patients report that caffeine can induce a Ménière's attack. Some migraine patients find that caffeine may abort a migraine.

Introduction

Chemically, *caffeine* is 1,3,7-trimethylxanthine. This molecule can cross the blood-brain barrier and is sufficiently similar to adenosine that it can bind to adenosine receptors and block the effect of endogenous adenosine. Adenosine inhibits arousal and thus has a soporific effect, so blocking adenosine receptors (by caffeine) has the opposite effect—namely, increasing alertness and reducing fatigue [1]. Caffeine also affects several neurotransmitter systems, including acetylcholine, noradrenaline, and dopamine [2].

Caffeine and the Vestibular System

In 2022, Ghahraman and colleagues published a review of the effects of caffeine on the auditory and vestibular systems [3]. While it seems logical to hypothesize that the overall alerting effect of caffeine is responsible for effects on the vestibular system [4], as of this writing, the mechanism by which caffeine influences the vestibular system is incompletely understood [3].

Briefly, as far as the vestibular system is concerned, caffeine has measurable effects on some parameters of ocular motor function in normal subjects. Caffeine decreases saccadic latency, increases peak saccadic velocity, and influences the peak velocity and amplitude of the fast phases of optokinetic nystagmus [5]. Caffeine has also been reported to improve adaptive tracking performance [4].

Caffeine does not appear to have significant measurable effects on cervical vestibular evoked myogenic potentials [6–11], ocular vestibular evoked myogenic potentials [6], caloric testing [6, 12], or posturography [13, 14] in normal subjects.

Very limited data suggest that caffeine may potentiate the beneficial effects of medications for motion sickness (promethazine, meclizine, scopolamine) [15].

The studies cited above pertain to the effects of caffeine on vestibular tests *in normal subjects.* Whether caffeine influences otovestibular test results in patients with specific diseases has not been studied. Consequently, it is unknown whether caffeine can influence the accuracy of otovestibular tests in diagnosing disease.

Caffeine and the Auditory System

Physiological studies report conflicting results on the effect of caffeine on the cochlea. For instance, some literature reports that caffeine can shorten outer cochlear hair cells [16, 17], thereby enhancing cochlear responses, while other studies report that caffeine suppresses cochlear responses [18].

Studies of brainstem auditory evoked responses sometimes report a reduction in response latencies [19], which is usually attributed to caffeine's ability to improve neural transmission in central auditory pathways.

M. Cherchi, *Otoneurology and Vestibular Medicine*, https://doi.org/10.1007/978-3-031-94841-1_161

Studies of higher-level auditory function are difficult to interpret because it is not easy to disentangle the effect of caffeine on the auditory system itself versus its effect on the cognitive processes needed to drive the measured behavioral responses.

Caffeine and Specific Otoneurological Diseases

The main clinical relevance of caffeine in otoneurology is for Ménière's disease and migraine-associated vertigo.

Anecdotally, some patients carrying a diagnosis of Ménière's disease report that caffeine intake increases the risk of having a Ménière's attack. Consequently, some literature recommends that management of Ménière's disease should include abstinence from caffeine [20, 21], though this recommendation is by no means universally accepted [22–24].

For migraine in general, reduction or elimination of caffeine is often recommended, though caffeine is somewhat of a "double-edged sword" for migraine [25, 26], in that caffeine withdrawal can trigger migraine [27], yet a spike in caffeine intake can sometimes abort migraine. For migraine-associated vertigo, as with other migraine variants, the recommendation of caffeine restriction is extrapolated from the broader literature on migraine [28], but it has not been studied systematically.

References

1. van Dam RM, Hu FB, Willett WC. Coffee, caffeine, and health. N Engl J Med. 2020;383(4):369–78.
2. Diukova A, Ware J, Smith JE, Evans CJ, Murphy K, Rogers PJ, Wise RG. Separating neural and vascular effects of caffeine using simultaneous EEG-FMRI: differential effects of caffeine on cognitive and sensorimotor brain responses. NeuroImage. 2012;62(1):239–49.
3. Ghahraman MA, Farahani S, Tavanai E. A comprehensive review of the effects of caffeine on the auditory and vestibular systems. Nutr Neurosci. 2022;25(10):2181–94.
4. Wilhelmus MM, Hay JL, Zuiker RG, Okkerse P, Perdrieu C, Sauser J, et al. Effects of a single, oral 60 mg caffeine dose on attention in healthy adult subjects. J Psychopharmacol. 2017;31(2):222–32.
5. Connell CJW, Thompson B, Turuwhenua J, Hess RF, Gant N. Caffeine increases the velocity of rapid eye movements in unfatigued humans. Psychopharmacology. 2017;234(15):2311–23.
6. Ledesma ALL, Souza MAd, Barreto C, Oliveira CA, Fayez Bahmad, editors. Effect of caffeine intake in vestibular function: a randomized., Triple-Blind, Controlled Trial 2016.
7. McNerney K, Coad ML, Burkard R. The influence of caffeine on calorics and cervical vestibular evoked myogenic potentials (cVEMPs). J Am Acad Audiol. 2014;25(3):261–7.
8. Sousa AM, Suzuki FA. Effect of caffeine on cervical vestibular-evoked myogenic potential in healthy individuals. Braz J Otorhinolaryngol. 2014;80(3):226–30.
9. Souza M, Costa K, Menezes PL. Effect of caffeine on vestibular evoked myogenic potential: a systematic review with meta-analysis. Braz J Otorhinolaryngol. 2018;84(3):381–8.
10. Tavanai E, Farahani S, Adel-Ghahraman M, Jalaie S, Kouti L, Shidfar F. Effects of caffeine on cervical vestibular evoked myogenic potential: a pilot study. Audit Vestib Res. 2015;24(1)
11. Tavanai E, Farahani S, Ghahraman MA, Soleimanian S, Jalaie S. Effects of caffeine on auditory- and vestibular-evoked potentials in healthy individuals: a double-blind placebo-controlled study. J Audiol Otol. 2020;24(1):10–6.
12. Felipe L, Simoes LC, Goncalves DU, Mancini PC. Evaluation of the caffeine effect in the vestibular test. Braz J Otorhinolaryngol. 2005;71(6):758–62.
13. Enriquez A, Sklaar J, Viirre E, Chase B. Effects of caffeine on postural stability. Int Tinnitus J. 2009;15(2):161–3.
14. McNerney KM, Coad ML, Burkard RF. The influence of caffeine on the sensory organization test. J Am Acad Audiol. 2014;25(6):521–8.
15. Estrada A, LeDuc PA, Curry IP, Phelps SE, Fuller DR. Airsickness prevention in helicopter passengers. Aviat Space Environ Med. 2007;78(4):408–13.
16. Skellett RA, Crist JR, Fallon M, Bobbin RP. Caffeine-induced shortening of isolated outer hair cells: an osmotic mechanism of action. Hear Res. 1995;87(1–2):41–8.
17. Slepecky N, Ulfendahl M, Flock A. Effects of caffeine and tetracaine on outer hair cell shortening suggest intracellular calcium involvement. Hear Res. 1988;32(1):11–21.
18. Bobbin RP. Caffeine and ryanodine demonstrate a role for the ryanodine receptor in the organ of Corti. Hear Res. 2002;174(1-2):172–82.
19. Dixit A, Vaney N, Tandon OP. Effect of caffeine on central auditory pathways: an evoked potential study. Hear Res. 2006;220(1–2):61–6.
20. Sanchez-Sellero I, San-Roman-Rodriguez E, Santos-Perez S, Rossi-Izquierdo M, Soto-Varela A. Caffeine intake and Meniere's disease: is there relationship? Nutr Neurosci. 2018;21(9):624–31.
21. Sharon JD, Trevino C, Schubert MC, Carey JP. Treatment of Meniere's disease. Curr Treat Options Neurol. 2015;17(4):341.
22. Hussain K, Murdin L, Schilder AG. Restriction of salt, caffeine and alcohol intake for the treatment of Meniere's disease or syndrome. Cochrane Database Syst Rev. 2018;12(12):CD012173.
23. Trinidade A, Robinson T, Phillips JS. The role of caffeine in otorhinolaryngology: guilty as charged? Eur Arch Otorrinolaringol. 2014;271(8):2097–102.
24. De Luca P, Cassandro C, Ralli M, Gioacchini FM, Turchetta R, Orlando MP, et al. Dietary restriction for the treatment of Meniere's disease. Transl Med UniSa. 2020;22:5–9.
25. Nowaczewska M, Wicinski M, Kazmierczak W. The ambiguous role of caffeine in migraine headache: from trigger to treatment. Nutrients. 2020;12(8):2259.
26. Alstadhaug KB, Andreou AP. Caffeine and primary (migraine) headaches-friend or foe? Front Neurol. 2019;10:1275.
27. Alstadhaug KB, Ofte HK, Muller KI, Andreou AP. Sudden caffeine withdrawal triggers migraine-a randomized controlled trial. Front Neurol. 2020;11:1002.
28. Mikulec AA, Faraji F, Kinsella LJ. Evaluation of the efficacy of caffeine cessation, nortriptyline, and topiramate therapy in vestibular migraine and complex dizziness of unknown etiology. Am J Otolaryngol. 2012;33(1):121–7.

Carbon Monoxide

162

Overview

Carbon monoxide (CO) is a colorless, odorless gas that binds to hemoglobin more avidly than oxygen, which impairs tissue oxygenation. CO intoxication is the most common cause of intentional and unintentional poisoning. CO intoxication, whether acute or chronic, can manifest with a variety of symptoms, including otovestibular ones such as hearing loss and/or disequilibrium. The hearing loss is sensorineural and can be unilateral, bilateral, or asymmetrical; this hearing loss is reversible in some cases. In the acute setting (typically an emergency department), treatment is with high-flow 100% oxygen.

Introduction

Carbon monoxide intoxication can manifest with a variety of symptoms. Neurologists are generally familiar with the selective vulnerability of the globus pallidus to carbon monoxide intoxication [1], but it can also manifest with otovestibular symptoms such as hearing loss and/or disequilibrium.

Epidemiology

Carbon monoxide intoxication is the leading cause of intentional and unintentional poisoning in the United States [2], accounting for 50,000 cases annually [3]. The fact that carbon monoxide is both colorless and odorless increases the risk of accidental intoxication.

Pathophysiological Mechanism of Disease

Carbon monoxide binds to hemoglobin more avidly than does oxygen, thereby diminishing the blood's capacity to deliver oxygen to tissues and resulting in ischemia. Other potentially contributory mechanisms include reoxygenation injury and direct toxic effects of carbon monoxide [4].

Clinical Presentation

Carbon monoxide intoxication can manifest with a variety of symptoms. Otoneurologically, it can manifest with sensorineural hearing loss and disequilibrium. "Dizziness" is among the most common symptoms of carbon monoxide intoxication, reported in 12.4% of cases [3]. Hearing loss is a less common manifestation of carbon monoxide intoxication, but it is important to recognize because the hearing loss is reversible in some cases [5].

Symptoms of carbon monoxide intoxication may become manifest after either acute or chronic exposure. The resulting sensorineural hearing loss can be unilateral or bilateral [6], and when bilateral, it can be asymmetrical [7]. The hearing loss may fluctuate [8], resembling Ménière's disease. The hearing loss may be reversible or irreversible [6]. Carbon monoxide intoxication can also present with isolated disequilibrium [9], or the auditory and vestibular symptoms may occur in combination [8].

Ocular Motor Examination

An experimental study by Tibbling [10] reported no effect of carbon monoxide on "vestibular nystagmus." Fielding and colleagues [11] reported that a victim of carbon monoxide intoxication with bilateral globus pallidus lesions exhibited a variety of saccadic abnormalities, including reduced gain in visually guided saccades, increased errors on antisaccades, higher anticipatory errors in memory-guided saccades and endogenously cued saccades, and shorter intersaccadic intervals in self-paced saccades. There are various reports of cerebellar lesions in carbon monoxide intoxication [4, 12–14], though no clear description of corresponding ocular motor correlates.

M. Cherchi, *Otoneurology and Vestibular Medicine*, https://doi.org/10.1007/978-3-031-94841-1_162

Testing: Auditory

The hearing loss associated with carbon monoxide intoxication is sensorineural. In some cases, the hearing loss is asymmetrical.

Figure 162.1, from Pillon [7], shows right greater than left-sided high-frequency sensorineural hearing loss.

The hearing loss in carbon monoxide intoxication can predominantly affect the low frequencies.

Seale and colleagues [8] document a case of carbon monoxide intoxication in which initially there is unilateral, predominantly low-frequency sensorineural hearing loss, which improved on a subsequent audiogram. Such fluctuation in low-frequency sensorineural hearing can resemble Ménière's disease.

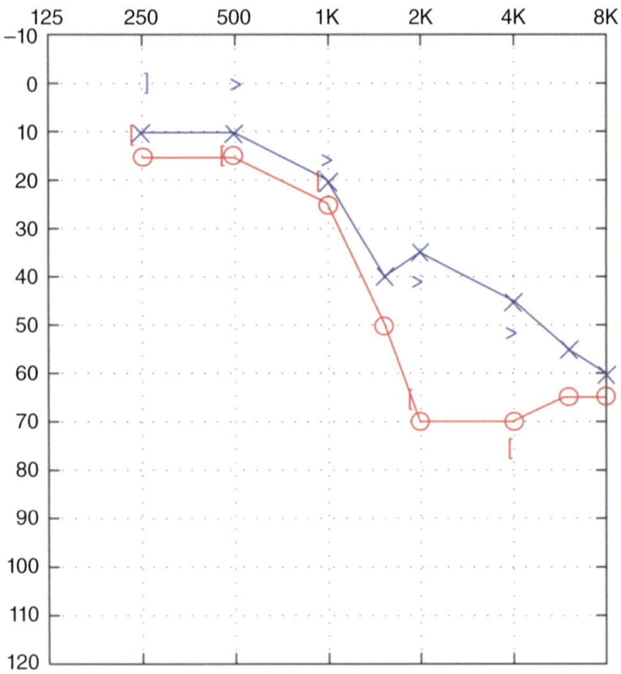

Fig. 162.1 Right greater than left-sided high-frequency sensorineural hearing loss in carbon monoxide intoxication. From Pillon [7]. Reproduced from Hindawi with permission under CC-BY 4.0

Testing: Other

Carbon monoxide binds to hemoglobin and forms carboxyhemoglobin, measurement of which can support a diagnosis of carbon monoxide intoxication. However, carboxyhemoglobin is not sensitive, as it can appear normal, especially in cases of chronic carbon monoxide exposure. Note that pulse oximetry does not distinguish oxyhemoglobin from carboxyhemoglobin, and thus cannot detect carbon monoxide intoxication.

Imaging

Carbon monoxide intoxication is often described as causing selective damage to the globus pallidus [15], but any distribution of supratentorial and infratentorial lesions is possible [16, 17].

Histopathology

Histopathological findings in the brain from carbon monoxide intoxication include necrosis in the globus pallidus and demyelination [4, 14, 18]. Specific findings in the labyrinth and vestibulocochlear nerve have not been described.

Differential Diagnosis

Since carbon monoxide intoxication can present with fluctuating low-frequency sensorineural hearing loss [8], Ménière's disease is on the differential diagnosis.

Treatment

Acute carbon monoxide intoxication is usually diagnosed in an emergency department, and in this setting, patients are treated with high-flow 100% oxygen. Toxicology consultation is reasonable. In some circumstances, hyperbaric oxygen therapy may also be appropriate, though this treatment modality is not widely available.

Prognosis

Carbon monoxide intoxication is a potentially reversible cause of sensorineural hearing loss [5], though there are no clear predictors of this.

References

1. Sharma S, Gupta R, Paul BS, Puri S, Garg S. Accidental carbon monoxide poisoning in our homes. Indian J Crit Care Med. 2009;13(3):169–70.
2. Thom SR, Keim LW. Carbon monoxide poisoning: a review epidemiology, pathophysiology, clinical findings, and treatment options including hyperbaric oxygen therapy. J Toxicol Clin Toxicol. 1989;27(3):141–56.
3. Shin M, Bronstein AC, Glidden E, Malone M, Chang A, Law R, et al. Morbidity and Mortality of Unintentional Carbon Monoxide Poisoning: United States 2005 to 2018. Ann Emerg Med. 2023;81(3):309–17.
4. Zhu L, Zhou C, Wu J, Fang S. Rare cerebellar lesions in a carbon monoxide poisoning case. Neurol Sci. 2014;35(1):117–9.
5. Shahbaz Hassan M, Ray J, Wilson F. Carbon monoxide poisoning and sensorineural hearing loss. J Laryngol Otol. 2003;117(2):134–7.
6. Mehrparvar AH, Davari MH, Mollasadeghi A, Vahidi MR, Mostaghaci M, Bahaloo M, et al. hearing loss due to carbon monoxide poisoning. Case Rep Otolaryngol. 2013;2013:940187.
7. Pillion JP. Sensorineural hearing loss following carbon monoxide poisoning. Case Rep Pediatr. 2012;2012:231230.
8. Seale B, Ahanger S, Hari C. Subacute carbon monoxide poisoning presenting as vertigo and fluctuating low frequency hearing loss. J Surg Case Rep. 2018;2018(8):rjy205.
9. Lakhani R, Bleach N. Carbon monoxide poisoning: an unusual cause of dizziness. J Laryngol Otol. 2010;124(10):1103–5.
10. Tibbling L. The influence of tobacco smoking, nicotine, CO and CO_2 on vestibular nystagmus. Acta Otolaryngol. 1969;68(1):118–26.
11. Fielding J, Lang W, White OB. Carbon monoxide poisoning: impact on ocular motility. Cogn Behav Neurol. 2010;23(4):256–61.
12. Velioglu M, Gumus T, Husmen G. Cerebellar lesions in the acute setting of carbon monoxide poisoning. Emerg Radiol. 2013;20(3):255–7.
13. Fan HC, Wang AC, Lo CP, Chang KP, Chen SJ. Damage of cerebellar white matter due to carbon monoxide poisoning: a case report. Am J Emerg Med. 2009;27(6):757:e5–7.
14. Mascalchi M, Petruzzi P, Zampa V. MRI of cerebellar white matter damage due to carbon monoxide poisoning: case report. Neuroradiology. 1996;38(Suppl 1):S73–4.
15. Jiang W, Wu Q, Zhou C, Zhao Z, Tan Y. Gray matter nuclei damage in acute carbon monoxide intoxication assessed in vivo using diffusion tensor MR imaging. Radiol Med. 2020;125(1):80–6.
16. Jeon SB, Sohn CH, Seo DW, Oh BJ, Lim KS, Kang DW, et al. Acute brain lesions on magnetic resonance imaging and delayed neurological sequelae in carbon monoxide poisoning. JAMA Neurol. 2018;75(4):436–43.
17. Kim DM, Lee IH, Park JY, Hwang SB, Yoo DS, Song CJ. Acute carbon monoxide poisoning: MR imaging findings with clinical correlation. Diagn Interv Imaging. 2017;98(4):299–306.
18. Lapresle J, Fardeau M. The central nervous system and carbon monoxide poisoning. II. Anatomical study of brain lesions following intoxication with carbon monixide (22 cases). Prog Brain Res. 1967;24:31–74.

Cocaine

163

Overview

Cocaine is extracted primarily from *Erythroxylum coca* and can be administered by inhalation, application to mucosal membranes, or injection (intravenously and intramuscularly). Pharmacologically, cocaine influences numerous neurotransmitter systems, including presynaptic monoamine reuptake, NMDA, and opioid receptors. Otovestibular effects are not well reported in the literature, but symptoms following cocaine use include hearing loss, tinnitus, and disequilibrium. Ocular motor examination may reveal opsoclonus. Instrumented vestibular studies may show an altered vestibulo-ocular reflex. MRI of the internal auditory canals may reveal intralabyrinthine hemorrhage.

Introduction

Cocaine is a sympathomimetic tropane alkaloid that occurs naturally in several hundred *Erythroxylum* species of shrub; the cocaine consumed by humans is largely extracted from *Erythroxylum coca* [1]. Cocaine can be ingested by chewing on the leaves of the plant. In industrialized countries, cocaine is usually delivered in a more concentrated form through various routes (inhaled, applied to mucosal membranes, injected intravenously or intramuscularly).

Epidemiology

It has been estimated that of adults aged 15–64 years, the proportion that have used cocaine in some form in the past year was 1% in South America, 1.2% in the European Union, and 2.1% in North America [1].

Pathophysiological Mechanism of Disease

Pharmacologically, cocaine functions primarily as a presynaptic monoamine reuptake inhibitor, thereby influencing the activity of dopamine, noradrenaline, and serotonin neurotransmitter systems [1].

Cocaine has other pharmacologic effects as well [1]. For example, cocaine can influence activity at NMDA, sigma-opioid receptors, and kappa-opioid receptors. Cocaine can also block voltage-gated sodium channels and maintain them in an inactive state, and this property enables cocaine to function as a topical anesthetic.

Clinical Presentation

Patell and colleagues [2] reported the case of a 32-year-old male who intranasally consumed 4 grams of cocaine daily for 2 days, then presented with an acute onset of bilateral sensorineural hearing loss [2].

Physical Examination

Physical examination of patients suffering from cocaine intoxication is usually dominated not by audiologic or vestibular symptoms, but rather by mania, myoclonus, cardiovascular symptoms, etc.

Ocular Motor Examination

Scharf reported the case of a 26-year-old woman who consumed cocaine intranasally and developed opsoclonus [3].

M. Cherchi, *Otoneurology and Vestibular Medicine*, https://doi.org/10.1007/978-3-031-94841-1_163

Testing: Auditory

Animal studies report that cocaine alters auditory evoked potentials and diminishes cochlear blood flow [4]; repeated exposure to cocaine results in decreased adrenergic innervation of the cochlea [5]. Comparable studies in humans have not been published.

Testing: Vestibular

Demer and colleagues [6] studied electronystagmography in nine cocaine users and nine healthy controls. They reported that saccades, tracking, vestibulo-ocular reflexes, and caloric testing were similar between the two groups. However, "Extreme visual enhancement of VOR [vestibulo-ocular reflex] gain by 4x magnified vision was significantly impaired in cocaine abusers as compared with controls" [6].

Imaging

Nicoucar and colleagues [7] reported the case of a 43-year-old male who consumed cocaine intranasally and then developed disequilibrium, unilateral tinnitus, and hearing loss. Brain MRI revealed an intralabyrinthine hemorrhage ipsilateral to the aural symptoms.

Treatment

The mainstay of treatment of cocaine-related symptoms is detoxification.

Notes

Cocaine is commonly consumed along with alcohol. Transesterification of both drugs produces cocaethylene, which has similar effects to cocaine, but has a longer half-life [8], is very toxic, and particularly epileptogenic.

Our discussion has focused on the otovestibular effects of cocaine. Cocaine has well-documented toxicity on multiple organ systems, including the central nervous system, cardiovascular, respiratory, renal, and hepatic systems.

References

1. Roque Bravo R, Faria AC, Brito-da-Costa AM, Carmo H, Mladenka P, Dias da Silva D, et al. Cocaine: an updated overview on chemistry, detection, biokinetics, and Pharmacotoxicological aspects including abuse pattern. Toxins (Basel). 2022;14(4):278.
2. Patell K, Al Armashi AR, Somoza-Cano FJ, Polanco VP, Ravakhah K. Rare case of sudden bilateral sensorineural hearing loss after cocaine abuse. Am J Ther. 2021;29(1):e137–e9.
3. Scharf D. Opsoclonus-myoclonus following the intranasal usage of cocaine. J Neurol Neurosurg Psychiatry. 1989;52(12):1447–8.
4. Shivapuja BG, Gu ZP, Saunders SS, Quirk WS. Acute effects of cocaine on cochlear function. Hear Res. 1993;69(1–2):243–50.
5. Shivapuja BG, Gu ZP, Liu SY, Saunders SS, Schoener EP. Effects of repeated cocaine injections on cochlear function. Brain Res. 1994;668(1–2):230–8.
6. Demer JL, Volkow ND, Ulrich I, Krajewski K, Davis CM, Porter FI. Eye movements in cocaine abusers. Psychiatry Res. 1989;29(2):123–36.
7. Nicoucar K, Sakbani K, Vukanovic S, Guyot JP. Intralabyrinthine haemorrhage following cocaine consumption. Acta Otolaryngol. 2005;125(8):899–901.
8. Pergolizzi J, Breve F, Magnusson P, LeQuang JAK, Varrassi G. Cocaethylene: when cocaine and alcohol are taken together. Cureus. 2022;14(2):e22498.

Marijuana

164

Overview

Marijuana (Cannabis sativa) contains many cannabinoid compounds, of which the best studied include tetrahydrocannabinol (THC), cannabidiol (CBD), and cannabinol (CBN). Cannabinoid receptors (CB1, CB2) are distributed throughout the central nervous system, including the audiologic and vestibular systems. Exogenously administered cannabinoids appear to have largely deleterious effects on vestibular and ocular motor function, and the majority of studies also report that cannabinoids can cause or exacerbate tinnitus.

Introduction

Marijuana (Cannabis sativa) contains over a hundred types of cannabinoid compounds. Those about which the most research data are available include tetrahydrocannabinol (THC), cannabidiol (CBD), and cannabinol (CBN) [1].

As of this writing, in the United States, marijuana is considered illegal under federal law, and in the medical context, cannabis is classified as a Schedule III drug. Many states have decriminalized the possession and use of marijuana. In 2023, the Department of Health and Human Services submitted a recommendation to the Drug Enforcement Agency that cannabis be reclassified from a Schedule III drug to a Schedule I drug.

Epidemiology

It has been estimated that 3–5% of the world population has at some point tried cannabis in some form, and that in the United States, roughly 1 in 7 to 1 in 8 people have done so [2]. It has further been estimated that 10% of people who have tried a cannabis product have a cannabis use disorder [3].

Pathophysiological Mechanism of Disease

There are several cannabinoid receptors, including the following:

- CB1, encoded by the CNR1 gene at locus 6q15 (OMIM 114610).
- CB2, encoded by the CNR2 gene at locus 1p36.11 (OMIM 605051).

Animal studies have shown that cannabinoid receptors are distributed throughout the central nervous system, including in the audiovestibular system [4–8].

There are numerous endogenous cannabinoids. Of the exogenous cannabinoids, THC is an agonist at CB1 and CB2 receptors, while CBD is a noncompetitive antagonist at CB1 and CB2 receptors [1].

The mechanisms by which cannabinoids influence audiovestibular function are unknown. However, binding cannabinoid receptors has downstream effects on multiple neurotransmitter systems, including modulation of multiple G-protein-coupled receptors, gamma aminobutyric acid, glutamate, and inflammatory pathways, many of which are present in audiologic and vestibular pathways. Thus, it should not be surprising that cannabinoids can have audiovestibular effects.

Clinical Presentation

Of people who seek medical care for complications from cannabis use, the most common otology-related complaints are disequilibrium, hearing loss, and tinnitus [9].

M. Cherchi, *Otoneurology and Vestibular Medicine*, https://doi.org/10.1007/978-3-031-94841-1_164

Ocular Motor Examination

Kibby and Halcomb [10] reported the case of a 25-year-old male who presented with 3 days of visual disturbance and dizziness following marijuana use. They documented gaze-evoked nystagmus.

Pradeep and colleagues [11] reported the case of a 19-year-old male with congenital nystagmus who reported that smoking marijuana improved his vision. The investigators compared performance on dynamic visual acuity testing after the patient had smoked marijuana versus after 3 weeks of abstinence from marijuana, and found that visual acuity in the right eye improved by 3 LogMAR lines in the left eye and 2 LogMAR lines in the right eye. They also reported that the magnitude of the congenital nystagmus improved at various degrees of gaze eccentricity after the patient had smoked marijuana, compared to after 3 weeks of abstinence.

Testing: Auditory

Spector [12] studied audiometry in 65 *chronic* marijuana users (61% males, age range 18–40 years, mean age 23 years) and 24 control subjects (41% males, mean age 28 years) and reported that, "No auditory differences were noted."

Though not evident on standard audiometry, the effects of cannabinoids on tinnitus are controversial (see below).

Testing: Vestibular

Spector [13] studied the *acute* effects of marijuana versus placebo by electronystagmography in 72 users (age range 21–40 years) and reported, "Results indicated no vestibular effect" on sinusoidal tracking, spontaneous nystagmus, positional nystagmus, and rotation of torsional swing.

Spector [12] studied electronystagmography in 65 *chronic* marijuana users (61% males, age range 18–40 years, mean age 23 years) and 24 control subjects (41% males, mean age 28 years) and reported that (1) a decrease in the maximum amplitude of torsional swing; (2) increased incidence of positional nystagmus; and (3) decrease slow-phase velocity "on right warm and left and right cold calorics but not in the left warm [caloric] test."

Schon and colleagues [14] reported the case of a 52-year-old male with clinically definite multiple sclerosis who had been suffering from dysconjugate elliptical pendular nystagmus. Using video oculography, they compared the effect of a placebo and two different doses of tetrahydrocannabinol, studying nystagmus and serum concentrations of tetrahydrocannabinol catabolites. They found a strong inverse correlation between serum concentrations of tetrahydrocannabinol catabolites and the magnitude of nystagmus.

Treatment of Symptoms Resulting from Cannabinoids

The mainstay of treatment for cannabis-related symptoms is detoxification.

Can Cannabinoids Treat Otovestibular Symptoms?

In the vestibular domain: Marijuana has been reported to help visual symptoms in one case report of congenital nystagmus [11] and two cases of pendular nystagmus related to multiple sclerosis [14, 15]. Despite these case reports, the overall evidence is stronger that cannabinoids have adverse effects on the vestibular and ocular motor systems.

In the auditory domain: Some tinnitus patients report symptom relief with cannabis [16], but the preponderance of the evidence is that cannabinoids have no therapeutic effect on tinnitus [17, 18], and some research suggests that cannabinoids may cause or worsen tinnitus [19].

Thus, we would agree with reviews that conclude the currently available evidence does not support a therapeutic role for cannabinoids in the management of otovestibular symptoms [20].

Notes

The reader should note that the above discussion is restricted to the relationship between cannabinoids and otovestibular symptoms. Cannabinoids have numerous effects on many other organ systems, including endocrine, hormonal, reproductive, genitourinary, cognitive, emotional, cardiovascular, gastrointestinal, hepatic, immunological, and others.

References

1. Guerra J, Naidoo V, Cacabelos R. Potential effects of cannabinoids on audiovestibular function: a narrative review. Front Pharmacol. 2022;13:1010296.
2. Anthony JC, Lopez-Quintero C, Alshaarawy O. Cannabis Epidemiology: a selective review. Curr Pharm Des. 2017;22(42):6340–52.
3. Connor JP, Stjepanovic D, Le Foll B, Hoch E, Budney AJ, Hall WD. Cannabis use and cannabis use disorder. Nat Rev Dis Primers. 2021;7(1):16.

4. Gong JP, Onaivi ES, Ishiguro H, Liu QR, Tagliaferro PA, Brusco A, Uhl GR. Cannabinoid CB2 receptors: immunohistochemical localization in rat brain. Brain Res. 2006;1071(1):10–23.
5. Chi DH, Kandler K. Cannabinoid receptor expression at the MNTB-LSO synapse in developing rats. Neurosci Lett. 2012;509(2):96–100.
6. Trattner B, Berner S, Grothe B, Kunz L. Depolarization-induced suppression of a glycinergic synapse in the superior olivary complex by endocannabinoids. J Neurochem. 2013;127(1):78–90.
7. Newsham-West D, Darlington CL, Smith PF. Potent effects of a selective cannabinoid receptor agonist on some Guinea pig medial vestibular nucleus neurons. Eur J Pharmacol. 1998;348(1):R1–2.
8. Baek JH, Zheng Y, Darlington CL, Smith PF. Cannabinoid CB2 receptor expression in the rat brainstem cochlear and vestibular nuclei. Acta Otolaryngol. 2008;128(9):961–7.
9. Phulka JS, Howlett JW, Hu A. Cannabis related side effects in otolaryngology: a scoping review. J Otolaryngol. 2021;50(1):56.
10. Kibby T, Halcomb SE. Toxicology observation: nystagmus after marijuana use. J Forensic Leg Med. 2013;20(4):345–6.
11. Pradeep A, Thomas S, Roberts EO, Proudlock FA, Gottlob I. Reduction of congenital nystagmus in a patient after smoking cannabis. Strabismus. 2008;16(1):29–32.
12. Spector M. Chronic vestibular and auditory effects of marijuana. Laryngoscope. 1974;84(5):816–20.
13. Spector M. Acute vestibular effects of marijuana. J Clin Pharmacol. 1973;13(5):214–7.
14. Schon F, Hart PE, Hodgson TL, Pambakian AL, Ruprah M, Williamson EM, Kennard C. Suppression of pendular nystagmus by smoking cannabis in a patient with multiple sclerosis. Neurology. 1999;53(9):2209–10.
15. Dell'Osso LF. Suppression of pendular nystagmus by smoking cannabis in a patient with multiple sclerosis. Neurology. 2000;54(11):2190–1.
16. Mavedatnia D, Levin M, Lee JW, Hamour AF, Dizon K, Le T. Cannabis use amongst tinnitus patients: consumption patterns and attitudes. J Otolaryngol. 2023;52(1):19.
17. Narwani V, Bourdillon A, Nalamada K, Manes RP, Hildrew DM. Does cannabis alleviate tinnitus? A review of the current literature. Laryngoscope Investig Otolaryngol. 2020;5(6):1147–55.
18. Zheng Y, Smith PF. Cannabinoid drugs: will they relieve or exacerbate tinnitus? Curr Opin Neurol. 2019;32(1):131–6.
19. Reyes Orozco F, Lin M, Clark B, Doherty JK, Oghalai JS, Voelker CCJ, Choi JS. Association between Cannabis use and Tinnitus in US adults. Otol Neurotol. 2023;44(6):611–8.
20. Smith PF, Zheng Y. Cannabinoids, cannabinoid receptors and tinnitus. Hear Res. 2016;332:210–6.

165 MDMA ("Ecstasy")

Overview

The compound *3,4-methylenedioxy-N-methamphetamine* is abbreviated *MDMA*, and commonly known as "*ecstasy*." The drug influences several neurotransmitter systems (serotonin, dopamine, norepinephrine) and endocrine systems (cortisol, prolactin, oxytocin). Its half-life is relatively short (11.0 ± 12.6 h in women and 7.4 ± 2.9 h in men), though some animal studies suggest longer-term neuroplastic effects. Users typically seek the drug's psychotropic effects (euphoria, feelings of intimacy, stimulation) and sexual effects (increased libido, enhanced erection/lubrication, intensity of orgasm), but the drug also has significant adverse effects (agitation, anxiety, tachycardia, hypertension, cardiac arrhythmias, hyperthermia, rhabdomyolysis, methemoglobinemia). The adverse effect of disequilibrium/dizziness is commonly reported; the mechanism is unclear, though hemodynamic effects (hypertension, tachycardia, cardiac arrhythmia) are plausible.

Introduction

The compound *3,4-methylenedioxy-N-methamphetamine* is abbreviated *MDMA* and known by a variety of street names, most popularly "ecstasy." Users typically seek the drug's psychotropic effects, such as "euphoria, a feeling of intimacy, altered visual perception, enhanced libido, and increased energy" [1], or its effects on sexual function, such as enhanced erection/lubrication and increased intensity of orgasm [2, 3]. The drug's adverse effects include agitation, anxiety, tachycardia, hypertension, cardiac arrhythmias, hyperthermia, and rhabdomyolysis [1].

Pharmacology

MDMA influences several neurotransmitter systems, functioning as an "indirect serotonin, dopamine, and norepinephrine agonist" [4]. It additionally has several endocrinological effects, producing "marked increases in cortisol, prolactin, and oxytocin" [5].

The half-life of MDMA is 11.0 ± 12.6 h in women and 7.4 ± 2.9 h in men [4]. Effects of MDMA "tend to be short lived" [6]. Some animal studies describe longer-term neuroplastic changes after repeated exposure [7], but whether such effects also occur in humans is unclear.

Adverse Effects

MDMA has been documented to provoke symptoms of disequilibrium [1, 4, 6, 8–10].

The mechanism by which MDMA might cause disequilibrium is unclear. Several of its hemodynamic effects, such as hypertension, tachycardia, and cardiac arrhythmia [1] may be contributory. A case report has described methemoglobinemia induced by MDMA [11], but most presentations are not so dramatic.

Ocular Motor Examination

As of this writing, there is almost no literature regarding ocular motor effects of MDMA. A single case report of infantile MDMA poisoning mentions "nystagmus" without further description [12].

M. Cherchi, *Otoneurology and Vestibular Medicine*, https://doi.org/10.1007/978-3-031-94841-1_165

References

1. Smith KM, Larive LL, Romanelli F. Club drugs: methylenedioxymethamphetamine, flunitrazepam, ketamine hydrochloride, and gamma-hydroxybutyrate. Am J Health Syst Pharm. 2002;59(11):1067–76.
2. Zemishlany Z, Aizenberg D, Weizman A. Subjective effects of MDMA ('ecstasy') on human sexual function. Eur Psychiatry. 2001;16(2):127–30.
3. Palamar JJ, Griffin-Tomas M, Acosta P, Ompad DC, Cleland CM. A comparison of self-reported sexual effects of alcohol, marijuana, and ecstasy in a sample of young adult nightlife attendees. Psychol Sex. 2018;9(1):54–68.
4. Pardo-Lozano R, Farre M, Yubero-Lahoz S, O'Mathuna B, Torrens M, Mustata C, et al. Clinical pharmacology of 3,4-meth ylenedioxymethamphetamine (MDMA, "ecstasy"): the influence of gender and genetics (CYP2D6, COMT, 5-HTT). PLoS One. 2012;7(10):e47599.
5. Dolder PC, Muller F, Schmid Y, Borgwardt SJ, Liechti ME. Direct comparison of the acute subjective, emotional, autonomic, and endocrine effects of MDMA, methylphenidate, and modafinil in healthy subjects. Psychopharmacology. 2018;235(2):467–79.
6. DeBattista C, Schatzberg AF. The black book of psychotropic dosing and monitoring. Psychopharmacol Bull. 2024;54(3):8–59.
7. Lanteri C, Doucet EL, Hernandez Vallejo SJ, Godeheu G, Bobadilla AC, Salomon L, et al. Repeated exposure to MDMA triggers long-term plasticity of noradrenergic and serotonergic neurons. Mol Psychiatry. 2014;19(7):823–33.
8. Baylen CA, Rosenberg H. A review of the acute subjective effects of MDMA/ecstasy. Addiction. 2006;101(7):933–47.
9. ter Bogt TFM, Engels RC. "Partying" hard: party style, motives for and effects of MDMA use at rave parties. Subst Use Misuse. 2005;40(9–10):1479–502.
10. Liechti ME, Kunz I, Kupferschmidt H. Acute medical problems due to ecstasy use. Case-series of emergency department visits. Swiss Med Wkly. 2005;135(43–44):652–7.
11. Verhaert LL. Methaemoglobinemia induced by MDMA? Case Rep Pulmonol. 2011;2011:494328.
12. Boucher A, Zine A, Jaziri F, Bernard N, Jeannoel P, Descotes J. Intoxication par ecstasy chez un nourrisson de 10 mois (Ecstasy poisoning in a 10-month-old infant). Arch Pediatr. 2009;16(10):1346–9.

166 Nicotine/Tobacco and Otoneurological Symptoms/Conditions

Overview

Nicotine is usually delivered by tobacco which can be inhaled or chewed. Nicotine and the numerous other compounds contained in tobacco can adversely affect every organ system. From the otoneurological perspective, nicotine can have effects on the ear and brain, exacerbating (and perhaps in some cases causing) hearing loss, tinnitus, disequilibrium, and migraine.

Introduction

There are now overwhelming scientific data that *nicotine*, particularly nicotine consumed through tobacco, harms health through its deleterious effects on every organ system in the body. As of 2022, diseases directly attributable to tobacco consumption accounted for seven million deaths worldwide annually; in the United States, it accounted for 480,000 deaths annually, which amounts to about 1300 deaths per day, and one out of every five deaths [1].

Other physicians can advise a patient on the specific relationships between smoking and health problems such as cardiovascular disease, lung problems, and cancer. Here we provide information on the fact that smoking can exacerbate a number of otoneurological conditions relevant to one's equilibrium and hearing.

Mechanism of Disease

Nicotine is a tertiary amine that binds to nicotinic cholinergic receptors; it has numerous downstream effects on several neurotransmitter systems, including dopamine, glutamate, GABA, and others [2].

Ocular Motor Examination

Nicotine exposure can cause spontaneous upbeat nystagmus [3, 4]. This can be subtle, and may only be visible under infrared video oculography.

Otoneurological Symptoms and Conditions Exacerbated by Nicotine and Tobacco

There is strong evidence that nicotine and tobacco increase the risk of *hearing loss* [5–14].

There is strong evidence that nicotine and tobacco increase the risk of developing or exacerbating *tinnitus* [15–22].

There is strong evidence that nicotine and tobacco adversely affect *ocular motor function* [23, 24], perhaps by interfering with the vestibulo-ocular reflex [25]. Nicotine is one of the few substances that can induce spontaneous upbeat nystagmus on the primary position of gaze [4, 26].

There is strong evidence that nicotine and tobacco adversely affect *postural stability* [3, 23, 24, 27].

There is good evidence that nicotine and tobacco increase the risk of *peripheral vestibular disease* [28].

There is good evidence that nicotine and tobacco can *interfere with the treatment* of a variety of vertiginous disorders [29].

There is strong evidence that nicotine and tobacco increase the risk of *migraine* [30–33]. The majority of studies conclude that smoking is a clear independent risk factor for the development, perpetuation, and worsening of migraine and its subtypes [30–34], with very few dissenting opinions [35–37]. Smoking is also a known risk factor for *stroke*, but in patients with a history of migraine and its subtypes, most studies find that the risk for stroke is increased even further [38–42], with very few dissenting opinions [43].

M. Cherchi, *Otoneurology and Vestibular Medicine*, https://doi.org/10.1007/978-3-031-94841-1_166

Quantification of Nicotine/Tobacco Use

In medical practice, it is common to quantify nicotine/tobacco exposure in "pack years," where one pack-year is equivalent to smoking one pack (20 cigarettes) per day for 365 days. While this is not an exact science, some approximate equivalents among different forms of tobacco are listed below [44].

There are online tools for calculating total pack-year exposure from different forms of nicotine/tobacco consumption, such as https://www.smokingpackyears.com (accessed 5/21/23).

The following are all approximately equivalent to 1 pack-year of nicotine/tobacco exposure:

- 1 pack of cigarettes per day for 1 year
- 20 cigarettes per day for 1 year
- 10 cigarillos per day for 1 year
- 8 pipe-bowls per day for 1 year
- 5 cigars/spliffs per day for 1 year
- 16 min of water pipe per day for 1 year
- 0.4 oz chewing tobacco per day for 1 year
- 11.3 g chewing tobacco per day for 1 year

Treatment

It is possible to undertake treatment for the individual consequences of nicotine/tobacco exposure (hearing loss, tinnitus, disequilibrium, migraine), but it is preferable to address the root cause of those symptoms, namely the nicotine/tobacco use itself.

Primary care physicians can offer a patient a variety of strategies to help cease using nicotine, including counseling, psychological treatment, and pharmacological treatment (transdermal or transmucosal nicotine, bupropion, varenicline). It usually takes several attempts to quit.

References

1. Centers for Disease Control and Prevention. Smoking and tobacco use: diseases and death 2022 [updated 29 July 2022]. Available from: https://www.cdc.gov/tobacco/data_statistics/fact_sheets/fast_facts/diseases-and-death.html.
2. Benowitz NL. Pharmacology of nicotine: addiction, smoking-induced disease, and therapeutics. Annu Rev Pharmacol Toxicol. 2009;49:57–71.
3. Pereira CB, Strupp M, Holzleitner T, Brandt T. Smoking and balance: correlation of nicotine-induced nystagmus and postural body sway. Neuroreport. 2001;12(6):1223–6.
4. Sibony PA, Evinger C, Manning KA. Tobacco-induced primary-position upbeat nystagmus. Ann Neurol. 1987;21(1):53–8.
5. Wang D, Wang Z, Zhou M, Li W, He M, Zhang X, et al. The combined effect of cigarette smoking and occupational noise exposure on hearing loss: evidence from the Dongfeng-Tongji cohort study. Sci Rep. 2017;7(1):11142.
6. Dawes P, Cruickshanks KJ, Moore DR, Edmondson-Jones M, McCormack A, Fortnum H, Munro KJ. Cigarette smoking, passive smoking, alcohol consumption, and hearing loss. J Assoc Res Otolaryngol. 2014;15(4):663–74.
7. Sung JH, Sim CS, Lee CR, Yoo CI, Lee H, Kim Y, Lee J. Relationship of cigarette smoking and hearing loss in workers exposed to occupational noise. Ann Occup Environ Med. 2013;25(1):8.
8. Ferrite S, Santana VS, Marshall SW. Interaction between noise and cigarette smoking for the outcome of hearing loss among women: a population-based study. Am J Ind Med. 2013;56(10):1213–20.
9. Tao L, Davis R, Heyer N, Yang Q, Qiu W, Zhu L, et al. Effect of cigarette smoking on noise-induced hearing loss in workers exposed to occupational noise in China. Noise Health. 2013;15(62):67–72.
10. Mohammadi S, Mazhari MM, Mehrparvar AH, Attarchi MS. Cigarette smoking and occupational noise-induced hearing loss. Eur J Pub Health. 2010;20(4):452–5.
11. Sharabi Y, Reshef-Haran I, Burstein M, Eldad A. Cigarette smoking and hearing loss: lessons from the young adult periodic examinations in Israel (YAPEIS) database. Isr Med Assoc J. 2002;4(12):1118–20.
12. Cruickshanks KJ, Klein R, Klein BE, Wiley TL, Nondahl DM, Tweed TS. Cigarette smoking and hearing loss: the epidemiology of hearing loss study. JAMA. 1998;279(21):1715–9.
13. Sumit AF, Das A, Sharmin Z, Ahsan N, Ohgami N, Kato M, Akhand AA. Cigarette smoking causes hearing impairment among Bangladeshi population. PLoS One. 2015;10(3):e0118960.
14. Pezzoli M, Lofaro D, Oliva A, Orione M, Cupi D, Albera A, et al. Effects of smoking on Eustachian tube and hearing. Int Tinnitus J. 2017;21(2):98–103.
15. Veile A, Zimmermann H, Lorenz E, Becher H. Is smoking a risk factor for tinnitus? A systematic review, meta-analysis and estimation of the population attributable risk in Germany. BMJ Open. 2018;8(2):e016589.
16. Bhatt IS. Prevalence of and risk factors for tinnitus and tinnitus-related handicap in a college-aged population. Ear Hear. 2018;39(3):517–26.
17. Kim HJ, Lee HJ, An SY, Sim S, Park B, Kim SW, et al. Analysis of the prevalence and associated risk factors of tinnitus in adults. PLoS One. 2015;10(5):e0127578.
18. Martines F, Sireci F, Cannizzaro E, Costanzo R, Martines E, Mucia M, et al. Clinical observations and risk factors for tinnitus in a Sicilian cohort. Eur Arch Otorrinolaringol. 2015;272(10):2719–29.
19. Mahboubi H, Oliaei S, Kiumehr S, Dwabe S, Djalilian HR. The prevalence and characteristics of tinnitus in the youth population of the United States. Laryngoscope. 2013;123(8):2001–8.
20. Shargorodsky J, Curhan GC, Farwell WR. Prevalence and characteristics of tinnitus among US adults. Am J Med. 2010;123(8):711–8.
21. Sindhusake D, Mitchell P, Newall P, Golding M, Rochtchina E, Rubin G. Prevalence and characteristics of tinnitus in older adults: the Blue Mountains hearing study. Int J Audiol. 2003;42(5):289–94.
22. Paschoal CP, Azevedo MF. Cigarette smoking as a risk factor for auditory problems. Braz J Otorhinolaryngol. 2009;75(6):893–902.
23. Uchida T, Hashimoto M, Suzuki N, Takegami T, Iwase Y. Smoking-induced body sway and its suppression by periodic saccades. Neurosci Lett. 1980;18(2):219–24.
24. Zingler VC, Denecke K, Jahn K, von Meyer L, Krafczyk S, Krams M, et al. The effect of nicotine on perceptual, ocular motor, postural, and vegetative functions at rest and in motion. J Neurol. 2007;254(12):1689–97.
25. Pereira CB, Strupp M, Eggert T, Straube A, Brandt T. Nicotine-induced nystagmus: three-dimensional analysis and dependence on head position. Neurology. 2000;55(10):1563–6.
26. Neveling R, Kruse KE. Über Nicotinnystagmus [On nicotine nystagmus]. Arch Ohren Nasen Kehlkopfheilkd. 1961;177(5):427–31.
27. Iki M, Ishizaki H, Aalto H, Starck J, Pyykko I. Smoking habits and postural stability. Am J Otolaryngol. 1994;15(2):124–8.

28. Wada M, Takeshima T, Nakamura Y, Nagasaka S, Kamesaki T, Kajii E, Kotani K. Association between smoking and the peripheral vestibular disorder: a retrospective cohort study. Sci Rep. 2017;7(1):16889.
29. Lin CY, Young YH. Effect of smoking on the treatment of vertigo. Otol Neurotol. 2001;22(3):369–72.
30. Hershey AD, Lipton RB. Lifestyles of the young and migrainous. Neurology. 2010;75(8):680–1.
31. Lopez-Mesonero L, Marquez S, Parra P, Gamez-Leyva G, Munoz P, Pascual J. Smoking as a precipitating factor for migraine: a survey in medical students. J Headache Pain. 2009;10(2):101–3.
32. Markush RE, Karp HR, Heyman A, O'Fallon WM. Epidemiologic study of migraine symptoms in young women. Neurology. 1975;25(5):430–5.
33. Rozen TD. A history of cigarette smoking is associated with the development of cranial autonomic symptoms with migraine headaches. Headache. 2011;51(1):85–91.
34. Robberstad L, Dyb G, Hagen K, Stovner LJ, Holmen TL, Zwart JA. An unfavorable lifestyle and recurrent headaches among adolescents: the HUNT study. Neurology. 2010;75(8):712–7.
35. Volans GN, Castleden CM. The relationship between smoking and migraine. Postgrad Med J. 1976;52(604):80–2.
36. Rasmussen BK. Migraine and tension-type headache in a general population: precipitating factors, female hormones, sleep pattern and relation to lifestyle. Pain. 1993;53(1):65–72.
37. Ulrich V, Olesen J, Gervil M, Russell MB. Possible risk factors and precipitants for migraine with aura in discordant twin-pairs: a population-based study. Cephalalgia. 2000;20(9):821–5.
38. Del Sette M. Migraine and cerebrovascular risk. Neurol Sci. 2010;31(Suppl 1):S127–8.
39. Kurth T. Migraine with aura and ischemic stroke: which additional factors matter? Stroke. 2007;38(9):2407–8.
40. MacClellan LR, Giles W, Cole J, Wozniak M, Stern B, Mitchell BD, Kittner SJ. Probable migraine with visual aura and risk of ischemic stroke: the stroke prevention in young women study. Stroke. 2007;38(9):2438–45.
41. Schurks M, Rist PM, Bigal ME, Buring JE, Lipton RB, Kurth T. Migraine and cardiovascular disease: systematic review and meta-analysis. BMJ. 2009;339:b3914.
42. Vargas BB, Dodick DW, Wingerchuk DM, Demaerschalk BM. Migraine with and without aura and risk for cardiovascular disease. Curr Atheroscler Rep. 2008;10(5):427–33.
43. Henrich JB, Horwitz RI. A controlled study of ischemic stroke risk in migraine patients. J Clin Epidemiol. 1989;42(8):773–80.
44. Wood DM, Mould MG, Ong SB, Baker EH. "Pack year" smoking histories: what about patients who use loose tobacco? Tob Control. 2005;14(2):141–2.

Phencyclidine

167

Overview

Phencyclidine (PCP) is an illegal drug that can be consumed by inhalation, ingestion, or intravenous injection. Acute intoxication can manifest with behavioral abnormalities (altered mental status, agitation, violent behavior, hallucinations, delusions), physical examination findings (ataxia, tachycardia, hypertension, fever), ocular findings (miosis), and ocular motor findings (various forms of nystagmus, dysconjugate gaze). The mechanism by which PCP exerts these effects is not understood.

Introduction

Phencyclidine, short for *phenyl-cyclohexyl-piperidine* hydrochloride *(PCP)*, was originally synthesized in the 1950s [1] by Park, Davis, and Co. [2] as an anesthetic for humans, but its medical application in humans was discontinued because of the postexposure behavioral and psychological disturbances. However, reports of the illicit use of PCP began to appear in 1965 [2].

PCP can be consumed by inhalation, ingestion, or intravenous injection [2].

PCP can have a variety of effects, and these are not predictable; it "may act as a stimulant, depressant, or hallucinogen, or as a combination of these" [3] and as an analgesic [1] and dissociative anesthetic [4, 5].

Clinical manifestations of acute PCP intoxication are often difficult to isolate since the drug is frequently consumed in conjunction with others [4, 5].

Epidemiology

Dominici and colleagues [4], citing the National Survey on Drug Use and Health, state that as of 2011, 6.1 million Americans age 12 years and older reported using PCP in their lifetime.

Pathophysiological Mechanism of Disease

Acute PCP intoxication has been associated with several eye movement abnormalities, particularly nystagmus, though the mechanism underlying this finding is unknown [6]. PCP is thought to have sympathomimetic and anticholinergic activity, yet some of its autonomic manifestations (e.g., diaphoresis, hypersalivation, bronchospasm) instead suggest either procholinergic or antiadrenergic activity [5].

Clinical Presentation

Acute PCP intoxication typically presents with some combination of behavioral abnormalities (altered mental status, agitation, violent behavior, hallucinations, and delusions) [5, 6].

Physical Examination

Acute PCP intoxication can present with a variety of physical examination findings, including ataxia, tachycardia, hypertension, and mild fever [5, 6].

Ocular Motor Examination

Ocular and ocular motor findings can include miosis [7], dysconjugate gaze [3], and various forms of nystagmus [1–9].

Barton and colleagues studied a series of 27 cases of acute PCP intoxication and reported nystagmus in 24 (89%) of them [6].

Dominici and colleagues studied a series of 184 cases of acute PCP intoxication and reported horizontal nystagmus in 118 (64.1%), vertical nystagmus in 90 (48.9%), and torsional nystagmus in 3 (2.6%) [4].

M. Cherchi, *Otoneurology and Vestibular Medicine*, https://doi.org/10.1007/978-3-031-94841-1_167

McCarron and colleagues [5] studied 1000 cases of acute PCP intoxication that presented to an emergency department. They reported that 574 cases (57.4%) exhibited "horizontal, vertical or rotary nystagmus… Some patients had two or three types of nystagmus simultaneously."

Herskowitz and Oppenheimer [8] reported the cases of two males (age 14 and 17) with acute PCP intoxication whose examinations showed, "On vertical or horizontal gaze, striking bursts of irregular, shuddery, jerk nystagmus occurred in the direction of gaze. Nystagmus was greatest in amplitude on upward gaze."

Pradhan studied experimental subjects in a controlled military laboratory setting and reported that, "The residual effects (e.g., numbness, light headedness, vertigo, ataxia and lateral nystagmus) persisted for at least 4 h postinjection" [2]. In contrast, Herskowitz and Oppenheimer reported the persistence of nystagmus as long as 4 days after acute intoxication [8].

Schwartz and Einhorn [3] studied a series of seven children with acute PCP intoxication and reported that 57% exhibited some form of nystagmus, and 27% exhibited dysconjugate gaze.

Testing: Vestibular

As of this writing, there were no published studies of instrumented vestibular testing in patients with acute PCP intoxication.

Treatment

Management of acute PCP intoxication may include haloperidol and benzodiazepines, which usually happens in an emergency room setting. After a patient has been stabilized, they are generally admitted to the hospital for subacute monitoring and care.

References

1. Mozayani A. Phencyclidine – effects on human performance and behavior. Forensic Sci Rev. 2003;15(1):61–74.
2. Pradhan SN. Phencyclidine (PCP): some human studies. Neurosci Biobehav Rev. 1984;8(4):493–501.
3. Schwartz RH, Einhorn A. PCP intoxication in seven young children. Pediatr Emerg Care. 1986;2(4):238–41.
4. Dominici P, Kopec K, Manur R, Khalid A, Damiron K, Rowden A. Phencyclidine intoxication case series study. J Med Toxicol. 2015;11(3):321–5.
5. McCarron MM, Schulze BW, Thompson GA, Conder MC, Goetz WA. Acute phencyclidine intoxication: incidence of clinical findings in 1,000 cases. Ann Emerg Med. 1981;10(5):237–42.
6. Barton CH, Sterling ML, Vaziri ND. Phencyclidine intoxication: clinical experience in 27 cases confirmed by urine assay. Ann Emerg Med. 1981;10(5):243–6.
7. Liden CB, Lovejoy FH Jr, Costello CE. Phencyclidine. Nine cases of poisoning. JAMA. 1975;234(5):513–6.
8. Herskowitz J, Oppenheimer EY. More about poisoning by phencyclidine ("PCP", "angel dust"). N Engl J Med. 1977;297(25):1405.
9. Tennant F. The rapid eye test to detect drug abuse. Postgrad Med. 1988;84(1):108–14.

Part XXVI

Diagnoses—The Role of Electromagnetic Fields and Radiation in Disequilibrium

168 Radiation and Its Otovestibular Effects

Overview

The management of some malignancies of the head and neck may include *radiation* therapy. Even with good planning, the temporal bone, and thus structures relevant to audiovestibular function, often fall within the radiation field. Radiation induces both short-term effects (perhaps through the mechanism of vasculitis) and long-term effects (by damaging DNA and thus disrupting cellular function, and probably through ischemia due to radiation-caused or radiation-accelerated atherosclerosis). Auditory effects—usually manifesting as hearing loss—may become apparent within months, and tend to be progressive (beyond what would be expected from age alone). Vestibular symptoms may present after some delay, sometimes years later. Audiometry will generally show predominantly sensorineural hearing loss, typically greater in the higher frequencies. Vestibular studies may reveal reduced or absent caloric responses, and delayed or absent vestibular evoked myogenic potential responses. The audiologic and vestibular deficits from radiation therapy are irreversible. Patients with auditory deficits should be referred to audiology regarding amplification (for hearing loss) and/or masking (for tinnitus). In patients with disequilibrium, it is medically reasonable to attempt a course of vestibular physical therapy, particularly if workup reveals vestibular weakness.

Introduction

Treatment with ionizing *radiation* plays an important role in the management of certain neoplasms of the head and neck. In some instances, radiation therapy is purposefully directed at neoplasms affecting audiovestibular structures, such as the treatment of a vestibular schwannoma with stereotactic radiosurgery (SRS) or fractionated stereotactic radiotherapy (FSRT). Neoplasms elsewhere in the head and neck may require fractionated radiotherapy (FRT) or intensity-modulated radiation therapy (IMRT) [1].

Radiation therapy probably works through multiple mechanisms.

> Deposition of radiation energy in tissue results in cell injury and death. Most of radiation's tissue effect is thought to be a result of the damage to DNA. This occurs both directly and indirectly. The latter, the predominant mechanism, involves ionization of surrounding water molecules to form free radicals, which, in turn, result in double- strand breaks in DNA. This injury may result in cell death during mitosis, induction of apoptosis (i.e., programmed cell death), recovery, and cell cycle arrest or terminal differentiation through activation of repair pathways that may also play a role in tumor suppression (e.g., activation of p53). The radio-response of a tissue depends on the inherent sensitivity of the cells, the kinetics of cell population, total dose, dose per fraction, and time-dose fractionation. Cells with fast turnover rate or higher mitotic activity exhibit more sensitivity to radiation, subjecting to cell death in attempting subsequent mitosis. This is the basis for therapeutic RT (i.e., relatively greater damage to highly reproductive tumor cells). Further, RT fractionation offers the potential for greater differential sparing of normal tissues and killing of tumor cells [1].

Radiation therapy is carefully planned with the goal of maximally targeting neoplastic tissue, while minimizing the effects on healthy tissues. However, even with optimal planning, "In treatment for malignancies of the brain, nasopharynx, paranasal sinuses, and several other sites in the head and neck, the temporal bone is included within the field receiving radiation" [2] and thus, audiovestibular structures may still lie in the path of the delivered radiotherapy.

Studies have shown that "The entire auditory-vestibular system is vulnerable to RT [radiotherapy] injury. Nearly half of all patients that have undergone RT for head and neck tumors demonstrate evidence of auditory or vestibular system pathology" [1].

Unlike some other aspects of oncologic care, there is no national-level consortium specifically tasked with the study of audiovestibular outcomes of cancer treatment. The LENT-SOMA ("Late Effects Normal Tissue Task Force – Subjective, Objective, Management, Analytic scales") framework was a proposal for tracking outcomes in multiple organ systems [3], including the ear, but was never validated, and has been criticized as inadequately granular for many purposes.

M. Cherchi, *Otoneurology and Vestibular Medicine*, https://doi.org/10.1007/978-3-031-94841-1_168

Consequently, the study of audiovestibular effects of cancer treatment has drawn on case reports and case series.

The audiologic effects of radiation therapy have been studied more thoroughly than the vestibular effects. As Frisch and colleagues state, "The scientific understanding of the effects of radiation on the vestibular apparatus is lacking compared to our understanding of its association with sensorineural hearing loss" [4].

Epidemiology

Taking into account the caveats cited above, Bhandare and colleagues [1], citing data from Kwong and colleagues [5], state that some degree of permanent hearing loss from radiation for treatment of head and neck cancers may be as high as 54%.

Many treatment regimens for head and neck cancers include platinum-based agents such as cisplatin, which has established ototoxicity. Most studies [6], though not all [5] conclude that the combination of radiation therapy and cisplatin has a synergistic effect with respect to adversely impacting hearing.

Bhandare and colleagues [1], citing data from Singh and Slevin [7], state that "vestibular dysfunction occurs in roughly 25% to 30% of patients treated with RT [radiation therapy] to the temporal bone." Drawing on data from several other studies [8–10], Bhandare and colleagues state that, "Caloric weakness has been reported in 9% to 36%" of patients treated with radiation involving the temporal bone.

Pathophysiological Mechanism of Disease

The mechanism by which radiation disrupts labyrinthine function is incompletely understood, however:

> The unique nature of the inner ear cell lines leads to unique manifestations of RT injury… Hair cells and the stria vascularis have been implicated as the two major sites of inner ear radiation toxicity [1].

Bhandare and colleagues [1], citing data from Leach [11], comment that:

> Histologic studies have shown that the greatest damage to inner ear is the result of injury to the vessels of stria vascularis… An autopsy study after a high dose of radiation demonstrated absence of the organ of Corti, macula of the utricle and the cristae of the semicircular canals [1].

Radiation can have different effects over time. It has been suggested that some of the short-term effects (typically within hours to days) of radiation may be attributable to vasculitis [12], while the longer-term effects are due to cellular dysfunction consequent to DNA damage from radiation.

Radiation is also known to provoke or accelerate atherosclerosis [13], and these effects may not become apparent for years. Most studies of this topic pertain to coronary artery disease and carotid artery disease. However, it is plausible that radiation has similar effects on smaller arteries, such as the labyrinthine artery, which is the only blood supply to the inner ear. Thus, labyrinthine ischemia (secondary to radiation-caused or radiation-accelerated atherosclerosis) is an additional mechanism by which radiation may have long-term effects on the audiologic and vestibular systems.

The longer-term effects of radiation are generally more difficult to study; as Gamble and colleagues comment, "The most obvious reason that we do not see many late sequelae of radiation of the ear is that most of these patients have a malignancy which causes their demise before enough time has elapsed for these changes to develop" [14].

Gamble and colleagues conducted temporal bone studies of irradiated guinea pigs and humans, and generalized that, "the vestibular apparatus may be more resistant to the effects of radiation than the cochlea" [14].

Clinical Presentation: Audiologic

Hearing loss has been reported to begin as early as 3 months after completing radiation therapy [15, 16]. Bhandare and colleagues [1], citing data of patients treated for nasopharyngeal carcinoma from Ho and colleagues [16] and Kwong and colleagues [5], state that the latency from completion of radiation therapy to onset of hearing loss is 1.5–2 years. The risk of progressive hearing loss (attributable to radiation and not to age) likely persists for at least 3–4 years [17], and observation of trends suggests that the trajectory of hearing loss likely extends further [16].

Clinical Presentation: Vestibular

Details are scanty regarding symptoms of disequilibrium in patients who have received radiation therapy in the head and neck. Often, the descriptions are limited to the level of case reporting; for instance, Leach states, "Vestibular symptoms were noticed even sooner, usually within a month or two but delayed for two and a half years in one patient who then suffered severe loss of balance" [11].

Young and colleagues [18] studied 113 nasopharyngeal carcinoma (NPC) patients who had received radiation therapy and complained of postirradiation "vertigo." Of these patients, 67 were men and 46 were women; their mean age at onset of NPC was 46 ± 8 years; the onset of "vertigo" occurred at a mean age of 56 ± 8 years, with an interval ranging from 1 to 27 years (mean 10 years). Given this inter-

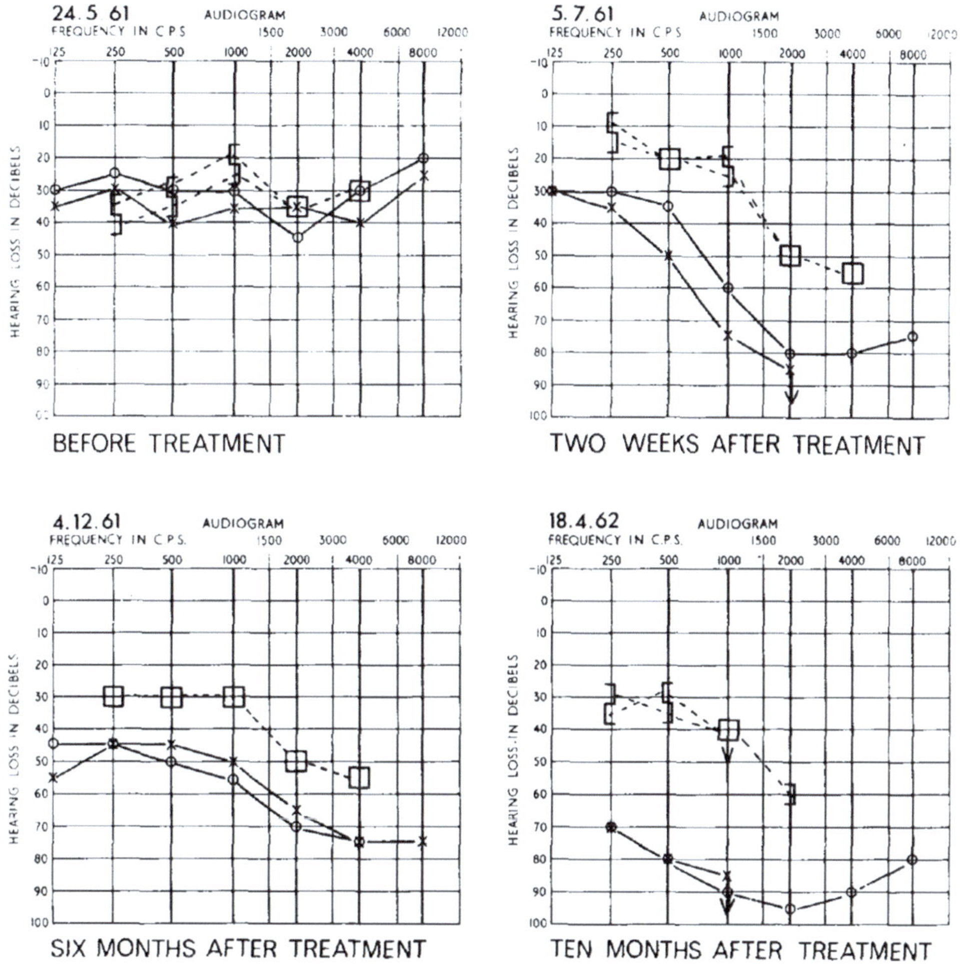

Fig. 168.1 Serial audiometry from prior to radiation therapy to 10 months after radiation therapy in a patient treated for nasopharyngeal carcinoma. (From Leach [11]. Reproduced with written permission from Cambridge University Press)

val, they concluded that "Vertigo is a late complication in irradiated NPC survivors."

Testing: In General

The degrees of affectation of audiologic and vestibular function may be different from each other. Jonahessen and colleagues studied audiometry and caloric testing in 33 patients who received unilateral radiation for brain tumor and reported that, "There was no correlation between the degree of reduced caloric response and sensorineural hearing loss" [10].

Testing: Auditory

Hearing loss from radiation tends to affect the high frequencies more than the low frequencies [5]. As we will discuss below, the hearing loss tends to be progressive.

Figure 168.1, from Leach [11], shows serial audiometry from prior to radiation therapy to 10 months after radiation therapy in a patient treated for nasopharyngeal carcinoma.

Testing: Vestibular

Johanessen and colleagues studied 33 patients who received unilateral radiation for brain tumor and reported that on electronystagmography, "Three patients [9%] had significantly reduced response to caloric stimulation of the vestibular inner ear of the irradiated side" [10].

Young and colleagues [18] studied 113 nasopharyngeal carcinoma (NPC) patients who had received radiation therapy and complained of postirradiation "vertigo." Of these patients, 67 were men and 46 were women; their mean age at onset of NPC was 46 ± 8 years; the onset of "vertigo" occurred at a mean age of 56 ± 8 years, with an interval ranging from 1 to 27 years (mean 10 years). They performed electronystagmography on these patients and reported that:

> Electronystagmography disclosed multidirection nystagmus in three patients and unidirection nystagmus in nine patients. The former consisted of three patients with radiation necrosis of the brain, whereas the latter included sudden deafness in five, intracranial relapse in one, and peripheral labyrinthine lesion in three. Abnormal eye tracking test was disclosed in 58 patients (51%) and abnormal optokinetic nystagmus test in 48 patients (42%). Excluding 5 patients who refused to perform caloric testing, 45 patients (42%) displayed absent ice-water caloric response in one or both ears, 39 patients (36%) had reduced caloric responses, and 24 patients (22%) had bilateral normal caloric responses [18].

Wu and colleagues [19] compared cervical vestibular evoked myogenic potentials (VEMPs) in two groups of nasopharyngeal carcinoma (NPC) patients (one group had received radiation therapy once, the other group had received radiation therapy twice) and healthy control subjects. They reported abnormalities in cervical vestibular evoked myogenic potential responses, ranging from delayed latencies (61%) to complete absence (5%). Interestingly, the delay in cervical vestibular evoked myogenic potentials increased after a second radiation treatment, as shown in Table 168.1. They speculated that "Delayed VEMPs in NPC survivors are possibly due to radiation-induced otitis media, a brainstem lesion, or neck fibrosis" [19].

Radiation can cause otitis media, and otitis media can, in turn, cause conductive hearing loss, which could alter VEMP responses. This means that there may be different contributors to VEMP abnormalities in patients who had received radiation, some from middle ear dysfunction (from otitis media) and some from damage to the labyrinth, vestibular nerve, or vestibular nuclei. In order to disentangle these effects, Yang and colleagues [20] compared cervical vestibular evoked myogenic potentials (VEMPs) in fourteen "radiation-induced otitis media (ROM)" nasopharyngeal carcinoma patients and 14 age- and gender-matched patients with chronic otitis media ("COM") who did not receive radiation therapy and did not have nasopharyngeal carcinoma. They reported that, "Most (89%) COM ears revealed normal VEMPs, and most (61%) ROM ears demonstrated delayed VEMPs, indicating that ROM is different from COM because of the larger affected areas such as retrolabyrinthine or brainstem involvement."

Table 168.1 Latencies of cervical vestibular evoked myogenic potentials in nasopharyngeal carcinoma patients (some radiated once, others radiated twice) versus healthy controls

	N	Latency I (ms)	Latency II (ms)
Control ears	20	14.9 ± 2.1	20.3 ± 1.3
Preirradiated nasopharyngeal carcinoma ears	10	13.7 ± 1.6	20.0 ± 1.8
Irradiated nasopharyngeal carcinoma ears (radiation once)	34	19.2 ± 4.3	25.3 ± 4.6
Irradiated nasopharyngeal carcinoma ears (radiation twice)	10	23.8 ± 2.5	29.8 ± 2.7

From Wu et al. [19]

Histopathology

Treatment of various head and neck cancers involves not only radiation but also chemotherapy (such as cisplatin), which is known to have significant ototoxicity, and in studies of such cases, it is difficult or impossible to distinguish the audiovestibular effects of the two treatment modalities.

Very few temporal bone studies have been published regarding patients who have received radiation therapy alone (i.e., with no chemotherapy). Hoistad and colleagues [2] reported on nine temporal bones from five patients who had received radiation therapy alone for various head and neck cancers. They reported:

> Histologic changes in the five patients (9 temporal bones) irradiated included one patient with a tympanic membrane thickened in the posterosuperior quadrant. Ossicles were normal except in one patient who had surgical changes caused by lateral resection of the temporal bone and mastoidectomy; this patient was missing all ossicles except the stapedial footplate. Another patient showed resorptive changes in the incus. Two bones demonstrated atrophy of the annular ligament but no obvious subluxation of the stapedial footplate. Mucosa of the middle ear was thickened in every irradiated temporal bone except one, which had been irradiated 9–10 months previously. One temporal bone demonstrated emptying of lacunae, fibrosis, and probable osteoradionecrosis near the internal carotid artery. Every irradiated mastoid cavity had fluid present and thickened mucosa. All of these temporal bones showed reactive compensatory fibrosis involving the mastoid cavity and near the facial nerve recess. In three of the five patients who had been irradiated, the round window membrane was thickened. We noted serous fluid in the utricle and saccule. The semicircular canals were normal. There was no endolymphatic or saccular hydrops. Atrophy of the stria vascularis and strial ligament was noted more commonly in the apical turns. All irradiated temporal bones showed decreased spiral ganglion cells (Fig. 3B). Outer hair cells were absent throughout all turns, and inner hair cells were present only sporadically in apical turns [2].

Schuknecht and colleagues [21] focused attention on the effect of radiation on the osseous structures in the temporal bone, but commented that the effect on soft tissues of the inner ear includes labyrinthitis.

Treatment

It is believed that the cell-death-related damage from radiation is permanent, so until regenerative therapies emerge, there is no direct treatment for the auditory and vestibular deficits arising from radiation therapy.

Patients with auditory deficits may consult with audiology regarding amplification (for hearing loss) and/or masking (for tinnitus).

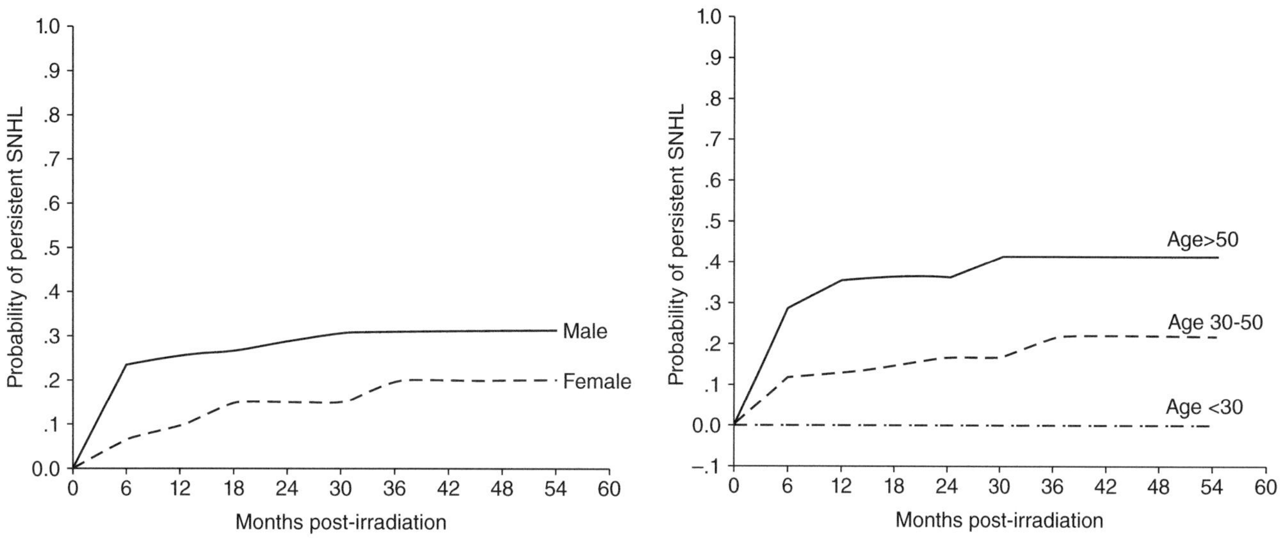

Fig. 168.2 The likelihood of developing persistent sensorineural hearing loss increases with time after radiation therapy, is higher in males (left panel), and is worse in higher age ranges (right panel). (From Kwong et al. [5]. Reproduced with written permission from Elsevier B.V)

If a patient has vestibular symptoms, it is medically reasonable to attempt a course of vestibular physical therapy. This should be especially encouraged if the workup reveals vestibular weakness.

Prognosis

In general, hearing loss following radiation therapy is progressive, and beyond what would be expected for age.

Ho and colleagues [16] present a case series showing the increasing proportion of ears with hearing loss up to 4.5 years after radiation therapy, and the progressively decreasing likelihood of hearing preservation at 4 kHz up to 54 months following radiation therapy.

Figure 168.2, from Kwong and colleagues [5], shows that the likelihood of persistent sensorineural hearing loss increases with time after radiation therapy, is higher in males, and is worse in higher age ranges.

It is unknown whether vestibular deficits from radiation therapy are progressive.

References

1. Bhandare N, Mendenhall WM, Antonelli PJ. Radiation effects on the auditory and vestibular systems. Otolaryngol Clin N Am. 2009;42(4):623–34.
2. Hoistad DL, Ondrey FG, Mutlu C, Schachern PA, Paparella MM, Adams GL. Histopathology of human temporal bone after cis-platinum, radiation, or both. Otolaryngol Head Neck Surg. 1998;118(6):825–32.
3. LENT SOMA tables. Radiother Oncol. 1995;35(1):17–60.
4. Frisch CD, Beadle BM, Gidley PW. Radiation effects on the temporal bone. In: Gidley PW, DeMonte F, editors. Temporal bone cancer. Cham: Springer International Publishing; 2018. p. 379–402.
5. Kwong DL, Wei WI, Sham JS, Ho WK, Yuen PW, Chua DT, et al. Sensorineural hearing loss in patients treated for nasopharyngeal carcinoma: a prospective study of the effect of radiation and cisplatin treatment. Int J Radiat Oncol Biol Phys. 1996;36(2):281–9.
6. Lambert EM, Gunn GB, Gidley PW. Effects of radiation on the temporal bone in patients with head and neck cancer. Head Neck. 2016;38(9):1428–35.
7. Singh IP, Slevin NJ. Late audiovestibular consequences of radical radiotherapy to the parotid. Clin Oncol (R Coll Radiol). 1991;3(4):217–9.
8. Chao WY, Tseng HZ, Tsai ST. Caloric response and postural control in patients with nasopharyngeal carcinoma after radiotherapy. Clin Otolaryngol Allied Sci. 1998;23(5):439–41.
9. Gabriele P, Orecchia R, Magnano M, Albera R, Sannazzari GL. Vestibular apparatus disorders after external radiation therapy for head and neck cancers. Radiother Oncol. 1992;25(1):25–30.
10. Johannesen TB, Rasmussen K, Winther FO, Halvorsen U, Lote K. Late radiation effects on hearing, vestibular function, and taste in brain tumor patients. Int J Radiat Oncol Biol Phys. 2002;53(1):86–90.
11. Leach W. Irradiation of the ear. J Laryngol Otol. 1965;79(10):870–80.
12. Borsanyi SJ, Blanchard CL. Ionizing radiation and the ear. JAMA. 1962;181:958–61.
13. Venkatesulu BP, Mahadevan LS, Aliru ML, Yang X, Bodd MH, Singh PK, et al. Radiation-induced endothelial vascular injury: a review of possible mechanisms. JACC Basic Transl Sci. 2018;3(4):563–72.
14. Gamble JE, Peterson EA, Chandler JR. Radiation effects on the inner ear. Arch Otolaryngol. 1968;88(2):156–61.
15. Wang LF, Kuo WR, Ho KY, Lee KW, Lin CS. A long-term study on hearing status in patients with nasopharyngeal carcinoma after radiotherapy. Otol Neurotol. 2004;25(2):168–73.
16. Ho WK, Wei WI, Kwong DL, Sham JS, Tai PT, Yuen AP, et al. Long-term sensorineural hearing deficit following radiotherapy in patients suffering from nasopharyngeal carcinoma: a prospective study. Head Neck. 1999;21(6):547–53.

17. Grau C, Overgaard J. Postirradiation sensorineural hearing loss: a common but ignored late radiation complication. Int J Radiat Oncol Biol Phys. 1996;36(2):515–7.
18. Young YH, Ko JY, Sheen TS. Postirradiation vertigo in nasopharyngeal carcinoma survivors. Otol Neurotol. 2004;25(3):366–70.
19. Wu CC, Young YH, Ko JY. Effect of irradiation on vestibular evoked myogenic potentials in nasopharyngeal carcinoma survivors. Head Neck. 2003;25(6):482–7.
20. Yang TL, Young YH. Radiation-induced otitis media—study of a new test, vestibular-evoked myogenic potential. Int J Radiat Oncol Biol Phys. 2004;60(1):295–301.
21. Schuknecht HF, Karmody CS. Radionecrosis of the temporal bone. Laryngoscope. 1966;76(8):1416–28.

Electromagnetic Fields

169

Overview

When patients ask about the possible effects of "*electromagnetic fields*" on their health, they are usually asking about magnetic fields generated by MRI scanners, electrical power lines, cellular telephones, and Bluetooth devices. Very strong magnetic fields (such as those present in an MRI scanner) can induce a magnetohydrodynamic force in the endolymph of the semicircular canals, causing deflection of the cupula with corresponding nystagmus. The much weaker magnetic fields generated by handheld devices do not appear to have any measurable effect on cochleo-vestibular function.

Introduction

Patients sometimes ask clinicians about the possible effects of "*electromagnetic fields* (EMFs)," and are usually referring either to the risks of using small devices (such as a cellular telephone or Bluetooth headset) or the risks of their home being near electrical power lines.

Although these inquiries regard "electromagnetic fields," and although electrical and magnetic phenomena are intimately related, the clinician should bear in mind that such questions really pertain to the *magnetic fields* generated by electrical devices. Magnetic fields influence the movement of electrons (i.e., electrical currents) and the orientation of atoms and compounds. Patients are generally *not* asking about *electrical* phenomena such as would occur in unintentional electrocution, or intentional treatment with transcranial direct current stimulation.

Pathophysiological Mechanism

There are several mechanisms by which magnetic fields could influence biological systems. Glover and colleagues categorize these into induced electrical currents, magnetohydrodynamics, and tissue magnetic susceptibility differences [1].

Very powerful magnetic fields, such as are used in MRI scanners, can generate a biologically meaningful Lorentz force that influences ionic currents in the endolymph of the labyrinth [2]. Such a magnetohydrodynamic force [1] in turn can exert actual physical force on the cupula of the semicircular canals. This is thought to be the mechanism by which MRI can induce nystagmus in normal individuals [3–6]. The strength of the magnetic field generated by a clinical MRI scanner is on the order of 1 to 3 Tesla.

Devices such as cellular telephones generate much weaker magnetic fields, sometimes referred to as "ELF" (extremely low frequency) magnetic fields [7, 8]. Experimental studies failed to detect any nystagmus induced by such weak fields [9].

Most research on this topic concludes that there is no evidence of any cochleo-vestibular effects of the electromagnetic fields generated by cellular telephones [10–13]. Skeptics cast doubt on the validity of such research, noting that some of the published studies were funded by the cellular telephone industry [14].

References

1. Glover PM, Cavin I, Qian W, Bowtell R, Gowland PA. Magnetic-field-induced vertigo: a theoretical and experimental investigation. Bioelectromagnetics. 2007;28(5):349–61.
2. Antunes A, Glover PM, Li Y, Mian OS, Day BL. Magnetic field effects on the vestibular system: calculation of the pressure on the cupula due to ionic current-induced Lorentz force. Phys Med Biol. 2012;57(14):4477–87.
3. Ward BK, Otero-Millan J, Jareonsettasin P, Schubert MC, Roberts DC, Zee DS. Magnetic vestibular stimulation (MVS) as a technique for understanding the normal and diseased labyrinth. Front Neurol. 2017;8:122.
4. Ward BK, Roberts DC, Della Santina CC, Carey JP, Zee DS. Magnetic vestibular stimulation in subjects with unilateral labyrinthine disorders. Front Neurol. 2014;5:28.

M. Cherchi, *Otoneurology and Vestibular Medicine*, https://doi.org/10.1007/978-3-031-94841-1_169

5. Ward BK, Roberts DC, Della Santina CC, Carey JP, Zee DS. Vestibular stimulation by magnetic fields. Ann N Y Acad Sci. 2015;1343:69–79.
6. Ward BK, Zee DS, Roberts DC, Schubert MC, Perez-Fernandez N, Otero-Millan J. Visual fixation and continuous head rotations have minimal effect on set-point adaptation to magnetic vestibular stimulation. Front Neurol. 2018;9:1197.
7. Cook CM, Saucier DM, Thomas AW, Prato FS. Exposure to ELF magnetic and ELF-modulated radiofrequency fields: the time course of physiological and cognitive effects observed in recent studies (2001–2005). Bioelectromagnetics. 2006;27(8):613–27.
8. Cook CM, Thomas AW, Prato FS. Human electrophysiological and cognitive effects of exposure to ELF magnetic and ELF modulated RF and microwave fields: a review of recent studies. Bioelectromagnetics. 2002;23(2):144–57.
9. Pau HW, Sievert U, Eggert S, Wild W. Can electromagnetic fields emitted by mobile phones stimulate the vestibular organ? Otolaryngol Head Neck Surg. 2005;132(1):43–9.
10. Balbani AP, Montovani JC. Mobile phones: influence on auditory and vestibular systems. Braz J Otorhinolaryngol. 2008;74(1):125–31.
11. Johansen C. Electromagnetic fields and health effects–epidemiologic studies of cancer, diseases of the central nervous system and arrhythmia-related heart disease. Scand J Work Environ Health. 2004;30(Suppl 1):1–30.
12. Kleinlogel H, Dierks T, Koenig T, Lehmann H, Minder A, Berz R. Effects of weak mobile phone – electromagnetic fields (GSM, UMTS) on event related potentials and cognitive functions. Bioelectromagnetics. 2008;29(6):488–97.
13. Mandala M, Colletti V, Sacchetto L, Manganotti P, Ramat S, Marcocci A, Colletti L. Effect of Bluetooth headset and mobile phone electromagnetic fields on the human auditory nerve. Laryngoscope. 2014;124(1):255–9.
14. Marino AA, Carrubba S. The effects of mobile-phone electromagnetic fields on brain electrical activity: a critical analysis of the literature. Electromagn Biol Med. 2009;28(3):250–74.

Part XXVII

Diagnoses—Presentations with Multiple Potential Causes

Multifactorial Disequilibrium

170

Overview

The phrase "*multifactorial disequilibrium*" (or "multifactorial dizziness") refers to the symptom of disequilibrium to which multiple etiologies are contributory. Humans maintain equilibrium through a system that involves multiple inputs (vestibular, visual, proprioceptive), multiple processing steps, and multiple outputs (ocular motor, somatomotor). This "many-to-many" arrangement has both a disadvantage (multiple vulnerabilities) and an advantage (redundancy). Redundancy can compensate for vulnerability up to a point, but beyond that threshold, compensation is no longer adequate, and the symptom of disequilibrium may become manifest. The clinical suspicion for multifactorial disequilibrium sometimes requires corroboration from consultation with several subspecialties. However, an otoneurologist or neurootologist should still evaluate these patients for discrete vestibular problems (such as benign paroxysmal positional vertigo, vestibular neuritis). Treatment and prognosis of multifactorial disequilibrium depend on which underlying pathologies are contributory. Vestibular rehabilitation therapy may be helpful.

Introduction

The phrase "*multifactorial disequilibrium*" (or "multifactorial dizziness") refers to the symptom of disequilibrium to which multiple etiologies are contributory. This idea has acquired an unfavorable reputation in neurology because it is sometimes viewed as a refuge when the diagnosis is unknown.

We agree that the diagnosis of multifactorial disequilibrium should not be a substitute for a reasonable attempt at using "Occam's razor" (the principle of parsimony) to seek a single underlying cause of this symptom. However, this heuristic (the principle of parsimony) is only that: it is a reasonable starting assumption, but it is not always correct. The reality is that clinicians do encounter cases of "one-symptom-from-many-diagnoses," and medical diagnostic thinking must accommodate this reality.

What Is the "Multi" in Multifactorial Disequilibrium?

In order to clarify the idea of multifactorial disequilibrium, it is helpful to take a high-level view of the organization by which the vestibular system maintains equilibrium. Neurological training encourages a "circuitry" conception of neurophysiology. We offer Fig. 170.1 as an illustration of this.

Many aspects of this *figure* will probably be familiar to trained physicians, while others may be less well known. We will highlight several points.

Overall: Many-to-Many Arrangement of Systems

This is a system in which there are multiple sensory inputs, multiple motor outputs, and multiple processing steps that get from the inputs to the outputs. In other words, this is a many-to-many arrangement of systems.

Input: Don't Forget Auditory Input

For didactic purposes, textbooks usually state that the vestibular system takes into account three sensory modalities as its inputs: (1) inner ear vestibular function; (2) vision; (3) the proprioceptive component of somatic sensation. These three are emphasized because they do, indeed, constitute the main inputs.

However, that view is also an oversimplification, because the brain takes into account other inputs as well, such as hearing. While not as well developed as in bats, humans actually do have a limited *echolocation* ability [1–7].

M. Cherchi, *Otoneurology and Vestibular Medicine*, https://doi.org/10.1007/978-3-031-94841-1_170

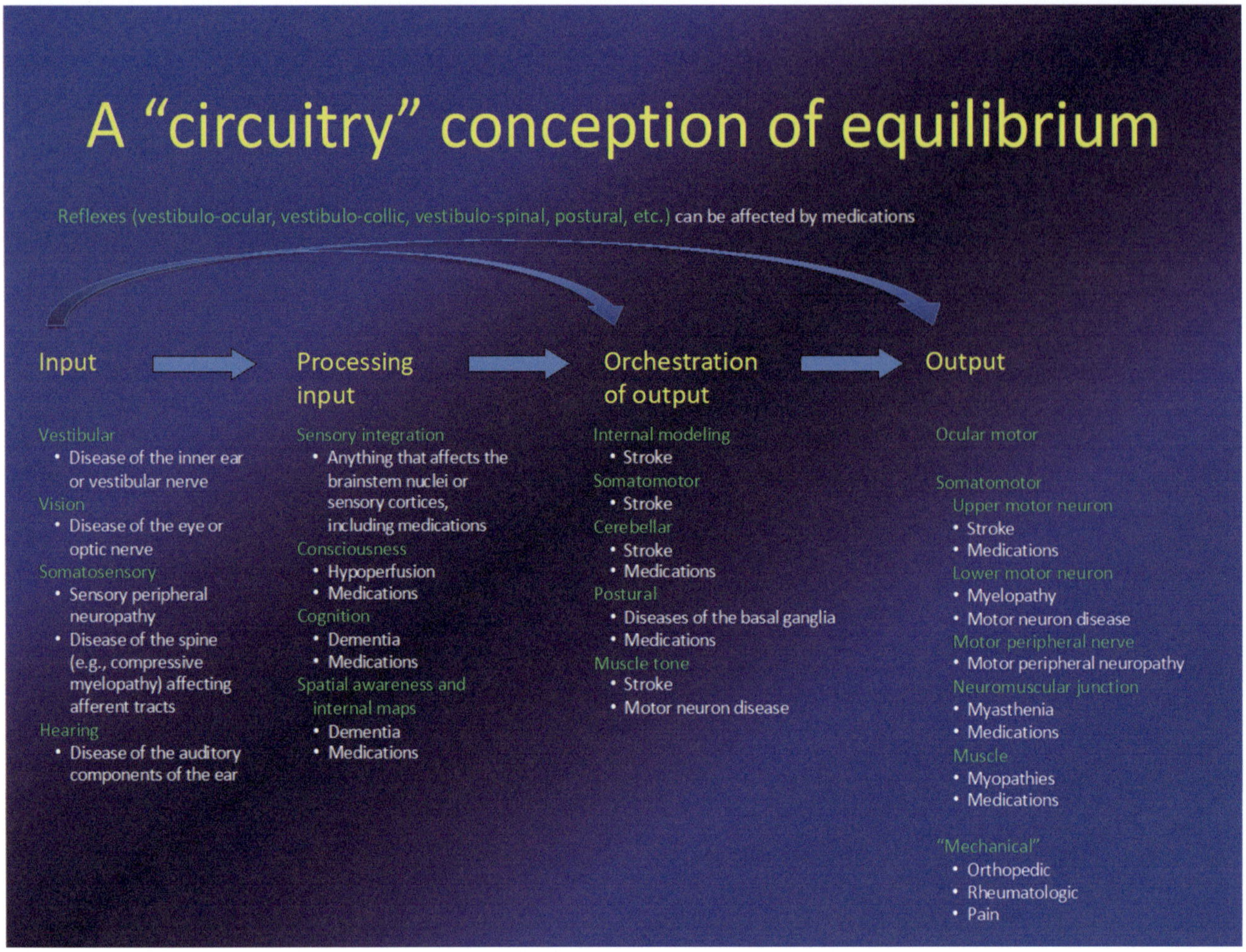

Fig. 170.1 A "circuitry" conception of equilibrium. (By Marcello Cherchi)

Auditory input enhances a person's spatial awareness and perception of movement [8]. Hearing loss is correlated with worsening of balance; conversely, hearing augmentation with hearing aids [9, 10] or cochlear implants [11–13] can improve a person's balance—though it remains unclear whether hearing amplification influences the clinically meaningful outcome of falls.

Processing Input: Sensory Integration

The brain does not only collect raw data from the various sensory input modalities. These inputs need to be interpreted and processed, and that processing may involve up-weighting or down-weighting a particular input, prioritizing the relevance of the inputs, arriving at an understanding of the environment (see "spatial awareness," below) and eventually combining those processed inputs into a coherent picture of a person's orientation in, and movement through, space. This interpretation, processing, and forming a coherent picture of one's environment, position, and movement is referred to as *sensory integration* [14, 15]. Sensory integration is probably also an important mechanism by which the brain "works around" sensory deficits (more on this below).

Processing Input: Spatial Awareness

Spatial awareness [16] has been an underappreciated aspect of the way a person develops and maintains a mental "map" of an environment. While this may be simply one component of a broader set of cognitive capacities, it likely influences a person's ability to negotiate an environment, and in this sense is relevant to "vestibular" function.

Orchestration of Output: Internal Modeling

At a high but nonconscious level, many of our motor functions—including the motor activities involved in maintenance of balance and ambulation—require us to make predictions about how our actions affect the world around us, and to "test" those predictions in real time. Computational biologists refer to this predictive task as *internal modeling*. This idea was originally proposed by Francis and Wonham [17], and further developed by researchers such as Wolpert and colleagues [18, 19].

An accessible example of internal modeling comes from what happens when you want to take a bottle of milk out of the refrigerator. If you see that the bottle of milk is full, you make a prediction that you will need to exert a certain amount of force to lift it up. If you believe that the milk bottle is empty when it is actually full, then your prediction of required force will be mistaken; the amount of force that you exert to lift the bottle will be greater than what is needed, and you will bump the milk bottle against the top of the refrigerator. This is diagrammed and discussed in detail by Wolpert and Ghahramani [18].

The ability to develop internal models is hard-wired, though the specific models themselves are built on data accumulated from real-world experience.

The underlying neuroanatomy of internal modeling is complex, and just beginning to be studied. Figure 170.2, from Scott [20], illustrates one proposal.

Consequences of the Many-to-Many Arrangement

We return to the "many-to-many" arrangement of this system, meaning that there are multiple sensory inputs and multiple motor outputs. This characteristic has advantages and drawbacks.

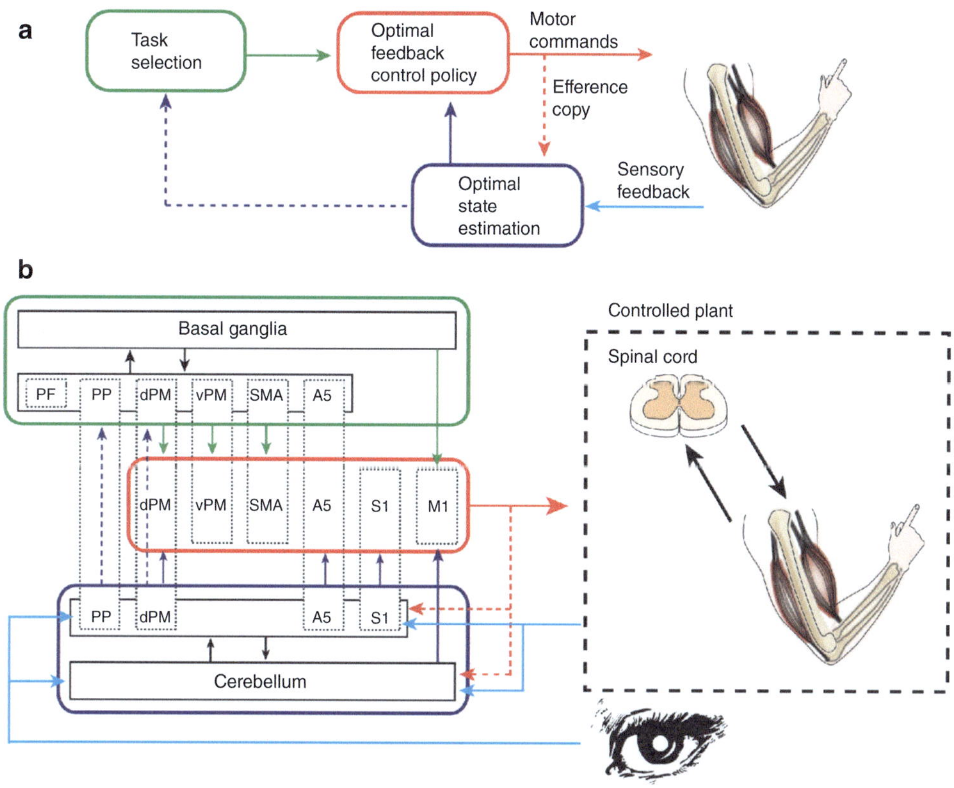

Fig. 170.2 Possible neuroanatomy underlying internal modeling. This diagrams optimal feedback control (OFC) and neural implementation. Panel (**a**): optimal feedback control policy converts state variables to motor commands. Optimal state estimation uses efference copy of motor commands and sensory feedback to estimate state variables. Task selection specifies the goal of the behavior which then defines the corresponding control policy. Panel (**b**): neural implementation of OFC-like voluntary control. The controlled plant is assumed to include the limb and spinal cord. The three basic processes, task selection, control policy, and state estimation, are each generated by highly distributed circuits. Dashed boxes denote individual cortical regions and some cortical regions are assumed to participate in two or even all three processes. Cortical connections are assumed amongst cortical regions involved in a given process. PF, PP, and S1 include multiple subdivisions not displayed for simplicity. Cortical abbreviations: PF prefrontal, PP posterior parietal, dPM dorsal premotor, vPM ventral premotor, SMA supplementary motor area, A5 area 5, S1 primary somatosensory, M1 primary motor. (From Scott [20]. Reproduced with written permission from Elsevier B.V)

An obvious drawback to this system is that there are many components, each of which is susceptible to a variety of pathologies. In other words, the fact that there are multiple components broadens its *vulnerability*.

In contrast, the multiplicity (of inputs and outputs) means that there is *redundancy*, which confers resilience against damage to any one modality. There is evidence that when confronted with diminished or unreliable input in one sensory modality, the brain may increase reliance on other modalities—a process sometimes referred to as "sensory reweighting" [21, 22] or "sensory substitution" [23]. For example, there is evidence that patients with vestibular loss begin to upweight their reliance on auditory cues for spatial awareness [24].

Returning to the "Multi" in Multifactorial Disequilibrium

The property of redundancy (in both sensory input and motor output) can partially offset the inherent vulnerability of this system—up to a certain point.

This is the point at which multisensory disequilibrium becomes relevant. While redundancy, sensory reweighting/substitution, and other strategies can compensate for sensory and motor deficits to a point, beyond that threshold, such compensation is no longer sufficient for adequate maintenance of equilibrium.

The relevance of multifactorial disequilibrium should now be clearer. A particular medical problem may interfere with only one component of the many-to-many arrangement described earlier, and the problem may be modest enough that, in isolation, the deficit that it imposes does not manifest as a clinically meaningful functional limitation that reaches a patient's awareness in the form of a disturbance of equilibrium.

But a human can be afflicted with multiple medical problems, of which each may interfere with a different component of the many-to-many arrangement. While each individual deficit by itself may not be sufficient to provoke disequilibrium, in aggregate, these deficits may start to interfere with the patient's equilibrium. Clinicians can envision plenty of circumstances in which this may occur.

An easy and common example would be poorly controlled diabetes mellitus. Diabetes can eventually lead to diabetic retinopathy, though this visual deficit in isolation may not cause disequilibrium. Diabetes can also lead to diabetic peripheral neuropathy, though this somatosensory deficit in isolation may not cause disequilibrium. Diabetes can also lead to diabetic autonomic neuropathy and problems such as orthostatic intolerance, though in isolation, this may not cause disequilibrium. However, if all three deficits are present simultaneously, then in aggregate they may cause disequilibrium. This is a circumstance in which one underlying disease (diabetes) causes multiple problems (retinopathy, peripheral neuropathy, autonomic neuropathy), each of which can contribute to disequilibrium.

There are also many examples of independently occurring diseases ultimately leading to multifactorial disequilibrium. This is more likely to happen with common diseases, simply because there is a higher likelihood that these will overlap (co-occur) in a single person. For example, a person who performs heavy labor outdoors may be susceptible to developing cataracts and arthritis. This person may also indulge in alcohol too frequently. In this scenario, there may be a visual deficit (from cataracts), a deficit in orchestration of output (alcoholic cerebellar degeneration) and a mechanical deficit (arthritis involving the spine, hips, knees, and ankles). Any one of these deficits in isolation may not have caused sufficient functional impairment for the person to seek care, but in aggregate, they may cause disequilibrium that brings the person to medical attention.

Clinical Presentation

One can imagine innumerable combinations and permutations of deficits in the "circuitry" of equilibrium described above, so there is no single presentation characteristic of multifactorial disequilibrium.

Physical Examination and Ocular Motor Examination

Findings on examination will depend on which pathologies are contributing to multifactorial disequilibrium.

Testing: Vestibular

Undertaking a screening otovestibular workup is nearly always reasonable for patients in whom multifactorial disequilibrium is suspected. Identification of an objective vestibular deficit can corroborate the suspicion for at least that component of the problem.

Testing: Other

Although patients with suspected multifactorial disequilibrium may initially be referred to otoneurology, the "multi" characteristic of this problem means that other specialists (and perhaps specialized testing) will be required, depending on the clinical scenario. For example, it may be appropriate to evaluate other sensory modalities, such as assessment of vision (consultation with ophthalmology) or somatosensory input (consultation with a specialist in neuropathy for possible electromyography and nerve conduction velocities). It may also be appropriate to evaluate motor output (consultation with neuromuscular neurology), check for basal ganglionic problems (consultation with a movement disorders specialist), or evaluate for mechanical limitations (consultation with an orthopedist or rheumatologist).

Differential Diagnosis

We have been repeating phrases such as, "each individual deficit in isolation may not cause disequilibrium, but in aggregate, multiple deficits can." In other words, we have been assuming that in patients with multifactorial disequilibrium, each individual deficit is modest.

However, one should bear in mind that even patients who truly have multifactorial disequilibrium are still vulnerable to more pronounced individual pathologies, and it is important to evaluate for these. For example, if the diabetic discussed earlier develops benign paroxysmal positional vertigo, or if the laborer discussed earlier develops vestibular neuritis, then these should be recognized and treated.

Treatment

Treatment of multifactorial disequilibrium will depend almost entirely on the pathologies identified.

If multiple deficits are found, but each is relatively modest, one should still consider a trial of vestibular rehabilitation therapy. Vestibular physical therapists are very skilled at training patients in strategies of sensory substitution, and diminishing fall risk. Occupational therapy consultation may also be appropriate.

Prognosis

The prognosis of multifactorial disequilibrium depends on the specific pathologies identified.

References

1. Tirado C, Gerdfeldter B, Karnekull SC, Nilsson ME. Comparing Echo-detection and Echo-localization in sighted individuals. Perception. 2021;50(4):308–27.
2. Flanagin VL, Schornich S, Schranner M, Hummel N, Wallmeier L, Wahlberg M, et al. Human exploration of enclosed spaces through echolocation. J Neurosci. 2017;37(6):1614–27.
3. Thaler L, Goodale MA. Echolocation in humans: an overview. Wiley Interdiscip Rev Cogn Sci. 2016;7(6):382–93.
4. Wallmeier L, Kish D, Wiegrebe L, Flanagin VL. Aural localization of silent objects by active human biosonar: neural representations of virtual echo-acoustic space. Eur J Neurosci. 2015;41(5):533–45.
5. Wallmeier L, Gessele N, Wiegrebe L. Echolocation versus echo suppression in humans. Proc Biol Sci. 2013;280(1769):20131428.
6. Schornich S, Wallmeier L, Gessele N, Nagy A, Schranner M, Kish D, et al. Psychophysics of human echolocation. Adv Exp Med Biol. 2013;787:311–9.
7. Schornich S, Nagy A, Wiegrebe L. Discovering your inner bat: echo-acoustic target ranging in humans. J Assoc Res Otolaryngol. 2012;13(5):673–82.
8. Campos J, Ramkhalawansingh R, Pichora-Fuller MK. Hearing, self-motion perception, mobility, and aging. Hear Res. 2018;369:42–55.
9. Negahban H, Bavarsad Cheshmeh Ali M, Nassadj G. Effect of hearing aids on static balance function in elderly with hearing loss. Gait Posture. 2017;58:126 9.
10. Rumalla K, Karim AM, Hullar TE. The effect of hearing aids on postural stability. Laryngoscope. 2015;125(3):720–3.
11. Weaver TS, Shayman CS, Hullar TE. The effect of hearing aids and cochlear implants on balance during gait. Otol Neurotol. 2017;38(9):1327–32.
12. Shayman CS, Earhart GM, Hullar TE. Improvements in gait with hearing aids and cochlear implants. Otol Neurotol. 2017;38(4):484–6.
13. Mazaheryazdi M, Moossavi A, Sarrafzadah J, Talebian S, Jalaie S. Study of the effects of hearing on static and dynamic postural function in children using cochlear implants. Int J Pediatr Otorhinolaryngol. 2017;100:18–22.
14. Peterka RJ, Loughlin PJ. Dynamic regulation of sensorimotor integration in human postural control. J Neurophysiol. 2004;91(1):410–23.
15. Peterka RJ. Sensorimotor integration in human postural control. J Neurophysiol. 2002;88(3);1097–118.
16. Ishikawa T. Spatial thinking, cognitive mapping, and spatial awareness. Cogn Process. 2021;22(Suppl 1):89–96.
17. Francis BA, Wonham WM. The internal model principle of control theory. Automatica. 1976;12(5):457–65.
18. Wolpert DM, Ghahramani Z. Computational principles of movement neuroscience. Nat Neurosci. 2000;3(Suppl):1212–7.
19. Wolpert DM, Ghahramani Z, Jordan MI. An internal model for sensorimotor integration. Science. 1995;269(5232):1880–2.
20. Scott SH. The computational and neural basis of voluntary motor control and planning. Trends Cogn Sci. 2012;16(11):541–9.
21. Asslander L, Peterka RJ. Sensory reweighting dynamics in human postural control. J Neurophysiol. 2014;111(9):1852–64.
22. Carver S, Kiemel T, Jeka JJ. Modeling the dynamics of sensory reweighting. Biol Cybern. 2006;95(2):123–34.
23. Sadeghi SG, Minor LB, Cullen KE. Neural correlates of sensory substitution in vestibular pathways following complete vestibular loss. J Neurosci. 2012;32(42):14685–95.
24. Vitkovic J, Le C, Lee SL, Clark RA. The contribution of hearing and hearing loss to balance control. Audiol Neurootol. 2016;21(4):195–202.

Presbyvestibulopathy

171

Overview

The term "*presbyvestibulopathy*" refers to disequilibrium that is only attributable to age-related reduction of vestibular function. Diagnostic criteria specify quantitative vestibular test results for this diagnosis. Histopathological studies have shown clear evidence of what appears to be age-related degeneration of most structures along the vestibular pathway (otoconia, vestibular hair cells, vestibular ganglion, vestibular nerve, vestibular nuclei), which is the likely mechanism underlying both the abnormal responses in quantitative vestibular function tests and the symptom of disequilibrium. A clinician should bear in mind that healthy aging is also associated with attrition of other sensoria (presbyopia, reduced proprioception), so in many if not most instances, reduced vestibular function is actually only one of several contributors to the symptom of disequilibrium, and thus multifactorial disequilibrium is more commonly encountered than "pure" presbyvestibulopathy. A trial of vestibular rehabilitation therapy is appropriate.

Introduction

The idea that the symptom of disequilibrium can arise exclusively from age-related deterioration of the vestibular system is not new [1], but a formal definition was proposed only relatively recently.

Agrawal and colleagues [2] authored a consensus paper for the Barany Society's classification committee proposing diagnostic criteria for *presbyvestibulopathy*, which they define as "Mild or incomplete vestibular losses attributable to the normative aging process." They purposefully choose this term to designate *incomplete* vestibular loss, analogous to terms such as "presbyopia" (age-related vision decline but not complete blindness) and "presbycusis" (age-related hearing loss but not complete deafness).

The specific diagnostic criteria they propose for presbyvestibulopathy are as follows:

A. Chronic vestibular syndrome (at least 3 months duration) with at least two of the following symptoms:
   1. Postural imbalance or unsteadiness
   2. Gait disturbance
   3. Chronic dizziness
   4. Recurrent falls
B. Mild bilateral peripheral vestibular hypofunction documented by at least one of the following:
   1. VOR [vestibulo-ocular reflex] gain measured by video-HIT [head impulse testing] between 0.6 and 0.8 bilaterally.
   2. VOR gain between 0.1 and 0.3 upon sinusoidal stimulation on rotatory chair (0.1 Hz, $V_{max} = 50°–60°/s$).
   3. Reduced caloric response (sum of bithermal maximum peak SPV [slow-phase velocity] on each side between 6° and 25°/s).
C. Age ≥60 years.
D. Not better accounted for by another disease or disorder.

Epidemiology

The epidemiology of presbyvestibulopathy per the Barany Society criteria is unknown. Many studies cite the increase in fall risk with age [3], but this is not a proxy for presbyvestibulopathy itself.

Pathophysiological Mechanism of Disease

If one takes the working definition of presbyvestibulopathy proposed by Agrawal and colleagues [2], the mechanism is reduced vestibular function due to the natural attrition that occurs in normal healthy aging. Data from histopathological studies support this idea (see below).

In addition, the neural plasticity mechanisms (which usually can help compensate to some extent for sensory deficits)

M. Cherchi, *Otoneurology and Vestibular Medicine*, https://doi.org/10.1007/978-3-031-94841-1_171

become less effective with age [4], so this "rescue mechanism" becomes progressively less available over time.

Clinical Presentation

Patients diagnosed with presbyvestibulopathy often describe an insidious onset of imbalance or unsteadiness, and because of the gradual onset, they may have difficulty pinpointing the start of symptoms. If a patient describes an apoplectic onset of disequilibrium, then other diagnoses should be considered.

Physical Examination

In patients meeting the Barany Society criteria [2] for presbyvestibulopathy and carrying no other diagnosis, physical examination is generally unrevealing. These patients perform poorly on unsighted tandem Romberg testing, but this is nonspecific since performance on that test declines with age in healthy individuals [5].

Testing: Vestibular

The only quantitative factors included in the diagnostic criteria proposed by Agrawal and colleagues [2] are video head impulse testing, caloric testing (on videonystagmography), and slow harmonic acceleration (on rotatory chair testing). Two of these tests are reasonably widely available (video head impulse testing, videonystagmography), while one is not (rotatory chair testing).

However, there is compelling evidence that responses on each clinical vestibular test deteriorate in normal healthy aging [6]:

- *Vestibulo-ocular reflexes*:
  - Vestibulo-ocular reflex (low end of the vestibular tuning frequency spectrum) of the horizontal semicircular canal, as measured on caloric testing [7–10].
  - Vestibulo-ocular reflex (middle of the vestibular tuning frequency spectrum), as measured on rotatory chair testing [11–13].
  - Vestibulo-ocular reflex gain (high-end of the vestibular tuning frequency spectrum), as measured on video head impulse testing [14–17].
  - Step velocity testing as measured on the rotatory chair testing [18].
- *Saccadic function*: Saccadic latency, velocity, and accuracy, as measured on oculography [19, 20].
- *Pursuit*: Smooth pursuit, as measured on oculography [21–25].
- *Optokinetic responses*:
  - Horizontal optokinetic responses, as measured on rotatory chair testing [11, 26].
  - Torsional optokinetic responses [27].
- *Otolith responses*:
  - Cervical vestibular evoked myogenic potentials' latencies, thresholds, and amplitudes [7, 9, 28–32].
  - Ocular vestibular evoked myogenic potentials' latencies, thresholds, and amplitudes [7, 32, 33].
  - Subjective visual vertical [34–38].
- *Postural responses*, as measured on computerized dynamic posturography [39, 40].

Despite the current limited definition of presbyvestibulopathy promulgated by the Barany Society [2], in clinical practice, we take into account the results of all vestibular tests when attempting to secure this diagnosis. We anticipate that future revisions of the diagnostic criteria for presbyvestibulopathy are likely to do the same.

Histopathology

Histopathological studies of the temporal bone and brain clearly document age-related changes in vestibular structures in otherwise healthy individuals, including the following:

- *Labyrinth*:
  - Reduced number of otoconia, degenerated otoconia [41].
  - Accumulation of lipofuscin inclusions in vestibular hair cells [41] and sustentacular cells [42].
  - Deformation of cilia [41, 42].
  - Decreasing hair cell counts [41, 43] of both type I and type II vestibular hair cells [44] and therefore decreasing hair cell densities.
- *Vestibular ganglion*: Decreasing number of nerve cells in Scarpa's ganglion [45].
- *Vestibular nerve*: Decreasing number of vestibular fibers [46].
- *Brainstem*: Loss of neurons in the vestibular nuclei [47].

Differential Diagnosis

To this point, we have discussed presbyvestibulopathy in the sense intended by the Barany Society's diagnostic criteria [2], which is to say that exclusively vestibular deficits account for disequilibrium ("Not better accounted for by another disease or disorder").

For purposes of research, it is desirable to isolate a single variable (in this case, "vestibular function") and observe its activity while keeping all other variables unchanged. In a complex biological system, this is rarely achievable.

Thus, we appreciate the Barany Society's proposal of a definition for presbyvestibulopathy as a theoretical construct. In practice, a clinician would rarely encounter an elderly person presenting with a chief complaint of disequilibrium who turns out to have only mild vestibular weakness and no other pathology at all. Even if an individual carries no other formal diagnoses, all sensory modalities undergo attrition with normal healthy aging. For example, it is well established that in normal healthy aging, there is a progressive decline in vision [48] and proprioception [49, 50].

In other words, in clinical practice, presbyvestibulopathy in isolation is probably rarely, if ever, encountered. The more likely scenario is that when an older person complains of disequilibrium, there are several factors contributing to that presentation [51], of which reduced vestibular function is only one. Thus, the otoneurologist and neuro-otologist should keep the likelihood of multifactorial disequilibrium in mind.

In such patients, it is also important to consider common vestibular diseases, such as benign paroxysmal positional vertigo, whose incidence increases with age.

Treatment

For a patient who meets the Barany Society's diagnostic criteria of presbyvestibulopathy [2], it is always reasonable to attempt a trial of vestibular rehabilitation therapy.

Prognosis

A process of age-related deterioration tends to be progressive, and this likely also applies to the clinical manifestations of presbyvestibulopathy.

References

1. Matheson AJ, Darlington CL, Smith PF. Dizziness in the elderly and age-related degeneration of the vestibular system. NZ J Psychol. 1999;28(1):10–6.
2. Agrawal Y, Van de Berg R, Wuyts F, Walther L, Magnusson M, Oh E, et al. Presbyvestibulopathy: diagnostic criteria consensus document of the classification committee of the Barany Society. J Vestib Res. 2019;29(4):161–70.
3. Zalewski CK. Aging of the human vestibular system. Semin Hear. 2015;36(3):175–96.
4. Paige GD. Senescence of human visual-vestibular interactions. 1. Vestibulo-ocular reflex and adaptive plasticity with aging. J Vestib Res. 1992;2(2):133–51.
5. Agrawal Y, Carey JP, Hoffman HJ, Sklare DA, Schubert MC. The modified Romberg balance test: normative data in U.S. adults. Otol Neurotol. 2011;32(8):1309–11.
6. Bermudez Rey MC, Clark TK, Wang W, Leeder T, Bian Y, Merfeld DM. Vestibular perceptual thresholds increase above the age of 40. Front Neurol. 2016;7:162.
7. Agrawal Y, Zuniga MG, Davalos-Bichara M, Schubert MC, Walston JD, Hughes J, Carey JP. Decline in semicircular canal and otolith function with age. Otol Neurotol. 2012;33(5):832–9.
8. Hajioff D, Barr-Hamilton RM, Colledge NR, Lewis SJ, Wilson JA. Re-evaluation of normative electronystagmography data in healthy ageing. Clin Otolaryngol Allied Sci. 2000;25(4):249–52.
9. Maes L, Dhooge I, D'Haenens W, Bockstael A, Keppler H, Philips B, et al. The effect of age on the sinusoidal harmonic acceleration test, pseudorandom rotation test, velocity step test, caloric test, and vestibular-evoked myogenic potential test. Ear Hear. 2010;31(1):84–94.
10. Peterka RJ, Black FO, Schoenhoff MB. Age-related changes in human vestibulo-ocular reflexes: sinusoidal rotation and caloric tests. J Vestib Res. 1990;1(1):49–59.
11. Baloh RW, Jacobson KM, Socotch TM. The effect of aging on visual-vestibuloocular responses. Exp Brain Res. 1993;95(3):509–16.
12. Peterka RJ, Black FO, Schoenhoff MB. Age-related changes in human vestibulo-ocular and optokinetic reflexes: pseudorandom rotation tests. J Vestib Res. 1990;1(1):61–71.
13. Chan FM, Galatioto J, Amato M, Kim AH. Normative data for rotational chair stratified by age. Laryngoscope. 2016;126(2):460–3.
14. Guerra Jimenez G, Perez FN. Reduction in posterior semicircular canal gain by age in video head impulse testing. Observational study. Acta Otorrinolaringol Esp. 2016;67(1):15–22.
15. Kim TH, Kim MB. Effect of aging and direction of impulse in video head impulse test. Laryngoscope. 2018;128(6):E228–E33.
16. Matino-Soler E, Esteller-More E, Martin-Sanchez JC, Martinez-Sanchez JM, Perez-Fernandez N. Normative data on angular vestibulo-ocular responses in the yaw axis measured using the video head impulse test. Otol Neurotol. 2015;36(3):466–71.
17. Mossman B, Mossman S, Purdie G, Schneider E. Age dependent normal horizontal VOR gain of head impulse test as measured with video-oculography. J Otolaryngol Head Neck Surg. 2015;44(1):29.
18. DiZio P, Lackner JR. Age differences in oculomotor responses to step changes in body velocity and visual surround velocity. J Gerontol. 1990;45(3):M89–94.
19. Irving EL, Lillakas L. Difference between vertical and horizontal saccades across the human lifespan. Exp Eye Res. 2019;183:38–45.
20. Warabi T, Kase M, Kato T. Effect of aging on the accuracy of visually guided saccadic eye movement. Ann Neurol. 1984;16(4):449–54.
21. Kato I, Sakuma A, Ogino S, Takahashi K, Okada T. Aging effects upon pursuit eye movements. Acta Otolaryngol Suppl. 1995;520(Pt 2):293–4.
22. Sakuma A, Ogino S, Takahashi K, Kato I. Aging effects upon smooth pursuit induced by step-ramp stimulation. Auris Nasus Larynx. 2000;27(3):195–200.
23. Sharpe JA, Sylvester TO. Effect of aging on horizontal smooth pursuit. Invest Ophthalmol Vis Sci. 1978;17(5):465–8.
24. Spooner JW, Sakala SM, Baloh RW. Effect of aging on eye tracking. Arch Neurol. 1980;37(9):575–6.
25. Zackon DH, Sharpe JA. Smooth pursuit in senescence. Effects of target acceleration and velocity. Acta Otolaryngol. 1987;104(3–4):290–7.
26. Lynch SJ, Nayak US, Isaacs B. Positional and optokinetic nystagmus in healthy old people. Age Ageing. 1985;14(2):122–4.
27. Farooq SJ, Gottlob I, Benskin S, Proudlock FA. The effect of aging on torsional optokinetic nystagmus. Invest Ophthalmol Vis Sci. 2008;49(2):589–93.

28. Ochi K, Ohashi T. Age-related changes in the vestibular-evoked myogenic potentials. Otolaryngol Head Neck Surg. 2003;129(6):655–9.
29. Janky KL, Shepard N. Vestibular evoked myogenic potential (VEMP) testing: normative threshold response curves and effects of age. J Am Acad Audiol. 2009;20(8):514–22.
30. Su HC, Huang TW, Young YH, Cheng PW. Aging effect on vestibular evoked myogenic potential. Otol Neurotol. 2004;25(6):977–80.
31. Welgampola MS, Colebatch JG. Vestibulocollic reflexes: normal values and the effect of age. Clin Neurophysiol. 2001;112(11):1971–9.
32. Bi X, Zhou HF, Su J, Zhang J, Wang MX. The normative values of vestibular evoked myogenic potential in different age-groups. Lin Chung Er Bi Yan Hou Tou Jing Wai Ke Za Zhi. 2016;30(13):1016–20.
33. Singh NK, Firdose H. Characterizing the impact of advancing age on 500 Hz tone-burst evoked ocular-vestibular evoked myogenic potentials. Eur Arch Otorrinolaringol. 2021;278:4259.
34. Baccini M, Paci M, Del Colletto M, Ravenni M, Baldassi S. The assessment of subjective visual vertical: comparison of two psychophysical paradigms and age-related performance. Atten Percept Psychophys. 2014;76(1):112–22.
35. Cakrt O, Slaby K, Kmet J, Kolar P, Jerabek J. Subjective visual and haptic vertical in young and elderly. J Vestib Res. 2016;25(5–6):195–9.
36. Fukata K, Amimoto K, Fujino Y, Inoue M, Inoue M, Takahashi Y, et al. The effects of aging on the subjective vertical in the frontal plane in healthy adults. J Phys Ther Sci. 2017;29(11):1950–3.
37. Toupet M, Van Nechel C, Grayeli AB. Subjective visual vertical tilt attraction to the side of rod presentation: effects of age, sex, and vestibular disorders. Otol Neurotol. 2015;36(6):1074–80.
38. Zakaria MN, Salim R, Tahir A, Zainun Z, Mohd Sakeri NS. The influences of age, gender and geometric pattern of visual image on the verticality perception: a subjective visual vertical (SVV) study among Malaysian adults. Clin Otolaryngol. 2019;44(2):166–71.
39. Peterka RJ, Black FO. Age-related changes in human posture control: motor coordination tests. J Vestib Res. 1990;1(1):87–96.
40. Peterka RJ, Black FO. Age-related changes in human posture control: sensory organization tests. J Vestib Res. 1990;1(1):73–85.
41. Walther LE, Westhofen M. Presbyvertigo-aging of otoconia and vestibular sensory cells. J Vestib Res. 2007;17(2–3):89–92.
42. Rosenhall U, Rubin W. Degenerative changes in the human vestibular sensory epithelia. Acta Otolaryngol. 1975;79(1–2):67–80.
43. Rauch SD, Velazquez-Villasenor L, Dimitri PS, Merchant SN. Decreasing hair cell counts in aging humans. Ann N Y Acad Sci. 2001;942:220–7.
44. Merchant SN, Velazquez-Villasenor L, Tsuji K, Glynn RJ, Wall C 3rd, Rauch SD. Temporal bone studies of the human peripheral vestibular system. Normative vestibular hair cell data. Ann Otol Rhinol Laryngol Suppl. 2000;181:3–13.
45. Richter E. Quantitative study of human Scarpa's ganglion and vestibular sensory epithelia. Acta Otolaryngol. 1980;90(3–4):199–208.
46. Rasmussen AT. Studies of the VIIIth cranial nerve of man. Laryngoscope. 1940;50:67–83.
47. Lopez I, Honrubia V, Baloh RW. Aging and the human vestibular nucleus. J Vestib Res. 1997;7(1):77–85.
48. Erdinest N, London N, Lavy I, Morad Y, Levinger N. Vision through healthy aging eyes. Vision (Basel). 2021;5(4):46.
49. Robbins S, Waked E, McClaran J. Proprioception and stability: foot position awareness as a function of age and footwear. Age Ageing. 1995;24(1):67–72.
50. Skinner HB, Barrack RL, Cook SD. Age-related decline in proprioception. Clin Orthop Relat Res. 1984;184:208–11.
51. Tuunainen E, Poe D, Jantti P, Varpa K, Rasku J, Toppila E, Pyykko I. Presbyequilibrium in the oldest old, a combination of vestibular, oculomotor and postural deficits. Aging Clin Exp Res. 2011;23(5–6):364–71.

Drop Attacks

172

Overview

The phrase "*drop attack*" refers to a clinical presentation in which there is no prodrome, there is an abrupt onset of falling to the ground, an abrupt offset of symptoms, and no loss of consciousness. The broad differential diagnosis includes otologic, neurologic, and cardiovascular etiologies. Undertaking a screening otovestibular workup is reasonable. If that is unremarkable, then depending on the history, referral to a sleep specialist or cardiologist may be warranted.

Definition

The phrase "*drop attack*" refers to a clinical presentation in which there is no prodrome, there is an abrupt onset of falling to the ground, an abrupt offset of symptoms, and no loss of consciousness. Patients use different descriptors, such as feeling "thrown" or "shoved" to the ground.

The absence of any prodrome or warning is an especially problematic feature of drop attacks, and can lead to injury.

Etiologies

Drop attacks have been reported in a broad range of diseases, with a differential diagnosis including the following:

Otologic

- Otolithic catastrophe of Tumarkin as a late stage of Ménière's disease [1, 2]
- Perilymphatic fistula [3]
- Peripheral pathologies, undifferentiated [4–6]
- Superior semicircular canal dehiscence [7]
- Vestibular paroxysmia [8]

Neurologic

- Arteriovenous fistula [9]
- Benign hereditary chorea [10]
- Cataplexy, even in the absence of narcolepsy [11]
- Chiari malformation [12, 13]
- Coffin-Lowry syndrome [14]
- Cyst of the lateral ventricle [15]
- Cyst of the third ventricle [16, 17]
- Epilepsy
  - Doose syndrome [18]
  - Lennox-Gastaut syndrome [19–21]
  - Rett syndrome [22]
  - Seizures, undifferentiated [23–25]
- Encephalitis [26]
- Migraine [2, 4]
- Startle [27]

Cardiovascular

- Cardiac valve abnormalities [28]

Cryptogenic

- One report documented "cryptogenic drop attacks: an affliction of women" [29]

Diagnostic Workup

Although the differential diagnosis listed above is impressive, practically, many of these patients do not present to an otoneurology or neuro-otology clinic. For example, the syndromic epilepsies (Doose syndrome, Lennox-Gastaut syn-

M. Cherchi, *Otoneurology and Vestibular Medicine*, https://doi.org/10.1007/978-3-031-94841-1_172

drome, Rett syndrome) have such dramatic clinical presentations that these patients are usually already under the care of an epileptologist who would be well aware of their association with drop attacks.

Consequently, from the otoneurological perspective, a reasonable workup may include the following:

- *Audiometry*. Low-frequency hearing loss may suggest Ménière's disease. Conductive hyperacusis may suggest semicircular canal dehiscence.
- *Cervical and ocular vestibular evoked myogenic potentials with thresholds*. An abnormally large response, with ipsilaterally reduced thresholds, may suggest semicircular canal dehiscence.
- *Other tests of unilateral vestibular function*, such as video head impulse testing and videonystagmography. These may suggest lateralizing pathology, such as a prior bout of vestibular neuritis.
- If any otovestibular test results are compatible with semicircular canal dehiscence, then consider a *temporal bone CT*.
- Consider *MRI of the brain and internal auditory canals without and with contrast*. This may detect structural otologic lesions, Chiari malformations, ventricular cysts, and cerebral malformations.

If all the above are unrevealing, then consider the following:

- Referral to a sleep disorders neurologist to be evaluated for cataplexy.
- If the history is ambiguous regarding loss of consciousness, then consider referral to a cardiologist.

Although migraine is on the differential diagnosis of drop attacks, we have never seen migraine presenting in this way. Such presentation must be exceedingly rare, and in any case, would be a diagnosis of exclusion; thus, we would be hesitant to treat empirically for presumptive migraine as an initial management strategy for drop attacks.

References

1. Baloh RW, Jacobson K, Winder T. Drop attacks with Meniere's syndrome. Ann Neurol. 1990;28(3):384–7.
2. Pyykko I, Pyykko N, Manchaiah V. Vestibular drop attacks in Meniere's disease and its association with migraine. Eur Arch Otorrinolaringol. 2020;277(7):1907–16.
3. Pace A, Iannella G, Rossetti V, Colizza A, Magliulo G. Vestibular drop attack: a potential origin from perilymphatic fistula? Med Hypotheses. 2020;140:109668.
4. Ishiyama G, Ishiyama A, Baloh RW. Drop attacks and vertigo secondary to a non-meniere otologic cause. Arch Neurol. 2003;60(1):71–5.
5. Ishiyama G, Ishiyama A, Jacobson K, Baloh RW. Drop attacks in older patients secondary to an otologic cause. Neurology. 2001;57(6):1103–6.
6. Lee H, Yi HA, Lee SR, Ahn BH, Park BR. Drop attacks in elderly patients secondary to otologic causes with Meniere's syndrome or non-Meniere peripheral vestibulopathy. J Neurol Sci. 2005;232(1–2):71–6.
7. Brantberg K, Ishiyama A, Baloh RW. Drop attacks secondary to superior canal dehiscence syndrome. Neurology. 2005;64(12):2126–8.
8. Kim S, Kong SK, Lee JM, Oh SJ. Drop attack treated by microvascular decompression of the superior vestibular nerve. Ear Nose Throat J. 2024;103:NP179.
9. Oh HJ, Yoon SM, Kim SH, Shim JJ. A case of pial arteriovenous fistula with giant venous aneurysm and multiple varices treated with coil embolization. J Korean Neurosurg Soc. 2011;50(3):248–51.
10. Rosati A, Berti B, Melani F, Cellini E, Procopio E, Guerrini R. Recurrent drop attacks in early childhood as presenting symptom of benign hereditary chorea caused by TITF1 gene mutations. Dev Med Child Neurol. 2015;57(8):777–9.
11. Egel RT, Lee A, Bump T, Javois A. Isolated cataplexy in the differential diagnosis of drop attacks: a case of successful clinical diagnosis and treatment. Case Rep Neurol Med. 2012;2012:757586.
12. Bardella L, Maleci A, Di Lorenzo N. Drop attack as the only symptom of type 1 Chiari malformation. Illustration by a case. Riv Patol Nerv Ment. 1984;105(5):217–22.
13. Straus D, Foster K, Zimmerman F, Frim D. Chiari drop attacks: surgical decompression and the role of tilt table testing. Pediatr Neurosurg. 2009;45(5):384–9.
14. Havaligi N, Matadeen-Ali C, Khurana DS, Marks H, Kothare SV. Treatment of drop attacks in coffin-Lowry syndrome with the use of sodium oxybate. Pediatr Neurol. 2007;37(5):373–4.
15. Bansal A, Kumar A, Sharma MC, Purkait S, Sharma BS. Lateral ventricular cystic choroid plexus papilloma presenting with recurrent drop attacks: a rare manifestation of a rare variant. Neurol India. 2015;63(4):619–21.
16. Pollack IF, Schor NF, Martinez AJ, Towbin R. Bobble-head doll syndrome and drop attacks in a child with a cystic choroid plexus papilloma of the third ventricle. J Neurosurg. 1995;83(4):729–32.
17. Zohrevandi B, Monsef Kasmaie V, Asadi P, Tajik H. Third ventricle colloid cyst as a cause of sudden drop attacks of a 13-year-old boy. Emergency. 2015;3(4):162–4.
18. Nickels K, Kossoff EH, Eschbach K, Joshi C. Epilepsy with myoclonic-atonic seizures (Doose syndrome): clarification of diagnosis and treatment options through a large retrospective multicenter cohort. Epilepsia. 2021;62(1):120–7.
19. Gresham J, Eiland LS, Chung AM. Treating Lennox-Gastaut syndrome in epileptic pediatric patients with third-generation rufinamide. Neuropsychiatr Dis Treat. 2010;6:639–45.
20. VanStraten AF, Ng YT. Update on the management of Lennox-Gastaut syndrome. Pediatr Neurol. 2012;47(3):153–61.
21. Verdian L, Yi Y. Cost-utility analysis of rufinamide versus topiramate and lamotrigine for the treatment of children with Lennox-Gastaut syndrome in the United Kingdom. Seizure. 2010;19(1):1–11.
22. Ueda K, Sood S, Asano E, Kumar A, Luat AF. Elimination of medically intractable epileptic drop attacks following endoscopic total corpus callosotomy in Rett syndrome. Childs Nerv Syst. 2017;33(11):1883–7.
23. Jea A, Vachhrajani S, Johnson KK, Rutka JT. Corpus callosotomy in children with intractable epilepsy using frameless stereotactic neuronavigation: 12-year experience at the Hospital for Sick Children in Toronto. Neurosurg Focus. 2008;25(3):E7.
24. Maehara T, Shimizu H. Surgical outcome of corpus callosotomy in patients with drop attacks. Epilepsia. 2001;42(1):67–71.
25. Sunaga S, Shimizu H, Sugano H. Long-term follow-up of seizure outcomes after corpus callosotomy. Seizure. 2009;18(2):124–8.

26. Vives-Rodriguez A, Sivaraju A, Louis ED. Drop attacks: a clinical manifestation of LGI1 encephalitis. Neurol Clin Pract. 2017;7(5):442–3.
27. Toelle SP, Wille D, Schmitt B, Scheer I, Thony B, Plecko B. Sensory stimulus-sensitive drop attacks and basal ganglia calcification: new findings in a patient with FOLR1 deficiency. Epileptic Disord. 2014;16(1):88–92.
28. Bossert T, Krakor R, Garbade J, Mohr FW. Fibroelastoma of the aortic valve causing drop attack. Eur J Cardiothorac Surg. 2003;24(4):639.
29. Stevens DL, Matthews WB. Cryptogenic drop attacks: an affliction of women. Br Med J. 1973;1(5851):439–42.

Coital Vertigo

173

Overview

Coital vertigo refers to disequilibrium typically beginning during or shortly after orgasm, lasting minutes to an hour. There is a paucity of literature on this subject, perhaps because patients are often embarrassed to report the circumstances of symptom onset. The differential diagnosis is thought to include otologic, neurologic, and cardiovascular etiologies.

Introduction

Disequilibrium occurring during or shortly after orgasm is referred to in the literature as *coital vertigo*. Symptoms usually last minutes (less commonly closer to an hour). This sometimes occurs due to the appearance of a perilymphatic fistula (in a patient with a prior history of otologic surgery) or due to other otologic diseases (such as superior semicircular canal dehiscence, Ménière's disease, benign paroxysmal positional vertigo). Other proposed mechanisms include cardiovascular (cardiac arrhythmia, hyperventilation, postorgasmic hypotension) and perhaps coital migraine (with migraine-associated vertigo). Workup for these possibilities is reasonable. The prognosis depends on the underlying cause.

Epidemiology

Coital vertigo is infrequently reported (perhaps because patients are embarrassed to report the circumstances of symptom onset), making it difficult to estimate incidence or prevalence.

Pathophysiological Mechanism of Disease

The literature suggests several possible mechanisms for coital vertigo.

- *Otologic Etiologies*:
  - Dawlatly [1] reported on four patients who had undergone various ear surgeries and developed episodes of disequilibrium during or immediately after sexual intercourse, attributed to the appearance of a perilymph fistula.
  - Lee and colleagues [2] reported on several patients in whom primary otologic disease appeared to be the mechanism of coital vertigo, including superior semicircular canal dehiscence (SSCD) and Ménière's disease.
  - Given the positions and changes of position that may occur during sexual activity, benign paroxysmal positional vertigo (BPPV) is a consideration.
- *Neurologic Etiologies:* Coital headache is fairly well described [3–6], and some such headaches are likely to be migraine [7], which in turn could manifest with migraine-associated vertigo (MAV).
- *Cardiovascular Etiologies:* Mann and colleagues [8] described a case of coital *syncope* and reviewed other reported cases, whose proposed mechanisms included cardiac arrhythmia, hyperventilation, and postorgasmic hypotension. Frank syncope is not usually experienced as disequilibrium, but can sometimes be confused therewith both by patients and clinicians.

Lee and colleagues [2] note that in the series of seven patients they described, two (29%) had no identifiable cause of coital vertigo.

M. Cherchi, *Otoneurology and Vestibular Medicine*, https://doi.org/10.1007/978-3-031-94841-1_173

Clinical Presentation

Lee and Kim [2] provided a case series of seven patients with coital vertigo. They reported the duration of disequilibrium to range from "a few minutes to 1 h." The patients we have encountered with this condition generally describe the symptom duration toward the shorter end of this range (a few minutes).

Lee and Kim [2] state that in their series of seven patients, 6 (86%) reported auditory symptoms, including tinnitus in 4 (57%), aural fullness in 2 (29%), autophony in 1 (14%), hearing impairment in 1 (14%), and hyperacusis in 1 (14%).

Physical Examination

Physical examination in patients complaining of coital vertigo is generally normal, though there are a few exceptions.

If coital vertigo is due to benign paroxysmal positional vertigo (BPPV), then the corresponding nystagmus may be observable on video oculography (VOG).

If coital vertigo is due to superior semicircular canal dehiscence (SSCD), then Valsalva-induced nystagmus may be observable on video oculography (VOG).

Differential Diagnosis

Given the mechanisms of disease described earlier, the differential diagnosis of coital vertigo includes the following:

- Development of a perilymphatic fistula in a previously operated ear
- Superior semicircular canal dehiscence (SSCD)
- Ménière's disease
- Benign paroxysmal positional vertigo (BPPV)
- Migraine-associated vertigo (MAV)
- Cardiovascular causes (arrhythmia, hyperventilation, postorgasmic hypotension)

Treatment

If a patient with coital vertigo reports a prior history of ear disease (particularly any kind of otologic surgery), then referral to an otolaryngologist is appropriate.

If there is no known prior history of otologic disease, then a medically reasonable workup includes the following:

- Audiometry to evaluate for Ménière's disease.
- Cervical and ocular vestibular evoked myogenic potentials (VEMP) with thresholds to screen for superior semicircular canal dehiscence (SSCD). If compatible with SSCD, then consider checking a temporal bone CT without contrast for confirmation.
- Electrocardiogram.

If the above are unrevealing, then referral to a cardiologist (e.g., to evaluate for cardiac arrhythmia) is medically reasonable.

Prognosis

The prognosis depends on the underlying cause, if determinable.

References

1. Dawlatly EE. Coital vertigo after ear surgery: when is sex safe? Am J Otolaryngol. 1998;19(4):278–82.
2. Lee SU, Kim HJ, Koo JW, Choi JY, Kim JS. Vertigo induced during coitus. Front Neurol. 2018;9:1187.
3. Martin EA. Headache during sexual intercourse (coital cephalalgia). A report on six cases. Ir J Med Sci. 1974;143(6):342–5.
4. Maynard P, Pace A. Primary headache associated with sexual activity: a review of the literature. Curr Pain Headache Rep. 2024;28:627.
5. Martinez JM, Roig C, Arboix A. Complicated coital cephalalgia. Three cases with benign evolution. Cephalalgia. 1988;8(4):265–8.
6. Ostergaard JR, Kraft M. Natural course of benign coital headache. BMJ. 1992;305(6862):1129.
7. Porter M, Jankovic J. Benign coital cephalalgia. Differential diagnosis and treatment. Arch Neurol. 1981;38(11):710–2.
8. Mann S, Craig MW, Gould BA, Melville DI, Raftery EB. Coital blood pressure in hypertensives. Cephalgia, syncope, and the effects of beta-blockade. Br Heart J. 1982;47(1):84–9.

Part XXVIII

Diagnoses—Otovestibular Symptoms Occurring in the Context of Other Conditions

Chronic Fatigue Syndrome and Otovestibular Symptoms

174

Overview

The etiology of *chronic fatigue syndrome* (CFS) is unknown. Some patients with CFS complain of disequilibrium, but the mechanism of this symptom is unclear. The usual proposals include "central vestibular" deficits and autonomic abnormalities. On rotatory chair testing, these patients may show low gain on slow harmonic oscillation and high gain on optokinetic testing. On computerized dynamic posturography, CFS patients may show low composite scores, visual dependence, and hip dominance. Most studies interpret these nonspecific findings to reflect central vestibular dysfunction, rather than peripheral vestibular disease. Most, though not all, studies report autonomic abnormalities in CFS patients.

Introduction

Chronic fatigue syndrome (CFS) is also referred to in some literature as myalgic encephalomyelitis. The underlying etiology of chronic fatigue syndrome is unknown [1].

Pathophysiological Mechanism of Disease

Since the etiology of CFS is not understood, the mechanism by which CFS might cause audiovestibular dysfunction is unknown. The usual proposals include autonomic dysfunction and some sort of central vestibular dysfunction.

Clinical Presentation

Furman commented that, "Patients with chronic fatigue syndrome (CFS) often complain of disequilibrium that is nonspecific" [2]. This matches our clinical experience of CFS patients who complain of a disturbance of equilibrium but have difficulty articulating a coherent description of this symptom.

Ocular Motor Examination

A case report described spontaneous nystagmus in one CFS patient [3].

Testing: Vestibular

Ash-Bernal and colleagues [4] performed a vestibular battery (caloric testing, rotatory chair testing, and computerized dynamic posturography) on 11 CFS patients and found "no predominant pattern of abnormalities," though they observed that:

- Rotatory Chair Test Findings:
  "Three [patients] had abnormally low earth vertical axis rotation (EVA) gains at the higher frequencies tested," and "As a group, the average EVA was significantly lower than normal in the 0.1–1.0 Hz range."
  - "In horizontal axis rotation, the CVS group had a higher than normal bias value for… optokinetic [responses]," and "Five of the 11 subjects had an abnormal OKN bias build up over the course of the run, equal to or actually exceeding the 60 degrees/s target velocity by as much as 14 degrees/s."
- "Patients typically performed below average in dynamic posturography."
- "One patient had abnormal caloric testing."

Ash-Bernal and colleagues concluded that, "Altogether, these results are more suggestive of central nervous system deficits than of peripheral vestibular disfunction" [4].

Furman [2] studied three patients and reported:

> None had abnormalities that suggested peripheral vestibular deficits, in that caloric responses were normal in each patient. However, all three patients had abnormal results on testing with dynamic posturography—one patient's results suggested disease of the vestibular system, and the others' results indicated nonspecific abnormalities more suggestive of CNS deficits.

M. Cherchi, *Otoneurology and Vestibular Medicine*, https://doi.org/10.1007/978-3-031-94841-1_174

> Additionally, one individual had positional nystagmus and evidenced asymmetry on rotation; both abnormalities of the vestibular system are nonspecific and nonlocalizing. Data from this small sample of patients suggests that… The abnormalities seen to date are more suggestive of CNS deficits than of peripheral vestibular deficits [2].

Palaniappan and Sirimanna [3] studied one CFS patient and reported spontaneous nystagmus and an asymmetrical vestibulo-ocular reflex.

Serrador and colleagues [5] compared computerized dynamic posturography results in 27 CFS patients and 22 matched controls. They reported that:

- "CFS patients had lower SOT [sensory organization testing] composite scores than healthy controls… indicating that they had worse overall balance."
- "Subjects with CRS had poorer scores on the SOT visual score. Low scores on this visual assessment indicate that CFS patients were unable to use visual information accurately."
- "CFS subjects had a lower strategy score (i.e., more non-ankle movement) than did the [healthy control] subjects (i.e., relatively more movement about the ankles)."

In short, CFS patients have poorer overall balance; they inappropriately rely on incorrect visual information; and they use a hip-dominant strategy.

Testing: Other

Some investigators have explored autonomic function in CFS patients. Most of these studies identify autonomic abnormalities, including hypotension [6]. For example:

- Bou-Holaigah and colleagues [7] studied tilt table testing in 23 CFS patients and 14 controls, and reported abnormal tilt table responses in 22 out of 23 CFS patients (96%) but only 4 out of 14 controls (29%).
- Freeman and Komaroff studied autonomic function in 23 CFS patients and concluded that, "Patients with CFS show alterations in measures of sympathetic and parasympathetic nervous system function" [8], including on measures of heart rate on standing and tilting, expiratory/inspiratory ratio of heart rate, Valsalva, and tilt table testing.
- De Becker and colleagues studied 21 CFS patients and 13 matched controls and found that, "There was a trend toward an increased heart rate during the cold pressor test" [9].
- Stewart and colleagues studied heart rate variability in head-upright tilt testing in 16 CFS patients (age 11–19 years) compared to 26 non-CFS patients being evaluated for syncope and 13 healthy control subjects. They concluded that, "CFS is associated with NMH [neurally mediated hypotension] during HUT [head upright tilt] in children. All indices of HRV [heart rate variability] are markedly depressed in CVS patients" [10].

Not all studies reach these conclusions. For example, Soetekouw and colleagues conducted autonomic testing on 37 CFS patients and 38 control subjects and concluded, "The findings of the study suggest that there are no gross alterations in cardiovascular autonomic function in patients with CFS" [11].

Treatment

There is no proven treatment for CFS. CFS frequently presents with symptoms involving multiple organ systems, manifesting with sleep disturbances, musculoskeletal symptoms, mood disorders, and cognitive deficits. Frequently, the care of a CFS patient is overseen by a rheumatologist; depending on a particular patient's symptoms, consultation with a sleep physician or a psychiatrist may be appropriate.

Regarding the vestibular symptoms of CFS, since some studies identify autonomic abnormalities (including hypotension), a number of trials have been undertaken to explore whether fludrocortisone may be helpful, but none of these studies demonstrated superiority of fludrocortisone over placebo [12–14].

References

1. Prins JB, van der Meer JW, Bleijenberg G. Chronic fatigue syndrome. Lancet. 2006;367(9507):346–55.
2. Furman JM. Testing of vestibular function: an adjunct in the assessment of chronic fatigue syndrome. Rev Infect Dis. 1991;13(Suppl 1):S109–11.
3. Palaniappan R, Sirimanna T. Peripheral vestibular dysfunction in chronic fatigue syndrome. Int J Pediatr Otorhinolaryngol. 2002;64(1):69–72.
4. Ash-Bernal R, Wall C 3rd, Komaroff AL, Bell D, Oas JG, Payman RN, Fagioli LR. Vestibular function test anomalies in patients with chronic fatigue syndrome. Acta Otolaryngol. 1995;115(1):9–17.
5. Serrador JM, Quigley KS, Zhao C, Findley T, Natelson BH. Balance deficits in chronic fatigue syndrome with and without fibromyalgia. NeuroRehabilitation. 2018;42(2):235–46.
6. Rowe PC, Calkins H. Neurally mediated hypotension and chronic fatigue syndrome. Am J Med. 1998;105(3A):15S–21S.
7. Bou-Holaigah I, Rowe PC, Kan J, Calkins H. The relationship between neurally mediated hypotension and the chronic fatigue syndrome. JAMA. 1995;274(12):961–7.
8. Freeman R, Komaroff AL. Does the chronic fatigue syndrome involve the autonomic nervous system? Am J Med. 1997;102(4):357–64.
9. De Becker P, Dendale P, De Meirleir K, Campine I, Vandenborne K, Hagers Y. Autonomic testing in patients with chronic fatigue syndrome. Am J Med. 1998;105(3A):22S–6S.

10. Stewart J, Weldon A, Arlievsky N, Li K, Munoz J. Neurally mediated hypotension and autonomic dysfunction measured by heart rate variability during head-up tilt testing in children with chronic fatigue syndrome. Clin Auton Res. 1998;8(4):221–30.
11. Soetekouw PM, Lenders JW, Bleijenberg G, Thien T, van der Meer JW. Autonomic function in patients with chronic fatigue syndrome. Clin Auton Res. 1999;9(6):334–40.
12. Peterson PK, Pheley A, Schroeppel J, Schenck C, Marshall P, Kind A, et al. A preliminary placebo-controlled crossover trial of fludrocortisone for chronic fatigue syndrome. Arch Intern Med. 1998;158(8):908–14.
13. Rowe PC, Calkins H, DeBusk K, McKenzie R, Anand R, Sharma G, et al. Fludrocortisone acetate to treat neurally mediated hypotension in chronic fatigue syndrome: a randomized controlled trial. JAMA. 2001;285(1):52–9.
14. Blockmans D, Persoons P, Van Houdenhove B, Lejeune M, Bobbaers H. Combination therapy with hydrocortisone and fludrocortisone does not improve symptoms in chronic fatigue syndrome: a randomized, placebo-controlled, double-blind, crossover study. Am J Med. 2003;114(9):736–41.

175 Fibromyalgia and Otovestibular Symptoms

Overview

Fibromyalgia (FM) is a rheumatologic disorder of unknown etiology that manifests with diffuse musculoskeletal pains and a variety of other somatic, constitutional, and cognitive/psychiatric symptoms. A substantial percentage of FM patients endorse auditory complaints (hearing loss, tinnitus) and vestibular complaints (disequilibrium and falls). From an otoneurological perspective, physical examination is normal. FM patients tend to have high-frequency (4–8 kHz) hearing loss beyond age-matched controls. Instrumented ocular motor testing of smooth pursuit may show reduced gain. Posturography shows poorer performance in all sensory organization testing conditions compared to healthy controls. Since these features do not distinguish whether otovestibular symptoms are due to FM or some other pathology, it is medically reasonable to undertake a screening otovestibular workup. Unless an alternative diagnosis is found, referral to audiology (regarding complaints of hearing loss and/or tinnitus) and a trial of vestibular rehabilitation therapy are reasonable. Of course, the FM itself should be managed by rheumatology.

Introduction

Fibromyalgia (FM), sometimes referred to as fibrositis in older literature, is a controversial clinical entity that manifests with widespread musculoskeletal pain and is often accompanied by other somatic symptoms, constitutional symptoms (such as fatigue), and cognitive/psychiatric symptoms [1]. Despite the prominent symptom of musculoskeletal pain and tenderness, there is no evidence of inflammation in the affected tissues, so more recent research has explored the possibility that FM is a form of central sensitization—in other words, a disorder of central pain regulation.

Problems with hearing and balance are among the many symptoms endorsed by FM patients, and some literature suggests that in some patients, otovestibular symptoms may be due to FM itself.

Epidemiology

Among rheumatological diseases, FM is relatively common. Queiroz [2] reviewed a large number of epidemiological studies on FM and reported that the global mean **prevalence** was 2.7%, ranging from a low of 0.4% in Greece to a high of 9.3% in Tunisia. Queiroz found that few studies explore the **incidence** of FM; of those that do, one study found an incidence rate of 5.83 new cases per 1000 person-years among females; another study found an incidence rate of 6.88 new cases per 1000 person-years among males, and 11.28 new cases per 1000 person-years for females. Globally, females are affected more than males by a ratio of 3:1.

Genetics

Study of the genetics of FM is still in a fairly preliminary phase [3]. Whether any specific genetic factors are driving otovestibular symptoms related to FM has not been studied.

Pathophysiological Mechanism of Disease

Even after acknowledging that FM is incompletely understood, it should be added that the mechanism by which FM might cause otovestibular symptoms is even more poorly understood. Rusy and colleagues [4] studied 12 children with FM using electronystagmography and rotatory chair testing and found no abnormalities. They concluded that,

M. Cherchi, *Otoneurology and Vestibular Medicine*, https://doi.org/10.1007/978-3-031-94841-1_175

"These findings suggest that central (brainstem) and peripheral vestibular (inner ear) mechanisms do not account for the complaints of dizziness in the pediatric patient with fibromyalgia. The common musculoskeletal abnormalities of fibromyalgia may affect their proprioceptive orientation, therefore giving them a sense of imbalance."

Clinical Presentation: Overall

Bayazit and colleagues [5] studied 24 female FM patients by history, physical examination, audiologic testing (audiometry, brainstem auditory evoked responses), and vestibular testing (bithermal caloric testing) and reported that, "Almost 50% of the patients had some sort of otologic symptoms" despite fairly normal otovestibular test results, and concluded that, "FM patients can complain otologic symptoms even though they do not have any clinically or audiologically detectable ear disease."

Clinical Presentation: Auditory Symptoms

Bayazit and colleagues [5] studied 24 female FM patients and reported that three (13%) complained of **hearing loss**, and four (17%) complained of **tinnitus**.

Mucci and colleagues [6] conducted a questionnaire-based study of 277 FM patients. They reported that 5.8% complained of "**tinnitus**."

Wolfe and colleagues [7] studied 1199 FM patients and found that 36.2% (95% confidence interval 33.4%–39.0%) complained of "hearing difficulties."

Le and colleagues [8] conducted a population-based retrospective cohort study of patients with new-onset FM. They compared 55,169 FM patients with 110,338 healthy controls and reported that, "The overall HL [**hearing loss**] risk in the FM group was 1.46-fold (95% confidence interval: 1.38–1.55) higher than that in the non-FM group."

Clinical Presentation: Vestibular Symptoms

Bayazit and colleagues [5] studied 24 female FM patients and reported that nine (38%) complained of "**dizziness**" and two (8%) complained of "**vertigo**."

Mucci and colleagues [6] conducted a questionnaire-based study of 277 FM patients in an attempt "to establish a prevalence rate of vestibular symptoms in FM." They reported that 63.6% of FM patients complained of "**dizziness**."

Rosenhall and colleagues [9] studied 168 FM patients, of whom 141 (83.9%) were women. They found that 72% of patients reported a complaint of "**vertigo**." They further analyzed patient descriptions, and commented that of all 168 patients, "Most of them (48% of all patients) had slight, diffuse dizziness causing no or only minor discomfort. Four percent of all patients complained of constant, severe dizziness. Eighteen percent complained of attacks of rotatory or nautical vertigo. Disturbance of gait without dizziness or vertigo was reported by 2% of patients."

Jones and colleagues [10] compared 34 FM patients (88% female) and 32 age-matched healthy controls through several questionnaires (Fibromyalgia Impact Questionnaire, Activities-specific Balance Confidence, report of falls) and a multifactor gait and balance assessment called the Balance Evaluation Systems Test (BESTest) developed by Horak and colleagues [11]. They concluded that "FM is associated with **balance problems** and **increased fall frequency**."

Jones and colleagues [12] compared 25 FM patients (88% female, mean age 50.8 ± 7.7 years) and 27 healthy control subjects (89% female, mean age 46.5 ± 10.9 years). They reported that, "FM patients reported significantly more falls over the past 6 months than HCs. Specifically, three (11%) of 27 HCs had fallen only once during the past 6 months, whereas 18 (72%) of 25 FM patients had **fallen** at least once, with 15 FM patients (60%) having fallen more than three times in the past 6 months."

Physical Examination

Aside from physical examination findings compatible with FM itself, there are no physical examination signs that are sensitive or specific for otovestibular symptoms in FM.

Ocular Motor Examination

Very few studies have specifically reported on positive bedside ocular motor findings in FM.

Rusy and colleagues [4] studied pediatric FM and specifically commented that the following bedside findings were normal: spontaneous nystagmus, oculocephalic reflex, dynamic visual acuity, head shaking nystagmus, Quix test, and the Dix-Hallpike maneuver.

In contrast, Bayazit and colleagues [5], who studied 24 female FM patients, reported that on the Dix-Hallpike maneuver, 21% had "rotary" nystagmus (without further description).

Testing: Audiologic

Gencer and colleagues [13] compared audiometric results in 35 female FM patients (mean age 48.1 years, range 30–65 years) and 25 healthy control subjects. They reported that, "At high frequencies, the mean air conduction threshold values and tympanometric values between the FM and control groups were statistically significant ($P < 0.05$ for both)." The frequencies at which pure tone hearing was statistically significantly worse in FM patients than in healthy controls were 4–8 kHz, as shown in Table 175.1.

Table 175.1 Comparison of pure tone audiometry (air conduction thresholds) results from 35 fibromyalgia patients and 25 healthy controls

Frequencies		Fibromyalgia (*n* = 35)	Control (*n* = 25)	*P*
250 Hz	Right	10.35 ± 5.31	10.02 ± 5.2	>0.05
	Left	10.25 ± 5.35	10.10 ± 5.61	>0.05
500 Hz	Right	11.35 ± 5.11	10.90 ± 5.2	>0.05
	Left	11.39 ± 6.22	11.01 ± 5.55	>0.05
1000 Hz	Right	13.34 ± 6.12	13.15 ± 7.1	>0.05
	Left	13.37 ± 6.99	13.32 ± 7.04	>0.05
2000 Hz	Right	15.01 ± 6.45	15.01 ± 6.98	>0.05
	Left	15.60 ± 7.50	15.36 ± 7.19	>0.05
4000 Hz	Right	19.80 ± 7.70	15.13 ± 7.20	**<0.05**
	Left	21.42 ± 7.72	15.06 ± 7.31	**<0.05**
6000 Hz	Right	25.76 ± 8.33	15.11 ± 7.22	**<0.05**
	Left	25.83 ± 8.67	15.62 ± 7.21	**<0.05**
8000 Hz	Right	30.22 ± 8.60	17.17 ± 8.41	**<0.05**
	Left	31.14 ± 9.15	17.12 ± 8.81	**<0.05**

From Gencer et al. [13]

Testing: Vestibular

Bayazit and colleagues [14] compared cervical vestibular evoked myogenic potentials (cVEMPs) in 24 FM patients (22 women, 2 men) and 21 healthy female controls. They reported statistically significant differences between the two groups; specifically, **the *n*23 latency was delayed, and the *p*13-*n*23 interpeak latency was prolonged, in FM patients** compared to controls. Figure 175.1 shows an example of the delayed *n*23 latency.

Rosenhall and colleagues [9] compared ocular motor performance in 168 FM patients, of whom 141 (83.9%) were women, and 90 healthy controls, of whom 48 were women. With respect to smooth pursuit eye movements, they reported that, "The mean velocity gain was significantly lower in the patient group than in the control group for both women and men ($p < 0.001$). Ninety-six of the 166 patients tested (58%) had pathological smooth pursuit eye movements. In all these cases the mean velocity gain of the 20 deg/s smooth pursuits was significantly reduced. In 24% of all the patients the smooth pursuits were severely abnormal with a mean velocity gain of ≤0.5."

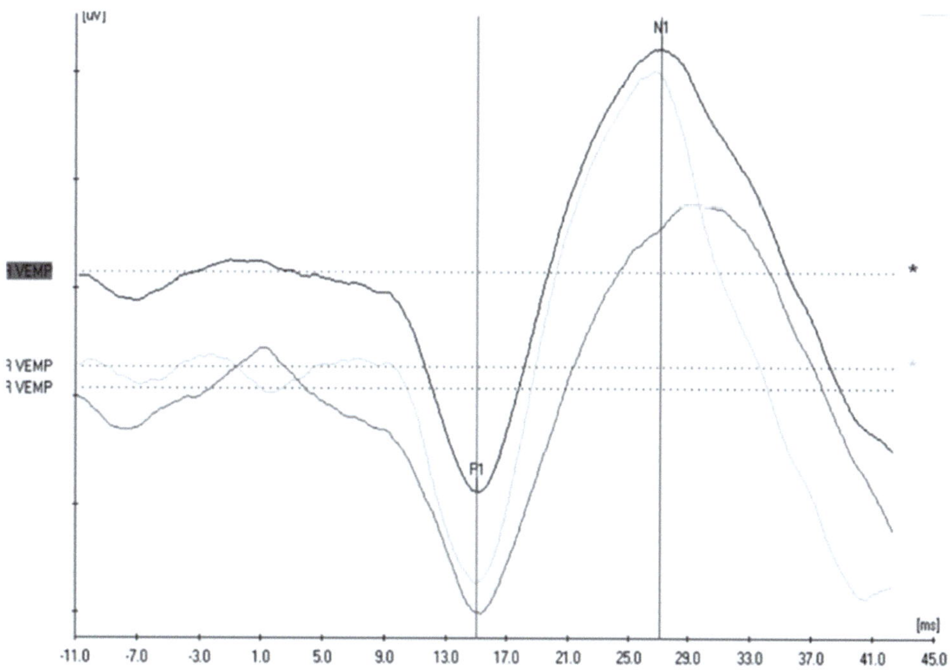

Fig. 175.1 Cervical vestibular evoked myogenic potentials show delayed *n*23 latencies in patients with fibromyalgia compared to healthy controls. (From Bayazit et al. [14]. Reproduced with written permission from Cambridge University Press)

Muto and colleagues [15] studied a case-control study of 117 women, of whom 67 carried a diagnosis of FM. The subjects completed several questionnaires and did (what from the authors' description sounds like) static posturography. They concluded that, "Individuals with FM had **impaired postural control** showing increased speed of oscillation of the center of gravity ($p = 0.0004$) and decreased balance self-efficacy ($p < 0.001$)."

Jones and colleagues [12] compared 25 FM patients (88% female, mean age 50.8 ± 7.7 years) and 27 healthy control subjects (89% female, mean age 46.5 ± 10.9 years) using computerized dynamic posturography (CDP). They reported that **on all six conditions of sensory organization testing, FM patients performed statistically significantly worse** than healthy controls.

Perez-se-Heredia-Torres and colleagues [16] compared performance on computerized dynamic posturography (CDP) in 20 women with FM and 20 matched healthy controls. They reported that on sensory organization testing, "Women with FM exhibited **balance deficiencies** and used different strategies for maintaining their balance in standing" compared to healthy controls.

Differential Diagnosis

In patients with FM, the characteristics of hearing and balance symptoms, and physical examination findings, cannot distinguish whether the etiology of the otovestibular symptoms is FM itself or some other disease. Consequently, the differential diagnosis (of the otovestibular symptoms) is broad, and a reasonable screening search for known otovestibular disorders should be considered.

Treatment

As of this writing, there were no specific trials regarding auditory and vestibular symptoms apparently associated with FM. If workup provides no evidence for an alternative explanation, then treatment of the fibromyalgia itself (under the guidance of a rheumatologist) is reasonable. If hearing loss and /or tinnitus are interfering with function, then referral to audiology is reasonable. Although vestibular rehabilitation therapy (VRT) has not been specifically studied for disequilibrium in FM, a trial of VRT is probably reasonable as it incurs no medical risk.

References

1. Arnold LM, Bennett RM, Crofford LJ, Dean LE, Clauw DJ, Goldenberg DL, et al. AAPT diagnostic criteria for fibromyalgia. J Pain. 2019;20(6):611–28.
2. Queiroz LP. Worldwide epidemiology of fibromyalgia. Curr Pain Headache Rep. 2013;17(8):356.
3. D'Agnelli S, Arendt-Nielsen L, Gerra MC, Zatorri K, Boggiani L, Baciarello M, et al. Fibromyalgia: genetics and epigenetics insights may provide the basis for the development of diagnostic biomarkers. Mol Pain. 2019;15:1744806918819944.
4. Rusy LM, Harvey SA, Beste DJ. Pediatric fibromyalgia and dizziness: evaluation of vestibular function. J Dev Behav Pediatr. 1999;20(4):211–5.
5. Bayazit YA, Gursoy S, Ozer E, Karakurum G, Madenci E. Neurotologic manifestations of the fibromyalgia syndrome. J Neurol Sci. 2002;196(1–2):77–80.
6. Mucci V, Demori I, Rapallo F, Molinari E, Losacco S, Marinelli L, et al. Vestibular disability/handicap in fibromyalgia: a questionnaire study. J Clin Med. 2022;11(14):4017.
7. Wolfe F, Rasker JJ, Hauser W. Hearing loss in fibromyalgia? Somatic sensory and non-sensory symptoms in patients with fibromyalgia and other rheumatic disorders. Clin Exp Rheumatol. 2012;30(6 Suppl 74):88–93.
8. Le TP, Tzeng YL, Muo CH, Ting H, Sung FC, Lee SD, et al. Risk of hearing loss in patients with fibromyalgia: a nationwide population-based retrospective cohort study. PLoS One. 2020;15(9):e0238502.
9. Rosenhall U, Johansson G, Orndahl G. Otoneurologic and audiologic findings in fibromyalgia. Scand J Rehabil Med. 1996;28(4):225–32.
10. Jones KD, Horak FB, Winters-Stone K, Irvine JM, Bennett RM. Fibromyalgia is associated with impaired balance and falls. J Clin Rheumatol. 2009;15(1):16–21.
11. Horak FB, Wrisley DM, Frank J. The balance evaluation systems test (BESTest) to differentiate balance deficits. Phys Ther. 2009;89(5):484–98.
12. Jones KD, King LA, Mist SD, Bennett RM, Horak FB. Postural control deficits in people with fibromyalgia: a pilot study. Arthritis Res Ther. 2011;13(4):R127.
13. Kapusuz Gencer Z, Balbaloglu O, Ozkiris M, Saydam L. Does fibromyalgia have an effect on hearing loss in women? Turk J Med Sci. 2017;47(6):1699–702.
14. Bayazit YA, Celenk F, Gunduz AG, Gunduz B, Ondag N, Meray J. Vestibular evoked myogenic potentials in patients with fibromyalgia syndrome. J Laryngol Otol. 2010;124(6):610–5.
15. Muto LH, Sauer JF, Yuan SL, Sousa A, Mango PC, Marques AP. Postural control and balance self-efficacy in women with fibromyalgia: are there differences? Eur J Phys Rehabil Med. 2015;51(2):149–54.
16. Perez-de-Heredia-Torres M, Huertas-Hoyas E, Martinez-Piedrola R, Palacios-Cena D, Alegre-Ayala J, Santamaria-Vazquez M, et al. Balance deficiencies in women with fibromyalgia assessed using computerised dynamic posturography: a cross-sectional study in Spain. BMJ Open. 2017;7(7):e016239.

Ehlers-Danlos Syndrome

176

Overview

The term "*Ehlers-Danlos syndrome* (EDS)" refers to a collection of diseases resulting from mutations in specific genes that encode connective tissues or specific enzymes. These can manifest with a rather broad array of phenotypes. Genetic testing can confirm the diagnosis. Some patients with EDS have hearing loss of various types. Some EDS patients may experience disequilibrium, for which plausible proximal mechanisms include orthostatic intolerance, superior semicircular canal dehiscence, and peripheral neuropathy.

Introduction

The first cases of what came to be known as *Ehlers-Danlos syndrome* (EDS) were originally described by Danish dermatologist, Dr. Edvard Lauritz Ehlers (1863–1937) [1] and French dermatologist, Henri-Alexandre Danlos (1844–1912) [2]. The eponymous designation of "Ehlers-Danlos syndrome" appears to have been originally proposed in 1936 by the English physician, Dr. Frederick Parkes Weber [3].

As additional cases emerged, it became apparent that there was significant phenotypic variability. With the advent of genetics, it is now understood that EDS comprises a range of distinct pathologies with different, though overlapping, clinical manifestations. Table 176.1 summarizes the types of Ehlers-Danlos syndrome in Online Mendelian Inheritance in Man as of this writing.

M. Cherchi, *Otoneurology and Vestibular Medicine*, https://doi.org/10.1007/978-3-031-94841-1_176

Table 176.1 Types of Ehlers-Danlos syndrome

Type	Other nomenclature	Inheritance pattern	OMIM disease	OMIM gene	Cytogenetic location	Gene(s)	Protein
Classical EDS, type 1	cEDS, EDS type I	AD	130000	120215	9q34.3	COL5A1	Type V collagen, alpha 1
Classical EDS, type 1	cEDS, EDS type I	AD	130000	120150	17q21.33	COL1A1 c.934C > T, p.(Arg312Cus)	Type I collagen, alpha 1
Classical EDS, type 2	cEDS, type II		130010	120190	2q32.2	COL5A2	Type 5 collagen, alpha 2
Hypermobile type EDS	clEDS, EDS type III	AD	130020	600985	6p21.3	TNXB	Tenascin XB, isoform 1
Classic-like EDS	EDS with TNXB deficiency	AR	606408	600985	6p21.3	TNXB	Tenascin XB, isoform 1
Vascular EDS	EDS type IV	AD	130050	120180	2q32.2	COL3A1	Type III collagen, alpha 1
Kyphoscoliotic EDS, type 1	EDS KSCL1, EDS type VIA	AR	225400	153454	1p36.22	PLOD1 (procollagen-lysine, 2-oxoglutarate 5-dioxygenase)	PLOD1 (procollagen-lysine, 2-oxoglutarate 5-dioxygenase)
Kyphoscoliotic EDS, type 2	EDS type VIB	AR	614557	614505	7p14.3	FKBP14	FK506 binding protein 14
Brittle cornea syndrome, type 1	BCS1	AR	229200	612078	16q24.2	ZNF469 (zinc finger protein 469)	ZNF469 (zinc finger protein 469)
Brittle cornea syndrome, type 2	BCS2	AR	614170	614161	4q27	PRDM5	PRDM5 (PR domain-containing protein 5)
Musculocontractural EDS type 1	mcEDS, D4ST1-deficient type EDS, EDS type VIB	AR	601776	608429	15q15.1	CHST14	CHST14 (carbohydrate sulfotransferase 14)
Arthrochalasic EDS type 1	aEDS, EDS type VIIA	AD	130060	120150	17q21.33	COL1A1	Type I collagen, alpha 1
Arthrochalasic EDS type 2	aEDS, EDS type VIIB	AD	617821	120160	7q21.3	COL1A2	Type I collagen, alpha 2
Cardiac-valvular EDS	cvEDS, EDS with COL1A2 deficiency	AR	225320	120160	7q21.3	COL1A2 (biallelic mutations that lead to COL1A2 NMD and absence of pro a2(I) collagen chains)	Type I collagen, alpha 2
Dermatosparexis EDS	EDS type VIIC	AR	225410	604539	5q35.3	ADAMTS2	A disintegrin-like and metalloproteinase with thrombospondin type 1 motif 2
Spondylodysplastic EDS, type 1	EDS SPD1	AR	130070	604327	5q35.3	B4GALT7	Beta-1,4-galactosyltransferase 7
Spondylodysplastic EDS, type 2	EDS SPD2	AR	615349	615291	1p36.33	B3GALT6	Beta-1,3-galactosyltransferase 6
EDS, spondylocheirodysplastic form, type 3	SCD-EDS	AR	612350	608735	11p11.2	SLC39A13	Solute carrier family 39 (zinc transporter), member 13
Periodontal EDS	EDS PD1	AD	130080	613785	12p13.31	C1R	Complement component 1, r subcomponent
Vascular EDS	EDS VASC	AD	130050	120180	2q32.2	COL3A1	Collagen type III, alpha 1
EDS with platelet dysfunction from fibronectin abnormality	EDS type X	AD	225310	135600	2q35	FN1	Fibronectin

AD autosomal dominant, *AR* autosomal recessive, *OMIM* online mendelian inheritance in man

Genetics

Most subtypes of Ehlers-Danlos syndrome result from mutations in genes encoding proteins for connective tissue structures or for enzymes.

Pathophysiological Mechanism of Disease: Hearing Loss

Weir and colleagues studied audiometry in 141 pediatric patients diagnosed with EDS, and found some form of hearing loss in 32 (22.7%). Of those 32 patients, the hearing loss was bilateral in 19 (59%) and unilateral in 13 (41%). Out of 282 ears (in the 141 EDS patients), pure conductive hearing loss was found in 25 (8.5% of ears), pure sensorineural hearing loss was found in 23 (8.2% of ears), and mixed hearing loss was found in only 3 (1.4% of ears). The authors reported that of all the EDS patients with hearing loss, 50% "received more than two diagnoses of otitis media throughout the course of their care," and "31.4% of the cohort with hearing loss had a diagnosis of Eustachian tube dysfunction." There are additionally several case reports of otosclerosis in EDS patients [4, 5].

Pathophysiological Mechanism of Disease: Disequilibrium

As far as symptoms of disequilibrium are concerned, there are several potential mechanisms by which EDS can cause disorders of equilibrium.

The most common mechanism appears to be autonomic dysfunction, manifesting primarily as orthostatic intolerance, particularly *postural orthostatic tachycardia syndrome* (POTS) [6–15].

Superior semicircular canal dehiscence (SSCD) has been reported in several cases of EDS, including unilateral [16] and bilateral [17, 18].

Peripheral neuropathy of various types may also be contributory [19–27].

In EDS patients with altered cervical mechanics, *cervicogenic vertigo* (CV) is also a reasonable consideration, though this has not been formally studied.

Clinical Presentation

The clinical presentation depends on the proximal mechanism of the symptom of disequilibrium.

Physical Examination

Physical examination findings will depend on the proximal mechanism of the symptom of disequilibrium.

It is reasonable for physical examination to include comparison of supine and standing pulse/pressure to seek evidence of postural orthostatic tachycardia.

Ocular Motor Examination

It is reasonable on infrared video oculography to assess for Valsalva-induced nystagmus which may elicit nystagmus if superior semicircular canal dehiscence is present.

Testing: Audiologic

If an EDS patient has hearing complaints, it is reasonable to check audiometry.

Testing: Vestibular

Consider checking cervical vestibular evoked myogenic potentials and ocular vestibular evoked myogenic potentials to evaluate for superior semicircular canal dehiscence (SSCD). Whether or not the patient has auditory symptoms, it is reasonable to check audiometry to assess for conductive hyperacusis (which supports a diagnosis of SSCD) and conductive hearing loss (which may interfere with the results of air-conducted vestibular evoked myogenic potentials).

Consider checking computerized dynamic posturography for an overall quantification of balance.

Testing: Other

Consider checking a tilt table test to evaluate for evidence of autonomic dysfunction.

Consider consultation with a medical geneticist regarding whether genetic testing is appropriate and, if so, which variants should be checked.

Imaging

From the otoneurological perspective, if screening tests (audiometry, cervical vestibular evoked myogenic potentials, ocular vestibular evoked myogenic potentials) are compatible with the presence of superior semicircular canal dehiscence, then consider a temporal bone CT without contrast.

Treatment

Care of an EDS patient tends to be multidisciplinary.

If a patient or their doctor suspects EDS, then consultation with a medical geneticist is sensible to determine whether genetic testing for diagnostic confirmation is appropriate.

Depending on the genetic subtype and clinical manifestations, the overall care of an EDS patient is often organized by a rheumatologist, often in consultation with other specialists, such as a dermatologist.

In EDS patients with hearing loss, referral to audiologist is reasonable.

In EDS patients with superior semicircular canal dehiscence, referral to an otolaryngologist is reasonable. Unfortunately, EDS patients are often poor surgical candidates due to poor tissue healing and, in some cases, vascular complications.

In EDS patients with postural orthostatic tachycardia syndrome, referral to a cardiologist is reasonable.

References

1. Ehlers EL. Cutis laxa. Neigung zu Haemorrhagien in der Haut, Lockering mehrerer Artikulationen [Lax skin: tendency to hemorrhages in the skin, loosening of several articulations]. Dermatologische Zeitschrift. 1901;8:173–4.
2. Danlos H-A. Un cas de cutis laxa avec tumeurs par comtusion chronique des coudes et des genoux (xanthome juvénile pseudo-diabetique de MM Hallopeau et Macé de Lépinay) [A case of cutis laxa with chronic comtusion tumors of the elbows and knees (pseudo-diabetic juvenile xanthoma of MM Hallopeau and Macé de Lépinay)]. Bulletin de la Societé francaise de dermatologie et de syphiligraphie. 1908;19:70–2.
3. Weber FP. Ehlers-Danlos syndrome. Proc R Soc Med. 1936;30(1):30–1.
4. Miyajima C, Ishimoto S, Yamasoba T. Otosclerosis associated with Ehlers-Danlos syndrome: report of a case. Acta Otolaryngol Suppl. 2007;559:157–9.
5. Thomas DM, Wright JL, Soucek S, Shalom AS. Ehlers-Danlos syndrome: aural manifestations and treatment. Am J Otolaryngol. 1996;17(6):432–3.
6. Celletti C, Borsellino B, Castori M, Censi F, Calcagnini G, Camerota F, Strano S. A new insight on postural tachycardia syndrome in 102 adults with hypermobile Ehlers-Danlos syndrome/hypermobility spectrum disorder. Monaldi Arch Chest Dis. 2020;90(2) https://doi.org/10.4081/monaldi.2020.1286.
7. Celletti C, Camerota F, Castori M, Censi F, Gioffre L, Calcagnini G, Strano S. Orthostatic intolerance and postural orthostatic tachycardia syndrome in joint hypermobility syndrome/Ehlers-Danlos syndrome, hypermobility type: neurovegetative dysregulation or autonomic failure? Biomed Res Int. 2017;2017:9161865.
8. De Wandele I, Rombaut L, Leybaert L, Van de Borne P, De Backer T, Malfait F, et al. Dysautonomia and its underlying mechanisms in the hypermobility type of Ehlers-Danlos syndrome. Semin Arthritis Rheum. 2014;44(1):93–100.
9. Hakim A, O'Callaghan C, De Wandele I, Stiles L, Pocinki A, Rowe P. Cardiovascular autonomic dysfunction in Ehlers-Danlos syndrome-hypermobile type. Am J Med Genet C Semin Med Genet. 2017;175(1):168–74.
10. Mathias CJ, Owens A, Iodice V, Hakim A. Dysautonomia in the Ehlers-Danlos syndromes and hypermobility spectrum disorders-with a focus on the postural tachycardia syndrome. Am J Med Genet C Semin Med Genet. 2021;187(4):510–9.
11. Miller AJ, Stiles LE, Sheehan T, Bascom R, Levy HP, Francomano CA, Arnold AC. Prevalence of hypermobile Ehlers-Danlos syndrome in postural orthostatic tachycardia syndrome. Auton Neurosci. 2020;224:102637.
12. Roma M, Marden CL, De Wandele I, Francomano CA, Rowe PC. Postural tachycardia syndrome and other forms of orthostatic intolerance in Ehlers-Danlos syndrome. Auton Neurosci. 2018;215:89–96.
13. Rowe PC, Barron DF, Calkins H, Maumenee IH, Tong PY, Geraghty MT. Orthostatic intolerance and chronic fatigue syndrome associated with Ehlers-Danlos syndrome. J Pediatr. 1999;135(4):494–9.
14. Wallman D, Weinberg J, Hohler AD. Ehlers-Danlos syndrome and postural tachycardia syndrome: a relationship study. J Neurol Sci. 2014;340(1–2):99–102.
15. Qarajeh R, Derbas LA, Abdallah M, Kureshi F. A 21-year-old woman with Ehlers-Danlos syndrome and persistent lightheadedness. Cleve Clin J Med. 2021;88(2):93–7.
16. Chung LK, Lagman C, Nagasawa DT, Gopen Q, Yang I. Superior semicircular canal dehiscence in a patient with Ehlers-Danlos syndrome: a case report. Cureus. 2017;9(4):e1141.
17. Preet K, Udawatta M, Duong C, Gopen Q, Yang I. Bilateral superior semicircular canal dehiscence associated with Ehlers-Danlos syndrome: a report of 2 cases. World Neurosurg. 2019;122:161–4.
18. Unterberger A, Miller J, Gopen Q, Yang I. Bilateral superior semicircular canal dehiscence concurrent with Ehlers-Danlos syndrome: a case report. Cureus. 2021;13(11):e19943.
19. Cazzato D, Castori M, Lombardi R, Caravello F, Bella ED, Petrucci A, et al. Small fiber neuropathy is a common feature of Ehlers-Danlos syndromes. Neurology. 2016;87(2):155–9.
20. Cook MK, Jordan M. Autoimmune small fiber neuropathy associated with Ehlers-Danlos syndrome treated with intravenous immunoglobulins. J Clin Neuromuscul Dis. 2021;22(3):160–3.
21. Fernandez A, Aubry-Rozier B, Vautey M, Berna C, Suter MR. Small fiber neuropathy in hypermobile Ehlers Danlos syndrome/hypermobility spectrum disorder. J Intern Med. 2022;292(6):957–60.
22. Galan E, Kousseff BG. Peripheral neuropathy in Ehlers-Danlos syndrome. Pediatr Neurol. 1995;12(3):242–5.
23. Igharo D, Thiel JC, Rolke R, Akkaya M, Weis J, Katona I, et al. Skin biopsy reveals generalized small fibre neuropathy in hypermobile Ehlers-Danlos syndromes. Eur J Neurol. 2023;30(3):719–28.
24. Pascarella A, Provitera V, Lullo F, Stancanelli A, Saltalamacchia AM, Caporaso G, Nolano M. Evidence of small fiber neuropathy in a patient with Ehlers-Danlos syndrome, hypermobility-type. Clin Neurophysiol. 2016;127(3):1914–6.
25. Schady W, Ochoa J. Ehlers-Danlos in association with tomaculous neuropathy. Neurology. 1984;34(9):1270–1.
26. Voermans NC, Drost G, van Kampen A, Gabreels-Festen AA, Lammens M, Hamel BC, et al. Recurrent neuropathy associated with Ehlers-Danlos syndrome. J Neurol. 2006;253(5):670–1.
27. Voermans NC, Knoop H, van Engelen BG. High frequency of neuropathic pain in Ehlers-Danlos syndrome: an association with axonal polyneuropathy and compression neuropathy? J Pain Symptom Manag. 2011;41(5):e4–6. author reply e-7

177 Pregnancy, Delivery, and Audiovestibular Symptoms

Overview

While there are no audiologic or vestibular disorders unique to *pregnancy* or parturition, these biological states may increase the risk of certain audiovestibular disorders.

Introduction

While there are no audiologic or vestibular disorders unique to *pregnancy* or parturition, these biological states may increase the risk of certain audiovestibular disorders. Among auditory disorders during pregnancy, sudden sensorineural hearing loss and Eustachian tube dysfunction are the usual considerations. Vestibular disorders during pregnancy are variegated, and can include otologic causes (benign paroxysmal positional vertigo, Ménière's disease, semicircular canal dehiscence), neurological causes (migraine-associated vertigo, unintended pharmacologic consequences of epidural injections), neurovascular causes (posterior reversible encephalopathy syndrome, vertebral artery dissection), and cardiovascular disorders (orthostatic intolerance, atrial myxoma).

Pathophysiological Mechanism of Disease for Audiologic Symptoms

Some degree of hearing fluctuation can occur in pregnancy, but it is often mild, transient, and self-limiting [1]. However, in some instances, this may manifest as sudden-onset sensorineural hearing loss (SSNHL) that is moderate to profound, and does not spontaneously resolve; this can occur antepartum or postpartum [2]. There are no clear treatment guidelines, but some authors advocate intravenous and transtympanic steroids [3], similar to the management of sudden sensorineural hearing loss outside of pregnancy. The mechanisms of sudden sensorineural hearing loss in pregnancy (and whether such mechanisms differ from the occurrence outside of pregnancy) are unknown.

Pathophysiological Mechanism of Disease for Vestibular Symptoms

Pregnancy may increase the risk of several conditions that may cause vestibular symptoms, including several otologic, neurologic, neurovascular, and cardiovascular disorders [4, 5].

Otologic Disorders

Some literature suggests that pregnancy increases the risk of benign paroxysmal positional vertigo (BPPV) [5, 6].

Some literature suggests that episodes of Ménière's disease can increase during pregnancy [5, 7, 8].

Semicircular canal dehiscence can present during gestation [9] or during labor [10].

A sensation of ear fullness during pregnancy is usually due to Eustachian tube dysfunction [11].

Neurological Disorders

In a minority of women, migraine may increase during pregnancy, including migraine-associated vertigo (MAV) [5].

A case report of a woman who developed disequilibrium following an epidural morphine injection for delivery speculated that, "the time of onset of the symptom coincided with the expected time of arrival of the morphine within intracerebral cerebrospinal fluid" [12].

M. Cherchi, *Otoneurology and Vestibular Medicine*, https://doi.org/10.1007/978-3-031-94841-1_177

Neurovascular Disorders

Peripartum or postpartum posterior reversible encephalopathy syndrome (PRES) can occur and manifest with vestibular symptoms [13]. It also often causes seizures and depressed consciousness.

Vertebral artery dissection can occur postpartum [14] or (similar to nonpregnant individuals) after chiropractic manipulation [15].

Cardiovascular Disorders

Hemodynamics change in several ways during pregnancy. The addition of placental circulation reduces maternal hemodynamic resistance, which can lower blood pressure. In addition, pregnancy may increase the risk of orthostatic intolerance, such as orthostatic hypotension [16–18].

Atrial myxoma can develop during pregnancy [19], and rarely can manifest with disequilibrium [20].

Clinical Presentation

The clinical presentation depends on the specific disorder.

Physical Examination

A pregnant person developing posterior reversible encephalopathy syndrome (PRES) may exhibit cerebellar findings.

Ocular Motor Examination

Ocular motor examination may reveal findings compatible with benign paroxysmal positional vertigo (BPPV) or semicircular canal dehiscence.

Testing: Auditory

In a pregnant patient complaining of hearing loss, audiometry may show sensorineural hearing loss.

Testing: Vestibular

One study of video head impulse testing (vHIT) during gestation reported that semicircular canal gain decreases beginning around the 20th week of gestation, then returns to baseline after pregnancy [21]. The authors speculated that this might be attributable to "volumetric changes probably given by hormonal actions."

If a pregnant patient develops semicircular canal dehiscence, then cervical vestibular evoked myogenic potentials (cVEMP) and ocular vestibular evoked myogenic potentials (oVEMP) may reveal corroborative findings (increased amplitudes and decreased thresholds ipsilateral to the lesion).

Imaging

In a pregnant patient with a history and physical examination compatible with semicircular canal dehiscence, then temporal bone CT (with shielding of the abdomen and pelvis) may confirm semicircular canal dehiscence.

If a pregnant patient with disequilibrium develops seizures or declining consciousness, then brain MRI may reveal findings compatible with posterior reversible encephalopathy syndrome (PRES).

Differential Diagnosis: Auditory

A pregnant individual complaining of hearing loss should consider a workup similar to that offered to nonpregnant persons, including an audiogram and perhaps otoacoustic emissions. If the overall scenario is compatible with sudden-onset sensorineural hearing loss, then referral to an otolaryngologist is appropriate.

In a pregnant person complaining of isolated aural fullness who has a normal physical examination, Eustachian tube dysfunction is a consideration. If decongestant strategies fail, then referral to an otolaryngologist is appropriate.

Differential Diagnosis: Vestibular

The workup of a pregnant person complaining of disequilibrium depends on several aspects of the clinical history and physical examination.

If a pregnant patient complains of brief bursts of positionally triggered disequilibrium and suggests benign paroxysmal positional vertigo (BPPV), and if this is confirmed by corroborative nystagmus on video oculography (VOG), then no further workup is necessary.

If a pregnant patient's episodes of disequilibrium are protracted and accompanied by aural symptoms (hearing loss, tinnitus, fullness), then Ménière's disease is a strong consideration.

If a pregnant patient complains of Valsalva-induced disequilibrium, or disequilibrium induced by loud noises (Tullio's phenomenon), and if video oculography (VOG) reveals Valsalva-induced nystagmus, then semicircular canal dehiscence is a consideration.

If a pregnant patient has a preexisting history of migraine (with or without vestibular symptoms), then migraine-associated vertigo (MAV) is a consideration, though it is always a diagnosis of exclusion.

If a pregnant patient develops symptoms following epidural injections, then a pharmacologic mechanism or cerebrospinal fluid leak should be considered.

If a pregnant patient with disequilibrium also develops seizures or progressive deterioration of consciousness, then PRES (posterior reversible encephalopathy syndrome) must be considered.

If a pregnant patient with disequilibrium also complains of neck pain, or if symptoms began during delivery (while pushing), then vertebral artery dissection is a consideration.

If a pregnant patient's symptoms have an orthostatic pattern, and if bedside examination confirms orthostatic hypotension, orthostatic hypertension, or postural orthostatic tachycardia, then no further workup is needed.

If a pregnant patient's symptoms and physical examination are compatible with cardiac embolization (including strokes) or cardiac failure, then an atrial myxoma is a consideration.

Treatment and Prognosis

Treatment and prognosis depend on the underlying disorder (irrespective of the gravid state).

References

1. Frosolini A, Marioni G, Gallo C, de Filippis C, Lovato A. Audio-vestibular disorders and pregnancy: A systematic review. Am J Otolaryngol. 2021;42(5):103136.
2. Zhang BY, Young YH. Sudden Deafness during Antepartum versus Postpartum Periods. ORL J Otorhinolaryngol Relat Spec. 2017;79(5):274–81.
3. Xie S, Wu X. Clinical management and progress in sudden sensorineural hearing loss during pregnancy. J Int Med Res. 2020;48(2):300060519870718.
4. Kirovakov Z, Kutsarov A, Todorov S, Penchev P. Vertigo during pregnancy: A narrative review of the etiology, pathophysiology, and treatment. Cureus. 2024;16(3):e55657.
5. Serna-Hoyos LC, Herrón Arango AF, Ortiz-Mesa S, Vieira-Rios SM, Arbelaez-Lelion D, Vanegas-Munera JM, et al. Vertigo in pregnancy: a narrative review. Cureus. 2022;14(5):e25386.
6. Çoban K, Yiğit N, Aydın E. Benign paroxysmal positional vertigo in pregnancy. Turk Arch Otorhinolaryngol. 2017;55(2):83–6.
7. Uchide K, Suzuki N, Takiguchi T, Terada S, Inoue M. The possible effect of pregnancy on Ménière's disease. ORL J Otorhinolaryngol Relat Spec. 1997;59(5):292–5.
8. Toshniwal V, Agrawal A, Toshniwal T, Toshniwal S, Khanke S, Bakshi S, et al. A systematic review of vertigo: negligence in pregnancy. Cureus. 2022;14(10):e29814.
9. Ogutha J, Page NC, Hullar TE. Postpartum vertigo and superior semicircular canal dehiscence syndrome. Obstet Gynecol. 2009;114(2 Pt 2):434–6.
10. Meehan T, Nogueira C, Rajenderkumar D, Shah J, Stephens D, Dyer K. Dehiscence of the posterior and superior semicircular canal presenting in pregnancy. B-ENT. 2013;9(2):165–8.
11. Swain SK, Pati BK, Mohanty JN. Otological manifestations in pregnant women – a study at a tertiary care hospital of eastern India. J Otol. 2020;15(3):103–6.
12. Goundrey J. Vertigo after epidural morphine. Can J Anaesth. 1990;37(7):804–5.
13. Ghia D, Cuganesan R, Cappelen-Smith C. Delayed angiographic changes in postpartum cerebral angiopathy. J Clin Neurosci. 2011;18(3):435–6.
14. Manasewitsch NT, Hanfy AA, Beutler BD, Antwi-Amoabeng D, Taha M, Elnaggar M, et al. Postpartum vertebral artery dissection: case report and review of the literature. Thromb J. 2020;18(1):30.
15. Monari F, Busani S, Imbrogno MG, Neri I, Girardis M, Ghirardini A, et al. Vertebral artery dissection in term pregnancy after cervical spine manipulation: a case report and review the literature. J Med Case Rep. 2021;15(1):530.
16. Wright L. Postural hypotension in late pregnancy. "the supine hypotensive syndrome". Br Med J. 1962;1(5280):760–2.
17. Serra-Serra V, Chandran R, Kyle PM, Redman CW. Cerebral hemodynamic changes during severe orthostatic hypotension in pregnancy. Acta Obstet Gynecol Scand. 1995;74(8):656–9.
18. Oxorn H. Postural hypotension in pregnancy. Can Med Assoc J. 1960;83(9):436.
19. Yuan SM. Cardiac myxoma in pregnancy: a comprehensive review. Rev Bras Cir Cardiovasc. 2015;30(3):386–94.
20. Shahbuddin HMA, Hussin SA, Wyh WI, Mamat AZ, Marzuki A, Yusof Z. Dizzy spells in pregnancy: successful antenatal removal of a huge left atrial myxoma with mitral valve obstruction. BMJ Case Rep. 2024;17(3):e259675.
21. Castillo-Bustamante M, Espinoza I, Briceño O, Vanegas JM, Tamayo MDM, Madrigal J. Vestibular findings on the video head impulse test (vHIT) in pregnancy: a cross-sectional study. Cureus. 2023;15(6):e41059.

Part XXIX

Diagnoses—Immunologically-Mediated Otovestibular Symptoms

178 Allergy and Otovestibular Symptoms

Overview

Otoneurologists and neuro-otologists sometimes encounter patients with both *allergic* symptoms and otovestibular symptoms whose examinations and workups are unrevealing, but the relationship between these symptoms is unclear. Studies report that, for example, patients diagnosed with allergic rhinitis report aural pruritis, subjective hearing loss and tinnitus, and vertigo more frequently than healthy control subjects, and audiologic testing also shows a greater frequency of abnormal results compared to healthy control subjects. This area has not been studied well, and there are no guidelines on how to proceed. Practically, when we encounter patients with both allergic symptoms and otovestibular symptoms whose otovestibular workups are normal, we will often suggest a trial of over-the-counter allergy medication. If that fails, we will refer for a formal allergy consultation, to identify whether there is a specific allergy whose treatment may be more targeted.

Introduction

The connection between *allergic* diatheses and otovestibular symptoms is poorly understood.

The older literature on this topic [1–7] is characterized by limited diagnostic techniques and treatments available at that time.

Allergy is common, and otovestibular symptoms are common, so these groups of symptoms will coincide simply by chance. Whether there is a greater-than-chance overlap of these two groups of symptoms is unclear, though some literature (reviewed below) suggests that there is.

The relevance in otoneurology and neuro-otology is that it is not rare for clinicians to encounter patients with allergic symptoms and otovestibular symptoms whose examinations and workups appear entirely normal. In this circumstance, it is reasonable to wonder whether there is a causative relationship (specifically, whether allergies can cause otovestibular symptoms) and whether treating for allergies might also alleviate otovestibular symptoms.

Pathophysiological Mechanism of Disease

Even if a greater-than-chance association can be demonstrated between allergic symptoms and otovestibular symptoms, this in itself does not explain whether there is a mechanistic relationship, much less what that relationship would be.

Dr. Jennifer Derebery and others [8–14] have advocated for the idea that Ménière's disease has an allergic basis. Even if this idea proves to be correct, it would still leave open the question of the broader relationship between allergy and otovestibular disease.

Clinical Presentation

Lasisi and Abdullahi [15] retrospectively reviewed records of 144 patients, consisting of 71 males and 73 females, mean age of 32 years (range 2–80 years), all diagnosed with nasal allergy. Of these 144 patients, they found that 95 (66%) complained of otovestibular symptoms. Of these 95 patients, 63 (66%) had itching of the external auditory canal, 55 (58%) complained of hearing loss, 39 (41%) complained of tinnitus, and 12 (13%) complained of vertigo, as shown in Table 178.1.

Of the 95 patients with otovestibular symptoms, the final diagnoses were shown in Table 178.2.

M. Cherchi, *Otoneurology and Vestibular Medicine*, https://doi.org/10.1007/978-3-031-94841-1_178

Table 178.1 Distribution of otovestibular symptoms in 95 patients with nasal allergies

Ear symptoms	Frequency (%)
Itching of external ear	63 (66.3%)
Hearing loss	55 (58%)
Tinnitus	39 (41%)
Vertigo	12 (12.6%)
Dull tympanic membrane	78 (82%)
Perforation of the tympanic membrane	4 (4.2%)
Peripheral vestibular signs of imbalance	11 (11.6%)
Positive position test	6 (6.3%)

From Lasisi and Abdullahi [15]

Table 178.2 Distribution of diagnoses among 95 nasal allergy patients with otovestibular symptoms

Clinical diagnosis	Frequency (%)
Idiopathic tinnitus	25 (26.3%)
Idiopathic SHL	17 (18%)
Idiopathic vertigo	13 (14%)
Benign paroxysmal positional vertigo	11 (12%)
Meniere's disease	9 (10%)
Vestibular hydrop	8 (8%)
Cochlear hydrop	6 (6%)
Autoimmune inner ear disease	6 (6%)
Total	95 (100%)

From Lasisi and Abdullahi [15]

Testing: Audiologic

Lasisi and Abdullahi [15] reported that of 75 nasal allergy patients whose audiometry had been documented, 43 (59%) were normal (0–25 dB), 17 (23%) had some sensorineural hearing loss, and 13 (18%) had conductive hearing loss. Of the 17 with sensorineural hearing loss, it was mild (26–45 dB) in 6 (35%), moderate (46–60 dB) in 7 (41%), and moderate-to-severe (61–70 dB) in 4 (24%).

Singh and colleagues [16] studied 30 patients diagnosed with allergic rhinitis, including 16 women (53%) and 14 men, with a mean age of 31 years (range 17–45 years), compared to 20 healthy controls. They studied pure tone audiometry, distortion product otoacoustic emissions, transient evoked otoacoustic emissions, and brainstem auditory evoked responses in all subjects.

They reported that the allergic rhinitis patients had statistically significant hearing loss at all frequencies compared to controls, as shown in Table 178.3.

They reported that allergic rhinitis patients were statistically significantly worse than controls on distortion product otoacoustic emissions, referring to all frequencies except the highest, as shown in Table 178.4.

Table 178.3 Audiometry results in patients with allergic rhinitis and healthy controls

Frequency (Hz)	Study group (mean ± standard deviation)	Control group (mean ± standard deviation)	Mean difference (95% confidence interval)	*p*
250	28.25 ± 7.55	10.38 ± 3.74	17.87 (14.20–21.54)	<0.001
500	28.67 ± 6.52	11.75 ± 2.58	16.91 (13.82–20.00)	<0.001
1000	26.67 ± 5.39	12.88 ± 4.68	13.79 (10.82–16.76)	<0.001
2000	24.75 ± 5.62	14.75 ± 3.33	10.00 (7.18–12.81)	<0.001
4000	28.83 ± 9.04	15.88 ± 4.38	12.95 (8.57–17.34)	<0.001
6000	29.08 ± 5.99	17.25 ± 1.97	11.83 (9.03–14.63)	<0.001
8000	33.25 ± 7.17	19.13 ± 2.60	14.12 (10.51–17.73)	<0.001
10,000	38.17 ± 14.35	23.50 ± 4.08	14.66 (8.02–12.31)	<0.001
12,000	45.33 ± 19.11	24.50 ± 7.03	20.83 (11.82–29.83)	<0.001
14,000	56.83 ± 22.73	26.88 ± 7.73	29.95 (19.32–40.59)	<0.001
16,000	68.58 ± 21.54	32.35 ± 11.32	36.33 (25.77–46.89)	<0.001

From Singh et al. [16]

Table 178.4 Distortion product otoacoustic emission results in patients with allergic rhinitis and healthy controls

Frequency (Hz)	Distortion product otoacoustic emissions (mean ± standard deviation)		*p*
	Study group	Control group	
1003	−0.55 ± 4.78	8.85 ± 6.03	<0.001
1409	−0.07 ± 3.85	8.75 ± 8.50	<0.001
2000	1.98 ± 4.67	10.53 ± 6.80	<0.001
2822	0.72 ± 3.91	6.53 ± 7.39	0.001
3991	1.03 ± 5.37	7.40 ± 6.29	<0.001
5649	1.62 ± 4.41	3.40 ± 3.02	0.122

From Singh et al. [16]

Table 178.5 Transient evoked otoacoustic emission results in patients with allergic rhinitis and healthy controls

Frequency (kHz)	Transient evoked otoacoustic emissions (mean ± standard deviation)		*p*
	Study group	Control group	
1	0.48 ± 1.64	4.43 ± 3.40	<0.001
1.5	1.53 ± 2.26	6.09 ± 5.39	0.02
2	0.82 ± 3.01	7.96 ± 5.51	<0.001
3	2.55 ± 2.97	10.09 ± 5.26	<0.001
4	0.40 ± 1.43	4.33 ± 4.74	<0.001

From Singh et al. [16]

Table 178.6 Brainstem auditory evoked response results in patients with allergic rhinitis and healthy controls

Parameter	Value (mean ± standard deviation)		*P*
	Study group	Control group	
Wave I latency	1.63 ± 0.13	1.52 ± 0.13	0.008
Wave II latency	3.67 ± 0.16	3.68 ± 0.14	0.875
Wave III latency	5.54 ± 0.24	5.53 ± 0.19	0.953
Wave I–III interpeak latency	2.03 ± 0.14	2.15 ± 0.12	0.003
Wave III–V interpeak latency	1.87 ± 0.17	1.85 ± 0.10	0.738
Wave I–V interpeak latency	3.90 ± 0.18	4.01 ± 0.16	0.038

From Singh et al. [16]

They reported that allergic rhinitis patients were statistically significantly worse than controls on transient evoked otoacoustic emissions, referring to all frequencies, as shown in Table 178.5.

They reported that allergic rhinitis patients were statistically significantly worse than controls on brainstem auditory evoked responses in Wave I latency, Wave I-III interpeak latency, and Wave I-V interpeak latency, as shown in Table 178.6.

The mechanism by which allergic rhinitis would provoke these otologic test results is unclear.

Testing: Vestibular

As of this writing, there were no recent publications that systematically studied vestibular tests in diagnosed allergy patients.

Testing: Other

Cisi and colleagues [17], and Domingues and colleagues [18] studied 75 patients with "imbalance due to peripheral origin" using skin prick testing for 13 inhalants and 5 foods. They observed that 25 (33%) of patients exhibited positive skin prick tests and at least one food allergen. Of these patients, four had positive skin prick tests even without any traditional allergic symptoms. They concluded that, "The prevalence of Type I hypersensitivity reactions to inhalants and food in the [studied] population was greater than in the general population."

Treatment

Since the relationship between allergy and otovestibular symptoms is unclear, it is unsurprising that there are no treatment trials.

Practically, when we encounter patients with both allergic symptoms and otovestibular symptoms whose otovestibular workups are normal, we will often suggest a trial of over-the-counter allergy medication. If that fails, we will refer for a formal allergy consultation, with the goal of identifying whether there is a specific allergy whose treatment may be more targeted.

References

1. Clemis JD. Allergic cochleovestibular disturbances. Trans Am Acad Ophthalmol Otolaryngol. 1972;76(6):1451–7.
2. Floberg LE, Koch H. Vestibular symptoms elicited through allergy. Acta Otolaryngol Suppl. 1954;116:94–9.
3. McLaurin JW. Desensitization by histamine (histamine azoprotein) in vertigo, periodic headaches and vasomotor (allergic) rhinitis; review of the literature and report of 102 personal cases. Laryngoscope. 1946;56:253–81.
4. Pisani S. Allergic etiology of vertiginous syndromes. Sem Med. 1958;112(21):831–8.
5. Powers WH, House WF. The dizzy patient–allergic aspect. Laryngoscope. 1969;79(7):1330–8.
6. Strandbygaard E. Vertigo attacks caused by allergy. Z Laryngol Rhinol Otol. 1955;34(6):371–6.
7. Wilson WH, Powers WH. Metabolic and allergic aspects of inner ear problems. J Otolaryngol Soc Aust. 1971;3(2):181–6.
8. Banks C, McGinness S, Harvey R, Sacks R. Is allergy related to Meniere's disease? Curr Allergy Asthma Rep. 2012;12(3):255–60.

9. Derebery MJ. The role of allergy in Meniere's disease. Otolaryngol Clin N Am. 1997;30(6):1007–16.
10. Derebery MJ, Berliner KI. Allergy and Meniere's disease. Curr Allergy Asthma Rep. 2007;7(6):451–6.
11. Derebery MJ, Berliner KI. Allergy and its relation to Meniere's disease. Otolaryngol Clin N Am. 2010;43(5):1047–58.
12. Derebery MJ, Valenzuela S. Meniere's syndrome and allergy. Otolaryngol Clin N Am. 1992;25(1):213–24.
13. Di Berardino F, Barozzi S, Cesarani A. Allergy and Meniere's disease: a review. Eur Ann Allergy Clin Immunol. 2005;37(8):299–300.
14. Weinreich HM, Agrawal Y. The link between allergy and Meniere's disease. Curr Opin Otolaryngol Head Neck Surg. 2014;22(3):227–30.
15. Lasisi AO, Abdullahi M. The inner ear in patients with nasal allergy. J Natl Med Assoc. 2008;100(8):903–5.
16. Singh S, Nagarkar AN, Bansal S, Vir D, Gupta AK. Audiological manifestations of allergic rhinitis. J Laryngol Otol. 2011;125(9):906–10.
17. Cisi ED, Moreira BRS, Ferreira de Mello J Jr. Hypersensitivity to inhalants and food in balance disturbances. Rev Laryngol Otol Rhinol (Bord). 2011;132(2):81–4.
18. Cisi ED, Moreira BRS, Ferreira de Mello J Jr. Hypersensitivity to inhalants and food in balance disturbances. Rev Laryngol Otol Rhinol (Bord). 2010;131(4–5):243–6.

Lupus

179

Overview

Systemic lupus erythematosus (SLE), sometimes referred to as simply "*lupus*," is an autoimmune disease of uncertain etiology that can involve nearly any organ system, and thus can have an extremely broad range of clinical presentations. SLE is more common in nonwhite populations and is much more common in women. SLE probably becomes manifest through a combination of genetic susceptibility and environmental/epigenetic factors. When SLE involves the central nervous system and specifically affects auditory or vestibular function, the mechanism of disease may involve deposition of immune complexes and cytotoxic damage. SLE patients may develop hearing loss that is usually sensorineural, typically involves the higher rather than the lower frequencies, and can be unilateral, asymmetrically bilateral, or symmetrically bilateral; the onset may be abrupt or gradual. SLE patients can develop other aural symptoms as well, such as tinnitus and aural fullness. While less common, SLE patients can also develop disequilibrium and ocular motor disorders.

Introduction

Systemic lupus erythematosus (SLE), sometimes referred to as simply "*lupus*," is an autoimmune disease of uncertain etiology that can involve nearly any organ system. Its remarkable clinical heterogeneity has led to reasonable questions regarding whether this is truly a single disease [1].

Over the centuries, the term "lupus" has been applied to what were probably a broad range of diseases, but in the early nineteenth century, the work of several scholars, including Moritz Kaposi, James H. Sequiera, H. Balean, and William Osler, led to the understanding of this as an autoimmune-mediated clinical entity with variable presentations [2].

Epidemiology

The incidence and prevalence of SLE have been difficult to determine for a variety of reasons; "Reports of the global incidence and prevalence of SLE vary widely, owing to inherent variation in population demographics, environmental exposures, and socioeconomic factors. Differences in study design and case definitions also contribute to inconsistent reporting" [3], and reported "variability may reflect true differences across populations, or result from methodological differences of studies" [4].

Epidemiological studies generally report SLE to be more common in nonwhite racial groups [4, 5] and is more common in women [5]. Although SLE occurs worldwide [5], there is a paucity of good-quality epidemiologic data for most countries [6]. Nevertheless, there appears to be considerable variability in geographic distribution, and considerable disagreement between studies; for example, Danchenko and colleagues report that, "There is a trend towards higher incidence and prevalence of SLE in Europe and Australia compared to the U.S.A. In Europe, the highest prevalence was reported in Sweden, Iceland and Spain" [4], while Rees and colleagues report North America to have the highest disease burden [7].

In the United States, analysis of data from the Centers for Disease Control suggests that in 2002–2009, the pooled incidence rate per 100,000 person-years was 5.1 (95% confidence interval 4.6–5.6) [8], and the pooled prevalence rate was 72.8 per 100,000 person-years (95% confidence interval 65.3–81.0) [9].

M. Cherchi, *Otoneurology and Vestibular Medicine*, https://doi.org/10.1007/978-3-031-94841-1_179

Genetics

The genetics of SLE are still being explored [10–21], and similar to a number of other autoimmune diseases, it seems likely that SLE becomes manifest when there is a confluence of genetic susceptibility and environmental/epigenetic factors.

Pathophysiological Mechanism of Disease

The etiology of SLE is unknown. The ability of SLE to involve the central nervous system probably results from "multiple contributors such as various immune effectors and the brain-intrinsic neuroimmune interfaces that are breached by the immune effectors. The postulated neuroimmune interfaces include the blood-brain barrier, blood-cerebrospinal fluid barrier, meningeal barrier, and glymphatic system," such that "the underlying pathophysiology can be classified as vasculitis and vasculopathy, APS [antiphospholipid syndrome], demyelinating syndrome, or autoimmune antibody-mediated encephalitis" [22].

As far as *auditory* disorders are concerned, many authors implicate the deposition of immune complexes [23], while others suggest a range of mechanisms; for instance, it has been speculated that, "The main mechanisms involved in inner ear damage include the autoimmune response, deposition of immune complexes in the vessels and, to a lesser extent, cytotoxic damage" [24].

As far as *vestibular* symptoms are concerned, there are scattered case reports and case series of patients with cerebellar syndromes. Among all patients with systemic lupus erythematosus, cerebellar ataxia appears to be rare; al-Arfaj and colleagues [25] report one out of 150 patients (***0.7%***); Singh and colleagues [26] report three out of 350 patients (***0.9%***); Cha and Kim [27] report a rate of ***2%***. Lupus-related cerebellar ataxia can occur at any age; Liao and colleagues [28] reported a case of an 11-year-old male whose presenting symptoms were disequilibrium and unsteady gait.

The mechanism of lupus-related cerebellar ataxia is unclear. Most authors postulate an autoimmune cerebellitis [29]. Some authors suggest cerebellar strokes secondary to lupus vasculopathy [30], while others dispute this [31]. There is a report of cerebellar ataxia secondary to lupus-related hydrocephalus [32].

Clinical Presentation: Auditory Alone

Abbasi and Yazdi [33] presented a case-control study of 45 SLE patients, in whom reported symptoms included hearing loss in 5 (11%), tinnitus in 3 (7%), and ear fullness in 1 (2%). They also performed audiometry on these patients and calculated that the relative risk for sensorineural hearing loss in SLE patients compared to controls was 3.7 (95% confidence interval 2.3–4.9).

Gomides and colleagues [34] reported on a series of 45 female SLE patients and found that 25 (56%) reported some auditory symptoms (hearing loss, ear fullness, tinnitus), and sensorineural hearing loss was identified in 7 (16%).

Sperling and colleagues [23] organized a self-directed questionnaire study of 84 SLE patients and found that 26 (31%) reported aural symptoms. Of these 84 patients, this included 13 (15%) who reported unilateral hearing loss with or without tinnitus, and 14 (17%) who reported bilateral hearing loss with or without tinnitus.

Caldarelli and colleagues [35] reported the case of a 51-year-old woman who presented with sudden sensorineural hearing loss on one side, and then 3 weeks later, on the other side, and subsequent workup led to a diagnosis of SLE.

Clinical Presentation: Vestibular Alone

Lupus-related cerebellar ataxia can present acutely [31, 36–38] or subacutely [39, 40] and can be rapidly progressive [41]. Cerebellar ataxia can occur in a patient with preexisting manifestations of systemic lupus erythematosus, but rarely can be the initial manifestation [42].

Clinical Presentation: Auditory and Vestibular

Maciaszczyk and colleagues [43] studied 35 unselected consecutive SLE patients and compared them to 30 healthy controls. Symptoms reported by the SLE patients included disequilibrium in 71%, tinnitus in 40%, hyperacusis in 17%, hearing loss in 17%, and ear fullness in 3%.

Batuecas-Caletrio and colleagues [44] studied 89 SLE patients. Of these, eight (9%) reported episodic vertigo.

Karatas and colleagues [45] studied 28 SLE patients and found that reported symptoms included tinnitus in 9 (32%), hearing loss in 2 (7%), and disequilibrium in 8 (29%), as shown in Table 179.1.

The temporal progression of hearing loss is variable. In some cases, the hearing loss presents acutely [35, 46], while in others it can progress over months [47].

Ocular Motor Examination

In rare cases, SLE can present with ocular motor abnormalities that are detectable on face-to-face examination.

Table 179.1 Auditory and vestibular symptoms reported by Karatas et al. [45]

	Tinnitus	Hearing loss	Vertigo	Asymptomatic
Tinnitus alone	2	–	–	–
Hearing loss alone	–	–	–	–
Vertigo alone	–	–	3	–
Tinnitus + hearing loss + vertigo	2	1	2	–
Tinnitus + hearing loss	3	1	–	–
Tinnitus + vertigo	2	–	3	–
Asymptomatic	–	–	–	9
Totals	9 (32%)	2 (7%)	8 (29%)	9 (32%)

Cakir and colleagues report a case of a 45-year-old woman who presented with acute-onset disequilibrium, ataxia, and diplopia, and was found to have internuclear ophthalmoplegia [48]. Galindo and colleagues [49] conducted a retrospective study of 268 SLE patients and identified internuclear ophthalmoplegia in four (1.5%).

Kunavarapu and colleagues [50] reported a case presenting with one-and-a-half syndrome.

Liao and colleagues [28] presented the case of an 11-year-old male whose presenting examination showed "bilateral vertical nystagmus."

Testing: Auditory

Abbasi and Yazdi [33] presented a case-control study of 45 SLE patients in whom they performed audiometry. They calculated that the relative risk for sensorineural hearing loss in SLE patients compared to controls was 3.7 (95% confidence interval 2.3–4.9).

Bowman and colleagues [51] prospectively studied 30 patients who had been hospitalized for an SLE exacerbation, and "found an ***8%*** incidence of substantial, previously undetected hearing loss without attributable cause."

Gomides and colleagues [34] reported on a series of 45 female SLE patients and found that 25 (56%) reported some auditory symptoms (hearing loss, ear fullness, tinnitus), and sensorineural hearing loss was identified in 7 (***16%***).

Jimenez-Alonso and colleagues [52] studied 91 consecutive SLE patients and found that 14 (***15%***) exhibited sensorineural hearing loss, "affecting mainly the high and middle frequencies," which was bilateral in 9 and unilateral in 5.

Kastanioudakis and colleagues [53] compared 43 SLE patients and 50 age-matched healthy controls. They found hearing loss in 9 (***23%***) of SLE patients, of whom eight had *sensorineural* hearing loss and one had *conductive* hearing loss. Of the nine patients with sensorineural hearing loss, the loss was bilateral and symmetrical in 1, bilateral but *asymmetrical* in four, and *unilateral* in 3. In one patient with conductive hearing loss, the loss was unilateral.

Andonopoulos and colleagues [54] compared 40 unselected consecutive SLE patients and compared them to 65 age-matched healthy controls, and divided them into age groups of 30–39, 40–49, 50–59, and 60–69 years. They found that, "A statistically significant decrease in hearing acuity at low frequencies (125–500 Hz) was found in the patients aged 16–59 compared with the controls, whereas in the first group of young patients (16–29) a similar decrease was observed in the high frequencies as well," and concluded that, "With regard to the high frequencies, our results suggest that young SLE patients exhibit a "premature aging" of the inner ear which eventually stabilizes. On the contrary, at low frequencies, there is a regular decrease in acuity in all groups (except for the elderly patients: 60–69), which may indicate a subclinical vestibular hydrops."

Maciaszczyk and colleagues [43] studied 35 unselected consecutive SLE patients and compared them to 30 healthy controls. They reported that, "Longer ABR [auditory brainstem evoked response] latency averages were observed in the group of SLE patients compared to control."

Testing: Vestibular

Karatas and colleagues [45] studied electronystagmography in 28 SLE patients and reported abnormalities in 50% of them.

Table 179.2 shows the specific electronystagmography abnormalities reported in 19 patients by Karatas and colleagues [45].

Table 179.2 Electronystagmography abnormalities in 28 patients with systemic lupus erythematosus. From Karatas et al. [45]

	Central vestibulopathy	Peripheral vestibulopathy	Nonspecific vestibulopathy	Mixed vestibulopathy	Total
Gaze nystagmus	–	–	–	1	1
Saccade nystagmus	–	–	–	–	–
Abnormal optokinetic nystagmus	1	–	–	1	2
Abnormal pursuit	1	–	–	–	1
Spontaneous nystagmus	–	–	–	–	–
Positional nystagmus	–	–	1	–	1
Directional preponderance	–	–	2	–	2
Absent caloric response	–	1	–	–	1
Reduced caloric response	–	9	–	2	11
Total	2	10	3	4	19

Testing: Other

Of the studies that compare audiologic findings with other clinical features or serum markers of disease activity, most find no clear correlation [44, 51, 54]. In contrast, Sperling and colleagues [23] report some correlation with average creatinine and C3 complement levels.

Imaging

In SLE patients with ocular motor abnormalities, brain MRI may identify discrete brainstem lesions [50].

In SLE patients with cerebellar ataxia, imaging may reveal cerebellar atrophy [27, 39, 40] or T2/FLAIR hyperintensity [29].

Differential Diagnosis

Since SLE can affect essentially every organ system, its differential diagnosis is extraordinarily broad.

It is rare for an otoneurologist or neuro-otologist to be the first clinician to make a diagnosis of lupus. The more common scenario is that a patient who already carries a diagnosis of SLE is referred for a complaint of disequilibrium or hearing loss, and the usual clinical query is whether the symptom is attributable to the already established diagnosis of SLE or whether it is due to independently occurring disease.

In this circumstance, it is medically reasonable to undertake a screening otovestibular workup to identify whether another treatable condition (separate from SLE) is present. If no other disease is identified, then management of otovestibular symptoms will be management of the SLE itself.

Treatment

The management of SLE patients is usually undertaken by a rheumatologist.

In lupus patients with cerebellar ataxia, management often includes steroids [25, 31, 37–40], usually intravenous [29], less commonly oral [30]. Other immunomodulators [38] such as azathioprine [25] or mycophenolate mofetil [30] are sometimes considered.

Prognosis

Prognosis for hearing loss in SLE patients is poor, particularly when it presents acutely [46].

References

1. Rivas-Larrauri F, Yamazaki-Nakashimada MA. Systemic lupus erythematosus: Is it one disease? Reumatol Clin. 2016;12(5):274–81.
2. Felten R, Lipsker D, Sibilia J, Chasset F, Arnaud L. The history of lupus throughout the ages. J Am Acad Dermatol. 2022;87(6):1361–9.
3. Barber MRW, Drenkard C, Falasinnu T, Hoi A, Mak A, Kow NY, et al. Global epidemiology of systemic lupus erythematosus. Nat Rev. Rheumatol. 2021;17(9):515–32.
4. Danchenko N, Satia JA, Anthony MS. Epidemiology of systemic lupus erythematosus: a comparison of worldwide disease burden. Lupus. 2006;15(5):308–18.
5. Pons-Estel GJ, Ugarte-Gil MF, Alarcon GS. Epidemiology of systemic lupus erythematosus. Expert Rev. Clin Immunol. 2017;13(8):799–814.
6. Tian J, Zhang D, Yao X, Huang Y, Lu Q. Global epidemiology of systemic lupus erythematosus: a comprehensive systematic analysis and modelling study. Ann Rheum Dis. 2023;82(3):351–6.
7. Rees F, Doherty M, Grainge MJ, Lanyon P, Zhang W. The worldwide incidence and prevalence of systemic lupus erythematosus: a systematic review of epidemiological studies. Rheumatology (Oxford). 2017;56(11):1945–61.
8. Izmirly PM, Ferucci ED, Somers EC, Wang L, Lim SS, Drenkard C, et al. Incidence rates of systemic lupus erythematosus in the USA: estimates from a meta-analysis of the Centers for Disease Control and Prevention national lupus registries. Lupus Sci Med. 2021;8(1):e000614.
9. Izmirly PM, Parton H, Wang L, McCune WJ, Lim SS, Drenkard C, et al. Prevalence of Systemic Lupus Erythematosus in the United States: Estimates From a Meta-Analysis of the Centers for Disease Control and Prevention National Lupus Registries. Arthritis Rheumatol. 2021;73(6):991–6.

10. Croker JA, Kimberly RP. Genetics of susceptibility and severity in systemic lupus erythematosus. Curr Opin Rheumatol. 2005;17(5):529–37.
11. Goulielmos GN, Zervou MI, Vazgiourakis VM, Ghodke-Puranik Y, Garyfallos A, Niewold TB. The genetics and molecular pathogenesis of systemic lupus erythematosus (SLE) in populations of different ancestry. Gene. 2018;668:59–72.
12. Hess EV, Farhey Y. Epidemiology, genetics, etiology, and environment relationships of systemic lupus erythematosus. Curr Opin Rheumatol. 1994;6(5):474–80.
13. Hess EV, Farhey Y. Etiology, environmental relationships, epidemiology, and genetics of systemic lupus erythematosus. Curr Opin Rheumatol. 1995;7(5):371–5.
14. James JA. Clinical perspectives on lupus genetics: advances and opportunities. Rheum Dis Clin North Am. 2014;40(3):413–32. vii
15. Lewkonia RM. The clinical genetics of lupus. Lupus. 1992;1(2):55–62.
16. Marion TN, Postlethwaite AE. Chance, genetics, and the heterogeneity of disease and pathogenesis in systemic lupus erythematosus. Semin Immunopathol. 2014;36(5):495–517.
17. Relle M, Weinmann-Menke J, Scorletti E, Cavagna L, Schwarting A. Genetics and novel aspects of therapies in systemic lupus erythematosus. Autoimmun Rev. 2015;14(11):1005–18.
18. Sestak AL, Fürnrohr BG, Harley JB, Merrill JT, Namjou B. The genetics of systemic lupus erythematosus and implications for targeted therapy. Ann Rheum Dis. 2011;70(Suppl 1):i37–43.
19. Sestak AL, Nath SK, Sawalha AH, Harley JB. Current status of lupus genetics. Arthritis Res Ther. 2007;9(3):210.
20. Shen N, Tsao BP. Current advances in the human lupus genetics. Curr Rheumatol Rep. 2004;6(5):391–8.
21. Yim YS, Wakeland EK. The genetics of lupus. Curr Opin Nephrol Hypertens. 2001;10(3):437–43.
22. Ota Y, Srinivasan A, Capizzano AA, Bapuraj JR, Kim J, Kurokawa R, et al. Central nervous system systemic lupus erythematosus: pathophysiologic, clinical, and imaging features. Radiographics. 2022;42(1):212–32.
23. Sperling NM, Tehrani K, Liebling A, Ginzler E. Aural symptoms and hearing loss in patients with lupus. Otolaryngol Head Neck Surg. 1998;118(6):762–5.
24. Di Stadio A, Ralli M. Systemic Lupus Erythematosus and hearing disorders: literature review and meta-analysis of clinical and temporal bone findings. J Int Med Res. 2017;45(5):1470–80.
25. Al-Arfaj HF, Naddaf HO. Cerebellar atrophy in systemic lupus erythematosus. Lupus. 1995;4(5):412–4.
26. Singh RR, Prasad K, Kumar A, Misra A, Padmakumar K, Malaviya AN. Cerebellar ataxia in systemic lupus erythematosus: three case reports. Ann Rheum Dis. 1988;47(11):954–6.
27. Cha JM, Kim HS. functional improvement after taking rehabilitation program in cerebellar ataxia in a patient with systemic lupus erythematosus: a case report. Brain Neurorehabil. 2020;13(2):e11.
28. Liao CH, Yang YH, Chiang BL. Systemic lupus erythematosus with presentation as vertigo and vertical nystagmus: report of one case. Acta Paediatr Taiwan. 2003;44(3):158–60.
29. Kristoff TJ, Shoskes A, Claytor B. Lupus cerebellitis presenting with ataxia: a case report. Neurohospitalist. 2022;12(1):117–20.
30. Ghosh K, Chatterjee A, Ghosh S, Chakraborty S. Cerebellar ataxia in a young patient: a rare path to lupus. J Neurosci Rural Pract. 2014;5(Suppl 1):S75–6.
31. Yaginuma M, Suenaga M, Shiono Y, Sakamoto M. Acute cerebellar ataxia of a patient with SLE. Clinical neurology and neurosurgery. 2000;102(1):37–9.
32. Ahmed HMA, El-Gohary R, Fayed F, El-Gendy H. Cerebellar ataxia and obstructive hydrocephalus, rare neurologic presentations in patients with systemic lupus erythematosus. Rheumatol Int. 2017;37(11):1917–30.
33. Abbasi M, Yazdi Z, Kazemifar AM, Bakhsh ZZ. Hearing loss in patients with systemic lupus erythematosus. Glob J Health Sci. 2013;5(5):102–6.
34. Gomides AP, do Rosario EJ, Borges HM, Gomides HH, de Padua PM, Sampaio-Barros PD. Sensorineural dysacusis in patients with systemic lupus erythematosus. Lupus. 2007;16(12):987–90.
35. Caldarelli DD, Rejowski JE, Corey JP. Sensorineural hearing loss in lupus erythematosus. Am J Otol. 1986;7(3):210–3.
36. Appenzeller S, Cendes F, Costallat LT. Cerebellar ataxia in systemic lupus erythematosus. Lupus. 2008;17(12):1122–6.
37. Chattopadhyay P, Dhua D, Philips CA, Saha S. Acute cerebellar ataxia in lupus. Lupus. 2011;20(12):1312–5.
38. Sy MCC, Reyes NGD, Zamora GT, Fernandez MLL. Cerebellar ataxia as a primary manifestation of neuropsychiatric systemic lupus erythematosus. BMJ Case Rep. 2021;14(2):e236825.
39. Iwasaki Y, Okamoto A, Shoda H, Takahashi Y, Fujio K, Kawahata K, et al. Subacute cerebellar ataxia and atrophy developed in a young woman with systemic lupus erythematosus whose cerebrospinal fluid was positive for antineuronal cell antibody. Lupus. 2012;21(3):324–8.
40. Manto MU, Rondeaux P, Jacquy J, Hildebrand JG. Subacute pancerebellar syndrome associated with systemic lupus erythematosus. Clin Neurol Neurosurg. 1996;98(2):157–60.
41. Casciato S, Mascia A, Quarato PP, D'Aniello A, Scoppetta C, Di Gennaro G. Subacute cerebellar ataxia as presenting symptom of systemic lupus erythematosus. Eur Rev Med Pharmacol Sci. 2018;22(21):7401–3.
42. Kutlubaev MA, Idrisova RF, Zakirova EN, Hardy TA. Cerebellar ataxia as a first manifestation of systemic lupus erythematosus. Acta Neurol Belg. 2020;120(5):1241–3.
43. Maciaszczyk K, Durko T, Waszczykowska E, Erkiert-Polguj A, Pajor A. Auditory function in patients with systemic lupus erythematosus. Auris Nasus Larynx. 2011;38(1):26–32.
44. Batuecas-Caletrio A, del Pino-Montes J, Cordero-Civantos C, Calle-Cabanillas MI, Lopez-Escamez JA. Hearing and vestibular disorders in patients with systemic lupus erythematosus. Lupus. 2013;22(5):437–42.
45. Karatas E, Onat AM, Durucu C, Baglam T, Kanlikama M, Altunoren O, et al. Audiovestibular disturbance in patients with systemic lupus erythematosus. Otolaryngol Head Neck Surg. 2007;136(1):82–6.
46. Riera JL, Del RMM, Musuruana JL, Cavallasca JA. Sudden sensorineural hearing loss in systemic lupus erythematosus and antiphospholipid syndrome: a clinical review. Curr Rheumatol Rev. 2020;16(2):84–91.
47. Khalidi NA, Rebello R, Robertson DD. Sensorineural hearing loss in systemic lupus erythematosus: case report and literature review. J Laryngol Otol. 2008;122(12):1371–6.
48. Cakir A, Duzgun E, Demir S, Aydin A. Jerky seesaw nystagmus with internuclear ophthalmoplegia as the presenting finding in systemic lupus erythematosus. J Fr Ophtalmol. 2014;37(10):e157–9.
49. Galindo M, Pablos JL, Gómez-Reino JJ. Internuclear ophthalmoplegia in systemic lupus erythematosus. Semin Arthritis Rheum. 1998;28(3):179–86.
50. Kunavarapu C, Kesavan RB, Pevil-Ulysee M, Mohan SS. Systemic lupus erythematosus presenting as "one-and-a-half syndrome". J Rheumatol. 2001;28(4):874–5.
51. Bowman CA, Linthicum FH Jr, Nelson RA, Mikami K, Quismorio F. Sensorineural hearing loss associated with systemic lupus erythematosus. Otolaryngol Head Neck Surg. 1986;94(2):197–204.
52. Jimenez-Alonso J, Gutierrez-Cabello F, Castillo JL, Sabio JM, Hidalgo-Tenorio C, Leon L, et al. Ear involvement in systemic lupus erythematosus patients: a comparative study. J Laryngol Otol. 2002;116:103–7.
53. Kastanioudakis I, Ziavra N, Voulgari PV, Exarchakos G, Skevas A, Drosos AA. Ear involvement in systemic lupus erythematosus patients: a comparative study. J Laryngol Otol. 2002;116(2):103–7.
54. Andonopoulos AP, Naxakis S, Goumas P, Lygatsikas C. Sensorineural hearing disorders in systemic lupus erythematosus. A controlled study. Clin Exp Rheumatol. 1995;13(2):137–41.

180 Rheumatoid Arthritis and Otovestibular Symptoms

Overview

Rheumatoid arthritis (RA) is a chronic autoimmune disorder of unknown etiology that can affect multiple organ systems. Some RA patients experience audiologic and vestibular symptoms, though the mechanism by which this occurs is unknown. Compared to healthy controls, among RA patients, there is a greater rate of hearing loss (predominantly sensorineural), delayed latencies in cervical vestibular evoked myogenic potentials, abnormal ocular motor findings (including caloric asymmetry), and abnormalities on computerized dynamic posturography. Such results are also found in many other diseases, so in an RA patient with otovestibular symptoms, it is medically reasonable to undertake a screening workup for more common causes of such findings. If no other cause is found, then treatment of RA (under the supervision of a rheumatologist) should be undertaken, though consultation with an audiologist (for amplification in RA patients with hearing loss) or a vestibular rehabilitation physical therapist (in RA patients with disequilibrium) is medically reasonable.

Introduction

Rheumatoid arthritis (RA) is a chronic autoimmune disorder of unknown etiology. The articular manifestations tend to be the most prominent, but the disease can affect multiple organ systems (cardiac, pulmonary, ophthalmologic, integumentary, and others).

Epidemiology

The global prevalence of RA has been estimated at 0.24% [1]. Compared to the global prevalence, the rates in the United States may be higher, with the overall age-adjusted prevalence in 2005 being 0.72% [2], and the prevalence appears to have increased from 2004–2014 [3]. RA affects females more than males.

Genetics

Dozens of genes have been associated with RA, particularly in genes encoding HLA (human leukocyte antigens). As with many autoimmune diseases, RA probably results from a combination of genetic predisposition and environmental exposures.

Pathophysiological Mechanism of Disease

The mechanism by which RA affects the audiovestibular system is unknown.

Clinical Presentation

Effat and Berty [4] conducted a questionnaire and examination-based study in 141 RA patients and 141 control subjects, and reported that otologic symptoms (otalgia, hearing loss, tinnitus, vertigo) "all had a significantly higher incidence in RA patients compared to control subjects."

Zonzini and colleagues [5] report a higher prevalence of falls in RA patients.

Ocular Motor Examination

As of this writing, there were no published studies regarding face-to-face ocular motor examination in RA patients.

M. Cherchi, *Otoneurology and Vestibular Medicine*, https://doi.org/10.1007/978-3-031-94841-1_180

Testing: Audiologic

Kakani and colleagues [6] studied audiometry in 25 RA patients and 25 healthy controls, and reported that six patients had mild sensorineural hearing loss and one had bilateral mixed hearing loss on audiometry.

Ozkiris and colleagues [7] conducted a prospective study of 81 RA patients and 81 healthy controls using low-frequency and high-frequency audiometry. They reported that, "The mean air conduction threshold values at high frequencies (4 kHz, 6 kHz, and 8 kHz) in the RA group were lower than control groups."

Yilmaz and colleagues [8] studied audiometry in 43 RA patients and 30 healthy controls. They reported that 12 RA patients (28%) and two controls (7%) had sensorineural hearing loss.

Testing: Cervical Vestibular Evoked Myogenic Potentials

Heydari and colleagues [9] studied cervical vestibular evoked myogenic potentials in 25 RA patients and 20 healthy controls. They reported that compared to controls, in RA patients, the mean peak latency of p13 was significantly higher in both ears, and the mean peak latency of n23 was significantly higher in the left ear. There were no significant differences in mean peak-to-peak amplitude and amplitude ratio between the groups.

Kent and colleagues [10] studied cervical vestibular evoked myogenic potentials in 34 patients with rheumatoid arthritis (RA), 24 patients with ankylosing spondylitis (AS), and 20 healthy controls. They reported that compared to healthy controls, "n1" latencies were prolonged on the right side in RA patients, and the amplitude asymmetry ratio was higher in RA patients.

Testing: Ocular Motor Function

Some, though not all, investigators report abnormal ocular motor findings in RA.

Ozkiris and colleagues [7] conducted a prospective study of 81 RA patients and 81 healthy controls using videonystagmography. They reported that, "VNG testing revealed central abnormalities in twenty patients (25%), peripheral abnormalities in five patients (6%), and mixed abnormalities in six patients (7%)."

Yilmaz and colleagues [8] studied electronystagmography in 43 RA patients and 30 healthy controls. They reported that, "the results of electronystagmography revealed central abnormalities in nine patients (20.9%), peripheral abnormalities in three (6.9%), and mixed abnormalities in three (6.9%). Smooth pursuit and saccade tracing impairments were significantly higher in patients with rheumatoid arthritis ($P > 0.05$). Canal paresis in patients with rheumatoid arthritis was significantly higher than that in the control group ($P = 0.039$)."

Kakani and colleagues [6] studied audiometry in 25 RA patients and 25 healthy controls, and reported that electronystagmography found *no abnormalities* in saccades or bithermal caloric stimulation in RA patients.

King and colleagues [11] studied several ocular motor functions in RA patients compared to controls. They reported *no statistically significant differences* between the two groups in gaze holding, vestibulo-ocular reflex gain or phase, optokinetic slow-phase velocity or quick-phase amplitude, optokinetic after-nystagmus slow-phase velocity or duration, or latency to the illusion of circular vection.

Testing: Computerized Dynamic Posturography

De Lorenzis and colleagues [12] studied computerized dynamic posturography in 30 ERA ("early rheumatoid arthritis") patients and 30 age- and sex-matched controls. They found that compared to controls, ERA patients exhibited lower composite equilibrium scores, lower somatosensory ratio, lower visual ratio, and lower vestibular ratios—in short, ERA patients performed more poorly across the board.

Testing: Other

A rheumatologist may monitor RA disease activity through blood tests, but these results do not appear to correlate with otovestibular manifestations of the disease.

Differential Diagnosis

The audiologic and vestibular abnormalities described above (high-frequency sensorineural hearing loss, abnormalities in cervical vestibular evoked myogenic potentials, and in ocular motor studies) are found in many other diseases as well.

In an RA patient complaining of audiologic or vestibular symptoms, it is medically reasonable to undertake a screening otovestibular workup to assess for more common causes of such symptoms. If this reveals no compelling evidence for an alternative explanation, then by exclusion, RA is reasonable to consider.

Treatment

Logically, treatment of audiologic and vestibular symptoms attributed to RA should probably be the same as treatment for RA itself, and is usually overseen by a rheumatologist.

However, in RA patients with hearing loss, it is reasonable to consult audiologist to be evaluated for amplification. Similarly, in RA patients with disequilibrium, it is reasonable to consider a trial of vestibular rehabilitation therapy in an attempt to reduce fall risk.

Prognosis

Curiously, of the studies that report otovestibular abnormalities in RA patients, most explicitly comment that these abnormalities did not correlate in any statistically significant way with disease duration [6–8].

References

1. Cross M, Smith E, Hoy D, Carmona L, Wolfe F, Vos T, et al. The global burden of rheumatoid arthritis: estimates from the global burden of disease 2010 study. Ann Rheum Dis. 2014;73(7):1316–22.
2. Myasoedova E, Crowson CS, Kremers HM, Therneau TM, Gabriel SE. Is the incidence of rheumatoid arthritis rising?: results from Olmsted County, Minnesota, 1955-2007. Arthritis Rheum. 2010;62(6):1576–82.
3. Hunter TM, Boytsov NN, Zhang X, Schroeder K, Michaud K, Araujo AB. Prevalence of rheumatoid arthritis in the United States adult population in healthcare claims databases, 2004-2014. Rheumatol Int. 2017;37(9):1551–7.
4. Effat KG, Berty A. Otological symptoms in patients with rheumatoid arthritis of the temporomandibular joint. Cranio. 2023;43(4):621–8.
5. Zonzini Gaino J, Barros Bértolo M, Silva Nunes C, de Morais BC, Sachetto Z, Davitt M, de Paiva ME. Disease-related outcomes influence prevalence of falls in people with rheumatoid arthritis. Ann Phys Rehabil Med. 2019;62(2):84–91.
6. Kakani RS, Mehra YN, Deodhar SD, Mann SB, Mehta S. Audiovestibular functions in rheumatoid arthritis. J Otolaryngol. 1990;19(2):100–2.
7. Özkırış M, Kapusuz Z, Günaydın İ, Kubilay U, Pırtı İ, Saydam L. Does rheumatoid arthritis have an effect on audiovestibular tests? Eur Arch Otorrinolaringol. 2014;271(6):1383–7.
8. Yilmaz S, Erbek S, Erbek SS, Ozgirgin N, Yucel E. Abnormal electronystagmography in rheumatoid arthritis. Auris Nasus Larynx. 2007;34(3):307–11.
9. Heydari N, Hajiabolhassani F, Fatahi J, Movaseghi S, Jalaie S. Vestibular evoked myogenic potentials in patients with rheumatoid arthritis. Med J Islam Repub Iran. 2015;29:216.
10. Kent AE, Gürses E, Karabekiroğlu F, Genç A. Balance impairment in patients with rheumatoid arthritis and ankylosing spondylitis. Curr Rheumatol Rev. 2023;20(4):426–34.
11. King J, Young C, Highton J, Smith PF, Darlington CL. Vestibulo-ocular, optokinetic and postural function in humans with rheumatoid arthritis. Neurosci Lett. 2002;328(2):77–80.
12. De Lorenzis E, Crudo F, Fedele AL, Fiorita A, Bruno D, Paludetti G, et al. Postural control and disability in patients with early rheumatoid arthritis. Clin Exp Rheumatol. 2021;39(6):1369–77.

181 Sarcoidosis and Otovestibular Symptoms

Overview

Sarcoidosis is an idiopathic immune disorder in which caseating granulomas appear in various organ systems. The disease is rare, and only about 5% of cases involve the central nervous system (neurosarcoidosis). The etiology is unknown. Of patients with neurosarcoidosis, almost one-third of patients have auditory and/or vestibular symptoms. The auditory manifestations include hearing loss and tinnitus; the hearing loss may be bilateral (usually asymmetrical) or unilateral, and is predominantly sensorineural; onset may be abrupt or gradual. The vestibular symptoms may be similarly abrupt or gradual in onset; instrumented vestibular testing usually identifies vestibular weakness (unilateral or bilateral). The auditory and vestibular symptoms may improve (though generally do not completely resolve) spontaneously; some clinicians elect treatment with immunosuppressants (such as corticosteroids), less commonly with immunomodulatory agents. For a neurosarcoidosis patient with otovestibular symptoms, an auditory and vestibular workup is reasonable; if brain imaging has not yet been performed, then MRI of the brain and internal auditory canals without and with contrast is reasonable. If otovestibular workup identifies vestibular weakness, then a trial of vestibular rehabilitation therapy is appropriate.

Introduction

The disease that eventually became known as *sarcoidosis* had been eponymously designated [1] Besnier-Boeck-Schaumann disease, in recognition of the original descriptions by French dermatologist Ernest Henri Besnier (1831–1909) [2, 3], Norwegian dermatologist Caesar Peter Moller Boeck (1845–1917) [4, 5], and Swedish dermatologist Jorgen Nielsen Schaumann (1879–1953) [6].

Epidemiology

Sarcoidosis is rare, with an estimated annual incidence of 11 to 36 cases per 100,000 people [7]. Of all patients with sarcoidosis, about 5% develop neurosarcoidosis [8, 9], which is to say, around 5.5 to 18 per one million people develop neurosarcoidosis. In the United States, the disease is significantly more common among African Americans than among Caucasians [10].

Genetics

It is unclear whether genetic susceptibility factors differ between neurosarcoidosis and other forms of sarcoidosis [11].

Pathophysiological Mechanism of Disease

Sarcoidosis is a systemic disease that can cause caseating granulomas in nearly every organ system. The etiology of sarcoidosis is unknown. As far as neurosarcoidosis is concerned, some investigators conclude that the auditory and vestibular symptoms are mostly attributable to vestibulocochlear nerve neuropathy [7], others propose a posterior fossa granulomatous meningitis [12], while others have documented labyrinthine involvement [8].

Clinical Presentation: Overall

Hooper and Holden [13] credit Colover [14] having been the first to describe vestibulocochlear nerve involvement by sarcoidosis.

M. Cherchi, *Otoneurology and Vestibular Medicine*, https://doi.org/10.1007/978-3-031-94841-1_181

Chen and McLeod [15] reviewed the records of 285 patients diagnosed with sarcoidosis, and identified 14 (4.9%) with neurosarcoidosis. Of the 14 patients with neurosarcoidosis, two (29%) had "deafness/vertigo."

Babin and colleagues report that "deafness and vertigo occur as the fourth most common manifestation of neural sarcoidosis" [8].

Clinical Presentation: Auditory

The cadence of hearing loss in neurosarcoidosis is variable. Colvin [7] reported that of 31 neurosarcoidosis patients with hearing loss, the loss was sudden in 15 (48%), "rapidly progressive" in 13 (42%), and "slowly progressive" in 3 (10%).

It has been reported that the subjective hearing loss in neurosarcoidosis can fluctuate [12, 16].

Colvin [7] studied 49 cases of neurosarcoidosis with audiologic symptoms, and reported that 30 (61%) complained of tinnitus; of these 30 patients, the tinnitus was unilateral in 11 (37%) and bilateral in 6 (20%).

Clinical Presentation: Vestibular

The cadence of disequilibrium in neurosarcoidosis is variable, ranging from apoplectic onset to gradual progression over several months. O'Reilly and Burrows [17] reported a case in which neurosarcoidosis manifested with "sudden… vertigo." Szmulewicz and Waterston reported the case of a "6-week history of progressive vertigo" [18]. Imran and colleagues reported a case in which "the vertigo had developed gradually over two months" [10]. Smith and colleagues reported a case presenting with "a 4-month history of progressive gait imbalance" [19].

Ocular Motor Examination

O'Reilly and Burrows [17] described a neurosarcoidosis patient who developed "sudden… vertigo" who, on examination, had spontaneous horizontal nonfatiguing nystagmus, suggestive of unilateral vestibular weakness.

Sidiropoulos and colleagues [20] described the case of a 49-year-old lady who over the course of 3 years, had gradually developed gait disturbances, truncal ataxia, and dysarthria. Physical examination showed bilateral rotatory nystagmus, dysmetric saccades, and palatal tremor. MRI showed an enhancing lesion in the right cerebellar peduncle extending into the pons, with hypertrophic olivary degeneration. Paratracheal lymph node biopsy confirmed sarcoidosis. This was a case in which neurosarcoidosis interrupted the Guillain-Mollaret triangle, manifesting with oculopalatal tremor, and thus is quite different from sarcoidosis involving the vestibulocochlear nerve.

Testing: Auditory

The hearing loss in neurosarcoidosis tends to be greater in the high frequencies on audiometry. The hearing loss is predominantly sensorineural (94%); a minority of patients (6%) have mixed hearing loss [7]. Colvin [7] reported that of 44 patients with hearing loss from neurosarcoidosis, the hearing loss was bilateral in 33 (75%) and unilateral in 11 (25%); of 24 patients with bilateral hearing loss, the loss was asymmetrical in 18 (75%) and symmetrical in 6 (25%).

Figure 181.1, from Babin and colleagues [8], displays a typical audiogram from a patient with deafness attributed to neurosarcoidosis (proven at autopsy).

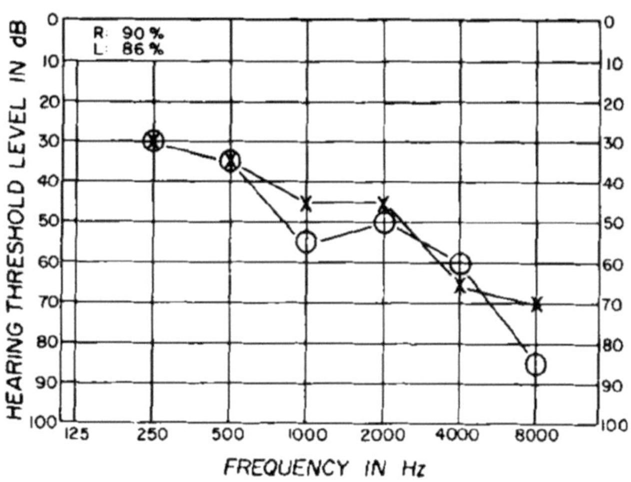

Fig. 181.1 Typical audiogram from a patient with sensorineural hearing loss attributed to neurosarcoidosis. From Babin et al. [8]. Reproduced with written permission from Sage Publications

Table 181.1 Vestibular test findings in neurosarcoidosis patients. From Colvin [7]

Vestibular test	Number of patients (percent)
Bithermal caloric testing ($n = 21$)	
Normal	2 (9)
Unilateral: Reduced response	4 (19)
Unilateral: Absent response	1 (5)
Bilateral: Reduced responses	6 (29)
Bilateral: Absent responses	3 (14)
Bilateral: Reduced response on one side and absent response on one side	4 (19)
Directional preponderance only*	1 (5)
Iced-water caloric testing ($n = 6$)	
Normal	0 (0)
Unilateral: Reduced response	2 (33)
Unilateral: Absent response	0 (0)
Bilateral: Reduced responses	3 (50)
Bilateral: Absent responses	1 (17)
Bilateral: Reduced response on one side and absent response on one side	0 (0)
Horizontal rotational chair testing ($n = 4$)	
Normal	0 (0)
Directional preponderance only	1 (25)
Bilateral: Reduced responses	2 (50)
Bilateral: Absent responses	1 (25)

Testing: Vestibular

Colvin reported that vestibular testing abnormalities were found in 23/24 (96%) of patients [7], with details shown in Table 181.1.

Testing: Other

Tissue biopsy is definitive.

Imaging

MRI of the brain without and with contrast is the imaging modality of choice to evaluate for neurosarcoidosis [10].

Histopathology

Babin and colleagues studied the temporal bones of a 32-year-old man who had been deaf since age 27 years and who had systemic sarcoidosis, including neurosarcoidosis. They described the following labyrinthine findings:

> A large portion of the crista of the superior semicircular canal had undergone degeneration characterized by eosinophilic coagulation. While the crista of the lateral semicircular canal was spared, a similar process had completely destroyed the posterior crista and maculae of the utricle and saccule. The latter was ruptured. These structures did not exhibit any inflammatory infiltrates. Both the perilymphatic and endolymphatic spaces of the labyrinth were variably eosinophilic. Few of the structures of the cochlea escaped destruction. The limbus and tectorial membrane remained intact. The fluid within all three scalae had an eosinic hue. A very few inflammatory round cells were present. The organ of Corti had almost completely degenerated in all turns, with the Reissner membrane adherent to it. The stria vascularis had also undergone degeneration. In some areas, often nearest the spiral vein, it was simply atrophic; in other areas, often close to the Reissner membrane, it was hypertrophic with loss of the usual layered architecture and notably devascularized. The number of cells in the spiral ganglion was only slightly reduced in contrast to the marked destruction elsewhere in the inner ear. There was a distinct, but not striking, infiltration of round cells in the modulus and bony spiral laminate without apparent loss of primary afferent nerve fibers. Examination of all areas of the internal auditory canal (IAC) revealed a striking lymphocytic infiltrate. In most sections, this took the form of an intramural perivascular infiltration. This infiltration extended the entire length of the IAC and involved all four nerves contained within it. In the cochlear nerve it was continuous with the infiltration within the modulus. In both the cochlear and vestibular nerves, loss of nerve fibers was apparent [8].

Differential Diagnosis

Neurosarcoidosis can manifest with the abrupt onset of hearing loss, tinnitus, and disequilibrium [21], thereby mimicking idiopathic *labyrinthitis*.

Since the hearing loss in neurosarcoidosis can fluctuate [12, 16] and be asymmetrical [7], the differential diagnosis includes *Ménière's disease*.

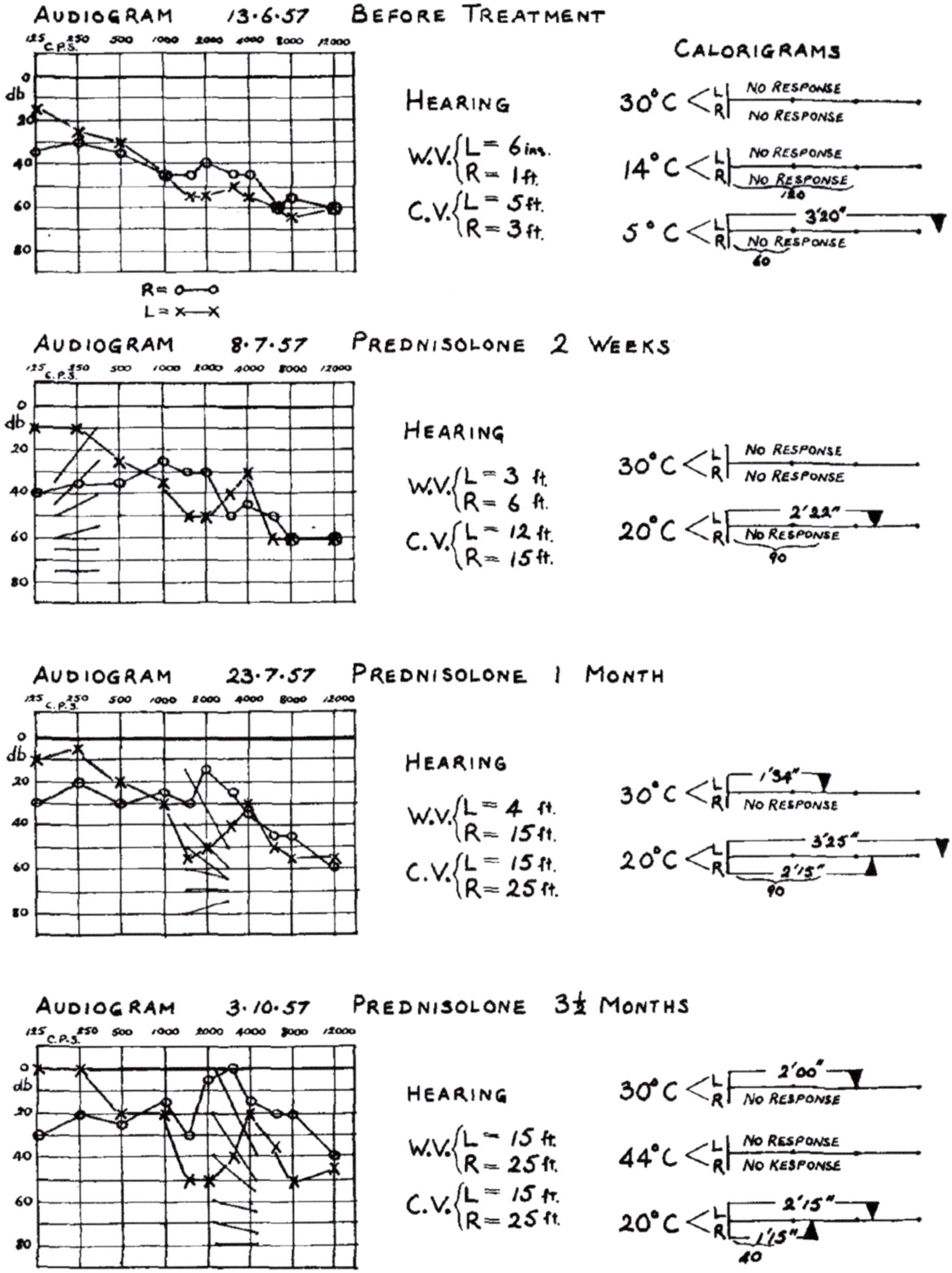

Fig. 181.2 Improvement in hearing and caloric responses in a neurosarcoidosis patient treated with prednisolone. From Gristwood [23]. Reproduced with written permission from Cambridge University Press

Szmulewicz and Waterston [18] described the case of a 57-year-old lady who presented with acute onset right ear symptoms (fullness, hearing loss, tinnitus) and disequilibrium; on examination, head impulse testing showed a reduced vestibulo-ocular reflex on the right. Four weeks later, she had another acute vertiginous episode and was found to have a bilaterally positive head impulse test. Lymph node biopsy confirmed sarcoidosis, but this presentation would be compatible with *bilateral sequential vestibular lesions* (labyrinthitis on the right, followed by vestibular neuritis on the left).

A patient with neurosarcoidosis involving the vestibular system is still vulnerable to developing more mundane vestibular pathology. For example, von Brevern and colleagues [9] report the case of a 30-year-old biopsy-proven neurosarcoid patient found to have bilateral vestibular weakness, who subsequently also developed bilateral posterior canal benign paroxysmal positional vertigo.

Treatment

Agari and colleagues [22] reported the case of a 69-year-old woman with bilateral sensorineural hearing loss and bilaterally absent caloric responses attributed to biopsy-proven sarcoidosis. She was treated with *prednisolone*, and after 2 months, "follow-up examination showed that her vestibular dysfunction and hearing loss had improved significantly."

Gristwood [23] documented improvement in audiologic and vestibular function (caloric responses) over 3.5 months of treatment with prednisolone, as shown in Fig. 181.2.

Imran and colleagues report on a neurosarcoidosis patient with disequilibrium that did not respond to prednisone but did improve with *methotrexate* [10].

Other, less studied immunomodulatory agents have been tried for neurosarcoidosis, such as mycophenolate mofetil, azathioprine, cyclophosphamide, cyclosporine, cladribine, and infliximab, though none have been scrutinized in a randomized controlled trial [10].

Prognosis

In some cases, the auditory and vestibular symptoms attributed to neurosarcoidosis improve spontaneously, though they generally do not completely resolve.

Final Comments

Neurosarcoidosis is rare, and thus should be low on the differential diagnosis of patients with new-onset auditory and/or vestibular symptoms. However, in a patient already diagnosed with sarcoidosis, the index of suspicion should be much higher, given that 5% of sarcoid patients develop neurosarcoidosis, and of those, about a third develop auditory and/or vestibular symptoms.

Given the differential diagnosis described earlier, workup should include auditory testing (audiometry, otoacoustic emissions) and vestibular testing (cervical vestibular evoked myogenic potentials, ocular vestibular evoked myogenic potentials, video head impulse testing, videonystagmography, and rotatory chair testing). If the hearing loss fluctuates, then electrocochleography is also reasonable. If brain imaging has not yet been performed, then an MRI of the brain and internal auditory canals without and with contrast should be considered. The clinician should keep in mind that neurosarcoidosis patients can develop more common otovestibular problems, such as benign paroxysmal positional vertigo.

By the time a sarcoidosis patient is referred to otoneurology, the disease is usually already being managed by a practitioner specializing in the organ system originally affected by the disease (e.g., pulmonology, dermatology). If otovestibular workup identifies vestibular weakness, then a trial of vestibular rehabilitation therapy is reasonable.

References

1. Scadding JG. The eponymy of sarcoidosis. J R Soc Med. 1981;74(2):147–57.
2. Besnier EH. Lupus pernio de la face. Ann Dermatol Syph. 1889;10:333–6.
3. Beeson BB. Ernest Besnier 1831–1909. Arch Derm Syphilol. 1929;20(1):95–9.
4. James DG. Caesar Peter Moeller Boeck (1845–1917). Boeck's sarcoidosis. JAMA. 1970;211(9):1537–8.
5. Boeck CPM. Multiple benign sarcoid of the skin. J Cutan Gen Urin Dis. 1899;17:543–50.
6. Schaumann JN. Recherches sur le lupus pernio et ses relations avec les sarcoïdes cutanées et sous-cutanées. Nordiskt Medicinskt Arkiv. 2009;49(17):1–81.
7. Colvin IB. Audiovestibular manifestations of sarcoidosis: a review of the literature. Laryngoscope. 2006;116(1):75–82.
8. Babin RW, Liu C, Aschenbrener C. Histopathology of neurosensory deafness in sarcoidosis. Ann Otol Rhinol Laryngol. 1984;93(4 Pt 1):389–93.
9. von Brevern M, Lempert T, Bronstein AM, Kocen R. Selective vestibular damage in neurosarcoidosis. Ann Neurol. 1997;42(1):117–20.

10. Imran TF, Nizami S, Eyzner I, Mirani N, Hossain T, Fede R, et al. Vertigo as a predominant manifestation of neurosarcoidosis. Case Rep Med. 2015;2015:397046.
11. Hardin A, Dawkins B, Pezant N, Rasmussen A, Montgomery C. Genetics of neurosarcoidosis. J Neuroimmunol. 2022;372:577957.
12. Hybels RL, Rice DH. Neuro-otologic manifestations of sarcoidosis. Laryngoscope. 1976;86(12):1873–8.
13. Hooper R, Holden H. Acoustic and vestibular problems in sarcoidosis. Arch Otolaryngol. 1970;92(4):386–91.
14. Colover J. Sarcoidosis with involvement of the nervous system. Brain. 1948;71(Pt. 4):451–75.
15. Chen RC, McLeod JG. Neurological complications of sarcoidosis. Clin Exp Neurol. 1989;26:99–112.
16. Brihaye P, Halama AR. Fluctuating hearing loss in sarcoidosis. Acta Otorhinolaryngol Belg. 1993;47(1):23–6.
17. O'Reilly BJ, Burrows EH. VIIIth cranial nerve involvement in sarcoidosis. J Laryngol Otol. 1995;109(11):1089–93.
18. Szmulewicz DJ, Waterston JA. Two patients with audiovestibular sarcoidosis. J Clin Neurosci. 2012;19(1):158–61.
19. Smith JH, Stovall KC, Coons S, Fife TD. Bilateral vestibular hypofunction in neurosarcoidosis: a case report. Ear Nose Throat J. 2011;90(1):E1–3.
20. Sidiropoulos C, Sripathi N, Nasrallah K, Mitsias P. Oculopalatal tremor, facial myokymia and truncal ataxia in a patient with neurosarcoidosis. J Clin Neurosci. 2014;21(12):2255–6.
21. Schick B, Brors D, Koch O, Schafers M, Kahle G. Magnetic resonance imaging in patients with sudden hearing loss, tinnitus and vertigo. Otol Neurotol. 2001;22(6):808–12.
22. Agari D, Koide R, Kashiyama T, Yoshida H, Naito R, Bandoh M. Neurosarcoidosis: a treatable cause of vestibular dysfunction. Lancet. 2007;369(9564):878.
23. Gristwood RE. Nerve deafness associated with sarcoidosis. J Laryngol Otol. 1958;72(6):479–91.

Scleroderma (Systemic Sclerosis) 182

Overview

Scleroderma, also called *systemic sclerosis*, is a rare multi-system connective tissue disease of unknown etiology. Some scleroderma patients complain of audiologic symptoms (hearing loss, tinnitus) and/or vestibular symptoms. Given the rarity of scleroderma, in some cases, these symptoms may be due to more common diseases that happen to be coincidental with scleroderma, while in other cases, these deficits may truly be the result of scleroderma itself. If no other cause is found, then referral to audiology and/or vestibular rehabilitation therapy is medically reasonable. The management of scleroderma is usually supervised by a rheumatologist.

Introduction

Scleroderma, also called *systemic sclerosis*, is a rare connective tissue disease of unknown etiology in which proliferative vascular lesions involving microvessels and small arteries result in collagen deposition and fibrosis in multiple organ systems, including the skin, gastrointestinal tract, lungs, kidneys, and, less commonly, the nervous system. Raynaud's phenomenon is sometimes the harbinger of scleroderma. Broad categories of scleroderma include diffuse systemic sclerosis and localized systemic sclerosis, the latter of which has lesser visceral involvement. Some patients with scleroderma develop hearing loss beyond what is expected for age; this hearing loss can be unilateral, bilateral (symmetric or asymmetric), sensorineural or mixed, gradual or abrupt. Scleroderma patients can also exhibit abnormalities on vestibular testing (such as on various ocular motor tests and static posturography) and can develop benign paroxysmal positional vertigo.

Epidemiology

Scleroderma is rare, with a reported annual incidence of 4–12 per one million individuals. It is more common among individuals aged 30–50 years [1]. Scleroderma affects women three [1] to four [2] times more frequently than men.

Genetics

The genetics of scleroderma are unknown. Most data implicate loci in the major histocompatibility complex and fibrillin-1, but other candidates include mutations in genes encoding for angiotensin-converting enzyme, nitric oxide synthase, CD19, and PTPN22.

Pathophysiological Mechanism of Disease

Scleroderma is a connective tissue disease of unknown etiology. Pathologically, it is characterized by marked proliferative vascular lesions involving small arteries and microvessels [3], collagen deposition in multiple organ systems [4], and diffuse connective tissue fibrosis that can involve the skin, gastrointestinal tract, lungs, heart, and kidneys [3], and less commonly, the central and peripheral nervous systems. Proposed pathophysiological mechanisms have included "endothelial dysfunction, fibroblast dysfunction, and dysregulation of the immune system" [3].

The mechanism by which scleroderma specifically causes audiologic and vestibular problems is unknown. Given the histopathologic changes described in a temporal bone study by Abou-Taleb and Linthicum [4], Deroee and colleagues [5] speculate, "It is possible that decreased blood flow due to

M. Cherchi, *Otoneurology and Vestibular Medicine*, https://doi.org/10.1007/978-3-031-94841-1_182

vasculitis is the cause of hearing loss," a theory also suggested by Rabelo and colleagues [1]. Contrastingly, Santarelli and colleagues [6] suggest that auditory neuropathy may be the mechanism of hearing loss in scleroderma.

Clinical Presentation

Raynaud's phenomenon is sometimes the initial manifestation of a disease process that eventually declares itself as scleroderma [4].

Amor-Dorado and colleagues [3] studied 27 scleroderma patients and reported that 19 (54%) complained of hearing loss.

Berrettini and colleagues [7] studied 37 scleroderma patients and found that 23 (62%) reported audiologic or vestibular complaints. Of those 23 patients, 11 (48%) complained exclusively of auditory symptoms (hearing loss or tinnitus), 4 (17%) complained exclusively of vestibular symptoms, and 8 (35%) complained of both auditory and vestibular symptoms.

Deroee and colleagues [5] reported the case of a 65-year-old man with scleroderma who suffered sudden sensorineural hearing loss bilaterally.

Testing: Audiologic

On audiometry, the hearing loss in scleroderma can be unilateral, bilateral (symmetric or asymmetric), sensorineural, conductive, or mixed, suggesting that both the inner ear and the middle ear can be affected.

Berrettini and colleagues [7] studied audiometry in 37 scleroderma patients and reported that 14 (38%) showed hearing loss. Of the 14 patients with hearing loss, only 1 (7%) had unilateral hearing loss, while the remaining 13 (93%) had bilateral hearing loss. Of the 14 patients with hearing loss, it was sensorineural in 10 (71%) and mixed in 4 (29%).

Kastanioudakis and colleagues [8] studied audiometry in 30 scleroderma patients and reported sensorineural hearing loss in 6 (20%) and mixed hearing loss in 1 (3%).

Amor-Dorado and colleagues [3] studied 35 scleroderma patients (limited systemic sclerosis) and reported that 27 (77%) showed hearing loss (beyond what was expected for age) on audiometry. Of these 27 patients with unexpected hearing loss, 24 (89%) had sensorineural hearing loss, which was symmetrical in 13 patients and asymmetrical in 11. Of the 27 patients with unexpected hearing loss, 3 (11%) had mixed (sensorineural and conductive) hearing loss. They also reported that scleroderma patients had abnormal tympanograms and stapedius reflexes compared to healthy controls.

Testing: Vestibular

Rabelo and colleagues [1] conducted a systematic literature review and found that studies reported vestibular abnormalities ranging from 11% to 63%, and that "the most common findings were changes in caloric testing, positional nystagmus, abnormal oculocephalic reflex, changes in clinical tests of sensory integration, and benign paroxysmal positional vertigo."

Amor-Dorado and colleagues [9] reported that 7 (17%) of 41 scleroderma patients exhibited oculographic abnormalities on vestibular testing. They further commented that the ocular motor findings were compatible with posterior canal benign paroxysmal positional vertigo in two scleroderma patients, and findings compatible with lateral canal benign paroxysmal positional vertigo in five scleroderma patients. Choi and colleagues [2] also reported a case of a scleroderma patient with direction-changing positional nystagmus compatible with lateral canal benign paroxysmal positional vertigo.

Berrettini and colleagues [7] conducted vestibular tests in 37 scleroderma patients and reported that 4 (11%) exhibited some abnormality. Of these, two exhibited positional nystagmus, one exhibited a "marked unilateral compensated vestibular deficit, and one exhibited bilateral caloric weakness.

Amor-Dorado and colleagues [3] reported that of 34 scleroderma patients who underwent vestibular testing, ten (29%) exhibited abnormalities in the oculocephalic reflex, nine (26%) exhibited head shaking nystagmus, three (9%) exhibited positional nystagmus in at least three positions, and 11 (31%) exhibited abnormal caloric asymmetry.

Amor-Dorado and colleagues [3] reported that 16 (46%) out of 34 scleroderma patients exhibited abnormalities on static posturography ("foam and dome test"). Similarly, Amor-Dorado and colleagues [9] reported that 20 (48%) of 42 scleroderma patients exhibited abnormalities on static posturography.

Testing: Other

Some investigators report the finding of a patulous Eustachian tube in scleroderma patients [5, 8].

Imaging

Scleroderma can cause intraparenchymal brain calcifications. Choi and colleagues [2] provide a clear example of intracerebral calcification from scleroderma, particularly involving the left cerebellar hemisphere.

Histopathology

Few temporal bone reports are available in patients with scleroderma. Abbou-Taleb and Linthicum [4] describe:

> The blood vessels in the stria [vascularis of the cochlea] were larger than normal and there was focal strial atrophy. The number of outer hair cells, dendrites, and ganglion cells was reduced by about 50 per cent in the area from the anterior basal turn to the hook. Blood vessels around the endolymphatic sac and in the internal auditory canal, mastoid cells, and bone marrow spaces had thickened walls and hyalinization. Perivascular fibrosis surrounded small arterioles and venules in some areas with a few ghost vessels. The amount of thickening of blood vessels around the ELS [endolymphatic sac] and the IAC [internal auditory canal] was approximately the same. There was no evidence of new bone formation or of hydrops [4].

Differential Diagnosis

The clinical history of the auditory and vestibular symptoms, and the abnormalities reported on audiologic and vestibular testing are neither specific nor sensitive for scleroderma. Given its rarity, if a scleroderma patient reports audiologic or vestibular symptoms, it is medically reasonable to undertake a screening audiologic and vestibular workup to seek more common causes of these symptoms.

Treatment and Prognosis

Scleroderma is usually managed by a rheumatologist.

Since the underlying cause of scleroderma is unknown, management is supportive rather than curative. As far as audiologic and vestibular symptoms are concerned, if a screening workup reveals no evidence for a more convincing explanation, then referral to audiology and/or vestibular rehabilitation therapy is medically reasonable.

References

1. Rabelo MB, Corona AP. Auditory and vestibular dysfunctions in systemic sclerosis: literature review. Codas. 2014;26(5):337–42.
2. Choi EJ, Lee DW, Park CW, Lee SH. A case of linear scleroderma involving cerebellum with vertigo. Korean J Audiol. 2012;16(2):87–90.
3. Amor-Dorado JC, Arias-Nunez MC, Miranda-Filloy JA, Gonzalez-Juanatey C, Llorca J, Gonzalez-Gay MA. Audiovestibular manifestations in patients with limited systemic sclerosis and centromere protein-B (CENP-B) antibodies. Medicine (Baltimore). 2008;87(3):131–41.
4. Abou-Taleb A, Linthicum FH Jr. Scleroderma and hearing loss: (histopathology of a case). J Laryngol Otol. 1987;101(7):656–62.
5. Deroee AF, Huang TC, Morita N, Hojjati M. Sudden hearing loss as the presenting symptom of systemic sclerosis. Otol Neurotol. 2009;30(3):277–9.
6. Santarelli R, Scimemi P, Dal Monte E, Genovese E, Arslan E. Auditory neuropathy in systemic sclerosis: a speech perception and evoked potential study before and after cochlear implantation. Eur Arch Otorrinolaringol. 2006;263(9):809–15.
7. Berrettini S, Ferri C, Pitaro N, Bruschini P, Latorraca A, Sellari-Franceschini S, Segnini G. Audiovestibular involvement in systemic sclerosis. ORL J Otorhinolaryngol Relat Spec. 1994;56(4):195–8.
8. Kastanioudakis I, Ziavra N, Politi EN, Exarchakos G, Drosos AA, Skevas A. Hearing loss in progressive systemic sclerosis patients: a comparative study. Otolaryngol Head Neck Surg. 2001;124(5):522–5.
9. Amor-Dorado JC, Barreira-Fernandez MP, Arias-Nunez MC, Gomez-Acebo I, Llorca J, Gonzalez-Gay MA. Benign paroxysmal positional vertigo and clinical test of sensory interaction and balance in systemic sclerosis. Otol Neurotol. 2008;29(8):1155–61.

Part XXX

Evaluation—Bedside Physical Examination

183 Dynamic Visual Acuity Testing of the Rotational Vestibulo-Ocular Reflex

Overview

The *rotational vestibulo-ocular reflex (rVOR)* maintains an image stable on the retina despite rotational movements of the head by rotating the eyes an equal amount but in the opposite direction of head movement. The middle frequency range of the vestibular tuning spectrum can be evaluated at the bedside with *dynamic visual acuity testing (DVA)*, sometimes also called the "dynamic illegible 'E' test" [1, 2]. This test is more sensitive for bilateral vestibular weakness than for unilateral vestibular weakness.

Introduction

There are several methods for assessing the rVOR at the bedside, including the dynamic visual acuity testing, the ophthalmoscope test, and the head impulse test. Dynamic visual acuity testing and the ophthalmoscope test assess the middle frequency range of the vestibular tuning spectrum, while the head impulse test assesses the high range. Assessment of the low range of the rVOR requires instrumented vestibular testing (such as rotatory chair testing).

Physiology and Neuroanatomy

The function of the vestibulo-ocular reflex (VOR) is to stabilize an image on the retina despite rotational and/or translational movements of the head [3]. Ideally the rotational VOR (rVOR) should rotate the eyes within the orbits by an amount that is equal in magnitude, but opposite in direction, to that of the head, thereby offsetting the head movement and maintaining the point of regard stable on the fovea (in other words, the gain of eye rotation to head rotation is 1.0).

The neuroanatomy of the rVOR is well understood [4]. Dynamic visual acuity testing (to be described in greater detail below) assesses the component of the rotational vestibulo-ocular reflex mediated by the lateral semicircular canals, also called the horizontal rVOR (hrVOR).

Note that the VOR can be assessed for any pair of coplanar semicircular canals; the relevant neuroanatomy is different for each canal [4].

M. Cherchi, *Otoneurology and Vestibular Medicine*, https://doi.org/10.1007/978-3-031-94841-1_183

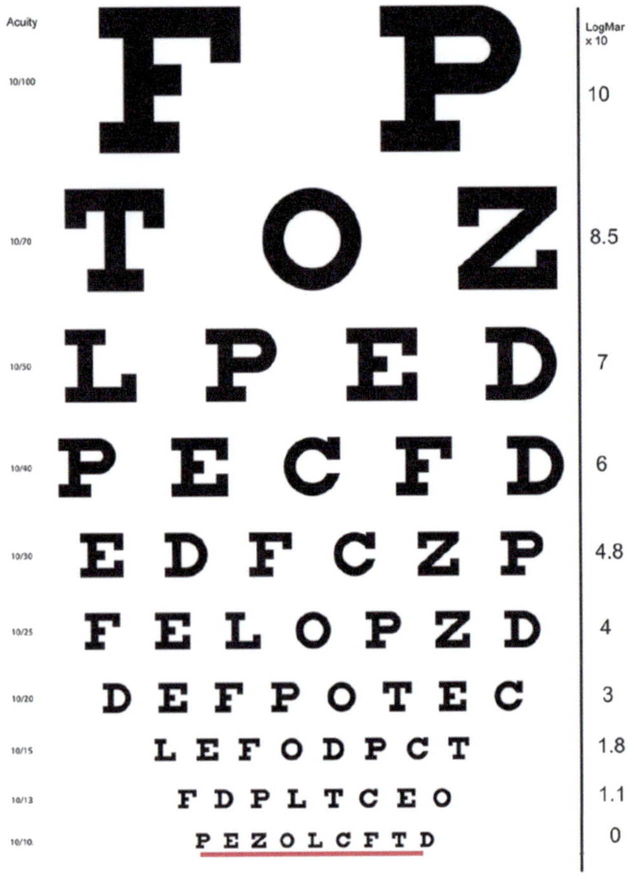

Fig. 183.1 Eye chart calibrated by LogMAR for dynamic visual acuity testing. From Hain and Cherchi [5]. Reproduced with written permission from Wolters Kluwer Health Limited

Equipment Needed

This test requires an eye chart that should be positioned about 12 feet away from the patient [1]. Ideally, the eye chart should be calibrated according to LogMAR (logarithm of the minimum angle of resolution), such as that shown in Fig. 183.1, from Hain and Cherchi [5].

How to Perform the Test

If the patient normally wears corrective lenses, then they should do so for this test as well.

First, the patient keeps their head still, looks at the eye chart—placed at a distance of about 12 feet from the patient [1]—and reads the farthest line down that they can.

Second, while the patient is still looking at the eye chart, the examiner passively oscillates the patient's head (rotating left to right around the vertical axis) at a frequency of about 1.5 Hz [6] to 2 Hz [7, 8]. While the patient's head is in motion, they again read the farthest line down that they can. Reading while the head is oscillating in this fashion cannot rely on visual tracking or smooth pursuit capabilities, which are too slow to process retinal slip at this frequency [1, 9, 10]; Rather, the eye movements are driven by the vestibulo-ocular reflex, which (when intact) is sufficiently rapid, and can respond to velocities up to 300 deg/sec [1].

Dynamic visual acuity testing is reported as "going from a LogMAR of [A] to [B]," where A is the LogMAR of the farthest line down that the patient can read with the head still, and B is the LogMAR of the farthest line down that the patient can read while the head is oscillating.

What This Test Assesses

Dynamic visual acuity testing as described above assesses the middle frequency range of the horizontal rVOR and, thus, the underlying neuroanatomical pathways of the vestibulo-ocular reflex. A lesion anywhere along this pathway can reduce or abolish the horizontal rVOR, and thus result in poor performance on dynamic visual acuity testing.

How to Interpret the Test Results

A patient with an intact horizontal rVOR may lose up to two lines on dynamic visual acuity testing [2, 10, 11].

If a patient has bilateral vestibular weakness or significant unilateral vestibular weakness, then they will lose three or more lines on dynamic visual acuity testing [7, 8, 12].

Dynamic visual acuity testing is more sensitive for bilateral vestibular weakness than for unilateral vestibular weakness, though it can detect unilateral vestibular weakness [13] that is sufficiently severe [1].

Limitations

This test requires that the patient have sufficient cervical rotational motion. Patients with cervical vertebral fusions, or significant cervical arthritis or cervicalgia, or other limitations on cervical range of motion, may be unable to perform this test.

Pitfalls

Note that the oscillatory motion in this test is passive [7]; the examiner moves the patient's head passively because active rotation (i.e., movement under the patient's volition) slightly improves gain [14, 15], though whether this meaningfully influences test results is unclear [16].

If a patient habitually uses corrective refractive lenses (whether spectacles or contact lenses), then they should also wear these during the ophthalmoscope test; otherwise, there may be false-positive results.

Diseases That May Be Diagnosed by This Test

Dynamic visual acuity testing of the horizontal rotational vestibulo-ocular reflex is particularly helpful in identifying bilateral vestibular weakness, of which there are many etiologies. It is less helpful in identifying unilateral vestibular weakness, which in turn may have many etiologies.

Additional Notes

There are computerized versions of dynamic visual acuity testing that may be more accurate [17, 18], but are not in wide clinical use.

References

1. Longridge NS, Mallinson AI. A discussion of the dynamic illegible "E" test: a new method of screening for aminoglycoside vestibulotoxicity. Otolaryngol Head Neck Surg. 1984;92(6):671–7.
2. Sargent EW, Goebel JA, Hanson JM, Beck DL. Idiopathic bilateral vestibular loss. Otolaryngol Head Neck Surg. 1997;116(2):157–62.
3. Rinaudo CN, Schubert MC, Figtree WVC, Todd CJ, Migliaccio AA. Human vestibulo-ocular reflex adaptation is frequency selective. J Neurophysiol. 2019;122(3):984–93.
4. Bronstein AM, Patel M, Arshad Q. A brief review of the clinical anatomy of the vestibular-ocular connections-how much do we know? Eye (Lond). 2015;29(2):163–70.
5. Hain TC, Cherchi M. Vestibular testing. Continuum. 2021;27(2):330–47.
6. Dannenbaum E, Paquet N, Hakim-Zadeh R, Feldman AG. Optimal parameters for the clinical test of dynamic visual acuity in patients with a unilateral vestibular deficit. J Otolaryngol. 2005;34(1):13–9.
7. Eggers SD, Zee DS. Evaluating the dizzy patient: bedside examination and laboratory assessment of the vestibular system. Semin Neurol. 2003;23(1):47–58.
8. Huh YE, Kim JS. Bedside evaluation of dizzy patients. J Clin Neurol. 2013;9(4):203–13.
9. Demer JL, Honrubia V, Baloh RW. Dynamic visual acuity: a test for oscillopsia and vestibulo-ocular reflex function. Am J Otol. 1994;15(3):340–7.
10. Petersen JA, Straumann D, Weber KP. Clinical diagnosis of bilateral vestibular loss: three simple bedside tests. Ther Adv Neurol Disord. 2013;6(1):41–5.
11. Kheradmand A, Zee DS. The bedside examination of the vestibulo-ocular reflex (VOR): an update. Rev Neurol (Paris). 2012;168(10):710–9.
12. Fife TD, Tusa RJ, Furman JM, Zee DS, Frohman E, Baloh RW, et al. Assessment: vestibular testing techniques in adults and children: report of the therapeutics and technology assessment Subcommittee of the American Academy of neurology. Neurology. 2000;55(10):1431–41.
13. Dannenbaum E, Paquet N, Chilingaryan G, Fung J. Clinical evaluation of dynamic visual acuity in subjects with unilateral vestibular hypofunction. Otol Neurotol. 2009;30(3):368–72.
14. Della Santina CC, Cremer PD, Carey JP, Minor LB. Comparison of head thrust test with head autorotation test reveals that the vestibulo-ocular reflex is enhanced during voluntary head movements. Arch Otolaryngol Head Neck Surg. 2002;128(9):1044–54.
15. Goebel JA, Isipradit P, Hanson JM. Manual rotational testing of the vestibulo-ocular reflex. Laryngoscope. 2000;110(4):517–35.
16. Furmen JM, Durrant JD. Head-only rotational testing in the elderly. J Vestib Res. 1998;8(5):355–61.
17. Herdman SJ, Tusa RJ, Blatt P, Suzuki A, Venuto PJ, Roberts D. Computerized dynamic visual acuity test in the assessment of vestibular deficits. Am J Otol. 1998;19(6):790–6.
18. Vital D, Hegemann SC, Straumann D, Bergamin O, Bockisch CJ, Angehrn D, et al. A new dynamic visual acuity test to assess peripheral vestibular function. Arch Otolaryngol Head Neck Surg. 2010;136(7):686–91.

MIX
Papier aus verantwortungsvollen Quellen
Paper from responsible sources
FSC® C105338

If you have any concerns about our products,
you can contact us on
ProductSafety@springernature.com

In case Publisher is established outside the EU,
the EU authorized representative is:
Springer Nature Customer Service Center GmbH
Europaplatz 3, 69115 Heidelberg, Germany

Printed by Libri Plureos GmbH
in Hamburg, Germany